Gender Trouble in the I

Stephanie Szitanyi

Gender Trouble in the U.S. Military

Challenges to Regimes of Male Privilege

Stephanie Szitanyi
Schools of Public Engagement
The New School
New York, NY, USA

ISBN 978-3-030-21227-8   ISBN 978-3-030-21225-4 (eBook)
https://doi.org/10.1007/978-3-030-21225-4

© The Editor(s) (if applicable) and The Author(s), under exclusive licence to Springer Nature Switzerland AG 2020
This work is subject to copyright. All rights are solely and exclusively licensed by the Publisher, whether the whole or part of the material is concerned, specifically the rights of translation, reprinting, reuse of illustrations, recitation, broadcasting, reproduction on microfilms or in any other physical way, and transmission or information storage and retrieval, electronic adaptation, computer software, or by similar or dissimilar methodology now known or hereafter developed.
The use of general descriptive names, registered names, trademarks, service marks, etc. in this publication does not imply, even in the absence of a specific statement, that such names are exempt from the relevant protective laws and regulations and therefore free for general use.
The publisher, the authors and the editors are safe to assume that the advice and information in this book are believed to be true and accurate at the date of publication. Neither the publisher nor the authors or the editors give a warranty, expressed or implied, with respect to the material contained herein or for any errors or omissions that may have been made. The publisher remains neutral with regard to jurisdictional claims in published maps and institutional affiliations.

Cover illustration: © DanielBendjy / Getty Images

This Palgrave Macmillan imprint is published by the registered company Springer Nature Switzerland AG.
The registered company address is: Gewerbestrasse 11, 6330 Cham, Switzerland

For Anyu and Olivia

Acknowledgments

I was frequently asked why I chose to engage in research related to the military while writing this book. For many, it was an unusual choice. After all, I have not served in the military, nor do I come from a military family. And yet, I have increasingly often felt like my life has been militarized even as a member of the American general public, a public falsely thought to be far removed from the institution that is "the military."

I am forever grateful for the host of incredible mentors who provided the resources for a "feminist awakening" to help connect those feelings of militarized existence with academic scholarship. Both Katie Verlin Laatikainen and Traci Levy were critical in this "aha!" moment during my undergraduate studies at Adelphi University, as was Paul Roe at Central European University. I wish to extend my particular thanks to Paul for his mentorship throughout my master's studies, and for introducing me to the works of both Cynthia Enloe and J. Ann Tickner (and feminist international relations theory, more broadly) for the first time in 2005 in his seminar, *Critical Security Studies*. Both Cynthia Enloe's and J. Ann Tickner's scholarship heavily inform the work in the forthcoming pages, and all my general musings related to gender and the military. Exposure to their scholarship has, in short, been life-changing.

This book project began as a master's degree thesis, and somehow later morphed into a doctoral dissertation at Rutgers University thanks to my doctoral chair and mentor, Mary Hawkesworth. I am forever grateful to Mary for allowing me to study under her, and for her utmost patience with me throughout my doctoral studies. I have learned a great deal from Mary

on all counts, but particularly from her diligent and detailed (but always only suggested) edits on dissertation chapter drafts, conference papers, and first publication submissions. I am also grateful to the other members of my dissertation committee, Susan J. Carroll, Mona Lena Krook, and my outside reader, Alison Howell.

I have been fortunate to engage in several formal and informal conversations around the topics presented in the book with some wonderful feminist scholars. My thanks to Melissa T. Brown, Terrell Carver, Maya Eichler, Cynthia Enloe, Megan MacKenzie, Melanie Richter-Montpetit, Audrey Reeves, and Laura Sjoberg, as well as countless others I shared ideas with at conferences or through email exchanges. I am also particularly grateful to Anca Pusca, with whom a serendipitous meeting resulted in finding a "home" for this book, to Katelyn Zingg for entertaining all my questions during the review and production phases, and to the two anonymous peer reviewers—their comments have undoubtedly strengthened this book in its final version.

The completion of this book would not have been possible without additional moral support from friends in all corners of the globe. I would particularly like to thank Sabrina Javor, Andrea Matolcsi, Randi Schnabel, Tanisha Johnson-Campbell, Aaron Hudson, Amanda Marziliano, Ashley Koning, Heather James, Leah Iannone, Dan Jacob, and John (Jack) Rahaim.

Finally, I am foremost indebted to my mother, Eva Szitanyi, as well as to my daughter, Olivia. My mother has passed on an impressive legacy of feminist activism that has undoubtedly impacted the person I've become and the research I engage in. She has sat and listened patiently to me read draft after draft of this book aloud, has provided copious amounts of coffee and snacks while I have worked through all hours of the night and, as Olivia's maternal grandmother, has heroically shepherded the responsibilities associated with parenting and child care almost solely for the last four years so that this book could come to fruition. Admittedly, she has, at times, been more dedicated to this project than I have, seeing the importance of it when I could not. More than anyone else, she has been this project's greatest fan, and for that I am eternally grateful. And to Olivia, for her patience with time and for giving up hours together each day in the final stages of this manuscript. I hope you open this book someday, read your name on this page, and know that my heart is always yours.

Contents

1 Introduction   1

2 The Masculine Warrior: Militarized Masculinities and Gender Regimes   29

3 The All-Volunteer Force: Patrolling Gendered Boundaries Through the Combat Ban   53

4 Violated Bodies: Combat Injuries and Sexual Assault in the U.S. Military   89

5 Military Museums and Memorial Sites: Disappearing Women in the Military   119

6 Gender and Military Recruitment Since the Lifting of the Combat Ban   147

7 Conclusion: The Challenge of Degendering the Military   175

Index   201

LIST OF FIGURES

Fig. 5.1	Exterior image of the *USS Midway* Museum	132
Fig. 5.2	Image of Dambuster missile	133
Fig. 5.3	Image of Eye of the Corps missiles	134
Fig. 5.4	Wide-angle photo of F9F Panther plane and description poster on flight deck of *USS Midway*	135
Fig. 5.5	Close-up photo of F9F Panther plane description poster on flight deck of *USS Midway*	136
Fig. 5.6	Image of Cougar plane with mannequin on flight deck of *USS Midway*	137

CHAPTER 1

Introduction

In January 2013, U.S. military officials announced the lifting of the long-standing combat ban, a policy that curtailed women service members' access to the most coveted positions within the organization for nearly five decades. As a result of the impending change, the military embarked on a series of gender-integrated unit experiments to ensure combat effectiveness would remain unimpaired. The results of those investigations would determine whether the four branches of the military under the Department of Defense—the Air Force, Army, Marine Corps, and Navy—would request exemptions toward preserving some 220,000 jobs as open to men only.

Of the four branches, the Marine Corps was the only branch to request exemptions. Specifically, the Corps argued adamantly against women's inclusion in ground combat, reconnaissance and intelligence units, as well as the prestigious Special Forces. The request was made based on the results of a Marine Corps experiment on gender integration, which indicated all-male units outperformed gender-integrated units in 69% of tasks (93 of 134)[1] including overall speed, firing accuracy, and casualty evacuation (Department of Defense 2015). Although the experiment was rife with methodological flaws, the timing of its release shortly after the Army's own announcement that three women had passed its grueling Ranger school training course counteracted any and all fanfare associated with women's ability to meet the necessary standards to serve side by side with their male counterparts.

The Marine Corps' vehement request for exemptions was ultimately denied in December 2015 when then Secretary of Defense, Ashton Carter, declared "there would be no exceptions...women will be allowed to drive tanks, fire mortars, lead infantry soldiers into combat. They'll be able to serve as Army Rangers, Green Berets, Navy SEALS, Marine Corps infantry, Air Force parajumpers and everything else previously open only to men" (Rosenberg and Phillips 2015, 2). As a result, the Marine Corps instituted new physical standards that were positioned as necessary for the branch to accommodate the more gender-inclusive policy. Headlined by Militarytimes. com as "New Marine Corps Fitness Standards for Combat Weed out Men, Women Alike," the article framed the new physical standards as beneficial for determining who the best performing Marines were, whether women or men (Baldor 2016). But the new standards seemed to overwhelmingly target women for exclusion. While only 40 out of 1500 (3%) male recruits were reported as not meeting the necessary standards, 6 out of every 7 females (86%) were failing the test, inhibiting their ability to choose from infantry, artillery, and other combat-related occupations (2).

The Marine Corps' sustained insistence that women are unqualified[2] for specific forms of combat service foregrounds an institutional conviction concerning the putative strength of the male body and a corollary belief in the inferiority of women's bodies. Despite the official elimination of the combat ban, the task of gender integration may be read critically as "gender trouble" for the branch in this environment; it challenges and threatens an institutional principle that views it necessary that masculinity be preserved as the cornerstone of military identity.

This tension between the institution's gendered identity and the behavior of service members' bodies in the above example is but one of several instances of "gender trouble" the contemporary U.S. military has had to face. Accounts of high percentages of sexual assault in both the military branches and the military academies alike, along with Congressional efforts to address institutional sexual violence, have flooded the media. After nearly two decades, the government officially abandoned its "Don't Ask, Don't Tell" (DADT) policy in September 2011,[3] a policy which mandated compulsory public heterosexuality and closeted homosexuality in the military. "Gender trouble" was also rife in the trial of Army Private First Class Bradley/Chelsea Manning, who was convicted of espionage, theft, and fraud for leaking classified documents to the online information source WikiLeaks. Manning's decision to leak

more than 750,000 classified documents was attributed by some to struggles with gender identity. Her incarceration in a male prison and initial denial of access to hormone treatment raised significant ethical issues concerning the military's treatment of enlisted transgender personnel.

How does the military cope with such episodes of "gender trouble" in an era of increased gender and sexual equality? And, perhaps more importantly, what do the ways in which it deals with these episodes tell us about the contemporary relationship between masculinity and military service in the United States?

To answer these questions, this book investigates challenges to the U.S. military's *gender regime*, that is organizational practices—both formal and informal—that structure gender relations and gender power within the institution (Connell 1987). These challenges—which the book refers to as "gender trouble"[4]—exist, primarily, in two forms; as gender-related (historical) policies previously set by the military that have been complicated by demands for gender equality, and by "deviant" manifestations of gender that challenge the institution's "existing gender logics of appropriateness" (Chappell and Waylen 2013, 603), and in so doing, resist conforming to its system of heteromale privilege. In the tension between institutional identity and demands for equality, the military faces a choice: to respond to calls to end gender and sex discrimination and shift its gender regime toward eliminating male, heterosexual privilege from all its operations (*degendering*), or instead, to reinforce specific forms of militarized masculinity as its primary organizing principle and value system (*regendering*).

How are we to know whether a gender regime is shifting? Recent repeals of institutional policies, namely "DADT" in 2011, the initial repeal of the transgender ban under the Obama administration in 2016,[5] and the ban against women in combat in 2016 may signal openings toward change. Indeed, open homosexuality and women's full inclusion in the most coveted roles of the military system may destabilize hegemonic and other militarized masculinity archetypes by expanding military identities to encompass multiple gendered subjects. But this may not be an outright cause for celebration. The elimination of long-standing, outdated policies, and equally, the creation of (seemingly) progressive, alternative policies should be considered with caution as they may not provide a full picture of the institution's intentions.

As this book suggests, some policy changes that appear to promote gender equality may in actuality be accompanied by measures that *reinforce* gender norms associated with militarized masculinities. Similarly, Chelsea Manning's pursuit of sex reassignment treatments while serving in

military prison, attention to sexual assault in the military branches and academies, and official recognition of cognitive and psychological injuries sustained in combat, trouble the military's power to contain how institutionally labeled "heroic" military bodies behave. As these examples suggest, *regendering* may involve changes in official narratives about gender and service, nation and sacrifice, valor and citizenship by providing new modes of inclusion that simultaneously (re)affirm the centrality of gender (read as masculinity) to the military as an organization. And as earlier policies make clear, the military can also silence expansive manifestations of gender within its own ranks, and mask open articulation of gender nonconformity and diverse sexual orientations among its troops, promoting conservative modes of male dominance within and outside of the military apparatus. Although the military has worked to generate explicit narratives that depict it as an inclusive organization, freed from past strictures barring women and gays from the armed services, closer examination of recurring tropes within these narratives indicate that the military continues to actively reinforce heteronormative masculinity in ways that marginalize not only women, but lesbian, gay, bisexual, and transgender military personnel as well.

Two bodies of literature—feminist international relations (IR) and feminist institutionalism—prove useful in uncovering potential gender regime shifts. The study presented in the book is grounded in both as they provide a robust understanding of the military's authority to shore up and convey a particular gender order. Gendered institutions—such as militaries—operate in accordance with norms that "construct and maintain power dynamics that favor men of the dominant race, ethnicity, and sexuality" (Hawkesworth 2012, 2018). They have been characterized as embodying male dominance, a masculinist culture, and homosociality—the establishment of intense bonds among men through the carefully orchestrated regulation of access to women (Belkin 2012; Brown 2012; Burke 2002; Enloe 1983, 1990, 2000; Cohn 2000; Francke 1997; Goldstein 2001; Kronsell 2012; Levy 1997; Lutz 2002; MacKenzie 2015). Charged with the nation's security, the military plays a crucial role in defining and upholding particular constructions of manhood and masculinized citizenship, while historically barring women, on the other hand, from the military, combat duty, and the revered valorization of that service.

Importantly, both feminist IR and feminist institutionalism help us move toward thinking of the U.S. military—if not militaries more generally—as not only a gendered institution, but rather a *gendering*[6]

institution[7] (Cohn 2000; Segal 1982, 1995), an active producer and communicator of social and cultural norms associated with gender. In gendered institutions, processes of gendering are often subtle, invisible, "sophisticated and hidden" (Sasson-Levy and Amram-Katz 2007, 107); they occur not necessarily through the implementation of policies themselves, but through discursive tools associated with the communication of those policies. As such, the book argues that the "discursive multiplicity" (106) associated with these simultaneously seen/veiled, visible/invisible, heard/silenced processes, and the often contradictory messages embedded in (sometimes simultaneous) processes of (re and de)gendering, requires tracing a diverse set of military public discourses "against the grain" (Ferguson and Turnbull 1999, 46),[8] that is interrogating gendered, social meanings (Millar in Woodward and Duncanson 2017, Carver 2002) behind those messages rather than accepting them at face value. This critical approach to "reading" the military's self-representational material foregrounds the dexterity of gendered power, revealing how the institution and its ideas on gender move to subtly reconstruct and reinforce "hierarchical gender differences" (Sasson-Levy and Amram-Katz 2007, 1) both internally to the organization and externally beyond the walls of the apparatus. In the case of the U.S. military, the hierarchical orders of value which the military projects, and which become embedded in society, subordinate the lives and contributions of ordinary individuals to the potent sacrifice of the manly warriors who risk everything for national defense (Sasson-Levy and Amram-Katz 2007). Ultimately, such careful feminist analysis provides a more nuanced understanding of the military, both as a gendered and gendering institution, that actively communicates social norms that become embedded in the lives of ordinary citizens—purely descriptive accounts of the military as an institution cannot suffice if the objective is to understand how military purposes and ideals are normalized and placed beyond question. Uncovering and understanding those nuances becomes a critical step for identifying possible pathways toward institutional change.

Both displays and representations of gender that reside within institutionally approved and prescribed structures "are profoundly influential on wider social understandings of what gender is" (Woodward and Duncanson 2017, 3), deeply reverberating outside the confines of the typically imagined institutional apparatus. How and what the institution communicates outwardly about its views on gender through presentations deemed acceptable and appropriate are particularly critical in non-conscription based militaries; the volunteer-soldier model of military service requires

that the military and society co-exist in an intimate relationship where one is reliant on the other for its existence and maintenance. In so doing, an era of growing sexual and gender equality forces the institution to consider what its relationship with masculinity is and how it positions its relationship with masculinity to the general public from which it draws individuals to serve. As such, "militaries [are] sources for understanding gender" (Woodward and Duncanson 2017, 3) and gendered power.

In exploring the possibility of the contemporary U.S. military's gender regime shifting away from promoting heteromale privilege, the goal of this book is twofold: to trace and analyze the nuances of gendered and gendering processes within the military, and to investigate how those are communicated by and through the institution with material which are, at least in part, created and used for the general public's consumption. Situated in postpositivist presuppositions associated with knowledge production,[9] it does so through a feminist interpretivist analysis, providing a "strategy to make gender visible" (Woodward and Duncanson 2017, 5). Feminist interpretivist methods illuminate gender and gendered relations where they might otherwise be overlooked and reveal how gender may be manipulated by institutions to achieve certain goals.

More specifically, the interpretivist methods deployed by the book[10] analyze visual, textual, archival, and cultural materials[11] disseminated and accessible to the public at large; they communicate the military's views on gender broadly, and masculinity specifically. In so doing, the military's cultural production—its words, arguments, justifications, and visual messages—is used to explore dimensions of meaning that exceed the intentions of individual speakers. Rather than taking words or images at face value, it explores tensions, ambiguities, inconsistencies, contexts, and subtexts that affect the messages conveyed. By using gender as an analytical category and tool for inquiry, the book brings into view not only the formal structures that produce inequalities, but the "rules, procedures, discourses, and practices" (Lovenduski 2005, 147) that contribute, maintain, or (re)produce those inequalities. As the forthcoming chapters convey, forms of masculinity dominate not only at the individual level, but as a prevailing gender (power) regime, an underlying logic that sustains judgments of what is normal, natural, and permissible. As a result, investigating the images and narratives the U.S. military circulates to explain its changing gender(ed) policies, it becomes possible to test claims about the "degendering" and "regendering" of the apparatus.

The messages that the military conveys about gender are complex, and at times, contradictory. In certain instances, these complexities are related to competing views and policies of various military organizations (Air Force, Army, Marines, Navy, Joint Chiefs of Staff, and the Department of Defense) that are contributing to emerging discourses. At other times, the contradictions reflect gaps between explicit commitments and actual practices. And often, these tensions are grounded in the clash of multiple forms of militarized masculinities *and* femininities vying for ascendancy, which challenge the contemporary form of hegemonic military masculinity.

As a result, readers should navigate the chapters of the book with a "feminist curiosity" (Enloe 2004, 220) searching underneath the surface of military messages for systematic deception or manipulation. As the book demonstrates, the military actively works to construct public discourses that favor and legitimate particular versions of masculinity—the male protector of hearth, home, and nation—and the feminine correlates—the protected wife, mother, daughter—that percolate from military ranks to the larger society. At times, the military as a whole uses those discourses to promote a hegemonic version of militarized masculinity—that is, in Western contexts, the ideal form of masculinity produced through (gendered) practices of military apparatuses, and primarily embodied by the white, heterosexual, able-bodied, male soldier—and at times different branches of the military purposefully promote varying versions of militarized masculinity (see Chap. 6 on social media military advertising). By insinuating that women, lesbian, gay, bisexual, and transgender troops are subpar or barriers to the efficient operations of the armed services, the military may practice a form of exclusion that relies upon selective, deliberate, and carefully orchestrated patterns of *inclusion*, complicating liberal feminist's "right to fight" rationale (Kennedy-Pipe in Woodward and Duncanson 2017). Those who create gender trouble for the established regime are allowed to serve but not on equal terms. Then, quite perversely, unequal performance is attributed to defects in those who diverge from hegemonic norms as the unequal conditions in the terms of their service are masked.

Whichever the case, these discursive maneuvers—which the book refers to as *gendered mechanisms of exclusion*—demonstrate how the cementing of certain masculinized narratives dialectically works to erase women's roles in the military from public consciousness, rendering military service performed by women invisible. At other times, discourses do note women's presence but simultaneously suggest feminine weakness, lack of ability, or lack of courage, thereby shoring up the image of heteromale service

members as the nation's ultimate defenders, reliable and trustworthy warriors who have earned a privileged status among citizens.

Ultimately, the book is interested in identifying "mechanisms for organizational transformation" (Carreiras in Woodward and Duncanson 2017, 115), which scholars of the feminist institutionalism paradigm have noted as one fruitful direction for future research. Doing so, however, first requires uncovering: if discourses produced by the military (and other gendered institutions, generally) actively work to reproduce inequalities within the organization despite external pressures that try to push it toward change (Britton and Logan 2008), exposing the processes through which the institution masks its work on reifying and maintaining inequality becomes critical for identifying pathways toward lasting change.

A Note on Butler and the Use of the Term "Gender Trouble"

The use of the term "gender trouble" up to this point in the introduction has likely conjured up Butlerian understandings of the term for the reader. It should, at least in part. To make clear, it is not gender itself—that is the stability of gender as a category, as Butler writes—that is being troubled in the cases contained in this book. Rather, it is the institution's gender regime of heteronormative masculinity that is being troubled through "troublemakers" whose performances reside outside the parameters of the institution's stipulated and promoted gender regime. I do, however, use the term deliberately to invoke Butler's (1990) theory of gender performativity and (re)production. While the military has historically been isolated from the pressures to change its gender regime, viewing gender through the lens of performativity provides "the ever-present possibility of military transformations and the inherent instability within militaries' gendered cultures and structures" (Woodward and Duncanson 2017, 4). More specifically, performativity positions gender as "never stable and always relational," (4) making transformation "inevitable" (4); it remains "open to any manipulation and alteration by individual agency, it could be done, undone, and maintained through daily interaction" (Yildirim et al. 2018, 664). Consequently, for the purposes of this study specifically, "gender trouble" refers to instances of gender nonconformity that contest the institutional gender regime,

exposing its weaknesses and vulnerabilities, threatening to undermine it or destabilize it. Although obedient displays of gender help to reinforce the gendered practices of the institution, "gender diversity and deviance challenges it" (Lorber 2005, 17). As such, the cases chosen for the book are those which I argue pose as challenges to the military's ability to maintain the gender regime of heteronormative warrior masculinity.

OTHER KEY TERMS OF THE BOOK

Gender Regime

Following R. W. Connell (1987), the book refers to the term *gender regime* as a system of principles and practices that determine how gender relations "play out" within an institution. As a social structure, gender relations materialize through an assemblage of operational policies and procedures, as well as cultural customs of the institution which, as I argue, establish and enable social hierarchies of privilege among groups of individuals associated with the organization (I return to this point below, and throughout the book, as it is an important one). Though these hierarchies generate "different and unequal masculinities" they are also "always defined in relation...to women" (Cockburn 2001, 16).[12]

Importantly, in the U.S. military—though, arguably, in the militaries of other contemporary Western democracies too—gender relations have habitually "played out" through the use of gender as an organizing tool, creating a bifurcated division and understanding of men and women, placing them in distinct and opposite categories (Lorber 2005). The result is a systemic regime of practices that attempt to neatly assign individuals to one of the two categories, "mak[ing] one category of people subordinate to the other" (11). In turn, this social bifurcation provides organizations with mechanisms for ordering and arranging its members, (socially) constructing and defining appropriate and inappropriate roles and behavior around gender difference. As Butler (1990) reminds us, both the possibility of maintaining and changing the regime rests on the ongoing, seemingly quotidian acts of gender (re)production and gendered relations between individuals within or outside of the parameters provided by the institution (see also Lorber 2000, 2005).

Most prominently, military policies and practices have used gender to organize groups into included and excluded categories, defining hard and fast boundaries of who is "in" and who is "out" based on biological understandings of gendered bodies' capabilities and capacities. Part and parcel to this is a belief that women are inherently more pacifistic due to biological functions, and are therefore inappropriate—or, at the very least, less appropriate—for military service (Goldstein 2001). Over time, these models of inclusion and exclusion have cemented and entrenched institutionally deemed acceptable/unacceptable corresponding versions of masculinities and counter femininities, gendered scripts which have literally determined which bodies are deemed (officially) eligible to serve in the military and in what forms. Among these are assumed characteristics of toughness/weakness, aggression/sensitivity, violence/nurturance, and domination/submission. In short, military gender regimes "shape their members' behaviours through the construction and reproduction of norms, and the development of rules and policies governing individual activities" (Woodward and Duncanson 2017, 2).

Besides, the prescribed parameters provided by military models of (gender) inclusion and exclusion can be found in nearly all core aspects of military life, from where bodies are allowed/not allowed to reside on military bases during boot camp and official service, setting pack load weight standards, deciding which sailors can or cannot serve on combat ships or submarines, or how women will deal with menstruation in remote combat environments. They are also found in policies related to where bodies are allowed/not allowed to be put to rest once they are deceased, thereby determining who is memorialized, valorized, and remembered. As the following chapter demonstrates, the binary organization of military bodies, through both formal institutional policies and socially constructed narratives, is one of the most fundamental characteristics associated with defining the military as a gendered institution.

Importantly, the impact of the tenets of the military's gender regime are felt not only by the individuals traditionally associated with the institution—members of the armed forces serving in the military—but also individuals "external" to the organization. However, any existing line between "internal" and "external" is often undefined and blurred, particularly in non-conscription based military models where the official institution relies on tactics that incentivize service for the public citizenry. Feminist scholars remind us that the tools of militarism and the militarization of nations often take place

outside of the four-wall confines of the official military apparatus, embedding themselves in societal values and mores; their ultimate power becomes most apparent when no longer visible at all, when citizens validate militaristic ideals without recognition, realization, or question. In thinking about institutional regimes then, the power of gender resides in the ability to dictate cultural scripts around acceptable versions of masculinity and femininity, both within the institution and the broader societies in which they exist. As the book argues, for the contemporary U.S. military, the accepted—indeed, heralded and valorized—performances of gender are those which adhere to a specific form of masculinity, that which is embodied by the white, heterosexual, heroic male warrior. I provide a review of some of these key rules and practices in the following chapter.

When bodies cannot or choose not to conform to the institutionally prescribed gender script, when they act outside of the allotted parameters and boundaries provided to them, their acts highlight the institutional regime's inability to contain those bodies. Doing so showcases the very vulnerability of both hegemonic militarized masculinity—and other forms of militarized masculinity—"illuminat[ing] the instability of the binary distinction" (Clark 2018, 1) that produced them in the first place. And while the military has historically been isolated from vectors that push it toward transformation, institutional anxieties proliferate; the inability to contain bodies within the limitations of the binary exacerbates the possibility of impending change (Woodward and Duncanson 2017).

Degendering and Regendering

What would be required for these "deviant" acts to push the institution to the brink of degendering its regime? The inclusionary/exclusionary model that I have outlined above may have us believe that the simple solution to degendering would be to remove barriers of discrimination, "adding" women where they have been "missing." Perhaps rightly so; the deliberate and systemic erasure of women—along with the erasure of their contributions—from histories the military has written remains a critically consistent attribute of gendered institutions specifically, and patriarchal systems more generally. But this somewhat simplistic institutionalist "resolution" misses important controversies both feminist IR theory and

postmodern feminist scholarship[13] more generally illuminate in relation to militarization and militarism. I return to a more detailed discussion of this in the following chapter, but it is worth noting that the most prominent of those controversies are important normative debates surrounding militarism and whether women's inclusion in military organizations is desirable in the first place (see Kennedy-Pipe 2017; Goldstein 2001; Duncanson 2013; Duncanson in Woodward and Duncanson 2017; Kronsell 2012).

Differently, for social constructionists, degendering refers to shifts that do "away with the binary gender divisions all together" (Lorber 2000, 83), and which entail finding ways for "render[ing] gender irrelevant" (Lorber 2000, 81). This may be accomplished, for example, by undercutting the structure of the gender binary—that is predicated on difference—by finding similarities between the two defined genders (see Sasson-Levy and Amram-Katz 2007). Alternatively, social constructionists assert that gender could also be rendered impertinent through a willingness on the military's part—seen thus far only to a limited degree—to talk about gender in ways that recognize and acknowledge soldiers' lived experiences prior to, during, and after military service as being intersected by a multitude of identity markers. In short, the work of degendering would at least, in part, be achieved through discursive tactics whereby discourses would "remove gender" in an effort that may "undermine[…] the role of gender in systems of power" (Berns 2001, 265).

For the purposes of this study, *degendering* refers to a shift away from, or elimination of, heteromale privilege as the defining structure for gendered relations in the military, and through which the warrior soldier archetype—"a key symbol of masculinity" (Morgan 1994, 165)—is dismantled. This may be accomplished, for example, through changes to policies, practices, and operations that remove gender as an organizing principle, or that blur divisions enabled by the binary. It may also be achieved by recognizing multiple genders, ultimately undermining the binary model of gendered relations and hegemonic militarized masculinity. As a result, any policy changes that result in including formerly excluded groups from serving in specific positions, or from service all together, may suggest a shift because it may, in theory, reconfigure the imaginary of how gender defines parameters for service. For example, the repeals of both "DADT," and the ban against women in combat may signal openings toward change. Indeed, open homosexuality and women's

full inclusion in the most coveted roles of the military system may destabilize (hegemonic) militarized masculinities by expanding military identities to encompass multiple gendered subjects.

Conversely, *regendering* refers to discursive mechanisms that reassert the centrality of gender for military institutional operations through the (heteronormative) binary division between masculinity and femininity, and as a result, reifies the warrior soldier at the pinnacle of the hierarchy of the military social order. Institutional regendering is often subtle and elusive; unlike degendering it may be thought of as "a more sophisticated and hidden process, harder to expose and change" (Sasson-Levy and Amram-Katz 2007, 2). It is also frequently complex and contradictory, making the task of identifying it as such far from straightforward. How can the military's implementation of more seemingly inclusive policies actually put military masculinity back at the fore of its institutional identity? As an example, the repeal of the combat ban would, on the surface, suggest an opportunity for women to be seen as equals to their male counterparts. But viewed differently, lifting the ban under the pretense that women be allowed to serve in the most coveted positions only if they meet standards previously set for those positions by the institution, standards set based on the expectations of "male bodies," reinforces the quintessential form of militarized masculinity. As such, the institution forces formerly excluded bodies to perform forms of masculinity, or meet archetypes of masculinity, that are institutionally valorized if they want access to being included in the most coveted positions of the military. In so doing, the inclusion of those bodies does not necessarily degenderize the institution. Rather, it reinforces previously established practices and structures predicated on masculinity. In other words, it regenders it.

Military officials often use discursive mechanisms to conceal their efforts toward institutional regendering; the power of multiplicity in these discursive maneuvers allows for a masking of the (re)confirmation of warrior masculinity as the hallmark of institutional identity through the concurrent presentation of superficially labeled gender-inclusive policies that seem to advance the cause toward gender equality or parity. At times, these discursive measures act to reframe policy debates and discussions toward narratives that have everything to do with gender. At other times, discourses completely ignore gender and instead resort to seemingly irrelevant or misplaced narratives, namely through rhetoric that is associated with securitization, medicalization, and even pathologization (see Chaps. 3 and 4).

A Note on Scope

What change would look like, and whether it can be achieved, is not necessarily something that feminist institutionalist and feminist IR scholars see eye-to-eye on. Military and defense institutions necessarily change over time—including changing policies that regulate how gender is "acted out" within the institution—but Joan Acker (1990) insists that the degree to which gender regimes can change is ultimately limited: "women's bodies cannot be adapted to hegemonic masculinity; to function at the top of male hierarchies requires that women render irrelevant everything that makes them women" (153). Differently, feminist IR scholars urge us to consider how hallmarks of military gender regimes, namely the formulation, ascendancy, and maintenance of particular militarized masculine archetypes are as "diverse and changing rather than as monolith and static—and as context specific rather than universally the same" (Eichler 2014, 90). Indeed, feminist scholarship on the military views "the ever-present possibility of military transformations and the inherent instability within militaries' gendered cultures and structures" (Woodward and Duncanson 2017, 4) as a productive vehicle for understanding productions of masculinity and femininity in larger societal contexts.

Nevertheless, any form of hegemonic militarized masculinity, or the prominence of multiple forms of militarized masculinities at any point in time, should not necessarily be automatically equated with the institution's gender regime. Instead, militarized masculinities and femininities may be viewed as one of many devices at the institution's disposal that may mirror and help support the overarching governing principle(s) of the gender regime. In the case of the contemporary U.S. military, the gender regime promotes heteromale privilege, and does so, in part, by valorizing certain forms of militarized masculinity, while simultaneously devaluing others.

Change, therefore, may come through the understanding of how processes of gendering occur within the military, specifically how masculinities and femininities are militarized. Pursuing this as a potential channel for change makes the elimination of personnel restrictions and policy changes such as Don't Ask Don't Tell, and the lifting of the combat ban on women important, not because it (officially) "adds" women where their presence was previously limited and denied, but because these policy changes may disrupt and contest "societal understandings of men's soldiering" (Eichler 2014, 90). In other words, these—and other—policy changes may provide a way toward disrupting, destabilizing, and ultimately, undoing the "link between militarism and masculinity" (Eichler 2014, 82).[14]

I return to the question of just what might be required to degender the contemporary U.S. military in the concluding chapter, but for now, the instances of gender trouble raised by the book are "trouble" for this very reason—they threaten to disrupt the link between militarism and masculinity. As a result, it is important to consider that the military may actively work to reestablish that gendered link when facing challenges to its gender regime, whether they be in the form of policy changes or otherwise. While the military has traditionally been thought of as well isolated from calls for change, prevalent or prominent form(s) of militarized masculinity should not be thought of as fixed or everlasting. Militarized masculinities have evolved over time, with some forms experiencing less prominence, allowing other forms to ascend to hegemonic status. Investigations of militarized masculinity/masculinities should, therefore, recognize complexities with their temporal and cultural scopes as they "become contested, redefined, or reaffirmed" (Eichler 2014, 84).

This book specifically investigates the possibility of changes to the military's gender regime—and with that, most coveted forms of militarized masculinity—contemporarily, that is starting from the abolition of the draft and with the subsequent creation of the All-Volunteer Force (AVF). Until the creation of the AVF, mandatory conscription-based military service "entrenched 'the male citizen soldier' version of militarized masculinity for over 200 years" (Eichler 2014, 84). As Chap. 3 showcases in detail, the removal of mandated military service (on men), and the creation of a volunteer-based public military in light of atrocities during the Vietnam War, ushered in an era of higher reliance on women's service. In so doing, the creation of the AVF marks an important moment in the history of the U.S. military for the possibility of shifting the forms of militarized masculinities and femininities approved by the military for its recruitment goals. An increased reliance on women to meet personnel targets opened up the possibility for "weakening the link between masculinity and the military" (Eichler 2014, 84). As such, the study presented begins there temporally, and covers nearly five decades of analysis to include the possibility of institutional change as a result of the eventual lifting of the combat ban on women.

The cases chosen for the book tackle gender trouble in relation to three critical and fundamental components of the warrior archetype's conceptualization—embodied as "heterosexual, white, of unrivalled physical and mental constitution, sexually potent, morally and nationally superior, and surpassing non-warriors in most respects" (Howard and Prividera 2015, 222), and which maintained hegemonic status for much of this period:

(1) notions of warriorhood through service that allows for the soldier's perpetuation of violence against a pre-defined enemy,
(2) service in environments which provide for the risk of bodily harm, mutilation, or death of the warrior through the theater of war (combat service),
(3) the construction of social memories associated with the masculine warrior's service through specific forms of valorization, memorialization, and remembrance.

Arguably, the military has faced true challenges to its gendered conceptualizations of masculinity in these core features of its hegemonic archetype; the increased presence of women in the military and increased proximity to war theaters in which service members are in combat front-line positions (Chap. 3), the engagement in the war theater being one that promotes the possibility for physical harm, wounding, and maiming (to the body, not the mind), if not death (Chap. 4), and the ability to valorize and memorialize that service, enabling a social memory of specific war stories to be told and generated among the general public, upholding and allowing for the continued prominence of the archetype (Chap. 5).

A true shift in the military's gender regime would produce a degendering of these core aspects of its most valorized form of militarized masculinity, allowing the essential archetype to be reimagined. As I have argued above and elsewhere, among the core aspects of the contemporary U.S. military's regime during this period of time are those which relate to combat service, harm and wounding, and the valorization and memorialization of service members. The cases covered in the book highlight the importance of narratives associated with these factors for the maintenance of the gender regime, and how they have been contemporarily challenged. Consequently, the first three empirical chapters of the book (Chaps. 3, 4, and 5) refute the assertion that contemporary policy debates, changes, and implementations have produced degendering of the U.S. military or shifts in the gender regime. Instead, these chapters should be read as examples of what degendering *is not* and how feminist qualitative analysis can be particularly useful in uncovering hidden regendering efforts.

Focusing on the military's regendering efforts also foregrounds the military's gender regime as being both heteronormative and racialized; its discursive production of regendering in moments of anxiety illustrates how the institution blatantly disregards diverse understandings of the lived experiences of its service members. Much of the material researched and analyzed for this project—again, the institution's *self-representations*, that

is materials that the institution uses to communicate its views on gender to the public—view and position "women" as a monolith, a group in which the intersectionality of lived experiences is generally not recognized, acknowledged, or accounted for. At times, the military's monolithic reference to, and representation of, women is based on discourses that revert them back to biological assumed capacities and capabilities, and at other times is based on socially constructed roles and behaviors deemed acceptable/unacceptable for women. As such, the book proceeds with an intersectional "analytical sensibility" (Cho et al. 2013, 795) but does not present a formal intersectional analysis. Racial "diversity," for example, is addressed to a limited degree in Chaps. 5 and 6 of the book, primarily asserting the continued prominence of the white male soldier as the hallmark of military masculinity, and consequently, service. Similarly, ongoing institutional anxieties associated with gender and displays of homosexual relationships by or between service members, and the fear of how those gender performances may destabilize coveted forms of militarized masculinities, can be found in Chaps. 5 and 6.

The systematic lack of acknowledgment for diverse lived experiences of service members beyond the completely oversimplified gender binary is perhaps not surprising. A true willingness on the part of the military to understand gender in ways that recognize and acknowledge soldiers' lives as being intersected by a multitude of identity markers would work toward degendering the institution. But as I suggest in the chapters that follow, the military has yet to show a concerted interest in doing so, and when it has, it has been in a highly limited fashion.[15]

THE MILITARY'S TROUBLING RESPONSE TO GENDER TROUBLE

Nonetheless, when confronting threats to its established gender regime, the military has many options. Thus, the response to particular instances of gender trouble is not to be thought of as foreordained. The following chapters do, however, illuminate certain patterns of responses to gender trouble over the past five decades, patterns which demonstrate the use of (gendered) mechanisms that position women as incompatible, or only selectively compatible, with the military apparatus, at best. Recurring military discourses render women's presence in the military invisible or characterize their contributions as negligible, thereby erasing

the very presence of women in the armed services from public consciousness. Other discourses acknowledge women's presence but frame their performance in negative terms, exaggerating "feminine weakness," lack of ability, or indeed, lack of courage, thereby shoring up the image of heteromale service members as the nation's ultimate defenders, reliable and trustworthy warriors—and as such, its best citizens.

Chapter 2, "The Masculine Warrior: Militarized Masculinities and Gender Regimes," explores critical feminist scholarship related to processes of militarism and militarization. It explores the pertinent role these processes play in the military's gender regime of heteromale privilege through their maintenance of the masculine warrior archetype. To examine whether the instances of gender trouble presented by the book work to destabilize the military's gender regime toward "change," the chapter first marks a baseline for comparison by providing a brief history of women's participation in the military to dispel the notion that the military is an exclusively male institution. Otherwise, any policy or process change that increases women's participation would deceptively be considered progress toward equality.

Instead, the chapter widens the scope of the military as a gendered institution to encompass other factors that uphold a gender regime that systemically promotes heteromale privilege. In doing so, it illustrates concerted efforts to shore up the image of the masculine warrior through (feminist understandings of) the subtle tools of militarism and militarization—that is, social processes that mutually link and reinforce the centrality of the military as an institution through which certain forms of masculinity are obtainable and achievable, and which are recognized as having a dominating influence on social, political, and economic affairs. Doing so, arguably, allows the institution to retain its ability to execute state policies through the organized use of force (war) without major contestation by way of the normalization around military values, ideology, and overall culture. The conversation around how these processes ground institutional discursive maneuvers on gender and the masculine warrior archetype provides a foundation for an examination of regendering efforts in the empirical chapters that follow thereafter.

Chapter 3, "The All-Volunteer Force: Patrolling Gendered Boundaries Through the Combat Ban," examines the paradox of "exclusionary inclusion" in the context of changing recruitment practices in the aftermath of the Vietnam War during the military's shift from conscription to the creation of an All-Volunteer Force. Occurring at the height of feminist activism in

the United States, this transformation coincided with considerable pressure on the military to degender military recruitment, opening all positions to women as well as men. Officially, military leaders pledged to make "incremental and deliberate" progress in the recruitment and inclusion of women, yet they also insisted that the combat ban be preserved, thereby curtailing women's access to the front—a precondition for promotion and career advancement—for nearly 50 additional years. I examine military officials' testimonies in Congressional hearings held between 1970 and 2015 to explore how "gender integration" could be pursued and justified on markedly unequal terms. I trace how the arguments to bar women from combat shifted over the course of five decades, appealing to biological, psychological, and social grounds for women's exclusion. Even as the grounds varied, the military cultivated anxieties about women's bodies and their fitness for combat duty—suggesting that women's military service was incompatible with the demands of national defense.

In advancing these exclusionary claims under the rubric of gender integration, military leaders shifted the terms of debate away from gender equality and evidence concerning women's abilities and aspirations to the military's "assessment of its needs," which did not require any evidentiary basis (Brown 2012). By using the language of gender integration, while simultaneously banning women from combat, the military limited the meaning of full integration. They assigned women to a narrow range of technologically advanced roles where "manpower" was insufficient (Stachowitsch 2012), limiting women's access to higher ranks through which they could attain leadership positions that afford power, respect, and prestige. The combat ban remained active until January 2016 despite intensive efforts by some political officials and diverse interest groups who sought a far more egalitarian approach to gender integration.

Chapter 4 explores how military discourses about war wounds work to exclude sexual violence from the theater of war. "Violated Bodies: Combat Injuries and Sexual Assault" takes recent attention to the high incidence of sexual assault within the military as its point of departure. It situates rape within a larger discussion about forms of violence committed on (gendered) bodies of soldiers, and examines how the U.S. military analyzes specific injuries to accredit "real war wounds" while marginalizing others. I contrast the military's treatment of wounds sustained in tours of combat with wounds caused by solider-on-soldier sexual assault. Whether those incidents of sexual violence occur in war zones or on military bases, neither the physi-

cal nor the psychological harms inflicted conform to notions of "military wounds." To probe this gendered hierarchy of harm, I analyze visual representations of sexual assault and combat injuries depicted in several recent documentaries, including the 2012 award-winning film, *The Invisible War*. My analysis shows how military discourses conjure specific gendered images of soldiering, which mask and silence sexual assault survivors. Instead of recognizing the injuries and trauma associated with rape, military officials engage in a sleight of hand, positioning the organization as a victim. By deploying a combination of securitization and medicalization discourses, military spokespersons frame the rape "epidemic" as a "cancer or plague" that threatens the very viability of the military as an institution. Heightened concern about the negative effects of rape on the military coexists with rampant unconcern about the well-being of soldiers raped while on duty.

Chapter 5, "Military Museums and Memorial Sites: Disappearing Women in the Military," analyzes war memorials, a common way that citizens in the United States consume the self-representations—images and narratives—produced and disseminated by the military. Through a study of military museums, most notably the *USS Midway* Museum in San Diego, California, and a "Segs4Vets" veteran's ceremony that occurred on the deck of the *Midway*, I provide a semiotic reading of military memorials that illuminates how gendered dimensions of public memory are produced and sustained. Although women have served in the Navy since 1908, and on naval ships since the 1970s, the exhibits on the *USS Midway* omit any reference to women's service. The public is invited to imagine a naval history completely devoid of women. By excluding women's military service from exhibits and remembrance ceremonies, military museums erase decades of women's participation in war and reify national defense as a thoroughly masculine domain. In this way, military museums contribute to specific notions of citizenship embedded in complex class, gender, and power relations. Structuring public memory in ways that "determine what is remembered (or forgotten), by whom, and for what end" (Gillis 1994, 3), these narratives of valor contribute to understandings of national belonging, linking membership to service in wars and subtly encoding citizenship as a male privilege.

In Chap. 6, "Gender and Military Recruitment Since the Lifting of the Combat Ban," I investigate whether military marketing and recruitment campaigns have become more inclusive in response to gender trouble over the past few years. The chapter provides a comparative analysis of online recruitment efforts in the four branches of the military between 2013 and 2015. Building

on Melissa T. Brown's (2012) analysis of print recruitment materials for the AVF, I examine how the gendered narratives of recruitment material have changed in light of the announcement in February 2013 of women's official inclusion in combat-related positions. Over the past 15 years, the military branches have increasingly utilized social media platforms to meet their recruitment goals. I explore both textual and visual self-representations of military service in the official recruitment Facebook pages of the Air Force, Army, Marines, and Navy, tracing key themes and modal tropes since the announcement of the end of the combat ban on women. In total, I categorized over 1000 Facebook posts to reveal trends across each individual branch, and across the military as a whole. My research demonstrates that women continue to be marginalized in online recruitment materials, frequently depicted in stereotypical roles or settings. Although the four branches vary in their online representations of women, they remain wedded to an overarching depiction of militarized masculinity in their recruitment materials—and a racialized depiction that foregrounds white men.

In the final chapter, I return to "The Challenge of Degendering the Military." I contrast the mechanisms of exclusion documented in the military's self-representations over the past five decades with the types of changes that would be required to recognize women's service and to integrate women fully in military operations. I begin by considering technological developments, such as the prominence of drone aircraft and other "unmanned" aerial strategies, which suggest forms of service no longer predicated upon military masculinity. Recourse to machines and robots removes the centrality of physical strength from warfare and raises the possibility that the military might move away from privileging men. In principle, high-tech modes of engagement should eliminate barriers for women that have been grounded in stereotypes about women's "weakness." Yet, my examination of the military's self-representations suggests that technological developments alone will not degender the military. Far more is at stake in the gender dynamics of the military than empirical questions about physical strength. The military's responses to women's continuing struggle to attain equal opportunities and equal treatment within the armed services suggest that far more must be changed to construct the U.S. military as a gender-inclusive organization. Despite the deployment of claims about gender integration, the U.S. military has incorporated women on markedly unequal terms. In the final chapter, I

identify the kinds of institutional transformation that would be required to "degender" the U.S. military.

Notes

1. According to the report, gender-mixed units performed better than all-male groups in two events, firing "hit & near miss with the M4" (Department of Defense 2015, 3) rifle, and firing accuracy with the M2 rifle.
2. On January 4, 2019, *Marine Corps Times* reported that for the first time in history, male and female recruits would be integrated in boot camp training, a temporary move necessitated "because the recruiting classes are typically much smaller in the winter months" (Snow 2019a quoting Military Official, 6). A Marine Official noted that the adjustment in boot camp strategy would be "look[ed] at how the company performs in this model as we continually evaluate how we make Marines" (Snow 2019a quoting Military Official, 4). The limited integration includes female and male recruits sharing living quarters, but continues the long-standing tradition of female recruits being led by female drill instructors in boot camp. The cohort completed boot camp on March 16th, 2019 (Snow 2019b).
3. The "DADT" policy was in effect from December 1993 until September 2011.
4. The term "gender trouble" is borrowed from Judith Butler's (1990) *Gender Trouble: Feminism and the Subversion of Identity* to refer to instances of gender performativity and production in the context of the U.S. military.
5. Though repealed by, at the time, President Barack Obama, the ban was reinstated by President Donald Trump in August of 2017 and portions went into effect in March 2018. The renewed ban under President Donald Trump initially banned all transgender individuals from serving openly in the military, then reversed course after federal courts stopped the policy, as written, from going into effect. Instead, the revised version of the policy declared "transgender persons with a history or diagnosis of gender dysphoria—individuals who the policies state may require substantial medical treatment, including medications and surgery—are disqualified from military service except under certain limited circumstances" (The White House 2018, 3). As of January 2019, in a 5–4 decision, the U.S. Supreme Court put several injunctions placed on the ban by the lower courts on hold, while fully maintaining one injunction. As of this writing, the Supreme Court had not yet decided whether or not it would directly hear a case on the policy.
6. As outlined later in this introduction, Acker (1990) provides five mechanisms of "gendering," that is processes through which gendering occurs.

Her mechanisms view gender as a "constitutive element" (147) in which construction, revision, and change is constant. The constant ebb and flow associated with this understanding of gender suggests the possibility of not only processes of gendering, but also degendering and regendering.

7. I use the term "institution" and "organization" somewhat interchangeably throughout the book. With indication of doing so, I also use the term "institution" to refer to the individual branches of the military, particularly in Chap. 6.
8. For exemplary examples of analyzing/reading discourses "against the grain" see Ferguson and Turnbull (1999) and Faust (2008). What is important for the book here is not necessarily the use of gender analysis to determine whether those presentations of gender are deemed appropriate/ inappropriate to the institution's ideal form of masculinity (sex gender paradox Butler 1990; Benhabib et al. 1995; Riley 1988; Scott in Ferree et al. 2000; Stern and Zalewski 2009), but rather to identify cases that the institution does deem inappropriate, cases that it may view as destabilizing/challenge the institution's ideas about ideal masculinity, and then to conduct gender analysis to investigate how it deals with those instances. Doing so, as the book argues, provides a more detailed and nuanced account of gendering processes of the institution, and a contemporary understanding of the relationship between masculinity and military service in the context of the United States.
9. For postpositivist approaches to the study of war and gender, and IR more generally, gender is more than a mere variable to be controlled for when testing universal generalizations. Instead, gender is viewed as both relational and as a system of power. Among the advantages of this understanding of gender for this project specifically is the use of postpositivist methods such as interpretivist tools, that allow for broader conceptualizations of militarization than those provided by mainstream IR.
10. These methods included historical analysis, discourse analysis, thick description, semiotics, and first-person observations.
11. These materials include Congressional hearings, documentary movies, museum and memorial sites, or recruitment marketing campaigns on social media platforms. These are material that are accessible and often created for the general public's consumption, providing a more nuanced understanding of how the military may convey messages related to gender to this audience.
12. See also Pateman (1998).
13. Following Goldstein (2001), the agenda for a degendered military institution is likely to look very different depending on the strand of feminism one refers to. Goldstein provides three strands of feminist theorizing on war and peace: liberal feminism, difference feminism, and postmodern

feminism. Liberal feminism argues that women face sex-based discrimination in the military, and should be allowed to serve in all capacities, equally to their male counterparts, and can meet standards of warriorhood just like men. Liberal feminists have argued that any future draft apply equally to women as it would to men, and that women be required to register for the Selective Service. Liberal feminists have been criticized for promoting that women enact forms of militarized masculinity in order to serve 'equally' to male counterparts while not advocating that men serving in the military do the same. Doing so, critics of Liberal Feminism argue, works to "prop up a male-dominated world instead of transforming it" (41). Difference feminism echoes the sentiments of standpoint feminist theory, suggesting that there are fundamental differences between men and women. Difference feminists argue "feminine" traits associated women are devalued in sexist systems instead of "valuing, celebrating, and promoting them" (41). Difference feminists do not necessarily agree on whether these traits are biological or attained through social forces, but view these differences positively. In the context of war and peace, Difference feminists view women as more pacifist and men more violent and, therefore, more appropriate for military service and waging war. Lastly, postmodern feminism questions the category of gender—whether it produces similarities or differences—rejecting gender's production through biology and instead urging for understandings of gender as social constructions, fluid and ever changing. Postmodern feminists (which, as Golstein points out, may typically include poststructuralists, postpositivists, postbehaviorists, and constructivists) view gender as being "everywhere" and some argue that gender is mapped onto the bodies of individuals through a series of "binary oppositions" (49). These binary categories (see Goldstein 2001 chart on p. 49) become particularly helpful for postmodern feminists in explaining socially constructed notions of masculinities and femininities across time and space. Critics of postmodern feminism, however, importantly point out that the fluidity and diversity associated with women's experience as a result of gender being imagined as socially constructed limits the strands ability to advance arguments that would require "women" to be viewed as a "meaningful category" (51).

14. Rejecting traits thought to be biologically associated with men and women's bodies, feminist IR, as Maya Eichler (2014) points out, urges scholars to never assume the link between masculinities and militarism as foreordained. See Enloe (2000) and Whitworth (2004).
15. In the cases presented by this book, the best indication (and that too is very limited) of a military willingness to think about the lived experiences as intersectional can be found in Chap. 6 Recruiting the Military during a Time of Transition.

References

Acker, J. 1990. Hierarchies, Jobs, Bodies: A Theory of Gendered Organizations. *Gender and Society* 4 (2): 139–158.
Baldor, L. C. 2016. New Marine Corps Fitness Standards for Combat Weed Out Men, Women Alike. Accessed September 20, 2016. Militarytimes.com.
Belkin, A. 2012. *Bring Me Men: Military Masculinity and the Benign Façade of American Empire, 1898–2001*. New York: Columbia University Press.
Benhabib, S., J. Butler, D. Cornell, and N. Fraser. 1995. *Feminist Contentions: A Philosophical Exchange*. New York: Routledge.
Berns, N. 2001. Degendering the Problem and Gendering the Blame: Political Discourse on Women and Violence. *Gender and Society* 15 (2): 262–281.
Britton, D.M., and L. Logan. 2008. Gendered Organizations: Progress and Prospects. *Sociological Compass* 2 (1): 107–121.
Brown, M.T. 2012. *Enlisting Masculinity: Gender and Recruitment of the All-Volunteer Force*. Oxford: Oxford University Press.
Burke, C. 2002. Women and Militarism. Accessed February 20, 2006. Available at http://wilpf.smilla.li/wpcontent/uploads/2012/10/Unknownyear_Women_and_Militarism.pdf.
Butler, J. 1990. *Gender Trouble: Feminism and the Subversion of Identity*. New York: Routledge.
Carreiras, H. 2017. Gendered Organizational Dynamics in Military Contexts. In *The Palgrave International Handbook of Gender and the Military*, ed. Rachel Woodward and Claire Duncanson, 543–559. London: Palgrave Macmillan.
Carver, T. 2002. Discourse Analysis and the "Linguistic Turn." *European Political Science* 2 (1): 50–53.
Chappell, L., and G. Waylen. 2013. Gender and the Hidden Life of Institutions. *Public Administration* 91 (3): 559–615.
Cho, S., K. Crenshaw, and L. McCall. 2013. Toward a Field of Intersectionality Studies: Theory, Applications, and Praxis. *Signs* 38: 785–810.
Clark, L.C. 2018. Grim Reapers: Ghostly Narratives of Masculinity and Killing in Drone Warfare. *International Journal of Feminist Politics* 20 (4): 602–623.
Cockburn, C. 2001. The Gendered Dynamics of Armed Conflict and Political Violence. In *Victims, Perpetrators or Actors: Gender, Armed Conflict and Political Violence*, ed. C. Moser and F. Clark. London: Zed Books.
Cohn, C. 2000. 'How Can She Claim Equal Rights When She Doesn't Have to Do As Many Pushups As I Do?' The Framing of Men's Opposition to Women's Equality in the Military? *Men and Masculinities* 3 (2): 131–151.
Connell, R.W. 1987. *Gender and Power: Society, the Person, and Sexual Politics*. Stanford: Stanford University Press.
Department of Defense. 2015. Marine Corps Force Integration Plan Red Team: Final Report. Accessed February 12, 2019. https://dod.defense.gov/Portals/1/Documents/wisr-studies/USMC%20-%20Center%20for%20

Strategic%20and%20International%20Studies%20Red%20Team%20analysis%20of%20Marine%20Corps%20research%20and%20analysis%20on%20gender%20integrat-1.pdf.
Duncanson, C. 2013. *Forces for Good? Military Masculinities and Peacebuilding in Afghanistan and Iraq.* Basingstoke: Palgrave Macmillan.
———. 2017. Anti-militarist Approaches to Gender and the Military. In *Palgrave International Handbook of Gender and the Military*, ed. Rachel Woodward and Claire Duncanson. London: Palgrave Macmillan.
Eichler, M. 2014. Militarized Masculinities in International Relations. *Brown Journal of World Affairs* 21 (1): 81–93.
Enloe, C. 1983. *Does Khaki Become You? The Militarization of Women's Lives.* Boston: South End Press.
———. 1990. *Bananas, Beaches, and Bases: Making Feminist Sense of International Politics.* Berkeley and Los Angeles: University of California Press.
———. 2000. *Maneuvers: The International Politics of Militarizing Women's Lives.* Berkeley: University of California Press.
———. 2004. *The Curious Feminist: Searching for Women in a New Age of Empire.* Berkeley and Los Angeles: University of California Press.
Faust, D.G. 2008. *This Republic of Suffering: Death and the American Civil War.* New York: Vintage Books.
Ferguson, K., and P. Turnbull. 1999. *Oh, Say, Can You See? The Semiotics of the Military in Hawai'i.* Minnesota: University of Minnesota Press.
Ferree, M.M., J. Lorber, and B.B. Hess. 2000. *Revisioning Gender (Gender Lens).* New York, NY: AltaMira Press.
Francke, L.B. 1997. *Ground Zero: The Gender Wars in the Military.* New York: Simon and Schuster.
Gillis, J.R. 1994. Memory and Identity: The History of a Relationship. In *Commemorations: The Politics of National Identity*, ed. John R. Gillis, 3–25. Princeton: Princeton University Press.
Goldstein, J. 2001. *War and Gender: How Gender Shapes the War System and Vice Versa.* Cambridge: Cambridge University Press.
Hawkesworth, M. 2012. *Political Worlds of Women: Activism, Advocacy, and Governance in the Twenty-First Century.* Boulder, CO: Westview Press.
———. 2018. *Political Worlds of Women: Activism, Advocacy, and Governance in the Twenty-First Century. Student Economy Edition.* Boulder, CO: Routledge.
Howard, J.W., III, and L.C. Prividera. 2015. Nationalism and Soldiers' Health: Media Framing of Soldiers' Returns from Deployments. In *A Communication Perspective on the Military*, ed. Erin Sahlstein Parcell and Lynne M. Webb, 217–276. New York: Peter Lang Publishing.
Kennedy-Pipe, K. 2017. Liberal Feminists, Militaries and War. In *The Palgrave International Handbook of Gender and the Military*, ed. Rachel Woodward and Claire Duncanson, 543–559. London: Palgrave Macmillan.

Kronsell, A. 2012. *Gender, Sex and the Postnational Defense: Militarism and Peacekeeping*. New York: Oxford University Press.

Levy, Y. 1997. How Militarization Drives Political Control of the Military: The Case of Israel. *Political Power and Social Theory* 11: 103–133.

Lorber, J. 2000. Using Gender to Undo Gender: A Feminist Degendering Movement. *Feminist Theory* 1 (1): 79–95.

———. 2005. *Breaking the Bowls: Degendering and Feminist Change*. New York: W. W. Norton & Company.

Lovenduski, J. 2005. *Feminizing Politics*. Cambridge: Polity Press.

Lutz, C. 2002. Making War at Home in the United States: Militarization and the Current Crisis. *American Anthropologist* 104 (2): 723–735.

MacKenzie, M. 2015. *Beyond the Band of Brothers*. Cambridge: Cambridge University Press.

Millar, K.M. 2017. Gendered Representations of Soldier Deaths. In *The Palgrave International Handbook of Gender and the Military*, ed. Rachel Woodward and Claire Duncanson, 543–559. London: Palgrave Macmillan.

Morgan, David H.J. 1994. Theater of War: Combat, the Military, and Masculinities. In *Theorizing Masculinities*, ed. Harry Brod and Michael Kaufman, 165. London: Sage.

Pateman, C. 1998. The Patriarchal Welfare State. In *Feminism, the Public and the Private*, ed. Joan Landes, 241–276. New York: Oxford University Press.

Riley, D. 1988. *'Am I That Name?' Feminism and the Category of 'Women' in History*. London: Palgrave Macmillan.

Rosenberg, M., and D. Phillips. 2015. All Combat Roles Now Open to Women, Defense Secretary Says. *The New York Times*. Accessed September 1, 2018. https://www.nytimes.com/2015/12/04/us/politics/combat-military-women-ash-carter.html.

Sasson-Levy, O., and S. Amram-Katz. 2007. Gender Integration in Israeli Officer Training: Degendering and Regendering the Military. *Signs* 33 (1): 105–133.

Scott, J.W. 2000. Some Reflections on Gender and Politics. In *Revisioning Gender*, ed. Myra Marx Ferree, Judith Lorber, and Beth B. Hess, 70–98. Oxford: AltaMira Press.

Segal, M.W. 1982. The Argument for Female Combatants. In *Female Soldiers – Combatants or Noncombatants?: Historical and Contemporary Perspectives*, 267–290. Westport: Greenwood Press.

———. 1995. Women's Military Roles Cross-Nationally: Past, Present, and Future. *Gender and Society* 9 (6): 757–775.

Snow, Shawn. 2019a. Male and Female Marine Platoons to Integrate at Recruit Training for the First Time. Accessed January 4, 2019. Marinetimes.com. https://www.marinecorpstimes.com/news/your-marine-corps/2019/01/04/male-and-female-marine-platoons-to-integrate-at-recruit-training-for-the-first-time/.

———. 2019b. Recruits with First Partially Gender-Integrated Marine Boot Camp Company Earn the Title Marine. Marinetimes.com. Accessed March 24, 2019. https://www.marinecorpstimes.com/news/your-marine-corps/2019/03/22/recruits-with-first-partially-gender-integrated-marine-boot-camp-company-earn-the-title-marine/.

Stachowitsch, S. 2012. Military Gender Integration and Foreign Policy in the United States: A Feminist International Relations Perspective. *Security Dialogue* 43 (4): 305–321.

Stern, M., and M. Zalewski. 2009. Feminist Fatigue(s): Reflections on Feminism and Familiar Fables of Militarisation. *Review of International Studies* 35 (3): 611–630.

The White House. 2018. Presidential Memorandum for the Secretary of Defense and the Secretary of Homeland Security Regarding Military Service by Transgender Individuals. Accessed March 24, 2019. https://www.whitehouse.gov/presidential-actions/presidential-memorandum-secretary-defense-secretary-homeland-security-regarding-military-service-transgender-individuals/.

Whitworth, Sandra. 2004. *Men, Militarism, and UN Peacekeeping*. Boulder, CO: Lynne Rienner Publishers, Inc.

Woodward, R., and C. Duncanson. 2017. An Introduction to Gender and the Military. In *The Palgrave International Handbook of Gender and the Military*, ed. Rachel Woodward and Claire Duncanson, 543–559. London: Palgrave Macmillan.

Yildirim, S., B. Ucaray-Mangitli, and Hakki Tas. 2018. Intimate Politics: Strategies and Practices of Female Mukhtars in Turkey. *British Journal of Middle Eastern Studies* 45 (5): 661–677.

CHAPTER 2

The Masculine Warrior: Militarized Masculinities and Gender Regimes

Introduction

The exploration of contemporary gender trouble as challenges to the institutionally established gender regime requires an historical perspective if the goal is to mark change and continuity. Specifically, it necessitates a baseline against which suggested and supposed change can be identified, situated, and scrutinized. One aspect of that baseline is the misperception that the military has been exclusively male; evidently, accepting that notion may prompt us to view any policy that increases the number of women serving in the military, or the capacities in which they can serve, as progressive and egalitarian. To avoid that inclination, this chapter begins with a brief history of women's service in the military. It then grounds the discursive mechanisms through which the military has been construed as an all-male organization despite the presence of women within the ranks. Following this, the next section expands the frame of gender trouble to include sexual orientation, tracing policies to exclude homosexuals from openly serving in the military, and which has cemented a gender regime that views heterosexuality as critical to its standard operating procedures. The final section of the chapter then considers the link between militarism and militarized masculinities, that is how militarized masculine archetypes, and gender regimes more broadly, become embedded in societies through the institution's projected messages on gender and military services. By examining the methods through which the military has sustained a gender regime of

heterosexual male dominance in the twentieth century, I lay a foundation for the analysis of more recent moves to counteract gender trouble in the contemporary era.

WOMEN IN AND FOR THE U.S. MILITARY

Although the U.S. military has historically often relied on publically depicting itself as an exclusively male institution, women have been instrumental in all wars conducted since the American Revolution (Kerber 1980). Women's participation has, however, been "built around understandings of appropriate gender roles" (Bailey 2009, 136) since the inception of the nation. During the Revolutionary War (1776–1783), the War of 1812, the Mexican War (1846–1848), and Civil War (1861–1865), women often disguised themselves as men to contribute to the war effort. Although women also served as nurses, cooks, saboteurs, and water bearers in these wars, their contributions were generally characterized as those of contract workers or volunteer services.

Women's military service was first officially recognized through the formation of the Army Nurse Corps in 1901 and the Navy Nurse Corps in 1908. Although women had served in various capacities and participated in activities related to Navy missions prior to that date, their service was not formally recognized by Congress until 1908. During the Civil War, for example, the Sisters of the Holy Cross treated wounded soldiers on the *USS Red Rover*, which became the Navy's first commissioned floating hospital. When Navy clerical workers were in short supply during the second decade of the twentieth century, the Navy began enlisting women specifically to "alleviate" the lack of personnel (Bailey 2009, 3).

During World War I, 21,480 nurses served at home and abroad, and 400 died in the war (Women in Military Service for America Memorial Foundation, Inc. 2014, 1). In accordance with prevailing notions about appropriate women's work, they served in caregiving capacities, taking care of wounded soldiers on the ground, in the air, as well as on battleships. In 1920, when the Army Reorganization Act was passed, military nurses were granted officer status with "relative rank" (Women in Military Service for America Memorial Foundation, Inc. 2014, 10), which ironically precluded them from having full rights and privileges. When women were demobilized after the end of World War I, the Army considered two proposals for women's service in the military. The first, the creation of a women's service corps, was instantly rejected. The second was a proposal put forth by Major Everett S. Hughes to fully integrate women into the Army, but it too was

rejected. As World War II broke out in Europe, and Americans weighed the consequences of intervention, the idea of a women's military corps emerged once more. As a result, the Women's Auxiliary Army Corps (WAAC) was established in May 1942. One year later, "auxiliary" was dropped from its title.

More than 150,000 women joined the Women's Army Corps by 1943, serving in various auxiliary and non-combat related positions. In addition, 60,000 army nurses served in military hospitals at home and overseas. Sixty-seven of these were captured as prisoners of war (POWs) in the Philippines and Japan and held captive for two and a half years. Overall, some 400,000 women served in the various branches of the military during World War II in non-combat positions.

Despite their valiant service in World War II, the full integration of the Women's Army Corps into the regular Army did not occur until 1948. According to military officials, the delay was due to women's official incorporation into the branch being seen as "a difficult sell" for the general public even though enlistment was "on a purely voluntary basis" and "no one was suggesting breeching the fundamental divide between men and women—that men fight and women don't" (Bailey 2009, 138). The military did, however, recognize the magnitude and importance of women's wartime service and proposed legislation to permit women to serve as regular and permanent members of the armed forces during periods of peace (Naval History and Heritage Command 2010, 4). Indeed, military officials urged Congress to recognize the utility of women in auxiliary positions both in periods of peace and war, arguing that women's service in a permanent capacity would assist with recruitment efforts for maintaining a standing military. Congress debated the legislation for two years (1946–1948), as many Congressmen and Senators remained unconvinced that women should have a permanent position in the armed forces (Frank 2013).

The passage of the 1948 Women's Armed Services Integration Act permitted women to serve in certain positions in the Air Force, Army, Marines, and Navy. The act, however, established a quota, restricting women's participation to 2% of the overall military force,[1] which approximated the number of women who had served during World War II. Women were excluded from Air Force and Navy carriers that were at risk of engaging in combat, and they were excluded from any position in which they might have command over men. The enactment of the 2% quota, along with other stipulations excluding women from combat, as well as the refusal to allow women to command men, appeared to act as an inclusive measure for the sake of having a prepared military for future wars, while simultaneously making women seem marginal to the military apparatus.

During the second half of the 1960s and the 1970s, women gained some ground within the military, although they remained restricted to non-combatant roles. The 2% restriction was, however, lifted in 1967 as the Vietnam War escalated. More than 11,000 military women were stationed in Vietnam during the conflict. Nearly all were volunteers; 90% served as military nurses, and the remainder worked as physicians, air traffic controllers, intelligence officers, clerks, and other positions in the U.S. Women's Army Corps, U.S. Navy, Air Force, Marines, and the Army Medical Specialist Corps. In 1972, at the same time that Congress debated the Equal Rights Amendment, the Navy initiated a pilot program to allow women to be posted on warships, specifically on the *USS Sanctuary*. The same year, both the Army and Navy Reserve Officer's Training Corps (ROTC) programs opened to women, as did all staff corps positions within the Navy. A year later, women were permitted to participate in aviation training, and in 1974, the first women graduated from Naval aviation training. Further training and academic inclusion came in 1975 when Congress authorized women to be admitted to military academies. The first cohort of women was admitted to the U.S. Naval Academy in 1976 (Naval History and Heritage Command 2010, 5). In the final years of the 1970s, the Navy—with approval from Congress—reinvigorated the combat exclusion by stipulating that women could only be assigned to non-combat ships.

Women made additional advances in the 1980s. The U.S. Naval Academy graduated its first female officer in 1980 (Naval History and Heritage Command 2010, 8). Women were selected for both the Air Force and Navy Test Pilot School, and qualified for jet training and as Enlisted Surface Warfare Specialists. Women became more visible on Navy ships; 193 women officers and 2185 sailors served on a total of 67 ships (8). In 1983, 37,000 women constituted 8% of the enlisted force of the Navy; the Department of Defense set a goal of increasing the number of women to serve in the Navy to 51,300 (9).

During the 1990s, combat aviation positions were opened to female aviators, the first woman sailor was permanently posted on a combat ship, and the combat vessel *USS Abraham Lincoln* was deployed to the Western Pacific for the first time with a mixed crew of men and women. Despite these openings, by 2014 women constituted only 15.1% (200,692) of the total Active Duty force of the military (United States Navy 2013), a percentage that has varied little since. Women comprised 18.8% of the 831,988 listed reservist and National Guard Troops. Although women

have been permitted to serve on submarines since January 2017, women have generally been assigned to fewer than 50% of the Navy's ships; only 135 ships had enlisted female sailors onboard by 2013 (United States Navy 2013).[2]

A review of the history of women's service illuminates one facet of the gendered nature of military institutions. Given long-standing projections of the military as an inherently male institution, the very presence of women comes as a surprise to many. But the number of women serving in the armed forces, the persistent underrepresentation of women in the armed forces, and the cultivation of lines of command that preclude women from supervising men, afford only a narrow glimpse of the military's gender regime. Women have been utilized and controlled in various ways by the military to serve the needs of male soldiers (see Enloe 1983). Even as women constitute a small percentage of military personnel, they have been fundamental to the continued operations of militaries on and off military bases, serving as "soldiers' wives, whores, servants, maids, and other camp followers" (Enloe 1983, 1). When considered in these expansive terms, women have been integral contributors to the effective functioning of militaries, even when officially absent—that is, when their roles are not officially acknowledged or are intentionally minimized. The careful management of the number of women allowed to serve in the military, along with their historical restriction to certain roles, helps illuminate how women's contributions in the military have been erased from public consciousness. But the military has devised multiple other mechanisms to sustain its gender regime. The next section explores some of those devices.

THE MILITARY AS A GENDERED INSTITUTION

The image of heteromale service members as the nation's ultimate defenders—its most reliable and trustworthy warriors—is intricately connected with constructions of women as weak and in need of protection. Carole Pateman (1998) has traced discourses that naturalize male power and female dependency to the long nineteenth century when nascent republics on both sides of the Atlantic created justifications for "equal citizenship" designed to assist in the overthrowing of monarchies. These justifications turned on a conceptualization of male autonomy or independence that was deemed inherent to white Euro-American men, but equally deemed absent in women and people of color. The conceptualization of autonomy— or independence—being specifically tied to the male biological body, according to Pateman (1998), was "related to the masculine capacity for

self-protection: the capacity to bear arms, the capacity to own property and the capacity for self-government" (248). States have used mandatory male military service, conscription, and militia duty as means to construct men as "bearers of arms" (248). Women, on the other hand, were "unilaterally disarmed" (248), barred from military service and from combat duty, as free, propertied men were assigned responsibility for the "protection of women and children" (248). Within the U.S. context, women were officially excluded from the colonial militias and from other rights of citizenship such as suffrage and office holding. Within the early American republican frame, citizenship itself became a gendered institution.

In contrast to narrow views that link gender to particular forms of male and female embodiment, the gendered institutions framework suggests that gender is "present in processes, practices, images and ideologies, and distributions of power in the various sectors of social life" (Acker 1992, 567). Gendered institutions can be understood as having been created and established by men, currently or previously dominated by men, or as institutions whose logic and operations are "interpreted from the standpoint of men in leading positions" (567). Gendered institutionalism's treatment underscores dialectic power structures within practices and principles of institutions; "[t]o say that an organization, or any other analytic unit, is gendered means that advantage and disadvantage, exploitation and control, action and emotion, meaning and identity are patterned through and in terms of a distinction between male and female, masculine and feminine" (Acker 1990, 146). In contrast to the notion that organizations are gender neutral (which informs liberal political theory) gender is seen as central to the functioning of an organization—fundamental and integral to its operating processes (Connell 1987; West and Zimmerman 1987).

Importantly, for feminist institutionalists, identifying an institution as "gendered" entails much deeper inquiry and analysis than that which simply concludes that women are absent from the institution. Instead, gendering is seen as produced through mechanisms that actively promote inequalities. No doubt, gendering does occur through establishing divisions which are specifically and consciously created to differentiate men and women. These divisions may, for example, pertain to how labor is divided, codes of dress and conduct, the allocation of physical spaces, or the distribution of power and authority. But gendering also occurs through more subtle—and often contradictory—vehicles such as the establishment of symbols, images, rituals, and traditions that "explain, express, reinforce,

or sometimes oppose" (Acker 1990, 146) those divisions of gender. Gendered symbolism may be visible in an organization's internal discourse and directives, public relations, celebrations and ceremonies, and coverage in print, social media, television, and radio. Similarly, gendering may also be found in scripted interactions between men and women within organizations that "enact dominance and submission" (146), and it surfaces through individual dress and deportment, and in regulations concerning appropriate clothing, language, and self-presentation. Whatever the modalities in which it is found, for feminist institutionalists, gender is seen as continuously constructed, contested, revised, and reestablished in the constant flow of social interactions and institutional regulations (Acker 1992). Yet, despite ongoing negotiation, a prevailing gender regime provides an underlying logic that sustains judgments of what is normal, natural, and permissible.

As the brief history of women's service in the military makes clear, for the better part of its existence the U.S. military has relegated the roles and duties associated with combat to men, entrenching a gender-based division of labor that has profound implications for individual performance and career advancement within the armed services. But the military has also deployed gender stereotypes to train men for the rigors of combat. Consider, for example, boot camp socialization, which is designed to achieve two primary purposes for newly enlisted service members (Belkin 2012; Burke 2002; Enloe 1983; Kier 1998; Levy 1997). Socialization to the military, for both men and women, involves intensive rituals designed to build comradery among the troops and obedience to the chain of command governing them. These rituals strip recruits of their individuality as they inculcate collective norms (Barkawi et al. 2002; Burke 2002; Enloe 1983; Kier 1998). Only those characteristics necessary for being a "good soldier" are permitted. Stereotypical masculine characteristics of "aggressiveness," "bravery," "endurance," and "discipline" are demanded, validated, and institutionally valorized, while stereotypical feminine characteristics such as "compassion" or "nurturing" are belittled and weeded out (Burke 2002, 13).[3] Military indoctrination teaches that "the good things are manly and collective; the despicable are feminine and individual" (Gilder quoted in Francke 1997, 155). Grueling exercises and ritual taunts teach recruits to do away with "feminine" traits like sensitivity or weakness. The end goal of this process is "to create a solidaristic group of male killers," which requires recruits to first and foremost "kill the woman in them" (Gilder quoted in Francke 1997, 155). By insistently

asserting a categorical binary between desirable traits coded "masculine" and undesirable traits coded "feminine," military training simultaneously indulges in gender symbolism and shores up belief in biological sex differences (Hooper 2001, 43). Individual soldiers are encouraged to function as a "unit" and feel loyal to their "brothers in arms," yet this fellow-feeling is far from egalitarian. Far from being grounded in a sense of community and cooperation, military rapport is based on a competitive hierarchy linked to a model of dominance, submissiveness, and a rigid constructed role of masculinity (Goldstein 2001; Kier 1998).

The language deployed in boot camp is riddled with negative depictions of women. During training exercises, any who lag behind or fall short in their performance are derided as "girls," "sissies," or other more derogatory invectives. In this way, basic training constructs the "Other" not just as the enemy who soldiers will confront in combat, but also as the *feminized* Other—women—who men define themselves against. As Cynthia Enloe (1983) has noted:

> To be masculine is to be not feminine. To prove one's manhood is imagined to be to prove (to oneself and to other men and women) that one is not 'a woman'. Consequently, experiencing military combat and identifying with the institution totally committed to the conduct of combat is, for those men trying to fulfill society's expectations, part and parcel of displaying and proving their male identity and thus qualifying for the privileges it bestows. (13–14)

Psychologically, boot camp establishes a connection between combat and manhood that provides for a "justification of the superiority of maleness in the social order" (Enloe 1983, 12–13). It ties violence to masculinity through "the widespread presumption that a man is unproven in his manhood until he has engaged in collective, violent, physical, struggles against someone categorized as 'the enemy'" or "the Other" (13), which may go some way toward explaining the frequency of gang rape within the military.

This socialization process constructs a model of aggressive hypermasculinity as the ideal to which new recruits aspire, and which can be achieved through "warrior" combat service. It also poses severe challenges for the socialization of women in the armed services. The male body has served as the norm in the design of military equipment as well as in the design of tests of physical strength and endurance. Backpacks, for example, have been developed to capitalize on upper body strength, which disadvantages women, who possess on average 25% less upper body strength (Cohn 2000).

Constructed for male bodies, backpacks geared to carry 100 pounds distribute considerable weight on the hip bones, which have caused disproportionate bone fractures in female troops (Bowman 2015). Although it has been proven that a slight alteration in backpacks would enable women to carry 100 pounds without endangering their hip bones, the military has refused to consider adoption of the redesigned packs. Further, numerous tests of physical strength also focus on upper body strength. By excelling in these tests, male recruits prove their "manhood," demonstrating that they are not "weak," they are not "girls," they are not inferior. Male recruits who do less well in these competitive exercises are taunted with obscenities typically reserved for women. Female recruits may excel in these physical tests, but even as they do so they hear the taunts directed at less successful male colleagues.

In this way, basic training itself defines feminine traits in opposition to masculine ones. If, as Burke (2002) suggests, soldiers—and by extension "real" men—are strong and brave and aggressive, then "real" women must be the opposite: weak, passive, and in need of protection. In this way, military training reinforces patriarchal notions about the strength and potency of the masculine soldier and contributes to the myth that women require protection. Gender integration in basic training, then, occurs within an institution founded on opposition to equality between the sexes, an institution whose training rituals routinely disparage all things feminine and undermine notions that women are capable of valor, strength, and military service (Enloe 1983), or on equal footing with their male counterparts. Basic training sets a tone: women exist only in relation to men, and that relation is one of stipulated inferiority regardless of rank.

Regulating and Policing Sexuality

Hegemonic militarized masculinity is often discussed in terms of a constellation of norms and values such as courage, strength, national honor, bravery, and service to country. Yet, sexuality is also an integral component of hegemonic masculinity within the military, and in society more generally. In an organization that is predominantly male, and that deploys troops on prolonged combat missions from which women have been historically banned, considerable effort has gone into ensuring that militarized masculinity—traits and characteristics traditionally associated with masculinity being attained through military service (Eichler 2014)—is profoundly coupled with "a certain kind of male heterosexual sexuality" (Acker 1990, 153).

Homosexuality was officially prohibited in the U.S. military until the early 1990s. Prior to that time, military personnel used physical exams and interviews to spot men with "effeminate characteristics" (Jain 2015, 12) and preclude their enlistment. Soldiers accused of homosexual behavior were discharged from the service. Beyond prohibiting male homosexual conduct, the military also went to great lengths to promote and facilitate normative heterosexuality. Officers were encouraged to marry, and the military issued handbooks for military wives that included detailed instructions concerning appropriate dress and conduct on base. The military also carefully regulated bars and clubs adjacent to military bases to facilitate sexual access to women free of venereal diseases. While on deployment, troops were allowed passes for "Rest and Recreation," which included transport to venues known for ample provision of sex workers. Thus, military policies shored up the "Madonna/whore" dichotomy, construing some women (mothers, wives, daughters) as pure and innocent and in need of protection, while positioning others as "available" for sexual use and underserving of respect.

As the presence of women in the military grew from 2% to 15% and women troops were integrated into military academies and basic training squadrons, they posed new challenges to the military's established mechanisms for sexual regulation. "Fraternization" between officers and enlisted personnel was prohibited, thereby setting certain limits on the sexual conduct of male officers and enlisted women, as well as female officers and enlisted men. Statistics concerning sexual harassment of women in the military, which indicate that more than 80% of enlisted women have been sexually harassed by male officers, demonstrate that bans on fraternization are clearly inadequate.[4] But even ineffectual fraternization bans do not regulate the sexual conduct of peers within the military. Thus, the presence of women within the military complicates military authorities' efforts to regulate the sexual behavior of military personnel. Where do women serving the military fall within the Madonna/whore schema? Does their very presence in the armed services negate their need for protection? Does their refusal to conform to stereotypes of weakness and subservience make them targets for assault? Does their very presence in the military jam the logic of male protection by showing that military men themselves constitute the prime threat to women troop's physical security?

If the growing number of women in the military has created one set of challenges to traditional military masculinity, gay and lesbian personnel have created another. Although homosexuals were barred from military

service for most of the twentieth century, under pressure from LGBTQ activists, President Bill Clinton converted the prohibition against gays and lesbians in the military to the "Don't Ask, Don't Tell" policy. Although the policy may have been preferable to an outright ban, for certain interest groups it maintained an institutionalization of different treatment for homosexual troops. In a nation that guarantees equal protection of the law, the "institutional and cultural privileging of a heterosexual masculine ideal" (Petersen 1998, 53) required some justification. Policy makers in Congress and the Department of Defense did not attempt to justify this policy with claims concerning homosexuals' lack of ability to excel in the military. Instead, they appealed to the military's ability to "accomplish its overall mission" (53). Articulating "prejudice-based arguments," policy makers asserted that enforced heterosexuality was necessary to maintain unit cohesion. They claimed that if homosexual soldiers were to serve openly, they would be the victims of harassment by their heterosexual counterparts—or worse, they would impair the operational readiness of heterosexual troops who would live in fear of sexual assault in the showers or in their bunks. In the case of leadership positions, policy makers posited that homosexual soldiers would not be able to command the respect necessary to keep their units functioning effectively. By allowing homosexuals to serve in the armed forces openly, the military not only risked the alienation of the general American public (which is imagined to be homophobic), but also risked creating a "perception of the military as a gay organization" (Snyder and Nyberg 1980, 81). Like the justifications for total exclusion of homosexuals from the military, military discourses on "Don't Ask, Don't Tell" manifested "political homophobia" (Currier 2010, 110), rhetoric circulated by political elites to justify differential treatment of some citizens. Insisting that national defense requires compulsory heterosexuality in the military, policy makers readily sacrificed the rights of homosexuals and gender nonconforming citizens to the putative needs of the collective.

Ultimately, the debates about "Don't Ask, Don't Tell" suggest that there are hard and fast lines that demarcate heterosexuality from homosexuality—whether in the military or beyond. They echo Aristotle's simplistic claim that "the male is he who mounts, and the female is she who is mounted." Heterosexuality is presumed to exist in pure opposition to homosexuality. Yet Aaron Belkin's (2012)[5] work highlights that opposition is neither firm nor secure conceptually or empirically. His research on military masculinity questions the fundamental and facile notions that heterosexuality/homosexuality exhaust the conceptual possibilities. In these

instances, investigating masculinity as a set of complex tensions, which play out on and through military personnel's bodies, forces them to "enter into intimate relationships with...unmasculine foils, *not* just to disavow them" (Belkin 2012, 24 emphasis added). As Belkin's (2012) series of case studies indicate, the production of masculine warriors requires individuals to embrace performances of *feminine and queer* acts, to engage in interactions with the *unmasculine* (24)—including male-on-male penetration and rape. The "impenetrability" associated with masculinity is constituted by penetrating another male, who in the act of being penetrated is rendered unmasculine or demasculinized, as the penetrator is viewed as the wholly masculine figure (80). The prevalence of male-on-male rape in the military is seldom addressed in public, but its frequent occurrence across the armed services raises fundamental questions about military efforts to regulate sexuality, much less enforce compulsory heterosexuality. They certainly trouble official military doctrines about appropriate sexual comportment and the myth of invariant heterosexuality.

Hegemonic Militarized Masculinity and the Warrior Archetype

In some cases, certain forms of militarized masculinity may reach hegemonic status, such as the masculine warrior in the contemporary U.S. military. The archetype's hegemonic status is a critical component of maintaining the institution's gender regime. Military and defense institutions—in the United States and in other contexts—have been studied as producers of hegemonic forms of masculinity, which establish and legitimate a hierarchy of power within the organization and the society in which they exist. Coined by R.W. Connell (1987, 1995), hegemonic masculinity is established in opposition to particular constructions of femininity and subordinate masculinities which are construed as inferior (see following paragraph). For Annica Kronsell (2005), hegemonic masculinity becomes visible in institutions through "a particular set of masculine norms and practices that have become dominant in specific institutions of social control" (281). Although Connell notes that determinate types of hegemonic masculinity surface within specific organizations, the content of hegemonic masculinity is specific and non-static.

Within the military, hegemonic masculinity currently combines strength and ingenuity, loyalty to the troops, principled obedience to the chain of

command, courage under fire, and ability to control emotions and carry out the mission even when facing death. Those who conform to "the masculine-warrior paradigm" (Magnusson 1998 quoted in Kronsell 2012, 46) help the nation achieve its national defense and foreign policy objectives, while also winning the respect of their peers and the admiration of their country. Although the masculine warrior paradigm vies in the twenty-first century with competing forms of hegemonic masculinity external to the institution (e.g. the hyperrational and inventive scientist, the super rich finance capitalist, or the glamorous Hollywood star), military masculinities—and this form of military masculinity in particular—continue to play an important role in shaping public norms of manliness and leadership. Indeed, R.W. Connell (1995) argues that the military plays a particularly prominent role in establishing definitions of hegemonic masculinity within American and European social and political cultures. While less concerned with tying specific forms of hegemonic masculinity to particular contexts than Connell, Joshua Goldstein (2001) suggests that military and defense organizations have consistently represented and been associated with particular gender stereotypes that hold impressive cultural cache. Further, Charles Tilly (1985) argues that the gender norms validated in military and defense institutions are particularly well suited for upholding pernicious gender stereotypes due to their close association with nation building activities and their prominence in international relations.

Militarism and Militarized Masculinities

Despite the long-standing prominence of the masculine warrior profile, men should not be thought of as inherently and naturally militaristic, and women conversely as naturally pacifistic.[6] Indeed, these stereotypical gendered characteristics persist despite women now being allowed to serve in the most violent positions available in the military (Eichler 2014). Importantly, feminist international relations asks us to interrogate and complicate the processes through which gendered identities become militarized, or are deemed unworthy of configuration through military service. Instead, at its most fundamental level, gendering by and of the institution results in the social linking and constructing of militarized masculinity in a variety of iterations.

The term militarized masculinity "refers to the assertion that traits stereotypically associated with masculinity can be acquired and proven through military service or action, and combat in particular" (Eichler

2014, 81). As I discuss later in this chapter, militarized masculinities are typically associated with characteristics of strength, bravery, violence, and aggression. Given the linkage between masculinity and military service "is constructed and maintained for the purposes of waging war" (Eichler 2014, 81), it is important that any investigation of militarized masculinities and the gender regimes they support not assume masculinity as being inherently tied to military service, or any relationship between the two as foreordained.

Central to the power of militarized masculinities' ability to privilege certain bodies and individuals over others is militarism, that is the political, social, and cultural promotion of the military institution—and the gendered identities valued by it, that is, it's mostly male personnel—as central to the society and state in which it exists. Militarism as a term is complex and multidimensional, stretching far beyond the mere defense spending and the formal military institution of a nation. It refers to supporting and endorsing a "central role" for the military; feminist international relations is interested in interrogating how militarism gives importance to military institutions. Moreover, when considered as a process, militarism reveals itself within societies both materialistically and ideologically. The material forms of militarism include "wars and direct military interventions, destabilization of other countries through proxy armies, foreign-sponsored coups, foreign and colonial occupation, military rule and abuse of human rights" (Burke 2002, 2). When considering militarism as an ideology, "militarism affects national governments of different political objectives with influence that has the means for becoming part of a social process which penetrates itself in all areas of a society" (2).

While we might expect to find variation in this degree of influence across cultures and over different periods of time, what is common is an attribute of militarism as "a dissemination of military values, symbols and language among the civilian population which promotes acceptance of hierarchies, nationalism which defines the 'other' as enemy, violence as a legitimate means of resolving conflicts, and strict division of proper masculine and feminine roles" (Burke 2002, 3). Institutional manifestation may also be pointed out in examining the military agency itself, as well as government monies and resources allocated for its upkeep. Militarism may also be seen as the result of the process of militarization[7] in which "military values, ideology and pattern of behavior achieve a dominating influence on the political, social, economic and external affairs of the state, and as a consequence the structural, ideological and behavioral

patterns of both the society and the government are 'militarized'" (Burke 2002, 2). At its most fundamental level, militarism's power reveals itself as a continued readiness on the part of states to execute their policies, if deemed necessary, through deliberate and organized use of physical force.

Theoretical formulations of militarism that highlight societal implications provide particular insight into how militarized versions of masculinity become established. Militarized masculinities are mutually reinforced by both society and the institution; ordinary members of a society play an active role in making militarized forms of masculinity valuable and desirable. "Significantly, militarized masculinities *shape, and are shaped by*, military practices, but also by state policies, security discourses, education programs, media debates, popular culture, family relations, personal identities, and more" (Eichler 2014, 84, emphasis added).

Militarization

Similarly, militarization's political and social dimensions demonstrate the institutional necessity for creating specific narratives of militarized masculinities that maintain the military as a tool of statecraft—that is, to wage war. Under these terms, militarization refers to the process in which "preparation for war is regarded as normal social activity, and is often considered a *desirable* (emphasis added) social activity at its height" (Mann 1987, 35). Militarization is further measured by the "extent to which the preparation of war and military spending is considered a routine and is reflected in the ascendancy of military thought over civilian political thought" (Levy 1997, 7). Moreover, political control over the military is measured by the extent to which the military's activities (military operations, budget, selection of weapons systems, mode of organization, mode of recruitment, etc.) are monitored by the main civilian state agencies, directly or through public participation (Levy 1997).

Consider Cynthia Enloe's (2002) eloquent illustration of militarization as,

> A step by step process by which a person or a thing gradually comes to be controlled by the military or comes to depend for its well-being on militaristic ideas. The more militarization transforms an individual society, the more that individual or society comes to imagine military needs and militaristic presumptions to be not only valuable but also normal. (3)

Many dimensions of everyday life are involved if military needs and concerns are to be accredited as legitimate. Ascendancy of military thinking saturates popular culture through film, fashion, and Humvees as consumption items, as well as through myths of national founding and narratives of national belonging. Building on Enloe's framework, Catherine Lutz (2002) suggests that,

> Militarization is simultaneously a discursive process, involving a shift in general societal beliefs and values in ways necessary to legitimate the use of force, the organization of large standing armies and their leaders, and the higher taxes or tribute used to pay for them. Militarization is intimately connected not only to the obvious increase in the size of armies and resurgence of militant nationalisms and militant fundamentalisms but also the less visible deformation of human potentials into the hierarchies or race, class, gender, and sexuality. (723)

These discursive understandings of militarization not only suggest that the power of the ideology is diffuse, but that it also does not emanate from a top-down structure. Rather, it is *found in and flows through* societal level structures that influence even those who are not part of the official military apparatuses, particularly ordinary citizens. The continuous flow of ideology through gendered military self-representations—images, narratives, values, and programs—by way of the manifold dimensions of social life and social memory produces a sort of normalcy that allows increasing militarization to go unnoticed and unquestioned. Kathy E. Ferguson and Phyllis Turnbull (1999) highlight that the state,

> acquires this status not simply through the perceptible presence of military objects and events, but through the social and economic insinuation of the military into other institutions, and the cultural imbrication of military codes, symbols, and values into daily life. These latter processes flag not the military per se but militarization as a dynamic, contested process of constituting a particular kind of order, naturalizing and legitimating the order, while simultaneously undermining competing possibilities for order. (2)

Implications of Feminist Understandings of Militarization

Feminist understandings militarization processes underscore an important point: when "looking" for militarization, we are unlikely to find "it" without engaging in and with interpretive analyses. Militarization is simultaneously

seen and unseen, and what is subtly hidden goes unquestioned. Evidently, these conceptualizations require a re-imagination of the militarization as that which moves through and beyond the four-wall confinement of the traditionally thought of military establishment. It suggests an ideology that is not restricted to the spatial boundaries of the military structures; rather, it is dispersed and embedded in many different levels of societal existence for continuous confirmation and reestablishment of the power of the ideology. As Ferguson and Turnbull (1999) point out, "the facts never speak for themselves; meaning does not dwell in objects but accrues through the narrative strategies by which the facts are recruited and made available for comprehension and contestation" (3). In this sense, understanding militarization from a feminist perspective requires "reading against the grain" (46) where simply looking at the military as an institution and its included apparatuses will not suffice if the objective is to understand the implications of the normalization of this (gendered) order in other levels of society, which is where militarization's eminent and dexterous power emerges.

In attempting to unearth the ascendancy of militarization of a particular nation and its society, feminist understandings remind us that the creation of these histories is typically a violent one in which many other histories are simultaneously silenced. These instances of violence often entail "the enforced movement of a variety of bodies across a variety of borders and the persistent transgression and frantic reinforcement of a range of critical boundaries" (Ferguson and Turnbull 1999, 3). They simultaneously reveal and hide themselves in "the material violence of displacement, uprooting, and resettlement; the discursive violence involved in reading a place through the lenses of one's own desire; the ontological violence of writing a particular kind of order onto bodies and spaces (Ferguson and Turnbull 1999).

Instead, the militarized present is illustrated in "the friction and slippage, inertia and momentum of the hazardous play of dominations" (Ferguson and Turnbull 1999, 3). In these uncontested histories of militarization, the historicity of organizations of sex, race, and class are inherently ignored and buried. In this regard, history is unceasingly being written to empower some and ignore others. As understandings shift away from economic and arms indicators, militarization emerges as a fluid, non-static process, one which fluctuates within and between different levels of society. Perhaps it is useful to think of it as a set of practices that embed themselves within society in waves. As Ferguson and Turnbull (1999) eloquently describe,

> waves wash up on beaches and then recede; some of the flotsam is carried up far enough to resist the suction of the water taking the rest back into the ocean. Things deposited on the shore are subject to continual suction, but some also become embedded in the sand. There is rarely a specific moment in the usual work of the ocean when the permanent embedding of an object can be determined; rather, that occurs over time. (Ferguson and Turnbull 1999, 28)

Consequently, it is neither possible nor fruitful to try to identify a single moment in which military order becomes embedded in a society in order to understand why it occurred, or why that particular military order is deemed necessary by the nation-state. It is the continuous reifying of normalcy in military order through unquestioned visible militarization and hidden unseen modes of militarization that are "rendered ordinary and unremarkable in establishment discourse" (34) that is so critical to the military institution's, and thus the nation-state's existence.

Conclusion

The military has established a tool kit of a wide spectrum of resources for sustaining its gender regime. Historically, the posturing of the institution as exclusively male—through the unacknowledged participation of women in the military, the insistent underrepresentation of women in the military, and the rejection of pathways for women rising above men in the military chain of command—merely scratches the surface in these efforts. Instead, feminist institutionalism highlights the multitude of ways in which institutions are gendered, and how the manipulation of gender relations within the institution may help it achieve concrete goals. Though undoubtedly an important part of the equation, gendered institutions are about much more than just gendered divisions of labor which promote women in traditionally "female roles" and in fewer numbers in comparison to male counterparts—this is but one symptom of institutional inequality. Instead, gendered institutions promote and enable inequalities in more covert and subtle ways, through symbols, images, rituals, and traditions. And importantly, feminist inquiry into critical understandings of militarism and militarization provide a conduit toward attending to the political and social dimensions through which the military establishes concrete narratives that tie desirable forms of masculinity with military service, and then disseminate and embed those messages—with equal subtlety—throughout the society from which it seeks recruits. Both feminist institutionalism and

feminist international relations theory research foreground just how instrumental women have been to the institution's ability to maintain specific social archetypes of (masculine) soldierhood.

The masculine warrior paradigm and profile has been the most prominent of these archetypes in the contemporary U.S. military, achieving hegemonic status since the inception of the All-Volunteer Force. What role were women to play in the newly imagined public military given societal notions of masculinity that were thought to be crumbling in the wake of the Vietnam War? And how would institutionally devised and promoted narratives about women's participation specifically, and gender relations more generally, satisfy calls for gender equality while preserving masculinity at the foundation of the institution's identity? I examine these questions in the following chapter.

Notes

1. It is interesting to note that legislation which eventually established the act was initiated by the military, not by members of Congress. After the successful participation of women during World War II, military officials urged Congress to recognize the utility of women in auxiliary positions both in periods of peace and war. Military officials argued that permitting women to serve in the military in a permanent capacity would assist recruitment efforts for maintaining a standing military during periods of peace. Prior to the introduction of the legislation, women were only permitted to serve the duration of a war, with an additional six months added for demobilization as a war wound down. Throughout 1946 and 1948, Congress debated the legislation, often unconvinced that women should have a permanent position in the military at all (Frank 2013). The enactment of a 2% quota, along with other stipulations excluding women from combat, female Marines not being able to serve on combat ships, female members of the Air Force not being allowed to serve on planes, and women not being allowed to command men acted to appear inclusive for the sake of having a prepared military for future wars, while also making women seem marginal to the military apparatus.
2. Women have done somewhat better in the military academies. At the Naval Academy, women made up 27% of the student population during the 2018–2019 academic year (U.S. News Best Colleges 2019b), compared to 26% at the Air Force Academy (U.S. News Best Colleges 2019a), and 22% at West Point Military Academy (U.S. News Best Colleges 2019c).

3. For studies on homosexuals in the military cross-nationally, see Elizabeth Kier (1998), who documents how basic training curricula involve a strategy of "breaking" the recruits and "molding" them into fighters (8).
4. In the three military academies (U.S. Military Academy at West Point, the U.S. Naval Academy, and the U.S. Air Force Academy) specifically, the Pentagon's Annual Report on Sexual Harassment and Violence at the Military Service Academies for academic year 2017–2018 reported a 50% increase in the number of unwanted sexual contacts since the previous report, while the number of reports on "unwanted sexual contacts" remained the same. The data is generated from an anonymous climate survey, the Service Academy Gender Relations Survey (SAGR), which approximately 12,000 students complete every two years (Department of Defense 2018).
5. Some of the content presented here related to Belkin's (2012) book, *Bring Me Men*, may have appeared in my published book review of his work. See Szitanyi (2012). "Aaron Belkin. Bring me Men: Military Masculinity and the Benign Façade of American Empire, 1898–2001." *International Feminist Journal of Politics*. Vol. 14 (4):572–574.
6. Indeed, interrogating, and complicating, how masculinities become militarized and how other gendered identities are declared worthy/unworthy of militarization instead of accepting them as fact remains central to the feminist international relations agenda on knowledge production.
7. As I discuss in the following section of the chapter, feminist international relations understandings of militarization have been paramount in observing and underscoring "the ways in which militarization structures everyday life far from military institutions or battlefields" (Lutz 2018, 1). This understanding of the term stems from anthropological perspectives and is differentiated from mainstream political science and international relations theory. For mainstream international relations theory, in particular, militarization has traditionally been operationalized as a monetary phenomenon that is studied in relation to fiscal measures such as annual defense budgets and spending figures per annum as a percentage of gross domestic product (GDP) (Slater and Nardin 1973). Discussions of militarization often focus on military capabilities in weapons and resources and in terms of arms capabilities, where the number of nuclear weapons one nation has is compared to those of another in suggesting a degree of threat it might pose. Large militaries and extensive arms capabilities are framed as necessary in order to maintain and secure the nation-state from outside threats. As is the case with feminist international relations, militarization is not thought to be fixed or static by mainstream international relations theory either, but very differently, fluctuations are measured and analyzed through increases or decreases in military expenditures and arms supplies (Slater and Nardin 1973).

References

Acker, J. 1990. Hierarchies, Jobs, Bodies: A Theory of Gendered Organizations. *Gender and Society* 4 (2): 139–158.
———. 1992. From Sex Roles to Gendered Institutions. *Contemporary Sociology* 21 (5): 565–569.
Bailey, B. 2009. *America's Army: Making the All-Volunteer Force*. Boston: Harvard University.
Barkawi, T., C. Dandeker, M.W. Petry, and E. Kier. 2002. Rights and Fights: Sexual Orientation and Military Effectiveness. *International Security* 27 (2): 181–201.
Belkin, A. 2012. *Bring Me Men: Military Masculinity and the Benign Façade of American Empire, 1898–2001*. New York: Columbia University Press.
Bowman, Tom. 2015. Can Female Marines Carry the Load and Kill the Enemy? NPR. Special Series: Back at Base. https://www.npr.org/2015/03/25/395279171/can-female-marines-carry-the-load-and-kill-the-enemy.
Burke, C. 2002. Women and Militarism. Accessed February 20, 2006. http://wilpf.smilla.li/wp content/uploads/2012/10/Unknownyear_Women_and_Militarism.pdf.
Cohn, C. 2000. 'How Can She Claim Equal Rights When She Doesn't Have to Do As Many Pushups As I Do?' The Framing of Men's Opposition to Women's Equality in the Military? *Men and Masculinities* 3 (2): 131–151.
Connell, R.W. 1987. *Gender and Power: Society, the Person, and Sexual Politics*. Stanford: Stanford University Press.
———. 1995. *Masculinities*. Cambridge, UK: Polity Press.
Currier, A. 2010. Political Homophobia in Postcolonial Namibia. *Gender & Society* 24 (1): 110–129.
Department of Defense. 2018. Annual Report on Sexual Harassment and Violence at the Military Service Academies: Academic Program Year 2017–2018. http://sapr.mil/public/docs/reports/MSA/APY_17-18/APY17-18_MSA_Report_FINAL.pdf.
Eichler, M. 2014. Militarized Masculinities in International Relations. *Brown Journal of World Affairs* 21 (1): 81–93.
Enloe, C. 1983. *Does Khaki Become You? The Militarization of Women's Lives*. Boston: South End Press.
———. 2002. *Maneuvers: The International Politics of Militarizing Women's Lives*. Berkeley: University of California Press.
Ferguson, K., and P. Turnbull. 1999. *Oh, Say, Can You See? The Semiotics of the Military in Hawai'i*. Minnesota: University of Minnesota Press.
Frank, L.T. 2013. *An Encyclopedia of American Women at War: From the Home Front to the Battlefields*. Santa Barbara, CA: ABC-CLIO.

Francke, L.B. 1997. *Ground Zero: The Gender Wars in the Military*. New York: Simon and Schuster.
Goldstein, J. 2001. *War and Gender: How Gender Shapes the War System and Vice Versa*. Cambridge: Cambridge University Press.
Hooper, C. 2001. *Manly States: Masculinities, International Relations, and Gender Politics*. New York: Columbia University Press.
Jain, P. 2015. Should the Homosexuals Be Allowed to Service in Armed Forces: A Critical Analysis. *ISOR Journal of Humanities and Social Sciences* 20 (2): 12–15.
Kerber, L.K. 1980. *Women of the Republic: Intellect and Ideology in Revolutionary America*. Chapel Hill: The University of North Carolina Press.
Kier, E. 1998. Homosexuals in the U.S. Military: Open Integration and Combat. *International Security* 23 (2): 5–39.
Kronsell, A. 2005. Gendered Practices in Institutions of Hegemonic Masculinity: Reflections from Feminist Standpoint Theory. *International Feminist Journal of Politics* 7 (2): 280–298.
———. 2012. *Gender, Sex and the Postnational Defense: Militarism and Peacekeeping*. New York: Oxford University Press.
Levy, Y. 1997. How Militarization Drives Political Control of the Military: The Case of Israel. *Political Power and Social Theory* 11: 103–133.
Lutz, C. 2002. Making War at Home in the United States: Militarization and the Current Crisis. *American Anthropologist* 104 (2): 723–735.
———. 2018. Militarization. *The International Encyclopedia of Anthropology*. Wiley. https://doi.org/10.1002/9781118924396.wbiea1304
Mann, M. 1987. The Roots and Contradictions of Modern Militarism. *New Left Review* 162 (2): 27–55.
Naval History and Heritage Command. 2010. Women in the United States Navy. Accessed January 28, 2014. http://www.history.navy.mil/special%20highlights/women/navywomen.pdf.
Pateman, C. 1998. The Patriarchal Welfare State. In *Feminism, the Public and the Private*, ed. Joan Landes, 241–276. New York: Oxford University Press.
Petersen, A. 1998. *Unmasking the Masculine – 'Men' and 'Identity' in a Sceptical Age*. London: SAGE Publications.
Slater, J., and T. Nardin. 1973. The Concept of Military-Industrial Concept. In *Testing the Theory of the Military-Industrial Complex*, ed. Steven Rosen, 27–60. Lexington: Lexington Books.
Snyder, W.P., and K.L. Nyberg. 1980. Gays and the Military: An Emerging Policy Issue. *Journal of Political and Military Sociology* 8: 71–84.
Szitanyi, S. 2012. Aaron Belkin. Bring Me Men: Military Masculinity and the Benign Façade of American Empire, 1898–2001. *International Feminist Journal of Politics* 14 (4): 572–574.
Tilly, C. 1985. War Making and State Making as Organized Crime. In *Bringing the State Back In*, ed. Peter B. Evans, Dietrich Rueschemeyer, and Theda Skocpol, 169–181. Cambridge: Cambridge University Press.

U.S. News Best Colleges. 2019a. United States Air Force Academy. Accessed March 25, 2019. https://www.usnews.com/best-colleges/united-states-air-force-academy-1369.
———. 2019b. United States Naval Academy. Accessed March 25, 2019. https://www.usnews.com/best-colleges/united-states-naval-academy-2101/overall-rankings.
———. 2019c. United States Military Academy. Accessed March 25, 2019. https://www.usnews.com/best-colleges/west-point-2893.
United States Navy. 2013. Women in the Navy Fact Sheet. Accessed January 28, 2014. http://www.public.navy.mil/bupers-npc/organization/bupers/WomensPolicy/Documents/Women%20in%20the%20Navy%20Fact%20Sheet_March%202013.pdf.
West, C., and D.H. Zimmerman. 1987. Doing Gender. *Gender and Society* 1 (2): 125–151.
Women in Military Service for America Memorial Foundation, Inc. 2014. Highlights in the History of Military Women. *Women in Military Service for America Memorial Foundation*. Accessed January 2, 2014. http://www.womensmemorial.org/Education/timeline.html.

CHAPTER 3

The All-Volunteer Force: Patrolling Gendered Boundaries Through the Combat Ban

Introduction

In the summer of 1968, then presidential candidate Richard Nixon made a campaign promise to the American public. As television screens across the nation projected images of flag-draped coffins returning U.S. casualties from the Vietnam War, Nixon declared the draft obsolete and vowed to create an all-volunteer, public military. Following Nixon's election, the Gates Commission[1] made recommendations concerning the feasibility of an All-Volunteer Force (AVF). An AVF, the Commission concluded, was in the nation's interest and would not only be beneficial to the institution, but was necessary for American society (The President's Commission on an All-Volunteer Armed Force 1970, iii).

Nixon's promise was more than a political tactic used to win an election. It was a response to an America where cultural understandings of masculinity and manhood were fragile and crumbling. Masculinity was thought to be in crisis from changing gender roles promoted by feminist activism, the unleashing of the civil rights movement, and from the loss of good paying, blue-collar jobs that gave men privilege and the economic ability to support a family. The ideal of the drafted "citizen-soldier" specifically was in disrepair as a result of U.S. troop participation in war atrocities that undermined notions of heroism and competence. The "citizen-soldier" was also challenged through competing displays of masculinity from young men who were anti-war, and whose displays of masculinity were made iconic through images of burning draft cards. A reimaging

of military service was necessary, and it would be used as a centerpiece for establishing a new blueprint on gender relations in American society.

That blueprint might have revised the military's established gender order. Recruiting a fully volunteer-based armed forces provided an opportunity to degender military ranks, increasing not only the overall number of women serving in the military, but also the types of positions in which they could serve. It might have produced a more egalitarian military, where women could serve side by side male counterparts in front-line combat roles, service seen as critical by the military for career advancement. Instead, women were conspicuously missing from the Commission's vision of a volunteer-based military (indeed, the terms "women recruits/volunteers," or "female recruits/volunteers" were mentioned only four and five times respectively in the Commission's 211-page report),[2] and women continued to be banned from combat roles for an additional 50 years. Indeed, the military's new blueprint on gender relations *regendered* the institution as a predominantly male voluntary force whose willing service to the nation strengthened notions of military masculinity.

The individual branches of the military recognized, however, that sustaining a volunteer-based military would require recruiting women in larger numbers to meet operational needs. Military officials vowed to increase the number of women under a volunteer military, but cautioned that women's increased "inclusion" be done carefully, rationally, and implemented through "deliberate and incremental" tactics. Gender integration would occur on terms that ensured women's service remained incompatible with combat service. Rationale for barring women from combat did not remain static over the 50-year span of the combat ban, but did consistently appeal to biological, psychological, and social grounds for women's exclusion. These arguments reflected persistent institutional anxieties about the limitations of the female body even as justifications for exclusion shifted from specific claims about the capabilities of women to the demands of national security. When the Department of Defense (DoD) announced the end of the combat ban in 2013,[3] it described its decision as the culmination of decades of "progress" and well-thought out policy implementation.

The incremental and deliberate progress referenced in the military's storyline can be viewed as a positive achievement of "full inclusion" only because the starting point was near total exclusion.[4] On the surface, the military's progressive claims about rescinding the combat ban appear to be qualified *degendering* of the institutional gender order. A closer look, however, indicates that the discursive practices used to justify

exclusion of women from combat positions across numerous decades shored up the masculinist identity of the institution. By suggesting that women's inclusion would require deliberate and incremental action, military leaders implied that there was something about women that necessitated careful control. The military's rhetoric in 2013 characterized the end of the combat ban as part of progressive modernization that recognized the equality of all groups participating in the institution. Yet this progressive narrative renders invisible the nature and scope of women's military service during the twentieth century and masks the *exclusionary* nature of the terms of their impending *inclusion*. It allowed continuing exclusion from certain roles to go unquestioned. And even today, it diverts attention away from the nuances and variation in tropes used throughout five decades to justify that exclusion.

To make the terms of *exclusionary inclusion* visible in the context of the military's claims about deliberate and incremental change, I first provide a short discussion of the Band of Brothers myth, which encapsulates institutional anxieties associated with women serving in combat. The chapter traces the changing policies that defined the terms of women's military service and then explores the justifications advanced to exclude women from combat roles from 1970 to 2013. Central to the military's narrative of deliberate and incremental inclusion of women are three discursive, gendered mechanisms of exclusion. The first, the myth of the *Band of Brothers*, focuses on arguments that suggest women lack the ability to be "natural soldiers," women threaten homosocial bonds necessary for unit cohesion, and for these reasons, women's exclusion is necessary for the security of the nation. Second, the military represents itself as a *rational comprehensive decision maker*, an *economic maximizer*, and a *modernizing force* through careful incremental policy change while depicting women through a host of misogynist stereotypes. Third, women's *bodies are positioned as leaky* through biological and psychological functions, and consequently a threat to the institution's ability to maintain the nation's security.

Testimonies given at Congressional hearings express the military's official position on women's inclusion in the AVF over the course of nearly five decades. The change from a draft-based military to a public, volunteer model raised multiple questions about recruitment tactics and the specific segments of the population that would be targeted for recruitment. Between 1970 and 2015, Congress held 70 hearings that involved significant discussions about whether women should be allowed to serve in combat capacities, in what fashion, and to what degree.[5] Discursive analysis of military testimonies in these Congressional hearings reveals shifting

grounds for women's exclusion over time, tracing the construction of women's presence as a threat to fighting units and to national security—a threat that has little to do with women's actual performance in military service.[6]

Debates surrounding women in combat have not occurred in a vacuum. In addition to providing a medium for interrogating the military's self-representation, Congressional hearings—and the testimonies of military officials at those hearings—offer a conduit for understanding how the various military branches represent themselves not only to the public, but to political officials and activist groups that may represent the public's interest. They also provide insight into the converging and diverging views of diverse interest groups deeply involved in efforts to shape policies pertaining to gender segregation and gender integration in the military.

Masculinization Through Band of Brothers and *Homo Economicus*

The newly minted, post-Vietnam volunteer-soldier was masculinized through the "Band of Brothers" myth and socialization model (Belkin 2012; Elshtain 1992; Enloe 1983, 1990; MacKenzie 2015). This form of military masculinity views unit cohesion as feasible only through male bonding. It depicts men fighting in combat as "exceptional, essential, and elite" (MacKenzie 2015, 198), bound by intense ties of loyalty. While models of masculinity establish specific parameters, boundaries, and profiles for manhood, their binary constructions simultaneously create the antithesis mode of (un)acceptable femininity. The Band of Brothers model specifically relied on circulating negative stereotypes and gendered myths about women to justify their continued exclusion from combat. Among the most prevalent of these myths, "that women are physically unfit for the demands of war, that the public cannot tolerate female casualties, and that female soldiers limit the cohesion of troops in combat" (MacKenzie 2015, 2) achieved "hegemonic status" in the initial decades of the AVF. Through the Band of Brothers mythos, gendered falsehoods about women's military service continue to circulate in public discourse even after the lifting of the combat ban and despite these claims being discredited by scholars and female soldiers alike.

The Band of Brothers myth is underpinned and supported by the military's self-representation as a rational decision maker (MacKenzie 2015). Part and parcel of the overarching narrative used by the military in defining

gender relations in the AVF, the image of the calculated decision maker casts the institution as a rational actor engaged in a cautiously deliberate, incremental process to integrate women. In the lead up to the creation of the AVF, this image of the military institution becomes paramount, as does projecting this image as central to its policy choices. At certain moments, the generic rationality ascribed to the military takes a particular form, stressing the role of economic maximizer. Military officials cast themselves as *homo economicus*, whose decisions are made based on economic self-interest and fiscal responsibility. This fiscally responsible subject surfaces as particularly dominant during periods of prospective military cuts. The integration of women is seen as incompatible with the *homo economicus* military. Instead, the exclusion of women can be altered only through deliberate, incremental inclusion according to military needs in periods of budgetary austerity. Fiscal constraint is as stern a task master as the homosociality required of the Band of Brothers. Both entrench a gendered order at the heart of the military's system of power.

Although the military characterizes itself as a rational actor during this period, Congressional testimonies simultaneously articulate a litany of *irrational* fears concerning the effects of women's presence on military operations—fears unsubstantiated by sustained research or by experiences of women serving in the military. Set against the background of the Band of Brother's image of masculinity, both the female body and the female mind are positioned as incompatible with military service. As the male body is imagined to be disciplined, self-contained, a steel trap, women's bodies are envisioned as weak, abject, porous, and "leaky" (Shildrick 1997). "Leakiness" emerges in two strands of argument that informed the women in combat debate. The first strand insists that biological differences between men and women impair women's physical ability to complete certain tasks. The second strand emphasizes potential negative consequences associated with the insertion of women into interpersonal processes of military groups.[7] The leaky body reduces the female subject to biological capacities, foregrounding women's (assumed) reproductive capabilities and menstrual processes as threats that must be contained lest they disrupt military order and operations. The leaky feminine body is also viewed as emotional and irrational, in many ways the opposite of the rational male subject embodied in the individual soldier and in the military institution more generally. Indeed, concerns regarding women's ability to serve adequately in the military have often emphasized the question of service in combat theaters, spaces thought to be more hazardous to the

unique "conditions" of womanhood. Within this reductive frame, menstruation and pregnancy are taken to be the definitive "conditions" of the feminized leaky body. The military's self-image as a rational decision maker is intricately tied to assumptions about women's erratic emotions and uncontained irrationality, which are fueled by male concerns, anxieties, and plain old "gut reactions" rather than empirical evidence.

The Leaky Body's Rejection for the Protection of the State

Within the parameters of assumptions associated with the leaky body (Shildrick 1997), placing women in combat roles is deemed an unacceptable "bodily practice." It is considered "unnatural" to put women *qua* mothers and daughters in harm's way. Indeed, sexual difference structures the masculinist logic of protection (Young 2003). Men go to war and are wounded or killed to protect their nation, symbolized by their women. While engaged in armed conflict, they envision home as a "natural state" of affairs in which women lovingly await their return, ready to greet them, care for them, rear children, and deal with the physical and psychological aftermaths of war. The notion and visual imagery of the Band of Brothers links military service to this "natural state." For those within the Band of Brothers, outsiders are seen "as inherent security threats" and "violence is deemed the most efficient way to solve the political problems" posed by such threats (MacKenzie 2015, 4). In 1982, Mady Wechsler Segal captured the gendered psychology underlying military masculinism:

> Let me offer an additional explanation for men's resistance to allowing women in combat units. I conjecture that there is a psychological differentiation between the real world and combat that enables some men to survive the enormous psychological stress of combat. One survives by preserving a mental picture of *the normal world back home* to which one will return from the horror world of combat. One is engaged in an elaborate game, albeit one with very high stakes, and when the game is over, one can *go home to an intact world*. One of the major components of the world back home is women, "our women", who are warm, nurturant and ultra-feminine. Women, at least, "our women," are not part of war. Indeed, one of the reasons for fighting is to protect our women and the rest of what is in the image of the world back home. If we allow these women into combat with us, then this psychological differentiation cannot be maintained and we lose this psychological defense. (Segal 1982, 278, emphasis added)

For the state, the idea of women in combat not only dissolves the romanticism of war, but may also be seen as defective and disorderly, or as a threat to the state's ability to provide national security for the maintenance of sovereignty.[8] Within such a frame, preservation of the state has its best chance through male-only military groups that exclude women from military roles. In other words, policies that exclude women from elite military roles are not actually about women, but rather about men (MacKenzie 2015).

Matters of the military are intrinsically connected to and bound up in notions of state sovereignty, which assign specific bodies to specific ends through a series of gendered practices. Bodies—both physical and material—are delineated in terms of insiders and outsiders, internal space and external space, subjects and "othered" subjects, protectors and the protected. Each dichotomy is gendered, socially defining appropriate/inappropriate manifestations of masculinities and femininities, pitting the dominant masculine against its inferior feminine counterpart. In its most obvious form, sovereignty establishes (gendered) borders; the state and its territories are thought to be feminine, while protection provided to her—most often in some form of armed forces—is masculine. For the purposes of the state, sovereignty dichotomizes "an orderly, internal space and an outside space of danger and disorder," a discursive maneuver that defines "violence as an intrusion upon the nation-state from an 'other' located outside of state boundaries" and from which national security must be provided (Wilcox 2013, 3). Women, specifically, are positioned as the ultimate "other," existing within the domestic sphere insulated from the dire world of armed conflict.

Rather than moving toward broadening the parameters for women's access to military service, the rational military actor is invested in these gendered dimensions of sovereignty. In its most fundamental form, the historical use of this narrative has allowed military officials to make the case that women should not be afforded the opportunity to serve equally to men. By specifically evoking concerns of national security, military officials position the women in combat ban as a necessity for maintaining the health of the sovereign state. Interest groups that argue for "equality between genders" allow the military to frame itself as a rational actor whose policy decisions are made in the best interest of the nation, as opposed to the (feminized) call for equality, which is seen to be based on emotion and sensitivity, and is therefore inferior. In the context of such gendered oppositions, women are framed as an existential threat to the state's existence. As their bodies attempt to prove their stability through conforming to the institution's masculinized standards, they are instead

framed as leaky through biological references that preclude them from ever being able to meet those standards. Hence, any integration of women into combat positions must proceed at a glacial pace.

EXCLUSIONARY TROPES IN THE COMBAT BAN

Examining the debates surrounding the creation of the AVF reveals the frequent recourse to notions of the military as a rational actor that must tread carefully in developing policies concerning women. As noted above, many of the arguments used in the "women in combat" discourse are made through reference to biological differences. On the surface, the biological argument is a simple one; because of physiological differences, women do not have the same strength or stamina as men, are therefore weaker, and less able to complete tasks associated with combat infantry. However, two essential assumptions underpin the narrative. First, that physical difference between men and women cannot be overcome, regardless of how much training women undergo, and second, that combat is a special form of activity, one which requires unique physical traits which women lack. These assumptions allow the narrative to link combat directly with manhood.

A closer look at this argument reveals the misogynist presuppositions that informed the combat exclusion policy. I highlight four strands that were used most frequently in the Congressional hearings between 1970 and 2015. These are tropes related to women's hormones and menstrual cycles, pregnancy and motherhood, sexual assault in the military, and medical differences in caring for women in the combat theater. While distinct, each trope framed women's bodies as leaky. While the military tried to present these reasons as rational causes to limit women's participation in combat roles, the claims typically relied on unfounded assertions based on gut feelings or inchoate emotions, rather than empirical evidence and calculations typical of rational actors.

Women in the Rollout of the AVF

Initially, the creation of the AVF acted as a strong factor in expanding the roles available to women serving in the military. Indeed, the success of the AVF was viewed by some in Congress as dependent on women. The expansion of women's service was established through a series of enacted

policies. However, when the Senate Armed Services Committee considered the issue of women in combat during hearings on reinstituting registration for Selective Services in 1979, confusion arose over the military's position on women's role in the military. Arguing for differential treatment between men and women's eligibility for the Selective Service program,[9] the Committee suggested "that it is not in the best interest of our national defense to register women for the Military Selective Service Act, which would provide needed military personnel upon mobilization or in the event of a peacetime draft for the armed forces" (United States Congress 1979, 5). By contrast, in that same year, hearings on the Defense Authorization Act of 1980 included several discussions focused on the need for a "fair and equitable reform of existing law providing for the registration and induction of persons for military service" (United States Congress 1980a, 2). If women were to be treated fairly in the military, the argument went, they should be equally eligible to be drafted during a time of war. Proponents of the measure believed it would provide them "with greater flexibility in determining assignments" (2). Unsurprisingly, views on whether women could serve in combat directly informed debates on whether women should be required to register for the Selective Service program. Military officials argued that women would have to be included in some form of Selective Service ensuring "we believe in equality" and "won't deny anyone the opportunity to serve their country" (12).

But despite the push to include women in a reformed version of the Selective Service program, the proposal would not allow women to serve in combat even if they were drafted. Instead, women would be more heavily utilized in support services, which in turn would make more men available for combat. Military and Congressional leaders argued that men would ultimately make or break the AVF as the essential soldiers to engage in combat environments. When challenged during a hearing on the Equal Rights Amendment Extension before the Subcommittee on the Constitution of the Committee on the Judiciary in 1978 on how he believed this would establish equality between men and women, former Chairman of the Joint Chiefs of Staff, Admiral Thomas Hinman Moorer of the U.S. Navy, suggested that equality would not be established when all could serve in all positions and in the same amounts, but rather "achieved when both men and women are asked to serve in proportion to the ability of the Armed Forces to use them effectively" (Moorer in United States Congress 1978, 217). Using women effectively would mean using

them in ways that could support men who would see combat and receive continued valorization for that service.[10]

Women's potential duties in the AVF were continuously questioned. They were frequently characterized as having raging hormones that caused them to be overly sentimental and to make decisions emotionally, not rationally. But the opinions among high ranking military officials on the topic often seemed fluid as they shifted throughout the 1970s (Bailey 2009). In the mid-1970s, for example, Army Chief of Staff, Bernard Rogers expressed disdain over the thought of women serving in combat: "women with rifles and fixed bayonets holding a forward position gives me heartburn" (Rogers quoted in Bailey 2009, 163), a guttural reaction to women serving on the front lines of the combat theater. However, by 1978, Rogers' position became one which saw women as "an integral part of the Army. They are not 'part-time soldiers—here in peace, gone in war.' Should war come, women will deploy with their units and, like male soldiers, share all risks...inherent in their specialty" (163). The Congressional decree to allow women into all military academies provided further pressure to integrate women into combat positions. Women who entered the academies' first coeducational classes performed better than their male counterparts, surviving their first year in the academies at a higher percentage, and graduating as officers and at the top of their class. However, as this first class of women graduated in 1980, the military began to question how to utilize them if they would not be allowed to serve in combat.

Conservatives remained adamant that women's military roles should be limited, preserving their exclusion from combat opportunities. A review of the health of the AVF at the end of the 1970s also placed women on a list of concerns that were labeled "problems of quality, of readiness, of size and training and capability" (Bailey 2009, 165). This concern was particularly articulated through a protective narrative of the military's conceptualization of masculinity. For James Webb,[11] one of the most outspoken critics of women in the military and military academies, the creation of the AVF was an abandonment of the "essential masculinity and rigorous nature" of the institution (Webb 1980, 3). Webb rejected any comparison between the integration of African American men and the integration of women into the armed forces as "unfounded bigotry" (3). Instead, Webb argued that women would never be combat ready soldiers, nor would they be able to lead male soldiers into combat because of biological differences, writing "despite what some would like to think, men and women are fundamentally different" (3).

Exploiting Women's Talents Versus Containing Threats

During the 95th Congress (summer of 1977) hearings by the Subcommittee on Priorities and Economy in Government of the Joint Economic Committee focused attention on the role of women in the Military. Four years after the initial rollout of the AVF, the subcommittee was especially "interested in the results of the actions taken regarding the employment, and utilization of women and the results in terms of economics, economic savings, costs and productivity gains or losses" (United States Congress 1977, 1). The subcommittee was interested in understanding why, just four years after the creation of the AVF, there seemed to be a lull in further increasing the number of women serving in the armed forces. Overall, the subcommittee viewed the gains attributed to the increasing involvement of women in the military as positive, but simultaneously acknowledged the "national security factor involved" (5).

Chaired by William Proxmire[12] (D-WI), the hearing framed the failure to increase women's numbers in the military (as compared to the initial increase experienced when the AVF was created), as careless, unsubstantiated prejudices. This framing was further extended in Senator Proxmire's opening remarks to characterize a painful lack of women in U.S. politics, pointing out that at the time of the hearings zero women were serving in the Senate, and only a few in the House. His remarks specifically suggested a notion of gendered institutionalism in which half of the population was being made completely invisible.

> I might conclude by saying that almost nobody in this society, particularly in the U.S. Senate, can speak by pointing to their own institution. Of the 100 Senators, there is not a woman who is a Senator. This institution is completely male. We are ignoring half of our intelligence, half of our population, half of our resources. The House has a little better situation but not much, with just a very few women in the House. Only 2 of our 50 Governors are women. So, throughout our society, we have certainly failed to provide the opportunity that we should provide that would serve our country so well. (Proxmire in United States Congress 1977, 3)

Overall, the committee members argued that the military was not taking full advantage of women's potential.

For the military branches, the question of women's qualifications for combat was central to the hearing discussions (United States Congress 1977, 5). In opening statements, Robert L. Nelson, Assistant Secretary of the Army for Manpower and Reserve Affairs, as well as Major General J. P. Kingston, Assistant Deputy Chief for Staff Personnel for the Army,

made comments about women engaging in sexual relations with male counterparts and becoming pregnant as a result of it, limiting the Army's ability to deploy them to combat zones when necessary.[13] For the Army, the view remained that "substantially more men than women [would be needed] to insure mission accomplishment...We must not institute policies which will drive men away from the recruiter and enlistment counselor" (Nelson in United States Congress 1977, 6). For the Air Force, the possibility of women serving in combat roles was a larger issue than just effectiveness. Referring to the combat exclusion policy being put in place by Congress, not by the military as an institution, Antonia Handler Chayes, Assistant Secretary of the Air Force for Manpower, Reserve Affairs, and Installations, questioned, "We have to ask ourselves, are we ready to ask women to serve as crew members on aircraft that may be shot down in hostile territory? Are we ready to require women to turn the keys to launch nuclear missiles?" (Handler Chayes in United States Congress 1977, 11). The linking of these two disparate concerns was suggestive of a misogynist militarized culture, which was at least partially established through the notion that women needed to be protected from war, and from themselves. The reference and questioning directed specifically at women indicated that both being shot down in hostile territory and launching nuclear missiles were activities deemed appropriate (only) for the hypermasculine warrior. The suggestion that women's possible service in combat capacities was a larger, general society issue in which appropriate roles in the military were attached to genders differently also made it possible for the institution to systematically mask the manipulation of women's exclusion.

While Senator Proxmire's urging to include more women in the military was admirable and may be considered an attempt to force the hand of the institution toward becoming more gender inclusive, many of the arguments and comments made during the hearing had strong sexist underpinnings. One such example was an argument which urged that more women be allowed to serve in the military simply for the sake of attracting more men to serve. When Senator Proxmire pointed out to the Army that their recruitment numbers were projected to fall 40% short of recruitment goals through 1985, he suggested that similar to colleges switching to coeducational campuses, research showed that the Army would have an easier time recruiting highly qualified males for service if there were more women in the service. "I know that one of the reasons why young men go to a particular college or did a few years ago when so many of them were segregated was because it was coeducational and young women for the

same reason. It would seem to me you have a better opportunity to attract male recruits in the Army if you had more women in the Army" (Proxmire in United States Congress 1977, 17).

Similarly, in an exchange between Senator Proxmire and Air Force Secretary Handler Chayes, Senator Proxmire returned to Secretary Handler Chayes' opening statement, which raised the issue of women at the helms of the switch for nuclear weapons.

Senator Proxmire: By raising this question, are you saying there are characteristics inherent in women that make them less responsible or less capable of making decisions than men? Why in the world should we hesitate to give women this responsibility any more than say appointing women as Assistant Secretaries of Defense or Assistant Secretary of the Air Force? (United States Congress 1977, 20).

Secretary Handler Chayes: There is no reason. The only issue here is the definition of 'combat.' I only raise it to indicate that currently it falls within the definition of 'combat,' and, because there is no clear-cut definition, interpretations do occur (20).

Senator Proxmire: I don't understand why you raised that as a question, a question apparently on your mind. Why do you ask that (20)?

Secretary Handler Chayes: The question already has been raised, not in my mind, Senator, but as I said previously the question is in the minds of many Congressmen and in the minds of many military personnel. I only raised it to illustrate the deep concern of military and congressional leadership. At present they seem to have an uncomfortable feeling about imposing this responsibility on women (20).

Senator Proxmire: What is your own personal feeling? Can you make a good, clear, strong recommendation one way or the other?...To wit, I want to get your views. I want to get your views (20).

Secretary Handler Chayes: My personal views (20)?

Senator Proxmire: Yes, indeed, your views on whether or not women should be in the position of turning the key on a nuclear missile (20).

Secretary Handler Chayes:	I guess I would be more comfortable with my views after the study. My view going in is really a wide open one—that is theoretically, I expect the study to offer a practical military viewpoint that will enhance my understanding (20).
Senator Proxmire:	I can't understand any difference here. Here—some people are that men have some different characteristics, certainly, there may be a difference in characteristic with respect to physical strength, that is the only one I can think of offhand. I don't know how people can argue that women are more emotional or less emotional, whatever. This is the thing that concerns me as to the reason why some people in the services seem to think women shouldn't be allowed to have this critical position. Certainly, there is no greater danger than in many other positions and probably a lot less (20).
Secretary Handler Chayes:	Senator, I think really it is the combat definition that is clouding the issue now. I think that that issue is going to be cleared (20).

This exchange reflects a particular institutional anxiety over the definition of combat in the new context of a changing landscape of war that was brought on by the Cold War. Evidently, it was not clear whether pressing a button on a nuclear weapon constituted combat. As I argue in the concluding chapter, this institutional anxiety is similar to an ongoing debate in the military more recently regarding drone warfare.

As the exchange between Senator Proxmire and Secretary Handler Chayes continued, the focus shifted to the question of whether women exhibited the necessary characteristics to launch a nuclear weapon.

Senator Proxmire:	Are there any characteristics of women which in your view would make them unacceptable in this position of being able to turn the key on a nuclear weapon? (United States Congress 1977, 21).
Secretary Handler Chayes:	I think you have to look at the definition of combat in the context of the statutes. In my view, there are no qualities which disqualify women from doing any job that a man can do except certain factors of strength. In my personal opinion, there are no quali-

	ties of emotions, or attitudes that would seriously disqualify them from any job (21).
Senator Proxmire:	Very good. I understand your position, then. That would apply in missile silos, too? Certainly, they have the strength to turn that key (21).
Secretary Handler Chayes:	And they need to be able to obey orders (21).
Senator Proxmire:	What was that (21)?
Secretary Handler Chayes:	I said they need to be able to obey orders and (21)—
Senator Proxmire:	Do you have any feeling there is any difference between men and women in obeying orders (21)?
Secretary Handler Chayes:	If anything, women tend to be more obedient. [Laughter] (21).
Senator Proxmire:	I want to introduce you to my wife sometime [Laughter] (21).

In assessing the military branches' concerns regarding integrating female enlistees into combat related positions, Senator Proxmire also raised an important differentiation between law and policy during the hearing. In an exchange with General Nelson from the Army, Senator Proxmire noted that combat positions would not necessarily have to be filled by men—and therefore recruitment efforts could be concentrated elsewhere as well—given that there was no law that said so. Though General Nelson admitted that "there is no law, but it is policy" (Nelson in United States Congress 1977, 27), he argued that whether women should be allowed to qualify for those positions would depend on results from Army administered stress management tests, suggesting that men and women handle high stress situations differently because of biological differences. Despite women performing better in basic training, as well as their first year in the military colleges, General Nelson did not view the integration of women into combat positions as a suitable way to meet AVF recruiting numbers, or as a way to increase the quality of the branch. When Senator Proxmire pointed out that the recruitment of women was also less costly in comparison to the recruitment of men, he pressed General Nelson on his personal views on women serving in the Army. General Nelson responded,

> Well, of course, that is difficult to answer. I have never been in the position of, say, as commander in the field or as a soldier. Let me say this, and I think it is important, that the combat role is one in which an immense amount of stress is found. In the testing we are doing, we are looking at that stress. Up

until very recently women were not given any combat training at all. They are now given 7 weeks basic training; the same basic combat training as men. This is being done because in today's world even those people in the relatively far rear areas could come under attack. We think it is very important for everyone in the Army to know the basics so that they can defend their position and defend themselves if they have to. And I think that is important. That is a little bit different than carrying the primary responsibility of direct combat with the hopeful result that you will engage and kill the enemy. The defense is a little different. But we have changed many things; for example, not long ago women were not allowed to throw a hand grenade. That is no longer true...I think that women not only handled themselves well, they did as well as the men. (Nelson in United States Congress 1977, 29)

This response too is suggestive of a context specific anxiety of women serving in certain roles in the Cold War era. Although General Nelson seemed to try to steer the conversation in that direction, Senator Proxmire remained adamant that biological differences did not exist between men and women in handling stress, and that the Army could change its policy on women serving in combat positions overnight, if it so chose to. He asserted, "I never heard of anybody making that contention, that women are less able to withstand stress than men...I think we ought to make the assumption, unless there is evidence that there is a difference, that some women can stand stress and some can't, some men can stand stress and some can't, there is no sex difference" (29). By the end of the exchange, General Nelson reframed the issue of women serving in combat; he shifted the matter from an issue of capability to a question of whether the general American public was ready to accept it.

Whether or not American society at large would be accepting of women in combat played a direct role in how the military branches recruited—or rather, chose not to recruit—women for non-traditional positions. In testimony provided by the Air Force, for example, it also became apparent that the branch was deliberately not recruiting women for non-traditional roles that had already been open to them. As suggested by the Brookings Institute Study, "Women in the Military," between 1973 and 1977, the Air Force could have opened up an additional 450,000 enlisted roles to women interested in entering the Air Force without overstepping the combat exclusion policy (referenced in United States Congress 1977, 38). Though the branches continued to argue that women in combat was an issue up for debate with the American public, the study concluded that the Air Force's gender demographics were more probably a result of the military branch's preference for maintaining a predominantly male institution.

The Leaky Body's Production of Time Lost

During the 1980s, Congress considered several bills designed to advance women's integration within the military. By the beginning of the decade, women comprised roughly 8% of the armed forces and were projected to comprise 13% by 1983. In 1987, when women made up only 10% of the overall armed forces, Senator William Dickinson (R-AL) introduced a bill to try to bring uniformity of interpretation of the definition of combat across the branches. In that same year, Senator Proxmire and Senator Cohen introduced a bill that pushed for opening all combat support positions within the Army, Navy, and Air Force to women.[14] But confusion around the military's intention to become more gender-inclusive became even more prominent during this decade as both Congressional and Military officials argued the AVF had been a failure and reverted back to whether or not reinstating the draft should be considered.

Two main tropes were used to criticize the abandonment of the draft. Because the AVF tried to recruit enlistees with pay increases and improved benefits, some argued that the AVF was astronomically more expensive than its conscription-based version. In arguing that considerable money could be saved if the costs associated with the AVF were eliminated, the military again positioned itself as a rational, fiscally responsible actor. That expenditures for "the active forces of the AVF [had] been within one and a half percent of the Congressionally authorized limits since 1974" (United States Congress 1980b, 156) dispelled the notion of the hyper-expensive AVF. Opponents of the AVF also argued that the AVF was weaker than the draft armed forces despite the DoD itself claiming in a 1978 report that the AVF was superior to the draft military. The DoD cited fewer disciplinary problems and higher levels of education among the ranks. In the 1980 hearing before the House Committee on the Judiciary regarding Judiciary Implications of Draft Registration, National Organization for Women (NOW) President, Eleanor Smeal, argued that opponents of gender-integrated forces offered arguments laden with sexist and racist bias. "Underneath the surface of these arguments are the racist and sexist attitudes, which pervade our society, coupled with undisguised economic exploitation" (Smeal in United States Congress 1980, 156).

Myths also persisted surrounding women's performance in the military in relation to the question of how a draft that excluded women would be effective. Some suggested that women were simply not needed; women, rather, needed "protection" from the draft. Yet others suggested that equal citizen-

ship depended on women's military service. "Omission from the registration and draft ultimately robs women of the right to first class citizenship and paves the way to underpaying women all the remaining days of our lives" (United States Congress 1980b, 157). Feminist groups argued that the masculinist justification of excluding women from service under a draft system had a larger implication, suggesting that it was acceptable to exclude them from decisions critical to the nation. Although women had been serving ably in the military for decades, the focus on the combat ban rendered this service invisible and diverted attention away from the question of what women could do "with the same training, benefits and salary as men" (159).

Proponents of reinstituting the draft could foresee that the move would regender the male-dominant institution, curtailing women's participation once again. It was well known that women served in much smaller proportions under the mandatory conscription model. Under the draft, for example, women comprised only 1% of the Army, compared to the Army's projection that women would constitute 13% by 1983 under a volunteer system. The military cited a range of reasons for the purported need for a registration system,[15] including "show[ing] the USSR that we mean business" (United States Congress 1980b, 156) and increasing our mobilization capabilities. But these arguments were far from persuasive. As Judy Goldsmith pointed out, a system of name gathering might "sound strong to Americans[16] who want to show that we are serious, in reality it proves nothing to the USSR which appreciates fully how little names on a list actually mean" (Goldsmith in United States Congress 1980, 156). The military's appeal to Cold War threats to national security offered a pretext for regendering the male-dominant institution, reversing gains in women's integration. Whether or not the military explicitly sought to use the draft as a vehicle to limit women's participation in the armed services, their pro-conscription narrative consistently drew attention away from potential negative implications of reinstating the draft on women's presence, focusing instead on national security in the context of a deepening Cold War. If the concern was truly about the ability to mobilize the most qualified individuals as quickly as possible in the case of a war, the military should have instead removed the "sex discriminatory restrictions" (156) from service in combat, which placed women as inferior, second class citizens within the military apparatus.

Intimately tied to arguments against allowing women into combat, and against allowing them to be drafted, was the issue of "time lost," time off taken by soldiers before their service contracts expire. For women, time lost was specifically framed in two contexts: one, in regard to their hor-

mones and menstrual cycles; the other, in regard to pregnancies and motherhood. Women's hormones were discussed in connection to their emotionalism and ability to make good decisions. Women's debilitating menstrual cycles were discussed through essentialist assertions that viewed all women in the military as a monolithic group, assuming that all women menstruate on a monthly basis and that all women deal with menstruation similarly. The nuances around these tropes, however, varied slightly during different eras. In the 1980s, panelists cited unsubstantiated generalities about women's emotionalism, argued that women were more likely to drop out of the military because of marriage, and cited menstruation as making women physically less able to complete tasks, suggesting that menstruation would cause female soldiers to lose more active time and mobilize more slowly than their male counterparts. This assertion marked women's bodies as leaky, incomplete, unstable, and unreliable, and an overall threat to the security of the sovereign nation.

Time lost by female soldiers from menstruation, pregnancies, and motherhood was further associated with hypothetical economic repercussions. The military framed itself as *homo economicus*, the rational actor concerned solely with the financial interests of the nation. Yet, their particular claims reflected the misogynist views of a hypermasculinized institution. These arguments suggested that major monies were spent and lost by having to invest in the recruitment of new soldiers to cover for women who took time off or left the forces. During a hearing on the performance of the military in Desert Storm, Air Force Chief of Staff Tony McPeak categorized pregnant soldiers as "nondeployable" (McPeak in United States Congress 1994, 103). Studies conducted by the NOW, however, showed "that there is little difference in the time lost by women and men and that, in fact, it appears that less time is lost by women. This is true even when pregnancy, the largest single factor in lost time for women, is included" (United States Congress 1980b, 160).[17] By contrast, in the 1990s, testimonies provided at Congressional hearings tied pregnancies established during a woman's term of service in the military with "policies that subsidize and encourage single parenthood" (United States Congress 1994b, 64). As the endemic nature of sexual violence in the military became more apparent during the decade, military officials argued that gender-integrated training increased the likelihood of sexual assault, and consequently pregnancies. Military officials often viewed this as a discrepancy in personnel policies, claiming "[s]exual assault misconduct is supposedly wrong, but pregnancy policies provide generous benefits to

unmarried servicewomen" (United States Congress 1997, 37).[18] Ultimately, forecasting high pregnancy rates positioned women as a national security threat that would detract from combat efficiency and readiness if gender-integrated training were permitted. Instead, the military encouraged women to serve in combat support positions, keeping the most prestigious roles for their male counterparts.

Combat and the Marginalization of Military Sexual Assault

The push to increase women's full military integration renewed in the 1990s as a result of feminist challenges of traditional gender assumptions, women's more prominent service in Operation Desert Storm, as well as political pressure on Pentagon officials after the sexual assault of women and men at the Tailhook Convention came to light (see Chap. 4). As part of that push, legislation was enacted that called for the elimination of barriers that stopped women from serving on combat-related aircraft and vessels of the Navy. Nonetheless, the Direct Ground Combat Definition and Assignment Rule Efforts identified six core reasons to maintain the combat exclusion policy. These included the suggestion that women lacked physical strength and stamina, problems related to living conditions, and lack of support from both Congress and the American public on the issue of women in combat. Moreover, this rule posited that including women "would not contribute to and could well detract from the readiness and effectiveness of the units" (United States Congress 1998, 6).

Reasoning that conflated combat service with sexual assault in the military took center stage during this period. Based on Congressional hearings, arguments most prominent against increasing women's access to combat positions during the 1990s were made through biological narratives that allowed Congressional proponents of the ban to argue that women were at increased risk of being sexually assaulted in combat situations just by being (assumed) women. By keeping women out of combat, the military was necessarily protecting them from their male counterparts. Framed in the shadows of the 1991 Tailhook Scandal, where Marines and Navy aviation officers sexually assaulted more than 80 women and several men during a military symposium in Las Vegas, several Congressional hearings directly tied the issue of women in combat[19] support roles to the Tailhook Sexual Assault Scandal. This framing allowed the military to argue that women's very livelihood and physical safety depended on keeping them out of combat environments altogether. In an exchange during the hearing, Senators raised

questions about an endemic culture of sexual violence in the military. Senators pressed military officials on whether any research had been conducted, data gathered, and analyzed on sexual assault and what each branch planned to do about it in order to ensure women's safety once they were officially allowed to serve in combat. In their response, military officials took an unexpected stance, suggesting that women's equality in the military would solve the problem of sexual harassment and assault. When women and men trained and served together, military men would treat women as equals. In this reversal, the military implied that it favored further integration of women, while masking their own responsibility for the low numbers of women in the military, and conveniently sweeping a major gendered form of violence under the table (see Chap. 4).

When further evidence of systematic sexual assault in the military emerged, Congressional representatives continued to question whether allowing women into combat roles would only make matters worse. During a 1997 Congressional hearing on Gender-Integrated Training before the Senate Subcommittee on Personnel of the Committee on Armed Forces, Subcommittee Chair Dirk Kempthorne [R-ID] argued that the 1996 Aberdeen Proving Ground sexual assault case in which groups of commissioned and non-commissioned officers sexually assaulted female trainees who were under their command may have occurred because "men and women were training together," which may have "contributed to the violations and the leadership failures that we saw at Aberdeen" (Kempthorne in United States Congress 1997, 2).

Medical Treatment and Care of Female Soldiers

U.S. participation in the war in Afghanistan and the war in Iraq in the aftermath of the September 11th attacks renewed the question of where to draw the line between combat and non-combat for the women deployed in those theaters of war. Between 2002 and 2013, Congress held 19 hearings concerning modes of injuries sustained in those conflicts and the adequacy of the military health care system to deal with those injuries, and held 9 hearings on sexual assault in the military. The impact of women's service in war settings—proximate to combat but not acknowledged as combat assignments—surfaced in these hearings, which routinely framed women veterans as distinct and different from their male counterparts. During this period, several hearings refer to "unique problems" (209) women faced returning home after tours of duty in the wars. In some

hearings, women were described as "female warriors," an honorific designation typically reserved for those who have served in combat and suffered severe wounds, including bodily maiming (see Chap. 4). By separating the discussion of the impact of war on women in the military, these hearings reinforced the perception that women's performance must always be measured as distinct from, but in comparison to, the male "norm," a subtle marginalization that removes women from the category of combatant. Discussions of post-deployment treatment programs for women create a double-edged sword that acknowledges women's service in combat theaters, while situating them in a peripheral space in need of separate study for conditions that deviate from those experienced by "ordinary" soldiers, who are presumed to be men.

But as Megan MacKenzie (2015) suggests, in certain instances, categorizing women's service in these wars as outside the boundaries of combat has a direct implication on services and treatments available to them as veterans. Despite the increased attention given to post-traumatic stress disorder (PTSD) in the context of wars in Afghanistan and Iraq, PTSD remains distinctly associated with "front-line combat," the definitive, masculinized warrior space, which grounds accolades and warrants honors for participants. This border has been leaky, however. Women have historically been asked to participate in combat missions, but have not received adequate training or the necessary services and treatment post-deployment to deal with their physical and psychological wounds. The partially acknowledged participation of women within combat spaces prior to the lifting of the combat ban had contradictory results. Rather than celebrate the valor of women troops, many in the military chose to refortify the boundaries of combat service, reaffirming the logic of masculine protection. In this way, PTSD as a diagnosis is reserved for those who are officially acknowledged as serving in combat spaces. How the association between combat and psychological injuries is treated by the military, however, generates contradictions in the logic behind modes of injury sustained through combat and their link to militarized manhood (see Chap. 4).

The 2013 Congressional hearings also examined strategies for implementing the newly lifted ban on women in combat, along with its potential impact. In preparation for the end of the combat ban, the military rolled out new rhetoric, referring to "gender-neutral," "gender-mixed," "gender-segregated," "gender-separated," "gender-integrated," and "gender-restrictive" units. As the terminology in the hearings shifted toward a suggested degendering of combat units, these new terms were offset by the familiar

gendered narratives that raised suspicion about the wisdom of rescinding the exclusion policy. For example, as the military branches were tasked with considering recommendations to limit women's participation in certain roles, Major General Bennet Sacolick, head of the Special Operations Units, argued that those decisions would be based solely on the demands of unit cohesion, thereby conjuring up images that sustain the mythic masculine stereotype of the Band of Brothers. To participate in special operations units—according to Sacolick—women would have to achieve a level of integration that would preserve unit cohesion, readiness, and morale—thus holding women responsible for the attitudes of men in their units. To further stack the deck against women's participation, the military insisted that they would have to deploy "in small self-contained teams for long periods of time in austere, geographically isolated locations…in very close quarters" with men (United States Congress 2014, 52). The deployment of equal numbers of women and men was never considered. Instead, United States Special Operations Command (SOCOM) suggested that likely social interactions within such gender-imbalanced units would have to be considered closely. The potential for a lack of "social cohesion" between men and women was of particular concern: if units did not "feel emotionally bonded with each other, […] task cohesion, referring to the mutual commitment among the individual team members in achieving the group objective" (53) could suffer.

Conclusions

Despite the opportunity to reimagine the gendered terms of military service, the need to reassert notions of masculinity that were thought to be crumbling in the United States in wake of the Vietnam War outweighed calls for gender equality. The narrative of deliberate and incremental inclusion allowed the military to reconfirm masculinity, albeit, in a new form, as integral to the profile of the new volunteer-soldier. It also allowed it to position itself as gender inclusive through the increased recruitment of women, while simultaneously masking the exclusionary practices around their integration. Anxieties ensuring women were kept away from combat service carried on for over 50 years; the discursive tools used to do so highlight the nuances of tropes and arguments for keeping women out of combat as they shifted over the decades. Most prominent of these were concerns over biological characteristics and capacities stereotypically attributed to female bodies. But perhaps more importantly, concerns over

women in combat were not really concerns about women at all. Rather, anxieties expressed by military officials often centered around how male soldiers would deal with women's full integration, and what impact allowing women into combat would have on mythical conceptions of unit cohesion and (hetero)male bonding in the Band of Brothers social order of loyalty and unity, showing the fundamental fragility of this masculine order. The Band of Brothers image was not only a male-only club, but one which gave life and limb for each other's protection. Women's integration into this social model was unfathomable as women were viewed as incompatible with processes of harm, injuring, or death produced by serving in war theaters. The potential for harm being inflicted on the body is a key way in which combat experience has been tied to militarized manhood. Limiting women's access to those roles has systemically stopped them for rising in the ranks and attaining leadership positions. But as the terms of "combat" change, and the number of casualties decrease as a result of advances in military technology, new "alternative" modes of harm may emerge, and I explore these in the next chapter. How harm and wounding is defined by the military and assigned to specific bodies in light of these changes, as Chap. 3 shows, might suggest the possibility for potential shifts to the military's system of heteromale privilege.

Appendix: Congressional Hearings for Chapter 3 (In Chronological Order)

Final report to the Congress of Secretary of Defense Melvin R. Laird before the House Armed Services Committee: Hearing Before the Committee on Armed Services, House of Representatives, 93rd Cong. 1, (1973).

To Provide Recognition to the Women's Air Force Service Pilots for Their Service During World War II by Deeming Such Service to Have Been Active Duty in the Armed Forces of the United States for Purposes of Laws Administered by the Veterans Administration: Hearing Before a Select Subcommittee of the Committee on Veterans' Affairs, House of Representatives, 95th Cong. 1, (1977).

The Role of Women in the Military. Hearings before the Subcommittee on Priorities and Economy in Government of the Joint Economic Committee. 95th Cong. 1, (1977). Second session, September 1, 1977.

Military Medical Health and Research: Hearings before a Subcommittee of the Committee on Government Operations, House of Representatives, 95th Cong. 2, (1978). Second session, July 26 and 27, 1978.

Selective Service System Plans Officer Personnel Management: Hearing before the Subcommittee on Manpower and Personnel of the Committee on Armed Services, United States Senate, 96th Cong. 1 (1979).

Judiciary Implications of Draft Registration—1980: Hearings before the Subcommittee on Courts, Civil Liberties, and the Administration of Justice of the Committee on the Judiciary, House of Representatives, 96th Cong. 2, (1980).

Drug Abuse in the Military—1981: Hearing before the Select Committee on Narcotics Abuse and Control, House of Representatives, 97th Cong. 1, (1981).

Department of Defense Authorization for Appropriations for Fiscal Year 1984: Hearings before the Committee on Armed Services, United States Senate, 98th Cong. 1 (1983).

Department of Defense Authorization and Oversight: Hearings on H.R. 5167, Department of Defense Authorization of Appropriations for Fiscal Year 1985, and Oversight of Previously Authorized Programs before the Committee on Armed Services, House of Representatives, 98th Cong. 2 (1984).

Department of Defense Authorization for Appropriations for Fiscal Year 1985: Hearings before the Committee on Armed Services, United States Senate, 98th Cong. 2 (1984).

Department of Defense Authorization for Appropriations for Fiscal Year 1987: Hearings before the Committee on Armed Services, United States Senate, 99th Cong. 2 (1986).

Department of Defense Authorization for Appropriations for Fiscal Years 1988 and 1989: Hearings before the Committee on Armed Services, United States Senate, 100th Cong. 1 (1987).

Nominations before the Armed Services Committee, United States Senate, 100th Cong. 1, (1987).

Department of Defense Appropriations for 1990: Hearings before the Department of Defense Subcommittee of the Committee on Appropriations, House of Representatives, 101st Cong. 1 (1989).

Merchant Mariners Fairness and Memorial Acts: Hearing before the Subcommittee on Merchant Marine of the Committee on Merchant Marine and Fisheries, House of Representatives, 101st Cong. 1, (1989).

Nominations before the Armed Services Committee, United States Senate, 101st Cong. 1, (1989).

Department of Defense Authorization for Appropriations for Fiscal Year 1991: Hearings before the Committee on Armed Services, United States Senate, 101 Cong. 2, (1990).

The Impact of the Persian Gulf War and the Decline of the Soviet Union on How the United States Does Its Defense Business: Hearings before the Committee on Armed Services, House of Representatives, 102nd Cong. 1, (1991).

Nominations before the Armed Services Committee, United States Senate, 102nd Cong. 1, (1991).

Gender Discrimination in the Military: Hearings before the Military Personnel and Compensation Subcommittee and the Defense Policy Panel of the Committee on Armed Services, House of Representatives, 102nd Cong. 2, (1992).

Implementation of the Repeal of the Combat Exclusion on Female Aviators: Hearing before the Military Personnel and Compensation Subcommittee of the Committee on Armed Services, House of Representatives, 102nd Cong. 2, (1992).

Nominations before the Armed Services Committee, Untied States Senate, 102nd Cong. 2, (1992).

Women in the Military: the Tailhook Affair and the Problem of Sexual Harassment: Report of the Military Personnel and Compensation Subcommittee and Defense Policy Panel of the Committee on Armed Services, House of Representatives, 102nd Cong. 2, (1992).

Assessment of the Plan to Lift the Ban on Homosexuals in the Military: Hearings before the Military Forces and Personnel Subcommittee of the Committee on Armed Services, House of Representatives, 103rd Cong. 1, (1993).

Department of Defense Authorization for Appropriations for Fiscal Year 1994 and the Future Years Defense Program: Hearings before the Committee on Armed Services, United States Senate, 103rd Cong. 1 (1993).

Hearings on National Defense Authorization Act for fiscal year 1994—H.R. 2401 and Oversight of Previously Authorized Programs before the Committee on Armed Services, House of Representatives, 103rd Cong. 1, (1993).

Hearings on National Defense Authorization Act for Fiscal Year 1994, H.R. 2401 and Oversight of Previously Authorized Programs: Hearings before Military Personnel and Compensation Subcommittee Hearings on Personnel Authorizations of the Committee on Armed Services, House of Representatives, 103rd Cong. 1, (1993).

Policy Concerning Homosexuality in the Armed Forces: Hearings before the Committee on Armed Services, United States Senate, 103rd Cong. 2, (1993).

Policy Implications of Lifting the Ban on Homosexuals in the Military: Hearings before the Committee on Armed Services, House of Representatives, 103rd Cong. 1, (1993).

Assignment of Army and Marine Corps Women under the New Definition of Ground Combat: Hearing before the Military Forces and Personnel Subcommittee of the Committee on Armed Services, House of Representatives, 103rd Cong. 2, (1994).

Department of Defense Authorization for Appropriations for Fiscal Year 1995 and the Future Years Defense Program: Hearings before the Committee on Armed Services, United States Senate, 103rd Cong. 2, (1994).

Morale, Welfare, and Recreation and Commissary Issues: Hearings before the Morale, Welfare, and Recreation Panel of the Committee on Armed Services, House of Representatives, 103rd Cong. 2, (1994).

Sexual Harassment of Military Women and Improving the Military Complaint System: Hearing before the Committee on Armed Services, House of Representatives, 103rd Cong. 2, (1994).

Hearing Before the Military Forces and Personnel Subcommittee of the Committee on Armed Services, "Desert Storm Mystery Illness/Adequacy of Care." House of Representatives, One Hundred Third Congress, Second Session, Hearing Held, March 15, 1994.

Nominations before the Armed Services Committee, United States Senate, 104th Cong. 1, (1995).

Department of Defense Appropriations for 1997: Hearings before a Subcommittee of National Security of the Committee on Appropriations, House of Representatives, 104th Cong, 2 (1996).

Extremist Activity in the Military/Committee on National Security, House of Representatives, 104th Cong. 2, (1996).

Army Sexual Harassment Incidents at Aberdeen Proving Ground and Sexual Harassment Policies within the Department of Defense: Hearing before the Committee on Armed Services, United States Senate, 105th Cong. 1 (1997).

Department of Defense Appropriations for Fiscal Year 1998: Hearings before a Subcommittee of the Committee on Appropriations, United States Senate, 105th Cong. 1 (1997).

Gender-Integrated Training and Related Matters: Hearing before the Subcommittee on Personnel of the Committee on Armed Services, United States Senate, 105th Cong. 1, (1997).

Department of Defense Authorization for Appropriations for Fiscal Year 1999 and the Future Years Defense Program: Hearings before the Committee on Armed Services, United States Senate, 105th Cong. 2 (1998).

Hearings on National Defense Authorization Act for Fiscal Year 1999— H.R. 3616 and Oversight of Previously Authorized Programs, before the Committee on National Security, House of Representatives, 105th Cong. 2, (1998).

Nominations before the Armed Services Committee, United States Senate, 105th Cong. 2, (1998).

Veterans' Employment Regarding Civilian Credentialing Requirements for Military Job Skills: Hearings before the Subcommittee on Benefits of the Committee on Veterans' Affairs, House of Representatives, 106th Cong. 1, (1999).

Issues Affecting Families of Soldiers, Sailors, Airmen, and Marines: Hearings before the Subcommittee on Personnel of the Committee on Armed Services, United States Senate, 108th Cong. 1, (2003).

Department of Defense Appropriations for Fiscal year 2005: Hearings before the Defense Subcommittee of the Committee on Appropriations, 108th Cong. 2, (2004).

Policies and Programs for Preventing and Responding to Incidents of Sexual Assault in the Armed Services: Hearing before the Subcommittee on Personnel of the Committee on Armed Services, United States Senate, 108th Cong. 1, (2004).

Status of the U.S. Army and U.S. Marine Corps in Fighting the Global War on Terrorism: Hearing before the Committee on Armed Services, United States Senate, 109th Cong. 1 (2005).

Department of Defense Appropriations for Fiscal year 2006: Hearings before the Defense Subcommittee of the Committee on Appropriations, United States Senate, 109th Cong. 1, (2006).

Department of Defense Appropriations for Fiscal Year 2007: Hearing before the Committee on Appropriations, United States Senate, 109th Cong. 1, (2006.)

Nominations before the Armed Services Committee, United States Senate, 109th Cong. 2 (2006).

Sexual Assault and Violence Against Women in the Military and at the Academies: Hearing before the Subcommittee on National Security, Emerging Threats, and International Relations of the Committee on Government Reform, House of Representatives, 109th Cong. 2, (2006).

Don't Ask, Don't Tell Review: Hearing before the Military Personnel Subcommittee of the Committee on Armed Services, House of Representatives, 110th Cong. 2 (2008).

Sexual Assault in the Military: Hearing before the Subcommittee on National Security and Foreign Affairs of the Committee on Oversight and Government Reform, House of Representatives, 110th Cong. 2, (2008).

Department of Defense Authorization for Appropriations for Fiscal Year 2010: Hearing before the Committee on Armed Services, United States Senate, 111th Cong. 1 (2009).

Nominations before the Senate Armed Services Committee: Hearings before the Committee on Armed Services, United States Senate, 111th Cong. 1, (2009).

Psychological Stress in the Military: What Steps Are Leaders Taking?: Hearing before the Military Personnel Subcommittee of the Committee on Armed Services, House of Representatives, 111th Cong. 1, (2009).

Sexual Assault in the Military: Prevention: Hearing before the Military Personnel Subcommittee of the Committee on Armed Services, House of Representatives, 111th Cong. 1 (2009).

Department of Defense Authorization for Appropriations for Fiscal Year 2011: Hearings before the Committee on Armed Services, United States Senate, 111th Cong. 2 (2010).

Ike Skelton National Defense Authorization Act for Fiscal Year 2011 2010.

Military Construction, Veterans Affairs, and Related Agencies Appropriations for 2011: Hearings before a Subcommittee of the Committee on Appropriations, House of Representatives, 111th Cong. 2, (2010).

Sexual Assault in the Military, part IV: Are We Making Progress?: Hearing before the Subcommittee on National Security and Foreign Affairs of the Committee on Oversight and Government Reform, House of Representatives, 111th Cong. 2, (2010).

Repeal of Law and Policies Governing Service by Openly Gay and Lesbian Service Members: Committee on Armed Services, House of Representatives, 112th Cong. 1, (2011).

Soldier and Marine Equipment for Dismounted Operations: Hearing before the Subcommittee on Tactical Air and Land Forces of the Committee on Armed Services, House of Representatives, 112th Cong. 1, (2011).

Department of Defense Appropriations for Fiscal Year 2014: Hearing before the Committee on Appropriations, 113th Cong. 1, (2013).

Nominations before the Armed Services Committee, United States Senate, 113th Cong. 1, (2013).

Review of Sexual Misconduct by Basic Training Instructors at Lackland Air Force Base: Hearing before the Committee on Armed Services, House of Representatives, 113th Cong. 1, (2013). Congress, First Session.

Pending Legislation Regarding Sexual Assaults in the Military. Hearing before the Comitee on Armed Services, United States Senate, 113th Cong. 1, (2013).

Women in Service Reviews: Hearing before the Subcommittee on Military Personnel of the Committee on Armed Services, House of Representatives, 113th Cong. 1, (2014).

Military Construction, Veterans Affairs, and Related Agencies for 2014: Hearing before the Subcommittee of the Committee on Appropriations, United States Senate, 113th Cong. 1, (2014).

Military Construction, Veterans Affairs, and Related Agencies for 2015: Hearing before the Committee on Appropriations, 113th Cong. 1, (2014).

Notes

1. Created by President Nixon on March 27, 1969, and chaired by former Secretary of Defense Thomas S. Gates, Jr., the Gates Commission relied heavily on a working group led by former University of Rochester President, Dean William H. Meckling, in its 100 hours of deliberation. The 15-member commission represented individuals from the academic, military, and business sectors, but most prominently also featured contemporary heavy hitters from the financial industries, suggesting that the question of whether or not an AVF would be possible was heavily based on economic principles. Both Milton Friedman (Paul Snowdon Russell Distinguished Service Professor of Economics at the University of Chicago at the time) and Alan Greenspan (Chairman of the Board for Townsend-Greenspan & Co. Economic Consultants) were members of the Commission. Of the 15 members who made up the Commission, 14 were men. The only woman, Jeanne Noble, on the commission, Jeanne Noble, was a professor at New York University, but perhaps more notably, the Vice President of the National Council of Negro Women and a former member of the National Advisory Commission on Selective Service. As an African American woman herself, Noble's participation on the Commission covered two "token" categories of representation.
2. As the Gates Commission report was made public, a discrepancy emerged between the Commission's recommendations and how those recommendations could be implemented. While female recruits were conspicuously missing from the report, the Army, which took on the largest burden in ensuring that the AVF would become a reality, was not as quick to dismiss the need for female volunteers in order to reach the numbers the Gates Commission proposed. The Army had been considering the need for increasing the number of women enlistees in order to diminish the number of draft calls as early as 1967.
3. On December 1, 2015, the Pentagon announced that there would be no exceptions made, and all combat roles, including Navy SEAL and secret operations positions, would become available to women starting January 1, 2016.
4. The combat exclusion, as well as the fluidity apparent in the changing and varied understandings of the term "combat," lies at the heart of the

narrative the military tells about women's participation in the armed forces. Not only did the interpretation of what "combat" referenced change over time, but it was interpreted differently across individual branches of the military. This allowed the military to limit women's access to roles at will. Thus the combat exclusion worked toward reaffirming military service as intrinsically tied to manhood. The shifting concept of "combat" connoted a specific kind of combatant who assumed all risks to protect, that is, a hypermilitarized, masculine warrior. Undergirding the combat exclusion were several misogynist assumptions including the idea that women are not "natural soldiers" and would never have the ability to obtain such status because of physiological "defects". It was believed that including women among combat troops would ultimately "ruin the bonds necessary for combat missions" (MacKenzie 2015, 15), undermining unit cohesion.

5. I arrived at this number by conducting a Boolean search of the HeinOnline Congressional Documents database, using the following search terms (women or female*) and (soldier* or military* or combat* or "armed forces" or "Marines" or "Marine Corps" or "Army" or "Navy" or "Air Force") and (combat). A majority of these discussions arose in Hearings on National Defense Authorization Acts between 1970 and 2015. Of these 70 hearings, 16 had titles specifically referring to women in the military or gendered policies. To focus my analysis of key themes in these debates, I eliminated any hearing in which women in the military were mentioned, but not in the context of the roles they should perform, the policies that affect their ability to serve, or the need to increase numbers of recruits. For example, the phrase "men and women who serve in the/or armed forces/military" appeared frequently in many of the Congressional Hearing records, but if this was the only way in which there was a mention, that hearing was not counted toward the total. The full list of the 70 Congressional hearings is listed in the Appendix.

6. My analysis includes hearings held in both the House and the Senate, including hearings conducted by the Senate Committee on Armed Services, the Senate Subcommittee on Manpower and Personnel of the Committee on Armed Services, the House Committee on Armed Services, the Senate Subcommittee on the Constitution of the Committee on the Judiciary, the House Subcommittee on Health of the Committee on Veterans' Affairs, the Senate Subcommittee on Personnel of the Committee on Armed Services, the House Committee on Veterans' Affairs, the House Subcommittee on National Security, Emerging Threats, and International Relations of the Committee on Government Reform, the House Military Personnel Subcommittee of the Committee on Armed Services, the House Subcommittee on Economic Opportunity of the Committee on Veterans' Affairs, the House Subcommittee on Disability Assistance and Memorial Affairs of the Committee on Veterans' Affairs, the House Subcommittee

on Hospitals and Health Care of the Committee on Veterans' Affairs, the House Committee on National Security, the House Military Readiness Subcommittee, the House Subcommittee on Civil Service of the Committee on Government Reform and Oversight, House of Representatives, Subcommittee on Human Resources and Intergovernmental Relations of the Committee on Government Reform and Oversight, the House Subcommittee on Benefits of the Committee on Veterans' Affairs, the House Military Personnel and Compensation Subcommittee of the Committee on Armed Services, the House Subcommittee on Education and Health of the Joint Economic Committee, the House Subcommittee on Courts, Civil Liberties, and the Administration of Justice of the Committee on the Judiciary, and the House Subcommittee on Benefits of the Committee on Veterans' Affairs.

7. These argument strands have persisted throughout the post-World War II period, although specific decades provide certain variation and nuance.
8. Feminist scholarship has theorized sovereignty as the "institution of power relations" in at least three key ways. First, sovereignty is thought to be a masculine institution and is represented by the state as a masculine body. Second, the space that the state embodies is in and of itself gendered through dichotomies of scientific knowledge production "between body and mind, culture and nature, men and women, and the private and public life of the state" (Wilcox 2013, 4). Third, the sovereign state is seen as a "protection racket," which requires men to protect women's bodies—even at the cost of mutilation and death in war.
9. The United States Selective Service program acts as a contingency system for a possible draft recall. As part of the program, all male citizens and all male immigrants are required to be registered between the ages of 18 and 25, and must register within 30 days of turning 18. Immigrants between the ages of 18 and 25 must register for the Selective Services within 30 days of arriving.
10. Narratives surrounding the issue even prior to the creation of the AVF used biologically determinist arguments to suggest that women's extended participation in the military would "weaken the combat efficiency" (United States Congress 1978, 347). Admiral Moorer suggested that not allowing women into combat was simply "commonsense" (342) and was particularly concerned with the psychological impact women serving in combat would have on male soldiers given that "men should have this protective attitude toward women. I would hope that in our society we never get to a point where we do not have the courtesies, the considerations, and the protective attitude toward women that we have today. I think that it would distract, in a heavy fire fight, the men in the unit who see around them large numbers of wounded and dead women. I do not think we should subject our men to that environment" (350).
11. Working at the time as part of the staff of the House Committee on Veterans Affairs, Webb was an outspoken critic of women in the military

academies. He wrote several articles on the topic, including "Women Can't Fight" which was published in 1979 in *The Washingtonian*. He went on to serve as the country's first Assistant Secretary of Defense for Reserve Affairs, as well as Secretary of the Navy before entering office as Senator for the State of Virginia in 2007.

12. William Proxmire served in the U.S. Senate, representing Wisconsin from 1957 and 1989.
13. Military data suggested, at the time, that as an aggregate across all the branches about 10% of female soldiers became pregnant during their time in the military (the highest numbers experienced by the Army were at 8%), making them ineligible for assignments overseas. In 1975, the military ended its policy requiring women who became pregnant during service to be discharged involuntarily; however, the policy was framed from an economic standpoint, pointing out that the policy had "resulted in a cost avoidance of $7.5 million—the estimated cost to replace those women had they been discharged" (The Role of Women in the Military 1977, 13).
14. During hearings on the Department of Defense (DoD) authorization for appropriations for fiscal years 1988 and 1989 before the Senate Committee on Armed Forces, Dr. David Armor, Principal Deputy Assistant Secretary of Defense and Acting Assistant Secretary, Force Management and Personnel, was asked about the DoD's position on the legislation. He argued that combat support positions were highly combat related and that if changes were to be made to the positions available to women they should be done by reviewing the combat exclusion policy within the statute. In essence, he argued that the proposed bill suggested "an abandonment of the principle [of keeping women out of combat] without addressing it." Reviewing the principle, he argued, should be taken up by Congress to consider, "Is it the will of the American people to allow women in combat positions? If it is, then I am sure the Congress will reflect that." Ultimately, Dr. Armor suggested that the proposed legislative measure was unnecessary given the services' ability of "reviewing occupations" and "open[ing] all kinds of jobs up," an argument which further promotes the military's narrative of rational and deliberate incremental integration and justifies limiting women's opportunity for advancement in the military.
15. Other commonly used arguments for reinstating the draft included mobilization in the case of a major war, a declining youth population would not be able to produce the necessary number of personnel for mobilization, a potential national youth training and work program would include military service in the program, and that considerable money could be saved in comparison to the costs associated with the AVF.
16. Following Eleanor Smeal, Judy Goldsmith was the President of the National Organization of Women (NOW) from 1982 to 1985.
17. Lost time accumulated by men is most usually associated with alcohol and drug usage.

18. Though more recent, research conducted by the Service Women's Action Network (SWAN) suggests that women service members' access to low or no cost reproductive care is highly limited. In fact, their research indicates a much higher rate of infertility among women serving, or who have served in the military, in comparison to civilian women. Their 2019 findings suggest that 30% of women in the military reported experience with infertility when actively trying to get pregnant (versus 12% in the general public). Women in the military who reported that they experienced infertility expressed that they believed it was caused by exposure to toxins during deployment, and in some instances, armor that was worn in the region of reproductive organs. SWAN's research also finds that allocation of financial resources to assist with reproductive healthcare depends on women service members' rank, with women service members being provided less access to preferred methods of birth control and higher rates of unwanted pregnancies. In the case of unwanted pregnancies, women are given very little choice to carry their pregnancies to term (unless the pregnancy is the result of rape, or carrying places the mother's life at risk) (Haring 2019) (see report—Service Women's Action Network 2018 "Access to Reproductive Health Care: The Experiences of Military Women 2018 Report").
19. In some hearings, concerns surrounding women's treatment in the military branches along with increasing revelations of sexual assault were compared to the integration of African Americans in the 1970s. As retired Air Force Major General Jeanne Holm explained during that same hearing "the purpose of harassment, whether in the military or civilian workplace, is always the same: to humiliate and degrade women, to make them feel they do not belong, and ultimately, to drive them out or to keep them out. It is most prevalent in communities that exclude women as in military combat fields" (13).

References

Bailey, B. 2009. *America's Army: Making the All-Volunteer Force*. Boston: Harvard University.
Belkin, A. 2012. *Bring Me Men: Military Masculinity and the Benign Façade of American Empire, 1898–2001*. New York: Columbia University Press.
Elshtain, J. 1992. *But Was It Just?: Reflections on the Morality of the Persian Gulf War*. New York: Doubleday.
Enloe, C. 1983. *Does Khaki Become You? The Militarization of Women's Lives*. Boston: South End Press.
———. 1990. *Bananas, Beaches, and Bases: Making Feminist Sense of International Politics*. Berkeley and Los Angeles: University of California Press.
Goldsmith, J. 1980. Testimony in United States Congress. House of Representatives. Committee on the Judiciary. "Judiciary Implication of Draft Registration – 1980." Hearings Before the Subcommon Courts, Civil Liberties, and the Administration of Justice, 96th Congr., 2nd Sess., 14 April and 22 May 1980. Washington, DC: U.S. Gov't. Print. Off., 1980 (Y4.J89/1:96/45).

Handler Chayes, A. 1977. Testimony in United States Congress. Hearings Before the Subcommittee on Priorities and Economy in Government of the Joint Economic Committee, "The Role of Women in the Military," 95th Cong., 1st Sess.

Haring, E. 2019. Military Women Need Better Reproductive Care at No Cost. *The Hill*, February 14. Accessed February 15, 2019. https://thehill.com/opinion/healthcare/430066-military-women-need-better-reproductive-care-at-no-cost

Kempthorne, D. 1997. Testimony in United States Congress. United States Senate. "Gender-Integrated Training and Related Matters: Hearing Before the Subcommittee on Personnel of the Committee on Armed Services," United States Senate, 105th Cong. 1.

MacKenzie, M. 2015. *Beyond the Band of Brothers*. Cambridge: Cambridge University Press.

McPeak, T. 1994. Testimony in United States Congress. Hearing Before the Military Forces and Personnel Subcommittee of the Committee on Armed Services, "Desert Storm Mystery Illness/Adequacy of Care." House of Representatives, One Hundred Third Congress, Second Session, Hearing Held, March 15, 1994.

Moorer, H. 1978. Testimony in United States Congress. Hearings Before the Subcommittee on the Civil and Constitutional Rights of the Committee on the Judiciary House of Representatives Equal Rights Amendment Extension Before the Subcommittee on the Constitution of the Committee on the Judiciary in 1978.

Nelson, R. L. 1977. Testimony in United States Congress. Hearings Before the Subcommittee on Priorities and Economy in Government of the Joint Economic Committee, "The Role of Women in the Military," 95th Cong., 1st Sess.

Proxmire, W. 1977. Testimony in United States Congress. Hearings Before the Subcommittee on Priorities and Economy in Government of the Joint Economic Committee, "The Role of Women in the Military," 95th Cong., 1st Sess.

Segal, M.W. 1982. The Argument for Female Combatants. In *Female Soldiers – Combatants or Noncombatants?: Historical and Contemporary Perspectives*, 267–290. Westport: Greenwood Press.

Service Women's Action Network. 2018. Access to Reproductive Health Care: The Experiences of Military Women 2018 Report. Accessed February 15, 2019. http://www.servicewomen.org/wp-content/uploads/2018/12/2018ReproReport_SWAN-2.pdf.

Shildrick, M. 1997. *Leaky Bodies and Boundaries: Feminism, Post-Modernism, and (Bio)ethics*. London: Routledge.

Smeal, E. 1980. Testimony in United States Congress. House of Representatives. Committee on the Judiciary. "Judiciary Implication of Draft Registration – 1980." Hearings Before the Subcommon Courts, Civil Liberties, and the Administration of Justice, 96th Congr., 2nd Sess., 14 April and 22 May 1980. Washington, DC: U.S. Gov't. Print. Off., 1980 (Y4.J89/1:96/45).

The President's Commission on an All-Volunteer Armed Force. 1970. *The Report of the President's Commission on an All-Volunteer Armed Force*. RAND Corporation.

United States Congress. 1977. Hearings Before the Subcommittee on Priorities and Economy in Government of the Joint Economic Committee, "The Role of Women in the Military," 95th Cong., 1st Sess.

———. 1978. Hearings Before the Subcommittee on the Civil and Constitutional Rights of the Committee on the Judiciary House of Representatives Equal Rights Amendment Extension Before the Subcommittee on the Constitution of the Committee on the Judiciary in 1978.

———. 1979. Senate Committee on Armed Services, "Requiring Reinstitution of Registration for Certain Persons Under the Military Selective Service Act, and For Other Reasons," Rept. 96–226, 96th Cong., 1st Sess.

———. 1980a. House of Representatives 4040—Department of Defense Authorization Act, Fiscal Year 1980. 96th Cong.

———. 1980b. House of Representatives. Committee on the Judiciary. "Judiciary Implication of Draft Registration – 1980." Hearings Before the Subcommon Courts, Civil Liberties, and the Administration of Justice, 96th Congr., 2nd Sess., 14 April and 22 May 1980. Washington: U.S. Gov't. Print. Off., 1980 (Y4.J89/1:96/45).

———. 1994b. House of Representatives. "Assignment of Army and Marine Corps Women Under the New Definition of Ground Combat": Hearing Before the Military Forces and Personnel Subcommittee of the Committee on Armed Services, 103rd Cong. 2.

———. 1997. United States Senate. "Gender-Integrated Training and Related Matters: Hearing Before the Subcommittee on Personnel of the Committee on Armed Services," United States Senate, 105th Cong. 1.

———. 1998. United States Senate. "Gender Issues: Information on DOD's Assignment Policy and Direct Ground Combat Definition." Report to the Ranking Minority Member, Subcommittee on Readiness, Committee on Armed Services. October 1998.

———. 2014. Women in Service Reviews: Hearing before the Subcommittee on Military Personnel of the Committee on Armed Services, House of Representatives, 113th Cong. 1.

Webb, J. 1980. The Draft: Why the Army Needs It. *The Atlantic*. Accessed October 8, 2019. https://www.theatlantic.com/magazine/archive/1980/04/the-draft/305456/.

Wilcox, L. 2013. Explosive Bodies and Bounded States. *International Feminist Journal of Politics* 16 (1): 66–85.

Women in Service Reviews: Hearing before the Subcommittee on Military Personnel of the Committee on Armed Services. 2014. House of Representatives, 113th Cong. 1.

Young, I.M. 2003. The Logic of Masculinist Protection: Reflections on the Current Security State. *Signs: Journal of Women in Culture and Society* 29 (1): 1–25.

CHAPTER 4

Violated Bodies: Combat Injuries and Sexual Assault in the U.S. Military

INTRODUCTION

Hegemonic militarized masculinity has been intimately connected to particular conceptions of bodily harm, those which have traditionally been associated with combat and the injuries sustained through enemy fire in times of war. Soldiers who accept the challenge of combat are defined as courageous, unsentimental patriots who face the possibility of death as a sacrifice for the greater good. Heralded for their valor in warfare, returning wounded veterans set "the standard of soldiering" (Eichler 2013, 257). Their heroic deeds are routinely conferred through the visible display of mutilations or missing limbs used for public consumption both in the media and through special commemoration ceremonies. The highest military distinctions, the *Purple Heart* and the *Medal of Honor*, are most often afforded to male service members for what is deemed courageous behavior in the combat war theater, accolades which set typically individual male soldiers apart from the Band of Brothers (see Chap. 3) collective. Death by combat is written into the history books.

Heroic conceptions of harm only account for a fraction of the visible manifestations of harm produced by war. Indeed, advances in both medical and military technology during Operation Enduring Freedom and Operation Iraqi Freedom have allowed soldiers to survive war experiences that would have otherwise caused death in previous conflicts. As deaths have decreased, deployment lengths and tours have increased,[1] as have the psychological injuries wars produce. These "alternative" forms of harm

have generated gender trouble; they complicate the military's ability to reify military service as being only for the physically strong, and have increased attention to the long-term consequences of combat experience among veterans, including unemployment, homelessness, drug and alcohol abuse, and suicide rates (Yuhl in Altinay and Peto 2016).

Injuries—beyond those valorized by medals—have marked the bodies of soldiers long before advancements in technology curtailed combat injuries and deaths. But not all wounds are seen as equal. Violence committed on bodies during military service does not in and of itself constitute wounding. Instead, selective military narratives on war wounds work to marginalize—and sometimes even exclude—certain forms of injury from the military's formulation of "war wounds." Most prominent of these exclusions are injuries—both physical and psychological—sustained through sexual assault. Pentagon statistics indicate that 6769 cases of sexual assault were reported for fiscal year 2017,[2] while studies conducted by organizations external to the military suggest that up to 33% of all service members—both men and women—may experience some form of sexual violence during their service in the military. Individuals who experience military sexual assault (MSA) often sustain powerful bodily injuries, through violent rapes, brutal beatings, broken bones, and/or head injuries. They often also suffer from psychological injuries, including post-traumatic stress disorder (PTSD), long after the act of sexual assault has ended. Descriptions of injuries sustained by survivors of sexual assault closely echo those of combat soldiers returning from tours of war.

Yet, the military selectively acknowledges and accredits some forms of injury, while diminishing, disappearing, and silencing others. Injuries sustained by active duty service members through sexual assault during military service—whether at home or abroad—are not considered military wounds. Instead, sexual assault is pathologized as a cancerous disease that infiltrates and spreads through the ranks, threatening the viability of the institution. Challenging stock notions of military heroism intricately tied to a duty to "protect and defend," wounds resulting from sexual assault by fellow soldiers or by commanding officers manifest gender trouble in the armed services. They are characterized as extrinsic to war making and are systematically silenced.

This chapter considers the following question: what role does gender play in the military's contemporary conceptualization of military wounding? It explores the military's perception of injury, wounds, and their relation to masculinity as an ideal for soldiering. Through a discourse analysis which juxtaposes wounding sustained through tours of combat in the war

theater against wounding experienced through solider-on-soldier sexual assault, I investigate both the physical and psychological effects of sexual assault that do not constitute and are not heralded as military wounds. Rather than recognizing the seriousness of sexual assault within the organization, the military tolerates these abusive practices. The selectivity of military narratives on war wounds marginalize, and sometimes exclude, certain forms of injury that do not conform to narrow constructions of "war wounds," discrediting pervasive forms of gendered harm (Weldes 1999).

Examination of a broad range of wounds incurred during military service deconstructs conventional conceptualizations of military wounding that are intimately connected to the heterosexual white male soldier's body. An adequate definition of military injury would not only include psychological injuries incurred through engagement in military operations and deployments, but other "invisible" wounds, such as those suffered through sexual assault within military ranks. Studying sexual violence, however, comes with its own set of methodological difficulties, particularly in gauging accurate numbers of incidents. After outlining these difficulties, I analyze unique aspects of gendered violence specific to the military. I draw attention to multiple discursive practices that create an institutional culture that silences survivors of gendered violence, and sexual assault in particular.

The military has routinely invoked three gendered mechanisms of exclusion that dissuade prosecution of sexual assault: (1) narratives that foreground the risks of retaliation against those who file complaints, (2) medicalized securitization narratives that turn the military itself into the victim of sexual assault, and (3) narratives that posit sexual assault as a violation of the sacred, familial-like bonds that soldiers supposedly share. Through discursive mechanisms that actively dissuade personnel from reporting cases of sexual assault, the military institutionalizes a culture of toleration for sexual assault. Such organizational toleration not only silences victims, but contributes to a sleight of hand that substitutes the military as the victim of allegations of sexual assault for the bodies of sexually violated military personnel as the chief concern at Congressional hearings. The documentary film *The Invisible War*[3] showcases specific narratives used in the military to ensure that sexual and psychological injuries are masked, silenced, and deemed worthy of the institution's attention when sexually violated personnel do step forward. In the final section of the chapter, I consider another set of unrecognized "invisible wounds"— proliferating mental health conditions associated with tours of war—to compare and contrast how the military grapples with these "alternative" processes of wounding.

Methodological Difficulties with Studying Sexual Violence

Although the utility of gendered violence (rape in particular) in international relations as a "tool to break down the enemy, to punish resistance to the military's goals, and to assert the military's strength and disciplinary power" (Nayak 2009, 153) has been well documented, less attention has been given to the endemic nature of rape among and within the personnel of militaries of sovereign nations. Complications with definitions and "measurements" have limited scholars' ability to subject the topic to systematic investigation despite sexual assault being "considered one of the most debilitating obstacles women have to face within organizational settings" (Carreiras 2006, 53). In spite of the Department of Defense's commitment to a zero-tolerance policy in the early 1980s, high-profile incidents such as the 1991 Tailhook Scandal and the 1996 sexual harassment incidents at the Army training facilities in Fort Leonard Wood, as well as the 2003 sexual assault scandal at the U.S. Air Force Academy suggest an ongoing, institutional problem that has transcended decades of (supposed) gender inclusive policies designed to produce a less masculinist, misogynist culture. Far from being a single instance of failed policy or inadequate intervention, persistent sexual violence in the military may be understood as a "tendency to reassert masculinity" (Carreiras 2006, 54), and remains an important topic for sustained inquiry.[4]

Despite challenges associated with investigations of MSA, studies suggest[5] that anywhere between 9.5% to 33% of women *report* attempted or completed rapes during their period of service (Bostock and Daley 2007; Coyle et al. 1996; Murdoch et al. 2007; Sadler et al. 2000; Skinner et al. 2000; Surís et al. 2007). Data suggests that nearly half of all female service members who served in Afghanistan and Iraq after the September 11th attacks experienced some form of sexual harassment. Moreover, it is thought that up to 40% of female soldiers—whether they report it or not—experience at least one incident of attempted or completed rape while completing their service in the military (MacKenzie 2015). The numbers of sexual assault cases vary across the branches of the military and higher numbers of veterans report sexual violations than active duty personnel. In the Air Force, for example, a large-scale study found that 9.5% of women had reported that "their most recent rape experience" (suggesting that these women had been raped multiple times, either in the military or in both the

military and civilian spheres) occurred during their military service (Bostock and Daley 2007, 927). In the Army, 10.5% of women reported rape while in military service (Murdoch et al. 2007). Among female veterans, 19.6% of women who have accessed Veterans' medical services after completing their term(s) in the military have reported and sought assistance to cope with having been raped (Coyle et al. 1996).

Generally, far less attention has been given to the prevalence of rape among male service members. As a result, research on MSA assumes women service members as the victims and male service members as the perpetrators. Very little data has been gathered about the gender of sexual assault perpetrators. The Department of Defense reported that in 2002 and 2003, 99% of alleged perpetrators were men and 91% of alleged victims were women (Department of Defense 2004). Studies conducted in the 1990s and the first decade of the twenty-first century estimated that between 1% and 12% of male soldiers experienced rape (Kimerling et al. 2007; Krinsley et al. 2003; Martin et al. 1998; Murdoch et al. 2004; Murdoch et al. 2007; Smith et al. 1999). These studies are ambiguous as to whether the referenced incidents of rape occurred during the individual's military service. Daniel W. Smith et al. (1999), for example, suggest a 12% lifetime prevalence rate (i.e. sexual assault committed against the individual at some point prior to the study) for male combat veterans diagnosed with post-traumatic stress syndrome (sample size: 129). In 92% of these cases, sexual assault occurred prior to the individual's tour of combat, but it is unclear whether sexual assault occurred when the individual was a civilian or military personnel.

It is true that studies of MSA face methodological difficulties. Studies by different authors lack consistency, differ in methodologies, sample ranges, and sizes, and definitions of sexual assault (Bostock and Daley 2007). Thus, it is important to consider that variation in reports of sexual assault may be due to variation in methodologies rather than to differences in incidence (Turchik and Wilson 2010). But regardless of unique research challenges, rape within the military is, in many ways, integrally related to military policies designed to regulate sexuality among the forces (Enloe 1983).[6] Historically, military leaders in the United States have provided male service members with opportunities to satisfy their supposed biologically driven heterosexual urges in ways that would not threaten effective military operations. Assuming that rape and illicit sexual encounters would proliferate if service members were not given a controlled outlet for their sexual cravings, military organizations have provided service members with prostitutes in

sites adjacent to military bases. In some instances, prostitutes were made available on bases themselves. Cynthia Enloe (1983) suggests that "it may be that there are aspects of the military institution and ideology, which greatly increase the pressure on militarized men to 'perform' sexually, whether they have a sexual 'need' or emotional feelings or not" (35) because they "are subjected to 24 hours a day pressures to conform to the standards of 'masculine' behavior" (35). Through socialization processes and through peer pressure, men in the military may be cajoled to engage in bonding rituals that are sexual in nature, including gang-raping prostitutes or other women made available to them. Although the military remains quite tight-lipped about its practices to provide outlets for the sexual energies of heterosexual male troops, evidence of these practices surface in regulations concerning sanitary hygiene in and around military bases and in zoning ordinances negotiated with local officials. In response to sexual assault among members of the military, however, tight lips give way to prevarications that disappear the military victims of sexual assault, and sometimes women in the military, more generally.

UNIQUE ASPECTS OF GENDERED VIOLENCE IN THE MILITARY

Undoubtedly, there are certain commonalities in military and civilian treatment of rape and sexual assault in the United States. Difficulties in gathering evidence and establishing proof, doubts about the veracity of those file reports, victim blaming, reluctance to bring charges, low prosecution rates, and even lower conviction rates are characteristic of the treatment of sexual assault in both spheres. Yet there are also important aspects of the treatment of sexual violence in the military specifically that are distinctive.

Three factors distinguish the treatment of sexual assault in the military from its treatment by civilian authorities. First, because the military services place such "emphasis on loyalty and community" (RAND review 2008, 11), military personnel who are sexually assaulted by their peers or their commanders experience different degrees of "shock and betrayal" (11) that somewhat resembles familial incest. Second, although reporting rates are lower among men, because men constitute 85% of military personnel, more than half of all reports of sexual assault involve male victims, which raises important questions about rape as a mechanism of "feminization" and as a means of social control within the military. Third, when service members are assaulted by perpetrators within their own units or

chain of command, they often remain vulnerable to repeated attacks, retaliation for filing reports, or other forms of abuse because they are unable to escape the specific posting to which they are assigned. Within the military, the perpetrator of sexual assault is often the victim's supervisor and superior, which decreases the likelihood that service members will report cases of sexual violence. Quite rightly, victims fear retaliation that could adversely affect their military careers. As Jennifer Mathers (in Cohn 2013) suggests, military personnel face critical obstacles when they choose to report sexual assault.

> But military women also face additional pressures specific to the armed forces. The military environment, with its strict hierarchies and chain of command, can make it more difficult for a woman to file a complaint and get the medical and psychological help that she needs because she has few other avenues to pursue if her superiors refuse to take her accusations seriously (or indeed if it was her commanding officer who assaulted her). Women who do report abuse are often pressured by other soldiers to withdraw the accusation to avoid damaging the reputation and career of "a good soldier," while those who persist in filing a complaint are likely to find themselves ostracized by their colleagues and passed up for promotion, and that their careers are effectively over. (Mathers in Cohn 2013, 143)

As Jennifer Mathers' (in Cohn 2013) reference to the "good soldier" narrative makes clear, the career accomplishments valorized by the military—particularly in the form of combat medals—are often deployed to silence rape victims. According to the military's value scheme, sexual assault is a minor infraction that ought not be allowed to sully a "good soldier's" reputation. By invoking the good soldier narrative, military personnel often allow perpetrators of sexual assault—almost exclusively male—to go unpunished. In addition to being used to dissuade victims from filing charges, the narrative "allows a defendant to cite unrelated, subjective factors during trial, such as a military record as a defense against sexual assault charges" (Tsongas and Turner 2014, 8).[7] These tactics are often particularly relevant during sentencing deliberations.

Victim and Survivor Discourses in Military Contexts

In addition to the unique distinctions between military and civilian sexual violence outlined above, the distinction between victim and survivor discourses take on additional meaning in military contexts surrounding wounds and wounding (Hannagan 2016). At issue is not

only how individuals who have experienced military sexual violence view or define themselves, but also how the military defines and represents them, inside and outside of the organizational apparatus. For members of groups who are marginalized within a particular institution or culture, being defined as a victim instead of a survivor can mean the difference between having their wounds acknowledged or ignored, and the difference between receiving or being denied necessary resources for rehabilitation after the experienced episode(s) of violence.

Within the military, the terms "victim" and "survivor" are used differently in the context of sexual assault and combat, discursively producing specific kinds of bodies, bodies that are gendered quite independently of military formations, and bodies that are further gendered through military training, deployment, and representation. As is, women often "do not think of themselves as survivors and do not see their own agency" (D'Costa in Ackerly, Stern, and True 2006, 140). When they do, they risk revictimization depending on how others within key institutions view them.

Through its treatment of troops who experience sexual assault, the military plays a powerful role in controlling the agency of personnel within the institution. Far from manifesting concern with the well-being of those who have been sexually violated, the military seems more concerned with discursive strategies "essential to the construction of a coherent national identity, the one created through privileged voices" (D'Costa in Ackerly, Stern, and True 2006, 142). While "neither victim nor survivor is a simple construct" (140), the military consciously—and problematically—attaches these concepts to some bodies and not to others in ways that silence sexually assaulted soldiers while glorifying soldiers wounded in combat, thereby shoring up militarized masculine identities and perpetuating gender inequities.

Selective Exclusions in Cultures of Silence

Whether survivors of sexual assault see themselves as such may be immaterial without a clear understanding of just how many individuals experience MSA, and how the institution may ensure that the public remains unaware of those actual figures. There are likely far more incidents of sexual assault occurring in the military in any given year than are reported. Low reporting rates are thought to occur for two primary reasons: service members' fear of retaliation stemming from reporting, and a strong lack of confidence among service members in the military justice system to punish

their perpetrators when they do come forward with reports of sexual assault. The combination of these two factors has created what Anu Bhagwati, former Executive Director of Service Women's Action Network (SWAN), calls "a culture of silence" (Bhagwati in United States Senate Armed Services Committee Hearing 2013), an institutional environment in which victims are unwilling to report these incidents, both in the military proper, and in the service academies. This silencing culture precludes the dissemination of truths about the endemic nature of MSA and limits victims' abilities to heal from their experiences.

Encouraging individuals who have experienced sexual assault in the military to speak out can be particularly sensitive and dangerous. Instead of facing potential forms of institutional retaliation—having charges filed against them, harassment and bullying, decrease in rank, or discharge—survivors remain silent. Through their silencing, the military shapes what is considered to be the "truth" about sexual assault; it conspicuously controls what narratives are used in the representation of the problem by creating an environment that dissuades victims from reporting. In so doing, the U.S. military uses "the power of dominant groups to define and address the concerns and interests of marginalized groups, mak[ing] it impossible for those groups to put their own needs on the public agenda" (D'Costa in Ackerly, Stern and True 2006, 131). By engaging in systematic silencing, the U.S. military maintains the power to define how the histories of soldiers experiencing sexual assault are told and projected to the general public. Practices of systemic silencing and manipulations of truth remain an essential means of preserving the masculinist identity of the military institution.

Sexual Assault as a Threat to the Institution (Not the Individual)

Official statements and discourses used by military officials about sexual assault have a peculiar way of engaging in this systemic silencing; they shift the focus from the individuals who experience sexual violence to the impact it has on the military. Although the military's organizational mission "relies on masculinist norms that perpetuate such violence" (Nayak 2009, 147), military spokespersons insist that sexual assault is an anomaly and not systemic to the organization. Far from being endemic to internal military culture and practices, they characterize sexual violence as something foreign that invades the military from the outside. Indeed, through

adopting language drawn from securitization and medical discourses, the military positions sexual violence as a "threat" to military institutions themselves. When political pressure concerning sexual violence escalates dramatically necessitating public relations management and an official response from the Pentagon, recourse to rhetoric that foregrounds a potent threat that must be contained enables the military to "not only [...] protect the image and representation of the military but also to allow the military to continue without interruption in its business of training people in warfare and potential attack on designated enemies" (Nayak 2009, 161).

Increased media attention to the endemic nature of military sexual violence in 2013 and 2014 provides a powerful example of the military's turn to securitization discourses to preserve the integrity of its institutions. As Alison Howell (2014) notes, "securitization theory seeks to understand the process by which threats are constructed. It studies how an issue gets elevated from 'normal politics' to a matter of security through the process of 'securitization,' or, the construction of an existential threat to a state or other referent object through securitizing speech acts" (964). In the case of sexual violence within the ranks, the military shifts its normal securitization stance from a focus on securing the nation against external forces that threaten the national security of the homeland to a new register that requires securitizing efforts against threats internal to the institution that may limit its operational effectiveness. It frames sexual assault as a source of "insecurity" for military institutions themselves, and turns to medical discourses to illuminate the nature of the "threat."

Statements made by branch generals to the Senate Armed Services Committee in the context of hearings on pending legislation regarding sexual assault portrayed sexual assault as a detriment to the institution's integrity, reputation, and ability to fulfill its purpose. The opening statement of General Robert Papp, Chief of Staff of the Coast Guard, before the Committee, provides a powerful example of this securitization narrative:

> Sexual assault is a virulent crime that devastates its victim. It also destroys unit discipline, it erodes cohesiveness and degrades our readiness...To execute our missions, all Coast Guard personnel must be bound by trust and mutual respect for one another. The crime of sexual assault not only damages the victim, it undermines morale, degrades readiness and damages mission performance. It is a deliberate act that violates law, policy and our Core Values of Honor, Respect and Devotion to Duty. (Papp in United States Congress 2013)

Responding to the Committee on behalf of the Army, General Raymond Odierno, Chief of Staff of the Army, also emphasized that "the credibility of the Armed Forces and the credibility of the Army [is] at stake" in efforts to contain the threat of sexual violence (Odierno). Similarly, in that same committee hearing, Navy Chief of Staff Admiral Jonathan Greenert suggested that sexual assault affected "the ability of the Navy to execute [its] mission. We must more effectively prevent and respond to sexual assault, or our readiness and credibility as a fighting force will suffer" (Greenert in United States Congress 2013). Other statements made articulate a fear that military sexual violence will "chip away" the public's faith in the institution.

Perhaps even more suggestively, military statements frequently depict sexual assault in medical terms, characterizing it as an "epidemic," "cancer," "disease," or "plague" that threatens the military as an institution. The use of medical metaphors as "shared language" (Howell 2014, 976) not only highlights affinities between institutions of war and medicine, but also suggests a "more deeply-rooted [...] ongoing shared strategic logic of warfare and medicine" (977). Although the intimate affinity between the two has most often played out externally on the battlefields of both inter- and intrastate wars (Howell 2014), medical metaphors in the context of sexual assault subtly position the military—not the individual—as the prime victim. As the violated bodies of raped women and men disappear, sexual assault is pathologized; it is reframed as a "disease" that the military must "sanitize" itself against. Consider the following statements made by various branch generals during the 2013 Senate Armed Services Committee Hearing:

> The violent crime of sexual assault *plagues* our society; it is unacceptable in any place. However, in the military it is especially repugnant because it breaks the sacred bond of trust between service members that is vital to readiness and our nation's security. (Papp in United States Congress 2013, 50 emphasis added)

> Sexual assault and harassment are like *a cancer* within the force—a cancer that left untreated will destroy the fabric of our force. (Odierno in United States Congress 2013, 15 emphasis added)

> Sexual assault is a crime against individual Marines that reverberates *within a unit like a cancer* undermining the most basic principle we hold dear—taking care of Marines. (Amos in United States Congress 2013, 33 emphasis added)

> Nothing saddens me more than knowing *this cancer exists in our ranks*, and that victimized *Airmen* (sic), on possibly the worst day of their lives, sometimes feel they cannot receive compassionate, capable support from our Air Force. (Welsh in United States Congress 2013, 44 emphasis added)

These comments displace the angst of the women and men who have been assaulted while on duty and foreground the pain of the military institutions. The emphasis on the impact of sexual assault on the military unit or institution as a whole substitutes a corporate body for an individual body as victim. This move both devalues the harms experienced by victims and survivors of sexual violence (Nayak 2009) and transforms the nature of the violation from sexual and physical injury to immaterial threat to the functioning of the military unit. The failure to duly acknowledge the victims of sexual assault "highlight[s]… the hypocrisy of military 'honor' [and] severely delimits democratic possibilities of all people mattering" (159).

But more than hypocrisy is at issue in this discursive maneuver. By referring to sexual assault as a "disease," "plague," or "cancer" that threatens the military, the site of therapeutic intervention is also altered. Rather than seeking to heal the wounds of the individual victims of sexual violence, the military gives priority to securing the health of military institutions. At a 2013 Senate Armed Services Committee Hearing, this shift plays out in the military's emphasis on strategies to cleanse the sullied organization of these contagious vectors of disease. Rather than addressing the concrete needs of rape victims, the military seeks to preserve the health of its chain of command. It actively resists proposals to remove the power of military commanders to handle reports of sexual assault in their units, and insists that existing complaint resolution mechanisms are essential for maintaining good order and discipline within their ranks. In addition, the discursive construction of sexual assault as a disease that threatens military institutions and governance depoliticizes sexual violence within the military. Questions of gender power as they relate to the use of rape to keep women "in their place," and to punish men who fail to conform to particular standards of military masculinity disappear. The need to investigate how gendering functions to produce and sustain relations of domination and subordination within military culture is obviated. Positioned as an anomalous infection that must be contained, sexual assault is characterized as a foreign body that bears no relation to the norms of the institution, and requires no systematic change of normal operating procedures for its eradication.

Selective Exclusions in Invisible Wounds

To illuminate the gendered nature of the silencing mechanisms the U.S. military has used in its past and current policies on sexual assault, I turn to testimonials of rape survivors documented in the highly acclaimed 2012 film, *The Invisible War*. By analyzing the testimonies of women and men who have attempted to report sexual assault, I trace how the military contains the "threat" of sexual violence by trying to make the problem disappear. I show how the chain of command uses doubt to impose a near insuperable burden of proof on those who report rape; how they force victims to shift from the harm of sexual violence itself to claims about collateral physical damage that is visible; how they engage in victim blaming, suggesting that victims are responsible for bringing sexual assault upon themselves; and how they refuse to recognize the wounds produced through sexual assault as war wounds, thereby making it more difficult for victims to heal and gain access to needed medical and psychological treatment. Through these various techniques, the military makes sexual violence within its ranks invisible, not only to those who experience sexual assault, but to the general public.

"Proving" Sexual Violence and Military Wounds

The Invisible War was produced, primarily, to document how the military as an institution handles reported cases of sexual assault. Through testimonials, the film shows how women and men who have been sexually assaulted, and attempt to file complaints, are ignored. A key to the recurrent dismissal of rape allegations turns on questions of evidence. As forms of intimate violence that pit the word of the victim against the word of the assailant(s), rape and other forms of sexual assault are notoriously hard to "prove." In the words of one soldier, "I reported it two different times to my squad leader and he told me that there was nothing he could do about it because I didn't have any proof" (*The Invisible War* 2012, 24:44).[8] The documentary suggests that in a calculated, systematic fashion, military investigators are told to ask female soldiers who make rape claims—and the alleged perpetrators—what the raped soldier was wearing at the time of the alleged assault and whether or not she has a (male) significant other, insinuating that the woman may have provoked her attack, or worse, was using sex as a means to ensnare a man in a relationship.

Testimonies provided in *The Invisible War* note that rape cases are systematically assigned to male investigators because the military is concerned that female investigators might provide overly sympathetic assessments. In addition to shoring up old myths about male objectivity and female sentimentality, this gendered practice also requires that women complainants discuss intimate, embarrassing, and painful details of their cases with male investigators who have been primed to manifest cold indifference. In an institution that prizes physical strength, women troops are forced to acknowledge their inability to fight off their assailant, thereby reinscribing their physical "inferiority." When soldiers report abuse, they are met with suspicion and with accusations of their untruthfulness. In the case of Rebecca Catagnus (U.S. Marine Corps), higher ranking officials told her that "I could choose to report it, if I wasn't, if you know, if they found out I, that what I was saying wasn't to be truthful, then I would be reduced in rank" (Catagnus in *The Invisible War* 2012, 24:20). By confronting complainants with disbelief and with threats of punishment, military investigators allow an old misogynist myth—that women are prone to file false rape accusations—to masquerade as objectivity. When reporting sexual violence, women are positioned as deviant subjects who are likely to report rape falsely. As Captain Debra Dickerson (U.S. Air Force) notes in the film, for military investigators, the presumption always favors the assailant, "If a man gets accused of rape, it's a set up. The woman is lying" (Dickerson in *The Invisible War* 2012, 24:15).

Within the military, sexual assault needs to be proven, but the kind of proof deemed compelling has nothing to do with the victim's word about non-consent, and little to do with the harms of sexual violence per se. For an institution prone to privilege combat as the criterion for establishing a body in pain, and accrediting only war wounds, visibly manifested in blood, gore, and mutilation, invisible wounds pose a unique challenge. Military investigators prefer visible evidence as proof of a real wound. In the absence of such brutal visuals, the military tends not to perceive any injury at all. As Christina Jones, who served in the U.S. Army, noted: "Even with the rape kit and everything and, and the person—my friend catching him raping me—they still don't believe me" (Jones in *The Invisible War* 2012, 24:37).

Even in cases where physical evidence is visible, the military has means to make it disappear. Hannah Sewell reported that she had been raped and went to the hospital to be treated for a range of injuries sustained during the sexual assault. Yet her case was closed without any charges brought against her assailant because her rape kit "went missing" (*The Invisible*

War 2012). Closing cases without bringing charges appears to be quite routine procedure for military investigators. Lack of evidence is repeatedly cited as the reason.

The Creation of Illegitimate, Self-Inflicted Wounds

The suggestion that sexual assault may be welcomed or self-inflicted is a pervasive narrative throughout *The Invisible War*. Women sexual assault victims are told by military investigators that they "were asking for it" (*The Invisible War* 2012) by wearing clothing other than those regulated by the military, or by flirting with soldiers, or by committing adultery. When victims are told that the act of sexual assault was their own fault, it is somewhat akin to declaring a wound self-inflicted, a status totally devoid of honor by military standards.

The victim blaming that women military personnel experience when they try to report sexual assault echoes messages sent by military authorities in response to other sex scandals. In 1991 during a national aviators' conference, women returning to their hotel rooms after an evening of convention partying were subjected to a "gauntlet" (*The Invisible War* 2012, 40:13). As they walked down a corridor where men positioned themselves against both walls, the women were manhandled, groped, mauled, taunted, and some were subjected to rape by an individual, others to gang rape. When Lieutenant Paula Coughlin (U.S. Navy) tried to report the incident to her superior, her report was summarily dismissed. As she recounts, "The next morning, I met with my boss for breakfast and I said, you know, 'what happened here?' and he said 'well that's what you get for walking down a hallway full of drunk aviators'" (Coughlin in *The Invisible War* 2012, 40:54).

Returning women veterans assigned to the prestigious Marines Barracks in Washington, D.C. encountered similar harassment and similar victim blaming when reporting cases of harassment. After serving in Operation Iraqi Freedom from 2009 to 2010, Lieutenant Ariana Klay (U.S. Marine Corps) received a completely non-supportive response to her experience of escalating sexual harassment.

> One of the first things I was told when I checked in was "don't wear any makeup because the Marines will think that they, that you want to sleep with them." And I thought that's just ridiculous (Klay in *The Invisible War* 2012,

45:12)...It got progressively worse and worse. They determined that I had welcomed the sexual harassment by wearing my regulation length uniform skirt and running in running shorts. (45:58)

In addition to being told that they had brought acts of sexual violence upon themselves, some women military personnel experienced retaliation when they reported incidents of sexual assault. Instead of bringing charges against the assailant, the military brought charges against the victim for infraction of military protocols. As Elle Helmer (U.S. Marine Corps) reported:

> The Colonel at one point said, "You know, Lieutenant Helmer boys, girls and alcohol just don't mix. We'll never really know what happened inside that office, only you and the Major know, and he's not talking. So, at this point, the investigation is closed for lack of evidence and we have reopened a new investigation against you for conduct unbecoming an officer and public intoxication. (Helmer in *The Invisible War* 2012, 48:33)

By turning rape into a matter of "she said, he said" altogether beyond the scope of any rational investigation, alleging that victims invited their attacks, blaming victims for being in the wrong place at the wrong time and, as such, responsible for putting themselves in danger, and insisting on forms of evidence that belie the nature of sexual violence, the military removes the injuries incurred in rape beyond the category of intelligible wound. For the military, wounds are sustained by soldiers in combat, who are fighting for a greater good—the annihilation of an enemy. Wounds are incurred through actions in keeping with the mission of the armed forces—to protect and defend. As such, the wounds of combat soldiers are wounds in which the nation can take pride. Injuries resulting from sexual violence fall below the threshold of military intelligibility. They threaten rather than sustain the military mission. They lack the graphic visuals associated with combat violence. And they reflect behavior altogether unbecoming military personnel, whether that be the violent behavior of the attacker or the reporting behavior of the violated. Little surprise then that the military is invested in making the problem of sexual violence in the ranks disappear.

Rape and Sexual Assault as Physical Wounds

Keenly aware of the military's reluctance to accept evidence of sexual violence, many veterans interviewed in *The Invisible War* (2012) devote intensive energy to documenting the physical injuries they sustained through sexual assault—injuries that are visible. Kori Cioca, who served in the Coastguard, described in detail her frequent harassment by her supervisor. She was often called by her supervisor at three o'clock in the morning to be picked up from a local bar and then subjected to verbal harassment on the return journey to base. When Kori first attempted to report the harassment, she is told: "Just because I didn't like somebody they weren't going to switch me away from this guy" (Cioca in *The Invisible War* 2012, 9:31). Sexual harassment, an illegal form of sex discrimination according to U.S. law, was thus transformed into a matter of personal taste, an issue of personality conflict. In the absence of any appropriate response from the chain of command, the harassment escalated from verbal harassment to physical violence, to rape:

> It was in the evening around taps, and he'd unlocked the door, and he'd come in, and he had an erection, and he tried to get me to touch him. And I took my right hand and I pushed him in the chest and started to yell for the other guys to kind of hear me to "hey, hey, hey!" Umm, he hit me across the left side of my face. I remember holding the closet thinking, "What just happened?" and my face hurt so bad. And when we went to the Command about it—me and this petty officer who saw my face—they just, they let it wait because they didn't want any kind of problems going on. A couple of weeks later, I needed the key to do my cleanup so I knocked on his door and he said "Yeah okay, come on in here, it's in here," and I said, "No, no, no, I'm going to wait out here," and he screamed at me and he made me come in and he grabbed my arm (long pause) and he raped me in his berthing area. (Cioca in *The Invisible War* 2012, 9:37)

Kori's jaw was broken during the first physical assault. Left untreated because her commanding officers wanted "no trouble," she endured intensive physical pain for months after the attack, forcing her to eventually consult a dentist.

> I was having the most horrible pain in my face that I went to the dentist 'cause I thought it was my teeth, or something was wrong. The doctor came in after the x-rays and asked me if I had been in a car accident. When he hit me in my face he dislocated my jaw and sent both of my disks forward in my

face so I don't have any disks where my, where they should be in my face. They told me I would probably need a partial bone replacement for where my bone had been laying on my nerve for so long it's starting to actually disintegrate. (Cioca in *The Invisible War* 2012, 15:18)

Although the dentist's diagnosis and the x-rays corroborated the validity of Kori's initial complaint and proved the existence of a physical wound, her injuries still remained below the military threshold of intelligibility. Indeed, when Cioca went to the Veterans Administration (VA) to get official x-rays to demonstrate that her jaw was injured during a sexual assault while on military duty, the VA ordered a back x-ray instead of a facial x-ray. Kori's conversation with her husband, Rob McDonald, after the appointment captures her frustration and disbelief when confronting such shabby treatment.

McDonald: How'd it go? What they say? (27:49)
Cioca: It went good. They, of course, they, they ordered the…they ordered a back x-ray instead of a face x-ray, but…they should have, they should know, should know what's wrong with me! Like, read my stuff and you'll see that it's my face. It's not my back, it's not my legs, my arms, it's my face! (27:52)
McDonald: So, the people who need the proof don't even know what your case is about (28:13).
Cioca: Like, I don't even know why we wasted the gas money, the trip, anything. It was completely a waste of time (28:17).

Hannah Sewell (U.S. Navy) also sustained severe physical injuries during her assault. A military hospital documented that the "main nerve in my spine was pinched in three places and my hips were rotated. I could barely walk. I had collapsed due to muscle spasms in my back because my back was injured during the rape" (Sewell in *The Invisible War* 2012, 20:12). Yet the hospital chose to focus on the physical injuries, largely ignoring that they were the product of rape.

The Invisible War illuminates the intricate techniques used by the military to disappear sexual assault within its ranks. By impugning victims' testimony, threatening dire effects of lodging complaints, deploying doubt to undermine evidence of sexual assault, and blaming victims for their

violation, military authorities press complainants toward silence or toward the production of physical evidence. But those who manifest visible physical injury are subjected to a new sleight of hand: their physical injuries are treated as self-generating and once again the rape disappears. Military investigative logic requires complainants to invest their efforts in documenting visible physical harms accruing from sexual violence, yet it affords no space for articulation of the nature of harm involved in the sexual violation itself. Soldiers' accounts of psychological pain, and their discussions of the harms associated with sexual penetration against their will, with the loss of physical integrity and autonomy are disappeared from the realm of relevant data for investigative officers.

Survivors of sexual assault often report that their treatment at the hands of investigative officers is akin to a "second rape." The authority of their experience and their veracity are questioned; they are subjected to brutal interrogation techniques and forced to recount their experiences of sexual degradation, humiliation, and terror. In the military, investigative tactics replicate many of these dehumanizing dimensions, but they also add an additional menace. In today's military, women must constantly prove that they are equal to their male counterparts and that they can meet standards derived from male bodies and male experiences. When women complain of sexual violence at the hands of their fellow service members, they are "feminized." Placed in the position of the victim, forced to emphasize the physical nature of their injuries, they are required to document their own "weakness," and their inability to protect themselves. Thus they are doubly disadvantaged, condemned to suffer the effects of sexual violence without any support from the military, and required to document their own physical deficiencies, providing evidence to a hostile military establishment that they do not belong in the military to begin with. Far from receiving recognition of the serious harm they have suffered at the hands of fellow members of the armed services, those who report sexual assault are perceived as a threat to the military institution itself.

Permanency of Wounds Through (Selective) Recognition Processes

As well documented in medical literature on MSA, the most commonly cited lasting impact of sexual assault is PTSD. Soldiers returning from combat missions in war zones are also treated for PTSD in high numbers. Yet women who are raped while serving in the military experience PTSD at a higher rate than men who have served in combat (*The Invisible War* 2012).

In depicting how sexual assault victims are systematically silenced, *The Invisible War* also raises questions about how military protocols for handling complaints of sexual assault work against the successful recovery of those who have been violated. Whether through threats of retaliation, demotion in rank, or frustrating interactions with the VA about appropriate medical treatment, the military ensures that sexual assault victims' physical and psychological injuries go ignored and are institutionally unrecognized. This lack of recognition by the military, or worse, the reclassification of the harm as self-inflicted, makes the process of healing nearly impossible. For how can one heal a wound that the military declares nonexistent? In the words of Amando Javier (U.S. Marine Corps), one of three male veterans who testify in the documentary, "It's really, really hard to forget. You know, up until now, I still...they live in my head, you know. I can hear them laugh, I can see their faces, I can see what they're doing to me" (Javier in *The Invisible War* 2012, 38:33). By refusing to acknowledge these acts of violence, the military simultaneously produces a sense of permanency to the wounds sustained by victims.

OTHER INVISIBLE WOUNDS AND THE EFFECTS OF NON-TREATMENT

In the case of sexual assault, military investigative techniques that disappear the harms of sexual violence ensure that perpetrators are not punished or brought to justice for their crimes. By consigning wounds incurred in sexual assault below the threshold of military visibility, these tactics also ensure that the wounds fester in the absence of any serious therapeutic intervention. But rape is not the only invisible wound that has escalated during recent military conflicts. Both the wars in Afghanistan and Iraq have provided evidence of increased incidences of PTSD and traumatic brain injuries.

The Frontline documentary, *The Wounded Platoon*, provides insight into the consequences of the military's failure to take invisible wounds seriously. Focusing on a unit of soldiers stationed in Fort Carson, a military base on the outskirts of Colorado Springs, Colorado, *The Wounded Platoon* documents the unprecedented increase in levels of crime in the city as a direct result of actions by soldiers who are unable to reintegrate into society upon their return from combat assignments. The documentary also showcases the strong stigma associated with combat soldiers—focusing exclusively on men—who request psychological assistance to deal

with the traumatic events of their tours. Returning soldiers experience depression, anxiety, paranoia, and hallucinations once being taken out of the war theater. In the case of Fort Carson, soldiers who sought help with these matters received hostile treatment. In the words of one soldier,

> It just seemed like if you came forward at Fort Carson, you were going to take it up the ass. It was very easy to convince people that I didn't need help. Like all I had to do was come in and say 'you know, ah I'm fine, and I just need to move on.' And they seemed to accept that and say, 'okay, you're good to go, see you later'. (Krebbs in The Wounded Platoon: Frontline Documentary 2010, 15:55)

The failure to take psychological harms seriously is emblematic of a masculinized warrior narrative that correlates mental health injuries with weakness.

As a result of this intensive stigmatization, almost all soldiers of third unit Platoon told doctors they did not need help. Instead, they engaged in dangerous behavior—including the consumption of drugs, attempts at suicide, and sleeping with guns and knives in their beds—in order to mitigate a heightened sense of paranoia upon their return. By the end of their first tour, four members of the Platoon were kicked out for failing drug tests, one was sent to jail for driving under the influence and fleeing, five were medically discharged, and five left Fort Carson because they chose not to reenlist.

Lieutenant Krebbs was sent back to Iraq for a second tour which lasted 15 months instead of 12. During his second deployment, he witnessed a traumatic road-side bombing that obliterated a fellow Platoon soldier. "The images of it—you know what I mean—the guy in the back seat was…basically just a black vest with boots in the floorboard of the Humvee. There was nothing, I could tell, of him left" (Krebbs in The Wounded Platoon: Frontline Documentary 2010, 38:55). After witnessing this incident, he began to behave irrationally for his environment. Krebbs began to go out on patrol with an unloaded weapon. When he realized that his actions had become a danger to his fellow Platoon members, he decided to report his behavior and seek help from a military psychologist on base. "I came clean, and I told that guy that, 'hey, I'm not loading my weapon anymore. I'm also having this problem, I'm also having this problem.' I wanted them to recognize that I had a problem and to say 'he can't go outside the wire anymore' (The Wounded Platoon: Frontline Documentary 2010). Instead, the military psychiatrist treating Craig declared him fit for patrol duties. "And then it was like, alright, after your hour session, go back. Put your boots on, let's go…you know what I mean? It blew my fucking mind. I shouldn't have been allowed to go outside of the wire" (The Wounded Platoon: Frontline Documentary 2010).

When members of the third Platoon returned home after their second, lengthened tour in Iraq induced by President George W. Bush's surge tactic, many found it extremely difficult to reintegrate into life on base. Several decided to leave the military rather than reenlist. Many attempted to "self-medicate," turning to drugs and alcohol to deal with the side-effects of chronic PTSD. It is during this reintegration period after their second tour of duty in Iraq that an extensive number of crimes and deaths start to occur in Colorado Springs, many of them caused by former members of the third Platoon. As documented by the film, "Since the Iraq war began, a total of eighteen soldiers from Fort Carson have been charged with murder, manslaughter or attempted murder committed at home in the United States, and thirty-six have committed suicide" (Narrator in The Wounded Platoon: Frontline Documentary 2010, 12:40). Three of 18 charged with murder, manslaughter, or attempted murder were convicted.

Lieutenant Jose Barco is one of the three men convicted from the Platoon. He was charged with two counts of manslaughter for a drive-by shooting in which he narrowly missed one person, and shot a pregnant woman in the leg. In his trial in front of a civilian court, Barco's lawyers argued emphatically that his actions were emblematic of poor decision-making resulting from the invisible wounds he had sustained from his tours in Iraq, wounds in the form of PTSD and a traumatic brain injury for which he never received adequate treatment. In asking for a minimum sentence, Barco's lawyers requested that the judge consider "the situation that Mr. Barco has with his medical condition, mental health, and his service to this country. It is important to know he has a diagnosis of PTSD, he also has some traumatic brain injury from the numerous explosions that he was subject to during his combat service in Iraq" (Defense attorney in The Wounded Platoon: Frontline Documentary 2010, 1:05:58).

In Barco's case, the judge was unimpressed by the PTSD diagnosis or by the fact that he had served his country in multiple deployments in Iraq. Barco was sentenced to two consecutive sentences of 30 and 22 years in prison. As the judge noted during his sentencing, "This was a stupid, angry, impulsive act. I expect young gang members who are at war with each other to do drive by shootings, somebody who's been in Iraq and who's a military veteran, I wouldn't expect that type of behavior from them. In fact, it brings considerable discredit upon the uniform that you wore to be engaged in that type of activity" (Schwartz in The Wounded Platoon: Frontline Documentary 2010, 1:06:36).

Drive-by shootings, domestic violence, attempted murders, and suicide are gendered crimes, particularly associated with men. *The Wounded Platoon* has the virtue of showing the very high costs of the military's failure to take invisible wounds seriously. But like the military itself, the film foregrounds male behavior, associates wounds only with combat, and asks no questions about the gendered nature of these untreated wounds or their effects. The lingering effects of sexual assault in the military remain unstudied.

Conclusions

Military discourses on war wounds invariably invoke specific gendered images of soldiering. Institutional insistence positions physical maiming, visible bodily injuries, and death by combat as intimately necessary for the maintenance of the masculine warrior profile. Alternative forms of injury pose gender trouble and are feminized and framed as weakness, or are systemically tolerated, silenced, or masked.

The insistence that only those wounds be acknowledged and heralded that reify militarized masculinity trouble the military's ability, as an institution, to publicly deal with alternative forms of wounding that do not fit the bill. Most recently, the heightened attention given to psychological injuries associated with tours in Operations Enduring Freedom and Iraqi Freedom have troubled the military's traditional association of harm and wounding with combat in the twenty-first century. When men step forward to assert the psychological impact of wars they are feminized through maneuvers that frame them as weak and overly sensitive. Similarly, male-on-male rape within the ranks raises calls to police sexualities and furthers notions about institutional homophobia, a counter image to the heterosexual male soldier that represents the military's gender order. When sexual assault in the military receives increased media attention or attention from advocacy groups, the military must move to counter the impact revelations of such scandals may have on the public's view of the institution.

The organizational move to deflect attention and concern for sexual assault survivors to concern for the institution is established through a combination of securitization and medicalization discourses. Presented as an anomaly, sexual assault within the ranks is characterized as a threat to the institution rather than a violation of individual bodies. Borrowing language from securitization theory, sexual violence is constructed as an external force that threatens the existence of the institution. This threat is described in medical terms as an "epidemic," "cancer," "disease," and a

"plague." Through medicalized securitization rhetoric, the military transforms gendered violence into containment tactics designed to protect military institutions.

Focusing on this medicalized discourse sheds particular light on how sexual assault in the military can be differentiated in important and marked ways from sexual assault in the civilian sphere. While many of the narratives that emerge from my analysis of the documentary *The Invisible War* may be comparable to dominant narratives of gendered violence in general—the physical nature of sexual assault, sexual assault needing to be proven through physical wounds in order to be believed, victims bringing attacks upon themselves through items of clothing—military officials' continuous discursive attempts to maintain the institution's masculinist culture provides a very different context for sexual assault. First, by focusing on the impact of sexual assault on the branch or overall institution, the military's representation of its public position on the topic produces a "devaluation of sexual violence victims and survivors" (Nayak 2009, 159). By referring to sexual assault as a "disease," "plague," or "cancer," not only are the experiences of victims silenced, but focus shifts to eradicating the disease. Second, statements which link sexual assault to a medical disease that threatens the existence of the military moves to depoliticize the problem; it removes responsibility and disassociates the systematic nature of the crime as being one that is specifically caused, tied to, and permitted by military culture.

And while sexual assault sustained through military service is not acknowledged as a military wound, unseen injuries sustained through combat do not fare any better. Soldiers struggling with the "invisible wounds" of war, that is psychological and physiological symptoms which do not necessarily mark their bodies visibly as a missing limb or other bodily maiming would, are represented through feminized discourses in which their experienced wounds are seen by the institution as a weakness which limits its ability to complete missions and continue with assigned foreign policy tactics. Discourses such as the "Good Soldier" narrative are accordingly used very differently depending on the situation; in the context of sexual assault, soldiers who are accused of committing rape against fellow soldiers often receive no punishment because they are otherwise argued to be valued, admirable members of the organization. In the context of crimes that are committed by soldiers who are psychologically wounded but were never diagnosed or never received proper treatment for PTSD and traumatic brain injuries, their acts of crime are seen as a disgraceful sullying of the honor of the uniform they wear and the country they serve.

Both invisible injuries stemming from combat tours and sexual assault within the ranks trouble the military's conceptualization of (honorable) wounding—visible forms of harm produced on bodies by enemy action. Although military and medical technology allow more and more soldiers to survive what had once been fatal injuries, and "invisible wounds" are fast becoming the most numerous incurred in theaters of war, the military has done little to expand the categories of bodies it valorizes, honors, and respects. Indeed, military museums and monuments tend to concentrate public attention on heroic notions of war wounds and the valiant male soldiers who suffer them. Chapter 5 demonstrates how this narrow focus is created and perpetuated.

Notes

1. 1.64 million U.S. Armed Services members deployed to Afghanistan and/or Iraq as part of Operation Enduring Freedom and Operation Iraqi Freedom respectively between October 2001 and 2008. Yet both utilized smaller numbers of U.S. troops and produced lower casualty rates of U.S. military killed or wounded than in other prolonged wars, including the Vietnam War and Korean War (RAND 2008, 2).
2. According to the Department of Defense Annual Report on Sexual Assault in the Military for fiscal year 2017, of the 6769 reports, 5864 involved military personnel. Of the 5864 cases, about 10% were related to incidents of sexual assault against a service member prior to joining the military. In total, 5277 cases were filed by service members during fiscal year 2017 for incidents of sexual assault that took place at some point in their military career. See Department of Defense (2018) report "Department of Defense Annual Report on Sexual Assault in the Military: Fiscal Year 2017." http://sapr.mil/public/docs/reports/FY17_Annual/DoD_FY17_Annual_Report_on_Sexual_Assault_in_the_Military.pdf.
3. The 2012 release of *The Invisible War* received considerable attention for several years after its release—both by the political and academic spheres—bringing public attention to the often unrecognized pervasiveness of sexual assault in the U.S. military. The film provides gripping depictions of soldiers—predominantly women, but also to a lesser extent, men—who have experienced sexual harassment and violence while serving in the military. The film has politicized the rape epidemic in the military and has been said to have produced a "greater influence on national policy, in a shorter amount of time, than nearly any other documentary in history" (Kooney 2013).
4. In response to modes of foreign policy and warfare that have been characterized as attempts to "feminize" the military, Carreiras (2006) suggests "the

prevalence of sexual harassment may be seen as the effect of pressure to reassert the masculinity of service members, in a period where the main function of the military is shifting from warfare to 'operations other than war' (OOTW) such as peacekeeping and humanitarian interventions" (54).
5. It should be noted that research on MSA has traditionally been conducted within the "medical and psychological communities" and in so doing, "takes up not only a particular way of talking about women and MSA, but...takes up a particular ontological position in which *agentic persons* are removed in favor of a focus on environmental conditions" (emphasis in original, Hannagan 2016, 3).
6. Building on Cynthia Enloe's observations in this area, a number of feminist scholars have investigated these issues in various empirical settings. For example, see Sandra Whitworth's (2004). *Men, Militarism, and UN Peacekeeping: A Gendered Analysis,* and Katherine HS Moon's (1997). *Sex Among Allies: Military Prostitution in U.S. – Korea Relations.*
7. The proposed FAIR Military Act would have limited the use of the "good soldier" narrative or defense during the trial of perpetrators of MSA (Tsongas and Turner 2014), but the bill did not progress past introduction during its initiation in the 113th Congress (2013–2014).
8. In-text citations after direct quotes and which are associated with the source *The Invisible War* connote timestamps at which individuals cited spoke the words quoted.

REFERENCES

Amos, J. 2013. Testimony. In United States Congress. Senate Armed Services Committee. Hearing on Pending Legislation Regarding Processes of Prosecution Against Perpetrators of Sexual Assault in the Military. Online: https://www.armed-services.senate.gov/imo/media/doc/pendinglegislation_sexualassaultsinmilitary_fullcomm_hearing_060413.pdf

Barco, J. 2010. Interview in The Wounded Platoon: Frontline Documentary. Public Broadcasting Station. Documentary Film.

Bostock, D.J., and J.G. Daley. 2007. Lifetime and Current Sexual Assault and Harassment Victimization Rates of Active-Duty United States Air Force Women. *Violence Against Women* 13 (9): 927–944.

Carreiras, H. 2006. *Gender and the Military: Women in the Armed Forces of Western Democracies.* New York: Routledge.

Catagnus, R. 2012. Interview in *The Invisible War.* Ro*co Films Educational. Documentary Film.

Cioca, K. 2012. Interview in *The Invisible War.* Ro*co Films Educational. Documentary Film.

Coughlin, P. 2012. Interview in *The Invisible War*. Ro*co Films Educational. Documentary Film.
Coyle, B.S., D.L. Wolan, and A.S. Van Horn. 1996. The Prevalence of Physical and Sexual Abuse in Women Veterans Seeking Care at a Veterans Affairs Medical Center. *Military Medicine* 161 (10): 588–593.
D'Costa, B. 2006. Marginalized Identity: New Frontiers of Research for IR? In *Feminist Methodologies for International Relations*, ed. Brooke A. Ackerly, Maria Stern, and Jacqui True, 19–41. New York: Cambridge University Press.
Defense Attorney. 2010. Interview in The Wounded Platoon: Frontline Documentary. Public Broadcasting Station. Documentary Film.
Department of Defense. 2004. Task Force Report on Care for Victims of Sexual Assault. Accessed November 4, 2013. http://www.defenselink.mil/news/May2004/d20040513SATFReport.pdf.
———. 2018. Department of Defense Annual Report on Sexual Assault in the Military. Accessed January 2, 2019. http://sapr.mil/public/docs/reports/FY17_Annual/DoD_FY17_Annual_Report_on_Sexual_Assault_in_the_Military.pdf.
Dickerson, D. 2012. Interview in *The Invisible War*. Ro*co Films Educational. Documentary Film.
Eichler, M. 2013. Women and Combat in Canada: Continuing Tensions Between 'Difference' and 'Equality. *Critical Studies on Security* 1 (2): 257–259.
Enloe, C. 1983. *Does Khaki Become You? The Militarization of Women's Lives*. Boston: South End Press.
Greenert, J. 2013. Testimony. In United States Congress. Senate Armed Services Committee. Hearing on Pending Legislation Regarding Processes of Prosecution Against Perpetrators of Sexual Assault in the Military. Online: https://www.armed-services.senate.gov/imo/media/doc/pendinglegislation_sexualassaultsinmilitary_fullcomm_hearing_060413.pdf
Hannagan, R.J. 2016. 'I Believe We Are the Fewer, the Prouder': Women's Agency in Meaning-Making After Military Sexual Assault'. *Journal of Contemporary Ethnography* 46 (5): 1–21.
Helmer, E. 2012. Interview in *The Invisible War*. Ro*co Films Educational. Documentary Film.
Howell, A. 2014. The Global Politics of Medicine: Beyond Global Health, Against Securitisation Theory. *Review of International Studies* 40 (5): 961–987.
Javier, A. 2012. Interview in *The Invisible War*. Ro*co Films Educational. Documentary Film.
Jones, C. 2012. Interview in *The Invisible War*. Ro*co Films Educational. Documentary Film.
Kimerling, R., K. Gima, M.W. Smith, A. Street, and S. Frayne. 2007. The Veterans Health Administration and Military Sexual Trauma. *American Journal of Public Health* 97 (12): 2160–2166.

Klay, A. 2012. Interview in *The Invisible War*. Ro*co Films Educational. Documentary Film.
Kooney, K. 2013. *The Invisible War*: A Timeline of Influence on U.S. Policy. Ro*co Films Educational. Email.
Krebbs, R. 2010. Interview in The Wounded Platoon: Frontline Documentary. Public Broadcasting Station. Documentary Film.
Krinsley, K.E., J.G. Gallagher, F.W. Weathers, C.J. Kutter, and D.G. Kaloupek. 2003. Consistency of Retrospective Reporting About Exposure to Traumatic Events. *Journal of Traumatic Stress* 16 (4): 399–409.
MacKenzie, M. 2015. *Beyond the Band of Brothers*. Cambridge: Cambridge University Press.
Martin, L., L. Rosen, D. Durand, R. Stretch, and K. Knudson. 1998. Prevalence and Timing of Sexual Assaults in a Sample of Male and Female U.S. Army Soldiers. *Military Medicine* 163: 213–216.
Mathers, J.G. 2013. Women and State Military Forces. In *Women and War*, ed. Carol Cohn, 124–145. Cambridge: Polity Press.
McDonald, R. 2012. Interview in *The Invisible War*. Ro*co Films Educational. Documentary Film.
Moon, Katharine H.S. 1997. *Sex Among Allies: Military Prostitution in U.S.-Korea Relations*. New York: Columbia University Press.
Murdoch, M., M.A. Polusny, J. Hodges, and N. O'Brien. 2004. Prevalence of In-service and Post-sexual Assault Among Combat and Noncombat Veterans Applying for Department of Veterans Affairs Posttraumatic Stress Disorder Disability Benefits. *Military Medicine* 169: 392–395.
Murdoch, M., J.B. Pryor, M.A. Polusny, and G.D. Gackstetter. 2007. Functioning and Psychiatric Symptoms Among Military Men and Women Exposed to Sexual Stressors. *Military Medicine* 172: 718–725.
Narrator. 2010. The Wounded Platoon: Frontline Documentary. Public Broadcasting Station. Documentary Film.
Nayak, M. 2009. Feminist Interrogations of Democracy, Sexual Violence, and the US Military. In *Theorizing Sexual Violence*, ed. Renee J. Heberle and Victoria Grace, 147–175. Oxon: Routledge.
Odierno, R. 2013. Testimony. In United States Congress. Senate Armed Services Committee. Hearing on Pending Legislation Regarding Processes of Prosecution Against Perpetrators of Sexual Assault in the Military. Online: https://www.armed-services.senate.gov/imo/media/doc/pendinglegislation_sexualassaultsinmilitary_fullcomm_hearing_060413.pdf
Papp, R. 2013. Testimony. In United States Congress. Senate Armed Services Committee. Hearing on Pending Legislation Regarding Processes of Prosecution Against Perpetrators of Sexual Assault in the Military. Online: https://www.armed-services.senate.gov/imo/media/doc/pendinglegislation_sexualassaultsinmilitary_fullcomm_hearing_060413.pdf

RAND Corporation. 2008. *Invisible Wounds of War: Psychological and Cognitive Injuries, Their Consequences, and Services to Assist Recovery*, ed. Terri Tanielian and Lisa H. Jaycox. Santa Monica, CA: RAND.

Sadler, A.G., B.M. Booth, B. Nielson, and B.N. Doebbeling. 2000. Health-Related Consequences of Physical and Sexual Violence: Women in the Military. *Obstetrics and Gynecology* 96 (3): 480–493.

Schwartz, L. 2010. Interview in The Wounded Platoon: Frontline Documentary. Public Broadcasting Station. Documentary Film.

Sewell, H. 2012. Interview in *The Invisible War*. Ro*co Films Educational. Documentary Film.

Skinner, K.M., N. Kressin, S. Frayne, T.J. Tripp, C.S. Hankin, D.R. Miller, et al. 2000. The Prevalence of Military Sexual Assault Among Female Veterans' Administration Outpatients. *Journal of Interpersonal Violence* 15 (3): 291–310.

Smith, D., B. Frueh, C. Sawchuck, and M. Johnson. 1999. Relationship Between Symptom Over-reporting and Pre- and Post-combat Trauma History in Veterans Evaluated for PTSD. *Depression & Anxiety* 10 (3): 119–124.

Surís, A., L. Lind, T.M. Kashner, and P.D. Borman. 2007. Mental Health, Quality of Life, and Health Functioning in Women Veterans: Differential Outcomes Associated with Military and Civilian Sexual Assault. *Journal of Interpersonal Violence* 22 (2): 179–197.

The Invisible War. 2012. Ro*co Films Educational. Documentary Film.

The Wounded Platoon: Frontline Documentary. 2010. Public Broadcasting Station. Documentary Film. Online. http://www.pbs.org/wgbh/pages/frontline/woundedplatoon/.

Tsongas, N., and M. Turner. 2014, June 16. What It Will Take to End Sexual Assault in the Military. *The Christian Science Monitor*. http://www.csmonitor.com/Commentary/Opinion/2014/0616/What-it-will-take-to-end-sexual-assault-in-the-military.

Turchik, J.A., and S.M. Wilson. 2010. Sexual Assault in the U.S. Military: A Review of the Literature and Recommendations for the Future. *Aggression and Violent Behavior* 15 (4): 267–277.

United States Congress. 2013. Senate Armed Services Committee. Hearing on Pending Legislation Regarding Processes of Prosecution Against Perpetrators of Sexual Assault in the Military. Online: https://www.armed-services.senate.gov/imo/media/doc/pendinglegislation_sexualassaultsinmilitary_fullcomm_hearing_060413.pdf.

Weldes, J. 1999. *Constructing National Interests: The US and the Cuban Missile Crisis*. Minneapolis: University of Minnesota Press.

Welsh, M. 2013. Testimony. In United States Congress. Senate Armed Services Committee. Hearing on Pending Legislation Regarding Processes of Prosecution Against Perpetrators of Sexual Assault in the Military. Online: https://www.armed-services.senate.gov/imo/media/doc/pendinglegislation_sexualassaultsinmilitary_fullcomm_hearing_060413.pdf

Whitworth, Sandra. 2004. *Men, Militarism, and UN Peacekeeping*. Boulder, CO: Lynne Rienner Publishers, Inc.

Yuhl, Stephanie E. 2016. Militarized US Women from the Wars in Iraq and Afghanistan: Citizenship, Homelessness, and the Construction of Public Memory in a Time of War. In *Gendered Wars, Gendered Memories: Feminist Conversations on War, Genocide and Political Violence*, ed. Ayse Gul Altinay and Andrea Peto, 159–178. New York: Routledge.

CHAPTER 5

Military Museums and Memorial Sites: Disappearing Women in the Military

INTRODUCTION

In times of gender trouble, U.S. military officials have both proclaimed commitments to equality and inclusion, and deployed various discursive tactics to reassert a heteromasculine gender order. Military spokespersons use language to assert certain "truths" about military life and operations, while skillfully disappearing, silencing, or masking inconvenient evidence that challenges those representations. The messages that the military disseminates and communicates, however, exceed spoken word, official pronouncements, and Congressional testimonies (see Chaps. 3 and 4). The military museum is one powerful example of a mode of military communication that shapes public understandings. Through their voluntary attendance at historical military sites, curated exhibits, and national museums and cemeteries, the public actively consumes carefully constructed narratives about military performance and national victories, shaping collective understandings of gender and war.

Like written and spoken word, physical spaces and historical sites retain and convey constructed narratives as uncontested truths. Transformed military spaces offer powerful lessons about the military and its role in

---

A version of this chapter was published in the *International Feminist Journal of Politics*. See Szitanyi, S. 2015. "Semiotic Readings of the USS Midway Museum," *International Feminist Journal of Politics*, 17 (2): 253–270.

© The Author(s) 2020
S. Szitanyi, *Gender Trouble in the U.S. Military*,
https://doi.org/10.1007/978-3-030-21225-4_5

society. Once solemn spaces designed to honor those who gave their lives for their country, military sites have more recently been commercialized to produce living history[1] through revitalized amusement park-like thrills in which visitors are urged to live the experiences provided by the site. Curators and exhibit designers are charged with creating interactive exhibits that are to attract diverse demographics, audiences that cut across ages, races, genders, and geographies of origin.

In the twenty-first century, the military's official public relations are on display in more than 400 institutionally related museums, monuments, and memorials[2] that span the United States and its territories.[3] As high "touch point" sites, these curated military installations offer the public opportunities to observe, interact with, and participate in selected facets of the military's gender order. In contrast to policy papers and Congressional testimony, military museums reach large audiences and offer visitors firsthand encounters with warfare, military technology, battle strategy, and hardships of life on the front, a sensory experience through which all five bodily senses are engaged. These "multimodal social semiotics" (Pang Kah Meng 2004, 31) construct messages through visual, audial, spatial, and written modalities. Indeed, visitors are active participants who (most likely unknowingly) shape the information they receive through a host of hands-on decisions they make while traversing these sites, and which, as a result, heighten their investment in military campaigns and outcomes.

Curated sites and exhibits are semiotic systems—systems of signs—that transmit messages through their contents. Within these built environments and historical spaces, museums, cemeteries, and memorials communicate meaning to their audiences through a wide spectrum of materials and modalities—exhibits, catalogs, books, recreational simulators, cafés, and even gift shops. Importantly, these "artifacts do not exist in a space of their own, transmitting meaning to the spectator, but on the contrary, are susceptible to a multiform construction of meaning, which is dependent on the design, the context of other objects, the visual and historical representation, the whole environment" (Smith 1989, 19). As museums orient their exhibits toward large target audiences, designers organize space and activities to provide visitors with a holistic gestalt that is experienced as entertainment and recreation even as it conveys a host of explicit and implicit messages.

In contrast to the static historical artifacts—such as press releases, recorded public statements, and written policy memoranda—covered thus far in this book, curated military spaces are prime sites for depicting change; they can be altered, reconfigured, and altogether redesigned.

Individual artifacts and entire exhibits can be added (to) or withdrawn (from), or spatially rearranged to reorient visitors' experience. As institutions and their associated policies show variation, or new (or recently declassified) evidence alters interpretations of the past, these exhibits and sites in which they exist can be "updated," restructured to enrich and provide temporally accurate and authentic accounts of military operations and outcomes. For this reason, military museums provide a particularly rich site for the investigation of shifts in military gender regimes; we may assume that any fundamental change to how the institution views gender and gender relations could be communicated to the public at large through these mediums, particularly those sites which attract a large number of visitors. This leads us to ask the following: given the contestations over gender in the military during the past half-century, have military museums altered their representations of women, homosexual, and transgender personnel in the military? Or have they actively reinforced messages that affirm heteromasculinity as the hallmark of military service, despite official policies that purport to be egalitarian and inclusive?

To explore these questions, the chapter provides a semiotic investigation of the *USS Midway* Aircraft Carrier, one of the longest serving vessels in the history of the U.S. Navy. Decommissioned in 1992, the *USS Midway* opened its doors to the public as a museum in June 2004 after being permanently docked in San Diego bay. By 2010, the museum had boasted its five millionth visitor, with an additional 1.4 million visitors to the *USS Midway* each year for eight consecutive years thereafter, making it "the most-visited floating ship museum in the world" (USS Midway Museum 2018). Thirty percent (30%) of the annual visitors to the *USS Midway* Museum visit from outside of the United States. As of 2019, the museum was listed as number one on Travelocity.com for a list of attractions in San Diego, California, and as the "5th most-popular U.S. museum of any type" on the site.[4] Additionally, it has been used as a filming set for a series of popular reality television shows in the U.S. including "American Idol," "The Bachelor," "Hell's Kitchen," and has been rented as a venue for "300 private events" (USS Midway Museum 2018) annually, including birthday parties and corporate events. The museum also markets educational services, providing classroom space aboard the aircraft carrier for science, technology, engineering, and math (STEM) field trips for over 50,000 elementary school-aged students each year. For younger children, the museum holds overnight stays aboard the aircraft twice a year, affording them the opportunity to "liv[e] an authentic life of an aircraft carrier sailor" (USS Midway Museum 2018).

In addition to acting as an educational resource for the public on the history of the aircraft carrier, large open spaces on the *USS Midway* also serve as the backdrop for 400 various military ceremonies—retirement, re-enlistment, memorial—each year, giving visitors the opportunity to "watch a young man or woman re-enlist on the flight deck or see an authentic change-of-command or inspirational retirement ceremony… illustrat[ing] America's values of strength, freedom, and peace for the entire family" (USS Midway Museum Website 2013). Through the exhibits, tours, and ceremonies held on the ship, the core of the *USS Midway* Museum's mission is officially characterized as "preserving and honoring the 225,000 young men and women who served aboard the *USS Midway*—and by extension all those who serve in uniform" (USS Midway Website until March 7th, 2013).[5]

Despite the general but explicit reference to honoring men and women in uniform, images of women were markedly absent from the *USS Midway* Museum and from an honoring ceremony of veterans held on the aircraft carrier during my visits to the museum.[6] The gender-inclusive narrative established by the museum's website coexists with a plethora of tributes to men and a lacuna for women. In spite of women's participation in certain combat-related and combat-support positions for more than a century (and officially allowed to perform in combat roles as of 2016), many routinely consumed messages about military service—explicit and symbolic—continue to leave women out of the picture. Through the erasure of women's performance in the armed services, war museums and memorials help ensure that war and the military remain male preserves in the public imagination. The *USS Midway* Museum illuminates this process of erasure.

As a contemporary Naval ship to be turned into a public memorial and museum—and with the shear annual public reach that it boasts—the *USS Midway* assumes a role greater than documenting the history of this specific aircraft carrier, where no women served on active duty.[7] As the mission statement notes, the museum seeks to honor "all those who serve in uniform," as such it assumes a representational responsibility to depict men's and women's roles within the Navy, capturing their crucial contributions to war efforts. By claiming to represent men and women in uniform while eliding women in the military, the *USS Midway* Museum contributes to a politics of erasure, a partitioning of the sensible that relegates women below the threshold of visibility.[8] Claiming to note women's presence through displays in which women are manifestly absent

constitutes a form of memorialization that replicates a long history of gendering practices in the military. This chapter illuminates those gendered dynamics, situating them in the context of the politics of social memory and showing how they help reproduce gendered hierarchies of citizenship.

Tracing the construction of social memory that validates certain forms of violence committed to defend the nation, while rendering other modes of violence invisible, the political potency of the militarized discourse lies not "in the accuracy or inaccuracy of its facts, but in its constitution of the ground on which to decide what counts as fact and what modes of comprehension to activate in understanding these facts" (Ferguson and Turnbull 1999, 45). In so doing, the military not so much hides but subtly conceals the incomplete truths it has created through a museum exhibit. Institutions within which military order embeds itself tell a specific history of militarization, a history in which many narratives are silenced. Historically, war stories, or "facts," were at least in part produced through the establishment of national cemeteries and processes of collective mourning over the dead at those sites. As I examine in the following section, these spaces produce selective and highly gendered military narratives about heroism associated with death through service. These narratives play an important part in gendered processes of militarization that become embedded in societies. Semiotics provides alternatives to the military's self-representation and probes for deeper understanding of how militarization as a discursive process has the power to create hierarchies among bodies, both internal and external to the military institution. As this chapter on the *USS Midway* demonstrates, narratives that simultaneously include and erase women in the military sustain a gendered order of citizenship, shoring up traditional notions of male service while rendering comparable service by women invisible.

While public memorials honoring women's service in the armed forces specifically do exist, they are few and far between. Katharine Millar (2015), for example, highlights that public representations of women's service related to combat, in particular, "is a severely limited universe" (3). Nonetheless, as of the writing of this book, the following memorials in the United States focus on women's military service specifically (to the best of my knowledge)—the Vietnam Women's Memorial in Washington, D.C., the Women in Military Service for America Memorial in Arlington Virginia, the Molly Marine Memorial in New Orleans, Louisiana, and the Sybil Ludington Memorial in Carmel, New York. In addition, the Las Cruces Women Veterans Memorial in Las Cruces, New Mexico, opened in spring of 2018 (Chavez 2018), as was a monument honoring the 6888

Central Postal Directory Battalion—the only all-black female military unit to serve during World War II—installed in Fort Levenworth, Kansas in late 2018 (Buffalo Soldier Educational and Historical Committee 2017). Further, it is worth noting that of the various branches of the military, the Army is the only one to have devoted a stand-alone museum to women's service specifically. That museum—the Army Women's Museum—is located in Fort Lee, Virginia.

What do women's only military heritage sites mean for feminist understandings of degendering? Arguably, a liberal feminist degendering agenda may point to these sites as new opportunities for inclusion; increased recognition of women's service through memorials and museums add and acknowledge women's stories where they have been omitted, critically rewriting histories that were previously established neglectfully. And while limited, the recognition that these spaces do (anemically) exist is an important one. Why not, then, engage in a semiotic investigation of these sites instead of the USS *Midway* where women did not serve during the commissioned history of the aircraft carrier?

I argue that the segregation of women's service and history as separate from those of the men who serve and served in those same branches reinforces a binary separation and understanding of women's service as being unequal and, as such, unhelpful in "rendering gender irrelevant" (Lorber 2000, 81). Indeed, it makes (dyadic constructions of) gender more palpable by an institutional side-stepping of social responsibility to educate publics appropriately. It avoids literal and figurative placements of women next to men in combat, visual representations that may assist in the public's (re)imagination of combat being open to and neutral of all differences. They fail to defuse constructed gendered relations between malehood and death through military service—"sidelining the exposure of female military personnel to violence…The separation between women and death is thus maintained" (Millar 2015, 13). While I do not provide a review of these memorials and monuments here (see Millar's 2015 excellent review of the Women in Military Service for America Memorial and the Vietnam Women's Memorial) I agree with Katharine Millar that the manner in which (at least some of the) memorials devoted to the depiction of women's military service only "perform their obligation to commemorate without constructing female soldiers—and, in many cases, female military personnel more generally—as legitimate subjects" (14). Instead, an incomplete, "partial commemoration" is constructed, through which the possibility of a "female warrior" is dissolved and "is virtually literally effaced" through silenced, half-true histories that have them "removed from the socially intelligible" (14) understandings of military

service and war. As in other cases of gender trouble, military memorials and heritage sites showcase women's military service and existence as service members as a "mis-fit," fundamentally lacking the ability "to be specifically and exhaustively categorized within the normative structures" (20), ultimately endangering the prescribed institutional gender order.

Gendering Memories of War Through Spatial Configurations

Known as the "science of signs," semiotics explores how meaning is communicated, not only through various groupings of texts, but also in verbal and non-verbal communications, theater and drama, and in diverse forms of aesthetic communications such as music, photography and architecture, and natural and built environments (Noth 1995). Interpreting military museums as curated spaces requires a rejection of the idea that space is "an empty container devoid of social history and relations" (Ferguson and Turnbull 1999, 111–112). Semiotic readings of spatial configurations offer a unique means to excavate structures of meaning within military cemeteries, museums, and memorials—spaces created to communicate certain meanings, but which also convey less explicit lessons. Military memorials offer specific narratives for consumption in a specific space. Efforts to ensure that institutionally accredited messages emerge and are absorbed at these sites may require the silencing and masking of alternative narratives and histories. In exploring how the military represents itself in these curated sites, semiotics calls attention not only to explicit messages, but also omissions, exclusions, and silences. By questioning what is hidden or systematically removed from certain histories, semiotics "locate the particular representational elements that both conceal and reveal the military's presence and power; illuminate the functions of these elements of representation as rituals of power; and expand discursive space so that other voices can be heard" (Ferguson and Turnbull 1999, xx).

In memorializing those who fight to defend their nation, war museums contribute to discourses on citizenship, in this case, through social constructions of memory "embedded in complex class, gender, and power relations that determine what is remembered (or forgotten), by whom, and for what end" (Gillis 1994, 3). By defining what counts as national service and who is allowed to perform it, "militarized practices of citizenship that define and administer bodies—social, physical, and environmental—

occlude alternative notions of citizenship and embodiment" (Ferguson and Turnbull 1999, xx). Beyond honoring particular feats as daring, narratives of valor—including those discussed in the previous chapter and associated with processes of (eligible) bodily harm and wounding (see Chap. 4)—contribute to understandings of national belonging, linking membership to service in wars, subtly encoding citizenship as a male terrain.

Gender structures the meaning of national service in wartime while memorials and other modes of memorialization work to numb grief associated with the deadly atrocities of war (Faust 2008). Soldiers have historically "turned to the resources of their culture, codes of masculinity, patriotism, and religion to prepare them" (5) for military adventures likely to end in death as to provide them with glory and feelings of heroism. During the Civil War, for example, in a context in which preparation for war was simultaneously preparation for death, conceptions of a "Good Death" subtly shifted from religious norms concerning life in service to one's God to life given in service to one's country. Condolence letters that military units sent home to loved ones of the deceased began to serve as testimonials of valiant military performance, patriotism, and manliness (24). Narratives of soldiers who "died a glorious death in defense of his Country" (24) became a marker of having lived "bravely and manfully" (25), blending "patriotism and piety… in what was at once a newly religious conception of the nation and a newly worldly understanding of faith" (26).

The creation of national cemeteries further anaesthetized the brutalities of war. During the Civil War, the first national cemetery was created with the aim of accounting for and honoring all soldiers who had perished in battle. Created after the Battle of Gettysburg, the Soldier's National Cemetery was consecrated by Abraham Lincoln, whose *Gettysburg Address* "signaled the beginning of a new significance for the dead in public life" (Faust 2008, 100). In so doing, the national soldier's cemetery made the death of the soldier a societal, public affair. More than a private loss to his family, the death of each soldier was a loss to the whole nation. It was no longer the responsibility of families to bury their military dead. Instead, a grateful nation assumed the responsibility to care for the mortal remains of those who died to defend it. In assuming the responsibility to account for and bury the dead, the nation-state created a dramatic shift in how those who died in combat would be remembered, creating a new index of equal citizenship. The physical configuration of the graves in the cemetery—the rows of aesthetically uniform tombs—framed the deaths of each

and every Union soldier as of equal importance to the nation. Regardless of a fallen soldier's rank, location of origin, site of war service, or personal misdemeanors, each warrior was equal in the eyes of the state (101).

The work of recording the deaths of Civil War soldiers and securing their burial often devolved upon women. Clara Barton was the first to make the case that in return for giving their lives to a cause that benefitted the nation, the nation should be responsible for identifying all unknown soldiers and providing a record of death for all soldiers who had died from the processes of war (Faust 2008, 101). Barton also suggested that women who undertook the national duty of accounting for and burying the dead performed an important service for their country. In calling attention to this service, Barton sought to establish an unprecedented contract between *women* and the state. As Faust notes, by having the state recognize women's labor in the war effort, Barton believed that women could "claim new rights of personhood and citizenship that derived from their wartime sacrifice" (230). Nevertheless, Barton understood that the state's recognition of women's wartime service would not entail equal citizenship. Dead soldiers, all of whom were men, would be valorized to the highest degree because of their ultimate sacrifice (Faust 2008, 230). Claims made to the state by others would never carry as much force.

As a new form of war memorialization, the national soldiers' cemetery established both equality (among the brave war dead) and inequality (differentiating between those who died in service of their nation and those who did not). This hierarchical citizenship was both legitimated through narratives celebrating the virtues of military sacrifice for the nation and rendered invisible as focus on the graves of fallen men eclipsed all recognition of the war service and sacrifices of women.

If memorialization of the massive losses in the American Civil War forged a new meaning of equal citizenship that helped to erase persistent gender and race inequalities that suffused the nation, World War II memorials helped normalize militarization in peace time. Militarization shores up particular aspects of state power, "not simply through the perceptible presence of military objects and events, but through the social and economic insinuation of the military into other institutions, and the cultural imbrication of military codes, symbols, and values into daily life. These latter processes flag not the military per se but militarization as a dynamic, contested process of constituting a particular kind of order, naturalizing and legitimating the order, while simultaneously undermining competing possibilities for order" (Ferguson and Turnbull 1999, 2).

As with the silencing culture of military sexual assault examined in the previous chapter, "facts" constructed by the military in war museums and military cemeteries silence many other stories (Ferguson and Turnbull 1999). National cemeteries become "a site for the production of the stories the state tells about why young men die in war" (Ferguson and Turnbull 1999, 109). Within these cemeteries, two spaces are typically produced: the burial space, which is created "with neat symmetrical rows of graves" and the memorial, "with its texts, maps, inscriptions, illustrations, instructions, and prayers" (109). The spatial configuration of graves and the memorial in relation to one another and in relation to other national cemeteries "constructs a barrier between violent death and heroic sacrifice, a symbolic barrier that pacifies death, sanitizes war, and enables future wars to be fought" (Ferguson and Turnbull 1999, 108–109). As codified space, national cemeteries silence egregious violence by carefully controlling the content of the narratives of violence crafted by the memorials' creators. According to Kathy E. Ferguson and Phyllis Turnbull, codified spaces are constructed to be "monoglossic, unidirectional, and generally self-referential; it is an order that can be written on the land without reference to any other logic" (114–115). Although thousands of individuals lie within the national cemeteries, these spaces have only one narrative to tell and do not have room for other, individual narratives to emerge. Gender and racial diversity are eclipsed as the patriotic narrative subsumes all the military dead under one monolithic, heroic subject. They envision honor established through service and death as uniformly applicable. Like formal equality, military honor incorporates each individual death within a larger conception of the collective good—a conception that has no space for gender, racial, ethnic, class, or sexual specificity. As militarization seeps into social memory, critical consciousness is also occluded. The state determines which campaigns constitute a collective good worth dying for as the "noble" causes of "the great wars" bleed into imperial expeditions, "humanitarian" interventions, and "coalitions of the willing."

Militarization, Masculinity, and Gendered Exclusions

The intricacy of this process is perhaps best illustrated by Kathy E. Ferguson and Phyllis Turnbull's (1999) discussion of the Fort DeRussy War Museum exhibit, created to narrate the military history of Hawaii through the creation of the fort during the American Civil War that was named after General Rene

Edward DeRussy. The political potency of the militarized discourse surrounding the DeRussy story is not "in the accuracy or inaccuracy of its facts, but in its constitution of the ground on which to decide what counts as fact and what modes of comprehension to activate in understanding these facts" (45). In so doing, the Army through a museum exhibit not so much hides, but subtly conceals the incomplete truths it has created. Institutions within which military order embeds itself tell a specific history of militarization, a history in which many narratives are silenced. That history is reestablished in those institutions unquestionably and taken as normal. In the case of Ferguson and Turnbull's (1999) study of Hawaii, we can view war museums as one example of such institutions. The narratives institutions tell, and simultaneously the silences they create, allow those histories to become further embedded in social discourses by the average citizen who consumes museums as modes of leisurely activity or education, naturally and without questioning. In the case of war museums, "the process of the inscription of meaning was/is not a simple ink-on-paper act that is accomplished with a single swift stroke. Rather, it happens as a series of new normalizing processes take over and are both contested and entrenched" (45).

In reading the spatial configurations of the DeRussy exhibit against the grain, Ferguson and Turnbull (1999) seek to "reactivate the motility of discourse and make space for fresh perceptions that call for careful attention to the dominant metaphors, recurrent figured of speech, authoritative grammars, and prevailing rhetorical moves that produce realist discourse" (46). They do so by paying particular attention to the exhibit's arrangement, which reveals its authority through a chronological order that works "along the long spine as a linear narrative" (47). The end of this linear line represents the objective of the exhibit; it is partially to create an image of the ultimate citizen through military service to the visitor, first in that the ultimate citizen is one who understands the militarized history of Hawaii as taught by the museum, and one who recognizes the glory that is attributed to those individuals who have served in the armed forces.

Charles Tilly (1985) has noted that democracies such as the United States are often born of war. As a violent, highly militarized process, the very birth of the nation sets stories in motion that tend to hide the violence by which laws are created, constitutions written, and subjects fixed within gender, racial, and class structures. Stories of valor under fire, courage through adversity, and sacrifice even of one's life for the sake of the nation are not only the hallmark of military service, but of memorialization. These narratives are

sedimented in monuments, images, and museums. They shape "social memory" (Mills 2007) in ways that require interrogation, particularly when the exhibits are deliberately designed and spatially configured by the military for representation of itself.[9]

Social memory is simultaneously intimately tied to "institutionalized forgetfulness" (Shapiro 1997, 22). More is at stake in particular memorializations than whether the information presented is accurate. In creating a space of military remembrance, museum designers have the power to decide what counts as fact and what does not, and which information will be consumed and which will be forgotten. Thus, Michael Shapiro suggests that "in order to situate the narrativized forms of forgetfulness in the present…one has to return to their points of emergence, to the presuppositions" (22). Perhaps one of the most fundamental presuppositions of military memorials is that they are male spaces, commemorations of men's willingness to defend, fight, and die for their countries. But memorials that naturalize men's role as defenders of hearth and home cover over the political work required to exclude women from national defense.

Military memorials that focus exclusively on male valor perpetuate gendered exclusions enacted by the state, while also figuring death in ways that mask the brutal violence of war. By naturalizing male military service and constructing narratives of a good death, a death worth having for one's country, war memorials contribute to the "management of memory" (Mills 2007, 28) in ways that support militarization and disregard women's valiant contributions to the services. In short, a semiotic interpretation of the USS *Midway* Museum challenges mainstream conceptualizations of militarization and provides a more sophisticated and nuanced understanding of how militarism becomes embedded in societal level institutions.

THE USS *MIDWAY* AND THE GENDERING OF SOCIAL MEMORY

An aircraft carrier of the U.S. Navy, the USS *Midway* was commissioned immediately after the end of World War II, named in honor of the Pacific Battle of the Midway between Japan and the United States. Construction on the ship commenced in October 1943 and was completed by March 1945 (USS Midway Museum 2018). The sheer magnitude of the 972-foot-long warship made it the largest ship in the world prior to 1955 (USS Midway Museum 2018). While in commission, the USS *Midway* was most well known for its participation in missions during the Vietnam War, as

well as its role in evacuating South Vietnamese during the final days of the conflict, its presence in the Arabian Sea during the Iran Hostage Crisis, as well as its participation in Operation Desert Storm. The *Midway* was finally decommissioned in April 1992 and opened to the public as a Naval museum on June 7, 2004 (USS Midway Museum 2018).

Although the military is not typically understood as either a producer of consumer goods or a generator of popular narratives, war museums and memorials produce gendered discourses that are consumed by all who visit these sites. When visiting military museums, military cemeteries, and military ceremonies open to the public, citizens consume images and narratives produced by multiple branches of the military, as well as civilians employed by them. Through the placement of objects, textual interpretations, visual imagery, and the content of exhibits and ceremonies, particular forms of military service celebrate specific kinds of masculinity, while simultaneously disappearing women.

In the case of war museums such as the *USS Midway*, "the process of the inscription of meaning was/is not a simple ink-on-paper act that is accomplished with a single swift stroke. Rather, it happens as a series of new normalizing processes take over and are both contested and entrenched" (Ferguson and Turnbull 1999, 2). Through spatial configurations and narratives of particular military feats (as well as critical omissions from those narratives) war museums produce a particular image of the military to be consumed by the citizens who visit them. The *USS Midway* Museum's exhibit and various activities on the ship such as the "Segs4Vets Veterans Honoring Ceremony" provide a powerful example of the inscription of gendered meaning. Through the pervasive depictions of maleness and masculinity and the blatant absence of depictions of women, the *USS Midway* Museum teaches visitors several lessons: by disappearing women from active duty in all ship exhibits, the museum represents Naval service and service on aircraft carriers as a male-only field. By celebrating the valor of men wounded in war, it valorizes and normalizes notions that male citizens deserve more respect because of the unique role they perform in the military.

Aboard the USS Midway

The sheer size of the *USS Midway* makes it difficult to miss if one is anywhere near the San Diego Bay park promenade. On approach to the ship, guests—during my visits—were greeted by banners that read "Live the Adventure, Honor the Legend," and "Battle Tested, Kid Approved" as they walked toward the

Fig. 5.1 Exterior image of the *USS Midway* Museum

ticket booths. While waiting for the ticket booths to open, visitors who stood in line conversed with male veterans who work or volunteer as docents on the ship. They spoke of service, sacrifice, and battle as they prepared visitors for the experience they were about to take part in (Fig. 5.1).

Upon entering the *Midway*, visitors were greeted in the main hangar where they were given the option of picking up audio headsets. A fanfare of patriotic music played over loud speakers to accompany visitors as they walked through the hall where they found various flight simulators available for riding in the fashion of an amusement park. Behind glass casings, guests could examine the original designs for the ship in its various phases of construction. All the while, missiles and bombs of various eras with names of "Dambusters" and "Eyes of the Corps" hung from the walls as militarist decoration (Figs. 5.2 and 5.3).

While the flow of the exhibit did not require visitors to explore the ship in any particular order, the flight deck was given prominence: signs throughout the main hall instructed visitors on how to proceed to reach it.

5 MILITARY MUSEUMS AND MEMORIAL SITES: DISAPPEARING WOMEN... 133

Fig. 5.2 Image of Dambuster missile

The flight deck is the most popular portion of the *USS Midway*'s exhibit, attracting children and adults alike who observe and often explore the insides of historic military airplanes and helicopters from numerous eras of military history. Boasting names of "hornet," "phantom," "panther," "cougar," "skywarrior," "skyhawk," and "viking," the flight deck was perhaps the most gendered portion of the exhibit. It blatantly projected notions of masculinity through its presentation of military strength, which was encapsulated in machinery and weapons that were further associated or conflated with animals or warriors, portrayed as powerful and deadly. The description of each airplane or helicopter on the flight deck provided visitors with a narrative of the life the machine lived, and its unquestionable (masculine) force. The F9F Panther, for example, was described as "the first carrier jet to see combat when the Korean War broke out in 1950. *Swift and packing a punch with bombs*, underwing rockets, and 20 mm cannons, the Panther flew *hazardous missions attacking targets deep in enemy territory* or in direct support of United Nations troops on the front line" (emphasis added) (Figs. 5.4 and 5.5).

Fig. 5.3 Image of Eye of the Corps missiles

Many of the airplanes on display had life-size cut-outs of fighter jet aviators—all male—who "stood guard" in front of the aircraft and greeted visitors as they passed by their airplanes. These cardboard service members were often used by visitors for photo opportunities during my visits to the museum; they posed beside them, using gestures that mimicked the stature and socially recognized symbols of military service. Aircraft and helicopters on display were brought back to life through the placement of manikins in key positions in the aircraft to demonstrate how it operated. The uniformed male manikins depicting the usage of the military machinery further normalized the masculine narrative that was initially established through words in the aircraft carrier's description (Fig. 5.6).

After viewing the flight deck, visitors were directed to a tour guided by male-veteran docents. The tour included the helm of the ship, where the Captain and his staff steered the boat, the Captain's sleeping quarters, and a display of decorated uniforms worn by Captains while stationed on the *USS Midway*. In another room, an automated manikin in Captain's uni-

Fig. 5.4 Wide-angle photo of F9F Panther plane and description poster on flight deck of *USS Midway*

form greeted visitors and spoke of the importance of service and sacrifice as he sat behind a desk writing letters to his loved ones back home, reinforcing gendered divisions of heroic service and familial sacrifice.

Following the guided portion of the tour, visitors were left to wander the lower levels of the ship on their own, where exhibits depicted the lives of average sailors. Meandering through a maze of corridors and halls, visitors observed sailors' living quarters, the mess hall (cafeteria), doctors' and dentists' offices, and emergency and operating rooms. In corridors, visitors found cardboard cut-outs of male service members from the Navy or Marines dressed in pristinely pressed uniforms with medals emblazing their chests. The figures portrayed in various duty assignments on the *USS Midway* included African American, Asian American, Latinos, and Caucasians, suggesting a multi-ethnic Band of Brothers, united in the service of their nation. The design of the exhibit implied that the traumas of racism and ethnocentrism that haunt the nation's streets are benignly transcended

Fig. 5.5 Close-up photo of F9F Panther plane description poster on flight deck of *USS Midway*

through military service in general and through service on all-male Naval aircraft carriers in particular.

The latest exhibits at the *USS Midway* Museum pledged to maintain the gendered norms of military masculinity. To mark the tenth anniversary of the opening of the museum, the *USS Midway* Museum constructed a state-of-the-art theater aboard the aircraft carrier to present a documentary film on the Battle of Midway, a decisive battle in the Pacific during World War II, during which the United States gained an impressive victory over Japan. Advertised on the museum's website in the lead-up to

Fig. 5.6 Image of Cougar plane with mannequin on flight deck of *USS Midway*

the exhibit, the film brought the Battle of Midway "to life in a gripping, emotional presentation!" (USS Midway Museum 2013). The preview video on the website linked memorialization to future military duty by connecting social memory not only to honoring those who have served the country, but to the duty to undertake military service:

> 2012 marks the 70th anniversary of the Battle of Midway, a battle whose legacy must be preserved for future generations of Americans. We must pass the baton of service and responsibility to our children and our grandchildren. We must instill a respect for service and sacrifice for America. These are the greatest ideals of the American spirit. Soon, 1 million USS Midway Museum visitors a year will relive the Battle of Midway… A holographic cast of actors will carry every guest into the Battle of Midway, a heart pounding blend of contemporary entertainment technology and timeless drama will inspire young and old alike. Every movie goer will walk in the footsteps of those who fought and sacrificed far from home. You will become part of a harrowing and inspirational odyssey that begins with the USS Yorktown at

the bottom of the sea, rises into the battle, and meets the battle's sailors, aviators, and even their worried and heartbroken families back home. This new visitor's experience represents an opportunity to cherish the American spirit of duty and devotion, an opportunity to strengthen the USS Midway Museum's mission to honor the legacy of those who serve our nation. You will serve America by educating and inspiring more than 1 million visitors a year for years to come. Together, we can keep the greatest generation's legacy alive by recognizing that the freedoms we enjoy today are built on a solid foundation, one that has been forged by those who have sacrificed for our nation in the past…the battle's legacy will stand as a beacon of education and inspiration for generations to come. (USS Midway Museum 2013)

The preview video put several narratives of militarized masculinity in play. The words "sacrifice" and "duty" were voiced as photos of male military service members or male veterans were shown. As the narrator mentioned worried families back home, images of a middle-aged, white woman with desperate tears in her eyes were presented, as she worried about the wellbeing of her deployed service member son. The preview reasserted and naturalized women's roles as soldier mothers and caretakers, and men's responsibility to enter the military, serve in combat, and heroically lose his life, if necessary, for the greater cause. Passing the "batons of service to our children and grandchildren" was visualized with an image of a male service member cradling a child, suggesting that military service promotes patriotism, nationalism, and even good, old-fashioned parenting. The preview urged visitors to share their experience at the museum, educating others about the critical link between male valor, military service, and a social order founded on freedom, an order that requires vigilant defense and as such makes militarization a rational and responsible policy.

Segs4Vets: Mobilizing America's Heroes

To further the overall mission of honoring the men and women who serve in uniform, the *USS Midway* flight deck also frequently acts as a venue for citizenship, re-enlistment, and retirement ceremonies. One organization that has used the physical space of the *USS Midway*, The Segs4Vets Veterans organization, holds annual honoring ceremonies for veterans. In these ceremonies, the organization dedicates Segway bikes to returning soldiers who have been disabled in combat. According to its website, Segs4Vets is

an unprecedented grassroots effort sustained and administered by volunteers representing grateful Americans, who passionately believe that when those serving our nation are sent into harm's way and suffer serious injury and permanent disability, they must have every resource and tool available to them, which will allow them to fulfill their dreams and live the highest quality of life possible. (Segs4Vets Website 2011)

By providing veterans with Segway bikes, the organization claims to "not draw attention to their disability, [but] provide a more healthful and psychological and physiological quality of life" (Segs4Vets Website 2011). In the advertising material for the honoring ceremony I attended on the flight deck of the *USS Midway* in March 2013, Segs4Vets encouraged the general public to "meet the Warriors" and "honor the sacrifices of these wonderful Men *and Women*, who have paid a great price, in their service to our country" (Segs4Vets Website 2011, emphasis added).

Like the *USS Midway* Museum's webpage, the Segs4Vets website attempts to establish a gender-inclusive narrative in which both men and women are seen as "warriors"—a title earned directly through military service and sacrifice for the protection of the nation. Eerily similar to the absence of women in the *USS Midway* Museum exhibits, the veterans' honoring ceremony, which I attended, honored 15 *male* veterans from various ethnic and racial backgrounds who had become disabled through serving in U.S. military engagements in Afghanistan and Iraq. The spatial configuration of the event helps illuminate how gendered hierarchies of citizenship are preserved and promoted. The ceremony took place at the end of the aircraft carrier in a high traffic area that all visitors were required to pass through. Precise rows of pristinely sanitized white chairs were set up, awaiting attendees to honor these former combatants. As bodies filled these chairs, they constituted the essential witnesses to acknowledge and valorize the service member's value, manifestly established through physical dismemberment in combat. In recognizing the courage of veterans, whose wounds so prominently mark their bodies, average citizens participate in what Susan Jeffords (1989) describes as "the spectacle" (9) of war. Drawing on the experience of the Vietnam War, Jeffords suggests that spectacular narratives created about the elements of warfare are "separated from [their] ostensible function" and "described only as their own display, their own theater" (9). Like any spectacle, war requires a "performer and audience" relationship (7). While news coverage brought the "performance" of soldiers on the battlefield into the living rooms of average citizens during the Vietnam era, contemporary veteran's ceremonies provide

a medium through which to applaud, honor, and valorize soldiers for their performances in war. As heroic symbols of service and sacrifice, veterans are put on display for popular consumption.

In the Segs4Vets ceremony, the fragmentation of body parts through violence inflicted on the physical corpus of the veteran were put on display for the audience members as part of the spectacle. Missing limbs were simultaneously made visible and erased through the mobility provided by the Segway bike. The technology of the Segway bike compensated the injured, feminized veterans for their physical injuries, while the adulation of the audience made the veteran whole, contributing to the reassembling and regeneration of masculinity. The damage inflicted by the "war machine" was erased as the masculine warrior re-emerged. Through such displays of the bodies of male soldiers, women are "effectively eliminated from the masculine narration of war and the society of which it is an emblem" (Jeffords 1989, 185–186). Masculinity becomes the dominant narrative through a careful construction of discourses of difference appropriate to military service. As the honored male veterans transcend differences grounded in rank, race, ethnicity, class, and (very recently) sexual orientation, war is constructed as an impermeable male terrain. Masculinity becomes the common denominator that connects veterans into a cohesive collective of the living and the dead who have served their nation in battle. The absence of women from these honoring ceremonies is not mere oversight, it is a symbolic erasure that preserves the U.S. military as male terrain.

Audience members are not silent participants in these veteran's ceremonies. On the contrary, their affirmation of wounded male warriors contributes not only to the consolidation of military masculinities, but also to the perpetuation of hierarchies of citizenship. For example, the Segs4Vets ceremony began with recitation of the "pledge of allegiance" and the performance of the national anthem. In the chairs provided for the public, visible modes of demarcation arose as the audience recited the pledge of allegiance. As average citizens placed their hands on their hearts, those in uniform—veterans, sailors currently serving in the Navy, as well as members of the San Diego Police Department—saluted the flag with fingertips at their temples. This simple, quotidian, unquestioned difference of gesture replicated a gendered division among the members of the collective that privileges head over heart. With the salute reserved for those whose relation to the state is established through military service, this small automated act feminizes non-service members and positions them as the heartfelt admirers who enhance the stature of those in uniform.

The hierarchy of citizenship that emerges in the context of honoring ceremonies places dismembered veterans at the pinnacle: the "ultimate citizen" is established through a narrative that celebrates heroism, sacrifice, and military power. All other groups of individuals are then placed lower in the hierarchy with rank established through specific relations to the disabled veterans. For example, families attending the ceremony constituted the ground on which the hierarchy is based, for it is to protect them that warriors engage in combat. Former soldiers of various eras are elevated in the hierarchy as military personnel who served their nation. Current enlisted personnel rise to the call of duty, carrying on the tradition of service and facing the same possibility of sacrifice of life and limb. Carrying weapons and dressed in their own uniforms, San Diego police stood together with the soldiers being honored as those who protect families within the homeland from internal threats, just as the soldiers protect the homeland from external enemies.

The hierarchy among citizens was established through the spatial arrangement of the ceremonial space. First and foremost, a deliberate physical division was created between audience and honorees. The rows of white chairs faced an elevated stage set with a podium and microphone, and bounded by the standing flag of the United States, and the respective flags of the various divisions of the military. Audience members were further divided from the valorized veterans through a row of aesthetic missiles that established a fence in front of the veterans. The missiles provided an unmistakable reminder of both the technologies and tragedies of war. Missiles are the tools through which the active military force protects the citizenry in the homeland, and they are the weapons that dismember and kill. The row of missiles stood as a barrier between military and civilian life, establishing a special realm for the disabled veterans who were honored because they bore the marks of war on their bodies. Their missing limbs are the spectacle that attests to the sacrifice required for one citizen to protect another. Their sacrifice legitimates the hierarchy among citizens of a democratic polity. And the exclusive place accorded male veterans shores up the notion that only men protect and defend, risking the ultimate sacrifice for their nation.

CONCLUSIONS

Like military memorials more generally, the spatial configuration of the *USS Midway* Museum claims to assert a simple truth: those who serve the nation deserve equal honor; those who sacrifice for their country deserve to live on in social memory. But careful attention to the imagery, metaphors,

figures of speech, and symbolism that undergirds these memorials reveals much more complicated social relations. Equal honor in death coexists with stark hierarchy in military life, and rigid prescriptions for gender hierarchy in the social order. Women's military service is not just elided in the *Midway* exhibits and the Segs4Vets ceremony; women are positioned as witnesses to the spectacle of male valor and as traditional caregivers who passively await news of the fate of their male protectors and defenders. Visitors who accept the truths presented in these fictive histories are invited to commit to a social order that privileges men, falsifies the historical record of both men and women, and celebrates willingness to die for a nation regardless of the moral merits of its cause.

The *USS Midway* exhibits and the veterans' honoring ceremony envision specific citizen subjects and citizen practices, weaving together notions of responsibility and honorable behavior, relationships between civilians, military personnel, and veterans; and defining military service and sacrifice as the pinnacle in a legitimate hierarchy of citizenship. Those who accept this militarized view of the world uncritically are condemned to a version of collective life in which the freedom of women is denuded and the lives of men and women are put at risk without adequate public debate about specific military practices or interventions.

This semiotic interpretation of the *USS Midway* Museum draws attention to the production of social memory as a site of subject production/ formation. The *USS Midway* Museum accredits a particular kind of citizen, replete with an overriding sense of duty to country, ready to bear arms unquestioningly for its defense, and prepared to sacrifice life and limb for its preservation. As the exhibits and the honoring ceremony make patently clear, they accredit a male citizen-soldier who earns his gender privilege through valiant military service. The *USS Midway* Museum endorses a version of twenty-first-century citizenship in the national security state, in which civilians and those who refuse military service are rightly subordinated to the privileges bestowed by a grateful nation upon those who engage in combat. The semiotic reading challenges the validity of the museum's account of military service, noting the critical erasure of women in uniform. It interrogates the meaning of equality grounded in military hierarchy and questions the profound costs of militarization, costs which justify death and physical dismemberment in pursuit of the nation's causes without any attention to their (im)moral aims.

As U.S. military engagements proliferate around the globe, the military's need to recruit carefully selected personnel to achieve its military

objectives continues to grow. Capitalizing on new technologies such as social media, the military has devised new methods to recruit personnel to its All-Volunteer Forces. Since the Obama administration's decisions to eliminate the ban against women in combat and against transgender service in the military, have the armed forces altered their recruitment efforts to target women and trans people? Have they transformed their recruitment videos and flyers to communicate more inclusive messages? Chapter 6 investigates contemporary recruitment practices in relation to these recent policy changes.

Notes

1. The term "living history" refers to a teaching/learning approach which promotes conveying knowledge about previous periods in history through the reenactment of those spaces and historical periods. Instead of reenacting specific moments in history, living history sites invite museum visitors, as an example, to survey what an aircraft carrier may have looked like during the period in which it was commissioned and in service. Exhibits are built with the intention of recreating the daily lives of sailors who may have served on that ship through showcasing daily activities they may have engaged in, uniforms they wore, and typical meals they may have eaten in those spaces. In so doing, living history sites require visitors to actively engage in and interact with curated exhibits with the goal of helping the visitor live the experience of serving on the ship in a previous time period.
2. It should be noted that this count includes sites that represent the military institution as a whole, as well as those that represent individual branches or units of the military.
3. This count is based on data from January 2017.
4. The *USS Midway* Museum's website actively incorporates and promotes reviews written on Travelocity.com by visitors. See current (as of February 2019) homepage of *USS Midway* Museum at https://www.midway.org/.
5. In addition to textual changes, the *USS Midway* Museum website has undergone a complete revamping in design. I estimate that the overall design changes to the site were made sometime between 2016 and 2018.
6. I visited the *USS Midway* a total of two times between 2012 and 2015. The veterans honoring ceremony took place during my first visit.
7. In an interview completed on March 6, 2013, *USS Midway* Exhibits and Curatorial Historian Karl Zingheim revealed to me that no women sailors served on the vessel. During sick leave and while stationed in ports, the *USS Midway* did, however, accommodate "female entertainers" onboard. Women were also onboard in 1975 when the *USS Midway* transported over

3000 men, women, and children who had been evacuated out of Saigon. As noted here, despite the fact that women did not serve on the ship as sailors, official materials including the 2013 version of the USS Midway Museum's website proclaimed the purpose of the exhibit as to honor *both the men and women* who served their country. Following my interview with Karl Zingheim, the language on the website of the USS Midway Museum was changed on March 7, 2013. I was unaware that this change would be made to the website—Karl Zingheim (2013) did not indicate that it would during our interview—but in email correspondence, Exhibits and Curatorial Director Duke Windsor apologized to me for the "typo." The previous version of the website (before the redesign) thereafter stated "The USS Midway Museum is dedicated to preserving and honoring the 200,000 young men who served aboard the USS Midway—and by extension all those who serve in uniform" (USS Midway Website as of March 7, 2013). As the chapter argues, the stakes in military memorialization are larger than documenting the lives of those who serve on any particular vessel.
8. The phrase, partitioning of the sensible, is borrowed from Jacques Rancière (2010, 140).
9. Mills (2007) suggests that social memory is concretized in everyday products and venues such as textbooks, ceremonies, statues, official holidays, as well as parks and monuments.

References

Buffalo Soldier Educational and Historical Committee. 2017. The 6888th Central Postal Directory Battalion Monument. Accessed February 15, 2019. http://www.womenofthe6888th.org/the-6888th-monument.

Chavez, J. 2018. Las Cruces Unveils Women Veterans Museum. ABC 7 News – KVIA. Accessed February 15, 2019. http://www.kvia.com/news/new-mexico/las-cruces-unveils-women-veterans-monument/714419838.

Faust, D.G. 2008. *This Republic of Suffering: Death and the American Civil War*. New York: Vintage Books.

Ferguson, K., and P. Turnbull. 1999. *Oh, Say, Can You See? The Semiotics of the Military in Hawai'i*. Minnesota: University of Minnesota Press.

Gillis, J.R. 1994. Memory and Identity: The History of a Relationship. In *Commemorations: The Politics of National Identity*, ed. John R. Gillis, 3–25. Princeton: Princeton University Press.

Jeffords, S. 1989. *The Remasculinization of America: Gender and the Vietnam War*. Bloomington: Indiana University Press.

Lorber, J. 2000. Using Gender to Undo Gender: A Feminist Degendering Movement. *Feminist Theory* 1 (1): 79–95.

Millar, K. 2015. Death Does Not Become Her: An Examination of the Public Construction of Female American Soldiers as Liminal Figures. *Review of International Studies* 41 (4): 757–779.

Mills, C. 2007. White Ignorance. In *Race and Epistemologies of Ignorance*, ed. S. Sullivan and N. Tuana, 11–38. Albany: SUNY Press.

Noth, W. 1995. *Handbook of Semiotics*. Bloomington: Indiana University Press.

Pang Kah Meng, A. 2004. Making History in From Colony to Nation: A Multimodal Analysis of a Museum Exhibition in Singapore. In *Multimodal Discourse Analysis: Systemic Functional Perspectives*, ed. Kay L. O'Halloran, 28–54. London: Continuum.

Rancière, J. 2010. *Dissensus: On Politics and Aesthetics*. Ed. and Trans. Stefen Corcoran. New York: Continuum.

Segs4Vets. 2011. The Segs4Vets Program. Accessed March 25, 2013. http://www.draft.org/Segs4Vets.aspx.

Shapiro, M.J. 1997. *Violent Cartographies: Mapping Cultures of War*. Minnesota: University of Minnesota Press.

Smith, C. 1989. Museums, Artefacts, and Meanings. In *The New Museology*, ed. P. Vergo, 6–21. London: Reaktion Books.

Szitanyi, S. 2015. Semiotic Readings of the USS Midway Museum. *International Feminist Journal of Politics* 17 (2): 253–270.

Tilly, C. 1985. War Making and State Making as Organized Crime. In *Bringing the State Back In*, ed. Peter B. Evans, Dietrich Rueschemeyer, and Theda Skocpol, 169–181. Cambridge: Cambridge University Press.

USS Midway Museum. 2013. Coming Attractions. Accessed March 20, 2013. http://www.midway.org/coming-attractions.

———. 2018. Museum Fact Sheet. 2018. Accessed September 9, 2019. https://www.midway.org/wp-content/uploads/2018/01/Midway_Fact_Sheet_2018.pdf.

Zingheim, K. Interviewed by: Szitanyi, S. Telephone Interview. (2013, March 6).

CHAPTER 6

Gender and Military Recruitment Since the Lifting of the Combat Ban

INTRODUCTION

"Who do we want to have serve, and how do we get them to sign up?" remain central questions for recruiters in all branches of the contemporary volunteer-based U.S. military. Recruiting "optimal" military personnel from the general public is a paramount goal. But who counts as an "optimal" recruit for the twenty-first-century armed forces? Inclusive rhetoric suggests that the military should be as diverse as the population itself—reflecting the rich racial, ethnic, class, gender, and sexual composition of the nation. Even citizenship is not a prerequisite as the military offers service as a pathway to citizenship for undocumented residents in the United States. Recruitment campaigns target particular demographics.[1] Recruitment offices in strip malls, for example, focus on high school graduates in low-income communities, whereas Reserved Officer Training Corps (ROTC) programs on college campuses concentrate on students from military families, as well as those seeking financing for their university studies. Each branch of the armed services devotes considerable sums to both market research and advertising to ensure that their carefully crafted materials reach prime audience segments.

In November of 2013, the online news source, *Politico*, ran an article titled "Army PR Push: 'Average-Looking' Women." The story uncovered internal Army communications between Colonel Lynette Arnhart and a group of analysts tasked with studying how to market women's integration into combat positions previously closed to them. The email, sent by

© The Author(s) 2020
S. Szitanyi, *Gender Trouble in the U.S. Military*,
https://doi.org/10.1007/978-3-030-21225-4_6

Arnhart, was intended to provide Army spokespeople guidance on how to frame women's integration into Army combat positions in materials created for "external communities," that is, the general public. Arnhart advised the Army to use "average-looking women" (quoted in Brannen 2013, 2) in public recruitment materials to convey the message that skills rather than looks were key to a successful career in the military. According to the *Politico* article, Arnhart noted, "In general, ugly women are perceived as competent while pretty women are perceived as having used their looks to get ahead" (3). Including "pretty women" in combat roles in recruitment materials, then, would send the wrong message, implying that their roles had more to do with flirting or sexual favors than with military prowess. Images of "ugly women," on the other hand, would make it clear that capability, competence, and aptitude were the qualities the Army sought in its recruits.

Although Arnhart's email to a team of public relations personnel was grounded in pernicious sexist stereotypes about working women, it indicates complexities, contradictions, and tensions embedded in the military's efforts to change its self-representations. Although women's participation in combat support roles and missions during Operation Enduring Freedom and Operation Iraqi Freedom could have provided ample images of courageous women in the military, these were not deemed suitable recruitment materials.

How, then, has the military altered its recruitment materials to reflect new opportunities for women following the end of the combat ban and for gay, lesbian, and transgender troops who may now serve openly in the military? This chapter investigates depictions of militarized masculinities and femininities in military recruitment material immediately after the Pentagon's announcement of the end of the ban against women in combat, from January 2013 to October 2015. The chapter builds on Melissa T. Brown's (2012) impressive study of military recruitment in *Enlisting Masculinity: The Construction of Gender in US Military Recruiting Advertising During the All-Volunteer Force*.[2] Brown (2012) examined print materials in military advertising campaigns between 1971 and 2007. Using an "interpretive textual approach, which…make[s] explicit the meanings encoded in the published words and images" (12), *Enlisting Masculinity* documents continuities and variations in the gendered images that populate recruitment materials over five decades. Using Brown's (2012) findings as a baseline, I expand the investigation of gendered and raced images in military recruitment materials to online media, which has now become a major vehicle for military recruitment.

Historical Depictions of the AVF Soldier in Military Advertising

Historically, the move from conscription to volunteer-based models of service has increased the number of women who serve in the militaries of many nations (see Segal 1995; Haltiner 2003; Caforio 2007; Iskra et al. 2002; Sandhoff et al. 2008). Those increases have occurred, in part, because the end of conscription has required military officials to anticipate a shortage of qualified recruits in advance, and to reconsider the number and types of roles available to women. In the United States, military officials devoted serious attention to advertising and marketing tactics that would succeed in recruiting the necessary number of men and women to fill the ranks of an All-Volunteer Force. In initial marketing materials for the All-Volunteer Force (AVF), women were depicted in two ways: as potential soldiers, and as props to lure men to join (Bailey 2009). In advertising strategies and recruitment tactics for the Army at the inception of the All-Volunteer Force, groundbreaking social market research was conducted on men to generate new conceptualizations of military masculinity, hoping that it would entice men to enlist in the aftermath of the Vietnam War (Bailey 2009). Men were the primary targets of these recruitment efforts and the depictions of women in these marketing campaigns reinforced a clear binary distinction between masculinity and femininity. To attract male recruits, for example, the Army presented eroticized images of "foreign" women, suggesting that overseas postings promised exotic romance rather than impending death. These ads portrayed military service in terms of individual freedom and adventure, promising professional and sexual opportunities. As Beth Bailey (2009) has noted, the message was explicit: "we'll make you an expert at whatever turns you on" (79).

Some ads did target women, promising new career opportunities. Yet when women and men appeared in ads together, the military roles assigned to them were markedly unequal. For example, in the initial 1971 rollout of the "Today's Army Wants to Join You" campaign, one flyer featured the headline, "We've got 300 good, steady jobs" (Bailey 2009, 162). Rows of Army personnel dressed in various uniforms suggested the kinds of jobs available for those who joined. Four women were included within the rows of uniformed soldiers: they wore only two uniforms. Two of the four women wore white uniforms which stood in stark contrast to the crowd of men in various shades of camouflage fatigues. The white uniforms suggested that the primary career available to women was that of "medical

aides." The emphasis on medical assignments was underscored by the difficulty in seeing the two women who were wearing camouflage uniforms. Surrounded by their male counterparts, the two women soldiers in fatigues were barely noticeable. The ad specified a range of gender-specific job titles: "Jobs for photographers, printers, truck drivers, teachers, typists, TV cameramen and repairmen, cooks, electricians, medical aides, meteorologists, motor and missile maintenance men" (Brown 2012, 72), thereby suggesting a traditional gendered division of labor. While a host of career possibilities awaited young men who enlisted, young women could expect relegation to caring roles. As the reference to "missile maintenance men" makes clear, combat-related positions were in no way open to women.

Configurations of masculinity have remained central to military advertising campaigns across the branches over several decades (Brown 2012). Each branch recruits from the general public and advertises in the public sphere based on its own internal culture and personnel needs. Specific models of masculinity have varied among the branches and over time (181). In periods when traditional conceptions of masculinity were challenged—such as after the Vietnam War and during the creation of the All-Volunteer Force—"the result was not the neutering of military service in recruiting appeals but the alteration of military masculinity" (178). As individual military branches engaged changing social conditions and expectations, they deployed distinctive articulations of masculinity to capture the imaginations of young recruits.

Between 1970 and 2007, for example, the Marine Corps circulated traditional depictions of masculinity that tied soldiering to a particular conception of the male warrior, "hard young men portrayed in martial contexts, either in a combat situation or on ceremonial display" (Brown 2012, 179). During this period, other branches appealed to economic incentives, traditional masculinized virtues such as courage, and unique training opportunities. The Air Force, Army, and Navy sought to attract recruits through narratives emphasizing economic incentives that framed economic benefits "in masculine terms…or contain other visual or textual elements that reinforce the masculinity of recruits" (179). In addition to economic incentives, the Army's recruitment strategy regularly deployed narratives of "character development and personal transformation" that celebrated personal qualities such as courage and strength, visually depicted through "weaponry and other martial visual markers" (179). By these means, the Army perpetuated traditional notions of warriorhood and

soldierhood accessible to unaggressive "regular guys" (179), a critical tactic for the branch tasked with recruiting the largest number of personnel. For the Navy, noneconomic pitches have routinely included the excitement of living on the sea and challenges within an environment that allows recruits to prove themselves in unique ways. After the attacks on the World Trade Towers and the Pentagon on September 11, 2001, however, "layering a warrior masculinity on top of other kinds of appeals" became visible in Navy recruiting campaigns, resurrecting the well-established image of the ultimate warrior, and "reaffirming the Navy's commitment to a strong form of masculinity" (179). The Air Force has not traditionally depicted service through martial images of masculinity. In addition to economic advantages available to young men through service, the Air Force "has offered, by association with the world's most advanced technology, the masculine advantages of mastery, dominance, and control" (180). In contrast to the other branches that often focus on "direct physical excitement" (180), the Air Force has enticed recruits through the excitement of technology, creating a video game-like environment that involves the conduct of missions from behind a joystick. This approach to recruiting for the Air Force is likely to continue as drone warfare begins to be more widely used.

Studying Online Military Recruitment Material

In the remaining section of this chapter, I explore online military advertising through Facebook posts, conducting a content analysis of more than 1000 posts between January 2013 and October 2015. Adapting the coding scheme developed by Brown (2012) to suit online material, I coded both verbal and visual materials, noting the appearance of actors, activities performed by individuals, uniforms worn, depictions of military hardware, and how individuals were grouped together in the advertising samples. By calculating frequencies of various images across the posts of all branches of the armed services, I identified modal representations of "optimal" military personnel. In addition to measuring frequencies, I analyzed the verbal and visual messages to capture key trends that emerge in the individual military branches' recruitment and marketing campaigns. In the following pages, I incorporate detailed descriptions of some of these posts to demonstrate the gendered narratives that competing branches of the military have circulated to identify worthy recruits for their organizations.

I chose to focus on Facebook rather than other social media platforms (including Twitter and Instagram) because each of the military branches has devoted far more effort to and secured more followers on this social media platform. As a platform for communication and consumption alike, Facebook's influence, both domestically and internationally, is undeniable. As of October 2015, Facebook had 1.55 billion active monthly users (Statista). By January 2015, of the 156.5 million users in the United States, 27 million fell within the military's target age group of 18–24.[3] Military Facebook posts—whether exclusively text-based, or those which include still photos or moving images in the form of videos—are dynamic conveyors of information about how the military branches view themselves and how they want potential recruits and the general public to view them. This content analysis provides powerful insight into the gender order that the military aspires to consolidate in the twenty-first century. Covering the period from the announcement of the planned end of the combat ban through the period in which the ban was actually lifted, the study documents how the various branches of the military perceive the changing role of women.

Online content—particularly that which is provided by social media platforms—is extensive and poses certain challenges for social science research. Megan MacKenzie (2015), for example, points out that online content is "unstable, instant, and edited" (166). Indeed, in contrast to print media, social media is often continuously edited. Facebook posts, in particular, can be posted, then edited, or even removed for whatever reason the site administrator deems necessary, often unbeknownst to the viewer of the page. Comments posted by page followers in response to the entries put up by the military branches may also be deleted. As with the changeability associated with military museum exhibits and spatial configurations, the constant availability of editing and removal of content from these pages allows the military to continuously (re)shape and (re)mold its self-representation to the public. In the context of a study designed to track change in gendered representations, however, I view MacKenzie's (2015) caution that "the content represents an unstable, edited, and potentially altered discourse" (166) less as a problem than as an opportunity. Each editing offers the possibility of introducing changes in the depiction of women in the military.

Any analysis of textual and visual representations must also grapple with ambiguity. My coding cannot capture how subjects depicted in the posts identify themselves. Battle fatigues can make it particularly difficult to identify the gender of military personnel. "Race" itself is a contested social

construct and cannot be imputed solely on the basis of skin tone. My categorization of subjects' gender and race, then, conforms to social conventions in the contemporary United States—as complicated and flawed as those may be. Nonetheless, the military carefully constructs its social media posts to communicate with the general public, using well-honed advertising techniques to convey specific messages about military service. Quantification of the images in these posts provides one measure of key elements in the recruitment messages that the military circulated to the general public over a two-year period following the announcement of an end to the combat ban on women.[4]

GENDERED DYNAMICS IN VIRTUAL MILITARY RECRUITMENT

Since the last year included in Melissa T. Brown's (2012) study, all four of the military branches under the Department of Defense have placed heavier emphasis on the use of social media platforms to meet their annual personnel recruitment targets. According to the Department of Defense (2015), all active components of the military branches have met or surpassed their recruitment targets since fiscal year 2010. Anxieties over the military's ability to meet recruitment and retention quotas increased in 2008 as President Bush's surge strategy escalated the number of U.S. troops deployed in Iraq and increasing numbers of soldiers incurred injuries. As public opinion began to favor a reduction in forces or a complete end to U.S. involvement, military officials began to worry that they might have another "Vietnam" on their hands that could negatively affect the public's view of their institution.

In 2011, *The New York Times* reported that the Army was "working hard to increase [their] social media" (Elliott 2011, 5) in addition to continuing to push advertising campaigns through television commercials, because they "fully recognize[d] that young people TiVo over commercials or are multitasking on their smartphones when the commercials are on" (Elliott 2011, 5). Part of the push toward social media was also enabled by a modernized approach to data-driven recruitment. In 2007, for example, the Marine Corps started using big data analyses gathered from Censuses to assess demographic shifts that would allow them to target specific population segments through paid Facebook advertising.

Table 6.1 provides an overview of the number and kinds of recruitment ads that each branch posted on its Facebook page from January 2013 through December 2015. The content of the ads developed by each branch is discussed in further detail below.

Table 6.1 Recruitment posts by branch

Military branch	Total # of posts	Institution	Position	Context			Gender					Gender and combat			
				Recruitment	Specific personnel	Technology	Women only	Men only	Women and men	Women only in combat	Men only in combat	Women only and military hardware	Men only and hardware	Women and men and combat	Women and men and hardware
Air Force	205 (100%)	29 (14.1%)	52 (25.4%)	81 (39.5%)	14 (6.8%)	27 (13.2%)	12 (5.6%)	79 (38.5%)	20 (9.8%)	0 (0%)	21 (10.2%)	4 (2.0%)	60 (29.3%)	3 (1.5%)	18 (8.8%)
Army	301 (100%)	103 (34.2%)	52 (17.3%)	12 (4.0%)	23 (7.6%)	35 (11.6%)	17 (5.65%)	90 (29.9%)	24 (8.0%)	1 (0.3%)	25 (8.3%)	4 (1.3%)	45 (15.0%)	3 (0.9%)	14 (4.7%)
Marine Corps	296 (100%)	112 (37.8%)	23 (7.8%)	114 (38.5%)	21 (7.1%)	19 (6.4%)	20 (6.8%)	184 (62.2%)	46 (15.6%)	1 (0.3%)	20 (6.8%)	7 (2.4%)	75 (25.3%)	1 (0.3%)	21 (7.1%)
Navy	277 (100%)	112 (40.4%)	38 (13.1%)	58 (20.1%)	19 (6.65)	14 (5.1%)	28 (10.1%)	65 (23.5%)	27 (9.7%)	1 (0.4%)	2 (0.7%)	3 (1.1%)	16 (5.8%)	1 (0.4%)	4 (1.4%)

Gendering of Technology in Air Force Posts

The U.S. Air Force Recruitment Facebook page was created on November 17, 2011. As of November 16, 2015, the page had 767,051 likes,[5] or followers. The page is described as the "Official U.S. Air Force Recruiting Facebook Page" (Air Force Facebook Page 2015). For the period covered in this study, 205 posts were coded from the Air Force's Facebook page. Of the 205 posts, 151 were posts that included still images/photos, and 53 were videos. One post was text-based only.

In terms of content, 29 posts conveyed messages about the Air Force as an institution, 52 posts about certain positions recruits can choose from when joining the Air Force, while 81 posts related to recruitment for the Air Force or general information on recruitment criteria. Fourteen posts featured specific members of the Air Force. Twenty-seven posts reflected technology the Air Force is currently using or developing. There were a total of 12 posts that depicted women only, 10 of which depicted a single woman, and 2 which depicted two or more women. Women were portrayed with men in 20 posts. However, one man or multiple men were depicted without women in 79 posts. Seven Air Force Posts were coded as in discernable for gender.

In posts in which women were depicted—whether alone, or with other women—not one post depicted women in the context of combat. Only four posts depicted women with military hardware. When portrayed along with men, three posts showed women with men engaging in some form of combat, but 18 posts that included both men and women depicted military hardware. When only men were included, 21 posts contained context related to combat, and 60 included military hardware.

As mentioned above, the content of a large number of the Air Force posts focused on specific positions in the branch, or conveyed information specific to recruiting individuals to join the Air Force. Take for example one such post which provided potential recruits with information on Survival, Evasion, Resistance and Escape (SERE) specialist positions.

The text displayed with the images of the post told viewers that these specialists "teach aircrew members everything they need to know…to be able to survive on their own in any environment under any conditions should their aircraft go down" (U.S. Air Force Facebook 2015). The photos included in the post depicted Caucasian males engaging in various activities in regard to this role; in the first one, a male Airman stood facing the camera, with three others with their backs turned toward the camera and packs on

their backs, with a helicopter in the far background. Looking at the images coupled with the text suggests that the male facing the camera is the SERE specialist. In the second image, the hatch of that helicopter is depicted open with the ground seen far below it, followed by an image of those men who the specialist was speaking to (assumingly) jumping from the helicopter with parachute packs on their backs. This depiction tells several different gendered stories about women's roles, on the individual level, branch level, and the overall military level. First, it suggests that men, not women, are Survival, Evasion, Resistance and Escape Specialists. Second, by including only male soldiers in these images, and by suggesting through the text that accompanies these photos that "[e]very single member of an aircrew" (U.S. Air Force Facebook 2015) must have the skills necessary to survive if their aircraft goes down, the Air Force suggests that women are not pilots, and further erases participation women have had—either historically, or contemporarily—in any duties or activities that require them to fly on military aircraft, images which would usually evoke connections to combat. Alternatively, this may also be read to suggest that although women do fly military aircraft, female lives are less important than those of their male counterparts, and therefore, it is less critical that they obtain the necessary training to ensure their survival. Finally, by including the sentences "[e]very single member of an aircrew in every military branch must be able to survive on their own…should their aircraft go down" (U.S. Air Force Facebook 2015) moves to produce this erasure not only in the context of the Air Force specifically, but throughout all branches of the military.

Many of the Air Force posts included in this study focused on providing visual representations of the military hardware used in the branch. For the Air Force, this appeared most often in the form of various aircraft. Posts during this period provide collages of multiple images of a variety of aircraft, and include pictures of unmanned aerial vehicles (drones). Oftentimes, no individuals are depicted in the images. All of them, however, convey messages of assault, conflict, and combat. The posts' textual reference to "collection of aviation technology" (U.S. Air Force Facebook 2015) however erases the damage inflicted in foreign nations by the aircraft and the people who operate them, whether that is done in-person or remotely.

The aircraft, gray in color and often depicted against the backdrop of a blue sky, appear smooth, seamless, clean, and sanitized, quite the opposite of the blood-sullied grounds they often produce through their missions. The presentation of the aircraft plays to elongate its smooth spears, and is sexualized as phallic objects (Cohn 1987). Sexualized discourses that

describe these technologies as "deep penetrating"—as is often done among military officials and defense "intellectuals" (Cohn 1987)—masculinize weaponry directly in relation to the male biological body. Doing so connotes specific messages of which genders are meant to "play" with these weapons, and which are not, designating appropriate roles for their service in the Air Force.

In some cases, Air Force posts also conveyed messages about the institution through narratives of community the recruits may experience once having successfully joined the branch. Take for example a picture collage with four different images that were added to the U.S. Air Force Recruiting Facebook page's "Life on Base" photo album on April 15, 2013. In the first, large image, a Caucasian airman is seen in the foreground with several other Caucasian airmen standing in line behind him, out of focus. They are looking toward the camera, saluting. In the second, smaller picture, a young boy has his back toward the camera and is bowling, while in the third and fourth pictures, aircraft are seen outside, as well as in a hangar.

By having the main, large photo in the Facebook post be that of a Caucasian airman, the "Life on Base" depiction suggests that men, such as the one featured in the images of the post, is the demographic norm of personnel in the Air Force. Indeed, only 22 of all of the Air Force posts examined in this study contained at least one person of color in the images used. Moreover, by having the text of the post say "When you're in the Air Force, you're part of a community that's a lot like the communities you'll find around the country" (U.S. Air Force Facebook 2013) the post not only disregards all other demographic groups that are not male and Caucasian from the military branch, but also from wider American society, reifying hierarchies of citizenship through the depiction of the image of the white, male soldier.

Racial and Gendered "Inclusion" in Army Posts

The GoArmy.com Facebook page was created on November 11, 2009. As of November 16, 2015, the GoArmy.com Facebook page had 1,042,634 likes, or followers. The page is described as "The Official Fan Page for http://www.goarmy.com. Learn more about the career opportunities available in the U.S. Army." For the period covered in this study, I coded 301 posts from the GoArmy.com Facebook page. Of the 301, 162 were posts that included still images/photos, and 136 were videos. Three posts were text-based only.

In terms of content, 103 posts conveyed messages about the Army as an institution, 52 posts about certain positions recruits can choose in the Army, and 12 posts related to recruitment for the Army or general information on recruitment criteria. Twenty-three posts featured specific members of the Army, while 35 posts related to technology the Army is developing or currently using. There were a total of 17 posts that depicted women specifically, 9 of which depicted one woman, and 8 which depicted two or more women. Women were portrayed with men in 24 posts. However, one man or multiple men were depicted without women in 90 posts. Sixteen Army posts were coded as not discernable.

In posts in which women were depicted—whether alone, with other women, or with men—only four depicted women in combat. Those same four posts also depicted women with military hardware. When depicted along with men, 3 posts depicted women with men engaging in some form of combat, and 14 posts that include both men and women depicted military hardware. When only men were included, 25 posts contained context related to combat, and 45 included military hardware. Two posts that included both men and women depicted combat scenarios, but in which women were depicted in either medical or scientific roles, while men were depicted on the front lines of a mission.

One of the main recruiting strategies discernable from a review of the GoArmy.com Facebook posts is the showcase of activities the branch is involved in outside of the military apparatus. These include activities at high schools, the Army National Hot Rod Association car races, Tough Mudder obstacle courses, and national leadership conferences. The Army's presence at these and similar venues not only showcases the multitude of ways in which it "sells" the institution to prospective recruits, but provides examples of how the institution interacts with civilians and becomes embedded in the larger general society.

One such montage of four photos of Army soldiers participating in an event off base illustrates this. In the first photo of the montage, a Caucasian, male Army soldier is standing next to two African American young men, and one African American young woman before a "U.S. Army" backdrop. In the second photo, a Caucasian, male Army soldier is standing on a stage in front of an audience of African American youth giving a talk. In the third photo, another Army soldier is standing in a classroom among a group of African American adults dressed in business attire and is giving a presentation during a roundtable discussion. Finally, in the fourth picture, a group of five African American Army soldiers stand beside one African

American civilian as the signage "U.S. Army—Army Strong" is seen behind them. The text of the post contextualizes the activities in the photo, notifying viewers that they are taking place at "HBCU (Historically Black Colleges and University) Atlanta Classic events," where members of the Army who are depicted in the photos are "speakers and youth mentors—engaging students, parents, community influencers, and educators on career and educational opportunities available in the Army and U.S. Army Reserve" (U.S. Army Facebook Page 2015). While the post is an example of the military recruiting specifically among a minority group—in this case, African American students who may be convinced to join the Army or Army Reserves—white male soldiers continue to represent over three-fourths of the total armed forces.

Increasingly, the Army has also used sponsorship opportunities at Tough Mudder 5K races to recruit for specific positions in the Army Reserves. These races, designed as extreme obstacle courses, are built to mimic military styled training courses to include "running through fire, steep inclines, ice water, mud and up to a dozen other military-style obstacles" (Agoglia 2013, 1). As a Facebook post depicts, during these races, the Army sets up a kiosk or booth area, where race participants can stop by before or after the race and pose in front of a U.S. Army logoed map of the obstacle course. In some cases, racers choose poses that are meant to reflect physical strength, for example, by flexing muscles and showing "toughness." The military-inspired nature of these obstacle course races glorifies the physicality of military training. And through the "Army Strong?" marketing campaign, the Army sponsorship of the races suggests that completing the race may be a good indication of being strong enough for serving in the Army by testing "their mental, emotional, and physical strength" (2) through the race. However, in juxtaposition to the above post, only Caucasian individuals are depicted in these photos, suggesting that African American and other minority bodies do not showcase "toughness" and are not seen as "Army Strong."

The use of gendered and raced bodies in specific contexts creates narratives that can be utilized at specific times, depending on messages the institution seeks to send. This is augmented through posts that the military institutions publish on national holidays, especially when those holidays are framed within the context of personal freedoms and the protection of the homeland. For example, in an Army Facebook post uploaded for the 4th of July celebration in 2014, and which includes a photo of a group of close to 30 soldiers wearing combat gear holding up a large American flag, the

soldiers all appear to be men, and are Caucasian. In the background, barren mountains are seen. Above the photo, the text reads "Today we celebrate freedom and the Soldiers who protect it" (U.S. Army Facebook Page 2014). Below the soldiers, superimposed on the image is the U.S. Army's logo, along with "Happy Independence Day 4 July 2014" in which the words "Independence Day" are placed in a yellow color that makes them stand out. By depicting male, Caucasian soldiers only within the context of a post that depicts messages of protection through combat service, the photo enables hierarchies of citizenship produced by the military to persist. The white male is hypermilitarized not only through being depicted in combat gear, standing in front of mountains that convey messages of possible deployments, but also through the text used in the post in which the right and honor to protect is given to these individuals specifically. All other groups of people are seen as lower on the ladder of hierarchy, whether that be other soldiers who are not afforded the caliber of honor linked to combat service or average citizens who are reminded of the need to have their freedoms protected during holidays such as the 4th of July.

Gendered Bodily "Transformations" in the Marines

The Marine Corps Recruiting Facebook Page was created on June 26, 2008. As of November 16, 2015, the Marine Corps Recruiting Facebook page had 4,267,897 likes, or followers. The page is described as the "The official Facebook page of Marine Corps Recruiting." For the period covered in this study, I coded 296 posts from the Marine Corps' Facebook page. Of the 296 posts, 199 were posts that included still images/photos, and 96 were videos.

In terms of content, 112 posts conveyed messages about the Marines as an institution, 23 posts about certain positions recruits can choose in the branch, 114 posts related to recruitment for the Marines or general information on recruitment criteria or text/images that are meant to encourage people to join. Twenty-one posts featured specific members of the Marine Corps. Six posts depicted Marines participating in specific missions, and 19 posts reflected technology the Marines are developing or currently using. There were a total of 20 posts that depicted women only, 14 of which depicted one woman, and 6 which depicted two or more women. Somewhat more often, women were depicted in images and videos alongside men, amounting to 46 posts. A large majority of the posts (184) depicted men only. The remaining 21 posts were coded as not discernible.

6 GENDER AND MILITARY RECRUITMENT SINCE THE LIFTING... 161

In posts which depict women—whether alone, or with other women—only one post displayed women in combat. Seven posts depicted women with military hardware. When shown with men, only one post depicted women with men engaging in some form of combat, but 21 posts that included both men and women depicted military hardware. When only men were included, 20 posts contained context related to combat, and 75 included military hardware.

Scenes of training exercises and boot camp initiations often appeared in the Marines Facebook posts. Conventionally referred to as Basic Training among the other branches, the Marine Corps refers to portions of its basic training program as "Basic Warrior Training," suggesting that recruits who make it through the Marine Corps basic training program epitomize the warrior persona often sold to recruits by the military. These posts exemplify how this narrative is particularly linked to maleness.

In one post, for example, the branch includes a photo of a drill sergeant standing with hands on hips on top of a wooden planked fence which appears to act as an obstacle on a training course. The drill sergeant appears to be male. Below him, we see four individuals on their stomachs, seemingly crawling on the course as they are dressed in full Marine combat gear and carry rifles in front of them. The individuals appear to be male, with one of the four appearing to be African American. Super imposed on the photo are the words "Basic Warrior Training" in the lower right-hand corner of the image. The text of the post, which appears above the image, notifies the viewer that "During the ninth week of recruit training, recruits undergo Basic Warrior Training where they learn basic field and combat skills" (U.S. Marine Corps Facebook Page 2014). The image, coupled with the text which appears above it, produces three layers of gendered contextualization. First, by only including men in the picture, it establishes that Marine recruits are more likely to be male. Second, it suggests that only male recruits are likely to engage in field duties and combat roles, and therefore, are privy to this training. Third, it suggests that only male recruits have the ability to become warriors by excluding the depiction of women in the photo, and finally, specifically links warriorhood to combat-related service. Looking through a feminist lens, the post signals to potential female recruits that the branch does not recognize women's service in combat, and, therefore, will never view them as ultimate warriors who otherwise would receive the highest of acclaim and honor for their service.

As detailed above, 20 out of nearly 300 posts depicted women only, whether only one woman, or multiple. Consider one such post. It shows a

photo of a woman in Marine Corps training apparel kneel on one leg as she carries a weighted bar across her shoulders. In the background of the photo we see several other women out of focus stand in rows behind her, staggered, engaging in the same activity. They are all wearing the same apparel. The activity takes place outside, as a dawning or dusking sky is visible. The text above the image informs viewers that the women depicted are engaging in basic training as they undergo a "transformation." The sentence reads, "To complete this transformation, recruits must have the desire and the ability to develop the mind and the body" (U.S. Marine Corps Facebook Page 2015).

The coupling of the image with the text tells a specific story about female Marine recruits' bodies. Though the term "recruits" in the text could refer to the gender of any recruit, the usage of a photo that only depicts women suggests that the context of the words used in the text above the photo is specific to the Marines' view of female recruits. By using the word "transformation," and referring to developing "the mind and the body," the branch is arguably suggesting that women's minds and bodies are not up to par with Marine Corps standards before starting basic training. Unlike the above post, basic training is framed as a "transformation," instead of "training," one which does not result in conferring skills that allows (female) recruits to engage in combat or attain warrior status.

But for the Marine Corps, combat is not only linked to warriorhood, it is also seen as a critical element of leadership. As previously indicated, over half of all of the posts reviewed for this study depicted men in visual representations. Consider one such image which shows a group of seven male Marines in combat uniforms carrying combat gear in their hands and on their bodies.

As they walk toward the direction of the camera, more Marines can be seen in the background, along with the top of a military helicopter. It appears from the photo that they have just disembarked from the helicopter.

Above the photo, the text reads, "Marines are imbued with 14 leadership traits that prepare them for battle. What are your leadership traits?" (U.S. Marine Corps Facebook Page 2015). The text contextualizes the photo to have viewers believe that the Marines depicted have just arrived for a combat-related deployment in a foreign nation in which they are likely to see battle. For potential recruits viewing the post, leadership qualities are argued to be necessary for preparation for battle. By including only male Marines in the photo, however, the post suggests that only men

are able to embody the necessary leadership traits, and are, therefore, the only ones to be prepared to succeed in combat. Through this association, the post quietly reaffirms military concerns over women serving in combat that echo those analyzed in Chap. 3, which position women in combat as a liability instead of an asset.

Gendered Adventure Narratives in the Navy

The Navy Life Facebook page was created on February 18, 2010. As of November 16, 2015, the U.S. Navy Life Facebook page had 383,036 likes, or followers. The page is described as "What's life like as a member of America's Navy? See the everyday challenges, duties and achievements of those who serve." For the period covered in this study, I coded 277 posts from the Navy life Facebook page. Of the 277, 242 were posts that included still images/photos, and 29 were videos. Six posts were text based only.

In terms of content, 112 posts conveyed messages about the Navy as an institution, 36 posts were about aircraft carriers, submarines, or other navy ships, 38 posts about certain positions recruits can choose in the Navy, and 58 posts related to recruitment for the Navy or general information on recruitment criteria. Nineteen posts featured specific members of the Navy, while the remaining posts related to technology the Navy is developing or currently using. There were a total of 28 posts that depicted women specifically, 18 of which depicted one woman, and 10 which depicted two or more women. Women were portrayed with men in 27 posts. However, one man or multiple men were depicted without women in 65 posts. Thirty-five Air Force posts were coded as in discernable.

In posts in which women were depicted—whether alone or with other women—one post depicted women in combat. Only three posts depicted women with military hardware. When depicted along with men, one post depicted women with men engaging in some form of combat, and four posts that included both men and women depicted military hardware. When only men were included, 2 posts contained context related to combat, and 16 included military hardware.

The Navy's recruitment strategy focused on depictions of adventure and travel. Several posts focused on providing information on a specific role or position in the branch. Take for example a post that conveyed information regarding the position of an Explosive Ordnance Disposal (EOD) Technician. The role requires individuals to defuse all explosive

devices including chemical, nuclear, biological, and improvised weapons. In the post related to this position, the role of the Explosive ordnance disposal technician is coupled with adventure in which the person depicted hangs from a helicopter high above an aircraft carrier with the ocean below them. The picture combined with the text above it tells potential recruits about the possible activities they may engage in as EOD technicians, depicting it as an adrenaline rush inducing role. Having the EOD technician hang above an aircraft carrier while being "out on the open seas" (U.S. Navy Facebook Page 2014) also furthers the notion of travel and adventure while being in the Navy. While the gender of the individual in the photo of the post is not discernable, Brown's (2012) study confirms that narratives of adventure have traditionally been linked to specific conceptualizations of masculinity.

Similarly, in a second post, an image depicts five sailors from behind as they stand aboard a Navy ship that is out at sea. They are seemingly conversing as the sun is either rising or setting in the distance. All five of the sailors depicted are male. The text of the post notifies us that the deck of the ship they are standing on is an amphibious assault ship. As in the post described above, by depicting the ship on open waters, the post conveys notions of adventure and travel. In this case, that narrative is specifically linked to masculinity through the depiction of male sailors only.

The photo coupled with the text notifies viewers of the post that individuals such as the men on the assault ship are protectors of ordinary citizens, working to protect them and the homeland as they sleep. By depicting only men in this context, the post suggests that the Navy views male sailors specifically as the protectors, and all other individuals as those who need protection. Moreover, by referring to the "rest of the world" and saying "Good night, everybody!" (U.S. Navy Facebook Page 2014) the post suggests that the Navy sees its male sailors not only as protectors of all other citizens of the United Nations, but as protectors of all other citizens worldwide.

In another post, viewers see two female sailors engaging in a game of soccer with several children on a sand and gravel pavement. The sailors are Caucasian and wear Navy uniform pants with casual sports shirts. The children are darker skinned and almost all wear what appears to be a school uniform. In the background, green trees and a few small buildings are seen. The text of the post notifies us that the scene depicted took place in Indonesia. The picture and text are posted in the context of a series of posts placed on the Facebook page by the Navy called "Around the World Wednesday" (U.S. Navy Facebook Page 2015). Each post depicts a picture

of sailors engaging in activities in a nation abroad. As observed above in the first two posts, the narrative of adventure and travel is continued. However, in this case, the usage of gender rewrites the narrative to suggest that male sailors and female sailors engage in different activities while abroad with the Navy. While above, male sailors are deployed abroad in order to be the protectors of the world, while female sailors fulfill nurturing roles for children in foreign lands. In notifying viewers that the sailors depicted in the photo are in Indonesia "in support of Pacific Partnership 2014…the largest annual multilateral humanitarian assistance and disaster relief preparedness missions conducted in the Asia-Pacific region" (U.S. Navy Facebook Page 2014), the specific depiction of female sailors instead of male sailors, combined with the textual description of why the Navy is in Indonesia, attempts to diminish potential colonial paradigms that may be evoked through this image. In this way, we see another example of how female bodies are used by the military through specific means for specific ends.

REGENDERING THE MILITARY OR DEGENDERING WOMEN?

My examination of more than 1000 Facebook posts reveals several trends. First, similar to Brown's (2012) findings, each military branch deploys different gendered narratives to market itself as it sees fit despite being part of one overarching institution. Each branch views itself as an individual and autonomous entity with its own culture and identity that must be conveyed to potential recruits. Where some of the branches draw attention to soldiers deployed on missions (e.g., Army), others emphasize current or future technology and the training opportunities provided through military service (Air Force).

Depictions of Combat

Perhaps not surprisingly, depictions of combat were most frequently seen on the Facebook sites of the Army and the Marine Corps. Unlike these two branches, the Air Force's Facebook posts focused more on technological capabilities being used, or those in development, by its personnel. The Navy, on the other hand, posted text and photos regarding travel and overall lifestyle experiences recruits could engage in by joining the Navy, romanticizing the international exposure component of possible combat deployments. Of the 277 Navy posts, 118 (almost 43%) did not show any images of Navy personnel; instead they showed pictures of various Navy ships or submarines in the waters or docked in a foreign country.

The combat-related posts provided by both the Army and the Marines were thoroughly gendered, sending a clear message about who was suitable to recruit to these highly coveted roles. Of the 39 combat-related posts on the Army's Facebook page, 30 posts contained images of exclusively male soldiers, shoring up the notion that only men should engage in combat activities. Consider, for example, this post: "U.S. Army Soldiers outperform under pressure because of their integrity, mental adaptability and personal courage," which ties the ability to serve in combat to depictions of physical strength, and exertion. These bodily abilities, which are framed as being specific to the biologically male body, are in turn tied to personal integrity. Of these 30 posts, nearly half (14) contained images that depicted only white males. Of the 24 combat specific posts on the Marines' Facebook page, 16 posts contained exclusively male images. Of the 16 posts, 6 contained images of white males only.

Depictions of Women Post-Combat Ban

Of the four branches of the military, only the Army made any mention of then Secretary of Defense Leon Panetta's announcement of the combat ban being lifted on January 24, 2013, or in subsequent days. The Army's mention, posted on the day of the announcement, read "ARMY TIMES: Senior Defense Officials say Pentagon Chief Leon Panetta is removing the military's ban on women serving in combat" (U.S. Army Facebook Page 2013). This text was coupled with a photo of white female soldiers seemingly on a foreign deployment talking to a group of male foreign teenagers. While the female soldiers are holding rifles in the photo, they are not depicted in the same way as male soldiers in combat. Those posts often show male soldiers actively using their weapons, driving tanks on patrol, or going on building raids. Although the combat ban has been lifted, representations of women such as this continue to reify the hypermasculinized binary that establishes suitable roles for men and women in the military.

As Brown (2012) found, the four branches of the military vary in their depictions of masculinity and in the roles they assign to women in their online recruitment materials. The Air Force frequently depicts women as medics. The Army displays women in ceremonies and ceremonial formations. Women soldiers and women Marines are shown helping civilians while on missions. The Navy showed female sailors playing with children. In the posts that include both women and men, sex segregation in military roles remains the norm. In the Air Force posts, for example, men were shown in combat roles or as pilots, while women were portrayed as medics or behind desks providing staff support.

Depictions of Military Families

Compared to the other branches of the military, the Navy provided the largest number of posts referring to, or depicting scenes of, military families. Some of these posts refer to the meeting of loved ones post-deployment, while others refer to scenes of community in depicting what life would be like for people who sign up to serve. These images are highly gendered, demonstrating that military conceptualizations of the "family" remain highly traditional with respect to sexual orientation.

In a comparative study of military family recruitment in Canada and the United States, Krystel Carrier et al. (2015) found that heteronormativity permeated recruitment and resource materials in both nations.[6] Both Canadian and U.S. materials used gender-neutral language when referring to military families and spouses, acknowledging that the gender of the military personnel deploying and the military spouse expected to provide the support was not fixed. Nonetheless, the gender-neutral language was illustrated with heterosexual couples, suggesting that marriage remains an affair between one man and one woman.

This study's content analysis also found a consistent pattern of gender-neutral language coupled with heterosexual partners in the Facebook posts depicting scenes of post-deployment reunion celebrations. Consider the following texts from Navy Facebook posts during this time period (2013–2015):

> Welcome home, Sailor! Who will be waiting for you at your homecoming? Tag them in the comments!
>
> There's nothing quite like a welcome home photo. #USNavy.
>
> It's beautiful to see Sailors reconnect with their loved ones.
>
> Tomorrow is World Kiss Day! 'Like' this photo to send a virtual kiss to your favorite Sailor.
>
> Family support is so important, especially to a Sailor. On Military Spouse Appreciation Day, we salute the spouses and families who support the Sailors of America's Navy.

The use of the term "Sailor" in these posts is applicable to both women and men serving in the Navy. Yet, in each of these instances the military personnel depicted in the image accompanying the text is male. Moreover, there is at least one perceived civilian in the photo as well. In each of these instances, the civilian is a female spouse or female child. It is also noteworthy that in each of these examples, both the military male and the civilian female are white. This suggests that the Navy not only views the military

family as heterosexual, but as gender traditional—the male serves in the military, and the female is the military spouse—and as "white."

It should be noted that the Navy website had several Facebook pages.[7] In addition to the Navy Life Facebook page, the Navy is the only branch that maintains a separate Facebook page for women, titled "Navy Women Redefined." According to the Navy, "Being a woman in today's Navy is as challenging as it is empowering. Unite with other female Sailors who have raised the bar and redefined what it means to be a woman in the Navy" (Navy.com 2015). Dedicated specifically to women in the Navy, this Facebook page provides some interesting language for visitors:

> What's it like being a woman in today's Navy? Challenging. Exciting. Rewarding. But above all, it's incredibly empowering. That's because the responsibilities are significant, the respect is well-earned and the lifestyle is liberating. Moreover, the chance to push limits personally and professionally is an equal opportunity for women and men alike. (Navy.com 2015)

> The idea that certain jobs are better suited for men and men alone is redefined in the Navy. Stereotypes are overridden by determination, by proven capabilities, and by a shared appreciation for work that's driven by hands-on skills and adrenaline. Here, women are definitely in on the action. And women who seek to pursue what some may consider male-dominated roles are not only welcome, they're wanted—in any of dozens of dynamic fields. (Navy.com 2015)

This language frames Navy women as subjects who redefine femininity, albeit the Navy's conceptualization of femininity. Indeed, the Navy invokes standard stereotypes about women to demonstrate the appeal of the Navy, casting naval service as a means to "override" traditional femininity by cultivating determination, proven capabilities, hands-on skills, and adrenaline—characteristics stereotypically associated with men. In other words, the Navy sells potential female recruits the possibility of "degendering," acquiring the accredited traits associated with "masculine" competence. Yet it offers women a version of equality and empowerment that requires assimilation to a male norm—a norm that remains elusive for women precisely because it privileges men. Far from "regendering" the military, women's participation simply reinforces male norms as definitive of military service.

Depictions of Futuristic Combat

Of all the branches, the Air Force Facebook page spent the most time posting information directly related to recruitment of personnel: 81 of the 205 posts involved pictures that foreground planes as emblematic of the Air

Force mission. Indeed, the Facebook page refers viewers to Flickr: "For more Air Force Photos, check out the U.S. Air Force on Flickr at http://www.flickr.com/photos/usairforce/" (U.S. Air Force Facebook Page 2014). These 81 posts included 8 that specifically provide a sample test question that recruits are likely to see on the Armed Services Vocational Aptitude Battery (ASVAB) entrance exam. Similar to Brown's (2012) finding of a lack of focus on martial masculinity, Facebook posts by the Air Force emphasized the opportunity to work with some of the greatest technology in the world. One post, for example, stated, "The U.S. Air Force is home to the world's most amazing collection of aviation technology. For more photos, video and stats on each aircraft download the new Tech Hangar app available for iPhone and Android" (U.S. Air Force Recruiting 2014). These posts particularly highlight the technological future of warfare. Recruits are encouraged to believe they might fly remotely piloted aircraft, or pilot "autonomous robots [that] will one day be able to enter dangerous environments without putting Airmen at risk" (U.S. Air Force Recruiting 2014). This technocentric message (attempts to) tie(s) masculinity to warfare conducted from afar, which carries the prestige of techno-culture without the risk of loss of limbs or bodily mutilation from serving in combat.

In several instances, Air Force Facebook posts portrayed women in technical positions and engaging in technical activities. These included posts which mentioned the importance of "learning basic electronics [as] the ground work for numerous Air Force careers" (U.S. Air Force Facebook Page 2014), being an air traffic controller, working in Aerospace maintenance, and the Air Force's Technical Training program. Nevertheless, posts related to piloting, calling in airstrikes on targets, combat air traffic controlling, and positions such as Special Missions Aviator, Air Force Special Operations Command, and Combat Rescue Officer remained the domain of male recruits. These posts refer to specific skills and capabilities that are needed to fulfill these roles. As an example,

> Taking control of one of the most advanced aircraft in the world—and pushing its performance to the limit—requires extraordinary skill and precision, and Air Force pilots make it look easy. While successfully completing their missions is paramount, the role of pilots as leaders and character models is just as important since they train and command crews in addition to flying. Air Force pilots deploy around the world to wherever there's a need as fighters, trainers, bombers, advisers and more. (U.S. Air Force Facebook Page 2014)

Although "pilot" is a gender-neutral term, the posts couple texts such as this with photo images of airmen, thereby suggesting that these positions

are exclusively for men. By indirection, the visual cues suggest that women exhibit a lack of the "extraordinary skill and precision" needed to excel. Moreover, including only men in piloting photos, and describing pilots as "leaders and character models," erases women's 30-year (official) history as pilots in the Air Force. As Brown (2012) suggested, the military's continuous reshaping of its cultures of masculinity in this era of technologically sophisticated weaponry, characterized by usage of drone warfare, may make the lifting of the combat ban on women irrelevant. Gender equality in combat positions may involve a degendering of women more than a regendering of the military institution itself.

Conclusions

Although the U.S. military has been actively debating when and under what conditions to regender its forces, contemporary recruitment efforts reinscribe male dominance within these institutions and endorse male norms as appropriate and unassailable. The content analysis of recruitment messages since the announcement of an end to the combat ban on women in January 2013 (and up until 2015) indicates that little has changed in the depiction of women. Women still remain marginalized in online recruitment materials, frequently depicted in stereotypical roles or settings. While the four military branches vary in their textual and visual representations of women, the selling point for each remains an overarching form of militarized masculinity, embodied in the image of the heterosexual, white, able, male.[8] For the Army and the Marine Corps, in particular, the achievement of this hegemonic form of masculinity is directly linked to combat service. For the Navy, international deployments, which may or may not result in combat, provide honor that is later recognized through homecoming ceremonies and reunions with families—routinely envisioned as heterosexual and traditional—featuring a male as the member of the military. And in the Air Force, militarized masculinity is intimately connected to technology, where recruits are lured through a new concept of combat, one which offers them the honor and prestige of fighting wars from behind a joystick.

The textual and visual evidence provided by these online recruitment materials makes it clear that announcements concerning the elimination of the combat ban on women and the allowance of open manifestations of gender and sexual variance among the troops are not sufficient to regender the military. The kinds of changes necessary to accomplish that objective are considered in the final chapter of this book.

Notes

1. As of the writing of this book, the policy, referring to both documented and undocumented residents, has been under a high degree of scrutiny by President Trump's administration.
2. The analysis provided here has also been shaped by the path breaking work of Beth Bailey (2009) in *America's Army: Making the All-Volunteer Force*.
3. I choose here to present data for 2015 because of the time scope of the study on military recruitment. But for those who may be interested, those numbers have continued to increase since 2015. As of January of 2018, Facebook had 1.8 billion active monthly users, with 214 million users specifically in the United States. Of those 214 million, 39.4 million are within the military's target age group range of 18–24 (Statista 2018).
4. U.S. military branches have multiple Facebook pages, which deliver content to various segments of their constituency. My content analysis focused on the official recruitment websites for each of the military branches (i.e., www.airforce.com, www.goarmy.com, www.marines.com, and www.navy.com/joining) viewing each post between January 24, 2013, and October 31, 2015. In total, I coded 1079 Facebook posts across the main Facebook page used for recruitment by the Air Force, Army, Marine Corps, and Navy. I was particularly interested in tracing how many times women were depicted in posts, whether alone, with other women, or with male counterparts. I was also interested in investigating how women were framed in those posts. A straightforward question guided the study: did the various branches of the military offer more gender-inclusive images of military performance after the announcement that the combat ban would be lifted? Accordingly, I coded each post for posting date, posting format (text, photo, or video), context (institution, mission, recruitment, position, personnel feature, technology), whether or not military personnel were depicted in the post (yes or no), number of military personnel depicted, gender (male, female, not discernable), race (white, not-white, not discernible), type of activity performed in photos and videos (personnel recruiting, boot camp, training, technical, medical, ceremony, standing in formation, religious, leisure, demonstration of skill), type of uniform worn in the post (civilian clothing, sports gear, training, combat, work, ceremonial), display of military hardware (yes or no), whether or not civilians were depicted in the post (yes or no), number and gender of civilians depicted (male, female, not discernable), race (white, not-white, not discernible), and type of activity performed by civilian(s) in photos or videos. In the final column of the coding sheet, I included notes about distinctive elements associated with each of the posts.
5. "Likes" referred to here are the number of likes, or followers of the actual branch's Facebook page, not a specific post. Once an individual has "liked" a page, any posts that are made to that page appear in the individual's newsfeed

as a way to "follow" content that is provided on the page. Individuals, who have chosen to "like" or, in other words, follow a military branch's Facebook page may choose to "unlike," or unfollow the page at any point in time.
6. It should be noted, that in the case of the United States, materials given to military families and/or spouses is not uniform across bases or branches of the military. While the Military One Source website produced by the Department of Defense acts as a general resource on all matters related to military and family support programs for all members of the military across all branches, based on the research conducted for this paper, we also found that each branch of the U.S. military had material specific to that branch. Moreover, the research also found material that was produced by individual bases. In summary, producing material across different levels of the military institution—material produced by the Department of Defense, each branch of the military, or individual bases within branches—may or may not produce consistent depictions and messages of what constitutes a "military family" or "military spouse."
7. This list includes US Navy Life, US Navy Chaplain Corps, US Navy Civil Engineer Corps, US Navy Health Care, US Navy Jag Corps, US Navy Nuclear Propulsion, Women Redefined, US Navy Cryptology and Technology, US Navy EOD, US Navy Air Rescue Swimmers, US Navy Diver, US Navy Athletes, Naval Reserve Officers Training Corps (NROTC), US Navy Reserve, US Navy Latinos, US Navy Events, and US Navy STEM.
8. Based on the Department of Defense demographics report for 2013, white Active Duty and Selected Reserve service members represented 71.5% (1,575,594) of the overall military force, with Black or African American Active Duty and Selected Reserve service members represented 16.5% (363,732). Female service members totaled 16.2% (358,156) and male service members 83.8% (1,846,680) of the overall military force.

References

Agoglia, John. 2013, September 13. Army Sponsors Tough Mudder Obstacles in an Attempt to Recruit 'Army Strong' Soldiers. *Athletic Business*. Accessed February 15, 2019. http://www.athleticbusiness.com/military/army-sponsors-tough-mudder-obstacles-in-an-attempt-to-recruit-army-strong-soldiers.html.

Bailey, B. 2009. *America's Army: Making the All-Volunteer Force*. Boston: Harvard University.

Brannen, K. 2013, November 19. Army PR Push: 'Average-looking Women'. *Politico*. Updated November 20, 2013. https://www.politico.com/story/2013/11/army-pr-push-average-looking-women-100065.

Brown, M.T. 2012. *Enlisting Masculinity: Gender and Recruitment of the All-Volunteer Force*. Oxford: Oxford University Press.

Caforio, G. 2007. Introduction. In *Social Science and the Military: An Interdisciplinary Overview*, ed. Guiseppi Caforio, 1–20. New York: Routledge.

Carrier, K., M. Eichler, and S. Szitanyi. 2015. The Lifting of Women's Combat Exclusion: A Comparison of the Shifting Gendered Politics of Military Families in Canada and the United States. *Paper Presented at the International Studies Association Annual Conference*, February 2015, New Orleans, Louisiana.

Cohn, C. 1987. Sex and Death in the Rational World of Defense Intellectuals. *Signs: Journal of Women in Culture and Society* 12 (4): 687–718.

Elliott, S. 2011, May 24. Army Seeks Recruits in Social Media. *The New York Times*. Accessed February 15, 2019. http://www.nytimes.com/2011/05/25/business/media/25adco.html.

Haltiner. 2003. The Decline of the European Mass Armies. In *Handbook of the Sociology of the Military*, ed. Giuseppe Caforio, 361–384. New York: Plenum Publishers.

Iskra, D., S. Trainor, M. Leithauser, and M.D. Segal. 2002. Women's Participation in Armed Forces Cross-Nationally: Expanding Segal's Model. *Current Sociology* 50 (5): 771–797.

MacKenzie, M. 2015. *Beyond the Band of Brothers*. Cambridge: Cambridge University Press.

Sandhoff, M., M. Segal, and D. Segal. 2008. Gender Issues in the Transformation to an All-Volunteer Force: A Transnational Perspective. In *The New Citizens Armies: Israel's Armed Forces in Comparative Perspective*, edited by Stuart Cohen, 111–113. New York: Routledge.

Segal, M.W. 1995. Women's Military Roles Cross-Nationally: Past, Present, and Future. *Gender and Society* 9 (6): 757–775.

Statista. 2018. Number of Facebook Users by Age in the U.S. as of January 2018 (in Millions). Accessed February 15, 2019. http://www.statista.com/statistics/398136/us-facebook-user-age-groups/.

U.S. Air Force Facebook Page. 2014. U.S. Airforce Recruiting. http://facebook.com/USAirForceRecruiting/.

U.S. Army Facebook Page. 2013. U.S. Army – GoArmy.com. http://facebook.com/goarmy.

U.S. Army Facebook Page. 2014. U.S. Army – GoArmy.com. http://facebook.com/goarmy.

U.S. Army Facebook Page. 2015. U.S. Army – GoArmy.com. http://facebook.com/goarmy.

U.S. Navy Facebook Page. 2015. "Forged by the Sea – America's Navy." http://facebook.com/americasnavy.

U.S. Marine Corps Facebook Page. 2014. Marine Corps Recruiting. http://facebook.com/marinecorps/.

U.S. Marine Corps Facebook Page. 2015. Marine Corps Recruiting. http://facebook.com/marinecorps/.

CHAPTER 7

Conclusion: The Challenge of Degendering the Military

The previous empiral chapters of this book have underscored instances of gender trouble that have failed to produce institutional degendering. What might be required to degender this markedly masculinist institution? How do the prospects for degendering fare in relation to regendering? To answer these questions, I return to the insights of feminist scholars and practitioners who have theorized how to "undo" gender, situating these transformative strategies in relation to the standard operating procedures within the military discussed throughout this book.

GENDERED INSTITUTIONS: TOOLS FOR "UNDOING" GENDER

As highlighted in the introductory chapter of this book, identifying which tool(s) would be necessary for degendering the military as an institution is, at least in part, a normative question. Defining the goal of degendering the institution may determine the approach(es) used. Taking Joshua S. Goldstein's (2001) outline of three different strands of feminist theorizing on gender and war as a frame—Liberal feminism, Difference feminism, and Postmodern feminism[1]—change may look entirely different for Liberal feminists as it might for Postmodern feminists. Liberal feminists and activists, for example, have argued that women can meet the same institutional standards as men, and should be allowed to serve and fight alongside men in the military fairly and equally, without being discriminated against based on sex. For Liberal feminists, the creation of

the All-Volunteer Force (AVF), the subsequent larger number of women recruited for a standing military, along with the increased roles women have been allowed to participate in—thought of as necessary to give women equal access to career advancement—are seen as having advanced these causes. Further change toward ensuring equality for women in the military would include requiring that all women register for the U.S. Selective Service Program, mandating women's participation in a draft if one were deemed necessary. For Liberal feminists, then, a degendering of the military would result in the right to participate—that is, a literal "inclusion" of women—in all aspects of the military apparatus from which they have been historically "excluded."

Differently, Feminist international relations (IR) has spent considerable time highlighting controversies associated with more nuanced understandings of militarization and militarism (see Chap. 2). The field's insight on these topics advance Postmodern feminist arguments on gender and the military to suggest that the increased inclusion of women in the armed forces, as argued by liberal feminists, should not be the goal. Indeed, for Feminist IR scholarship, the increased inclusion of women may act to further enable and mask the masculinist nature of the military. As such, any tool suggested as necessary or useful for degendering the military should be considered in relation to what a degendered version of the institution is imagined to be. But as I demonstrate through a review of tools for undoing gender, the contemporary U.S. military has shown an anemic willingness to use more inclusive understandings of gendered bodies and experiences, tactics that could work toward dismantling the warrior archetype.

In *Breaking the Bowls*, Judith Lorber (2005) outlines multiple contemporary approaches to "undo gender" (8). Among these are *gender mainstreaming* and *gender parity*. Other feminists who have advocated for the degendering of institutions have suggested tactics such as *gender visibility, gender balance, gender diversity, and gender freedom*. For Lorber, degendering is ultimately a call for "so-called gender normals" (13) to oppose and challenge culturally sanctioned gender performances prescribed by the binary system, through which women are positioned as unequal and inferior to men.

Gender mainstreaming focuses on systems of power that generate disadvantage for specific groups (Duncanson and Woodward 2015) and requires analysis of "the potential impact of policy decisions on women

and men" (Lorber 2005). It is used as a strategy to "ensur[e] that gender perspectives and attention to the goal of gender equality are central to all activities" (UN Women 2016, 3). Throughout the 50-year long debate on women's inclusion in military services during and after the establishment of the All-Volunteer Force, very little of that policy review included gender mainstreaming. While the core of that debate on the surface seemed to center on women, a more nuanced investigation of Congressional Hearings demonstrates that the narrative of "incremental and deliberate" progress toward inclusion privileged the impact women's inclusion might have on the institution, and the men who serve in it. As Chap. 3 highlights, this narrative placed a positive spin on the consistently small number of women (in comparison to men) serving in the military, while diverting attention away from the increasing roles women successfully performed in the military prior to the suspension of the combat ban.

More specifically, the Congressional hearings demonstrate that military officials justified women's exclusion from combat by positioning the military as a rational actor, concerned with efficient operation of the nation's defense. But the carefully crafted image of *homo economicus* circulated during the first decade of the AVF gave way to misogynist arguments that imagined women to be "leaky bodies" that threatened the operations of the institution. In the 1980s, anxieties surrounding "time-lost" by women generated exaggerated claims about negative effects of "hormones," menstruation, pregnancies, and motherhood. In the 1990s, grounds for continued exclusion included fears concerning sexual assault when women served side by side men in combat. In the early twenty-first century, the military generated worries about the different treatment women's bodies would require when returning from combat zones. Even as women performed valiantly in multiple deployments on the front lines in the post 9/11 era, binary understandings of gender remained salient, and were used by the military to manipulate discourses about women's bodies, foregrounding fertility and fragility to justify the continuance of the combat ban.

Institutional response to Congressional hearings on military sexual assault showed similar difficulties with respect to *gender mainstreaming*. Shifting attention from the individual women and men who have experienced sexual assault by their fellow service members, military testimony provided to Congressional committees used securitization and medicalization discourses to position the individual military branches themselves as the victims of the "sexual assault epidemic." Military officials perform an

extraordinary sleight of hand when they characterize sexual assault in the military as a "plague" or "cancer" that threatens the very survival of the organization; they engage in a systematic masking and erasing of victims' wounds. Doing so ignores the physical and psychological horrors of individuals attacked and refocuses attention to the military as a whole. Diverting the focus of Congressional investigations away from the physical and psychological harms of the individuals, the military refocused attention onto the defense institutions, eliding critical questions about whether perpetrators of sexual violence are brought to justice by the military, or whether the military cultivates a culture hostile to gender equality. Although debates surrounding gender inclusion in the military seem to give due attention to their impact on women service members, they typically place the institution, and its male soldiers specifically, at the center of concern.

Gender parity requires equal numbers of women and men in all ranks within institutions and, arguably, favors liberal feminists' "right to fight" advocacy agenda. On a macro level, gender parity in the U.S. military would require, at a bare minimum the same number of women and men to serve in the individual branches, or in the institution as a whole. Yet, as of April 2019, women represent less than 20% of the active duty members of the military (Department of Defense 2019). Gender parity would require that the same number of *women and men* be included in all boot camp and training cohorts, leadership positions, deployment units, and military occupations—those which have traditionally been viewed as military positions for women, as well as those which have historically been closed-off to women. In line with postmodern feminism's agenda, to eliminate male privilege the military would also have to forego using gender as an organizing principle for evaluation and assessment, eliminating the "idealized masculine model" as the unquestioned standard of comparison (Lorber 2005). To move toward gender parity, the military would have to systematically transform its handling of sexual assault and sexual harassment, affording women the opportunity to rise in the ranks toward leadership positions free from threats of retaliation, demotion, or dismissal from the service when they report cases of sexual assault.

Similarly, *gender balancing* strategies would require the military to recruit equal numbers of men and women for each military occupation. Toward that end, the military would have to reconsider how it depicts men and women in military marketing campaigns, and in military museums and memorials. Indeed, the military would have to transcend binary con-

structions of gender and rethink how inclusion of LGBTQ troops would alter the way in which the military interacts with citizens in recruitment, training, and promotion. Evidence drawn from recent recruitment materials suggest that neither gender parity nor gender balancing is a priority for the U.S. military. Less than 1% of the recruitment images posted by the Air Force, Army, Marines and Navy during the period of study covered in Chap. 6 depicted women in combat roles—a far cry from the percentage of women troops deployed in recent theaters of combat. The profound underrepresentation of women (and racial minorities) in military recruitment materials is one powerful indicator of the military's continuing role in shoring up gendered and raced hierarchies of citizenship. Rather than fostering gender inclusion and gender and sexual variance, the military entrenches the white, heterosexual male as the ideal soldier and courageous citizen.

Gender visibility illuminates "the extent to which societies, cultures, groups, and individuals are gendered" (Lorber 2005, 10) and how particular gender orders are maintained by various institutions of power. New York Democratic Senator Kristen Gillibrand's tireless work on military sexual assault reform, through the proposed Military Justice Improvement Act, has made *gender visible* in the context of the abysmal treatment of those who complain of sexual assault and sexual harassment, where retaliation against the victim supplants punishment of the sexual assailant. Documentary films such as *The Invisible War* demonstrate how the military renders sexual assault victims and their injuries invisible. In direct contrast to the norms of gender visibility, the military typically acts to ensure that unequal treatment goes unnoticed, unacknowledged, and unredressed.

The military's institutional gender order becomes starkly visible in its treatment of war wounds, which consigns those who suffer some forms of injury to second-class status (Chap. 4). Wounding sustained from tours of combat that do not visibly mark the bodies of soldiers through noticeable maiming or the missing of limbs are feminized through narratives of inadequacy. In crimes committed by veterans grappling with post-traumatic stress disorder (PTSD) and traumatic brain injuries, their acts of crime are seen as a disgraceful sullying of the honor of the uniform they wear and the country they serve.

And as Chap. 5 documented, certain soldiers are dishonored by having their service erased from military histories disseminated to the public. Military museums, memorials, and ceremonies broadcast gendered stories of valor, celebrating and commemorating an idealized male heterosexual

warrior, while eliding the service records of gay, gender nonconforming, and women troops. As the case study of the USS *Midway* demonstrated, women were erased from naval history, as a century of women's valiant and diverse contributions were made to disappear. Through the semiotic systems established in war museums and memorials, masculinity is re-inscribed as the essence of the armed forces and circulated unproblematically to the public.

Gender diversity recognizes the flawed demarcation of human bodies into male and female and celebrates the multiple modes of embodiment and sexuality that occur in nature, which are structured by "cultures, religions, ethnicities, social class, sexualities, bodies, and other major characteristics" (Lorber 2005, 12). Gender diversity in the U.S. military would at a bare minimum acknowledge and honor the service of a multitude of embodied subjects at the intersections of gender, race, sexual orientation, or other identity markers, emphatically rejecting notions of hegemonic masculinity. Generally, most institutional material reviewed and analyzed for this book addressed women service members—when they are addressed—as a monolithic group, disregarding intersected variation of lived experiences before, during, and after their service in the military. At other times, documents such as the Gates Commission report that urged Nixon to move forward with creating an All-Volunteer Force mention "women" only in passing, or leave them out altogether. And more recently, rather than moving toward inclusion, recruitment materials show that the military routinely fails to depict gender diversity among its troops. Similar to Brown's (2012) study of print military advertisements from 1970 to 2007, the content analysis presented in Chap. 6 found that various forms of masculinity remained the hallmark of recruitment materials. For the Air Force, service was tied to techno-masculinity, for the Army and Marines, service was tied to warrior masculinity established through frontline combat roles, and for the Navy, service was tied to adventurous masculinity often established through travel opportunities available with the branch. Further, depictions of military life overwhelmingly embed notions of traditional, white, heterosexual armed forces and families. Images of families reuniting post-deployment show white male soldiers greeted by white female civilian spouses and children. Although the texts in these posts use gender neutral language, the visual images reify heteronormativity, "whiteness," and traditional gender roles as fundamental to military identity.

Gender freedom grants "gender rebels" the right to "liv[e] outside the binary sex/gender system, against the grain of gender norms and expecta-

tions" (Lorber 2005, 12–13) freely and without repercussions from institutional rules. "Don't Ask, Don't Tell," a policy which had prohibited lesbian, gay, and bisexuals from openly serving in the military, was repealed in 2011. Yet few LGBTQ troops are featured in military recruitment materials, and certain high-profile cases, such as the demonization of Lynndie England and Chelsea Manning, suggest that gender nonconformity—in both hetero and homosexual formats—continues to be treated by the military as deviant and unacceptable. Rather than challenging the military's heteronormative and binary gender system, instances of gender trouble are quickly criminalized and marked for punishment by gender conforming soldiers who participate in courts martial.

Although the spectrum of feminist theorizing, scholarship, and activism has advanced multiple strategies to "undo" gender, the U.S. military has, seemingly, not made use of them, showing instead a discernible preference toward reinforcing hegemonic white heteromasculinity as the cornerstone of military service. Despite the U.S. military's gender regime "hav[ing] proven exceptionally resilient in adapting to technological changes" (Daggett 2015, 362), technological developments that facilitate warfare by robots, drones, and other mechanical means may now provide an opening for change by removing the centrality of physical strength from institutionally defined conceptualizations of "combat." Indeed, cyborg and robotic versions of soldiering could eliminate gender and gender binarism altogether, fulfilling Lorber's (2000) call for "rendering gender irrelevant" (81). Could autonomous technology and these "unmanned" military tactics provide an alternative path, a disruption of the dyadic understanding of gender and warfare that, as a result, unsettles military masculinity? After all, high-tech modes of engagement could, in principle, eliminate barriers for women, grounded in gender stereotypes about women's "weakness." The final section of the book considers whether this possibility is likely to have palpable degendering or regendering effects.

The Disorienting Puzzle of Drone Warfare

Questions surrounding what the increased use of drone warfare might mean for gendered understandings of state warcraft have advanced, and equally, further complicated our knowledge about how militarized masculinities work (in Western contexts) (Clark 2018). Scholarship on drone warfare—along with other artificial intelligence technologies—has been studied as producers of hypermasculinity, particularly in the context of

both the U.S. and British militaries (Holmqvist 2013; Kunashakaran 2016), as "posthuman bodies" that blur lines around the relationship between people and technology (Wilcox 2017), and as disruptive modalities that produce a series of disorientations for the intelligibility of war (Daggett 2015; Manjikian 2014), each directly or indirectly addressing the potential implications they may have on contemporary understandings of militarized masculinities (Clark 2018).

Despite these important contributions, Lindsay C. Clark (2018) highlights inconsistencies in accounts of gendered discourses and narratives drones produce in relation to ideas on militarized masculinity/masculinities. Like discourses presented in earlier chapters, understanding and articulating drone warfare as gendered seems to be quite "messy"; that is, confusing, complex, and sometimes even contradictory. Drone warfare may, for example, bolster forms of "masculine" service while simultaneously highlighting "feminine" skills needed to be a drone operator. Similarly, instead of making gender unnecessary or irrelevant for techno-based modes of service, it seems "technological advances [may] create a new soldier who is a more virile and a more deadly male" (Manjikian 2014, 56) while simultaneously feminizing said individual through military service that is ridiculed for being conducted from thousands of miles away from the traditionally imagined war theater. In turn, military discourses that try to respond and position drone warfare as falling in line with the institutional gender regime are even messier; they are slippery and often contested.

How are we then to deal with the "disorienting" (Daggett 2015) (gender) "trouble" (Clark 2018 paraphrasing Haraway 2017), that is, messiness produced by these seemingly "inherently irresolvable" (Clark 2018, 9) inconsistencies and contradictions? Is the feminist scholarly toolkit, perhaps, too limited in providing useful mechanisms for analyzing how drone warfare may or may not challenge the military's gender regime?

Not necessarily. I offer one productive way forward through a feminist lens, as that which considers the "disorientation" provided by drones' seeming inability to "fit" within "existing feminist engagements" (Clark 2018, 9) as situational, standpoint based, and dependent on vantage and positionality. This is not to say that the theoretical interventions made by scholars such as Manjikian (2014), Daggett (2015), Wilcox (2017), and Clark (2018) have not advanced our understanding of this phenomena—they undoubtedly have—but it is to highlight the complexity of relations

between and among actors that drone warfare exposes. In other words, when we position drone warfare as disorienting, it necessitates a pause in the midst of our intellectual pursuit to ask "disorienting for whom?" The manner in which drone warfare is gendered for drone operators, for example, may be very different from how it is gendered for targeted individuals on the receiving end of those operations, and the gendering of (power) relations between the two groups may equally be different than the gendering of relations between drone operators and other service members with whom they serve in the same military organization. This may, in essence, seem obvious, and if so, further engagement toward untangling the "(cob)web" (Clark 2018, 3) like complexities of these power relations seems necessary and worthy.

The remainder of this chapter is therefore particularly interested in how drone warfare may create gender trouble, and consequently, a disorientation for the military or military officials specifically responsible for the creation and dissemination of coherent messages related to the institution's preferred dyadic understandings of gender, and which uphold the masculine warrior archetype for the general public. In turn, I am also equally interested in how drone warfare may be disorienting for the general public, and how it may deviate from its current (gendered) understandings of masculinity, military service, and war. The forthcoming analysis and comments, therefore, focus on these groups specifically.

"Unmanning" Warfare: Techno Challenges to Militarized Masculinity

Given feminist insight on the gendered nature of drone warfare, I return to the question of whether techno-based versions of warfare may provide a conduit for degendering the military's gender regime of heteromale privilege. In the introductory chapter of this book, I provided an overview of gender regimes as an amalgamation of institutional procedures, policies, and practices that dictate how gender(ed) relations "play out" (Connell 1987). I argued that as social constructions, gender relations within the military establish systems of social hierarchies of privilege and status among individuals "associated" with the organization. Through insight from feminist international relations' understanding of processes of militarism and militarization, I argued that individuals thought to be far removed from the military apparatus, that is members of the public at

large, who are not and have not served in the military, are also associated with the organization as (gendered) military values and mores are actively communicated by and disseminated to the general public from which the volunteer-based military sources individuals to serve. In particular, I highlighted that in gender regimes, social hierarchies of privilege are created by using gender as a binary organizing and ordering tool through which (biological) women and men are put in oppositional categories, and through which appropriate and inappropriate roles and behaviors are structured and defined around gender difference. Finally, I argued that the contemporary masculine warrior archetype has been paramount in helping the military bolster and maintain its gender regime of heteromale privilege through its continuous binary calibration and contradistinction of less masculine and feminized counterimages of soldiering.

Crucial for the archetype's pernicious dexterity to outlive calls for institutional change has been the romanticism associated with war. As I have suggested elsewhere in this book,[2] the romanticism of war in the West has historically been associated with (white, heterosexual) male soldiers going off to battle to face the possibility of giving their lives in the ultimate sacrifice of death. Those who pay the ultimate sacrifice return home in flag-draped coffins, inextricably intertwining and cementing images of war, masculinity, and patriotism. Heroism, whether marked by bodily harm, physical maiming, or death, is narrated as bravery—the courage to face the enemy to protect family and nation. Celebrated in folktales, novels, poetry, film, and oral histories, such heroism is seldom associated with life-long and painful challenges to overcome the psychological and physical scars of war. In short, the romanticism of war and, with that, the calibration of the (hetero) masculine warrior as heroic through visual and discursive maneuvers have provided an intelligibility to war.

What happens when this intelligibility is disrupted? Arguably, the kinds of warfare enabled by advances in military technology are at great remove from such romanticized notions of heroism; while historically, the U.S. military has been able to adapt discursive frames of militarized masculinities to include advances in technology,[3] warfare by way of drones and other autonomous aerial vehicles may produce institutional gender trouble through the blurring of a series of social parameters that would otherwise make war intelligible for both service members and the general public. In so doing, the increased reliance on drone technology in the conduct of war may problematize the military's ability to devise a narrative of service that positions it as fitting into the warrior archetype profile, fail-

ing to provide "a reliable mapping device" though which "readaptation" could be achieved (Daggett 2015, 362).[4] I consider the potential these blurred parameters have for disrupting the link between heralded heteronormative masculinity and military service in relation to the masculine warrior profile as a pathway to degendering the military below.

The (Disruptive) Destruction of the Masculine Warrior?

Concretely, the parameters of (gendered) relations allotted between soldiers and drone technology have trouble fitting, and are mismatched, with the parameters of (gendered) relations provided by the warrior archetype in soldiers and war. On top of failing to fit the gendered mold of the romanticism of war, techno-based war forms actively feminize drone operators through a variety of discursive narratives and tropes. In short, "drone warfare remains slippery and illegible" (Daggett 2015, 362) for both the institution and the public, producing an uncomfortability, which, as I show later in this final chapter, forces the military to concertedly use ways to try to (re)establish and stabilize warriorhood in and through these more technology-based forms of warfare.

War by way of drones deviates from the intelligibility provided by the masculine warrior paradigm producing a series of gendered anxieties related to three fundamental components of the archetype's profile, core elements of the institutionally defined "ultimate soldier," and which I have covered through the cases in the empirical chapters of this book:

- The physical presence in, or proximity to,[5] a war theater in which service members are on the front lines of engaging in conflict with a pre-defined enemy (Chap. 3)
- The engagement in the war theater being one that allows for and promotes the possibility for physical harm, wounding, and maiming (to the body, not the mind), if not death (Chap. 4)
- The ability to valorize and memorialize that heralded form of service, enabling a social memory of specific war stories to be generated and disseminated among the general public, upholding and allowing for the continued prominence of the archetype (Chap. 5)

Following both Cara Daggett (2015) and Mary Manjikian's (2014) work on the intelligibility/unintelligibility of war drone warfare produces, I argue that the blurring, or "confusion of boundaries" (Haraway

1985/1991, 133),[6] of and around these specific defining components of the masculine warrior paradigm—the most coveted form of militarized masculinity in the contemporary U.S. military[7]—produces "queer moments of disorientation" (Daggett 2015, 361), primarily for the general public, but also for military officials and personnel, making war overall uneasy and unintelligible. I consider each of these archetypal components in turn below.

Proximity to the Enemy: The Blurring of the War Theater

Since 2013, "manned" and "unmanned" drones have been used with increasing frequency for military pursuits.[8] Drone operators—most usually located thousands of miles away from the physical combat zone—target sites and individuals of interest from behind a joystick with which they launch attacks. Through the drones' live feeds, operators are visually placed in the very environment in which life or death is determined for targeted individuals, but from which they are far removed (physically) and able to avoid the possibility of any bodily harm.

Unlike traditional images of combat, drone warfare is cloaked, unseen, and most often, invisible. The results of drone strikes and intelligence gathered through drone-led missions are seldom shared outside control rooms. The silent and invisible virtual reality of drone warfare might well challenge hegemonic narratives of military masculinity by "unsettl[ing] the cross-referential nature of military-masculinity that refers to and derives meaning from" physical combat (Bayard de Volo 2016, 52). Instead of traveling long distances to a combat theater that is defined through the (supposed) location of a pre-defined enemy (individual or group), drone command centers are situated in remote regions of the United States, whose locations are typically unknown to the public. Drone operators—men and women—sit behind computer screens thousands of miles away from their targets; their "boots" never touch the "ground" of lengthy deployments typically associated with the traditional imaginary of "combat." The physicality of war, which traditionally associates stamina and toughness—specifically imagined as attributes of the heteromale body—as necessary for the act of killing, is infantilized, relating drone operators' engagement to that of playing a video game—easy, comfortable, lacking danger or the possibility for trauma, and in fact, enjoyable.

Drone operators often endure scorn from their peers who are tasked with engaging in traditional forms of warfare, institutional colleagues who label "virtual warriors" (see below) feminized and infantilized derogatory terms such as "cowardly," as they are positioned by peers as being thoroughly insulated from danger and risk of "real" combat.

Techno-mental Injuries: The Blurring of Harm and Death

Despite this spatial divide that is thought to remove the possibility for trauma, drone operators have increasingly stepped forward to discuss their experiences, and to bring attention to the negative psychological impact their roles have on them, impact which the military has largely ignored.[9] In some cases, drone operators who have attempted to seek help for mental trauma associated with their job responsibilities have been retaliated against and threatened to have their security clearances revoked if they did (Press 2018). At the core of this institutional response is a confusion and anxiety for both the military and the public around (alternative) understandings of "combat" drone warfare may necessitate: should launching a drone be considered a combat mission? Are drone operators combat soldiers? And are drone operators eligible for the same honor, valor, and respect as front-line ground combat troops? If so, what does the lifting of the ban on women in combat really mean for gender equality in the military and how will this play out among average citizens? Developments in drone warfare are likely to shape the future of military operations. In certain ways, it is also likely to determine how the general public views the future nature of warfare, and thereby, the (in)significance of the lifting of the combat ban on women.

Similarly, the nature of war injury changes for those military service members (typically those in the Air Force) who operate weaponized drones and other forms of autonomous aerial vehicles. The dire physical injuries and mortal wounds valorized in military medals, museums, and memorials (see below) will give way to the very real psychological harms associated with anonymous delivery of death across long distances—psychological harms that appear to affect men and women comparably in the form of PTSD. Shortages of drone pilots increase burnout rates among soldiers who are overworked (Martin 2011). Drone operators and pilots report ongoing sleep disturbances and uneasy consciences. Indeed, killing through drones results in a queering of death by war (Daggett 2015) that has trouble fitting the mold of previously institutionally prescribed (accept-

able) forms of militarized masculinities; the spatial divide between the killer and the killed in drone warfare positions the drone operator as "the definitional opposite of bravery" (Greenwald 2012, 10). They do not face any danger or risk of bodily injury or death. Those who seek medical help for sleep disturbances, post-traumatic stress disorder, or the urge to abandon post, are treated harshly enough by military command to secure their silence. The psychological injuries incurred in drone warfare receive the same hostile treatment associated with sexual assault and PTSD. Falling outside the category of war wounds acknowledged and valorized by military authorities, drone operators who seek medical assistance may be threatened with psychiatric discharge, a form of punitive retaliation for modes of suffering that does not register on the military's roster of recognized wounds (Martin 2011).

(Lacking) Spectacle of War: The Blurring of Recognized and Valorized Service

Drone warfare certainly does not position drone sensors and operators as heroes—they need not manifest heroism or bravery to complete missions. Drone pilots and sensors say little about their jobs to spouses, friends, and family. Their deeds are unsung. Valorization for missions completed through the use of drones does not afford the same recognition as those completed through traditionally imagined theaters of combat. Medal and award recognition programs for drone missions have met heavy resistance—both in the United States, and in other Western Democracies, including the United Kingdom—as they are deemed inappropriate for drone operators who are positioned as not having anything risked in their roles (Tilghman 2016).[10]

The hidden and remote locations of drone operators also sever visual ties between male soldiering and the combat zone. Drone missions are executed by women, who work side by side with their male counterparts. Because drone operators are unseen during missions, however, the gender of the combatant becomes invisible and irrelevant. Women who work as drone operators or sensors enter a sphere of operations in which physical embodiment is systemically unrecognized—yet this lack of recognition is shared equally by men in these roles. If drone warfare becomes the modal form of future combat, this high-tech tactical engagement will be peculiarly disembodied—for both American military personnel and the public from which the military institution must draw people to serve—

rendering the elimination of the ban against women in combat insignificant as both male and female drone pilots disappear from visible theaters of war. Rather than marking a major moment in women's military history, elimination of the combat ban will signify a minor blip in the technological transformation of war.

Perhaps more importantly, drone warfare troubles the public's experience of war; it provides a confusing and "uniquely disorientating [technology]" (Daggett 2015, 364), effectively blurring their "[location] along traditional gendered maps that orient killing in war" (364). As a result, the diminished articulation of a nation "being at war" limits the public's ability to valorize and mourn those facing injuries and death fighting in that war. In so doing, the use of drones in warfare holds the potential to transform military-civilian relations. As earlier chapters have demonstrated, much of the military's hold on the public imagination is related to the gendered spectacle of war—powerful images of courageous men risking everything to protect their homeland. A critical component of that spectacle is the act of killing and the experience of death. For the public, drone warfare fundamentally blurs the line of defining when a nation is "at war," limiting the public's ability to valorize those who "fight" the wars and mourn their injuries and deaths. In short, under ordinary circumstances, drone warfare lacks the basic elements of spectacle necessary for war to be intelligible for a public: it is invisible and disembodied (specifically for the public of the nation using drone warfare as a foreign policy tactic), altering the public's visceral experience of war and its consequences. Its visible effects include shattered homes, piles of rubble, cities of ruin—images that may make viewers question the validity of war, especially when breaking news indicates that the drones missed military targets and destroyed hospitals, schools, and wedding parties. Under such circumstances, the anonymity of the drone operators may be their most prized possession—far more valuable than military medals or commendations. The occasional grainy image on the nightly news does not evoke the positive or negative passion generated by watching soldiers engage in real-time broadcasts of fire fights. Mutilated limbs and flag-draped coffins are replaced with virtual reality screens, which wash away the brutality of war. Drone warfare literally refigures combat; the war becomes invisible to those not conducting it.

The Military's (Continued) Troubling Response to Gender Trouble

Despite the potential changes military technology holds for disrupting military masculinity and military-civilian relations, there is reason to believe that they will not necessarily produce such degendering effects as military officials work diligently to find discursive ways "to fit the new technology into existing gender stereotypes rather than allowing for the development of new ones" (Manjikian 2014, 52). But despite this possibility, the military works to build a new narrative around this form of service that produces a new version of militarized masculinity, one that is built upon the idea of "a highly developed masculine sensibility" (53). Specifically, military officials attempt to regender drone operators by positioning them as "cyberwarriors" "making the argument that the soldier who utilizes internet-based command, communications and control technology now 'commands' the domain called the internet, just as other types of soldiers command the air, the sea and the land or terrain" (Convertino et al. 2007 quoted in Manjikian 2014, 53).

The military performs multiple functions in contemporary society; war making is only one. The military provides a means of capacity building, educational opportunity, upward mobility, and a pathway to power, particularly for low-income youth. It offers comradery, honor, and a disciplined life that many find powerfully attractive. The military serves as a potent symbol of U.S. power, which is leveraged in complex ways during international crises. Developments in military technology alone will not alter these complex military roles and relations.

WHAT'S AT STAKE? THE REGENDERING AND REMILITARIZATION OF AMERICAN SOCIETY

In light of my analysis on drone warfare and what I have argued to be an inability to adequately degender the military, that is destabilize its gender regime of heteromale privilege, where are we to go from here? What are the implications for this military order persisting, both for the members currently serving in the institution, and, more generally, the public at large, the public from which the institution relies on drawing individuals to serve?

As I have shown throughout this book, for feminist scholarship interested in the military as a gendered institution, questions surrounding societal implications of militarism and processes of militarization have been

paramount (Belkin 2012; Brown 2012; Burke 2002; Enloe 1983, 1990, 2000, 2002; Cohn 2000; Ferguson and Turnbull 1999; Francke 1997; Goldstein 2001; Hooper 2001; Kronsell 2012; Levy 1997; Lutz 2002). Processes of militarization serve as systems of power that are felt widely by citizens of communities otherwise thought to be far removed from the actual military apparatus itself. As such, investigations of the military's changing gender representations as have been promoted within the general public remain paramount as gender remains "present in the processes, practices, images and ideologies, and distributions of power in the various sectors of social life" (Acker 1992, 567).

The U.S. military remains a vastly gendered institution. Despite recent policy changes to eliminate women's exclusion from combat and to permit openly gay, lesbian, bisexual, and trans personnel to serve, the military persists as an organization that embraces notions of male superiority and preserves an institutional gender order that privileges men—heterosexual men, in particular. Activists who have sought to degender the military suggest that the gender hierarchy established within the military has multiple effects on society at large. The military endorses binary constructions of gender, heteronormativity, and presumptions of male strength that marginalize women, gays, and gender nonconformers despite its occasional adoption of inclusive, egalitarian rhetoric. The manifold images of the military that circulate through recruitment materials, museums, and memorials suggest that military masculinity is benevolent and protective, eliding the brutal gendering practices in basic training, pervasive sexual harassment, and sexual assault within the institution.

Although these and other forms of "gender trouble" have opened up the possibility for the institution to adopt meaningful degendering change, the military has proven to be far more likely to regender, shoring up hegemonic conceptions of military masculinity in response to challenges to the prevailing gender regime. The traced gendered nuances of narratives surrounding the combat exclusion policy since the inception of the All-Volunteer Force, specifically between 1970 and 2015, revealed an overarching narrative of women's increasing inclusion in the military though an "incremental and deliberate" process and which "culminated" in the lifting of the combat ban. That pervasive narrative revealed positivist arguments made by military officials to justify keeping women out of combat for as long as they did. It allowed the military to position itself as a rational actor, first through the image of *homo economicus* during the first decade after the creation of the AVF, and thereafter, through arguments

that positioned women as "leaky bodies" that threatened the operations of the institution. In the 1980s, anxieties surrounded time-loss produced by women through hormones and menstruation and pregnancies and motherhoods. In the 1990s, fear of increases in sexual assault by placing women side by side men in combat were cited, while the 2000s were filled with arguments on differentiated treatment women's bodies would require when returning from combat zones. A critical reading of Congressional hearings from this period underscores how the military manipulates ideas about women's bodies and uses their bodies by specific means for specific ends to justify the continuance of policies to the general public.

Indeed, narratives surrounding soldiers' bodies and their "appropriate" usage have been paramount to the establishment and maintenance of the masculine warrior profile. These narratives, most prominently, have depicted bodies worthy of honor and accolade as those which sustain bodily maiming and mutilation through tours of combat. But the usage of military bodies in narratives of institutional self-representations surpasses simple depictions of the physicality associated with service. Juxtaposing multiple narratives around soldiers' bodies unearths preferences by the military of some bodies over others, the recognition of certain forms of wounding over others, creating a systemic grouping of "invisible wounds." These wounds not only lack visibility or visual tangibility on the corpus itself, but become invisible through the silencing and shaming of victims, and by the masking of injuries through discourses manipulated by the military.

Uncovering the ways in which the military silences and masks certain forms of wounding provides a particularly useful way to explore the relation between wounding as being necessary for the perpetuations of masculinity for the establishment of the ideal warrior. Gendering by the institution determines the acceptable contours for definitions of "wounds" and "wounding," which wounds are acknowledged/ignored, which wounds are expressed/silenced, and which wounds are to be heralded by the public/masked or erased from the public. For military sexual assault in particular, the military engages in a systematic masking and erasing of victims' wounds. This erasure is primarily established through a combination of securitization, medicalization, and pathologization discourses. Securitized discourses highlight important ways in which military sexual assault is markedly different from sexual assault in the civilian sector. Moreover, both securitization and medicalization rhetoric allow the mili-

tary to redirect concern for the individual victims of sexual assault to itself as it positions the military as on the receiving end of the "plague" and "cancer" thought to be infiltrating and threatening the very livelihood of the institution. Doing so ignores the physical and psychological horrors of individuals attacked and refocuses attention to the military as a whole. It also draws attention away from whether perpetrators of sexual violence are brought to justice by the military. Wounding sustained from tours of combat that do not visibly mark the bodies of soldiers through noticeable maiming or the missing of limbs are feminized through narratives of inadequacy. Crimes committed by veterans grappling with PTSD and traumatic brain injuries are seen as a disgraceful sullying of the honor of the uniform they wear and the country they serve.

Dishonor associated with certain groups serving in the military is often met with their removal from military histories. Silenced histories, as uncovered through semiotic techniques, provide for an understanding of militarization's converging and diverging vectors of power, established primarily through traversing the daily lives of average citizens. Military museums, memorials, and ceremonies are fruitful sites for this analytic endeavor; the power to determine which histories are broadcasted and which are silenced resides with museum curators and exhibit directors who make deliberate decisions regarding the design and spatial configurations of exhibits, which often force visitors to move through the space in specific ways. As such, the military organization maintains an ability to carefully orchestrate narratives on gender and service for the general public to consume at these curated sites. War museums and memorials help ensure that war remains a male preserve in the public psyche.

Contemporary modes of online recruiting, and the materials used for recruiting on social media, seem to show a tendency to do the same. Online recruitment material in the form of Facebook posts for the four branches since the initial announcement of the rescinding of the combat exclusion policy in 2013 up until October of 2015 shows that other than one post made on the Army's Facebook recruitment page, none of the branches made mention of the repeal of the combat ban on women. In fact, depictions of women during the time studied were very limited overall. In instances when women were depicted whether alone, with other females, or with male counterparts, less than 1% of posts, across all four of the branches, depicted women in combat roles or scenarios. Similar to Brown's study of print military advertisements from 1970 to 2007, depictions of masculinity through the use of men's bodies specifically remained

the hallmark of recruitment materials. For the Air Force, service is tied to techno-masculinity, for the Army and Marines, service is tied to warrior masculinity established through front-line combat roles, and for the Navy, service is tied to adventurous masculinity often established through travel opportunities made available with the branch.

Ultimately, institutional anxieties associated with these, and other, instances of "gender trouble" produce a doubling down on reaffirming its preferred gender order. They expose the vulnerability of its continued, and adamant, understanding of gender as dyadic, a binary system that often showcases its inability to fit with the needs of the society in which it exists, demonstrating a narrow and limited ability to respond responsibly and adequately to "queer" performances of gender that do not fit the "neat" structures of the binary. True degendering, then, ultimately requires destabilizing the institution's perceived need for the binary, removing—or at the very least—"reduc[ing] the salience of dichotomous hierarchical gender norms" (Eichler 2014, 88 see also Lorber 2005), which should help in "detach[ing] soldiering from violent and aggressive notions of masculinity" (88).

As a result, to suggest that contemporary military policy changes, and even the rescinding of some of those policies, represent a period of degendering in the U.S. military's history is to provide false hope. Demands to degender the military involve higher stakes than idle calls for an innovative "social experiment." As the cases in this book make clear, the masculinist ideals embedded in military institutions and operations seep beyond the boundaries of military bases, influencing beliefs, values, fashions, interpersonal relations, and public imaginaries of appropriate gender behavior and aspiration. In manifold subtle ways, military masculinities—and the masculine warrior in particular—re-inscribe the white, heterosexual, male soldier as the embodiment of the patriot, the proper citizen, the natural leader, and the nation's honor. By comparison, women, gay, lesbian, bisexual, and gender nonconforming individuals are second class—and their continued marginalization within the military ensures that they remain as such.

Critics of militarization point out that the influence of military values is pervasive, permeating celebrations of national holidays, school textbooks, sporting events, films and television, video arcades, museums and memorials, highway names, welcome announcements on airplanes, philanthropic endeavors, clothing and jewelry, fitness programs that replicate obstacle courses, training rituals that mimic "boot camps," popular music,

and colloquial language. The gendered hierarchies that military institutions normalize and naturalize structure daily life within militarized cultures. Although the military is not the only institution that shores up heteronormativity and male domination, its pervasive presence consolidates male power and helps to embed the "legal, bureaucratic, and socially institutionalized binary gender system" (Lorber 2005, xv).

The demand to degender the military, then, is a demand to make good on the promise of equal citizenship, to dismantle the hallmarks of militarized masculinity, and to challenge the ongoing militarization of American society. It is a call to view all subjects and their lives equally, and a call to dissolve hierarchies of citizenship established at the intersections of gender, race, class, and sexual orientation. It would interrogate the societal implications of militarism that have allowed a culture of violence to thread throughout layers of society and become unimposingly natural. Degendering would require releasing the tenants of masculinity upon which the institution was built, allowing for, and, indeed, mandating "new ways of recruiting, training, and motivating soldiers that do not rely on the privileging of masculinity and denigration of femininity" (Eichler 2014, 89). It would entail erasing standards of performance that were once predicated upon the abilities of the male corpus, and reenvisioning the modalities of operation the military uses. In short, a degendering of the U.S. military would be a reimaging of how wars are conducted, along with the very foreign policy tools at the disposal of nation states. Given the centrality of the military to the perpetuation of male power and its well-honed mechanisms for regendering, degendering this institution will not be easy, but it is a worthy project for those committed to equal citizenship and gender justice.

Notes

1. Goldstein (2001) includes "ecofeminism" as part of "difference feminism" "because it begins from radical connectedness" (47).
2. In particular, see Chap. 4, "Violated Bodies: Combat Injuries and Sexual Assault in the U.S. Military."
3. In particular, see Chap. 6, "Gender and Military Recruitment Since the Lifting of the Combat Ban."
4. Daggett (2015) offers a similar argument suggesting discursive measures have attempted to secure and stabilize the "disorientation" produced by death through drones but that they have not yet been able to do so "reliably" (361).

5. See Daggett (2015) on the importance of the distance/proximity frame (referred to as distance-intimacy binary by Daggett) as one of two main axes of reference that makes war intelligible, and through which drone warfare as a form of war becomes unintelligible through its inability to "fit" traditional axes of framing.
6. Though used by Haraway's Manifesto of Cyborgs (1985/1991), I borrow this phrase to hone in on Haraway's eloquent explanation of "fracturedness." Unintelligible matters, such as the atrocities associated with war, are arguably unintelligible because they are fractured, requiring calibrated parameters or boundaries, or "axes" as Daggett (2015) writes, through which they are discursively made to make sense.
7. It is also worth considering how other forms of militarized masculinity fare if drone warfare has difficulty calibrating with the military's hegemonic version of militarized masculinity. Does drone warfare bring forth the possibility of another form of military service, one which rejects gendered conceptualization of service from ascending above the masculine warrior through the relationship it produces between the soldier and machine? The previous chapter, for example, suggests that the Air Force, in particular, has been heavily promoting and advancing a technology-driven depiction of military service in its social media recruitment material. In theory, technological developments should give us cause for optimism. Cyborgs—as originally imagined by Haraway (1985/1991)—may provide the possibility for removing gender, or rendering gender irrelevant from warfare all together. But Manjikian (2014) argues that military technology and narratives are building "super soldiers" instead of cyborgs. Through this narrative, drone operators are positioned as "controlling technology" as opposed to being controlled or dependent on it. Instead, the relationship between the soldier and technology mimics narratives associated with "ways in which man has traditionally utilized and subdued nature and thus, in this way reinforcing a binary distinction between that which dominates and that which is dominated" (Manjikian 2014, 56 referencing Kaplan 1994). Perhaps not surprisingly then, my assessment of Air Force recruiting posts on Facebook echoes a favoritism for positioning techno-based military service as that which is most suitable for male recruits and which can be mapped onto understandings of masculinity. In particular, the technocentric messages produced and disseminated by the branch work diligently to disassociate notions of military valor, honor, and prestige as being that which can primarily be achieved as a result of sacrifices (bodily maiming, injury, or death) made in or in proximity to combat environments by recalibrating (acceptable) performances of masculinity as being those which display control over robotic technologies that enter combat war theaters on their behalf. Militarized masculinity conducted from afar, as a way to try to advance claims that valor and prestige can be achieved without necessitating the risk of loss of limbs or bodily mutilation from serving in combat.

8. Though used with increased frequency, drone warfare should not be thought of as new. As a tool of foreign policy, the United States has used contemporary versions of unarmed drones since 2000 specifically in relation to the surveillance of the nation of Afghanistan, and weaponized unmanned drones since the 9/11 attacks on the United States, and the resulting, subsequent policies of the George W. Bush administration.
9. The Air Force has, however, worked toward providing mental health support to drone operators since 2017. In particular, the branch has brought in physicians and psychologists to work together with chaplains counseling operators. See McCammon (2017).
10. In February of 2013, then Secretary of Defense, Leon Panetta, created the "Distinguished Warfare Medal"—more colloquially known as the "Nintendo Medal"—to honor service members participating in combat operations from remote stations. The medal was particularly favored by the Air Force, the primary military branch to engage in and operate drone and other autonomous aerial forms of warfare for the military. The Air Force hoped that recognizing the service of drone operators through the medal program would help retain them despite serious "burn out" rates. Instead, the medal was eradicated just a few weeks later at the request of the new, incoming Secretary of Defense, Chuck Hagel. After a subsequent two-year review, the Pentagon rolled out a new form of recognition for drone operations, a quarter-inch sized "R" ("Remote") pin that could be added to any non-combat related award medal (Tilghman 2016).

References

Acker, J. 1992. From Sex Roles to Gendered Institutions. *Contemporary Sociology* 21 (5): 565–569.

Bayard de Volo, L. 2016. Unmanned? Gender Recalibrations and the Rise of Drone Warfare. *Politics & Gender* 12 (1): 50–77.

Belkin, A. 2012. *Bring Me Men: Military Masculinity and the Benign Façade of American Empire, 1898–2001*. New York: Columbia University Press.

Brown, M.T. 2012. *Enlisting Masculinity: Gender and Recruitment of the All-Volunteer Force*. Oxford: Oxford University Press.

Burke, C. 2002. Women and Militarism. Accessed February 20, 2006. http://wilpf.smilla.li/wp content/uploads/2012/10/Unknownyear_Women_and_Militarism.pdf.

Clark, L.C. 2018. Grim Reapers: Ghostly Narratives of Masculinity and Killing in Drone Warfare. *International Feminist Journal of Politics* 20 (4): 602–623. Published Online.

Cohn, C. 2000. 'How Can She Claim Equal Rights When She Doesn't Have to Do as Many Pushups as I Do?' The Framing of Men's Opposition to Women's Equality in the Military? *Men and Masculinities* 3 (2): 131–151.

Connell, R.W. 1987. *Gender and Power: Society, the Person, and Sexual Politics*. Stanford: Stanford University Press.

Daggett, C. 2015. Drone Disorientations: How 'Unmanned' Weapons Queer the Experience of Killing in War. *International Feminist Journal of Politics* 17 (3): 361–379.

Department of Defense. 2019. DoD Personnel, Workforce Reports & Publications – Table of Active Duty Females by Rank/Grade and Service. Published April 2019. Accessed June 27, 2019. https://www.dmdc.osd.mil/appj/dwp/dwp_reports.jsp.

Duncanson, C., and R. Woodward. 2015. Regendering the Military: Theorizing Women's Military Participation. *Security Dialogue* 47 (1): 3–21.

Eichler, M. 2014. Militarized Masculinities in International Relations. *Brown Journal of World Affairs* 21 (1): 81–93.

Enloe, C. 1983. *Does Khaki Become You? The Militarization of Women's Lives*. Boston: South End Press.

———. 1990. *Bananas, Beaches, and Bases: Making Feminist Sense of International Politics*. Berkeley and Los Angeles: University of California Press.

———. 2000. *Maneuvers: The International Politics of Militarizing Women's Lives*. Berkeley: University of California Press.

———. 2002. *Maneuvers: The International Politics of Militarizing Women's Lives*. Berkeley: University of California Press.

Ferguson, K., and P. Turnbull. 1999. *Oh, Say, Can You See? The Semiotics of the Military in Hawai'i*. Minnesota: University of Minnesota Press.

Francke, L.B. 1997. *Ground Zero: The Gender Wars in the Military*. New York: Simon and Schuster.

Goldstein, J. 2001. *War and Gender: How Gender Shapes the War System and Vice Versa*. Cambridge: Cambridge University Press.

Greenwald, G. 2012. Bravery and Drone Pilots. *Salon*. Accessed September 20, 2016.

Haraway, D.J. 1985/1991. A Cyborg Manifesto: Science, Technology, and Socialist-Feminism in the Late 20th Century. In *Simians, Cyborgs and Women: The Reinvention of Nature*, 149–181. New York: Routledge.

———. 2017. *Staying with the Trouble: Making Kin in the Chthulucene*. Old Saybrook, CT: Tantor Media.

Holmqvist, C. 2013. Undoing War: War Ontologies and the Materiality of Drone Warfare. *Millennium – Journal of International Studies* 41 (3): 535–552.

Hooper, C. 2001. *Manly States: Masculinities, International Relations, and Gender Politics*. New York: Columbia University Press.

Kaplan, L.D. 1994. Woman as Caretaker: An Archetype that Supports Patriarchal Militarism. *Hypatia* 9 (2): 123–133.

Kronsell, A. 2012. *Gender, Sex and the Postnational Defense: Militarism and Peacekeeping*. New York: Oxford University Press.

Kunashakaran, S. 2016. Un(wo)manned Aerial Vehicles: An Assessment of How Unmanned Aerial Vehicles Influence Masculinity in the Conflict Arena. *Contemporary Security Policy* 37 (1): 31–61.

Levy, Y. 1997. How Militarization Drives Political Control of the Military: The Case of Israel. *Political Power and Social Theory* 11: 103–133.

Lorber, J. 2000. Using Gender to Undo Gender: A Feminist Degendering Movement. *Feminist Theory* 1 (1): 79–95.

———. 2005. *Breaking the Bowls: Degendering and Feminist Change*. New York: W. W. Norton & Company.

Lutz, C. 2002. Making War at Home in the United States: Militarization and the Current Crisis. American Anthropologist 104 (2): 723–735.

Manjikian, M. 2014. Becoming Unmanned: The Gendering of Lethal Autonomous Warfare Technology. *International Feminist Journal of Politics* 16 (1): 48–65.

Martin, Rachel. 2011, December 18. Report: High Levels of 'Burnout' in U.S. Drone Pilots. NPR. Accessed March 21, 2019. https://www.npr.org/2011/12/19/143926857/report-high-levels-of-burnout-in-u-s-drone-pilots.

McCammon, S. 2017, April 24. The Warfare May Be Remote But the Trauma Is Real. NPR. Accessed March 27, 2019. https://www.npr.org/2017/04/24/525413427/for-drone-pilots-warfare-may-be-remote-but-the-trauma-is-real.

Press, Eyal. 2018. The Wounds of the Drone Warrior. *The New York Times*. Accessed March 27, 2019. https://www.nytimes.com/2018/06/13/magazine/veterans-ptsd-drone-warrior-wounds.html.

Tilghman, A. 2016, January 6. DoD Rejects 'Nintendo Medal' for Drone Pilots and Cyber Warrior. *Military Times*. Accessed February 15, 2019. https://www.militarytimes.com/2016/01/06/dod-rejects-nintendo-medal-for-drone-pilots-and-cyber-warriors/.

UN Women. 2016. Gender Mainstreaming. Accessed February 15, 2019. http://www.un.org/womenwatch/osagi/gendermainstreaming.htm.

Wilcox, L. 2017. Drones, Swarms, and Becoming-Insect: Feminist Utopias and Posthuman Politics. *Feminist Review* 116 (1): 25–45.

Index[1]

A

Air Force, 1, 2, 7, 21, 31, 32, 47n1, 64, 68, 69, 83n5, 92, 99, 102, 150, 151, 155–157, 163, 165, 166, 168–170, 171n4, 180, 187, 194, 196n7, 197n9

All-Volunteer Force (AVF), 15, 18, 21, 47, 53–76, 149–151, 176, 177, 180, 191

Army, 1, 7, 21, 30–32, 62, 64, 65, 67–70, 82n2, 83n5, 85n13, 92, 93, 99, 102, 129, 147–150, 153, 157–160, 165, 166, 170, 171n4, 180, 193, 194

Army Nurse Corps, 30

Army Reorganization Act, 30

B

Band of Brothers, 55–58, 75, 76, 89, 135

Brown, Melissa T., viii, 4, 19, 21, 148, 150, 151, 153, 164–166, 169, 170, 180, 191, 193

Butler, Judith, 8, 22n4, 23n8

C

Combat, 1, 2, 4, 10, 13, 16, 19, 20, 31, 32, 34–37, 41, 47n1, 53–76, 89–113, 122–124, 126, 133, 138, 139, 141, 142, 147–170, 177, 179–181, 186–189, 191–194, 196n7

Combat ban on women, 14, 15, 21, 170, 193

Content analysis, 151, 152, 167, 170, 171n4, 180

D

Degendering, 3, 6, 11, 15, 16, 23n6, 54, 74, 124, 165–170, 175–195

[1] Note: Page numbers followed by 'n' refer to notes.

Discourse
  medicalization, 13, 20, 111, 177, 192
  pathologization, 192
  securitization, 20, 98, 111, 177
"Don't Ask, Don't Tell," 2, 14, 39

E
Eichler, Maya, viii, 14, 15, 24n14, 37, 41–43, 89, 194, 195
Enloe, Cynthia, vii, viii, 4, 7, 33, 35–37, 43, 44, 56, 93, 94, 114n6, 191
Exclusion, 2, 7, 10, 19, 21, 32, 39, 54–56, 60, 62, 64, 68, 72, 75, 82–83n4, 85n14, 90, 101–108, 125, 128–130, 177, 191, 193
  of women, 55, 57

F
Feminization/feminized, 36, 58, 59, 94, 107, 111, 112, 113n4, 140, 179, 184, 187, 193

G
The Gates Commission, 53, 82n1, 82n2
Gender
  balance, 176, 179
  binary understandings of, 177
  bodies, 10, 19, 176
  deviance, 9
  diversity, 9, 176, 180
  equality, 3, 6, 13, 19, 47, 75, 170, 177, 178, 187
  freedom, 176, 180
  hierarchies, 20, 123, 139, 195
  inclusion, 10, 157–160, 178, 179
  institution/institutions, 5
  integration, 1, 2, 19, 21, 37, 54, 56
  mainstreaming, 176, 177
  mechanisms of exclusion, 7, 55, 91
  nonconformity, 4, 8, 181
  normals, 176
  order, 4, 54, 111, 119, 120, 125, 152, 179, 191, 194
    as an organizing principle, 178
  parity, 176, 178, 179
  performances, 17, 176
  performativity, 8, 22n4
  perspectives, 177
  regime, 3, 6, 8, 9, 14–18, 29–47, 121, 181–184, 190, 191
  relations, 3, 9, 46, 47, 54, 57, 121, 183
  (re)production, 8
  trouble, 2, 3, 7, 8, 15, 17, 22n4, 29, 30, 90, 111, 119, 125, 175, 181, 183, 184, 190, 191, 194
  undo, 175, 176, 181
  visibility, 176, 179

H
Hegemonic militarized masculinity, 11, 13, 37, 40–46, 89
Heteronormative masculinity, 4, 8, 185
Homo economicus, 56–58, 71, 177, 191

I
Inclusion, 1, 3, 4, 7, 10, 12, 13, 32, 54, 55, 57, 75, 119, 124, 176, 177, 179, 180, 191
  gender, 10, 157–160, 178, 179
"Incremental and deliberate," 19, 54, 177, 191
Injury, 4, 19, 20, 73, 74, 89–113, 139, 140, 153, 179, 187–189, 192, 196n7
Intersectional/intersectionality, 17, 24n15
The Invisible War, 101–108

M

MacKenzie, Megan, viii, 4, 56, 59, 74, 83n4, 92, 152
Marine Corps, 1, 2, 102–104, 108, 150, 153, 160–162, 165, 170, 171n4
Marines, 2, 7, 21, 22n2, 31, 32, 47n1, 72, 83n5, 99, 103, 135, 160–163, 166, 180, 194
Masculinities/masculinity, 2–4, 6, 7, 9–16, 18, 23n8, 24n13, 24n14, 36, 37, 40–43, 47, 48n6, 53, 56, 57, 59, 62, 75, 90, 114n4, 126, 128–131, 133, 140, 149–151, 164, 166, 169, 170, 180, 183, 184, 192–195, 196n7
Militarism, 10, 12, 14, 18, 24n14, 29, 41–43, 46, 130, 176, 183, 190, 195
Militarization, 10, 12, 18, 23n9, 42–46, 48n6, 48n7, 123, 127–130, 138, 142, 176, 183, 191, 193–195
Militarized femininities/femininity, 7, 14, 15, 148
Militarized masculinities/masculinity, 3, 7, 11, 13–17, 21, 24n13, 29–47, 89, 111, 138, 148, 170, 181–190, 195, 196n7
Military/militaries
  as *homo economicus*, 57, 71, 191
  injuring/injury, 19, 73, 89–113, 179
  as a rational actor, 57, 59, 60, 177, 191
  self-representations, 5, 21, 44, 56, 123, 148
  sexual assault, 72–73, 89–113, 128, 177, 179, 192
  wounding/wounds, 20, 89–113, 193
Military Justice Improvement Act, 179
Military sexual assault (MSA), 72–73, 90, 92, 93, 96, 102, 107, 114n5, 114n7, 128, 177, 179, 192

N

Navy, 2, 7, 20, 21, 30–33, 61, 69, 72, 82n3, 83n5, 85n11, 99, 121, 122, 135, 140, 150, 151, 163–168, 170, 171n4, 180, 194
Navy Nurse Corps, 30

P

Post-Traumatic Stress Disorder (PTSD), 74, 90, 107, 108, 110, 112, 179, 187, 188, 193

R

Regendering, 3, 4, 6, 11, 13, 16, 18, 23n6, 70, 165–170, 175, 181, 190–195

S

Semiotic/Semiotics, 20, 23n10, 120, 121, 123–125, 130, 142, 180, 193
Sexual assault
  as a cancer, 99, 100, 111, 112, 178, 193
  epidemic, 177
  in the military, 4, 72, 73, 97, 111, 112, 113n2, 178
  as a plague, 99, 100, 112, 178, 193
  sexual violence, 19, 95, 98–101, 112
  victims, 103, 108, 179
Sexual violence, 2, 19, 71, 73, 90, 91, 94–105, 107, 108, 111, 112, 178, 193

T

Techno-masculinity, 180, 194
Traumatic Brain Injury/Injuries, 108, 110, 112, 193

U

USS Midway Museum, 20, 121, 122, 130–132, 136–139, 141, 142, 143n4, 144n7
  See also USS Midway Aircraft Carrier

W

War
  as statecraft, 43
  war wounds, 19, 47n1, 90, 91, 101, 102, 111, 113, 188
Women
  access, 19, 59, 72, 76, 83n4
  in auxiliary positions, 31, 47n1
  bodies, 2, 14, 19, 24n14, 55, 57, 60, 71, 84n8, 177, 192
  in combat, 3, 12, 56–61, 68, 72, 74, 76, 85n14, 143, 148, 158, 161, 163, 179, 187, 189, 193
  combat ban on, 14, 15, 21, 153, 170, 187, 193
  enlisted women, 32, 33, 38, 68, 150
  exclusion, 54–56, 64, 177, 191
  exist in relation to men, 37
  as incompatible, 17, 19, 54, 57, 76
  as inferior to men, 176
  integration of, 57, 60, 62, 67, 73
  as leaky bodies, 55, 57, 60, 71, 177, 192
  in the military, 16, 29, 38, 46, 62, 64, 68, 71, 73, 74, 83n5, 84n11, 86n18, 94, 119–143, 148, 152, 166, 176
  military corps, 31
  military occupations for, 178
  military service, 19, 20, 30, 55, 56, 70
  official inclusion, 21
  participation in the military, 18, 84n10
  participation in war, 20
  protection of, 34
  recruitment of, 19, 67, 75
  representation of, 17
  service, 15, 20, 21, 30, 31, 33, 35, 54, 60, 73, 74, 161
  service members, 1, 86n18, 93, 180
  troops, 38, 74, 102, 179, 180
  underrepresentation of, 33, 46, 179
  war-time service, 31, 126, 127
  weakness, 7, 18, 21, 181
Women's Armed Services Integration Act, 31
Women's Army Corps, 31, 32
Women's Auxiliary Army Corps (WAAC), 31
The Wounded Platoon, 109, 110
Wounding, 16, 76, 90, 91, 95, 111, 113, 126, 179, 185, 192, 193

CPSIA information can be obtained
at www.ICGtesting.com
Printed in the USA
LVHW081910250621
691162LV00009B/517

DICTIONNAIRE
DES
JARDINIERS
ET DES
CULTIVATEURS,
PAR
PHILIPPE MILLER.

TOME CINQUIEME.

DICTIONNAIRE
DES
JARDINIERS
ET DES
CULTIVATEURS,
PAR
PHILIPPE MILLER:
Traduit de l'Anglois fur la VIIIe. Edition;

Avec un grand nombre d'Additions de différens genres ;
Par MM. le Préfident DE CHAZELLES,
le Confeiller HOLANDRE, &c.
NOUVELLE ÉDITION,
*Dans laquelle on a rectifié un très-grand nombre d'endroits de
l'Édition de Paris, afin de rendre la Traduction Françoife
conforme à l'Original Anglois ; & de plus, on y a ajouté les
noms Anglois des Plantes, & plufieurs nouvelles Notes.*

TOME CINQUIEME.

A BRUXELLES;
Chez BENOIT LE FRANCQ, Imprimeur-Libraire ;
rue de la Magdelaine.

M. DCC. LXXXVIII.

DICTIONNAIRE

DES

JARDINIERS.

MEA

MEADIA. *Catesb. Carol.* 3. P. 1. *Dodecatheon. Lin. Gen. Plant.* 183. [*Virginian Bears-Ear.*] Oreille d'ours de Virginie.

Caractères. Cette plante a une petite enveloppe de plusieurs feuilles, qui renferme plusieurs fleurs; chacune de ces fleurs a un calice persistant, & formé par une feuille découpée en cinq segmens longs & réfléchis; la corolle est monopétale, & divisée en cinq parties; son tube est plus court que le calice, & son sommet réfléchi en arrière : cette fleur a cinq étamines courtes, obtuses, postées dans le tube, & terminées par des anthères à pointe de flèche, qui sont jointes en un bec, avec un germe conique, qui soutient un style mince plus long que les étamines, & couronné par un stigmat obtus : le calice se change dans la suite en une

Tome V.

MEA

capsule ovale, oblongue, & a une cellule qui s'ouvre au sommet, & qui contient beaucoup de petites semences.

Ce genre de plante est rangé dans la première section de la cinquieme classe de LINNÉE, qui renferme celles dont les fleurs ont cinq étamines & un style.

Le titre de ce genre lui a été donné par M. MARC CATESBY, Membre de la Société Royale en l'honneur du Docteur MEAD, à qui les Sciences ont beaucoup d'obligations; mais comme il n'étoit pas un grand Botaniste, LINNÉE n'a pas voulu qu'aucune plante portât son nom : en conséquence, il l'a changé en celui de *Dodecatheon*, nom donné par PLINE à une espece de *Primevere*, qui a des racines jaunes & des feuilles à-peu-près semblables à celles de la *Laitue* de jardin.

A

MEA

Nous n'avons qu'une espece de ce genre, qui est la

Meadia (Dodecathcon.) *Catesb. Hist. Carol. app. 1. tab. Trew. Ehret t. 12.*

Meadia, Auricula ursi Virginiana, floribus Boraginis instar rostratis, Cyclaminum more reflexis. Pluk. Alm. 62. tab. 79. fol. 6; Oreille d'ours de Virginie, dont la fleur a un bec comme celle de la Bourache, & des pétales réfléchis comme ceux du Cyclamen.

DODECATHEON ou *Oreille d'ours de Virginie. Linn. Syst. Plant. t. 1. pag. 414.*

Cette plante croît naturellement dans la Virginie & dans quelques autres parties de l'Amérique septentrionale, d'où elle a été envoyée par M. BANISTER, il y a plusieurs années, au Docteur COMPTON, Evêque de Londres, dans le jardin duquel j'ai vu cette plante pour la première fois en l'année 1709 : elle a péri ensuite, & cette espece a été perdue pour l'Angleterre : mais on en a reçu d'autres quelques tems après, & on l'a beaucoup multipliée.

Cette plante a une racine jaune & vivace, de laquelle sortent au printems plusieurs feuilles d'environ six pouces de longueur, sur deux & demi de largeur, d'abord érigées, mais qui se couchent ensuite sur la terre, sur-tout si les plantes sont fort exposées au soleil : du centre de ces feuilles sortent, suivant la force des racines, deux, trois ou quatre tiges de huit ou neuf pouces de hauteur, lisses, nues, & terminées par une ombelle de fleurs, sous laquelle est située une enveloppe à plusieurs feuilles : chaque fleur est soutenue par un pédoncule long, mince, & recourbé de maniere qu'elle pend vers le bas. Elle est monopétale ; sa corolle est profondément découpée en cinq segmens en forme de lame, & réfléchis vers le haut comme ceux du *Cyclamen* ou *Pain de pourceau* ; elle a cinq étamines courtes placées dans le tube, & surmontées par des antheres à pointe de flèche, & jointes ensemble autour du style, qui forme une espece de bec. Ces fleurs sont d'une couleur pourpre tirant sur celle de fleur de *Pêcher* ; elles ont un germe oblong, & placé dans le fond du tube, qui se change ensuite en une capsule ovale, renfermée dans le calice, & sur l'extrémité de laquelle le style reste fixé. Cette capsule s'ouvre au sommet, lors de sa maturité, & laisse sortir les semences qui sont attachées au style. Cette plante fleurit au commencement du mois de Mai, ses semences mûrissent en Juillet, & bientôt après les tiges & les feuilles périssent ; de sorte que les racines restent dans l'inaction jusqu'au printems suivant.

Culture. Cette plante se multiplie par les rejettons que sa racine pousse assez librement, quand elle se trouve dans un sol léger, humide, & à l'ombre. Le meilleur tems pour enlever les racines & détacher les rejettons, est le mois d'Août, après que les feuilles & les tiges sont flétries, afin qu'elles

MEA

puiffent être bien établies avant l'approche des gelées : on peut auffi la multiplier par fes graines qu'elle produit en abondance : on les répand auffi-tôt qu'elles font mûres, fur une plate-bande humide & à l'ombre ; les plantes poufferont au printems, alors on les tiendra conftamment nettes : on les arrofera, fi le tems eft fec, & on ne les expofera pas au foleil : car tandis que ces plantes font jeunes, elles font fort fenfibles à la chaleur, & j'en ai vu périr un grand nombre en deux ou trois jours, parce qu'elles croiffoient au plein foleil. On ne doit pas les transplanter avant que leurs feuilles foient détruites ; mais alors on peut les enlever avec précaution, & les planter dans des plates-bandes à l'ombre, dont le fol foit humide & defferré : on les place à huit pouces environ de diftance entr'elles, ce qui leur fuffira, parce qu'elles ne doivent y refter qu'une année : au bout de ce tems, lorfqu'elles feront affez fortes pour fleurir, on pourra les transplanter dans les plates-bandes du parterre, à l'ombre, où elles feront un bel ornement, tant quelles feront couvertes de fleurs.

On a d'abord regardé cette plante comme délicate, & en conféquence on la plaçoit dans des fituations chaudes. ce qui l'a fait fouvent périr ; mais l'expérience a fait connoître depuis, qu'elle eft affez dure pour n'être point endommagée par les froids les plus rigoureux de ce pays, & qu'elle ne

MED

réuffit point dans un fol fort fec & trop expofé au foleil.

MEDEOLA. *Lin* (en. Plant. 411. [*Climbing African Afparagus.*]

Caractères. La fleur n'a point de calice ; la corolle a fix pétales oblongs, ovales, égaux, étendus & tournés en arriere ; la fleur à fix étamines auffi longues que la corolle, & terminées par des antheres courbées, & trois germes cornés, qui fe terminent en ftyles couronnés par des ftigmats recourbés : ces germes fe changent dans la fuite en une bande ronde, divifée en trois parties, & à trois cellules, qui contiennent chacune une femence en forme de cœur.

Ce genre de plantes eft rangé dans la troifieme fection de la fixieme claffe de LINNÉE, qui renferme celles dont les fleurs ont fix étamines & trois ftyles.

Les efpeces font :

1°. *Medeola Afparagoïdes, foliis ovato lanceolatis alternis, caule fcandente ;* Afperge d'Afrique, avec des feuilles ovales, en forme de lance, alternes, & ayant une tige grimpante.

Afparagus Africanus fcandens, Myrti folio. Hort. Pifs. 17 ; Afperge grimpante d'Afrique à feuilles de Myrte.

Laurus Alexandrina ramofa, foliis è fummitate caulium prodeuntibus. Herm. Lugd. B. 679. *f.* 631.

2°. *Medeola angufti folia, foliis lanceolatis alternis, caule fcandente ;* Medeola avec des feuilles en forme de lance, & alternes, & une tige grimpante,

A 2

MED

Asparagus Africanus, *Myrti folio angustiori*. *Hort. Piss.* 17 ; Asperge grimpante d'Afrique, avec une feuille de myrte plus étroite.

3°. *Medeola Virginiana*, *foliis verticillatis*, *ramis inermibus*. *Lin. Sp. Plant.* 339 ; Medeola avec des feuilles verticillées & des branches unies.

Medeola foliis stellatis, *lanceolatis*, *fructu baccato*. *Gron. Virg.* 39.

Lilium, *sive Martagon pusillum*, *floribus minutissimè herbaceis*. *Pluk. Alm.* 410. *tab.* 328, *fol.* 4 ; le Lys ou petit Martagon, avec de fort petites fleurs herbacées.

Asparagoïdes. La premiere espece croît naturellement au Cap de Bonne-Espérance ; elle a une racine composée de plusieurs bulbes ou nœuds oblongs, qui se joignent au sommet comme ceux des Renoncules : de ces racines sortent deux ou trois tiges fermes & sarmenteuses, qui se divisent en branches, & s'élevent à quatre ou cinq pieds de hauteur, si elles rencontrent quelque soutien, ou qu'elles puissent s'attacher, sans quoi elles rampent sur la terre. Ces tiges sont garnies de feuilles ovales, en forme de lance, terminées en pointe aiguë, alternes, sessiles, d'un vert clair en-dessous, & foncé au-dessus ; ses fleurs naissent sur les côtés des tiges, quelquefois simples, & d'autrefois au nombre de deux, sur un pedoncule mince & court : elles sont composées de six pétales égaux, étendus, & d'un blanc sale ; & de six étamines aussi longues que la corolle, & terminées par

des antheres inclinées. Dans leur centre est placé un germe à trois cornes, porté sur un style court, & couronné par trois stigmats épais & recourbés. Ce germe se change en une baie ronde & à trois cellules, qui renferment chacune une semence en forme de cœur. Cette plante fleurit au commencement de l'hiver, & ses semences mûrissent dans le printems.

Angusti-folia. La seconde espece, qui est aussi originaire du Cap de Bonne-Espérance, d'où ses semences m'ont été envoyées, a une racine semblable à celle de la premiere ; mais ses tiges sont moins grosses, plus élevées, & moins divisées en branches : ses feuilles sont plus longues, plus étroites, & d'une couleur grisâtre ; ses fleurs sortent des parties latérales des branches au nombre de deux ou trois sur chaque pédoncule ; elles sont d'un blanc herbacé, de la même forme que celles de l'espece précédente, & elles paroissent dans le même tems : mais celle-ci n'a point encore produit de fruits dans ce pays. Comme elle ne varie jamais en la multipliant de semences, on ne peut douter qu'elle ne soit une espece distincte.

On multiplie ces deux especes par les rejettons de leurs racines, de maniere qu'on peut se passer de leurs graines, quand on en possede une fois quelques plantes : d'ailleurs ces semences restent ordinairement long tems dans la terre, & les plantes qu'elles produisent ne fleurissent qu'au bout de deux

MED

áns, au-lieu que les rejettons donnent des fleurs dans l'année suivante. Le tems le plus propre pour transplanter ces racines, est le mois de Juillet, lorsque leurs tiges sont entiérement flétries, parce qu'elles commencent à pousser vers la fin d'Août, continuent à croitre tout l'hiver, & se flétrissent au printems. Ces racines doivent être plantées dans des pots remplis d'une bonne terre de jardin potager, & peuvent rester en plein air jusqu'aux fortes gelées; alors on les porte dans un endroit abrité, parce qu'elles sont trop délicates pour pouvoir résister sans abri aux froids de nos hivers. Si on les place dans une bonne serre, elles profiteront & fleuriront très-bien; mais elles ne produiront point de fruits, à moins qu'elles ne soient dans une serre de chaleur tempérée. Pendant l'hiver, lorsque ces plantes sont en vigueur, il faut les arroser fréquemment & légerement; mais lorsque leurs tiges commencent à se flétrir, on leur donne très-peu d'humidité, sans quoi elles pourriroient, parce qu'alors elles sont dans un état d'inaction : pendant ce tems, on les place de façon qu'elles jouissent du soleil du matin, & on ne leur donne que très-peu ou point d'eau; mais lorsque leurs tiges poussent, on les remet à une exposition chaude, & on les arrose souvent & légerement.

Les fleurs de ces plantes n'ont pas grande apparence, on ne les cultive pas pour leur beauté; mais comme leurs tiges sont

MED 5

grimpantes, & leurs feuilles vigoureuses en hiver, elles augmentent la variété dans la serre.

Virginiana. La troisieme espece est originaire de l'Amérique septentrionale. LINNÉE l'a réunie à ce genre où je l'ai laissée moi-même, quoique, si je m'en souviens bien, ses caractéres ne s'accordent pas exactement avec ceux des autres; car sa fleur n'est ni polypétale ni découpée en beaucoup de segmens, mais elle a seulement cinq étamines. je ne puis cependant assurer ce que j'avance, parce que je n'ai point vu cette plante depuis quelques années : celle-ci a une racine foible & écailleuse, de laquelle sort une simple tige de huit pouces environ de hauteur, & garnie de feuilles en-spirale à une petite distance de la terre; mais au sommet, il y a deux feuilles opposées, entre lesquelles naissent trois foibles pédoncules inclinés vers le bas, & qui soutiennent chacun une fleur d'une couleur pâle & herbacée avec une pointe de pourpre : elles paroissent en Juin, mais je n'ai jamais vu leurs fruits.

Cette plante est assez dure pour rester en plein air; mais elle ne se multiplie pas beaucoup ici. Comme elle ne produit point de semences en Europe; on ne peut la propager que par ses rejettons.

MEDICA. *Tourn. Inst. R. H.* 410. *Tab.* 231. *Medicago. Tourn. Inst.* 412. *Lin. Gen. Plant.* 805. Cette plante prend le nom de *Medica* parce que, suivant PLINE quand DARIUS fils D'HYS-,

A 3

MED

TASPE amena son armée en Grece, il avoit avec lui une grande quantité de graines de cette espece, qu'il fit semer pour nourrir son bétail ; ce qui a répandu cette plante dans la Grece. [*Medick or Lucerne.*] Luserne.

Caracteres. Le Calice de la fleur est en cloche, & formé par une feuille découpée en cinq pointes égales ; la corolle est papilionnacée ; l'étendart est ovale, entier, & son bord est réfléchi ; les deux ailes sont oblongues, ovales, & fixées par un appendice à la carène, qui est oblongue, & divisée en trois parties obtuses & réfléchies vers l'étendard ; la fleur a dix étamines, dont neuf sont jointes presque jusqu'à leur sommet ; son germe est oblong, comprimé, recourbé, posté sur un style court, & couronné par un petit menat, qui est, ainsi que les étamines, enveloppé par la carène & l'étendard. Le germe se change, quand la fleur est passée en un légume comprimé, & en forme de croissant, qui renferme plusieurs semences en forme de rein.

Ce genre de plantes est rangé dans la troisieme section de la dix-septieme classe de LINNÉE, qui renferme celles à fleurs papilionnacées, qui ont dix étamines, divisées en deux corps : il a joint aussi le *Medicago* de TOURNEFORT à ce genre, & n'en a fait qu'un seul sous le titre de MEDICAGO. Mais TOURNEFORT distingue le caractere du *Medicago* & du *Medica*, en ce que

l'enveloppe du légume de ce dernier est comprimée & recourbée, & que celle du *Medicago* est torse comme une vis. Le titre de *Medica* ayant été anciennement appliqué à la Luserne, je l'étendrai aux especes dont les légumes sont semblables : je renverrai les autres au genre du *Medicago*.

1°. *Medica sativa, pedunculis racemosis, leguminibus contortis, caule erecto, glabro.* Lin. Sp. 1096. *Hort. Cliff.* 377. Roy. *Lugd.-B.* 281. *Crantz. Austr.* pag. 434. *Neck. Gallob. p.* 317. *Pollich. Pal. n.* 712. *Pal. it.* 1. *p.* 370. *Kniph. cent.* 8, *n.* 67. *sub Medicago.* Luserne avec des pédoncules branchus, des légumes tordus, une tige lisse & droite.

Medicago sativa. Lin. Syst. Plant. *t.* 3. *p.* 574. *Sp.* 5.

Medica major, erectior, floribus purpurascentibus. J. B. 2. 382; la plus grande espece de Luserne à fleurs pourpre, communément appelé Luserne ; & par les François, Foin de Bourgogne.

Fœnum Burgundicum. lob. ic. 2. *p.* 36.

2°. *Medica falcata, pedunculis racemosis, leguminibus lunatis, caule prostrato.* Flor. Suec. 620. 677. *Dalib. Paris.* 229. *Crantz. Austr. p.* 434. *Neck. Gallob.* 317. *Pollich. pal. n.* 713. *Kniph. cent.* 11. *p.* 67. *Fl. Dan. t.* 233. *sub Medicago* ; Luserne avec des pédoncules branchus, des légumes en forme de lune, & des tiges trainantes.

Medica sylvestris, floribus croceis. J. B. 2. 383 ; Luserne sau-

MED

vage, à fleurs couleur de safran.

Medicago falcata. Lin. Syft. Plant. t. 3 p. 574. Sp. 6.

Trifolium fylveftre luteum, filiquâ curvatâ. Bauh. Pin. 330. Falcata. Riv. t. 84.

Lens major repens. Tabern. p. 502 Hall. R.

Medica flavo flore. Clus. Hift. 2. p. 243.

3°. *Medica radiata, leguminibus reni-formibus, margine dentatis, foliis ternatis. Hort. Cliff. 377. Hort. Ups. 230. Roy. Lugd.-B. 381. Gron. Orient. 231. Kniph. cent. 10. n. 58. fub Medicago ;* Luferne avec des légumes en forme de rein, dentés au bord, & des feuilles à trois lobes.

Medicago radiata. Lin. Syft. Plant. t. 3. p. 573. Sp. 3.

Trifolium filiquâ falcatâ. Bauh. Pin. 330.

Medicago annua, trifolii facie. Tourn. Inft. R. H. 412. Luferne annuelle, qui reffemble au Trefle.

Lunaria radiata Italorum. Lob. ic. 2. p. 38.

4°. *Medica Hifpanica, caule herbaceo procumbente, foliis pinnatis, leguminibus ciliato-dentatis ;* Luferne à tige trainante & herbacée, avec des feuilles ailées, & des légumes à dents garnies de poils.

Medicago Vulnerariæ facie Hifpanica. Tourn. Inft. R. H. 412 ; Luferne d'Efpagne, qui reffemble à la *Vulnéraire*, appelée par les Anglois *Doigt de Dame*.

5°. *Medica Italica, caule herbaceo proftrato, foliis ternatis, foliolis cunei-formibus, fupernè ferratis, leguminibus margine integerrimis ;* Luferne avec une ti-

MED 7

ge herbacée & couchée fur terre, des feuilles à trois lobes en forme de coin, & fciés à l'extrémité, & des légumes dont les bords font entiers.

6°. *Medica Cretica, caule herbaceo proftrato, foliis radicalibus integerrimis, caulinis, pinnatis, leguminibus dentatis ;* Luferne avec une tige herbacée & couchée fur terre, des feuilles radicales entieres, celles de la tige ailées, & les légumes dentés.

Luferne de Candie, ayant l'apparence de *Vulnéraire* ou *Doigt des Dames*.

7°. *Medica arborea, leguminibus lunatis, margine integerrimis, caule arboreo. Hort. Cliff. 376. Hort. Ups. 230. Roy. Lugd.-B. 382. Kniph. cent. 5. n. 55. fub Medicago ;* Luferne avec des légumes en forme de lune, dont les bords font entiers, & une tige d'arbre.

Medicago trifolia, frutefcens, incana. Tourn. Inft. R. H. 412 ; Luferne en arbre, velue & à trois feuilles, ou le *Cytifus Virgilii*.

Medicago arborea. Lin. Syft. Plant. t. 3. p. 573. Sp. 1.

Cytifus incanus, filiquis falcatis. Bauh. Pin. 389.

Cytifus Marante. Lob. ic. 2. p. 46.

Sativa. La premiere efpece a une racine vivace, & des tiges annuelles, qui s'élevent dans une bonne terre à la hauteur d'environ trois pieds, & font garnies à chaque nœud de feuilles à trois lobes, en forme de lance, d'un pouce & demi de longueur fur fix lignes de largeur, un peu élevées vers leur extrémité, d'un vert

MED

foncé, & placées alternativement fur les tiges : fes fleurs
croiffent en épis de deux ou
trois pouces de longueur, &
font portées fur des pédoncules de deux pouces de longueur, qui fortent des aiffelles
de la tige ; elles font papilionnacées, d'une belle couleur
pourpre, & elles produifent
des légumes comprimés , &
en forme de croiffant, qui renferment plufieurs femences en
forme de rein. Cette plante fleurit dans le mois de Juin,& fes femences mûriffent en Septembre.

Elle donne les variétés fuivantes :

Luferne à fleurs violettes.

Luferne à fleurs d'un bleu
pâle.

Luferne à fleurs panachées.

Ces différences qui fe trouvent dans la fleur, font des
accidens produits de femences ; c'eft - pourquoi on ne
doit point les regarder comme
des efpeces diftinctes : cependant comme celles à fleurs
d'un bleu pâle, & celles
à fleurs panachées ne deviennent jamais auffi fortes que
celles à fleurs pourpres, les
cultivateurs doivent les diftinguer, parce qu'elles font d'un
moindre rapport.

On croit que cette plante a été
apportée originairement de la
Médie ; les Efpagnols l'appellent *Alfafa* ; les François, *Luferne* ou *grand Trefle* ; & plufieurs
Ecrivains fur la Botanique la
nomment *Fœnum Burgundiacum*,
ou *foin de Bourgogne* : mais il
y a lieu de douter qu'elle foit
la *Medica* de VIRGILE, de CO-
LUMELLE, de PALLADIUS, &

MED

d'autres anciens Auteurs d'A-
griculture, qui n'avoient pas
befoin de vanter la qualité de ce
fourrage, & de donner des inftructions fur fa culture dans
les pays qu'ils habitoient.

Mais quoiqu'elle ait été fi
recommandée par les anciens,
& cultivée avec tant d'avantage par nos voifins, en France
& en Suiffe, depuis plufieurs
années; cependant elle n'a pas
encore été reçue jufqu'à préfent dans ce pays avec autant
d'empreffement qu'on pourroit
le defirer : on ne l'y cultive point
en grande quantité, quoiqu'on
foit certain qu'elle réuffiroit
auffi bien en Angleterre que
dans aucun autre pays : on peut
la couper très-fouvent ; elle eft
extrêmement dure, & réfifte
au froid le plus vif de notre
climat. On peut en donner une
preuve ; car fes femences écartées en automne ont produit
des plantes qui ont réfifté aux
froids les plus rudes, & font
devenues très-fortes l'année
fuivante.

Les femences de cette plante
ont été apportées de France
en Angleterre vers l'année
1650 ; mais elle n'y a point
réuffi, foit faute de foins dans
la culture, foit par trop d'at-
tachement aux anciens ufages ;
& pour n'avoir pas voulu effayer quelques expériences, elle
a été entierement négligée dans
notre Ifle. J'ignore la véritable
raifon du peu de fuccès de
cette premiere tentative ; mais
il eft certain qu'elle eft prefque abandonnée aujourd'hui.
Cependant j'efpere que les inftructions que je donnerai dans

cet article fur la culture de cette plante précieufe, encourageront les habitans de ce pays à faire de nouveaux effais, pour fe procurer cette efpece de fourrage, qui croît également dans les pays les plus chauds & les plus froids, avec cette différence feulement que, dans tous les pays chauds, tels que l'Amérique Efpagnole, où cette plante forme la plus grande partie des pâturages néceffaires à la nourriture du bétail, on la coupe chaque femaine, au-lieu que dans les pays froids on ne la fauche guere plus que quatre ou cinq fois dans l'année; & il eft très-vraifemblable que cette plante peut être d'une grande utilité aux habitans des Barbades, de la Jamaïque, & autres Ifles chaudes de l'Amérique où la nourriture du bétail eft la chofe la plus néceffaire: car (d'après le rapport du Pere Feuillé) cette plante réuffit très-bien dans l'Amérique Efpagnole, & particulierement aux environs de Lima, où on la coupe chaque femaine, pour la porter fur les marchés; cette efpece de fourrage étant la feule qu'on cultive dans ce pays.

Cette plante eft auffi très-commune dans le Languedoc, la Provence, le Dauphiné, & fur tous les rivages du Rhône, où elle produit abondamment, & où on peut la faucher cinq ou fix fois dans l'année: les chevaux, les mulets, les bœufs, & autres animaux domeftiques l'aiment beaucoup, fur-tout quand elle eft verte; le bétail noir la préfere lorfqu'elle eft

feche. Cependant l'excès de ce fourrage eft regardé comme très-dangereux; il eft excellent pour donner beaucoup de lait aux vaches & aux chevres: on prétend auffi qu'il vaut mieux que tout autre pour les chevaux; les moutons, les chevres, &c. s'en nourriffent volontiers quand l'herbe eft jeune & tendre.

Les inftructions données par tous ceux qui ont écrit fur cette plante, font fi imparfaites, que, fi on les fuivoit dans ce pays, on s'en trouveroit fort mal; car plufieurs veulent que l'on mêle fes femences avec de l'Avoine ou de l'Orge, comme cela fe pratique pour le Trefle: mais par cette méthode, elles pouffent rarement bien; & quand elles réuffiffent, les plantes filent, & deviennent fi foibles parmi ces autres efpeces, qu'elles font une année entiere à recouvrer leur force, quand elles peuvent en revenir; d'autres confeillent de les femer fur un fol bas, riche & humide, qui eft le plus mauvais terrein, après celui de glaife, parce que, dans de pareils terres, les racines de cette plante pourriffent en hiver, & qu'en une ou deux ans la récolte entiere eft détruite.

Le fol dans lequel cette plante réuffit le mieux dans ce pays, eft une terre légere, feche & fablonneufe; mais elle doit être bien labourée, & exactement débarraffée de toutes les racines des herbes nuifibles, qui furmonteroient les jeunes plantes, & en arrêteroient les progrès.

Le meilleur tems pour femer

MED

ce fourrage eft vers le milieu d'Avril, lorique le tems eft fixé au beau ; car fi on le feme dans une terre fort humide, ou par un tems pluvieux, les fe— mences crevent & périflent, comme il arrive fouvent à plu- fieurs efpeces de plantes ligneu- fes : c'eft-pourquoi il faut tou- jours obferver de les femer dans une faifon feche ; & quand il furvient de la pluie une fe- maine ou dix jours après, les plantes paroiflent bientôt au- deflus de la terre.

La méthode que je confeil- lerois pour femer cette efpece, feroit de bien labourer & her- fer la terre, de maniere qu'elle foit très-meuble ; de tirer une rigolle dans toute la largeur du terrein, d'un demi-pouce environ de profondeur, dans laquelle on répandroit la fe- mence fort clair avec une tré- mie attachée à une charrue à rigolle, que l'on recouvriroit de fix lignes d'épaifleur avec la même terre ; on creuferoit enfuite une autre rigolle à deux pieds & demi environ de la premiere, & l'on continueroit ainfi fur toute la piece de terre, en laiflant toujours la même diflance entre les rangs, & en femant fort clair dans les ri- golles. Par cette méthode, un âcre de terre exigera à-peu- près fix livres de femences ; fi l'on en emploie davantage, & que les plantes croiffent bien, elles fe trouveront fi rappro- chées & fi ferrées, qu'elles fe nuiront & fe détruiront dans l'efpace d'une ou deux années ; au-lieu qu'en leur donnant af- fez de place, elles acquerront

MED

une grofleur confidérable, & leurs racines deviendront très- fortes. J'ai mefuré la couronne d'une racine qui m'apparte- noit ; elle s'eft trouvée avoir dix-huit pouces de diamètre, & j'ai coupé deflus près de quatre cents rejettons en une feule fois, ce qui eft une ré- colte extraordinaire ; & cela fur un fol fec, graveleux, & de mauvaife qualité, qui n'a- voit point été engraiflé depuis plufieurs années : & cette ra- cine avoit au moins quatorze années ; ce qui prouve que, fi cette plante étoit bien culti- vée, elle dureroit long–tems, & feroit toujours auffi bonne que fi elle étoit femée nouvel- lement : fes racines pénetrent toujours profondément dans la terre, pourvu que le fol foit fec, & qu'elles ne rencontrent pas un gravier dur à un pied au-deflous de la furface ; ce- pendant elles y pénétreroient encore, & s'y enfonceroient, ainfi que je l'ai obfervé, après en avoir enlevé quelques-unes qui avoient plus de quatre pieds de longueur, & qui avoient pénétré plus de deux pieds dans un gravier auffi dur que le roc, qu'on ne pouvoit déflerrer fans pioche & pince de fer, en em- ployant beaucoup de force.

Ce qui me détermine à con- feiller de femer cette plante en rangs, c'eft afin qu'elle puiffe avoir affez de place pour croî- tre, & qu'on ait la facilité de travailler la terre pour détruire les mauvaifes herbes, & aug- menter les progrès des plan- tes ; ce qui fe fait très-aifément après chaque récolte, avec une

MED

houe Hollandoise. Au moyen de ce houage, elles repousseront mieux, en moins de tems, & feront beaucoup plus fortes que dans les endroits où ce travail n'aura pu être fait. Aussi-tôt que les plantes poussent, il faut houer la terre entr'elles avec une houe ordinaire à main, & détruire en même tems quelques plantes dans les endroits où elles sont trop serrées, afin que les autres puissent acquérir de la force. On répete ce travail deux ou trois fois, tandis qu'elles sont jeunes, & suivant les progrès qu'elles ont faits, en choisissant toujours un tems sec pour mieux détruire les mauvaises herbes, qui reprendroient racine, si on le faisoit par un tems humide.

Par le moyen de ce traitement, ces plantes auront acquis deux pieds & plus de hauteur au commencement d'Août, tems auquel les fleurs commenceront à paroître ; alors il faudra les couper, pour la premiere fois, dans un tems sec, sur-tout si on veut en faire du Foin : il sera nécessaire de les remuer souvent, afin qu'elles soient plutôt seches & plutôt enlevées ; car si elles séjournoient long-tems sur les racines, elles les empêcheroient de repousser. Quand la récolte est enlevée, on remue la terre entre les rangs avec la houe Hollandoise, & l'on en ameublit la surface, pour faire pousser les plantes en peu de tems. Au milieu de Septembre, leurs branches auront atteint la hauteur de quatre pouces ; alors on pourra les laisser brouter par les moutons, car elles ne seroient pas bonnes à être coupées dans cette saison : il ne faut pas non plus laisser ces branches sur les plantes, parce qu'elles périroient aux approches de la gelée, tomberoient sur les racines, & les empêcheroient de pousser au printems suivant ; mais les moutons ne doivent pas rester trop long tems dessus, de peur qu'ils n'endommagent les couronnes des racines.

La meilleure méthode est de les laisser manger jusqu'en Novembre, c'est-à-dire, jusqu'à ce que ces plantes aient cessé de pousser, en observant cependant que le grand bétail ne passe pas dessus dans la premiere année, parce que ces racines étant jeunes, seroient en danger d'être détruites par les pieds des bestiaux qui les fouleroient, ou qui les arracheroient ; mais les moutons au contraire rendent service aux racines, & engraissent la terre, pourvu cependant qu'ils ne paturent pas les couronnes de trop près.

Au commencement de Février, il faut remuer la terre entre les racines avec une houe, pour les faire repousser, & toujours avec précaution, pour ne pas blesser les couronnes, dont les boutons sont alors très-gonflés & prêts à s'ouvrir. Par ce moyen, si le sol est chaud, les branches acquerront dans peu de tems cinq à six pouces de hauteur, & on pourra les laisser pâturer jusqu'après la premiere semaine d'Avril, si l'on manque de fourrage ; après quoi on les laissera croître pour

une récolte, qui fera en état d'être coupée au commencement de Juin, & qu'on tâchera d'enlever le plutôt qu'il fera possible, afin de pouvoir labourer la terre avec la houe Hollandoise, & la préparer à donner au milieu du mois de Juillet une seconde récolte, que l'on conduira comme la premiere : après cela on y introduira les troupeaux pendant l'automne ; & comme alors les racines auront coulé profondément dans la terre, il y aura peu de risque qu'elles soient endommagées par le grand bétail : mais il faudra toujours observer de ne pas le laisser dessus, quand elles auront cessé de pousser, de peur qu'il ne mange les couronnes au-dessous des boutons ; ce qui leur feroit beaucoup de tort, & peut-être les détruiroit. Au moyen de cet arrangement, on peut toujours se procurer par année deux récoltes, & faire pâturer deux fois cette plante. Dans des étés favorables, il est possible de faucher trois fois, & d'y mettre deux fois les troupeaux qui amélioreront beaucoup le terrein, sur-tout s'il est sec & stérile ; car sans ce secours, l'herbe y croîtroit peu dans les années seches, où le fourrage est le plus nécessaire & le plus rare : dans ce cas, la *Luserne* est propre à servir de nourriture aux animaux, au moins un mois plutôt que l'herbe ordinaire ou le *Trefle* ; car j'ai vu cette plante à huit pouces de hauteur le 10 de Mars, tandis que l'herbe, dans le même endroit, avoit à peine un pouce de longueur.

Je suis entierement convaincu que le froid ne fait aucun tort à cette plante ; car dans l'hiver de 1729 à 1730, j'avois quelques racines de cette espece qui avoient été arrachées en Octobre, & qui resterent sur la terre en plein air jusqu'au commencent de Mars : ces racines ayant été replantées alors, elles repousserent très-vigoureusement bientôt après ; pendant même qu'elles étoient sur la terre, elles poussoient des fibres, & commençoient à produire des branches à la couronne D'un autre côté, je suis très-persuadé que l'humidité détruit ces racines ; car j'en ai semé plus d'un âcre dans une piece de terre humide, pour essayer si cette plante y réussiroit ; elle y a très-bien poussé, a beaucoup fleuri pendant l'été ; mais en hiver, les grandes pluies ont commencé à faire pourrir les racines vers le bas, & la plupart ont été détruites avant le printems.

Quelques personnes ont conseillé dernierement de semer la Luserne à la volée, & de faire usage d'une grande herse, pour arracher & détruire les mauvaises herbes qui poussent naturellement parmi les plantes : mais ce conseil a été donné sans beaucoup de réflexion & il faut espérer qu'il ne sera pas suivi par les personnes prudentes, que l'on prie de jetter un coup d'œil sur quelques-unes de ces terres ainsi cultivées pendant trois ou quatre années ; je ne doute

MED

doute pas qu'elles ne foient convaincues que cette méthode eft très-mauvaife, pour peu qu'elles aient égard à la proprété & au produit.

Les meilleurs cantons d'où l'on peut tirer la graine de Luferne, font la Suiffe & les parties feptentrionales de la France ; car celle que l'on fait venir de ces contrées réuffit mieux en Angleterre que celle des pays plus chauds : mais on la recueilleroit auffi bien ici, & en auffi grande abondance, fi l'on étoit affez curieux pour laiffer croître la première pouffée ; pour cela il faudroit réferver une petite quantité de plantes fur lefquelles on laifferoit parvenir les femences à leur maturité : elles fe perfectionnent ordinairement vers le commencement de Septembre, alors on les coupe, & on les fait fecher dans une grange ouverte, & où l'air puiffe paffer librement, mais où elles foient à l'abri de l'humidité ; car fi ces femences y étoient expofées, elles germeroient dans leurs enveloppes, & ne pourroient plus fervir à rien. Quand elles font tout-à-fait feches, on les bat, on les nettoie entierement de leurs enveloppes, & on les conferve dans un endroit fec, jufqu'au tems où l'on doit en faire ufage. Ces graines, recueillies en Angleterre, font bien préférables à toutes celles qu'on apporte des pays étrangers, ainfi que je l'ai obfervé, d'après plufieurs expériences. Les plantes élevées de femences du pays font devenues beaucoup plus fortes que celles qui

Tome V.

MED

ont été produites par des graines de France, de Suiffe & de Turquie, qui avoient été femées dans le même tems, fur le même terrein, & à la même expofition.

Je fuis porté à croire que cette plante n'a pas encore réuffi en Angleterre, parce qu'elle a été femée avec du grain : elle ne profite point du tout de cette maniere ; car quoiqu'elle foit fort dure quand elle eft devenue groffe, cependant quand elle commence à pouffer, le voifinage des autres plantes la gêne tellement, qu'elle réuffit rarement bien ainfi : c'eft-pourquoi il eft indifpenfable de la femer feule, & d'avoir le plus grand foin de la tenir nette de mauvaifes herbes, jufqu'à ce qu'elle ait acquis de la force, après quoi elle fe défend affez bien : peut-être auffi l'a-t-on femée dans une mauvaife faifon, ou par un tems humide, qui a fait pourrir fes graines ; ce qui a découragé les Cultivateurs, & les a empêchés de faire de nouveaux effais : mais, quoi qu'il en foit, j'ofe affurer qu'en fuivant la méthode que je viens d'indiquer, cette plante réuffira auffi bien en Angleterre qu'aucune des autres qu'on y cultive ; qu'elle produira une récolte beaucoup plus abondante qu'aucune autre efpece de fourrage, & qu'elle fubfiftera beaucoup plus long-tems : car fi la terre eft bien labourée après chaque récolte, & que l'on faffe pâturer la derniere pouffée, comme il a été prefcrit, les plantes refteront en vigueur pendant quarante

B

14 MED

annnées, & même davantage, sans avoir befoin d'être renouvellées, pourvu qu'on ne les laiffe pas monter en femences; ce qui les affoibliroit plus que fi on les coupoit quatre fois dans une année. Le fourrage que cette plante fournit doit être mis dans des granges bien fermées, parce qu'il eft trop tendre pour pouvoir être confervé en meule en plein air, comme l'autre Foin. Il fe conferve ainfi pendant trois ans, quand il a été bien féché auparavant. Dans les pays étrangers, on compte qu'un âcre de terre donne affez de ce fourrage par année pour l'entretien de trois chevaux.

Des perfonnes dignes de foi m'ont affuré qu'elles avoient cultivé cette plante en Angleterre, & que trois âcres de ce fourrage leur avoient nourri dix chevaux de charette, depuis la fin d'Avril jufqu'au commencement d'Octobre, fans aucun autre Foin, quoiqu'ils aient travaillé conftamment: ainfi, le meilleur ufage que l'on puiffe faire de cette l'erbe, eft de la couper & de la donner en verd au bétail. Par tout où cela eft pratiqué, on obferve qu'en finiffant de faucher le champ, la partie qui a été coupée la première, eft en état d'être fauchée une feconde fois; de forte que la récolte peut fe continuer dans le même champ, depuis le milieu d'Avril jufqu'à la fin d'Octobre. Dans les années favorables, & lorfque les étés font pluvieux, on peut obtenir fix récoltes par an; mais dans les ·tems les plus fecs, on en a

MED

toujours trois ou quatre. Quand la Luferne commence à fleurir, on la coupe; car fi on la laiffoit plus long-tems, fes tiges deviendroient dures, fes feuilles périroient, & le bétail ne la mangeroit pas auffi volontiers. Lorfque l'on en cultive un grand terrein, on devroit en couper une partie avant que les fleurs paroiffent, afin de n'avoir pas tout à récolter en même tems.

Quand on convertit la Luferne en Foin, elle exige beaucoup de travail; car fes tiges étant fort fucculentes, elles ont befoin d'être fouvent retournées, & d'être expofées à l'air pendant quinze jours, & même plus long-tems, avant d'être affez feches pour être renfermées. Comme elle doit être plus travaillée que le *Sainfoin*, lorfqu'elle eft coupée, on devroit la tranfporter fur quelques prairies, afin que la terre qui eft à nud dans les intervalles des rangs, ne foit point emportée & mêlée avec le foin à chaque ondée, & que d'ailleurs les plantes puiffent plutôt repouffer; mais il eft moins avantageux de réduire cette plante en Foin, que de la faire manger en verd pour toute forte de bétail, & furtout pour les chevaux, qui l'aiment beaucoup, & auxquels elle eft très-bonne de toute manière; car ils travaillent autant en ne mangeant que de ce fourrage, que s'ils étoient nourris avec l'Avoine & le Foin fec ordinaire.

Falcata. La feconde efpece croît naturellement dans la France Meridionale, en Efpa-

MED

gne, en Italie, & autres contrées. Elle a été regardée comme une variété de la premiere ; mais je l'ai souvent élevée de femences, & ne l'ai jamais vu varier. Les tiges de celle-ci font plus petites, moins élevées, & toujours inclinées vers la terre ; fes feuilles n'ont pas la moitié de la largeur de celles de la précédente ; fes fleurs naiffent en épis courts & ronds, & font de couleur de fafran. Cette plante fleurit vers le même-tems que la premiere, & fes femences mûriffent à la fin de l'été. On peut la multiplier aifément par fes graines ; fa racine eft vivace, & fubfifte plufieurs années ; mais on la cultive rarement dans d'autres pays.

Radiata. La troifieme efpece, qui eft originaire de l'Italie, eft une plante annuelle, qui pouffe plufieurs tiges minces, branchues, d'un pied & demi de longueur, couchées fur la terre, & garnies de feuilles à trois lobes ovales, en forme de lance & entiers ; fes fleurs fortent fimples, fur des pédoncules minces aux côtés des branches ; elles font petites, jaunes, & femblables, pour la forme, à celles de l'efpece précédente ; elles produifent des légumes gros, plats, & en forme de croiffant, dont les bords font divifés en dentelures, terminés par de beaux poils ; chaque légume renferme quatre ou cinq femences en forme de rein. Cette plante fleurit dans les mois de Juin & de Juillet, & fes femences mûriffent en automne.

MED 15

Hispanica. La quatrieme efpece naît fans culture en Efpagne ; elle eft auffi annuelle : fes tiges, longues d'un pied & demi, trainent fur la terre, & font garnies de feuilles ailées, & compofées de deux paires de petits lobes un peu blanchâtres, & alternes fur les nœuds ; fes pédoncules, longs & minces, foutiennent chacun à leur fommet quatre ou cinq fleurs de couleur d'or, auxquelles fuccedent des légumes comprimés en forme de croiffant, de moitié moins larges que ceux de la troifieme efpece, & qui ont auffi des dents velues. Cette plante fleurit & perfectionne fes femences vers le même tems que la précédente.

Italica. La cinquieme naît fur les rivages de la mer, dans plufieurs parties de l'Italie ; elle eft auffi annuelle : fes tiges font courbées, herbacées, d'un pied environ de longueur, & garnies de feuilles à trois lobes en forme de coin, & fciés vers leur extrémité ; fes fleurs fortent fur des pédoncules minces aux nœuds de la tige : ils ont à peu-près un pouce de longueur, & foutiennent chacun cinq ou fix fleurs d'un jaune pâle, qui font remplacées par des légumes épais en forme de croiffant, dont les bords font en tiges, & qui contiennent chacun trois ou quatre petites femences en forme de rein. Cette plante fleurit & produit fes femences vers le même tems que les deux précédentes.

Cretica. La fixieme efpece

croît ſpontanément dans les Iſles de l'Archipel ; elle eſt annuelle , & pouſſe de ſes racines pluſieurs feuilles oblongues d'environ deux pouces & demi de longueur, étroites à leur bâſe , mais larges vers leur extrémité , où elles ſont arrondies & couchées ſur la terre : du milieu de ces feuilles ſortent des tiges minces, d'un pied à-peu-près de longueur, qui produiſent des branches foibles , & garnies de feuilles aîlées & blanches ; celles du bas des tiges ſont compoſées de deux paires de lobes égaux en longueur, & terminés par un lobe impair ; mais celles du haut ont trois lobes : ſes fleurs naiſſent aux extrémités des tiges ; elles ſont petites, jaunes, & de la même forme que celles des .autres eſpeces, & ſont remplacées par des légumes comprimés, en forme de croiſſant, & dentés ſur leurs bords, qui contiennent trois ou quatre ſemences en forme de rein. Cette plante fleurit & perfectionne ſes ſemences vers le même tems que les autres.

Les Botaniſtes conſervent ces eſpeces dans leurs jardins : on les ſeme ſur une terre ouverte & fraiche, où elles doivent reſter, parce que ces plantes ne ſouffrent pas aiſément la tranſplantation, à moins qu'elles ne ſoient très-jeunes. Comme elles étendent leurs branches ſur la terre, il faut les ſemer au moins à deux pieds & demi de diſtance ; & quand elles pouſſent, elles n'exigent plus aucun autre ſoin que d'ê-

tre tenues nettes de mauvaiſes herbes : elles commencent à fleurir dans le mois de Juin, & leurs fleurs ſe ſuccedent juſqu'à l'automne ſur les tiges & branches, qui s'étendent continuellement. Les premieres fleurs ſont les ſeules qui produiſent de bonnes ſemences ; celles qui viennent enſuite n'ont pas le tems. de perfectionner les leurs avant les premiers froids.

Arborea. La ſeptieme eſpece croît naturellement dans les Iſles de l'Archipel, en Sicile, & dans les parties chaudes de l'Italie ; elle s'éleve à la hauteur de huit ou dix pieds, avec une tige d'arbriſſeau couverte d'une écorce griſe, & diviſée en pluſieurs branches, couvertes, étant jeunes, d'un duvet blanc ; elles ſont garnies à chaque nœud de feuilles à trois lobes, ſupportées par des pétioles d'un pouce environ de longueur ; chaque nœud en produit deux ou trois, de ſorte que les branches en ſont fortement couvertes. Ces lobes ſont petits, en forme de lance, & blancs en-deſſous ; ces feuilles durent toute l'année : les fleurs naiſſent ſur des pédoncules qui ſortent aux côtés des branches ; elles ſont d'un jaune brillant ; chaque pédoncule en ſoutient quatre ou cinq, qui ſont remplacées par des légumes comprimés en forme de croiſſant, & qui renferment chacun trois ou quatre ſemences en forme de rein.

Cette plante fleurit durant une grande partie de l'année, & même toute l'année, quand

MED

les hivers font favorables, ou quand elle eſt abritée ; de maniere qu'elle eſt rarement ſans fleurs : mais celles de pleine terre commencent à fleurir en Avril, & continuent juſqu'en Décembre.

Les fleurs qui paroiſſent de bonne heure en été, perfectionnent leurs ſemences en Août ou au commencement de Septembre, & les autres mûriſſent ſucceſſivement, juſqu'à ce que le froid les arrête.

On peut multiplier cette plante, en la ſemant ſur une couche de chaleur modérée, ou ſur une plate-bande chaude dans une terre légere, au commencement du mois d'Avril. Quand les plantes pouſſent, on les débaraſſe ſoigneuſement de toutes mauvaiſes herbes & on les laiſſe ſans les remuer juſqu'au mois de Septembre ſuivant, quand elles ſont ſemées ſur une terre ordinaire ; mais celles qui ſont ſur une couche chaude doivent être tranſplantées dans des pôts vers le milieu de l'été, tenues à l'ombre, juſqu'à ce qu'elles aient formé de nouvelles racines, & placées enſuite dans un endroit où elles ſoient à l'abri des vents violens : on les y laiſſera juſqu'à la fin d'Octobre, pour les mettre alors ſous un châſſis vitré ordinaire, où elles ſoient à couvert des fortes gelées ; car les plantes qui ont été élevées délicatement, ſont ſujettes à ſouffrir dans les tems rudes, ſur-tout tandis qu'elles ſont jeunes. Au mois d'Avril ſuivant, on peut les tirer des pots,

pour les mettre en pleine terre dans les places qui leur ſont deſtinées, ſur une terre légere, & à une expoſition chaude, où elles ſupporteront très-bien le froid de nos hivers ordinaires & continueront à produire des fleurs durant la plus grande partie de l'année : on les eſtime ſur-tout, parce qu'elles conſervent leurs feuilles pendant tout l'hiver.

Celles qui ont été ſemées ſur une plate-bande en plein air, peuvent auſſi être tranſplantées au mois d'Août ſuivant de la même maniere ; mais il faut les enlever avec précaution, & conſerver, s'il eſt poſſible, une bonne motte de terre à leurs racines : on les arroſe & on les tient à l'ombre juſqu'à ce qu'elles aient pris racine ; après quoi il ſuffira de les tenir toujours nettes, & de retrancher les branches luxurieuſes, pour qu'elles ne s'étendent point trop ; mais on ne doit jamais les tailler de bonne heure au printems, ni trop tard en automne ; car ſi la gelée ſurvenoit auſſi-tôt après, leurs branches ſeroient détruites, & ſouvent la plante entiere périroit.

On a toujours conſervé ces plantes dans les ſerres, parce qu'on les croyoit trop tendres pour pouvoir réſiſter en plein air aux gelées de l'hiver ; mais j'en ai eu quelques-unes de grandes, qui ont ſubſiſté pendant pluſieurs années à une expoſition chaude, ſans aucune couverture ; elles étoient même beaucoup plus fortes,

MED

& ont mieux fleuri que celles de la ferre. Il eft cependant prudent d'en mettre une ou deux à couvert, de peur que, dans les hivers rigoureux qui furviennent quelquefois en Angleterre, celles de pleine terre ne foient détruites.

On peut la multiplier par boutures, qu'il faut planter en Avril dans une planche de terre légere : on les arrofe, & on les tient à l'ombre, jufqu'à ce qu'elles aient pris racine, après quoi on les expofe à l'air, & on les laiffe dans cette planche jufqu'au mois de Juillet ou d'Août : alors elles auront pouffé de bonnes racines, & on pourra les tranfplanter à demeure dans les places qui leur font deftinées, en obfervant, comme il a été dit ci-devant, de les arrofer & tenir à l'ombre jufqu'à ce qu'elles aient formé de nouvelles racines : on les éleve enfuite en tiges droites, en les fixant à des bâtons ; car fans cela elles font fujettes à fe courber & à croître irrégulierement : quand leurs tiges ont atteint la hauteur qu'on veut leur donner, on peut leur former des têtes régulieres, en les taillant chaque année, pour les tenir en bon ordre.

Cette plante croit en grande abondance dans le Royaume de Naples, où les chevres s'en nourriffent, & donnent un lait dont les habitans font une grande quantité de fromages : on la trouve auffi dans les Ifles de l'Archipel. Les Turcs fe fervent du bois de cet arbriffeau, pour faire des poignées à leurs fabres, & les **Caloyers** de Patmos en font des lits.

Plufieurs perfonnes ont fuppofé, comme il a été obfervé ci-deffus, que cet arbriffeau étoit le *Cytife* de VIRGILE & de COLUMELLE ; & fur ce que quelques anciens Ecrivains d'Agriculture font mention de cette plante comme étant extraordinaire, & digne d'être cultivée pour fourrage, elles en recommandent la culture en Angleterre.

Cette plante peut être très-utile en Candie, en Sicile, à Naples, & dans d'autres pays chauds : mais je fuis perfuadé qu'elle ne réuffiroit jamais dans notre climat, de maniere à devenir d'un avantage réel ; car les fortes gelées la détruifent, ou au moins l'endommagent fi confidérablement, qu'elle ne peut recouvrer fa premiere verdure avant le milieu ou la fin du mois de Mai ; d'ailleurs comme fes branches ne fouffrent pas d'être coupées plus d'une fois dans un été, qu'elles ne font pas d'une longueur confidérable, & que fes tiges deviennent fort ligneufes, & en rendent la taille très-penible, elle ne vaut pas la peine & les frais qu'elle occafionneroit pour la cultiver ; je ne penfe pas même qu'on doive en faire l'effai, parce que nous avons beaucoup de plantes qui lui font préférables : mais dans des pays chauds, fecs & remplis de rochers, elle doit être très-avantageufe, parce qu'il y a peu d'autres plantes qui puiffent réuffir dans de pareils fols ; elle y fubfifte plufieurs

MED

années, & y profite très-bien.

Quoiqu'on ne puisse la cultiver en Angleterre pour fourrage , cependant la beauté de son feuillage , qui dure toute l'année , & la succession continuelle de ses fleurs, doivent lui mériter une place dans tous les beaux jardins : quand elle est mêlée avec d'autres arbrisseaux du même crû, elle donne une variété très-agréable.

Comme beaucoup de personnes paroissent curieuses de savoir quel est le vrai *Cytise* dont les anciens font mention , j'ai pris la peine de rapporter ce qu'ils en ont dit , & la description qu'ils en ont donnée, afin que l'on puisse juger quel peu de fond on doit faire sur ces Auteurs, pour nous décider dans cette question.

THÉOPHRASTE dit que le *Cytise* est si ennemi des autres plantes, qu'il les fait périr, en leur retranchant leur nourriture, & que son organisation intérieure est si dure & si compacte, qu'il ressemble beaucoup à l'*Ebene*. ARISTOMAQUE l'Athénien (comme on peut le voir dans PLINE) dit comme VARRON & COLUMELLE , & probablement d'après eux, que le *Cytise* est fort propre à la nourriture des moutons , & quand il est sec, pour maintenir en santé & engraisser les porcs, de même que le grain; mais qu'il rassasie plus vite les quadrupedes , & engraisse le bétail si promptement , qu'il ne se soucie plus d'orge.

Aucune nourriture ne produit une plus grande quantité ni un meilleur lait , & ce four-

MED 19

rage vaut mieux que toute autre chose dans les maladies du bétail : de plus , étant donné sec ou dans une décoction d'eau mêlée de vin , aux Nourrices dont le lait vient à manquer, il aide beaucoup à les rétablir, rend les enfans plus forts, & les fait tenir plutôt sur leurs pieds. Cette plante est aussi très-bonne en verd pour les enfans , ou sèche, en l'humectant un peu.

DÉMOCRITE & ARISTOMAQUE disent qu'il n'y a rien de meilleur pour les abeilles , & qu'elles ne manquent jamais de nourriture , quand elles trouvent assez de *Cytise*.

Quand il ne survient point de pluie après avoir semé le *Cytise* , dit COLUMELLE , il faut l'arroser quinze jours de suite.

On le seme, suivant les anciens , après les équinoxes , & il parvient à sa perfection en trois années : on le fauche dans l'équinoxe du printems, car il fleurit tout l'hiver.[DAL.] On travaille ce Foin à bon marché ; un seul garçon ou une vieille femme suffit pour cet ouvrage.

Le *Cytise* paroît blanc à la vue ; & pour le décrire en un mot, c'est un arbrisseau plus gros que le *Trefle*.

En hiver, quand il est humecté , dix livres de ce fourrage suffisent pour un cheval, & on en donne une moindre quantité aux autres animaux. Quand il est sec, il contient plus de substance , & il en faut moins pour la nourriture des bestiaux.

Cet arbrisseau a été trouvé

B 4

MED

dans l'Isle de Cythnus , d'où il a été transplanté dans toutes les Cyclades , & ensuite dans les villes de la Grece , où il a donné une grande augmentation de fourrage.

Il ne craint ni le chaud ni le froid ; il résiste à la grêle & à la neige. HYGINUS ajoûte qu'il est à l'abri des injures de l'ennemi , parce que son bois n'est d'aucune valeur.

GALIEN dit aussi , (dans son livre *de Antid.*) « Le *Cytise* » est un arbrisseau ; en Mysie, » & dans la partie la plus voi- » sine de notre province, il » y a une étendue de terrein » appelée Brotton , toute cou- » verte de *Cytises*, & tout le » monde convient que les abeil- » les ramassent une très–gran- » de quantité de miel sur les » fleurs de ces arbrisseaux. Cet- » te plante est ligneuse,& s'éle- » ve à la hauteur d'un Myrte ». GALIEN ajoûte encore , que sept de ses feuilles seulement, mêlées avec de l'eau chaude , comme celles de la Mauve, aident à faire la digestion.

CORNARUS écrit avec trop d'assurance, que le *Cytise* n'a jamais été transporté en Allemagne , ou qu'il y a péri de-puis long-tems , sur ce que dit *Pline*, qu'il étoit fort rare en Italie de son tems; mais il ne me persuadera pas que ce qui est rare en Italie, ne puisse pas croître en Allemagne. (*Io. Bruh.*)

STRABON pense autrement que DIOSCORIDE , PLINE & GALIEN. Il veut absolument que le *Cytise* soit un arbre , & il le compare au *Balsamum*,

MED

arbre odoriférant; ce qui a été cause sans doute que CORNA-RUS a prétendu que cette plan-te approchoit beaucoup de l'ar-brisseau.

C'est ce que dit PLINE , qui assûre que son bois n'a aucune valeur : c'est-pourquoi il ajoûte qu'il produit des branches li-gneuses , qui ne sont ni ten-dres ni molles comme celles des herbes.

Mais VIRGILE fait entendre qu'il n'est ni arbre ni arbris-seau , en disant :

" . . .*Non , me pascente , capellæ,* " *Florentem Cytisum , & salices* " *carpetis amaras* ,,

Buc. Eclog. I.

" *Sic Cytiso pastæ distendunt ubera* " *vaccæ* ,,

Eclog. IX.

" *Nec Cytiso saturantur apes,nec fronde* " *capellæ* ,,

Eclog. X.

VIRGILE, dis-je , indique très-clairement dans ces vers qu'il n'est ni arbre ni arbrisseau ; car les chevres n'en mangent point, & ne le pourroient pas, quoiqu'elles soient accoutumées à brouter les arbres à fleurs. Ce que dit CORNARUS n'a au—cun fondement , que cette plante doit nécessairement pro-duire des branches ligneuses, d'après ce que PLINE avance, que le bois n'est d'aucune va-leur. Le contraire est plutôt établi *que le bois n'est d'aucune valeur*, parce qu'il produit des branches souples & remplies de jus, dont les chevres ne peu-vent se rassasier.

THÉOCRITE dit au contraire, que le Cytise est une nourri-ture fort agréable aux chevres

MED

Ηαὶξ τὸν κύτισον, ὁ λύκος τὴν αἶγα
διώκει.

„ Capra Cytifum , lupus capellam
 fequitur „

que VIRGILLE a ainfi imité :
„ Torva Leæna Lupum fequitur , Lupus ipfe capellam;
„ Florentem Cytifum fequitur lafciva
 „ capella „.

AMATUS, pour éluder cette difficulté, conclut que le *Cytife* tient de l'arbre & de l'arbriffeau, parce que PLINE le diftingue en différens genres : comme arbre, il le met au genre féminin, & comme arbriffeau au genre mafculin ; ce qui ne vaut pas la peine d'y faire attention.

COLUMELLE met le *Cytife* au genre féminin, & THÉOCRITE, ainfi que d'autres, au mafculin, comme ROBERT CONSTANTIN *dans fon Lexique Grec*, qui l'appelle ἀμφόφυλλον & THÉOCRITE nomme cet arbriffeau κύλιλον & d'autres κύτισον & τῆλις, de Cythnus ou de Cythifa nom d'une Ifle citée par SERVIUS.

Dans quelques manufcrits de DIOSCORIDE, on trouve les fauffes dénominations de *Telinen triphyllon ;* & dans d'autres, *Lotum grandem.*

La defcription que donne DIOSCORIDE du *Cyrife*, n'eft pas affez exacte pour y reconnoître le vrai *Cytife.*

Dans plufieurs efpeces de *Cytifes*, il eft difficile de décider quel eft le veritable, fpécifié par les anciens.

Les plus inftruits croient que c'eft celui que MARANTHUS a repréfenté, & qui eft le même que notre *Medica*, que l'on a rangé fous le genre de *Cytifus*, avant que TOURNEFORT

ait établi celui de *Medicago*, à caufe que les capfules reffemblent à celles du *Medica* ou de la *Luferne.*

Cette plante croît en abondance dans l'Abbruze, où les chevres qui s'en nourriffent, donnent beaucoup de lait, dont on fait une grande quantité de fromage. J'ai eu des femences & des échantillons de la plante, qui m'ont été envoyés par des perfonnes de la plus grande habileté en Botanique, qui m'ont affuré que cette plante étoit généralement regardée par toùs les gens inftruits du pays, comme étant celle dont VIRGILE fait mention.

Le *Trifolium fruticans*, fuivant DODONÉE, ou *Polemonum*, fuivant d'autres, eft improprement appellé *Cytifus* par plufieurs.

Quelques perfonnes prétendent que le *Trifolium candidum Dodonæi*, eft le *Cytife* de COLUMELLE. (*Voy.* fur cela *Lib. Hift. n. 9. 17.* des *herbes à trois feuilles.*)

TRAGUS écrit que l'on doit rejetter l'opinion de ceux qui veulent faire paffer le *Trifolium pratenfe* pour un *Cytife ;* d'autres, que le *Trifolium candidum* de DODONÉE, le *Rectum Melilotum vulgarem*, ne font autres que le *Cytife* des Anciens, fuivant que le dit DODONÉE; mais leur opinion eft mal fondée.

RUELLIUS écrit, qu'il craignoit que MARCELLUS n'ait pris le *Cytife* pour le *Medica.*

MEDICAGO. *Lin.Gen.Plant.* 805. *Medica. Tourn. Inft. R. H.* 410. *tab. 231.* [*Snail Trefoil.*] LUSERNE.

MED

Caractères. Le calice de la fleur est cylindrique , érigé & formé par une feuille decoupée au bord en cinq segmens aigus & égaux ; la corolle est papillonnacée ; l'étendard est ovale & érigé , & les bords font réfléchis ; les ailes font oblongues , ovales, & fixées à la carène par un appendice ; la carène est ovale , divisée en deux parties , obtuse & réfléchie. La fleur a dix étamines, dont neuf font jointes, & l'autre féparée , & qui font toutes terminées par de petites antheres; fon germe est oblong , porté , fur un ftyle court, enveloppé avec les étamines par la carène , & couronné par un fort petit ftigmat : ce germe fe change enfuite en un légume long , comprimé , tordu en fpirale , & qui renferme plufieurs femences en forme de rein.

Ce genre de plantes est renfermé dans les mêmes claffe & fection que le *Medica.*

Les efpeces font :

1°. *Medicago marina , pedunculis racemofis , leguminibus cochleatis , fpinofis , caule procumbente tomentofo. Hort. Cliff.* 378. *Roy. Lugd.-B.* 381. *Sauv. Monfp.* 186. *Gron. Orient.* 230. *Pal. it.* 3. *p.* 590. *Kniph. cent.* 4. *n.* 46; Luferne avec des pédoncules branchus , des légumes épineux & en forme de limaçon , ayant une tige velue & trainante.

Trifolium cochleatum maritimum tomentofum. Bauh. Pin. 329. *Clus. Hift.* 2. *p.* 242.

Medica marina. Lob. icon. 38; Medica marin ou Luferne.

Medicago incana. Riv. 205.

2°. *Medica fcutellata , leguminibus cochleatis , inermibus , ftipulis dentatis , caule angulofo diffufo , foliolis oblongo-ovatis , acutè dentatis ;* Luferne avec des légumes unis , & en forme de limaçon , des ftipules dentées , une tige angulaire & étendue , & des feuilles petites, oblongues , ovales , & à dents aiguës.

Medicago polymorpha. B. ; variété. *Lin. Syft. Plant. t.* 3. *p.* 575. *Sp.* 9.

Medica fcutellata. J. B. 2.384; Luferne communément appelée Limace.

Trifolium cochleatum , fructu latiore. Bauh. Pin. 329.

3°. *Medicago tornata , leguminibus tornatis , inermibus , ftipulis acutè dentatis, foliolis ferratis ;* Luferne avec des légumes tordus & unis , des ftipules à dents aiguës , & des folioles fciées.

Medica tornata minor lenis. Park. Theat. 1116 ; Luferne à plus petits fruits , tordus & liffes.

Medicago polymorpha. Y. variété. *Lin. Syft. Plant. t.* 3. *p.* 576. *Sp.* 9.

4°. *Medicago intertexta , leguminibus cochleatis , fpinofiffimis , aculeis utrinque tendentibus.* Luferne avec des légumes épineux en forme de limaçon , & dont les épines s'étendent de chaque côté.

Medicago polymorpha. V. cinquieme variété. *Lin. Syft. Plant. t.* 3. *p.* 577. *Sp.* 9.

Medicago magno fructu , aculeis furfùm & deorfùm tendentibus Tourn. Inft. R. H. 411 ; Luferne à gros fruits , dont les épines font dirigées vers le haut & vers le bas, communément appelé hériffon.

MED

*Medica echiniata , spinofa ,
echinis magnis, utrinque turbinatis , cum spinulis reflexis. Raii.
Hift.* 962.

5°. *Medicago laciniata , leguminibus cochleatis, spinofis , foliolis acutè dentatis tricufpidifque ;*
Luferne avec des légumes épineux , en forme de limaçon ,
& dont les lobes ont des dentelures aiguës , & font terminées en trois pointes.

*Medicago polymorpha ;*dernière
variété. *Lin. Syft. Plant. t.* 3.
p. 579. *Sp.* 9.

Medica cochleata dicarpos , capfulâ rotundâ spinofâ , foliis eleganter diffectis.H. L. B. ; Trefle
en forme de limaçon , ayant
un double fruit , une capfule
ronde & épineufe , & des feuilles agréablement découpées.

Trifolium echinato capite. Dodard. Mem. 1. p. 123.

Il y a plufieurs autres efpeces de ce genre , qui croiffent naturellement dans les parties chaudes de l'Europe , &
que l'on conferve fouvent dans
les Jardins de Botanique , pour
la variété : mais comme on ne
les admet pas dans les collections d'agrément, je n'en parlerai pas ici.

Marina. La premiere , qui croit
fans culture fur les rivages de
la mer méditerranée , eft une
plante vivace , dont les tiges
font velues, trainantes, d'environ un pied de longueur, &
divifées en plufieurs petites
branches , garnies à chaque
nœud de petites feuilles à trois
lobes , velues , & fupportées
par de courts pétioles : les
fleurs fortent fur les côtés &
aux extrémités des branches en
petites grappes ; elles font d'un
jaune brillant , & font remplacées par des fruits ronds ,
petits , en forme de limaçon ,
remplis de duvet , & armés de
quelques courtes épines : ces
fleurs paroiffent dans les mois
de Juin & de Juillet , &
font remplacées par des femences qui mûriffent en Septembre. Cette plante fe multiplie par fes graines, qu'il faut
femer au printems dans une
plate - bande chaude de terre
feche, où les plantes doivent
refter ; quand elles ont pouffé ,
on peut en tranfplanter deux
ou trois dans de petits pots ,
pour les mettre à l'abri des
froids de l'hiver, qui détruifent
fouvent celles de pleine terre ,
quoiqu'elles puiffent fupporter
les gelées des hivers ordinaires,
quand elles fe trouvent, dans
un fol fec & une fituation
abritée. Celles qui reftent en
place n'exigent aucune autre
culture, que d'être éclaircies
où elles font trop ferrées , &
d'être tenues nettes de mauvaifes herbes. On peut auffi
multiplier cette efpece par boutures , que l'on plante , en Juin
& Juillet , dans une plate-bande
à l'ombre , & qu'on couvre de
vitrages , pour en exclure l'air.
Ces boutures prendront racine
dans l'efpace de fix femaines,
& pourront enfuite être tranfplantées dans une plate-bande ,
comme les plantes élevées de
femence.

Scutellata. La feconde efpece
eft une plante annuelle , qui
croit naturellement dans les parties chaudes de l'Europe ; mais
en Angleterre on la cultive

MED

souvent dans les jardins, à cause de la singularité de son fruit, qui est contourné en forme de limaçon, & qui, à mesure qu'il mûrit, devient d'un brun foncé, & a l'apparence, à une certaine distance, d'un limaçon qui pâture sur la plante. Cette espece a des branches traînantes; ses fleurs sont d'un jaune pâle, & sont produites sur les parties latérales des branches; elles paroissent dans les mois de Juin & de Juillet, & leurs semences mûrissent en automne. On multiplie cette espece au moyen de ses graines, qu'on seme au milieu d'Avril, dans les places où les plantes doivent rester : on les éclaircit quand elles sont trop serrées, & on les tient nettes de mauvaises herbes; c'est en cela que consiste leur culture.

Tornata. La troisieme est aussi une plante annuelle qui croît dans les mêmes contrées que la précédente; elle a des branches traînantes, & des fleurs jaunes comme la seconde : mais son fruit est beaucop plus long & plus tordu; de sorte qu'il a la forme d'un tonneau que l'on nomme pipe, étant moins large à chaque bout qu'au milieu. On conserve souvent cette espece dans les jardins, pour la variété; elle peut être multipliée & traitée de la même maniere que la seconde.

Intertexta. La quatrieme est une plante annuelle qu'on cultivoit beaucoup plus autrefois dans les jardins Anglois, qu'on ne le fait aujourd'hui; ses tiges, ses feuilles & ses fleurs sont semblables à celles des deux

MEL

especes précédentes; mais son fruit est beaucoup plus gros, & fortement armé de longues épines en forme d'hérisson; ce qui lui en a fait donner le nom, Comme ces épines sont hérissées en tout sens, il est difficile de manier ce fruit sans se blesser. On la multiplie par ses graines comme la seconde, & les plantes exigent le même traitement; elles fleurissent dans le mois de Juin, & leurs semences mûrissent en Septembre.

Laciniata. La cinquieme croît naturellement en Syrie; elle est aussi annuelle, & les tiges traînent comme celles de la précédente; leurs lobes sont en forme de coin, fortement dentés sur leurs bords, & terminés au sommet par trois pointes aigues : ses fleurs sont d'un jaune pâle, & le fruit a la forme d'un limaçon; mais il est petit, & armé de plusieurs épines foibles. Cette plante fleurit vers le même tems que l'espece précédente, & peut être cultivée de la même maniere.

MEDICINIER *ou* Pignon d'Inde. *Voyez* Jatropha cur-cas.

MELAMPYRUM. *Tourn. Inst. R. H.* 173. *tab.* 78. *Lin. Gen. Plant* 660. Μελαμπυρον, de μίλας, noir, & πυρὸς Froment. [*Cowwheat.*] Bled de vache.

Caracteres. Le calice de la fleur est persistant, & formé par une feuille tubulée, divisée en quatre segmens sur ses bords; la corolle est labiée, ou pourvue d'un tube recourbé & comprimé au bord; la levre

MEL

supérieure est en forme de casque, comprimée & découpée au sommet ; la levre inférieure est unie, érigée & divisée en trois segmens obtus & égaux : la fleur a quatre étamines en forme d'alène, & recourbés au-dessous de la levre supérieure ; deux de ces étamines sont plus courtes que les autres, & elles sont terminées par des antheres oblongues ; dans le centre est placé un germe à pointe aigue, qui soutient un style & couronné par un stigmat obtus ; le calice se change, quand la fleur est passée, en une capsule oblongue, à pointe aigue, & à deux cellules, qui renferment deux semences grasses & ovales.

Ce genre de plantes est rangé dans la seconde section de la quatorzieme classe de LINNEÉ, qui comprend celles dont les fleurs ont deux étamines longues & deux courtes, avec des semences renfermées dans une capsule.

Les especes sont :

1°. Melampyrum pratense, floribus fœcundis lateralibus, conjugationibus remotis, corollis clausis. Flor. Suec. 513. 548. Reyg. Ged. 1. p. 159. Crantz. Austr. p. 304. Deneck. Gallob. p. 265. Scop. carn. ed. 2. n. 758. Pollich. pal. n. 586. Mattusch. Sil. n. 561 ; Melampyrum avec des fleurs fructueuses, postées à une certaine distance les unes des autres sur les côtés, & des corolles fermées.

Melampyrum foliis imis integerrimis, mediis dentatis, floralibus hastatis. Hall. Helv. n. 308.

Melampyrum luteum lati-folium.

MEL   25

C. B. P. 234 ; Bled de vache, jaune, à feuilles larges.

Melampyrum foliis lanceolatis, florum paribus remotis. Fl. Lapp. 240.

2°. Melampyrum cristatum, spicis quadrangularibus, bracteis cordatis, compactis, denticulatis, imbricatis. Flor. Suec. 510, 545. Crantz. Austr. p. 300. Scop. carn. ed. 2. n. 757. Pall. it. 1. p. 20. Pollich. pal. n. 584. Kniph. cent. 11. n. 71 ; Bled de vache avec des épis quadrangulaires, des bractées en forme de cœur, & comprimées, & des dents imbriquées.

Melampyrum foliis integerrimis, floribus spicatis, bracteis duplicatis, cristatis. Hall. Helv. n. 311.

Melampyrum luteum Linariæ folio. Bauh. prodr. 112.

Melampyrum angusti-folium, cristatum. Pluk. alm. 249. t. 99. f. 2.

Melampyrum luteum angustifolium. C. B. P. 234 ; Melampyrum jaune, à feuilles étroites ; Bled de vache.

3°. Melampyrum arvense, spicis conicis laxis, bracteis dentato-setaceis, coloratis. Flor. Suec. 511 ; 546. Crantz. Austr. p. 301. Deneck. Galiob. p. 265. Pollich. pal. n. 585 ; Melampyrum avec des épis lâches, & de forme conique, des bractées dentées, & garnies de pailles rudes & colorées.

Melampyrum lanuginosum Bæticum. Bauh. pin. 234.

Melampyrum purpurascente comâ. Bauh. pin. 234 ; Melampyrum avec des sommets pourpres, ou le Bled rouge.

Triticum vaccidum. Dod. pemp. 541 ; Bled de vache.

26 MEL MEL

4°. *Melampyrum nemorofum* , *floribus fœcundis lateralibus* ; *bracteis dentatis cordato - lanceolatis* ; *fummis coloratis* , *fterilibus* , *calycibus lanatis. Flor. Suec.* 512. 547. *Gort. Ingr.* 97. *Reyg. Ged.* 1. *p.* 159. *Crantz. Auftr. p.* 302. *Mattusch. Sil. n.* 460. *Oed. Flor. Dan. t.* 305. *Kniph. cent.* 11. *n.* 72 ; Bled de vache avec des fleurs fructueufes & latèrales, des bractées dentées en forme de cœur & de lance, des fommets ftériles & colorés, & des calices laineux.

Melampyrum fylvaticum. Riv. f. 81. *Hall. R.*

Melampyrum comá cœrulea. C. B. P. 234 ; Melampyrum avec des fommets bleus.

Parietaria fylveftris. 1. *Clus. Hift.* 2. *p.* 44 ; Pariétaire des bois.

On cultive rarement ces plantes dans les jardins. La premiere efpece croît naturellement dans plufieurs parties de l'Angleterre; on trouve la feconde en abondance dans les Comtés de Bedfort & de Cambridge. La quatrieme croît dans les parties feptentrionales de l'Europe. La troifieme nait fpontanément dans des terres fablonneufes en Norfolk ; mais elle n'y eft pas bien commune. Dans la Frife occidentale , & en Flandres , on en voit beaucoup parmi les bleds. CLUSIUS prétend qu'elle gâte le pain dans ces pays, & le rend noir , & que ceux qui en font ufage font ordinairement attaqués d'une pefanteur de tête, comme s'ils avoient mangé de l'*ivraie*: mais M. RAY affure qu'il a fort fouvent mangé de ce pain,

& qu'il ne lui a jamais trouvé de goût défagréable , quoiqu'il foit cependant regardé comme mal-fain par les gens du pays, fans qu'ils aient jamais cherché à féparer cette graine du bled. TABERNŒMÒNTANUS déclare qu'il en a fouvent mangé fans en éprouver aucun mal ; que ce pain eft fort bon , & que la graine eft une nourriture excellente pour le bétail, & particuliérement pour engraiffer les bœufs & les vaches ; ce qui doit engager à la cultiver.

Les graines de ces plantes doivent être femées en automne, auffi - tôt qu'elles font mûres, fans quoi il eft rare qu'elles croiffent dans la premiere année. Quand les plantes pouffent, on les débaraffe avec foin des mauvaifes herbes qui les environnent ; & dès qu'elles commencent à montrer leurs fleurs, on les emploie à nourrir le bétail ; mais il ne faut donner aux beftiaux qu'un petit canton à pâturer à la fois , parce qu'ils fouleroient aux pieds & détruiroient les autres plantes.

Les troifieme & quatrieme efpeces font un bel effet avec leurs fommets pourpre & bleus, pendant les mois de Juillet & d'Août. Toutes ces plantes font annuelles.

MELANTHIUM. [*Starrflower.*] Fleur étoilée.

Caractères. La fleur a un calice, fi ce n'eft pas une corolle, compofé de fix pétales oblongs, ovales, étendus & perfiftants ; elle a fix étamines minces, érigées, inférées au—

MEL

deſſus des onglets, & terminées par des antheres angulaires : le germe, qui eſt rayé & globulaire, ſoutient trois ſtyles courbés, diſtincts, & couronnés par des ſtigmats obtus ; ce germe ſe change enſuite en une capſule ovale, & à trois cellules, unies en-dedans, qui renferment pluſieurs ſemences ovales & applaties.

Ce genre de plantes eſt claſſé dans la troiſieme ſection du ſixieme ordre de LINNÉE, qui a pour titre, *Hexandria Trigynia*, avec celles dont les fleurs ont ſix étamines & trois ſtyles.

Les eſpeces ſont :

1°. *Melanthium Virginicum, petalis unguiculatis. Lin. Sp. Plant.* 483 ; Fleur étoilée avec des pétales à onglets.

Melanthium foliis linearibus, integerrimis, longiſſimis ; floribus paniculatis. Gron. Virg. 59.

Aſphodelo affinis Floridana, ramoſo caule, floribus Ornithogali, obſoletis. Pluk. tab. 434. f. 8.

2°. *Melanthium Sibericum, petalis ſeſſilibus. Amœn. Acad. 2. p. 349. t. 11 ;* Fleur étoilée de Sibérie, avec des pétales ſeſſiles.

Ornithogalum ſpicis florum longiſſimis, ramoſis. Flor. Siber. p. 45.

3°. *Melanthium punctatum, petalis punctatis, foliis cucullatis. Amœn. Acad. 6 ;* Melanthium avec des pétales ponctués, & des feuilles en capuchon.

Melanthium Capenſe. Lin. Syſt. Plant. t. 2. p. 127. Sp. 3.

Virginicum. La premiere eſpece croît naturellement dans la Virginie & dans quelques autres parties de l'Amérique

MEL 27

ſeptentrionale ; mais comme elle eſt peu remarquable, on ne la cultive guere que dans les jardins de Botanique ; ſes tiges à fleurs s'élevent à la hauteur de ſix ou huit pieds, & ſe diviſent vers le haut en deux ou trois feuilles linéaires : ſes fleurs, qui ſont compoſées de ſix pétales étendus, & d'une couleur ſombre & uſée, produiſent rarement des ſemences en Angleterre.

En plantant les racines de cette eſpece dans une platebande de terre légere, & pas trop ſeche, elles profiteront, & produiront des fleurs dans ce pays ; mais on la multiplie difficilement.

Sibericum. La ſeconde, qui eſt originaire de la Sibérie, eſt actuellement rare en Angleterre ; mais quand on l'a une fois obtenu, on peut l'y multiplier, en plantant ſes racines bulbeuſes dans une plate-bande, à l'expoſition de l'orient.

Punctatum. La troiſieme, qu'on rencontre au Cap de BonneEſpérance, eſt trop tendre pour réuſſir ici en plein air ; mais ſi on place ſes racines dans une plate - bande, qu'on les couvre d'un châſſis en hiver, & qu'on les traite comme celles de l'*Ixia*, elles profiteront & fleuriront annuellement.

MELASTOMA. *Lin. Gen. Plant.* 481. *Groſſularia. Sloan. Hiſt. Jam. Plum. Sp.* 18. [*The American Gooſeberry-tree.*] Groſeiller d'Amérique.

Caracteres. Le calice de la fleur eſt perſiſtant, & formé par une feuille gonflée comme une veſſie, & obtuſe ; la corolle a

MEL

cinq pétales ronds, & insérés dans le bord du calice : la fleur a dix étamines courtes, & terminées par des antheres oblongues, érigées & un peu courbées ; sous le calice est placé un germe rond, qui soutient un style mince, & couronné par un stigmat recourbé & denté ; ce germe devient ensuite une baie à cinq cellules, couronnée par le calice, & qui renferme plusieurs petites semences.

Ce genre de plantes est rangé dans la premiere section de la dixieme classe de LINNÉE, intitulée, *Decandria Monogynia*, qui comprend celles dont les fleurs ont dix étamines & un style.

Les especes sont :

1°. *Melastoma Plantaginis folio*, *foliis denticulatis, ovatis, acutis. Lin. Sp. Plant.* 389 ; Groseiller d'Amérique, avec des feuilles ovales, à pointe aiguë & dentée.

Grossularia Americana, Plantaginis folio amplissimo. Plum. Sp. 18.

2°. *Melastoma acinodendron*, *foliis denticulatis, sub-trinerviis, ovatis, acutis. Lin. Sp. Plant.* 558. *Lin. Syst. Plant. t.* 2. *p.* 284. *Sp. 1* ; Groseiller d'Amérique, à feuilles ovales, & garnies de dents aiguës, avec trois veines.

Acinodendrum Americanum pentaneuron, foliis crassis, hirsutis, adambitum rarioribus serris. Pluk. mant. 4. *t.* 159. *f.* 1.

Grossularia alia Plantaginis folio, fructu rariore violaceo. Plum. Sp. 18.

Grossulariæ fructu arbor maxi-

MEL

ma, non spinosa. Sloan. Jam. 164. *Hist.* 2. *p.* 84. *t.* 196. *f.* 1.

3°. *Melastoma hirta, foliis denticulatis, quinque-nervibus, ovato-lanceolatis, caule hispido. Lin. Sp.* 390. *Syst. Plant. t.* 2. *p.* 285. *Sp. 4* ; Groseiller d'Amérique, à feuilles dentées, & en forme de lance, ayant cinq veines, & une tige épineuse.

Grossularia fructu non spinoso, Malabathri, foliis longâ & rufâ lanugine hirsutis, fructu majore cæruleo. Sloan. Jam. 165. *Hist.* 2. *p.* 85. *t.* 197. *f.* 2. *Rai. dendr.* 74.

Arbuscula Jamaïcensis, quinque-nerviis, minutissimè dentatis foliis, & caule pubescentibus. Pluk. Alm. 40. *t.* 264. *f.* 1. *vel potiùs.* 265. *f.* 1.

Grossularia Plantaginis folio angustiori hirsuto. Plum. Sp. 18. *ic.* 141.

4°. *Melastoma holosericea, foliis integerrimis, trinerviis, oblongo-ovatis, subtùs tomentosis, racemis brachiatis, spicis bipartitis. Lin. Sp.* 559. *Syst. Plant. t.* 2. *p.* 286 ; Melastoma avec des feuilles entieres, oblongues, ovales, & cotonneuses en-dessous, & des épis de fleurs divisés en deux parties.

Arbor racemosa Brasiliana, folio Malabathri. Breyn. cent. tab. 2. *& 4.*

Acinodendrum Americanum, ampliori folio trinervi, inferiùs albâ lanugine incano, maximo, utrinque glabro. Pluk. mant. 4. *t.* 250. *f.* 2.

Muiua. Marcgr. Bras. 117. *Burm. Ind.* 104.

5°. *Melastoma grossularioïdes, foliis lanceolatis, utrinque glabris, nervis tribus ante basim coëuntibus. Hort.*

MEL

Hort. Cliff. 162 ; Melaſtoma avec des feuilles en forme de lance, unies ſur les deux ſurfaces, ayant trois veines qui ſe joignent avant d'atteindre la bâſe.

Groſſularia fructu non ſpinoſo, Malabathri foliis oblongis, floribus herbaceis, racemoſis, fructu nigro. Sloan. Cat. 165.

6°. *Melaſtoma bicolor, foliis lanceolatis, nervis tribus longitudinalibus, ſubtùs glabris, coloratis. Hort. Cliff.* 162 ; Melaſtoma avec des feuilles en forme de lance, ayant trois veines longitudinales, unies & colorées en-deſſous.

7°. *Melaſtoma Malabathrica, foliis lanceolato-ovatis, quinque-nervibus, ſcabris. Flor. Zeyl.* 171 ; Melaſtoma avec des feuilles en forme de lance, ovales, & à cinq veines rudes.

Melaſtoma-quinque nervia hirta major, capitulis ſericeis villoſis. Burm. Zeyl. 155. *tab.* 73.

Kedali.Rheed.Mal. 4. *p.* 87.42. *Fragarius niger. Rumph. Amb.* 4. *p.* 137. *ſ.* 72.

8°. *Melaſtoma lævigata, foliis oblongo-ovatis, minutiſſimè dentatis, infernè ſericeis, quinque-nervibus, floribus racemoſis* ; Melaſtoma à feuilles oblongues, ovales, légerement dentées ſur les bords, ſoyeuſes en-deſſous, & à cinq nervures, avec des fleurs diſpoſées en paquets longs.

Groſſulariæ fructu arbor maxima, non ſpinoſa, Malabathri folio maximo, inodora, flore racemoſo albo. Sloan. Cat. Jam. 165.

9°. *Melaſtoma petiolata, foliis denticulatis, ovatis, acuminatis, infernè nitidiſſimis, petiolis longiſ-*

Tome V.

MEL

ſimis ; Melaſtoma avec des feuilles ovales, dentées ſur leurs bords, terminées en pointe aigue, fort luiſantes en-deſſous, & ſupportées par de longs pétioles.

10. *Melaſtoma umbellata, foliis cordatis, acuminatis, integerrimis, infernè incanis, floribus umbellatis* ; Melaſtoma avec des feuilles entieres, en forme de cœur, à pointe aigue, blanches en-deſſous, ayant des fleurs qui croiſſent en ombelle.

Sambucus, Barbadenſis dicta, foliis ſub-incanis. Pluk. Phyt. tab. 221. *folio* 6.

11°. *Melaſtoma racemoſa, foliis oblongo-cordatis, acuminatis, denticulato-ſerratis, floribus racemoſis, ſparſis* ; Melaſtoma avec des feuilles oblongues, en forme de cœur, à pointe aigue, ayant des dents en forme de ſcie, & des fleurs éparſes dans la longueur des épis.

12°. *Melaſtoma verticillata, foliis ovato-lanceolatis, quinque-nervibus, ſubtùs aureis, floribus verticillatis, caule tomentoſo* ; Melaſtoma à feuilles ovales, en forme de lance, & fortifiées par cinq nervures de couleur d'or en deſſous, avec des fleurs verticillées, & une tige laineuſe.

13°. *Melaſtoma acuta foliis lanceolatis, acutis, denticulatis, infernè incanis, trinervibus, floribus racemoſis* ; Melaſtoma à feuilles en forme de lance, aiguës, dentées ſur leurs bords, blanches en-deſſous, & à trois nervures, avec des fleurs en paquets.

14°. *Melaſtoma glabra, foliis ovato-lanceolatis, acuminatis, integerrimis utrinque glabris, tri-*

C

MEL

nervibus, floribus racemosis; Melastoma à feuilles entieres, ovales, en forme de lance, terminées en pointe aiguë, à trois veines, & unies sur les deux surfaces, avec des fleurs en paquets longs.

Arbor Surinamensis, Canellæ folio utrinque glabro. Phyt. tab. 249. fol. 5.

15°. *Melastoma quinque-nervia, foliis ovatis, quinque-nervibus scabris, floribus racemosis alaribus*; Melastoma à feuilles ovales, rudes, & à cinq veines, avec des fleurs en paquets sur les côtés des branches.

16°. *Melastoma octandra, foliis lanceolatis, trinervibus glabris, marginibus hispidis*; Melastoma à feuilles unies, en forme de lance, à trois veines, & armées de piquants velus sur les bords.

17°. *Melastoma aspera, foliis ovatis, quinque-nervibus, glabris, marginibus hispidis*; Melastoma avec des feuilles ovales, unies, & à cinq veines, avec des bords velus & épineux.

18°. *Melastoma scabrosa, foliis ovato-lanceolatis, scabris, acuminatis, quinque-nervibus, floribus racemosis*; Melastoma à feuilles en forme de lance, ovales, à pointes aiguës, & à cinq veines, & des fleurs disposées en paquets longs.

Le titre de ce genre lui a été donné par le Professeur BURMAN, d'Amsterdam, dans son *Thesaurus Zeylanicus*: quelques-unes des plantes qui le composent, ont été appelées *Sambucus*, d'autres *Christophoriana*, & plusieurs *Acinodendron* par le Docteur PLUKNET; mais le Chevalier Sir HANS SLOANE,

MEL

& le P. PLUMIER les ont nommées *Grossularia*, d'où je les ai appelées Groseillers, qui est le nom sous lequel quelques espèces sont connues en Amérique.

Plantaginis folio. La premiere s'éleve à la hauteur de quatre ou cinq pieds; sa tige & ses branches sont couvertes de poils longs & bruns; ses feuilles, opposées sur ses branches, ont cinq pouces de longueur sur deux de largeur, sont couvertes d'un duvet brun, & fortifiées par cinq nervures qui s'étendent d'une extrémité à l'autre, mais dont les trois intérieures se joignent avant d'atteindre la base, & par des petites côtes transversales; son fruit croît aux extrémités des branches, sous la forme d'une baie bleue, charnue, & aussi grosse qu'une noix muscade.

Acinodendron. La seconde espece s'éleve sous la forme d'un grand arbre, garni de plusieurs branches courbées, & couvertes d'une écorce brune; ses feuilles sont opposées, unies, entieres, de plus de cinq pouces de longueur sur deux de largeur au milieu, & fortifiées par trois veines profondes qui coulent à travers; les deux surfaces de ces feuilles sont d'un vert clair & uni; elles sont fortement dentées sur leurs bords, & terminées en pointe aiguë: les fruits, qui croissent en épis clairs aux extrémités des branches, fort éloignés les uns des autres sur leurs épis, sont de couleur violette.

Hirta. La troisieme s'éleve à la hauteur de vingt pieds, avec

un tronc gros , & couvert d'une écorce brune ; fes feuilles font fort larges , d'un brun foncé en‑deffus , & d'un brun jaunâtre en‑deffous , douces au toucher , & couvertes d'un duvet mou ; fes tiges font garnies de poils rudes , & fes feuilles placées par paires fur les branches. Cette efpece d'arbre produit un bel effet quand on le regarde d'une certaine diftance.

Holofericea. La quatrieme s'éleve rarement à plus de huit ou dix pieds de hauteur ; fes feuilles ont environ quatre pouces de longueur ; elles ont trois veines, qui fe joignent avant d'atteindre la bâfe ; elles font entieres, fatinées en‑deffous , d'un vert clair au‑deffus , & placées par paires fur les branches.

Groffularioïdes. La cinquieme s'éleve à plus de fept ou huit pieds de hauteur , & fe divife en plufieurs branches couvertes d'une écorce unie , de couleur pourpre ; elles font minces , & garnies de feuilles en forme de lance , de cinq pouces de longueur fur deux de largeur au milieu , unies fur les deux furfaces , entieres fur leurs bords , & terminées en pointe aiguë : fes fleurs, qui naiffent en paquets longs & pendans , font de couleur herbacée , & ont de longs ftyles , qui s'étendent à une grande diftance au‑delà des corolles , & perfiftent après la fleur: fon fruit eft petit & noir , lorfqu'il eft mur.

Bicolor. La fixieme s'éleve à quatre ou cinq pieds de hauteur , & fe divife en plufieurs branches minces , unies , & garnies de feuilles en forme de lance , de trois pouces de longueur fur quinze lignes de largeur , d'un vert luifant au‑deffus , blanches en‑deffous , fillonnées par trois veines longitudinales , qui fe réuniffent avant d'atteindre la bâfe , entieres & alternes fur les branches : fes fleurs fortent en panicules clairs aux extrémités des branches ; elles font petites , blanches , & ont de longs tubes ; elles font remplacées par de petits fruits de couleur pourpre.

Malabathrica. La feptieme a une tige angulaire de fix ou fept pieds de hauteur , qui pouffe des branches oppofées , & garnies de feuilles en forme de lance , ovales , rudes , placées par paires , velues , d'un vert foncé en‑deffus , & d'un vert pâle en‑deffous : fes fleurs, qui fortent aux extrémités des branches ; deux ou trois enfemble , font larges , d'une couleur de rofe tirant fur le pourpre, & placées'dans des calices grands & velus ; elles font remplacées par des fruits ronds de couleur pourpre, couronnés par les calices , & remplis d'une chair pourpre , qui environne les femences.

Lævigata. La huitieme s'éleve à la hauteur de vingt pieds , avec une groffe tige droite , couverte d'une écorce grife , & divifée au fommet en plufieurs branches angulaires & garnies de feuilles oblongues , ovales, de près d'un pied de longueur fur fix pouces de largeur au milieu , d'un vert foncé

en-deſſus, ſoyeuſes en-deſſous, avec cinq fortes côtes longitudinales, dentées ſur leurs bords, & oppoſées : ſes fleurs ſortent aux extrémités des branches en paquets longs & lâches : elles ſont blanches, & produiſent des fruits ronds de couleur pourpre, & remplis d'une chair qui renferme les ſemences.

Petiolata. La neuvieme s'éleve à la hauteur de près de trente pieds, avec une tige forte, droite, couverte d'une écorce griſe, & diviſée au ſommet en pluſieurs branches angulaires applaties, & garnies de feuilles ovales, dentées ſur leurs bords, de ſept pouces de longueur, ſur preſque cinq de largeur, portées par paires oppoſées ſur de fort longs pétioles, d'un vert luiſant en-deſſus, d'une couleur d'or pâle ſatinée en-deſſous, & fortifiées par cinq fortes côtes longitudinales, & un grand nombre de plus petites, tranſverſales : ſes fleurs ſont produites en panicules lâches aux extrémités des branches ; elles ſont blanches, & remplacées par des fruits de couleur pourpre, & à-peu près de la même groſſeur que ceux de la précédente.

Umbellata. La dixieme a une tige d'arbriſteau de dix à douze pieds de hauteur, couverte d'une écorce velue, & diviſée vers ſon ſommet en pluſieurs branches garnies de feuilles en forme de cœur, terminées en pointe aiguë, de cinq pouces de longueur ſur trois de largeur vers leur bâſe, entieres ſur leurs bords, d'un vert foncé

au-deſſus, & blanches en-deſſous, à cinq veines longitudinales, avec pluſieurs plus petites, tranſverſales, oppoſées & ſupportées par des pétioles velus, & de deux pouces & demi de longueur : ſes fleurs ſont produites aux extrémités des branches en eſpece d'ombelle ; elles ſont d'une couleur de roſe pâle, larges & placées dans des calices velus ; elles produiſent des fruits noirs, & un peu plus gros que les baies de *Sureau.*

Racemoſa. La onzieme s'éleve à la hauteur de huit à neuf pieds, en une tige d'arbriſſeau, couverte d'une écorce d'un brun foncé, & diviſée au ſommet en pluſieurs branches écartées, & garnies de feuilles oblongues en forme de lance, de ſix pouces de longueur ſur trois de largeur vers la bâſe, terminées en pointe aiguë, diviſées ſur leurs bords en dentelures très-fines, unies ſur les deux ſurfaces, & d'un vert foncé.

Verticillata. La douzieme a une tige d'arbriſſeau de cinq ou ſix pieds de hauteur, diviſée en pluſieurs petites branches couvertes d'une écorce velue, cotonneuſe, & de couleur de fer rouillé : ces branches ſont garnies de feuilles ovales en forme de lance, d'un pouce & demi de longueur ſur neuf lignes de largeur au milieu, d'un vert foncé en-deſſus, & d'une couleur de fer rouillé en-deſſous, avec cinq veines longitudinales oppoſées, & ſeſſiles aux branches : ces fleurs ſont produites en têtes verticillées aux nœuds des tiges ;

MEL

elles font petites, de couleur tirant fur le pourpre, & des fruits petits & noirs leur fuccedent.

Acuta. La treizieme eft un arbriffeau qui ne s'élève guere qu'à trois pieds de hauteur, & fe divife vers le bas en plufieurs branches minces, & garnies de feuilles en forme de lance, terminées en pointe aiguë, de cinq pouces de longueur fur un & demi de largeur au milieu, fciées fur leurs bords, & d'un vert foncé en-deffous, avec trois veines longitudinales, oppofées & fupportées par de courts pétioles : fes fleurs naiffent en paquets clairs aux extrémités des branches; elles font blanches, & remplacées par de petits fruits de couleur pourpre.

Glabra. La quatorzieme a une tige d'arbriffeau de huit à neuf pieds de hauteur, & divifée vers le fommet en plufieurs branches minces, liffes, & garnies de feuilles ovales en forme de lance, de fept pouces de longueur fur trois de largeur, terminées en pointe aiguë, entieres fur leurs bords, unies fur les deux furfaces oppofées, & à trois veines longitudinales : fes fleurs fortent aux extrémités des branches, en panicules lâches & produifent de petits fruits de couleur pourpre.

Quinque-nervia. La quinzieme s'élève à la hauteur de cinq ou fix pieds avec plufieurs tiges d'arbriffeau, qui fe divifent en plufieurs branches courbées, & garnies de feuilles ovales de la longueur de trois pouces

MEL 33

fur prefque autant de largeur, traverfées par cinq veines longitudinales, rudes, d'un vert foncé en-deffus, d'un vert pâle en-deffous, dentées fur leurs bords, fupportées par des pétioles très-velus, quelquefois oppofées, & d'autres fois alternes fur les branches : fes fleurs naiffent fur les côtés des tiges en paquets fort lâches; elles font petites, d'une couleur herbacée, & font remplacées par de petits fruits de couleur pourpre, & remplis de petites femences.

Octandria. La feizieme a une tige d'arbriffeau de fept à huit pieds de hauteur, & divifée en plufieurs branches liffes, & garnies de feuilles en forme de lance, de quatre pouces environ de longueur fur quinze lignes de largeur au milieu, liffes fur les deux furfaces, d'un vert foncé, garnies de trois veines longitudinales, & dont les bords font fortement garnis de poils hériffés & piquans : fes fleurs font difpofées en paquets lâches aux extrémités des branches; elles font petites, de couleur pourpre, & font remplacées par de petits fruits noirs.

Afpera. La dix-feptieme reffemble à la précédente en plufieurs parties; mais fes feuilles font ovales, d'un peu plus de deux pouces de longueur fur quinze lignes de largeur, garnies de cinq veines longitudinales, liffes fur les deux furfaces, d'un vert foncé, oppofées, & fupportées par de courts pétioles : fes fleurs croiffent en paquets lâches aux extrémités

C 3

des branches ; elles font plus groffes que celles de l'efpece précédente , & de la même couleur : les bords des feuilles de cette efpece font fortement garnis de poils piquans, comme ceux de la précédente.

Scabrofa. La dix—huitieme s'éleve avec une tige d'arbriffeau à la hauteur de huit à neuf pieds , & fe divife en branches oppofées , ainfi que les feuilles , qui ont fept pouces de longueur fur trois de largeur, & qui font rudes en-deffus, entieres fur leurs bords , terminées en pointe aiguë , d'un vert tendre fur les deux furfaces, & fupportées par de courts pétioles : fes fleurs, qui naiffent en panicules larges , & clairs aux extrémités des branches , font petites , blanches , '& remplacées par des fruits ronds , petits , & de couleur pourpre.

Culture. Toutes les efpeces font originaires des parties chaudes de l'Amérique ; mais on en connoît encore un grand nombre d'autres , dont la plupart ont été trouvées à la Jamaïque par le Docteur HOUSTOUN , qui a envoyé les femences de plufieurs en Europe. Quelquesunes ont réuffi; mais prefque toutes les plantes qu'elles ont produites ont été détruites par le rude hiver de 1740 ; depuis ce tems, on n'en a point envoyé de nouvelles en Europe.

La diverfité du feuillage de ces plantes forme un coup-d'œil très-agréable ; plufieurs d'entr'elles ont des feuilles fort larges ; & la plupart font de differentes couleurs fur les deux furfaces; le deffous eft blanc ,

de couleur d'or ou brun , & leur furface fupérieure offre différentes teintes de vert, de maniere qu'elles ont une belle apparence dans les ferres pendant toute l'année ; d'ailleurs ces plantes n'ont rien d'utile qui puiffe les faire rechercher , & ce n'eft que la fingularité de leur feuillage qui leur fait donner une place dans toutes les collections des curieux.

On voit aujourd'hui très-peu de ces plantes dans les jardins de l'Europe, ce qui peut, être occafionné par la difficulté de les tranfporter d'Amérique dans des pots , & de ce que leurs femences, qui font très-petites , fe deffechent bientôt, quand on les a tirées de la chair, & réuffiffent rarement. La meilleure méthode pour obtenir ces plantes eft de mettre les fruits entiers dans du fable fec, auffi-tôt qu'ils font mûrs, & de les envoyer en Angleterre par les premiers vaiffeaux. Dès qu'on les reçoit, il faut les tirer hors du fable , mettre les femences dans des pots remplis d'une terre légere , & les plonger dans une couche de tan de chaleur modérée. Lorfque les plantes pouffent , & qu'elles font en état d'être enlevées , on les place féparément chacune dans de petits pots remplis de terre légere ; on les replonge dans une autre couche de tan , & on les traite enfuite comme l'*Annona*, dont le Lecteur peut confulter l'article.

MELESE. *Voyez* LARIX.

MELIA. *Lin. Gen. Plant.* 473. *Azedarack. Tourn. Inft. R.H.* 616. *tab.* 387. [*The Bead-tree.*]

MEL

Arbre à Chapelet ; Lilas des-Indes, Faux Sycomore.

Caractères. Le calice de la fleur est petit, érigé & formé par une feuille divisée à son sommet en cinq pointes obtuses ; la corolle est composée de cinq pétales longs, étroits, en forme de lance, & étendus ; elle a un nectaire cylindrique, formé par une feuille aussi longue que les pétales, & découpée en deux parties sur ses bords : la fleur a dix petites étamines insérées au sommet du nectaire, & terminées par des antheres qui ne paroissent pas au-dessus ; son germe est conique, & soutient un style cylindrique, & couronné par un stigmat obtus & denté ; ce germe se change ensuite en un fruit mou & globulaire, qui renferme un noyau rond, sillonné par cinq rainures rudes, & à cinq cellules, qui contienoent chacune une semence oblongue.

Ce genre de plantes est rangé dans la premiere section de la dixieme classe de LINNÉE, intitulée *Decandria Monogynia*, qui renferme celles dont les fleurs ont dix étamines & un style.

Les especes sont :

1°. *Melia Azedarach*, *foliis bipinnatis*. Flor. Zeyl. 162. Gron. Orient. 133. Kniph. cent. 2. n. 44 ; Lilas des Indes à feuilles doublement ailées.

Azedarach. Dod. Pempt. 848. Burm. Zeyl. 40. Herm. Lugd.-B. 652. Raii Hist. 1546 ; Faux Sycomore ou Lilas des Indes.

Melia foliis decompositis. Hort. Cliff. 161. Roy. Lugd.-B. 462.

MEL 35

Arbor Fraxini folio, flore cœ-ruleo. Bauh. Pin. 415.

Pseudo-Sycomorus. Cam. Epit. 181.

B. *Melia semper virens.* Lin. Syst. Plant. t. 2. p. 272. Variété

Azedarach semper virens & florens. Tourn. Inst. 616.

Azadirachta Indica, foliis ramosis minoribus, flore albo, sub-cæruleo, purpurascente, majore. Comm. Hort. 1. p. 147. t. 76.

2°. *Mel'a azadirachta, foliis pinnatis.* Hort. Cliff. 161. Fl. Zeyl: 161. Roy Lugd.-B. 462 ; Lilas des Indes à feuilles ailées.

Olea Malabarica Fraxini folio. Pluk. Alm. 269. t. 147. f. 1.

Azedarach foliis falcato-serratis. Burm. Zeyl. 40. t. 15.

Azadirachta Indica folio Fraxini, Breyn. ic. 21. t. 15.

Arbor Indica, Fraxino similis, Oleæ fructu. Bauh. pin. 416.

Aria Bepou. Rheed. Mal. 4. p. 107. f. 52.

Azedarach. Burm. Fl. Ind. 101.

Azedarach. La premiere espece, qui croit naturellement en Syrie, a été portée en Espagne & en Portugal, où elle est à présent aussi commune que si elle en étoit originaire ; dans les pays chauds, elle devient un grand arbre qui s'étend au-dehors en plusieurs branches, garnies de feuilles ailées, & composées de trois ailes plus petites, dont les lobes sont entaillés & dentés sur leurs bords ; elles sont d'un vert foncé en-dessus, & plus pâle en-dessous : les fleurs sortent sur les côtés des branches en longs paquets clairs ; elles sont composées de cinq pétales longs, étroits, en

C 4

MEL

forme de lance, & de couleur bleue, & font remplacées par des fruits oblongs de la groffeur d'une petite *Cerife*, d'abord verts, mais qui deviennent d'un jaune plus pâle en mûriffant, & qui renferment une noix fillonnée par cinq rainures profondes, & à quatre ou cinq cellules, qui contiennent chacune une femence oblongue. Cette efpece produit des fleurs en Juillet, mais elle ne donne pas fouvent des femences en Angleterre ; elle perd fes feuilles en automne, & en pouffe de nouvelles au printems. On dit que la chair qui environne la noix eft vénéneufe, & qu'en la mêlant avec de la graiffe, elle empoifonne les chiens qui la mangent. Les Catholiques Romains percent ces noix pour en faire des chapelets.

On a introduit ces plantes, il y a quelques années, dans les Ifles de l'Amérique, où, fuivant ce que l'on m'a affuré, elles reftent en fleurs, & produifent des fruits durant la plus grande partie de l'année. On m'a envoyé quelques-uns de ces fruits fous le titre de *Lilas des Indes*, & les plantes qu'ils ont produites ont été les mêmes que celles qui viennent de la Syrie.

On multiplie cette efpece par fes graines, qu'on peut faire venir de l'Italie ou de l'Efpagne, où ces arbres produifent annuellement des fruits mûrs. On feme ces baies dans des pots remplis de terre fraiche & légere ; on les plonge dans une couche de tan de chaleur tempérée ; & fi les femences

MEL

font bonnes, elles poufferont dans l'efpace d'un mois ou de fix femaines. Quand ces plantes ont paru, on les arrofe fréquemment, & on leur donne beaucoup d'air, en foulevant les vitrages chaque jour : au mois de Juin, on les expofe en plein air, dans une fituation bien abritée, pour les durcir avant l'hiver ; & en Octobre, on place les pots qui les contiennent fous un chaffis de couche chaude, où les plantes puiffent jouir de l'air dans les tems doux, & être à l'abri des fortes gelées. Pendant l'hiver, on les arrofe légerement, mais pas trop fouvent ; car lorfque leurs feuilles font tombées, il ne leur faut que très-peu d'humidité.

Au mois de Mars fuivant, on tire les plantes des pots de femences, pour les divifer ; on les met chacune féparément dans de petits pots remplis d'une terre fraiche & légere, & on les plonge dans une couche de chaleur modérée, pour les aider à prendre racine, & hâter leur accroiffement; mais il ne faut pas les laiffer trop filer. Au mois de Juin, on les expofe au grand air comme auparavant : pendant les trois ou quatre premiers hivers, on les tient à l'abri du froid ; mais lorfqu'elles font devenues affez groffes & ligneufes, elles fupportent le plein air contre une muraille expofée au midi. La bonne faifon pour les mettre en pleine terre, eft le mois d'Avril : alors on les tire des pots, en coupant feulement l'extérieur de la motte de terre ; on

MEL

les met dans les trous qui ont été creufés pour les recevoir, & l'on ferre exactement la terre autour de leurs racines : quand le tems eft fec, on les arrofe deux fois par femaine, jufqu'à ce qu'elles aient pouffé de nouvelles fibres ; mais il faut les planter dans un fol fec, fans quoi il eft à craindre que les fortes gelées ne les détruifent.

Azadirachta. La feconde efpece croît naturellement en Amérique, où elle devient un grand arbre ; fa tige eft épaiffe, fon bois eft d'un jaune pâle, & fon écorce eft de couleur pourpre foncé, & d'une faveur fort amere ; fes branches s'étendent au loin en tout fens, & font garnies de feuilles blanches, ailées, & compofées de cinq ou fix paires de lobes oblongs, à pointe aiguë, & terminées par un lobe impair ; ils font fciés fur leurs bords, d'un vert clair, & d'une odeur défagréable ; fes feuilles font fupportées par de longs pétioles, quelquefois oppofés & quelquefois alternes : fes fleurs, qui naiffent en panicules longs & branchus fur les parties latérales des branches, font petites, blanches, & placées dans de petits calices découpés en cinq fegmens aigus ; elles font remplacées par des fruits ovales de la groffeur d'une petite olive, d'abord verts, enfuite jaunes, & pourpre lorfqu'ils font mûrs. La chair qui environne la noix eft huileufe, âcre & amere ; la noix eft blanche, & de la même forme que celle de l'efpece précé-

MEL

dente. Cet arbre croît dans des terreins fablonneux de l'Amérique, & dans l'Ifle de Céylan, où il eft toujours vert. Il produit des fleurs & des fruits deux fois l'année.

Cette efpece eft à préfent fort rare en Angleterre, ainfi que dans les Jardins Hollandois, où elle étoit plus commune, il y a quelques années : on la multiplie par fes graines, ainfi que la premiere ; mais comme elle eft beaucoup plus tendre, il faut la tenir conftamment dans une couche de tan, tandis que fes plantes font jeunes. En été, on peut les placer fous un châffis ; mais en hiver, on les tient dans la ferre chaude, où on les traite comme les autres plantes qui viennent des mêmes contrées : quand elles ont acquis de la force, on peut les conduire plus durement, en les tenant en hiver dans une ferre fèche, en les expofant pendant deux ou trois mois de l'été, en plein air, dans une fituation chaude & abritée ; mais on ne doit pas les laiffer dehors trop longtems, & on les arrofe légerement en hiver. Au moyen de ce traitement, ces plantes produiront des fleurs annuellement ; & comme elles confervent leurs feuilles toute l'année, elles fervent d'ornement dans les ferres chaudes.

La premiere efpece, à laquelle on donne ordinairement le nom de *Zizyphus alba*, en Portugal & en Efpagne ; de *Pfeudo-Sycomorus*, en Italie, eft la même que la plupart des Botaniftes appellent *Azedarach* ;

mais LINNÉE, a changé ce nom en celui de *Melia*, qui a été donné par THÉOPHRASTE à une espece de *Frêne*.

MÉLIANTHE ou PIMPRE-NELLE D'AFRIQUE. *Voyez* ME-LIANTHUS.

MÉLIANTHUS. *Tourn. Inst. R. H.* 430. *tab* 215. *Lin. Gen. Plant:* 712 μιλίανθος de μιλι du miel, & ανθος; une fleur. [*Honey-flower.*] Fleur à miel, Mélianthe ou Pimprenelle d'Afrique.

Caractères. Le calice de la fleur est large, coloré, inégal, & divisé en cinq segmens, dont les deux supérieurs sont oblongs & érigés; l'inférieur, court, & semblable à un sac, & ceux du milieu en forme de lance, & opposés, la corolle a quatre pétales étroits; en forme de lance, réfléchis à leurs pointes, étendus au-dehors, & figurés comme le calice en deux levres jointes à leurs côtés; la fleur a un nectaire formé par une feuille placée dans le segment inférieur du calice fixé au réceptacle, court, comprimé sur le côté, & découpé sur ses bords; elle a quatre étamines érigées & en forme de lance, dont les deux inférieures sont un peu plus courtes que les deux autres, & qui sont toutes terminées par des antheres oblongues, & en forme de cœur : dans le centre est placé un germe quarré, qui soutient un style érigé, & couronné par un stigmat divisé en quatre parties. Ce germe se change dans la suite en une capsule quadrangulaire, & divisée par des cloisons en plusieurs cellules,

dont chacune renferme une semence presque globulaire, fixée au centre de la capsule.

Ce genre de plantes est rangé dans la seconde section de la quatorzieme classe de LINNÉE, qui comprend celles dont les fleurs ont deux étamines longues, & deux plus courtes, & dont les semences sont renfermées; dans les calices.

Les especes sont :

1°. *Melianthus major, stipulis solitariis, petiolo adnatis. Hort. Cliff.* 492. *Hort. Ups.* 181. *Roy. Lugd.-B.* 402. *Kniph. cent* 12. *n.* 70 ; Mélianthe avec des stipules simples qui crbissent près du pétiole.

Melianthus Africanus. H. L. B. 414; le plus grand Mélianthe d'Afrique.

2°. *Melianthus minor, stipulis geminis distinctis. Hort. Cliff.* 492. *Roy. Lugd.-B.* 402. *Kniph. cent.* 8. *n.* 68. *Fabric. Helmst.* 420 ; le plus petit Mélianthe, avec deux stipules distinctes.

Melianthus Africanus minor, fœtidus. Com. Rar. Pl. 4. *tab.* 4.

Melianthus Hysiquanensis minor, fœtidus. Raii. Dendron. 120.

Major. La premiere espece croît naturellement au Cap de Bonne-Espérance, d'où elle a été portée en Hollande en l'année 1672 ; elle a une racine ligneuse & vivace, qui s'étend au loin de tous côtés, & qui produit plusieurs tiges ligneuses de quatre ou cinq pieds de hauteur, & herbacées vers le sommet, où elles sont garnies de feuilles larges & ailées, qui embrassent les tiges de leur bâse. Ces feuilles ont une stipule large & simple,

MEL

fixée sur le côté supérieur du périole, avec deux oreilles à la bâfe, qui embraffent auffi la tige ; elles font formées par quatre ou cinq paires de lobes fort larges, terminées par un lobe impair, & profondément découpées fur leurs bords en fegmens aigus, & de couleur grife ; entre ces lobes regne une bordure à double feuille ou aîle fur le côté fupérieur de la côte du milieu ; de maniere qu'elle joint la bâfe des lobes enfemble. Cette bordure eft auffi profondément dentée : les fleurs font placées en longs épis entre les feuilles vers le fommet des tiges ; elles font d'un brun de chocolat, & de la forme des fleurs en gueule ; mais elles different de celles de cette claffe, en ce qu'elles ont quatre pétales ; elles font remplacées par des capfules oblongues, quarrées, & divifées par une cloifon centrale en quatre cellules, qui renferment chacune une femence ronde. Cette plante fleurit dans le mois de Juin ; mais elle ne produit point de femences en Angleterre, à moins que l'année ne foit très-chaude.

On confervoit autrefois cette plante dans les ferres, comme les autres efpeces exotiques tendres ; mais en la plaçant dans un fol fec, & à une expofition chaude, elle fupporte très-bien le froid de nos hivers ordinaires. Dans les tems de gelée, les fommets des tiges font quelquefois détruits, mais les racines fubfiftent & repouffent au printems fuivant, de forte qu'on rifque

MEL 39

peu de la perdre. Celles qui croiffent en plein air, fleuriffent toujours mieux que celles qu'on tient renfermées dans la ferre, parce qu'elles filent moins, ce qui empêche communément les plantes de fleurir ; car il eft très-rare de voir produire des fleurs à celles qui fe trouvent placées dans les ferres, où elles pouffent foiblement, & périffent bien-tôt après ; de forte que, malgré que les tiges deviennent ligneufes, elles ne font cependant pas de longue durée : mais les racines s'étendent beaucoup, quand elles ont affez de place pour cela, & pouffent annuellement une grande quantité de rejettons. Quand les plantes croiffent en pleine terre, la plupart de ces tiges ou rejettons, qui ne font pas endommagés par la gelée, fleuriffent ordinairement au printems fuivant : ainfi, la méthode la plus fûre pour les faire fleurir, eft de couvrir leurs branches dans les tems de gelée avec des rofeaux ou des nattes, pour empêcher le froid de détruire leurs fommets, & encore mieux de les placer contre une muraille à une bonne expofition, & fur des décombres fecs, dans lefquels elles ne pafferont pas fi vigoureufement que dans une bonne terre, où elles deviendront moins fucculentes, & feront par conféquent moins fujettes à être endommagées par le froid. Si l'hiver eft rude, on attache les tiges contre la muraille, & on les couvre pour les conferver, fans quoi elles périffent fouvent juf-

MEL

qu'à la racine, & produifent rarement des fleurs.

On peut multiplier cette efpece par rejettons ou par marcottes, que l'on choifit dans les jeunes branches de côté, & que l'on tient couchées depuis le mois de Mars jufqu'en Septembre, en obfervant de choifir celles qui font garnies de fibres. Lorfqu'elles ont pouffé des racines, elles n'exigent plus aucun autre foin que d'être tenues nettes de mauvaifes herbes : on les multiplie auffi par boutures, qu'on peut planter pendant tout l'été, & qui prennent très-bien racine, fi on les arrofe & fi on les tient à l'ombre : on peut les tranfplanter enfuite dans les places qui leur font deftinées.

Minor. La feconde efpece, qu'on rencontre auffi dans les environs du Cap de Bonne-Efpérance, d'où elle a été apportée en Europe, s'élève avec des tiges rondes, molles & ligneufes, à la hauteur de cinq ou fix pieds, & pouffe deux ou trois branches latérales, garnies de feuilles ailées, comme celles de la précédente, mais de moitié moins larges, & qui ont deux ftipules diftinctes qui adherent à leurs périoles. Ces feuilles font d'un vert foncé en-deffus, & blanchâtres en-deffous : fes fleurs fortent des parties latérales des tiges en panicules lâches & pendans, qui foutiennent chacun fix ou huit fleurs de la même forme que celles de la première efpece, mais plus petites, & dont les pétales ont leur bâfe verte, & leur extrémité de cou-

MEL

leur de *Safran* ; en-dehors, dans la partie gonflée des pétales, eft une tache d'un beau rouge. Ces fleurs ont deux étamines longues & deux plus courtes, qui font toutes terminées par des antheres jaunes ; elles produifent des capfules quarrées, & plus courtes que celles de la précédente, dans lefquelles font renfermées quatre femences ovales dans des cellules féparées. Cette plante fleurit dans le même tems que la première.

Cette efpece n'étend pas fes racines comme la précedente ; auffi ne fe multiplie-t-elle pas avec autant de facilité : mais fes boutures étant plantées fur une vieille couche chaude, dont la chaleur eft éteinte, & couvertes de cloches, pour en exclurre l'air, prennent racine affez aifément ; elles peuvent être mifes dans des pots, & abritées en hiver fous un châffis de couche ordinaire, pendant un an ou deux, jufqu'à ce qu'elles aient acquis de la force ; enfuite on les plante dans une plate-bande chaude, & on les traite comme l'efpece précédente. Au moyen de ce traitement, je les ai vu fleurir beaucoup mieux qu'aucune de celles qui avoient été foignées plus délicatement. Ces plantes perfectionnent leur femences dans les années favorables.

MELILOT. *Voy.* TRIFOLIUM MELILOTUS.

MELILOTUS. *Voyez* TRIGONELLA.

MELINET. *Voyez* CERINTHE.

MELINET ORIENTAL *V*, ONOSMA ORIENTALIS. L.

MEL

• MELISSA. *Tourn. Inst. R.H.*
193. *tab.* 91. *Lin. Gen. Plant.*
647 ; ainsi appelée de μίλι
miel , parce que les abeilles en
recueillent fur cette plante ;
elle est auffi nommée Meliffo-
phyllon φιλλον & μίλι feuille
de miel. [*Baume*] Meliffe.

Caractéres. Le calice de la
fleur , qui est ouvert , angulaire , & en forme de cloche , a
fon bord en deux levres , dont
la fupérieure est découpée en
trois parties ouvertes & réfléchies , & l'inférieure est courte,
aiguë , & divifée en deux fegmens ; la corolle est labiée ,
& fon tube est cylindrique ;
fes levres font ouvertes ; la
fupérieure est courte , érigée ,
fourchue , ronde , & découpée
à l'extrémité ; l'inférieure est
divifée en trois parties , dont
celle du milieu est la plus large : la fleur a quatre étamines
en forme d'alêne , dont deux
font auffi longues que la corolle , & les deux autres n'ont
que la moitié de cette longueur ; elles font terminées par
de petites antheres , qui fe joignent par paires ; fon germe est
divifé en quatre parties, qui foutiennent un style mince , auffi
long que la corolle, placé, avec
les étamines fous la levre fupérieure , & couronné par un
stigmat mince , divifé en deux
parties , & réfléchie. Le germe
fe change , quand la fleur est
paffée , en quatre femences ,
nues & poftées dans le calice.

Ce genre de plantes est rangé
dans la premiere fection de la
quatorzieme claffe de LINNÉE ,
qui renferme celles dont les

MEL 41

fleurs ont deux étamines longues , & deux plus courtes ,
& dont les femences font nues.

Les efpeces font :

1°. *Meliffa officinalis* , *racemis
axillaribus verticillatis , pedicellis
fimplicibus. Lin. Sp. Plant.* 592.
Scop. carn. ed. 2. *n.* 739. *Kniph.
cent.* 4. *n.* 48. *Sabb. Hort.* 3. *f.*
61. *Blackw. f.* 27. *Regn. bot.* ;
Méliffe avec des paquets de
fleurs verticillées , qui fortent
des côtés des tiges , & font fou-
tenus par des pédoncules fimples & petits.

Meliffa hortenfis. C. B. P.
229 ; Méliffe de jardin ou Baume commun. Citronnelle.

*Apiaftrum five Meliffophyllum.
Lob. ic.* 227.

2°. *Meliffa Romana* , *floribus
verticillatis , feffilibus , foliis hirfutis* ; Méliffe avec des fleurs
verticillées & feffiles , & des
feuilles velues.

*Meliffa Romana molliter hir-
futa & grave-olens. H. R. Par.* ;
Méliffe Romaine, avec des feuilles garnies d'un duvet mou ,
ayant une odeur forte.

3°. *Meliffa grandi-flora* , *pedunculis axillaribus dichotomis ,
longitudine florum. Lin. Sp. Plant.*
592. *Knip. cent.* 7. *n.* 56 ; Méliffe dont les pédoncules qui
fortent des aiffelles de la tige ,
font fourchus , & de la longueur des fleurs.

*Thymus racemis lateralibus ,
fparfis ; corollis calyce triplò longioribus. Scop. carn. ed.* 1. *p.* 457.
n. 3. *ed.* 2. *n.* 732.

Calamintha magno flore. C.B.P.
229 ; Calament à grandes fleurs.

*Calamintha montana praeftantior.
Befl. Fyft. Aeft.* 7. *f.* 1.

4°. *Meliffa Calamintha* , *pe—*

MEL

dunculis axillaribus dichotomis , longitudine foliorum. *Lin. Sp. Plant.* 593. *Mat. Med.* 153. *Crantz. Auftr.* p. 285. *Kniph. cent.* 4 ; Méliffe avec des pédoncules fourchus, qui fortent des aiffelles de la tige, & font auffi longs que les feuilles.

Thymus Calamintha. Scop. carn. ed. 2. *n.* 733.

Calamintha vulgaris & officinarum Germaniæ. C. B. P. 228 ; Calament commun des boutiques d'Allemagne.

Calamintha montana. Dod. pempt. 98. *Riv. f.* 46. *Blackw. f.* 166.

5°. *Meliffa Nepeta, pedunculis axillaribus dichotomis , folio longioribus , caule decumbente. Lin. Sp. Plant.* 593 ; Méliffe avec des pédoncules fourchus, & plus longs que les feuilles qui fortent des aiffelles de la tige, & une tige tombante.

Calamintha Pulegii odore fivè Nepeta. Bauh. Pin. 228 ; Calament à odeur de Pouliot, ou Pouliot fauvage.

6°. *Meliffa Cretica, racemis terminalibus , pedunculis folitariis breviffimis. Lin. Sp. Plant.* 593 ; Baume avec des épis de fleurs qui terminent les tiges, & qui croiffent fur des pédoncules fimples & très-courts.

Calamintha incana, Ocymi foliis. C. B. P. 228 ; Calament blanc, à feuilles de Bafilic.

Calamintha, Pulegii odore minor. Barr. ic. 1166.

7°. *Meliffa Majorani-folia , foliis ovatis glabris , floribus verticillatis feffilibus, pedunculis folitariis breviffimis ;* Baume à feuilles ovales & unies, à fleurs verticillées & feffiles, & à tiges fimples & très-courtes.

Calamintha Romana , Majoranæ folio , Pulegii odore. Eocc. mus. ; Calament Romain à feuilles de Marjolaine , & à odeur de Pouliot.

8°. *Meliffa fruticofa, ramis attenuatis, virgatis , foliis fubtùs tomentofis. Lin. Sp. Plant.* 593 ; Baume en arbriffeau avec des branches minces comme des verges, & des feuilles velues en-deffous.

Calamintha Hifpanica frutefcens, Mari folio. Tourn. Inft. 194 ; Calament d'Efpagne en arbriffeau , à feuilles de Marum.

Calamintha montana incana minor. Moris. Hift. 3. *p.* 413. *n.* 6.

Officinalis. La premiere efpece croît naturellement fur les montagnes , près de Geneve, & dans quelques parties de l'Italie : on la cultive ici comme une plante médicinale, & pour la cuifine ; elle a une racine vivace & une tige annuelle , quarrée , branchue , de deux ou trois pieds de hauteur, & garnie à chaque nœud de feuilles placées par paires , de deux pouces & demi de longueur fur près de deux de largeur à leur bâfe , plus étroites vers leur extrémité , & découpées fur leurs bords ; celles du bas ont de longs pétioles : fes fleurs naiffent en petits paquets lâches aux aiffelles de la tige , fur des pédoncules minces & verticillés ; elles font labiées, leur levre fupérieure eft érigée & fourchue , l'inférieure eft divifée en trois parties , & celle du milieu eft ronde , & découpée à fon extrémité. Ces fleurs font blanches , & paroiffent en Juillet. La plante entiere ré-

MEL

pand une odeur agréable, qui approche un peu de celle du *Citron*.

On la regarde comme cordiale, céphalique, & propre à guérir les maux de tête & de nerfs. On extrait de cette herbe une eau simple; on s'en sert aussi en guise de thé, & on lui attribue sous cette forme des effets salutaires. Cette plante donne une variété à feuilles panachées.

On la multiplie aisément, en divisant ses racines : on fait cette opération au mois d'Octobre, afin que les rejettons puissent avoir le tems de s'enraciner avant les gelées. On peut diviser ces racines en trèspetites parties; car il suffit que chacune ait trois ou quatre boutons : on les plante à deux pieds de distance dans des planches d'une terre commune de jardin, où elles s'étendront & s'entrelaceront; la seule culture qu'elles exigent est d'être tenues nettes de mauvaises herbes, & d'être débarassées en automne de toutes les tiges mortes : on laboure la terre entre les plantes dans cette saison (1).

Romana. La seconde espece se trouve aux environs de Rome, & dans plusieurs autres endroits de l'Italie; elle a une racine vivace, & une tige an-

(1) Les feuilles de la Mélisse qu'on emploie de préférence aux autres parties de cette plante, ont une odeur agréable, qui approche de celle du citron & une saveur un peu âcre & balsamique. Cette plante contient une petite quantité d'huile éthérée,

MEL 43

nuelle comme la précédente : ses tiges sont minces, ses feuilles beaucoup plus courtes que celles de la premiere; la plante entiere est velue, & d'une odeur forte & désagréable; ses fleurs sont verticillées, sessiles, & plus petites que celle de la

d'une odeur suave, un principe résineux, actif, & assez abondant, & une substance gommeuse presque inerte quand elle est separée des autres principes.

Cette plante, qu'on regarde comme une des plus efficaces & amies de l'homme, agit en agaçant, en discutant, & en secouant un peu. *Hermann*, qui lui attribue les propriétés les plus merveilleuses, pense que ses parties les plus volatiles peuvent réparer les pertes du fluide spiritueux du sang & des nerfs : elle tient un rang distingué entre les médicamens céphaliques, nervins, stomachiques, carminatifs, uterins, &c., & convient sur-tout dans le vertige. la foiblesse de mémoire, l'épilepsie, l'apoplexie, le relâchement de l'estomac, la cardialgie, les affections hystériques & hypocondriaques, la mélancolie, la suppression des regles & des vuidanges, les fleurs blanches, l'asthme humide, les palpitations; & enfin dans toutes les maladies qui reconnoissent pour cause une foiblesse dans le genre nerveux.

On emploie cette plante dans les bouillons, & en infusion théiforme; mais sa préparation la plus ordinaire, est son eau distillée, simple ou composée. L'eau de Mélisse simple s'ordonne dans les potions cordiales & anti-hystériques, & l'eau de Mélisse composée est employée de préférence dans les maladies du cerveau & des nerfs.

La Mélisse entre dans la composition du syrop d'Armoise dans le Catholicon simple, &c.

44 **MEL**

premiere ; elle fleurit vers le même tems : on la cultive rarement dans les jardins ; mais elle peut être traitée de la même maniere que la précédente.

Grandi-flora. La troisieme est originaire des montagnes de la Toscane & de l'Autriche : on la conserve dans plusieurs Jardins anglois, pour la variété ; elle a une racine vivace & une tige annuelle, haute d'environ un pied, & garnie à chaque nœud de deux feuilles opposées, d'un pouce & demi de longueur sur neuf lignes de largeur, sciées sur leurs bords, & d'un vert luisant en-dessus ; des aisselles de la tige sortent des pédoncules solitaires, de six lignes de longueur, qui se divisent en deux autres plus petits, dont chacune soutient deux fleurs séparées. Ces fleurs sont grandes, de couleur pourpre, & de la même forme que celles des autres especes. Cette plante fleurit en Juin, & ses semences mûrissent en Août. On peut la multiplier comme la premiere, & traiter les plantes de la même maniere.

Calamintha. La quatrieme est le Calament commun des boutiques, qui croît sans culture dans plusieurs parties de l'Angleterre ; mais on l'admet rarement dans les jardins : elle a une racine vivace, de laquelle sortent plusieurs tiges quarrées, velues, d'environ un pied de longueur, & garnies à chaque nœud de deux feuilles de la largeur environ de celles de la *Marjolaine*, un peu dentées sur leurs bords, & d'une odeur fort pénétrante : ses fleurs

naissent en têtes verticillées sur les côtés des tiges, soutenues par des pédoncules qui se divisent par paires, & de la même longueur que les feuilles. Ces pédoncules supportent plusieurs petites fleurs bleuâtres, qui paroissent en Juillet, & sont remplacées chacune par quatre semences, petites, rondes & noires. Cette herbe est d'usage en Médecine ; elle est plus chaude, & renferme plus de parties subtiles & volatiles que la *Menthe* ; elle excite l'urine, nettoie le foie, & adoucit la toux. Cette espece peut être plantée dans les jardins, & traitée de la même maniere que la premiere.

Nepeta. La cinquieme naît spontanément en Angleterre : ses tiges sont plus longues, & inclinées vers la terre ; ses feuilles, qui sont plus larges & plus dentées sur leurs bords, ont une odeur forte comme celle du *Pouliot* ; les têtes verticillées des fleurs sont plus rapprochées les unes des autres que celles de la quatrieme : mais en toutes autres choses, elles se ressemblent parfaitement.

Cretica. La sixieme, qu'on rencontre dans la France méridionale & en Italie, n'est pas d'une si longue durée que l'espece précédente ; car elle subsiste rarement plus de deux ou trois ans ; ses tiges sont minces, un peu ligneuses, & garnies de feuilles rondes, blanches, petites, & opposées à chaque nœud : ses fleurs sont produites en têtes verticillées vers le haut des tiges, qui sont terminées par un épi lâche ;

MEL

che; elles font petites, blanches, & de la même forme que celles des autres efpeces; elles paroiffent dans le mois de Juin, & font remplacées par des femences, qui mûriffent en automne, & qui produifent fans foin une quantité fuffifante de jeunes plantes, quand on leur permet de fe répandre.

Majorani-folia. La feptieme, qui croît naturellement en Italie, eft une plante bis annuelle, dont les tiges, longues d'environ huit pouces, font inclinées vers la terre, & garnies de feuilles rondes, à-peu-près de la grandeur de celles de la *Marjolaine*, & d'un vert clair : fes fleurs naiffent en têtes ferrées & verticillées fur la partie haute des tiges, chacune fur un pédoncule féparé; elles font groffes, & d'un pourpre brillant; elles paroiffent dans les mois de Juillet & Août, & leurs femences mûriffent en automne : on la multiplie par fes graines, qu'on met en terre auffi-tôt qu'elles font mûres, pour que les plantes pouffent au printems fuivant : mais fi l'on ne les feme qu'au printems, elles ne germent qu'un an après. On peut auffi la multiplier par boutures, qui prennent aifément racine, en les plantant en été, & en les tenant à l'ombre. Ces plantes réfiftent au froid de l'hiver, fi elles font placées dans une plate-bande chaude : mais pour en conferver l'efpece, il faut mettre une plante ou deux dans des pots, & les enfermer en hiver fous un châffis.

Fruticofa. La huitieme eft ori-

Tome V.

MEL 45

ginaire de l'Efpagne ; elle a des tiges minces d'arbriffeau de neuf pouces environ de longueur, qui pouffent des petites branches latérales, oppofées, & garnies de petites feuilles blanches, à pointe ovale, & placées par paires; elles reffemblent beaucoup à celles du *Marum*; fes fleurs font difpofées en épis verticillés aux extrémités des tiges; elles font petites & blanches; elles paroiffent en Juillet, & leurs femences mûriffent en automne. La plante entiere a une odeur forte de *Pouliot*, & elle eft d'auffi peu de durée que la feptieme efpece. On peut la multiplier par femences ou par boutures, comme la précédente, & la traiter de même.

MELISSE ou CITRONELLE. *Voyez* MELISSA OFFICINALIS.

MELISSE DES BOIS. *Voyez* MELITIS MELISSOPHYLLUM.

MELISSE DES MOLDAVES, *ou* LA MOLDAVIQUE. *V.* DRACOCEPHALUM MOLDAVICA.

MELISSE DES MOLUQUES, *ou* LA MOLUQUE. *V.* MOLUCELLA LÆVIS.

MELITTIS. *Lin. Gen. Plant. ed. non. n.* 789. [*Greater Dead-Nettle.*] La grande Ortie morte.

Caracteres. Le calice de la fleur eft érigé, conique, en forme de cloche, & avec deux levres, dont la fupérieure eft grande & découpée, & l'inférieure eft courte, & divifée en deux parties : la corolle eft labiée; fon tube eft plus long que le calice, les levres font plus épaiffes; la fupérieure eft ronde, unie & érigée, & l'in-

D

MEL

férieure eft divifée en trois parties étendues & obtufes : la fleur a quatre étamines en forme d'alêne, & fituées fous la levre fupérieure, dont deux font un peu plus longues que les autres, & qui font toutes terminées par des antheres divifées en deux parties obtufes, & placées en travers ; fon germe, qui eft obtus, velu, & divifé en quatre parties, foutient un ftyle mince, couronné par un ftigmat partagé en deux parties aiguës ; la fleur eft remplacée par quatre femences qui mûriffent dans le calice.

Ce genre de plantes eft rangé dans la première fection de la quatorzieme claffe de LINNÉE, intitulée *Didynamia Gymnofpermia*, qui comprend celles dont les fleurs ont deux étamines longues & deux courtes, & qui produifent quatre femences nues qui mûriffent dans le calice.

Nous n'avons qu'une efpece de ce genre, qui eft :

1°. *Meiitis Meliffophyllum. Hort. Cliff.* 309. *Roy. Lugd.-B.* 309. *Sauv. Monfp.* 150. ; la plus grande *Ortie* morte.

Meliffophyllum. Hall. Helv. n. 244. *Riv. Mon. f.* 21.

Meliffa Fuchfii. Cam. Epit. 99. 30 ; Melilfe des bois.

Lamium montanum, Meliffæ folio. Bauh. Pin. 231.

Cette plante croît naturellement dans quelques bois de l'Angleterre occidentale, & dans le pays de Galles, en Allemagne, & près de Montpellier ; elle a une racine vivace, qui pouffe au printems trois ou quatre tiges, & quelquefois

un plus grand nombre, fuivant l'âge & la force de la plante. Ces tiges s'élevent à la hauteur d'un pied & demi ; elles font quarrées, & garnies de feuilles femblables à celles de l'*Ortie* morte commune, mais plus larges, plus rudes, & fupportées par de plus longs pétioles ; elles font oppofées deux à deux fur chaque nœud : fes fleurs naiffent aux nœuds des tiges, précifément au deffus des pétioles des feuilles ; elles font de la même forme que celles de l'*Ortie* morte, & la levre fupérieure eft érigée ; elles paroiffent dans le mois de Mai, & font alors un très-bel effet. Si la faifon n'eft pas chaude, ces fleurs confervent leur beauté plus de trois femaines. Comme elles produifent rarement de bonnes femences dans les jardins, on ne les multiplie guere qu'en divifant leurs racines ; mais quand elles font deftinées à fervir d'ornement, il ne faut enlever ces racines que tous les trois ans, & ne pas les divifer en trop petites parties, fans quoi elles ne fleuriroient pas dans la première année. Le meilleur tems pour faire cette opération, eft le commencement du mois d'Octobre, afin qu'elles puiffent avoir le tems de pouffer de nouvelles fibres avant les gelées. Elles exigent un fol marneux, & l'expofition du levant, où les plantes profiteront & produiront des fleurs en abondance.

MELO. *Tourn. Inft. R. H.* 104. *tab.* 32. *Cucumis. Lin. Gen. Plant,* 969. Cette plante prend

MEL

fon nom de μᾶλω une pomme, parce que fon fruit a quelque reffemblance avec une pomme. [*The Melon.*] Melon.

Caractéres. Cette plante a des fleurs mâles & des fleurs femelles fur le même pied : les fleurs mâles ont un calice en cloche, & formé par une feuille dont le bord eft terminé par cinq poils rudes en forme d'alène ; la corolle eft monopétale, en cloche, fixée au calice, & découpée fur fes bords, en cinq fegmens veinés, & rudes ; la fleur a trois étamines courtes, inférées dans le calice, & jointes enfemble ; deux d'entr'elles ont des pointes divifées en deux parties : les antheres font linéaires, & montent & defcendent au-dehors des étamines auxquelles elles adherent : les fleurs femelles n'ont ni étamines ni antheres, mais feulement un germe gros, ovale, & fitué au deffous de la fleur, qui foutient un ftyle court, cylindrique, & couronné par trois ftigmats épais & boffus : ce germe devient enfuite un fruit ovale, & à plufieurs cellules remplies de femences ovales, terminées en pointe aiguë, plâtes, & renfermées dans une chair molle.

LINNÉE a joint ce genre au *Colocynthus*, à l'*Anguria*, & au *Cucumis*, dont il n'a fait que des efpeces. Cet ordre peut être admis en fuivant fon fyftème; mais ceux qui regardent le fruit comme un caractere diftinctif du genre, conviendront qu'ils doivent être féparés; & quoique ces plantes puiffent être réunies dans un fyftème bota-

MEL 47

nique, on ne peut le faire dans un ouvrage de la nature de celui-ci.

On cultive dans différentes contrées un grand nombre de variétés de ce fruit ; & dans ce pays même, on en multiplie beaucoup qui ne font pas de grande valeur, fur-tout les perfonnes qui en font commerce, & qui n'en cherchent que la groffeur & le nombre. Je ne ferai mention ici que de très-peu de variétés ; car la plupart, & fur-tout le Melon commun, ne méritent pas la culture.

Cantaloup. Le Melon *Cantaloup* eft l'efpece la plus eftimée ; fon nom lui vient du lieu d'où il a été tiré, qui eft à quatre milles de Rome, où le Pape a une maifon de campagne. Ce fruit y eft cultivé depuis long-tems ; mais il a été originairement apporté de cette partie de l'Arménie qui confine à la Perfe, où il eft en fi grande abondance, qu'on en donne pour un écu autant qu'un cheval peut en porter.

Lorfque la chair de ce Melon eft bien mûre, elle eft délicieufe, & ne nuit en aucune maniere aux eftomacs les plus délicats; les Hollandois en font fi friands, qu'ils n'en cultivent guere d'autres ; & pour le diftinguer, ils le nomment fimplement *Cantaloup*, fans y joindre le titre de Melon, qu'ils donnent aux autres efpeces.

L'écorce de ces fruits eft très-rude, & couverte de boutons ou verrues; ils font d'une groffeur médiocre, plutôt ronds que longs, & la chair de la

D 2

plupart eft couleur d'orange, quoiqu'il y en ait quelques-uns d'une chair verdâtre : mais ces derniers ne font jamais auffi bons que ceux à chair rouge.

Romain. Plufieurs eftiment le Romain ; ce fruit eft bon quand il eft bien conditionné, que la plante eft en bon état, & la faifon feche ; il eft plus précoce que le *Cantaloup*, & l'on doit choifir cette efpece de préférence, quand on veut en avoir de bonne heure.

Succado. Le *Succado* eft auffi un bon fruit, & auffi printanier ; mais lorfqu'il vient dans la faifon ordinaire, il n'eft pas fi bon que le *Cantaloup*.

Zatte. Le *Zatte* eft encore un très-bon Melon ; mais il n'eft guere plus gros qu'une forte orange, & un peu applati aux deux extrémités ; fon écorce reffemble à celle du *Cantaloup* ; mais fon extrême petiteffe le rend peu propre à être cultivé.

Portugal. Le *Petit Portugal* eft un affez bon fruit ; & comme d'ailleurs il eft très-fécond, on l'admet plutôt que les autres dans les jardins ; les perfonnes fur-tout qui préférent la quantité à la qualité, le cultivent de préférence ; il eft auffi très-précoce.

Galloway Noire. C'eft le plus précoce de toutes les efpeces de Melons : il fut apporté de Portugal par Mylord Galloway, il y a plufieurs années ; mais il eft à préfent rare en Angleterre, parce qu'il y a dégénéré, en le mêlant avec les autres efpeces : il mûrit plutôt qu'aucun autre Melon printanier. Ce fruit eft très-bon lorfqu'il parvient naturellement à fa maturité.

Ces variétés fuffifent pour fatisfaire les curieux ; car il y en a très-peu d'autres qui méritent la peine d'être cultivées, & ceux qui aiment ce fruit ne s'attachent qu'au feul *Cantaloup.* Quand on veut en avoir de printaniers, on doit préférer ceux qui viennent d'être rapportés, & avoir foin de les cultiver féparément des *Cantaloups* ; car s'ils y étoient mêlés, il feroit impoffible de conferver ces efpeces parfaitement pures : auffi les Jardiniers Hollandois & les Allemands ont grand foin de ne planter près des *Cantaloups* aucune autre efpece de Melon. On peut en dire autant des Concombres, & des Citrouilles ou Callebaffes, dont la pouffiere fécondante peut altérer la qualité des *Cantaloups* ; & je fuis convaincu, par une longue expérience, qu'ils ont raifon à cet égard. J'ai obfervé que plufieurs perfonnes, faute de cette précaution, ont diminué la bonté de leurs fruits, fans en favoir la raifon, & l'ont imputé à ce qu'elles n'avoient pas changé de femences, qu'elles croyoient s'être altérées à la longue ; car de tems en tems il eft néceffaire d'en faire venir d'autres d'un lieu éloigné : mais peu de gens prennent la peine de choifir eux-mêmes les femences des bonnes efpeces, & s'en rapportent à d'autres pour cela ; ce qui les a fouvent trompés, ainfi que je l'ai été moi-même.

On ne doit pas fe fervir des

MEL

semences de Melons, qu'elles
n'aient trois ans : mais passé
six ans, il ne faut plus en
faire usage ; car quoiqu'elles
poussent au bout de dix ou
douze années, cependant les
fruits qu'elles donnent ont rarement la chair aussi épaisse que
ceux qui proviennent de semences plus fraîches. Il en est
de même des graines légeres
qui nagent sur l'eau, lorsqu'on
vient de les recueillir. J'ai quelquefois essayé ces dernieres,
après les avoir gardées trois ans;
mais pas un des fruits qui en
ont été produits, ne s'est trouvé
avoir la chair aussi épaisse &
aussi ferme que ceux des semences lourdes prises dans le
même fruit, quoiqu'elles aient
été semées & cultivées sur la
même couche, & avec le même
soin.

Culture. Après avoir parlé du
choix des especes & des semences, je vais donner la méthode de les cultiver pour en
obtenir une grande quantité
de bons fruits; mais cette méthode sera bien différente de
celle qu'on emploie aujour-
d'hui en Angleterre. Plusieurs
personnes y trouveront certainement des défauts ; mais c'est
celle de tous les bons Jardiniers de Hollande & d'Allemagne, où l'on mange une très-
grande quantité de fort bons
Cantaloups. Je ne donne d'ailleurs cette méthode, qu'après
avoir éprouvé, par une longue
suite d'expériences, qu'elle est
la meilleure de toutes.

On voit souvent des gens
qui se vantent d'avoir des Melons printaniers ; mais ces fruits

MEL

49

ne valent pas mieux que des
Courges, quoiqu'ils occasionnent beaucoup de frais & de
peines pour se les procurer un
peu plutôt ; & quand ils parviennent à leur grosseur, la
tige est communément torse, ce
qui empêche les sucs de monter jusqu'aux fruits, & les
fait avorter. Pour les colorer, & achever de les mûrir,
on les couvre d'une bonne
épaisseur d'herbe nouvellement
fauchée, pour les faire fermenter : mais ces fruits ainsi forcés, ont la chair mince, sans
eau & sans saveur ; de sorte
qu'après quatre mois de travail,
& beaucoup de dépenses en
fumier &c., on obtient à peine trois ou quatre paires de
Melons assez mauvais, & plus
propres à être jettés que mangés.
Ainsi, je conseillerai toujours
de ne faire mûrir ces fruits
qu'au milieu ou à la fin de
Juin, ce qui est assez tôt pour
notre climat; mais depuis ce
tems jusqu'à la fin de Septembre, on peut en avoir en
abondance, s'ils sont bien traités. J'en ai eu jusqu'au milieu
d'Octobre, lorsque l'automne
s'est trouvé favorable.

Pour s'en procurer aussi long-
tems, il faut en semer en deux
ou trois saisons différentes.
Les premiers doivent être semés vers le milieu ou à la fin
de Février, si l'année est précoce ; car sans cela, il faut
différer jusqu'à la fin de ce mois;
le succès dépend d'élever les
plantes en vigueur, & la chose
n'est pas aisée, si le tems devient mauvais, après qu'elles
ont commencé à pousser, par-

D 3

MEL

ce qu'on ne peut leur donner beaucoup d'air frais : ainfi, il faut éviter de les femer trop tôt.

Dans la faifon que j'ai indiquée, on peut les femer fur le haut d'une couche de *Concombres*, s'il y en a , finon l'on ramaffe du crotin nouveau de cheval, que l'on met en monceau , pour le faire fermenter, & que l'on remue , pour lui communiquer une chaleur égale , comme on le fait pour les couches de *Concombres* : on conduit & on éleve ces plantes comme celles des *Concombres* , jufqu'à ce qu'elles foient placées à demeure : c'eft-pourquoi j'invite le Lecteur à recourir à cet article, pour éviter les répétitions.

La feconde faifon pour femer des Melons, eft à - peuprès le milieu du mois de Mars : ces deux femis font deftinés à fournir des plantes propres à être mifes fous des vitrages ; car celles qu'on veut planter fous cloches ou fous des châffis de papiers huilés, ne doivent être femées qu'en Avril ; fi on le fait plutôt, les plantes filent & allongent leurs rameaux hors des cloches , avant qu'il foit poffible de les découvrir, parce qu'il furvient fouvent de fortes gelées au milieu du mois de Mai , & que dans ce cas les branches qui font hors des cloches , & qui ne font pas couvertes de nattes , fouffrent beaucoup de la gelée : d'ailleurs fi les plantes ont affez pouffé pour remplir les cloches, & n'ont pas la liberté de s'étendre , elles feront étouffées ,

MEL

& fouffriront de la chaleur & du foleil pendant le jour. J'ai femé le 3 de Mai, fur une couche chaude , des *Cantaloups* qui n'ont point été tranfplantés , mais qu'on a feulement recouverts avec des papiers huilés , & j'ai recueilli une bonne quantité de très-bons fruits, qui ont commencé à mûrir à la fin d'Août , & fe font fuccédés jufqu'à la fin d'Octobre.

Couches. Voici la méthode de faire les couches fur lefquelles doivent refter les plantes : il faut toujours les placer dans une fituation chaude , & à l'abri du froid & des vents violens , fur-tout de ceux de l'orient & du nord, qui font généralement fâcheux au printems ; de maniere que , fi les couches y étoient expofées , il feroit difficile de donner de l'air aux jeunes plantes : il faut auffi les parer du vent du fud-oueft , qui eft fouvent impétueux en été & en automne , & qui non-feulement dérange les branches, mais les endommage auffi beaucoup : c'eft-pourquoi la meilleure expofition qu'on puiffe choifir pour ces couches, eft au midi, ou un peu inclinées à l'orient, & abritées à une certaine diftance par des arbres fur les autres côtés. Cette place doit être fermée d'un bon enclos de rofeaux, qui valent mieux pour cet effet qu'aucune autre chofe , parce qu'ils parent mieux les vents que ne feroient des murailles qui les renvoient fur les couches ; mais ces enclos de rofeaux doivent être éloignés des couches ;

MEL

afin qu'ils ne donnent point d'ombrage durant une partie de la journée : on y pratique une porte aſſez large pour le paſſage d'une brouette , afin de pouvoir y tranſporter du fumier , de la terre , &c., & on la tient fermée , pour empêcher d'entrer tous ceux qui n'y ont point à travailler; car ſouvent des ignorans viſitent les couches , donnent mal-à-propos de l'air aux plantes , & quelquefois même les laiſſent à découvert , ou ferment les vitrages quand ils doivent être ouverts, ce qui fait beaucoup de tort aux jeunes plants.

On prépare enſuite la terre pour les plantes , & c'eſt en cela que les Jardiniers Hollandois & Allemands ſont fort experts. Le mélange ordinaire eſt un tiers de terre graſſe , un tiers d'écurement des foſſés ou d'étangs , & un tiers de fumier fort conſommé & réduit en terreau; le tout doit être bien mêlé , & mis à part une année avant de s'en ſervir : on le remue ſouvent , pour l'ameublir & le bien façonner.

La compoſition qui réuſſit le mieux en Angleterre , eſt de deux tiers de terre graſſe & légere , avec un tiers de fumier de vache bien réduit en terreau, en mêlant & façonnant le tout enſemble une année avant de s'en ſervir , de maniere que l'hiver & l'été puiſſent paſſer deſſus , & en obſervant de la remuer ſouvent , & de ne pas y laiſſer croître de mauvaiſes herbes : on trouvera cette façon auſſi bonne que toute autre. Comme ces plantes réuſſiſſent

mieux lorſqu'elles ſont tranſplantées jeunes, il faut amaſſer une quantité de fumier proportionnée aux couches que l'on veut faire, en comptant quinze bonnes brouettes pour chaque châſſis : on le remue deux ou trois fois pour le préparer, comme il a été dit à l'article des *Concombres ;* & quinze jours après , lorſqu'il eſt en état d'être employé, on creuſe la couche pour l'y placer. Cette couche doit être plus large que le châſſis , & d'une longueur proportionnée à leur nombre; quant à la profondeur , elle doit être creuſée ſuivant que le ſol eſt ſec ou humide. Dans une terre ſeche, elle ne doit pas avoir moins d'un pied ou d'un pied & demi ; car plus elles ſont profondes , & mieux elles réuſſiſſent, pourvu qu'il n'y ait rien à craindre de l'humidité. En mettant le fumier dans la couche, on doit le bien mêler, & ſuivre en tout la méthode qui a été indiquée pour les *Concombres.* Lorſque cette couche eſt faite, on place les châſſis deſſus, pour en attirer l'humidité, & on ne la couvre de terre que trois ou quatre jours après , lorſqu'on s'apperçoit qu'elle eſt au dégré de chaleur qu'elle doit avoir ; car les couches nouvellement faites ſont quelquefois ſi ardentes, qu'elles brûleroient la terre qui ſe trouveroit deſſus , & alors il vaudroit mieux ôter cette terre brûlée, dans laquelle les plantes ne profiteroient jamais.

Dès que la couche eſt parvenue au dégré de chaleur qui lui eſt néceſſaire , on la cou-

MEL

vre de terre, feulement à l'épaiffeur de deux pouces, excepté au milieu de chaque châffis, où les plantes doivent êtres placées ; car il faut élever dans cet endroit une butte de quinze pouces au moins de hauteur, terminée en cône tronqué. Deux ou trois jours après que l'on aura mis la terre fur la couche, elle fera affez échauffée pour recevoir les plantes ; alors on les tranfplante le foir, &, s'il eft poffible, lorfqu'il regne un peu de vent : on enleve foigneufement les plantes avec un tranfplantoir, de peur de déranger leurs racines ; car fi elles étoient endommagées, elles feroient longtems à reprendre, & refteroient prefque toujours languiffantes. Le Melon eft plus délicat & plus difficile à tranfplanter que le *Concombre*, fur-tout le *Cantaloup*, qui eft long-tems à prendre vigueur, s'il n'eft pas tranfplanté auffi-tôt que paroît fa troifieme feuille, que les Jardiniers appellent rude. Ainfi, lorfqu'il arrive que les couches ne peuvent pas être prêtes à les recevoir pour ce tems, il faut mettre chaque plante dans un petit pot, tandis qu'elles font jeunes, & les plonger dans la couche chaude où elles doivent être placées, ou bien dans quelques couches de *Concombres*, pour les faire avancer. Lorfque la couche eft en état, on les tire des pots en motte, & fans leur donner aucune fecouffe. On préfere cette derniere méthode pour les *Cantaloups*, parce qu'il ne doit y avoir qu'une feule plan-

MEL

te fous chaque châffis ; & en s'y prenant ainfi, on eft affuré qu'elle réuffira, fans avoir befoin d'en mettre plufieurs enfemble, comme on a coutume de le faire pour les Melons ordinaires. Lorfque les plantes font placées fur le fommet des buttes de terre, on les arrofe légerement, ce qui doit être réitéré une ou deux fois après, jufqu'à ce qu'elles aient pouffé de bonnes racines, après quoi elles exigent rarement d'être arrofées ; car trop d'humidité moifit le pied, le pourrit jufqu'à la racine, & l'empêche de produire de bons fruits.

Quand les plantes font bien enracinées dans les nouvelles couches, on y met une plus grande quantité de terre, en commençant autour des buttes où font les plantes, pour procurer aux racines le moyen de s'étendre ; & en y mettant de la terre de temps en temps, on la preffe le plus qu'il eft poffible. Lorfque toute cette terre eft placée, elle doit avoir au moins un pied & demi d'épaiffeur fur toute la couche ; mais il faut avoir foin d'élever les châffis de maniere que les vitrages ne foient pas trop près des plantes, de peur qu'elles ne foient brûlées par le foleil.

Lorfque les pieds de Melons ont pouffé quatre feuilles, il faut pincer le fommet de la plante, en obfervant de ne pas l'écorcher, ou le couper net avec la ferpette, afin que la plaie fe referme plutôt. Cette opération les fait trocher, & pouffer des branches latérales,

MEL

qui produiront du fruit. Ainsi, lorsqu'il y a deux, & même un plus grand nombre de ces branches, on les pince aussi, pour leur en faire pousser d'autres, que les Jardiniers appellent coulans, & qui servent à couvrir la couche. La maniere de traiter les Melons étant à-peu-près la même que celle qu'on emploie pour les concombres, je ne répéterai point ici ce que j'ai dit ailleurs ; j'observerai seulement que les Melons exigent beaucoup plus d'air, & moins d'arrosemens que le *Concombre*, & que l'eau qu'on leur donne doit être répandue à une certaine distance du pied.

Si les plantes réussissent bien, elles couvriront toute la couche, & s'étendront jusqu'aux cadres en cinq ou six semaines de temps; alors il faudra creuser la terre entre les couches, ou autour de la couche, s'il n'y en a qu'une; y faire une tranchée de quatre pieds environ, aussi profonde que la couche, & y mettre jusqu'à cette hauteur du fumier chaud, qu'on presse & qu'on foule aux pieds : on le couvre ensuite avec la même terre que celle de la couche, jusqu'à l'épaisseur d'un pied & demi, & même davantage, & on la serre autant qu'il est possible. Au moyen de cela, cette couche se trouvera avoir douze pieds de largeur, ce qui lui est absolument nécessaire; car les racines des plantes s'étendront & rempliront entierement cet espace : sans cette précaution, il est ordinaire de voir les branches se flétrir avant que le fruit soit

MEL 53

parvenu à sa grosseur, parce que les racines, ne pouvant plus s'étendre, se ramassent sur le côté des couches, dans le tems que le fruit commence à paroitre ; & faute de nourriture, les extrémités des branches se dessèchent bientôt par l'action du soleil & de l'air; ce dont on s'apperçoit dans peu, par le dépérissement des feuilles, qui se fannent pendant la chaleur du jour. Dans ce cas, les plantes vont toujours en déclinant, les fruits ne peuvent plus prendre d'accroissement ; & s'ils parviennent à leur maturité, ils n'ont que très-peu de chair, & font farineux & de mauvais goût; au-lieu que les plantes bien conditionnées, & auxquelles on a donné le supplément de nourriture qui leur est nécessaire, se conservent vertes & vigoureuses, jusqu'à ce que les gélées les détruisent, & fournissent une seconde récolte de fruits, qui parviennent quelquefois à une bonne maturité : mais les premiers fruits font toujours excellens, & d'une grosseur plus considérable que les Melons ordinaires ; leurs feuilles font fort larges, & d'un vert foncé qui annonce la plus grande vigueur.

En élargissant les couches, comme il vient d'être dit, on se procure un nouvel avantage, en ce que le fumier que l'on remet sur les côtés, réchauffe celui de la couche, & fait un très-grand bien aux plantes, qui commencent alors à pousser leurs fruits, sur-tout lorsque la saison est encore

MEL

froide, comme cela arrive souvent dans ce pays vers le mois de Mai. Lorsque les plantes ont rempli les châssis, & demandent plus d'espace, on élève les cadres avec des briques de trois pouces d'épaisseur, pour donner la liberté aux branches de couler dessous. Si les plantes sont fortes, ces branches s'étendront à sept ou huit pieds de chaque côté, ce qui exigeroit plus de place, & obligeroit à retrancher une plante sur chaque couche ; car lorsque les branches sont trop touffues, les fruits se nouent rarement bien, ou tombent quand ils ont atteint la grosseur d'un œuf : c'est pourquoi les châssis destinés à contenir des Melons, doivent avoir au moins six pieds de largeur.

Il n'y a point de partie du jardinage dans laquelle les praticiens different plus que dans la culture des Melons, parce qu'on ne trouve dans les livres aucunes regles sûres, d'après lesquelles on puisse se diriger : c'est pourquoi je vais exposer en peu de mots ce qui est nécessaire pour y réussir.

J'ai déja parlé du pincement des plantes, quand elles commencent à pousser, pour se procurer des branches latérales, appelées par les Jardiniers, coulans. On réitere cette opération sur toutes celles qui se montrent, parce que c'est sur ces branches que le fruit doit être produit : mais lorsqu'il y en a un nombre suffisant, il ne faut plus les arrêter, mais attendre que le fruit se montre : il poussera bientôt en abon-

dance ; alors on examinera avec soin les branches trois fois la semaine, pour reconnoître les fruits : on choisira sur chaque coulant celui qui est le plus près du pied, qui a le plus gros pédoncule, & qui annonce devoir devenir le plus fort ; on retranchera tous les autres qui peuvent se trouver sur le même coulant, & l'on coupera aussi l'extrémité du coulant au troisieme nœud au-dessus du fruit, pour arrêter la sève, & nourrir le fruit.

Quelques Jardiniers ont coutume, pour faire nouer le fruit, d'enlever quelques fleurs mâles, dont la poussiere fécondante est mûre, & de les poser sur les fleurs femelles qui sont au sommet des fruits ; ils secouent avec les doigts cette poussiere séminale sur les pistiles des fleurs femelles, pour aider la nature, & faire gonfler promptement le germe du fruit. Cette pratique paroît nécessaire, lorsque les plantes sont élevées sous des vitrages où le vent n'a point d'entrée, & ne peut par conséquent transporter cette poussiere fécondante de la fleur mâle sur la fleur femelle.

En retranchant tous les autres fruits, on procure la totalité de la sève & de la nourriture à celui que l'on a laissé, qui avorteroit, si l'on en conservoit un plus grand nombre : en ne réservant qu'un fruit sur chaque coulant, il s'en trouvera autant que la plante pourra en nourrir ; car si on en laissoit plus de huit sur chacune, ils seroient petits & mal conditionnés. J'en ai vu quelques

MEL

fois quinze ou vingt sur une seule plante de Melon ordinaire; mais ils n'étoient parvenus qu'à une grosseur médiocre, quoiqu'ils n'eussent pas besoin d'autant de nourriture que le *Cantaloup*, dont l'écorce est très-épaisse; & la chair en étoit fort mince. Après avoir pincé trois nœuds au-dessus des fruits, il faut visiter souvent les plantes, pour retrancher les nouveaux coulans qui pourroient naître sur les branches, ainsi que les nouveaux fruits; ce qu'il est nécessaire de recommencer souvent, jusqu'à ce que les fruits réservés soient parvenus à une grosseur suffisante pour attirer toute la séve & la nourriture des plantes, dont la vigueur commence alors à diminuer : on les arrose, après avoir fait cette opération à quelque distance des tiges, pour faire arrêter & grossir les fruits.

Il est nécessaire de tenir les vitrages soulevés, pour donner de l'air aux plantes; car sans cela le fruit n'arrêteroit pas; & si la saison est fort humide, on les enleve même tout-à-fait, sur-tout dans les soirées, pour y admettre les rosées, pourvu qu'il y ait un peu de vent : mais il ne faut pas laisser les couches sans vitrages pendant la nuit entiere, de peur que le froid ne devienne trop vif. Dans les tems chauds, ces plantes peuvent être découvertes depuis dix heures du matin jusqu'au soir.

Lorsque les plantes se font étendues au-delà des châssis, si le tems devient froid, on couvre les branches qui débor-

MEL 55

dent avec des nattes; car si ces rejettons étoient endommagés, l'accroissement des fruits seroit retardé, & les plantes souffriroient beaucoup. Les arrosemens doivent être faits dans les allées où les racines se font étendues : au moyen de cette attention, les plantes feront des progrès rapides, & les tiges, étant toujours seches, se conserveront en bon état; mais on ne doit les arroser qu'une fois la semaine, par un tems très-sec & chaud, & il est nécessaire de leur donner dans ce moment le plus d'air qu'il est possible.

Après avoir traité de la culture des Melons que l'on éleve sous des châssis, je vais parler de la maniere de conduire ceux que l'on plante sous des cloches ou glaces à main. Les plantes qu'on veut disposer ainsi, doivent être élevées comme les précédentes.

Vers la fin d'Avril, si la saison est avancée, on pourra faire les couches; alors il faut se pourvoir d'une quantité de fumier chaud, proportionnée au nombre de cloches que l'on veut employer, en comptant six ou huit fortes brouettes de fumier pour chaque cloche.

Quand on ne fait qu'une couche, il faut la creuser de quatre pieds & demi de largeur, & lui donner une longueur proportionnée au nombre des cloches, qui doivent être placées à quatre pieds l'une de l'autre; car lorsque les plantes sont trop rapprochées, leurs branches s'entrelacent, & couvrent si fort la couche, que le fruit ne peut nouer. En creu-

MEL

fant la foffe, on réferve trois ou quatre pieds de largeur à chaque côté, & l'on proportionne fa profondeur à la fécherefe ou à l'humidité du fol ; mais, comme on l'a déja obfervé ci-deffus, la couche fera d'autant meilleure, qu'elle fera plus profonde. On doit auffi avoir la même attention pour mêler le fumier; & quand il eft placé dans la couche, il faut élever un monceau de terre d'un pied & demi de hauteur, à chaque place où les plantes doivent être mifes, & l'on ne répand fur le refte de la couche que quatre pouces d'épaiffeur de terre, ce qui fuffira pour empêcher l'évaporation du fumier : on met enfuite les cloches fur le fommet, & on les preffe de façon que la terre des buttes puiffe s'échauffer, & être en état de recevoir les plantes que l'on y placera, comme il a été dit ci-deffus, deux ou trois jours après, fi la couche a le dégré de chaleur qui lui eft néceffaire. Lorfque les plantes font dans des pots où elles avancent également bien, on fe contente d'en mettre un feul fous chaque cloche; mais fans cela, il faut placer dans chaque endroit deux plantes, dont on retranche enfuite la plus foible, quand toutes deux réuffiffent : on les arrofe auffi-tôt qu'elles font en place, pour faire pénétrer la terre entre leurs racines, & on les tient à l'ombre, jufqu'à ce qu'elles aient pouffé de nouvelles fibres. Si les nuits font fraîches, il fera néceffaire de couvrir les cloches avec des

MEL

nattes, pour conferver la chaleur de la couche.

Lorfque l'on a deffein de faire plufieurs couches, on les place à huit pieds de diftance l'une de l'autre, afin qu'il refte un efpace fuffifant entre chacune, dans lequel les racines puiffent s'étendre, ainfi qu'on l'a déja obfervé plus haut.

Quand les plantes font bien enracinées, on pince leurs fommets & les rejettons, & on les conduit comme celles des châffis. Pendant la chaleur du jour, on fouleve le côté des cloches oppofé au vent, pour y introduire l'air ; car fans cela elles fileroient, & s'affoibliroient : ce qu'il faut prévenir avec foin ; car fi les coulans ne font pas affez vigoureux, ils ne font point en état de nourrir leurs fruits.

Lorfque les plantes ont atteint le côté des cloches, & que le tems eft favorable, on pofe les cloches fur trois briques, & on les éleve ainfi à deux pouces au-deffus de la furface, pour laiffer paffer les branches, & leur donner la liberté de s'étendre ; alors on couvre toute la couche avec de la terre, jufqu'à la hauteur d'un pied & demi, & on la foule avec les pieds le plus qu'il eft poffible. Si les nuits font froides, on étend des nattes fur les couches, afin que le froid ne nuife point aux tendres rejettons des branches; mais comme ces *Cantaloups* craignent l'humidité, il fera néceffaire d'établir des cercles en arcades, pour foutenir ces nattes. Cette méthode eft la feule qu'on puiffe em-

ployer pour faire réuffir cette efpece en Angleterre, où les faifons font fi variables & fi incertaines, que j'ai perdu, par des pluies du mois de Juin, plufieurs couches de ces Melons, qui étoient dans le meilleur état.

Si le tems devient froid, il eft néceffaire de creufer autour des couches des tranchées de la même profondeur, & de les remplir de fumier chaud, qu'on éleve à la même hauteur que celui des couches, comme il a été dit pour les couches à châffis; & quand on peut fe procurer beaucoup de ce fumier, on creufe encore l'intervalle qui fépare les couches, on le remplit de même, & on le recouvre d'un pied & demi de terre, qu'on foule exactement. Cette opération procurera une nouvelle chaleur aux couches, & fera paroitre le fruit bientôt après.

Il faut arrofer ces plantes avec beaucoup de précaution, en prenant garde de ne pas mouiller les pieds; & lorfqu'on pince les coulans, & que l'on ôte les fruits fuperflus, pour faire profiter ceux que l'on réferve, il faut le faire légerement; enfin il faut fuivre exactement tout ce qui a été prefcrit au fujet de la culture des Melons placés fous les châffis, en obfervant toujours de les couvrir avec des nattes dans les tems pluvieux & durant les nuits froides. Si l'on fuit, fans s'en écarter, les regles qui viennent d'être données, on rifquera peu de voir manquer ces fruits, & les branches con-

ferveront leur vigueur, jufqu'à ce que les froids de l'automne les détruifent.

Plufieurs perfonnes ont élevé, depuis peu, des Melons fous des châffis de papiers huilés, qui ont bien réuffi en beaucoup d'endroits : mais en fuivant cette méthode, on doit faire en forte que ces châffis foient éloignés des plantes, fans quoi leurs branches deviendront foibles, fileront, & donneront rarement du fruit en abondance. Ainfi, lorfqu'on fe propofe de faire ufage de ces couvertures, je confeille d'élever les plantes fous des cloches, comme il vient d'être dit ci-deffus, jufqu'à ce que leurs branches foient devenues affez longues pour ne pouvoir plus y être contenues : alors on fe fervira de papiers huilés au lieu de nattes ; ce qui vaudra beaucoup mieux, fi l'on s'y prend avec difcernement.

Le papier qu'on emploie pour cela doit être fort, & pas trop foncé en couleur : on le frotte d'huile de lin, qui fe deffechera bientôt, quand il aura été collé fur les châffis, & on ne s'en fervira qu'après que l'odeur fera diffipée, parce que cette odeur pourroit être trèsnuifible aux plantes.

Lorfque les fruits font arrêtés, on continue à retrancher tous ceux qui fe trouvent de trop, ainfi que les coulans foibles qui uferoient la féve : on retourne légerement deux fois la femaine les fruits réfervés, pour les expofer de tous côtés à l'air & au foleil ; car fi on les laiffoit toujours fur la terre

MEL

dans la même position, le côté qui la toucheroit deviendroit tendre & blanchâtre, faute de ce secours.

Ces plantes exigent un peu d'arrosement dans les tems secs; mais on doit le faire dans les allées, à quelque distance du pied des plantes, & tout au plus une fois par semaine, ou chaque dix jours. En suivant cette méthode, la terre doit être bien humectée; au moyen de cela, on avancera l'accroissement du fruit, & on en rendra la chair épaisse : mais ce qu'il faut le plus observer, c'est de ne pas trop arroser les plantes, parce que l'humidité leur est très-nuisible, & de leur donner en tous tems le plus d'air libre qu'il est possible, lorsque la saison le permet.

Lorsque ces fruits sont tout-à-fait mûrs, on doit avoir attention de les couper à tems ; car si on les laissoit quelques heures de plus sur la plante, ils perdroient beaucoup de leur délicatesse : pour cela il faut les visiter au moins deux fois par jour ; on les coupe dès le matin, avant que le soleil les ait échauffés : mais si l'on est forcé de les cueillir plus tard, on les tient dans de l'eau de source ou dans de la glace pilée, pour les rafraîchir avant de les manger. Ceux qui sont cueillis le matin doivent être conservés dans un lieu frais, jusqu'à ce qu'on les serve sur la table. On reconnoît que ces fruits sont mûrs, par l'odeur qu'ils exhalent lorsqu'on en a rompu le pédoncule ; car il ne

MEL

faut jamais attendre que les *Cantaloups* changent de couleur; ce qui n'arrive que lorsqu'ils sont trop mûrs.

La méthode que l'on vient de donner pour les *Cantaloups*, sera également bonne pour toutes les autres especes, ainsi que l'expérience me l'a prouvé. La coutume ordinaire de ne mettre que trois ou quatre pouces de terre d'épaisseur sur les couches, expose les plantes à se flétrir avant que les fruits soient parvenus à leur maturité, parce que les racines gagnent bientôt le fumier, & s'étendent dans les côtés de la couche, où leurs tendres fibres sont exposées à l'air & au soleil, ce qui fanne les feuilles pendant la chaleur du jour ; & il seroit alors nécessaire de les couvrir avec des nattes, pour prévenir leur dépérissement, & de les arroser plus souvent, pour les conserver, quoique cela soit très-préjudiciable à leurs racines : au lieu qu'en couvrant les couches d'une largeur & d'une épaisseur de terre suffisante, les plantes supportent jusqu'à l'automne les plus grandes ardeurs du soleil dans notre climat, sans avoir besoin d'humidité, & sans que leurs feuilles puissent en souffrir. Je recommande toujours de ne conserver les semences que des fruits les mieux conditionnés, les plus fermes, & du meilleur goût, de les laisser deux ou trois jours dans leur jus, sans les laver, & de ne conserver que celles qui se précipitent au fond de l'eau.

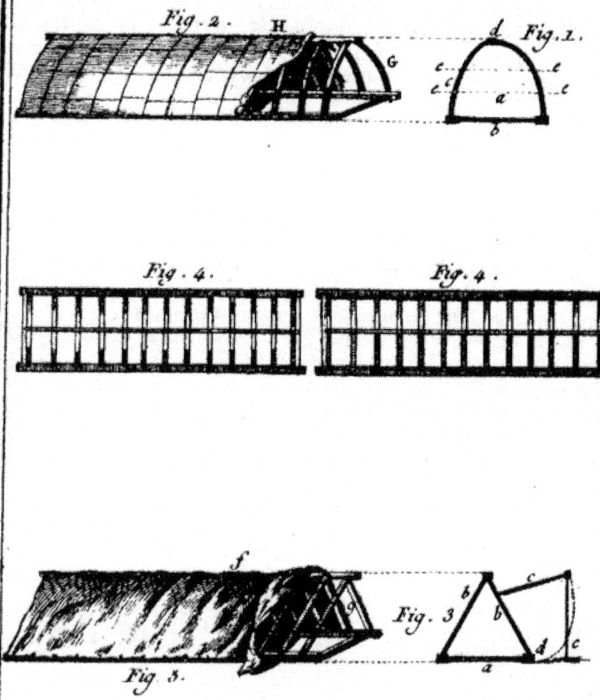

Tom. 5   Page. 69

Fig. 2.   H   d *Fig. 1.*
                              G

Fig. 4.   *Fig. 4.*

Fig. 3.

Fig. 5.

Deux especes de Cadre garnis de papiers Huilés pour la Couverture des Melons.

*Explication des Planches qui re-
préfentent deux fortes de cadres
garnis de papiers huilés, pour
couvrir les Melons.*

La premiere figure reffemble à la couverture d'un charriot ou fourgon; elle a à fa bâfe un cadre de bois, auquel font fixées les extrémités des demi-cercles. Ce cadre doit avoir cinq ou fix pieds de largeur; s'il étoit plus étroit, il ne pourroit pas couvrir toute la couche; & s'il étoit plus large, on ne pourroit le remuer qu'avec beaucoup de peine. A montre la largeur, B le cadre de bois à la bâfe, C l'arc de cercle, D une lame mince de bois attachée au-deffous des cercles, pour les contenir dans leur pofition.

La diftance entre chaque cercle ne doit pas être de plus d'un pied, & il doit y avoir de chaque côté deux ficelles fixées aux cercles d'une extrémité à l'autre du châffis, aux places marquées EEEE, pour contenir le papier huilé, & l'empêcher d'être enfoncé par l'eau des pluies. Chaque cadre ne doit pas avoir plus de dix pieds de longueur, qui eft celle de trois châffis, ce qui fuffira pour couvrir trois plantes. On proportionne le nombre des châffis à celui des plantes qu'on veut cultiver.

Seconde figure. La feconde figure repréfente deux de ces cadres unis; G montre le profil du cadre; H le papier retourné, pour faire voir la façon de le placer fur les cadres.

Troifieme figure. La troifieme figure repréfente une autre ef-

pece de cadre en forme de toit. A montre la bâfe, BB les deux côtés inclinés, C un des côtés qui peut être foulevé, pour donner de l'air aux plantes, D la place où il fe ferme, & E le bois qui le foutient. Il fera très-utile de faire ces volets avec des charnieres, & de les placer alternativement fur chaque côté, pour pouvoir donner de l'air du côté oppofé au vent, & même ouvrir les deux côtés à la fois, lorfque les plantes demandent beaucoup d'air, & qu'il fait un tems chaud. La figure cinquieme eft conftruite de même; G repréfente fon profil, & F la couverture de papier. Ces fortes de cadres peuvent être faits avec des lattes droites ou avec des morceaux de fapin, afin qu'ils foient plus légers; mais la bâfe du cadre à laquelle elles font attachées doit être plus forte. Quelques perfonnes qui ont fait l'effai de ces deux manieres, trouvent la derniere plus commode pour donner de l'air aux plantes, parce que dans la premiere on ne peut en donner qu'en foulevant plus ou moins le cadre entier d'un côté; & lorfque la faifon devient chaude, on eft obligé d'ôter les cadres tout-à-fait, & de laiffer couler les plantes dehors. Lorfque les cadres font peints avec la compofition fuivante, ils fe confervent long-tems; pour cela, il faut prendre fix livres de poix, une demi-pinte d'huile de lin, une livre de poudre de briques: on mêle le tout enfemble, & l'on s'en fert quand il eft chaud.

MEL

Lorsque cette espece d'enduit est sec, il devient fort dur, & forme un ciment que l'humidité ne peut pénétrer. Cette composition est aussi la meilleure dont on puisse se servir pour conserver les charpentes exposées à l'humidité.

Lorsque ces cadres sont parfaitement secs, on colle les papiers dessus. Le meilleur papier pour cet usage est celui qui vient de Hollande, & dont on fait des enveloppes : il est fort, & lorsqu'il est huilé, il devient transparent, & laisse passer les rayons de la lumiere. Quand l'empoix dont on s'est servi pour le coller est bien sec, on le frotte d'huile de lin, qui le pénetre aussi-tôt. Ainsi, il n'est pas nécessaire de l'enduire des deux côtés : on laisse secher l'huile avant de l'exposer à l'humidité, sans quoi le papier se déchireroit. En collant le papier sur le cadre, il faut avoir soin de le rendre fort uni, & de le bien coller sur toutes les traverses du cadre, ainsi qu'aux ficelles, pour empêcher que le vent ne le souleve & ne le déchire.

Ce qui est dit de ces cadres à l'article de la culture des Melons, suffira pour donner l'idée de l'usage que l'on doit en faire. On observera seulement ici de ne pas tenir ces couvertures trop près des plantes, de peur qu'elles ne filent & ne s'affoiblissent, & afin qu'elles aient assez d'air, à proportion de la chaleur de la saison. Ces couvertures de papiers huilés sont absolument nécessaires pour la culture des Melons, & encore meilleures pour couvrir les boutures de plantes exotiques, ainsi que pour beaucoup d'autres usages.

Comme ce papier ne dure guere plus d'une année, il faut le renouveler à chaque printems ; mais quand les cadres sont bien construits, & qu'on les met à l'abri de l'humidité lorsqu'on ne s'en sert plus, ils peuvent durer plusieurs années, sur-tout si on a le soin de les placer sur des rouleaux de paille, pour les garantir de l'humidité. Cette paille peut encore servir à couvrir les plants d'*Asperges* pendant l'hiver.

MELOCACTUS. ⎫ *Voyez*
MELOCARDUUS. ⎭ CACTUS.

MELOCHIA. [*Jews Mallow.*] Mauve de Juif, espece de Bette-rave d'Egypte.

Caractères. Le calice de sa fleur est persistant, & formé par une feuille découpée jusqu'au milieu en cinq segmens ; la corolle est composée de cinq pétales larges & étendus : les étamines sont enveloppées dans le tube du germe, & ont cinq antheres : la fleur a un germe rond, avec cinq styles en forme d'alêne, érigés, persistans, & couronnés par des stigmats simples. A cette fleur succedent des capsules rondes à cinq angles, à deux cornes & à cinq cellules, dans chacune desquelles est renfermée une semence angulaire & applatie.

Ce genre de plantes est rangé dans la premiere section de la seizieme classe de LINNÉE, intitulée *Monadelphia Pentandria*, les

MEL

les fleurs de cette claſſe ayant leurs étamines & leurs ſtyles réunis en un tube , & celles de cette ſection n'ayant que cinq étamines.

Les eſpeces ſont :

1°. *Melochia pyramidata , floribus umbellatis , oppoſitis foliis , capſulis pyramidatis , pentagonis , angulis acutis , foliis nudis. Hort. Cliff.* 343. *Fl. Zeyl.* 245. *Læſl. it.* 255 ; Melochia avec des fleurs en ombelles , & oppo-ſées aux feuilles , des capſules pyramidales à cinq angles , & des feuilles nues.

Althæa Braſiliana fruteſcens , incarnato flore , fagopyri ſemine. Pluk. Phyt. tab. 131. *f.* 3.

2°. *Melochia tomentoſa , floribus umbellatis axillaribus , capſu-lis pyramidatis , pentagonis , an-gulis mucronatis , foliis tomentoſis. Lin. Sp.* 943. *Jacq. Amer.* 193; Melochia avec des fleurs en ombelles aux aiſſelles de la tige , des capſules pyramidales à cinq angles , & des feuilles cotonneuſes.

Abutylon herbaceum procum-bens , Betonicæ folio. flore pur-pureo. Sloan. 97. *Hiſt. Sp.* 220. *t.* 139. *f.* 1.

3°. *Melochia depreſſa , floribus ſolitariis , capſulis depreſſis pen-tagonis : angulis obtuſis , ciliatis. Flor. Leyd.* 348 ; Melochia avec des fleurs ſolitaires , des capſules comprimées , à cinq angles , obtuſes , & garnies de poils.

Abutylon Americanum , Ribe-fii foliis , flore carneo , fructu pentagono aſpero. Houſt. MSS.

4°. *Melochia concatenata , ra-cemis confertis terminalibus , cap-ſulis globoſis , ſeſſilibus. Flor. Zeyl.*

Tome V.

MEL   61

247 ; Melochia avec des épis en grappes qui terminent les tiges , & des capſules globulai-res & ſeſſiles.

Alcea Indica , floſculis parvis , faſciculatim ramulis adfixis. Pluk. Alm. 26. *t.* 9. *f.* 5.

5°. *Melochia ſupina , floribus capitatis , foliis ovatis ſerratis , cau-libus procumbentibus. Lin. Sp.* 944 ; Melochia avec des fleurs en tête , des feuilles ovales & ſciées , & des tiges traî-nantes.

Alcea ſupina puſilla , Geranii exigui maritimi folio & facie , Maderaſpatenſis , fructu in ſummo caule glomerato , pericarpio duro. Pluk. Phyt. tab. 132. *f.* 4.

Pyramidata. La premiere eſ-pece croît naturellement dans le Bréſil , comme une herbe ſauvage & commune : ſa tige eſt un peu en arbriſſeau , & s'éleve à la hauteur de quatre ou cinq pieds ; ſes fleurs naiſ-ſent en ombelles ſur les côtés de la tige , & oppoſées aux feuilles; elles ſont d'une cou-leur de chair pâle , & ſont rem-placées par des capſules pyra-midales à cinq angles & à cinq cellules , qui renferment cha-cune une ſemence angulaire.

Tomentoſa. La ſeconde eſ-pece croît ſans culture à la Jamaïque , & dans quelques au-tres parties chaudes de l'Amé-rique ; elle a une tige traînan-te , herbacée , & garnie de feuil-les cotonneuſes de la même forme que celles de la *Bétoine:* ſes fleurs ſont produites en om-belles aux aiſſelles de la tige ; elles ſont de couleur pourpre , & produiſent des capſules py-ramidales & à cinq angles.

E

Depreſſa. La troiſieme eſpece, qui a été découverte à la Havane par le Docteur HOUSTOUN, s'eleve à la hauteur de cinq ou ſix pieds, avec une tige d'arbriſſeau, garnie de feuilles angulaires ſemblables à celles du *Groſſeiller* en buiſſon : ſes fleurs ſont ſolitaires ſur les parties latérales de la tige ; elles ſont de couleur de chair, & de la même forme que celles de la petite Mauve à fleurs, & ſont remplacées par des capſules rudes à cinq angles, qui renferment cinq ſemences ſemblables à celles de la *Mauve*.

Concatenata. La quatrieme croit naturellement dans les deux Indes ; elle a une tige herbacée, & terminée par pluſieurs paquets oblongs de fleurs, auxquels ſuccedent des capſules angulaires & à cinq cellules, qui renferment chacune une ſimple ſemence.

Supina. La cinquieme, qu'on rencontre dans les Indes orientales, eſt une plante annuelle, dont les tiges ſont traînantes, couchées ſur la terre, & garnies de petites feuilles ſemblables à celles de la *Bétoine* ; ſes fleurs & ſes fruits ſortent en grappes aux extrémités des branches.

Culture. On conſerve toutes ces plantes dans les jardins de Botanique, pour la variété ; mais comme elles ont peu de beauté, on les cultive rarement dans ceux d'agrément : on les multiplie par leurs graines qu'on ſeme ſur une couche chaude. Quand les plantes pouſſent, on les traite ſuivant la méthode qui a été preſ-

crite pour le *Sida*, dont le Lecteur doit conſulter l'article, pour éviter les répétitions.

Les premiere & troiſieme eſpeces ſont des arbriſſeaux que l'on peut, avec quelque ſoin, conſerver dans une ſerre chaude pendant l'hiver, au moyen de quoi on en obtiendra des ſemences, qui parviennent rarement en maturité dès la premiere année, à moins que les plantes ne ſoient avancées de bonne heure au printems, & que l'été ne ſoit chaud. Les trois autres eſpeces perfectionnent généralement les leurs la même année qu'elles ont été ſemées.

MELON. *Voyez* MELO.

MELON D'EAU ou PASTEQUE. *Voyez* CUCURBITA CITRULLUS. ANGURIA.

MELON - CHARDON. *Voyez* CACTUS MELOCACTUS.

MELONGENA. *Tourn. Inſt. R. H.* 151. *tab.* 65. *Solanum. Lin. Gen. Plant.* 224. [*Mad Apple, or Egg-Plant.*] Mayenne, Melongene, Aubergine.

Caracteres. Le calice de la fleur eſt perſiſtant, & formé par une feuille profondément découpée en cinq ſegmens aigus & étendus ; la corolle eſt monopétale, & diviſée en cinq parties entierement ouvertes & réfléchies ; la fleur a cinq étamines en forme d'alêne, & terminées par des antheres oblongues & réunies : dans ſon centre eſt placé un germe oblong qui ſoutient un ſtyle mince, couronné par un ſtigmat obtus. Ce germe ſe change dans la ſuite en un fruit ovale, oblong, à une cellule couverte

d'une pulpe charnue , & remplie de femences plates & arrondies.

Ce genre de plantes eft rangé dans la feptieme fection de la feconde claffe de TOURNEFORT , qui renferme les herbes à fleurs en forme de roue , & monopétales dont le pointal fe change en un fruit mou. LINNÉE a joint ce genre, & le *Lycoperficon* de TOURNEFORT au *Solanum* , dont il n'a fait que des efpeces : mais comme le fruit de cette plante n'a qu'une cellule, elle doit être féparée du *Solanum*, dont le fruit en a deux, & dont il y a déja un fi grand nombre d'efpeces connues, qu'il n'eft pas néceffaire d'y en ajouter d'autres, qui peuvent en être féparées par leurs différentes propriétés. LINNÉE place celle-ci dans la première fection de la cinquieme claffe.

Les efpeces font :

1°. *Melongena ovata , caule inermi herbaceo foliis oblongo-ovatis , tomentofis , integris , fruêtu ovato* ; Aubergine à tige unie & herbacée, avec des feuilles longues, ovales, velues, entieres, & un fruit ovale.

Melongena fruêtu oblongo violaceo. Tourn. Inft. 151 ; Melongene avec un fruit oblong & violet.

Solanum Melongena. Lin. Syft. Plant. tom. 1. p. 515. Sp. 19.

2°. *Melongena teres , caule inermi herbaceo , foliis oblongo-ovatis , tomentofis , fruêtu tereti* ; Aubergine à tiges unies & herbacées , avec des feuilles oblongues, ovales, velues, & un fruit recourbé & cylindrique.

3°. *Melongena fruêtu incurvo* ; Melongene avec un fruit recourbé.

4°. *Melongena fpinofa , foliis finuatis , laciniatis , fruêtu tereti , caule herbaceo* ; Aubergine avec une tige épineufe & herbacée, des feuilles découpées & finuées, & un fruit cylindrique.

Solanum Pomiferum , fruêtu fpinofo. J. B. 3. 619 ; Morelle portant Pomme, dont le fruit eft épineux.

Ovata. La premiere efpece croît naturellement en Afie, en Afrique, & en Amérique, ou les habitans fe nourriffent de fon fruit : on la cultive en Efpagne, dans les jardins, comme un fruit bon à manger, fous le nom de Barenkeena. Les Turcs en font le même ufage, & l'appellent *Badinjan* ; les Italiens , *Melanzana* ; & les habitans des Ifles Britanniques en Amérique , *Brown-John* , ou *Brown-Jolly.* Cette plante eft annuelle, & fa tige, qui eft herbacée & un peu ligneufe, s'éleve à près de trois pieds de hauteur, & pouffe des branches latérales & garnies de feuilles oblongues, ovales, de fept à huit pouces de longueur fur quatre de large, cotonneufes, légerement finuées, fans être dentées, placées fans ordre, & fupportées par des pétioles fort épais : fes fleurs fortent fimples fur les côtés des branches ; elles ont un calice épais, charnu, & formé par une feuille profondément découpée en cinq fegmens aigus, entierement ouverte, & armée de piquans en-dehors : ces fleurs font

monopétales , & divisées sur leurs bords en cinq parties étendues en forme d'étoile , & un peu réfléchie ; elles sont bleues, & les sommets qui sont unis ensemble dans leur centre , sont de couleur jaune ; à ces fleurs succede un fruit ovale , charnu, à-peu-près aussi gros qu'un œuf de cygne , & de la même forme d'un pourpre foncé d'un côté, & blanc de l'autre : les fleurs paroissent dans les mois de Juin & de Juillet , & les fruits mûrissent en Septembre.

Il y a plusieurs variétés de cette espece, une à fruits blancs, appelée par quelques-uns *Plante à œufs* ; une à fruits jaunes ; & une troisieme à fruits d'un rouge pâle. Toutes ces variétés sont généralement constantes, les semences de chacune produisant le même fruit : mais comme elles ne different que dans les couleurs de ces fruits, je ne crois pas devoir les donner comme des especes distinctes (1).

Teres. La seconde differe de la premiere par la forme de son fruit, qui a communément huit à neuf pouces de longueur, & qui est cylindrique & droit; mais en toutes autres choses les plantes sont les mêmes : cependant comme elle ne varie jamais quand on la multiplie dans les jardins , il n'y a aucun doute qu'elle ne soit

plutôt une espece distincte. Il y a encore deux variétés de celle-ci, l'une à fruits pourpre , & l'autre à fruits blancs ; mais la derniere est la plus commune en Angleterre.

Incurva. La troisieme differe des deux précédentes par la forme de ses feuilles, qui sont profondément sinuées sur leurs bords : son fruit est oblong , recourbé, de couleur jaunâtre, & plus gros à son extrémité que dans aucune autre partie.

Spinosa. La quatrieme , dont les semences m'ont été envoyées des Indes , est très-différente des précédentes ; ses tiges & ses feuilles sont armées d'épines très-fortes, & ces feuilles sont plus larges, & profondément dentées sur les côtés : ses fleurs sont plus larges, & d'un bleu plus foncé ; son fruit est long , cylindrique & blanc.

Culture. On mange ces fruits dans la plupart des contrées méridionales, où on les regarde comme délicats ; mais on croit qu'ils provoquent à l'amour.

On multiplie ces plantes par leurs graines , qu'il faut semer en Mars , sur une couche de chaleur tempérée; quand elles poussent , on les transplante sur une autre couche chaude, à quatre pouces de distance ; on les arrose, & on les tient à l'ombre jusqu'à ce qu'elles aient formé de nouvelles racines : on leur donne ensuite beaucoup d'air dans les tems chauds , sans quoi elle fileroient & deviendroient foibles : il faut aussi les arroser

---

(1) Les feuilles de cette plante sont annodines & résolutives : on peut les employer en forme de cataplasme , comme celles de la Mandragore, sur les hémorrhoïdes, les cancers, les brûlures, &c.

souvent, fi l'on veut qu'elles faffent des progrès : mais quand elles font devenues affez fortes pour remplir le châffis, ce qui arrive ordinairement vers le milieu ou la fin de Mai, on les tranfplante dans une piece de terre riche, ou dans les plates-bandes du parterre, à deux pieds de diftance entr'elles, en confervant, lorfqu'on les enleve, une motte de terre à leurs racines, autant qu'il eft poffible ; car fans cela elles font fujettes à manquer : on les arrofe beaucoup, & on les tient à l'ombre jufqu'à ce qu'elles foient bien enracinées, après quoi elles n'exigeront plus aucun autre foin que d'être tenues nettes de mauvaifes herbes, & arrofées dans les tems fort fecs.

Le fruit paroît au mois de Juillet : fi alors la faifon eft fort feche, on arrofe très-fouvent ces plantes, pour faire groffir les fruits, & en augmenter le nombre. Ces fruits mûriffent vers la fin d'Août : on doit avoir foin de conferver les femences de chaque efpeçe féparément ; mais on cueille les fruits qu'on veut manger, avant qu'ils foient tout-à-fait mûrs.

On ne conferve ces plantes dans les Jardins Anglois, que par curiofité ; car on en mange rarement le fruit dans ce pays, à moins qu'il ne s'y trouve quelques Italiens ou Efpagnols qui ont coutume d'en manger chez eux.

MÉLONIERE.
On appelle ainfi une partie d'un jardin potager, ou un endroit uniquement deftiné à la culture des Melons. Cet efpace doit être à découvert au fud-eft, mais abrité du côté de l'oueft, du nord-oueft, & du nord-eft, par des murs, des paliffades ou des haies ; mais les haies font préférables à toutes les autres efpeces de clôtures. La Meloniere doit être auffi dans un lieu fec, car il n'y a rien de plus nuifible aux Melons que l'humidité. Comme le printems eft très-fouvent pluvieux, fi le fol fe trouve humide, on ne pourra faire les élévations de terre que bien tard. Il faut difpofer la Meloniere de façon qu'elle foit auffi près du fumier qu'il eft poffible, afin d'épargner aux Jardiniers la peine de le tranfporter bien loin. Il fera auffi très-commode d'avoir dans le voifinage un baffin rempli d'eau, pour arrofer les Melons dans le befoin : mais l'eau ne manque guere en Angleterre.

Pour ce qui eft de l'étendue du terrein, il doit être proportionné au nombre de levées de terre qu'on veut faire ; ce que l'on peut aifément calculer, en donnant douze pieds de longueur à chaque levée, & en plaçant les trous à quatre pieds de diftance ; mais le mieux eft de prendre affez de terrein pour n'être pas gêné.

Il faut entourer la Meloniere d'une haie de joncs, & la tenir toujours fermée pendant que les Melons croiffent ; car fi on les montre à toutes les perfonnes qui fe promenent dans les jardins, & dont la plupart font curieufes de manier

les tiges, pour en voir le fruit, on s'en trouvera mal, rien ne faisant autant de tort à ces plantes que de déranger souvent leurs feuilles.

La pratique ordinaire, dans la plupart des jardins des Gentilshommes, est d'enclorre un petit espace de terrein ou de murailles ou de palissades, pour l'employer à cet usage; mais cette méthode n'est point bonne, parce que les Melons ne réussissent guere plus de trois ans dans la même place, à moins qu'on n'en change la terre, & qu'on n'y en mette de la nouvelle; ce qui est toujours très-dispendieux. Ainsi, la meilleure maniere est d'avoir une assez grande quantité de roseaux, qu'on peut transporter où l'on veut, & de changer tous les ans les Melons de place. Si le terrein est assez grand pour être divisé en trois ou quatre pieces, on pourra transporter la haie, jusqu'à ce que tout le canton ait été occupé, après quoi on retournera à l'endroit où l'on avoit commencé, qui fera alors aussi bon qu'une terre neuve; & comme, par cette méthode, on laisse toujours en place un côté de la clôture, la peine est moindre que si l'on étoit obligé de transporter la totalité de la haie à une grande distance. Ces clôtures de roseaux sont préférables, pour cet usage, aux haies ordinaires & aux palissades.

MELOPEPO. V. Cucurbita.

MELOTHRIA. Lin. Gen. Plant. 48. Le nom que porte ce genre lui a été donné par Linnée, dans le *Hortus Cliffortianus*. Quelques Auteurs l'ont placé sous celui de *Cucumis*, & d'autres sous celui de *Bryonia*; mais Linnée a fort éloigné cette plante de tous ces genres, parce qu'elle n'a que trois étamines. Van-Royen l'a rapprochée de la *Bryone*, parce qu'elle a des fleurs mâles & hermaphrodites. [*Small Creeping Cucumber.*] Petit Concombre.

Caracteres. Le calice de la fleur est formé par une feuille en forme de cloche, & légerement découpée sur ses bords en cinq parties. Dans les fleurs hermaphrodites, ce calice reste sur l'embryon : la fleur mâle est monopétale, & en forme de roue; elle a un tube de la longueur du calice : dans le centre de la fleur hermaphrodite est placé un pointal, qui soutient un style cylindrique, & accompagné par trois étamines coniques, insérées dans le tube de la fleur, & étendues à la même longueur; les fleurs mâles ont trois étamines, terminées par des antheres doubles, presque rondes, & comprimées; le pointal des fleurs hermaphrodites devient ensuite une petite baie ovale, & à trois divisions, dans lesquelles sont renfermées des semences petites & plates.

Nous n'avons qu'une espece de cette plante, qui est la

Melothria pendula. Lin. Hort. *Cliff.* 490. *Hort. Ups.* 15. *Gron. Virg.* 10. *Roy. Lugd.-B.* 528. *Kniph. Orig. cent.* 4. *n.* 49; petit Concombre rampant.

MEL

Cucumis minima fruĉtu ovali, nigro, lævi. Sloan. Hiſt. 1. *p.* 227. *t.* 142. *f.* 1 ; le plus petit Concombre à fruit liſſe, noir & ovale.

Bryonia, olivæ fruĉtu, minor. Plum. Spec. 3. *ic.* 66. *f.* 2.

Cette plante croît ſauvage dans les bois de la Caroline & de la Virginie, ainſi que dans pluſieurs Iſles de l'Amérique ; elle rampe ſur la terre, & pouſſe des branches minces, garnies de feuilles angulaires, qui reſſemblent un peu à celles du Melon, mais beaucoup plus petites. Ces branches pouſſent à chacun de leurs nœuds des racines qui pénetrent dans la terre, & fourniſſent par ce moyen plus de nourriture aux plantes, dont les tiges s'étendent à une grande diſtance de tous côtés, & couvrent un très-grand eſpace : ſes fleurs ſont fort petites, de la même forme que celles du Melon, & d'une couleur de ſoufre pâle. En Amérique, ſon fruit eſt de la groſſeur d'un pois, & d'une figure ovale ; il devient noir en mûriſſant. Les habitans de ces contrées font quelquefois mariner ce fruit, quand il eſt encore vert.

Il eſt beaucoup plus petit en Angleterre, & ſi caché par les feuilles, qu'on le trouve difficilement. Cette plante ne réuſſit pas en plein air ici. On la multiplie en ſemant ſes graines ſur une couche chaude. Quand on permet à ces plantes de s'étendre, elles couvrent bientôt toute la ſurface d'une grande couche : les ſemences des fruits mûrs qui s'écartent, pouſſent

MEN

de nouvelles plantes, quand la terre où elles ſont tombées eſt employée à d'autres couches chaudes ; elles n'exigent aucun autre ſoin que d'être arroſées. On conſerve cette eſpece dans quelques jardins, pour la variété, mais elle n'eſt d'aucun uſage.

MEMBRANE (une), eſt une peau ligneuſe, qui ſépare les ſemences, dans les legumes des plantes

MÉNIANTE *ou* TREFFLE D'EAU. *V* MENYANTHES TRIFOLIATA.

MÉNISPERMUM. *Tourn. Inſt. R. H.* 1705. *Lin. Gen. Plant.* 1131. [*Moon-Seed.*] Semence étoilée.

Caraĉteres. Cette plante a des fleurs mâles & des femelles ſur différens pieds : les fleurs mâles ont des calices compoſés de deux feuilles courtes & linéaires ; la corolle a quatre pétales ovales & étendus au-dehors, & en-dedans huit pétales ovales concaves, plus petits que les extérieurs, & rangés en quatre enchaînures : la fleur a pluſieurs étamines cylindriques, plus longues que les pétales, & terminées par des antheres courtes, obtuſes, & à quatre lobes ; les fleurs femelles ont une corolle & un calice ſemblable à ceux des mâles, huit étamines avec des antheres tranſparentes & ſtériles, & deux germes ovales & courbés, qui ſoutiennent chacun un ſtyle ſolitaire, recourbé & couronné par un ſtigmat diviſé en deux parties. Ce germe ſe change enſuite en deux baies rondes en forme de reins, &

E 4

MEN

à une cellule qui renferme une grosse semence de même forme.

Ce genre de plantes est rangé dans la dixieme section de la vingt-deuxieme classe de LINNÉE, qui comprend celles qui ont des fleurs mâles & femelles sur différens pieds, & dont les fleurs mâles ont douze étamines.

Les especes sont :

1°. *Menispermum Canadense*, *foliis peltatis, subrotundis, angulatis. Hort. Cliff.* 140. *Hort. Upf.* 91. *Gron. Virg.* 153. *Gmel. Sib.* 3. *p.* 107. *n. 86* ; Menisperme à feuilles rondes, angulaires & en forme de bouclier.

Menispermum Canadense scandens, umbilicatis foliis. Tourn. Act. Par. 1705 ; Menisperme grimpant du Canada, à feuilles en forme de nombril, ou la Coque du Levant.

Hedera monophyllos Virginiana, Convolvuli foliis. Pluk. Alm. 181. *t. 36. f. 1.*

Cissampelos. Rupp. Jen. 67. Gmel. R.

2°. *Menispermum Virginicum*, *foliis cordatis, peltatis, lobatis. Flor. Virg.* 40 ; Menisperme à feuilles en forme de cœur & de targe, & découpées en lobes.

Menispermum folio hederaceo. Hort. Elth. 223. *tab.* 178. *f.* 219 ; Menisperme à feuilles de Lierre.

3°. *Menispermum Carolinum*, *foliis cordatis, subtus villosis. Lin. Sp. Plant.* 340 ; Menisperme à feuilles en forme de cœur, & velues en-dessous.

Canadense. La premiere espece croît naturellement dans le Canada, & dans plusieurs autres parties de l'Amérique septen-trionale, où elle pousse dans les bois ; elle a une racine épaisse & ligneuse, de laquelle sortent des tiges grimpantes, qui deviennent ligneuses, s'élevent à la hauteur de douze ou quatorze pieds, & se roulent autour des plantes voisines, pour se soutenir. Ces tiges sont garnies de feuilles larges, rondes & unies, dont les pétioles sont placés presqu'au milieu du dessous des feuilles, & dont le dessus forme un creux, & ressemble à un nombril : ses fleurs sortent en paquets lâches sur les côtés des tiges ; elles sont de couleur herbacée, petites, & composées de deux rangs de pétales oblongs & ovales, & d'étamines fort courtes : les fleurs mâles ont dix étamines terminées par des antheres simples, & les fleurs femelles ont, dans leur centre, deux germes qui se changent en baies, & dont chacune contient une semence en forme de rein. Cette plante fleurit dans le mois de Juillet, & ses semences mûrissent en automne.

On peut multiplier aisément cette espece, en marcottant ses branches qui auront de bonnes racines pour l'automne suivant: alors elles seront en état d'être séparées des vielles plantes, & d'être transplantées dans les places où elles devront rester ; mais il faut leur donner un soutien, car leurs branches sont foibles & minces, comme dans leur pays originaire ; elles grimpent sur les arbres, & s'élevent à une hauteur considérable : on peut les placer de

MEN

même ici dans des quartiers déferts, où elles profiteront mieux que dans une fituation ouverte.

Virginicum. La feconde efpece différe de la premiere par la forme de fes feuilles, qui font angulaires, & quelquefois figurées en cœur : leurs pétioles adherent à leur bâfe ; ainfi, elles n'ont point la forme de nombril en-deffus. Les tiges de celle-ci deviennent ligneufes, & s'élevent prefque à la même hauteur que celles de la premiere ; fes fleurs & fes baies ne different point de celles de la précédente : on la multiplie auffi de la même maniere.

Carolinum. La troifieme eft originaire de la Caroline, d'où fes femences ont été envoyées en Angleterre : plufieurs perfonnes ont penfé qu'elle étoit la même que la feconde, de laquelle elle differe par fes branches, qui ne deviennent pas ligneufes : fes tiges font herbacées ; fes feuilles font entieres & velues, & n'ont que la moitié de la largeur de celles de la précédente : la plante eft auffi moins dure ; car dans les hivers rigoureux, celles qui font expofées en plein air, font quelquefois détruites, au–lieu que celles de la feconde ne périffent jamais par le froid. Cette efpece ne produit point de fleurs en Angleterre, à moins que l'année ne foit fort chaude.

On peut multiplier cette plante, en divifant fes racines, qui s'étendent au - dehors fur les côtés, & que l'on coupe tous les deux ans. Le meilleur tems pour faire cette opération, eft le printems, un peu avant qu'elles

commencent à pouffer. Ces racines doivent être mifes à une expofition chaude & dans un fol léger ; car dans une terre forte qui retient l'humidité en hiver, elles font fujettes à pourrir : on les plante contre une muraille expofée au midi ou à l'oueft, de maniere qu'on puiffe y attacher leurs tiges, pour les empêcher de ramper. Dans cette fituation, ces plantes fleuriront fouvent, & en les couvrant pendant les fortes gelées, leurs tiges pourront être préfervées des injures du froid.

Cette efpece a peu de beauté ; cependant comme on la cultive dans plufieurs jardins pour la variété, j'en ai fait mention ici.

MENTHA. *Tourn. Infl. R. H.* 188. tab. 89. *Lin. Gen. Plant.* 633. Mℓ*θη* Déeffe, fuivant les Anciens. Les Poëtes lui donnent auffi le nom de bonne odeur, de maniere que quand on trouve cette expreffion dans leurs ouvrages, le nom de la plante eft fous–entendu. *Mentha* vient auffi de *Mens*, en latin, l'efprit, parce que cette plante le reconforte. [*Mint.*] Menthe.

Caracleres. La corolle eft labiée, monopétale, érigée & poftée fur un calice perfiftant, tubulé, monophylle, & découpé en cinq fegmens égaux : le tube de la corolle eft un peu plus long que le calice ; les levres font découpées en quatre parties prefqu'égales, dont la fupérieure eft un peu plus large & dentée. La fleur a quatre étamines en forme d'aléne, érigées & fixées à une certaine diftance les unes des autres,

MEN

dont les deux plus voisines sont
les plus longues ; elles sont
terminées par des antheres rondes, & dans le fond du tube
est placé un germe à quatre
pointes, qui soutient un style
mince, érigé & couronné par
un stigmat étendu & divisé en
deux parties. Ce germe se change, quand la fleur est passée,
en quatre semences nues, &
postées dans le calice.

Ce genre de plantes est rangé
dans la premiere section de la
quatorzieme classe de LINNÉE,
qui comprend celles dont les
fleurs ont deux étamines longues, & deux plus courtes, &
dont les semences mûrissent
dans le calice.

Les especes sont :

1°. *Mentha viridis, floribus
spicatis, foliis oblongis, serratis.
Hort. Upsal. 168* ; Menthe avec
des fleurs en épis, & des feuilles oblongues & sciées.

*Mentha angusti-folia, spicata.
C. B. P. 227* ; Menthe en épis,
& à feuilles étroites, communément appelée *Menthe à lance.*

2°. *Mentha glabra, floribus
spicatis, foliis longioribus, glabris,
superné minimè serratis* ; Menthe
avec des fleurs en épis, & des
feuilles plus longues, unies,
& fort légerement sciées vers
leur pointe.

*Mentha angusti folia, spicata,
glabra. Rand.* ; Menthe en épis,
à feuilles unies & étroites.

3°. *Mentha candicans, foliis
lanceolatis, serratis, subtùs incanis, floribus spicatis, hirsutissi-
mis* ; Menthe avec des feuilles
en forme de lance, sciées &
velues en-dessous, avec des
fleurs en épis, & très-velues.

*Mentha sylvestris candicans,
odore sativi. Doody. Raii Syn.
App.* ; Menthe sauvage de couleur blanche, de la même
odeur que celle de jardin.

4°. *Mentha sylvestris, spicis
confertis, foliis serratis, tomentosis, sessilibus. Hort. Cliff. 306* ;
Menthe avec des fleurs en épis,
disposées en grappes, & des
feuilles cotonneuses, sciées, &
sessiles aux tiges.

*Mentha sylvestris longiori folio.
C. B. P. 227* ; Menthe sauvage
à plus larges feuilles. *Menthastre.*

5°. *Mentha aquatica, spicis
crassioribus, foliis ovato lanceolatis, serratis, subtùs tomentosis,
petiolatis* ; Menthe avec des
épis plus épais, & des feuilles ovales en forme de lance,
sciées, cotonneuses en-dessous, & supportées par des pétioles.

*Menthastri aquatici genus hirsu-
tum, spicâ latiore. J. B, 3. 222* ;
Menthe d'eau, velue, à plus
gros épis. Menthe aquatique.
Le Pouliot.

6°. *Mentha Piperita, spicis crassioribus interruptis, foliis lanceolatis, acutè serratis* ; Menthe
avec des épis de fleurs plus
épais, & éloignés, & des
feuilles en forme de lance, &
fortement sciées.

*Mentha fervida nigricans,
Piperis sapore. Rand. Hort. Chel.
cat.* ; Menthe tirant sur le noir,
& chaude, ayant un goût de
Poivre, communément appelée
Menthe à Poivre.

7°. *Mentha crispa, floribus
spicatis, foliis cordatis, dentatis,
undulatis, sessilibus. Hort. Cliff.
306. Hort. Ups. 168. Mat.
Med. 147. Ejusph. cent. 11.*

MEN

n. 75 ; Menthe avec des fleurs en épis, & des feuilles en forme de cœur, dentées, ondées, & seffiles aux tiges.

Mentha crispa Danica, sive Germanica speciosa. Mor. Hist. 3. p. 367 ; Menthe Danoise ou d'Allemagne, frisée.

8°. *Mentha Rotundi-folia, spicis confertis, foliis ovatis, rugosis, sessilibus* ; Menthe avec des épis rapprochés, & des feuilles ovales, rudes & seffiles.

Menthastrum folio rugoso rotundiori, spontaneum, flore spicato, odore gravi. J. B. 3. 219 ; Menthe sauvage avec une feuille plus ronde & rude, & des fleurs en épis, qui ont une odeur forte.

9°. *Mentha rubra, spicis confertis, interruptis, foliis oblongo-ovatis, acuminatis, dentatis, sessilibus*; Menthe avec des épis de fleurs interrompus, des feuilles oblongues, ovales, à pointe aiguë, dentées & seffiles.

Mentha rotundi-folia, rubra, Aurantii odore. Mor. Hist. 3. 369 ; Menthe rouge à feuilles rondes, & à odeur d'Orange, communément appelée *Menthe d'Orange.*

. 10°. *Mentha Chalepensa, foliis oblongis, dentatis, utrinque tomentosis, sessilibus, spicis tenuioribus* ; Menthe à feuilles oblongues, dentées, velues sur les deux surfaces, & seffiles, produisant des épis de fleurs fort étroits.

Menthastrum Chalepense, angustifolium, raro florens. Boërh. Ind. alt. 1. *p.* 185 ; Menthe sauvage d'Alep, avec des feuilles étroites, mais qui fleurit rarement.

11°. *Mentha palustris, floribus capitatis, foliis ovatis, serratis, petiolatis, staminibus corollá longioribus. Hort. Cliff.* 306. *Fl. Suec.* 482. 517. *Roy. Lugd. B.* 325. *Dalib. Paris.* 177. *Crantz. Austr. p.* 332. *Pollich. pal. n.* 552. *Neck. Gallob. p.* 250. *Kniph. cent.* 11. *n.* 74 ; Menthe avec des fleurs en tête, des feuilles ovales, sciées & pétiolées, & des étamines plus longues que la corolle.

Mentha aquatica. Lin. Syst. Plant. tom. 3. *p.* 43. *Sp.* 7.

Mentha rotundi-folia palustris, sive aquatica major. C. B. P. 227 ; la plus grande Menthe d'eau ou de marais à feuilles rondes.

Sisymbrium sylvestre. Dalech. Hist. 677.

12°. *Mentha nigricans, floribus capitatis, foliis lanceolatis, serratis, subpetiolatis. Lin. Sp. Plant.* 576 ; Menthe avec des fleurs disposées en tête, & des feuilles en forme de lance, avec de fort courts pétioles.

Mentha fervida nigricans latifolia. Rand. ; Menthe à Poivre, & noirâtre, à larges feuilles.

13°. *Mentha arvensis, floribus verticillatis, foliis ovatis, acutis, serratis, staminibus corollá brevioribus. Lin. Sp. Plant.* 577. *Hort. Cliff.* 307. *Fl. Suec.* 481 ; 516. *Roy. Lugd.-B.* 326. *Dalib. Paris.* 178. *Scop. carn. ed.* 2. *n.* 740. *Pollich. pal. n.* 553. *Mattusch. Sil. n.* 429. *Flor. Dan. t.* 512. ; Menthe avec des fleurs verticillées, des feuilles ovales, aiguës & sciées, & des étamines plus courtes que la corolle.

Mentha arvensis, verticillata, hirsuta. J. B. 3. 2. 217 ; Menthe des champs, velue & verticillée.

Calamentha arvensis verticillata. Bauh. pin. 229 ; Calament des boutiques.

14°. *Mentha exigua, floribus verticillatis, foliis ovatis, dentatis, staminibus corollâ longioribus* ; Menthe avec des fleurs verticillées, des feuilles ovales & dentées, & des étamines plus longues que les pétales.

Mentha aquatica exigua. Traj. lib. 1. c. 6. ; la plus petite Menthe d'eau.

Calamentha aquatica Belgarum & Matthioli. Lob. ic. 505.

15°. *Calamentha Gentilis floribus verticillatis, foliis ovatis, marginibus ciliatis, staminibus corollam æquantibus* ; Menthe à fleurs verticillés & à feuilles ovales, dont les bords sont velus, avec des étamines égales à la corolle.

Mentha verticillita rotundiori folio, odore Ocimi. Dale. ; Menthe verticillée, avec des feuilles plus rondes, à odeur de basilic. Beaume, où menthe des jardins.

16°. *Mentha hirsuta, floribus verticillatis, foliis-ovatis, serratis, hirsutis, staminibus corollâ longioribus* ; Menthe à fleurs verticillées, avec des feuilles ovales, sciées & velues, & des étamines plus longues que la corolle.

Mentha aquatica, sivè Sisymbrium hirsutius. J. B. 3. 2. 224. Moris. Hort. p. 370 ; Menthe d'eau ou Sisymbrium velu.

Sisymbrium hirsutum. Raü Angl. 3. p. 233.

17°. *Mentha verticillata, floribus verticillatis, foliis, lanceolatis, acutis, serratis, rugosis, staminibus corollam æquantibus* ; Menthe à fleurs verticillées, avec des feuilles en forme de lance, à pointe aiguë, & sciées, & des étamines égales à la corolle.

Mentha verticillata, longiori acuminato folio, odore aromatico. Rand. Hort. Chel. Cat. ; Menthe verticillée avec une feuille plus longue, & à pointe aiguë, ayant une odeur aromatique.

Il y a plusieurs autres variétés de ce genre, qui naissent spontanément en Angleterre, & dont je possede douze & plus dans ma collection ; mais je soupçonne que quelques-unes ne sont que des variétés accidentelles, occasionnées par la différence du sol & de l'exposition où elles ont été trouvées. Je ne les ai pas toutes dénombrées ici ; mais je crois que celles dont j'ai fait mention sont des especes distinctes, parce que je les ai cultivées pendant plus de trente années, & que je ne les ai jamais vu varier. J'en ai élevé plusieurs de semences, & les ai toujours trouvées semblables à celles sur lesquelles les graines avoient été recueillies.

Viridis. La premiere espece est celle que les Jardiniers cultivent, pour la porter sur les marchés, parce qu'on s'en sert en Médecine, ainsi que pour la cuisine. On la connoît généralement sous le nom de *Menthe* à lance ; mais quelques personnes lui donnent celui de

MEN

Menthe au cerf. PARKINSON & GERARD l'appellent *Menthe* romaine. Cette plante eſt ſi bien connue, qu'il n'eſt pas néceſſaire d'en donner la deſcription. On en connoît deux variétés, l'une à feuilles friſées, & l'autre à feuilles panachées ; mais je les ai obtenues toutes deux de l'eſpece commune. Comme quelques perſonnes les conſervent dans leurs jardins pour la variété, j'en fais mention ici (1).

Cette plante eſt fort eſtimée

---

(1) On emploie en Médecine pluſieurs eſpeces de ce genre ; mais comme leurs principes ſont toujours les mêmes, & qu'elles jouïſſent toutes de propriétés ſemblables, ce que je dirai de cette premiere eſpece peut auſſi être appliqué aux autres, qui peuvent lui être ſubſtituées dans la plupart des cas.

La Menthe eſt un des meilleurs remedes ſimples que fournit le regne végétal ; ſon odeur eſt balſamique, forte & pénetrante, & ſon goût eſt chaud & un peu amer, elle fournit par l'analyſe, outre un principe ſpiritueux très-volatil, une quantité conſidérable d'huile eſſentielle très-active, de réſine très-chaude, & de matiere gommeuſe preſque inerte, lorſqu'elle eſt dégagée des autres principes.

Les propriétés ſtomachiques, carminatives & utérines de cette plante ſont très-marquées ; on s'en ſert avec beaucoup de ſuccès dans les affections venteuſes & hyſtériques, les vices de digeſtion occaſionnés par le relâchement de l'eſtomac, les fleurs blanches, le flux de ventre invétéré, l'aſthme humide, les engorgemens catharreux de la poitrine, les pâles couleurs, les ſuppreſſions des regles, &c. On la prépare en infuſion dans l'eau ou le vin, depuis une pincée juſqu'à

---

pour les maux d'eſtomac, le défaut d'appétit, & pour les vomiſſemens : on en extrait une eau ſimple, un eſprit ; on en fait un ſyrop compoſé, & une huile préparée dans les boutiques.

Glabra. La ſeconde eſpece a des feuilles plus unies & plus étroites que celles de la premiere ; mais en toutes autres choſes, ces deux eſpeces ſe reſſemblent, de maniere qu'on les cultive ſouvent dans les jardins, pour l'uſage, ſans diſtinction.

Candicans. La troiſieme croît naturellement en Angleterre ; ſes feuilles ſont plus courtes & plus larges au milieu que celles des précédentes ; leurs dentelures ſont plus aiguës, & leur ſurface inférieure eſt cotonneuſe & fort blanche ; ſes tiges ſont plus diviſées vers le ſommet, & terminées par un plus grand nombre d'épis ; qui ſont interrompus vers leur partie baſſe. Cette eſpece répand une odeur ſemblable à celle de la *Menthe* de jardin.

Sylveſtris. La quatrieme a des feuilles plus longues & plus larges qu'aucune des précédentes ; elles ſont cotonneuſes & blanches, & leurs dentelures ſont plus éloignées & fort aiguës ; elles ſont ſeſſiles & velues. Les épis de fleurs ſont plus minces, & réunis en nom-

---

deux : ſon eau diſtillée eſt auſſi d'un grand uſage, ainſi que ſon extrait, qu'on emploie dans les mêmes circonſtances & à la même doſe que l'extrait d'Abſynthe.

Les ſyrops de Menthe ſont très-utiles dans les maladies de poitrine.

bre au haut d'une tige velue. Cette espece est le *Menthaftrum*, ou la *Menthe* fauvage des boutiques, qui entre dans les trochifques de *Myrrhe*.

Aquatica. La cinquieme croît naturellement dans des endroits humides de plufieurs parties de l'Angleterre : on lui donne le nom de *Menthe* fauvage en épis, ou *Menthe* d'eau : fes tiges font plus courtes que celles des précédentes, & velues, ainfi que les feuilles qui font ovales, en forme de lance, fciées fur leurs bords, & d'une couleur pâle : fes fleurs croiffent en épis courts & épais aux extrémités des tiges ; leurs étamines font plus courtes que la corolle.

Piperita. La fixieme naît auffi, fans culture, dans quelques parties de l'Angleterre : je l'ai trouvée fur les bords de la riviere qui coule entre Mitcham & Croydon en Surry ; fes tiges font liffes & de couleur pourpre ; fes feuilles font plus petites que celles de la *Menthe* commune ; elles font en forme de lance, fciées fur leurs bords, & d'un vert plus foncé qu'aucune de celles des précédentes ; leurs côtes du milieu & leurs nervures font pourpre, & un peu velues en-deffous ; les épis de fleurs font plus courts & plus épais que ceux de la *Menthe* commune ; elles font éloignées & interrompues au bas ; leur couleur eft pourpre foncé, & leurs étamines font plus longues que la corolle. La plante entiere a une faveur chaude & mordante comme le *Poivre*, & une

odeur agréable. On tire de cette plante, par la diftillation une eau qui eft auffi eftimée que celle de la *Menthe* commune, qui fert aux mêmes ufages : on la regarde comme un remede excellent contre la pierre & la gravelle.

Crifpa. La feptieme efpece a été apportée du Dannemarck, où l'on croyoit qu'elle croiffoit naturellement ; mais LINNÉE la donne comme étant originaire de la Sibérie : fes tiges font velues, & à-peu-près auffi hautes que celles de la *Menthe* commune ; fes feuilles font en forme de cœur, profondément dentées fur leurs bords, ondées, frifées, feffiles, & de couleur verte : fes fleurs font pourpre, & naiffent en épis interrompus aux extrémités des tiges ; leur calices font découpés prefque jufqu'au fond, & le ftyle eft divifé en deux parties, & porté au-delà de la corolle.

Rotundi-folia. La huitieme, qui fe trouve dans plufieurs parties de l'Angleterre, s'éleve à-peu-près à la même hauteur que la *Menthe* ordinaire, avec une tige forte, quarrée, velue ; branchue au-dehors vers le fommet, & garnie de feuilles ovales, rudes, feffiles, d'un vert foncé, & crénelé fur leurs bords : les épis de fleurs, courts & ferrés, croiffent en grappes fur le haut des tiges ; ces fleurs font d'un blanc herbacé, & les étamines font étendues au-dehors, au-delà de la corolle.

Rubra. La neuvieme, à laquelle on donne communément le nom de *Menthe d'Orange*, à caufe de fon odeur, qui appro-

che de celle de l'écorce *d'O- range*, s'éleve à-peu près à la même hauteur que la *Menthe* commune, avec une tige droi- te, lisse & moins branchue; ses feuilles font beaucoup plus lar- ges, & leurs dentelures font profondes, & terminées en pointe aiguë: ses épis de fleurs font interrompus, & croissent en grappes sur le sommet des tiges. Ses fleurs font d'une couleur pâle, & leurs étamines font plus courtes que les co- rolles. On cultive ordinaire- ment cette plante dans les jar- dins, à cause de son odeur agréable.

Chalepensa. La dixieme se trouve dans les environs d'A- lep; assez dure pour profiter en plein air en Angleterre: ses tiges font minces, quarrées, de couleur pourpre vers le bas, & velues vers le haut: elle pousse rarement des branches; mais elle est garnie de feuilles oblongues, dentées, velues sur les deux surfaces, & sessiles: ses épis de fleurs font simples, & fort minces; ses fleurs ne paroissent pas souvent: mais quand elles se montrent, ce n'est que sur la fin de l'été; ses racines font fort rampantes, & la seule méthode d'en obtenir des fleurs, est de les resserrer dans des pots, pour les em- pêcher de s'étendre.

Paluftris. La onzieme, qui naît sans culture dans les fos- sés en Angleterre, est commu- nément connue sous le nom de *Menthe* d'eau; ses tiges ve- lues, & d'environ un pied de hauteur, poussent vers leur ex- trémité des branches garnies de feuilles ovales, sciées & sup- portées par de longs pétioles: ses fleurs sortent en épis ronds aux extrémités des branches; elles font de couleur pourpre, & leurs étamines font plus lon- gues que les corolles. La plante entiere a une odeur très-forte, qui approche de celle du *Pou- liot.* Cette espece est quelque- fois d'usage en Médecine: on la croit plus chaude que la *Men- the* de jardin; & on la regarde comme propre à chasser les vents de l'estomac, & à guérir la colique.

Nigricans. La douzieme, qu'on rencontre encore dans des fos- sées en Angleterre, a des tiges de couleur pourpre, lisses, courtes, & chargées de bran- ches; ses feuilles font petites, en forme de lance, de cou- leur foncée, légerement sciées sur leurs bords, & supportées par des courts pétioles: ses fleurs font aussi de couleur pour- pre, & disposées en têtes ron- des aux extrémités des tiges; leurs étamines font plus lon- gues que la corolle. Cette es- pece a une saveur chaude & mordante, mais cependant moins âcre que celle de la *Menthe* à *Poivre*, à laquelle on la sub- stitue quelquefois. Il y a une variété de cette espece à odeur de *Pouliot.*

Arvensis. La treizieme croît naturellement dans les terres labourables de plusieurs parties de l'Angleterre; mais on la cul- tive peu dans les jardins: c'est le *Calament* d'eau de boutiques, dont cependant on se sert ra- rement en Médecine. Les tiges de cette plante font velues,

hautes d'environ un pied , garnies de feuilles ovales, & terminées en pointe aiguë : fes fleurs font difposées en fort groffes têtes, & verticillées autour des tiges; elles font petites , de couleur pourpre , & leurs étamines font plus courtes que la corolle : la plante a une odeur forte comme celle du *Pouliot*.

Exigua. La quatorzieme croît dans les lieux aquatiques de plufieurs parties de l'Angleterre ; elle a des tiges foibles , trainantes , d'un pied & demi de longueur , & garnies de feuilles petites , ovales , dentées fur leurs bords , & fupportées par de longs pétioles, fes fleurs croiffent en groffes têtes verticillées autour des tiges ; elles font de couleur pourpre , & leurs étamines font plus longues que la corolle.

Gentilis. La quinzieme fe trouve en abondance fur les bords de la route, entre Bocking & Gosfild en Effex; fes tiges font beaucoup plus petites , & moins longues que celles de la précédente ; fes feuilles font plus courtes , plus rondes, & très-peu dentées fur leurs bords ; mais leurs dentelures font garnies de poils : les têtes verticillées de fes fleurs font plus petites , & la plante entiere a une odeur de *Bafilic*.

Hirfuta. La feizieme croît naturellement dans les foffés & fur les bords des rivieres , dans plufieurs parties de l'Angleterre ; fes tiges font velues , quarrées, & de plus d'un pied de hauteur ; fes feuilles font ovales , fciées , & fort velues :

fes fleurs font difpofées en groffes têtes verticillées vers l'extrémité des tiges ; elles font de couleur pourpre , & leurs étamines font plus longues que la corolle. Cette plante répand une odeur plus agréable que celle de la *Menthe* d'eau commune , & c'eft pour cela qu'on l'appelle *Menthe* douce aquatique , pour la diftinguer. On la trouve dans la plupart des Pharmacopées au nombre des efpeces médicinales ; mais on s'en fert peu à préfent en Médecine.

Verticillata. La dix-feptieme naît fpontanément fur les bords de la riviere Medway, entre Rochefter & Chatham ; elle s'éleve à la hauteur d'environ deux pieds, avec des tiges minces , velues , & garnies de feuilles en forme de lance , terminées en pointe aiguë , & fciées fur leurs bords : fes tiges font garnies de groffes têtes de fleurs verticillées prefque dans toute leur longueur , de forte que chaque tige porte fouvent dix ou douze de ces têtes verticillées. Ces fleurs font de couleur pourpre , & leurs étamines font egales aux corolles. Cette plante répand une odeur aromatique fort agréable.

Culture. Toutes les efpeces de *Menthe* peuvent être aifément multipliées, en divifant les racines au printems, ou par boutures, que l'on peut planter pendant tout l'été dans un fol humide : on les arrofe, fi la faifon eft feche, jufqu'à ce qu'elles aient pouffé des racines ; mais après cela , elles n'exigent plus aucun autre foin que d'être tenues nettes de mauvaifes

MEN

ses herbes : on les plante en planches de quatre pieds environ de largeur, entre lesquelles on laisse un sentier de deux pieds de chaque côté. Ces plantes doivent être placées à quatre ou cinq pieds de distance entr'elles, parce qu'elles s'étendent beaucoup par leurs racines ; c'est aussi pour cette raison qu'on ne doit pas les laisser plus de trois ans sans les transplanter : car si elles restent plus long-tems en place, leurs racines s'entrelacent de manière qu'elles sont bientôt attaquées de pourriture, & se détruisent les unes les autres.

Quelques personnes aiment beaucoup la salade de *Menthe* en hiver & au printems : pour s'en procurer, elles enlevent des racines avant Noël, les couvrent d'un pouce environ de terre fine, les mettent à l'abri avec des nattes ou des vitrages ; & un mois après, la *Menthe* pousse, & est bientôt bonne à manger.

Quand on veut recueillir cette plante pour l'usage de la Médecine, on doit le faire par un tems fort sec, & précisément quand elle est en fleurs : car si on la laissoit plus long-tems, elle ne seroit ni aussi belle ni aussi bonne ; & si on la coupe par un tems humide, elle devient noire, & perd de ses qualités : il faut la suspendre dans un endroit à l'ombre, & la laisser secher, jusqu'à ce qu'on en fasse usage. Quand cette espece se trouve dans un sol fertile, elle donne trois récoltes par année; mais celles que l'on fait après le mois de

Tome V.

Juillet, sont rarement bonnes. Ainsi toutes les branches qui poussent après ce tems, doivent rester en place jusqu'à la Saint-Michel : alors on les coupe ; & après avoir enlevé toutes les mauvaises herbes qui se trouvent sur la planche, on y répand un peu de terre meuble & riche, pour faire pousser vigoureusement les racines au printems suivant.

Comme l'eau distillée de toutes les especes de *Menthe* est regardée comme saine & cordiale, je pense qu'on pourroit la substituer aux mauvaises liqueurs spiritueuses avec lesquelles le menu peuple s'enivre ; car l'eau de la *Menthe à Poivre* est aussi chaude pour l'estomac qu'aucune de ces mauvaises liqueurs dont on fait usage ; & en la mêlant avec d'autres herbes agréables & aromatiques, on pourroit certainement se procurer une liqueur beaucoup plus agréable au goût, & plus saine que celles que l'on vend communément.

MENTHA CATARIA. *Voy.* NEPETA CATARIA. L.

MENTHASTRE ou *Menthe Sauvage. Voyez* MENTHA SYLVESTRIS.

MENTHE AQUATIQUE. *Voy.* MENTHA AQUATICA.

MENTHE-COQ. *Herbe au Coq* ou *Coq des Jardins. Voyez.* TANACETUM BALSAMITA.

MENTHE *des Jardins* ou *Baume. Voy.* MENTHA GENTILIS.

MENTHE *frisée Voyez* MENTHA CRISPA.

MENTHE *à Poivre. Voy.* MENTHA PIPERITA.

MENTZELIA. *Plum. nov.*

F

78 MEN

Gen. Plant. 40. tab. 6. Lin. Gen. Plant. 595.

Cette plante a été ainsi nommée par le P. PLUMIER, qui l'a découverte dans les établissemens françois en Amérique, en l'honneur de MENTZELIUS, Médecin de l'Electeur de Brandebourg, qui a publié un Index des Plantes, en latin, en grec, & en allemand.

Caractères. Le calice de la fleur est étendu, découpé en cinq parties, & placé sur un germe long & cylindrique : la corolle a cinq pétales, étendus & un peu plus longs que le calice ; la fleur a plusieurs étamines érigées, garnies de poils rudes, & terminées par des antheres simples ; du germe long & cylindrique, situé sous la fleur, s'éleve un style à poils rudes de la longueur des pétales, & couronné par un stigmat simple. Ce germe se change dans la suite en une capsule longue & cylindrique, & a une cellule qui renferme plusieurs petites semences.

Ce genre de plantes est rangé dans la premiere section de la treizieme classe de LINNÉÉ, qui comprend celles dont les fleurs ont plusieurs étamines & un style.

Nous n'avons qu'une espece de ce genre, qui est la

Mentzelia aspera. Hort. Cliff. 492. PLUMIER l'appelle Mentzelia foliis & fructibus asperis. Nov. Gen. Plant. 41. ic. 174. f. 1 ; Mentzelia à feuilles & fruits piquans.

Onagra Americana, folio Betonicæ, fructu hispido. Tourn. Inst. 302.

Cette plante croit en abon-

MEN

dance à la Vera-Cruz, d'où ses semences, qui ont été apportées en Angleterre par le Docteur HOUSTOUN, ont réussi dans le Jardin de Botanique de Chelséa.

Elle est annuelle, & s'éleve avec une tige mince, lisse, roide, & un peu ligneuse, à la hauteur de plus de trois pieds, & pousse, de distance en distance, des branches torses, qui coulent l'une dans l'autre, & sont garnies de feuilles en forme de pointes de hallebarde, alternes sur les branches, supportées par de courts pétioles, & couvertes de piquans courts, qui s'attachent aux habits de ceux qui s'y frottent. Ces branches se séparent aisément des plantes, & s'attachent aux habits, comme les semences du Metilot : ses fleurs naissent simples aux nœuds de la tige ; elles sont placées sur un germe cylindrique d'un pouce de longueur, étroit à sa base, & plus large vers le haut : au sommet, sort un calice qui s'étend comme celui de l'Onagra, & les pétales de la fleur s'ouvrent & s'épanouissent sur le calice ; ils sont d'un jaune pâle, & plus long que le calice. Dans le milieu s'éleve un grand nombre d'étamines, & sur le germe est un style simple, aussi long que la corolle, & couronné par un stigmat simple. Ce germe se change ensuite en une capsule longue, cylindrique, & armée de piquans, ainsi que les feuilles, qui s'attachent aussi aux habits de ceux qui en approchent. Cette capsule n'a qu'une cel-

MEN

lule remplie de petites femences. Comme cette plante eft annuelle, & qu'elle périt auffitôt que fes femences font mûres, on doit les répandre fur une couche chaude dans le commencement du printems, afin que les plantes puiffent faire des progrès rapides, fans quoi elles ne produiroient point de femences mûres dans ce pays ; quand elles font parvenues à un pouce de hauteur, on les met chaçune féparément dans un pot d'un fou, rempli d'une terre riche & légere : on les plonge dans une couche chaude de tan, & on les tient à l'ombre, jufqu'à ce qu'elles aient formé de nouvelles racines ; après quoi on les arrofe fouvent, & on leur donne de l'air chaque jour, à proportion de la chaleur de la faifon, & de la couche où elles font. Six femaines après qu'elles ont été tranfplantées, fi elles ont fait de bons progrès, leurs racines auront rempli les pots, & il fera néceffaire de leur en donner de plus grands, que l'on remplira de terre riche & légere, & que l'on replongera dans la couche de tan de la ferre, où on leur donnera de l'air, & on les arrofera dans tous les jours chauds. Au moyen de ce traitement, ces plantes s'éleveront à la hauteur de trois pieds, & produiront des femences mûres à la fin d'Août ou au commencement de Septembre.

MENYANTHES. C'eft le *Paluftre trifolium* ou *l'herbe des Marais*. [*Bog-Bean*.] Treffle d'eau ou le Menianthe.

MER 79

Cette plante eft commune dans les lieux marécageux de différentes parties de l'Angleterre ; mais comme on ne la cultive jamais dans les jardins, je n'en parlerai pas d'avantage. J'obferverai feulement que l'on fait aujourd'hui un grand cas de cette plante, parce qu'on la regarde comme un remede excellent pour guérir les rhumatifmes, prévenir la goutte & plufieurs autres défordres. On lui donne fouvent fur les marchés le nom d'*Herbe* ou *Treffle de Marais* ; elle croît en abondance dans des lieux marécageux de plufieurs parties de l'Angleterre, où ceux qui en fourniffent les marchés vont la cueillir (1).

MERCURIALE. *Voy.* MERCURIALIS. I..

MERCURIALE *à trois femences. Voyez* ACALYPHA.

MERCURIALIS. *Tourn. Inft. R. H.* 534. *tab.* 308. *Lin. Gen. Plant.* 998.

Cette plante prend fon nom de MERCURE, parce que les Anciens avoient imaginé que ce Dieu avoit mis cette plante en ufage. [*Mercury.*] Mercuriale.

Caraêteres. Cette plante a des fleurs mâles & des fleurs femelles fur différens pieds : les fleurs mâles ont des calices

(1) La racine de Treffle d'eau n'a aucune odeur, mais elle eft d'une médiocre amertume ; fes propriétés médicinales ne différent point de celles de la Fraxinelle & de la Gentiane, quoiqu'on la regarde, dans certains pays, comme un fpécifique contre le fcorbut & les fievres intermittentes.

F 2

étendus & découpés en trois segmens concaves ; elles n'ont point de corolle, mais seulement neuf ou douze étamines érigées, velues & couronnées par des antheres globulaires & naines : les fleurs femelles qui n'ont point non plus de corolle, sont pourvues de deux nectaires en forme d'aléne & à pointe aiguë, & d'un germe large, simple, & séparé de ses voisins par un sillon. Ces germes sont ronds, comprimés, & ont un sillon épineux à chaque côté, ils soutiennent deux styles réfléchis, épineux, & couronnés par des stigmats aigus & réfléchis. Ces germes se changent dans la suite en une capsule presque ronde, de la forme d'un scrotum, & à deux cellules, qui renferment chacune une semence arrondie.

Ce genre de plantes est rangé dans la huitieme section de la vingt-deuxieme classe de LINNÉE, qui comprend celles dont les fleurs mâles & les fleurs femelles croissent sur différentes plantes, & dont les fleurs mâles ont chacune neuf étamines.

Les especes sont :

1°. *Mercurialis annua, caule brachiato, foliis glabris. Hort. Cliff.* 461. *Hort. Ups. 298. Mat. Med. 216. Roy. Lugd.-B. 263. Dalib. Paris. 302. Scop. carn. ed. 2. n. 1226. Neck. Gallob. p. 410*; Mercuriale à tige branchue & à feuilles unies.

Mercurialis spicata & testiculata mas & femina. C. B. P. 121; Mercuriale avec des fleurs en épis & testiculées, dont les fleurs mâles & les femelles naissent sur des plantes différentes,

& à laquelle on donne le nom de Mercuriale de France.

Mercurialis mas. Dod. Pempt. 658. *Mercurialis spicata, sivé femina. Dod. Pempt. 658.*

2°. *Mercurialis perennis, caule simplicissimo, foliis scabris. Hort. Cliff.* 461, *Fl. Suec. 823; 913. Roy. Lugd.-B. 203. Dalib. Paris. 302. Neck. Gallob. p. 510. Pollich. pal. n. 93*; Mercuriale avec une tige simple & des feuilles rudes.

Mercurialis montana spicata & testiculata. C. B. P. 122; Mercuriale de montagne ou Mercuriale de marais, avec des fleurs en épis & en forme de testicule, mâle & femelle sur différentes plantes, ou la Mercuriale vivace.

Cynocrambe mas. Cam. Epit. 999. *Cynocrambe femina. Cam. Epit. 999.*

3°. *Mercurialis tomentosa, caule sub-fruticoso, foliis tomentosis. Hort. Cliff.* 461. *Roy. Lugd.-B. 203. Sauv. Monsp. 128. Gouan. Monsp.* 507; Mercuriale à tige de sous-arbrisseau, avec des feuilles cotonneuses.

Mercurialis fruticosa incana, spicata & testiculata. Tourn. Inst. R. H. 534; Mercuriale blanche en arbrisseau, avec des fleurs en épis, & testiculées.

Phyllon testiculatum. Bauh. Pin. 122. *Phyllon spicatum. Bauh. Pin. 122.*

Annua. La premiere espece, à laquelle on donne communément le nom de *Mercuriale de France*, d'où elle a peut-être été portée en Angleterre ; car quoiqu'elle soit à présent devenue une herbe commune & sauvage dans les jardins & sur les tas de fumiers, cependant

MER

on la trouve rarement à quelque diftance des habitations. Cette plante eft annuelle, & fa tige, qui eft branchue & haute d'environ un pied, eft garnie de feuilles en forme de lance d'environ un pouce & demi de long, dentées fur leurs bords, de couleur pâle ou d'un vert jaunâtre. Les plantes mâles ont des épis de fleurs herbacées, qui croiffent aux extrémités des tiges, & tombent bientôt après; mais les plantes femelles ont des fleurs tefticulées, qui naiffent fur les côtés des tiges, & font remplacées par des femences qui produifent une grande quantité de plantes des deux fexes. Les feuilles & les tiges de cette efpece font d'ufage en Médecine : on les regarde comme apéritives & émollientes (1),

(1) Cette plante eft très-laxative & émolliente, & d'un ufage affez fréquent en Médecine : on s'en fert peu intérieurement, quoiqu'elle foit regardée comme purgative ; mais on la fait entrer dans prefque toutes les décoctions émollientes & les lavemens laxatifs : on en prépare un miel qui fert aux mêmes ufages, à la dofe de deux ou trois onces. Quelques perfonnes font néanmoins cuire une poignée de feuilles de cette plante dans un bouillon de veau, qu'ils prennent pour fe lâcher le ventre. On en prépare un fyrop fimple & un fyrop compofé, qui porte le nom de Syrop de Longue-Vie : on prefcrit le premier à la dofe d'une ou deux onces, pour lâcher le ventre : on attribue au fecond la propriété de purifier le fang & de fortifier les digeftions. La Mercuriale entre encore dans la compofition du Lénitif, du Catholicon, &c.

MER   81

Perennis. La feconde efpece croît fous les haies & dans les bois de plufieurs parties de l'Angleterre; elle a une racine vivace qui rampe dans la terre ; fes tiges font fimples, fans branches, de dix à douze pouces de hauteur, & garnies de feuilles rudes qui naiffent par paires fur chaque nœud, d'un vert foncé, & dentées fur leurs bords : les fleurs mâles font difpofées en épis fur des plantes différentes de celles qui produifent les femences.

Cette efpece eft vénéneufe, & l'on en a eu depuis peu plufieurs preuves. Des perfonnes du peuple en ayant mangé les feuilles dans un tems fec où les légumes étoient très-rares, s'en trouverent très-incommodées.

Tomentofa. La troifieme croît naturellement dans la France méridionnale, en Efpagne & en Italie ; elle s'éleve à la hauteur d'un pied & demi, avec une tige branchue d'arbriffeau, garnie de feuilles ovales, placées par paires, & couvertes d'un duvet blanc fur les deux furfaces : les fleurs mâles croiffent en épis courts aux côtés des tiges, fur des plantes differentes de celles qui produifent les fruits ; elles font blanches & tefticulées. Si l'on donne à cette efpece le tems de répandre fes graines, les plantes poufferont au printems fuivant; mais fi on ne les met en terre que dans cette faifon, elles paroîtront rarement dans la même année. Cette plante exige une fituation chaude, & un fol fec & rempli de décombres, dans lequel elle durera trois ou qua-

F 3

MES

tre années ; mais les fortes gelées la détruisent souvent.

MERISIER. CERISIER. *Voyz*
Cerasus vulgaris.

MERVEILLE DU PÉROU.
Voyez Mirabilis.

MESEMBRYANTHEMUM.
Dill. Gen. 9. *Hort. Elth.* 179.
Ficoïdes. Tourn. Act. R. Par.
1705. [*Fig Marygold.*] Figue
d'Inde. Ficoïde.

Caracteres. Le calice de la
fleur est persistant, étendu, &
formé par une feuille découpée
au sommet en cinq parties aiguës ; la corolle a un pétale
divisé presque jusqu'au fond en
plusieurs segmens linéaires, dif-
posés en plusieurs rangs, mais
joints ensemble à leur bâse ;
dans le milieu sont arrangées
un grand nombre d'étamines
velues, & terminées par des
antheres inclinées : sous la
fleur est placé un germe obtús
& à cinq angles, qui soutient
quelquefois cinq styles, & souvent dix ou plus, qui sont réfléchis & couronnés par des
stigmats simples. Ce germe se
change dans la suite en un fruit
rond, charnu, avec autant de
cellules qu'il y a de styles, lesquelles sont remplies de petites semences.

Ce genre de plantes est rangé
dans la quatrieme section de la
douzieme classe de Linnée, qui
comprend celles dont les fleurs
ont pour le moins vingt étamines insérées dans le calice, &
cinq styles.

Les especes sont :

1°. *Mesembryanthemum nodiflorum*, *foliis alternis*, *teretiusculis*, *obtusis*, *basi ciliatis. Hort.
Upsal.* 129. Mesembryanthe-

mum à feuilles cylindriques,
obtuses, velues & alternes.

Ficoïdes Neapolitana, *flore candido. H. L.* ; Figue de Naples
à fleurs blanches ou Kali d'E-
gypte.

*Kali Crassulæ minoris foliis.
Bauh. Pin.* 289. *Moris. Hist.* 2.
p. 610. *S.* 5. *t.* 33. *f.* 7.

2° *Mesembryanthemum crystallinum*, *foliis alternis*, *ovatis*, *papillosis*, *undulatis. Hort. Cliff.*
216. *n.* 1. *Hort. Upf.* 127. *Roy.
Lugd.-B.* 281. *Kniph. cent.* 4.
n. 50 ; Mesembryanthemum à
feuilles ovales, obtuses, ondées
& alternes.

*Mesembryanthemum crystallinum,
Plantaginis folio undulato. Dill.
Elth.* 231. *t.* 180. *f.* 221.

Ficoïdes Africana, *folio Plantaginis undulato*, *micis argenteis
asperso. Tourn. Act. R. Par.* 1705 ;
Ficoïde avec des feuilles ondées de *Plantain*, marquées de
taches argentées, commun-é-
ment appelée Ficoïde à diamant, ou Plante de diamant,
ou Glaciale.

3°. *Mesembryanthemum geniculi-
florum*, *foliis semi-teretibus*, *papillosis*, *distinctis*, *floribus sessilibus*, *axillaribus. Lin. Gen. Plant.*
481 ; Mesembryanthemum avec
des feuilles à moitié cylindriques, couvertes de boutons
distincts, & des fleurs sessiles
aux aisselles des tiges.

*Mesembryanthemum ramis un-
dique papillosis*, *folio crassioribus.
Hort. Cliff.* 218. *n.* 22. *Roy.
Lugd.-B.* 285.

Ficoïdes Capensis, *folio tereti*,
flore albido. Pet. Gaz. 78. *fol.* 3 ;
Ficoïde du Cap à feuilles cylindriques, & à fleurs blanchâtres.

MES

4°. *Mesembryanthemum nocti-florum , foliis semi - cylindricis , impunctatis , distinctis , floribus pedunculatis , calycibus quadrifidis.* Lin. Sp. Plant. 481 ; Mesembrianthemum avec des feuilles presque cylindriques , des fleurs sur des pédoncules , & des calices divisés en quatre parties.

Ficoïdes Africana , erecta , arborescens , lignosa , flore radiato , primò purpureo , dein argenteo , interdiù clauso , noctu aperto. Boerh. Ind. Alt. 1. 290; Ficoïde droite , ligneuse , & en arbre , avec une fleur radiée , d'abord pourpre, ensuite argentée , fermée pendant le jour , & ouverte la nuit.

5°. *Mesembryanthemum splendens , foliis semi - teretibus , impunctatis , recurvis , distinctis , congestis , calycibus terminalibus , digiti - formibus.* Lin. Sp. 689 ; Mesembryanthemum avec des feuilles à demi—cylindriques , sans taches, récourbées , distinctes , & en grappe , dont le calice , qui termine , est en forme de doigt.

Ficoïdes Capensis frutescens , foliis teretibus , confertis , glaucis flore albo. Bradl. Suec. 1. p. 7. f. 6.

6°. *Mesembryanthemum umbellatum , foliis subulatis , scabrido punctatis , connatis , apice patulo , caule erecto , corymbo trichotomo.* Lin. Sp. Plant. 481. Kniph. cent. 3. n. 63 ; Mesembryanthemum à feuilles en forme d'alêne réunies , & à taches rudes , avec une tête érigée & un corymbe de fleurs aux triples divisions de la tige.

Mesembryanthemum frutescens , floribus albis , umbellatis. Dill. Elth. 277. t. 208. f. 266.

MES 83

Ficoïdes Africana erecta , tereti-folia , floribus albis , umbellatis. Par. Bat. 166. Bradl. Suec. 4. p. 12 ; Ficoïde érigée avec une feuille cylindrique , & des fleurs blanches disposées en ombelles.

7°. *Mesembryanthemum calami-forme , acaule , foliis subteretibus , adscendentibus , impunctatis , connatis , floribus octogynis.* Lin. Sp. Plant. 481; Mesembryanthemum sans tige , avec des feuilles presque cylindriques réunies à leur bâse , & des fleurs garnies de huit styles.

Mesembryanthemum acaule , calami-forme. Dill. Elth. 229. t. 186. f. 228.

Ficoïdes Capensis humilis , Cepeæ folio , flore stamineo. Bradl. Suec. p. 10. fol. 19 ; Ficoïde du Cap , avec une feuille d'Oignon , & des fleurs avec des étamines.

8°. *Mesembryanthemum Tripolium , foliis alternis , lanceolatis , planis , impunctatis ; caulibus laxis simplicibus ; calycibus pentagonis.* Hort. Cliff. 217. n. 11. Hort. Ups. 128. Roy. Lugd.-B. 283. Kniph. cent 8. n. 70 ; Mesembryanthemum avec des feuilles en forme de lance, sans marque de points, une tige simple & foible , & un calice à cinq angles.

Ficoïdes Africana procumbens , Tripolii folio , flore argenteo. Hort. Chels. ; Ficoïde d'Afrique tombante , avec une feuille de Tripolium & une fleur argentée.

9°. *Mesembryanthemum Belli-di—florum , acaule , foliis trique-tris , linearibus , impunctatis , apice trifariàm dentatis.* Hort. Cliff. 218. n. 13. Roy. Lugd.-B. 283. Knorr. Del. 1. tab. G. 5. a. n. 1. ; Me-

F 4

sembryanthemum sans tige, à feuilles étroites, triangulaires, unies & dentées à leurs pointes.

Ficoïdes Capensis humilis, folio triangulari, in summitatem dentato, flore minore purpurascente. Bradl. Suec. p. 9. tab. 18. Ficoïde du Cap, avec une feuille triangulaire, découpée à son extrémité, & une plus petite fleur de couleur pourpre.

Mesembryanthemum Bellidiflorum. Dill. Elth. 244. t. 189. f. 233.

10°. *Mesembryanthemum subulatum, acaule, foliis subulatis triquetris, dorso superne serratis* ; Mesembryanthemum sans tige, à feuilles triangulaires & en forme d'alène, dont la partie du dos est sciée vers le sommet.

11°. *Mesembryanthemum deltoïdes, foliis triquetris, dentatis, impunctatis.* Hort. Cliff. 218. Roy. Lugd.-B. 284; Mesembryanthemum avec des feuilles triangulaires, dentées, sans marques de points, & en forme de delta.

Ficoïdes Africana, folio triangulari, crasso, brevi, glauco, ad tres margines aculeato. Boerh. Ind. Alt. 1. 290 ; Ficoïde d'Afrique, avec une feuille épaisse, grise, triangulaire, & garnie de poils rudes sur les trois bords.

Ficus Aizoïdes Africana, erecta, folio triangulari, breviusculo, fimbriato, floribus roseis odoratis. Volk. Hesp. 223. f. 224.

12°. *Mesembryanthemum caulescens foliis deltoidibus, lateralibus, minime dentatis* ; Mesembryanthemum garnie de tiges avec des feuilles en forme de delta, dont les côtés sont un peu dentés.

Ficoïdes Africana, folio triangulari, glauco, brevissimo, crassissimo, margine non spinoso. Boerh. Ind. Alt. 1. 290 ; Ficoïde d'Afrique, avec des feuilles fort épaisses, courtes, grises, triangulaires, & sans épines sur leurs bords.

13°. *Mesembryanthemum barbatum, foliis sub-ovatis papulosis distinctis, apice barbatis.* Hort. Cliff. 216. Hort. Ups. 127. Roy. Lugd.-B. 282. Kniph. cent. 10. n. 60 ; Mesembryanthemum avec des feuilles presqu'ovales, ayant des vessies barbues & distinctes à leur pointe.

Ficoïdes, seu Ficus. Aizoïdes Africana, folio variegato aspero, ad apicem stella spinosa armato. Boerh. Ind. Alt. 1. p. 291 ; Ficoïde d'Afrique avec une feuille rude & panachée, dont la pointe est armée d'épines en forme d'étoile.

14°. *Mesembryanthemum stellatum, caulibus decumbentibus, foliis teretibus, papillosis, apice herbatis* ; Mesembryanthemum avec des tiges inclinées, des feuilles cylindriques, couvertes de vessies, dont les pointes sont barbues, & en forme d'étoile.

Ficoïdes Capensis frutescens, folio tumido, extremitate stellata; flore purpureo. Bradl. Suec. Dec. 1. tab. 6 ; Ficoïde en arbrisseau du Cap, à feuilles à pointes étoilées & gonflées, & à fleurs pourpre.

Ficus Aizoïdes, folio tereti, in villos radiatos abeunte, flore rubro. Volk. Hesp. 222. t. 124. f. 6.

15°. *Mesembryanthemum hispidum, foliis cylindricis, papulosis, distinctis, caule hispido.*

MES

Lin. Sp. Plant. 482 ; Mesembryanthemum avec une tige épineuse , des feuilles réfléchies & cylindriques, & des vessies charnues.

Ficus Aizoïdes tereti-folia, foliis cryfallino rore eleganter confperfis ; floribus diluè rofeis. Volk. Hefp. 221.

Ficoïdes Afra , fruticofa , caule lanugine argenteá ornato , folio tereti , parvo , longo, guttulis argenteis quasi fcabro , flore violaceo. Boerrh. Ind. Alt. 1. 291 ; Ficoïde d'Afrique , en arbrisseau, ayant des tiges ornées d'un duvet argenté , avec des feuilles longues , petites , cylindriques , & marquées de taches semblables à des gouttes argentées, & une fleur violette.

16°. *Mefembryanthemum villofum , caule foliifque pubefcentibus. Hort. Cliff.* 217 ; Mesembryanthemum , dont les tiges & les feuilles font couvertes d'un duvet doux.

17°. *Mefembryanthemum fcabrum , foliis fubulatis , diftinctis , fubtùs undiquè punctato-muricatis , calycibus muticis. Hort. Cliff.* 219. *n.* 20. *Hort. Ups.* 120 ; Mesembryanthemum avec des feuilles en forme d'alène , diftinctes & rudes en-dessous , & des calices garnis de paille.

Ficoïdes Afra , folio triangulari, viridi , longo , afpero , flore violaceo. Boerrh. Ind. Alt. 290 ; Ficoïde avec une feuille longue , verte , rude , & triangulaire , & une fleur violette.

Mefembryanthemum purpureum fcabrum , ftaminibus collectis. Dill. Elth. 260. *t.* 197. *f.* 251.

18°. *Mefembrianthemum uncinatum , articulis caulinis terminatis*

MES   85

in folia conata , acuminata , fubtiùs dentata. Hort. Cliff. 218. *n.* 16. *Roy. Lugd.-B.* 284 ; Mesembryanthemum dont les nœuds des tiges font terminés par des feuilles en pointe aiguë , jointes à leurs bàfes , & dentées en-dessous.

Ficoïdes Afra , folio triangulari , glauco, perfoliato , breviffimo , apice fpinofo. Boerrh. Ind. Alt. 290 ; Ficoïde d'Afrique , avec une feuille courte , triangulaire , couleur de vert de mer , & trouée , dont les fommets font épineux, communément appelée Ficoïde de Chiendent.

19°. *Mefembryanthemum perfoliatum , foliis majoribus , apicibus tri-acanthis. Hort. Elth.* 252 ; Mesembryanthemum avec de larges feuilles trouées , dont les fommets font garnis de trois épines.

Ficoïdes Africana frutefcens , perfoliata , folio triangulari , glauco, punctato, cortice lignofo , candido , tenui. Tourn. Act. Par. 1705 ; Ficoïde d'Afrique en arbrisseau , avec une feuille triangulaire , grife , marquée de points , enfilée dans le difque , ayant une écorce ligneufe , mince & blanche , communément appelée Ficoïde à corne de cerf.

20°. *Mefembryanthemum fpinofum , foliis tereti-triquetris , punctatis , diftinctis , fpinis ramofis. Hort. Cliff.* 216. *n.* 3. *Roy. Lugd.-B.* 281 ; Mefembryanthemum avec des feuilles cylindriques , triangulaires , marquées de points , diftinctes , & armées d'épines branchues.

Ficoïdes Africana , aculeis longiffimis & foliolis nafcentibus ex foliorum alis. Tourn. Act. R. Par.

86 MES

1705 ; Ficoïde d'Afrique avec de longues épines, & de plus petites feuilles, qui s'élevent des aîles des grandes.

21°. *Mesembryanthemum tuberosum, foliis subulatis, papillosis, distinctis, apice patulis, radice capitatá. Hort. Cliff.* 216. *n.* 4. *Roy. Lugd.-B.* 282 ; Mesembryanthemum avec des feuilles en forme d'alène, couvertes de tubercules, ayant une racine à tête.

Mesembryanthemum frutescens, radice ingenti tuberosá. Dill. Elth. 275. *t.* 207. *f.* 264.

Ficoïdes Africana, folio triangulari recurvo, floribus umbellatis obsoleti coloris, externè purpureis. Tourn. Act. Par. 1705 ; Ficoïde d'Afrique avec une feuille triangulaire & recourbée, & des fleurs en ombelle d'une couleur usée, & pourpre au-dehors.

22. *Mesembryanthemum tenuifolium, foliis subulatis, semi-teretibus, glabris, distinctis, internodis longioribus. Hort. Cliff.* 220. *n.* 26. *Hort. Upf.* 128. *Roy. Lugd.-B.* 286 ; Mesembryanthemum avec des feuilles en forme d'alène, à demi cylindriques, unies & distinctes, dont les nœuds sont à une distance plus grande.

Ficoides Capensis humilis, teretifolia, flore coccineo. Bradl. Suec. p. 13. *t.* 9 ; petite Ficoïde du Cap, avec une feuille cylindrique & une fleur écarlate.

23°. *Mesembryanthemum stipulaceum, foliis subtriquetris, compressis, incurvis, punctatis, distinctis, congestis, basi marginatis. Lin. Sp.* 693. *Hort. Cliff.* 220. *n.* 29. *Roy. Lug.-B.* 287 ; Mesembryanthemum avec des feuilles triangulaires, recourbées, comprimées, & marquées de points, distinc-

tes, dont les bâses sont bordées, & en grappes ou rapprochées.

Mesembryanthemum, frutescens, flore purpureo rariore. Hort. Elth. tab. 209.

24°. *Mesembryanthemum crassifolium, foliis semi-cylindricis, impunctatis, connatis, apice triquetris, caule repente semi-cylindrico. Hort. Cliff.* 217. *n.* 9. *Roy. Lugd.-B.* 283. *Knorr. Dell.* 2. *f. M.* 4 ; Mesembryanthemum avec une tige cylindrique & rampante, & des feuilles à moitié cylindriques, unies, réunies à leur bâse, & dont les extrémités sont triangulaires.

Ficoïdes Africana reptans, folio triangulari, viridi, flore saturatè purpureo. Bradl. Suec. p. 16. *tab.* 38 ; Ficoïde d'Afrique rampante, avec une feuille verte & triangulaire, & une fleur d'un pourpre foncé.

25°. *Mesembryanthemum falcatum foliis sub-acinaci-formibus, incurvis, punctatis, distinctis, ramis teretibus. Hort. Cliff.* 219. *n.* 19. *Roy. Lugd.-B.* 285. *Knorr. Dell.* 2. *f. M.* 4 ; Mesembryanthemum avec des feuilles en forme de coutelas recourbé, marquées de points, distinctes, & des branches cylindriques.

Ficoïdes Afra, folio triangulari, ensi-formi, brevissimo, flore dilutè purpurascente, filamentoso. Bradl. Suec. Dec. 5. *tab.* 42 ; Ficoïde d'Afrique avec une feuille triangulaire, courte, & en forme de cimeterre, ayant une fleur d'un pourpre pâle.

26°. *Mesembryanthemum glomeratum, foliis teretiusculis, compressis, punctatis, distinctis, caule paniculato, multifloro. Lin. Sp.* 694;

MES

Mesembryanthemum avec des feuilles cylindriques, comprimées, tachetées de points, distinctes, & une tige en panicule, qui produit plusieurs fleurs.

Mesembryanthemum falcatum minus, flore carneo minore. Hort. Elth. tab. 213. *f.* 274.

27°. *Mesembryanthemum edule, foliis æquilateri-triquetris, acutis, strictis, impunctatis, connatis, carinâ sub-serratis, caule ancipiti. Lin. Sp.* 695; Mesembryanthemum avec des feuilles équilatérales, aigues, & sans points, jointes à leurs bâses, & dont la carêne est sciée, avec une tige en forme de coutelas.

Mesembryanthemum falcatum majus, flore amplo luteo. Dill. Elth. 283. *t.* 212. *f.* 272.

Ficoïdes, seu Ficus Aizoïdes Africana major, procumbens, triangulari folio, fructu maximo eduli. H. L. 244; la plus grande Ficoïde d'Afrique rampante, à feuilles triangulaires, qui produit un gros fruit bon à manger, ou Figuier des Hottentots.

28°. *Mesembryanthemum bicolorum, foliis subulatis, levibus, punctatis, distinctis, caule frutescente, corollis bi-coloribus. Lin. Sp. Plant.* 695; Mesembryanthemum à feuilles en forme d'alêne, marquées de points, distinctes & lisses, ayant une tige d'arbrisseau, & des corolles de deux couleurs.

Ficoïdes Capensis frutescens, folio tereti, punctato, petalis luteis. Bradl. Suec. 1. *p.* 8. *tab.* 7; Ficoïde du Cap, en arbrisseau, avec une feuille cylindrique marquée de points, & de pétales jaunes.

29°. *Mesembryanthemum acina-*

MES 87

ci-forme, foliis acinaci-formibus, impunctatis, connatis, angulo carinali scabris, petalis lanceolatis. Lin. Sp. 695. *Hort. Cliff.* 219 *n.* 18. *Roy. Lugd.-B.* 284. *Kniph. cent.* 10. *n.* 59; Mesembryanthemum avec des feuilles rudes, triangulaires, & sans points, jointes à leurs bâses, & dont la carêne est rude, avec des pétales en forme de lance.

Ficoïdes Africana, folio longo, triangulari, incurvo, caule purpureo. Tourn. Act. Par. 1705; Ficoïde d'Afrique à feuilles longues, triangulaires, & recourbées, & à tige pourpre.

30° *Mesembryanthemum loreum, foliis semi-cylindricis, recurvis, congestis, basi interiore gibbis, connatis, caule pendulo. Lin. Sp.* 694; Mesembryanthemum avec des feuilles à moitié cylindriques, recourbées & rapprochées à leurs bâses avec une tige pendante.

Mesembryanthemum loreum. Hort. Elth. tab. 200. *f.* 255.

31°. *Mesembryanthemum serratum, foliis subulatis, triquetris, punctatis, distinctis, angulo carinali retrorsûm serratis. Lin. Sp.* 696. *Hort. Cliff.* 218. *n.* 15. *Roy. Lugd.-B.* 284; Mesembryanthemum avec des feuilles en forme d'alêne, & triangulaires, ayant des points, distinctes, & l'angle de la carêne sciée.

Mesembryanthemum serratum, flore acetabuli-formi, luteo. Hort. Elth. tab. 192. *f.* 238.

32°. *Mesembryanthemum tuberculatum, acaule, foliis semi-cylindricis, connatis, externè tuberculatis. Hort. Cliff.* 219; Mesembryanthemum sans tige, avec

MES

des feuilles à moitié cylindriques, chargées de tubercules au-dehors, & jointes à leur bâse.

Ficoïdes Afra, folio triangulari, longo, fucculento, caulibus rubris. Boerrh. Ind. Alt. 290; Ficoïde d'Afrique à feuilles longues, triangulaires, fucculentes, & à tige rouge.

33°. *Mefembryanthemum veruculatum, foliis triquetro-cylindricis, acutis, connatis, arcuatis, impunctatis, diftinctis. Hort. Cliff.* 220. *n.* 24. *Hort. Upf.* 128. *n.* *Roy. Lugd.-B.* 285. *Kniph. cent.* 10. *n.* 62; Mefembryanthemum avec des feuilles triangulaires & cylindriques, jointes à leurs bâfes, courbées, & non marquées de points, diftinctes.

Mefembryanthemum foliis veruculi-formibus, floribus mellinis, umbellatis. Dill. Elth. 268. *t.* 203. *f.* 259.

Ficoïdes Afra arborefcens, folio tereti, glauco, apice purpureo, craffo. Boerrh. Ind. Alt. 291; Ficoïde d'Afrique en arbre, avec une feuille grife, cylindrique, & dont le fommet eft épais, & de couleur pourpre.

34°. *Mefembryanthemum glaucum, foliis triquetris, acutis, punctatis, diftinctis, calycinis foliolis ovato-cordatis. Lin. Sp.* 696. *Hort. Cliff.* 220. *n.* 27. *Roy. Lugd.-B.* 283; Mefembryanthemum avec des feuilles aiguës, triangulaires, marquées de points, diftinctes, ayant les folioles des calices ovales & en forme de cœur.

Ficoïdes Afra, caule lignofo, erecta, folio triangulari, enfi-formi, fcabro, flore luteo magno. Boerrh. Ind. Alt. 289. *Bradl. Suec.* 4. *p.*

15. *f.* 37; Ficoïde d'Afrique avec une tige ligneufe & droite, une feuille triangulaire, & en forme de coutelas & rude, & une groffe fleur jaune.

35°. *Mefembryanthemum corniculatum, foliis triquetro-femi-cylindricis, fcabrido-punctatis, fupra bafim lineâ elevatâ connatis. Lin. Sp.* 697; Mefembryanthemum qui produit beaucoup de branches garnies de feuilles triangulaires, à moitié cylindriques, rudes, ponctuées, & jointes à leurs bâfes.

Ficoïdes Afra, folio triangulari, longiffimo, marginibus obtufioribus, flore amplo, intùs pallidè luteo, extùs lineâ rubrâ longâ picto. Boerrh. Ind. Alt. 289; Ficoïde d'Afrique, avec une feuille longue & triangulaire, dont les bords font plus obtus, & une groffe fleur d'un jaune pâle en-dedans, & marquée d'une longue raie rouge au-dehors.

36°. *Mefembryanthemum expanfum, foliis planiufculis, lanceolatis, impunctatis, patentibus, diftinctis, oppofitis, alternatifque, remotis. Lin. Sp.* 697. *Knorr. Dell.* 2. *t.* M. 3; Mefembryanthemum avec des feuilles unies, en forme de lance, fans marque de points, diftinctes, oppofées, alternes, & placées à une certaine diftance les unes des autres.

Mefembryanthemum tortuofum, foliis Sempervivi expanfis. Dill. Elth. 234. *t.* 182. *f.* 223.

Ficoïdes Capenfis, folio lato acuto, flore albo, intùs luteo. Pet. Gaz. t. 78. *f.* 10.

Ficoïdes Africana procumbens, foliis planis, conjugatis, lucidis. Bradl. Suec. 3. *p.* 7. *f.* 16.

MES

Ficoïdes Africana humi fufa,
folio triangulari, longiori, glauco,
flore flavefcente. Tourn. Acad.
R. Par. 1705 ; Ficoïde d'Afrique, rampante, avec une plus
longue feuille grife, triangulaire, & une fleur jaunâtre.

37°. *Mefembryanthemum mi-*
cans, foliis fubulatis, triquetris,
pánctatis, diftinctis, caule fcabro.
Lin. Sp. 696 ; Mefembryanthemum avec des feuilles triangulaires en forme d'alène, dif-
tinctement tachetées, ayant
une tige rude.

Mefembryanthemum micans,
flore phæniceo, filamentis atris.
Hort. Elth. tab. 215. *f.* 282.

38°. *Mefembryanthemum tor-*
tuofum, foliis planiufculis,
oblongo-ovatis, fubpapillofis, con-
fertis, connatis, calycibus triphyl-
lis, bicornibus. Lin. Sp. 697.
Kniph. cent. 8. n. 69 ; Mefembrianthemum avec des feuilles
unies, oblongues, ovales, &
réunies à leur bâfe, ayant un
calice à trois feuilles avec deux
cornes.

Ficoïdes Capenfis, procumbens,
Oleæ folio, flore albo, medio
croceo. Bradl. Suec. Dec. 2. p. 7.
tab. 16 ; Ficoïde du Cap, rampante, à feuilles d'Olivier,
avec une fleur blanche, &
de couleur de fafran dans le
milieu.

39°. *Mefembryanthemum ringens,*
fub acaule, foliis ciliato-dentatis,
punctatis. Lin. Hort. Cliff. 218 ;
Mefembryanthemum avec une
tige courte, & des feuilles
dentées, velues & ponctuées.

Ficoïdes Capenfis humilis, folio
triangulari, prope fummitatem den-
tato, flore luteo. Bradl. Suec.
Dec. 2. p. 8. tab. 17 ; petite

MES

Ficoïde du Cap, avec une
feuille triangulaire & dentée
vers l'extrêmité, & à une
fleur jaune, communément appelée Ficoïde à gueule de
chien.

Ringens caninum. Linn. Syft.
Pl. Sp. 40.

40°. *Mefembryanthemum roftra-*
tum, acaule, foliis femi-cylin-
dricis, connatis, externè tubercu-
latis. Lin. Sp. 696. *Hort.*
Cliff. 219. *n.* 23. *Roy. Lugd.-B.*
285. *Kniph. cent.* 10. *n.* 61 ;
Mefembrianthemum fans tige,
à feuilles à moitié cylindriques, jointes à leurs bâfes,
& tuberculées fur le dehors.

Mefembryanthemum roftrum ar-
deæ referens. Dill. Elth. 240.
t. 186. *f.* 229.

Ficoïdes Afra, folio triangu-
lari, enfi-formi, craffo, brevi, ad
margines laterales multis majori-
bus fpinis aculeato. Martyn.
cent. 30. *tab.* 30 ; Ficoïde d'Afrique, avec une feuille triangulaire, courte, épaiffe, & en
forme de cimeterre, dont les
bords font garnis de plufieurs
groffes épines, communément
appelée Ficoïde à gueule de
chat.

41°. *Mefembryanthemum dola-*
bri-forme, acaule, foliis dolabri-
formibus, punctatis. Hort. Cliff.
219. *n.* 17. *Roy. Lugd.-B.* 284.
Kniph. cent. 1. *n.* 51 ; Mefembryanthemum avec des feuilles
en forme de hache, & ta-
chées.

Ficoïdes Capenfis humilis,
foliis cornua cervi referentibus,
petalis luteis, nocti-flora. Bradl.
Suec. 1 *p.* 11. *tab.* 10 ; Ficoïde
nain du Cap, avec des feuilles
femblables aux cornes d'un

cerf, des pétales jaunes, & une fleur qui s'ouvre la nuit.

42°. *Mesembryanthemum difforme, acaule, foliis difformibus, punctatis, connatis. Prod. Leyd.* 287 ; Mesembryanthemum avec des feuilles difformes, ponctuées & rapprochées.

Ficoïdes Afra, foliis latissimis, crassissimis, lucidis, difformibus. Boerrh. Ind. Alt. 292 ; Ficoïde d'Afrique, avec des feuilles fort larges, épaisses, luisantes, & difformes.

43°. *Mesembryanthemum lucidum, acaule, foliis lingui-formibus, lucidis, emarginatis* ; Mesembryanthemum sans tige, ayant des feuilles luisantes en forme de langue, & dentées au sommet.

Ficoïdes Afra acaulos, foliis latissimis, crassis, lucidis, conjugatis, flore aureo amplissimo. Tourn. Acad. R. scient. 1705 ; Ficoïde d'Afrique, sans tige, à feuilles très-larges, grasses, luisantes, & disposées par paires, avec une fort grosse fleur jaune.

44°. *Mesembryanthemum linguiforme, acaule, foliis lingui-formibus, altero margine crassioribus, impunctatis. Lin. Sp.* 699. *Hort. Cliff.* 219. *n.* 8. *Hort. Ups.* 128. *Roy. Lugd.-B.* 282. *Knorr. Del.* 1. *f.* 5. *f. G.* 6. *n.* 2. ; Mesembryanthemum sans tige, avec des feuilles en forme de langue, très-grasses, & des bords plus épais que ceux de la précédente, & sans taches.

Ficoïdes Afra, acaulos, foliis latissimis, lucidis, conjugatis, flore aureo amplo, pedunculo bre-

vi. *Boerrh. Ind. Alt.* ; Ficoïde d'Afrique, sans tiges, avec des feuilles très-larges, épaisses, luisantes, placées par paires, & une grosse fleur dorée sur de courts pedoncules.

45°. *Mesembryanthemum albidum, acaule, foliis triquetris, integerrimis* ; Mesembryanthemum sans tige, avec des feuilles triangulaires & entieres.

Mesembryanthemum foliis robustis, albicantibus. Hort. Elth. 243 ; Mesembryanthemum avec des feuilles fortes & blanchâtres.

46°. *Mesembryanthemum pugioni-forme, foliis alternis, confertis, subulatis, triquetris, longissimis, impunctatis. Hort. Cliff.* 216. *n.* 2. *Hort. Ups.* 129. *Roy-Lugd.-B* 281 ; Mesembryanthemum avec des feuilles alternes, en forme d'alêne, triangulaires, fort longues, & sans taches.

Mesembryanthemum, folio pugioni-formi, flore aureo, stamineo. Dill. Elth. 280. *t.* 210. *f.* 269.

Ficus Capensis, Caryophylli folio, flore aureo specioso. Bradl. Suec. Del. 2. *p.* 5. *tab.* 14 ; Ficoïde du Cap, à feuille de Girofflier, avec une belle fleur de couleur d'or.

Ces plantes sont presque toutes originaires du Cap de Bonne-Espérance ; leurs semences ont été d'abord envoyées en Hollande, dans beaucoup de jardins curieux, d'où elles se sont ensuite répandues dans plusieurs parties de l'Europe. On leur donnoit, dans l'ancienne Botanique, le nom de *Chrysanthemum* : depuis, HERMANN

MES 91

& Tournefort les ont appelées *Ficoïdes*, à caufe de leurs capfules, qui reffemblent affez à de petites *Figues* ; mais enfuite on a fini par les nommer *Mefembryanthemum*, terme qui fignifie une fleur qui s'épanouït au milieu du jour, ce qui arrive à la plupart des efpeces. Comme il y en a trois ou quatre qui s'ouvrent vers le foir, & font fermées pendant tout le jour, quelques perfonnes les ont féparées des autres, & leur ont donné le nom de *Nycterianthemum* : cependant comme les caracteres de toutes ces plantes font abfolument les mêmes, on ne doit pas les féparer.

La plupart des plantes de ce genre ont de belles fleurs, qui paroiffent dans des tems différens ; les unes fleuriffent au commencement du printems, d'autres pendant l'été, quelques-unes en automne, & d'autres enfin pendant l'hiver : plufieurs de celles-ci produifent des fleurs en telle quantité, que les plantes en font entierement couvertes ; elles ont toutes des feuilles épaiffes & fucculentes : quelques-unes les ont graffes ; mais leur forme varie fuivant les différentes efpeces, & elles font une variété agréable, même lorfqu'elles ne font pas en fleurs.

Comme une defcription détaillée de toutes ces plantes augmenteroit inutilement le volume de cet Ouvrage, puifque leurs titres fuffifent pour les faire reconnoître, je n'en parlerai pas davantage ;

je me bornerai à donner la maniere de les cultiver.

Toutes ces efpeces font vivaces, à l'exception des deux premieres, qui font annuelles.

Les efpeces vivaces fe multiplient aifément par bouture, qu'on peut planter pendant tout l'été : celles qui ont des tiges & des branches d'arbriffeau, prennent aifément racine, lorfqu'elles font placées dans une planche de terre légere, & couvertes de nattes ou de vitrages ; mais fi l'on fe fert de vitrages, il faut les tenir à l'ombre chaque jour, lorfque le foleil eft chaud. Ces boutures n'ont pas befoin d'être coupées plus de cinq ou fix jours avant d'être plantées. Pendant ce tems, on les tient dans une chambre feche, & pas trop expofée au foleil, afin que les parties coupées puiffent fe fecher avant qu'on les plante, fans quoi elles feroient en danger de fe pourrir. On peut les planter à trois pouces environ de diftance, en obfervant de preffer fortement la terre autour, fans engager aucune feuille, qui, étant remplie d'humidité, communiqueroit fa pourriture à la tige, & la détruiroit ainfi : c'eft-pourquoi, lorfqu'on a détaché des boutures fur les vieilles plantes, on retranche autant de feuilles qu'il eft néceffaire pour rendre les tiges nues dans une longueur fuffifante.

Quand elles font plantées, on les arrofe un peu, pour raffermir la terre, mais avec modération, parce qu'une trop

MES

grande humidité leur eft fort contraire : fi on les tient à l'ombre lorfque le foleil eft ardent, depuis dix heures du matin jufqu'à trois ou quatre heures, on empêchera la terre de fe deffecher trop vite, & les boutures n'auront befoin d'être arrofées qu'une fois la femaine. S'il furvient une pluie légere, on enlevera les vitrages & les couvertures, pour les y expofer ; mais on les mettra foigneufement à l'abri des fortes ondées : au moyen de ce traitement, ces boutures auront pouffé de bonnes racines au bout de fix femaines; alors on les enlevera avec foin, & on les mettra féparément dans de petits pots remplis d'une terre légere & fablonneufe, que l'on placera dans une fituation abritée, & à l'ombre, après les avoir un peu arrofées, pour affermir la terre fur les racines : on les laiffera ainfi pendant huit ou dix jours, pour leur faire pouffer de nouvelles fibres ; mais après ce tems, on les tranfportera à une expofition abritée, & plus expofée au foleil, où elles refteront jufqu'à l'automne.

Pendant l'été, on peut les arrofer deux fois la femaine, & même trois fois dans les tems chauds ; mais il ne faut pas leur donner trop d'eau : & en automne, lorfque le foleil a moins d'activité, on ne les arrofe plus qu'une fois dans le même efpace de tems ; car fi on les arrofoit trop fouvent, elles deviendroient fucculentes ; leurs branches & leurs feuilles fe rempliroient d'une

humidité fi abondante, que les premieres gelées de l'automne les détruiroient : lorfqu'au contraire on les tient plus feches, leur accroiffement eft plus lent ; mais elles deviennent affez dures pour réfifter aux petites gelées. Il faut auffi empêcher que leurs racines ne pénetrent pas par les trous des pots, car dans ce cas les plantes pouffent avec trop de force ; & lorfqu'on veut enlever ces pots, leurs racines fe déchirent, leurs feuilles & leurs branches fe fannent, & les plantes périffent, ou ne fe rétabliffent que long-tems après. Pour prévenir cet accident, on souleve les pots chaque quinze jours ; & lorfque les racines ont commencé à pouffer par les ouvertures, on les coupe auffi-tôt.

Les efpeces qui croiffent aifément, doivent être changées trois fois dans l'été, afin qu'on puiffe retrancher leurs racines, & redreffer leurs branches ; mais il ne faut jamais leur donner de la trop bonne terre, pour les raifons qui viennent d'être expofées. Celle qui leur convient le mieux, eft une terre nouvelle, fans fumier ni autres engrais, à laquelle on ajoûte encore du fable ou des décombres, fi elle eft trop forte. La quantité de ce fable ou de ces décombres doit être proportionnée à la qualité de la terre, qu'il faut rendre affez légere, pour empêcher l'humidité d'y féjourner.

Nous allons paffer à préfent au traitement des efpeces, dont les

MES

les tiges & les feuilles font très-fucculentes.

Les boutures de celles-ci doivent être féparées des plantes dix ou quinze jours avant de les planter, afin que leurs parties bleffées aient le rems de fe deffecher; leurs feuilles baffes doivent être auffi retranchées dans une longueur fuffifante. Comme ces efpeces ne s'élevent pas à une grande hauteur, il fuffira de dégarnir leurs tiges dans la longueur d'un pouce & demi : il faut les placer fous des vitrages, pour qu'elles foient à couvert de l'humidité, & les arrofer beaucoup moins: au refte, le même traitement leur convient.

Comme les racines de ces efpeces ne s'étendent pas autant que celles des précédentes, il ne faut les changer que deux fois l'année tout au plus : il est auffi néceffaire de les tenir dans de petits pots; on doit leur donner une terre légere & fans engrais, & ne les pas trop arrofer pendant l'été: en hiver, elles n'ont befoin que de très-peu d'eau.

Si l'on tient en hiver ces efpeces fucculentes fous des châffis où l'on puiffe leur donner beaucoup d'air libre dans les tems doux, & les abriter de la gelée, elles profiteront beaucoup mieux que fi elles étoient traitées plus délicatement.

Les efpeces en arbriffeau n'ont befoin que d'être placées fous des châffis ordinaires, pour les mettre à couvert des gelées & de l'humidité; car plus on les conduit durement, plus elles produifent de fleurs. Quelques-

Tome V.

unes d'entr'elles font fi durés, qu'on pourroit les laiffer en pleine terre contre une muraille, à une bonne expofition, en les plantant dans une terre fèche & ftérile. Ces dernieres fleuriroient beaucoup mieux que fi elles étoient tenues à couvert.

Nodiflorum. La premiere efpece croît naturellement en Égypte, où on la brûle pour en retirer les cendres, qui font propres à faire du favon dur, & du verre de la meilleure qualité : elle est annuelle, & ne perfectionne pas fes femences en Angleterre. Si on la tient dans la ferre ou dans une couche chaude, fes tiges deviennent longues & minces, & ne produifent pas beaucoup de fleurs ; au-lieu que celles qu'on éleve fur des couches chaudes, & qu'on expofe enfuite en plein air, fleuriffent affez aifément : mais elles ne perfectionnent point leurs femences.

Comme cette plante croît dans la Caroline méridionale, auffi bien que dans fon pays natal, elle peut devenir très-utile à cette Colonie, fi l'on parvient à en perfectionner la culture.

Cryftallinum. La feconde efpece, qui est auffi annuelle, est originaire du Cap de Bonne-Efpérance : on la multiplie, à caufe de la fingularité de fes feuilles & de fes tiges, qui font entierement couvertes de tubercules tranfparens & remplis d'humidité, qui réfléchiffent la lumiere, & les rendent brillantes comme la glace, lorfqu'elles font expofées au foleil ; ce qui a fait donner à cette ef-

G

94 MES

pece le nom de *Glaciale* ou *Fi-coïde de Diamant*.

Celle-ci fe multiplie par fes graines, qu'il faut femer fur une couche chaude dans le commencement du printems. Lorfque les plantes fortent de la terre, on les remet fur de nouvelles couches chaudes, pour les avancer; & quand elles y ont pris racine, on les arrofe légerement, parce que l'humidité les pourriroit. Lorfqu'elles font devenues affez fortes, on les met féparément dans de petits pots remplis d'une terre fraîche, légere, & fans fumier, & on les plonge dans une couche chaude de tan, en obfervant de les tenir à l'abri de la chaleur du jour, jufqu'à ce qu'elles aient formé de nouvelles racines : alors on leur donne beaucoup d'air frais chaque jour dans les tems chauds, pour les empêcher de filer. Vers la fin du mois de Juin, on peut en accoutumer quelques-unes à fupporter le plein air, & les ôter enfuite des pots, pour les tranfplanter dans des plates-bandes chaudes, où elles profiteront & étendront leurs branches à une grande diftance fur la terre : mais comme ces plantes ne donneront pas beaucoup de fleurs, on en confervera quelques-unes dans de petits pots, que l'on placera fur les tablettes d'une ferre, ou fous des vitrages, en empêchant les racines de fortir au travers des trous des pots, afin qu'elles y foient refferrées, & qu'elles puiffent, par ce moyen, donner beaucoup de fleurs, dont on obtiendra chaque année de très-bonnes femences.

MES

MESPILUS. Μέσπιλος, en Grec. *Tourn. Infl. R. H.* 641. tab. 410. *Lin. Gen. Plant.* 549. [*The Medlar.*] Neflier.

Caracteres. Le calice de la fleur perfifte ; il eft formé par une feuille découpée en cinq fegmens, étendus & concaves : la corolle eft compofée de cinq pétales ronds & concaves, & inférés dans le calice ; le nombre des étamines eft différent dans plufieurs efpeces ; il s'en trouve depuis dix jufqu'à vingt & plus ; elles font auffi inférées dans le calice, & terminées par des antheres fimples. Le germe, qui eft placé fous la fleur, foutient un certain nombre de ftyles, depuis trois jufqu'à cinq, couronnés par des ftigmats à têtes ; le germe fe change dans la fuite en une baie ronde ou ovale, furmontée par le ftyle, & dans laquelle font renfermées quatre ou cinq femences dures.

Ce genre de plantes eft rangé dans la quatrieme fection de la douzieme claffe de LINNÉE, qui comprend celles dont les fleurs ont vingt étamines inférées dans le calice, & cinq ftyles.

Les efpeces font :

1º. *Mefpilus fylveftris, inermis, foliis lanceolatis, dentatis, acuminatis, fubtùs tomentofis, calycibus acuminatis* ; Neflier fans épines à feuilles en forme de lance, dentées, pointues, & cotonneufes en deffous, avec des calices à pointe aiguë.

Mefpilus folio Laurino major, fructu minori, rariori fubftantiá. Hort. Cath. ; le plus grand Neflier, avec une feuille de Laurier, & un fruit plus petit, & moins garni de chair.

MES

2°. *Mespilus Germanica*, *inermis*, *foliis lanceolatis*, *integerrimis*, *subtùs tomentosis*, *calycibus acuminatis*. *Hort. Cliff.* 189. *Hort. Ups.* 129. *Mat. Med.* 127. *Roy. Lugd.-B.* 270 ; Nefflier sans épines, avec des feuilles en forme de lance, entieres, & velues en-dessous, & des calices en pointe aigue.

Mespilus Germanica, *folio Laurino*, *non serrato*, *sivè Mespilus sylvestris*. *C. B. P.*; Nefflier d'Allemagne à feuilles de Laurier, dont les bords sont entiers, ou Nefflier ordinaire.

Mespilus. *Dod. Pempt.* 801.

3°. *Mespilus Pyracantha*, *spinosa*, *foliis lanceolato-ovatis*, *crenatis*, *calycibus fructûs obtusis*. *Hort. Cliff.* 189. *Virg. Cliff.* 44. *Roy. Lugd.-B.* 271. *Scop. Carn.* *n.* 596 ; Nefflier épineux, avec des feuilles en forme de lance, ovales & crenelées, avec des calices obtus sur les fruits.

Mespilus Aculeata, *Amygdali folio.* *Tourn. Inst.* 642. *Duham. arb.* 7 ; Nefflier épineux à feuilles d'Amandier, appelé Pyracantha, Buisson ardent ou Epine toujours verte.

Oxyacantha Dioscoridis. *S. spinâ acutâ*, *Pyri folio.* *Bauh. Pin.* 454. *Rai. Hist.* 1459.

Uva Ursi. *Dalech. Hist.* 164.
Rhamnus 3. *Dioscoridis.* *Lob. ic.* 2. *p.* 182.

4°. *Mespilus cordata*, *foliis cordato ovatis*, *acuminatis*, *acutè serratis*, *ramis spinosis*, *Fig. Plant. tab.* 179; Nefflier à feuilles en forme de cœur, ovales, à pointe aiguë, & fortement sciées, avec des branches épineuses.

5°. *Mespilus Amelanchier*, *inermis*, *foliis ovalibus*, *serratis*, *cauliculis hirsutis.* *Lin. Sp. Plant.* 478. *Jacq. Austr. t.* 300. *Dœrr. Nass. p.* 262; Nefflier sans épines, à feuilles ovales, & sciées, & à tiges velues.

Mespilus folio rotundiori, *fructu nigro*, *subdulci.* *Tourn. Inst.* 642 ; Nefflier à feuilles plus rondes, qui produit un fruit noir & douçâtre, communément appelé Amelanchier.

Mespilus inermis, *foliis ovalibus*, *serratis*, *acutis.* *Hort. Cliff.* 189. *Roy. Lugd.-B.* 271.

Pyrus, *foliis ovatis*, *serratis*, *subtùs tomentosis*, *coalescentibus.* *Hall. Helv. n.* 1095.

Sorbus Amelanchier, *foliis ovalibus*, *serratis*, *carinatis*, *fructu globoso*, *multi-loculari.* *Crantz. Austr. p.* 90.

Mespilus floribus pentagynis; *racemis terminalibus*; *foliis ovatis*, *obtusis*, *serratis*; *caule inermi.* *Scop. carn. ed.* 1. *p.* 584. *ed.* 2. *n.* 595.

Pyrus Amelanchier. *Du Roi. Harpk.* 2. *p.* 219.

Alni effigie, *lanato folio*, *minor.* *Bauh. Pin.* 452.

Vitis Idœa III. *Clus. Hist.* 1. *p.* 75.

6°. *Mespilus Canadensis*, *foliis ovato-oblongis*, *glabris*, *serratis*, *caule inermi.* *Lin. Sp. Plant.* 478 ; Nefflier à feuilles ovales, oblongues, unies & sciées, avec des branches sans épines.

Mespilus inermis, *foliis subtùs glabris*, *obverse—ovatis.* *Flor. Virg.* 54. *Duham. arb.* 9 ; Nefflier sans épines, à feuilles ovales & obverses, & unies en-dessous.

7°. *Mespilus Cotoneaster*, *inermis*, *foliis ovatis*, *integerrimis*,

96 MES

jubtus tomentosis, Hort. Cliff. 189; Neflier sans épines, à feuilles ovales, entieres & cotonneufes en-deffous.
Mefpilus folio fub-rotundo, fructu rubro. Tourn. Inft. R. H. 642; Neflier à feuilles rondes & fruits rouges, communément appelé Coignaffier nain ou Coroneafter.
Mefpilus foliis ovato-acuminatis, integerrimis, fubtus lanatis, bacca globofa. Cranız, Auftr. p. 81. Coroneafter folio rotundo, non ferrato. Bauh. Pin. 452.
Chamæ-Mefpilus Gefneri, Cluf. Hift. 1. p. 60.
Chamæ-Mefpilus Cordi. Bauh. Pin. 452.

8°. *Mefpilus Chamæ-Mefpilus, inermis, foliis ovalibus, acutè ferratis, glabris, floribus capitatis, bracteis deciduis linearibus, Lin. Sp. Plant.* 479; Neflier fans épines, à feuilles unies, ovales & fciées, avec des fleurs à têtes, & des bractées linéaires qui tombent.
Coroneafter folio oblongo ferrato, C.B.P. 452; Coignaffier bâtard, avec une feuille oblongue & fciée.

Cratægus foliis ovalibus, acutè ferratis, glabris, caule inermi. Jacq. Vind. 243. *Auftr.* t. 231.
Cratægus foliis ovalibus utrinquè glabris, plicatis, indivifis, ferratis, Hort. Cliff. 497. *Vir. Cliff.* 43. *Roy. Lugd.-B.* 272.
Sorbus fruticofa, foliis oblongo-ovalibus, ferratis, fructu ovato bi-loculari, Cranız, Auftr. p. 83. *Hift.* 1. *f.* 3.
Coroneafter forte Gefneri, Cluf. 1. *f.* 3.

9°. *Mefpilus Orientalis, foliis ovatis, craffis, integerrimis, fubtus

---

tomentofis, floribus umbellatis axillaribus;* Neflier à feuilles ovales, épaiffes, entieres &, velues en-deffous, avec des fleurs difpofées en ombelles aux aiffelles de la tige.
Chamæ-Cerafus Idæa. Alp. Exot. 5 ; Cerifier nain du mont Ida.

10°. *Mefpilus Arbuti-folia, inermis, foliis lanceolatis, crenatis, fubtus tomentofis, Hort. Cliff.* 189. *Roy. Lugd.-B.* 271. *Du Roi, Harpk.* 3. *p.* 418 ; Neflier de Virginie, à feuilles d'Arboufier, fans épines, en forme de lance, crenelées, & velues en deffous.

Sorbus Virginiana, folio Arbuti. H. L. 578.
Cratægus virginiana, foliis Arbuti. T. Mill. f. 109.
Sorbus aucuparia Virginiana, foliis Arbuti. Bryn. Prodr. 1. *p.* 15.

11°. *Mefpilus Virginiana, inermis, foliis oblongo-ovatis, fubtus tomentofis, fructu ovato, pedunculis longiffimis;* Neflier uni de Virginie, avec des feuilles oblongues, ovales & cotonneufes en-deffous, produifant un fruit ovale fur de longs pédoncules.

Sylveftris. La premiere efpece croît naturellement en Sicile, où elle devient un grand arbre : elle s'éleve avec une tige plus droite & des branches plus érigées que celles du *Neflier hollandois;* fes feuilles font plus étroites, & point fciées fur leurs bords : fes fleurs font plus petites que celles du Neflier hollandois, & le fruit a la forme d'une poire.
Germanica. La feconde efpece, à laquelle on donne gé-

néralement le nom de *Nefflier Hollandois*, ne s'élève jamais avec une tige droite ; mais elle pousse des branches courbes & difformes à une petite hauteur de terre ; ses feuilles sont fort larges, entieres & velues en-dessous : ses fleurs sont très-grosses, ainsi que le fruit, qui est rond, & presque de la forme d'une *Pomme.* Comme ce fruit est le plus gros de tous ceux de ce genre, on cultive cette espece dans les jardins. On en connoît cependant une autre, dont le fruit est encore plus gros : on le nomme le *Nef-flier de Nottingham* ; sa saveur est plus forte & plus piquante que celui-ci, dont il me paroît n'ê-tre qu'une variété : aussi n'en ai-je point fait mention comme d'une espece distincte (1).

Amelanchier. La cinquieme naît sans culture en Autriche, en Italie & en France, parti-culierement près de Fontaine-bleau : elle s'élève avec plu-sieurs tiges minces à la hau-teur d'environ trois pieds, & pousse de petites branches la-térales, couvertes d'une écorce d'un pourpre foncé, sans épi-nes, & fortement garnies de feuilles ovales, d'environ neuf lignes de longueur sur six de large, & légérement sciées sur leurs bords ; les petites bran-

ches latérales qui produisent les fleurs, sont fort velues & co-tonneuses, ainsi que les pédon-cules & le dessous des feuilles ; mais leur surface supérieure est unie & verte ; les fleurs sor-tent en paquets aux extrémités des branches ; elles ont cinq pétales longs & étroits, & en-viron dix étamines. Ces fleurs sont remplacées par de petits fruits, qui deviennent noirs en mûrissant. Les Jardiniers don-nent à cette espece le nom de *Coing* de la Nouvelle-Angleter-re : on en connoît une variété qui croît naturellement dans l'Amérique Septentrionale. Com-me ses feuilles sont en forme de coin, & sans dentelures sur leurs bords, on a pensé qu'elle pourroit bien être une espece distincte.

Canadensis. La sixieme, qui est originaire du Canada, est aussi un arbrisseau qui s'élève rarement au-dessus de la hau-teur de cinq pieds, & se divise en plusieurs branches lisses & couvertes d'une écorce pour-pâtre : ses feuilles sont placées sur des pétioles longs & min-ces ; elles ont un pouce & demi de longueur sur un de large, unies aux deux surfaces, & un peu sciées sur leurs bords : les fleurs qui naissent en petits paquets aux extrémités des bran-ches, sont à - peu - près de la grosseur de celles de l'*Aubépine* commune, & elles produisent de petits fruits, qui prennent une teinte de pourpre en mû-rissant.

Cotoneaster. La septieme es-pece se trouve sur les monta-gnes des Pyrénées, &dans d'au-

---

(1) Les feuilles & les fruits du Nefflier sont astringens : on s'en sert quelquefois dans les anciens cours de ventre ; mais ces fruits doivent être bien mûrs, ou confits au sucre. Les Nefles entrent dans la composition du syrop de Myr-rhe, de Mésue.

G 3

tres parties froides de l'Europe ; elle s'éleve avec une tige unie d'arbriffeau, à la hauteur d'environ quatre pieds, & fe divife en quelques petites branches couvertes d'une écorce pourpre, & garnies de feuilles ovales entieres, d'un peu plus d'un pouce de longueur, fur à-peu-près neuf lignes de large, & fupportées par de fort petits pétioles. Les fleurs fortent au nombre de deux ou trois enfemble fur les côtés des tiges ; elles font petites, de couleur pourpre, & fefliles ; elles paroiffent dans le mois de Mai, & produifent de petits fruits ronds, & d'un rouge brillant, lorfqu'ils font mûrs.

Chamæ-Mefpilus. La huitieme efpece, qu'on rencontre dans les contrées feptentrionales de l'Europe, a une tige unie de quatre ou cinq pieds de hauteur, qui pouffe des branches minces, couvertes d'une écorce pourpâtre, & garnies de feuilles ovales, unies, d'environ deux pouces de longueur, fur un & demi de largeur, & divifées fur leurs bords en dentelures, dont les pointes font tournées vers le haut ; elles ont des pétioles longs & minces, & font d'un vert jaunâtre fur les deux furfaces : les fleurs croiffent quatre ou cinq enfemble en une tête ferrée aux aiffelles de la tige ; elles font d'une couleur pourpâtre, & entr'elles fortent des bractées de même couleur, qui tombent quand les feuiiles commencent à fe flétrir : le fruit eft petit & rouge, lorfqu'il eft mûr.

Orientalis. La neuvieme ef-

pece croît naturellement fur le mont Ida, dans l'Ifle de Crète, où les pauvres Bergers fe nourriffent de fon fruit. Cet arbriffeau a une tige unie, d'environ huit pieds de hauteur, & divifée en plufieurs branches unies, & garnies de feuilles de deux pouces & demi de longueur fur près de deux de large, d'une fubftance épaiffe, & d'un vert foncé au-deffus, mais cotonneufes en-deffous, & & fupportées par de courts pétioles : fes fleurs fortent aux côtés de la tige, fur des branches courtes & foibles, au nombre de cinq ou fix réunies en un paquet ferré ; elles font de couleur pourpre, & leur corolle eft un peu plus longue que le calice, qui eft velu & découpé en cinq fegmens obtus : le fruit eft gros, rond, & d'un beau rouge, lorfqu'il eft mûr.

Arbuti-folia. La dixieme efpece, qui fe trouve dans l'Amérique Septentrionale, où elle s'éleve rarement au-deffus de cinq pieds de hauteur, pouffe quelques branches droites & garnies de feuilles en forme de lance, dont les bords font crenelés, & qui font cotonneufes en – deffous : fes fleurs naiffent en petits paquets fur les côtés & aux extrémités des branches, & font remplacées par des fruits ronds un peu comprimés, & de couleur pourpre, lorfqu'ils font mûrs.

Virginiana. La onzieme efpece eft originaire des mêmes contrées que la précédente ; elle s'éleve à la hauteur de fix ou huit pieds, & pouffe des branches latérales garnies de

MES

feuilles oblongues, ovales, entieres & cotonneuses en-dessous : ses fleurs naissent en petits paquets sur de longs pédoncules ; elles ont chacune cinq petits pétales blancs, étroits & rétrécis à leurs bâses, & produisent des fruits ovales & de couleur bleue, lorsqu'ils sont mûrs. Les habitans de l'Amerique les mangent, quand les autres especes de fruits sont rares ; mais ils ne sont pas fort agréables au goût.

Culture. Toutes ces especes sont assez dures pour résister en plein air dans ce pays ; & l'on trouve beaucoup de ces plantes dans nos jardins, où elles font un bel effet pendant qu'elles sont en fleurs ; & en automne, quand leur fruit est mûr, elles produisent une variété agréable ; de sorte que des bois plantés sans ordre, dans différentes parties d'un jardin, avec ces especes d'arbrisseaux, font un charmant coup-d'œil : d'ailleurs leurs fruits servent de nourriture aux oiseaux & aux animaux sauvages.

On greffe ordinairement les especes Américaines, dans les pépinieres, sur l'Epine blanche commune : mais les plantes ainsi multipliées n'acquierent jamais la moitié de la hauteur de celle qu'on multiplie d'une autre maniere, de sorte qu'on devroit toujours choisir les plantes qui n'ont point été greffées, & qui croissent sur leurs propres racines.

Plusieurs personnes ne peuvent se déterminer à les faire venir de semences, parce qu'elles ne croissent pas les pre-

MES   99

mieres années, & que l'on est trop long-tems à les attendre ; mais si l'on met ces semences à terre en automne, aussi-tôt que les fruits sont mûrs, elles pousseront au printems suivant. En tenant constamment nettes les plantes qui en proviennent, & en les arrosant dans les tems secs, elles feront de grands progrès : deux ans après, on les place à demeure, parce qu'étant transplantées jeunes, elles réussissent beaucoup mieux que quand elles sont plus âgées, pourvu que la terre qui leur est destinée soit bien labourée, & débarassée de mauvaises herbes & des racines inutiles. Le meilleur tems pour les transplanter est l'automne, quand leurs feuilles sont tombées ; on les tient constamment nettes de mauvaises herbes, & on laboure, chaque hiver, la terre entr'elles, pour hâter leur accroissement : il suffira de les nettoyer trois ou quatre fois pendant l'été.

Les greffes du *Mespilus* & du *Cratægus* prennent les unes sur les autres ; elles réussissent aussi sur le *Coignassier* & le *Poirier*, & ces deux derniers prennent aussi sur le *Nesslier* ; de sorte que ces arbres ont une grande affinité entr'eux, & pourroient être placés avec plus de convenance sous le même genre que le *Poirier* & le *Pommier*, qui ne prennent pas l'un sur l'autre : mais quoique le *Poirier* réussisse sur l'Epine blanche, il n'est cependant pas prudent de faire usage de ces sujets ; car ils rendent généralement le fruit petit, cassant & pier-

G 4

MIC

reux : de forte que les fruits qui proviennent de cette ef-pece de greffe, ne font pas bons, à moins que ce ne foit peut-être quelques *Poires* fort tendres & fondantes.

METHONICA DE MALA-BAR. *Voyez* GLORIOSA.

MEUM. *Voy.* ATHAMANTA MEUM.

MICOCOULIER. *Voy.* CELTIS. *L.*

MICROPUS. *Lin. Gen. Plant.* 892. *Gnaphalodes. Tourn. Inft. R. H.* 439. *tab.* 261. [*Baftard Cudweed.*] Gnaphalium bâtard.

Caractères. Cette plante a des fleurs femelles & hermaphro-dites renfermées dans les mêmes calices doubles ; dix fleurs her-maphrodites compofent le dif-que ; elles ont un pétale en forme d'entonnoir, érigé & dé-coupé en cinq parties au fom-met, cinq courtes étamines hé-riffées & terminées par des an-theres cylindriques, & un ger-me ufé, qui foutient un ftyle court, mince, & couronné par un ftigmat ufé : dans le même calice font cinq petites fleurs qui occupent la circonférence ; elles ont chacune un germe ovale, comprimé & caché fous les écailles du calice intérieur, & un ftyle à leur côté qui eft hériffé, & fe tourne vers les fleurs hermaphrodites. Ce ftyle eft couronné par un ftigmat à pointe aiguë, & divifé en deux parties. Les fleurs femelles ont chacune une femence fimple & ovale, renfermée dans de peti-tes feuilles du calice ; mais les fleurs hermaphrodites font fté-riles.

Ce genre de plantes eft rangé dans la quatrieme fection de la dix-neuvieme claffe de LINNÉE, qui comprend celles dont les fleurs font compofées de fleu-rons femelles & fertiles dans le rayon, & de fleurons her-maphrodites ftériles dans le difque.

Nous n'avons qu'une efpece de ce genre dans les Jardins Anglois.

Micropus fupinus, caule pro-ftrato, foliis geminis. Hort. Up-fal. 275. *Prod. Leyd.* 145 ; Mi-cropus ou Gnaphalium bâtard à tige traînante.

Gnaphalodes Lufitanica. Tourn. Inft. R. H. 439 ; Gnaphalium bâtard de Portugal.

Micropus foliis floralibus op-pofitis, fructibus muricatis. Gouan. Illuftr. p. 74.

Gnaphalium fupinum, echina-to femine. Pluk. Alm. 171. *t.* 187. *f.* 6. *Rai. Suppl.* 191.

Pfeudo-Gnaphalium fupinum, femine echinato. Moris. Hift. 3. p. 93.

Cette plante eft annuelle, & croît naturellement en Por-tugal, fur les bords de la mer ; fes racines pouffent plufieurs branches traînantes, de fix ou huit pieds de longueur, gar-nies de petites feuilles ovales & argentées, dont les bâfes embraffent les tiges : fes fleurs fortent des aiffelles des tiges en petites grappes ; elles font fort petites, blanches, & por-tées dans un double calice, dont l'intérieur eft fi large, qu'il cache prefque les fleurs. Cette plante fleurit dans les mois de Juin & de Juillet, & fes fe-mences mûriffent en automne : on la conferve fouvent dans les jardins, pour la beauté

de fes feuilles argentées. Si on feme fes graines en automne, ou fi on leur permet de s'écarter, les plantes pouf-feront au printems, & n'exigeront aucun autre foin, que d'être tenues nettes de mauvaifes herbes, & éclaircies où elles feront trop ferrées ; mais fi on ne les met en terre qu'au printems, elles croiffent rarement la premiere année.

MICROSCOPE (un) eft un inftrument de dioptrique, qui groffit les petits objets, & au moyen duquel on peut diftinguer toutes leurs parties.

Cet inftrument peut être d'une très-grande utilité à ceux qui recherchent, dans le tiffu intime des plantes, la maniere dont la végétation s'opere. On peut, par fon moyen, obferver de près les plus petits vaiffeaux des plantes, & leurs parties les plus cachées, pour en reconnoître les fonctions, ainfi que les petites parties de fleurs, qui ne font pas vifibles à l'œil nud.

MIGNARDISE ou PETIT ŒILLET DE JARDIN. V. DIANTHUS BARBATUS. L.

MILIUM. Tourn. Inft. R. H. 514. tab. 298. Lin. Gen. Plant. 73 ; ainfi appelée de mille, à caufe de la multitude de fes grains. [Millet.] Millet.

Caractères. Cette efpece de plante a une fleur dans chaque bafle : la bafle s'ouvre en deux valves ovales & à pointe aiguë ; la corolle eft divifée en deux parties, & plus petite que le calice ; la fleur a trois étamines fort courtes, velues, & terminées par des fom-

mets oblongs, & un germe rond, qui foutient deux ftyles velus, & couronnés par deux ftigmats en forme de vergettes ; le germe fe change enfuite en une femence ronde, & couverte par le pétale de la fleur.

Ce genre de plantes eft rangé dans la feconde fection de la troifieme claffe de LINNÉE, qui renferme les plantes dont les fleurs ont trois étamines & deux ftyles.

Les efpeces font :

1°. Milium Panicum, paniculâ laxâ flaccidâ, foliorum vaginis pubefcentibus ; Millet avec des panicules lâches & pendants, dont les gaînes des feuilles font velues.

Milium femine luteo & albo. C. B. P. 26 ; Millet à femences jaunes & blanches.

Panicum Miliaceum. Lin. Syft. Plant. tom. 1. p. 160. Sp. 23.

2°. Milium fparfum, paniculâ fparfâ, erectâ, glumis ariftatis ; Millet avec un panicule lâche & érigé, & une bafle barbue.

Milium paniculâ amplâ, fparfâ. Houft. MSS. ; Millet avec un grand panicule érigé & épars.

Panicum Capillare. Lin. Syft. Plant. tom. 1. p. 160. Sp. 24.

Milium effufum, floribus paniculatis, difperfis, muticis. Flor. Suec. 55. Dalib. Paris. 23. Pollich. Pal. n. 67 ; Millet avec des fleurs en panicule, difperfè & fans barbe.

Gramen fylvaticum, paniculâ miliaceâ, fparfâ. C. B. P. 8. Theatr. 141. Moris. Hift. 3. S 8. t. 5. f. 10 ; Gramen des boi. avec un panicule femblable à celui du Millet. Faux Milled; à Oifeaux.

MIL

4°. *Milium confertum, floribus paniculatis confertis*. Prod. Leyd. 57 ; Millet avec des panicules de fleurs disposées en paquets.

Gramen paniculatum Alpinum latifolium, paniculâ Miliaceâ fparfâ. Scheu. Gr. 134 ; Gramen des Alpes en panicule, à larges feuilles, avec un panicule épars comme celui du Millet.

Panicum. La premiere espece croît naturellement dans les Indes ; mais on la cultive à présent dans plusieurs parties de l'Europe, comme une graine bonne à manger ; elle s'éleve à la hauteur de trois ou quatre pieds avec une tige de roseau canelée, & à chaque nœud de laquelle sort une feuille semblable à celle du Roseau, jointe au sommet de la gaîne, qui embrasse & couvre ce nœud de la tige au-dessous de la feuille. Cette gaîne est couverte de poils mous ; mais la partie de la feuille qui est étendue, n'en a point. Cette feuille a plusieurs petits sillons longitudinaux, qui partent tous de la côte du milieu. Le sommet de la tige est terminé par un panicule large & lâche, qui pend d'un côté ; elle a une fleur pleine de barbe, à laquelle succede une semence petite & ronde, qu'on apprête souvent en bouillie.

Il y a deux variétés de cette plante, l'une à semences blanches, & l'autre à semences noires, mais qui se ressemblent en toute autre chose.

Elle a été rangée par LINNÉE sous le titre de *Panicum*; mais comme elle est plus générale-ment connue sous son ancien nom, je crois devoir le lui continuer.

Sparfum. La seconde espece, qui a été trouvée à la Vera-Cruz, a une tige plus mince que celle de la précédente ; elle s'éleve à la hauteur d'environ trois pieds : les gaînes qui l'environnent n'ont point de filets, mais elles sont canelées ; ses feuilles sont plus courtes que celles de la premiere, le panicule est érigé, les barbes & les cosses sont plus courtes.

Les deux autres especes croissent sans culture dans les forêts, & ne sont jamais cultivées dans les campagnes : ainsi elles n'exigent aucune autre description.

Le Millet commun a été originairement apporté des pays orientaux, où on le cultive beaucoup, & d'où l'on nous en envoie tous les ans une certaine quantité, parce que beaucoup de personnes le mangent avec plaisir ; mais on le multiplie fort peu en Angleterre, si ce n'est dans de petits jardins, par curiosité, & pour nourrir la volaille : il y mûrit ordinairement bien.

Culture. On seme cette graine au commencement d'Avril, sur un sol chaud & sec ; mais pas trop épaisse, parce que ces plantes se divisent en plusieurs branches, & qu'elles exigent beaucoup de place. Quand elles commencent à pousser, on les débarasse de mauvaises herbes qui peuvent se trouver parmi elles ; mais ensuite

MIL

elles prennent le deſſus , empêchent de pouſſer ces mauvaiſes herbes. Ces graines mûriſſent en Août ; alors on les coupe & on les bat comme les autres graines. Quand elles commencent à mûrir , il faut les mettre à l'abri des oiſeaux, qui dévoreroient bientôt toute la récolte.

MILLEFEUILLE *V.* Achillea.

MILLEPERTUIS. *Voyez* Hypericum.

MILLERIA. *Houſt. Gen. Nov. Martyn. cent.* 4. *Lin. Gen. Plant.* 881. [*Milleria.*] Millerie.

Caractères. La fleur eſt compoſée de pluſieurs fleurons hermaphrodites , & d'un fleuron femelle , qui ſont tous renfermés dans un calice commun , formé par une feuille perſiſtante,& diviſée en trois parties. Les fleurons hermaphrodites ont une corolle tubulée , érigée , & découpée au bord en cinq parties , cinq étamines velues , avec des antheres érigées & linéaires, jointes dans leur milieu au côté de l'étamine , & de la longueur de la corolle , avec un germe oblong & étroit, qui ſoutient un ſtyle mince , couronné par deux ſtigmats étroits, obtus & étendus. Ces fleurons ſont ſtériles. Le fleuron femelle eſt monopétale : la corolle, qui s'étend au-dehors ſur un côté en forme de langue , eſt dentée au ſommet ; elle a un germe gros & triangulaire , qui ſoutient un ſtyle mince , couronné par deux ſtigmats longs & hériſſés. Ce germe ſe change enſuite en une ſemence oblongue , obtuſe,

MIL 103

triangulaire , & renfermée dans le calice.

Ce genre de plantes eſt rangé dans la quatrieme ſection de la dix-neuvieme claſſe de Linnée , qui comprend celles dont les fleurs ſont compoſées , & dont les fleurons hermaphrodites ſont ſtériles, & les femelles fructueuſes.

Les eſpeces ſont :

1°. *Milleria quinque flora , foliis cordatis , pedunculis dichotomis. Hort. Cliff.* 425. *Roy. Lugd.B.* 182 ; Millerie à feuilles en forme de cœur , & à tiges fourchues.

Milleria annua , erecta , major , foliis conjugatis , floribus ſpicatis , luteis.Houſt.MSS. ; le plus grand Millerie droit & annuel , avec des feuilles diſpoſées par paires , & des épis de fleurs jaunes.

Milleria maculata , foliis infimis cordato-ovatis , acutis , rugofis ; caulinis , lanceolato-ovatis , acuminatis; Millerie dont les feuilles du bas ſont ovales , rudes , en forme de cœur , & à pointe aiguë , & celles du haut ovales , en forme de lance, & pointues.

Milleria annua , erecta , ramofior , foliis maculatis , profundiùs ferratis. Martyn. Dec. 5 ; Millerie droit , annuel , & branchu , avec des feuilles tachetées & profondément ſciées.

3°. *Milleria biflora , foliis ovatis , pedunculis ſimpliciſſimis. Hort. Cliff.* 425. *t.* 25. *Hort. Ups.* 275. *Roy. Lugd.B.* 182. *Lœfl. it.* 239 ; Millerie avec des feuilles ovales & des pédoncules ſimples.

Milleria annua , erecta , minor , foliis Parietariæ , floribus ex ſo

104 MIL   MIL

liorum alis. Houft. MSS. ; le plus petit Millerie droit & annuel, avec des feuilles de Pariétaire, & des fleurs qui s'uniffent aux aîles des feuilles.

4°. *Milleria triflora, foliis ovato-lanceolatis, acuminatis, trinerviis, pedunculis alaribus, trifloris* ; Millerie avec des feuilles ovales, en forme de lance, à pointe aiguë & à trois veines, des pédoncules qui fortent des aîles des feuilles, & trois fleurs.

Milleria annua, erecta, foliis Parietariæ longioribus, floribus ex foliorum alis. Édit. Prior; Millerie annuel & érigé, avec une plus longue feuille de Pariétaire, & des fleurs placées aux aîles des feuilles.

Quinque-flora. La première efpece a été découverte à Campêche par le Docteur WILLIAM HOUSTOUN, qui a envoyé, en 1731, fes femences en Europe. Comme les caracteres qui la diftinguent étoient différens de tous ceux des autres genres de la claffe à laquelle elle appartient, on en a conftitué un genre particulier fous ce titre.

Elle s'éleve à la hauteur de quatre, cinq ou fix pieds, avec une tige branchue, & garnie de feuilles en forme de cœur, d'environ quatre pouces de longueur fur trois de largeur vers leur bâfe, terminées en pointe à l'extrémité, légerement fciées fur leurs bords, & fortifiées de chaque côté de la côte du milieu par deux nervures qui s'écartent d'abord, fe joignent près de leur bâfe, & fe rencontrent à la pointe, qui eft ordinairement oblique au pé-

tiole. Ces feuilles font d'un vert léger, velues & oppofées; leurs pétioles ont à-peu-près un pouce de longueur, & font ornés de chaque côté par un prolongement de la feuille en forme d'aile. Les tiges fe divifent en fourche vers le haut, & les pédoncules qui fortent des divifions, fe fous-divifent encore par paires, & font terminés par des épis clairs de fleurs jaunes, compofées de quatre ou cinq fleurons hermaphrodites, ftériles, & d'un demi-fleuron femelle, auquel fuccede une femence fimple, oblongue, angulaire, & enveloppée dans le calice de la fleur. Cette plante fleurit dans les mois de Juillet & Août, & fes femences mûriffent en automne.

Sparfum. La feconde efpece, qui a été découverte à Campêche par ROBERT MILLAR, en l'année 1734, a quelque reffemblance avec la première : mais fes tiges s'élevent à fix ou fept pieds de hauteur; fes branches s'étendent fort loin de tous côtés ; fes feuilles ont fept pouces de longueur fur quatre & demi de largeur vers leur bâfe, & font terminées en pointe aiguë ; elles font plus profondément fciées fur leurs bords, & ont plufieurs taches larges, noires, & écartées en-deffus; leur furface eft plus rude, & d'un vert plus foncé que celles de la première efpece; les feuilles du haut font longues, & en forme de lance; les pédoncules font branchus, & s'étendent plus au-dehors : les épis de fleurs font plus courts que ceux de la précédente.

MIL

Biflora. La troisieme, que le Docteur HOUSTOUN a encore trouvée à Campêche, est aussi une plante annuelle, qui s'éleve au-dessus de deux pieds de hauteur, avec une tige herbacée, & divisée en branches un peu au-dessus de sa racine; elle forme trois ou quatre tiges minces & nues presque jusqu'au sommet, où elles ont deux feuilles ovales en forme de lance, opposées, d'environ deux pouces de longueur sur neuf lignes de largeur vers leur bâse, terminées en pointe, velues, rudes, légerement dentelées sur leurs bords, fortifiées par trois veines longitudinales, & supportées par des pétioles nuds & d'environ un pouce de longueur : ces fleurs naissent en petites grappes aux pétioles des feuilles; leur calice commun est composé de trois feuilles orbiculaires, serrées ensemble, dans chacune desquelles sont placées deux fleurons hermaphrodites stériles, & un demi-fleuron femelle fructueux, auquel succede une semence ronde, angulaire, & renfermée dans le calice. Cette plante fleurit & perfectionne ses semences vers le même tems que la précédente.

Confertum. La quatrieme espece a été découverte à Campêche par ROBERT MILLAR; elle est annuelle, & s'éleve à la hauteur de trois ou quatre pieds, avec une tige droite, & garnie dans toute sa longueur de feuilles ovales en forme de lance, & de quatre pouces à-peu-près de longueur sur presque deux de largeur à leur bâse; elles ont trois veines longitudinales, & vers le sommet il y en a deux de plus, qui s'écartent de la côte du milieu, & se rejoignent à la pointe; la surface supérieure de ces feuilles est d'un vert foncé & unie, & l'inférieure est d'un vert pâle; elle sont dentées sur leurs bords. Les fleurs, qui naissent en petites grappes aux aîles des feuilles, renferment chacune trois fleurons hermaphrodites, & un femelle; ils sont portés sur de courts pédoncules, & ont des calices semblables à ceux de la précédente, mais beaucoup plus petits. Cette espece fleurit & perfectionne ses semences plus tard qu'aucune des précèdentes, de sorte qu'elles ne mûrissent qu'autant que les plantes sont poussées de bonne heure au printems.

Les semences de celle-ci doivent être répandues au commencement du printems sur une couche de chaleur modérée. Quand les plantes ont poussé, & qu'elles ont atteint la hauteur de deux pouces, on les met chacune séparément dans des pots remplis d'une terre riche & légere : on les plonge dans une couche de tan de chaleur modérée; on les tient à l'ombre, jusqu'à ce qu'elles aient formé de nouvelles racines, & on les arrose souvent. Lorsqu'elles sont bien enracinées, on leur procure beaucoup d'air, en soulevant les vitrages de la couche chaque jour dans les tems chauds, & on les arrose

MIM

souvent pendant les chaleurs, parce qu'elles ont befoin d'une abondante humidité. Au moyen de ce traitement, elles s'éleveront à une hauteur confidérable, un mois après qu'elles auront été tranfplantées: c'eftpourquoi il faudra les mettre alors dans de grands pots, & les placer dans la ferre chaude, où on les plongera dans la couche de tan, afin qu'elles puiffent avoir affez de place pour croître, fur-tout les premiere & feconde efpeces, qui deviennent ordinairement hautes & pouffent beaucoup de branches, quand elles font bien traitées; mais comme les autres ne s'élevent guere qu'à trois ou quatre pieds de hauteur, & n'étendent pas leurs branches fort loin, elles exigent moins de place.

Au milieu de Juillet, ces plantes commenceront à fleurir, & leurs femences mûriront un mois ou fix femaines après. Lorfque ces graines commencent à devenir d'un brun foncé, il faut les recueillir fans différer, parce qu'alors elles fe répandent d'elles-mêmes, & tombent par la moindre fecouffe. Ces plantes continuent à fleurir jufqu'à la Saint-Michel, & même plus tard, fi la faifon eft favorable; mais elles périffent aux approches des premiers froids de l'automne.

MILLET *Voyez* MILIUM.

MILLET DES INDES. *Millet noir* ou *Sorghum. Voyez* HOLcus SORGHUM.

MILLET NOIR, *grand. Voy.* HOLCUS.

MIMOSA: *Tourn. Inft. R. H.*

MIM

605. *tab.* 375. *Lin. Gen. Plant.* 597. [*The fenfitive Plant.*] Plante fenfitive. Acacia véritable.

Caractères. Le calice de la fleur eft petit & formé par une feuille découpée au fommet en cinq parties; la corolle eft monopétale, en forme d'entonnoir, & a cinq pointes: la fleur a plufieurs étamines longues, velues, & terminées par des antheres penchées, avec un germe oblong, qui foutient un ftyle court, mince, & couronné par un ftigmat difforme. Ce germe fe change, quand la fleur eft paffée, en un légume long, noueux, & divifé par plufieurs partitions tranfverfales, qui renferment des femences comprimées de différentes formes. Quelques-unes des efpeces ont plufieurs fleurs mâles, femelles & hermaphrodites mêlées enfemble.

Ce genre de plantes a été joint par LINNÉE, à l'*Acacia* de TOURNEFORT, & à l'*Inga* de PLUMIER. Il l'a placé dans la premiere fection de fa vingttroifieme claffe, qui renferme celle squi ont des fleurs mâles, femelles & hermaphrodites fur la même plante, avec plufieurs étamines & un ftyle.

Les efpeces font:

1°. *Mimofa punctata inermis, foliis bipinnatis, fpicis, erectis, floribus decandris, inferioribus caftratis, corollatis, caule erecto, t ereti. Lin. Sp.* 1502; Plante fenfitive, fans épines, avec des feuilles ailées, des fleurs à dix étamines, dont celles du bas des épis n'en ont point, ayant une tige érigée & cylindrique.

Mimofa frutefcens inermis, fili-

MIM

quis compreſſis , falcatis , umbellatis , pedunculo longiſſimo. *Brown. Jam.* 253.

Æſchynomene mitis prima. Comm. Hort. 1. *p.* 61. *f.* 31.

2º. *Mimoſa plena inermis , foliis bipinnatis , ſpicis pentandris , inferioribus plenis. Hort. Upſal.* 145 ; Mimoſa ſans épines , avec des feuilles doublement ailées , des épis de fleurs à cinq étamines , & ceux du bas à fleurs doubles.

Mimoſa non ſpinoſa , paluſtris et herbacea, procumbens , flore luteo,pleno. Houſt. MSS. : Mimoſa de marais , & herbacée , traînante , & ſans épines , avec une double fleur jaune.

3º. *Mimoſa Fernambucana,inermis , decumbens , foliis bipinnatis , ſpicis cernuis , pentandris , inferioribus caſtratis. Hort. Upſal.* 145 ; Mimoſa ſans épines , avec des tiges penchées , des feuilles doublement ailées, des épis courbées , & des fleurs à cinq étamines , dont celles du bas n'en ont point.

Mimoſa inermis , foliis duplicato-pinnatis , ſiliquis linearibus , glabris. Hort. Cliff. 209.

Mimoſa Americana pigra , ſiliquis longis , anguſtis , Allium olentibus. Pluk. Alm. 552. *t.* 307. *f.* 3.

Mimoſa ſpuria Fernambucana , dicta Mimoſa Italica. Zan. Hiſt. 151 ; Mimoſa bâtarde , appelée Mimoſa d'Italie , ou Acacia de Fernambour.

4º. *Mimoſa pudica , aculeata , foliis pinnatis ;* Senſitive piquante & à feuilles ailées.

Mimoſa herbacea , procumbens & ſpinoſa , caule tereti & villoſo , ſiliquis articulatis. Houſt. MSS. ; Senſitive herbacée &

traînante , avec des épines , une tige cylindrique & velue , & des légumes noueux.

5º. *Mimoſa pudica , foliis ſubdigitatis , pinnatis , caule aculeato , hiſpido. Lin. Sp.* 1501 ; Senſitive avec une feuille ailée , une tige épineuſe & penchée , de petits légumes en grappes , & des enveloppes ou coſſes épineuſes.

Mimoſa ſpinoſa 3. *ſiliquis parvis echinatis. Breyn. cent.* 40. *t.* 18.

Mimoſa humilis fruteſcens & ſpinoſa , ſiliquis conglobatis. Plum. Cat. ; Senſitive épineuſe en arbriſſeau bas , avec des légumes en grappes , communément appelée la plante Senſitive baſſe.

Æſchinomene ſpinoſa , flore globoſo albido , ſiliculis articulatis , echinatis. Comm. Hort. 1. *p.* 57. *f.* 29.

6º. *Mimoſa quadri-valvis , aculeata , foliis bipinnatis , caule quadrangulo , aculeis recurvis , leguminibus quadri-valvibus. Lin. Sp. Plant.* 1508 ; Senſitive épineuſe , avec des feuilles doublement ailées , une tige quarrée , des épines recourbées , & des légumes à quatre valves.

Mimoſa herbacea procumbens & ſpinoſa , caule quadrangulo , ſiliquis quadri-valvibus. Houſt. MSS. Senſitive traînante , épineuſe & herbacée , avec une tige quadrangulaire , & des légumes à quatre valves.

7º. *Mimoſa ſenſitiva , foliis conjugatis , pinnatis , partialibus , bijugis , intimis minimis , caule aculeato. Lin. Sp. Plant.* 1501 ; Senſitive à feuilles conjuguées ,

ailées, & à deux paires de lobes, dont les intérieurs sont plus petits, avec une tige épineuse.

Mimosa spinosa prima , sivè Brasiliana lati-folia , siliquis radiatis. Breyn. cent. 1. 31. Trew. Ehrt. f. 95 ; premiere plante sensitive épineuse , à larges feuilles du Bresil , à légumes radiés. Sensitive veritable.

8°. *Mimosa asperata , caule fruticoso, foliis bipinnatis , aculeatis , aculeis geminis , siliquis radiatis , hirsutis. Fig. Plant. tab. 182. fol. 3.* ; Sensitive avec une tige en arbrisseau , des feuilles à doubles ailes, & épineuses, dont les épines croissent par paires, & des légumes velus & en rayons.

Æschynomene spinosa quarta, sivè foliolis Acaciæ angustioribus, frondibus validissimas spinas habentibus. Breyn. cent. 1. 43. t. 19 ; quatrieme Æschynomene épineuse , avec des folioles d'Acacia très-étroites, & armées de fortes épines.

Æschynomene spinosa quinta. Commel. Hort. 1. p. 59. f. 30.

9°. *Mimosa viva inermis , foliis conjugatis , pinnatis , partialibus , quadri-jugis , sub-rotundis , caule inermi herbaceo. Lin. Sp. 1500* ; Sensitive avec une tige rampante , herbacée, & sans épines , des feuilles ailées & conjuguées, & des fleurs globulaires qui sortent des aisselles des tiges.

Mimosa herbacea non spinosa, minima , repens. Sloan. Hist. Jam. 2. p. 58 ; Sensitive herbacée, très-petite, rampante , & sans épines.

10°. *Mimosa Nilotica , spinis stipularibus , patentibus foliis bipinnatis , partialibus , extimis , glandula interstinctis , spicis globosis , pedunculatis. Hasselq. it. 475. Mat. Med. 221. Fabric. Helmst. 327. Blackw. f. 377.* ; Mimosa avec des feuilles à doubles ailes , des épis globulaires, des pédoncules aux fleurs , & des glandes aux pétioles.

Acacia Ægyptiaca. Hern. Mex. 866 ; Acacia d'Egypte.

11°. *Mimosa Farnesiana , spinis stipularibus distinctis , foliis bipinnatis , partialibus , octo-jugis , spicis globosis , sessilibus. Hort. Upsal. 146.*

Acacia Indica , foliis Scorpioïdis leguminosæ , siliquis fuscis , teretibus , resinosis. G. L. ; Acacia des Indes, avec des légumes cylindriques, bruns & résineux. Casse ou Acacia de Farnese.

12°. *Mimosa cornigera , spinis stipularibus geminis connatis, foliis bipinnatis. Hort. Cliff. 208. Roy. Lugd.-B. 470. Jacq. Amer. 266* ; Acacia avec deux épines jointes à leur bâse , & des feuilles doublement ailées.

Acacia cornuta Indiæ Orientalis. Seb. Thes. 1. p. 113. t. 70. f. 13.

Acaciæ similis Mexicana , spinis cornu similibus. Comm. Hort. 1. p. 209. f. 107 ; le grand Acacia cornifere, ou l'Acacia cornu.

13°. *Mimosa unguis—cati ; spinosa , foliis bigeminis obtusis. Hort. Cliff. 207. Roy. Lugd.-B. 470* ; Acacia épineux à quatre feuilles obtuses.

Acacia quadrifolia , siliquis cincinnatis. Plum. ic. 4.

Mimosa fruticosa , foliis ovatis binato-binatis ,

M I M

binatobinatis, seminibus compressis, atro-nitentibus, flosculis rubellis adnatis. Brown. Jam. 252.

Acacia arborea major spinosa, pinnis quatuor majoribus sub-rotundis, siliquis variè intortis. Sloan. Jam. 152. Hist. 2. p. 56.

Acaciæ quodam modo accedens, sivè Ceratiæ & Acaciæ media, Jamaïcensis, spinosa, bi-geminatis foliis, flosculis stamineis, atro-nitente fructu, siliquis intortis. Pluk. Phyt. 1. f. 6. ; Acacia avec des feuilles disposées par paires, des tiges épineuses, de petites fleurs à étamines, un fruit d'un brun clair, & des siliques torses, communément appelé Griffe de Chat.

14°. *Mimosa arborea inermis, foliis bipinnatis, pinnis dimidiatis, acutis, caule arboreo.* Lin. Sp. 1503 ; Acacia en arbre & sans épines, avec des feuilles doublement aîlées, dont les lobes sont à pointe aiguë.

Acacia non spinosa Jamaïcensis, foliis latâ basi in metæ formam fastigiatis. Pluk. Alm. 6. t. 251. f. 2.

Acacia arborea maxima, non spinosa, pinnis majoribus, flore albo, siliquâ contortâ, coccineâ, ventricosâ, elegantissimâ. Sloan. Jam. 157.

15°. *Mimosa purpurea, inermis, foliis conjugatis, pinnatis, foliis infimis minoribus.* Lin. Sp. 500 ; Acacia pourpre & sans épines, avec des feuilles conjuguées, aîlées, & plus petites vers le bas.

Acacia Americana frutescens, non aculeata, flore purpurascente. Plum. Cat. ; Acacia en arbrisseau, d'Amérique, sans épines, & à fleurs pourpre.

Tome V.

M I M 109

16°. *Mimosa Houstoniana, inermis, foliis bipinnatis, glabris, pinnis tenuissimis, siliquis latis, villosis.* Fig. Pl. 5 ; Acacia sans épines, avec des feuilles unies, & doublement aîlées, dont les lobes sont fort étroits, & les siliques larges & velues.

Acacia Americana, non spinosa, flore purpureo, staminibus longissimis, siliquis planis, villosis, pinnis foliorum tenuissimis. Houst. MSS. ; Acacia d'Amérique sans épines, avec une fleur pourpre, de très-longues étamines, des siliques plates & velues, & des lobes très-étroits aux feuilles.

Gleditsia inermis. Lin. Sp. Plant. 1509. edit. 3.

17°. *Mimosa lutea aculeata, foliis bipinnatis glabris, floribus globosis pedunculatis, aculeis longissimis* ; Acacia épineux, avec des feuilles unies & doublement aîlées, des fleurs globulaires, postées sur des pédoncules, & de fort longues épines.

Acacia spinosa, foliorum pinnis tenuissimis, glabris, floribus globosis, luteis, spinis longissimis. Houst. MSS. ; Acacia épineux avec des feuilles fort étroites & unies, des fleurs rondes & jaunes, & des épines très-longues.

18°. *Mimosa glauca, inermis, foliis bipinnatis, partialibus, sex-jugis, pinnis plurimis, glandulâ inter infima.* Lin. Sp. Plant. 1502 ; Acacia sans épines, avec des feuilles doublement aîlées, dont les aîles sont séparées, & ont de petits lobes entr'elles, & des glandes vers le bas.

Acaciæ similis Americana non

H

fpinofa, *floribus globofis*, *albis*, *ramofis. Kigg. Beaum.* 3.

Acacia non fpinofa, *flore albo*, *foliorum pinnis latiufculis*, *glabris*, *filiquis longis*, *planis. Houft. MSS. Trew. Ehret. f.* 36 ; Acacia à fleurs blanches & fans épines, ayant les lobes des feuilles larges & liffes, & des filiques longues & plates. Senfitive pareffeufe.

19°. *Mimofa anguftiffima*, *inermis*, *foliis bipinnatis*, *pinnis anguftiffimis*, *glabris*, *leguminibus tumidis* ; Acacia fans épines, avec des feuilles doublement ailées, dont les lobes font unis & très-étroits, & les légumes gonflés.

Acacia non fpinofa, *floribus globofis*, *albis*, *foliorum pinnis tenuiffimis*, *glabris*, *filiquis ad fingula grana tumidis. Houft. MSS.* ; Acacia fans épines, avec des fleurs globulaires & blanches, des lobes unis & très-étroits, & des filiques noueufes.

20°. *Mimofa Campeachiana*, *fpinofa*, *foliis bipinnatis*, *pinnis anguftis*, *fpinis fingulis cornu bovinum per longitudinem fiffum referentibus* ; Acacia avec des feuilles doublement ailées, dont les lobes font étroits, des épines féparées en forme de corne de bœuf, & divifées dans leur longueur.

Acacia fpinofa, *tenui-folia*, *fpinis fingulis cornu bovinum per longitudinem fiffum referentibus. Houft. Cat.* ; Acacia épineux, à petites feuilles, dont les épines font fimples, en forme de corne de bœuf, que l'on auroit fendue en deux.

21°. *Mimofa cinerea*, *fpinis*

folitariis, *foliis bipinnatis*, *floribus fpicatis. Flor. Zeyl.* 215 ; Acacia avec des épines folitaires, des feuilles doublement ailées, & des fleurs en épis.

Acacia fpinofa, *tenui-folia*, *filiquis latis*, *fpinis minimis*, *recurvis*, *folitariis. Houft. Cat.* Acacia épineux, à feuilles étroites, avec des légumes larges, & de petites épines recourbées & folitaires.

22°. *Mimofa lati-folia*, *inermis*, *foliis conjugatis*, *pinnis terminalibus oppofitis*, *lateralibus alternis. Lin. Sp.* 1499 ; Acacia fans épines, avec des feuilles conjuguées, dont les lobes du haut font oppofés, & ceux de côté alternes.

Acacia non fpinofa, *juglandis folio*, *flore purpurafcente. Plum. Sp.* 17. *ic.* 9. ; Acacia fans épines, à feuilles de Noyer, & à fleur pourpre.

23°. *Mimofa circinalis*, *aculeata*, *foliis conjugatis*, *pinnatis*, *pinnis æqualibus*, *ftipulis fpinofis. Lin. Sp.* 1499 ; Acacia épineux, avec des feuilles conjuguées & ailées, dont les lobes font égaux, & des ftipules épineufes.

Acacia foliis amplioribus, *filiquis circinatis. Plum. Sp.* 17. *ic.* 5 ; Acacia à larges feuilles & des filiques roulées.

24°. *Mimofa Fagi-folia*, *inermis*, *foliis pinnatis bijugis*, *petiolo marginato. Lin. Sp.* 1498. *Jacq. Amer.* 264. *t.* 164 ; Acacia à feuilles de Hêtre, & fans épines, dont les ailes ont quatre lobes, & dont les pétioles font ailés.

Arbor filiquofa, *Faginis foliis*, *Americana*, *floribus comofis. Pluk.*

M I M

Phyt. tab. 141. *fol.* 2 ; Poix doux. *Faba dulcis. Jacq.*

Punctata. La premiere espece croît naturellement dans la plupart des Isles des Indes Occidentales , où elle a été trouvée dans des lieux chauds & humides, sous la même latitude que la Virginie ; elle s'élève à la hauteur de six ou sept pieds, avec des tiges droites, branchues & ligneuses vers la racine, quoiqu'elles ne soient point vivaces, au moins dans notre pays, où elles ne passent pas l'hiver, dans quelque situation qu'on puisse les mettre. Ces tiges sont lisses & garnies de feuilles doublement ailées , & composées de quatre à cinq paires de lobes longs & ailés, qui ont environ vingt paires de folioles rangées dans la longueur de la côte du milieu. Ces feuilles sont lisses, rondes à leur extrémité, d'un beau vert en-dessus, & pâles en-dessous. Les petites feuilles se rapprochent quand on les touche ; mais leurs pétioles ne se courbent pas comme ceux des plantes plus tendres. On donne à cette espece le nom de Sensitive , pour la distinguer : ses fleurs naissent sur de longs pédoncules, qui sortent des ailes des pétioles ; elles sont disposées en têtes globulaires, qui penchent vers le bas , & de couleur jaune : celles qui ont des pétales renferment dix étamines ; celles qui sont au-dessus des épis, sont remplacées par des légumes d'un pouce & demi de longueur, sur un quart de pouce de largeur , d'un brun foncé , lorsqu'ils sont mûrs , & qui renferment trois ou quatre semences luisantes , noires , & comprimées.

Plena. La seconde a été découverte par le Docteur Houstoun , à la Vera-Cruz, où il l'a trouvée dans des eaux croupissantes : ses tiges sont fort larges & plates ; elles nàgent sur la surface des eaux comme les mauvaises herbes des étangs, mais dans les endroits où l'eau est desséchée & tarie : ses tiges sont rondes & érigées, ce qui leur arrive aussi lorsqu'on les cultive dans des jardins , de sorte qu'on les croiroit différentes , si l'on n'en étoit pas prévenu.

Cette espece , cultivée dans les jardins, ressemble beaucoup à la premiere : mais ses tiges ne sont jamais si droites ; les ailes des feuilles sont plus longues & plus horisontales : ses têtes de fleurs sont plus grosses ; les étamines sont plus longues, & les fleurs du bas des épis , qui n'ont point d'étamines , sont un peu plus grosses. Les légumes de cette espece sont plus courts , & beaucoup plus larges que ceux de la premiere. Cette plante est aussi annuelle dans ce pays ; elle a été, depuis peu, rencontrée , par un de mes amis, dans des campagnes marécageuses de l'Isle des Barbades , d'où il m'en a envoyé des semences, avec une grande branche de la plante, dans une bouteille de verre remplie d'eau imprégnée de sel , qui l'a conservée dans le même état où elle étoit quand on l'avoit cueillie avec ses fleurs & ses légumes.

H 2

Fernambucana. La troisieme croît aussi, sans culture, dans toutes les Isles des Indes Occidentales, où on lui donne le nom de *Senfitive pareſſeuſe*, parce que ſes feuilles n'ont point de ſentiment, quand on les touche: ſes tiges ne s'élevent guere qu'à la hauteur de deux pieds & demi; elles ſont liſſes, & garnies de feuilles doublement ailées, & compoſées de trois ou quatre paires d'ailes plus courtes; ſes lobes ſont beaucoup plus étroits que ceux des eſpeces précédentes; ſes têtes de fleurs ſont plus petites, & ſes légumes plus longs & plus étroits que ceux de la ſeconde.

Cette eſpece ſubſiſte en hiver, au moyen d'une chaleur modérée.

Pudica. La quatrieme, qui a été découverte à la Vera-Cruz par le Docteur HOUSTOUN, a une tige ligneuſe, inclinée vers la terre, & de deux ou trois pieds de hauteur, qui pouſſe latéralement pluſieurs branches armées d'épines courtes & jaunes, placées ſous les pétioles des feuilles, & hériſ—fées dans toute leur longueur de poils aigus : ſes pétioles, qui ont trois pouces de longueur, ſont terminées par quatre à cinq feuilles ailées, réunies en un point à leur baſe, & écartées à leur extrémité, comme les doigts d'une main-ouverte. Ces ailes ont environ deux pouces de longueur, & ſont formées par un grand nombre de petits lobes étroits & diſpoſés par paires dans la longueur de la côte du milieu, qui eſt cou-

verte en deſſous, ainſi que les tiges, de poils hériſſés : ſes fleurs, qui ſont d'un jaune pâle, réunies en têtes globulaires, ſortent des ailes des pétioles ſur des pédoncules aſſez longs, & ſont remplacées par des légumes petits & pointus, qui contiennent deux ou trois ſemences d'un noir luiſant.

Pudica foliis ſub-digitatis. La cinquieme eſt la plus commune de toutes dans les Isles des Indes Occidentales, ainſi que dans les jardins anglois. Les Marchands de graines vendent ſouvent ſes ſemences ſous le nom de *Plante humble :* ſes racines ſont compoſées d'un grand nombre de fibres entrelacées, deſquelles ſortent pluſieurs tiges ligneuſes & inclinées vers la terre, à moins qu'on ne leur fourniſſe un ſoutien. Ces tiges ſont armées d'épines courtes, recourbées, & garnies de feuilles ailées, compoſées de quatre & quelquefois de cinq ailes, dont les bâſes ſe joignent en un point où elles ſont inſérées aux pétioles, & s'étendent vers le haut comme les doigts de la main. Ces ailes ſont plus courtes que celles de l'eſpece précédente, & les tiges ne ſont pas garnies de poils : les fleurs qui naiſſent aux aiſſelles de la tige, ſur de courts pédoncules, ſont de couleur jaune, & rapprochées en petites têtes globulaires ; elles produiſent des légumes courts, plats & noueux, qui renferment chacun deux ou trois ſemences orbiculaires, bordées & comprimées. Ces légumes ſont diſpoſés en grappes ſerrées, &

M I M

font couverts de poils aigus.

Quadrivalvis. La fixieme efpece croit naturellement à la Vera-Cruz, d'où le Docteur HOUSTOUN a envoyé fes femences ; elle a une racine rampante & vivace, qui s'étend & fe multiplie confidérablement dans les fables, où on la trouve fauvage : fes tiges font minces, à quatre angles aigus, & fortement armées d'épines courtes & recourbées : fes feuilles font fupportées par des pétioles longs, piquans, & écartés les uns des autres fur les branches ; elles font compofées de deux paires d'ailes, placées à un pouce environ de diftance : les ailes font courtes, & leurs petits lobes font étroits, & moins rapprochés les uns des autres que dans plufieurs autres efpeces ; les pédoncules fortent des ailes des pétioles, & foutiennent de petites têtes globulaires de fleurs pourpre, auxquelles fuccedent des légumes quarrés de deux pouces de longueur, à quatre valves, & à quatre cellules, qui contiennent chacune plufieurs femences angulaires.

Cette efpece étend fi fort fes racines, qu'elle ne peut produire autant de fleurs & de femences que la plupart des autres. Les plantes que l'on fe procure en divifant ces racines font toujours foibles ; de forte que la meilleure maniere de la multiplier eft par femences, lorfqu'on peut s'en procurer. C'eft une des efpeces dont les pétioles tombent ou fe retirent lorfqu'on les touche.

Senfitiva. La feptieme, qu'on

M I M 113

rencontre auffi à la Vera Cruz, d'où le Docteur HOUSTOUN a également envoyé fes femences, s'éleve à la hauteur de fept ou huit pieds, en tiges minces, ligneufes, & armées d'épines courtes & recourbées ; fes feuilles naiffent fur des pétioles piquans, qui foutiennent chacun deux paires d'ailes, dont les extérieures, qui ont deux lobes réunis à leurs bafes, font arrondies en-dehors, & droites intérieurement ; ce qui leur donne la forme de cifeaux à tondre les moutons. Ces paires de lobes extérieurs font beaucoup plus larges que les intérieurs ; ils ont près de deux pouces de longueur fur un de largeur au milieu : de l'endroit où ils font inférés à la tige, fortent de petites branches qui produifent trois ou quatre têtes globulaires chargées de fleurs d'un pourpre pâle, & placées de côté fur de courts pédoncules : la tige principale porte vers fon extrémité, dans la longueur d'un pied, plufieurs têtes de fleurs ; elle eft terminée, ainfi que les branches, par d'autres têtes de fleurs, auxquelles fuccedent des légumes larges, plats & noueux, qui s'ouvrent en quatre valves, & renferment une, deux, & quelquefois trois femences orbiculaires & comprimées. Les feuilles de cette efpece fe meuvent très lentement lorfqu'on les touche ; mais les pétioles tombent, ou fe renverfent promptement, lorfqu'on les preffe.

Afperata. La huitieme a encore été trouvée par le même Docteur HOUSTOUN, à la Vera-

H 3

Cruz ; elle a une tige d'arbriſſeau droite, de cinq pieds environ de hauteur, velue & armée d'épines fortes, courtes, larges, blanches, placées de chaque côté, quelquefois oppoſées, & d'autres fois alternes. Les feuilles ſont compoſées de cinq ou ſix paires d'ailes oppoſées, & rangées dans la longueur d'une forte côte ; entre chaque paire ſont placées deux épines courtes & fortes, dont les pointes ſe préſentent à chaque côté ; les petites feuilles ou lobes qui compoſent ces ailes, ſont extrêmement étroites, & placées très-près les unes des autres : les fleurs ſortent vers le ſommet de la tige ſur des courts pédoncules ; elles ſont d'un pourpre clair, & rapprochées en têtes globulaires ; les tiges ſont auſſi terminées par des têtes plus petites de fleurs ſemblables, qui ſont ſuivies par des légumes noueux d'environ deux pouces de longueur ſur un quart de pouce de largeur, qui s'écartent par le haut en forme de rayons, & ſont réunis par leurs bâſes au pédoncule, au nombre de cinq ou ſix. Ces légumes ſe ſéparent à chaque articulation, laiſſant les deux membranes élevées ; & les ſemences, qui ſont comprimées & quarrées, tombent de chaque nœud. Ces légumes ſont d'abord velus, & deviennent liſſes en mûriſſant.

Cette plante eſt vivace, & peut être conſervée en hiver dans la ſerre chaude. Cette méthode eſt la ſeule qu'on puiſſe employer pour en obtenir des ſemences mûres. Cette eſpece

fleurit rarement dans la premiere année : ſes pétioles ne tombent pas quand on les touche ; mais les petites feuilles des ailes ſont ſenſibles.

Viva. La neuvieme croît naturellement à la Jamaïque ; elle a des tiges traînantes & herbacées, qui pouſſent des racines à chaque nœud : ces racines pénetrent dans la terre, & s'étendent à une grande diſtance. La même choſe lui arrive en Angleterre, lorſqu'elle eſt placée dans la couche de tan. J'ai poſſédé ici une ſeule plante de cette eſpece, qui s'eſt étendue à près de trois pieds en quarré pendant un été ; ſes branches étoient ſi ſerrées & ſi épaiſſes, qu'elles couvroient toute la ſurface de la couche : mais lorſqu'on lui donne ainſi la liberté de s'étendre, elle produit rarement des fleurs ; ſes tiges, qui n'ont point d'épines, ſont toutes garnies de feuilles ailées, compoſées de deux paires d'ailes courtes, dont les petits lobes ſont étroits, liſſes & placés ſur de courts pétioles. Les feuilles de cette eſpece ſe retirent ſortement ſi peu qu'on les touche ; de ſorte que dans les endroits où elle s'eſt développée, on peut, avec un bâton, tracer ſur ſes feuilles, telle figure que ce ſoit ; & cette figure reſtera viſible juſqu'à ce que les feuilles ſe ſoient redreſſées. Les fleurs de cette plante naiſſent aux ailes des pétioles, ſur des pédoncules nuds, & d'environ un pouce de longueur ; elles ſont d'un jaune pâle, rapprochées en une petite tête globulaire, & produiſent des

MIM

lègumes courts, plats & noueux, qui renferment trois ou quatre femences rondes & comprimées.

Culture. On multiplie toutes ces plantes par leurs graines, qu'il faut femer, au commencement du printems, fur une bonne couche chaude. Si ces graines font fraîches, les plantes paroîtront au bout de quinze jours ou trois femaines, & exigeront enfuite d'être traitées avec beaucoup de foin. Il ne faut pas trop les arrofer ni les laiffer s'affoiblir en filant; mais il eft néceffaire de leur donner de l'air frais dans tous les tems, lorfque la faifon eft tempérée : environ quinze jours ou trois femaines après que les plantes auront paru, elles feront en état d'être transplantées, furtout fi la couche dans laquelle elles ont été femées, a été tenue a un dégré de chaleur convenable; alors on prépare, pour les recevoir, une nouvelle couche chaude, qui doit être faite une femaine avant d'en faire ufage, de façon que fa trop grande chaleur foit diffipée avant de la charger de terre, & que cette terre foit échauffée pour y mettre les plants : on enleve enfuite les plantes, en confervant leurs racines entieres, & on les place tout de fuite dans la nouvelle couche à trois ou quatre pouces de diftance, en preffant un peu la terre fur leurs racines : on les arrofe légerement, pour les joindre à la terre; on les tient à l'ombre jufqu'à ce qu'elles aient pouffé de nouvelles fibres, & l'on conferve les vitrages de la couche pendant les nuits, pour y con-

MIM

ferver la chaleur. Lorfque ces plantes font enracinées, on les arrofe fréquemment, mais avec moderation; on leur donne de l'air chaque jour, à proportion de la chaleur du tems, pour les empêcher de filer, & on les tient conftamment à un dégré de chaleur modérée, fans quoi elles ne feroient prefque point de progrès. Environ un mois après, fi ces plantes font affez fortes, on les enleve avec précaution, en confervant à leurs racines autant de terre qu'il eft poffible, & on les met, chacune féparement, dans de petits pots remplis d'une bonne terre de jardin potager; on les plonge dans une bonne couche chaude de tan; on les tient à l'ombre jufqu'à ce qu'elles aient formé de nouvelles racines, & on les traite enfuite de la même maniere que les autres plantes tendres & exotiques des pays très-chauds.

Comme les efpeces droites & hautes s'élevent bientôt affez pour atteindre les vitrages de la couche, fur-tout lorfqu'elles font de bons progrès, il faut les transplanter dans de larges pots, & les placer dans la ferre chaude, où elles avanceront beaucoup, fi on les y tient plongées dans la couche. La premiere efpece fleurit fouvent ici, fi on l'éleve de bonne heure au printems, & fi l'on hâte fes progrès fur plufieurs couches. J'en ai obtenu deux ou trois fois des femences mûres; mais on ne peut en efpérer de pareilles que dans les années chaudes.

Les efpeces vivaces fubfiftent

en hiver dans des serres chaudes, & donnent des fleurs & des semences mûres l'été suivant. Quelques - unes peuvent être multipliées par marcottes, que l'on fépare des vieilles plantes, lorfqu'elles ont pris racine : je les ai auffi multipliées quelquefois par boutures ; mais les plantes élevées de femences font préférables à toutes les autres.

Ces plantes n'exigent aucune culture particuliere & différente de celle que l'on emploie pour les autres efpeces des pays chauds ; mais il faut toujours les tenir à un dégré de chaleur convenable, & ne pas trop les arrofer, fur-tout dans les tems froids : on doit auffi éviter de les tenir trop feches, parce que plufieurs de ces efpeces, qui périffent naturellement dans des lieux humides, ont befoin de fréquens arrofemens : il faut auffi empêcher que leurs racines ne filtrent & ne s'échappent dans la couche de tan, par les trous des pots ; ce qui les feroit pouffer très-vigoureufement, & les expoferoit à périr, lorfqu'en les enlevant enfuite, on viendroit à rompre ces racines. Pour prévenir cet accident, il faut foulever de tems en tems les pots qui font plongés dans le tan ; & fi quelques racines commencent à paffer par les trous, on les coupe tout près. Lorfque ces racines font entrelacées & trop ferrées, on les tire hors des pots, pour les raccourcir, & on les met dans d'autres de même grandeur, ou de plus grands, fi les plantes l'exigent, en évitant cependant de leur

en donner de trop grands, dans lefquels elles ne pourroient profiter.

Au milieu du mois de Juin, on peut ôter des pots les efpeces rampantes, & les planter à une expofition chaude. Si on les couvre de cloches, elles fubfifteront ainfi pendant l'été ; mais elles ne deviendront pas fort groffes, & feront bientôt détruites par les premiers froids de l'automne. Ceux qui n'ont pas la commodité des ferres chaudes ou des couches de tan, peuvent élever ces plantes fur des couches chaudes ordinaires dans le printems : quand elles ont acquis de la force, on peut les mettre en pleine terre, comme nous venons de le dire, afin d'en jouir pendant l'été : mais celles ci font peu de progrès, & ne parviennent pas à la même perfection que les autres ; elles ne confervent non plus leur fenfibilité, lorfqu'elles font entierement expofées à l'air.

Il eft inutile d'entretenir le Lecteur de plufieurs contes que les Voyageurs rapportent au fujet de ces plantes, parce qu'ils n'ont pas même le mérite de la vraifemblance ; je me bornerai à faire mention de ce que quarante ans d'obfervations m'ont fourni l'occafion de remarquer. Ces plantes font plus ou moins fenfibles lorfqu'on les touche & qu'on les preffe, fuivant que l'air où elles croiffent eft plus ou moins chaud ; car les plantes qui font tenues dans les ferres chaudes, rétréciffent leurs feuilles auffi-tôt qu'elles font touchées ou avec la main

ou avec un corps quelconque, & même lorsqu'elles sont exposées au contact du vent. Quelques-unes de ces especes rétrécissent seulement leurs petites feuilles placées sur la côte du milieu, d'autres montrent leur sensibilité, non seulement dans cette partie, mais aussi dans leurs pétioles, qui se courbent vers le bas au moindre attouchement. Les premieres sont appelées *Sensitives*, & les secondes *Plantes humbles*; mais lorsqu'elles sont placées dans des lieux plus froids, elles se resserrent beaucoup moins & ne se meuvent pas avec autant de vitesse que celles qui sont tenues à une plus grande chaleur. Celles que l'on expose entierement en plein air ont trèspeu de sensibilité, sur-tout dans les tems froids, elles ne se ferment pas non plus la nuit, comme celles qui se trouvent dans une situation chaude.

J'ai aussi observé que ce n'est pas la lumiere qui les fait ouvrir, comme plusieurs personnes l'ont prétendu; car dans les plus grands jours de l'été, elles sont généralement repliées à cinq ou six heures du soir, quoique le soleil ne se cache que deux ou trois heures après; & quoique les vitrages de la serre où elles sont placées soient couverts avec des volets qui en excluent la lumiere au milieu du jour, cependant si l'air de la serre est chaud, les feuilles de ces plantes continueront à se tenir ouvertes, comme je l'ai remarqué plusieurs fois : j'ai aussi trouvé souvent leurs feuilles tout-à-fait étendues au point du jour ; de sorte qu'il est évident que la lumiere n'est pas cause de leur extension.

J'ai encore remarqué que les plantes auxquelles on procure une forte chaleur pendant l'hiver, conservent leur vigueur & leur sensibilité ; mais que celles qu'on tient à une chaleur tempérée, ne se meuvent que très peu, ou même point du tout, lorsqu'on les touche.

Si on touche quelques-unes des feuilles au haut de la plante, elles excitent la sensibilité des inférieures, qu'elles frottent en tombant, de façon qu'en se touchant l'une l'autre, elles se replient toutes. Lorsque l'air de la serre où sont ces plantes est à un dégré de chaleur convenable, elles se relevent entierement en huit ou dix minutes. Je les ai souvent examinées dans le tems qu'elles se redressoient, & j'ai remarqué que cela s'opéroit par des mouvemens de vibration semblables à ceux du battant d'une cloche.

Quelques-unes de ces especes sont si sensibles, que la plus petite goutte d'eau les met en mouvement ; mais les autres ne se remuent pas sans un frottement plus considérable.

Les racines de toutes les especes ont une odeur très-forte & désagréable, qui ressemble à l'odeur putride d'un égoût.

J'ai lu dans quelques relations, que les feuilles & les branches de ces plantes ont une qualité vénéneuse, & que les Indiens en tirent par expression un poison lent, dont la racine même de la plante est le remede spécifique ; mais je ne

118 MIM

puis ni certifier la vérité de cette assertion, ni la nier, n'ayant jamais fait aucune expérience pour constater la propriété de ces plantes. Si elles renferment un poison mortel, comme on le dit, la sensibilité dont elles sont douées sert peut-être à avertir les hommes de ne point y toucher, & les épines dont quelques-unes d'entr'elles sont armées, suffisent pour écarter les animaux: aussi je n'ai jamais entendu dire qu'aucun animal les ait attaquées pour s'en nourrir.

Toutes ces plantes sont originaires de l'Amérique; ainsi elles sont restées inconnues jusqu'à la découverte de ce nouveau Continent. J'en ai envoyé des semences, il y a quelques années, à la Chine, où elles ont réussi & ont beaucoup excité l'admiration.

Les *Acacias* ressemblent si fort au *Mimosa* par leurs caracteres, que LINNÉE les a compris dans le même genre; & comme son système est généralement adopté, je suivrai ici son exemple.

Nilotica. La dixieme espece est l'*Acacia*, ou l'arbre dont on tire le vrai *Suc d'Acacia* & la *Gomme Arabique.* Quoiqu'il soit originaire de l'Egypte, on le trouve dans plusieurs parties de l'Amerique, d'où ses semences m'ont été envoyées en Angleterre. Ces graines ont fort bien réussi dans plusieurs jardins des environs de Londres.

Cet arbre parvient à une grande hauteur dans son pays natal; mais il ne s'éleve guere qu'à huit ou dix pieds en Angleterre, où il fleurit sou-

MIM

vent en automne, mais sans donner aucunes semences.

Farnesiana. La onzieme espece est la plus commune à la Jamaïque, à la Barbade, & dans toutes les autres contrées chaudes de l'Amérique: la bonne odeur & la beauté de ses fleurs la font multiplier dans la plus grande partie de l'Europe. Quoiqu'elle soit originaire d'un pays très-chaud, elle s'est néanmoins habituée en Italie, en Portugal & en Espagne, où elle est devenue fort commune.

Les Jardiniers Italiens, qui apportent ici annuellement des *Orangers*, y joignent toujours aussi de jeunes plants de cette espece, sous le nom de *Gazia*; mais comme ils sont trop tendres pour résister dans nos serres communes, il y en a très-peu qui réussissent. J'ai élevé quelques plantes de cette espece, qui sont parvenues au-dessus de seize pieds de hauteur; elles ont produit un grand nombre de fleurs dans les mois de Juillet & Août: mais elles avoient été tenues en hiver dans des serres chaudes, & pendant l'été dans des caisses de vitrages, pour les mettre à couvert de l'humidité & du froid; car elles ne pourroient pas fleurir en plein air dans ce pays: leurs fleurs sont d'un jaune clair, & répandent une odeur douce & agréable. On lui donne le nom d'*Arbre à éponge*, dans les Indes Occidentales, & de *Cassie*, dans la France méridionale.

Cornigera. La douzieme est à présent très rare en Angleterre; car on ne la trouve que dans

quelques jardins curieux : elle est armée d'épines diſpoſées par paires extrêmement groſſes, courbes & blanchâtres ; mais je ne me ſouviens pas de l'avoir jamais vue en fleurs. A en juger par des échantillons deſſechés que j'ai reçus de Campêche, & qui étoient chargés de pluſieurs fleurs, elle ne paroiſſent pas être d'une grande beauté, & les arbres qui les produiſent, n'ont pas grande apparence, même dans leur pays natal : leurs branches ſont toujours difformes, & peu garnies de feuilles, même dans leur plus grande vigueur. Ces arbres ſont même dépouillés pendant pluſieurs mois de ſuite, de ſorte que cette eſpece n'eſt remarquable que par la forme extraordinaire de ſes épines, dont les branches & les tiges ſont armées, & qui reſſemblent aux cornes de quelques animaux, par leur tiſſu, & leurs ſinguliers contours.

Circinalis. La vingt-troiſieme eſpece a été apportée des Iſles de Bahama, par M. CATESBY, en 1726 ; ſes ſemences ſont plates, moitié d'un beau rouge, & moitié d'un noir foncé : elles ſont formées dans des légumes longs & tors, qui s'ouvrent d'un côté, lorſque les ſemences ſont mûres, & les laiſſent ſuſpendues en-dehors, après un fil mince, pendant quelque tems ; ce qui produit un bel effet. Les feuilles de cet arbre, qui ſont horiſontalement placées en dehors, ſe diviſent en pluſieurs ramifications ; les lobes qui les compoſent, ſont ronds & très-ré-

gulierement diſpoſés : ſes fleurs n'ont point encore paru en Angleterre ; mais on les connoît, d'après un deſſin fait ſur la plante, dans le pays : elles paroiſſent être jolies.

Unguis cati. La treizieme eſpece a été apportée de la Jamaïque dans le jardin de Botanique de Chelſéa : elle a quatre lobes larges à chaque feuille ; ſes épis ſont courts, fermes & courbes : ſes ſemences ſe forment dans des légumes tors, comme ceux de la précédente. Cette plante eſt bien décrite dans l'*Hiſtoire naturelle de la Jamaïque* du Chevalier Sir HANS SLOANE. Les habitans de l'Amérique lui donnent le nom de *Doctor-long*, qui auſſi eſt celui ſous lequel ſes ſemences ſont ſouvent apportées en Angleterre.

Culture. La plupart des autres eſpeces ci-deſſus mentionnées ont été recueillies par le Docteur HOUSTOUN, à la Jamaïque, à la Vera-Cruz & à Campêche, d'où il a envoyé les ſemences d'une grande partie en Europe. Pluſieurs croiſſent à préſent dans le jardin de Botanique de Chelſéa, & quelques-unes y ont donné des fleurs & des fruits en abondance.

Ces eſpeces ſont toutes tendres, & doivent être conſervées dans des ſerres chaudes en hiver, & ſeulement un peu expoſées en plein air pendant l'été dans une ſituation chaude

On les multiplie, en les ſemant ſur des couches chaudes au printems : peu de tems après, les plantes paroiſſent au-deſſus de la terre ; & cinq

ou six semaines après, elles sont en état d'être transplantées. On prépare alors une nouvelle couche assez chaude, & l'on se pourvoit d'une quantité suffisante de petits pots de la valeur d'un sou, qu'on remplit d'une terre fraîche, légere & sablonneuse, & qu'on plonge dans cette couche, qui ne doit pas être de fumier ; car si elle étoit faite avec du fumier chaud de cheval, il faudroit la couvrir de terre jusqu'au niveau des pots, dont le fond seroit posé sur le fumier, & alors les racines des plantes souffriroient de la trop grande chaleur, au-lieu que les lits de tan ont rarement une chaleur aussi violente.

Deux ou trois jours après, & aussi-tôt que la terre des pots est échauffée, on enleve avec soin les jeunes plantes de dessus la premiere couche ; on en met quatre ou cinq dans chaque pot ; on les arrose légerement, pour affermir la terre, & on les tient à l'ombre, en couvrant les vitrages avec des nattes, jusqu'à ce qu'elles aient formé de nouvelles racines ; on leur donne ensuite de l'air, en soulevant les vitrages à proportion de la chaleur du tems, & de la force des plantes.

Nilotica. Farnesiana. Cornigera. Les dix, onze & douzieme especes étant trop tendres, surtout lorsqu'elles sont jeunes, exigent une couche chaude de tan ; & à mesure qu'elles grandissent, il faut les mettre dans de plus grands pots. La terre dans laquelle on les pla-

ce, doit être un peu plus légere & plus chargée de sable que pour les autres especes ; mais il faut toujours éviter de leur donner de trop grands pots, qui leur seroient aussi nuisibles qu'aux *Orangers*, & de ne pas trop les arroser, surtout en hiver. La dixieme espece, qui est plus dure, peut rester dans une serre chaude ordinaire, lorsqu'elle est devenue ligneuse : elle n'a besoin que d'une chaleur modérée en hiver ; & en été, on peut la placer en plein air, lorsque le tems est chaud, sans cependant l'y exposer tout-à-fait, au moins pendant les quatre ou cinq premieres années, & jusqu'à ce qu'elle soit devenue très-ligneuse : car ces plantes sont fort tendres, & se conservent très-difficilement dans ce climat. La serre chaude où elles sont placées pendant l'hiver doit être tenue au-dessus du point de la température fixée pour les dernieres serres dans les thermometres ; mais en été, ces plantes ont besoin d'être arrosées fréquemment, quoique cependant toujours avec modération. La onzieme espece est un très-bel arbre ; la douzieme perd ses feuilles précisément avant que les nouvelles commencent à paroître ; de sorte que cet arbre n'est dépouillé qu'un mois ou six semaines pendant le printems. Plusieurs personnes, qui n'étoient point prévenues de cela, voyant leurs arbres dégarnis de feuilles, les ont cru morts, & les ont arrachés. Je fais cette observation, afin d'enga-

MIM

ger à être plus circonspect, lorsqu'on arrache des arbres que l'on croit morts, & pour qu'on ait assez de patience pour les conserver encore durant l'été suivant ; car j'ai vu plusieurs plantes qne l'on regardoit comme perdues, reprendre vigueur au mois de Juillet suivant, & d'autres qui étoient détruites jusqu'à la surface de la terre, repousser de leurs racines.

Les trois especes d'*Acacia* armées d'épines en forme de cornes, font souvent dépouillées de feuilles pendant deux ou trois mois, & paroissent comme mortes ; mais leur feuillage se renouvelle aux approches de l'automne, qui est la saison de leur plus grande vigueur : il faut les exposer en plein air, & à l'abri des vents violens pendant deux mois de l'été, pour les débarrasser des insectes qui les infectent ; & en hiver on leur procure une chaleur modérée.

Toutes les autres especes dont il a été question, se multiplient par leurs graines, qui mûrissent rarement en Angleterre, & que l'on doit par conséquent se procurer de l'Amérique, & particulierement de Campêche, où l'on en trouve une grande quantité d'especes, dont la plupart ne font connues des Botanistes que depuis très-peu de tems. Pour les transplanter en Angleterre, il est nécessaire de les dépouiller de leurs légumes dans le moment qu'on les recueille, & de les envelopper de papiers, dans lesquels on met des feuilles de tabac, ou des herbes venimeuses, pour préserver les femences des insectes, fans quoi elles feroient rongées & détruites avant d'arriver en Angleterre, parce que ces insectes percent ces semences, & y déposent leurs œufs, & que les vers qui en sortent s'en nourrissent, & détruisent les germes ; ce qui est souvent arrivé aux femences qui m'ont été envoyées de l'Amérique.

Plusieurs de ces *Acacias* font très-délicats tandis qu'ils sont jeunes ; mais après deux ou trois ans, ils deviennent assez forts pour supporter le plein air en été, quoique très - peu puissent resister en hiver dans une ferre, à moins qu'elle ne soit échauffée artificiellement dans les grands froids.

MIMULUS. *Lin. Gen. Plant.* 761. *Cynorrhynchium. Mitch.* 3. [*Mimulus.*] Le Masque ou la Monavie.

Caracteres. Le calice de la fleur est oblong, prismatique, & formé par une feuille persistante ; la corolle est monopétale, labiée & érigée ; son tube est de la longueur du calice, & son bord est divisé au sommet en deux parties réfléchies fur le côté ; la levre inférieure est large, & divisée en trois segmens, dont celui du milieu est le plus petit ; le palais est convexe, & separé en deux segmens : la fleur a quatre étamines minces, dont deux font plus longues que les autres, & qui font toutes terminées par des antheres en forme de rein, & divisées en deux parties ; son germe est conique, & soutient un style

MIM

mince, couronné par un stigmat ovale, divisé en deux parties, & comprimé : ce germe devient ensuite une capsule ovale & à deux cellules remplies de petites semences.

Ce genre de plantes est rangé dans la seconde section de la quatorzieme classe de LINNÉE, qui renferme celles dont les fleurs ont deux étamines longues & deux courtes, & dont les semences sont contenues dans une capsule.

Nous n'avons qu'une espece de ce genre en Angleterre.

Mimulus ringens, erectus, foliis oblongis, linearibus, sessilibus. Hort. Upsal. 176. tab. 2. Act. Ups. 1741. p. 82 ; Monavie érigée, avec des feuilles oblongues, linéaires & sessiles.

Euphrasia Floridana, Lysimachiæ glabræ, siliquosæ foliis, quadrato caule, ramosior. Pluk. Amalth. 83. t. 393. f. 3.

Lisimachia galericulata, sive Gratiola elatior, non ramosa. Gron. Virg. 69.

Gratiola Canadensis latifolia, flore magno cæruleo. Boerh. Lugd.-B. 2. p. 265.

Digitalis perfoliata glabra, flore violaceo minore. Mor. Hist. 2. p. 479 ; Digitale unie & perforée, avec une petite fleur violette. Le Masque ou la Monavie.

Cette plante croît naturellement dans l'Amérique septentrionale, où on la trouve dans les terres humides ; elle a une racine vivace & une tige annuelle qui périt en automne : la tige est quarrée, d'un pied & demi de hauteur, & garnie à chaque nœud de deux feuilles oblongues, lisses, plus larges

à leur base, où elles se joignent presqu'autour de la tige, & terminées en pointe aiguë ; la partie basse de la tige pousse deux ou trois branches courtes, & son sommet est orné à chaque nœud de deux fleurs qui sortent des ailes des feuilles, tout autour de la tige : ces fleurs ont un calice oblong, recourbé, à cinq angles, & découpé au sommet en cinq parties, du milieu duquel s'éleve la fleur, avec un long tube recourbé, qui s'ouvre au sommet en deux levres, dont la supérieure est érigée, & légerement découpée à l'extrémité en deux parties ; l'inferieure est inclinée vers le bas, & divisée en trois foibles segmens. Ces fleurs sont de couleur violette, & n'ont point d'odeur ; elles paroissent dans le mois de Juillet, & sont remplacées par des capsules oblongues & à deux cellules, remplies de petites semences, qui, dans les années chaudes, mûrissent en automne.

Culture. Cette plante est fort dure au froid ; mais elle exige un sol mou, marneux, & pas trop exposé au soleil. On peut la multiplier en divisant ses racines en automne, sans cependant les réduire en trop petites parties : on se la procure aussi par semences, qni doivent être mises en terre en automne, aussitôt qu'elles sont mûres ; car si on ne les seme qu'au printems, elles croissent rarement dans la même année. On peut les placer sur une plate-bande exposée aux rayons du soleil du matin, & l'on peut les disperser ensuite dans tout le parterre.

MIR

MIRABILIS. *Lin. Gen. Plant.*
215. *Jalapa. Tourn. Inst. R. H.*
129. *tab.* 50. [*Marvel of Peru,*
or Four o'Clock Flower.] Merveille du Pérou, ou Belle-de-
Nuit.

Caractères. Le calice a cinq
petites feuilles ovales, & en
forme de lance ; il est érigé,
gonflé & persistant ; la corolle,
qui est monopétale, & en forme
d'entonnoir, a un tube mince,
posté sur le nectaire, étendu
au-dessus, & découpé en cinq
segmens obtus. La fleur a cinq
étamines minces, adhérentes à
la corolle, égales, penchées,
& terminées par des antheres
rondes, avec un germe rond en
dedans du nectaire, qui soutient un style mince, couronné
par un stigmat globulaire. Ce
germe devient ensuite une noix
ovale & à cinq angles, dans
laquelle est renfermée une semence.

Ce genre de plantes est rangé
dans la premiere section de la
cinquieme classe de LINNÉE, qui
comprend celles dont les fleurs
ont cinq étamines & un style.

Les especes sont :

1º. *Mirabilis Jalapa, floribus*
congestis, terminalibus, erectis.
Lin. Sp. Plant. 252 ; Belle-de-
nuit avec des paquets de fleurs
érigées qui terminent les tiges.

Mirabilis. Hort. Cliff. 53. *Hort.*
Ups. 43. *Fl. Zeyl.* 85. *Rumph.*
Amb. 5. *p.* 253. *f.* 89. *Osb. it.*
225. *Blackw. t.* 404. *Kniph. cent.*
6. *n.* 61. 62. *Knorr. Del. 1. t.* 3.
Sabb. Hort. 2. *f.* 11.

Nyctage. Roy. Lugd.-B. 417.
Cold. Noveb. 29.

Solanum Mexicanum, flore
magno. Bauh. Pin. 168.

Admirabilis Peruviana. Clus.
Hist. 2. *p.* 87 ; Merveille du
Perou.

2º. *Mirabilis dichotoma, floribus*
sessilibus, axillaribus, erectis, so-
litariis. Amœn. Acad. 4. *p.* 267 ;
Belle-de-nuit à fleurs solitaires,
érigées & sessiles.

Mirabilis Jasmini Rosa. Clus.
Hist. 2. *p.* 90.

Jalapa officinarum. Mart. cent.
1. *f.* 1 ; Faux Jalap des Boutiques.

Solanum Mexicanum, flore par-
vo. Bauh. Pin. 168. *Prodr.* 91.

3º. *Mirabilis longi-flora, flo-*
ribus congestis, terminalibus, lon-
gissimis, nutantibus, foliis sub-vil-
losis. Act. Holmens. 1755. *p.* 176.
t. 6. *f.* 1. *Amœn. Acad.* 4. *p.* 268.
Kniph. cent. 7. *n.* 59; Belle-de-
nuit à longues fleurs du Pérou,
dont les fleurs rapprochées en
paquets & penchées, terminent
les tiges, & ont des feuilles
velues en-dessous.

Alzoyati, Mirabilis Mexicana.
Hern. Mex. 120. *f.* 2.

Jalapa. La premiere espece
est la *Merveille du Pérou*, qu'on
cultive depuis plusieurs années
dans les jardins anglois, comme plante d'ornement. On en
connoît plusieurs variétés, qui
ne different que par la couleur
de leurs fleurs, & dont deux
conservent constamment leurs
différences ; l'une a des fleurs
pourpre & blanches, d'autres
sont d'un blanc uni, & la plupart sont panachées en deux
couleurs. Toutes ces variétés
se trouvent souvent sur la
même plante, & quelquefois
sur différens pieds ; les unes ont
des fleurs rouges & jaunes; d'autres portent ces deux variétés,

& ont en même tems des fleurs unies; d'autres, au lieu de fleurs unies, ont encore des fleurs panachées : mais je n'ai jamais vu les femences de l'efpece pourpre & de la blanche produire la jaune & la rouge, ni cette derniere fe changer en la premiere. J'ai cultivé ces deux efpeces pendant plus de quarante années ; & quoiqu'elles ne changent pas l'une dans l'autre, cependant, comme elles ne different des autres que par la couleur de leurs fleurs, je ne les ai pas données comme des efpeces diftinctes.

Dichotoma. La feconde efpece eft fort commune dans les Ifles des Indes Occidentales, où on lui donne le nom de *Fleur de quatre heures*, parce qu'elle s'ouvre vers cette heure. Je n'ai jamais vu de variétés de cette efpece : fes fleurs font d'un rouge pourpâtré, & de moitié moins groffes que les autres ; fes tiges font épaiffes, gonflées & noueuffes ; & comme fes feuilles font auffi plus petites, & fon fruit plus rude, on ne peut pas douter qu'elle ne foit une efpece diftincte : d'ailleurs je n'ai jamais remarqué aucune altération dans les plantes de cette efpece élevées de femences, après les avoir cultivées pendant plufieurs années.

TOURNEFORT ayant été informé, par le P. PLUMIER, que la racine de cette plante étoit le *Jalap des Boutiques*, a conftitué ce genre, & lui a donné ce titre ; mais le Docteur HOUSTOUN ayant appris le contraire dans l'Amérique Efpagnole, a

porté en Angleterre un deffin de la plante à *Jalap*, fait par un Efpagnol, ainfi que deux ou trois racines de la véritable efpece, qu'il a plantées dans un jardin à la Jamaïque, mais qui, malheureufement, ont été détruites par les cochons, depuis fon départ. Depuis ce tems, on a été entierement convaincu que le vrai *Jalap* étoit un *Convolvulus*, auquel RAY avoit donné, plufieurs années avant, le nom de *Jalap* des Indes Occidentales, fans qu'il paroiffe fur quoi il s'étoit fondé. Quelques années après, j'ai reçu de l'Amérique Efpagnole, trois femences de *Jalap*, dont une a produit une grande plante à racines bulbeufes, auffi groffes que celles du *Jalap* qu'on apporte en Angleterre ; mais la plante n'a pas produit de fleurs pendant les trois années qu'elle a fubfifté, & elle a péri dans l'hiver de 1739 à 1740. Depuis ce tems, je n'ai pu me procurer de nouvelles femences ; cependant je fuis entierement convaincu que le vrai *Jalap* eft une efpece de *Convolvulus* : au furplus les racines de la *Belle-de-Nuit* font purgatives ; & quand on les donne à double dofe, elles produifent le même effet que le *Jalap*.

Longi-flora. J'ai reçu du Méxique la troifieme efpece, il y a quelques années ; fes femences m'ont d'abord été envoyées de Paris par M. LE MONIER, de l'Académie Royale des Sciences, & enfuite de Madrid, par le Docteur HORTEGA : les tiges de celle-ci rampent fur la terre,

MIR

terre, fi on ne leur fournit pas un foutien ; elles ont environ trois pieds de longueur, & fe divifent en plufieurs branches garnies de feuilles en forme de cœur, & oppofées ; les feuilles, ainfi que les tiges, font velues, vifqueufes, & s'attachent aux doigts de ceux qui les touchent : les fleurs qui naiffent aux extrémités des branches, font blanches, & ont des tubes fort longs & minces ; elles répandent une odeur agréable & mufquée, & elles reffemblent à celles des autres efpeces qui fe ferment pendant tout le jour, & s'ouvrent au coucher du foleil. Les femences de cette efpece font plus groffes que celles des autres, & auffi raboteufes que celles de la feconde.

Les deux variétés de la premiere efpece ornent beaucoup les jardins pendant les mois de Juillet, Août & Septembre ; & fi le tems continue à être doux, leurs fleurs fe fuccedent fouvent jufqu'à la fin d'Oɛobre. Ces fleurs ne s'épanouiffent point avant la foirée, tandis que le tems continue chaud ; mais lorfqu'il fait frais, & que le foleil eft caché par des nuages, elles reftent ouvertes prefque tout le jour : elles font fi nombreufes aux extrémités des branches, que lorfqu'elles font épanouies, les plantes paroiffent en être entierement couvertes : les unes font unies, & d'autres panachées fur la même plante ; ce qui produit un bel effet. On multiplie ces efpeces par femences, pour le choix defquelles on devroit

MIR 125

avoir foin de ne conferver aucune des plantes dont les fleurs font unies ; & quand on veut n'avoir que des efpeces panachées, on a foin de retrancher toutes les fleurs de couleur unie, fur les pieds de fleurs panachées que l'on deftine à produire des femences.

On répand ces femences en Mars fur une couche de chaleur modérée : quand les plantes paroiffent, on leur donne beaucoup d'air dans les tems doux, pour les empêcher de filer ; & quand elles ont atteint la hauteur de deux pouces, on les tranfplante fur une autre couche de chaleur très-modérée, ou bien on les met chacune féparément dans un petit pot rempli de terre légere, & on les plonge dans une couche de chaleur tempérée. Cette derniere méthode eft la plus fûre, parce qu'alors on ne court aucun rifque en les tirant des pots, pour les mettre dans les plates-bandes, & qu'elles confervent toutes leurs mottes & leurs racines. On n'a pas befoin de les tenir à l'ombre, au-lieu que celles qu'on tranfplante de la feconde couche dans les plates-bandes, ne confervent que peu de terre à leurs racines, & ont alors befoin d'être couvertes avec foin, fans quoi elles périffent fouvent.

Quand elles font fur la feconde couche chaude, il faut les tenir à l'ombre, jufqu'à ce qu'elles aient formé de nouvelles racines, leur donner enfuite beaucoup d'air, pour les empêcher de filer, & dans

Tome V. I

le mois de Mai, les accoutumer par dégrés à supporter l'air ouvert. Si, au commencement du mois de Juin, la saison est favorable, on les transplante dans les plates-bandes du parterre, en laissant entr'elles une distance suffisante; & quand elles sont bien enracinées, elles n'ont plus besoin d'aucun soin. On peut les semer au commencement d'Avril, sur une plate-bande chaude, où les plantes croîtront très-bien, mais où elles fleuriront tard.

Comme les semences de ces plantes mûrissent très-aisément chaque année, on ne se donne guere la peine de conserver leurs racines; cependant, en les tirant hors de terre en automne, & en les mettant dans du sable sec pendant l'hiver, à l'abri des gelées, on peut les replanter au printems: elles deviendront, par cette méthode, beaucoup plus grosses, & fleuriront plutôt que les plantes de semences. Si l'on couvre aussi ces racines avec du tan pendant l'hiver, pour empêcher la gelée d'y pénétrer, on pourra les laisser dans les plates-bandes, pourvu néanmoins que le sol en soit sec. Si l'on plante dans de gros pots les racines qu'on a tirées de la terre, & qu'on les plonge dans une couche chaude, sous un châssis profond, on hâtera considérablement leurs progrès, & elles s'éleveront à la hauteur de quatre ou cinq pieds, ainsi que j'en ai souvent fait l'expérience. Ces plantes fleurissent de très-bonne heure, & font un agréable effet, lorsqu'elles

sont entremêlées avec d'autres.

Les deux autres especes exigent le même traitement; mais comme la seconde n'est pas tout-à-fait si dure que les autres, à moins qu'on n'ait hâté son accroissement de très-bonne heure; elle ne fleurit que fort tard, & ne perfectionne point ses semences.

MIROIR DE VENUS. *Voyez* CAMPANULA SPECULUM, ET CAMPANULA HYBRIDA.

MIRTE. *Voyez*. MYRTUS.

MIRTILE. *V.* VITIS IDÆA.

MISEREON. BOIS GENTIL, *ou la* LAURÉOLE FEMELLE. *Voyez* DAPHNE MÉZERÉON. L.

MITELLA. *Tourn. Inst. R. H.* 241. *tab.* 126. *Lin. Gen. Plant.* 496, ainsi appelé de *Mitella*, Latin, qui signifie une petite mitre, parce que les capsules de cette plante ressemblent à une mitre d'Evêque, [*Bastard American Sanicle.*]Sanicle bâtard d'Amérique, ou petite Mitre.

Caracteres. Le calice de la fleur est figuré en cloche, & formé par une feuille persistante; & divisé en cinq parties. La corolle est composée de cinq pétales terminés en plusieurs pointes, velus, & insérés dans le calice, ainsi que dix étamines, qui sont en forme d'alêne, plus courtes que la corolle, & terminées par des antheres rondes: la fleur a un germe rond, divisé en deux parties, & un style peu apparent, couronné par deux stigmats obtus: le calice se change dans la suite en une capsule ovale, & a une cellule qui s'ouvre en deux valves, & qui contient un grand nombre de petites semences.

MIT

Ce genre de plantes est rangé dans la seconde section de la dixieme classe de LINNÉE, qui comprend celles dont les fleurs ont dix étamines, & deux styles ou stigmats.

Les especes sont:

1°. *Mitella diphylla*, *scapo diphyllo.* Lin. Gen. Nov. 29. Hort. Cliff. 167. Roy. Lugd.-B. 459. Kniph. cent. 1. n. 59; petite Mitre avec des tiges de fleurs garnies de deux feuilles.

Mitella Americana, *florum petalis fimbriatis.* Tourn. Inst. 242; petite Mitre d'Amérique avec des fleurs dont les pétales sont frangés.

Cortuza Americana altera, *floribus minutim fimbriatis.* Mentz. Pug. t. 10.

Cortuza Americana, *spicato flore*, *petalis fimbriatis.* Herm. Par. 130.

Sanicula, *sivè Cortusa Indica*, *flore spicato fimbriato.* Dodart. Mem. 299.

2°. *Mitella nuda*, *scapo nudo.* Amæn. Acad. 2. p. 252. Amm. Act. Petrop. 2. p. 352; petite Mitre à tige nue.

Mitella scapo nudo, *corollarum petalis fimbriatis.* Gmel. Sib. 4. p. 175. t. 63. f. 2.

Diphylla. La premiere espece croît naturellement dans les bois de plusieurs parties de l'Amérique Septentrionale; elle a une racine vivace, de laquelle sortent plusieurs feuilles en forme de lance, & angulaires, dont quelques-unes sont obtuses, & d'autres terminées en pointe aiguë; elles sont dentées sur leurs bords, d'un vert luisant, un peu velues, & supportées par de longs pétioles:

MOL

les tiges des fleurs s'élevent immédiatement de la racine; elles ont vers leur bâses deux ou trois feuilles angulaires, & au milieu deux petites feuilles à angles aigus, & opposées. Ces tiges s'élevent à huit ou neuf pouces de hauteur, & sont terminées par un épi lâche de fleurs petites & blanches, dont les pétales sont garnis de franges sur leurs bords; elles paroissent au commencement de Juin, & produisent des capsules rondes, & remplies de petites semences.

Nuda. La seconde espece, qui naît sans culture dans les parties septentrionales de l'Asie, est d'un crû plus bas que la premiere, & ne s'éleve guere qu'à cinq ou six pouces de hauteur; ses feuilles sont moins angulaires, & ses tiges sont toujours nues & sans feuilles; ses épis de fleurs sont plus courts & plus serrés.

Culture. Ces deux especes se multiplient, en divisant leurs racines en automne: on les plante à l'ombre dans un sol mou & marneux.

MITELLA MAXIMA. *Voyez* BIXA.

MOLDANIQUE *ou* MÉLISSE DES MOLDAVES. *Voyez* DRACO-CEPHALUM MOLDAVICA. *L.*

MOLÉNE *ou* BOUILLON—BLANC MALE. *Voyez* VERBASCUM THAPSUS. *L.*

MOLLE *on* MASTIC DES INDES, *ou* POIVRIER DU PÉROU. *Voyez* SCHINUS.

MOLLUGO. *Lin. Gen. Plant.* 99. [*Mollugo.*] Grateron. Morsgeline.

Caracteres. Le calice de la

I 2

fleur est formé par cinq petites feuilles oblongues, colorées en-dedans, & persistan-
tes : la corolle est composée
de cinq pétales ovales, plus
courts que le calice : la fleur
a trois étamines hérissées, postées près du style, & terminées par des antheres simples ;
son germe est ovale, & a
trois sillons ; il soutient trois
styles fort courts, & couronnés par des stigmats obtus. Ce
germe devient, quand la fleur
est passée, une capsule ovale,
& à trois cellules remplies de
semences en forme de rein.

Ce genre de plantes est rangé dans la troisieme section de
la troisieme classe de LINNÉE,
qui comprend celles dont les
fleurs ont trois étamines &
trois styles.

Les especes sont :

1°. *Mollugo verticillata, foliis
verticillatis, cunei-formibus, acutis, caule sub-diviso decumbente,
pedunculis unifloris. Hort. Upsal.
24. Kniph. Orig. 8. n. 71* ; Grateron à feuilles verticillées, en
forme de coin, & aiguës, avec
des branches trainantes & divisées, & des pédoncules qui
soutiennent une seule fleur.

*Mollugo foliis sæpiùs septenis
lanceolatis. Gron. Virg.* 14.

. *Alsine spergula Mariana, latiore folio, floribus ad nodos pediculis curtis circà caulem insidentibus, calycibus eleganter punctatis.
Pluk. Mant. 9. t. 332. f. 5.*

*Alsine procumbens, Gallii facie.
Ehret. Pict. t. 6. f. 3.*

2°. *Mollugo quadrifolia, foliis
quaternis obovatis, paniculâ di-
chotomâ. Hort. Cliff. 28* ; Grateron qui produit à chaque

nœud quatre feuilles ovales,
& qui a un panicule à la division des branches.

*Hernaria Alsines folio. Tourn.
Inst. 507* ; Herniaire à feuilles
de Mouron.

*Mollugo tetraphylla. Lin. Sp.
Plant. 1. p. 89.*

*Polycarpon tetraphyllum. Lin.
Syst. Plant. tom. 1. p. 247.*

*Triclis tristemon, foliis conju-
gatis. Hall. Gætt.* 25.

*Anthyllis marina Alsine-folia.
Bauh. Pin. 282.*

*Anthyllis Alsine-folia, polygo-
noïdes major. Barr. Rar. 103. f. 534.*

Il y a deux ou trois especes
de ce genre qui sont rarement
admises dans les jardins ; &
dont, pour cette raison, je ne
ferai point mention ici.

Ces deux especes sont an-
nuelles. La premiere, qui est
originaire des pays chauds, est
moins dure que la seconde ;
elles sont toutes deux trainantes, & leurs tiges sont couchées sur la terre. La premiere,
qui s'étend à huit ou neuf pouces de chaque côté, est garnie à
chaque nœud de six ou sept feuilles petites & étendues en forme
d'étoile : ses fleurs sont petites,
comme celles du *Mouron* ; chaque pédoncule en soutient une ;
elles sont remplacées par des
capsules ovales, & remplies de
petites semences qui poussent
au printems suivant, sans aucun soin, si on leur permet
de se répandre ; mais quand il
arrive qu'elles tombent sur la
terre dont on se sert pour couvrir une couche chaude, les
plantes sont plus précoces, &
deviennent plus fortes que celles de plein air.

MOL

On conferve ces plantes dans quelques jardins, pour la variété, quoiqu'elles ne foient pas fort belles.

MOLUCCA. *Voyez* MOLUC-CELLA.

MOLUCCELLA. *Lin. Gen. Plant.* 643. *Molucca. Tourn. Inst. R. H.* 187. *tab.* 88. Cette plante prend fon nom des Isles Moluques, où elle a été trouvée. [*Molucca Bahn.*] Meliffe des Moluques, *ou* la Moluque.

Caractères. La fleur a un calice gros, perfiftant, & formé par une feuille profondément dentée & ouverte fur fes bords : la corolle eft labiée ; fon tube eft court ; la levre fupérieure eft érigée, concave, & entiere, & l'inférieure eft divifée en trois parties, dont celle du milieu eft plus longue que les autres ; la fleur a quatre étamines placées fous la levre féperieure, dont deux font plus courtes que les autres, & qui font toutes terminées par des antheres fimples ; le germe eft en quatre parties, & foutient un ftyle placé avec les étamines, & couronné par un ftigmat divifé en deux portions ; ce germe fe change dans la fuite en quatre femences angulaires, convexes, & poftées dans le calice.

Ce genre de plante eft rangé dans la premiere fection de la quatorzieme claffe de LINNÉE, avec celles dont les fleurs ont deux étamines longues & deux courtes, & qui produifent des femences nues renfermées dans le calice.

Les efpeces font :

1°. *Molucella lævis*, *calycibus*

MOL 129

campani-formibus, *fub-quinque-dentatis* ; *denticulis æqualibus. Lin. Sp.* 821. *Roy. Lugd.-B.* 314. *Hort. Ups.* 172. *Gron. Orient.* 75. *Kniph. cent.* 11. *n.* 77. *Sabb. Hort.* 3. *t.* 45. *Regn. Bot.* ; Méliffe des Moluques avec des calices en forme de cloches, découpés en cinq dentelures égales.

Molucca lævis. Dod. Pempt. 92 ; Méliffe des Moluques, unie, ou la Moluque.

Meliffa Moluccana odorata. Bauh. Pin. 229.

2°. *Molucella fpinofa*, *calycibus ringentibus*, *octo-dentatis. Lin. Sp.* 821. *Roy. Lugd.-B.* 314. *Hort. Ups.* 172. *Sabb. Hort.* 3. *t.* 46 ; Méliffe des Moluques, dont les calices font en mafque, & découpés en huit dents.

Molucca fpinofa. Dod. Pempt. 92 ; Méliffe des Moluques, épineufe.

Meliffa Moluccana fœtida. Bauh. Pin. 229.

Lævis. La premiere efpece s'élève avec une tige quarrée, à trois pieds de hauteur, & s'étend au-dehors en plufieurs branches liffes, qui fortent par paires, & font garnies de feuilles rondes profondément entaillées fur leurs bords, fupportées par de longs pétioles oppofés, liffes, & d'un vert clair fur les deux furfaces : fes fleurs naiffent en têtes verticillées aux pétioles de feuilles ; elles ont de fort grands calices étendus & découpés en cinq parties ; immédiatement au-deffous des fleurs, fortent deux paquets de longues épines, un à chaque côté de la tige ; chaque paquet en contient cinq ou fix, qui s'élevent de la mê-

I 3

me pointe. Les fleurs font petites, & placées dans le fond de larges calices ; de maniere qu'elles ne font pas vifibles de quelque diftance : elles font d'un blanc tirant fur le pourpre, & de la même forme que toutes les labiées ; leur levre fupérieure eft entiere, & creufée en forme de cuiller, & l'inférieure eft découpée en trois fegmens, dont celui du milieu eft le plus long. Quand la fleur eft paffée, le germe fe change en quatre femences angulaires, en forme de maffue, & renfermées dans le calice. Cette plante fleurit dans le mois de Juillet ; mais fes femences ne mûriffent point en Angleterre, à moins que l'année ne foit chaude & feche : fon odeur paroît défagréable à quelques perfonnes, & plaît à d'autres.

Spinofa. La feconde efpece a des tiges quarrées, liffes, & de couleur tirant fur le pourpre, qui s'élevent à la hauteur de quatre pieds, & jettent des branches au-dehors de la même maniere que la premiere ; fes feuilles font plus petites, fupportées par de courts pétioles, d'une couleur plus foncée, & plus fortement dentées fur leurs bords : les calices des fleurs, qui font moins larges, font découpés en huit fegmens, terminés chacun par une pointe aiguë ; fes fleurs font femblables à celles de la précedente, ainfi que les femences : mais cette efpece eft moins dure que la premiere.

La premiere croît naturellement dans plufieurs parties de la Syrie, & la feconde dans les Ifles Moluques, d'où ce genre a pris fon nom; elles font toutes deux annuelles, & périffent auffi-tôt que leurs femences font mûres. Comme elles font originaires des pays chauds, elles perfectiônnent rarement leurs femences en Angleterre, quand on ne les feme qu'au printems. Ainfi, il vaut mieux les élever dans des pots en automne : on les tient fous le vitrage d'une couche chaude en hiver, de maniere qu'elles puiffent avoir de l'air dans les tems doux ; mais il eft effentiel de les conferver feches, fans quoi elles font très fujettes à être attaquées de pourriture, quand elles reftent trop longtems couvertes pendant les gelées. Au printems, on peut tirer les plantes hors des pots, avec leur motte entiere, & les mettre dans une plate-bande chaude, à l'abri des vents violens : on les arrofe un peu, pour fixer la terre à leurs racines; après quoi elles n'exigent plus aucun autre foin que d'être tenues nettes de mauvaifes herbes. Il faut les foutenir avec des baguettes, pour empêcher qu'elles ne foient rompues par les vents. Les plantes ainfi confervées pendant l'hiver fleuriffent à la fin du mois de Juin, & peuvent donner de bonnes femences.

MOLUQUE *ou* MÉLISSE DES MOLUQUES. *Voyez* MOLUCCELLA LÆVIS *L.*

MOLY. *Voyez* ALLIUM.

MOMIE *ou* CIRE A GREFFER. On appelle de ce nom une efpece de cire compofée d'une livre de poix noire & d'un quar-

MOM

teron de térébenthine ordinaire : on met ce mélange dans un pot de terre que l'on place sur du feu en plein air. En faisant cette espece de cire, il faut avoir à la main un couvercle, afin de pouvoir l'éteindre lorsqu'elle s'enflamme ; ce qu'il faut faire plusieurs fois, afin que le feu fasse évaporer ses parties volatiles.

On connoît que cette cire a assez bouilli, quand, en en versant un peu sur une assiette plate d'étain; elle s'épaissit aussitôt : on répand alors cette poix fondue dans un autre pot ; on y ajoûte un peu de cire commune, qu'on mêle avec la premiere, & l'on conserve cette matiere pour l'usage.

Le Docteur AGRICOLA donne la maniere de l'employer.

Quand vous voulez, dit-il, couvrir des racines avec cette cire, il faut la faire fondre, la laisser refroidir un peu, & tremper les deux bouts de la racine que vous voudrez planter (car le Docteur la propose pour les morceaux de racines ou d'arbres qu'on veut planter) l'un après l'autre, mais pas trop avant dans la cire; mettez les ensuite dans l'eau, plantez-les dans la terre, le petit bout en bas, de maniere que le gros bout puisse paroitre un peu au-dessus, pour profiter de l'avantage de l'air; & comprimez bien la terre autour de ces racines, afin qu'elles n'en reçoivent pas trop d'humidité, car cela les feroit pourrir.

Le même auteur donne la méthode suivante, pour faire une momie propre aux plantes exotiques.

Prenez trois livres de térébenthine de Venise ; faites-la fondre dans un pot de terre fort, sur un feu lent ; quand elle est tout-à-fait fondue, ajoutez-y une demi-livre de gomme copale, bien pulvérisée & tamisée, & remuez toujours avec un petit bâton, en augmentant le feu par dégrés. Ce mélange se dissoudra insensiblement ; on laissera ensuite la térébenthine s'évaporer, & le mélange s'épaissir : lorsqu'il aura acquis une consistance suffisante, on pourra en former de petits bâtons semblables à ceux de cire d'Espagne, que l'on conservera pour l'usage.

Ce Docteur prétend que cette momie est un excellent vulnéraire pour les plantes, parce qu'elle n'est pas sujette à se corrompre comme d'autres substances gommeuses; elle empêche toute pourriture entre la tige & la racine, & par ce moyen le calus se forme bien plutôt, & s'étend sur toutes les parties; de sorte que la tige se trouve entierement unie avec la racine, & lui donne de la force & de la vigueur.

Voici la maniere de faire la momie végétale, suivant le même Auteur.

Remplissez un grand chaudron ou pot de terre aux trois quarts, avec de la poix noire ordinaire ; ajoutez y un peu de résine fine, ou de la poix sulfurisée, avec un peu de cire jaune ; faites fondre ce mélange, jusqu'à ce qu'il devienne liquide ; ôtez le ensuite de des-

I 4

132 MOM MOM

ſus le feu, & placez-le ſur la terre, juſqu'à ce qu'il ceſſe de fumer : lorſqu'il ſera refroidi, vous pourrez, avec une broſ-ſe, en frotter les inciſions fai-tes pour la greffe ordinaire ou celle en écuſſon, &c. &c. &c.

Méthode du même Auteur, pour faire la Momie pourpre aux arbres des jardins & des forêts.

Prenez trois livres de téré-benthine ordinaire, & quatre livres de poix commune ; fai-tes fondre la térébenthine ſur le feu ; &, après avoir mis la poix en poudre, jettez-la dans la térébenthine : quand le tout ſera mêlé & épaiſſi, ôtez-le du feu, & conſervez-le pour l'uſage.

On peut former, avec cette compoſition, de petits bâtons ſemblables à ceux de cire d'Eſ-pagne, ou la conſerver dans de petits pots : on la fait fon-dre ſur un feu lent, quand on veut s'en ſervir, & on y trempe une petite broſſe, pour en frot-ter les greffes.

Le même Auteur donne la recette ſuivante pour faire la momie noble ou la cire à greffer.

Prenez deux livres de poix pure, qu'on appelle poix vierge de Ratisbonne ; ajoutez-y une demi-livre de térébenthine ; mê-lez ces deux drogues enſemble dans un pot de terre ; placez-le ſur le feu, afin que les par-ties les plus déliées de la téré-benthine s'évaporent, ſans quoi ce mêlange nuiroit beaucoup aux arbres & aux racines : fai-tes la même épreuve que pour la premiere compoſition, afin

de connoître ſi elle eſt aſſez cuite, après quoi vous y ajou-terez encore une demi-livre de cire vierge, & une demi-once de myrrhe & d'aloës pul-vériſés. Quand toutes ces dro-gues ſeront bien mêlées, vous en formerez des bâtons, ou vous la conſerverez dans des pots de fayance.

Le tems le plus propre pour faire uſage de cette compoſi-tion ſur les racines, eſt dans les mois de Septembre, d'Oc-tobre, de Novembre, quoiqu'on réuſſiſſe aſſez bien dans tous les tems de l'année ; mais l'au-tomne eſt la ſaiſon la plus fa-vorable. AGRICOLA dit que la ſeule différence conſiſte en ce que tout ce qui eſt planté au printems, pouſſe aux mois de Juin & de Juillet, & que ce qui eſt planté en automne, ne commence à pouſſer qu'au mois d'Avril.

Le même Auteur fait men-tion de beaucoup de merveil-les opérées par ſes momies. Ceux qui voudront ſatisfaire leur curioſité à ce ſujet, pourront recourir à ſon traité.

MOMORDICA. Tourn. Inſt. R. H. 103. 29. 30. Lin. Gen. Plant. 1090. [Male Balſam Ap-ple.] Pomme de Merveille.

Caractères. Cette plante a des fleurs mâles & des fleurs fe-melles ſur le même pied ; les fleurs mâles ont un calice éten-du, & formé par une feuille ; la corolle eſt monopétale, & adhérente au calice : la fleur a trois étamines, dont deux ont des antheres diviſées en deux parties, & des oreilles à chaque côté, & la troiſieme a

MOM

une anthere fimple avec des oreilles. Toutes ces étamines font jointes en un corps. Les fleurs femelles ont un calice & une corolle femblables à ceux de la fleur mâle; elles font poftées fur le germe, & ont trois courts filamens fans antheres. Le germe, qui foutient un ftyle cylindrique, divifé en deux parties, & couronné par trois ftigmats oblongs & boffus, fe change dans la fuite en un fruit oblong, qui s'ouvre avec élafticité, & montre trois cellules membraneufes, remplies de femences comprimées.

Ce genre de plantes eft rangé dans la dixieme fection de la vingt-unieme claffe de LINNÉE, qui renferme celles qui ont des fleurs mâles & des fleurs femelles placées fur le même pied, & dont les étamines font jointes enfemble.

Les efpeces font :

1°. *Momordica Balfamina, pomis angulatis, foliis glabris, patenti-palmatis. Hort. Cliff.* 451. *Hort. Ups.* 293. *Roy. Lugd.-B.* 262. *Kniph. cent.* 7. *n.* 60, Pomme de Merveille, avec des fruits angulaires & ondés, & des feuilles unies & étendues en forme de main.

Momordica vulgaris. Tourn. Inft. R. H. 103; Pomme de Merveille, commune.

Balfamina rotundi-folia repens, five Mâs. Bauh. Pin. 306.

Charantia. Dod. Pempt. 670.

2°. *Momordica Charantia, pomis angulatis, tuberculatis, foliis villofis, longitudinaliter palmatis. Hort. Cliff.* 451. *Fl. Zeyl.* 351. *Roy. Lugd.-B.* 262; Pomme

MOM 133

de Merveille avec des fruits angulaires & couverts de tubercules, & de feuilles velues, allongées, & en forme de main.

Momordica Zeylanica pampineâ fronde, fructu longiore. Tourn. Inft. R. H. 103; Pomme de Merveille de Céylan, à feuilles de Vigne, & à fruit plus long.

Balfamina cucumerina Indica, fructu majore flavefcente. Comm. Hort. 1. *p.* 103. *f.* 54.

Amera Indica. Rumph. Amb. 5. *p.* 410. *f.* 151.

Pandipavel. Rheed. Mal. 8. *p.* 17. *f.* 9.

3°. *Momordica Zeylanica, pomis ovatis, acuminatis, tuberculatis, foliis glabris, palmatis, ferratis*; Pomme de Merveille, avec des fruits ovales, couverts de tubercules, & terminés en pointe aiguë,& des feuilles unies en forme de main, & fciées.

Momordica Zeylanica, pampineâ fronde, fructu breviori. Tourn. Inft. 103; Pomme de Merveille, de Céylan, avec une feuille de Vigne, & un fruit plus court.

Pavel. Rheed. Mal. 8. *p.* 18. *t.* 10. *Satb. Hort.* 1. *f.* 61.

4°. *Momordica Elaterium, pomis hifpidis, cirrhis nullis. Lin. Sp. Plant.* 1010. *Mat. Med.* 208. *Kniph. cent.* 8. *n.* 72. *Sabb. Hort.* 1. *f.* 64. *Regn. Bot.*; Concombre fauvage, avec un fruit piquant, & fans vrilles aux branches.

Cucumis fylveftris, afininus dictus C. B. P. 314; Concombre fauvage, appelé Concombre des ânes: c'eft l'Elaterium de Boerrhaave.

Balfamina. La premiere efpece croît naturellement en

134 **MOM**

Afie , & les feconde & troi-fieme , dans l'Ifle de Ceylan. Ces plantes font annuelles , & periffent bientôt après que leurs fruits font mûrs ; elles ont , comme les *Concombres* & les *Melons* , des tiges trainantes , qui s'étendent à trois ou qua-tre pieds de longueur , & pouf-fent des branches latérales, ar-mées de vrilles , au moyen defquelles elles s'attachent à routes les plantes voifines , pour fe défendre contre les vents. Ces branches font garnies de feuilles femblables à celles de la *Vigne*. Les feuilles des pre-miere & troifieme efpeces font unies, profondément découpées en plufieurs fegmens , & éten-dues en forme de main ; mais celles de la feconde font plus longues & velues. Le fruit de la premiere , qui eft ovale & terminé en pointe aiguë , a plu-fieurs angles profonds , & des tubercules aigus placés fur les bords : il devient rouge , ou de couleur pourpre en mûrif-fant ; il s'ouvre avec élaftici-té , & jette au loin fes femen-ces (1).

Charantia. Le fruit de la fe-conde eft beaucoup plus long que celui de la premiere , mais moins profondément canelé; fes tubercules font répandus fur

toute fa furface , & ne font pas fi aigus que ceux de la précédente. Ce fruit devient jaune en mûriffant , & lance fes femences avec élafticité.

Zeylanica. Le fruit de la troi-fieme eft court & pointu, com-me celui de la premiere, mais moins gonflé au milieu ; fes angles font moins profonds , & toute fa furface eft fortement garnie de tubercules aigus. Ce fruit devient d'une couleur d'o-range foncée en mûriffant , & jette fes femences de la même maniere.

Elaterium. La quatrieme , à laquelle on donne le nom de *Concombre* fauvage , a un fruit qui jette impétueufement fes femences avec un jus vifqueux , dans lequel elles font renfer-mées , lorfqu'on le touche à fa maturité , ce qui la fait auffi quelquefois nommer *Noli me tangere*. Cette plante croît na-turellement dans quelques par-ties chaudes de l'Europe ; mais en Angleterre, on la cultive dans les jardins, pour fon fruit qui eft d'ufage en Médecine, ou plutôt pour la fécule du jus du fruit , qui eft l'*Elaterium* des Boutiques.

Cette plante a une racine groffe , charnue , & prefque femblable à celle de la *Brionne*, de laquelle fortent au printems plufieurs tiges épaiffes , rudes & traînantes , qui fe divifent en plufieurs branches , & s'é-tendent à trois pieds de tous côtés. Ces branches font gar-nies de feuilles épaiffes , ru-des, prefque en forme de cœur , de couleur grife , & foutenues fur de longs pétioles : fes fleurs

_____

(1) Cette plante a la réputation d'être un des meilleurs vulnéraires connus. L'huile d'amandes douces dans laquelle on a fait infufer fes graines, paffe fur-tout pour un ex-cellent remede contre la piquûre des tendons , les bleffures de toutes efpeces , les hemorrhoides , les ul-ceres de la matrice , la gerfure des mammelles , &c.

naissent aux aisselles de la tige ; elles sont mâles & femelles , & naissent en différens endroits sur la même plante ,
comme celles du *Concombre* commun ; mais elles sont beaucoup plus petites , d'un jaune
pâle , & verdâtres au fond ;
les fleurs mâles sortent sur des
pédoncules courts & épais ,
& les femelles sont placées sur
les jeunes fruits , qui , après
que les fleurs sont fannées, croissent jusqu'à la longueur d'un
pouce & demi , & se gonflent
comme les *Concombres.* Ces
fruits sont gris comme les feuilles , & couverts de piquans
courts : ils ne changent point
de couleur en mûrissant , comme la plupart des autres fruits
de cette classe ; mais si l'on
entreprend de les cueillir , ils
quittent les pédoncules , & jettent leurs semences & leur jus
avec une grande force ; de sorte
que , partout où il y a de ces
plantes , si on laisse le fruit
jusqu'à sa maturité, les semences s'écartent dans tous les environs à une grande distance,
& produisent une grande quantité de nouvelles plantes au
printems suivant.

Quand on veut faire usage
du fruit, il faut toujours le
recueillir avant qu'il soit mûr ;
car si l'on attend ce moment,
on perd la plus grande partie de son jus, qui est la seule
chose qui soit utile. Celui qui
reste mêlé avec le parenchyme
du fruit, n'est pas, à beaucoup
près, aussi bon. L'*Elaterium* fait
du jus pur, est beaucoup plus
blanc, & conserve sa vertu
beaucoup plus long-tems que

celui qu'on obtient par expression.

Culture. Les trois premieres
especes sont annuelles : on seme
leurs graines sur une couche
chaude au commencement du
mois de Mars ; quand les plantes poussent, on les transplante
sur une nouvelle couche chaude, comme on le pratique pour
les *Concombres* & les *Melons* ,
en mettant deux plantes de la
même espece sous chaque vitrage : on les arrose, on les
tient à l'ombre, jusqu'à ce qu'elles aient formé de nouvelles
racines ; on les traite ensuite
comme les *Concombres*, en permettant à leurs branches de
s'étendre sur la terre de la
même maniere, & on les tient
nettes de mauvaises herbes.

Avec ce traitement, pourvu
qu'on ne leur donne pas trop
d'humidité, & qu'on ne les expose pas trop au plein air, elles produiront leurs fruits en
Juillet, & leurs semences mûriront en Août & Septembre,
tems auquel il est nécessaire de
les cueillir aussi-tôt qu'on voit
le fruit s'ouvrir.

On conserve ces plantes dans
les jardins des curieux, à cause
de la singularité de leurs fruits ;
mais comme elles occupent
beaucoup de place sur les couches chaudes, qu'elles exigent
beaucoup de soins, qu'elles ont
peu de beauté, & qu'elles ne
sont pas d'un grand usage, on
ne les cultive guere en Angleterre, à moins que ce ne
soit dans les jardins de Botanique, pour la variété.

Quelques personnes mettent
ces plantes dans des pots, fixent

leurs tiges à des baguettes, pour les empêcher de ramper sur la terre, & les placent dans une serre chaude, où elles produisent leurs fruits assez bien, si elles sont traitées avec intelligence : de cette maniere, elles font plus d'effet que quand on laisse traîner leurs branches comme celles des *Concombres* & *Melons*. Cependant quand on les laisse ramper sur la terre, qui est la maniere naturelle dont elles croissent, elles profitent beaucoup mieux, & produisent plus de fruits ; car quoi-qu'elles soient armées de vril-les, ce n'est cependant pas pour grimper, mais seulement pour se fixer à tous les soutiens voisins, & se garantir des secousses des vents, qui brisent souvent leurs branches, quand elles y sont exposées.

La quatrieme espece s'éleve aisément des semences, qui poussent une grande quantité de plantes au printems suivant, quand on leur donne le tems de se répandre, ou quand on les seme sur une planche de terre légere ; les plantes poussent environ un mois après, & peuvent être transplantées ensuite dans un terrein ouvert en rangs éloignés de trois ou quatre pieds, & à une distance égale entr'elles. Si l'on fait cette opération avec soin, tandis qu'elles sont encore jeunes, on ne court aucun risque de les voir manquer ; & quand elles ont pris racine, il suffit de les débarrasser des herbes inutiles qui les environnent. Si la terre dans laquelle elles sont plantées est seche, leurs racines subsisteront pendant trois ou quatre années, à moins qu'il ne survienne un hiver très-rude, qui les feroit périr.

MONARDA. *Lin. Gen. Plant.* 34. *Leonurus. Tourn. Inst. R. H.* 187. *tab.* 87. [*Monarda*].

Caracteres. Le calice est tu-bulé, cylindrique, & formé par une feuille canelée & dé-coupée sur ses bords en cinq parties égales : la corolle est monopétale & labiée ; elle a un tube cylindrique plus long que le calice, & divisé au sommet en deux levres, dont la supérieure est étroite, en-tiere & érigée, & l'inférieure, qui est large, & divisée en trois parties réfléchies, a son segment du milieu long & étroit, & les latéraux obtus : la fleur a deux étamines hérissées, de la longueur de la levre supérieure, dans laquelle elles sont enveloppées, & ter-minées par des antheres com-primées & érigées ; dans le fond du tube est placé un germe à quatre pointes, qui soutient un style mince, enve-loppé avec des étamines, & couronné par un stigmat aigu, & divisé en deux parties. Ce germe se change ensuite en quatre semences nues, ren-fermées dans le calice.

Ce genre de plantes est ran-gé dans la premiere section de la seconde classe de LINNÉE, qui comprend celles dont les fleurs ont deux étamines & un style.

Les especes sont :

1°. *Monarda fistulosa*, ca-pitulis terminalibus, caule obtus-angulo. *Hort. Upsal.* 12. Mat.

MON

Med. p. 39. *Kniph. Orig. cent.* 2. *n.* 47. *Fabric. Helmſt.p.* 95 ; Monarde avec des têtes de fleurs qui terminent les tiges, & dont les angles ſont obtus.

Monarda floribus capitatis, caule obtuſo. Vir. Cliff. 3. *Roy. Lugd·B.* 313. *Hort.Cliff.* 11. *Riv.Mons.* 58·

Leonurus Canadenſis, Origani folio. Tourn.Inſt. R. H. 187 ; Leonurus de Canada, à feuilles d'Origanum. Monarde du Canada.

Origanum fiſtuloſum Canadenſe. Corn. Canad. 13. *f.* 14.

Monarda mollis. Amœn. Acad. 3. *p.* 399.

Clinopodium majus Virginienſe, foliis minùs hirtis, acutioribus, floribus fiſtuloſis. Moris. Hiſt. 3. *p.* 474 ; Variété.

2°. *Monarda didyma, floribus capitatis, ſub—didynamis, caule acutangulo. Lin. Sp. Plant.* 32. *Kniph. Orig, cent* 2. *n.* 46 ; Monarde avec des fleurs rapprochées en têtes, dont les étamines ſont preſque en deux corps, & une tige aiguë & angulaire.

Monarda caule acutè angulato, capitulis terminalibus. Hort. Cliff. 495. *Cold. Noveb.* 7.

Monarda floribus capitatis, verticillatiſque, caule acutangulo, foliis lanceolato-ſerratis, glabris. Butt. Cun. 226. *Trew. Ehret.* 32. *f.* 66 ; Monarde à fleurs recueillies en tête, & verticillées, avec une tige à angles aigus, & des feuilles unies, ſciées, & en forme de lance, communément appelées Thé d'Oſwego. Monarde de Penſylvanie.

3°. *Monarda punɛtata, floribus verticillatis, corollis punɛtatis. Hort. Upſal.* 12. *Hort. Cliff.* 495. *Gron. Virg.* 9. *Roy.*

MON 137

Lugd.-B. 313. *Sabb. Hort. Rom.* 3. *f.* 86. 87 ; Monarde à fleurs verticillées, dont les pétales ſont ponɛtuées.

Clinopodium Virginianum anguſti-folium, floribus amplis luteis, purpurâ maculâ notatis, cujus caulis ſub-quovis verticillo decem, vel duodecim foliolis rubentibus eſt circumcinɛtus. Baniſt. Raii Sup. 300 ; Clinopode de Virginie, à feuilles étroites, avec de groſſes fleurs jaunes tachetées de pourpre, & des tiges garnies de dix ou douze petites feuilles rougeâtres ſous chaque tête verticillée de fleurs.

F.ſtuloſa. La premiere eſpece, qui croît naturellement dans le Canada & dans quelques autres parties Septentrionales de l'Amérique, a une racine vivace, compoſée de fortes fibres, qui s'étendent au loin de tous côtés : ſes tiges s'élèvent à la hauteur d'environ trois pieds ; elles ſont velues, & ont des angles obtus ; elles pouſſent vers le haut deux ou quatre branches latérales, garnies de feuilles oblongues, larges à leur bâſe, mais terminées en pointe aiguë, velues, un peu dentées ſur leurs bords, poſtées ſur des pétioles courts & velus, & oppoſées : la tige & les branches ſont terminées par des têtes de fleurs pourpre, qui ont une enveloppe compoſée de cinq feuilles à pointe aiguë au milieu : ces fleurs ont chacune deux étamines plus longues que les corolles, avec un ſtyle de la même longueur, & couronné par un ſtigmat diviſé en deux parties ; elles paroiſſent dans

le mois de Juillet , & pro- duisent des femences qui mû- riffent en automne.

Didyma. La feconde efpece eft originaire des mêmes contrées que la premiere : les habitans de l'Amérique font infufer fes feuilles en guife de *Thé* , & lui donnent pour cette raifon le nom de *Thé d'Ofwego* , fous lequel nom elle a été apportée en Angleterre ; elle a une ra- cine vivace , & une tige an- nuelle qui périt en automne. Les tiges de cette efpece font liffes & ont quatre angles aigus ; elles s'élevent à la hau- teur d'environ deux pieds , & font garnies de feuilles unies , ovales , en forme de lance , dentées fur leurs bords , oppo- fées , fupportées par de fort courts pétioles , & qui répan- dent , lorfqu'on les froiffe , une odeur fort agréable & ra- fraîchiffante : les tiges pouffent vers leur fommet deux ou quatre petites branches laté- rales , garnies de petites feuilles de la même forme que celles de la précédente : les fleurs naiffent en groffes têtes verti- cillées aux extrémités' des ti- ges , & fouvent il y a une tête plus petite de fleurs ver- ticillées , qui croît à un nœud au-deffous de la groffe tête , & une pareille qui s'élève au- deffus , fur un pédoncule nud.

Ces fleurs font d'un rouge brillant ; elles ont deux levres , dont la fupérieure eft longue , étroite & entiere , & l'infé- rieure eft découpée en trois parties ; elles ont chacune deux étamines plus longues que la corolle , & terminées par des

antheres comprimées , & plu- fieurs ont encore deux étamines plus courtes & fans antheres. Cette plante fleurit dans le mois de Juillet ; mais dans les années pluvieufes , ou quand elle fe trouve placée dans un fol humide , elle continue à produire de nouvelles fleurs jufqu'au milieu ou la fin de Septembre.

Culture. On peut multiplier ces deux efpeces , en divifant leurs racines : la premiere ne fe multiplie pas auffi vite que la feconde ; mais comme elle produit une grande quantité de femences , on peut y fuppléer par ce moyen. Quand on feme fes graines en automne , auffi- tôt qu'elles font mûres , les plantes pouffent au printems fuivant ; mais fi on ne les met en terre qu'au printems , il eft rare qu'elles paroiffent dans la même année.

Lorfqu'elles ont pouffé , & qu'elles font en état d'être en- levées , on les tranfplante dans une plate-bande , à l'ombre , à neuf pouces environ de dif- tance ; & lorfqu'elles ont for- mé de nouvelles racines , el- les n'exigent plus aucun au- tre foin que d'être tenues net- tes de mauvaifes herbes. En automne , on les tranfplante à demeure dans les plates-ban- des , où elles fleuriront dans l'été fuivant , & produiront des femences mûres : leurs racines dureront plufieurs années , & pourront être divifées chaque deux ans , pour les multiplier. Cette efpece exige une terre molle & marneufe , & une fi- tuation peu expofée au foleil.

MON

La feconde efpece perfectionne rarement fes femences en Angleterre ; mais elle fe multiplie affez promptement par fes racines rampantes , ainfi que par boutures , qui prennent auffi aifément racine que celles de la *Menthe* , lorfqu'on les plante au mois de Mai , fur une plate - bande à l'ombre ; mais comme fes racines s'étendent beaucoup , il n'eft pas néceffaire de fe fervir d'autre moyen pour la multiplier.

Cette efpece demande un fol humide & léger. Si on la plante à l'expofition du levant, elle reftera plus long-tems en fleurs que dans une fituation plus chaude ; elle produit un agréable effet dans les jardins : fes feuilles ont une odeur fraîche & agréable , & bien des perfonnes aiment à fe fervir de fes jeunes feuilles en guife de *Thé.*

Punctata. La troifieme naît fans culture dans l'Amérique Septentrionale ; elle eft bis-annuelle , car fes racines periffent dans la feconde année , lorfqu'elle a perfectionné fes femences ; elle a des tiges quarrées d'environ deux pieds de hauteur , qui pouffent, depuis leur bâfe jufqu'au fommet , des branches garnies de feuilles en forme de lance , & difpofées en grappes à chaque nœud. Outre ces feuilles , on en voit encore deux autres , qui font larges & oppofées , & plufieurs petites à chaque côté de la tige ; les plus larges ont environ deux pouces & demi de longueur fur neuf lignes de largeur , & font légerement dentées fur leurs bords. Vers le haut de la tige fortent les fleurs en groffes têtes verticillées , dont chacune a une enveloppe compofée de dix ou douze petites feuilles , d'un rouge pourpre en-deffus. Ces fleurs font larges , de la même forme que celles des autres efpeces, d'un jaune fale , & tachetées de pourpre ; elles ont chacune deux longues étamines , placées fous la levre fupérieure, & terminées par des antheres comprimées & divifées en deux parties ; à ces fleurs fuccedent quatre femences nues , & renfermées dans le calice. Cette plante fleurit en Juillet ; & fi l'été eft favorable , fes femences mûriffent quelquefois en automne.

On la multiplie par fes graines , qu'on feme fur une plate-bande de terre légere à l'expofition du levant où les plantes leveront fort aifément. Quand elles font en état d'être tranfplantées , on peut les placer dans une plate-bande à l'ombre , comme celles de la premiere efpece. Si , par hafard , elles pouffent des tiges de fleurs dans la premiere année , il faut les couper , pour fortifier les racines & leur faire pouffer des jets latéraux : car fi on les laiffe fleurir , il eft à craindre qu'elles ne périffent pendant l'hiver. On enleve ces plantes en automne , pour les mettre dans des plates-bandes ouvertes du parterre, où elles fleuriront dans l'été fuivant. Dans les années feches , il faut les arrofer beau-

MOR

coup ; car sans cela, elles ne deviennent pas aussi belles, & ne produisent pas de bonnes semences.

MONARDE *Voy.* MONARDA.

MONAVIE *ou* LE MASQUE. *Voyez* MIMULUS.

MONBIN. *Voyez* SPONDIAS LUTEA. *L.*

MONTIA. *Voyez* HELIO-CARPOS.

MORÆA. *Lin. Gen. Plant.* 60. [*Morea.*] la Morée.

Caracteres. La gaîne de la fleur a deux valves ; la corolle est composée de six pétales, dont les trois supérieurs sont érigés & divisés en deux parties, & les trois inférieurs sont étendus : la fleur a trois étamines courtes, & terminées par des antheres oblongues ; le germe, qui est placé au-dessous de la fleur, soutient un style mince, couronné par un stigmat érigé & divisé en trois parties ; ce germe devient ensuite une capsule à trois angles, qui forment trois sillons, & à trois cellules remplies par plusieurs semences rondes.

Ce genre de plantes est rangé dans la premiere section de la troisieme classe de LINNÉE, intitulée, *Triandria Monogynia*, qui comprend celles dont les fleurs ont trois étamines & un style.

Les especes sont :

1°. *Moræa vegeta*, *spathâ uni-florâ*, *foliis gladiolatis* ; Morée avec une fleur dans chaque spathe, & des feuilles en forme d'épée.

Moræa foliis canaliculatis. Lin. Sp. 59; Morée à feuilles canelées.

2°. *Moræa juncea*, *spathâ bi-florâ*, *foliis subulatis* ; Morée avec deux fleurs dans chaque spathe, & des feuilles en forme d'alêne.

Moræa foliis subulatis. Lin. Sp. 59 ; Morée à feuilles en forme d'alêne.

Ces deux plantes sont originaires du Cap de Bonne Espérance, d'où leurs semences m'ont été envoyées. Ces graines ont réussi dans le jardin de Chelséa, & les plantes qu'elles ont produites ont donné plusieurs fois des fleurs. Ces fleurs different de toutes celles des autres genres de la même classe : je l'ai nommée *Moræa*, en l'honneur de ROBERT MORE, Ecuyer de Shrewsbury, qui est très-versé en Botanique, ainsi que dans beaucoup d'autres parties de l'Histoire Naturelle.

Vegeta. La premiere espece a, comme l'*Iris à feuilles de jonc*, des racines fibreuses, desquelles sortent plusieurs feuilles en forme d'épée, de cinq ou six pouces de longueur, sur un demi-pouce de largeur au milieu, mais plus étroites vers les deux extrémités, d'un vert foncé, & placées l'une sur l'autre à leur bâse, comme celles de l'*Iris* ; sa tige qui sort de la racine entre les feuilles, & s'éleve à près de huit pouces de hauteur, est garnie d'une petite feuille à chaque nœud, & terminée par une fleur couverte d'une spathe à deux valves. Cette fleur est d'un blanc sale ; chacun de ses pétales est tacheté d'un rouge pourpre vers sa partie supérieure, & d'une tache

tache grande, belle, & jaune à l'onglet; dans fon centre font trois étamines minces, terminées par des antheres oblongues, & un ftyle couronné par un ftigmat oblong & divifé en deux parties. Ces fleurs paroiffent dans le mois de Juin, & leurs femences mûriffent à la fin de Juillet.

Juncea. La feconde efpece a une racine bulbeufe, un peu comprimée fur les côtés, & couverte d'une peau unie & foncée èn couleur; de cette racine s'élevent trois ou quatre feuilles en forme d'alène, & d'un vert pâle, dont quelques-unes ont cinq pouces de longueur, & d'autres fept ou huit, fur environ fix lignes de largeur, & qui font terminées en trois angles : les pédoncules des fleurs s'élevent à-peu-près à la hauteur de fix pouces, & font généralement courbés au nœud du bas; ils font garnis à chaque nœud d'une petite feuille, dont la bâfe embraffe prefque la tige, & terminés par deux fleurs entourées d'une fpathe fanñée : ces fleurs font de couleur d'orange ; leurs pétales font larges vers le haut, & joints enfemble à leurs bâfes; elles paroiffent en Juin, & leurs femences mûriffent à la fin de Juillet.

Culture. La premiere efpece fe multiplie par femences, ou en divifant fes racines; la feconde, auffi par femences, ou par le moyen de fes rejettons. Le meilleur tems pour tranfplanter & féparer ces rejettons, ainfi que pour divifer les racines de la premiere, eft le

Tome V.

mois d'Août, afin qu'elles puiffent pouffer de nouvelles fibres avant l'hiver. Cette faifon eft auffi la meilleure pour mettre leurs graines en terre. Si on les répand dans de petits pots, & qu'on les tienne dans une couche de vieux tan, fous un vitrage ordinaire en hiver, on ne court pas le rifque de les voir manquer. Les plantes exigent auffi le même abri en hiver ; car comme elles font trop tendres pour profiter en plein air dans ce pays, & qu'elles font fujettes à filer dans les ferres, on eft obligé de les mettre fous des châffis, de maniere qu'elles puiffent y jouir de beaucoup d'air en hiver, quand le tems eft doux, & être auffi à l'abri des gelées & des fortes pluies : par ce moyen, elles fleuriront & perfectionneront leurs femences beaucoup mieux que de toute autre maniere. En été, il faut les tenir entierement en plein air jufqu'au mois d'Octobre, & les remettre enfuite à couvert.

MORELLE CERISETTE *ou* AMOMOM. *Voyez* SOLANUM PSEUDOCAPRICUM.

MORELLE A FRUIT NOIR. *Voy.* SOLANUM NIGRUM. *L.*

MORELLE GRIMPANTE DE MALABAR. *Voy.* BASELLA.

MORELLE GRIMPANTE *ou* VIGNE-VIERGE. *Voyez* SOLANUM DULCAMARA.

MORELLE MORTELLE *ou* POISON. *Voyez* ATROPA *L.*

MORELLE *ou* RAISIN D'Amérique. *V.* PHYTOLACCA. *L.*

MORGELINE. *Voy.* ALSINE MEDIA. *L.* SPERGULA. *L.* MOLLUGO.

K

MOR

MORINA. *Tourn. Cor.* 48. *tab.* 480. *Lin. Gen. Plant.* 39. *Diototheca. Vaill. Mém. Acad.* 1722. [*The Morina.*] la Morin.

Caraêleres. Le calice de la fleur est double; celui qui est placé sous le fruit est tubulé, cylindrique, persistant, & formé par une feuille découpée sur ses bords; celui de la fleur est tubulé, divisé en deux parties, persistant, & aussi formé par une feuille; la corolle est monopétale, son tube est long, élargi vers le haut, & un peu courbé; son sommet est divisé en deux levres, dont la supérieure est petite & découpée en deux segmens, & l'inférieure est divisée en trois parties égales & obtuses, dont celle du milieu s'étend au-delà des deux autres. La fleur a deux étamines hérissées, situées près du style, & terminées par des antheres en forme de cœur, & érigées: le germe, qui est globulaire, est placé sous la fleur; il soutient un style mince, plus long que les étamines, & couronné par un stigmat en forme de targe. Ce germe se change, quand la fleur est passée, en une semence simple, couronnée par le calice.

Ce genre de plantes est rangé dans la premiere seêtion de la seconde classe de LINNÉE, qui comprend celles dont les fleurs ont deux étamines & un style.

Nous ne connoissons encore qu'une espece de ce genre, qui est la

Morina Orientalis. Hort. Cliff. 14.

Morina Orientalis Carlinæ folio. Tourn. Cor. it. 3. p. 131. f.

132; la Morin Orientale à feuilles de Carline.

Morina Persica. Lin. Syst. Plant. tom. 1. p. 73.

Cette plante a été découverte dans le Levant par TOURNEFORT, qui lui a donné ce nom, en l'honneur du Doêteur MORIN, Médecin de Paris.

Elle croît naturellement près d'Erzerum en Perse, d'où elle a été portée dans le Jardin Royal de Paris &`en Angleterre; mais toutes ces plantes ont été détruites par le rude hiver de 1740, à l'exception d'une seule qui se trouvoit dans le jardin de M. DUHAMEL.

Sa racine, qui est épaisse & cylindrique, s'enfonce profondément dans la terre, & pousse plusieurs fibres épaisses & aussi grosses que le doigt: sa tige s'éleve presque à la hauteur de trois pieds; elle est lisse, de couleur pourpre vers le bas, velue & verte au sommet, & garnie à chaque nœud de trois ou quatre feuilles épineuses, comme celles de la *Carline*, de quatre ou cinq pouces de longueur sur un pouce & demi de largeur, d'un vert luisant en-dessus, un peu velues en-dessous, & armées d'épines sur leurs bords: ses fleurs naissent aux aîles des feuilles tout autour de la tige; elles ont des tubes fort longs, étroits au fond, larges au sommet, & un peu courbés; leur bords sont évasés & divisés en deux grosses levres, dont la supérieure est dentelée à son extrémité & plus ronde, & l'inférieure est découpée en trois segmens obtus; sous la levre

MOR

font fixées deux étamines courbées & couronnées par des fommets jaunes. Ces fleurs paroiffent dans le mois de Juillet, & ne produifent jamais de femences ; quelques-unes font blanches, & d'autres de couleur tirant fur le pourpre, fur la même plante.

On multiplie cette plante au moyen de fes graines, qu'il faut femer en automne, auffitôt qu'elles font mûres, fans quoi elles ne pouffent pas dans le premier été : car j'ai fouvent obfervé qu'en ne les mettant en terre qu'au printems, elles ne germent qu'au bout de quatorze ou quinze mois. On les feme dans les places où elles doivent refter, parce que ces plantes pouffent des racines qui pénétrent très - profondément dans la terre ; & que, quand on vient à les rompre, en les tranfplantant, elles profitent rarement après. On peut les femer fur des plates-bandes d'une terre légere, en marquant les endroits où elles ont été placées, afin qu'on ne les dérange point ; car lorfqu'on a labouré la terre où elles fe trouvent, il arrive fouvent qu'elles ne pouffent que dans l'année fuivante, quoiqu'elles aient été femées en automne. La terre où on les a femées, doit être tenue nette de mauvaifes herbes. Quand les plantes commencent à pouffer, on les éclaircit dans les endroits où elles font trop ferrées, en laiffant entr'elles environ huit pouces d'intervalle, & on les tient conftamment nettes. Au printems, un peu avant qu'elles

MOR 143

commencent à pouffer de nouvelles feuilles, on laboure légerement la terre autour de leurs racines, & on en répand un peu de la nouvelle fur la furface, pour les ranimer.

En automne, ces plantes périffent jufques fur terre, & pouffent de nouvelles feuilles au printems fuivant ; mais elles ne produifent de fleurs qu'à l'âge de trois ans : après ce tems, elles fleuriffent chaque été, & leurs racines durent plufieurs années, pourvu qu'elles ne foient pas dérangées ou détruites par de trop grands froids.

MORINGHA. *Voyez* GUILANDINA MORINGA. *L.*

MORS DU DIABLE, *Succife* ou *Scabieufe des bois. Voyez* SCABIOSA SUCCISA. *L.*

MORT AUX RATS. *Voyez* HAMELLIA PATENS. *L. &* HAMAMELIS VIRGINIANA. *L.*

MORUS. *Tourn. Inft. R. H.* 589. *tab.* 363. *Lin. Gen. Plant.* 936, de μαυρος, noir, parce que fon fruit eft ordinairement de cette couleur. [*The Mulberry Tree.*] Mûrier.

Caractères. Cet arbre a des fleurs mâles placées à quelque diftance des fleurs femelles fur la même tige ; les fleurs mâles font recueillies en chatons longs & cylindriques ; elles font apétales, & ont quatre étamines en forme d'alêne, érigées, plus longues que le calice, & terminées par des antheres fimples : les fleurs femelles font raffemblées en têtes rondes ; elles n'ont point de pétales, mais feulement un germe en forme de

K 2

MOR

cœur, qui foutient deux ftyles longs , rudes , réfléchis , & couronnés par des ftigmats fimples ; le calice de celles-ci fe change dans la fuite en un fruit large , charnu, fucculent , & compofé de plufieurs tumeurs ou baies , qui renferment chacune une femence ovale.

Ce genre de plantes eft rangé dans la quatrieme fection de la vingt-unieme claffe de LINNÉE , qui comprend celles qui ont des fleurs mâles & femelles fur le même pied , & dont les fleurs mâles ont quatre étamines.

Les efpeces font :

1°. *Morus nigra , foliis cordatis. Hort. Cliff.* 441. *Hort. Upf.* 283. *Mat. Med.* 201. *Roy. Lugd.-B.* 211. *Dalib. Paris.* 290. *Kniph. cent.* 3. *n.* 64. *Regn. bot.* ; Mûrier à feuilles en forme de cœur.

Morus fruclu nigro. C. B. P. 459 ; Mûrier à fruits noirs , ou le Mûrier commun.

2°. *Morus laciniata , foliis palmatis , hirfutis* ; Mûrier à feuilles en forme de main , & velues.

Morus fruclu nigro minori , foliis eleganter laciniatis. Tourn. Infl. R. H. ; Mûrier à petit fruit noir , & à feuilles élégamment découpées.

3°. *Morus rubra , foliis cordatis , fubtus villofis , amentis cylindricis. Lin. Sp. Plant.* 986 ; Mûrier à feuilles en forme de cœur , & velues en-deffous , avec des chatons cylindriques.

Morus foliis cordatis , fcabris , villofis , amentis cylindricis. Du Roi Harpk. 1. *p.* 430.

Morus foliis fubtus tcmentofs ,

MOR

amentis longis , dioïcis. Gron. Virg. 146.

Morus Virginienfis arbor , Loti arboris inftar ramofa , foliis ampliffimis. Pluk. Phyt. tab. 246. *fol.* 4. *Duham. Arb.* 7 ; Mûrier de Virginie , branchu comme l'Alifier , & à feuilles très-larges.

4°. *Morus alba , foliis obliquè cordatis , lævibus. Hort. Cliff.* 441. *Hort. Ups.* 283. *Roy. Lugd.-B.* 211. *Dalib. Par.* 290. *Gmel. it.* 3. 374. *Scop. cam. ed.* 2. *n.* 1176. *Du Roi Harpk.* 1. *p.* 473 ; Mûrier avec des feuilles obliques en forme de cœur , & liffes.

Morus fruclu albo. C. B. P. 459 ; Mûrier à fruits blancs.

Morus candida. Dod. Pempt. 810.

5°. *Morus tinctoria , foliis obliquè-cordatis , acuminatis , hirfutis* ; Mûrier à feuilles obliques en forme de cœur , hériffées & terminées en pointe aiguë.

Morus fruclu viridi , ligno tinctorio. Sloan. Hift. Jam. 2. *p.* 3 ; Mûrier à fruit vert , & dont le bois eft teint en couleur de foufre , ou bois fuftique.

Fuftick-Wood. Raü. Dendr. 666.

6°. *Morus papyri-fera , foliis palmatis , fruclibus hifpidis. Lin. Sp. Plant.* 986 ; Mûrier à feuilles en forme de main , & à fruits épineux.

Morus fativa , foliis Urticæ mortuæ , cortice papyri-fera. Kæmp. Aman. 471 ; Mûrier cultivé , à feuilles d'Ortie morte , ayant une écorce propre à faire du papier.

7°. *Morus fTartarica , foliis*

MOR

ovato-oblongis , utrinquè æqualibus , inæqualiter ferratis. Flor. Zeyl. 337 ; Mûrier à feuilles ovales , oblongues , liffes fur les deux furfaces . & fciées inégalement.

Tinda parva. Hort. Mal. 1. p. 87. *fol.* 49.

Morus Indica. Lin. Syft. Plant. tom. 4. p. 135. *Sp.* 5.

Tinda parva Rheed. Mal. 1. p. 87.

8°. *Morus Zanthoxylum , foliis ovato-oblongis , acuminatis , obliquis , ramis aculeatis ;* Mûrier à feuilles ovales , oblongues , à pointe aiguë , & obliquement placées fur les pétioles , avec des branches épineufes.

Zanthoxylum aculeatum , Carpini foliis , Americanum , cortice cinereo. Pluk. Phyt. 239. *fol.* 3 ; Zanthoxylum épineux d'Amérique à feuilles de Charme , ayant une écorce cendrée.

Morus quæ Tata-iba. Plum. ic. 199. *f.* 204.

Tata-iba. Marcgr. Bras. 119.

Nigra. La première efpece eft le Mûrier noir commun, qu'on cultive pour fon fruit. Cet arbre croît naturellement en Perfe , d'où il a été d'abord porté dans les parties méridionales de l'Europe , & enfuite dans toutes les autres contrées où les hivers ne font pas fort rigoureux ; car dans le nord de la Suede, il ne fubfifte pas en plein air : on le plante contre des murailles dans plufieurs parties de l'Allemagne , où on le traite comme on fait ici pour le *Pêcher* & les autres fruits tendres.

Cet arbre produit les deux

MOR 145

fexes : les fleurs mâles ou les chatons fe trouvent fur le même pied avec les fruits ; mais il arrive fouvent que quelques-uns de ceux qu'on élève de femences , n'ont que des fleurs mâles , & ne produifent point de fruits, de forte que ceux qui plantent ces arbres , pour en recueillir les fruits, ne doivent jamais choifir des tiges de femences,à moins qu'on ne leur ait vu produire du fruit dans les pépinieres,où l'on doit toujours marquer ceux qui font fructueux. Il arrive auffi quelquefois que des arbres élevés de marcottes ne produifent que des fleurs mâles ; car j'ai fouvent obfervé que quelques-unes de groffes branches de ceux-ci ne donnoient que des chatons, pendant que les autres étoient chargées de fruits : de maniere que fi l'on ne choifit pas ces branches fructueufes pour les marcotter, on courra le même rifque qu'avec les arbres élevés de femences : il ne faut pas non plus marcotter les branches qui fortent près des racines des vieux arbres; car celles-ci ne donnent des fruits que plufieurs années après qu'elles font plantées , quoique les arbres d'où elles proviennent foient extrêmement fructueux. J'en ai vu quelques-uns qui n'ont donné que des chatons pendant plufieurs années , & qui enfuite ont produit des fruits. J'ai obfervé la même chofe fur les *Novers* ; & mon ami le Chevalier RATHGEB m'a dit avoir fait la même obfervation fur le *Lentifque* & le *Térébinthe.*

K 3

Comme les vieux Mûriers font non seulement plus fructueux que les jeunes, & que leurs fruits font encore beaucoup plus gros & plus favoureux, lorfqu'on poffede quelques-uns de ces vieux arbres dans un jardin, il faut les choifir de préférence pour marcotter, & auffi prendre en même tems les branches les plus fructueufes. Ces branches pouffent des racines dans une année ; au bout de ce tems, on les fépare des vieux arbres : mais comme les branches les plus fructueufes font fouvent très-éloignées de la terre, & qu'on ne peut les y courber affez pour les y fixer, on fe fert de caiffes ou de paniers remplis de terre, qu'on éleve à leur hauteur. La meilleure méthode pour multiplier cette efpece eft la bouture. Ces boutures prennent très-aifément racine, fi elles font bien choifies & conduites avec intelligence, & elles forment de bons arbres par la fuite. Pour faire ces boutures, il faut choifir des branches de l'année précédente, & conferver un nœud du bois de deux ans à leur partie baffe. On ne raccourcit point ces boutures, mais on les plante dans toute leur longueur, & on laiffe deux ou trois boutons au-deffus de la terre. La meilleure faifon pour les planter eft le mois de Mars, lorfque les fortes gelées font paffées ; on les place dans une terre riche & légere, que l'on comprime bien autour ; on les couvre de vitrages, pour leur faire pouffer plutôt des racines ; &

quand on n'a pas cette facilité, on couvre la terre qui les environne avec de la mouffe, pour l'empêcher de fe deffécher : au moyen de cette méthode, les boutures n'exigeront que très-peu d'eau, & réuffiront beaucoup mieux qu'en les arrofant fouvent. Lorfque ces boutures ont bien pouffé, & produit de bonnes branches, on peut les tranfplanter au printems fuivant dans une pépiniere, où on les dreffe régulierement en tiges, en enfonçant contre chacune des piquers auxquels on attache les tiges principales, en retranchant la plupart des branches laterales, & en n'en laiffant que deux ou trois pour retenir la féve ; car lorfque ces tiges font entierement dépouillées de leurs branches de côté, toute la féve monte au fommet, & les têtes deviennent trop fortes pour que les tiges puiffent les foutenir.

Au bout de quatre années de féjour dans la pépiniere, ces plantes pourront être placées à demeure. Cette tranfplantation s'exécute alors avec plus de fûreté que fi elles étoient plus jeunes, ou qu'elles fuffent devenues d'une groffeur plus confidérable.

Si l'on plante ces boutures dans une planche entierement expofée au foleil, il fera prudent de difpofer au-deffus des cerceaux, pour pouvoir les couvrir avec des nattes pendant la chaleur du jour, jufqu'à ce qu'elles aient pris racine : mais après cela, on les expofera au foleil le plus qu'il fera poffible ; ce qui leur fera faire de grands

MOR

progrès, pourvu que la terre soit couverte de mousse ou de terreau, pour l'empêcher de se dessécher : au moyen de cela, le soleil durcira les branches, & les rendra plus propres à résister aux premieres gelées de l'automne ; au-lieu que, si elles se trouvent placées plus à l'abri, elles croîtront plus vigoureusement : mais étant plus succulentes, les premieres gelées d'Octobre détruiront souvent leurs sommets ; & si l'hiver suivant est rude, elles périront jusqu'à la racine, & quelquefois entierement. J'ai essayé deux ou trois fois de planter des boutures de Mûriers sur une couche chaude, où elles ont très-bien réussi.

Cette idée m'est venue, en observant des baguettes de Mûriers, qui ayant été coupées pour servir de fourches, & placées dans une couche chaude, pour soutenir des branches de *Concombres*, avoient pris racine, quoiqu'elles eussent été séparées de l'arbre depuis longtems. Ainsi, si l'on est pressé de multiplier ces arbres, on peut en planter des boutures sur une couche de chaleur modérée, où elles prendront racine beaucoup plutôt qu'en pleine terre.

Cet arbre se plaît dans une terre riche & légere, telle que celle qui se trouve ordinairement dans les vieux jardins potagers des environs de Londres, où il y a une grande profondeur de bonne terre. On voit dans quelques-uns de ces jardins des arbres très-vieux, qui sont encore fort sains & fructueux, & dont les fruits sont

MOR 147

gros & plus savoureux que ceux des plus jeunes arbres. Je n'ai encore vu aucun de ces arbres planté dans une terre forte, dans des lieux bas, ou sur de la glaise, la craie, ou le gravier, qui se soit conservé sain & fructueux ; leurs tiges & leurs branches sont au contraire toujours couvertes de mousse, & le peu de fruits qu'ils produisent sont petits, de mauvais goût, & mûrissent tard.

En plantant ces arbres à l'abri des vents impétueux du midi & du nord ouest, leurs fruits se conserveront mieux ; mais les plantations ou les bâtimens qui les garantissent de l'action des vents, doivent être assez éloignés pour ne point les priver de l'aspect du soleil ; car s'ils sont trop à l'ombre, la rosée du matin séjourne long-tems sur leurs fruits, & les fait pourrir sur l'arbre. Quand ils sont ainsi abrités, il n'est plus nécessaire de les émonder ni de retrancher aucunes branches, quand même elles se croiseroient, parce que le fruit est toujours produit sur le jeune bois.

Laciniata. La seconde espece est originaire de la Sicile, d'où l'on m'a envoyé quelques semences, qui ont produit un grand nombre de plantes. Celle-ci differe totalement, par ses feuilles, du *Mûrier* commun, & je ne doute point qu'elle ne soit une espece distincte. Cet arbre est aussi moins élevé que le précédent, & son fruit, qui est petit & sans saveur, ne vaut pas la peine d'être cultivé. Plusieurs des arbres de cette espece que j'ai élevés, ont pro-

K 4

duit du fruit pendant deux ou trois ans dans le jardin de Chel-séa.

Alba. Le *Mûrier blanc* est communément cultivé pour ses feuilles, qui servent à nourrir les vers à soie en France, en Italie, &c. Quoique les Persans fassent toujours usage des feuil-les du *Mûrier noir* commun, pour leurs vers, un Gentil-homme, digne de foi, qui s'est servi des deux especes de feuilles, m'a assuré que les vers nourris avec les feuilles de *Mûrier noir* donno'ent une soie bien meilleure que celle des vers nourris avec la feuille du *Mû-rier blanc*; mais il observe qu'il ne faut jamais donner aux vers des feuilles du *Mûrier noir*, quand ils ont mangé pendant quelque tems les feuilles du *Mûrier blanc*, parce que ce chan-gement de régime les fait très-souvent périr.

On ne doit pas laisser trop grandir les arbres destinés à nourrir des vers à soie : mais il vaut beaucoup mieux les te-nir en haies; & au-lieu de cueillir les feuilles l'une après l'autre, on doit cueillir toutes les jeunes branches avec leurs feuilles; ce qui est beaucoup plus expéditif, & n'occasionne pas autant de dommage aux arbres.

Le *Mûrier blanc* est aussi dur que le *Mûrier noir*, & peut être multiplié comme lui par semen-ces ou par marcottes; mais la premiere méthode est la plus prompte, & celle qu'on doit employer de préférence, lors-qu'on veut se procurer une grande quantité d'arbres. On

peut tirer ces graines de la Fran-ce Méridionale & de l'Italie. On les seme en Angleterre sur une couche de chaleur modé-rée, sur laquelle on dispose des cerceaux, pour la couvrir de nattes. On fait cette opération à la fin du mois de Mars; on recouvre les graines de terre légere, jusqu'à l'épaisseur de trois lignes; on les arrose sou-vent dans les tems fort secs; on les tient à l'ombre pendant la chaleur du jour, & on les couvre dans les nuits froides. Au moyen de ce traitement, les plantes pousseront au bout de cinq ou six semaines; & comme elles sont très-délicates, quand elles commencent à paroî-tre, il faut les garantir avec soin des gelées du matin, qui sur-viennent souvent dans le mois de Mai. Durant l'été suivant, il suffit de les tenir nettes de mauvaises herbes : mais il faut en avoir soin dans le premier hiver, & sur-tout les couvrir en automne; car sans cela les premieres gelées les détruiroient jusqu'à la racine. Au mois de Mars suivant, on les transplante en pépiniere, & deux ou trois ans après, on les place à de-meure.

Il y a deux ou trois varié-tés de cet arbre, qui different par la forme de leurs feuilles, & par la grosseur & la cou-leur de leurs fruits; mais com-me toutes ces plantes ne sont utiles que par leurs feuilles, il faut toujours préférer celles qui ont de plus fortes tiges & des feuilles plus larges.

Rubra. La troisieme espece, qui est le *Mûrier* de Virginie

MOR

à larges feuilles , & à branches noires , est moins commune en Angleterre que les précédentes. On voit un grand arbre de cette espece dans le jardin de l'Evêque de Londres , à Fulham , où il existe depuis plusieurs années. On m'a assuré qu'il n'avoit jamais produit de fruits, quoiqu'il ait été couvert, il y a quelque tems , d'un grand nombre de chatons semblables à ceux du *Noisetier*; ce qui a déterminé RAY à lui donner le nom de *Corylus* : mais ce peut être un arbre mâle, qui ne produit point de fruits , ce qui arrive souvent dans les autres especes de *Mûriers*. Les feuilles de celui-ci ressemblent un peu à celles du *Mûrier* commun , mais elles sont moins rudes.

On n'a pas encore multiplié cet arbre en Angleterre ; & quoiqu'il ait été greffé sur des *Mûriers* noirs & blancs , je n'ai point appris que ces greffes aient réussi, ce qui me feroit penser qu'il n'est pas de ce genre. Comme cet arbre est fort grand , il ne peut être marcotté ; ce seroit cependant la maniere la plus sûre de le perpétuer. Il est fort dur , & résiste très-bien en plein air aux froids de notre climat. Ceux qui aiment la variété dans les arbres & arbrisseaux , désirent fort pouvoir se procurer cette espece.

Tinctoria. La cinquieme est l'arbre dont le bois sert aux Teinturiers : il est plus connu sous le nom de *Fustique* , appliqué au bois , que par on fruit , qui n'est pas fort

MOR 149

estimé. Il croît naturellement dans presque toutes les Isles de l'Amérique , & en plus grande abondance à Campêche que par-tout ailleurs. On exporte ce bois de la Jamaïque , où on le trouve plus communément que dans aucune autre des Isles-Britanniques. Cet arbre, dans son pays natal , s'éleve au-dessus de soixante pieds de hauteur ; son écorce est d'un brun clair , & quelquefois sillonnée ; son bois est ferme , solide , & d'un jaune brillant : il pousse de tous côtés plusieurs branches couvertes d'une écorce blanche, & garnie de quatre feuilles de quatre pouces de longueur , larges à leur bâse , découpées au pétiole , où elles sont arrondies , & plus larges d'un côté que de l'autre ; de maniere qu'elles paroissent placées obliquement sur les pétioles : leur largeur diminue par dégrés vers l'extrémité qui se termine en pointe aiguë ; elles sont rudes comme celles du *Mûrier* commun , d'un vert foncé , & supportées par de courts pétioles. Vers l'extrémité des jeunes branches sortent les chatons courts, & de couleur pâle , herbacée ; le fruit , qui sort sur de courts pédoncules dans d'autres parties des mêmes branches, est de la grosseur d'une grosse *Noix Muscade*, de forme ronde , couvert de protubérances comme la *Mûre* commune , vert en-dedans & au dehors , & d'une saveur douce & sucrée, lorsqu'il est mûr.

Cet arbre est trop délicat pour réussir dans ce pays, à

150 MOR

moins qu'on ne le conferve dans une ferre chaude. On voit dans le jardin de Chelféa plufieurs de ces plantes qui ont été élevées avec des femences envoyées de la Jamaïque par WILLIAM WILLIAMS, Écuyer, avec plufieurs autres efpeces curieufes qui naiffent fpontanément dans cette Ifle. Les femences de cette efpece pouffent aifément fur une couche chaude : lorfque les plantes font en état d'être enlevées, on les met chacune dans un petit pot rempli de terre fraîche & légere; on les plonge dans une couche chaude de tan, & on les tient à l'ombre, jufqu'à ce qu'elles aient formé de nouvelles racines : on les traite enfuite comme les autres plantes qui viennent des mêmes contrées, en les tenant toujours dans la couche de tan de la ferre, où elles feront de grands progrès. Ces plantes confervent leurs feuilles, durant une grande partie de l'année, dans la ferre chaude.

Papyri-fera. La fixieme efpece croît fans culture à la Chine & au Japon : on la trouve auffi dans la Caroline Septentrionale, d'où fes femences m'ont été envoyées. Les Habitans du Japon font du papier avec fon écorce, & cultivent ces arbres pour cet ufage fur les collines & les montagnes, comme nous cultivons ici les *Ofiers* : ils coupent les jeunes branches en automne, pour fe fervir de leur écorce. On a élevé plufieurs de ces plantes par femences, il y a quelques années, dans le jar-

MOR

din du Duc de NORTHUMBER-LAND, qui a eu la bonté de m'en donner une. Cette efpece profite très-bien en plein air, fans aucun abri, ainfi que plufieurs autres des mêmes contrées, qui croiffent fur des montagnes. Elle produit des branches très-fortes & vigoureufes : mais elle ne paroît pas être d'un crû fort élevé; car fes branches pouffent fur les côtés depuis la racine jufques vers le haut : fes feuilles font larges, quelques-unes font entieres, d'autres profondément découpées en trois lobes, & plufieurs en cinq, fur-tout tandis que les arbres font jeunes; elles fe divifent en forme de main, & font d'un vert foncé, rudes au toucher en-deffus, mais d'un vert pâle & un peu velues en-deffous; elles tombent aux approches des premieres gelées de l'automne, comme celles du *Mûrier* commun. KOEMPFER donne la defcription de fon fruit; il eft un peu plus gros qu'un pois, couvert de poils longs, de couleur pourpre, & compofé de protubérances; il devient d'un pourpre noir en mûriffant, & il eft rempli d'un jus doux.

On peut multiplier cet arbre, en marcottant les branches, comme on le pratique pour le *Mûrier* commun, ou bien par boutures, comme il a été dit ci-devant pour le *Mûrier noir.*

Tartarica. La feptieme efpece croît naturellement dans l'Inde, où elle devient un grand arbre : fon écorce eft molle, épaiffe, jaunâtre, & remplie

MOR

d'un fuc laiteux & aftringent, comme celle du *Figuier* ; fes branches fortent de tous côtés, & font garnies de feuilles oblongues, ovales, & placées fur de courts pétioles : tous les côtés de ces feuilles font égaux ; mais leurs bords font inégalement fciés : elles font rudes ; & d'un vert foncé en-deffus, pâles en-deffous, & alternes fur les branches : fes fleurs naiffent en têtes rondes aux pétioles des feuilles, de chaque côté des branches, & font d'un blanc herbacé : les fleurs mâles ont quatre étamines ; les fleurs femelles produifent un fruit rond, d'abord vert, enfuite blanc, & d'un rouge foncé, lorfqu'il eft mûr. J'ai reçu de Bombay des femences de cette efpece, qui ont réuffi dans le jardin de Chelféa. Cette plante eft trop délicate pour pouvoir fubfifter hors des ferres chaudes en Anterre ; car en ayant élevé un grand nombre, après qu'elles eurent acquis de la force, j'en plaçai quelques-unes dans différentes fituations, où elles étoient à l'abri des gelées : mais aucune ne réfifta à l'hiver. Je n'ai fauvé que celles qui étoient reftées plongées dans la couche de tan de la ferre ; elles ont été traitées comme les autres plantes tendres, & légerement arrofées pendant l'hiver. Au moyen de ce traitement, ces plantes ont profité, & confervé leurs feuilles pendant toute l'année.

Zanthoxylum. La huitieme efpece fe trouve à la Jamaïque & dans les Ifles de Bahama,

MOU 151

d'où fes femences m'ont été envoyées : on vend fon bois, & on l'emploie aux mêmes ufages que celui de la cinquieme efpece, dont les Botaniftes ne l'ont pas trop bien diftinguée : celle-ci ne parvient pas à une groffeur auffi confidérable que la cinquieme ; fes branches font plus minces, fes feuilles font plus étroites, plus rondes à leur bâfe, fciées fur leurs bords, & terminées en pointe aiguë : du pètiole de chaque feuille fortent deux épines aiguës, qui, dans les plus vieilles branches, ont jufqu'à deux pouces de longueur. Le fruit a la même forme que celui de la cinquieme efpece, mais il eft plus petir.

MOSCATELLE. TUBÉREU-SE ou RACINE CREUSE. *Voyez* ADOXA.

MOURON. *Voy.* ANAGALLIS. L. ALSINE.

MOURON CORNU ou OREILLE DE SOURIS DES BLÉS. *Voyez* CERASTIUM DICHOTO-MUM.

MOURON PORTANT BAIES. *Voyez* CUCUBALUS. L.

MOURON MARITIME. *V.* GLAUX.

MOURON JAUNE SAUVAGE. *Voyez* LYSIMACHIA NEMORUM.

MOUSSE. *Voyez* MUSCUS.

MOUSSERONS ou CHAMPIGNONS. MUSHROOMS. *Voyez* FUNGUS. *Suppl.*

Plufieurs perfonnes penfent que les Champignons naiffent de la putréfaction du fumier, de la terre, &c. &c. ou on les trouve. Quoique cetteopinion foit affez généralement reçue parmi ceux qui ne fe

donnent pas la peine de réfléchir, cependant les Natura-
listes les regardent comme de
véritables plantes, quoiqu'on
n'ait pas encore découvert
parfaitement leurs fleurs & leurs
semences : mais comme on cultive les Champignons dans les
environs de Londres, & qu'un
grand nombre de personnes en
font beaucoup de cas, je vais
indiquer la méthode que sui-
vent les Jardiniers pour les multiplier.

Il ne sera pas hors de propos de donner une description
de la véritable espece ; car il
y en a beaucoup de mal-sains,
qu'on a recueillis souvent par
ignorance, & dont on a éprouvé des effets funestes.

Le vrai *Champignon* ou *Mouf-
seron* est rond comme un bouton, des qu'il commence à paroître. Le dessus & la tige sont
très-blancs, quand on les
ouvre, & le dessous est de
couleur de chair livide : la
partie charnue est fort blanche
intérieurement. Quand on laif-
se les *Champignons* sans les
cueillir, ils acquierent une
grandeur considérable, & s'é-
tendent tellement, qu'ils deviennent presque plats ; la partie inférieure, qui est d'abord
rouge, devient alors presque
noire.

Quand on veut cultiver des
Champignons, si l'on n'a point
de couches qui en produisent,
on va les chercher dans quelques riches pâturages, aux
mois d'Août & de *Septembre*,
parce que c'est dans cette faison qu'ils se montrent : quand
on en a trouvé, on creuse la

terre autour de leurs racines ;
on la trouve très-souvent remplie de petits boutons blancs,
qui sont ou des rejettons ou
de jeunes *Mousserons*. On enleve ces *Champignons* avec soin,
en conservant une bonne motte
de terre à leurs racines. Comme on ne peut les trouver que
dans la saison où ils naissent
naturellement, on en cherche
aussi dans de vieux fumiers,
& surtout dans ceux où il y a
beaucoup de litiere que la pluie
n'a pas encore pénétrée : on en
trouve encore quelquefois en
fouillant dans les vieilles couches. Ce frai de *Champignons* a
l'apparence d'une terre blanche ; il pousse de longues fibres,
qui les font distinguer aisément.
On peut aussi s'en procurer en
choisissant du fumier rempli
de litiere qui n'aura pas encore
fermenté : on mêle ce fumier
avec de la terre forte, & on
le met à couvert de la pluie.
Plus on en exclut l'air & plutôt le frai se développe. Il ne
faut pas serrer beaucoup ce
mélange, pour le faire fermenter, parce qu'on détruiroit par-
là le frai. Ce frai paroîtra au
bout de deux mois, sur tout si
le monceau est bien couvert
de vieux chaume ou de litiere
long-tems exposée à l'air, de
maniere qu'elle ne fermente
plus quand le frai sera formé
dans le tas de fumier : alors
vous pourrez le transporter
dans des couches. Ces couches
doivent être faites de fumier
mêlé de beaucoup de litiere
qui n'ait pas été mise en tas
pour fermenter. Le fumier qui
a été répandu sur la terre pen-

dant un mois, & même plus long-tems, est le meilleur : on place cette couche sur un terrein sec, & on pose le fumier sur la surface de la terre. La largeur de cette couche, mesurée à sa base, doit être de deux pieds & demi ou trois pieds, & sa longueur proportionnée à la quantité de *Champignons* que l'on désire : on entasse le fumier jusqu'à l'épaisseur d'un pied, & on le couvre de quatre pouces de terre forte; on met encore dix pouces d'épaisseur de fumier, & pardessus une autre couche de terre, en rapprochant en talus les deux côtés de la couche, & en l'élevant de cette maniere, jusqu'à ce qu'il y ait trois lits de fumier, & autant de terre.

Quand cette couche est ainsi disposée, on la couvre de litiere ou de vieux chaume, pour empêcher la pluie d'y pénétrer & y conserver l'humidité. On peut la laisser dans cet état pendant huit ou dix jours; & après ce tems, elle sera en état de recevoir le frai. La chaleur de cette couche doit être modérée; si elle étoit trop forte, elle détruiroit le frai, ainsi que l'humidité. Quand on a trouvé du frai, il faut le tenir sec, jusqu'à ce qu'on en fasse usage; car plus il est sec, mieux il réussit. J'avois laissé, pendant quelques mois, une grande quantité de ce frai près du fourneau de ma serre, où il étoit devenu si sec, qu'il paroissoit n'être plus propre à rien; & cependant ce frai a produit plutôt & en plus grande quantité que tout autre.

La couche ayant acquis le dégré de chaleur qui lui est nécessaire pour recevoir le frai, on enleve la litiere, & on nivelle les côtés; on couvre alors toute la couche de terre riche, légere & seche, jusqu'à l'épaisseur d'un pouce, & l'on enfonce le frai dans cette terre, en plaçant les mottes à quatre ou cinq pouces de distance : on couvre ces mottes avec la même terre, jusqu'à l'épaisseur d'un peu plus d'un demi-pouce, & on met par-dessus assez de litiere, pour que la pluie ne puisse y pénétrer, & que la couche ne se desseche point. Quand on fait ces couches au printems ou en automne, on doit toujours choisir une température douce; de cette maniere, le frai prend bien plutôt, & les champignons paroissent au bout d'un mois; mais les couches faites en été, & par un tems chaud, ou en hiver quand il fait très froid, ne donnent des *Champignons* que beaucoup plus tard.

Le grand secret dans l'entretien de ces couches, est de les tenir dans un état convenable d'humidité, de ne leur pas donner trop de fraîcheur : en été, on peut les découvrir, afin qu'elles reçoivent les pluies douces dans les tems favorables; lorsqu'il fait sec, on les arrose un peu de tems en tems, mais avec modération; en hiver, on les tient aussi seches qu'il est possible, & assez couvertes pour empê-

MOU

cher le froid d'y pénètrer.
Lorfqu'il gele ou qu'il fait très-
froid, on y met de la litiere
prife fur un fumier, ce qui
les avance beaucoup ; mais il
n'en faut pas pour toute cette
litiere à la fois. On commence
par un fimple lit de paille feche ; & toutes les fois qu'on
obferve que cette litiere diminue, on en remet de la
nouvelle, & l'on augmente
fon épaiffeur, fuivant que le
tems eft plus ou moins froid.
Si l'on obferve exactement tout
ce qui vient d'ètre dit, on
aura beaucoup de *Champignons*
pendant toute l'annèe. Ceux
qui croiffent fur de pareilles
couches font beaucoup meilleurs que ceux qu'on ramaffe
dans les champs.

Si le frai prend bien, une
couche ainfi foignée fera bonne
pendant plufieurs mois, & produira une grande quantité de
Moufferons. On pourra y prendre du frai pour en garnir d'autres. On conferve ce frai dans
un endroit fec, jufqu'à ce qu'il
foit tems de s'en fervir ; ce
qui ne pourra être qu'après
cinq ou fix femaines, afin qu'il
ait le tems de fe deffecher
avant de le planter dans la
nouvelle couche ; car il ne
réuffiroit pas bien fans cette
précaution.

Quelquefois il arrive que
des couches faites de cette maniere ne produifent point de
Moufferons avant fix mois : mais
on ne doit point pour cela
les détruire ; car j'en ai vu
qui, après un certain tems,
ont produit beaucoup, & ont
continué longtems à donner
des *Champignons*.

MUN

MOUTARDE SENEVÉ. *V.*
SINAPIS NIGRA. *L.*

MOUTARDE BLANCHE.
Voyez SINAPIS ALBA. *L.*

MOUTARDE BATARDE.
Voyez ARABIS.

MOUTARDE BATARDE
DE MITHRIDATE. *Voy.* BIS-
CUTELLA. *l.*

MOUTARDE DES INDES
ou ÉTRANGERE. *Voyez* CLEO-
ME. *L.*

MOUTARDE DE HAIE,
VELAR, TORTELLE ou HER-
BE AU CHANTRÉ. *Voyez* ERY-
SIMUM. *L.*

MOUTARDE CYLINDRI-
QUE. *Voyez* TURRITIS. *L.*

MOXA DES CHINOIS. *V.*
ARTEMISIA VULGARIS. *L.*

MUCILAGE. On nomme
ainfi une fubftance vifqueufe
& gluante qui fe trouve autour
des femences.

MUCILAGINEUX fe dit des
fubftances vifqueufes & gluantes.

MUFLE DE VEAU. *Voyez*
ANTHIRRINUM. *L.*

MUGUET ou LYS DES VAL-
LÉES. *Voyez* CONVALLARIA MA-
JALIS. *L.*

MUGUET PETIT ou CAIL-
LE-LAIT. *Voyez* GALLIUM. *L.*

MULTI-SILIQUOSUS fe dit
des plantes dont chaque fleur
eft remplacée par plufieurs filiques diftinctes, longues,
minces, & quelquefois courbes, qui s'ouvrent d'elles-mêmes, quand les femences font
parvenues à leur maturité, &
les laiffent tomber. De ce genre
font les *Pieds d'Ours*, les *Co-
lombines*, la *Joubarbe* ordinaire,
le *Nombril de Vénus*, &c. &c.

MUNTINGIA. *Plum. Gen.*

MUN

Nov. 41. *tab.* 6. *Lin. Gen. Pl.*
575. Cette plante n'a pas d'autre nom.

Caractères. Le calice de la fleur est découpé en cinq segmens jusqu'au fond; la corolle est composée de cinq pétales en forme de cœur, étroits à leurs bâses, insérés dans le calice, & étendus comme une rose; la fleur a un grand nombre d'étamines, terminées par des antheres rondes. Dans son centre est placé un germe rond, sans style, mais couronné par un stigmat divisé en plusieurs parties. Ce germe se change, quand la fleur est passée, en un fruit mou à une cellule, couronné par le stigmat comme un nombril, & rempli de petites semences.

Ce genre de plantes est rangé dans la premiere section de la treizieme classe de LINNÉE, qui renferme celles dont les fleurs ont plusieurs étamines & un stigmat. Suivant le système de TOURNEFORT, il doit être rangé dans la huitieme section de la vingt-unieme classe, qui contient les arbres & arbrisseaux, avec une fleur en rose, dont le calice devient un fruit, & dont les semences sont dures.

Nous n'avons qu'une espece de ce genre, qui est le

Muntingia Calabura. Jacq. Hist. tab. 107. *p.* 166.

Muntingia folio sericeo molli, fructu majori. Plum. Nov. Gen. 41. *ic.* 205; Muntingia avec une feuille molle & soyeuse, & un gros fruit.

Muntingia pedunculis unifloris. Hort. Cliff. 203.

Muntingia fruticosa, villosa, foliis serratis, oblongis, uno latere brevioribus. Brown. Jam. 245.

Calabura alba. Pluk. Alm. 75. *Mant.* 34. *t.* 152. *f.* 4.

Loti arboris folio angustiori, rubi-floro, fructu polyspermo umbilicato. Sloan. Jam. 162. Hist. 2. *p.* 80. *t.* 194. *f.* 1. Raii dendr. 32.

Mallam - Toddali. Rheed. mal. 4. *t.* 40. Jean Reichard, dans le *Syst. Plant.* de LINNÉE, prétend que cette plante de RHEEDE n'est point le *Muntingia.*

Le nom de *Muntingia* a été donné à ce genre par le P. PLUMIER, en l'honneur du Docteur MUNTING, Professeur de Botanique à Groningue en Hollande, qui a publié un volume in-folio de Botanique, intitulé : *Phytographia curiosa,* dans lequel il y a plusieurs figures de plantes gravées sur des planches de cuivre. Le même Auteur a publié deux autres ouvrages de plantes (in-4°.), dont l'un, qui a pour titre *Aloïdarum,* traite de plusieurs especes d'*Aloës.* Le titre du second est, *De Herbá Britanniá Antiquorum.*

Cette plante est dessinée par le Chevalier HANS SLOANE, dans son *Histoire de la Jamaïque,* sous le titre de *Loti arboris folio angustiori, rubi-floro, fructu polyspermo umbilicato.* 2. *p.* 80 : elle s'éleve à la hauteur de trente pieds & plus dans son pays originaire ; elle pousse vers son sommet plusieurs branches couvertes d'une écorce lisse, d'un pourpre foncé, & garnies de feuilles de trois pouces environ de longueur sur neuf lignes de largeur à leur bâse, où elles sont

MUN

arrondiés en forme de cœur près du pédoncule, terminées en pointe aiguë, fort laineufes en-deffous, unies au-deffus, d'un vert luifant, légerement fciées fur leurs bords, & alternes : fes fleurs naiffent aux aîles des pétioles fur de longs pédoncules ; elles font compofées de cinq pétales en forme de cœur, blancs, étendus & femblables à ceux de la Ronce, & de plufieurs étamines de moitié moins longues que les pétales, & terminées par des antheres globulaires ; dans le centre eft placé un germe rond, & couronné par un ftigmat à plufieurs pointes, qui fe change, quand la fleur eft paffée, en un fruit charnu, ombiliqué, auffi gros que celui de l'*Aubépine*, & d'une couleur de pourpre lorfqu'il eft mûr. Ce fruit renferme plufieurs petites femences dures & angulaires. Cette efpece a produit des fleurs & des fruits en Angleterre.

M. Robert Millar a envoyé de la Jamaïque les femences de cette plante, qui ont réuffi dans quelques jardins Anglois.

On la multiplie par fes graines, qu'il faut femer dans des pots remplis d'une terre riche & légere, & les plonger dans une couche de tan d'une chaleur modérée, en obfervant de foulever les vitrages dans les tems chauds, pour leur donner de l'air. Ces graines reftent fouvent une année en terre avant de germer ; dans ce cas, il faut les tenir conftamment nettes de mauvaifes herbes, & les laiffer dans la couche chaude

MUN

jufqu'à la Saint-Michel : alors on peut les mettre dans la ferre chaude, & les plonger dans la couche de tan, entre les grandes plantes, où elles refteront pendant l'hiver : on les arrofe de tems en tems durant cette faifon. Quand la terre paroît feche, & au commencement de Mars, on retire les pots de la ferre chaude, pour les placer fous les châffis d'une nouvelle couche de tan, ce qui fera pouffer les plantes bientôt après.

Quand elles ont atteint la hauteur d'environ deux pouces, on les enleve hors des pots avec beaucoup de précaution ; on les met chacune féparément dans de petits pots remplis d'une terre riche & légere, on les replonge dans la couche chaude, & on les tient à l'ombre, jufqu'à ce qu'elles aient formé de nouvelles racines, après quoi on les arrofe exactement, & on leur donne beaucoup d'air dans les tems chauds. Ces plantes peuvent refter dans cette couche jufqu'aux premieres nuits froides de l'automne ; alors on les met dans la ferre chaude, & on les plonge dans la couche de tan, où elles veulent être tenues chaudement en hiver, fur-tout tandis qu'elles font jeunes ; il faut auffi leur donner fouvent de l'eau, mais peu à la fois, de peur que les tendres fibres de leurs racines ne fe pourriffent. Il fera prudent de tenir ces plantes dans la ferre pendant toute l'année, en leur donnant beaucoup d'air dans les tems chauds ; mais à mefure qu'elles acquierent de la force,

MUR

force, elles deviennent plus robustes, & peuvent être exposées au-dehors pendant deux ou trois mois de l'été, & conservées en hiver dans une serre seche, à une chaleur modérée.

MURAILLES. Les murs sont absolument nécessaires dans les jardins, pour faire mûrir les fruits qui sont trop délicats pour se perfectionner dans notre climat sans ce secours. On les construit avec differens matériaux ; dans quelques pays on se sert de pierres, & dans d'autres de briques, suivant la facilité que l'on a de se les procurer.

Des tous les matériaux propres à faire des murs de jardin, la brique est le meilleur. Les murailles ainsi construites sont non seulement plus propres, mais aussi plus chaudes & plus favorables aux fruits ; outre cela, on a plus de facilité d'y enfoncer des clous pour retenir les arbres, & ces clous n'ont pas besoin d'être aussi forts, parce que les joints qui séparent les briques sont beaucoup moins larges que ceux qui se trouvent entre les moëllons des murs construits en pierres. Ces murailles étant couronnées de pierres de taille, & fortifiées de distance en distance, par des colonnes ou pilastres qui séparent les arbres & brisent l'effort des vents, produisent un très-agréable effet.

En quelques endroits de l'Angleterre, on construit des murs fort commodes en briques & en pierres ; dans certains pays, les briques n'ont point assez de résistance pour pouvoir être

Tome V.

MUR 157

employées seules, & ne sont jamais aussi durables que la pierre : aussi quelques personnes, pour rendre leur construction plus solide, ont bâti des doubles murs, dont l'extérieur est de pierres, & l'intérieur de briques, ou un mur de pierres revêtu de briques. Quand on suit cette méthode, il faut avoir grand soin de bien lier les briques avec les pierres, sans quoi les unes se sépareront des autres, sur-tout quand, après de grandes pluies, il survient une forte gelée, qui fait gonfler le mortier, & tomber les briques qui servent de revêtement.

Dans les endroits où les murs sont bâtis entierement en pierres, il faut y joindre des treillages, pour pouvoir palisser plus commodément les branches des arbres. Le bois de ces treillages ne doit point avoir plus d'un pouce & demi d'épaisseur sur deux & demi de large : on croise ces lattes l'une sur l'autre à quatre pouces de distance ; car si on les rapprochoit davantage, il seroit difficile d'y arranger comme il faut les branches des arbres. Comme ce treillage sera fixé contre la muraille, les arbres n'en seront éloignés que de deux pouces ; & au moyen de cela, le fruit mûrira mieux que s'il étoit plus près du mur. Ainsi, il est absolument nécessaire de revêtir de treillages les murs construits en pierres ; car sans cela les fruits acquerroient difficilement le dégré de perfection qu'ils doivent avoir, & on ne pourroit pas y fixer les branches des arbres.

L

MUR

On a essayé de donner différentes formes aux murailles; quelques-uns les ont tracées en demi-cercles, d'autres en angles plus ou moins grands, & plus inclinés du côté du nord, pour s'opposer aux vents froids; mais aucune méthode n'a aussi bien réussi que de les faire droites & perpendiculaires. La vérité de ce que j'avance, a été confirmée par un exemple frappant. A Goodwood, en Suffex, maison de campagne du Duc de RICHMOND, au milieu de deux murs exposés au midi, il y avoit deux segmens de cercle dans lesquels on avoit planté des arbres fruitiers de la même espece que ceux qui couvroient la partie de la muraille tracée en ligne droite. Ces premiers arbres n'ont produit que de mauvais fruits, & les arbres eux-mêmes ont péri en peu d'années, par la nielle qui les attaquoit à chaque printems; lorsque les branches de ceux qui couvroient la partie droite du mur, venoient à s'étendre sur la partie cintrée, elles brouissoient, & périssoient aussi bientôt après.

Lorsqu'on eut arraché ces arbres, on les remplaça par de la vigne; mais les raisins qu'elle produisit ne mûrirent point, ou ne furent bons qu'un mois après les autres de la même espece qui se trouvoient contre la partie droite du mur; de maniere qu'on fut encore obligé d'arracher cette vigne, & on mit en place des figuiers, dont les fruits furent aussi très-mauvais. Ainsi, il est complettement démontré que les arbres

MUR

fruitiers ne peuvent réussir contre des murailles concaves, parce qu'il y regne toujours de forts courans d'air, qui rendent ces situations beaucoup plus froides que celles qui sont sans aucun abri.

J'ai vu aussi, dans le jardin de M. LE COUR, en Hollande, des murs bâtis en angles de différentes formes; mais ils n'ont pas mieux réussi que les cercles, & je n'y ai pas trouvé un seul arbre qui fût en bon état, & qui produisit du fruit.

Différentes personnes ont proposé plusieurs autres plans pour bâtir les murs, afin d'accélérer la maturité des fruits. Il y a eu entr'autres un livre fort ingénieusement écrit, sous le titre de *Murs à fruits perfectionnés*, en les couchant en talus. L'Auteur a fait voir, par bien des calculs, qu'en cette position le mur reçoit une plus grande quantité de rayons du soleil, qu'étant perpendiculaire; d'où il a conclu que les murs bâtis ainsi sont préférables aux autres pour accélérer la maturité des fruits; il s'est même donné la peine de calculer les différentes inclinaisons que les murs doivent avoir dans les différens climats, afin d'y recevoir un plus grand nombre de rayons du soleil. Quoique cette théorie semble être démontrée, cependant les expériences n'ont point réussi; car comme on est obligé de construire ces murs contre des levées de terre, les exhalaisons ou vapeurs qui s'élevent de cette terre, font perdre l'avan-

MUR

tage qu'on pourroit retirer de l'augmentation des rayons du soleil : d'ailleurs ces murs en talus étant plus exposés aux rosées froides de la nuit, les fruits sont fort retardés dans leur accroissement, & les fleurs sont bien plus sujettes à être détruites par les gelées du matin. Si l'on ajoute encore que ces murailles inclinées sont bien plus exposées aussi au vent & à la pluie, on trouvera, après avoir comparé leurs avantages avec les inconvéniens qui en résultent, que les murs perpendiculaires leur sont de beaucoup préférables ; car ce ne sont pas tant les rayons les plus forts du soleil que le fruit demande, qu'une continuation de chaleur modérée, & sur-tout l'aspect du soleil levant, dont les rayons dissipent de bonne heure l'humidité de la nuit ; & pour cela, les murs perpendiculaires sont préférables aux murs en talus ; parce qu'ils ont le matin les rayons du soleil directs, tandis qu'ils ne tombent qu'obliquement sur les talus ; aussi les murailles exposées à l'est sont bien meilleures que celles qui regardent le midi, & les fruits y mûrissent bien plutôt.

D'autres veulent qu'on noircisse les murs, ou qu'on les peigne en brun. Ils supposent que le noir, absorbant une plus grande quantité de rayons, conserve par conséquent plus long-tems la chaleur : mais cela est plus vrai dans la théorie que dans la pratique ; car, quoiqu'il faille avouer qu'un mur noir est plus chaud au toucher qu'un mur ordinaire,

MUR 159

cependant comme le fruit en est toujours un peu éloigné, il ne profite pas beaucoup de de cette chaleur, tandis que la chaleur réfléchie accélere la maturité des fruits : c'est-pourquoi je conseillerois de faire des essais de toutes ces méthodes, avant de les mettre en pratique, & de ne pas croire sur parole ceux qui proposent de nouveaux moyens, malgré leur ton affirmatif ; car quelquefois leurs systêmes n'ont pour bâse que des principes mal fondés, ou ne sont appuyés que sur une seule épreuve. Les personnes qui ont conseillé de bâtir les murs en talus, ont également imaginé de les noircir, d'après les mêmes principes ; mais il faut éviter d'introduire de pareilles méthodes, jusqu'à ce qu'on ait fait des expériences suffisantes pour en assurer l'utilité.

Quand on veut faire la dépense de bâtir des murs solides, on trouvera qu'ils réussissent mieux que ceux qui sont construits légerement, non seulement par rapport à la durée, mais aussi par la chaleur qu'ils procurent. Ainsi, un mur de deux briques d'épaisseur réussira mieux qu'un d'une brique & demie ; & si on le revêt encore d'une couche de mortier, pour remplir & fermer les joints, il sera d'une plus longue durée, & l'air ne pénétrera pas si aisément au travers.

Dans la pratique actuelle du jardinage, on entoure rarement les jardins de murailles, ce qui est certainement bien fait ; car, par cette méthode,

MUR

on se conserve la vue de la campagne, & on évite de plus une dépense considérable, dans laquelle on est entraîné, non seulement pour la construction de ces murailles, mais encore pour l'entretien des arbres, sans qu'il en revienne ni beaucoup de profit, ni de plaisir : car, quand on plante beaucoup d'arbres contre ces murs, ils sont rarement bien soignés, & ne produisent par conséquent qu'une petite quantité de fruits mal nourris & de mauvais goût : c'est pourquoi il faut restrein— dre l'étendue des murs à la quantité des fruits dont on a besoin pour sa consom— mation : mais comme il est tou— jours nécessaire d'entourer un potager de murailles, pour mettre en sûreté les légumes, ces murailles fourniront une surface suffisante pour la quan— tité d'arbres nécessaire, parce qu'un potager est toujours pro— portionné au nombre de ceux qui composent la famille du pro— priétaire. Cependant si l'éten— due des murs du jardin pota— ger ne suffit pas, on en peut bâtir d'autres qui traversent le jardin ; & lorsque le terrein est assez étendu, on peut en— core en élever qui croisent les premiers, en laissant entr'eux au moins quatre-vingts ou cent pieds d'intervalle; & com— me le jardin potager doit être éloigné de l'habitation, on peut en cacher les murs par des plantations d'arbres, qui ser— viront encore à abriter les ar— bres fruitiers.

Le meilleur aspect pour des arbres fruitiers, en Angleterre,

MUR

est celui de l'est & du sud-est. Les arbres profitent ainsi de l'aspect du soleil du matin, & sont moins exposés aux vents d'ouest & du sud-ouest, qui sont les plus nuisibles aux fruits en Angleterre. Quelques per— sonnes condamnent cependant cette méthode d'exposer les murs au sud-est, à cause des nielles du printems : mais plu— sieurs années d'expérience & d'observations m'ont appris que les murs exposés au sud-ouest sont aussi exposés à ces niel— les que ceux de tout autre as— pect; & je crois qu'en se don— nant la peine d'observer, pen— dant sept années consécutives, quels sont les murs les plus ex— posés à la nielle, on trouvera que ceux du sud-est en sont aussi peu gâtés que ceux de tout autre aspect. Ainsi lors— qu'on établit un jardin potager, il faut élever à cette exposition un mur aussi étendu que la si— tuation du terrein le permet.

Après l'aspect dont nous ve— nons de parler, vient celui du midi, & ensuite celui du sud— est, qui est préférable à celui du sud-ouest, pour les raisons qui viennent d'être énoncées : mais comme dans la plupart des jardins il y a des murs ex— posés au sud-ouest & à l'ouest, on peut les garnir avec des espèces d'arbres fruitiers qui n'exigent pas beaucoup de cha— leur. Les murs qui regardent le nord, ne sont bons que pour des *Poires* à cuire, des *Prunes*, des *Cerises-Morelles* propres à êtres conservées, ou quelques *Cerises-Ducs*, qui, mûrissent plus tard que les autres, ser—

vent à fournir la table, jusqu'à ce que les *Pêches* & les *Brugnons* soient mûrs.

Les personnes curieuses de bons fruits font construire un treillage qui éloigne du mur les branches de deux pouces, & qui sert à les attacher. Cette méthode est excellente, parce que les fruits, se trouvant à une distance convenable du mur, n'en sont point endommagés, & en reçoivent la chaleur réfléchie : on évite aussi par-là de dégrader les murailles, en y enfonçant des clous; ce qui fait tomber le mortier, & forme des retraites pour les chenilles & les autres insectes qui détruisent les fruits.

Ces treillages peuvent être différens, suivant les especes d'arbres qu'ils doivent soutenir. Ceux qu'on destine aux *Pêchers*, aux *Brugnons*, & aux *Abricotiers*, qui produisent pour l'ordinaire leurs fruits sur les jeunes branches, doivent être construits avec des lattes éloignées de trois ou quatre pouces en quarré : mais pour toutes les autres especes dont les fruits naissent sur le vieux bois, les mailles peuvent avoir cinq ou six pouces, & huit ou neuf pouces pour la vigne, dont on place les branches à une bien plus grande distance que celles d'aucune autre espece d'arbres.

On peut employer dans ces treillages toutes sortes de bois : mais on se sert communément de sapin, & sur-tout du sapin jaune, qui peut durer plusieurs années, s'il est bien sec lorsqu'on le met en œuvre, &

si l'on a soin de le bien conduire ; mais le plus durable de tous les bois pour cet usage, est le chêne, sur-tout celui qui a été coupé en hiver. Cependant si l'on veut faire ces treillages avec économie, on peut acheter des lattes de frène, & s'en servir de la même maniere que pour les espaliers des plates-bandes, avec cette différence seulement que chaque quatrieme latte doit être forte, & fixée au mur avec des crochets de fer, pour soutenir le tout ; & comme il faut les placer plus près les unes des autres qu'on ne le pratique ordinairement pour les espaliers, les lattes droites & fortes ne doivent pas être à plus de trois ou quatre pieds de distance l'une de l'autre, on attache solidement avec des clous les lattes horisontales sur les lattes perpendiculaires, & on les fixe de maniere qu'elles ne puissent pas se déplacer. Les autres piquets ou lattes plus minces, qui sont placées à côté des plus grosses, peuvent être attachées avec du fil de fer. On palisse les branches au treillage avec des *Osiers*, du *Chanvre*, ou quelqu'autre lien mou ; mais on ne doit pas embrasser la branche dans cette ligature, pour la fixer ensuite au treillage avec un clou.

On ne doit pas dresser ces treillages avant que les arbres soient grands, & qu'ils portent beaucoup de fruits ; jusqu'à ce tems, on peut élever les jeunes arbres, & disposer leurs branches contre quelques lattes minces, faites de *Frène* ou d'au-

L 3

tres bois : au moyen de cela , le treillage fera neuf lorfque les arbres commenceront à porter du fruit ; il durera beaucoup plus long-tems : aulieu qu'en le mettant en place avant que les arbres foient plantés , il eft prefque pourri quand il doit commencer à fervir. Lorfqu'on a le projet d'employer des treillages , il faut mettre plufieurs crampons de fer dans le mur , en le bâtiffant, à la diftance qu'on veut donner aux piquets perpendiculaires dont nous avons parlé ; car fi on les chaffe dans le mur après qu'il eft fait , on arrache le mortier & on le dégrade. En conftruifant un mur autour d'un jardin potager, on enduit exactement la face contre laquelle on veut planter les arbres , & on la rend auffi unie qu'il eft poffible , de maniere que les pilaftres n'aient pas plus de trois ou quatre pouces de faillie : on laiffe entre chaque pilaftre quatorze pieds de diftance , lorfqu'on veut planter contre ce mur des *Péchers* & des *Brugnons*. Chaque arbre fe trouvant , par cette difpofition , placé exactement entre deux pilaftres , produit un effet plus agréable ; mais quand on doit y mettre des *Abricotiers* , des *Pruniers* ou des *Cerifiers* , les pilaftres ne doivent être qu'à dix pieds de diftance , & alors on plante les arbres contre les pilaftres mêmes , de deux l'un , afin qu'ils aient affez de place pour s'étendre ; & comme la faillie de ces pilaftres obligera d'avancer auffi le treillage , les branches des arbres

feront placées d'une maniere uniforme ; mais quand les pilaftres ne failliffent que de quatre pouces du côté du jardin , il faut leur donner plus d'épaiffeur en-dehors , afin que le mur en foit mieux foutenu.

L'épaiffeur ordinaire des murailles de jardin , fi elles font conftruites en briques , eft de treize pouces ; ce qui fait une brique & demie : mais en général elle doit être proportionnée à la hauteur : car fi elles ont douze ou quatorze pieds d'élévation , & même davantage , comme il arrive fouvent , il faut mettre au moins deux briques & demie pour la fondation , conferver la même épaiffeur à-peu-près jufqu'à un pied au deffus de la terre , & la diminuer enfuite de deux pouces de chaque côté ; ce qui reduira le mur à l'épaiffeur de deux briques. A cinq ou fix pieds au-deffus de la terre , on peut encore diminuer cette épaiffeur jufqu'à une brique & demie , & continuer ainfi jufqu'au haut. Dans ces murs , les pilaftres doivent être plus forts que dans les murs ordinaires : il faut auffi les faire plus bas ; car leur hauteur donne plus de prife aux vents violens , & les expofe à être renverfés. Si les pilaftres ne failliffent pas en-dehors de la longueur d'une brique , & de fon épaiffeur en-dedans , on donnera une plus grande force aux murailles , en les plaçant à dix ou douze pieds d'intervalle.

Il n'eft cependant pas néceffaire d'élever ces murs au—

MUR

deffus de neuf ou dix pieds , à moins qu'on ne veuille y placer des *Poiriers* , qui s'étendent beaucoup , & demandent un grand efpace : mais comme il n'y a que quelques efpeces de *Poires* d'hiver qui exigent le fecours d'une muraille , on n'éleve que la partie où on veut les planter. Les *Péchers* & *Brugnons* n'ont befoin que de dix pieds de hauteur. Toutes les fois qu'on les fait monter plus haut , les arbres fe dégarniffent par le bas , & n'y produifent point de fruits ; & quoique les *Abricotiers* , les *Pruniers* & les *Cerifiers* s'élevent à une hauteur plus confidérable , cependant , fi on les plante à une diftance convenable , & fi l'on conduit leurs branches horifontalement depuis le bas , ils ne garniront pas fi-tôt un mur de dix pieds de hauteur. La vigne peut être tenue auffi baffe qu'aucune autre efpece d'arbres fruitiers. Lorfqu'on la plante contre une muraille peu élevée , il faut la traiter comme on le fait dans les vignobles , en coupant la plus grande partie du bois qui a porté fruit l'année précédente , pour faire place aux jeunes rejettons qui doivent en donner l'année fuivante. Ces rejettons ont rarement plus de trois pieds de longueur.

Si l'endroit où l'on veut planter des *Poiriers* eft expofé au fud-eft , pofition où les fruits mûriront très-bien , on éleve alors les murailles au moins à quatorze pieds. Comme ces arbres s'étendent confidérablement lorfqu'ils font greffés fur des fauvageons , il ne faut pas les tailler ni les arrêter dans leur accroiffement , parce qu'on les empêcheroit de porter des fruits , en leur faifant pouffer un grand nombre de branches gourmandes , qui font toujours ftériles , & on ne doit jamais les mêler avec d'autres arbres fruitiers plus petits , parce qu'alors les murs paroitroient dégarnis , & il y auroit des arbres plantés à une double diftance des autres. Ainfi , de tous les arbres qui ont befoin du fecours d'une muraille pour mûrir leurs fruits , il n'y en a aucun à qui il en faille une plus élevée qu'au *Poirier* , fi ce n'eft cependant le *Figuier* , que l'on peut planter contre un pareil mur dans les endroits vuides , quoiqu'on puiffe mettre auffi cette efpece d'arbre derriere les murs d'un office ou des écuries , qui font des endroits convenables , parce que les domeftiques ne font pas fort curieux de ce fruit ; & en le plaçant ainfi dans un lieu fréquenté , il eft moins à craindre que les fruits n'en foient dévorés par les oifeaux.

MURAILLES CHAUDES *ou* PROPRES A ÊTRE ÉCHAUFFÉES.

Je vais donner à préfent quelques inftructions pour conftruire des murailles chaudes propres à hâter la maturité des fruits , telles qu'on les fait aujourd'hui affez communément en Angleterre.

Dans quelques endroits , cette conftruction exige beaucoup de dépenfes , & la maniere dont elles font difpofées ,

MUR

entraîne une forte confommation de matieres combuftibles ; mais quand elles font bâties avec jugement, la premiere depenfe eft beaucoup moindre, & les frais de chauffage ne font pas fi confidérables, puifqu'il ne fera néceffaire d'y faire du feu que pendant trois ou quatre mois, en commençant vers le milieu ou la fin de Janvier, & en ceffant à la fin de Mai, tems auquel il fuffit de fermer exactement les châffis tous les foirs, & pendant les mauvais tems. Une demi-heure de foleil fur ces vitrages, dans cette faifon, fuffit pour échauffer l'air qui y eft renfermé, & pour faire mûrir nos fruits d'Europe.

Quelques perfonnes plantent de la vigne & d'autres arbres fruitiers à côté des ferres, & y font entrer quelques-unes de leurs branches, afin d'avoir des fruits précoces ; mais cette méthode eft affez mauvaife, quand la ferre eft deftinée à la culture des *Ananas*, à qui il faut une plus grande chaleur qu'à tous les autres fruits, de maniere qu'ils ne peuvent jamais bien réuffir enfemble. Quand on laiffe entrer une quantité fuffifante d'air pour l'accroiffement des autres fruits, les *Ananas* périffent faute d'une chaleur convenable ; & d'un autre côté, quand la ferre eft échauffée convenablement pour les *Ananas*, la chaleur eft trop forte pour les autres fruits. La *Vigne*, comme on l'a déja dit, doit être plantée contre un mur féparé, parce qu'elle exige plus d'air, lorfqu'elle

MUR

commence à pouffer, que toute autre efpece d'arbres fruitiers.

La hauteur ordinaire des murailles chaudes eft à peu-près de dix pieds ; ce qui fuffit pour toutes les efpeces de fruits qui peuvent être forcés : car les arbres qu'on foumet à une chaleur artificielle, ne font jamais auffi vigoureux que ceux qui reftent toujours expofés en plein air : & quand on n'a pas une affez grande étendue de mur pour laiffer repofer une partie de ces arbres de deux années l'une, ils s'affoibliffent bientôt, & périffent en peu d'années. Une muraille deftinée à fournir des fruits précoces pour l'agrément d'une famille ordinaire, ne doit pas avoir moins de quatre-vingts ou cent pieds de longueur. Ainfi, quand on veut avoir ces fruits dans leur grande perfection, & des arbres qui confervent leur vigueur pendant plufieurs années, la muraille doit avoir trois fois cette longueur. Comme on n'en emploie qu'un tiers chaque année, les arbres qui garniffent les deux autres parties, auront toujours deux ans pour recouvrer leur vigueur. Ils acquerront ainfi une plus grande quantité de bois à fruits, & ces fruits feront plus beaux & en plus grand nombre que fur les arbres que l'on force chaque deux ans. Comme les vitrages font conftruits de maniere à pouvoir être tranfportés, la dépenfe des murs plus longs ne fera pas bien confidérable.

Les fondations de ces murs doivent avoir quatre briques

MUR

& demie d'épaiſſeur, pour ſou-
tenir les tuyaux de cheminée;
autrement, ſi une partie de
ces tuyaux poſoit ſur les briques, & l'autre ſur la terre,
leur bâſe feroit inégale, & ils ſe
dérangeroient bientôt; & lorſqu'il ſe forme des fentes dans
ces cheminées, la fumée s'é-
chappe à travers, & cela les
empêche de tirer. Si cette fumée pénetre dans le vitrage,
elle nuit beaucoup aux fruits,
& leur communique un goût
déſagréable. Il ſuffit de conſerver cette épaiſſeur au mur juſqu'à ſix pouces au-deſſus de
la terre où doit être poſée la
bâſe & la fondation de la premiere cheminée. Ces ſix pouces d'élévation ſuffiſent pour la
mettre au deſſus de l'humidité:
on peut enſuite diminuer l'é-
paiſſeur de ces murs, & les
réduire à trois briques & demie
d'épaiſſeur. Ainſi, le mur doit
avoir par-derriere deux briques
d'épaiſſeur, ce qui eſt abſolument néceſſaire pour jetter la
chaleur ſur le devant, & pour
l'empêcher de ſe perdre à travers. Le mur de face, c'eſt-à-
dire, celui contre lequel les
arbres ſont placés, ne doit
avoir que quatre pouces d'é-
paiſſeur; par ce moyen, les
cheminées auront neuf pouces
de diametre, & on pourra les
couvrir avec des tuiles de douze pouces de longueur, qui
ne doivent poſer que d'un
pouce & demi de chaque côté.
Les fours où l'on allume le
feu, ſe pratiquent par-derriere,
& leur nombre doit être proportionné à la longueur du
mur. On donne ordinairement

MUR 165

à chaque tuyau depuis quarante
juſqu'à cinquante pieds de longueur: mais je ne conſeille
pas de les faire plus longs; car
lorſque les fours ſont plus éloignés de leur extrémité, il faut
y faire de plus grands feux
pour échauffer les murs, ce
qui occaſionne une trop grande
chaleur dans le voiſinage des
ces foyers. On couvre ces
fours, pour empêcher le vent
& la pluie d'y pénétrer; car
ſans cela les feux ne brûleroient pas également. Quelques perſonnes élevent par-
deſſus des hangards en bois;
mais il vaut mieux les conſtrui-
re en briques; & les couvrir
de tuiles. Ceux qui ſont en
bois ſe pourriſſent en peu de
tems, exigent des réparations
annuelles, & ſont expoſés aux
dangers des incendies. Comme
il eſt néceſſaire que les fours
ſoient placés au-deſſous des
fondations de la premiere cheminée, il faut pratiquer des
marches pour deſcendre dans
le hangard, & parvenir à l'em-
bouchure du four. Ainſi, ces
hangards doivent avoir au
moins huit pieds dans-œuvre;
les marches en occuperont à-
peu-près quatre, & il en reſtera
autant pour ſe remuer, faire
le feu, & ôter les cendres.
Quand la longueur des murs
exige deux fours, on les pratique au milieu du même hangard, ce qui épargne beaucoup de frais. En donnant à
ce hangard dix pieds de longueur ſur ſix de largeur, on
a plus de place pour ſoigner
les feux. On place les marches
à une de ſes extrémités, de

MUR

maniere que la porte ne se trouve pas vis-à-vis l'ouverture des fours, afin que le feu brûle plus également ; car lorsqu'elle est placée en face, le vent y pénetre sans obstacle, le fait brûler avec trop de violence, & consomme en peu de tems les matieres qu'on emploie pour l'entretenir.

Comme ces fours peuvent être construits de la même maniere que ceux dont on a déja donné l'idée pour les serres, je n'en dirai rien ici ; j'observerai seulement que, quand les deux fours sont joints ensemble, il faut qu'ils soient séparés par une muraille de trois briques au moins d'épaisseur, sans quoi cette cloison seroit bientôt détruite, & la moindre ouverture qui s'y formeroit, fournissant un passage à la fumée d'un tuyau à l'autre, les empêcheroit de tirer.

Le tuyau inférieur qui reçoit immédiatement la fumée, devant avoir deux pieds & demi de profondeur, il est nécessaire que le mur de derriere ait au moins deux briques & demie d'épaisseur jusqu'au haut de ce tuyau ; on peut ensuite réduire son épaisseur à deux briques, largeur qu'il faut conserver jusqu'à son extrémité. Le second tuyau, qui doit retourner au dessus du premier, doit avoir deux pieds de diametre, le troisieme un pied & demi, & le quatrieme un pied. Ces quatre tuyaux, avec leur couverture, s'éleveront à huit pieds de hauteur, de maniere qu'il y aura à-peu-près deux pieds au-dessus, pour y

fixer les châssis, & pour le chaperon du mur. Ces quatre tuyaux suffiront pour échauffer l'air renfermé dans les vitrages ; car la fumée aura perdu sa chaleur en les traversant. Quand on construit ces murs, on doit avoir soin d'y engager, de distance en distance, quelques crochets ou crampons de fer qu'on laisse saillir de deux pouces, & qui servent à soutenir le treillage. Ces crampons doivent être assez longs pour se prolonger dans le mur de derriere ; celui de devant n'ayant que quatre pouces d'épaisseur, ne seroit pas assez fort pour supporter le treillage : mais il faut observer de ne pas les faire passer à travers les tuyaux, parce qu'on ne pourroit pas les nettoyer ; de sorte que la meilleure maniere est de les placer immédiatement au-dessous des tuiles qui servent de couvertures aux tuyaux, & à trois ou quatre pieds de distance, ce qui suffira, pourvu que ces crochets soient assez forts. Comme il est nécessaire que les tuyaux soient bien enduits de terre forte en-dedans, on doit aussi plâtrer le dessous des tuiles qui les couvre jusqu'au niveau des crochets, afin que ces tuyaux n'offrent aucune inégalité ; car sans cela, la suie s'attacheroit aux crochets, & boucheroit à la longue le passage de la fumée. Il conviendra aussi de couvrir ces tuyaux du côté du treillage avec des sacs de houblon, ou quelqu'autre toile grossiere, comme on l'a déja dit pour les

MUR

serres chaudes, afin de fermer toutes les issues, de maniere que la fumée ne puisse trouver aucun passage; sans cette précaution, la fumée pénetre souvent, sur-tout quand les murs sont aussi minces qu'il est nécessaire qu'ils soient ici. Cette couverture fortifiera aussi les parois de ces tuyaux, & réunira tout l'ouvrage. A chaque extrémité de ces tuyaux, on pratiquera de petites arcades dans le mur de derriere, de maniere qu'on puisse les nettoyer, & en ôter toute la suie, lorsqu'il sera nécessaire de le faire; ce qui donnera beaucoup moins de peine que d'ouvrir ces tuyaux en face : on n'endommagera pas non plus les arbres fruitiers, en s'y prenant ainsi, & on ne gâtera pas les tuyaux, comme on le feroit si on les ouvroit par devant.

Les plates-bandes qui se trouvent en face de ces murailles chaudes, doivent avoir quatre pieds de largeur, ce qui suffit pour les talus des vitrages : on peut y semer un rang de pois nains, pour en avoir de bonne heure, ou un rang de féves naines, qui y réussiront également bien, & qui ne nuiront point aux arbres, si on ne les plante pas trop près. On éleve sur le bord de cette plate-bande un petit mur de quatre ou six pouces au-dessus du niveau sur lequel on place les châssis des vitrages, pour les garantir de la pourriture, & qui retiendra la terre de la plate-bande.

Les vitrages qu'on destine à couvrir ces murs, doivent être divisés en deux rangées.

MUR 167

Comme il est nécessaire qu'ils s'étendent depuis le bas presque jusqu'au haut du mur, ils auroient plus de douze pieds de longueur, s'ils étoient d'une seule piece. Lorsqu'ils ont plus de six pieds de longueur, ils font trop lourds pour être changés, sur-tout si les cadres font d'une force proportionnée au poids du verre. On doit faire ces cadres de maniere que celui du haut puisse glisser sur celui du bas, & en pratiquant d'un côté trois petits trous dans le bois qui soutient les cadres à un pied de distance l'un de l'autre, on pourra baisser les vitrages supérieurs d'un ou de trois pieds, suivant le volume d'air qu'on voudra y introduire, & les arrêter dans cette position au moyen d'une cheville de fer qu'on passera dans ces trous. Le rang inférieur des vitrages peut être construit de maniere qu'on puisse les ôter aisément; mais comme il est absolument nécessaire qu'ils soient en talus, & que le rang supérieur glisse par-dessus, on ne peut pas les faire monter : il n'y a d'ailleurs aucune nécessité de les remuer, parce qu'il vaut mieux laisser entrer l'air par le haut que par le bas.

Les pieces de bois qui soutiennent les châssis, doivent être fixées en bas dans la traverse placée sur le petit mur; & en haut, par les crampons de fer qu'on a eu la précaution d'engager dans le mur en le bâtissant. Ces pieces de bois doivent être en sapin, qui ne plie pas comme le chêne & d'autres especes, lorsqu'il est

dans une pareille pofition. Ces bois doivent être forts, fans quoi ils dureroient peu, furtout fi l'on eft obligé de les changer tous les ans : on cloue au haut de ces pieces une planche forte, au-deffous de laquelle les châffis doivent gliffer. L'ufage de cette planche eft d'empêcher que le vent n'enleve le rang fupérieur des châffis, & que la pluie ne pénetre jufqu'aux arbres : c'eft-pourquoi elle doit joindre le plus exactement qu'il eft poffible, & couvrir auffi à-peu-près deux pouces de la partie fupérieure des châffis, pour faire couler l'eau fur les vitrages, & les tenir fermes.

La largeur de ces châffis doit être de trois pieds ou un peu plus, fuivant que la divifion de la longueur du mur le permet. Il eft indifférent qu'ils foient un peu plus larges, pourvu que leur pefanteur n'empêche pas de les remuer. Si on les fait trop larges pour qu'un homme puiffe les embraffer, on ne les tranfporte pas commodément. Les barres ou lattes qui foutiennent le verre, doivent être placées en longueur ; car fi elles étoient en travers, elles arrêteroient l'eau des pluies, qui, pénétrant par le joint & tombant fur les plantes, leur feroit beaucoup de tort, fur-tout aux arbres, lorfqu'ils font en fleurs.

Le plomb qui réunit les vitrages, doit être large & exactement joint, pour fermer tout paffage à l'humidité.

A chaque extrèmité de ces rangs de vitrages, il reftera un efpace entre les châffis & le mur, que l'on doit bien boucher pour empêcher l'air d'y pénétrer ; car fans cela cette ouverture deviendroit fort nuifible aux arbres. Quelques perfonnes y mettent des planches ; mais fi on le fermoit avec un châffis vitré, difpofé de maniere qu'on puiffe l'ouvrir en partie, pour laiffer entrer l'air de tems en tems, cela feroit bien plus avantageux. Quand le vent fouffle directement contre le châffis de face, on pourroit ouvrir à chaque bout une de ces petites vitres, pour tempérer la chaleur, qui eft fouvent trop forte fous ces vitrages.

Les efpeces de fruits qu'on plante ordinairement pour les forcer, font les *Cerifiers*, les *Pruniers*, les *Abricotiers*, & les *Brugnons* : mais ces derniers réuffiffent rarement ; & comme ils font de peu de durée, ils ne valent pas la peine d'être plantés contre des murs chauds. Pour ce qui eft de la *Vigne*, je fuis d'avis qu'on la place féparément ; car comme elle exige plus d'air que les arbres, lorfqu'elle commence à pouffer, elle ne pourroit pas réuffir avec eux fous le même vitrage, au-lieu que les autres profperent tous dans le même endroit, parce qu'ils demandent à-peu-près le même dégré de chaleur.

Les efpeces les plus propres à être plantées contre ces murs, font : *Cerifiers.*

Le Cerifier de Mai printannier, & le Mayduc.
  Pruniers.

La Mirabelle ; le Damas noir

MUR

ou le Maroc précoce ; le gros Damas violet de Tours ; & le Drap d'or.

Péchers.

Le Muscat rouge ; la Magdelaine rouge ; le Montauban ; le Newington printannier ou précoce ; la Violette hâtive.

Brugnons.

La Muscade précoce de Fraichild ; l'Ebruge.

Abricotier.

Le Masculin.

Comme ces especes font les plus printannieres, elles sont aussi les plus propres à être plantées contre ces murs, quoiqu'elles ne soient pas aussi bonnes que quelques autres especes de fruits. Cependant comme elles mûrissent trois semaines ou un mois avant les autres, elles se perfectionnent bientôt, étant accélérées par une chaleur artificielle.

En preparant la plate-bande pour y planter ces arbres fruitiers, il faut avoir le même soin que pour ceux qu'on place contre des murs en plein air. Ainsi je n'en dirai rien ici, & je renvoie le Lecteur aux endroits de cet ouvrage où il trouvera toutes les instructions nécessaires à ce sujet. On palisse les branches qui commencent à pousser ; mais on ne place point le treillage que les arbres ne soient en état de produire beaucoup de fruits. Jusqu'à ce tems, on peut les soutenir avec des pesseaux à l'ordinaire, & l'on attend, pour les forcer, qu'ils soient devenus assez vigoureux, c'est-à-dire, au moins de quatre ou cinq ans, suivant les progrès qu'ils

MUR 169

auront faits. Si on les force trop jeunes, ils s'affoiblissent si fort, qu'ils ne poussent que très-rarement des branches bien nourries dans la suite : d'ailleurs la petite quantité de fruits que les jeunes arbres produisent, n'indemniseroit point des frais qu'ils exigeroient ; la dépense en bois ou en charbon, & les soins étant les mêmes pour de petits arbres capables de produire au plus six ou sept fruits, que pour ceux qui en donnent trois ou quatre douzaines.

Ainsi, plus on donne de tems à ces arbres pour se fortifier avant de les forcer, plus ils sont en état de dédommager de la peine & des dépenses.

La meilleure méthode est de ne faire ni cadres, ni treillage, ni aucune autre chose en bois, avant que les arbres soient assez avancés pour être forcés ; car si l'on place ces ouvrages aussi-tôt que le mur est bâti, comme on le pratique quelquefois, ils seront à moitié pourris avant qu'ils puissent être d'aucun usage : mais en suivant ma méthode, on doit avoir grand soin de ne pas endommager les arbres en plaçant le treillage.

Quand ces arbres auront acquis assez de force pour produire une certaine quantité de fruits, la partie qu'on veut forcer au printems suivant, doit être taillée au commencement de l'automne. On coupe les petites branches entierement, ou on les raccourcit beaucoup, parce qu'elles périroient presque toutes, lorsqu'elles se trou-

MUR

veroient exposées à une chaleur artificielle : & quoique quelques-unes soient bien chargées de boutons à fleurs, si elles sont foibles, elles n'auront pas la force de les nourrir ; de sorte que les fleurs épuisant toute la séve, les branches périssent bientôt après. Les branches plus fortes doivent aussi être raccourcies à une longueur convenable, comme nous l'avons dit pour les arbres en plein air, avec cette seule différence que les arbres qu'on destine à être forcés, ne doivent pas avoir leurs branches si longues, parce que la chaleur les affoiblit beaucoup ; & comme tous les boutons à fruits réussissent sous les châssis, parce qu'ils sont à couvert des injures de l'air, il en faut laisser beaucoup moins. Les branches doivent être fixées régulierement au treillage, à une distance convenable l'une de l'autre, afin que celles qui doivent pousser au printems suivant, ne s'ombragent point mutuellement. J'ai conseillé de tailler ces arbres au commencement de l'automne, afin que les branches qu'on a conservées, puissent attirer à elles toute la séve, & qu'étant bien remplies de sucs de la séve en hiver, elles soient plus disposées à la végétation, lorsque les feux sont allumés.

On commence à allumer les feux vers le milieu ou à la fin de Janvier, suivant que le tems est plus ou moins favorable ; car si l'on fait fleurir les arbres trop tôt, il est à craindre qu'ils ne réussissent pas, à cause des grands froids qui peuvent survenir encore : c'est-pourquoi la méthode la plus sure est de ne commencer à allumer le feu que vers la fin de Janvier, parce qu'on aura besoin de donner de l'air aux arbres, quand ils seront en fleurs, ce qui est impraticable dans le mauvais tems. - Les arbres qui fleuriront vers le milieu de Fevrier, donneront des fruits assez tôt. Les *Cerises* mûriront au commencement d'Avril, les *Abricotiers* au mois de Mai, & bientôt après suivront les *Prunes*, les *Pêches* & les *Brugnons*.

Quelques personnes plantent sur les plates-bandes, au-devant des arbres frutiers, des *Fraisiers*, qui souvent réussissent très-bien ; mais quand on adopte cette méthode, il faut avoir attention d'empêcher ces plantes de ramper sur la platebande ; car elles épuiseroient toute la substance de la terre, & feroient beaucoup de tort aux arbres. Ainsi, quand on désire avoir des *Fraises* précoces, je conseille de mettre ces plantes dans des pots ou séparément, à une bonne distance, sur une plate-bande à l'ombre, dans une terre forte, une année avant de les forcer, & d'arracher pendant ce téms tous les fils qui poussent, pour fortifier la racine principale & la préparer à produire du fruit. A la Saint-Michel, on peut les transplanter, avec de grosses mottes de terre à leurs racines, dans les plates-bandes, au-devant des arbres qu'on veut forcer au printems sui-

MUR

vant, pour qu'elles aient le tems de pouſſer des racines avant cette ſaiſon. Si on les arroſe lorſque les boutons à fleurs commencent à paroître, elles produiront une grande quantité de *Fraiſes*, qui mûriront vers la fin d'Avril ou au commencement de Mai. Lorſque ces plantes ont donné leurs fruits, il eſt bon de les enlever auſſi-tôt, afin qu'elles ne privent pas les arbres de la nourriture qui leur eſt néceſſaire.

Fraiſes précoces. J'inſérerai ici une autre pratique dont on uſe aſſez communément pour ſe procurer des *Fraiſes* dans le commencement du printems, quoique cela n'appartienne pas proprement à cet article. On élève les *Fraiſiers* dans des pots ou dans des plates-bandes comme on l'a dit plus haut, pendant un an au moins. Vers le commencement de Fevrier, on leur prépare une couche de chaleur tempérée ; & d'une grandeur proportionnée au nombre de plantes qu'on veut forcer. Les vitrages qu'on deſtine à les couvrir, peuvent être de la même forme que ceux que l'on emploie pour les couches chaudes ordinaires ſur leſquelles on plante des *Concombres* printanniers. On couvre cette couche avec de la terre forte & neuve, à huit pouces d'épaiſſeur, & l'on y place les *Fraiſiers* enlevés en motte, en laiſſant entr'eux une diſtance ſuffiſante. Comme il faut toujours les dépouiller des fils qui pouſſent, ils ne s'étendront pas

MUR   171

beaucoup ſur la couche, juſqu'à ce que leurs fruits ſoient paſſés ; alors on les arroſe légerement, pour comprimer la terre autour de leurs racines, & on répete cet arroſement à meſure que la terre ſe deſſeche, pour leur faire produire de nouveaux fruits. Pendant les nuits froides, on couvre les vitrages avec des nattes, pour conſerver une chaleur convenable dans les couches ; mais durant la journée, lorſque le tems eſt favorable, on leve les châſſis, pour y admettre l'air ; car ſi l'on pouſſoit trop ces plantes, ſur-tout lorſqu'elles commencent à fleurir, elles ne produiroient pas beaucoup de fruits. Si le froid continue long-tems, & que la chaleur des couches diminue, on place autour du fumier chaud, pour en renouveller la chaleur, en obſervant toujours de ne pas l'employer trop chaud, de peur qu'il ne brûle les racines des plantes. Si ces *Fraiſiers* ſont vigoureux & en état de porter du fruit ; ſi on les tranſplante avec de bonnes mottes, & ſi l'on entretient avec ſoin la chaleur de la couche, on aura une grande quantité de fruits à la fin ou au commencement de Mai, & les plantes continueront à en donner de nouvelles, juſqu'à ce que celles de pleine terre puiſſent leur ſuccéder.

Les eſpeces les plus propres à être forcées, ſont les *Fraiſes* écarlate ou celles des Alpes ; car les *Haut-Bois* ou *Caprons* ſont trop rampans pour cela.

MUR

Pour revenir à nos murailles chaudes, tout ce que j'ai inféré ici touchant la maniere de forcer les fruits, n'a été que pour les faire mûrir plutôt qu'on ne peut les avoir contre un mur ordinaire; mais dans quelques endroits de l'Angleterre, où la plupart de nos meilleures especes de fruits se perfectionnent rarement, il seroit fort à propos de construire de semblables murailles, pour se procurer les bonnes especes de *Péches* & de *Prunes*, qui n'y peuvent mûrir autrement. Ces murs feroient surtout utiles dans les cantons où le chauffage est à bon marché. Les murs étant bâtis, la dépense pour le reste feroit peu considérable. Je ne conseillerois cependant pas de faire les frais des vitrages, à moins que ce ne soit pour une petite longueur de muraille, mais de se servir de canevas ou de papiers huilés, qui rempliront le même objet : car, comme il ne feroit pas nécessaire de couvrir ces arbres avant le commencement de Mars, tems auquel on allumeroit les feux, avant qu'ils soient en fleurs, le tems est souvent assez chaud pour pouvoir les découvrir vers midi, & les exposer au soleil; car lorsqu'on les tient trop couverts, leurs branches filent, & leurs feuilles pâlissent. Comme le but de cette méthode n'est que de faire fleurir ces arbres trois semaines ou un mois plutôt, il n'est pas nécessaire de les échauffer par de grands feux, ni de les tenir trop couverts.

MUR

Au lieu de canevas, on peut se servir plus utilement de papiers huilés, que l'on emploie comme il a été dit pour les *Melons*, en collant ensemble autant de feuilles qu'il en faut pour couvrir les châssis : quand la colle est seche, on les attache sur ces cadres, & & on les enduit d'huile au moyen d'une brosse; ce qui donnera la transparence au papier, & le rendra propre à l'usage auquel il est destiné. Ce papier durera une saison, & il n'en coûtera pas beaucoup pour le réparer : c'est pourquoi il faut le préférer au canevas. Toutes les plantes réussiront mieux sous ce papier que sous le canevas, ou toute autre couverture qui n'admet point aussi bien les rayons de la lumiere. Les châssis qui doivent porter ces papiers huilés, n'ont pas besoin d'être aussi forts que ceux qui soutiennent des vitrages; & comme ils ne restent que trois mois exposés aux injures de l'air, ils peuvent durer long-tems, si l'on a soin de les mettre à couvert aussi-tôt qu'on n'en a plus besoin. Dans l'espace de trois mois, c'est-à-dire, depuis le commencement de Mars jusqu'à la fin de Mai, les arbres feront bien feuillés, & les jeunes branches auront fait assez de progrès pour être en état de protéger les fruits. Il ne faut pas ôter ces couvertures tout d'un coup, mais accoutumer par dégrés les arbres au plein air, sans quoi le changement feroit peut-être trop subit, & pourroit faire
tomber

MUR

tomber les fruits, fur-tout s'il furvient des nuits froides.

Les perfonnes qui voudront adopter cette méthode, pourront fe procurer les meilleures efpeces de fruits dans les parties feptentrionales de l'Angleterre, où, fans un pareil fecours, il eft impoffible de les faire mùrir ; & comme la houille eft fort commune dans ces cantons, la dépenfe du feu fera peu confidérable. Je fuis fort étonné que ceux qui habitent cette partie feptentrionale, ne fuivent point ce procédé ; car ils favent bien que dans leur voifinage il y a de pareilles murailles, qui, à la vérité, ont été élevées plutôt par la curiofité que pour l'ufage ; & ces murs, pour la plupart, font fi mal conftruits, que l'on confomme quatre fois plus de matieres combuftibles qu'il n'en faut en fuivant la méthode que je viens d'indiquer. Quand la chaleur n'eft pas également diftribuée par tout le mur, quelques arbres en ont trop, & d'autres pas affez.

Certaines gens conftruifent leurs murs de maniere que la plus grande partie de la chaleur fe porte fous la plate-bande contre la racine des arbres, parce qu'ils penfent que la chaleur eft auffi néceffaire qu'aux branches : mais c'eft une erreur ; le feu doit néceffairement nuire aux racines des arbres, en deffechant l'humidité de la terre, & en brûlant leurs fibres délicates, qui fe trouvent à fa portée. On doit donc rejetter cette

Tome V.

pratique, & élever toujours le premier tuyau d'un pied, ou au moins de quelques pouces audeffus du niveau de la plate-bande, fuivant que la terre eft feche ou humide, au-lieu de l'enfoncer fous la terre, où il ne ferviroit qu'à la deffecher, au-lieu d'échauffer l'air autour des arbres, ce que l'on doit uniquement chercher par cette chaleur artificielle. On fait quelquefois entrer une branche de vigne dans la ferre, & cette branche produit des fruits auffi promptement que fi l'arbre entier avoit été forcé, tandis que toutes les autres branches du même arbre expofées en plein air, n'en font point du tout accélérées, quoiqu'elles reçoivent la nourriture du même pied ; ce qui prouve, d'une maniere évidente, qu'il n'eft pas néceffaire d'échauffer les racines des arbres, pour en obtenir plutôt des fruits, ou pour hâter leur maturité.

J'ai auffi entendu parler de murs conftruits pour forcer des fruits, avec une ouverture depuis le haut jufqu'en bas, de maniere qu'ils formoient une double muraille de diftance en diftance, pour y faire du feu : mais cette méthode eft peu avantageufe ; car fi les murs font ouverts en haut pour laiffer fortir la fumée, la chaleur doit s'échapper ; & fi cette fumée ne fait pas trois ou quatre tours dans des tuyaux de briques, la chaleur fe diffipera par le haut, fans rendre le moindre fervice aux arbres.

Quand on a planté contre

M

des murailles les meilleures especes de fruits, si l'on veut faire mûrir parfaitement ces fruits, il faut remettre les couvertures sur les arbres, en cas que l'automne se trouve froid & pluvieux ; & en faisant un peu de feu pour dessecher l'humidité, on empêchera le fruit de moisir, & l'on avancera sa maturité. Si l'on suit cette méthode, il faut ôter les couvertures lorsque le tems le permet, afin que le fruit jouisse du plein air, sans quoi il seroit insipide & d'une mauvaise qualité. Quoique dans mes instructions précédentes pour forcer les fruits, j'aie conseillé de laisser reposer les arbres deux ou trois ans, pour qu'ils recouvrent leur vigueur, cependant il ne faut pas l'entendre des arbres que l'on se contente d'avancer pour perfectionner leurs fruits ; car comme il ne faut pas faire allumer les feux avant le commencement de Mars, ces arbres n'en seront point affoiblis, parce qu'ils seront accoutumés au plein air longtems avant la maturité de leurs fruits, & ils auront le tems de perfectionner leurs boutons pour l'année suivante. Ainsi, on peut forcer ces arbres tous les ans, sans leur nuire beaucoup, pourvu qu'on les traite avec soin.

En forçant des arbres fruitiers, quelques personnes placent des thermometres sous les châssis, pour mieux régler la chaleur : mais il faut alors les suspendre à l'ombre ; car si, au printems, ils res-toient seulement une heure exposés aux rayons du soleil, l'esprit-de-vin se raréfieroit, & s'éleveroit jusqu'au haut du tube, tandis que, sous le châssis, l'air ne seroit que tempéré : mais comme l'usage principal de ce thermometre est de régler le feu, il sert à peu de chose pendant le jour ; car une heure de soleil sur ces châssis échauffera assez l'air pour perfectionner les fruits d'Europe, sans chaleur artificielle ; ce qui fait qu'on n'a pas besoin d'allumer le feu pendant le jour, à moins que le tems ne se trouve fort mauvais ; & si, par les feux de la nuit, l'air est échauffé au point tempéré marqué sur le thermometre botanique, les fruits réussiront beaucoup mieux que dans une plus grande chaleur.

Aux environs de Londres, quelques personnes s'occupent à élever des fruits printaniers, pour fournir les marchés ; ce qu'ils operent par la seule chaleur du fumier, n'ayant point de murs chauds dans leurs jardins.

Voici la maniere dont elles s'y prennent. Après avoir mis en tas une grande quantité de fumier nouveau, comme on le pratique pour les couches chaudes, lorsque ce fumier a acquis une chaleur convenable, on le place derriere la muraille contre laquelle sont plantés les arbres fruitiers, en lui donnant quatre pieds d'épaisseur au bas ; mais en diminuant par dégrés cette épaisseur jusqu'à un pied ou dix

pouces vers le haut : on comprime ce fumier légerement avec la fourche, pour empêcher la chaleur de se dissiper trop tôt ; mais il ne faut pas le battre trop, de peur qu'il ne fermente pas : on unit sa surface autant qu'il est possible, afin que l'eau de la pluie puisse s'écouler facilement; & en le couvrant de chaume, comme on le fait quelquefois, on l'empêche de se pourrir trop tôt, & l'on conserve plus long-tems sa chaleur. On ne place point ce fumier dans le tems où l'on allume ordinairement les feux, mais un peu plus tard, c'est-à-dire, vers le milieu de Fevrier. Ce fumier conserve sa chaleur pendant un mois ou cinq semaines ; après ce tems, on en prépare d'un autre, & l'on enleve le premier, ou bien on le mêle avec le nouveau, qui renouvellera sa chaleur, & la conservera jusqu'à la derniere saison : on couvre le mur avec du papier huilé, comme nous l'avons dit, & l'on traite les arbres de la même maniere ; mais il faut avoir plus de soin d'ôter ces châssis, lorsque le tems le permet, sans quoi la fumée du fumier occasionneroit une grande humidité qui pénetreroit à travers le mur, & nuiroit beaucoup aux arbres, sur-tout quand ils sont en fleurs.

Quelques Jardiniers, par cette méthode, ont forcé de longs murs garnis de vieux arbres, qui ont produit une grande quantité de fruits tous les ans, & qui ont rapporté au-delà de la dépense : mais comme il est difficile en plusieurs endroits de se procurer une quantité suffisante de fumier, les murs chauds y sont moins dispendieux.

J'ai vu construire des murailles en bois, pour forcer des arbres fruitiers, au moyen du fumier : mais ces murailles de bois ne valent rien ; car l'odeur & les exhalaisons du fumier, qui passent à travers les fentes des planches, nuit beaucoup aux arbres : d'ailleurs ces planches étant toujours humides, tant que le fumier l'est lui-même, les arbres en souffrent beaucoup ; & comme ces planches se pourrissent en peu d'années, elles sont plus coûteuses que les murs, sans donner le même produit.

MURE DE RONCE. *Voyez* Rubus Cæsius. *L.*

MURIER. *Voyez* Morus.

MURIER NAIN. *Voyez* Rubus Chamæmorus.

MUSA. *Plum. Nov. Gen.* 24. *tab.* 34. *Lin. Gen. Plant.* 1010. [*The Plaintaine-tree.*] Bananier ou Figuier d'Adam.

Bihai, troisieme espece, est remis sous le nom d'*Héliconia.*

Caracteres. Cette plante a des fleurs mâles & femelles, & même quelques fleurs hermaphrodites sur le même pédoncule ou dans la même grappe ; elles sont produites sur une simple tige ou spadix : les fleurs mâles sont placées sur la partie haute du poinçon, & les femelles vers le bas ; elles sont rassemblées en grappes, qui ont chacune une enveloppe qui tombe. Ces fleurs

176 MUS

font labiées ; les pétales conftituent la levre fupérieure , & le nectaire l'inférieure ; elles ont fix étamines en forme d'alêne , dont cinq font fituées dans le pétale , & la fixieme dans le nectaire : celle-ci eft une fois plus longue que les autres , & terminée par une anthere linéaire ; les autres n'ont point d'antheres : le germe, qui eft fous la fleur , eft long , & a trois angles obtus ; il foutient un ftyle érigé , cylindrique , & couronné par un ftigmat rond. Ce germe fe change dans la fuite en un fruit oblong , triangulaire , charnu, couvert d'une peau épaiffe , & divifé en trois parties.

Ce fruit fe nomme *Bannane* ou *Figue Bannane* , & la grappe qui raffemble ces fruits, eft appelée *Régime.*

Ce genre de plantes eft rangé dans la premiere fection de la vingt troifieme claffe de LINNÉE , qui renferme celles qui ont des fleurs mâles, femelles & hermaphrodites fur la même tige. PLUMIER le range dans la claffe de TOURNEFORT , avec les fleurs irrégulieres de plufieurs pétales; & GARCIN le place parmi les plantes à *Fleurs de lys.*

Les efpeces font :

1º. *Mufa parafidiaca , fpadice nutante , floribus mafculis perfiftentibus. Lin. Sp. 1477. Burm. Ind.* 217 ; Bananier avec un fpadix penché , & des fleurs mâles qui perfiftent.

Mufa fructu Cucumerino longiore. Plum. Nov. Gen. 24 ; Bananier avec un plus long fruit en forme de Concombre , communément appelé arbre de Bananier.

Mufa racemo fimpliciffimo. Hort. Cliff. 467. *Hort. Ups.* 301. *Fl. Zeyl.* 368. *Roy. Lugd.-B.* 10. *Haffelq. it.* 491. *Gron. Orient.* 324.

Mufa Cliffortiana. Lin. Muf. 1. *f.* 1. *Trew. Ehret. f. 18* 19. 20.

Mufa. Cluf. Exot. 229. *Rumph. Amb.* 5. *p. 125. f.* 60.

Ficus Indica , fructu racemofo , folio oblongo. Bauh. Pin. 508.

Palma humilis, longis latifque foliis. Bauh. Pin. 107.

Bata. Rheed. Mal. 1. *p.* 17. *f.* 12. 13. 14.

2º. *Mufa fapientum , fpadice nutante , floribus mafculis deciduis. Lin. Sp.* 1477 ; Bananier avec un épi penché , & des fleurs mâles qui tombent.

Mufa fructu Cucumerino breviore. Plum. Nov. Gen. 24 ; Bananier avec un fruit plus court en forme de Concombre , nommé fimplement Bananier.

Mufa fpadice nutante , fructu breviore oblongo. Brown. Jam. 363.

Mufa fpadice nutante , fructu breviore obtufè angulato. Lin. Syft. Plant. tom. 4. *p.* 295. *Sp.* 2.

Mufa caudice maculato , fructu recto-rotundo , breviore , odorato , Sloan. Jam. 192. *Hift.* 2. *p.* 147. *Trew. Ehret.* 4. *f.* 21. 22. 23.

Mufæ affinis altera. Bauh. Pin. 580.

Ficus Indica racemofa , foliis venuftè venofis , fructu minore. Pluk. Alm. 145.

Paradifiaca. On cultive communément la premiere efpece dans les Ifles des Indes Occidentales, où fon fruit fert à la nourriture des Negres , qui le mangent en guife de pain. Quelques Blancs le préferent auffi à prefque toute autre

MUS

nourriture , & fur-tout au pain de *Yams* & à la *Caſſave*.

Cette plante s'éleve avec une tige molle & herbacée , à quinze ou vingt pieds de hauteur , & même davantage. La partie baſſe de la tige eſt ſouvent auſſi groſſe que la cuiſſe , mais plus mince par dégrés juſqu'au ſommet , où les feuilles ſortent ſur chaque côté. Ces feuilles ont ſouvent plus de ſix pieds de longueur ſur deux de large , la côte du milieu eſt fort charnue , & donne origine à un grand nombre de nervures tranſverſales, qui s'étendent juſqu'aux bords. Ces feuilles ſont minces & tendres ; de ſorte que , quand elles ſont expoſées en plein air , le vent, qui a beaucoup de priſe ſur elles, les déchire ordinairement ; elles ſortent de la tige principale , qu'elles enveloppent de leurs bâſes. Quand elles commencent à paroître , elles ſemblent être roulées ; mais à meſure qu'elles s'élevent au deſſus de la tige, elles s'étendent entierement , & ſe penchent en arriere. Comme elles ſortent roulées , ainſi qu'on vient de le dire, leur accroiſſement vers le haut eſt ſi prompt, qu'on pourroit preſque le ſuivre à l'œil nud ; car ſi l'on tire une ligne horiſontale à leur extrémité , on verra qu'en une heure de tems elles ſe ſont élevées à un pouce au deſſus.

Quand la plante eſt parvenue à ſon entiere hauteur, les épis des fleurs paroiſſent dans le centre des feuilles ; ils ont ſouvent à-peu-près quatre pieds de longueur, & ſont inclinés ſur le côté : les fleurs ſortent en grappes, celles du bas ſont les plus larges , & les autres diminuent de largeur à meſure qu'elles ſont plus voiſines de l'extrémité. Chaque paquet ou grappe eſt couvert d'une gaîne d'une belle couleur pourpre en-dedans , & qui tombe quand les fleurs s'ouvrent. Le haut de l'épi eſt garni de fleurs mâles ou ſtériles, qui ne produiſent point de fruits, & celles de la ſeconde eſpece tombent avec leurs enveloppes. Le fruit de cette plante a huit ou neuf pouces de longueur ſur plus d'un pouce de diametre ; il eſt un peu recourbé, à trois angles, d'abord vert, & d'un jaune pâle lorſqu'il eſt mûr ; ſa peau eſt rude, & recouvre une chair molle d'une ſaveur douce & agréable. La tige du fruit ou le régime eſt ſouvent aſſez gros pour peſer plus de quarante livres.

On coupe toujours le fruit de la premiere eſpece, qui eſt la *Banane*, avant ſa maturité ; on le fait cuire ſous la cendre, & on le mange en guiſe de pain ; ſes feuilles ſervent de ſerviettes & de nappes, & on s'en ſert encore pour nourrir les cochons.

Sapientum. La ſeconde eſpece , à laquelle on donne communément le nom de *Bananier*, differe de la premiere par ſes tiges marquées de raies & de taches d'un pourpre foncé. Son fruit, que l'on nomme *Figue Banane*, eſt plus court, plus droit, & plus rond : les fleurs mâles tombent, la chair

est plus molle & d'un goût plus sucré ; aussi le mange-t-on toujours au dessert, & il est rare qu'on en fasse le même usage que du précédent, ce qui fait qu'on ne le cultive pas en si grande abondance.

Culture. Ces deux plantes ont été portées des Isles Canaries en Amérique. On croit qu'elles avoient été transportées dans ces Isles de la côte de Gui-née, où elles croissent natu-rellement. On les cultive aussi en Egypte, & dans plusieurs autres pays chauds, où elles acquierent tout leur dévelop-pement environ dix mois après qu'elles ont été plantées, & donnent des fruits mûrs. Quand leurs tiges sont coupées, leurs racines poussent plusieurs re-jettons qui produisent aussi du fruit dix mois après ; de sorte qu'en les coupant dans des tems différens, ces fruits se succedent sans interruption pendant toute l'année.

En Europe, on conserve quelques-unes de ces plantes dans les jardins des curieux qui ont des serres chaudes assez grandes pour les conte-nir, dans plusieurs desquelles serres elles ont perfectionné leurs fruits assez bien : mais comme elles s'elevent à une hauteur considérable, & que leurs feuilles sont grandes, elle exigent plus de place dans la serre qu'on ne voudroit leur en donner. On les multiplie par les rejettons qui sortent des racines de celles qui ont produit des fruits ; & quand les jeunes plantes sont gênées dans leur crû, elles poussent

aussi des rejettons, qu'il faut enlever soigneusement, en y conservant quelques fibres : on les plante dans des pots remplis d'une terre riche & legere, & on les plonge dans la couche de tan de la serre.

On peut enlever ces rejet-tons dans tous les tems de l'été ; mais il est toujours plus avantageux de les détacher, tandis qu'ils sont encore très-jeunes, parce que leurs racines étant devenues grosses, elles ne poussent pas si aisément de nouvelles fibres, & que les plantes se pourrissent souvent, quand on coupe, en les en-levant, la partie épaisse de leurs racines.

Il faut arroser beaucoup ces plantes pendant l'été ; car la surface de leurs feuilles étant fort étendue, elles per-dent beaucoup d'humidité par la transpiration dans les tems chauds. En hiver, on les arrose très-legerement ; mais on ré-pete souvent cette opération.

Les pots dans lesquels ces plantes sont placées doivent être proportionnés à leur gros-seur ; car leurs racines s'é-tendent ordinairement fort loin. La terre qu'on leur donne doit être riche & legere, & le dégré de chaleur auquel elles profitent le mieux, est le même que celui qui convient aux *Ananas.* Au moyen de ce trai-tement, plusieurs des plantes que j'ai possédées, ont perfec-tionné leurs fruits, & se sont élevées à la hauteur d'environ vingt pieds.

La méthode la plus sûre pour faire porter du fruit à

MUS

ces plantes dans notre climat, est, après qu'elles ont crû pendant quelque tems dans des pots, & qu'elles ont poussé de bonnes racines, de les enlever avec la motte de terre, & de les planter dans la couche de tan de la serre chaude, en observant de mettre un peu de vieux tan contre leurs racines, pour que leurs fibres puissent pénétrer; bientôt après, ces racines s'étendront à plusieurs pieds de tous côtés, & les plantes feront beaucoup plus de progrès que celles qui sont gênées dans des pots ou dans des caisses. Quand la couche a besoin d'être renouvelée avec du nouveau tan, il faut en laisser une assez grande quantité de vieux autour de leurs racines, non-seulement pour ne point les endommager en l'enlevant, mais encore pour empêcher que le tan nouveau ne les brûle. Ces plantes ne font des progrès qu'autant qu'elles sont bien arrosées. En hiver, on donne à chacune environ deux pintes d'eau, deux fois la semaine; mais en été, il leur en faut au moins quatre pintes chaque deux jours. Si leurs tiges de fleurs paroissent au printems, on pourra espérer de leur voir perfectionner leurs fruits; mais quand ces tiges poussent plus tard, les plantes périssent quelquefois avant que leurs fruits ne soient mûrs. Les serres chaudes dans lesquelles elles sont placées, doivent avoir au moins vingt pieds de hauteur, sans quoi il n'y aura pas assez de place pour l'étendue de leurs feuilles; car lorsqu'elles sont en vigueur, ces feuilles ont souvent huit pieds de longueur & deux de largeur, & les tiges ont quatorze pieds jusqu'à la division des feuilles; de maniere que quand les serres n'ont pas assez de hauteur, les feuilles sont gênées, & l'accroissement des plantes est fort retardé : d'ailleurs quand les feuilles se penchent contre les vitrages, & qu'elles croissent avec vigueur, ces vitrages courent risque d'être brisés; car j'ai vu, dans cette circonstance, ces feuilles casser les vitrages d'une serre, & sortir de deux ou trois pouces au-dessus dans une seule nuit.

J'ai eu des régimes de fruits de la premiere espece, qui ont mûri parfaitement en Angleterre, & qui pesoient plus de quarante livres : mais ce fruit n'est pas assez bon pour engager à faire la dépense de le cultiver dans ce pays.

On préfere la seconde espece à la premiere dans les pays chauds où on la cultive, parce que son fruit est beaucoup plus agréable au goû'. Les régimes de celle-ci sont moins gros que ceux de la premiere espece, & le fruit n'en est pas si long : il devient d'un jaune foncé à mesure qu'il mûrit; son goût ressemble un peu à celui d'une *Figue* farineuse. Des personnes qui ont résidé en Amérique, & qui y ont mangé de ces fruits, ont trouvé que ceux qui croissent en Angleterre leur étoient peu inférieurs en qualité. Je pense

M 4

180 MUS MUS

que les habitans de ces contrées ne font tant de cas de ces fruits, que parce qu'ils n'en ont pas beaucoup d'autres qui leur soient préférables ; mais ils seroient peu recherchés en Europe, quand même on pourroit les y avoir dans leur plus grande perfection.

MUSC ou KETMIE D'AMÉRIQUE. *Voyez* HIBISCUS ABELMOSCHUS.

MUSC ou GRAPPE DE JACINTHE. *Voyez* MUSCARI.

MUSCARI. *Tourn. Inst. R. H.* 347. *tab.* 180. *Hyacinthus. Lin. Gen. Plant. Ed. nov. n.* 461. [*Musk, or Grape Hyacinth.*] Musc, vulgairement appelé Grappe de Jacinthe.

Caractères. La fleur n'a point de calice ; la corolle est monopétale, en forme de cruche, & réfléchie sur ses bords : la fleur a trois nectaires sur le sommet du germe, & six étamines en forme d'alène, plus courtes que la corolle, & dont les anthères sont réunies ; dans son centre est placé un germe rond à trois angles, qui soutient un style simple, couronné par un stigmat obtus. Ce germe se change dans la suite en une capsule ronde à trois angles, & à trois cellules remplies de semences rondes.

LINNÉE a joint ce genre à la *Jacinthe,* qui est placée dans la première section de la sixieme classe, avec les plantes, dont les fleurs ont six étamines & un style.

Les especes sont :

1°. *Muscari botryoïdes,* corollis globosis, uniformibus, foliis canaliculato-cylindricis, strictis ; Muscari avec des corolles globulaires & uniformes, & des feuilles cylindriques & en forme de gouttieres serrées.

Muscari arvense, Junci-folium cæruleum minus. Tourn. Inst. 348 ; petit Muscari bleu des champs, à feuilles de Jonc, communément appelé Grappe de Jacinthe.

Hyacinthus botryoïdes. Lin. Syst. Plant. tom. 2. *p.* 80. *Sp.* 12.

Hyacinthus foliis gramineis, spicâ ovatâ, floribus globosis fæcundis. Hall. Helv. n. 1246.

Hyacinthus botryoïdes vernus minor, latifolius, cæruleus, inodorus. Bauh. Hist. 2. *p.* 572.

Hyacinthus botryoïdes purpureus III. Clus. Hist. 1. *p.* 181.

2°. *Muscari comosum,* corollis angulato-cylindricis, summis sterilibus, longiùs pedicellatis ; Muscari avec des corolles angulaires & cylindriques qui sont stériles au sommet de l'épi, où elles ont de plus longs pédoncules.

Muscari arvense, latifolium purpurascens. Tourn. Inst. 347 ; Muscari pourpre des champs, à larges feuilles, communément appelé Jacinthe à beau poil.

Hyacinthus comosus. Lin. Syst. Plant. tom. 2. *p.* 79. *Sp.* 11. *Scop. carn.* 2. *n.* 423. *Pollich. pal. n.* 342. *Jacq. Austr. t.* 126. *Kniph. cent.* 2. *n.* 34.

Hyacinthus spicâ longissimâ, floribus supremis sterilibus, erectis, inferioribus fæcundis, patulis. Hall. Helv. n. 1247.

Hyacinthus corollis globosis, summis pedunculatis, foliis ensiformibus. Sauv. Monsp. 17.

MUS

Hyacinthus. Cam. Epit. 798.

3°. *Muſcari racemoſum , corol-
liis ovatis , ſummis feſſilibus ,
foliis laxis* ; Muſcari avec des
corolles ovales , dont les ſommets ſont ſeſſiles , & les feuilles
moins ſerrées.

*Muſcari obſoletiore flore. Cluſ.
Hiſt.* 1. *p.* 178 ; Muſcari à
fleurs de couleur uſée , communément appelé Jacinthe de
Muſc.

*Hyacinthus racemoſus. Lin.
Syſt. Plant. tom.* 2. *p.* 80. *Sp.*
13. *Sauv. Monſp.* 17. *Jacq.
Auſtr. t.* 187.

*Hyacinthus foliis carinatis ,
ſpicâ ovatâ , floribus globoſis.
Hall. Helv. n.* 1245.

Allium caninum exiguum. Trag.
750.

4°. *Muſcari monſtroſum corollis ſub-ovatis* ; Muſcari avec
des corolles preſque ovales.

*Hyacinthus paniculâ cæruleâ ,
C. B. P.* 42. ; Jacinthe bleue en
panicule , appelée Jacinthe plu-
macée.

*Hyacinthus monſtroſus. Lin.
Syſt. Plant. tom* 2. *p.* 79. *Sp.* 10.

5°. *Muſcari Orchioïdes , corollis ſexpartitis , petalis tribus
exterioribus brevioribus* ; Muſcari avec des corolles diviſées
en ſix parties , & dont les trois
pétales extérieurs ſont les plus
courts.

*Hyacinthus Orchioïdes , Afri-
canus, major, bifolius, maculatus ,
flore ſulphureo , obſoleto , majore.
Breyn. Prod.* 3. 24 ; la plus
grande Jacinthe d'Afrique , reſ-
ſemblant à l'Orchis , à deux
feuilles tachetées , ayant une
grande fleur d'une couleur de
ſoufre uſée.

Hyacinthus Orchioïdes. Lin.

MUS 181

Syſt. Plant. tom. 2. *p.* 80. *Sp.*
14. *Jacq. Hort. t.* 178.

*Orchis anguſtifolia maculata.
Buxb. cent.* 3. *p.* 10. *tom.* 16.

Botryoïdes. La premiere eſpece crcît naturellement dans
les vignes & les terres labourées , en France , en Italie &
en Allemagne : quand elle eſt
une fois établie dans un jardin , il n'eſt pas aiſé de la détruire , car ſes racines ſe multiplient conſidérablement ; &
ſi on lui laiſſe écarter ſes ſemences , tout le terrein en eſt
bientôt rempli. Il y a trois variétés de cette eſpece , l'une
à fleurs bleues , la ſeconde à
fleurs blanches , & la troiſieme
à fleurs cendrées : la premiere
a une petite racine ronde &
bulbeuſe , de laquelle ſortent
pluſieurs feuilles de ſix pouces environ de longueur ,
étroites , recourbées ſur leurs
bords , & en forme de gouttieres ; du centre de ces feuilles
s'éleve une tige nue , & garnie vers ſon ſommet d'un épi
ſerré de fleurs bleues en forme de cruches , ſeſſiles au pédoncule , & qui répandent une
odeur d'un empois nouveau ,
ou de noyaux de prunes frais.
Cette plante fleurit en Avril ,
& ſes ſemences mûriſſent à la
fin de Juin.

Comoſum. La ſeconde eſpece
eſt originaire de l'Eſpagne &
du Portugal , d'où ſes racines
& ſes graines m'ont été envoyées ; elle a une racine bulbeuſe auſſi groſſe qu'un *Oignon*
médiocre , de laquelle ſortent cinq ou ſix feuilles d'un
pied de longueur ſur neuf
lignes de largeur à leur bâſe ,

mais plus étroites par dégrés jufqu'à la pointe : la tige de fleurs, qui s'élève à un pied environ de hauteur, eft nue dans la moitié de fa longueur vers le bas ; mais le haut eft garni de fleurs pourpre, cylindriques, angulaires, poftées fur des pédoncules de fix pouces de longueur ; elles font placées horifontalement : la tige eft terminée par une touffe de fleurs dont les corolles font ovales, & qui font ftériles, parce qu'elles n'ont ni ftyle ni germe. Cette efpece fleurit à la fin d'Avril ou au commencement de Mai ; elle donne une variété à fleurs blanches, & une autre à fleurs bleues ; mais la pourpre eft la plus commune.

Racemofum. La troifieme a des racines groffes, ovales & bulbeufes, d'où s'élevent plufieurs feuilles de huit ou neuf pouces de longueur, fur fix lignes de largeur, un peu recourbées fur leurs côtés, terminées en pointe obtufe, & qui fe roulent les unes fur les autres à leur bâfe : la tige de fleurs fort du milieu de ces feuilles ; elle eft nue vers le bas, & garnie en haut de petites fleurs rapprochées en épis, dont les corolles font ovales, en forme de cruche, réfléchies fur leurs bords, d'une couleur de pourpre cendrée, ou de couleur ufée, comme fi elles étoient fanées, & d'une odeur agréable de mufc. Ces tiges n'ont que fix pouces de hauteur, & les fleurs n'ont pas grande apparence ; mais quand elles font nombreufes, elles parfument l'air à une diftance confidérable. Cette plante fleurit en Avril, & fes femences mûriffent en Juillet.

Il y a deux variétés de cette efpece, dont la premiere a des fleurs de même couleur que celles de la précédente fur le bas de l'épi, mais plus larges, & tirant fur le pourpre. Les fleurs du haut font jaunes, & d'une odeur fort agréable.

Les Jardiniers Hollandois donnent à cette plante le nom de *Tibcadi Mufcari.* Comme celle-ci eft regardée comme n'étant qu'une variété de la troifieme, je ne l'ai pas mife au nombre des efpeces. Il y en a une autre à très-grandes fleurs, qui a été nouvellement obtenue de femences en Hollande. Les Fleuriftes vendent fa racine une guinée.

Monftrofum. La quatrieme a une racine groffe & bulbeufe, de laquelle fortent plufieurs feuilles unies, d'un pied de longueur, fur fix lignes environ de largeur à leur bâfe, & terminées en pointe obtufe. Les tiges des fleurs s'élevent à la hauteur d'un pied & demi ; elles font nues vers la bâfe dans la longueur de fept ou huit pouces : mais au-deffus commencent les panicules de fleurs qui terminent les tiges. Ces fleurs naiffent fur des pédoncules d'un pouce & plus de longueur, qui foutiennent chacun trois, quatre ou cinq fleurs, dont les corolles font découpées en filamens minces comme des poils, & font d'un bleu pourpâtre : mais comme elles n'ont ni étamine ni germe,

elles ne produifent jamais de femences. Cette plante fleurit dans le mois de Mai : quand fes fleurs font paffées, les tiges & les feuilles périffent jufqu'à la racine, qui en repouffe de nouvelles au printems fuivant.

Orchioïdes. La cinquieme efpece fe trouve au Cap de Bonne-Efpérance, d'où j'en ai reçu des femences, qui ont réuffi dans le jardin de Chelféa. Les plantes qu'elles ont produites ont fleuri pendant plufieurs années : elle a une racine blanche, bulbeufe, & de la groffeur d'une noifette, qui ne produit ordinairement que deux feuilles, & quelquefois trois. Quand les racines font fortes, ces feuilles ont cinq ou fix pouces de longueur, & un pouce de largeur au milieu ; elles font terminées en pointe aiguë, d'un vert luifant, & marquées de plufieurs taches ou protubérances fur leur furface fupérieure. La tige de fleurs qui s'éleve au milieu de ces feuilles, jufqu'à la hauteur de fix ou fept pouces, eft ronde, unie, nue dans la longueur de trois pouces, & terminée par un épi de fleurs de couleur de foufre pâle, mais qui n'ont point de pédoncules : la corolle eft monopétale, d'une forme irréguliere, & découpée au fommet en fix parties ; les étamines font prefque de la longueur de la corolle, & poftées autour du ftyle, qui eft également long. Ces fleurs paroiffent dans le mois de Mars, mais elles produifent rarement de bonnes femences ici.

Culture. Les quatre premieres efpeces font fort dures, & profitent en plein air ; elles n'exigent point d'autre culture que les autres fleurs dures à racine bulbeufe : l'on enleve ces racines chaque deux ou trois ans, pour féparer leurs bulbes ; car comme quelques-unes des autres efpeces fe multiplient affez confidérablement, fi on les laiffe venir en gros paquets, elles ne fleuriffent pas fi bien que fi elles étoient féparées. Le meilleur tems pour les tirer de terre, eft auffi-tôt après que leurs tiges & leurs feuilles font flétries : on les fépare enfuite fur une natte dans une chambre feche & à l'ombre, où on les tient pendant quinze jours, pour les faire fecher ; après quoi on peut les conferver dans des caiffes comme les autres racines bulbeufes jufqu'à la Saint-Michel, qui eft le tems de les replanter dans les plates-bandes du parterre, où elles doivent être traitées comme les *Jacinthes* communes & dures.

La premiere efpece ne doit point être admife dans les jardins à fleurs, parce que fes racines fe multiplient fi fort, qu'elles deviennent embarraffantes.

Comme la feconde a peu de beauté, on n'en conferve que quelques plantes pour la variété : elle eft fi dure, qu'elle profite dans tous les fols & à toutes les expofitions.

La troifieme mérite une place dans les jardins, à caufe de la bonne odeur de fes fleurs,

184 **MUS**

sur-tout la variété à fleurs jaunes, appelée *Tibcady*.

La quatrieme doit aussi être placée dans les plates - bandes ordinaires du parterre, où elle augmentera la variété : ainsi, on ne doit point du tout la méprifer.

On les multiplie toutes aifément par leurs rejettons, que la plupart de leurs racines pouffent en grande abondance ; de forte qu'on eft rarement obligé de les femer, à moins que ce ne foit pour acquérir de nouvelles variétés.

Comme la cinquieme efpece eft trop tendre pour réuffir en plein air dans ce pays, il faut planter fes racines dans des pots remplis d'une terre riche & légere ; & en automne, les mettre fous un châffis de couche chaude, où elles puiffent être à l'abri du froid : mais elles exigent autant d'air qu'il eft poffible en tems doux ; car lorfqu'elles font placées dans une ferre, leurs feuilles filent, deviennent longues & étroites, leurs tiges reftent toujours foibles, & ne fleuriffent jamais bien. Ces fleurs fe conservent un mois quand elles ne filent point ; mais elles périffent prefque toujours dans une ferre.

Ces racines doivent être tranfplantées en Juillet ; celles dont les tiges & les feuilles périffent pendant l'été, veulent être placées en plein air, & très-peu arrofées, quand leurs feuilles font fanées.

MUSCIPULA. *Voy.* SILENE.
MUSCOSUS MOUSSEUX *ou* COUVERT DE MOUSSE. Ce

MUS

mot exprime quelquefois le coton ou duvet qui couvre les plantes ou les fruits.

MUSCUS. [*Mofs.*] La Mouffe eft une plante qui autrefois n'étoit regardée que comme une excroiffance produite par la terre, les arbres, &c. Cependant les plantes que l'on connoît fous ce nom ne font pas moins parfaites que les autres, quoiqu'elles foient plus grandes & mieux développées ; elles ont des racines, des branches, des fleurs, des femences, quoiqu'on ne puiffe les multiplier par leurs graines, par quelque méthode que ce foit.

Les Botaniftes les diftinguent en plufieurs genres, fous chacun defquels font placées plufieurs efpeces : mais comme ces plantes ne font d'aucun ufage & n'ont point de beauté, ce n'eft pas la peine d'en parler ici.

Elles fleuriffent principalement dans des pays froids & en hiver ; elles font fouvent fort nuifibles aux arbres fruitiers, qui croiffent dans des fols froids & ftériles, contre lefquels elle s'attachent fi étroitement, qu'elles les privent d'air entierement : le feul remede, dans ce cas, eft d'arracher une partie des arbres, de labourer la terre entre ceux qui reftent ; & au printems, lorfque la terre eft humide, on ratiffe ces arbres avec un inftrument de fer, de forme circulaire ; on enleve toute la mouffe qui les couvre, & on a foin de l'emporter. Cette opération, qu'on réitere deux ou

MYA

trois fois , ainſi que le labour , peut entierement détruire toute la mouſſe des arbres : maisſi l'on ne retranche pas une partie de ces arbres, & ſi l'on ne cultive pas bien la terre, il ſera inutile de ratiſſer la mouſ-ſe , parce qu'alors la cauſe ſub-ſiſtant toujours , elle ſe repro-duira en peu de tems.

MYAGRUM. *Tourn. Inſt. R. H.* 211. *tab.* 99. *Lin. Gen. Plant.* 713. [*Godl of Pleaſure.*] la Ca-meline.

Caractères. Le calice de la fleur eſt compoſé de quatre feuilles oblongues , ovales & colorées ; la corolle a quatre pétales ronds, obtus , & pla-cés en forme de croix ; la fleur a ſix étamines auſſi longues que la corolle , dont quatre ſont cependant un peu plus longues que les autres , & qui ſont toutes terminées par des antheres ſimples ; dans le cen-tre , eſt placé un germe ova-le , qui ſoutient un ſtyle mince & couronné par un ſtigmat ob-tus. Ce germe ſe change , quand la fleur eſt paſſée , en un lé-gume turbiné , court , en forme de cœur , à deux valves , avec un ſtyle rigide au ſommet , & qui renferme des ſemen-ces rondes.

Ce genre de plantes eſt rangé dans la première ſection de la quinzieme claſſe de LINNÉE , qui comprend celles dont les fleurs ont quatre étamines lon-gues , & deux plus courtes, avec des ſemences renfermées dans de petits légumes courts.

Les eſpeces ſont:

1°. *Myagrum ſativum , ſili-culis ovatis , pedunculatis , po-*

MYA 185

lyſpermis. Hort. Cliff. 328. *Fl. Suec.* 541. 464. *Roy. Lugd.-B.* 330. *Dalib. Paris.* 193. *Neck. Gallob. p.* 273. *Gmel. Tub. p.* 194. *Pollich. pall. n.* 602. *Mat-tuſch. Sil. n.* 473. *Kniph. cent.* 11. *n,* 78 ; Cameline avec des ſiliques ovales , ſoutenues par des pédoncules , & qui ren-ferment pluſieurs ſemences.

Alyſſum ſativum. Scop. carn. ed. 2. *n.* 794.

Camelina ſativa. Crantz. Auſtr. p. 18 ; la Cameline.

Myagrum ſylveſtre. Bauh. Pin. 109. *Dill. Giſſ. p.* 134.

Alyſſon ſegetum , foliis auri-culatis , acutis. Tourn. Inſt. R. H. ; Herbe à l'enragé , qui croît dans les bleds , avec des feuilles oreillées , & à pointe aiguë , communément appellée *Or de plaiſir* en Angleterre.

2°. *Myagrum Alyſſum, ſiliculis cordatis , pedunculatis , polyſpermis, foliis denticulatis , obtuſis* ; Mya-grum avec des ſiliques en forme de cœur, poſtées ſur des pédon-cules , & qui renferment plu-ſieurs ſemences , & des feuilles dentelées & obtuſes.

Alyſſon ſegetum , foliis auri-culatis acutis, fructu majori. Tourn. Inſt. 217 ; Herbe à l'enragé à feuilles oreillées qui produit un plus gros fruit.

3°. *Myagrum rugoſum, ſili-culis globoſis , compreſſis , punc-tatis , rugoſis. Hort. Cliff.* 328 ; Myagrum avec de petites ſili-ques globulaires , comprimées, & marquées de points rudes.

Rapiſtrum arvenſe , folio auri-culato, acuto. Tourn. Inſt. 211; eſ-pece de Moutarde des champs , avec une feuille à oreilles poin-tues.

MYA

4°. *Myagrum perenne, filiculis biarticulatis, dispermis, foliis extrorsùm sinuatis, denticulatis.* Hort. Ups. 182. Scop. carn. ed. 2. n 795. Jacq. Austr. t. 414; Myagrum avec des filiques courtes à deux nœuds, qui renferment deux femences, & dont les feuilles extérieures font finuées & dentelées.

Crambe foliis lanceolatis, dentato-finuatis. Hort. Cliff. 340. Roy. Lugd.-B. 329. Gort. Gelz. 404.

Rapistrum monospermum. C. B. P. 95; espece de Moutarde à une femence. Rapistre monosperme.

5°. *Myagrum perfoliatum, filiculis ob-cordatis, fub-fessilibus, foliis amplexicaulibus.* Hort. Ups. 182. Hort. Cliff. 328. Roy. Lugd.-B. 330. Sauv. Monf. 77. Kniph. cent. 10. n. 64; Myagrum avec de petites filiques presqu'en forme de cœur, prefque fessiles, & des feuilles amplexicaules.

Myagrum loculo fœcundo conico, fterili, biloculari. Hall. Helv. n. 524.

Myagrum monospermum, latifolium. C. B. P. 109. Prodr. 52. t. 51. Moris. Hist. 2. p. 257. S. 3. t. 21. f. antepenult.; Myagrum à larges feuilles, avec une femence dans chaque filique.

Sativum. La premiere espece croit naturellement dans les champs femés en bled, dans la France méridionale & en Italie; je l'ai auffi trouvée dans les bleds du parc de Eafthamfted, maifon de campagne de GUILLAUME TRUMBULL, Ecuyer; mais elle n'eft pas com-

mune dans ce pays. Cette plante, qui eft annuelle, s'éleve avec une tige droite, à la hauteur d'environ un pied & demi, & poufle vers fon fommet deux ou quatre branches latérales, érigées, liffes & remplies d'une moëlle fpongieufe: les feuilles du bas ont trois ou quatre pouces de longueur; elles font de couleur pâle, ou d'un vert jaunâtre, & ont des oreilles à leur bâfe: celles des tiges, qui font plus étroites à mefure qu'elles font plus voifines du fommet, font entieres, & embraffent prefque les tiges de leurs bâfes: fes fleurs croiffent en épis clairs ou lâches aux extrémités des branches, fur des pédoncules d'un pouce de longueur, & font compofées de quatre petits pétales jaunâtres placés en forme de croix; à ces fleurs fuccedent des capfules ovales, bordées, couronnées au fommet par le ftyle de la fleur, & à deux cellules remplies de femences rouges.

Alyffum. La feconde espece, qui eft auffi une plante annuelle, differe de la premiere, en ce qu'elle a une tige plus haute, des feuilles beaucoup plus longues, plus étroites, régulierement dentelées fur leurs bords, & terminées en pointe obtufe: fes fleurs font auffi plus larges, mais de la même forme & de la même couleur; fes capfules font plus groffes, & en forme de cœur.

Ces deux plantes fleuriffent dans les mois de Juin & de Juillet, & leurs femences mûriffent en Septembre.

Rugofum. La troifieme efpece croît naturellement fur le bord des terres labourées, dans la France Méridionale & en Italie : elle eft annuelle ; fes feuilles baffes ont cinq ou fix pouces de longueur ; elles font velues, fucculentes, garnies d'oreilles à leurs bâfes, & terminées en pointe aiguë ; fes tiges, qui s'élevent à la hauteur d'un pied & demi, font velues, caffantes, garnies de branches vers leur fommet, comme les deux précédentes, & terminées par des épis courts & lâches de petites fleurs pâles, auxquelles fuccedent de petites capfules rudes, rondes, & comprimées à l'extrémité. Cette plante fleurit en Juillet, & fes femences mûriffent en automne.

Perenne. La quatrieme efpece, qu'on rencontre auffi parmi les bleds en France & en Allemagne, eft encore une plante annuelle ; fes feuilles baffes font larges, dentelées & velues ; fes tiges pouffent des branches vers le bas, & font garnies de feuilles de quatre pouces de longueur fur deux de largeur, velues & dentelées inégalement ; fes tiges font terminées par des épis fort longs & lâches de fleurs jaunes, qui font remplacées par des légumes courts & à deux nœuds, qui renferment chacun une femence ronde. Cette plante fleurit à-peu-près dans le même tems que la précédente.

Perfoliatum. La cinquieme eft originaire de la France méridionale & de l'Italie ; elle a une tige liffe, branchue, & de plus de deux pieds de hauteur ; fes feuilles baffes ont cinq ou fix pouces de longueur ; elles font unies, fucculentes, & un peu dentelées ; celles du haut embraffent prefque les tiges de leurs bâfes : fes fleurs naiffent en épis longs & lâches ; elles font jaunes & feffiles à la tige, & elles produifent des légumes en forme de cœur, comprimés & divifés, par une partition longitudinale, en deux cellules, qui contiennent chacune une femence ronde. Cette plante fleurit dans le même tems que la précédente.

Culture. En laiffant écarter les femences de toutes ces plantes en automne, elles poufferont fans aucuns foins & n'exigeront que d'être éclaircies & nettoyées de mauvaifes herbes. Celles qui pouffent en automne, perfectionnent toujours leurs femences, au-lieu que celles du printems manquent quelquefois.

MYOSOTIS. *Dill. Gen.* 3. *Lin. Gen.* 180. [*Moufe-ear.*] Oreille de Souris.

Caractères. Le calice de la fleur eft oblong, érigé, découpé en cinq pointes, & perfiftant. La corolle, qui eft en forme de foucoupe, a un tube court, cylindrique, & divifé fur fes bords en cinq fegmens obtus, dont l'évâfement eft fermé par cinq petites écailles qui fe joignent & débordent. La fleur a cinq étamines courtes, placées dans le cou du tube, & terminées par de petites antheres, & quatre ger-

188 M Y O

mes qui foutiennent un ftyle mince de la longueur du tube, & couronné par un ftigmat obtus. Ces germes fe changent, quand la fleur eft paffée, en quatre femences ovales, renfermées dans le calice.

Ce genre de plante eft rangé dans la premiere fection de la cinquieme claffe de LINNÉE, intitulée *Pentandria Monogynia*, avec celles dont les fleurs ont cinq étamines & un ftyle.

Les efpeces font :

1°. *Myofotis Virginica, feminibus aculeato-glochidibus, foliis ovato-oblongis, ramis divaricatis*. Lin. Sp. 189 ; Oreille de Souris avec des femences épineufes, des feuilles oblongues & ovales, & des branches étendues.

Myofotis feminibus hifpidis, foliis lanceolato-ovatis. Gron. Virg. 19.

Cynogloffum Virginianum, flore & fructu minimo. Mor. Hift. 3. tab. 10. fol. 9. ; Cynogloffe de Virginie, avec une petite fleur & de petites femences.

2°. *Myofotis lappula, feminibus aculeis glochidibus, foliis lanceolatis, pilofis. Flor. Suec. 150. 158. Dalib. Par. 57. Pollich. pall. n. 182. Gmel. it. 1. p. 117. Kniph. cent. 11. n. 79. Flor. Dan. t. 692* ; Oreille de Souris avec des femences épineufes, & des feuilles velues & en forme de lance.

Lithofpermum feminibus echinatis. Hort. Cliff. 46. Roy Lugd.-B. 405.

Cynogloffum minus. C. B. P. 257 ; la plus petite Cynogloffe.

M Y O

Cynogloffa minor montana ferotina altera. Col. Ecphr. 179. 180. Haller.

3°. *Myofotis Apula, feminibus nudis, foliis hifpidis, racemis foliofis. Lin. Sp. 189* ; Oreille de Souris avec des femences nues, des feuilles piquantes, & des tiges branchues & feuillées.

Echium luteum minimum. C. B. P. 254 ; la plus petite Viperine jaune.

Echioides lutea minima, Apula campeftris. Col. Ecphr. 1. p. 184. f. 185.

Anchufa lutea minima. Lob. ic. 312.

Lithofpermum feminibus lævibus, corollis vix calycem fuperantibus, foliis lanceolatis. Roy. Lugd.-B. 405. Sauv. Monfp. 62.

Il y a encore une ou deux autres efpeces de ce genre, qui croiffent naturellement en Angleterre ; mais comme on les admet rarement dans les jardins, je n'en fais pas mention. Celles dont il vient d'être queftion ne font guere cultivées que dans les jardins de Botanique ; car elles ont peu de beauté, & ne font d'aucun ufage. Ceux qui défirent les conferver, doivent les femer en automne, fur une planche de terre ouverte, ou dans une plate-bande de terre légere ; au printems, on éclaircit les plantes, quand elles font trop ferrées, & on les tient nettes de mauvaifes herbes : c'eft en cela que confifte toute leur culture. Si on leur laiffe écarter leurs femences elles fe propagent fans aucun foin.

MYO-

MYR

MYOSURUS. [*Mouse-tail.*]
Queue de Souris.

Cette plante reſſemble beaucoup à la *Renoncule*, dans le genre de laquelle elle eſt rangée par quelques Botaniſtes : ſes fleurs ſont extrêmement petites, & produiſent des épis longs & minces ſemblables à des queues de ſouris, ce qui a fait donner ce nom à la plante; elle croît ſans culture ſur des terres humides, dans différentes parties de l'Angleterre, où elle fleurit à la fin d'Avril, donne des ſemences mûres un mois après, & périt enſuite. Comme on ne la cultive guere dans les jardins, je n'en parlerai pas davantage.

MYRICA. *Lin. Gen. Plant. 981. Gale. Tourn. act. R. Scient.* **1706.** [*The Candleberry Myrtle, Gale, or Sweet-Willow.*] le Myrte à chandelle ou Arbre de Cire, Gale ou Saule doux; appellé par quelques-uns, *Myrtus Brabantica*, Myrte Hollandois, ou Piment royal.

Caracteres. Les fleurs mâles naiſſent ſur des plantes différentes de celles qui produiſent les femelles ; les fleurs mâles ſont raſſemblées en un chaton lâche, oblong, ovale, & imbriqué à chaque côté ; ſous chaque écaille eſt placée une fleur en forme de croiſſant, ſans pétales, mais à quatre ou ſix étamines courtes, minces, & terminées par de grands ſommets jumeaux & à deux lobes : les fleurs femelles n'ont ni corolles ni étamines, mais ſeulement un germe ovale, qui ſoutient deux ſtyles minces & couronnés par des ſti-

Tome V.

MYR 189

gmats ſimples. Ce germe ſe change dans la ſuite en une baie à une cellule, qui renferme une ſimple ſemence.

Ce genre de plantes eſt rangé dans la quatrieme ſection de la vingt-deuxieme claſſe de LINNÉE, qui comprend celles dont les fleurs mâles ont quatre étamines, & qui ſont ſur des pieds différens de ceux qui produiſent le fruit.

Les eſpeces ſont :

1°. *Myrica gale, foliis lanceolatis, ſub-ſerratis, caule fruticoſo. Lin. Sp. Plant.* 1024. *Mat. Med.* 211. *Gort. Ingr.* 159. *Flor. Dan. f.* 327. *Kniph. cent.* 9. *n.* 70. 71 ; Piment royal à feuilles en forme de lance, & ſciées, & à tige d'arbriſſeau.

Myrica foliis lanceolatis, fructu ſicco. Fl. Lapp. 373. *Fl. Suec.* 817. 907. *Hort. Cliff.* 455. *Roy. Lugd.-B.* 527. *Dalib. Paris.* 300. *Rhus Myrti-folia Belgica. Bauh. Pin.* 414.

Chamæleagnus. Dod. Pempt. 780. *App.*

Gale frutex odoratus Septentrionalium. J. B. 1. *p.* 2. 225 ; la Gale d'Occident ou Saule odorant Septentrional & en arbriſſeau. Le Piment royal.

2°. *Myrica ceri-fera, foliis lanceolatis, ſub-ſerratis, caule arboreſcente. Kalm. Fabric. Helmſt.* 410 ; Arbre de Cire avec des feuilles en forme de lance & ſciées, & une tige d'arbriſſeau.

Myrtus Brabantiæ, ſimilis Carolinienſis baccifera, fructu racemoſo, ſeſſili, monopyreno. Pluk. Phyt. tab. 48. *fol.* 9. *Catesb. Car.* 1. *p.* 69. *f.* 69 ; Myrte de Hollande, ſemblable à celui de Caroline qui produit des

N

MYR

baies disposées en paquets & sessiles. Arbre de Cire.

3°. *Myrtus Caroliniensis, foliis lanceolatis serratis, caule suffruticoso*; Arbre de Cire à feuilles en forme de lance & sciées, & à tige d'arbrisseau.

Myrtus Brabantiæ, similis Caroliniensis humilior, foliis latioribus & magis serratis. Catesb. Car. vol. 1. *p.* 13; le plus petit Myrte du Brabant, semblable à celui de Caroline, & à feuilles plus larges & plus profondément sciées, ou le Myrte à chandelle.

4°. *Myrtus Aspleni-folia, foliis oblongis, alternatim sinuatis. Hort. Cliff.* 456. *Gron. Virg.* 153. *Cold. Nov.* 224; Piment royal à feuilles oblongues, ovales & sinuées alternativement.

Gale Mariana, Asplenii folio. Pet. Muf. 773; Piment royal du Maryland, à feuilles de Scolopendre.

Liquidambar peregrinum. Lin. Syst. Plant. tom. 4. *p.* 171. *Sp.* 2. *Duham. Arb.* 1. *p.* 366.

Myrti Brabantiæ affinis Americana, foliorum laciniis Asplenii modo divisis. Pluk. Alm. 250. *t.* 100. *f.* 6. 7.

5°. *Myrica Querci-folia, foliis oblongis, oppositè sinuatis, glabris. Hort. Cliff.* 456. *Roy. Lugd.-B.* 527. *Burm. Ind. t.* 98. *f.* 1; Piment royal avec des feuilles oblongues, unies, & dont les sinuosités sont opposées.

Cariotrage Matodendros Africana, Botryos amplioribus foliis densis. Pluk. Amalth. 65.

Laurus Africana minor, Querci folio. Hott. Amst. 2. *p.* 161;

MYR

petit Laurier d'Afrique à feuilles de Chêne.

6°. *Myrica hirsuta, foliis oblongis, oppositè sinuatis hirsutis;* Myrica avec des feuilles oblongues & velues, dont les sinuosités sont opposées.

7°. *Myrica cordi-folia, foliis sub-cordatis, serratis, sessilibus. Hort.Cliff.*456.*Roy.Lugd.-B.*527; Myrica à feuilles sciées, presque en forme de cœur,& sessiles.

Alaternoïdes, Ilicis folio crasso, hirsuto. Walth. Hort. 3. *f.* 3.

Tithymali facie planta Æthiopica, Ilicis aculeato folio Pluk. Alm. 373. *t.* 319. *f.* 7.

Myrica foliis sub-cordatis, integris, sessilibus. Burm. Afr. 263. *t.* 98. *f.* 3.

Gale Capensis, Ilicis cocciferæ folio. Pet. Muf. 774; Piment royal du Cap, à feuilles de Chêne de Kermès.

Gale. La premiere espece croît naturellement dans les marais de plusieurs parties de l'Angleterre particulierement dans les pays septentrionaux & au couchant, ainsi que dans le parc de Windsor, & près de Turnbridge-Wells; elle s'éleve, avec plusieurs tiges d'arbrisseau, à près de quatre pieds de hauteur, & se divise en plusieurs branches minces, garnies de feuilles roides en forme de lance, d'un pouce & demi environ de longueur, sur six lignes de largeur au milieu; d'un vert tendre ou jaunâtre, unies, un peu sciées à leur extrémité, & alternes sur les branches. Lorsque ces feuilles sont froissées, elles répandent une odeur agréable. Les fleurs mâles, ou chatons, sont pro-

MYR

duites fur les parties latérales des branches, & croiffent fur des plantes différentes de celles qui produifent les femelles. Ces dernieres font remplacées par de petites baies raffemblées en grappes, qui renferment chacune une femence. Cet arbriffeau fleurit dans le mois de Juillet, & fes femences mûriffent en automne.

Quelques perfonnes font ufage des feuilles de cette efpece en guife de thé ; mais on les croit nuifibles au cerveau. Il y a peu d'années qu'un favant Médecin a donné un Traité, pour prouver que cet arbriffeau étoit le vrai the ; mais il n'a réuffi qu'à montrer fon peu de connoiffance.

Comme il croît naturellement dans des marais, & qu'il ne feroit pas poffible de le faire réuffir dans un terrein fec, on le cultive rarement dans les jardins.

Ceri-fera. La feconde efpece croît fans culture dans l'Amérique Septentrionale, dont les Habitans tirent de fes baies une efpece de cire, qui leur fert à faire des chandelles. La maniere de la recueillir & de la préparer, a été indiquée par M. CATESBY, dans fon *Hiftoire de la Caroline*.

Celle-ci naît fpontanément dans des marais & terres humides, où elle s'éleve avec plufieurs tiges d'arbriffeau à la hauteur de huit ou dix pieds, & pouffe plufieurs branches garnies de feuilles roides, en forme de lance, de trois pouces environ de longueur, fur un de largeur au milieu, unies,

MYR 191

entieres, ayant à peine des pétioles, d'un vert jaunâtre & luifant, mais plus pâle en-deffous, alternes, affez voifines des branches, & qui répandent une odeur fort agréable, quand elles font froiffées. Les chatons font produits fur des plantes différentes de celles qui portent les fruits ; ils ont environ un pouce de longueur, & font ériges. Les fleurs femelles fortent fur les côtés des branches en paquets longs, & produifent des baies rondes, petites, & couvertes d'une efpece de farine. Cet arbriffeau fe plait dans une terre molle & humide, où il fait beaucoup de progrès, & il réfifte au plein air fans aucun abri.

Carolinienfis. La troifieme efpece fe trouve dans la Caroline, & s'éleve à la même hauteur que la précédente ; fes branches font moins fortes, & couvertes d'une écorce grifâtre ; fes feuilles font plus courtes, plus larges, & fciées fur leurs bords ; mais en toute autre chofe, elle reffemble à la feconde ; les baies de celle-ci fervent auffi au même ufage.

On multiplie ces efpeces par leurs graines, qui pouffent au printems fuivant, lorfqu'on les feme en automne ; mais qui ne germent qu'une année après, lorfqu'on ne les met en terre que dans cette derniere faifon. Ces plantes doivent être arrofées dans les tems fecs, & mifes à l'abri des gelées, tandis qu'elles font jeunes ; mais lorfqu'elles ont acquis de la force, elles réfiftent très-bien au froid de notre climat.

Aspleni-folia. La quatrieme naît fans culture aux environs de Philadelphie, d'où on en a apporté plufieurs plantes en Angleterre. Celles qui ont été plantées dans un fol humide, ont très-bien réuffi. Les racines de quelques-unes de ces plantes rampent, & pouffent des rejettons en abondance, auffi-bien que dans leur pays natal.

Cette efpece s'éleve avec des tiges minces d'arbriffeau, à la hauteur d'environ trois pieds ; elles font velues, divifées en plufieurs branches minces, & garnies de feuilles de trois ou quatre pouces de longueur fur fix lignes de large, alternes, découpées prefque jufqu'à la côte du milieu, fort reffemblantes à celles de la *Scolopendre*, d'un vert foncé, velues en-deffous, & feffiles aux tiges : les fleurs mâles ou chatons naiffent fur les côtés des branches, entre les feuilles ; elles font ovales & érigées : mais je ne puis donner la defcription de fes fruits, parce que je ne les ai jamais vus.

On peut multiplier cette plante par boutures, ou par les rejettons qui fortent de fa racine. Si on la plante dans un fol humide & léger, elle fupportera le froid auffi-bien que les deux précédentes.

Querci-folia. Les cinquieme & fixieme efpeces font originaires du Cap de Bonne-Efpérance ; elles ne different des autres qu'en ce qu'elles ont des feuilles fort unies & luifantes, & que celles des au-

tres font velues. J'ignore fi elles font réellement des efpeces diftinctes ; mais comme elles m'ont été envoyées de Hollande comme telles, & que les plantes confervent toujours leurs différences, je les donne ici pour deux efpeces féparées.

Elles s'élevent avec des tiges minces d'arbriffeau à la hauteur d'environ quatre pieds, & fe divifent en branches plus petites, liffes dans une efpece, velues dans l'autre, & fortement garnies de feuilles d'un pouce & demi de longueur, fur prefque un pouce de largeur. Quelques-unes de ces feuilles ont deux, d'autres trois dentelures profondes fur leurs bords, & font oppofées. Dans une efpece, elles font unies & luifantes ; & dans l'autre, velues, & d'un vert plus foncé ; mais elles font toutes feffiles, & terminées en pointe obtufe où elles font encore découpées : entre ces feuilles fortent quelques chatons de forme ovale, & qui tombent. Toutes les plantes que j'ai vues étoient des plantes mâles ; ainfi, je ne puis donner aucune defcription du fruit : elles confervent leurs feuilles durant toute l'année ; mais comme elles font trop tendres pour fubfifter pendant l'hiver en plein air dans ce pays, il faut les conferver dans la ferre pendant cette faifon : elles ne produifent point de femences ici, & on ne peut les y multiplier que par marcottes, qui ne prennent pas racine fort aifé-

MYR

ment ; ce qui est cause que ces plantes ne font pas communes à présent en Angleterre, d'autant plus que les boutures ne poussent des racines que très-difficilement : car j'en ai fait plusieurs fois l'essai, & aucune de ces tentatives ne m'a réussi. Les Jardiniers Hollandois n'ayant pas été plus heureux, ces plantes font aussi rares chez eux qu'en Angleterre.

Quand les marcottes sont placées, on tord à chaque nœud la partie de la branche qui est couchée en terre, comme on le pratique pour les Œillets : mais on n'emploie pour cela que les jeunes branches ; car les vieilles ne poussent point de racines. Ces marcottes restent souvent deux ans en terre avant d'être assez enracinées pour pouvoir être transplantées ; car il ne faut pas les séparer avant qu'elles aient formé de bonnes racines, sans quoi elles font fort sujettes à manquer.

Quand elles font détachées des vieilles plantes, on les met chacune séparément dans de petits pots remplis d'une terre molle, riche & marneuse, & on les place sous un châssis ordinaire, où on les tient à l'ombre au milieu du jour, pour leur faire pousser plus aisément de nouvelles racines. On peut ensuite les tenir en été dans une situation chaude, & les renfermer en automne dans une serre, où on les traitera de la même maniere que les autres plantes des mêmes contrées. La meilleure sai-

MYR 193

son pour marcotter les branches, est, comme je l'ai déjà dit, le mois de Juillet ; & un an après elles feront en état d'être enlevées.

Cordi-folia. La septieme espece, qui a été trouvée au Cap de Bonne-Espérance, a une tige foible d'arbrisseau de cinq ou six pieds de hauteur, qui pousse plusieurs branches longues, minces, & fortement garnies, dans toute leur longueur, de petites feuilles en forme de cœur, sessiles aux branches, légerement dentelées, & ondées sur leurs bords : ses fleurs sortent entre les feuilles en paquets ronds ; mais toutes les plantes que j'ai vues jusqu'à présent n'avoient que des fleurs mâles, avec un nombre indéterminé d'étamines, qui toutes étoient renfermées dans une enveloppe commune & écailleuse. Ces fleurs paroissent dans le mois de Juillet, & n'ont point grande apparence. Les feuilles de cette espece se conservent vertes toute l'année.

On la multiplie de la même maniere que les deux especes précédentes ; mais comme elle prend difficilement racine par marcotte, elle n'est pas commune dans les jardins de l'Europe ; elle exige le même traitement que les deux précédentes.

MYROBALANUS. *Voyez* Spondias purpurea. Spondias lutea. Phyllanthus embica. *L.*

MYRRHE DU CANADA. *Voyez* Sison Canadense. *L.*

MYRRHIS. *Voyez* Chæ-

MYR

ROPHYLLUM. SCANDIX. SI—
SON. *L.*

MYRTE. *Voyez* MYRTUS. *L.*
MYRTE HOLLANDOIS *ou*
PIMENT ROYAL. *Voyez* MY—
RICA. *L.*

MYRTILLE *ou* AIRELLE.
Voyez VACCINIUM MYRTIL-
LUS. *L.*

MYRTUS. *Tourn. Inst. R. H.*
640. *tab.* 409. *Lin. Gen. Plant.*
543. [*Myrtle.*] Myrte.

Caractères. Le calice de la
fleur est formé par une feuille découpée sur ses bords en
cinq pointes aiguës ; il est persistant , & placé sur le germe :
la corolle a cinq pétales larges , ovales, & insérés dans
le calice ; la fleur a un grand
nombre de petites étamines
aussi insérées dans le calice ,
& terminées par de petites antheres. Le germe , qui est placé sous la fleur , soutient un
style mince & couronné par
un stigmat obtus. Ce germe
se change dans la suite en une
baie ovale à trois cellules , &
couronnée par le calice ;
chaque cellule contient une
ou deux semences en forme
de rein.

Ce genre de plantes est rangé dans la premiere section
de la douzieme classe de LIN-
NÉE, qui renferme celles dont
les fleurs ont environ vingt
étamines & un style.

Les especes sont :

1°. *Myrtus communis ; foliis
ovatis , pedunculis longioribus ;*
Myrte avec des feuilles ovales & de longs pédoncules aux
fleurs.

*Myrtus lati-folia romana. C.
B. P.* 468 ; Myrte romain à

larges feuilles *ou* Myrte commun à larges feuilles.

*Myrtus communis romana. Lin.
Syst. Plant. tom.* 2. *p.* 477. *Sp.*
1 ; premiere Variété.

2°. *Myrtus Belgica , foliis
lanceolatis , acuminatis ;* Myrte
avec des feuilles en forme de
lance, & terminées en pointe
aiguë.

*Myrtus lati-folia Belgica. C.
B. P.* 469; Myrte Hollandois
à larges feuilles.

*Myrtus communis Belgica. Lin.
Syst. Plant. tom.* 2. *p.* 477. *Sp.*
1 ; sixieme variété.

3°. *Myrtus acuta , foliis lanceolato-ovatis , acutis ;* Myrte
avec des feuilles en forme de
lance , ovales & à pointe aiguë.

*Myrtus sylvestris , foliis acu-
tissimis. C. B. P.* 469 ; Myrte
sauvage, avec des feuilles à
pointe fort aiguë.

*Myrtus communis Lusitanica.
Lin. Syst. Plant. tom.* 2. *p.* 477 ;
cinquieme variété. *Sp.* 1.

4°. *Myrtus Bætica foliis ovato-
lanceolatis , confertis ;* Myrte
avec des feuilles ovales en
forme de lance , & rapprochées en paquets.

Myrtus lati-folia Bætica. 2.
*vel foliis Laurinis , confertim
nascentibus. C. B. P.* 469 ; second Myrte d'Espagne à larges feuilles de Laurier , disposées en paquets , communément appelé Myrte à feuilles
d'Oranger.

*Myrtus communis Bætica. Lin.
Syst. Plant. p.* 477. *Sp.* 1. ;
quatrieme variété.

5°. *Myrtus Italica , foliis ovato-
lanceolatis acutis , ramis erectiori-
bus ;* Myrte à feuilles ovales
en forme de lance, & à pointe

MYR

aiguë , avec des branches érigées.

Myrtus communis Italica. C. B. P. 468. *Lin. Syſt. Plant. tom.* 2. *p.* 477. *Sp.* 1 ; troiſieme variété. Myrte commun d'Italie , appelé Myrte érigé.

6°. *Myrtus Tarentina , foliis , ovatis , baccis rotundioribus* ; Myrte à feuilles ovales , & à baies plus rondes.

Myrtus minor vulgaris. C. B. P. 469 ; le plus petit Myrte commun , appelé Myrte commun à feuilles de Buis , ou le Myrte de Tarente.

Myrtus communis Tarentina. Lin. Syſt. Plant. tom. 2. *p.* 476. *Sp.* 1. ; ſeconde variété.

7°. *Myrtus minima , foliis lineari-lanceolatis , acuminatis* ; Myrte à feuilles linéaires , en forme de lance , & à pointe aiguë.

Myrtus foliis minimis & mucronatis. C. B. P. 469 ; Myrte avec des feuilles plus petites , & à pointe aiguë , communément appelé Myrte à feuilles de Romarin.

Myrtus communis mucronata. Lin. Syſt. Plant. tom. 2. *p.* 477. *Sp.* 1. , ſeptieme variété.

8°. *Myrtus Zeylanica , pedunculis multi-floris , foliis ovatis , ſub-petiolatis. Lin. Sp. Plant.* 472 ; Myrte avec pluſieurs fleurs ſur chaque pédoncule , & des feuilles ovales ſur de courts pétioles.

Myrtus foliis ovatis , acuminatis , obtuſiuſculis. Fl. Zeyl. 182.

Myrtus Zeylanica odoratiſſima , baccis niveis , monococcis. H. L. 434 ; Myrte de Ceylan , très-odorant , avec des baies blan-

MYR 195

ches comme la neige , qui renferment une ſeule ſemence.

Mirtoïdes foliis ovatis , Hort. Cliff. 489. *Roy. Lugd.-B.* 535.

Communis. La premiere eſpece eſt le *Myrte* commun à larges feuilles , qui eſt une des plus dures que nous ayons : ſes feuilles ont un pouce & demi de longueur ſur un de largeur ; elles ſont d'un vert luiſant , & ſupportées par de courts pétioles : ces fleurs , qui ſont plus larges que celles des autres eſpeces , naiſſent ſur de longs pédoncules aux côtés des branches , & ſont remplacées par des baies ovales , & d'un pourpre foncé , qui renferment trois ou quatre ſemences dures & en forme de rein. Cette eſpece fleurit en Juillet & en Août , & ſes baies mûriſſent en hiver. Quelques-uns donnent à cette eſpece le nom de *Myrte* fleuriſſant , parce qu'elle produit une plus grande quantité de fleurs , qui ſont auſſi plus larges que celles des autres.

Belgica. La ſeconde a des feuilles beaucoup plus petites , plus pointues , & plus rapprochées ſur les branches que celles de la précédente ; elles ſont d'un vert foncé , & leur côte mitoyenne eſt de couleur pourpre en-deſſous : ſes fleurs ſont plus petites , & ont de plus courts pédoncules que celles de la premiere ; elles paroiſſent un peu plus tard en été , & perfectionnent rarement leurs baies en Angleterre.

Le *Myrte* à doubles fleurs eſt , je crois , une variété de celui-ci ; car ſes feuilles ,

N 4

196 MYR

le port de la plante , & la groffeur de fes fleurs s'accordent mieux avec cette efpece qu'avec aucune autre.

Acuta. La troifieme fe trouve dans la France Méridionale & en Italie ; fes feuilles font beaucoup plus petites que celles de la feconde , d'un peu moins d'un pouce de longueur fur fix lignes au plus de largeur , ovales , en forme de lance , terminées en pointe aiguë , d'un vert trifte , & feffiles aux branches : fes fleurs font plus petites qu'aucune des autres , & naiffent aux aîles des feuilles vers l'extrémité des branches ; fes baies font petites & ovales.

Bætica. La quatrieme efpece a une tige & des branches plus fortes qu'aucune des précédentes , & s'éleve à une plus grande hauteur ; fes feuilles font ovales , en forme de lance , difpofées en paquets autour des branches , & d'un vert foncé : fes fleurs font d'une groffeur médiocre , & fortent éparfes & en petit nombre entre les feuilles ; fes baies font ovales , & plus petites que celles de la premiere : mais elles mûriffent rarement en Angleterre. Des Jardiniers , les uns donnent à cette efpece le nom de *Myrte à feuilles d'Oranger* ; d'autres celui de *Myrte à feuilles de Laurier.* Celle-ci n'eft pas fi dure que la précédente.

Italica. La cinquieme eft le *Myrte* commun d'Italie, qui a des feuilles ovales , en forme de lance , & terminées en pointe aiguë ; fes branches & fes feuilles font plus érigées que celles des précédentes , ce qui la fait nommer par les Jardiniers *Myrte érigé* ; fes fleurs font moins groffes , & leurs pétales font marqués de pourpre à leur pointe , lorfqu'ils font fermés ; fes baies font petites , ovales , & de couleur pourpre. Il y a dans cette efpece une variété à baies blanches, qui n'offre d'ailleurs aucune autre différence : je crois auffi que le *Myrte*, qui produit la Noix Mufcade , n'eft qu'une variété de celui-ci ; car j'ai élevé de femences plufieurs de ces plantes, qui étoient fi femblables au *Myrte* Italien , qu'on avoit peine à les diftinguer.

Tarentina. La fixieme efpece , à laquelle on donne communément le nom de *Myrte à feuilles de Buis* , a des feuilles ovales , petites , feffiles aux branches , d'un vert luifant , & terminées en pointe obtufe : fes branches font foibles , & pendent fouvent vers le bas, fi on les laiffe croitre fans les tailler ; elles font couvertes d'une écorce grifâtre : fes fleurs font petites , & paroiffent tard en été ; fes baies font petites & rondes.

Minima. La feptieme efpece, qu'on appelle *Myrte à feuilles de Romarin* , ou *Myrte à feuilles de Thym* , a des branches érigées & des feuilles feffiles aux branches : ces feuilles font petites , étroites , terminées en pointe aiguë , & d'un vert luifant ; elles répandent une odeur agréable , quand elles font froiffées : fes fleurs , qui

font petites, se montrent plus tard que celles des autres, & produisent rarement des baies en Angleterre.

Il y a d'autres variétés de ces *Myrtes* que l'on multiplie dans les jardins pour en faire commerce ; mais comme elles ne sont que des produits accidentels occasionnés par la culture, il est inutile d'en faire mention ici. Celles que je viens de rapporter me paroissent être réellement des espèces distinctes ; car, après les avoir élevées presque toutes de semences, je n'ai jamais observé que les légères altérations qu'on y observe quelquefois, rendissent à les rapprocher les unes des autres. *Zeylanica*. La huitième, qui est originaire de l'île de Céylan étant beaucoup plus délicate qu'aucune des autres, on ne peut la conserver pendant l'hiver en Angleterre, sans le secours d'une chaleur artificielle : sa tige est forte, érigée, couverte d'une écorce lisse & grise, & divisée vers le haut en plusieurs branches minces, roides, & garnies de feuilles ovales, opposées, de deux pouces environ de longueur sur un pouce & un quart de largeur, terminées en pointe, d'un vert luisant, & portées sur de fort courts pétioles : ses fleurs naissent aux extrémités des branches sur un pédoncule commun, qui, se divisant en plusieurs autres, fournit à chacune un pédoncule fort mince. Ces fleurs ressemblent beaucoup à celles du *Myrte* italien ; elles paroissent toujours dans les mois de Décembre & de Janvier ; mais elles ne produisent jamais de baies en Angleterre.

Culture. Pour procéder avec ordre dans les détails relatifs à la culture de ces plantes, je commencerai par la méthode qu'on doit suivre pour traiter & multiplier les espèces communes : je parlerai après de la derniere, qui exige une culture différente ; & comme on multiplie les variétés de l'espèce commune, pour en faire commerce, je donnerai les noms sous lesquels elles sont connues, afin que les curieux puissent les distinguer.

Il y a deux espèces de *Myrtes à Noix Muscade*, dont l'une a des feuilles plus larges que l'autre.

Le *Myrte à Nid d'Oiseau*, le *Myrte à Noix Muscade rayé* ou *panaché*, celui *à feuilles de Romarin panachées*, un autre *à feuilles de Buis panachées*, & le *Myrte à larges feuilles panachées*.

Toutes ces plantes peuvent être multipliées par boutures. Au mois de Juillet, on choisit quelques branches droites, jeunes, vigoureuses, & de six ou huit pouces de longueur ; on enleve les feuilles de leur partie baffe dans la longueur d'environ trois pouces ; l'on tord le bout, qui doit être mis en terre ; &, après avoir rempli de terre riche & légere un pot d'une grandeur proportionnée à la quantité de boutures qu'on veut y mettre, on les y plante à deux pouces environ de dif-

tance, & l'on comprime fortement la terre tout autour ; l'on place ensuite ce pot sous le châssis d'une couche ordinaire ; on le plonge dans du vieux fumier ou du vieux tan, pour empêcher la terre de se secher trop vîte ; on le couvre avec des nattes pendant la chaleur du jour ; on lui donne de l'air à proportion de la chaleur de la saison, & on l'arrose chaque deux ou trois jours, suivant que la terre du pot l'exige. Au moyen de ce traitement, les boutures prendront racine en six semaines de tems. Quand elles commenceront à pousser des branches, on les accoutumera par dégrés au plein air, auquel on les exposera tout-à-fait vers la fin d'Août ou au commencement de Septembre, en les plaçant à l'abri des vents : on les laissera ainsi jusqu'au milieu ou à la fin d'Octobre, & on les enfermera alors dans la serre, en les plaçant dans l'endroit le plus frais, afin qu'elles puissent jouir de l'air toutes les fois que le tems sera doux ; car il suffit de les tenir à couvert de grands froids, à l'exception cependant des *Myrtes à feuilles d'Oranger*, & de ceux à *Noix de Muscade panachés*, qui sont un peu plus tendres que les autres, & qui ont besoin d'une situation plus chaude.

Il faut les arroser souvent pendant l'hiver, mais légerement. Quand quelques-unes de leurs feuilles paroissent flétries, on les ôte aussi-tôt, & l'on tient les pots nets de mauvaises herbes, qui détruiroient

les jeunes plantes, si elles s'étendoient dessus.

Si l'on place ces pots pendant l'hiver sous un châssis ordinaire de couche chaude où ils puissent être à l'abri du froid & avoir de l'air dans les tems doux, les jeunes plantes réussiront mieux que dans une serre ; pourvu qu'elles ne soient point exposées à trop d'humidité, & que l'on ne les couvre pas beaucoup ; ce qui les feroit moisir, & leur feroit perdre leurs feuilles.

Au printems suivant, on tire ces plantes des pots avec précaution, en conservant une motte de terre à leurs racines : on les met chacune séparément dans de petits pots remplis d'une terre riche & légere ; on les arrose exactement pour fixer la terre à leurs racines, & on les tient sous un châssis, jusqu'à ce qu'elles aient formé de nouvelles racines ; après quoi on les accoutume au plein air, auquel on les expose tout-à-fait au mois de Mai pour tout l'été, en les plaçant dans une situation abritée, où elles puissent être à couvert de grands vents.

Pendant l'été, il faut les arroser souvent, sur-tout quand elles sont dans de petits pots, dont la terre se desseche promptement dans cette saison : c'est-pourquoi elles ne doivent être exposées qu'au soleil du matin ; car si elles en recevoient toute la chaleur pendant le jour, l'humidité de la terre contenue dans ces petits pots seroit bientôt dissipée, & les plantes seroient par-là beau-

coup retardées dans leur accroissement.

Au mois d'Août suivant, on examine les pots, pour voir si les racines ne sortent pas par les trous dont leur fond est percé; en ce cas, on les remet dans de plus grands pots, qu'on remplit également de terre riche; on coupe les racines qui se sont roulées autour des pots; on desserre la terre de l'extérieur des mottes avec les mains, & l'on en retranche même une partie, afin que les racines puissent plus aisément pénétrer dans la nouvelle terre; on les arrose ensuite, & on les place à l'abri des vents : alors on peut tailler ces plantes, pour leur faire prendre une figure réguliere; & quand elles sont inclinées à avoir des tiges courbes, on les redresse, en les fixant contre des baguettes minces & droites.

On les assujettit ainsi tandis qu'elles sont jeunes; mais une fois qu'elles ont acquis de la force, elles se maintiennent droites sans aucun secours, & leurs branches pourront être taillées de maniere à prendre des formes rondes ou pyramidales, telle qu'on le jugera à propos, ou qu'il sera nécessaire de le faire, pour pouvoir les conserver dans une serre, où l'espace est ordinairement resserré, ce qui les rend aussi plus agréables : mais comme les plantes ainsi disposées ne produisent point de fleurs, il ne faudroit point tailler l'espece à doubles fleurs, parce que c'est dans ses fleurs

que consiste sa plus grande beauté; d'ailleurs on peut laisser croître naturellement une plante ou deux de chaque espece, pour les laisser fleurir & se procurer des bouquets : mais cela gâte beaucoup celles qui ont toujours été abritées, & dont les branches sont ordinairement taillées.

A mesure que ces plantes grossissent, il faut leur donner tous les ans de plus grands pots, en proportionnant toujours leur capacité au volume des racines; car si ces pots étoient trop grands, les plantes ne feroient que de très-foibles progrès, & même périroient tout-à-fait : c'est-pourquoi, en les tirant des premiers pots, on doit ôter la terre de leurs racines, & les desserrer légerement en-dedans, afin qu'elles ne soient pas trop rapprochées : on les remet ensuite dans les mêmes pots, pourvu qu'ils ne soient pas trop petits; on en remplit le fond & les côtés avec de la nouvelle terre, & on les arrose abondamment, pour fixer cette terre aux racines : ce qu'il faut souvent répéter, parce qu'elles ont besoin de beaucoup d'humidité dans toutes les saisons, & sur-tout dans les tems chauds.

La meilleure saison pour changer ces plantes, est en Avril ou en Août; car si on le fait beaucoup plutôt, leur accroissement devient plus lent, & elles ne peuvent plus repousser de nouvelles racines assez tôt : si au contraire on le fait plus tard, c'est-à-dire, en automne, les premiers froids

les empêchent de prendre racine. Il n'est cependant pas prudent non plus de les changer
dans les grandes chaleurs de
l'été, parce qu'il faudroit alors
les arroter trop souvent, &
les tenir à l'ombre, sans quoi
elles seroient sujettes à lan-
guir pendant un tems confidérable ; d'ailleurs, elles ne
pourroient être placées avec
les autres plantes exotiques,
& servir d'ornement dans les
différentes parties du jardin.
Dès que ces plantes sont re-
mises en pots, on les tient à
couvert, jusqu'à ce qu'elles
aient formé de nouvelles racines, c'est-à-dire, pendant
trois semaines ou un mois, si
la saison est seche & chaude.

En Octobre, lorsque les
nuits commencent à être froides, on enferme ces plantes
dans la serre; mais si l'automne
est favorable, comme cela arrive souvent, elles peuvent
rester à l'air jusqu'au commencement de Novembre ; car si
on les mettoit trop tôt dans
la serre, & que l'automne fût
chaud, elles pousseroient de
nouvelles branches foibles,
qui seroient en danger de se
moisir en hiver, lorsque, par
les grands froids, on est obligé
de tenir les fenêtres exactement fermées : c'est-pourquoi
il faut toujours les laisser dehors tant que la saison le permet, les sortir au printems,
avant qu'elles aient commencé
à pousser des branches ; &
pendant qu'elles sont dans la
serre, leur donner autant d'air
qu'il est possible dans les tems
doux.

J'ai vu les trois premieres
especes plantées en plein air,
dans des situations chaudes,
& sur un sol sec, où elles
ont très-bien supporté le froid
de nos hivers pendant plusieurs
années. On ne les couvroit,
pendant les plus fortes gelées,
qu'avec deux ou trois nattes,
& on mettoit sur la surface
de la terre, autour des racines, un peu de terreau, pour
empêcher la gelée d'y pénétrer. Dans Cornwall, & dans
le Comté de Devon, où les
hivers sont plus doux que dans
la plupart des autres parties
de l'Angleterre, on voit de
grandes haies de *Myrtes* plantées depuis plusieurs années,
qui profitent très-bien & sont
vigoureuses ; quelques-unes
ont plus de six pieds de hauteur. Je crois que si l'espece
à doubles fleurs étoit mise en
pleine terre, elle supporteroit
le froid aussi bien que les précédentes, parce qu'elle est originaire de la France Méridionale. Cette derniere, & celle
à feuilles d'Oranger, ont plus de
peine à prendre racine par bouture ; mais en les plantant vers
la fin de Juin, en ne choisissant que des branches tendres,
& en plongeant les pots dans
une vieille couche de tan qui
ait perdu sa plus grande chaleur, elles prendront très-facilement racine, ainsi que je
l'ai souvent éprouvé, pourvu
que l'on ait soin de couvrir
les vitrages chaque jour. L'espece *à feuilles d'Oranger*, &
celles à feuilles panachées,
étant un peu plus tendres que
les autres, doivent être mises

MYR

dans la ferre un peu plutôt en automne, & placées plus loin des fenêtres.

La huitieme espece est à présent rare en Europe, & on la trouve dans très peu de jardins. LINNÉE, dans les premieres éditions de ses Ouvrages, a séparé cette plante des *Myrtes*, & lui a donné le nom de *Myrtine* : mais dans son *Species Plantarum*, il l'a rejointe à ce genre, auquel, suivant son système, elle appartient spécialement ; car le nombre de ses pétales, de ses étamines, & de ses styles, s'accorde avec ceux du *Myrte* ; mais elle en differe par les parties de la fructification, cette espece n'ayant qu'une semence dans chaque fruit, & le *Myrte* en ayant quatre ou cinq.

Cette plante est très-rare dans nos jardins, parce que ses semences ne mûrissent point en Europe, & qu'on ne peut la multiplier que par marcottes ou par boutures. Les marcottes sont ordinairement deux ans avant de pousser des racines, & souvent les boutures manquent. On préfere cependant cette derniere méthode, qui réussit quand on s'y prend dans une saison convenable, & que l'on y apporte tous les soins nécessaires : c'est-pourquoi il faut planter ces boutures dans le mois de Mai, après avoir choisi les branches de l'année précédente, au bout desquelles on laisse un peu de bois de deux ans, & on les plante dans de petits pots remplis d'une terre molle & marneuse. On préfere toujours les

MYR

petits pots aux grands, pour cette opération ; on les plonge dans une couche de tan de chaleur très-modérée ; & en couvrant chaque pot d'une cloche de verre, malgré les vitrages qui sont au-dessus, les boutures prendront plutôt racine : il faut aussi les tenir à l'ombre pendant la chaleur du jour, & les arroser legerement toutes les fois que la terre des pots se trouve seche, mais ne pas leur donner trop d'humidité. Les boutures qui réussissent auront pris racine vers le mois de Juillet ; alors on les accoutumera à supporter le plein air, auquel il sera prudent de les exposer entierement au milieu du même mois, afin qu'elles puissent acquérir de la force avant l'hiver : mais on ne doit pas les transplanter avant le printems. En automne, on met ces pots dans une serre tempérée, & durant l'hiver, on les arrose legerement : au printems suivant, on les enleve des pots avec précaution ; on les plante chacune séparément dans de petits pots remplis de terre légere de jardin potager, & on les plonge dans une couche de chaleur modérée, pour les avancer & leur faire pousser de nouvelles racines : on les endurcit ensuite par dégrés ; & en Juillet, on les place en plein air, dans une situation abritée, où elles peuvent rester jusqu'à la fin de Septembre, pour être mises alors dans la serre chaude.

Cette plante ne peut subsister pendant l'hiver, en An-

MYR

gleterre, dans une serre ordinaire: mais en la mettant dans une serre de chaleur moderée, elle fleurira durant cette saison : on peut la tenir en plein air dans une situation chaude, durant les mois de Juillet, Août & Septembre.

MYRTUS BRABANTICA. *Voyez* MYRICA.

MYRTUS PIMENTA. *Voy.* CARYOPHYLLUS PIMENTA.

N

NAP

NAPEL. *Voyez* ACONITUM NAPELLUS.

NAPÆA. *Lin. Gen. Plant.* 748. *Malva H. L.* [*Napea.*] Mauve de Virginie, Nymphe des Bois.

Caractères. Cette plante a des fleurs mâles & des fleurs hermaphrodites sur des racines différentes ; les fleurs mâles ont des calices en forme de cruche, persistans, & formés par une feuille découpée au sommet en cinq segmens : les corolles ont cinq pétales oblongs, joints à leur bâse, mais étendus & divisés au sommet : ces fleurs ont plusieurs étamines velues, réunies vers le bas en une espece de colonne cylindrique, & terminées pas des antheres rondes & comprimées. Les fleurs hermaphrodites ont un pareil calice, une corolle, & des étamines semblables à celles des mâles ; mais elles ont encore un germe conique, qui soutient un style cylindrique, divisé au sommet en dix parties, couronnées par des stigmats simples. Ce germe se change dans la suite en un fruit ovale, enveloppé par le calice, & di-

visé en dix cellules qui renferment chacune une semence en forme de rein,

Ce genre de plantes est rangé dans la douzieme section de la vingt-deuxieme classe de LINNÉE, qui comprend celles à fleurs mâles & femelles sur différens pieds, & dont les fleurs ont plusieurs étamines jointes par leur bâse au style, & qui forment ensemble une colonne. Comme les plantes de ce genre ont des fleurs mâles & hermaphrodites sur des pieds différens elles ne different point de toutes les especes de plantes malvacées auxquelles elles appartiennent proprement, les fleurs étant monopétales, les étamines & les styles étant joints à leur bâse en forme de colonne ; ce qui constitue les caracteres essentiels de cette classe.

Les especes sont :

1°. *Napæa dioïca, pedunculis involucratis, angulatis, foliis scabris, floribus dioïcis. Flor. Virg.* 102. *Amœn. Acad.* 3. *p.* 18. *Fabric. Helmst.* 282. *Trew. in. Nov. Act. A. N. C. tom.* 1. *t.* X ; Mauve de Virginie, avec

NAP

des pédoncules enveloppés & angulaires, des feuilles rudes & des fleurs mâles & hermaphrodites sur différens pieds.

Napæa scabra. Lin. Syst. Plant. tom. 4. p. 282. Sp. 2.

Abutilon folio profundè dissecto, pedunculis multi-floris, mas & fœmina. Ehret. Pict. 7 & 8 ; Abutilon avec une feuille profondément divisée, & plusieurs fleurs mâles & femelles sur chaque pédoncule.

Althæa magna, Aceris folio, cortice Cannabino, floribus parvis, semina rotatim in summitate caulium, singula singulis cuticulis rostratis cooperta ferens. Banist. Virg. 1928.

2°. *Napæa hermaphrodita, pedunculis nudis lævibus, foliis glabris, floribus hermaphroditis. Kniph. cent. 8. n. 73* ; Mauve de Virginie, avec des pédoncules nuds & lisses, des feuilles unies & des fleurs hermaphrodites.

Napæa lævis. Lin. Syst. Plant. t. 4. p. 282. Sp. 1.

Sida foliis palmatis, laciniis lanceolato - attenuatis. Hort. Cliff. Hort. 346. Upsal. 198. Roy. Lugd.-B. 348.

Althæa Ricini folio Virginiana. H. L. ; Mauve de Virginie à feuilles de Ricin.

Malva Virginiana, Ricini-folio. Herm. Lugd.-B. 22. f. 23.

Dioïca. La premiere espece a des racines vivaces, composées de plusieurs fibres épaisses & charnues, qui pénetrent profondément dans la terre, & se réunissent au sommet en une grosse tête, de laquelle sortent un grand nombre de feuilles rudes, velues, de près

d'un pied de diametre, profondément découpées en six ou sept lobes, & irrégulierement dentelées sur leurs bords; chaque lobe a une forte côte, & toutes ces côtes se réunissent en un centre au pétiole : les pétioles sont gros & longs; ils sortent immédiatement de la racine, & s'écartent en-dehors de tous côtés : les tiges de fleurs s'élevent à la hauteur de sept ou huit pieds, & se divisent en plusieurs branches garnies à chaque nœud d'une feuille de la même forme que celles du bas, mais qui sont d'autant moins grandes, qu'elles sont plus voisines du sommet, où elles ont rarement plus de trois lobes, qui sont divisés jusqu'au pétiole. Au sommet de cette tige fort de côté, à chaque nœud, un long pédoncule, divisé en branches vers son extrémité, & qui soutient plusieurs fleurs blanches, tubulées au fond, où les segmens de la corolle sont joints. La corolle est divisée en cinq segmens oblongs; dans son centre s'éleve la colonne à laquelle sont réunies les étamines par leur bâse. Dans les fleurs hermaphrodites, le style est joint à la colonne. Les fleurs hermaphrodites sont remplacées par des fruits comprimés, orbiculaires, contenus dans le calice, & divisés en cinq cellules, qui renferment chacune une semence en forme de rein : mais les fleurs mâles sont stériles. Cette espece fleurit dans le mois de Juillet, & ses semences mû

riſſent en automne ; bien—
tôt après, la tige périt, mais
les racines ſubſiſtent pluſieurs
années.

Hermaphrodita. La ſeconde
a auſſi une racine vivace, qui
coule ſouvent ſur la ſurface
de la terre ; elle pouſſe des
tiges liſſes, hautes d'environ
quatre pieds, & garnies de
feuilles unies, alternes, ſupportées par des pétioles longs
& minces, profondément découpées en trois lobes, terminées en pointe aiguë, &
irrégulierement ſciées ſur leurs
bords : celles du bas de la
tige ont près de quatre pouces de longueur ſur preſqu'autant de largeur ; mais elles diminuent par dégrés à meſure
qu'elles ſont plus voiſines du
ſommet : à la bâſe des feuilles
ſort le pédoncule de la fleur,
qui a trois pouces de longueur,
& qui eſt diviſé à ſon extrémité
en trois autres plus petits,
dont chacun ſoutient une fleur
blanche de la même forme que
celles de la premiere eſpece,
mais plus petite, dont la colonne, formée par les étamines, eſt plus longue, &
dont les étamines ont leurs antheres étendues en-dehors, &
au-delà de la corolle.

Culture. Ces deux plantes ſe
trouvent dans la Virginie, &
dans d'autres parties de l'Amérique Septentrionale. On peut
tirer de leur écorce une eſpece
de chanvre, ainſi que de celle
de pluſieurs malvacées. Dans
quelques-unes, qui croiſſent
naturellement dans les Indes,
ces fibres ſont ſi fines, qu'on
en fabrique un fil très-délié,

qui ſert à faire de très-belles
toiles.

Ces deux eſpeces ſe multiplient aiſément par leurs graines, qu'on ſeme au printems
ſur une terre commune, où les
plantes leveront facilement,
& n'exigeront aucun autre
ſoin que d'être tenues nettes
de mauvaiſes herbes. En automne, elles pourront être
tranſplantées dans les places
où elles doivent reſter. Elles
ſe plaiſent dans un ſol riche
& humide ; mais comme elles
y croiſſent avec vigueur, il
ne faut pas manquer de leur
donner beaucoup de place.

On peut multiplier encore
la ſeconde, en diviſant ſes racines rampantes, en automne :
mais comme ces plantes ne
ſont pas fort belles, il ſuffit
d'en avoir une ou deux de
chaque eſpece, dans un jardin, pour la variété.

NAPUS. *Voyez* BRASSICA &
RAPA.

NARCISSE. *V.* NARCISSUS.

NARCISSE D'AUTOMNE.
Voyez AMARILLIS LUTEA. *L.*

NARCISSE DE CONSTAN-
TINOPLE. *Voyez* NARCISSUS
TAZETTA.

NARCISSE DE MATHIOLE.
V. PANCRATIUM ILLYRICUM. *L.*

NARCISSO – LEUCOIUM.
Voyez GALANTHUS.

NARCISSUS. *Lin. Gen. Plant.*
364. Cette plante prend ſon nom
de ναρκη, un engourdiſſement
ou ναρχὸς ſtupeur, parce que
l'odeur de ſa fleur cauſe, à
ce qu'on dit, un aſſoupiſſe-
ment & une eſpece de ſtupi-
dité. PLUTARQUE nous dit que
cette plante fut conſacrée aux
Dieux

NAR

Dieux infernaux. Les Poëtes avancent que NARCISSE, fils de CÉPHISE & de la Nymphe LYRIOPE, étoit d'une si grande beauté, que, s'étant approché d'une fontaine pour y boire, & ayant apperçu sa belle image dans l'eau, il en devint amoureux, & s'étant consommé en vains désirs, il fut transformé en une fleur de ce nom. [*The Daffodil.*] Narcisse, Asphodèle, Jonquille.

Caractères. Les fleurs sont renfermées dans une gaîne oblongue & comprimée, qui se sépare, s'ouvre sur un côté, & se fane : elles ont un calice cylindrique en forme d'entonnoir, & formé par une feuille épanouïe sur ses bords ; la corolle est composée de six pétales ovales au-dehors du nectaire, & insérés à l'extérieur & au-dessus de la bâse de son tube : la fleur a six étamines en forme d'alène, fixées au tube de la corolle, plus courtes que le nectaire, & terminées par des antheres oblongues ; son germe est presque rond, obtus, à trois angles, & placé au-dessous du réceptacle ; il soutient un style long, mince, & couronné par un stigmat divisé en trois parties. Ce germe se change ensuite en une capsule obtuse, presque ronde, à trois angles & à trois cellules remplies de semences globulaires.

Ce genre de plantes est rangé dans la premiere section de la sixieme classe de LINNÉE, qui renferme celles dont les fleurs ont six étamines & un style.

Tome V.

NAR

Les especes sont :

1°. *Narcissus pseudo-Narcissus, spathâ uniflorâ, nectario campanulato, erecto, crispo, æquante petalum ovatum.* Lin. Sp. Plant. 414. Scop. carn. 2. n. 395. Leers. Herb. n. 243. Dœrr. Nass. p. 158 ; Narcisse avec une fleur dans chaque spathe, dont le nectaire est érigé en forme de cloche, & égal aux pétales qui sont ovales.

Narcissus sylvestris pallidus. calyce luteo. C. B. P. 52 ; Narcisse champêtre & de couleur pâle, avec un calice jaune, ou Asphodèle commun d'Angleterre. Faux Narcisse, ou Narcisse des bois.

Bulbocodium vulgatius. Bauh. Hist. 2. p. 593.

2°. *Narcissus Poeticus, spathâ uni-florâ, nectario rotato, brevissimo, scariofo, crenulato.* Hort. Ups. 74. Scop. carn. ed. 2. n. 394. Knorr. 1. s. N. 4. Kniph. cent. 7. n. 62 ; Narcisse de Poëte avec une fleur dans chaque spathe, & un nectaire fort court, en forme de roue, & découpé sur ses bords.

Narcissus albus, circulo purpureo. C. B. P. 48 ; Narcisse blanc, ayant un cercle pourpre dans le milieu de la fleur, ou Narcisse de Poëte.

Narcissus medio purpureus. Dod. Pempt. 223.

3°. *Narcissus incomparabilis, spathâ uni-florâ, nectario campanulato erecto, petalo dimidio breviore* ; Narcisse avec une seule fleur dans chaque spathe, & un nectaire érigé & en forme de cloche n'ayant que moitié de la longueur des petales.

Narcissus incomparabilis, flore

O

pleno, partim flavo, partim croceo. H. R. Par.; Narciffe im-
comparable à doubles fleurs,
partie presque jaunes, & partie couleur de safran.

4º. Narciffus medio - luteus,
fpathâ bi-florâ, neclario campanulato, breviffimo, floribus nutantibus; Narciffe avec deux fleurs
dans chaque fpathe, un nectaire très-court & en forme de
cloche, & des fleurs penchées.

Narciffus medio luteus vulgaris.
Parck; Narciffe commun, dont
la fleur eft jaune dans le milieu, appelé Prime-vere incomparable.

5º. Narciffus albus, fpathâ uni-
flora, neclario campanulato breviffimo, petalis reflexis; Narciffe avec une feule fleur dans
chaque fpathe, un neclaire
fort court, & en forme de
cloche, & des pétales ré-
fléchis.

Narciffus albus, folius reflexis,
calice brevi aureo. H. R. Par.;
Narciffe à fleurs blanches &
à feuilles refléchies, avec un
calice court & de couleur d'or.

6º. Narciffus Bulbocodium,
fpathâ uni-florâ, neclario turbi-
nato petalis majori,genitalibus decli-
natis. Lin.Sp.Plant 417; Narciffe
à feuilles de Jonc, ayant une
feule fleur dans chaque fpathe,
un neclaire turbiné, plus grand
que les pétales, & des étamines penchées.

Pfeudo-Narciffus Junci folio 2,
flavo flore. Clus. Hifl. 66; communément appelé Narciffe ou
Cotillon à panier, ou la Trompette de Medufe.

7º. Narciffus ferotinus, fpathâ uni-florâ, neclario breviffimo
fex-partito. Lin. Sp. Plant. 290.

Læfl. it. 19; Narciffe avec une
fleur dans chaque fpathe, &
un neclaire fort court divifé
en fix parties.

Narciffus autumnalis minor.
Clus. Hifp. 251; le plus petit
Narciffe d'automne.

Narciffus ferotinus. Clus. Hifl.
1. p. 162.

8º. Narciffus Tazetta, fpathâ
multi-florâ, neclario campanulato,
foliis planis. Hort Upfal. 74;
Narciffe avec plufieurs fleurs
dans chaque fpathe, un nectaire en forme de cloche, &
des feuilles entieres.

Narciffus luteus polyanthos
Lufitanicus. C. B. P.; Narciffe
jaune de Portugal, avec plufieurs fleurs, communément
appelé Narciffus Polyanthus.

Narciffus lati-folius, flore pror-
fûs albo. 1. 2. Clus. Hifl. 1. p.
155; Narciffe de Conftantinople.

9º. Narciffus Jonquilla, fpathâ multi-florâ, neclario campanulato brevi, foliis fubulatis.
Hort. Upfal. 75; Jonquille
avec plufieurs fleurs dans chaque fpathe, un neclaire court
& en forme de cloche, &
des feuilles en forme d'alêne.

Narciffus Junci-folius minor.
Er. 2. Clus. Hifl. 1. p. 159.

Narciffus Junci - folius luteus
minor. C. B. P. 51; le plus petit Narciffe jaune, à pétales
de Jonc, appelé Jonquille.

Ces efpeces font toutes celles que j'ai vues dans les jardins Anglois, quoiqu'il y ait
dans chacune un grand nombre de variétés qui different
affez les unes des autres, pour
qu'il foit difficile de diftinguer
à laquelle elles appartiennent.

Comme je les ai toutes exactement observées, je tâcherai, autant que cela sera possible, de les présenter chacune sous leur véritable espece, tant celles à fleurs doubles que celles qui passent pour les meilleures.

Pseudo-Narcissus. La premiere est le Narcisse Anglois commun, qui croît naturellement sur les bords des bois & des champs, dans plusieurs parties de l'Angleterre, elle a une racine grosse & bulbeuse, de laquelle sortent cinq à six feuilles plates, d'environ un pied de longueur sur un pouce de largeur, d'une couleur grisâtre, & un peu creusées dans le milieu, comme la quille d'un bateau : sa tige, qui s'éleve à la hauteur d'un pied & demi, a deux angles aigus & longitudinaux, & produit à son sommet une fleur simple, renfermée dans une spathe mince, qui se déchire & s'ouvre sur le côté, pour laisser sortir la fleur, qui se fane ensuite, & reste sur le haut de la tige : la fleur est monopétale ; la corolle est découpée en six segmens presque jusqu'au fond, où ils se réunissent, & qui s'étendent vers le haut. Dans le milieu est placé un nectaire en forme de cloche, appelé par les Jardiniers, *godet*, qui est de la même longueur que la corolle, & érigé ; la fleur penche sur un côté de la tige ; la corolle est d'une couleur de soufre pâle, & le nectaire est jaune. Cette plante fleurit au commencement d'Avril ; quand ses fleurs sont passées, le germe se change en une

capsule presque ronde, & à trois cellules remplies de semences rondes & noires, qui mûrissent en Juillet : elle se multiplie fortement par les rejettons de sa racine.

Les variétés de cette espece sont :

Narcisse à pétales blancs, avec un godet d'un jaune pâle.

Narcisse à pétales jaunes, avec un godet doré.

Narcisse commun, double & jaune.

Narcisse à fleurs doubles, avec trois ou quatre godets l'un dans l'autre.

Je crois que le *Narcisse* de JEAN TRADESCANT peut encore être placé ici.

Poeticus. La seconde espece croit sans culture dans la France Méridionale & en Italie ; sa racine est bulbeuse, mais plus petite & plus ronde que celle de la précédente : ses feuilles sont plus longues, plus étroites & plus plates ; les tiges ne s'élevent pas plus haut que les feuilles, qui sont grises : au sommet de la tige sort une fleur, en ouvrant sa spathe ; cette fleur penche d'un côté, & sa corolle est découpée en six segmens arrondis à leur extrémité, d'un blanc de neige ; & entierement étendus ; dans son centre est placé un très-court nectaire ou godet, frangé sur ses bords, & qui a un cercle d'un pourpre brillant. Ces fleurs répandent une odeur agréable ; elles paroissent dans le mois de Mai, & sont rarement suivies de semences : mais cette plante se multiplie assez promtement par les rejettons.

Le *Narciffe blanc à fleurs doubles* eft la feule variété de ce-
dui-ci , quoique , dans quelques
livres , on ait fait mention de
plufieurs autres.

Incomparabilis. La troifieme
efpece croît naturellement en
Efpagne & en Portugal , d'où
l'on m'a envoyé fes racines :
les bulbes de celle-ci reffemblent à ceux de la premiere ;
fes feuilles font plus longues ,
& d'un vert plus foncé ; fes
tiges de fleurs s'élevent à une
hauteur plus confidérable ; les
fegmens de la corolle font plus
ronds , plus étendus , & plus
applatis que ceux de la premiere ; le neƈtaire ou godet ,
qui en occupe le centre , n'a
que la moitié de la longueur
de la corolle , & il eft bordé
d'une frange de couleur d'or.
Cette plante fleurit dans le
mois d'Avril ; mais elle produit
rarement des femences dans ce
pays ; elle varie plus qu'aucune
des autres : car une feule racine m'a donné les variétés
fuivantes.

Cette racine avoit produit
la premiere année des fleurs
très-doubles , connues fous le
nom de *Narciffe incomparable* ;
les fix fegmens extérieurs du
pétale étoient blancs , & plus
longs qu'aucun des autres ; le
milieu étoit fort garni d'autres
pétales plus courts , dont quelques-uns étoient blancs , d'autres jaunes , & recueillis en
forme globulaire. Quelques-
unes de ces racines produifirent l'année fuivante des fleurs
moins doubles que celles de la
premiere année ; elles n'avoient
point de pétales blancs en-de-

dans ; mais les plus grands pétales étoient de couleur de foufre , & les autres jaunes : ces
fleurs dégénérerent enfuite en
femi-doubles , & devinrent enfin des fleurs fimples avec un
godet moitié moins long que
le pétale. Comme elles ont
continué ainfi pendant plufieurs années , on peut con-
clure que ces variétés n'é-
toient que des accidens de femences.

Medio luteus. La quatrieme
efpece eft originaire de la France Méridionale & de l'Italie :
on la trouve auffi dans quelques parties de l'Angleterre ;
mais il eft probable que ces
dernieres proviennent de quelques racines qui ont été jet-
tees hors des jardins avec des
immondices. Les racines de
cette efpece font plus rondes ,
& moins groffes que celles de
la premiere ; fes feuilles font
longues , de couleur grife , &
plus liffes ; fes tiges font de
la même longueur que les feuilles , & ont communément une
fleur dans chaque fpathe ; mais
quelquefois , quand les racines
font fortes , & en ont deux , la
fleur penche vers le bas , les
fegmens de la corolle font un
peu ondés fur leurs bords , le
neƈtaire ou godet eft court &
bordé de jaune. Cette plante
fleurit dans le mois de Mai :
l'odeur de fes fleurs n'eft pas
fort agréable ; & comme elles
ne font pas bien belles , on les
cultive peu dans les jardins ,
fur-tout depuis que les plus
belles efpeces fe font fort
multipliées. Je n'ai jamais vu
aucune variété de celle-ci , &

n'ai jamais observé aucune altération dans ses fleurs.

Albus. La cinquieme ressemble un peu à la quatrieme ; mais ses fleurs sont plus blanches, les segmens de la corolle sont réfléchis, & le bord du nectaire ou godet est d'un jaune doré. Celle-ci a quelqu'affinité avec la seconde espece.

Bulbocodium. La sixieme se trouve en Portugal, d'où j'en ai reçu les racines ; ses bulbes sont petites ; ses feuilles sont fort étroites, & à-peuprès semblables à celles du *Jonc,* mais un peu plus comprimées, sillonnées par une rainure longitudinale sur un côté, & de huit ou neuf pouces de longueur ; la tige est mince, cylindrique, de six pouces de hauteur, & terminée par une fleur qui est d'abord renfermée dans une spathe ; la corolle, qui a à peine six lignes de longueur, est découpée en six segmens aigus ; le nectaire ou godet, dont la hauteur est de plus de deux pouces, est fort large au bord, mais plus étroit vers la bâse, & ressemble un peu, par sa forme, au panier que portent les Dames, ce qui en a fait donner le nom à cette fleur. Celle-ci paroit dans le mois d'Avril ; mais elle n'est point suivie de semences en Angleterre. Je ne connois point de variétés de cette espece.

Serotinus. La septieme, qu'on rencontre en Espagne, a une petite racine bulbeuse, & produit un petit nombre de feuilles étroites, sa tige, qui est

noueuse & de neuf pouces environ de hauteur, soutient à son extrémité une seule fleur, qui est d'abord renfermée dans une spathe ; la corolle est découpée en six segmens étroits & blancs ; le nectaire ou godet est jaune. Cette fleur paroît sur la fin de l'automne ; ses racines, qui sont délicates, sont souvent détruites en Angleterre par les fortes gelées ce qui fait qu'elles sont rares dans ce pays.

Tazetta. La huitieme espece naît spontanément en Portugal & dans les Isles de l'Archipel ; elle donne un plus grand nombre de variétés que toutes les autres.

Comme ces fleurs font un bel ornement dans les jardins, & qu'elles paroissent de bonne heure au printems, les Fleuristes Hollandois, Flamands & François, ont pris beaucoup de peines pour les perfectionner ; de sorte qu'à présent les Catalogues imprimés des Hollandois contiennent plus de trente de ses variétés, dont les principales sont rapportées ci-après.

Narcisses à pétales jaunes, avec des godets ou nectaires de couleur d'orange, jaune ou de soufre.

1. Le grand Alger.
2. Le Bouquet des Dames.
3. La grande Cloche.
4. La Royale Dorée.
5. Le Sceptre doré.
6. Le Triomphant.
7. Le Très-beau.
8. L'Etoile dorée.
9. Le Mignon.
10. Le Zelandier.
11. La Madouse.

12. Le Soleil doré.

*Les suivantes ont des pétales blancs,
avec des godets ou nectaires
jaunes, ou de couleur de soufre.*

1. L'Archiduchesse.
2. Le Bouquet triomphant.
3. La nouvelle Dorothée.
4. La Passe-Bozelman.
5. Le Superbe.
6. Le grand Bozelman.
7. La Czarine.
8. Le grand Monarque.
9. Le Czar de Moscovie.
10. La Surpassante.

Quelques-unes ont des pétales & des godets blancs ; mais elles ne sont pas si estimées que les autres, à l'exception de deux variétés, qui ont de gros paquets de petites fleurs blanches, que leur odeur suave fait rechercher. Il y en a encore une à fleurs très-doubles & très-odorantes, dont les pétales extérieurs sont blancs, ainsi que ceux du milieu dans quelques-unes ; & dans d'autres, de couleur d'orange. Cette variété est celle qui fleurit la première au printems. On lui donne généralement le nom de *Narcisse de Chypre* ; mais elle paroît être une espece distincte des autres. Celle-ci, comme la plupart des autres fleurs doubles, ne produit jamais de semences, aussi ne la multiplie-t on que par ses rejettons. C'est le plus beau de tous les *Narcisses*, quand on le fait fleurir dans une chambre sur des carasses de verre remplies d'eau ; mais lorsqu'il est planté en pleine terre, si on ne le couvre pas avec des nattes pendant les gelées, pour garantir ses boutons des impressions

du froid, il fleurira rarement ; car ses feuilles poussent de bonne heure en automne, & ses boutons de fleurs paroissent vers Noël ; ils sont tendres ; & s'il survient une forte gelée, quand ils sont hors de terre, ils périssent ordinairement : mais quand ils sont mis exactement à l'abri du froid, ils fleurissent en Fevrier, & souvent même dans le mois de Janvier, si le tems est favorable.

Jonquilla. La neuvieme espece est la *Jonquille*, que tout le monde connoît si bien, qu'il est inutile d'en donner la description. Nous avons une grande & une petite Jonquille à fleurs simples, & l'espece commune à fleurs doubles, qui est la plus estimée.

Je vais commencer par indiquer la maniere d'élever les *Narcisses polyanthus* par semences, pour en obtenir de nouvelles variétés.

Faute d'avoir suivi cette pratique, nous avons été obligés de tirer annuellement ces racines des pays étrangers, à un très-haut prix, à cause des grands envois que l'on en faisoit en Angleterre, au-lieu que, si nous étions aussi industrieux que nos voisins pour les multiplier, nous pourrions bientôt les égaler, & même les surpasser dans la perfection des especes de fleurs ; comme on peut le voir par la grande quantité d'*Œillets*, d'*Auricules*, de *Renoncules*, &c. qui ont été élevées de semences en Angleterre, & qui surpassent en beauté presque toutes les fleurs de même espece dans la plus

grande partie de l'Europe.

Il faut avoir grand foin, en ramaffant les femences, de ne recueillir que celles des fleurs de bonne efpece, & fur-tout de celles qui ont plufieurs fleurs fur une tige, qui s'élevent très-haut, & ont de beaux godets ; par ce moyen, on pourra efpérer d'en obtenir de bonnes fleurs : mais quand on ne ramaffe que des femences ordinaires, on fe prépare beaucoup de peines & de dépenfes mal-à-propos, puifqu'on ne peut en obtenir que des fleurs communes & peu eftimables.

Quand on s'eft procuré de bonnes femences, on fe pourvoit de caiffes ou de terrines peu profondes, telles qu'on les fait exprès pour élever des plantes de femence, & dont le fond doit être percé de trous, pour laiffer écouler l'humidité.

Vers le commencement du mois d'Août, on les remplit d'une terre nouvelle, légere & fablonneufe ; car cette faifon eft la plus propre pour femer la plupart des fleurs à racine bulbeufe : on nivelle exactement la furface de cette terre ; on y repand les femences fort épaiffes, & on crible par-deffus une terre légere, jufqu'à l'épaiffeur d'environ fix lignes : on met ces caiffes ou terrines dans un lieu où elles foient expofées au foleil, feulement depuis fon lever jufques vers dix heures, & on les laiffe ainfi jufqu'au commencement d'Octobre ; alors on les place à une expofition plus chaude, & on les pofe fur des briques, afin que l'air

puiffe circuler plus librement par-deffous, & diffiper le trop d'humidité.

On les expofe auffi au plein foleil ; mais on les garantit avec foin des vents du nord & de l'eft. Si la gelée devient rude, il eft néceffaire de les couvrir, pour prévenir leur deftruction. On peut les laiffer dans cette fituation jufqu'au commencement d'Avril ; alors les plantes auront commencé à pouffer.

On les débarraffera auffi avec foin de mauvaifes herbes ; on les arrofera fouvent, fi le tems eft fec : on mettra de tems en tems les caiffes ou terrines à l'ombre dans leur premiere pofition, & on les couvrira au milieu du jour ; car la chaleur du foleil du midi eft trop forte pour les jeunes plantes.

A la fin de Juin, lorfque leurs feuilles font flétries, on enleve la furface de la terre où fe trouve alors le poil cotonneux des femences, qui endommageroit beaucoup les jeunes racines, fi on les laiffoit ; mais il faut avoir foin de ne pas creufer affez profondément pour toucher les racines, qu'on fortifiera enfuite en criblant par deffus de la nouvelle terre légere, jufqu'à l'épaiffeur d'environ fix lignes : on repete cette opération au mois d'Octobre, quand on remet les caiffes au foleil.

Pendant l'été, quand le tems eft pluvieux, & que la terre des caiffes paroît fort humide, on les met au foleil, jufqu'à ce qu'elle foit deffechée ;

NAR

car fi les racines avoient trop
d'humidité, tandis qu'elles font
dans l'inaction, il feroit à craindre qu'elles ne fuffent attaquées de pourriture : c'eft-
pourquoi il ne faut jamais leur
donner d'eau après que leurs
feuilles font tombées ; mais
feulement les placer à l'ombre
comme il a été dit ci-deffus.

C'eft en cela que confiftent
tous les foins qu'elles exigent dans les deux premieres
faifons , jufqu'à ce que leurs
feuilles foient mortes ; mais
dans le fecond été, après qu'elles ont été femées, il faut enlever ces racines, en paffant
la terre des caiffes à travers
un crible fin, & les planter
à trois pouces de diftance entr'elles, & à trois pouces de
profondeur dans des planches
de terre nouvelle & légere
qu'on aura préparées d'avance.

Ces planches doivent être
plus ou moins élevées au-deffus du niveau du fol, fuivant
que le terrein eft plus fec ou
plus humide : s'il eft fec, trois
pouces fuffiront ; & fi au contraire il eft humide, on les
éleve à fix ou huit pouces ,
en les arrondiffant un peu,
pour laiffer écouler l'humidité.
Si les planches font dreffées
en Juillet, qui eft le meilleur
tems pour tranfplanter les racines , les mauvaifes herbes
y pousferont bientôt en abondance ; alors on houe légerement la terre pour les détruire, fans enfoncer affez profondément pour toucher quelques-unes des racines. Cette
opération fe répete auffi fouvent qu'il eft befoin d'arracher

NAR

les mauvaifes herbes qui repouffent, en obfervant toujours de la faire par un tems
fec , afin de les détruire entierement. Vers la fin d'Octobre,
lorfque les planches font bien
nettoyées, on crible deffus un
peu de terre riche & légere,
jufqu'à l'épaiffeur d'environ un
pouce ; ce qui déterminera ,
par la premiere pluie d'hiver,
les racines à pouffer par le
bas , & leur fera faire de
grands progrès au printems.

Si le froid devient rude en
hiver, on couvre les planches
avec du vieux tan, des cendres de charbon de terre, ou
même avec du chaume de pois ,
ou quelqu'autre couverture légere, pour empêcher la gelée
de pénétrer jufqu'aux racines ,
qui pourroient en être endommagées fortement, tandis
qu'elles font encore fort jeunes.

Au printems, lorfque les
plantes commencent à paroître au-deffus de la terre, on
en remue légerement la furface, pour faire périr les mauvaifes herbes : mais en faifant
ce travail, il faut prendre bien
garde de ne pas endommager
les plantes. Si la faifon devient
feche, on les arrofe légerement de tems en tems, pour
renforcer les racines.

Quand les feuilles font flétries, on nettoie les planches
de toutes mauvaifes herbes ,
& l'on y crible un peu de
terre neuve & légere, comme
il a été dit ci-deffus, ce qui
doit être reiteré dans le mois
d'Octobre : mais les racines
ne doivent refter que deux ans

dans ces planches. Comme, après ce tems, elles auront acquis affez de volume pour exiger plus de place, on les enlevera auffi-tôt que leurs feuilles feront fanées, & on les placera dans de nouvelles planches profondément labourées, afin que les fibres des racines puiffent y pénétrer. Ces racines doivent être plantées à fix pouces de diftance, & à fix pouces de profondeur dans la terre.

En automne, & avant que les gelées fe faffent fentir, on répand fur les planches du tan pourri, pour empêcher la gelée d'y pénétrer. Si l'hiver eft rude, il eft prudent d'y mettre une plus groffe épaiffeur de tan, & d'en répandre auffi dans les fentiers, pour les préferver du froid, ou bien on les couvre de paille ou de chaume de pois: fans ces précautions, elles pourroient être détruites entierement. Au printems, auffi-tôt que le danger des fortes gelées eft paffé, on enleve les couvertures, & on tient les planches nettes de mauvaifes herbes pendant tout l'été fuivant. A la Saint-Michel on répand de la nouvelle terre par-deffus, & on les recouvre de tan, en continuant ainfi, jufqu'à ce qu'elles fleuriffent; ce qui a lieu généralement dans la cinquieme année: alors on marque toutes celles qui paroiffent bonnes, pour les enlever auffi tôt après la chûte de leurs feuilles, & les planter à une plus grande diftance, dans de nouvelles planches préparées; mais on laiffe

dans l'ancienne celles qui n'ont point encore fleuri, & dont on fait peu de cas. Ainfi, en enlevant les racines qui ont été marquées, il faut avoir foin de ne pas déranger celles qui doivent refter,& de remettre les planches de niveau, en criblant de la nouvelle terre par-deffus, comme on l'a fait auparavant, pour fortifier les racines; car il arrive fouvent dans les plantes de femence de cette efpece, que la premiere fois qu'elles fleuriffent, elles ne font pas à beaucoup près, auffi belles qu'elles le font dans la feconde année: c'eft pour cette raifon qu'on n'en doit rejetter aucune qu'elles n'aient fleuri deux ou trois fois, pour s'affurer de leur véritable valeur.

Après avoir donné des inftructions pour femer & traiter ces racines, jufqu'à ce qu'elles aient acquis affez de force pour fleurir, je vais en donner pour les planter, & les traiter de maniere à leur faire produire de belles & groffes fleurs.

Toutes les efpeces de *Narciffes* qui produifent beaucoup de fleurs fur une tige, doivent être placées dans une fituation à l'abri du froid & des grands vents, fans quoi elles font fujettes à être endommagées par le froid en hiver, ou à avoir leurs tiges rompues, quand elles font en fleurs: car quoique leurs tiges foient généralement affez fortes, cependant le nombre de fleurs qui fe trouvent fur chacune, rend leurs têtes très-lourdes, fur-tout lorfqu'après la pluie, elles font

NAR

chargées d'eau, & les grands vents les flétrissent bientôt, lorsqu'elles y sont exposées ; de sorte qu'une plate-bande, à l'abri d'une haie & à l'exposition du sud-est, est préférable à toute autre pour ces especes de fleurs.

Les premiers rayons du soleil levant dessécheront l'humidité que la nuit dépose sur ces fleurs, & les rendront plus belles & mieux épanouies que si elles avoient été plantées au soleil de l'après-midi. Elles se conserveront beaucoup mieux ainsi, qu'étant exposées à la fureur des vents du couchant & du sud-ouest, qui leur sont souvent fort nuisibles.

Quand on a choisi une exposition convenable, on prépare une terre qui leur soit propre ; car si la nature du sol est très-forte & de mauvaise qualité, il sera prudent d'élever des plates-bandes avec une nouvelle terre, en enlevant l'ancien sol jusqu'à la profondeur d'environ trois pieds. Celui qui convient le mieux à ces fleurs, est une terre nouvelle, légere & marneuse, à laquelle on a encore ajouté un peu de fumier de vache pourri. Ces différentes matieres étant exactement mêlées, & la vieille terre étant enlevée, comme il a été dit ci-dessus, on met au fond une couche de fumier ou de tan de quatre ou cinq pouces d'épaisseur : on en nivelle exactement la surface ; on la couvre de deux pieds de bonne terre préparée ; & après l'avoir bien dressée, on trace des li-

NAR

gnes à des distances égales de sept ou huit pouces, où les racines doivent être plantées. On plante ces racines, en observant de les placer droites, & les têtes en haut, & on les recouvre ensuite de huit pouces environ, avec la terre préparée. En faisant cette opération, il faut avoir bien soin de ne pas déplacer les racines : on dresse ensuite la surface de la plate-bande, & on rend les côtés droits, pour qu'elle soit plus agréable.

Le meilleur tems pour planter ces racines, est la fin d'Août ou le commencement de Septembre ; car si on les tient trop long-tems hors de terre, leurs fleurs deviennent trop foibles : il faut aussi observer la nature du sol où on les plante, & si le terrein est humide ou sec, afin de se régler en conséquence pour former les plates-bandes ; car si le sol est très-fort & la terre humide, il faut alors y rapporter une terre légere, & élever les plates-bandes à six ou huit pouces, & même à un pied au-dessus du niveau, sans quoi les racines seroient en danger de périr par trop d'humidité ; mais si au contraire le sol est sec, & la terre naturellement légere, on y en mêle une un peu plus forte, & on n'éleve les plates-bandes que de trois ou quatre pouces au-dessus du terrein ; car si elles étoient trop hautes, les racines souffriroient beaucoup dans les sécheresses du printems, & les fleurs ne seroient pas si belles. Il arrive aussi que, dans les

hivers très-rudes, les plates-bandes fort élevées au-dessus du terrein sont plus exposées au froid, à moins qu'on ne remplisse les sentiers avec du tan pourri ou de la litiere.

Pendant l'été, la seule culture que ces fleurs exigent, est d'être tenues nettes de mauvaises herbes, & d'être débarrassées de leurs feuilles, quand elles sont entierement fanées; mais on ne doit jamais enlever ces feuilles avant qu'elles soient tout-à-fait détruites, comme on le fait quelquefois, parce que cela affoiblit beaucoup leurs racines.

Vers le milieu d'Octobre, si les semences des mauvaises herbes ont commencé à croître sur les plates-bandes, on houe légerement la surface de la terre par un tems sec, pour les détruire; & avant que les gelées arrivent, on couvre la terre de deux pouces de tan pourri, pour empêcher le froid d'y pénétrer, après quoi elles n'exigeront plus aucun soin jusqu'au printems. Lorsque leurs feuilles paroissent au-dessus de la terre, on remue légerement la surface des plates-bandes avec une petite truelle, sans endommager les plantes : on unit la terre avec les mains, et on enleve toutes les mauvaises herbes, qui repousseroient bientôt, de maniere qu'elles seroient désagréables à la vue, si on les y laissoit dans cette saison, & épuiseroient le suc de la terre. Avec ce traitement, ces racines fleuriront très-bien. Quelques-unes paroîtront en Mars, d'autres en

Avril; & si on les laisse, elles conserveront leur beauté durant un mois entier, & feront un très-agréable effet dans les parterres.

Quand les fleurs sont passées, & les feuilles détruites, il faut remuer la surface de la terre, pour empêcher les mauvaises herbes de croître; & en mettant un peu de fumier très-consommé par-dessus les plates-bandes, la pluie en fera entrer les sels dans la terre; ce qui disposera ces racines à bien fleurir l'année suivante. Pendant l'été, elles n'exigent aucun autre soin que d'être tenues nettes de mauvaises herbes. Au mois d'Octobre, on remue la surface des plates-bandes; on enleve les mauvaises herbes avec un rateau, & on met de la terre nouvelle par-dessus, de l'épaisseur d'environ un pouce, pour remplacer celle qui a été enlevée par les houages. Au printems suivant, on traite ces plantes comme il a été prescrit pour l'année précédente.

Ces plantes ne doivent être levées de terre que tous les trois ans, si l'on veut qu'elles fleurissent & se multiplient fortement, parce que la premiere année que ces racines sont transplantées, elles ne fleurissent jamais aussi fort que la seconde & la troisieme, & elles ne se multiplient pas autant, quand on les leve de terre trop souvent; mais si on les laisse plus long-tems sans les remuer, le grand nombre de rejettons affoiblit les grosses bulbes, qui ne donnent plus

enfuite que de très petites
fleurs : c'eft-pourquoi, quand
on les enleve, il faut retrancher tous les petits rejettons
qu'on place dans une planche
en pépiniere à part, & planter les groffes bulbes, pour
fleurir. Si on veut les remettre dans la même plate-bande
où elles étoient avant, il faut
enlever toute la terre jufqu'à
la profondeur de deux pieds,
& la remplacer avec une nouvelle, compofée comme il a
été dit ci-deffus. La même opération fera néceffaire, fi on
les plante dans un autre endroit : c'eft la pratique conftante des Jardiniers Hollandois,
qui ont peu de place pour
changer leurs racines. Tous
les ans, ils renouvellent la
terre de leurs plates-bandes,
de maniere que la même place
eft conftamment occupée par
les mêmes fleurs : mais ils enlevent leurs racines chaque année, parce que, leur objet
étant d'en faire commerce,
plus elles font rondes, & plus
elles ont de valeur, ce qu'on
ne peut obtenir qu'en ôtant
tous les ans les rejettons; car
fi on les laiffe deux ou trois
ans fans les détacher, ils deviennent gros, fe preffent les
uns contre les autres, &
les côtés intérieurs s'applatif-
fent. Ainfi, lorfqu'on les deftine à être vendues, il faut les
enlever annuellement, auffi-
tôt que leurs feuilles font flétries, conferver les groffes
bulbes hors de terre, jufqu'au
milieu ou à la fin d'Octobre,
& planter les rejettons au commencement de Septembre, ou

même plutôt, afin qu'ils puiffent acquérir de la force, &
produire des fleurs l'année fuivante; mais lorfque ces racines ne fervent qu'à orner un
parterre, on ne doit les lever
que tous les trois ans : au
moyen de cela, les rejettons
s'accumulent, forment de gros
paquets, & produifent un grand
nombre de tiges à fleurs, qui
font un bien plus bel effet que
s'il n'y avoit qu'une fimple fleur
à chaque racine, comme il arrive quand on les leve tous les ans.
Les efpeces communes de
Narciffe doivent toujours être
placées dans les grandes plates-
bandes d'un parterre, où elles
font une variété agréable,
quand elles font en fleurs, &
entremêlées avec d'autres racines bulbeufes. Ces efpeces
font fort dures, & profitent
dans prefque tous les fols &
dans toutes les fituations ; ce
qui les rend fort propres à
orner des jardins champêtres.
On peut les planter à l'ombre
des arbres, où elles profiteront pendant plufieurs années,
fans être enlevées, & produiront annuellement au printems
une grande quantité de fleurs,
qui feront un charmant effet,
avant que les feuilles des arbres
commencent à paroître.
Jonquille. On plante les *Jon-
quilles* dans des planches ou
plates-bandes féparées des autres racines, parce qu'elles doivent être enlevées chaque
année, fans quoi elles font
fujettes à s'allonger, & à devenir minces ; ce qui les empêche de bien fleurir dans la
fuite. C'eft auffi ce qui arrive

NAS

quand on les tient plufieurs années de fuite dans le même fol. Pour éviter cet inconvénient , ils eft prudent de les tranfporter fouvent d'une partie du jardin dans une autre , ou au moins de renouveller fréquemment la terre. Cette méthode eft la plus fûre pour conferver ces fleurs dans leur perfection.

Le fol qui convient le mieux aux racines, eft une terre marneufe , pas trop legere ni trop forte , fraiche , exempte de toutes racines d'arbres & d'herbes nuifibles , & fans aucun mélange de fumier ; car on a obfervé qu'une terre riche leur eft rarement bonne pendant long-tems , qu'elle fait pouffer les racines par le bas, & les rend longues & minces.

Ces fleurs font très-eftimées, à caufe de leur charmante odeur : mais cette odeur eft fi forte , que fort peu de Dames peuvent la fupporter ; elle eft d'une telle activité, qu'elle les fait fouvent tomber en foibleffe , fur-tout fi ces fleurs font renfermées dans une chambre : c'eft-pourquoi il ne faut jamais les placer trop près des habitations , ni les garder dans les appartemens où l'on reçoit compagnie.

NARD ASPIC *ou* LAVANDE MASLE. *Voyez* LAVENDULA SPICA. *L.*

NARD CELTIQUE. *V.* VALERIANA CELTICA. *L.*

NASITOR *ou* CRESSON ALENOIS. *Voyez* LEPIDIUM SATIVUM. *L.*

NASTURTIUM. *Voyez* LEPIDIUM.

NAT

NASTURTIUM INDICUM. *Voy.* TROPÆOLUM MINUS. *L.*

NATUREL , eft ce qui n'a point été altéré , & fe montre tel que la nature le produit.

NATURE. On prend ce mot en différens fens. M. BOYLE , dans fon traité fur les différens fens attachés à ce mot , nous en donne huit acceptions diverfes.

1°. *Nature* fe dit du fyftême du Monde , de la machine de l'Univers , ou de l'affemblage de tous les êtres créés.

Dans ce fens , nous difons que DIEU eft l'Auteur de la Nature : en parlant du foleil , nous difons , le Pere de la Nature , parce qu'il échauffe la terre & la rend féconde ; nous l'appelons l'Œil de la Nature , parce qu'il éclaire l'Univers. S'il eft queftion d'un phénix , d'une licorne , d'un fatyre , nous difons qu'il n'eft point de ces êtres dans la Nature.

2°. Dans un fens moins étendu , le mot *Nature* comprend les différentes claffes d'êtres créés & non créés , corporels & fpirituels. Dans ce fens , nous difons la Nature humaine , c'eft-à-dire , tous les hommes qui font doués d'une ame raifonnable ; la Nature des Anges, la Nature divine.

3°. Le mot *Nature* , pris dans un fens plus reftreint , comme l'effence d'une chofe ou l'attribut qui conftitue fon être. Par exemple , nous difons : La nature de l'ame eft de penfer.

4°. On emploie encore particulierement le mot de *Nature* , pour exprimer l'ordre &

NEF

le cours fixe des choses matérielles, l'enchaînement des causes secondes, ou les loix que Dieu, par sa volonté, a imposées aux corps. Dans ce sens, nous disons, le jour & la nuit se succedent naturellement l'un à l'autre. La Physique est l'étude de la Nature, & la Nature a rendu la respiration nécessaire à la vie.

5°. La *Nature* signifie aussi les différens pouvoirs appartenans à un corps, & sur-tout à un corps animé. Dans ce sens, nous disons, la Nature est forte, la Nature est foible, elle est épuisée.

6°. Dans un sens plus stricte, on se sert de ce mot pour désigner la Providence, le principe de toutes choses; & cette puissance qui est répandue dans tout l'Univers, qui se meut & agit dans tous les corps, leur donne une certaine propriété, & produit certains effets. Dans ce sens, la *Nature* signifie la qualité ou vertu que Dieu a donnée à ses créatures, soit animales, soit végétales.

En parlant de l'action de la *Nature*, on ne veut dire autre chose, sinon que les corps agissent les uns sur les autres d'une maniere conforme aux loix générales du mouvement que le Créateur a établies.

NAVET. *Voy.* RAPA NAPUS.

NAVETTE *ou* NAVET SAUVAGE. *Voy.* BRASSICA GONGYLODES. 1. NAPUS SYLVESTRIS.

NÉBULEUX, signifie couvert de nuages, brouillard & tems couvert.

NEFLIER. *V.* MESPILUS. *L.*

NEI

NUGUNDO. *Voyez* VITEX NEGUNDO. *L.*

NEIGE. *Snow. Angl.* On définit la neige un météore formé dans la région moyenne de l'air, des vapeurs élevées par l'action du soleil, ou par le feu souterrain. Ces vapeurs étant congelées dans l'air, deviennent plus dures, augmentent en gravité spécifique, & se précipitent sur la terre en floccons.

La neige qui tombe, peut être proprement attribué au froid de l'atmosphere, au travers duquel elle passe. Quand l'atmosphere est assez échauffé pour la fondre avant qu'elle arrive à nous, nous l'appelons *pluie*; & si elle n'est pas dissoute, on l'appelle *neige*. La neige est fort utile à la terre, qu'elle fertilise; elle préserve les grains & autres végétaux des gelées rigoureuses, & sur-tout des vents froids & pénétrans.

On croit que la neige abonde en particules salines & fertiles, autant & même plus que la pluie. On pense qu'étant plus pesante, elle pénetre bien plus avant dans la terre que la pluie, & devient par-là plus avantageuse aux plantes. C'est par cette raison que plusieurs personnes entassent la neige autour de leurs arbres de forêts, sur-tout lorsque le sol est naturellement chaud & brûlant.

Suivant M. LE CLERC, quelques nuages qui devoient se tourner en pluie, en sont quelquefois empêchés par le froid, & se transforment en une sub-

stance que nous appelons *Neige* ; elle est formée de particules aqueuses, puisqu'après sa dissolution elle se change en eau. Ainsi, nous concevons aisément que la Neige, composée de particules aqueuses, condensées par le froid & rassemblées en floccons, de maniere qu'il reste entr'elles des interstices, n'est point transparente comme l'eau, parce que les particules, durcies par le froid, se trouvent rassemblées confusément, & que la lumiere ne peut traverser leurs pores tortueux.

Quand il arrive que la région de l'air qui se trouve au-dessous des nuages est très-froide, les gouttes d'eau qui la traversent se gelent en tombant, & parviennent à nous en grains que nous nommons *Grêle*. Ces grains sont plus gros ou plus petits selon la grosseur des gouttes de pluie dont ils sont formés, & diverses causes leur donnent aussi des formes différentes.

Le Docteur GREW, dans son *Discours sur la nature de la Neige*, observe que plusieurs de ces parties sont d'une forme réguliere, & que la plupart sont, pour ainsi dire, autant d'étoiles d'une glace parfaite & transparente, sur chaque pointe desquelles en sont placées d'autres qui ont des angles semblables. Parmi celles-ci, il y en a plusieurs autres irrégulieres, qui ne sont telles, que parce qu'elles ne sont que des fragmens, & qu'elles ont perdu leurs pointes régulieres ; d'autres aussi, par dif-

férens vents, semblent avoir été dissoutes & gélées de nouveau en forme irréguliere, de maniere qu'il semble que le corps de la neige soit un amas de glaces semblables à celle qui pend aux gouttieres dans quelques circonstances. Cet effet a lieu, lorsqu'un nuage se dissolvant en pluie, les gouttes rencontrent un air plus froid, qui les change en glace, & leur donne cette forme anguleuse qu'on remarque dans la neige ; mais si ces floccons traversent ensuite une région plus tempérée, ou s'ils sont agités en différens sens par des vents opposés, leurs angles les plus déliés se résolvent en eau, ou sont rompus par le frottement qu'ils éprouvent ; ce qui dérange leur régularité, & nous fait voir cette neige sous la forme de gros floccons (1).

Quoique la neige soit véritablement de la glace, elle est néanmoins d'une grande

_____

(1) La forme anguleuse & étoilée qu'on remarque dans la neige, est un produit de la crystallisation de l'eau, qui, comme tous les autres corps qui se réunissent en une masse solide, après avoir été séparés par un fluide en molécules très-petites, affecte une forme qui lui est propre, & se montre sous celle de petits octaëdres groupés en forme d'étoiles. On distingue fort bien ces petits crystaux, lorsqu'on examine un floccon de neige à travers une bonne loupe. On peut en voir de semblables dans ces *Dendrites* ou ramifications qu'on observe contre les vitres des appartemens par un tems de forte gelee,

NEP

légereté, parce qu'elle offre à l'air une furface fort confidérable, en comparaifon de la petite quantité de matiere qu'elle contient. C'eft ainfi que l'*Or*, quoique le plus pefant des métaux, étant réduit en lames fort minces, peut, par ce moyen, devenir affez léger pour flotter dans l'air.

NENUFAR *ou* NYMPHEA. *Voyez* NYMPHÆA. *L.*

NEPETA. *Lin. Gen. Plant.* 629. *Cataria. Tourn. Inft. R. H.* 202. *tab.* 95. [*Cat's-mint*, *or Nep.*] Herbe aux Chats.

Caracteres. Le calice de la fleur eft tubulé, cylindrique, & découpé fur fes bords en cinq parties aiguës : la corolle eft labiée & monopétale ; elle a un tube recourbé, cylindrique, & ouvert au fommet : la levre fupérieure eft érigée, ronde, & dentelée à fon extrémité ; la levre inférieure eft large, concave, entiere, & fciée fur fes bords : la fleur a quatre étamines en forme d'alêne, fituées fous la levre fupérieure, dont deux font plus courtes que les autres, & qui font toutes terminées par des antheres inclinées : dans le fond du tube, eft placé un germe divifé en quatre parties, qui foutient un ftyle mince & couronné par un ftigmat partagé en deux portions aiguës ; ce germe fe change dans la fuite en quatre femences ovales, placées dans le calice.

Ce genre de plantes eft rangé dans la premiere fection de la quatorzieme claffe de LINNÉE, qui renferme celles dont les

fleurs ont deux étamines longues & deux courtes, & qui produifent des femences nues, renfermées dans le calice.

Les efpeces font :

1°. *Nepeta Cataria*, *floribus fpicat's*, *verticillis fub-pedicellatis*, *foliis petiolatis*, *cordatis*, *dentato-ferratis. Lin. Sp. Plant.* 796. *Mat. Med.* 146. *Reyg-ged.* 2. *p.* 100. *Scop. carn. ed.* 2. *n.* 743. de *Neck. Gallob. p.* 249. *Pall. it.* 1. *p.* 25. *Pollich. pal. n.* 549. *Mattufch. Sil. n.* 426 ; Herbe aux Chats, avec des fleurs en épis, dont les têtes verticillées ont des pédoncules très-courts, & des feuilles pétiolées, en forme de cœur, & dentelées en forme de fcie.

Cataria, *foliis cordatis*, *verticillis fpicatis. Hall. Helv. n.* 246.

Mentha Cataria vulgaris & major. Bauh. 228.

Cataria major vulgaris. Tourn. Inft. R. H. 202. la plus grande Herbe aux Chats, commune.

Cataria Herba, *Dod. Pempt.* 99.

2°. *Nepeta minor*, *floribus fpicatis*, *fpicis interruptis*, *verticillis pedicellatis*, *foliis fub-cordatis*, *ferratis*, *petiolatis* ; Herbe aux Chats, produifant des fleurs en épis, avec des têtes verticillées interrompues, poftées fur des pédoncules, & des feuilles fciées prefque en forme de cœur, & fupportées par des pétioles.

Mentha Cataria minor. Bauh. Pin. 228.

Cataria minor vulgaris. Tourn. Inft. R. H. 202 ; la plus petite Herbe aux Chats, commune.

3°. *Nepeta angufti-folia*, *floribus fpicatis*, *verticillis fub-feffilibus*,

NEP

bus , foliis cordato-oblongis , serratis, sessilibus ; Herbe aux Chats , à fleurs en épis, dont les têtes verticillées sont presque sessiles aux tiges , avec des feuilles oblongues , en forme de cœur, sciées & sessiles.

Cataria angusti-folia major. Tourn. Inst. R. H. 202 ; la plus grande Herbe aux Chats , à feuilles étroites.

4°. Nepeta paniculata , floribus paniculatis , foliis oblongo-cordatis , acutis, serratis, sessilibus ; Herbe aux Chats , à fleurs paniculées, avec des feuilles oblongues, en forme de cœur , aiguës, sciées & sessiles.

Cataria quæ Nepeta minor, folio Melissæ Turcicæ. Boërh. Ind. Alt. 174 ; la plus petite Herbe aux Chats , à feuilles de Mélisse de Turquie.

5°. Nepeta Italica , floribus sessilibus verticillato-spicatis , bracteis lanceolatis longitudine calycis , foliis petiolatis. Lin. Sp. Plant. 798. Jacq. Hort. t. 112 ; Herbe aux Chats , dont les fleurs croissent en épis verticillés , & sessiles à la tige , avec des bractées en forme de lance de la longueur du calice , & des feuilles pétiolées.

Cataria minor. Tourn. Inst. R. H. 202 ; la plus petite Herbe aux Chats des Alpes.

Mentha Cataria minor Alpina. Bauh. Pin 228. Prodr. 110.

6°. Nepeta violacea , verticillis pedunculatis corymbosis , foliis petiolatis, cordato-oblongis , dentatis. Lin. Sp. Plant. 797. Scop. carn. ed. 2. n. 744. Pall. it. 1. p. 154. ; Herbe aux Chats , avec des têtes verticillées ,

Tome V.

NEP   221

rondes , & postées sur des pédoncules , & des feuilles oblongues en forme de cœur , & dentelées.

Nepeta montana purpurea major, sparsâ spicâ. Barr. ic. 601 Boc. Mus. 2. p. 46. f. 36.

Cataria Hispanica , Betonicæ folio angustiori , flore cæruleo. Tourn. Inst. R H. 202 ; Herbe aux Chats d'Espagne , avec des feuilles étroites de Bétoine , & une fleur bleue.

7°. Nepeta tuberosa , spicis sessilibus terminalibus , bracteis ovatis coloratis , foliis summis sessilibus. Hort. Cliff. 311. Roy. Lugd.-B. 316. Gouan. Illustr. 36. Kniph. cent. 9. n. 72 ; Herbe aux Chats , avec des fleurs en épis & sessiles, ayant des bractées ovales & colorées, & dont les feuilles du haut sont sessiles aux tiges.

Mentha tuberosâ radice. Bauh. Pin. 227.

Cataria Hispanica , supina , Betonicæ folio , tuberosâ radice. Tourn. Inst. R. H. 202 ; Herbe aux Chats d'Espagne , avec une tige penchée, une feuille de Bétoine , & une racine tubéreuse.

8°. Nepeta hirsuta , floribus sessilibus verticillato-spicatis , verticillis tomento-obvolutis. Hort. Cliff. 311. Roy. Lugd.-B. 316 ; Herbe aux Chats , avec des fleurs en épis verticillés & sessiles , dont les têtes verticillées sont couvertes de duvet.

Horminum spicatum Lavendulæ flore & odore. Boc. Plant. Sic. 48. tab. 25 ; Orvale en épis , à odeur & à fleurs de Lavande.

9°. Nepeta Virginica , foliis

lanceolatis, capitulis terminalibus, staminibus flore longioribus. Lin. Sp. Plant. 571 ; Herbe aux Chats , avec des feuilles en forme de lance , des tiges terminées par des têtes de fleurs, & des étamines plus longues que les corolles.

Clinopodium foliis lanceolatis, capitulis terminalibus. Hort. Cliff. 305. Gron. Virg. 65.

Clinopodium Amaraci folio , floribus albis. Pluk. Alm. 110. t. 85. f. 2 ; Basilic des Champs , à feuilles de Marjolaine , & à fleurs blanches.

Clinopodium , floro albo , ramosius , angustioribus foliis glabris , Virginianum. Moris. Hist. 3. p. 374. S. 11. f. 8.

10°. *Nepeta Orientalis , floribus spicatis , verticillis crassioribus , foliis cordatis , obtusè dentatis , petiolatis* ; Herbe aux Chats , avec des fleurs en épis , dont les têtes verticillées sont fort épaisses , & des feuilles en forme de cœur , à dents obtuses & pétiolées.

Cataria Orientalis , Tenerii folio , Lavendulæ odore , verticillis florum crassissimis. Tourn. Cor. Inst. 13 ; Herbe aux Chats, Orientale , à feuilles de Germandrée , & à odeur de Lavande , avec des têtes de fleurs fort épaisses.

11°. *Nepeta procumbens , floribus verticillatis , bracteis ovatis , hirsutis , foliis cordato-ovatis , crenatis , caule procumbente* ; Herbe aux Chats , avec des fleurs verticillées , des bractées ovales & velues , des feuilles ovales , en forme de cœur & crenelées , & une tige traînante.

Cataria. La première espece

est l'Herbe aux Chats , qui croît naturellement sur les bords des chemins & des haies , dans plusieurs parties de l'Angleterre ; elle a une racine vivace , de laquelle sortent plusieurs tiges branchues , quarrées , de deux pieds de hauteur , & garnies à chaque nœud de deux feuilles en forme de cœur , opposées , supportées par de longs pétioles , sciées sur leurs bords , & velues en-dessous : ses fleurs croissent en épi au sommet des tiges ; au-dessous des épis sont placées deux ou trois têtes de fleurs verticillées , qui ont de fort courts pédoncules : ces fleurs sont blanches , & ont deux levres, dont la supérieure est érigée , & l'inférieure un peu réfléchie & dentelée à la pointe ; elles produisent toutes quatre semences ovales & noires , qui mûrissent dans le calice.

Toutes les parties de cette plante répandent une odeur forte , qui tient de celle de la Menthe & de celle du Poillot. On lui donne le nom d'*Herbe aux Chats* , parce que ces animaux l'aiment beaucoup , surtout quand elle est fannée ; car alors ils se roulent dessus , la déchirent en morceaux , & la mâchent avec grand plaisir. M. Ray rapporte , qu'ayant transplanté quelques-unes de ces plantes dans son jardin , elles furent bientôt détruites par les chats : mais celles qui poussèrent de semences dans le même jardin , n'en furent point endommagées ; ce qui vérifie l'ancien proverbe : *Si vous les plantez , les Chats le*

NEP

mangeront ; fi vous les femez, les Chats n'y toucheront pas. Je l'ai fouvent éprouvé moi-même, & l'expérience a toujours réuffi; car ayant tranfplanté une de ces plantes dans une partie éloignée du jardin, à la diftance de deux pieds de pareilles plantes venues de femences, ces dernieres ne furent point touchées, au lieu que la premiere fut déchirée en morceaux, & détruite entierement par les chats : d'ailleurs, j'ai toujours obfervé que, quand ces plantes croiffoient beaucoup enfemble, elles n'étoient jamais endommagées par ces animaux. Cette efpece fleurit en Juin & Juillet, & fes femences mûriffent en automne : elle eft d'ufage en Médecine (1).

La feconde efpece eft originaire de l'Italie & de la France Méridionale; fes tiges font minces, & leurs nœuds font plus éloignés; fes feuilles font plus étroites, & la plante entiere eft plus blanche que la premiere efpece ; fes épis de fleurs font divifés en têtes verticillées, dont les plus baffes font à deux pouces de diftance, d'autres à un pouce, & celles du haut à fix pouces. Ces différences font

---

(1) L'Herbe aux Chats ne differe point de la Menthe fauvage, quant à fes propriétés médicinales : on l'emploie avec le Marrhube blanc, la Matricaire, & dans les décoctions & autres préparations anti-hiftériques. elle a auffi la réputation d'être un bon remede aperitif, & de pouvoir être employée avec fuccès contre la jauniffe.

NEP

perfiftantes ; car j'ai toujours vu les femences de chaque efpece produire les mêmes plantes.

Angufti-folia. Les tiges de la troifieme font moins branchues que celles des deux précédentes; elles font plus minces, & leurs nœuds font plus éloignés; fes feuilles font petites, étroites, prefqu'en forme de cœur, blanches, fciées, fur leurs bords, & fupportées par de courts pétioles : fes épis de fleurs font plus interrompus que ceux de la feconde, & leurs têtes verticillées font poftées fur des pédoncules : elle croit fans culture en Italie.

Paniculata. La quatrieme efpece, qu'on rencontre en Sicile, s'élève avec une tige forte & quarrée prefque à trois pieds de hauteur; les nœuds du bas font à quatre ou cinq pouces de diftance; fes feuilles font longues, étroites, en forme de cœur, profondément découpées ou fciées fur leurs bords, & feffiles : fes fleurs croiffent en panicules dans la longueur des tiges; elles font d'un pourpre pâle, & paroiffent à-peu-près dans le même tems que celles des autres efpeces.

La cinquieme fe trouve fur les Alpes ; fes tiges n'ont guere plus d'un pied & demi de hauteur, & pouffent très-peu de branches; fes fleurs font verticillées en forme d'épis, placées à une certaine diftance les unes des autres, & feffiles; fes feuilles font courtes, ovales, en forme de cœur, & fupportées par des pétio-

les : la plante entière est blanche, & d'une odeur forte.

Violacea. La sixieme, qu'on rencontre en Espagne, a des tiges d'environ deux pieds de hauteur, qui poussent quelques branches minces sur les côtés ; ses feuilles sont en forme de cœur, & dentelées sur leurs bords : ses fleurs naissent en têtes rondes & verticillées ; elles sont de couleur bleue, & portées sur des pédoncules. Il y a une variété de cette espece à fleurs blanches.

Tuberosa. La septieme, qui est originaire du Portugal, a une racine épaisse & noueuse, de laquelle sortent deux tiges, souvent inclinées vers la terre d'environ deux pieds & demi de longueur, & qui poussent deux branches latérales opposées : ses feuilles sont oblongues, crenelées sur leurs bords, sessiles aux tiges, & d'un vert foncé ; le sommet de cette tige, dans plus de la longueur d'un pied, est garni de têtes de fleurs verticillées, & éloignées de deux pouces les unes des autres vers le bas, mais plus rapprochées vers le haut, presque sessiles aux tiges, & protégées par des bractées ovales, petites & colorées : ses fleurs sont de couleur bleue, & de la même forme que celles des autres especes. Il y a une variété de celle-ci, dont les tiges sont érigées, & qui ne differe des autres qu'en cela.

Hirsuta. La huitieme naît spontanément en Sicile ; ses tiges s'élevent à près de deux pieds de hauteur, & poussent vers le bas des branches garnies de feuilles en forme de cœur, obtuses, un peu dentelées, & supportées sur de longs pétioles : les tiges sont terminées par de longs épis de fleurs verticillées, séparées, sessiles, & enveloppées d'un duvet blanc. Ces fleurs sont blanches, & paroissent en Juillet.

Virginica. La neuvieme se trouve dans l'Amérique Septentrionale ; elle a une racine vivace, de laquelle sortent plusieurs tiges quarrées de deux pieds de hauteur, & garnies de feuilles velues, qui ressem- un peu à celles de la *Marjolaine*, mais plus larges : ses fleurs sont verticillées, dans la longueur & à l'extrémité de la tige, en grosses têtes rondes ; elles sont d'une couleur de chair pâle, & leurs étamines sont plus longues que la corolle. Cette plante fleurit en Juillet.

Orientalis. La dixieme espece a été découverte dans le Levant par M. de TOURNEFORT, qui a envoyé ses semences à Paris : ses tiges sont fortes, & s'élevent à près de trois pieds de hauteur ; ses feuilles sont en forme de cœur, hachées en dentelures, émoussées sur leurs bords, & portées sur de courts pétioles : ses fleurs croissent en épis verticillés au sommet des tiges ; les têtes verticillées sont fort grosses, rapprochées, & terminées en pointe obtuse. Ces fleurs sont d'une couleur de chair pâle ; la plante entiere est blanche, & répand une odeur forte.

NEP

Procumbens. La onzieme croît naturellement dans les rochers de l'Isle de Candie, où les habitans l'emploient aux mêmes usages que la *Germandrée aquatique*; ses tiges. quarrées, & d'un pied de longueur, traînent sur la terre, & poussent quelques branches minces sur les côtés; ses feuilles ressemblent fort à celles du *Mentastrum* à feuilles rondes, & sont sessiles à la tige : ses fleurs croissent en têtes verticillées; elles sont grosses, rondes, sessiles à la tige, & entourées de bractées ovales & velues; les corolles sont blanches, & paroissent à peine hors de leurs calices. Les racines de cette espece subsistent rarement au-delà de deux années; mais comme les semences mûrissent bien, en leur donnant le tems de se répandre, les plantes se renouvelleront chaque printems.

Culture. Toutes ces especes sont fort dures, & ne craignent point les gelées : on les multiplie aisément par leurs graines; car celles qui tombent naturellement, produisent des plantes sans aucun soin ; & en les semant au printems ou en automne, elles réussissent également, sans exiger aucune autre culture, que d'être éclaircies & tenues nettes de mauvaises herbes. Si on les seme sur un sol sec & de mauvaise qualité, elles ne deviendront pas fortes ; mais elles subsisteront plus longtems, & seront plus belles que si elles étoient placées dans une terre riche, où elles sont plus suc-

NER

culentes, & ont une odeur moins forte.

NERFS (les) sont des veines ou cordons longs, qui coulent au travers ou dans la longueur des feuilles.

NERIUM. *Lin. Gen. Plant.* 262. *Nerion. Tourn. Inst. R. H.* 604. *tab.* 374. [*The Oleander, or Rose Baye.*] Laurier-Rose.

Caracteres. Le calice de la fleur est persistant, & divisé en cinq segmens aigus ; la corolle est monopétale, & en forme d'entonnoir ; son tube est cylindrique, son bord ou limbe est large & découpé en cinq segmens larges, obtus & obliques : la fleur a un nectaire qui termine le tube, avec des antheres à pointes étroites réunies ensemble & terminées par un long filet ; son germe est oblong, divisé en deux parties ; & son style, qui est à peine visible, est couronné par un stigmat simple. Ce germe se change dans la suite en deux légumes longs, coniques, terminés en pointe aiguë, & remplis de semences oblongues, posées l'une sur l'autre en écailles de poisson, & couronnées de duvet.

Ce genre de plantes est rangé dans la premiere section de la cinquieme classe de LINNÉE, qui comprend celles dont les fleurs ont cinq étamines & un style.

Les especes sont :

1°. *Nerium Oleander, foliis lineari-lanceolatis, ternis. Hort. Cliff.* 76. *Hort Ups.* 53. *Flor. Zeyl.* 108 *Roy. Lugd.-B.* 412 ; Laurier-Rose à feuilles linéaires, en forme de lance, &

P 3

NER

placé par trois autour des tiges.

Nerion floribus rubescentibus. C. B. P. 464; Laurier - Rose à fleurs rouges.

Rhododendrum. Ded. Pempt. 851.

Areli. Rheed. Mal. 9. *p.* 1. *f.* 1. 2.

2°. *Nerium Indicum, foliis linearibus, rigidis;* Laurier-Rose des Indes, à feuilles étroites & rudes.

Nerium Indicum angusti-folium, floribus odoratis simplicibus. H. L. 447; Laurier-Rose des Indes, à feuilles étroites, produisant des fleurs simples, d'une odeur agréable.

3°. *Nerium lati-folium, foliis lanceolatis, longioribus, flaccidis;* Laurier-Rose à plus longues feuilles, en forme de lance, & molles.

Nerium Indicum lati-folium, floribus odoratis plenis. H. L. 449. *f.* 447; Laurier-Rose des Indes, à larges feuilles & à fleurs doubles, d'une odeur agréable, communément nommé Laurier-Rose à fleurs doubles.

Oleander. La premiere espece croît naturellement dans la Grece, & dans plusieurs autres contrées voisines de la mer Méditerrannée, toujours sur les bords des rivieres & des ruisseaux. On en connoît deux variétés, l'une à fleurs blanches, & l'autre à fleurs rouges; du reste, elles ne different en rien, & peuvent être regardées comme étant la même espece, quoique celle à fleurs blanches croisse rarement sans culture ailleurs que dans l'Isle de Candie.

NER

Elle s'éleve, avec plusieurs tiges, à la hauteur de huit ou dix pieds : ses branches sortent par trois autour des tiges principales : leur écorce est unie, & de couleur pourpre dans celle à fleurs rouges ; mais la blanche a une écorce d'un vert clair; les feuilles, pour la plupart, sont disposées par trois autour des tiges, sur de fort courts pétioles : leurs pointes sont dirigées vers le haut; elles ont trois ou quatre pouces de longueur sur neuf lignes de largeur au milieu, & sont d'un vert foncé, fort roides, & terminées en pointe aiguë; les fleurs qui naissent aux extrémités des branches, en gros paquets lâches, sont, dans la premiere variété, de couleur pourpre brillant ou cramoisi; & dans l'autre, d'un blanc sale; elles ont des tubes courts & évasés au sommet, où elles sont découpées en cinq segmens obtus & roulés vers le bas, ce qui les rend obliques au tube : le nectaire est déchiqueté en filets capillaires; il est placé à l'ouverture du tube, en - dedans duquel sont situées cinq étamines, & un germe qui en occupe le fond, & qui se change dans la suite en un légume brun, cylindrique, double, d'environ quatre pouces de longueur, qui s'ouvre longitudinalement sur un côté, & renferme des semences oblongues, placées l'une sur l'autre en forme d'écailles de poisson. Cet arbrisseau fleurit dans les mois de Juillet & Août : dans les années chaudes, ses fleurs

NER

font remplacées par des légumes ; mais fes femences mûriffent rarement en Angleterre.

Quand les étés font chauds & fecs, ces plantes font un agréable effet ; car alors elles fleuriffent fortement : mais dans les années froides & humides, fes fleurs périffent fouvent fans s'ouvrir. La variété à fleurs blanches eft plus tendre que celle à fleurs rouges ; & fi le tems n'eft pas favorable, quand fes fleurs paroiffent, elles fe pourriffent, & n'ont point d'apparence, à moins qu'on ne la tienne fous des vitrages (1).

Indicum. La feconde efpece, qui croît fans culture dans les Indes, s'éleve à la hauteur de fix ou huit pieds, avec des tiges d'arbriffeau, couvertes d'une écorce brune, & garnies de feuilles de trois ou quatre pouces de longueur, fur trois lignes de largeur au plus, d'un vert clair, avec leurs bords réfléchis, quelquefois oppofées, & quelquefois alternes, & fouvent difpofées par trois autour des branches : fes fleurs naiffent en paquets laches aux extrémités des branches ; elles font d'un rouge pâle, & ont

____

(1) Le fuc de cet arbriffeau, par fa violente caufticité, eft un véritable poifon : auffi ne s'en fert on jamais intérieurement, quoique cette plante ait été recommandée par CESALPIN en infufion dans le vin, contre la morfure des ferpens. Les feuilles de cet arbriffeau, deffechées & reduites en poudre, forment un puiffant fternutatoire, qui peut être utile dans quelques circonftances.

NER 227

une odeur de *Mufc* agréable. Cette efpece fleurit en même tems que la précédente ; mais fes fleurs s'ouvrent rarement ici, à moins qu'elles ne foient placées dans une caiffe de vitrage aérée, où elles foient à l'abri du froid & de l'humidité.

Lati-folium. La troifieme efpece croît naturellement dans les deux Indes : elle a d'abord été apportée dans les Ifles Britanniques de l'Amérique, du continent Efpagnol. Les Habitans de ces Ifles lui donnent le nom de *Rofe de la mer du fud.* Sa beauté, & la bonne odeur de fes feuilles, engagerent les habitans à la cultiver. Ils en avoient formé des haies dans plufieurs endroits ; mais les beftiaux les ayant broutées pendant une difette de fourrage, elles ont été prefque toutes détruites ; de forte qu'à préfent on n'en voit plus que dans quelques jardins, où elles produifent le plus bel effet pendant une trèsgrande partie de l'année ; car, dans ces pays chauds, elles font rarement fans fleurs. Cette efpece a été regardée par quelques perfonnes qui n'avoient qu'une connoiffance fuperficielle des plantes, comme n'étant qu'une variété de la commune ; mais ceux qui les ont cultivées l'une & l'autre, ont dû remarquer que la première fubfifte, pendant tout l'hiver, en plein air, dans une expofition chaude, & que celle-ci ne peut être confervée en Angleterre fans le fecours d'une ferre chaude, & qu'elle ne

P 4

fleurit pas, si elle n'est tenue dans une caisse de vitrage en été. La troisieme espece n'a été connue ici qu'au commencement du dernier siecle ; mais la premiere se multiplie dans les jardins anglois depuis près de deux-cents ans. Les semences de la premiere espece n'ont jamais produit de plantes semblables à celles de la troisieme, malgré l'assurance positive de quelques personnes dépourvues de connoissances.

Les feuilles de cette espece ont six pouces de longueur sur un pouce de largeur au milieu ; elles sont d'une texture beaucoup plus mince que celles de la premiere, & leurs extrémités sont généralement resléchies ; elles sont d'un vert clair, & placées irregulierement sur les branches, quelquefois par paires, d'autres fois alternes, & souvent par trois autour des branches : ses fleurs naissent en très-gros paquets aux extrémités des branches, sur de longs pédoncules ; elles ont trois ou quatre rangs de pétales placés en dedans l'un de l'autre, & sont plus ou moins doubles, & beaucoup plus larges que celles de l'espece commune : leur odeur est celle de l'*Aubépine* ; elles sont d'un rouge léger ou couleur de pêche, & la plupart joliment panachées, d'un rouge plus foncé, ce qui les rend très-agréables. Elles paroissent ordinairement dans les mois de Juillet & Août ; & si on les tient dans une serre chaude, elles conservent leur beauté jusqu'à la Saint-Michel.

Comme les fleurs de cette espece sont doubles, elles ne produisent point de semences : mais à présent nous connoissons celle à fleurs simples ; car la seconde est certainement distincte.

Culture. On croit que toutes les especes de *Laurier-Rose* ont une qualité vénéneuse ; & cette opinion paroît fondée ; car lorsqu'on rompt leurs jeunes branches, elles répandent un suc laiteux, & les grosses étant brûlées, ont une odeur fort délagréable : mais ce genre de plantes a été confondu avec le *Chamærhododendros* de TOURNEFORT, par plusieurs Auteurs, qui ont appliqué au *Nerium* les qualités pernicieuses de l'autre. C'est ainsi que le Miel de Trébisonde, qu'on regarde comme fort mal-sain, etoit cru recueilli par les abeilles sur les fleurs du *Nerium* ; tandis qu'elles le prennent sur celles du *Chamærhododendros*, ainsi que TOURNEFORT l'a très-bien observé : mais la ressemblance de leurs noms dans la langue grecque, est cause que ces deux plantes ont souvent été confondues.

On multiplie toujours ici ces plantes par marcottes ; car quoique les boutures prennent quelquefois racine, cependant cette méthode n'est point sûre, & l'on s'en tient à la premiere. Comme elles sont fort sujettes à produire des rejettons de leurs racines, on préfere ces racines, pour en faire de marcottes ; car les vieilles branches ne poussent point de racines. Quand on veut les

coucher, on fait une fente à un de leurs nœuds, comme on le pratique pour les Œillets, ce qui les aide à prendre racine. Si l'on fait cette opération en automne, & si on les arrose à propos, elles auront produit des racines au bout d'un an : alors on les enleve avec une truelle ; & si elles sont bien enracinées, on les détache de la vieille plante, & on les met chacune séparément dans de petits pots remplis d'une terre molle & marneuse. Celles de l'espece commune n'exigent aucun autre soin que d'être placées à l'ombre & légérement arrosées, suivant la saison, jusqu'à ce qu'elles aient formé de nouvelles fibres ; mais les deux autres doivent être plongées dans une couche de chaleur très-modérée, pour hâter leurs progrès, & leur faire prendre racine, en observant de les tenir à l'ombre pendant la chaleur du jour. Quand les plantes de l'espece commune ont acquis des racines, on peut les placer dans une situation abritée avec d'autres plantes exotiques dures ; mais à la fin d'Octobre, on les met ou dans la terre ou sous un châssis de couche chaude, de maniere qu'elles soient à l'abri des gelées de l'hiver, & qu'elles puissent jouïr de l'air dans tous les tems doux.

Cette espece est si dure, qu'elle pourroit subsister en plein air dans les hivers doux, étant placée à une exposition chaude ; mais comme elle est sujette à être détruite par les fortes gelées, la meilleure méthode est de tenir les plantes dans des pots ou des caisses, quand elles sont grandes, afin de pouvoir les abriter en hiver, & les exposer au plein air en été, dans une situation chaude & abritée. Pendant l'hiver, elles peuvent être mises avec les Myrtes & autres plantes exotiques plus dures, afin qu'elles aient autant d'air qu'il est possible dans les tems doux, & qu'elles soient seulement à couvert des fortes gelées : car si elles étoient tenues trop chaudement en hiver, elles ne fleuriroient pas bien ; & quand elles n'ont point assez d'air, les extrémités de leurs branches se moisissent : ainsi, plus elles sont traitées durement, sans être exposées aux fortes gelées, & mieux elles reussissent.

Les deux autres especes exigent un traitement différent, sans lequel elles n'ont aucune apparence. Quand les jeunes plantes ont formé de nouvelles racines, on les accoutume par dégrés à supporter le plein air, auquel on les expose entierement en Juillet, pour les y laisser jusqu'au mois d'Octobre, pourvu que le tems continue à être doux ; mais elles doivent toujours être tenues dans une situation abritée ; & dès que les premiers froids approchent, on les met sous un abri : car lorsque leurs feuilles sont endommagées par la gelée, elles deviennent d'un jaune pâle, & ne recouvrent pas leur couleur ordinaire avant l'automne suivant. Ces

plantes étant confervées dans une bonne ferre en hiver, deviendront plus fortes que celles qui font traitées plus délicatement : mais au mois de Mai, quand les boutons commencent à paroitre, il faut les placer dans une caiffe de vitrage ouverte, où elles puiffent être à couvert des injures du tems, & avoir beaucoup d'air dans les tems chauds. Avec ce traitement, les fleurs s'ouvriront, & conferveront long-tems leur beauté : lorfqu'elles font entierement épanouïes, il y a peu d'autres fleurs qui puiffent leur être comparées, foit pour le coup-d'œil, foit pour l'odeur, qui approche de celle de l'*Epine blanche*. Ces bouquets de fleurs font très gros, quand les plantes font fortes.

NERPRUN *ou* NOIRPRUN. BOURG-EPINE. *Voyez* RHAMNUS CATHARTICUS. L. HIPPOPHAE. *L.*

NERPRUN DE MALABAR. *Voy.* LAWSONIA SPINOSA.

NEZ COUPÉ *ou* FAUX PISTACIER. *Voyez* STAPHILEA PINNATA. *L.*

NICOTIANA. *Tourn. Inft. R. H.* 117. *tab.* 41. *Lin. Gen. Plant.* 220.

Cette plante porte le nom de JEAN NICOT, Confeiller de FRANÇOIS II, Roi de France, qui, en l'année 1560, étant Ambaffadeur à la Cour de Portugal, acheta ces femences d'un Hollandois venant de l'Amérique, & les envoya à la Reine de France, CATHERINE DE MÉDICIS, qui les fit femer. Ces graines produifirent des plantes & d'autres femences.

Les Indiens l'appellent *Tabac*, parce qu'elle croît dans une des Ifles Caribbes, nommée *Tabaco* ou *Tabago*.

La plus petite efpece eft connue par quelques-uns fous le nom de *Hyofciamus*, parce qu'une partie de fes caracteres s'accorde avec ceux de cette plante. On lui donne auffi la dénomination de *Priapeia*. [*Tobacco.*] Nicotiane ou Tabac.

Caracteres. Le calice de la fleur eft perfiftant, & formé par une feuille découpée en cinq fegmens aigus : la corolle eft monopétale, & en forme d'entonnoir ; elle a un tube long, évafé fur fes bords, & terminé en cinq pointes aiguës : la fleur a cinq étamines en forme d'alêne, auffi longues que le tube, un peu penchées, & terminées par des antheres oblongues ; fon germe, qui eft ovale, foutient un ftyle mince, couronné par un ftigmat découpé. Ce germe fe change enfuite en une capfule ovale, fillonnée par une rainure à chaque côté, & à deux cellules, qui s'ouvrent au fommet, & font remplies de femences rudes, fixées à la cloifon.

Ce genre de plantes eft rangé dans la premiere fection de la cinquieme claffe de LINNÉE, qui renferme celles dont les fleurs ont cinq étamines & un ftyle.

Les efpeces font :

1°. *Nicotiana latiffima, foliis ovato-lanceolatis, rugofis, femiamplexicaulibus ;* Tabac avec des feuilles ovales, rudes, & en for-

NIC

me de lance, qui embraffent les tiges à moitié.

Hyofcyamus Peruvianus. Ger. 357; Tabac ou Jufquiame du Pérou.

2°. *Nicotiana Tabacum, foliis lanceolato-ovatis, feffilibus, decurrentibus, floribus acutis. Lin.* Sp. *Plant.* 258. *Mat. Med.* 64. *Blacw. t.* 146. *Kniph. cent.* 4. *t.* 55. *Ludw. Eft. t.* 167. *Knorr. Del.* 1. *t. T.* 11. *Sabb. Hort.* 1. *t.* 89; Tabac à feuilles ovales, en forme de lance, coulantes fur les tiges, & feffiles.

Nicotiana foliis lanceolatis. Hort. *Cliff.* 56. Hort. *Ups.* 45. *Roy. Lugd.-B.* 423.

Nicotiana major lati-folia. C. B. P. 169.; Le plus grand Tabac à larges feuilles.

Blennochoes. Reneal. Spec. 37. *t.* 38.

3°. *Nicotiana angufti-folia, foliis lanceolatis, acutis, feffilibus, calicibus acutis, tubo floris longiffimo. Plat.* 185; Tabac avec des feuilles en forme de lance, aiguës & feffiles, des calices à pointe aiguë, & un fort long tube aux fleurs.

Nicotiana major angufti-folia. C. B. P. 170; le plus grand Tabacà feuilles étroites.

4°. *Nicotiana fruticofa, foliis lineari-lanceolatis, acuminatis, femi-amplexicaulibus, caule fruticofo;* Tabac avec des feuilles linéaires, en forme de lance, & à pointe aiguë, qui embraffent les tiges à moitié, & à tige d'arbriffeau.

Nicotiana major anguftiffimo folio, perennis. Iufs.; le plus grand Tabac à feuilles très-étroites & vivaces.

5°. *Nicotiana alba, foliis ova-*tis, *acuminatis, femi-amplexicaulibus, capfulis ovatis, obtufis;* Tabac à feuilles ovales, dont les pointes font aiguës, & qui embraffent les tiges à moitié avec des capfules ovales & obtufes.

Nicotiana major lati-folia, floribus albis, vafculo brevi. Martyn. Dec. 5; le plus grand Tabac à fleurs blanches, avec de larges feuilles, & une capfule courte.

6°. *Nicotiana ruftica, foliis petiolatis, ovatis, integerrimis, floribus obtufis. Lin. Sp.* 258. *Blackw.* 437. *t. Kniph. cent.* 3. *n.* 65. *Sabb. Hort.* 1. *t.* 90; Tabac à feuilles pétiolées, ovales & entieres, & à fleurs obtufes.

Nicotiana minor. C. B. P. 170; le plus petit Tabac, communément appelé Tabac anglois. Nicotiane ou Herbe à la Reine.

Pachyphylla. Reneal. Spec. 40.

7°. *Nicotiana rugofa, foliis ovatis, rugofis, petiolatis;* Tabac à feuilles ovales, ridées & pétiolées.

Nicotiana minor, foliis rugofioribus, amplioribus. Vaill.; le plus petit Tabac à feuilles plus larges & plus ridées.

8°. *Nicotiana paniculata, foliis petiolatis, cordatis, integerrimis, floribus paniculatis, obtufis, clavatis, Lin. Sp. Plant.* 259. *Kniph. cent.* 2. 48; Tabac avec des feuilles en forme de cœur, entieres, & portées fur des pétioles, avec des fleurs en panicules, obtufes, & des tubes en forme de maffue.

Nicotiana minor, folio cordiformi, tubofloris prælongo. Feuill. obf. 1. *p.* 717. *tab.* 10; le plus

petit Tabac, avec une feuille en forme de cœur, & un fort long tube à la fleur.

9°. *Nicotiana glutinofa, foliis petiolatis, cordatis, integerrimis, racemofis, floribus fecundis ringentibus, calycibus inæqualibus. Lin. Sp. Plant.* 259 ; Tabac à feuilles en forme de cœur, entieres & pétiolées, avec des pédoncules branchus, des corolles labiées, & des calices inégaux.

10°. *Nicotiana humilis, foliis ovato-lanceolatis, obtufis, rugofis, calycibus breviffimis. Plat.* 185 ; Tabac avec des feuilles ovales, rudes, obtufes, & en forme de lance, & des calices très-courts.

Nicotiana humilis, Primulæ Veris folio. Houft. MSS.; Tabac nain, à feuilles de Primevere.

Nicotiana pufilla. Lin. Syft. Plant. t. 1. p. 504. *Sp.* 7.

Latiffima. La Premiere efpece eft celle qu'on femoit autrefois le plus communément en Angleterre, & qui a toujours été prife pour le Tabac commun à larges feuilles de Gaspar Bauhin & autres ; mais elle en eft très-différente : fes feuilles ont plus d'un pied & demi de longueur, fur un pied de largeur ; leurs furfaces font fort rudes & glutineufes : quand elle croît fur un fol riche & humide, elle s'élève à plus de dix pieds de hauteur ; la bâfe des feuilles embraffe la tige à moitié ; le haut de cette tige fe divife en plufieurs branches, qui font terminées par des paquets lâches de fleurs érigées avec de longs tubes,

& d'un pourpre pâle. Cette plante fleurit dans les mois de Juillet & Août, & fes femences mûriffent en automne : c'eft l'efpece de Tabac que l'on porte ordinairement au marché dans des pots, pour orner les boutiques & les balcons de Londres. Quelques perfonnes lui donnent le nom de *Tabac d'Oroenoko.*

Tabacum. La feconde efpece eft le *Tabac* à larges feuilles de Gaspard Bauhin ; les tiges de celui-ci s'élevent rarement à plus de cinq ou fix pieds, & fe divifent en un plus grand nombre de branches que la premiere efpece : fes feuilles ont environ dix pouces de longueur fur trois & demi de largeur ; elles font unies, terminées en pointe aiguë, & feffiles aux tiges : fes fleurs font plus larges, & d'un pourpre plus brillant que celles de là precédente ; elles paroiffent & perfectionnent leurs femences dans le même tems. Cette plante eft connue par quelques-uns fous le nom de *Tabac de bonne odeur.*

Angufti-folia. La troifieme efpece s'éleve avec une tige droite & branchue, à quatre ou cinq pieds de hauteur : fes feuilles du bas ont un pied de longueur fur trois ou quatre pouces de large ; celles des tiges font beaucoup plus étroites, & diminuent à mefure qu'elles font plus voifines du fommet ; leur pointe eft fort aiguë ; elles font feffiles aux tiges, & fort glutineufes : fes fleurs naiffent en paquets lâches aux extrémités des tiges ; elles ont

NIC

de longs tubes , & font de couleur pourpre , ou d'un rouge brillant ; elles paroissent dans le même tems que celles des especes precedentes , & leurs semences mûrissent en automne. (1).

(1) Ce n'est point ici le lieu de traiter des bons ou mauvais effets qui résultent de l'usage habituel que nous faisons du Tabac. Beaucoup d'Auteurs en ont parlé, & l'on trouve dans quelques - uns de savantes dissertations a ce sujet, qui sont aussi inutiles que toutes celles qu'on pourroit faire contre le luxe & la bonne chere.

Le Tabac est narcotique, âcre & irritant ; l'habitude que nous avons d'en faire usage , ne nous a pas tellement habitués à son action , que nous ne puissions éprouver encore ses puissans effets, en le prenant sous une forme différente de celle à laquelle nous sommes accoutumés. C'est ainsi que ceux qui n'usent que de Tabac en poudre, éprouvent une violente ivresse, des vomissemens, & même des convulsions , lorsqu'ils viennent à le fumer pour la premiere fois. L'usage intérieur du Tabac est toujours dangereux . & quoiqu'il soit un des plus puissans purgatifs & émétiques que nous connoissions, ce n'est qu'avec beaucoup de prudence & dans des cas extraordinaires qu'on doit se déterminer à l'employer ainsi. On peut l'administrer avec un peu plus de sûreté en lavemens , en le faisant infuser à la dose d'une once. Il produit quelquefois de cette maniere d'excellens effets, sur - tout dans les maladies comateuses, l'apoplexie, &c. L'eau simple qu'on retire des feuilles de cette plante, étant en quelque sorte dépouillée du principe vireux & narcotique , peut être donnée avec que'que succès dans les affections catharra-

Fruticosa. La quatrieme s'éleve avec des tiges fort branchues , à la hauteur d'environ cinq pieds ; les feuilles du bas de la tige ont un pied de longueur sur un & demi de large à la base , où elles l'embrassent à moitié ; elles ont environ trois pouces de largeur au milieu, & sont terminées en pointes longues & aiguës. Les tiges se divisent en plusieurs petites branches, terminées par des paquets lâches de fleurs, teintes d'un pourpre brillant, & auxquelles succedent des capsules a pointe aiguë. Cette plante fleurit à-peu-près dans le même tems que la précédente; mais si on la place dans une serre chaude, elle subsiste pendant l'hiver. Les semences de cette espece m'ont été envoyées pour celle du *Tabac* du Brésil.

Alba. La cinquieme croît naturellement dans les bois de l'Isle de Tabago, d'où ses semences m'ont été envoyées par le feu Docteur ROBERT MILLAR ; elle s'éleve à la hauteur d'environ cinq pieds ; sa tige ne pousse pas autant de branches que la tige de la précédente; ses feuilles sont ovales , & de quinze pouces environ de longueur sur deux de largeur au milieu ; mais elles deviennent plus étroites à mesure qu'elles approchent du sommet, & elles embrassent les tiges de leur base à moitié : ses

les & les engorgemens du poumon. Les feuilles fraiches du Tabac sont vulnéraires & détersives : on s'en sert pour mondifier les ulceres sordides.

fleurs, qui croissent en paquets plus serrés que celles de la précédente, sont blanches, & produisent des capsules courtes, ovales & obtuses. Cette plante fleurit & perfectionne ses semences vers le même tems que la quatrieme.

Rustica. La sixieme est communément appellée *Tabac Anglois*, parce qu'étant la plus dure de toutes, elle est la premiere qui ait été introduite sous ce nom en Angleterre. Ses semences mûrissent très-aisément; & quand on leur permet de se répandre, elles produisent des plantes qui poussent sans aucun soin par-tout où elles se trouvent; de sorte qu'elle est devenue une herbe sauvage dans plusieurs endroits; mais elle a été apportée originairement de l'Amérique sous le nom de *Petum.* DODONÆUS, TABERNEMONTANUS, & autres, l'ont appellée *Hyoscyamus luteus*, à cause de l'affinité qu'elle paroît avoir avec la *Jusquiame*; mais ses fleurs sont tubulées, & non labiées comme celles de cette derniere plante, & ses capsules ne s'ouvrent pas en couvercle au sommet comme celles de la *Jusquiame* : ses tiges s'élevent rarement à plus de trois pieds de hauteur; ses feuilles sont ovales, unies, alternes sur les tiges, & portées sur de courts pétioles : ses fleurs croissent en petits paquets desserrés sur le sommet des tiges; elles sont de couleur herbacée, & ont des tubes courts évasés, & découpés en cinq segmens obtus; elles paroissent dans le mois de Juil-

let, & sont remplacées par des capsules rondes, remplies de petites semences qui mûrissent en automne.

Rugosa. La septieme espece s'éleve avec une tige forte, à la hauteur d'environ quatre pieds; ses feuilles ont la même forme que celles de la précédente; mais elles sont plus fortement sillonnées sur leurs surfaces, deux fois plus larges, d'un vert plus foncé, & portées sur de plus longs pétioles : ses fleurs sont aussi plus larges que celles de la sixieme, & de la même forme. Cette plante est certainement différente de la précédente; car les ayant sémées. l'une & l'autre pendant plus de trente ans, je ne les ai jamais vu varier.

Paniculata. La huitieme a été trouvée dans la vallée de Lima, par le P. FEUILLÉE, en l'année 1710; & depuis peu, ses semences ont été envoyées du Pérou à Paris par DE JUSSIEU le jeune. La tige de cette plante s'éleve au-dessus de trois pieds de hauteur, & se divise vers son sommet en plusieurs branches paniculées, rondes, & un peu velues; ses feuilles sont en forme de cœur, de quatre pouces environ de longueur, sur trois de largeur, & portées sur de longs pétioles : ses fleurs, qui sortent en panicules lâches aux extrémités des branches, ont des tubes d'un pouce environ de longueur, en forme de massue, & dont les bords sont légerement découpés en neuf segmens obtus & réfléchis; elles

font d'un vert jaunâtre , & produifent des capfules rondes, & remplies de fort petites femences. Cette plante fleurit à-peu-près dans le même tems que les autres efpeces.

Glutinofa. Les femences de la neuvieme ont été envoyées du Pérou avec celles de la précédente , par DE JUSSIEU le jeune : fa tige , qui eft ronde, & haute de près de quatre pieds , pouffe deux ou trois branches vers le bas ; fes feuilles font larges , en forme de cœur , un peu ondées , gluantes & portées fur de longs pétioles : fes fleurs croiffent en épis defferrés au fommet de la tige ; elles ont des tubes courts, ouverts, & courbés prefque comme les fleurs labiées ; elles font d'un pourpre pâle , & leur calice eft inégalement découpé, un des fegmens étant deux fois plus large que les autres.

Humilis. La dixieme efpece a été découverte à la Vera-Cruz par le feu Docteur HOUSTOUN, qui en a envoyé les femences en Angleterre : elle a une racine épaiffe & conique, qui pénetre profondément dans la terre ; du haut de cette racine, fortent, fix ou fept feuilles ovales, en forme de lance , étendues fur la terre, à-peu-près auffi larges que celles de la *Primevere* commune, mais d'un vert plus foncé ; fa tige s'éleve à un pied environ de hauteur , & pouffe des branches qui forment trois ou quatre divifions, à chacune defquelles eft placée une petite feuille. Ces branches font ter-

minées par un épi clair de fleurs qui font petites, tubulées, d'un vert jaunâtre , & ont des calices fort courts , & découpés fur leurs bords en cinq fegmens aigus ; fes capfules font petites , ovales , & divifées en deux cellules remplies de petites femences.

Culture. Toutes ces efpeces , à l'exception des fixieme , feptieme & huitieme, exigent la même culture. Comme elles font trop délicates pour pouvoir être femées en pleine terre, il faut les élever fur une couche chaude, comme on le dira ci-après.

On repand leurs graines dans le mois de Mars fur une couche de chaleur modérée. Quand les plantes ont pouffé , & qu'elles font devenues affez fortes pour être enlevées, on les tranfplante dans une couche tempérée, à quatre pouces de diftance entr'elles , en obfervant de les arrofer & de les tenir à l'ombre, jufqu'à ce qu'elles aient formé de nouvelles racines ; après quoi on leur donne de l'air à proportion de la chaleur de la faifon , fans quoi elles fileroient , deviendroient très-foibles , & feroient moins en état de fupporter l'air ouvert. Quoiqu'il foit néceffaire de les arrofer fouvent , il ne faut néanmoins leur donner que très peu d'eau à la fois, tandis qu'elles font jeunes ; mais quand elles ont acquis une certaine force, on doit les arrofer fouvent & en abondance. Ces plantes doivent refter dans cette couche jufqu'au milieu du mois de

Mai : alors, fi elles ont bien réuffi, elles fe toucheront, & il fera néceffaire de les accoutumer à fupporter le grand air par dégrés ; après quoi on les enlevera foigneufement, en confervant une groffe motte de terre à chaque racine, & on les plantera dans un fol riche & léger en rangs éloignés de quatre pieds, & à trois pieds entr'elles dans les rangs : on les arrofe avec foin, jufqu'à ce qu'elles foient bien enracinées ; mais elles n'exigeront plus aucun autre foin que d'être tenues nettes de mauvaifes herbes, jufqu'à ce qu'elles commencent à montrer leurs tiges de fleurs, dont il faut alors couper les fommets, afin que les feuilles foient mieux nourries. Par cette méthode, elles deviendront plus larges, & d'une fubftance plus épaiffe. Au mois d'Août, qui eft le tems où elles auront acquis toute leur longueur, on les recueillera, pour les employer à l'ufage auquel elles fo t deftinées ; car fi on les laiffoit plus long-tems, celles du bas commenceroient à dépérir. Tout ceci doit s'entendre des plantes qu'on cultive pour l'ufage ; car celles qu'on ne deftine que pour l'ornement, doivent être placées dans les plates-bandes du parterre, où on les laiffera parvenir à leur hauteur entiere. Elles continueront à fleurir depuis le mois de Juillet, jufqu'à ce que les gelées les détruifent.

Les trois petites efpeces de Tabac font ordinairement confervées dans les jardins de Botanique, pour la variété ; mais elles font rarement cultivées pour l'ufage. On trouve la premiere fur des tas de fumier dans différentes parties de l'Angleterre. Les fixieme & feptieme font dures, & peuvent être élevées : en les femant en Mars fur une terre légere, elles y poufferont, & pourront être tranfplantées enfuite dans telle partie du jardin que ce foit, où elles profiteront fans aucun foin.

La derniere efpece, étant un peu plus tendre que les autres, doit être femée de bonne heure au printems fur une couche chaude. Quand les plantes pouffent, on les tranfporte fur une autre couche de chaleur modérée, où on les arrofe exactement, en leur donnant beaucoup d'air dans les tems chauds ; lorfqu'elles ont acquis affez de force, on les tranfplante féparément dans des pots, que l'on plonge dans une couche tempérée, pour les faire avancer. Vers le milieu de Juin, on peut tirer quelques plantes hors des pots, & les placer dans une terre riche ; mais il fera prudent d'en tenir une ou deux dans des pots, qu'on placera dans la terre chaude, pour en conferver l'efpece, en cas que le mauvais tems empêche les autres de perfectionner leurs femences [a].

---

[a] La culture du Tabac étant deven un objet très-important dans les Pays Bas, dans l'Allemagne, la Hongrie &c, l'on a cru

NICOTIANE.

NIE

NICOTIANE *ou* TABAC.
Voyez NICOTIANA.
NIELLE *ou* TOUTES EPICES.
Voyez NIGELLA ARVENSIS.

rendre un service au public en communiquant la note suivante qui vient de bonne main, & n'est que le pur résultat de l'expérience.

NOTE.

Sur la Culture & la Préparation du Tabac, en Virginie & en Maryland.

Pour les semailles du Tabac, de même que pour celles de toute autre plante, il ne faut se servir que d'une semence choisie.

En Virginie & Maryland, on seme les graines de Tabac sur des couches ; d'où on les transplante a la premiere pluie dans un terrain préparé en mottes comme une Houblonniere ; & c'est ce qui arrive ordinairement vers la fin d'Avril. Un terrain vierge & un peu humide, est celui où cette plante croit avec le plus de force. Elle amaigrit extrêmement le sol.

Un mois après qu'elles ont été transplantées, elles s'élevent a la hauteur d'un pied ou plus. Si elles poussent en haut trop vite on les étête, afin de mieux fournir leurs feuilles de sucs : pour la même raison on les dépouille des feuilles inférieures qui sont trop près de la terre, en ne laissant sur la tige que depuis huit jusqu'a douze feuilles ; on ôte avec beaucoup d'attention la vermine de celles-ci, & on a le même soin de sarcler tout le terrain planté & d'arracher tous les jets de Tabac qui poussent par le pied ou par la tige de la plante. Cette opération se fait regulierement une ou deux fois par semaine.

En moins de trois mois après la transplantation, les plantes ont ac-

Tome V.

NIE   237

NIELLE DES BLÉS *ou* COMPAGNON DES BLÉS. *V.* AGROSTEMA GITHAGO.
NIELLE. [*Mildew*] *Angl.*

quis toute leur croissance, & elles font alors de 4 ou 5 pieds de hauteur, & souvent davantage.

La tige est droite, velue, gluante & de la grosseur du poignet : les feuilles font alternes, d'un verd pâle jaunâtre, & fort grandes vers le pied de la plante. Quelque tems avant de les cueillir, on les étête de nouveau, pour faire refluer les sucs dans les feuilles, qui deviennent alors d'un verd foncé. On connoit a ces signes que le Tabac est mur.

On coupe les plantes à quelques doigts de la terre a mesure qu'elles mûrissent, & on les renversé sur leur motte de terre pour le reste de la journée ce qui fait faner les feuilles. Le soir on les amoncele & on les laisse suer une nuit. Si les plantes font fort chargées de sucs, on les expose de nouveau au soleil pendant le jour suivant afin de mieux faire mûrir & épaissir ces sucs, & ensuite on les porte sous des hangars, construits de maniere que l'air y entre de toutes parts, mais non pas la pluie. On les y pend, chaque plante séparément, & on les laisse sécher pendant 4 à 5 semaines. Si la saison est froide on se sert de feu pour les sécher. Le Tabac de Maryland qu'on destine pour la pipe, est presqu'entierement seché par le moyen du feu : il devient jaunâtre, & est le plus cher de tous.

Après l'entier dessechement des plantes, on les retire des hangars par un tems humide ; car autrement elles tomberoient en poussiere. On les étend sur des claies en monceaux, on les couvre, & on les laisse suer une semaine ou deux, selon leur qualité & celle de la saison, les visitant très-frequem-

Q

NIE

C'est une maladie qui survient aux plantes, & qu'on croit être occasionnée par une rosée qui tombe deffus. Cette

ment pour examiner le degré de leur chaleur, & pour ouvrir & retourner les monceaux afin d'empêcher qu'aucune partie ne s'échauffe trop. Cette fermentation pourroit aller jufqu'à l'inflammation ; mais une trop forte effervefcence détruit la qualité du fuc & des fels, & fait pourrir le Tabac. C'eft la partie la plus difficile de la préparation de cette plante ; elle n'admet aucune regle générale, & dépend entièrement de l'expérience & d'une pratique d'habitude. Un negre qui y eft accoutumé, fait diftinguer le degré convenable de chaleur en poufiant fa main dans un monceau, cent fois mieux que ne feroit un phyficien avec fon thermometre.

Cette opération ainfi que la fermentation étant complettement achevées, on depouille les tiges de leurs feuilles, feparant les feuilles du fommet de celles d'en bas, en deux ou trois claffes. Celles-ci étant entierement fechées de nouveau, on les met par couches régulieres dans les Barrils ou Boucauts refpectifs, pofant par-deffus à plufieurs reprifes à mefure qu'on les remplit une forte planche ronde, qu'on comprime chaque fois avec un lévier qui fait l'effet d'un poids de deux, trois ou quatre mille livres de compreffion. On foutient que cette maniere d'Emballage très-compacte, eft un des points les plus effentiels pour la bonne confervation du Tabac. On envoye quelquefois le plus fin Tabac en forme de carottes, & avant que de les mettre dans cette forme, les feuilles font dépouillées de leurs groffes fibres.

On fe fert de machines pour rendre le Tabac qui compofe ces carottes extrêmement compacte.

NIE

rofée y féjournant, faute de foleil pour l'attirer, corrode, détruit, gâte, par fon acrimonie, la fubftance intérieure des

On a foin de faire ces opérations, c'eft-à-dire de remplir les Boucauts & de former les carottes, dans un tems humide, quand le Tabac feché eft plus fouple.

Le Tabac ainfi préparé eft envoyé au marché ; mais avant que de pouvoir être vendu, il doit fubir l'examen des officiers publics, inftitués pour cela, & nommés *Infpecteurs de Tabac*, qui en déterminent la qualité. Tout Tabac non convénablement preparé, ou qui a été mouillé en chemin, & qui par ces caufes ou d'autres a fermenté de nouveau dans les Boucauts ou dans les Carottes, eft condamné au feu & perdu pour le propriétaire. Les Colonies Américaines ont des Loix utiles, ou plutot néceffaires pour regler tous ces objets. C'eft par la ftricte obfervance de ces loix, que leur Tabac s'eft perfectionné & que le commerce qu'on en fait s'eft étendu au point où on le voit.

D'après des notices que j'ai acquifes à Londres en 1777, il refulte que l'année avant la rupture des colonies avec la mere-patrie, le produit du Tabac de deux Provinces, de la Virginie & du Maryland, étoit comme il fuit. Elles envoyoient à la Grande-Bretagne pour la fomme annuelle de 768,000 liv. fterling. Le prix moyen de ce Tabac étoit à huit liv. fterling, par Boucaut de douze à quatorze cent liv. pefant chacun ; ce qui fait 96 mille Boucauts d'exportation. De cette quantité on comptoit qu'environ 13,500 Boucauts fe confommoient dans les Royaumes Britanniques, lefquels payoient vingt-fix liv. fterling un fch. par Boucaut de droits à l'Etat : en tout 351,675 liv. fterling. Les autres 82, 500 Boucauts étoient exportés en

plantes, & empêche la circulation de la féve nutritive ; ce qui fait flétrir les feuilles , & endommage beaucoup les fleurs

& les fruits : D'autres croyent que la *Nielle* eft plutôt une fubftance concrete, qui exfude à travers les pores des feuilles.

Cependant ce que les Jardiniers appellent communément

d'autres pays de l'Europe par les Négotians Anglois. On ajoute que cette feule branche de commerce employoit alors environ 330 vaifleaux & 4000 mariniers.

Les Américains qui ont fourni la plupart des notices précédentes, ajoutent les fuivantes.

1³. Que le bon Tabac complettement préparé & emballé de la maniere qui eft fpécifié ci-deffus , ne fouffre plus ni fueur ni fermentation, à moins de quelque accident extraordinaire.

2°. Que le Tabac mal-préparé, non fuffifamment feché & fermenté , de même que quand il n'eft pas affez comprimé & compacte dans le Boucaut, s'échauffe & fermente de nouveau, & fe pourrit enfuite.

3°. Que le Tabac d'une deuxieme recolte , c'eft-à-dire, les jets & rejettons qui pouffent des tiges après que la premiere plante a été coupée, eft toujours mauvais & hors d'état de fe conferver par aucune préparation ; par conféquent fon exportation chez l'étranger, foit pur, foit mélangé, eft conftamment prohibée par les loix.

4°. Plus un Tabac eft cultivé dans un fol gras & humide , plus il eft abondant en huiles & en fels âcres , & plus auffi il demande un defféchement & une fermentation longs & foignés. Une préparation fuffifante pour un Tabac o dinaire ne l'eft pas pour celui-ci , car il fermente de nouveau & fe corrompt enfuite.

5°. Un bon Tabac bien préparé , mais qui a été mouillé par accident dans le Boucaut , fermente de nouveau & fe pourrit en-

fuite, auffi loin que l'humidité a pénétré.

6°. Par cette nouvelle fermentation , les feuilles fe moififfent , perdent leur odeur & leur goût, deviennent blanches , & fe corrompent au point de n'être plus d'aucun ufage , fi ce n'eft comme engrais.

7°. Dans un fol très-riche & humide , la plante de Tabac monte jufqu'au delà de fix pieds d'élevation, & fes feuilles s'étendent de tous côtés à un diamètre qui n'eft guere moindre que fa hauteur. Une plante de cette efpece contient tant de fucs gras, tant de fels âcres, qu'il eft difficile de la préparer, enforte qu'elle puiffe fe conferver longtems fans nouvelle fermentation.

8°. Le Tabac le plus fin & le plus délicat, eft celui qui croit dans un fol modérément riche & léger, vers l'arriere de la Virginie & du Maryland, près des montagnes d'Allegany ; mais le produit en eft beaucoup moindre que dans les prairies humides & fur les bords des rivières plus près de la mer.

9°. Si le fol eft trop léger & fablonneux, & par conféquent trop chaud & fec, la plante fe brûle & produit fort peu.

10°. Au refte, un très-grand dégré de chaleur eft néceffaire, tant pour la culture que pour la préparation du Tabac : la chaleur des mois de Juin, Juillet & Août en Virginie, eft ordinairement d'environ 100 dégrés de Fahrenheit, ou de 30 dégrés de Reaumur, fouvent plus, rarement moins : cette Province eft comprife entre 36 & 40 dégrés de latitude feptentrionale

Nielle, est un insecte qu'on trouve souvent en grande quantité, & qui se repaît de l'exsudation des plantes.

Il y en a qui disent que la *Nielle* est une vapeur épaisse & gluante, que les plantes, les fleurs, & la terre même, dans un tems tranquille & calme, exhalent au printems & en été, quand il n'y a ni assez de soleil pour l'attirer à une hauteur considérable, ni assez de vent pour la dissiper ; & cette vapeur, restant près de la surface de la terre, se condense dès que le frais du soir commence à se faire sentir, tombe sur les plantes, en bouche les pores par sa substance épaisse & gluante, empêche leur transpiration, & arrête la séve qui doit monter pour nourrir leurs fleurs & leurs fruits.

On dit encore que la *Nielle* est une rosée corrosive & rongeante, qui naît des vapeurs que la terre exhale, & qui, après s'être élevée à une certaine hauteur, retombe ensuite sur les tendres boutons qui commencent à s'ouvrir, les infecte par son acrimonie, & empêche la circulation de la séve nutritive dans les vaisseaux qui lui sont propres ; d'où naît la flétrissure des feuilles, & l'altération que les fleurs & les fruits éprouvent.

Quelques personnes observent, que les lieux les plus sujets à la *Nielle* sont ceux qui sont environnés de clôtures, ou masqués par des montagnes, & sur-tout ceux qui sont exposés au levant. La raison qu'ils apportent pour prouver qu'a l'exposition du levant les terres sont plus sujettes à la *Nielle*, c'est que le soleil attire ces vapeurs à lui de la même manière que le feu dans une chambre attire l'air, & qu'après les avoir mises en mouvement, sans avoir assez d'activité pour les élever sous forme de nuages, jusqu'à la moyenne région de l'air, il les attire cependant jusqu'à ce qu'il soit descendu au-dessous de l'horison, & qu'alors ces rosées tendent vers la terre d'où elles sont sorties, en se portant à l'ouest, & en frappant à angle droit les corps exposés à l'est.

Mais je pense que ce qui fait que les plantes exposées à l'est, sont plus sujettes à la *Nielle*, c'est que dans cette position, elles sont aussi plus exposées aux vents secs qui bouchent leurs pores, & arrêtant leur transpiration, produisent l'épaississement de la séve sur la surface des feuilles. Cette séve, qui est naturellement sucrée, attire les insectes, qui, y trouvant une nourriture convenable, y déposent leurs œufs, & s'y multiplient si vîte, qu'ils couvrent en peu de tems toute la surface des plantes, corrodent les vaisseaux, & empêchent par-là la circulation de la séve. Il est très-probable que les excrémens de ces insectes entrent dans les vaisseaux des plantes, & qu'en se mêlant avec la séve, ils peuvent causer une infection dans toutes leurs parties ; car on voit que toutes les fois qu'un arbre a beaucoup souffert de la Nielle, il ne recouvre pas sa vigueur

avant deux ou trois ans, & quelquefois même il ne fe rétablit jamais entierement.

On croit encore que ce qui eft caufe que les vallons fourniffent plus d'humidité que les montagnes, c'eft que la rofée s'éleve de la terre & des arbres pendant le jour, comme nous l'avons dit. On en donne pour preuve ces brouillards que l'on voit bien plus fouvent dans les vallons que fur les collines. Cette humidité que le foleil attire, refte fufpendue près de la terre, à moins que le vent ne favorife fon élévation : après le coucher du foleil, elle retombe fur les plantes, pénetre celles dont l'écorce eft encore tendre, bouche les pores que la chaleur avoit ouverts, & arrête ce mouvement de la fève qui, dans les végétaux, nourrit les fleurs & les branches.

On a remarqué que cette *Nielle*, dans les *Cérifiers* à grandes feuilles, comme le *Cœur noir* ou le *Cœur blanc*, attaque leurs fommets, quand les jeunes branches qui naiffent à la Saint-Jean commencent à pouffer ; que cette *Nielle* arrête tellement leur accroiffement, que ces arbres pouffent par le bas ; & au fommet des jeunes branches, on voit plufieurs petits moucherons qui fe nourriffent de cette rofée. On peut auffi faire très-aifément cette obfervation fur les feuilles du *Chéne* & de l'*Erable*.

Les uns penfent que la *Nielle* & la bruïne ou rouille ne font qu'une feule maladie ; mais d'autres prétendent que la

Nielle eft très diftincte des bruïnes ou rouilles. Ces derniers difent que les bruïnes naiffent de la condenfation des exhalaifons graffes & humides qui fortent, dans un été chaud & fec, des fleurs des plantes & de la terre même. Ces exhalaifons étant condenfées en une matiere graffe & glutineufe, par la fraicheur & le calme de l'air, retombent enfuite fur la terre ; une partie refte fur les feuilles du *Chéne* & d'autres arbres qui ont leurs feuilles unies, & qui, par cette raifon, n'abforbent pas fi aifément l'humidité que celles de l'Orme & d'autres feuilles rudes.

Les autres parties de la *Nielle* reftent fur les épis & les tiges de froment, & les tachent de couleurs différentes de celle qui leur eft naturelle. Cette *Nielle*, devenue une fubftance graffe & glutineufe, par la chaleur du foleil, refferre fi étroitement les épis, qu'elle en empêche l'accroiffement, & les rend fort légers à la moiffon.

Quelques-uns croient que les *Nielles* font la nourriture principale des abeilles, parce qu'étant douces & fucrées, elles peuvent être aifément changées en miel.

NIELLE, BRUÏNE ou ROUILLE. *Blight*, en anglois.

Comme il n'y a rien de plus nuifible aux arbres fruitiers que les *Nielles*, rien auffi n'exige plus notre attention que de tâcher de les garantir de cette maladie.

Pour remédier à ce mal, il eft néceffaire de connoître d'abord

Q 3

242　**N I E**

la vraie caufe ; car quoique plufieurs perfonnes curieues aient
tenté de l'expliquer , cependant très-peu font parvenues
à découvrir la vérité , fi ce
n'eft le Docteur HALES , qui,
dans fon favant livre, intitulé
Statique des Végétaux, nous a
donné quelques expériences
exactes fur l'accroiffement &
l'afpiration des plantes , avec
les différens effets que l'air
produit fur les végétaux. En
joignant à cela des obfervations
faites avec foin , nous pour-
rons parvenir à découvrir la
caufe des *Nielles* ; mais ici , je
ne puis paffer fous filence les
caufes que plufieurs de nos
Ecrivains modernes fur le Jardinage ont attribuées aux *Nielles* , & quelles font les différentes méthodes qu'ils ont pref-
crites pour parvenir à les détruire & à empêcher la perte
des fruits.

Quelques-uns ont penfé que
les Nielles font ordinairement
occafionnées par des infectes
dont un vent d'orient apporte
les œufs en grande quantité
d'un endroit éloigné , & les
dépofe fur la furface des feuilles & des fleurs des arbres à
fruits. Pour prévenir ce défordre, on confeilloit de brûler
de la litiere humide , de maniere que la fumée fût portée
par le vent fur les arbres , &
on imaginoit pouvoir détruire
les infectes : d'autres confeillent l'ufage du *Tabac* réduit en
poudre , ou d'arrofer les arbres
avec une eau dans laquelle on
a fait infufer des tiges de *Tabac* pendant douze heures. Ils
prétendent qu'on peut détrui-

N I E

re les infectes par ce moyen ;
& rendre aux arbres leur premiere vigueur.

Du poivre en poudre répandu fur les fleurs des arbres à
fruits, a auffi été recommandé
comme fort utile en pareil cas ;
d'autres indiquent, comme le
meilleur de tous les remedes ,
d'ôter les feuilles des arbres ,
quand elles font ridées & fanées, & de couper les plus
petites branches , lorfqu'elles
produifent des rejettons courbés & défigurés : ils veulent
auffi que l'on arrofe les arbres
avec un arrofoir ou une pom-
pe à main.

Ces conjectures fur les Nielles, quelques fpécieufes qu'elles paroiffent d'abord , feront
trouvées peu conformes à la
vérité , quand on aura examiné cette matiere avec attention.

Mais voyons d'abord ce que
des obfervations exactes &
des expériences fuivies nous
apprennent fur la vraie caufe
de cette maladie.

1°. Les Nielles font fouvent
occafionnées par un vent fec
d'orient, qui a continué pendant plufieurs jours de fuite ,
fans pluie ou fans rofée , &
durant lequel la tranfpiration
des fleurs étant arrêtée, on les
voit changer de couleur, fe
faner, & périr bientôt. S'il arrive que ce vent continue long-
tems , & que les feuilles des
arbres en foient auffi affectées,
leur tranfpiration s'épaiffit, devient gluante , adhere à la furface des feuilles , & devient
une nourriture pour ces petits
infectes, que l'on trouve tou-

jours dévorant les feuilles & les tendres branches des arbres fruitiers, toutes les fois que cette Nielle a lieu : mais ces infectes ne font pas la premiere caufe des Nielles, comme quelques perfonnes l'ont imaginé, quoiqu'il faille convenir que lorfqu'ils fe jettent fur un arbre où ils trouvent une nourriture qui leur eft propre, ils s'y multiplient fortement, & contribuent beaucoup à augmenter ce défordre ; de forte que, quand la faifon leur eft convenable, & qu'on n'a pas pris un grand foin pour prévenir leurs ravages, on ne peut imaginer combien d'arbres fouffrent de cette infection.

Le meilleur remede que j'aie connu jufqu'à préfent pour guérir cette pefte, & qui m'a toujours réuffi, eft de nettoyer & d'arrofer légerement les arbres de tems en tems avec de l'eau ordinaire, c'eft-à dire, fans aucun mélange. Plutôt on fait cette opération, quand on craint ce danger, mieux on réuffit. Si les rejettons les plus tendres paroiffent être fort infectés, on les lave avec un drap de laine, jufqu'à ce qu'on ait enlevé, s'il eft poffible, toute la matiere glutineufe qui s'oppofoit au paffage de leurs parties volatiles. En plaçant auffi près de ces arbres quelques terrines ou cuves larges & plates, remplies d'eau, les émanations de cette eau, s'attachant à leurs branches, les humecteront, & les tiendront dans un état de foupleffe qui leur fera fort avantageux : mais

cette opération ne doit être faite que dans la matinée, afin que l'humidité puiffe être diffipée avant l'approche du froid de la nuit, fur tout quand il y a quelque apparence de gelée. Il ne faut pas non plus mettre cette pratique en ufage lorfque le foleil eft trop chaud ; car on courroit rifque de voir brûler les tendres rejettons.

Une autre caufe de la Nielle du printems, eft une forte gelée blanche, à laquelle fuccede une chaleur vive. Cette circonftance eft une de celles qui font périr les fruits avec le plus de promptitude. Le froid de la nuit flétrit les parties tendres des fleurs, &, le foleil dardant enfuite fes rayons fur les efpaliers, l'humidité répandue en pétits globules fur les fleurs fait l'office d'autant de lentilles ou verres ardens, qui brulent non-feulement les fleurs qui viennent d'éclorre, mais encore les autres parties des plantes.

Malgré tout ce qui vient d'être dit, les Nielles ne font fouvent qu'un affoibliffement ou maladie intérieure des arbres. Cette propofition paroitra démontrée à ceux qui voudront fe donner la peine de confidérer que, parmi les arbres qui garniffent un efpalier, qui jouiffent tous du même afpect, & de l'influence de l'air & du foleil, qui peuvent les rendre également fains, cependant il arrive trèsfouvent que plufieurs d'entr'eux different des autres confidérablement en force & vigueur ; & comme nous voyons fou-

Q 4

vent que ces arbres foibles font continuellement niellés, pendant que les plus vigoureux, dans la même expofition, échappent à ce fléau, il eft naturel d'attribuer cette différence à leur bonne conftitution. Ainfi, cette foibleffe dans les arbres doit procéder, ou de ce qu'ils manquent d'une nourriture fuffifante pour les maintenir dans une parfaite vigueur, ou de quelque mauvaife qualité du fol dans lequel ils croiffent, ou peut être de quelques vices dans le tronc ou dans la greffe, ou enfin d'un mauvais traitement dans la taille, &c. Toutes ces caufes peuvent produire dans les arbres des défordres dont ils guériffent difficilement. Si la Nielle provient d'une foibleffe de l'arbre, on doit s'efforcer d'en découvrir la caufe. Cette foibleffe peut être occafionnée, comme nous l'avons dit, par un défaut dans la taille ; ce qui n'eft que trop ordinaire : car on voit fouvent des *Péchers* dont on laiffe étendre les branches dans toute leur longueur, pour les faire parvenir en peu d'années jufqu'au haut des murs. Ces branches, au lieu de porter du fruit, font fi foibles, qu'à peine elles peuvent produire des fleurs ; & le peu de vigueur qu'elles poffèdent étant bientôt abattue, les fleurs tombent, & fouvent les branches fe flétriffent, en partie ou même en totalité ; alors on attribue cet accident à la Nielle, quoiqu'il ne provienne que d'une mauvaife taille, qui a épuifé totalement l'arbre.

D'autres perfonnes laiffent croître leurs arbres comme ils y font naturellement difpofés, fans arrêter les rejettons, ou fans retrancher les branches gourmandes, dont deux ou trois fuffifent pendant un été, pour épuifer la plus grande partie de la nourriture des arbres : & comme on retranche enfuite ces branches en totalité dans la taille d'hiver, toute la force de l'arbre n'a été employée qu'à nourrir des branches inutiles ; & celles qui doivent porter du fruit, font devenues fi foibles, qu'elles ne font plus en état de fe conferver. Le remede à ce mal fera donné dans l'article de la taille des *Péchers*, &c. *Voy.* l'article *Taille des Arbres*.

Mais fi la foibleffe des arbres provient d'un défordre intérieur, le mieux eft de les arracher fur le champ ; & après avoir renouvelé la terre, d'en replanter d'autres à leur place.

Si le fol eft un gravier ou fable chaud & brûlant, on pourra prefque toujours le regarder comme la caufe du mal qui a eu lieu lorfque les racines des arbres fe font allongées au-delà de la terre des plates-bandes. Dans ce cas, il fera beaucoup plus prudent de les ôter & de les remplacer par de la *Vigne*, des *Figuiers*, des *Abricotiers*, ou quelques autres efpeces de fruits qui puiffent bien réuffir dans un pareil fol, plutôt que d'être trompé annuellement dans fes efpérances, parce qu'il eft prouvé, par une expérience conftante, que les *Abricotiers* abforbent l'humidité avec une

plus grande force que les *Pê-chers* & les *Pavies*, & que par conséquent ils font plus en état de raffembler les particules nutritives de la terre que les autres, qui exigent un fol riche & capable de leur fournir une nourriture abondante, fans beaucoup de difficulté. Nous voyons fouvent les *Pêchers* réuffir à merveille dans de pareilles places, fur-tout fi on les conduit avec art, tandis que les feps de *Vigne* & les *Figuiers*, qui tranfpirent fort lentement, s'abreuvent d'une humidité fi abondante, que leurs fruits perdent ce goût agréable & fucré dont ils font remplis, lorfqu'ils croiffent dans un terrein fec, ce qu'on peut attribuer aux principes aériens raffinés qu'ils récueillent, lorfqu'ils font dans un état d'infpiration : & comme ces arbres n'aiment point à tirer de la terre beaucoup de fucs humides, & qu'ils réuffiffent mieux dans un fol fec & aride que dans un terrein gras & fertile, on devroit toujours affortir les efpeces de fruits à la nature du fol, & ne pas prétendre les forcer tous à réuffir dans la même terre.

Une autre efpece de Nielle dont il eft très difficile de préferver les arbres à fruits, eft celle qui eft occafionnée par les fortes gelées du matin, qui, lorfqu'elles furviennent dans le tems que les arbres font en fleurs, ou tandis que le fruit eft encore fort jeune, occafionnent la chute des fleurs & des fruits, & quelquefois endommagent les extremités

des rejettons & les feuilles.

La feule méthode jufqu'à préfent connue pour prévenir ce mal, eft de couvrir les murailles d'efpaliers foigneufement avec des nattes, des canevas, des rofeaux, &c. : on attache ces couvertures de maniere qu'elles ne puiffent être agitées par le vent ; on les laiffe la nuit, & on les ôte chaque jour, fi le tems le permet. Ce moyen eft le plus propre pour parvenir au but qu'on fe propofe, quoique plufieurs perfonnes l'aient négligé, dans l'idée qu'il n'étoit pas d'une fort grande utilité : mais fi elles n'en ont pas obtenu le fuccès qu'elles en attendoient, c'eft qu'elles s'y font mal prifes, en laiffant les arbres trop long-tems couverts ; ce qui aura rendu les plus jeunes branches & les feuilles trop tendres pour fupporter le plein air, lorfqu'elles y ont été expofées enfuite, foit parce qu'on aura expofé les arbres trop vite à l'air, après les avoir tenus long tems couverts.

Ceux qui ont fait ufage de ces couvertures avec intelligence, les ont toujours trouvées fort utiles, & ont fouvent confervé leurs fruits, tandis qu'ils ont été détruits dans les jardins voifins. Quoique les foins que cette précaution exige puiffent paroître onéreux, cependant on trouvera que cette peine n'eft pas fort grande, & qu'on en fera amplement dédommagé, fi l'on fixe ces couvertures au haut de la muraille, & qu'on y place des poulies, pour pouvoir les

relever & les baisser à son gré.

Il y a une autre espece de *Nielle* qui se montre quelquefois plus tard dans le printems, & qui endommage souvent, en Avril ou en Mai, des vergers & des plantations entieres. Nous ne connoissons aucun remede contre ce mal : on l'appelle la *Nielle* de feu ; elle détruit en peu d'heures non-seulement les fruits & les feuilles, mais aussi une partie des branches, & souvent même des arbres entiers.

On croit que cet accident provient de quelques bouffées de vapeurs transparentes & flottantes, qui prenant différentes formes, & souvent celles d'un hémisphere ou d'un demi-cylindre, dans leurs surfaces inférieures ou supérieures, rendent convergens les rayons du soleil, qui embrasent & consument plus ou moins les plantes & les arbres qui sont exposés, à proportion de leur intensité.

Le savant Boerhave, dans sa *Théorie de la Chymie*, s'exprime ainsi : « Ces nuages » blancs, qui paroissent pen- » dant l'été, sont pour ainsi » dire autant de miroirs qui » occasionnent une chaleur ex- » cessive. Ces nuées sont quel- » quefois rondes, quelquefois » concaves, &c. Quand notre » hémisphere en est couvert, » le soleil, en y dardant ses » rayons, doit produire une » chaleur violente, puisque » plusieurs de ces rayons, qui, » sans ces nuages, ne par- » viendroient peut-être jamais » jusqu'à la terre, se divergent » alors, & se réfléchissent jus- » qu'à nous. Le soleil étant » d'un côté, les nuages d'un » autre, ces derniers font par- » faitement l'office des verres » ardens, & occasionnent le » phénomene du tonnerre.

» J'ai vu quelquefois, con- » tinue-t-il, une espece de » nuage concave, rempli de » grêle & de neige, qui, tant » qu'il s'est trouvé sur l'hori- » zon, a produit une chaleur » extrême, parce que sa con- » densation lui faisoit réfléchir » beaucoup plus fortement les » rayons du soleil ; mais dès » que cette nuée étoit passée, » il survenoit un froid rude ; » & aussi-tôt que la grêle étoit » tombée, on sentoit revenir » une chaleur modérée : donc » des nuées concaves & gla- » cées produisent, par leur » grande reflexion, une cha- » leur vigoureuse ; & en tom- » bant, un froid excessif ».

D'après cela, comme le Docteur HALES l'observe, nous voyons que les Nielles peuvent être occasionnées par la réflexion des nuages, ainsi que par la réfraction des vapeurs épaisses & transparentes dont il vient d'être question.

Nous ne connoissons aucun moyen, ainsi que nous l'avons déja observé, qui puisse prévenir ou remédier à cet accident : mais comme il est plus fréquent dans les plantations où les vapeurs stagnantes de la terre & la transpiration abondante des arbres sont renfermées, & ne peuvent être dissipées par les vents, & où

NIG

on les voit souvent, dans un temps calme, monter en si grande abondance, qu'on les apperçoit de l'œil nud, mais encore mieux avec des télescopes à réflexion, de maniere à rendre obscur & vacillant un objet clair & distinct · & comme on voit aussi que les plantations dans lesquelles les arbres sont éloignés, & qui ne sont point environnées de collines & de forêts, ne sont point sujettes à de pareils inconvéniens, cela doit nous engager à placer nos jardins potagers & nos vergers dans des lieux plus convenables, à donner une plus grande distance entre les arbres, & à choisir dès situations ouvertes & saines, afin que l'air puisse circuler plus librement entre les arbres, pour dissiper ces vapeurs, avant qu'elles soient trop rassemblées, & les empêcher de nuire à ces plantations : d'ailleurs les fruits qui naissent à l'air libre sont toujours d'un goût plus agréable que ceux qu'on recueille dans un lieu renfermé & environné d'un air plus épais, parce qu'étant souvent dans un état d'aspiration, ils se nourrissent de ces vapeurs nuisibles, & deviennent cruds & de mauvais goût, comme sont presque tous ceux qu'on recueille en Angleterre.

NIGELLA. *Tourn. Inst. R. H.* 258. *tab.* 134. *Lin. Gen. Plant.* 606, ainsi appelé, comme si c'étoit *Nigrella*, de la couleur de ses semences, parce qu'elles sont la plupart noires. On la nomme aussi *Mélan-*thum, de κίλας noire, & de ἄνθος une fleur; c'est-à-dire, *fleur noire*, quoique la fleur ne soit point de cette couleur; elle s'appelle encore *Melaspermum* de σπέρμα, semence, & de μέλας, noire. *Fleur de Fenouil* ou *le Diable dans un Buisson.* En Anglois, *Fennel Flower, or Devil in a bush*, en françois, *la Nielle.*

Caracteres. La fleur n'a point de calice, mais seulement un perianthe feuillé; la corolle a cinq pétales ovales, obtus, unis, étendus, rétrécis à leur bâte; la fleur a huit nectaires placés en cercle, chacun desquels a deux levres, dont l'extérieure est la plus large, & l'inférieure est divisée en deux parties unies convexes, marquée de deux points; celle de ces parties, qui est la plus interne, est plus courte, plus étroite, & terminée par une ligne ovale. Cette fleur a un grand nombre d'étamines en forme d'alène, plus courtes que les pétales; & terminées par des antheres obtuses, comprimées & droites : dans quelques unes sont cinq, & dans d'autres dix germes oblongs, convexes, érigés & terminés par des styles longs, roulés, persistans, & pourvus de stigmats fixés longitudinalement. Ces germes se changent dans la suite en autant de capsules oblongues, comprimées, divisées par un sillon, mais reunies en-dedans, & remplies de semences rudes & angulaires.

Ce genre de plantes est rangé dans la cinquieme section de la treizieme classe de LINNÉE, qui renferme celles dont

les fleurs ont plusieurs étamines & cinq styles.

Les espèces sont :

1°. *Nigella arvensis*, *pistillis
quinis*, *petalis integris*, *capsulis
turbinatis.* Lin. Sp. Plant. 534.
Scop. carn. 2. n. 657. Pollich.
p al. n 514. Mattusch. Sil. n. 387.
Œder. Nass. p. 159; Nielle ayant
cinq pistiles , des pétales entiers , & des capsules turbinées.

Nigella arvensis cornuta. C. B.
P. 145 ; la Nielle des champs,
ayant des cornes , ou toute
épiée.

Melianthum sylvestre alterum.
Cam. Epit. 552.

2°. *Nigella Damascena*, *floribus involucro folioso cinctis.* Hort.
Cliff. 215. Hort. Upsf. 157. Roy.
Lugd.-B. 481. Blackw. 158.
Knaph. cent. 10. n. 66 ; Nielle
ayant des fleurs entourées d'une
enveloppe feuillée.

Nigella angusti-folia, *flore majore simplici cæruleo.* C. B. P.
145 ; Nielle à feuilles étroites , avec des fleurs séparées,
plus grosses , & bleues ; Che-
veux de Vénus , ou Nigelle de
Damas.

Melianthum sylvestre. Matth.
Diss. 529.

3°. *Nigella sativa*, *pistillis
quinis , capsulis muricatis sub-
rotundis , foliis sub-pilosis.* Hort.
Upsf. 154. Mat. Med. 139.
Bahm. Lips. 173. Ludw. Ect.
t. 83. Knaph. cent. 7. n. 63 ;
Nielle ayant cinq pistiles , des
capsules épineuses & presque
rondes , & des feuilles un peu
velues.

*Nigella flore minore simplici can-
do.* C. B. P. 145 ; Nielle
avec une plus petite fleur blanche & séparée.

Melianthum sativum. Cam.
Epit 551.

4°. *Nigella Cretica* , *pistillis
quinis corollâ longioribus , petalis
integris* ; Nielle de Crète à cinq
pistiles plus longs que la corolle , & à pétales entiers.

*Nigella Cretica lati-folia odo-
rata.* Park. Theat. 1376 ; Nielle
de Crète à larges feuilles , &
d'une odeur agréable.

5°. *Nigella lati-folia*, *pistillis
denis corollâ brevioribus* ; Nielle
avec dix pistiles plus courts
que la corolle.

Nigella alba simplici flore. Alp.
Exot. 261 ; Nielle avec une
fleur simple & blanche.

6°. *Nigella Hispanica* , *pistillis
denis corollam æquantibus.* Hort.
Upsal. 154. Sauv. Monsp. 253 ;
Nielle avec dix pistiles de même
longueur que la corolle.

Nigella lati-folia, *flore majore
simplici cæruleo.* C. B. P. 145.
Prodr. 75. Moris. Hist. 3. p. 516.
S. 12. t. 18. f. 8. ; Nielle à
larges feuilles , ayant une grosse
fleur simple & bleue.

7°. *Nigella Orientalis*, *pistillis
denis corollâ longioribus.* Hort.
Cliff. 215. Hort. Upsf. 153. Roy.
Lugd-B. 481 ; Nielle ayant dix
pistiles plus longs que la corolle.

Nigella Chalepensis lutea , *cor-
niculis longioribus.* Moris. Hist.
3. p. 516. S. 12. t. 18. f. 10.
Raii. App. 525.

Nigella Orientalis, *flore flavescente , semine alato , plano.*
Tourn. Cor. 19 ; Nielle du Levant , à fleurs jaunâtres , & à
semences unies & ailées.

Arvensis. La première espèce
croît naturellement parmi les
Bleds dans la France , en Italie
& en Allemagne ; mais on la

NIG

conferve rarement dans les jardins : elle s'éleve avec des tiges minces à un pied environ de hauteur, & pouffe quelquefois des branches vers le bas ; mais fouvent elles font fimples, & feulement gärnies de quelques feuilles très-finement découpées, & un peu reffemblantes à celles de l'*Anet* ; chaque tige eft terminée par une fleur formée par cinq pétales pointus & difpofés en forme d'étoile ; elles font d'un bleu pâle, & n'ont point d'enveloppe feuillée au deffous d'elles ; elles font remplacées par des capfules garnies de cinq cornes peu longues, inclinées en différens fens au fommet, & remplies de femences rudes & noires. Il y a une variété de cette efpece à fleurs blanches, & une autre à fleurs doubles (1).

Damafcena. La feconde, qui croit en Efpagne & en Italie parmi les bleds, s'éleve à la hauteur d'un pied & demi, avec une tige droite, branchue, & garnie de feuilles beaucoup plus longues & plus belles que celles de la premiere : fes fleurs font larges, d'un bleu pâle, & pourvues d'une longue enveloppe ; à ces fleurs fuccedent des capfules plus groffes, gonflées, & armées de cornes au fommet. Il y a auffi dans cette efpece une variété à fleurs fimples & blanches, & une autre à fleurs doubles, que l'on feme dans les jardins, pour fervir d'ornement.

Sativa. La troifieme efpece fe trouve dans l'Ifle de Candie ; elle s'éleve à-peu-près à la même hauteur que la précédente : fes feuilles ne font pas auffi agréablement découpées que celles de la feconde ; mais elles font un peu velues : au fommet de chaque tige eft une fleur compofée de cinq pétales blancs, légerement découpés à leur extrémité en trois pointes, à laquelle fuccede une capfule oblongue, gonflée, armée de cinq cornes à fon fommet, & remplie de femences de couleur pâle.

Cretica. La quatrieme efpece, qu'on rencontre encore dans l'Ifle de Candie, s'éleve à la hauteur d'environ un pied, avec des tiges branchues & garnies de feuilles plus courtes & plus larges que celles des autres efpeces ; chaque branche eft terminée par une fleur fans enveloppe, compofée de cinq pétales ovales, & de cinq piftiles plus long»

---

(1) On regarde la graine de Nielle comme fortifiante, chaude, difcuffive, céphalique, carminative, anthelminthique, emménagogue, utérine, &c. Ses principes font la gomme, la réfine, & une petite quantité d'huile ethérée : on s'en fert avec quelque fuccès, à la dofe d'un gros, en infufion vineufe, dans les affections catharrales de la poitrine, l'afthme, contre le vertige, la céphalalgie, le coryfa, les pâles couleurs, les obftructions des regles, &c. On emploie auffi ces graines dans les epithêmes fecs, contre le rhumatifme, l'hemi-crânie, & toutes les affections catharrales de la tête ; elles entrent dans la compofition du fyrop d'Armoife, dans l'électuaire de baie de Laurier dans les trochifques de Cipres, &c.

que la corolle ; sa capsule n'est pas fort gonflée : elle a aussi cinq cornes minces au sommet, & les semences sont d'un brun clair & jaunâtre.

Lati-folia. La cinquieme est aussi originaire de l'Isle de Candie ; elle s'éleve à la hauteur d'un pied avec une tige branchue, & garnie de feuilles semblables à celles des *Pieds d'Alouette* ; ses fleurs ont cinq pétales larges, ovales, & entiers, dix pistiles plus courts que la corolle, un grand nombre d'étamines vertes, & des filamens bleus, & des capsules semblables à celles de la derniere espece.

Hispanica. La sixieme s'éleve à un pied & demi de hauteur ; ses feuilles les plus basses sont joliment découpées, & celles des tiges ont des segmens plus larges : ses fleurs sont plus grosses que celles des autres especes, & d'un plus beau bleu ; leurs pistiles sont égaux à la corolle : les capsules sont cinq cornes, & sont d'une texture plus ferme que celle d'aucune des autres. Cette plante croît naturellement dans la France méridionale & en Espagne ; elle donne une variété à fleurs doubles.

Orientalis. La septieme se trouve aux environs d'Alep, dans des campagnes ensemencées en Bled ; sa tige est haute d'un pied & demi, & garnie de feuilles longues & agréablement découpées : ses fleurs, qui naissent aux extrémités des branches, sont composées de cinq pétales jaunâtres, dont les bâses ont huit nectaires,

entre lesquels s'élevent un grand nombre d'étamines, & une quantité inégale de germes : dans quelques-unes, il ne s'en trouve que cinq, & dans d'autres huit ou neuf; ils sont oblongs, comprimés, réunis ensemble sur le côté intérieur, & terminés en cornes; ils s'ouvrent longitüdinalement, & contiennent plusieurs semences minces, comprimées, & bordées tout autour.

On multiplie communément la variété de cette espece à fleurs doubles, pour servir d'ornement dans les parterres ; mais celle à fleurs simples n'est cultivée que dans les jardins de Botanique, pour la variété.

Culture. On peut multiplier toutes ces plantes, en semant leurs graines sur une terre légere où elles doivent rester ; car elles ne réussissent pas aisément quand elles sont transplantées. Ainsi, quand on veut qu'elles soient entremêlées parmi les autres fleurs annuelles, dans les plates-bandes d'un parterre, il faut les semer en touffes à des distances convenables. Quand les plantes poussent, on les éclaircit, & on n'en laisse que trois ou quatre dans chaque touffe : la seule culture qu'elles exigent est d'être tenues nettes de mauvaises herbes ; elles produiront leurs fleurs dans le mois de Juillet, & leurs semences mûriront en Août : alors on les recueille, on les fait sécher, on nettoie séparément celles de chaque espece & on les conserve dans un endroit sec.

Le meilleur tems pour se-

NIT

mer ces graines eſt le mois d'Août ; auſſi-tôt qu'elles ſont mûres, il leur faut un ſol ſec & une expoſition chaude, où elle ſubſiſtent pendant l'hiver, & fleuriſſent fort l'année ſuivante. En les ſemant en différens tems, les fleurs peuvent ſe ſuccéder durant la plus grande partie de l'été.

Ces plantes ſont annuelles, & périſſent auſſi tôt que leurs graines ſont mûres. Si on leur permet de s'écarter ſur les plates-bandes, elles pouſſent ſans aucun ſoin.

NIGELLASTRUM. *Voyez* AGROSTEMA.

NINZIN & GINZENG. *Voy.* PANAX.

NIRURY. *Voyez* PHYLLANTHUS NIRURI. *L.*

NISSOLIA. *Voyez* LATHYRUS.

NITRE. C'eſt une eſpece de ſel imprégné par l'air d'une abondance d'eſprits qui le rendent volatil.

M. LE CLERC en parle ainſi :

On fait une grande quantité de *Nitre* en Egypte ; mais il n'eſt pas auſſi bon parce qu'il eſt brun, & rempli de pierres & de nœuds.

On fait le nitre à-peu-près de la même maniere que le ſel commun, excepté qu'on ſe ſert pour ce dernier d'eau de mer, & que l'on prend l'eau du Nil pour le *Nitre*.

Quand le Nil ſe retire, les puits reſtent remplis pendant quarante jours ; & auſſi-tôt que le *Nitre* devient ferme, on ſe hâte de l'enlever, de peur qu'il ne ſe fonde de nouveau ; on le met en monceaux, &

NIT
251

il ſe conſerve fort bien dans cet état.

Le *Nitre* de Memphis ſe pétrifie : on en voit pluſieurs carrieres auprès de cette Ville : on en fait des vâſes, & quelquefois on le fait fondre avec du ſoufre, pour le mêler avec du charbon de terre.

Les Egyptiens emploient ce *Nitre* pour les ouvrages auxquels ils veulent procurer une longue durée.

Le *Nitre*, pour être bon, doit être leger, friable, & preſque de couleur pourpre. Il n'y a guere de différence entre le *Nitre* naturel & l'artificiel ; le premier ſe raffine de lui-même, & l'autre eſt purifié par l'art. Tout *Nitre* eſt une eſpece de ſel, &, à proprement parler, il ne differe guere du ſel commun, qu'en ce qu'étant bien purifié, il eſt plus acide, plus leger, & qu'il s'enflamme aiſément.

La raiſon de cette différence, dit le même Auteur, ſemble être, 1°. Que les angles ou les deux extrémités des particules oblongues du *Nitre* ſont plus courts que les angles des particules ſalines.

2°. Que les particules du *Nitre* étant plus fines & plus remplies de pores, quand elles ſont pénétrées par le feu, elles entrent auſſi-tôt en mouvement, juſqu'à ce qu'elles ſe briſent & s'enflamment.

3°. Le *Nitre* eſt plus leger que le ſel commun, parce que celui-ci contient plus de matieres homogènes ſous le même volume que le *Nitre*.

Le Docteur LISTER dit,

qu'ayant examiné les particules du *Nitre* à travers le microscope, il y avoit remarqué six angles, des côtés parallélogrammes, & une pointe pyramidale.

Quelques Auteurs pensent que les sels nitreux sont destinés par la Nature principalement à l'accroissement des plantes.

D'autres pensent différemment, & disent, que lorsque le *Nitre* touche les plantes, il les détruit beaucoup plus qu'il ne les nourrit. Cependant ils avouent que le *Nitre* & d'autres sels rendent la terre plus légere, en divisent les parties concretes, la disposent par ce moyen à être imbibée par l'eau, & s'insinuent dans les plantes pour concourir à leur accroissement.

On peut observer comment l'humidité agit sur tous les sels, & avec quelle facilité ils se liquéfient & coulent avec elle. Quand leurs particules sont divisées, & qu'elles ont abandonné les corps auxquels elles étoient attachées, il faut que ces corps se dissolvent immédiatement après.

La pierre la plus dure, si elle a quelques particules de sel mêlées avec le sable dont elle est composée, après avoir été exposée à un air humide, se dissout, & tombe en poussiere en peu de tems. Cet effet est bien plus prompt dans la terre commune & la glaise, quand elle est dure, parce qu'elles ne sont pas d'une consistance aussi solide & aussi compacte que la pierre.

Quelque propre que soit la terre à la production du *Nitre*, on ne peut espérer d'en tirer beaucoup, à moins que ses parties ne soient divisées. C'est par cette raison qu'on la bêche, qu'on la laboure, qu'on la herse, & qu'on en brise les mottes ; c'est aussi de cette maniere que le *Nitre*, le *sel de mer*, & d'autres sels avancent la végétation.

Une personne rapporte, qu'habitant une campagne dans le voisinage d'une salpétriere, où l'on apporte le salpêtre des pays étrangers, pour le raffiner & le rendre propre à faire de la poudre à canon, il observa que ce bâtiment se trouvoit placé de maniere que la fumée du *Nitre* venoit frapper la plupart des arbres de son verger ; & que, malgré l'opinion de quelques personnes qui croyoient que ces vapeurs dussent être nuisibles à ces arbres, il éprouva cependant un effet contraire ; car son verger lui fournit tous les ans une récolte abondante de fruits, tandis que ses voisins n'en avoient que très-peu ou presque point, quoique son verger ne fut pas moins exposé, par sa situation naturelle, aux bruines & aux mêmes vents que les autres, d'où il a conclu que les vapeurs nitreuses, se mêlant avec l'air qui entouroit son jardin, empêchoient la nielle, & faisoient périr les chenilles.

Le Lord BACON, dans son *Histoire Naturelle*, recommande l'usage du *Nitre* pour la conservation de la santé. Plusieurs Cultivateurs habiles ne l'ont
pas

NIV

pas moins recommandé pour l'avancement des végétaux, pourvu que la quantité en fut bien proportionnée.

Il est certain que l'air contient beaucoup de particules salines, parce qu'étant continuellement exposé aux émanations de la terre & de la mer, il doit aussi recevoir une très-grande quantité de particules salines, qui sont de nature différente, suivant les lieux d'où elles sont tirées (1).

NIVEAU (un). [*a Level.*]
C'est un instrument de Mathématique, qui sert à tirer une ligne parallele à l'horison. Il est utile, non-seulement dans la Maçonnerie, mais encore pour mesurer la hauteur des terres entre deux lieux différens, pour la conduite des eaux, & pour saigner les marais.

---

(1) Il ne faut pas être bien versé dans les connoissances chymiques, pour reconnoître combien les principes établis dans cet article sont peu conformes à ceux de la Chymie moderne sur la nature du Nitre. Les progrès que nous avons faits à cet égard sont dus à une connoissance plus parfaite de l'acide nitreux, qui, combiné avec l'alkali fixe végétal, jusqu'au point de saturation, forme le Nitre commun, & différens sels nitreux particuliers, lorsqu'il est uni à d'autres bâses.

Je n'entrerai point dans un plus grand détail à ce sujet, parce que ce que je pourrois ajouter seroit déplacé ici. Le Lecteur pourra consulter, s'il est curieux d'en savoir davantage, le *Dictionnaire de Chymie de* MACQUER, & d'autres ouvrages modernes, où il trouvera amplement de quoi se satisfaire.

Tome V.

NIV

Un *Niveau d'eau* montre la ligne horisontale, d'après ce principe, que l'eau prend toujours son niveau.

L'instrument le plus simple pour cet usage, est un long canal de bois, dont les deux côtés sont paralleles à sa bâse; de sorte qu'étant également rempli d'eau, sa surface fera voir la ligne de niveau.

On fait aussi ce Niveau avec deux gobelets attachés aux deux extrémités d'un tuyau de trois ou quatre pieds de longueur, & d'un pouce à peu-près de diametre, au moyen duquel l'eau se communique d'un gobelet dans l'autre. Ce tuyau étant mobile sur un tuyau formé par une boule & un creux, lorsqu'il est placé de maniere que les deux gobelets sont remplis d'eau à une hauteur égale, leurs surfaces marquent la ligne du Niveau.

Au lieu de ces gobelets, on peut faire cet instrument avec deux cylindres de verre de trois ou quatre pouces de longueur, fixés à chaque extrémité du tuyau avec de la cire ou du mastic; alors le tuyau étant rempli d'eau commune ou colorée, ce liquide montera dans les cylindres, & fixera la ligne de niveau, la hauteur de l'eau, par rapport au centre de la terre, étant toujours la même dans les deux cylindres. Ce Niveau est très-commode pour niveler de petits espaces.

Si vous voulez niveler quelque piece de terre dont vous pouvez voir les deux extrémités, étant placé dans le centre, posez avec de guidons votre ni-

R

veau dans le milieu, soit que
ce soit un Niveau d'eau, ou
tout autre instrument ; élevez-
le assez pour que vous puissiez
voir un demi-pied ou un pied
au-dessus de la plus haute partie de votre terrein ; mettez un
jalon au milieu de l'espace, de
maniere que son extrémité soit
de niveau avec les guidons ;
placez un autre jalon sur la
partie la plus élevée du terrein,
de maniere que son extrémité
soit de niveau avec l'extrémité
de celui du milieu ; tournez
alors votre Niveau du côté du
guidon ; placez ensuite dans la
partie la plus basse du terrein
un autre jalon, qui soit de niveau avec les deux autres ;
& vous aurez alors trois points
de niveau ; tenez ensuite votre
Niveau exactement sur le jalon
du milieu, & tournez-le jusqu'à ce qu'il forme des angles
droits avec les trois jalons ;
après quoi, fixez-en deux autres à chaque côté, qui soient
de niveau avec les trois premiers, & vous aurez cinq points
en deux lignes, qui formeront
un Niveau exact.

Si le terrein est étendu, vous
pouvez placer encore deux
autres rangs de jalons à coté
du Niveau ; mais cinq points
suffisent dans un petit espace.

Cette opération étant terminée, ôtez le Niveau, & regar-
dez les têtes de deux de vos
jalons, en en faisant placer
d'autres entr'eux, jusqu'à ce
que vous en ayez autant que
vous le jugerez nécessaire. On
peut se servir pour cela d'une
regle, laquelle étant placée de
niveau avec la tête du jalon,

vous servira à regarder au-
dessus des autres, pour établir
dans les intervalles tous les
points de niveau nécessaires.

Le terrein étant ainsi marqué avec des jalons dont toutes
les têtes sont de niveau, de
façon cependant que ces piquets soient au-dessus de la partie la plus élevée du terrein :
dans de certains endroits, le
piquet du milieu, & ceux qui
sont dans le rang de traverse,
feront la ligne de niveau que
l'on veut donner au terrein :
on abaissera alors les éminen-
ces, & l'on haussera les parties
basses, jusqu'à ce que tout soit
de niveau avec la ligne du milieu. Si le terrein se trouve fort
inégal, il faut d'abord mesurer
une ligne au-dessus de la tête
de tous ces piquets, prendre
ensuite le Niveau au milieu de
chacun, & par le moyen de la
regle de trois, proportionner votre terrein au milieu des piquets.

Par exemple, si vous avez
un vallon de dix perches de
longueur, & de deux pieds
de profondeur, & une colline
longue de cinq perches, à combien de pieds faut-il creuser
ces cinq perches, pour remplir le vallon ? On peut résou-
dre cette question par la regle
de trois inverse : de cette maniere, cinq est à deux comme
dix est à quatre.

5 — 2 — 10.
2

5)20(4.
Ainsi, on doit creuser quatre pieds dans cette éminence,
pour niveler le vallon.
Si vous avez à diminuer de

NOL

quatre pieds, le sommet d'une montagne sur deux perches de longueur, vous commencez par enlever ces quatre pieds.

Placez au sommet de la montagne un piquet qui s'élève de deux ou trois pieds au-dessus de la surface ; mettez en un autre de la même hauteur dans l'endroit creusé, & un troisieme à trois perches de distance du dernier, de maniere que sa tête se trouve de niveau avec la tête des deux autres : le piquet du milieu doit être enfoncé d'un pied de profondeur.

A la distance de six perches, mettez-en un autre comme auparavant, & enfoncez le de deux pieds dans la terre ; placez ensuite un autre piquet à la distance de neuf perches, & enfoncez-le de trois pieds ; vous pourrez de même mettre encore d'autres piquets à des distances égales, qui vous dirigeront, & vous empêcheront de vous tromper.

NOIRPRUN ou NERPRUN. *Voyez* RHAMNUS CATHARTICUS *L.*

NOISETIER ou AVELINIER. *Voyez* CORYLUS AVELLANA. *L.*

NOISETIER MAGIQUE. *V.* HAMAMELIS. *L.*

NOIX D'ACAJOU. *Voyez* ANACARDIUM.

NOIX DE CHOCOLAT. *Voyez* CACAO.

NOIX MÉDICINALE D'AMÉRIQUE. *Voyez* JATROPHA MULTIFIDA ET CURCAS. *L.*

NOIX DE BEN. *Voyez* GUILANDINA MORINGHA.

NOIX DE TERRE. *Voyez* ARACHIS.

NOLANA. *Royen. Lin. Gen.*

NOL

Plant. 193. ; Cette plante n'a pas de nom Anglois.

Caractéres. Le calice de la fleur est formé par une feuille turbinée à sa base, divisée en cinq segmens aigus, en forme de cœur, & persistante ; la corolie est en cloche, plissée, étendue, & deux fois plus large que le calice : la fleur a cinq étamines en forme d'alène, érigées & terminées par des antheres à pointe de flèche, & cinq germes ronds qui entourent un style cylindrique érigé, & couronné par un stigmat à tête. La baie succulente intérieure du réceptacle se change en quatre cellules qui renferment les semences.

Ce genre de plantes est rangé dans la premiere section de la cinquieme classe de LINNÉE, qui comprend les plantes dont les fleurs ont cinq étamines & un style.

Nous ne connoissons à présent qu'une espece de ce genre, qui est :

Nolana prostrata. Lin. Sp. 202. *Dec.* 1. *tab.* 2 ; Nolana rampant.

Atropa foliis geminatis, caly—cibus polycarpis, caule humi fuso. Gouan. Monsp. 82 ; Moreile mortelle, ayant deux feuilles à chaque nœud, des calyces de fleurs avec plusieurs femences, & une tige traînante.

Walkeria. Ehret. Act. Angl. 1764. *n.* 53. *p.* 130. *f* 10.

Zwingera. Act. Helv. 5. *p.* 267. *f.* 1.

Neudorffia Peruviana repens, flore cærulco. Adans. Plant. Fam. 219.

Cette plante croît naturel-

lement en Egypte, d'où ses semences m'ont été adressées par M. Forchal, un de ceux qui ont été envoyés par le feu Roi de Danemark pour faire des découvertes au Levant.

Cette plante est annuelle, & pousse des tiges traînantes couchées sur la terre, & divisées en plusieurs branches garnies de feuilles ovales, unies, en forme de lance, & portées sur de courts pétioles; elles sortent simples à quelques nœuds, par paires à d'autres, & souvent au nombre de trois ou quatre ensemble aux nœuds du haut : ses fleurs sont produites seules aux fourches des branches, sur de longs pédoncules; elles sont de la même forme que celles du *Cerisier* d'hiver, & ont des tubes courts d'un beau bleu, dont les bords sont entièrement épanouïs : à ces fleurs succèdent quatre semences nues, placées dans le calice. Cette plante fleurit en Juillet, & ses semences mûrissent au commencement de Septembre.

Il faut répandre ses graines en Mars sur une couche chaude : quand les plantes qui en proviennent sont en état d'être enlevées, on les met chacune séparément dans un petit pot rempli de terre légere, & on les plonge dans une nouvelle couche chaude, pour hâter leur progrès, sans quoi elles ne perfectionneront pas leurs semences dans ce pays; mais quand leurs fleurs s'ouvrent, ce qui a lieu dans le mois de Juillet, il faut leur donner beaucoup d'air dans les tems chauds, pour empêcher ces fleurs de tomber sans produire de semences. Au moyen de ce traitement, les plantes continueront à fleurir jusqu'à ce que les premieres gelées les détruisent, & donneront des semences mûres au commencement de Septembre.

NOLI TANGERE ou Balsamine jaune. *Voy.* Impatiens noli tangere.

NOMBRIL DE VÉNUS. *Voyez* Cotyledon umbilicatum. L.

NOYER. *Voy.* Juglans.

NOYER DE MALABAR. *Voy.* Justicia adathoda.

NUMMULAIRE ou Herbe aux Ecus. *Voyez* Lysimachia nummularia. L.

NUX AVELLANA. *Voyez* Corylus.

NUX JUGLANS. *Voyez* Juglans.

NUX VESICARIA. *Voy.* Staphilodendron.

NYCTANTHES. *Lin. Gen. Plant.* 16. *Jasminum. Raii Meth. Plant.* [*Arabian Jasmine.*] Jasmin d'Arabie.

Caracteres. Le calice de la fleur est cylindrique, persistant, & formé par une feuille découpée en huit ou dix segmens aigus : la corolle, qui est en forme de soucoupe, & monopétale, a un tube cylindrique plus long que le calice, & découpé sur ses bords en huit ou dix segmens étendus : la fleur a deux petites étamines en forme d'alène, situées au fond du tube, & terminées par des antheres érigées, &

un germe rond enfoncé, qui soutient un style simple de la longueur du tube, & couronné par un stigmat érigé, & divisé en deux parties. Il se change, quand la fleur est passée, en une baie ronde & à deux cellules, qui renferment chacune une semence grosse & ronde.

Ce genre de plantes est rangé dans la premiere section de la seconde classe de LINNÉE, qui comprend celles dont les fleurs ont deux étamines & un style.

Les especes sont :

1°. *Nyctanthes Sambac, caule volubili, foliis sub-ovatis, acutis.* Hort. Ups. 4. Hort. Cliff. 5. Fl. Zeyl. 12. Roy Lugd.-B. 398. ; Jasmin d'Arabie, avec une tige tournante, & des feuilles aiguës & presqu'ovales.

Syringa Arabica, foliis Mali Aurantii. Bauh. Pin. 398.

Jasminum sivè Sambac Arabum, folio acuminato. Till. Pis. 87. f. 31.

Jasminum Arabicum. Clus. Cur. 3 ; Jasmin d'Arabie.

Flos Manoræ. Rumph. Amb. 5. p. 52. f. 31.

2°. *Nyctanthes hirsuta, petiolis pedunculisque villosis.* Lin. Sp. Plant. 6 ; Jasmin des Indes, avec des feuilles pétiolées, & des pédoncules velus aux fleurs.

Jasminum Indicum bacciferum, flore albo majore, noctu olente. Com. Hort. Mal. ; Jasmin des Indes portant baies, avec une fleur grande & blanche, qui répand de l'odeur pendant la nuit.

Rava-Pon. Rheed. Mal. 4. p. 99. t. 48. Raii. Hist. 1702. Burm. Ind. 4.

Sambac. La premiere espece, qui croît naturellement dans les Indes, a été autrefois transportée dans les Isles de l'Amérique, où on la cultive comme plante d'ornement : elle s'éleve en tiges foibles & penchées à la hauteur de quinze ou vingt pieds, & pousse plusieurs petites branches garnies de feuilles ovales, unies, de trois pouces environ de longueur sur à-peu-près deux de largeur, d'un vert clair, opposées, placées sur de courts périoles, & terminées en pointe aiguë : ses fleurs naissent latéralement vers les extrémités des branches & des rejettons, sur de courts pédoncules, qui en soutiennent toujours trois, dont les deux inferieures sont opposées, & celle du milieu plus longue. Ces fleurs ont des calices cylindriques, courts & découpés presque jusqu'au fond en huit segmens étroits & étendus tout-à-fait à plat : elles sont d'un blanc pur, & répandent une odeur très-agréable, qui ressemble à celle de la fleur d'Orange, mais plus douce. Lorsque ces fleurs sont ouvertes, elles tombent à la moindre secousse, & souvent d'elles-mêmes, pendant la nuit ; de sorte que la terre au-dessous, lorsqu'elles sont entierement fleuries, s'en trouve couverte tous les matins : elles prennent bien-tôt ensuite une couleur de pourpre. Cette plante produit des fleurs durant une grande partie de l'année, lorsqu'on la tient dans une serre de chaleur convenable.

Il y a une variété de cette

efpece à fleurs très-larges, doubles, & d'une odeur extrêmement fuave : elle croît naturellement fur la côte de Malabar, où les femmes enfilent ces fleurs pour les mettre autour de leur cou en guife de collier. Cette plante étoit cultivée, il y a quelques années, dans les jardins de Hampton-Court ; mais elle y a péri, avec plufieurs autres très-rares, par l'ignorance du Jardinier. Depuis lors on ne la trouvoit plus en Europe, excepté dans les feuls jardins du Grand Duc de Tofcane, où on la fait garder à vue, afin d'empêcher que l'on n'en prenne ni boutures ni marcottes pour la multiplier : mais j'ai reçu dernierement une plante de cette efpece de la côte de Malabar, avec plufieurs autres fort curieufes qui m'ont été apportées par le Capitaine QUICK : elle eft à préfent en fi bon état, & fleurit fi bien, que j'efpere pouvoir bientôt la multiplier confidérablement ; ce qui fera une grande acquifition pour les jardins anglois.

LINNÉE a pris l'efpece de Jafmin appelée *Gardenia*, pour celui-ci ; mais depuis que ma plante a fleuri, on a reconnu qu'elle eft une variété du *Nyctanthes* : fes fleurs prennent une couleur pourpre avant de tomber, & celles du *Gardenia* fe changent en couleur de buffle : en outre, le *Nyctanthes* eft une plante foible, qui eft toujours penchée, au lieu que le *Gardenia* vient en tige droite. Le même Auteur fe trompe

encore beaucoup, en la confondant avec la plante de RUMPHIUS ; car ces deux plantes different en beaucoup de chofes, comme il paroît par le deffin qu'en a donné BURMANN. Si LINNÉE l'avoit comparée à ce deffin, en faifant attention à la defcription donnée de cette plante dans le jardin de Pife, il n'auroit pas confondu ces deux efpeces, & ne les auroit pas regardées comme n'en formant qu'une feule.

Hirfuta. La feconde eft originaire des Indes, où elle s'éleve à la hauteur de trois pieds, & fe divife en plufieurs branches garnies de feuilles larges, ovales, unies, d'un vert luifant, & portées fur des pétioles velus ; elles fortent fans ordre dans la circonférence des branches : fes fleurs font produites aux aîles des feuilles fur les côtés des branches, foutenues par des pédoncules longs & velus, qui en ont chacun fept ou huit ; elles répandent une très-bonne odeur : mais elles ont des tubes plus longs que ceux de l'efpece précédente. Ces fleurs s'ouvrent dans la foirée, & tombent au matin ; ce qui a fait donner à cette efpece, par plufieurs, le nom d'*Arbor triftis*. Elle eft fort rare à préfent en Europe.

Culture. Les Jardiniers q font un commerce d'Orangers, nous apportent fouvent de l'Italie les plantes de la premiere efpece : mais elles font toujours greffées fur des tiges de *Jafmin* commun ; ce qui eft caufe qu'elles ne croiffent pas auffi vite, & qu'elles deviennent

désagréables à la vue, quand elles font parvenues à une certaine grandeur : leurs tiges font auffi fort fujettes à repouffer de leurs racines des rejettons, qui attirent toute la nourriture de la greffe, & la font périr, fi l'on n'a pas conftamment le foin de les arracher : c'eft-pourquoi la meilleure méthode eft de les multiplier par marcottes ou par boutures. Les marcottes font plus fures ; car fi on ne traite les boutures avec le plus grand foin, elles ne prennent point racine ; & comme les tiges de cette efpece fe plient aifément, il eft facile de les coucher dans des pots remplis de terre douce & marneufe, que l'on plonge dans une couche chaude de tan. En faifant cette opération au printems, ces boutures auront des racines pour l'automne fuivant ; & fi on les arrofe avec foin, elles pourront alors être féparées des vieilles plantes, pour être mifes chacune féparément dans de petits pots : on les replongera dans une couche de tan, & on les tiendra à l'ombre, jufqu'à ce qu'elles aient formé de nouvelles racines.

Si l'on veut les multiplier par boutures, il faut les planter depuis le mois de Mai jufqu'au mois d'Août, dans des pots remplis d'une terre douce & marneufe, & les plonger dans une couche de tan de chaleur tempérée. Ces pots doivent être affez larges pour contenir chacun dix à douze boutures. Si on les couvre exactement avec des cloches à Melon, pour en exclure l'air extérieur, elles prendront bientôt racine : il faut auffi les tenir à l'ombre pendant la chaleur du jour, & les arrofer légèrement, lorfque la terre eft deffechée. De cette maniere, ces boutures auront pouffé des racines pour le mois d'Août ; alors on pourra les tranfplanter dans des pots féparés, & les traiter enfuite comme les marcottes.

Ces plantes peuvent être confervées à un dégré de chaleur modérée ; mais en les plongeant dans la couche de la ferre chaude, elles profiteront beaucoup mieux, & produiront une grande quantité de fleurs. Comme elles confervent leurs feuilles toute l'année, elles font un bel effet dans les ferres pendant toutes les faifons.

La feconde efpece exige le même traitement ; mais comme elle eft beaucoup plus difficile à multiplier, elle eft très-rare en Europe. On a apporté de Florence, depuis quelques années, deux ou trois de ces plantes, qui, ayant été confiées à une perfonne peu habile, ont été bientôt détruites.

NYMPHÆA. *Tourn. Inft. R. H.* 260. *tab.* 137. 128. *Lin. Gen. Plant.* 579, ainfi nommée, parce qu'elle croît dans l'eau, que les Poëtes ont imaginé être la réfidence des Nymphes. [*The Water Lily.*] Nenufar ou Lys d'Eau.

Caracteres. Le calice de la fleur eft perfiftant, & compofé de quatre ou cinq feuilles colorées. La corolle a plufieurs

pétales plus petits que le calice, pôftés fur le côté du germe, la plupart dans une enchaînure fimple : la fleur a un grand nombre d'étamines courtes, unies, recourbées, & terminées par des fommets oblongs comme des fils, & placés latéralement ; fon germe eft gros, ovale, fans ftyle, mais avec un ftigmat orbiculaire, uni, en forme de bouclier, rapproché, perfiftant, & dont le bord eft crenelé ; le germe fe change dans la fuite en un fruit dur, ovale & charnu, avec un cou rude & étroit, couronné au fommet, & divifé en dix ou quinze cellules remplies de chair, garnies de plufieurs femences rondes.

Ce genre de plantes eft rangé dans la premiere feftion de la treizieme Claffe de LINNÉE, qui comprend celles dont les fleurs ont un grand nombre d'étamines, & un feul ftyle.

Les efpeces font :

1°. *Nymphæa lutea, foliis cordatis, integerrimis, calyce petalis majore, pentaphyllo. Flor. Lap.* 218. *Fl. Suec.* 426. 469. *Hort. Cliff.* 203. *Roy. Lugd.-B.* 480. *Dalib. Paris.* 150. *Gmel. Sib.* 4. *p.* 142. *Scop. carn.* 2. *n.* 639. *Pollich. pal. n.* 508 ; Nenufar avec des feuilles entieres & en forme de cœur, dont les calices ont cinq feuilles plus grandes que les pétales.

Nymphæa lutea major. C. B. P. 193.; le plus grand Lys d'eau, jaune.

Nymphea lutea. Cam. Epit. 635.

2°. *Nymphæa alba, foliis cordatis integerrimis, calyce quadrifido. Lin. Sp. Plant.* 510. *Mat.*

Med. p. 135. *Gmel. Sib.* 4. *p.* 183. *t.* 72. *Crantz. Auftr. p.* 142. *Pollich. pal. n.* 509. *Mattufch. Sil. n.* 380. *Scop. carn. n.* 640. *Fl. Dan. t.* 602. *Blackw. t.* 498. 499 ; Nenufar blanc, avec des feuilles entieres & en forme de cœur, ayant un calice à quatre feuilles.

Nymphæa alba. Cam. Epit. 634.

Nymphæa alba major. C. B. P.; le plus grand Lys d'eau, blanc.

Nymphæa alba minor. Hort. Aichft. Vern. ord. 7. *t.* 3. *f.* 1 ; Variété.

Il y a quelques autres efpeces de ce genre, qui font originaires des pays chauds : mais comme elles ne peuvent être cultivées dans nos climats fans beaucoup de difficulté, je n'en parlerai point ; car elles ne réuffiroient pas, à moins qu'on n'imaginât un moyen d'avoir une eau dormante dans la ferre chaude, où l'on pourroit y placer ces plantes : mais comme l'humidité que cette eau répandroit feroit très-contraire aux autres plantes, il faudroit avoir une ferre particuliere, qui fût uniquement deftinée aux plantes aquatiques.

Les deux efpeces ci-deffus croiffent dans des eaux ftagnantes de plufieurs parties de l'Angleterre ; elles ont de groffes racines qui s'enfoncent dans la terre, & leurs feuilles, qui s'étendent & flottent fur l'eau, font larges, rondes, & en forme de cœur : les fleurs s'elevent entre les feuilles, & nagent fur la furface de l'eau. L'efpece blanche a une odeur douce & foible ; les fleurs paroiffent en Juillet, & font remplacées

N Y M

par des capsules groffes, rondes, & remplies de femences noires & luifantes, qui mûriffent vers la fin d'Août, & tombent alors au fond de l'eau.

Le meilleur moyen de multiplier ces plantes, eft de fe procurer quelques-unes de leurs capfules toutes prêtes à mùrir & à s'ouvrir, & de les jetter dans des canaux ou eaux dormantes. Ces femences s'y enfonceront, & les plantes qui paroîtront au printems fuivant, produiront de groffes fleurs en

N Y M

Juillet. Ces plantes fe multiplient fi confidérablement , quand elles font une fois établies dans un lieu, qu'en peu d'années elles couvrent toute la furface de l'eau.

J'ai vu cultiver ces plantes dans de petits jardins, en fe fervant de grands baquets remplis d'eau ; elles y réuffiffoient très-bien , & produifoient chaque année une grande quantité de fleurs : mais comme ces baquets doivent être doublés en plomb, peu de perfonnes veulent en faire la dépenfe (1).

O

O C Y

O C Y

OBELISCOTHECA. *Voy.* RUDBECKIA.

OBIER. *Voy.* VIBURNUM OPULUS. *L.*

OCHRUS. *Voyez* PISUM. OCHRUS. *L.*

OCRE. *V.* PISUM OCHRUS. *L.*

OCULUS CHRISTI. *Voy.* HORMINUM SYLVESTRE.

OCYMUM. *Tourn. Inft. R. H.* 203. *tab.* 96. *Lin. Gen. Plant.* 651. [*Bafil.*] Bafilic.

·Caractères. Le calice de la fleur eft court, perfiftant, & formé par une feuille divifée en deux levres, dont la fupérieure eft unie, en forme de cœur, & partagée en deux parties, & l'inférieure eft découpée

en quatre fegmens aigus : la corolle eft labiée, monopétale , & renverfée ; elle a un tube

---

(1) Les racines du Nymphæa font d'un ufage affez fréquent en Médecine : mais elles ont peu de vertus , & on pourroit fort bien s'en paffer ; car nous avons une multitude de fimples qu'on pourroit leur fubftituer avec beaucoup d'avantages. Ces racines font fans odeur ; leur faveur eft un peu aftringente & amere, & elles ne peuvent agir que bien foiblement fur la membrane de l'eftomac : leur vertu tempérante n'eft pas mieux fondée, quoiqu'on les emploie avec confiance dans les fievres ardentes , les infomnies , dans l'ardeur d'urine , l'inflammation des vifceres , &c.

262 O C Y

court & étendu ; la levre droite est large , & divisée en quatre parties obtuses & égales ; la levre réfléchie est longue , étroite & sciée : la fleur a quatre étamines placées dans la levre inférieure , & réfléchies, dont deux font un peu plus longues que les autres , & qui font toutes terminées par des antheres en forme de croissant. Le germe, qui est divisé en quatre parties, soutient un style mince , placé avec les étamines , & couronné par un stigmat partagé en deux segmens. Ce germe se change, quand la fleur est passée , en quatre femences nues, renfermées dans le calice.

Ce genre de plantes est rangé dans la premiere section de la quatorzieme classe de Linnée , qui comprend celles dont les fleurs ont deux étamines longues & deux courtes, & dont les femences n'ont point de péricarpe.

Les especes font :

1º. Ocymum Basilicum , foliis ovatis glabris, calycibus ciliatis. Hort. Cliff. 315. Hort. Ups. 168. Roy. Lugd. B. 322. Mat. Med. 154. Kniph. cent. 4. n. 56. Regn. Bot. ; Basilic avec des feuilles ovales & unies, & des calices velus.

Ocymum caryophyllatum majus. C. B. P. 226 ; le plus grand Basilic, ou Basilic commun à odeur de Cloux de Giroffle.

2º. Ocymum minimum , foliis ovatis integerrimis. Hort. Upsal. 169 ; Basilic à feuilles ovales & entieres.

Ocymum foliis ovatis incanis.

Hort. Cliff. 315. Roy. Lugd. B. 322.

Ocymum minimum. C. B. P. 226 ; le plus petit Basilic , communément appelé Basilic en buisson.

3º. Ocymum medium hirsutum , foliis ovato-lanceolatis , acuminatis , dentatis ; Basilic velu , à feuilles ovales , en forme de lance , dentelées, & à pointe aiguë.

Ocymum medium vulgatius & nigrum. J. B. 3. p. 2. 247 ; Basilic commun d'une moyenne grandeur , & noir.

4º. Ocymum Americanum , foliis ovato-oblongis , serratis , bracteis cordatis , reflexis , concavis , spicis filiformibus. Lin. Sp. Plant. 833 ; Basilic à feuilles ovales , oblongues, & sciées, dont les bractées font en forme de cœur , réfléchies & concaves , avec des épis fort minces.

5º. Ocymum Campechianum , foliis lanceolatis , subtùs incanis, petiolis longissimis villosis , floribus pedunculatis ; Basilic à feuilles en forme de lance , & blanches en dessous, avec des pétioles fort longs , & velus aux feuilles, & des pédoncules aux fleurs.

Ocymum Campechianum odoratissimum. Houst. M S S. ; Basilic de Campèche très-odorant.

6º. Ocymum frutescens , racemis fœcundis lateralibus , caule erecto. Lin. Sp. Plant. 832 ; Basilic avec des épis de fleurs fructueux, qui naissent sur les côtés de la tige , & une tige droite.

Ocymum Zeylanicum , perenne , odoratissimum, lati-folium. Burm.

OCY

Zeyl. 174. tab. 80. fol. 1 ; Bafilic vivace de Ceylan, à larges feuilles, & à odeur douce.

Ocymum gratiffimum. Lin. Syft. Plant. tom. 3. p. 93. Sp. 3.

Les trois premieres especes, qui croiffent naturellement dans les Indes & en Perfe, donnent beaucoup de variétés, qui different par la grandeur, la forme & la couleur de leurs feuilles, ainfi que par leur odeur : mais comme ces différences font accidentelles, je n'en parlerai point ici, étant convaincu, par des expériences réitérées, que les femences de chaque plante produifent toujours plufieurs variétés.

La premiere efpece s'éleve avec une tige branchue à la hauteur d'un pied & demi; ces feuilles font larges, ovales & unies; fa tige eft velue & quarrée; fes feuilles font placées par paires & oppofées, & fes branches fortent de la même maniere; fa tige eft terminée en épis de fleurs, placés fans ordre, & de cinq ou fix pouces de longueur; les branches en produifent de femblables, quoique plus courts & placés de même. Toutes les parties de cette plante ont une odeur forte de *Cloux de Giroffle.*

Les variétés de cette plante font :

1. *Ocymum* à feuilles de Bafilic, de couleur pourpre, & garnies de franges.

2. *Ocymum* à feuilles de Bafilic, vertes & frangées.

3. *Ocymum* vert, dont les feuilles font garnies de cloux.

4. *Ocymum* à larges feuilles de Bafilic.

Minimum. La feconde efpece eft une plante baffe & touffue, de fix pouces de hauteur au plus, garnie de branches vers le bas, & qui forme une tête ronde; fes feuilles font petites, ovales, unies, & oppofées fur de courts pétioles : fes fleurs, qui naiffent en grand nombre vers le fommet des branches, font plus petites que celles de la précédente, & produifent rarement des femences mures en Angleterre.

Voici quelques variétés de cette efpece.

1. Le plus petit *Bafilic*, à feuilles d'un pourpre noir.

2. Le plus petit *Bafilic*, à feuilles panachées.

Medium. La troifieme efpece eft le *Bafilic* commun, dont on fait ufage en Médecine, ainfi que dans la cuifine, fur-tout chez les François, qui en mettent beaucoup dans leurs potages & leurs ragoûts. Cette plante s'éleve à dix pouces de hauteur, & pouffe vers le bas des branches difpofées par paires, & oppofées; fes tiges & fes branches font quarrées; fes feuilles font ovales, en forme de lance, terminées en pointe aiguë, & dentelées fur leurs bords. Toutes les parties de cette plante font velues, & répandent une odeur forte de Cloux de Giroffle, qui déplaît à bien des gens, mais qui plaît à d'autres. Cette efpece entre dans la compofition de l'eau Brionnée (1).

_____

(1) Toutes les parties de cette

Les variétés de cette espece font :

1. Le *Bafilic* commun, à feuilles très-vertes, & à fleurs violettes.

2. Le *Bafilic* à feuilles frifées, avec des épis de fleurs courts.

3. Le *Bafilic* à feuilles étroites & à odeur de Fenouil.

4. Le *Bafilic* moyen à odeur de citron.

5. Le *Bafilic* à feuilles garnies de cloux.

6. Le *Bafilic* à feuilles de trois couleurs.

Americanum. La quatrieme efpece croît naturellement dans les Indes : elle s'éleve, avec des tiges branchues & de couleur pourpre, à la hauteur d'un pied & demi, en forme cylindrique ou en pyramide ; fes feuilles font courtes, veloutées, de forme ovale & oblongue, terminées en pointe obtufe, dentelées fur leurs bords, & placées fur des pétioles affez longs : fes tiges font terminées par trois épis de fleurs, dont celui du milieu eft le plus long. Ces épis font longs & minces : fes fleurs naiffent fur de courts pédoncules ; fous chaque paquet de

plante font très-nervines, céphaliques, cardiaques, carminatives, utérines, & peuvent être employées avec fuccès en infufion vineufe dans les différentes maladies contre lefquelles on preferit la Méliffe.

Les femences de Bafilic entrent dans la compofition de la poudre de Guttete, dans le Tryfera de Nicolas d'Alexandrie, dans la poudre de Diamofchi, de Mefué, &c.

fleurs font deux petites bractées oppofées, en forme de cœur, concaves & réfléchies. Les fleurs font petites : dans quelques plantes, elles font de couleur pourpre ; mais en général toujours blanches : leurs calices font unis, & divifés en cinq parties fur leurs bords ; le ftyle de la fleur eft plus long que la corolle. Cette plante répand une odeur agréable, forte & aromatique.

Campechianum. La cinquieme efpece s'éleve avec une tige droite à la hauteur d'environ deux pieds, & pouffe vers fon fommet deux & quelquefois quatre branches oppofées, & garnies de feuilles en forme de lance ; leurs pétioles ont deux pouces de longueur, & font velus : les fleurs font difpofées dans des épis en bâle au fommet des tiges, fur des pédoncules qui en foutiennent chacun trois ; elles font à-peu près de la grandeur de celles du *Bafilic* commun, & blanches. Toute cette plante a une odeur forte & aromatique.

Frutefcens. La fixieme efpece eft originaire de l'Ifle de Ceylan ; elle s'éleve à la hauteur d'environ un pied, avec une tige branchue & garnie de feuilles linéaires en forme de lance, & dentelées : fes fleurs croiffent en épis, garnies de bâles aux extrémités des tiges, comme celles du *Bafilic* ordinaire. Toutes les parties de cette plante répandent une odeur d'anis.

Culture. La plupart de ces efpeces font annuelles, & fe multiplient par leurs graines,

qui doivent être femées en Mars fur une couche de chaleur tempérée. Lorfque ces plantes ont pouffé, on les tranfporte fur une autre couche tempérée, on les arrofe, & on les tient à l'ombre, jufqu'à ce qu'elles aient formé de nouvelles racines; après quoi, on leur donne beaucoup d'air dans les tems doux, pour les empêcher de filer & de s'affoiblir. Dans le mois de Mai, on les enleve avec une motte de terre à leurs racines; on les tranfplante dans des pots, ou dans les plates-bandes, & on les tient à l'ombre jufqu'à ce qu'elles aient pouffé de nouvelles fibres: elles n'exigent plus enfuite aucun autre foin que d'être tenues nettes de mauvaifes herbes, & d'être arrofées dans les tems chauds.

Quoiqu'on ne multiplie ordinairement ces plantes que par femences, cependant, quand on a quelques efpeces particulieres venues de graines, & qu'on veut les conferver, on peut les propager par boutures, qu'on plante, pendant les mois de Mai ou de Juin, fur une couche de chaleur tempérée; on les abrite pendant environ dix jours, jufqu'à ce qu'elles aient pouffé des racines; & au bout de trois femaines, elles feront en état d'être enlevées, & d'être placées dans des pots, ou dans les plates-bandes avec les plantes de femences.

Ces plantes perfectionnent leurs graines dans le mois de Septembre; alors on choifit dans chaque efpece les plus belles tiges, pour en conferver les femences féparément jufqu'au printems fuivant.

On nous apporte ordinairement au printems les femences de ces plantes de la France Méridionale ou de l'Italie, parce qu'elles mûriffent rarement en plein air dans ce pays; mais quand on eft curieux de fe procurer celles de quelques variétés, il faut placer ces plantes dans des caiffes de vitrage aërées, ou dans une ferre chaude en automne, lorfque le tems commence à être froid & humide; & en les arrofant à propos, elles perfectionneront très bien leurs femences en Angleterre.

La cinquieme efpece, qui eft beaucoup plus tendre qu'aucune des autres, a été découverte à Campêche par le Docteur GUILLAUME HOUSTOUN: il faut répandre fes graines fur une couche chaude au commencement du printems; lorfque les plantes ont pouffé, on les tranfporte fur une autre couche fort tempérée, pour les faire avancer; quand elles ont acquis de la force, on les met chacune féparément dans des pots, & on les place fur une couche de chaleur modérée ou dans la ferre, en leur procurant beaucoup d'air dans les tems chauds. Quand elles font à couvert du froid & de l'humidité, elles perfectionnent affez bien leurs femences en Angleterre.

ŒIL DE BŒUF. *Voy.* ANTHEMIS TINCTORIA, BUPH-

266 ŒNA

TALMUM , & CHRYSANTHE-MUM.

ŒIL DE BOURIQUE. Voyez DOLICHO-URENS.

ŒIL DE CHRIST ou AS-TE Voy. ASTER AMELLUS.

ŒIL D'OISEAU. Voy. A-DON...

ŒILLET. Voy. DYANTHUS CARYOPHYLLUS.

ŒILLET D'INDE. Voyez T...

ŒILLET MARIN ou STA-TI... Voy. STATICE

ŒNANTHE. Tourn. Inst. R. ... 166. Lin. Gen. Plan. , de οἶνος de οὖν un sep de ... , & ἄνθος, une fleur. Les Anciens appeloient Œnan-the toutes les plantes qui fleu-rissent en même tems que la Vigne , ou dont les fleurs ont la même odeur [Water Drop-wort.] Filipendule aquatique ou l'Œnanthe.

Caractères. La fleur de cette plante est ombellée : l'ombel-le principale n'a que peu de rayons ; mais les ombelles particulieres en ont plusieurs courts ; l'enveloppe principale est composée de plusieurs feuil-les simples , plus courtes que l'ombelle ; les plus petites om-belles ont plusieurs petites feuilles : les rayons de l'om-belle principale sont irrégu-liers ; les fleurs du disque sont hermaphrodites , & composées de cinq pétales en forme de cœur , courbes & presqu'égaux: celles des rayons sont males , & ont cinq larges pétales iné-gaux , & divisés en deux par-ties , & cinq étamines simples , terminées par des antheres ron-des. Le germe , qui est situé

ŒNA

sous la fleur hermaphrodite , soutient deux styles en forme d'alène , persistant , & couron-nés par des stigmats obtus. Ce germe se change dans la suite en un fruit ovaie , divisé en deux parties , & qui renferme deux semences presqu'ovales , convexes d'un côté , & unies sur l'autre.

Ce genre de plantes est ran-gé dans la seconde section de la cinquieme classe de LINNÉE , qui comprend celles dont les fleurs ont cinq étamines & deux styles.

Les especes sont :

1°. Œnanthe crocata foliis omnibus multi-fidis, obtusis, sub-æqualibus. Hort. Cliff. 99. Flor. 237. 251. Roy. Lugd.-B. 107. Blackw. f. 575. Jacq. Hort. Val. 3. f. 55 , Œnanthe dont toutes les feuilles se terminent en plusieurs pointes obtuses & presque égales.

Œnanthe Chærophylli foliis. Bauh. Pin. 162.

Œnanthe succo viroso , Cicutæ facie , Lobelii. J. B. 3. p. 2. 193 , Œnanthe à suc de Ciguë.

2°. Œnanthe fistulosa , stolo-nifera , foliis caulinis , pinnatis , fili-formibus , fistulosis. Lin. Sp. Plant. 254. Crantz. Austr. 201. Neck. Gallob. 150. Pollich. pal. n. 290. Mattusch. Sil. n. 199. Kniph. cent. 5. n. 60. Dœrr. Nass. p. 160 ; Filipendule aqua-tique , dont les feuilles des ti-ges sont étroites , fistuleuses , & ailées.

Œnanthe aquatica. C. B. P. 162 ; Filipendule aquatique.

Œnanthe foliis caulinis , fistu-losis , teretibus. Hort. Cliff. 99.

ŒNA

Flor. Suec. 236. 250. *Roy. Lugd.-B.* 108.

Œnanthe aquatica triflora. Moris. Hist 3. *p.* 269. *S.* 9. *t.* 7. *f.* 8. *Juncus odoratus. Dodon. Cer.* 242.

3°. *Œnanthe Pimpinelloides, foliolis radicalibus cuneatis, fissis caulinis integris, linearibus, longissimis, canaliculatis. Hort. Cliff* 99. *Roy. Lugd-B.* 108. *Sauv. Monsp.* 259. *Jacq. Austr. f.* 394. *Crantz. Austr. p.* 201. *Scop. carn.* 2. *n.* 364. *Pollich. pal. n.* 291 : Filipendule aquatique, dont les feuilles du bas sont ovales & découpées, & celles des tiges entieres, étroites, très longues, & canelées.

Œnanthe Apii folio. C. B. P. 162 ; Filipendule aquatique à feuilles d'Ache.

Œnanthe aquatica Pimpinellæ saxifragæ divisura. Pluk. Alm. 268. *f.* 49.

4°. *Œnanthe prolifera, umbellularum pedunculis marginalibus, longioribus, ramosis, masculis. Hort. Ups.* 63. *Jacq. Hort. vol.* 3. *f.* 62 ; Filipendule aquatique, dont les pédoncules sur le bord des ombelles sont plus longs, branchus, & portent des fleurs mâles.

Œnanthe prolifera Apula. C. B. P. 193 ; Filipendule fertile de la Pouille.

5°. *Œnanthe globulosa, fructibus globosis. Hort. Cliff.* 99. *Roy. Lugd.-B.* 108 ; Filipendule à fruits ronds.

Œnanthe foliis bipinnatis, fructibus globosis. Gouan. Illustr. 18. *f.* 9.

Œnanthe Lusitanica, semine crassiori globoso. Tourn. Inst. 313 ; Filipendule de Portugal, avec une semence plus charnue, & globulaire.

ŒNA 267

Crocata. La premiere espece est fort commune sur les bords de la Tamise aux environs de Londres, ainsi que sur les bords des grands fossés & des rivieres dans differentes parties de l'Angleterre. Cette plante, qui s'éleve communément à quatre ou cinq pieds de hauteur, pousse des tiges fortes & noueuses, & remplies d'une séve jaunâtre & fétide, qui s'écoule quand on les rompt : ses feuilles ressemblent un peu à celles de la Ciguë ordinaire ; mais elles sont d'un vert plus clair ; ses racines se divisent en quatre ou cinq grosses parties de forme cylindrique, qui, quand on les sépare, ressemblent beaucoup à celles des Panais. Quelques personnes, faute de les connoître, en ayant mangé, ont été empoisonnées avec toute leur famille.

Cette plante est une des plus vénimeuses que nous connoissions : le suc qui en découle ressemble d'abord à du lait, & prend ensuite une couleur de safran : pour peu que l'on en avale, tout ce que ce jus touche se crispe ; immédiatement après survient une inflammation, à laquelle succede une terrible gangrene ; &, ce qui est pis encore, on ne connoît aucun antidote contre ce poison : c'est pourquoi on doit avoir grand soin de reconnoître cette plante, pour l'éviter ; car, comme nous venons de le dire, son usage est certainement funeste.

Les qualités pernicieuses de cette plante ont fait penser

ŒNO

à quelques perſonnes qu'elle pourroit bien être la *Cigue* des anciens : mais, ſuivant WEPFER, le *Sium alterum Olmaſtri facie* de LOBEL, eſt ce que les Anciens appelloient *Cituta*, ainſi qu'on peut le voir dans le livre même de WEPFER, qui a pour titre, *De Cituta*.

Fiſtuloſa. La ſeconde eſpece eſt fort commune dans les terres humides & ſur les bords des rivieres, dans differentes parties de l'Angleterre : on la regarde comme auſſi dangereuſe que la premiere.

Comme toutes ces eſpeces croiſſent naturellement dans des lieux humides, lorſqu'on veut les cultiver, il faut les ſemer en automne, auſſi-tôt que leurs graines ſont mûres, ſur un ſol humide, où elles pouſſeront & profiteront très-bien. Elles n'exigent aucun autre ſoin que d'être tenues nettes de mauvaiſes herbes.

ŒNOTHERA. *Lin. Gen. Plant.* 424. *Onagra. Tourn. Inſt. R. H.* 302. *tab.* 156. [*Tree Primroſe.*] Primevere en arbre. L'Herbe aux Anes.

Caracteres. Le calice de la fleur eſt formé par une feuille ; ſon tube eſt long, cylindrique, & découpé ſur ſes bords en quatre ſegmens aigus & tournés en arriere ; la corolle a quatre pétales en forme de cœur, inſérés en longueur dans les diviſions du calice : la fleur a huit étamines en forme d'alène, recourbées, fixées dans le tube du calice, & terminées par des antheres oblongues & penchées ; ſon germe, qui eſt

cylindrique & placé ſous le tube du calice, ſoutient un ſtyle mince, couronné par un ſtigmat épais, & diviſé en quatre parties obtuſes & réfléchies. Ce germe ſe change, quand la fleur eſt paſſée, en une capſule cylindrique quarrée, & à quatre cellules remplies de petites ſemences angulaires.

Ce genre de plantes eſt rangé dans la premiere ſection de la huitieme claſſe de LINNÉE, dans laquelle ſont compriſes celles dont les fleurs ont huit étamines & un ſtyle.

Les eſpeces ſont :

1°. Œnothera biennis , *foliis ovato-lanceolatis, planis, caule muricato, ſubvilloſo. Vir. Cliff.* 33. *Hort. Ups.* 94. *Gron. Virg.* 254. *Roy. Lugd.-B.* 251. *Hall. Helv. n.* 994 ; Herbe aux Anes, à feuilles unies, ovales, & en forme de lance, avec une tige rude & velue.

Lyſimachia lutea corniculata. Bauh. Pin. 245. 516. *Moris. Hiſt.* 2. *p.* 271.

Hyoſciamus Virginianus. Alp. Exot. 325. *f.* 324.

Onagra lati-folia. Tourn. Inſt. 382 ; Primevere en arbre, à larges feuilles, ou le Jambon.

2°. Œnothera anguſti-folia, *foliis lanceolatis, dentatis, caule hiſpido* ; Herbe aux Anes, à feuilles en forme de lance & dentelées, avec une tige épineuſe.

Œnothera parvi-flora. Linn. Syſt. Plant. tom. 2. *pag.* 147. *Sp.* 2.

Onagra anguſti folia, caule rubro, flore minore. Tourn. Inſt. R. H. 302 ; Primevere en arbre à feuilles étroites, avec une tige

ŒNO

tige rouge & une plus petite fleur.

3°. *Œnothera glabra* , *foliis lanceolatis* , *planis* , *caule glabro* ; Herbe aux Anes , à feuilles unies & en forme de lance , avec une tige liffe.

4°. *Œnothera molliffima* , *foliis lanceolatis undulatis. Vir. Cliff.* 33. *Gron. Virg* 42. *Roy. Lugd. B.* 251. *Kniph. cent.* 4. *n.* 57 ; Herbe aux Anes , à feuilles ondées , & en forme de lance.

Œnothera foliis lineari-lanceolatis dentatis , *floribus è medio caule. Hort. Ups.* 144.

Onagra Bonarienfis villofa , *flore mutabili. Hort. Elth.* 297. *t.* 219. *f.* 286 ; Primevere en arbre de Buenos-Ayrès , velue & à fleurs changeantes.

5°. *Œnothera pumila* , *foliis radicalibus ovatis* , *caulinis lanceolatis* , *obtufis* , *capfulis ovatis* , *fulcatis.* ; Herbe aux Anes , dont les feuilles radicales font ovales , & celles des tiges en forme de lance & obtufes , avec des capfules ovales & fillonnées.

Lyfimachia Marylandica parva , *foliis anguftis* , *acutis. Raii Suppl.* 416.

Les autres efpeces qui étoient autrefois placées dans ce genre , fe trouvent à préfent fous le titre de JUSSIÆA & LUDWIGIA , auxquels je renvoie le Lecteur.

Les trois premieres croiffent naturellement dans la Virginie , & autres parties de l'Amérique Septentrionale , d'où leurs femences ont été envoyées en Europe au commencement du feizieme fiecle ; mais elles font à préfent devenues fi communes en Angleterre , qu'elles femblent en être originaires.

Biennis. La premiere a une racine longue , épaiffe & cylindrique , qui pénetre profondément dans la terre ; de cette racine naiffent plufieurs feuilles obtufes , qui s'étendent fur la terre ; entr'elles fortent des tiges qui s'élevent à la hauteur de trois ou quatre pieds ; elles font d'un vert pâle , un peu velues , à-peu-près de la groffeur du doigt , remplies de moëlle , & garnies de feuilles longues , étroites , feffiles , & placées fans ordre : fes fleurs naiffent dans toute la longueur de la tige , aux ailes des feuilles ; elles ont un tube étroit , de plus de deux pouces de longueur , & qui fort du haut du calice , qui eft découpé en quatre fegmens aigus & réfléchis vers le bas : la corolle de la fleur eft divifée en quatre fegmens larges , obtus , qui s'ouvrent dans la foirée. Plufieurs perfonnes lui donnent le nom de *Primevere de Nuit.*

Ces plantes commencent à fleurir vers la Saint-Jean , & continuent à donner de nouvelles fleurs à mefure que les tiges avancent en hauteur ; de forte que ces fleurs fe fuccedent fur la même tige jufqu'à l'automne.

Angufti-folia. La feconde a des tiges rouges , remplies de protubérances rudes , mais moins élevées que celles de la premiere ; fes feuilles font étroites , & fes fleurs plus petites.

Glabra. La troifieme differe de la premiere , en ce que fes

tiges font plus courtes ; fes feuilles plus étroites, fes fleurs plus petites ; & de la feconde, en ce que fes tiges font liffes & d'un vert pâle. Ces différences font conftantes. Ainfi, cette efpece eft certainement diftincte.

Molliffima. La quatrieme fe trouve à Buenos Ayrès ; elle a une tige d'arbriffeau de deux pieds de hauteur, velue, & garnie de feuilles étroites, en forme de lance, terminées en pointe aiguë, feffiles, & un peu ondées fur leurs bords : fes fleurs fortent aux ailes des feuilles dans la longueur des tiges, comme dans les autres efpeces, elles font d'abord d'un jaune pâle ; mais à mefure qu'elles fe flétriffent, elles fe changent en une couleur d'orange ; elles font plus petites que celles des précédentes, & ne s'épanouiffent que dans la foirée ; leurs capfules font minces, cylindriques & velues. Cette plante fleurit en même tems que la troifieme.

La cinquieme efpece eft originaire du Canada, d'où fes femences ont été envoyées à Paris il y a quelques années. Cette plante eft vivace ; fa racine eft fibreufe ; fes feuilles baffes font petites, ovales, & feffiles à la terre ; fa tige eft mince, d'un pied de hauteur, & garnie de petites feuilles en forme de lance, d'un vert clair, terminées en pointe émouffée, & feffiles aux tiges : fes fleurs font produites aux ailes des feuilles comme celles des autres ef-

peces ; elles font petites, d'un jaune brillant, paroiffent en même tems que celles de la précédente, & font remplacées par des capfules courtes, ovales, fillonnées, & remplies de petites femences.

Culture. Les trois premieres efpeces font des plantes fort dures, qui fe multiplient d'elles-mêmes, fans aucun foin, quand elles font une fois établies dans un jardin, & qu'on en laiffe écarter les femences ; elles font bis-annuelles, & périffent auffi-tôt que leurs graines font mûres. Il faut les femer en automne ; car celles que l'on conferve jufqu'au printems, levent rarement dans la premiere année. Quand les plantes pouffent, on les éclaircit, & on les débarraffe des mauvaifes herbes : c'eft en cela que confiftent tous les foins qu'elles exigent jufqu'en automne ; alors on les met à demeure dans les places qui leur font deftinées ; mais comme leurs racines s'enfoncent confidérablement dans la terre, il faut avoir grand foin de ne pas les couper en les enlevant ; elles profitent dans prefque tous les fols & à toutes les expofitions ; elles fleuriffent même dans les petits jardins de Londres, mieux que la plupart des autres plantes.

La quatrieme efpece eft devenue affez commune dans les jardins anglois ; en lui laiffant écarter fes femences, les plantes pouffent au printems fuivant, & n'exigent aucun autre foin que d'être tenues nettes de mauvaifes herbes, &

OLD

éclaircies où elles font trop ferrées. Si on les conferve dans des pots, & fi on les place en automne dans une Serre, elles fubfiftent en hiver; mais comme elles produifent des fleurs & des femences en plein air, on les garde rarement plus long-tems.

La cinquieme efpece eft vivace, & peut être multipliée par femences ou par la divifion de fes racines. La premiere méthode fe pratique au printems; mais on les feme dans des pots, & on les tient en hiver fous un châffis de couche. Les plantes paroîtront au printems; & quand elles feront en état d'être enlevées, on pourra les placer chacune féparément dans de petits pots, pour pouvoir les mettre en hiver fous un châffis ordinaire. Il fuffira de tenir les autres dans une plate-bande à une bonne expofition, où elles fupporteront très-bien le froid de nos hivers, & où elles produiront des fleurs & des femences en abondance dans l'été fuivant. Il fera peu néceffaire de divifer leurs racines, parce que les plantes de femences feront beaucoup plus fortes, & fleuriront mieux que celles de rejettons.

OIGNON. *Voyez*. CEPA. *L.*
OIGNON DE MER *ou* SQUILLE. *Voy.* SCILLA. *L.*

OLDENLANDIA. *Plum. Nov. Gen. 42. tab. 36. Lin. Gen. Plant.* 143. [*Oldenlandia.*] Oldenlande.

Caraéteres. Le calice de la fleur eft perfiftant, placé fur le germe, & découpé en cinq parties; la corolle a quatre petales ovales, étendus, & une fois plus longs que le calice: la fleur a quatre étamines terminées par de petites antheres, & un germe rond placé en-deffous, qui foutient un ftyle fimple, & couronné par un ftigmat dentelé; ce germe devient enfuite une capfule globulaire, & à deux cellules remplies de petites femences.

Ce genre de plantes eft rangé dans la premiere feétion de la quatrieme claffe de LINNÉE, où fe trouvent celles dont les fleurs ont quatre étamines & un ftyle.

Nous n'avons qu'une efpece de ce genre dans les jardins anglois.

1°. *Oldenlandia corymbofa, pedunculis multi-floris, foliis lineari-lanceolatis. Lin. Sp. Plant.* 119; Oldenlande produifant plufieurs fleurs fur chaque pédoncule, & des feuilles linéaires en forme de lance.

Oldenlandia humilis Hvffopifolia. Plum. Nov. Gen. ; Oldenlande nain à feuilles d'Hyffope.

Cette plante a été découverte en Amérique par le P. PLUMIER, qui l'a ainfi nommée en l'honneur de HENRI-BERNARD OLDENLAND, Allemand, Difciple du Doéteur HERMANN de Leyde, & très-curieux en Botanique.

Les femences de cette plante ont été envoyées en Angleterre par M. ROBERT MILLAR, qui les avoit recueillies à la Jamaïque: elle eft baffe & annuelle; elle s'éleve rarement au-deffus de trois ou quatre pouces, & fe divife en plu-

fieurs branches, qui s'étendent près de la terre, & font
garnies de longues feuilles é-
troites & oppofées : du milieu
des feuilles & des aîles fort
une tige de fleurs d'un pouce
ou un peu plus de longueur,
& divifée en trois ou quatre
plus petits pédoncules, qui
foutiennent chacun à leur extrémité une petite fleur blanche.

Culture. Il faut femer au
commencement du printems les
graines de cette plante fur une
couche chaude ; quand elles
ont pouffé, on met les plantes fur une autre couche chaude, ou dans de petits pots que
l'on plonge dans une couche
de tan d'une chaleur tempérée ; on les arrofe, & on les
tient à l'ombre jufqu'à ce qu'elles aient formé de nouvelles racines, après quoi on leur procure beaucoup d'air dans les
tems chauds, & on les rafraîchit fouvent avec de l'eau : au
moyen de ce traitement, ces
plantes fleuriront en Juin, &
leurs femences mûriront en
Juillet ; à mefure que leurs
branches s'allongeront, elles
produiront de nouvelles fleurs
jufqu'à l'automne, tems auquel
les plantes périront. En leur
permettant de répandre leurs
graines dans des pots, il en
paroîtra bien-tôt d'autres qui
fubfifteront pendant l'hiver,
pourvu qu'elles foient placées
dans une ferre chaude, & elles fleuriront de bonne heure
au printems fuivant.

OLEA. *Tourn. Inft. R. H.*
598. *tab.* 370. *Lin. Gen. Plant.*
20. de bais, [*The Olive.*]
L'Olivier.

Caracteres. Le calice de la
fleur eft petit, tubulé, & formé par une feuille découpée
fur fes bords en quatre fegmens. La corolle eft compofée d'un pétale en forme d'entonnoir tubulé, & découpé en
quatre parties tout-à-fait ouvertes : la fleur a deux étamines courtes, & terminées par
des antheres érigées ; fon germe eft rond, & foutient un
ftyle court, couronné par un
ftigmat épais & divifé en deux
parties. Ce germe fe change
dans la fuite en un fruit ou
baie ovale & liffe, & à une
cellule, qui renferme un noyau
oblong & ovale.

Ce genre de plantes eft rangé dans la premiere fection de
la feconde claffe de LINNÉE,
avec celles dont les fleurs ont
deux étamines & un ftyle.

Les efpeces font :

1°. *Olea Gallica, foliis li-*
neari-lanceolatis, fubtùs incanis ;
Olivier à feuilles linéaires, en
forme de lance, & blanches
en-deffous.

Olea Europæa. Lin. Syft. Plant.
t. 1. p. 19.

Olea fruêtu oblongo minori.
Tourn. Inft. R. H. 599; Olivier à petit fruit oblong, communément appelé Olive de Provence.

Olea. Dod. Pempt. 821. *Du-*
ham. Arb. 2. p. 57.

2°. *Olea Hifpanica, foliis*
lanceolatis, fruêtu-ovato ; Olivier
à feuilles en forme de lance,
produifant un fruit de la forme d'un œuf.

Olea fruêtu maximo. Tourn.
Inft. R. H. 599; Olivier pro-

OLE

duifant le plus gros fruit , nommé Olive d'Efpagne.

3°. *Olea fylveftris , foliis lanceolatis , obtufis , rigidis , fubtùs incanis* ; Olivier à feuilles en forme de lance , obtufes , roides , & blanches en-deffous.

Olea Sylveftris , folio duro , fubtùs incano. C. B. P. 470 ; Olivier fauvage , à feuilles fermes & blanches en-deffous.

4°. *Olea Africana , foliis lanceolatis , lucidis , ramis teretibus* ; Olivier à feuilles en forme de lance , & luifantes , avec des branches cylindriques.

Olea Afra , folio longo , lato , fupra atro-viridi fplendenti , infrâ pallidè viridi. Boerrh. Ind. Alt. 2. 218 ; Olivier d'Afrique , à feuilles longues , larges , luifantes , d'un noir verdàtre en-deffus , & pâle en-deffous.

5°. *Olea Buxi-folia , foliis ovatis , rigidis , feffilibus* ;. Olivier à feuilles ovales , roides & feffiles aux branches.

Olea Afra , folio Buxi craffo , atro-viridi , lucido , cortice albo , fcabro. Boerrh. Ind. Alt. 2. 218 ; Olivier d'Afrique , à feuilles épaiffes de Buis , d'une couleur fombre & luifante , avec une écorce rude & blanche , communément nommé Olivier à feuilles de buis.

Gallica. La premiere efpece eft celle que les habitans de la France Méridionale cultivent principalement , parce qu'elle produit la meilleure huile , qui forme une grande branche de commerce pour la Provence & le Languedoc ; c'eft auffi le fruit de cette efpece que l'on eftime le plus quand il eft mariné. Il y a quelques variétés de cet arbre ; la premiere s'appelle *Olive picholine* ; il y en a une autre qui produit un fruit d'un vert foncé , une autre à fruit blanc ; & enfin une derniere avec un fruit plus petit & plus rond ; mais comme toutes ces variétés ne font que des accidens de femences , je ne les ai point mifes au nombre des efpeces diftinctes.

L'Olivier devient rarement un grand arbre ; cet arbre a ordinairement deux ou trois tiges qui fortent de la même racine , s'élevent à vingt ou trente pieds de hauteur , & pouffent , dans prefque toute leur longueur , des branches latérales , couvertes d'une écorce grife , & garnies de feuilles roides d'environ deux pouces & demi de longueur fur un & demi de largeur au milieu , mais plus étroites par dégrés vers les deux extrémités , d'un vert vif en deffus , blanches en deffous , & oppofées : fes fleurs font difpofées en petits paquets aux ailes des feuilles ; elles font petites , blanches , & ont de gros tubes qui s'ouvrent au fommet. Ces fleurs produifent un fruit ovale , qui , dans les pays chauds , mûrit en automne.

Hifpanica. La feconde efpece eft principalement cultivée en Efpagne , où ces arbres s'élevent à une hauteur plus confidérable que ceux de l'efpece précédente ; fes feuilles font beaucoup plus larges , & moins blanches en-deffous : fon fruit eft prefque deux fois plus gros que celui de l'*Olivier* de

Provence ; mais fon goût eft plus âcre, & l'huile que l'on en tire eft trop forte, & n'eft point eftimée en Angleterre.

Sylveftris. La troifieme efpece eft l'*Olivier* fauvage, qui croit naturellement dans les bois de la France Méridionale, en Efpagne & en Italie ; auffi ne la cultive-t-on jamais. Ses feuilles font beaucoup plus courtes & plus roides que celles des autres ; fes branches font fouvent armées d'épines, & fon fruit eft petit, & n'eft d'aucun ufage.

Africana. Les quatrieme & cinquieme efpeces croiffent naturellement au Cap de Bonne-Efpérance ; la quatrieme s'éleve à la même hauteur de la premiere, à laquelle elle reffemble un peu : mais l'écorce en eft plus rude, les feuilles moins longues, & d'un vert luifant en-deffus. Comme cet arbre ne produit point de fruits en Europe, je ne puis en donner aucun détail.

Buxi-folia. La cinquieme efpece, qui s'éleve rarement au-deffus de quatre ou cinq pieds de hauteur, pouffe plufieurs branches depuis fa racine jufqu'au fommet, en forme de buiffon. Ces branches font cylindriques, & couvertes d'une écorce grife ; les feuilles font ovales, fort roides, & plus petites que celles des autres efpeces. Cet arbre n'a point produit de fruits en Angleterre.

Culture. Les curieux confervent toutes ces efpeces dans leurs jardins, mais elles font trop tendres pour profiter en plein air dans le voifinage de Londres, où on les plante quelquefois contre des murailles, en les abritant un peu pendant les fortes gelées. Elles s'y entretiennent affez bien ; mais dans le Comté de Devon, quelques-uns de ces arbres ont crû en plein air pendant plufieurs années, & n'ont été que rarement endommagés par la gelée : cependant les étés n'y font pas affez chauds pour leur faire perfectionner leurs fruits. J'ai vu plufieurs de ces arbres plantés contre des murailles chaudes, à la maifon de Cambden, près de Kenfington ; ils y ont fort bien réuffi, tant que leurs fommets ne fe font point élevés au-deffus des murailles : mais auffi-tôt qu'ils les ont furpaffées, ils ont toujours été détruits pendant les hivers. En 1719, ils ont produit une grande quantité de fruits, qui font parvenus à une groffeur affez confidérable pour pouvoir être marinés : mais depuis ce tems, ces fruits font reftés petits.

L'*Olivier* a été regardé par les Anciens comme un arbre maritime, & ils penfoient qu'il ne pouvoit profiter, pour peu qu'il fût éloigné de la mer ; mais l'expérience nous a appris qu'il réuffiffoit très-bien dans tous les pays où l'air eft d'une chaleur convenable. Cependant cette efpece d'arbre fupporte mieux les vapeurs & les brouillards de la mer que la plupart des autres.

Dans le Languedoc & la Provence, où l'on cultive beaucoup d'*Oliviers*, on les multiplie avec de groffes branches qu'on détache de leurs racines. Comme

OLE

ils font fouvent endommagés en hiver par les fortes gelées, ils pouffent plufieurs tiges de leurs racines, qu'on enleve au moyen d'une hache, quand ils font devenus affez forts, en y confervant quelques racines. On fait cette opération au printems, quand le danger des gelées eft paffé ; on les plante, en les enterrant à la profondeur de deux pieds, & l'on couvre la furface avec un peu de litiere ou de terreau, pour empêcher le foleil & le vent d'y pénétrer & de fecher la terre. Quand ces tiges ont pouffé de nouvelles fibres, on laboure la terre avec foin, & l'on détruit les mauvaifes herbes qui les environnent.

Cet arbre croît prefque dans tous les fols ; mais quand il eft planté dans une terre riche & humide, il devient plus grand & a plus d'apparence que dans un mauvais terrein : cependant le fruit en eft moins eftimé que celui qui provient d'un foi maigre, & l'huile n'en eft pas fi bonne. La terre de craie eft la meilleure pour ces arbres, & l'huile qu'on extrait de leurs fruits eft plus eftimée, plus fine, & fe conferve plus long-tems.

Dans les pays où l'on eft curieux de faire de la bonne huile, on eft fouvent obligé de fe procurer des bâtons de l'efpece d'*Oliviers* communs, pour les planter. Quand *ils* ont pouffé de bonnes racines, on les greffe avec les meilleures efpeces. En Languedoc, on cultive principalement le *Cormeau*, *l'Ampallant* & le *Moureau*, qui font trois variétés de

OLE 275

la premiere efpece ; mais en Efpagne, ils emploient toujours la feconde, parce qu'ils cherchent la groffeur du fruit, & une plus grande quantité, qu'ils préferent à la qualité. Si la culture de ces arbres étoit bien entendue par les habitans de la Caroline, & qu'ils y donnaffent tous leurs foins, elle pourroit leur procurer une branche de commerce d'une grande valeur ; car on ne peut pas douter du fuccès, les étés étant affez chauds pour mûrir ce fruit, & le conduire à fa plus grande perfection.

En Angleterre, on ne conferve ces arbres que par curiofité, comme on les place en hiver dans les Serres, pour en augmenter la variété. Je vais donner ci-après une méthode pour les multiplier & les traiter convenablement.

On peut multiplier ces plantes, en marcottant leurs jeunes branches, comme on le pratique pour les autres arbres : on les laiffe pendant deux ans fans y toucher ; mais comme après ce tems, elles auront pouffé des racines, on pourra les féparer alors des vieilles plantes, & les mettre dans des pots remplis d'une terre fraiche & légere, ou en pleine terre à une expofition chaude.

La meilleure faifon pour les tranfplanter, eft le commencement d'Avril, en choififfant pour cela, s'il eft poffible, un tems humide. Celles que l'on met dans des pots doivent être tenues à l'ombre dans la Serre, jufqu'à ce qu'elles

S 4

OLE

äient formé de nouvelles racines ; mais on doit répandre
du terreau autour des racines
de celles qui font en plein air,
pour empêcher la terre de fecher trop vite , & on les arrofe de tems en tems, en obfervant cependant de ne jamais
leur donner trop d'humidité ,
qui feroit pourrir les tendres
rejettons de leurs racines, &
détruiroit les arbres. Quand les
plantes font bien enracinées,
on peut expofer celles des pots
en plein air avec les autres
efpeces exotiques dures, & les
placer en hiver dans la Serre,
où elles doivent être traitées
comme les *Myrtes*, & les autres
arbres & arbriffeaux moins tendres : mais celles de pleine terre
n'exigent aucun foin jufqu'en
hiver ; alors on répand du
terreau fur la terre autour de
leurs racines , pour empêcher
la gelée d'y pénétrer profondément ; & fi le froid devient
confidérable, on les couvre de
nattes pour les abriter : & ,
afin qu'elles ne foient point
endommagées, on a foin d'enlever les couvertures, quand
la gelée eft paffée, de peur
que les feuilles & les tendres
branches ne fe moififfent faute
d'air ; ce qui feroit auffi nuifible aux arbres, & peut-être
plus que s'ils avoient été expofés à la gelée ; car en mettant beaucoup de terreau, ou
en les laiffant trop long-tems
couvertes, leur écorce eft fouvent endommagée, & elles ne
fe rétabliffent de long-tems :
au-lieu que la gelée ne détruit
que les tendres rejettons, fans
faire de tort au corps de l'ar-

OLE

bre, ni aux plus groffes branches, qui repouffent au printems fuivant.

Tous les ans , au printems,
on nous apporte de l'Italie ces
arbres avec les *Orangers* , les
Jafmins, &c. ; de maniere qu'on
peut s'en procurer aifément ;
ce qui vaut mieux que de les
élever de marcottes dans ce
pays , car cette méthode eft
fort longue & ennuyeufe ; &
fouvent, dans ceux que l'on
apporte ici, on en trouve qui
ont des tiges fi groffes , que
ceux qu'on multiplieroit en Angleterre ne pourroient en acquérir de femblables dans moins
de dix à douze ans. On fait
tremper leurs racines dans
l'eau pendant vingt-quatre heures , & après les avoir exactement nettoyées, on les plante dans des pots remplis de
terre fraîche, légere & fablonneufe ; on les plonge dans une
couche de chaleur modérée ;
on les met à l'abri de la trop
grande ardeur du foleil pendant
la chaleur du jour , & on les
arrofe toutes les fois que la
terre des pots fe trouve feche.

Par cette méthode , elles
commenceront à pouffer fix femaines ou deux mois après :
alors on leur donnera de l'air
à proportion de la chaleur de
la faifon ; & quand elles feront
bien enracinées, on les accoutumera par dégrés au plein
air , auquel on les expofera ,
en les plaçant à couvert des
grands vents. On les laiffera
ainfi jufqu'au mois d'Octobre
fuivant , pour les retirer a-
lors dans la ferre, comme il a
été dit ci deffus. Quand , au

moyen de ce traitement, ces plantes ont poussé de fortes racines, & formé d'assez bonnes têtes, on peut les tirer des pots avec leurs mottes, les mettre en pleine terre à une exposition chaude, & les traiter comme on l'a dit ci-dessus, en parlant des jeunes plantes : elles produiront toutes des fleurs au bout de deux ou trois ans, & même du fruit dans les années très-chaudes, pourvu qu'elles soient saines. L'*Olivier de Lucques*, & celui *à feuilles de Buis* étant les plus durs de tous, il faut les préférer pour les mettre en pleine terre ; mais la première espece forme de plus grands arbres.

OLIVIER. *Voy.* OLEA.

OLIVIER DE BOHEME ou OLIVIER SAUVAGE. *Voy.* ÉLEAGNUS. *L.*

OLIVIER SAUVAGE DES BARBADES *Voy.* BONTIA. *L.*

OMPHALODES. *Voy.* CYNOGLOSSUM LUSITANICUM.

ONAGRA. *Voy.* ŒNOTHERA. *L.*

ONOBRICHIS. *Voy.* HEDYSARUM. *L.*

ONONIS. *Lin. Gen. Plant.* 772. ANONIS. *Tourn. Inst. R. H.* 408. *tab.* 229. [*Rest-harrow*, *Cammock*, *Pettywin.*] Arrête-Bœuf ou Bugrande.

Caractéres. Le calice de la fleur est découpé en cinq segmens étroits, & terminés en pointe aiguë, dont les supérieurs sont un peu élevés & arqués, & les inférieurs penchés sur la carène ; la corolle est papillonnée ; l'étendard est en forme de cœur, abaissé sur les côtés, & plus large

que les ailes ; les ailes sont ovales & courtes ; la carène est pointue, & plus longue que les ailes : la fleur a dix étamines jointes ensemble, & terminées par des antheres simples ; son germe, qui est oblong & velu, soutient un style simple, & couronné par un stigmat obtus. Ce germe se change dans la suite en un légume gonflé, & à une cellule qui renferme des semences en forme de rein.

Ce genre de plantes est rangé dans la troisieme section de la dix-septieme classe de LINNÉE, intitulée : *Diadelphia Decandria*, qui comprend celles dont les fleurs ont dix étamines jointes en deux corps.

Les especes sont :

1°. *Ononis spinosa*, *floribus sub-sessilibus solitariis lateralibus*, *caule spinoso. Hort. Cliff.* 359. *Kniph. cent.* 5. *n.* 61. *Regn. bot ;* Arrête-Bœuf avec des fleurs solitaires & sessiles aux côtés des branches, & une tige épineuse.

Anonis spinosa flore purpureo. C. B. P. 389 ; Anonis épineux à fleurs pourpre, quelquefois appelé *Cammock* ou *Petty-win ;* & dans quelques pays, *Genet françois. L'Anonis épineux.*

Anonis. Fuchs. Hist. 60. *Riv. t.* 69.

2°. *Ononis mitis*, *floribus sub-sessilibus solitariis lateralibus*, *ramis inermibus. Hort. Cliff.* 359. *Fl. Suec.* 622. *Roy. Lugd.-B.* 375 ; Ononis avec des fleurs solitaires & sessiles aux tiges, & des branches sans épines.

Anonis spinis carens purpurea. C. B. P. 389 ; Arrête-Bœuf pourpre & sans épines.

Anonis mitior. 1. *Clus. Hist.* 99.

3°. *Ononis repens , caulibus diffusis , ramis erectis , foliis superioribus solitariis , stipulis ovatis. Lin. Sp.* 1006 ; Ononis avec des tiges diffuses & érigées , dont les feuilles du haut font folitaires , & les stipules ovales.

Ononis maritima procumbens , foliis hirfutis pubefcentibus. Pluk. Alm. 33 ; Arrête-Bœuf maritime & rampant , avec des feuilles velues & hériffées.

4°. *Ononis tridentata, foliis ternatis, carnofis , fub-linearibus , tridentatis , fruticofa , pedunculis bi-floris. Lin. Sp. Plant.* 718 ; Ononis en arbriffeau , avec des feuilles à trois lobes , charnues, étroites , & à trois dents , & dont chaque pédoncule foutient deux fleurs.

Anonis Hispanica frutefcens , folio tridentato carnofo. Tourn. Infl. 408 ; Arrête-Bœuf d'Espagne en arbriffeau , avec une feuille charnue & à trois dents.

5°. *Ononis fruticofa , floribus paniculatis, pedunculis fub-trifloris, ftipulis vaginalibus , foliis ternatis, lanceolatis , ferratis. Hort. Cliff.* 358. *Roy. Lugd. - B.* 376 ; Ononis en arbriffeau , avec des fleurs en panicules , qui croiffent trois enfemble fur chaque pédoncule , des stipules en forme de gaîne , & des feuilles à trois lobes en forme de lance , & fciées.

Anonis purpurea frutefcens , non fpinofa. Dodart. Mem. 57. *f.* 57.

Anonis purpurea verna præcox frutefcens , flore rubro amplo. Moris. Hift. 2. *p.* 170. *Duham. Arbr.* 1. *t.* 58 ; Arrête-Bœuf printanier , précoce & en arbrif-

feau , qui produit une fleur groffe & pourpre.

6°. *Ononis natrix , pedunculis unifloris , ariftatis , foliis ternatis, ovatis , ftipulis integerrimis. Hort. Cliff.* 358. *Roy. Lugd. - B.* 375. *Sauv. Monfp.* 189 ; Ononis avec une feule fleur fur chaque pédoncule , terminée par une barbe ou paille , des feuilles ovales & à trois lobes , & des stipules entieres.

Anonis vifcida , foliis petiolatis , ternatis, petiolis uni-floris. Hall. Helv. n. 358.

Anonis natrix. Scop. carn. ed. 2. *n.* 878.

Ononis lutea , Cam. épit. 445. *Natrix. Riv. tetr.* 68. *Lob. ic.* 2. *p.* 28.

Anonis vifcofa , fpinis carens , lutea major. C. B. P. 389 ; Arrête-Bœuf vifqueux & fans épines , produifant une grande fleur jaune.

7°. *Ononis vifcofa , pedunculis unifloris , ariftatis , foliis fimplicibus infimis , ternatis. Lin. Sp.* 1009. *Ger. Prou.* 486 ; Ononis avec une feule fleur fur chaque pédoncule , terminée par une barbe , dont les feuilles du bas font fimples , & à trois lobes.

Anonis lutea vifcofa lati-folia minor , flore pallido. Barr. rar. 840. *f.* 1239.

Anonis annua erectior , lati-folia glutinofa Lufitanica. Tourn. Infl. 409 ; Arrête-Bœuf de Portugal , à larges feuilles , vifqueux , érigé & annuel.

8°. *Ononis minutiffima , floribus fub-feffilibus lateralibus , foliis ternatis glabris , ftipulis fetaceis, calycibus ariftis corollâ longioribus. Lin. Sp. Plant.* 1007. *Jacq.*

ONO

Auſtr. t. 240 ; Ononis avec des fleurs ſeſſiles aux côtés des tiges, des feuilles à trois lobes, des ſtipules garnies de pailles, & dont les barbes du calice ſont plus longues que la corolle.

Anonis ſpinoſa lutea minor. Bauh. Pin. 389.

Anonis flore luteo parvo. H. R. Par. ; Arrête-Bœuf à petites fleurs jaunes.

9° *Ononis criſtata, pedunculis uniſloris prœlongis, ramis inermibus, foliis ternatis, glabris, vaginis acutè dentatis ;* Ononis avec une fleur placée ſur un long pédoncule, des branches ſans épines, des feuilles liſſes & à trois lobes, & des gaînes à dents aiguës.

Ononis Criſta. Lin. Syſt. Plant. tom. 3. *p.* 429. *Sp.* 15.

Anonis glabra inermis, pedunculis uniſloris prœlongis, vaginis criſtatis. Allion. ; Arrête-Bœuf uni & ſans épines, ayant une fleur ſur un long pédoncule, & une gaîne à crête.

10°. *Ononis Ornithopodioides, pedunculis bi-floris ariſtatis, leguminibus linearibus, cernuis. Prod. Leyd.* 376 ; Ononis avec deux fleurs ſur chaque pédoncule, terminées par une barbe, & auxquelles ſuccedent des légumes linéaires & penchés.

Anonis ſiliquis Ornithopodii. Boerrh. Ind. Alt. 2.34 ; Arrête-Bœuf, avec des légumes de Pied d'oiſeau.

Fœnum Grœcum Siculum, ſiliquis Ornithopodii. Tourn. Inſt. 409.

11°. *Ononis rotundi-folia, fruticoſa, pedunculis tri-floris, ca-*

ONO   279

lycibus triphyllo-bracteatis, foliis ternatis ſubrotundis. Hort. Cliff. 358. *Roy. Lugd.-B.* 376. *Jacq. Auſt. App. f.* 49 ; Ononis avec des pédoncules à trois fleurs, qui ſortent aux côtés des branches, des calices avec des bractées à trois feuilles, & des feuilles preſque rondes & à trois lobes.

Anonis foliis ſub-rotundis, ſerratis, ternatis, petiolis multifloris. Hall. Helv. n. 357.

Cicer ſylveſtre, lati-folium, triphyllum. C. B. P. 347 ; Pois chiche à trois feuilles larges.

Cicer ſylveſtre tertium. Dod. Pempt. 525.

Cicer ſylveſtre verius. Lob. ic. 2. *p.* 73.

12°. *Ononis mitiſſima, floribus feſſilibus ſpicatis, bracteis ſtipularibus, ovatis, ventricoſis, ſcarioſis, imbricatis. Lin. Sp.* 1007 ; Ononis avec des fleurs ſeſſiles & en épis, ayant des bractées en forme de ſtipules, ovales, gonflées, & en écailles imbriquées.

Anonis Alopecuroïdes, mitis, annua, purpuraſcens. Hort. Elth. 28. *tab.* 24 ; Arrête-Bœuf en queue de Renard, uni, annuel & pourpre.

Anonis purpurea ſpicata, erecta, annua, lati-folia, ſiliquis rectis lenti-formibus. Moris. Hiſt. 2. *p.* 169. *S.* 2. *f.* 17.

13°. *Ononis Alopecuroïdes, ſpicis folioſis, foliis ſimplicibus ovatis, obtuſis, ſtipulis dilatatis. Lin. Sp. Plant.* 1 08. *Kniph. cent.* 8. *n.* 75 ; Ononis en épis feuillés, avec des feuilles ſimples, ovales & obtuſes, & des ſtipules étendues.

Ononis floribus spicatis. Hort.
Cliff. 358. Roy. Lugd.-B. 376.
Anonis Sicula Alopecuroides.
Tourn. Inst. 408 ; Arrête-Bœuf
de Sicile en queue de Renard.
Anonis purpurea spicata Alo-
pecuroides major. Boerrh. Lugd.-B.
2. p. 33.

14°. *Ononis anil , foliis ter-*
natis , ovatis , petiolis longissimis ,
leguminibus hirsutis ; Ononis avec
des feuilles ovales & à trois
lobes, placées sur de fort longs
pétioles, & des légumes hérissés.
Anonis Americana , folio la-
tiori , subrotundo. Tourn. Inst. R.
H. 409 ; Arrête Bœuf d'Amérique , à feuilles plus larges ,
& presque rondes.

15°. *Ononis decumbens, foliis*
ternatis , lineari-lanceolatis , caule
decumbente , floribus spicatis ala-
ribus , leguminibus glabris ; Ononis avec des feuilles à trois
lobes , linéaires , & en forme
de lance , une tige traînante ,
des fleurs en épis qui sortent
des ailes des feuilles , & des
légumes unis.
Anonis Americana , angusti-
folia , humilior & minùs hirsuta.
Houst. MSS. ; le plus petit
Arrête-Bœuf d'Amerique , à
feuilles étroites , & moins hérissées.

Spinosa. La premiere espece
est une mauvaise herbe qu'on
rencontre communément dans
plusieurs parties de l'Angleterre , mais qu'on cultive rarement dans les jardins ; elle
a une racine forte & rampante qui s'étend au loin dans la
terre , & qu'on détruit très-
difficilement : ses tiges s'élèvent à la hauteur d'un pied &
demi ; elles sont minces , de

couleur pourpre , velues , &
garnies sur les côtés de petites
branches armées de piquans
très aigus : ses fleurs sortent
seules aux côtés des branches ;
elles sont papilionnacées , de
couleur pourpre , & produisent de petits siliques qui renferment une ou deux semences en forme de rein. Cette
plante fleurit durant une grande partie de l'été , & ses semences mûrissent en automne.
Sa racine est une des cinq racines apéritives ; son écorce
est regardée comme propre à
guérir les rétentions d'urine ,
& les obstructions du foie &
de la rate. Il y en a une variété à fleurs blanches (1).

Mitis. La seconde espece ,
qui croît naturellement dans
plusieurs parties de l'Angleterre , a été regardée par plusieurs personnes comme une
variété de la premiere ; mais
je les ai élevées toutes deux
de semences , & j'ai toujours
remarqué qu'elles conservoient
leurs différences : les tiges de
celle-ci sont velues & plus
diffuses que celles de la précédente ; ses feuilles sont plus
larges & plus rapprochées des
branches ; ses tiges sont moins
droites , & n'ont point d'épi-

---

(1) On regarde cette plante
comme un très-bon remede apéritif
& diurétique ; on fait infuser l'é-
corce de sa racine, ou l'on emploie son eau distillée contre la
jaunisse, la gravelle, les suppres-
sions des regles, des urines & des
hémorroides , &c. La décoction de
ses feuilles est détersive : on en prépare des gargarismes ,dont on fait
usage dans le scorbut, les ulceres
des amygdales , &c.

ONO

nes : ſes fleurs & ſes légumes reſſemblent à ceux de la premiere. Il y a auſſi une variété à fleurs blanches de cette eſpece.

Repens. La troiſieme, qu'on rencontre ſur les bords de la mer dans pluſieurs parties de l'Angleterre, a une racine rampante, de laquelle ſortent pluſieurs tiges velues d'environ deux pieds de longueur, couchées ſur la terre de tous côtés, & garnies de feuilles velues & à trois lobes ; celles du bas des tiges ſont larges & ovales, & celles du haut plus petites & plus étroites : ſes fleurs reſſemblent à celles de la premiere eſpece ; elles ſortent ſimples ſur les côtés des tiges, & ſont d'une couleur de pourpre brillant ; ſes légumes ſont courts, & renferment chacun deux ou trois ſemences. Cette plante fleurit dans le mois de Juillet, & ſes graines mûriſſent en automne.

Tridentata. La quatrieme ſe trouve en Eſpagne & en Portugal ; elle s'élève avec des tiges d'arbriſſeau à la hauteur d'un pied & demi, & ſe diviſe en pluſieurs branches minces remplies de nœuds, & garnies de feuilles étroites à trois lobes, épaiſſes, charnues, & ſupportées par de courts pétioles : ſes fleurs naiſſent aux extrémités des branches en panicules lâches ; quelques pédoncules ſoutiennent deux fleurs, & d'autres n'en ont qu'une ; elles ſont de couleur pourpre, & paroiſſent dans le mois de Juin : leurs ſemences mûriſſent en Septembre.

ONO 281

Fruticoſa. La cinquieme, qui eſt originaire des Alpes, eſt un très bel arbriſſeau nain, qui s'élève, avec des tiges ligneuſes, à la hauteur de deux pieds, & ſe diviſe en pluſieurs branches garnies de feuilles à trois lobes, étroites, ſciées ſur eurs bords, & ſeſſiles : ſes fleurs ſont produites en panicules aux extrémités des branches, ſur de longs pédoncules, qui en ſoutiennent chacun trois ; elles ſont de couleur pourpre ; la ſtipule eſt une eſpece de gaîne qui embraſſe le pédoncule de la fleur. Cet arbriſſeau fleurit à la fin de Mai & au commencement de Juin ; ſes fleurs ſont remplacées par des légumes gonflés, d'un pouce environ de longueur, velus, & dans leſquels ſont renfermées trois ou quatre ſemences en forme de rein, qui mûriſſent dans le mois d'Août.

Natrix. La ſixieme eſpece naît ſpontanément dans la France Méridionale & en Eſpagne ; elle a une racine vivace & une tige annuelle, qui s'élève à la hauteur de deux pieds, & pouſſe vers le bas des branches courtes & latérales, garnies de feuilles à trois lobes, ovales, velues & gluantes : ſes fleurs ſortent en épis aux extrémités des tiges ; elles ſont groſſes, d'un jaune brillant, & poſtées ſur de longs pédoncules, qui s'étendent au-delà des feuilles, & penchent vers le bas : elles paroiſſent à la fin de Juin, & produiſent des légumes gonflés, d'un pouce de longueur, & renferment trois ou quatre ſemences brunes &

en forme de rein, qui mûrissent en Septembre.

Viscosa. La septieme croît sans culture en Portugal, d'où l'on m'en a envoyé les semences : elle est annuelle ; sa tige forte, herbacée, velue, & d'un pied & demi de hauteur, pousse dans toute sa longueur des branches fort garnies de feuilles gluantes à trois lobes , dont celui du milieu est le plus large , & ovale, & les deux de côté sont longs, étroits, arrondis à leur extrémité, & dentelés sur leurs bords ; les pédoncules des fleurs , qui sortent seuls des aisselles des tiges , soutiennent chacun une fleur d'un jaune pâle, & érigée dans le milieu du pédoncule , qui déborde la fleur. Cette plante fleurit en Juillet, & ses semences mûrissent en automne.

Minutissima. La huitieme espece , qui croît naturellement dans la France Méridionale & en Italie, est une plante annuelle, dont les tiges s'élevent à neuf pouces environ de hauteur, & poussent une ou deux branches latérales vers le bas ; ses feuilles sont petites , à trois lobes, ovales, postées sur de longs pétioles, & dentelées sur leurs bords : ses fleurs sortent seules aux aisselles de la tige ; elles sont petites, jaunes , & fort rapprochées de la tige ; elles ont des stipules aiguës & hérissées sous le calice ; leurs légumes sont fort courts, gonflés, & renferment deux ou trois semences en forme de rein. Cette plante fleurit en Juillet, & ses semences mûrissent en automne.

Cristata. La neuvieme espece se trouve sur les Alpes ; elle a une racine vivace, de laquelle sortent plusieurs tiges minces , traînantes , d'environ six pouces de longueur , & garnies de petites feuilles à trois lobes , ovales, dentelées sur leurs bords, & postées sur de courts pétioles : ses fleurs naissent seules vers le sommet de la tige , sur des pédoncules longs & minces , qui sortent des aîles des feuilles, & soutiennent chacun une fleur jaune ; la gaîne embrasse la bâse du pédoncule , & a des dents aiguës. Ces fleurs paroissent en Juin , & les semences mûrissent en automne.

Ornithopodioïdes. La dixieme espece est une plante annuelle, qu'on rencontre en Sicile ; ses tiges s'élevent à neuf pouces de hauteur , & poussent vers le bas une ou deux branches garnies de petites feuilles à trois lobes , & postées sur de courts pétioles : ses fleurs sortent aux côtés des branches, postées sur de courts pédoncules , qui en soutiennent chacun deux petites de couleur jaune , auxquelles succedent des légumes noueux, comprimés , comme ceux du *Pied d'Oiseau*, & qui renferment quatre ou cinq semences en forme de rein. Cette plante fleurit dans le mois de Juillet, & ses semences mûrissent en automne.

Rotundi-folia. La onzieme espece , qui est originaire des Alpes & des montagnes de la Suisse, s'éleve à la hauteur d'un pied & demi, avec une tige

simple, & garnie de feuilles ovales à trois lobes, poſtées ſur de longs pétioles : les pédoncules des fleurs ſortent des aîles des feuilles ; ils ſont longs, minces, & ſoutiennent chacun trois fleurs d'un jaune pâle, que remplacent des légumes courts & gonflés, dans chacun desquels ſont renfermés deux ou trois ſemences. Cette eſpece fleurit en Juin, & ſes ſemences mûriſſent en Septembre.

Mitiſſima. La douzieme eſpece a pouſſé dans de la terre qui avoit été apportée de la Barbade ; mais elle ne paroît pas être originaire de cette Iſle, car elle s'éleve aiſément de ſemences ici en plein air, & perfectionne ſes graines en automne ; elle ne réuſſiroit pas même à un plus grand dégré de chaleur : elle a une tige droite, d'un pied & demi de hauteur, qui pouſſe de petites branches latérales, garnies de feuilles rondes à trois lobes, ſciées ſur leurs bords, & poſtées ſur de courts pétioles : ſes fleurs croiſſent en épis courts & feuillés aux extrémités des branches ; elles ſont petites & d'une couleur pourpre pâle : elles paroiſſent dans le mois de Juillet, & produiſent des légumes courts & gonflés, qui contiennent deux ou trois ſemences en forme de rein, leſquelles mûriſſent en automne.

Alopecuroïdes. La treizieme ſe trouve en Portugal, en Eſpagne & en Italie ; elle eſt annuelle, & s'éleve à la hauteur d'un pied, avec des tiges droites, branchues, & garnies de feuilles ſimples & ſeſſiles, dont les plus larges ſont ovales, & d'un pouce de longueur ſur neuf lignes de largeur : celles du haut ſont étroites, terminées en pointe obtuſe, & légerement dentelées ſur leurs bords : ſes fleurs croiſſent en épis feuillés & fort rapprochés aux extrémités des tiges ; leurs calices ſont velus ; elles ſont groſſes, de couleur pourpre, & paroiſſent en Juillet ; à ces fleurs ſuccedent des légumes cylindriques, & d'un pouce environ de longueur, qui renferment chacun quatre ou cinq ſemences en forme de rein. Cette plante a pluſieurs noms dans les différens livres de Botanique.

Anil. La quatorzieme eſpece, qui croît naturellement dans les Iſles de l'Amérique, eſt une plante annuelle, qui s'éleve à la hauteur de deux pieds, avec une tige branchue, & garnie de feuilles à trois lobes, ovales, & poſtées ſur des périoles fort longs & velus : ſes fleurs ſortent en épis lâches aux extrémités des branches ; elles ſont groſſes, d'un jaune pourpâtre, & produiſent des légumes très-gonflés & velus, qui renferment chacun cinq ou ſix groſſes ſemences en forme de rein. Cette eſpece fleurit dans les mois de Juillet & Août, & ſes ſemences mûriſſent en automne. On faiſoit autrefois avec cette plante de l'*Indigo*, qui, à ce que je crois, étoit d'une moindre valeur que celui qu'on fait avec l'*Anil* ; auſſi ne la cultive-t-on plus depuis pluſieurs années dans quelques parties des Iſles.

Decumbens. La quinzieme a été découverte par le Docteur HOUSTOUN, à la Vera-Cruz, dans la Nouvelle-Espagne, d'où il a envoyé ses semences en Angleterre. Cette plante est vivace, & pousse de ses racines plusieurs branches fortes, étendues, inclinées vers la terre, & garnies de feuilles étroites, à trois lobes, & très-velues : ses fleurs sortent en panicules clairs aux extrémités des branches ; elles sont jaunes, & produisent des légumes unis, gonflés, & de six lignes environ de longueur, qui renferment chacun deux ou trois semences en forme de rein. Cette plante fleurit en Juillet, & ses semences mûrissent quelquefois ici en automne.

Culture. Les trois premieres espèces ne sont jamais cultivées dans les jardins, parce qu'elles sont des herbes embarrassantes, lorsqu'elles sont une fois établies dans les champs ; leurs racines s'étendent & se multiplient considérablement ; elles sont si dures & si fortes, que la charrue peut à peine les rompre, & ce n'est qu'avec beaucoup de difficultés qu'on parvient à défricher un terrein où elles sont une fois établies.

Les quatrieme & cinquieme especes sont des plantes basses & en arbrisseaux, qui se multiplient par semences : la quatrieme est trop tendre pour profiter en plein air en Angleterre, à moins qu'elle ne soit plantée dans une situation chaude, & couverte pendant les fortes gelées. En semant ces deux espe-

ces en Avril sur une planche de terre légere, les plantes pousseront en Mai ; alors on les tiendra nettes de mauvaises herbes, & l'on en arrachera quelques-unes, par un tems humide, dans les endroits où elles sont trop serrées, pour les transplanter à quatre ou cinq pouces de distance. Celles de la quatrieme espece doivent être placées sur une plate-bande chaude & abritée ; & celles de la cinquieme, dans une plate-bande à l'ombre, où elles profiteront très-bien, aussi-tôt qu'elles seront enracinées ; mais il faut les tenir nettes de mauvaises herbes jusqu'à l'automne suivant, pour les transplanter alors à demeure. Celles qu'on a laissé croître dans la planche de semis, doivent être aussi traitées de la même maniere. Comme elles ne réussissent point dans des pots, il faut toujours les mettre en pleine terre, où la cinquieme fleurira beaucoup, & poussera souvent une quantité de rejettons de ses racines ; mais la quatrieme est plus sensible au froid. Ces plantes fleurissent dans la seconde année, & font un bel effet pendant qu'elles sont en fleurs : la cinquieme produit des semences en abondance.

La sixieme espece se multiplie par ses graines, qu'il faut semer claires en rigoles sur une terre légere. Quand les plantes poussent, on les tient nettes de mauvaises herbes jusqu'à l'automne, & on les enleve alors avec précaution, pour les placer dans les plates-bandes où elles

ONO

où 'elles doivent rester. Elles fleuriront dans la seconde année, & perfectionneront leurs semences ; leurs racines subsistent plusieurs années, & sont fort dures au froid.

Les septieme, huitieme & onzieme especes sont des plantes dures & annuelles qu'on multiplie par semences. On répand ces graines dans les places où les plantes doivent rester; elles n'exigent aucun autre soin que d'être éclaircies où elles sont trop serrées, & tenues nettes de mauvaises herbes.

La neuvieme espece est une plante dure & vivace : mais comme elle a peu d'apparence, on ne la cultive guere, si ce n'est dans les jardins de Botanique, pour la variété. Elle pousse chaque année des semences écartées, & profite dans tous les sols & à toutes les expositions.

La quatorzieme est annuelle : on répand ses graines au printems sur une couche de chaleur modérée ; quand les plantes sont en état d'être enlevées, on les remet sur une autre couche tempérée, pour les faire avancer, & on les traite de la même maniere que le *Souci* de France & l'*Œillet d'Inde*. Dans le mois de Juin, on les enleve avec une bonne motte de terre, & on les plante dans des plates-bandes ouvertes, où elles profiteront, fleuriront dans le mois suivant, & perfectionneront leurs semences en automne, si on les tient à l'ombre jusqu'à ce qu'elles soient bien enracinées.

La quinzieme est délicate,
Tome V.

ONO 285

& doit être semée sur une bonne couche chaude au printems. Quand les plantes sont assez fortes pour être enlevées, on les met chacune séparément dans de petits pots remplis d'une terre marneuse & légere ; on les plonge dans une couche chaude de tan, & on les tient à l'ombre jusqu'à ce qu'elles aient formé de nouvelles racines ; après quoi on les traite de la même maniere que les autres plantes tendres qui viennent des mêmes contrées. On les met en été dans la couche de tan de la serre chaude, où elles donneront des fleurs dans l'été suivant ; mais elles ne perfectionnent pas souvent leurs semences en Angleterre.

ONOPORDUM. *Lin. Gen. Plant. 834. Vaill. Act. Par. 1718. Carduus. Tourn. Inst. R. H. 440. tab. 253.* [*Woolly Thistle.*] Chardon cotonneux.

Caracteres. Le calice commun est rond, en forme de cloche, imbriqué, & formé par un grand nombre d'écailles terminées par des épines. La fleur est composée de plusieurs fleurons hermaphrodites, en forme d'entonnoir, égaux, uniformes, pourvus chacun d'un tube étroit, gonflé au bord, & découpé en cinq parties ; ils ont cinq étamines courtes, velues, & terminées par des antheres cylindriques, & un germe ovale, couronné de duvet, qui soutient un style mince, couronné par un stigmat. Ce germe se change ensuite en une simple semence couronnée de duvet, & placée dans le calice.

T

OPH

Ce genre de plantes est rangé dans la premiere section de la dix-neuvieme classe de LINNÉE, avec celles dont les fleurs ne sont composées que de fleurons hermaphrodites & fructueux.

Les especes sont :

1°. *Onopordum Acanthium*, *calycibus squarrosis*, *foliis ovato-oblongis*, *sinuatis. Lin. Sp. Plant.* 827. *Pollich. pal. n.* 772. *Neck. Gallob.* 340. *Mattusch. Sil. n.* 593. *Dœrr. Nass. p.* 161 ; Chardon cotonneux avec des calices raboteux, & des feuilles ovales, oblongues & sinuées.

Acanos Spina. Scop. carn. ed. 2. *n.* 1013.

Spina alba, *tomentosa*, *latifolia*, *sylvestris. Bauh. Pin.* 382. *Carduus tomentosus*, *Acanthi folio*, *vulgaris. Tourn. Inst. R. H.* 441. ; Chardon commun & cotonneux à feuilles d'Acanthe. *Pet d'Ane* ou *Epine blanche. Acanthium. Dod. Pempt.* 721.

2°. *Onopordum Illyricum*, *calycibus squarrosis*, *foliis lanceolatis*, *pinnati-fidis. Lin. Sp. Plant.* 1158. *Gouan. Monsp.* 424. *Jacq. Hort. f.* 148 ; Chardon cotonneux, avec des calices raboteux, & des feuilles en forme de lance, ailées & terminées par plusieurs épines.

Spina tomentosa altera spinosior. Bauh. Pin. 382.

Carduus tomentosus, *Acanthi folio angustiori. Tourn. Inst. R. H.* 441 ; Chardon cotonneux, à feuilles étroites d'Acanthe.

Acanthium Illyricum. Lob. ic. 1. *Barr. ic.* 501.

3°. *Onopordum Arabicum, calycibus imbricatis. Hort. Ups.* 249. *Jacq. Hort. f.* 149 ; Chardon cotonneux, avec des calices imbriqués.

Carduus tomentosus, *Acanthi folio*, *altissimus*, *Lusitanicus. Tourn. Inst.* 441. Le plus grand chardon cotonneux de Portugal, à feuilles d'Acanthe.

Carduus tomentosus, *Acanthium dictus*, *Arabicus. Pluk. Alm.* 85. *t.* 154. *f.* 5.

Acanthium altissimum Lusitanicum. Moris. Hist. 3. *p.* 153.

4°. *Onopordum Orientale*, *calycibus squarrosis*, *foliis oblongis*, *pinnato-sinuatis*, *decurrentibus*, *capite magno* ; Chardon cotonneux, avec des calices rudes, des feuilles oblongues, ailées, sinuées & coulantes, & une grosse tête de fleurs.

Carduus tomentosus, *Acanthi folio*, *Aleppicus*, *magno flore. Tourn. Inst. R. H.* 441 ; Chardon cotonneux d'Alep, à feuilles d'Acanthe, avec une grande fleur.

5°. *Onopordum acaulon*, *subacaule. Lin. Sp.* 1159 ; Chardon cotonneux, dont les têtes sont sessiles à la terre, ou sans tige.

Onopordon acaulon, *flore albicante. D. Jussieu. Vaill. Mem.* 1718 ; Chardon cotonneux, sans tige, dont la fleur est blanchâtre.

Il y a quelques autres especes de ce genre, que l'on conserve dans les jardins de Botanique, ainsi que plusieurs variétés, qui different par la couleur de leurs fleurs : mais comme ces plantes sont rarement admises dans d'autres endroits, il est inutile de les décrire ici.

Acanthium. La premiere espece croît naturellement dans des lieux incultes de plusieurs parties de l'Angleterre ; elle

ONO

est bis-annuelle, & pousse dans la première année plusieurs feuilles larges, cotonneuses, sinuées sur leurs bords, & piquantes, qui s'étendent sur la terre & résistent à l'hiver. Au printems suivant, la tige sort du milieu des feuilles, & s'éleve jusqu'à cinq ou six pieds de hauteur sur les tas de fumier, ou dans une bonne terre : elle se divise vers son sommet en plusieurs branches garnies d'ailes feuillées, qui coulent dans toute leur longueur, dentelées, & dont chaque dent est terminée par une épine : les tiges sont couronnées par des têtes écailleuses de fleurs pourpre, qui paroissent en Juin, & produisent des semences oblongues, angulaires, & couronnées d'un duvet velu, au moyen duquel le vent les tranporte à une grande distance ; de sorte que, par-tout où ces plantes perfectionnent leurs semences, elles s'y multiplient beaucoup, & deviennent fort embarrassantes.

Illyricum. La seconde espece se trouve en Espagne, en Portugal, & dans le Levant ; sa tige devient plus haute que celle de la précédente ; ses feuilles sont beaucoup plus longues & plus étroites, & leurs dentelures sont régulieres, & terminées par des épines aiguës ; ses têtes de fleurs sont plus grosses, & les épines du calice sont plus longues que celles de la premiere.

Arabicum. La troisieme s'éleve à la hauteur de neuf ou dix pieds ; ses tiges se divisent en plusieurs branches ; ses feuilles sont plus longues que celles des autres especes : ses têtes de fleurs sont grosses & de couleur pourpre ; le calice a des écailles couchées l'une sur l'autre comme celles d'un poisson. Cette plante croît naturellement en Espagne & en Portugal.

Acaulon. La cinquieme a plusieurs feuilles oblongues, ovales, & cotonneuses, qui s'étendent sur la terre ; du milieu de ces feuilles sort une tête de fleurs sessiles à la terre ou sans tige, plus petite que les têtes de la précédente, & dont les fleurs sont blanches. On cultivoit autrefois quelques-unes de ces plantes pour la table : mais c'étoit avant que les jardins anglois fussent fournis d'autres especes qui leur sont bien préférables ; car à présent il est très-rare que l'on en fasse usage. Elles n'exigent aucune culture ; car en laissant écarter leurs semences, ces plantes se reproduisent sans aucun soin.

ONOSMA. *Lin. Gen.* 187. Cette plante n'a pas d'autre Nom en Anglois.

Caracteres. Le calice de la fleur est persistant, & formé par une feuille érigée & découpée en cinq segmens ; la corolle est en forme de cloche & monopétale ; elle a un tube court & enflé sur ses bords qui sont divisés en cinq parties nues & trouées ; la fleur a cinq étamines courtes, terminées par des antheres en pointe de fleche, & de la longueur de la corolle ; son germe est divisé en quatre par-

ONO

ties ; il foutient un ftyle min-ce , couronné par un ftigmat ebtus , & fe change dans la fuire en quatre femences placées dans le calice.

Ce genre de plantes eft rangé dans la première feҫtion de la cinquieme claffe de Lin-née , intitulée : *Pentandria Monogynia* , avec celles dont les fleurs ont cinq étamines & un ftyle.

Les efpeces font :

1°. *Onofma fimpliciffima , foliis confertiffimis , lanceolato-linearibus pilofis*. Lin. Sp. 196 ; Onofma avec des feuilles linéaires , velues , en forme de lance , & rapprochées.

Echium Creticum. Alp. Exot. 130. *t.* 129 ; Vipérine de Crète.

2°. *Onofma Orientalis , foliis lanceolatis , hifpidis , fruɕibus pendulis. Lin. Sp.* 196 ; Onofma avec des feuilles en forme de lance , & velues , & des fruits pendants.

Cerinthe Orientalis , Amœn. Acad. 4. p. 267 ; Melinet O-riental.

3°. *Onofma Echioïdes , foliis lanceolatis hifpidis , fruɕibus erectis. Lin. Sp* 196. *Jacq. Auftr.* f. 295 ; Onofma avec des feuilles velues & en forme de lance , & des fruits érigés.

Anchufa lutea , minor. C. B. P. 255 ; la plus petite Bugloffe jaune.

Symphytum foliis lingulatis , hifpidis. Hall. Helv. n. 601.

Cerinthe foliis , lanceolato-linearibus , hifpidis , femaribus quaternis diflinɕis , fruɕibus erectis. Hort. Cliff. 48. *Roy. Lugd.-B.* 408.

Cerinthe Echioïdes. Scop. carn. 2. n. 197. *Kniph. cent.* 3. n. 26.

OPH

Les premiere & feconde efpeces font généralement des plantes bis-annuelles, qui périffent auffi-tôt après qu'elles ont perfeɕionné leurs femences ; quelquefois cependant elles fubfiftent pendant trois ou quatre années , quand elles croiffent dans des crevaffes de murailles ou des fentes de rochers , parce qu'alors elles font gênées dans leur crûe , qu'elles font moins remplies d'humidité , plus fermes , & par conféquent moins fufceptibles des impreffions du froid. Si ces trois efpeces pouvoient être placées fur de vieilles murailles ou dans des décombres , comme il arrive quelquefois qu'elles s'y trouvent quand leurs femences y font portées par les vents , on les conferveroit beaucoup mieux qu'en les tenant dans une bonne terre. Pour faire croître ces plantes dans ces décombres ou fur des murailles , il faut les femer auffi-tôt que leurs graines font mûres , & laiffer les tiges des vieilles plantes par-deffus , pour les abriter du foleil & hâter leurs progrès. Quand elles font bien établies dans ces endroits , elles s'y confervent & s'y multiplient d'elles-mêmes.

Comme on ne cultive guere ces plantes que dans les jardins de Botanique , il n'eft pas néceffaire d'en dire davantage ; elles fleuriffent au commencement du printems , & perfeɕionnent leurs femences dans le mois de Juin.

OPALE. *Voyez* Acer Opa-lus.

OPHIOGLOSSUM. *Tourn.*

OPH

325. [*Adder's Tongue.*] Langue de Serpent, Herbe fans couture.

Cette plante croît naturellement dans des près humides ; mais comme on ne peut la conferver long-tems dans les jardins, elle n'y eft point admife. (1).

OPHRYS. *Tourn. Inft. R. H.* 437. *tab.* 250. *Lin. Gen. Plant.* 902. [*Twyblade.*] double feuille. *Bifolium.*

Caractères. Elle a une tige fimple, avec une fpathe en forme de gaîne : la fleur n'a point de calice ; la corolle eft compofée de cinq pétales oblongs, qui fe joignent en montant, de maniere qu'ils forment un arc : le pétale du bas eft divifé en deux parties ; le nectaire eft pendant & en forme de carène le long des pétales. Cette fleur a deux courtes étamines poftées fur le piftile, avec des antheres érigées, & fixées au bord intérieur du nectaire ; fon germe eft oblong, tordu, & placé fous la fleur avec un ftyle qui adhere au bord intérieur du nectaire, & qui eft couronné par un ftigmat ufé. Ce germe devient enfuite une capfule ovale, à trois angles, obtufe, & a une cellule qui s'ouvre en trois petites valves, & montre des femences femblables à de la pouffiere, dont elle eft remplie.

Ce genre de plantes eft rangé dans la premiere fection de

_____

(1) On ne fait pas un grand ufage de cette plante ; cependant on la regarde comme vulnéraire, étant employée tant à l'exterieur, qu'intérieurement.

OPH 289

la vingtieme claffe de LINNÉE, qui comprend celles dont les fleurs ont deux étamines jointes au ftyle. Cet Auteur a réuni auffi à ce genre plufieurs efpeces d'*Orchis.*

Les efpeces font :

1°. *Ophrys nidus avis, bulbis fibrofo-fafciculatis, caule vaginato, aphyllo, nectarii labio bifido. Lin. Sp. Plant.* 1339. *Gmel. Sib.* 1. *p.* 25. *n.* 24. *Crantz. Auftr. p.* 475. fub *Epipactide.Scop. carn. ed.* 2. *n.* 1131. *Pollich. pal. n.* 853. *Mattufch. Sil. n.* 662. *Dœrr. Naff. p.* 162 ; Ophrys avec une racine bulbeufe garnie de fibres en paquets, une tige en forme de gaîne, & un nectaire, dont la levre eft divifée en deux parties.

Neottia bulbis fafciculatis, nectarii labio bifido. Act. Ups. 1740. *p.* 33. *Fl. Suec.* 442. 815. *Dalib. Paris.* 277.

Epipactis aphylla, flore inermi, labello bicorni. Hall. Helv. n. 1290. *f.* 37.

Orchis abortiva fufca. Bauh. Pin. 86.

Nidus avis. Lob. ic. 195. *Riv. Hex. f.* 7.

Ophrys bi-folia. C. B. P. 87 ; Double-feuille ou Ophrys commun.

2°. *Ophrys cordata, bulbo fibrofo, caule bifolio, foliis cordatis. Lin. Sp. Plant.* 946. *Scop. carn.* 2. *n.* 1133 ; Ophrys avec une racine bulbeufe, garnie de fibres, & une tige à deux feuilles, & en forme de cœur.

Ophrys foliis cordatis. Fl. Lapp. 247. *Fl. Succ.* 739. 809. *Act. Ups.* 1740. *p.* 29. *Gmel.* 1. *p.* 25.

Epipactis foliis binis, cordatis,

T 3

labello bifido, posticè bidentato. Hall. Helv. n. 1292. t. 22. f. 4.

Bifolium minimum. Bauh. Hist. 3. p. 534.

Ophrys minima. C. B. P. 87 ; le plus petit Ophrys.

Ophrys minima, floribus purpureo croceis, Mentz. Pug. t. 9. f. 3 ; Variété à fleurs, d'un pourpre couleur de safran.

3°. Ophrys spiralis, bulbis aggregatis, oblongis, caule subfolioso, floribus spirali-fœcundis, nectarii labio indiviso, crenulato. Act. Ups. 1740. p. 32. Dalib. Paris 277 ; Ophrys avec des bulbes oblongues & en grappes, une tige feuillée, des fleurs en spirale & fécondes, & une levre non divisée, & crenelée au nectaire.

Orchis spiralis, alba, odorata. J. B. 2. 769 ; Orchis blanc, odorant, & en spirale, appelé Triple Trace des Dames.

Epipactis bulbis cylindricis, spicâ spirali, labello crenulato. Hall. Helv. n. 1294. f. 38.

Triorchis alba odorata minor. Bauh. Pin. 84.

Satyrium odoriferum. Brusf. Herb. 1. p. 105.

Testiculus odoratus. Lob. ic. 186.

Triorchis, sivè Tetrorchis alba odorata, major. Bauh. Pin. 84 ; Variété plus grande & odorante.

Orchiastrum æstivum palustre, album odoratum. Mich. Gen. 30. f 26 ; seconde variété qui croit dans les marais.

Epipactis foliis plerisque ex lineari lanceolatis. Gmel. Sib. 2. p. 13. t. 3. f. 1.

4°. Ophrys monorchis, bulbo globoso, caule nudo, nectarii labio trifido. Act. Ups. 1740 ; Ophrys à bulbes globulaires, avec une tige nue, & dont la levre du nectaire est divisée en trois parties.

Herminium radice globosâ. Fl. Lapp. 317.

Orchis odorata moschata, sivè monorchis. C. B. P. 84 ; Orchis jaune & odorant, ou Orchis musqué.

Monorchis. Mich. Gen. 39. f. 26. Rupp. Gen. 421. t. 2.

Orchis lutea, hirsuto-folia. Bauh. Pin. 84 ; Variété à fleurs jaunes, & à feuilles hérissées.

Triorchis lutea, folio glabro. Bauh. Pin. 84 ; seconde variété à feuilles unies & à fleurs jaunes.

Triorchis lutea altera. Bauh. Pin. 84 ; troisieme Variété.

Monorchis bifolia, flore pallidè virente, Prussica. Mentz. Pug. t. 5. f. 5 ; quatrieme variété à deux feuilles, & à fleurs d'un vert pâle.

5°. Ophrys Antropophora, bulbis subrotundis, scapo folioso, nectarii labio lineari, tri-partito, medio elongato, bifido. Lin. Sp. Plant. 948 ; Ophrys avec des bulbes rondes, une hampe ou pédoncule feuillé, & dont la levre du nectaire est étroite, & divisée en trois parties, le segment du milieu étant étendu au dehors & découpé en deux portions.

Neottia bulbis sub-rotundis, nectarii labio quadrifido. Act. Ups. 1740. p. 32. Dalib. Paris. 277.

Orchis flore nudi hominis effigiem repræsentante, fœminâ. C. B. P. 82 ; Orchis dont la fleur est femelle, & représente l'effigie d'un homme nud.

OPH

Orchis Anthropophora Orcades. Col. Ecphr. 1. p. 320.

6°. Ophrys Infecti-fera, bulbis sub rotundis, scapo folioso, nectarii labio sub-quinque-lobo. Lin. Sp. Plant. 948. Gouan. Monsp. 473. Pollich. pal. n. 858 ; Ophrys avec des bulbes presque rondes, un pédoncule feuillé, ayant la levre du nectaire divisée presque en cinq lobes.

Orchis muscam referens, major. C. B. P. 83 ; le plus grand Orchis mouche.

Cypripedium bulbis sub-rotundis, foliis oblongis caulinis. Act. Ups. 1740. p. 26. Fl. Suec. 737. 818.

7°. Ophrys arachnites, bulbis sub-rotundis, caule folioso, nectarii labio trifido ; Ophrys à bulbes presque rondes, avec une tige feuillée, & dont la levre du nectaire est divisée en trois portions.

Orchis fucum referens, major, foliolis superioribus candidis, & purpurascentibus. C. B. P. 83 ; Orchis ressemblant à une abeille, dont les folioles supérieures sont blanches & pourpre.

Orchis arachnites. Lob. ic. 135.

8°. Ophrys sphegodes, bulbis sub-rotundis, caule sub-folioso, nectarii labio trifido, hirsuto ; Ophrys avec des bulbes presque rondes, une tige un peu feuillée, & dont la levre du nectaire est velue, & divisée en trois parties.

Orchis sivè Testiculus sphegodes, hirsuto flore. J. B. 2. 727 ; Satyrion en forme d'abeille, avec des aîles vertes, & une fleur hérissée.

Nidus avis. La premiere es-

OPH 291

pece croît naturellement dans les bois, & quelquefois dans les pâturages humides de plusieurs parties de l'Angleterre : sa racine est composée de plusieurs fibres fortes, desquelles sortent deux feuilles ovales, veinées, de trois pouces de longueur sur deux de largeur, & jointes à leur bâse ; entr'elles s'éleve une tige nue de huit pouces environ de hauteur, & terminée par un épi lâche de fleurs herbacées, semblables à des moucherons, & composées de cinq pétales, avec une levre longue & divisée en deux parties. Ces fleurs ont une houpe ou étendard qui en occupe le sommet, & deux aîles sur les côtés ; elles sont placées sur un germe qui se gonfle dans la suite, & devient une capsule qui s'ouvre, en mûrissant, en six parties, & qui est remplie de petites semences semblables à de la poussiere. Cette plante ne souffre point de culture ; mais on peut la transplanter dans un jardin, en la plaçant à l'ombre, où elle subsistera pendant plusieurs années, si les racines ne sont pas remuées : elle fleurit dans le mois de Mai ; mais on ne peut la multiplier. Le meilleur tems pour enlever ses racines, est dans les mois de Juillet & d'Août, lorsque ses feuilles se flétrissent ; car plus tard il seroit difficile de les trouver (1).

_____

(1) Quoique cette plante ne soit pas d'un usage bien frequent, on la regarde cependant comme propre à déterger les anciens ulceres.

T 4

Cordata. La feconde efpece fe trouve dans les parties feptentrionales de l'Angleterre ; mais elle croît rarement dans le midi de notre Ifle : elle a une petite bulbe garnie de plufieurs fortes fibres ; elle pouffe vers le bas deux petites feuilles en forme de lance : fa tige s'éleve à la hauteur d'environ quatre pouces, & fe termine par un épi de petites fleurs herbacées, & de la même forme que celles de la précédente.

Spiralis. La troifieme naît fpontanément fur des montagnes de craie, dans plufieurs parties de l'Angleterre ; elle a une racine oblongue, en grappe, & bulbeufe, de laquelle fort une tige fimple de fix pouces de hauteur, & garnie de deux feuilles oblongues vers le bas, mais ordinairement nues au deffus : fes fleurs font petites, blanches, & raffemblées en épis lâches fur le fommet de la tige : elles ont une odeur de mufc, & paroiffent en Août.

Cette efpece fe trouve auffi fur des pâturages humides, dans les provinces feptentrionales de l'Angleterre. Je l'ai encore rencontrée en grande abondance fur le terrein d'Enfield, affez près de la ville de Londres.

Monorchis ; *Anthropophora.* Les quatrieme & cinquieme efpeces naiffent fur les montagnes de craie, près de Northfleet, en Kent, ainfi que fur les montagnes de Causham, près de Reading : elles ont des racines rondes & bulbeufes, defquelles fortent quelques feuilles oblongues ; leurs tiges s'élevent à un pied & demi de hauteur, & font garnies de quelques feuilles étroites : leurs fleurs croiffent en épis lâches fur le fommet de la tige ; les unes font de couleur de fer rouillé, & d'autres de couleur herbacée : la levre du nectaire eft divifée en trois parties, dont celle du milieu eft étendue au dehors, plus longue que les autres, & divifée en deux parties. Le fommet de ces fleurs eft en forme de capuchon : elles paroiffent dans le mois de Juin, & reffemblent à un homme nud.

Sphegodes. La huitieme efpece qu'on rencontre fur des pâturages arides, dans plufieurs parties de l'Angleterre, eft ordinairement connue fous le nom d'*Orchis d'Abeille Bourdon* ou d'*Orchis Mouche.* Il y en a deux ou trois variétés, que l'on trouve en Angleterre, & plufieurs autres en Efpagne & en Portugal : celle-ci a une racine ronde & bulbeufe ; fes feuilles reffemblent à celles du *Plantin* à feuilles étroites ; fa tige s'éleve à la hauteur de fix ou fept pouces, & a deux ou trois feuilles érigées, qui l'embraffent en forme de gaîne. Au fommet de cette tige fortent deux ou trois fleurs fans ergots, mais avec des houpes pourpre & des aîles ; le nectaire eft large, formé comme le corps d'une petite abeille, & de couleur de fuie foncée, avec deux ou trois lignes couant à travers, & d'une couleur

---

Dans quelques contrées, les payfans l'appliquent fur leurs bleffures, après l'avoir fimplement écrafée.

plus foncée ou plus claire, qui paroissent plus brillantes ou plus pâles, suivant la position de la fleur. Ces fleurs paroissent au commencement du mois de Juin. Il y a quelques variétés de cette plante qui different par la couleur & la grosseur de leurs fleurs.

Toutes ces especes peuvent être conservées dans les jardins, mais elles ne s'y multiplient point. Le meilleur tems pour enlever leurs racines est, ainsi que nous l'avons déja dit, précisément dans le moment où leurs tiges périssent, parce qu'alors elles sont dans l'inaction, & qu'on peut aisément trouver leurs racines. Dans ce tems, leurs bulbes sont tout-à-fait formées pour fleurir l'année suivante, & ne peuvent plus se rétrécir ; mais quand on les enleve dans d'autres tems de l'année, où leur végétation est encore en vigueur, leurs bulbes n'étant pas tout-à-fait mûres, se dessechent, & périssent souvent, ou si elles en échappent, elles ne recouvrent pas leur premiere vigueur avant deux ans.

Quand on les transplante dans un jardin, chaque espece doit être placée dans le sol qui lui convient. Celles qui croissent naturellement dans des pâturages humides doivent être plantées dans des plates-bandes humides & à l'ombre ; d'autres qui croissent dans les bois, peuvent être mises sous des arbres dans les quartiers déserts ; mais celles qu'on trouve sur des montagnes de craie, doivent avoir un terrein de

craie préparé pour les recevoir dans une situation ouverte. Ces plantes étant ainsi distribuées, ne doivent plus être touchées : on se contente de les tenir nettes de mauvaises herbes. Moins la terre sera remuée, plus elles feront de progrès, & plus long-tems on les conservera.

OPULUS. *Voy.* VIBURNUM.

OPUNTIA. *Tourn. Inst. R. H.* 239. *tab.* 122. *Tuna. Hort. Elth.* 295. *Cactus. Lin. Gen. Plant.* 539. Cette plante est appelée *Opuntia*, d'après le passage de *THÉOPHRASTE*, qui dit qu'elle se trouve aux environ d'*Opuntium* ; Raquette. Figue d'Inde. Cardasse.

Caracteres. La corolle de la fleur est composée de plusieurs pétales, obtus, concaves, placés circulairement, & posés sur le germe : la fleur a un grand nombre d'étamines en forme d'alène, insérées dans le germe, plus courtes que les pétales, & terminées par des antheres oblongues & érigées. Le germe, qui est placé sous la fleur, soutient un style cylindrique de la longueur des étamines, & couronné par un stigmat divisé en plusieurs parties ; il se change dans la suite en un fruit charnu, avec un nombril, & a une cellule remplie par plusieurs semences rondes.

Ce genre de plantes est rangé dans la seconde section de la sixieme classe de TOURNEFORT, qui renferme les herbes avec une fleur en rose, dont le pistile & le calice deviennent un fruit, avec une capsule.

294 O P U

Le Docteur LINNÉE le place dans la premiere section de la douzieme classe, où se trouvent les plantes, dont les fleurs ont plus de dix-neuf étamines intérées ou dans le calice ou aux pétales de la fleur.

Les especes sont :

1°. *Opuntia vulgaris, articulis ovatis, compressis, spinis setaceis.* Hall. Helv. n. 1099 ; Figue d'Inde, avec des nœuds ovales & comprimés, & des épines hérissées.

Cactus Opuntia. Lin. Syst. Plant. tom. 2. p. 470. Sp. 18.

Opuntia vulgò herbariorum. J. B. 1. 154 ; Opuntia commun ou Figuier d'Inde.

Ficus Indica, folio spinoso, fructu majore. Bauh. Pin. 458.

Opuntia ficus Indica, articulis ovato-oblongis, spinis setaceis ; Figuier d'Inde, dont les nœuds sont longs, ovales, & hérissés de fines épines.

Opuntia folio oblongo media. Tourn. Inst. R. H. 239 ; Figuier d'inde moyen, à feuilles oblongues.

Cactus ficus Indica. Lin. Syst. Plant. t. 2. p. 470. Sp. 19

3°. *Opuntia Tuna, articulis ovato-oblongis, spinis subulatis* ; Figuier d'Inde avec des nœuds oblongs, ovales, & garnis d'épines en forme d'aléne.

Cactus Tuna. Lin. Syst. Plant. t. 2. p. 470. Sp. 20.

Opuntia major, validissimis spinis munita. Tourn. Inst. R. H. 239 ; le plus grand Figuier d'Inde, armé de très-fortes épines.

Tuna major, spinis validis, flavicantibus, flore gilvo. Dill. Elth. 396. t. 295. f. 238.

4°. *Opuntica elatior, articulis* ovato-oblongis, spinis longissimis, nigricantibus* ; Figuier d'Inde, avec des nœuds ovales, oblongs, & garnis d'épines très-longues & noirâtres.

Tuna elatior spinis validis, nigricantibus. Hort. Elth. tab. 194 ; le plus grand Figuier d'Inde, armé de très-fortes épines noirâtres.

5°. *Opuntia maxima articulis ovato-oblongis, crassissimis, spinis inæqualibus* ; Figuier d'Inde avec des nœuds oblongs, ovales, épais, & garnis d'épines inégales.

Opuntia maxima, folio spinoso, latissimo & longissimo. Tourn. inst. 240 ; le plus grand Figuier d'Inde, avec des feuilles très-larges, très-longues & épineuses.

6°. *Opuntia Cochenillifera, articulis ovato-oblongis, sub-inermibus* ; Figuier d'Inde, avec des nœuds ovales, oblongs, & presque sans épines.

Tuna mitior, flore sanguineo Cochenillifera. Dill. Elth. 399. t. 297. f. 388.

Opuntia maxima folio oblongo, rotundo majore, spinulis mollibus & innocentibus obsito, flore striis rubris variegato. Sloan. Cat. Jam. 194 ; le plus grand Figuier d'Inde, avec une feuille plus grande, ronde & oblongue, garnie de petites épines molles, & point dangereuses, ayant une fleur panachée de raies rouges, communément appelé le *Figuier d'Inde à Cochenille.*

Ficus Indica major lævis, vermiculos proferens. Pluk. Alm. 146. t. 281. f. 383.

Cactus Cochenillifer. Lin. Syst. Plant. tom. 2. p. 471. Sp. 21.

OPU

7°. *Opuntia Curaffavica*, articulis cylindrico-ventricofis, compreffis, fpinis fetaceis ; Figuier d'Inde avec des nœuds cylindriques, gonflés & comprimés, hériffés de fines épines. *Caɛtus Curaffavicus. Lin. Syft. Plant. tom.* 2. *p.* 471. *Sp.* 22.

Caɛtus tereti-compreffus articulatus, ramofus. Hort. Cliff. 182. *Hort. Ups.* 120. *Roy. Lugd. B.* 280. *Opuntia minima, Americana, fpinofiffima. Bradl. Suec.* 1. *p.* 5. *f.* 4.

Ficus Indica, *fivè Opuntia minor caulefcens arbufculæ in modum, ramis cineritiis, fpinofiffima. Pluk. Alm.* 147. *t.* 281. *f.* 3.

Ficus Indica, *feu Opuntia Curaffavica minima. Hort. Amft.* 1. 107 ; Figuier d'Inde , ou le plus petit *Opuntia* de Curaffao, fouvent appelé *Pinpillow*. Peloton garni d'épingles.

8°. *Opuntia fpinofiffima*, articulis longiffimis , tenuibus , compreffis, fpinis longiffimis , confertiffimis, glacilibus , albicantibus , armatis. *Houft. MSS.* ; Figuier d'Inde, qui pouffe beaucoup de tiges, avec des nœuds très-longs, minces, comprimés, armés d'épines très-longues , fort rapprochées, fines & blanches, auquel les jardiniers donnent le nom de *Jufle au Corps de Robinfon Crufoe.*

9°. *Opuntia phyllanthus*, prolifer , enfi-formis , compreffus , ferrato - repandus ; Figuier d'Inde, avec des nœuds comprimés, en forme d'épée , dont les dents fe tournent en arriere. *Caɛtus phylanthus. Lin. Syft. Plant. t.* 2 *p.* 471. *Sp.* 23.

Caɛtus foliis enfi - formibus , obtufè ferratis. Hort. Cliff. 183. *Lugd-B.* 281.

Caɛtus mitis minor , farmento flexili-rotundo , frondibus longis , compreffis , crenatis , ad crenas floridis. Brown. Jam. 237.

Ficus , fivè Opuntia non fpinofa , Scolopendriæ folio finuato. Raii dendr. 21.

Cereus Scolopendriæ folio brachiato. Hort. Elth 73. *tab.* 64 ; Cierge à feuilles branchues de Scolopendre.

Toutes ces plantes font originaires de l'Amérique , quoique la premiere fe trouve quelquefois fur les bords des routes , aux environs de Naples , en Sicile & en Efpagne ; mais il eft vaifemblable qu'elles y ont été anciennement apportées de l'Amérique.

Vulgaris. Cette efpece eft depuis long-tems dans les jardins anglois ; fes nœuds ou branches font ovales ou rondes, comprimées fur les deux côtés, & plates ; elles ont de petites feuilles qui fortent aux nœuds fur leurs furfaces , ainfi que fur leurs bords du haut , & qui tombent en peu de tems. A ces mêmes nœuds il y a trois ou quatre courtes épines hériffées qui ne paroiffent point , à moins qu'on ne les regarde de très-près , mais fi on les manie , elles s'infinuent dans la chair , en fe féparant de la plante , & deviennent très-embarraffantes, & fouvent difficiles à retirer. Les branches de cette efpece s'étendent fur la terre, où elles pouffent de nouvelles racines , & finiffent par couvrir un terrein confidérable : elles ne s'élevent jamais en hauteur ; elles font charnues & herbacées tandis

qu'elles font jeunes : mais à mesure qu'elles vieilliffent, elles fe deflechent, & deviennent d'une texture dure & ligneufe. Les fleurs fortent fur les bords fupérieurs des branches dures, & quelquefois elles font produites fur leurs côtés ; elles font poffées fur l'embryon de leur fruit, & font compofées de plufieurs pétales ronds, concaves & étendus ; elles font d'un jaune pâle, & renferment un grand nombre d'éramines attachées à l'embryon du fruit, & terminées par des antheres oblongues ; dans le centre eft placé un ftyle, couronné par un ftigmat à plufieurs pointes. Quand les fleurs font paffées, l'embryon fe gonfle, & devient un fruit oblong, dont la peau ou enveloppe eft garnie de petites épines en faifceau : l'intérieur du fruit eft charnu, de couleur de pourpre ou rouge, & renferme plufieurs femences noires. Cette plante fleurit dans les mois de Juillet & d'Août ; mais fon fruit ne mûrit en Angleterre que dans les années très chaudes.

J'ai eu quelques branches de cette efpece, qui m'ont été envoyées par M. PIERRE COLLINSON, Membre de la Société Royale, qui m'a affuré les avoir reçues de Terre-neuve, où elle croît naturellement. Comme ce pays eft très-feptentrional, il eft inconcevable comment cette plante peut y fupporter le froid ; car, quoiqu'elle fubfifte en plein air en Angleterre dans une fituation chaude & dans un fol fec,

cependant les hivers durs la détruifent généralement, fi l'on ne la met point à l'abri des gelées.

Ficus Indica. La feconde efpece a des branches oblongues, ovales, plates, plus érigées que celles de la premiere, & armées d'épines longues, hériffées, rapprochées en faifceaux à chaque point, & étendues en forme d'étoile : fes fleurs croiffent fur l'embryon du fruit, & fortent aux bords fupérieurs des feuilles comme celles de la précédente ; mais elles font plus larges & d'un pourpre plus foncé : la peau extérieure du fruit eft auffi armée de plus longues épines.

Cette efpece eft la plus commune à la Jamaïque, & c'eft fur fon fruit que l'on répand la femence de cochenille, ainfi que fur l'efpece fauvage, appelée *Sylvefter.* On m'a envoyé quelques-unes de ces plantes de la Jamaïque avec ces infectes vivans deffus. Le Docteur HOUSTOUN, qui me les avoit adreffées, faifoit alors l'hiftoire de ces infectes : mais dans ce moment, il a été attaqué d'une maladie qui l'a conduit au fombeau. La cochenille a vécu fur ces plantes pendant trois ou quatre mois, & a péri enfuite.

Quand on mange le fruit de cette plante, il teint l'urine en couleur de fang.

La troifieme efpece a des branches plus fortes que la feconde ; elles font armées de plus grandes épines en forme d'alène, blanchâtres, & en faifceaux comme celles des

autres : ses fleurs sont grosses, d'un jaune brillant, & le fruit a la même forme que celui de la seconde.

Elatior. La quatrieme espece s'éleve beaucoup plus haut qu'aucune des précédentes : ses branches sont plus grosses, plus épaisses, d'un vert plus foncé, & armées d'épines fortes & noires, qui sortent en faisceaux comme celles des autres ; mais les paquets d'épines sont placés à une plus grande distance : ses fleurs sont produites sur les bords du haut des branches ; elles sont plus petites que celles des autres especes, & d'une couleur pourpârre, ainsi que les étamines : le fruit a la même forme que celui de la premiere ; mais il ne mûrit pas ici.

Maxima. La cinquieme est la plus grande de toutes celles que nous connoissons : ses branches ont plus d'un pied de longueur sur huit pouces de largeur ; elles sont fort épaisses, d'un vert foncé, & armées de quelques épines courtes & hérissées ; les plus vieilles branches deviennent presque comme des cierges, & très-fortes. Je n'ai point encore vu les fleurs de cette espece, quoique j'aie des plantes qui ont plus de dix pieds de hauteur.

Cochenillifera. On a toujours regardé la sixieme comme étant la plante sur laquelle on nourrit la cochenille : elle a des branches oblongues, lisses, vertes, érigées, de huit ou dix pouces de hauteur, & presque dépourvues d'épines ; car on en apperçoit à peine de quelque distance quelques-unes qui sont molles & peu dangereuses au toucher. Les fleurs de cette plante sont petites, de couleur pourpre, & placées sur l'embryon du fruit, de même que celles des autres : mais elles ne s'étendent pas autant ; elles ne paroissent que sur la fin de l'automne, & le fruit tombe en hiver, sans parvenir ici à quelque dégré de maturité. On cultive cette plante dans les campagnes de la nouvelle Espagne, pour nourrir la cochenille ; mais elle croît naturellement dans la Jamaïque, où l'on y découvriroit vraisemblablement la véritable *Cochenille*, si des personnes habiles vouloient se donner la peine de faire cette recherche.

Curaſſavica. La septieme espece est supposée croître naturellement à Curaſſao : elle a des nœuds cylindriques, gonflés, & fortement armés d'épines minces & blanches : ses branches s'étendent en-dehors de tous côtés ; & quand elles ne sont point soutenues, elles tombent sur la terre, & se séparent souvent aux différens nœuds qui poussent des racines, & forment par ce moyen de nouvelles plantes. Cette espece produit rarement des fleurs en Angleterre. On l'appelle *Pinpillow* dans les Indes Occidentales, parce que ses branches ressemblent beaucoup à un peloton garni d'épingles.

Spinoſiſſima. La huitieme espece m'a été envoyée de la Jamaïque par le Docteur

HOUSTOUN, qui l'y a trouvée en grande abondance : mais il n'a jamais vu de fleurs ni de fruits sur aucune des plantes ; elle n'en a point produit non plus en Angleterre : ses branches, qui ont des nœuds beaucoup plus longs que ceux des autres, sont aussi plus étroites & plus comprimées ; ses épines sont fort longues, minces, & d'un brun jaunâtre ; elles sortent en paquets sur toute la surface des branches, en se croisant : ce qui les rend plus dangereuses à manier ; car, pour peu qu'on les touche, elles s'attachent à la main, se séparent des branches, & pénetrent dans la chair.

Phyllanthus. La neuvieme espece, qui est originaire du Brésil, a des branches fort minces, régulierement dentelées sur leurs bords, comme la *Scolopendre*, d'un vert clair, en forme de sabre, lisses & sans épines : ses fleurs sortent des côtés & à l'extrémité des branches sur les embryons, comme celles des autres especes ; elles sont d'un jaune pâle, & les fruits qui leur succedent, sont de la même forme que ceux de la premiere : ils mûrissent rarement en Angleterre.

Culture. Toutes ces especes, excepté la premiere, sont trop tendres pour profiter en plein air dans ce pays, & l'on ne peut en conserver beaucoup pendant l'hiver, sans chaleur artificielle ; car, lorsqu'elles sont placées dans une serre, elles deviennent d'un jaune pâle, & leurs branches se rétrécissent & se pourrissent souvent aux premieres chaleurs du printems.

On les multiplie toutes par leurs branches, qu'on peut séparer aux nœuds pendant tout l'été : on les tient dans un lieu sec & chaud, pendant une quinzaine de jours, pour faire secher leurs parties blessées, sans quoi l'humidité qu'elles absorberoient par-là, les feroient bientôt pourrir, comme cela arrive à la plupart des autres plantes succulentes. Le sol dans lequel on les place, doit être composé d'un tiers de terre fraîche & légere de pâturage, d'un tiers de sable de mer, & d'un tiers formé par un mélange de tan pourri & de décombres de chaux à parties égales : on mêle exactement ces differentes matieres ; on les tient en tas pendant trois ou quatre mois avant de s'en servir, & l'on retourne ce mélange au moins une fois par mois, afin que toutes les parties puissent être exactement unies ensemble ; après quoi, on le passe à travers un gros crible, pour en séparer les plus grosses pierres & les mottes ; mais il ne faut point cribler cette terre trop fin, faute que l'on commet assez communément : on met ensuite quelques petites pierres au fond des pots, pour aider l'écoulement de l'humidité, ce qui doit être observé pour toutes les plantes succulentes ; car, si l'eau y séjournoit, elle feroit pourrir leurs racines, & les détruiroit bientôt.

Ce qu'on vient de dire doit s'entendre de la premiere

espece ; mais pour les autres, il faut mettre les pots dans une couche de chaleur modérée, qui leur fera prendre racine très-facilement : on les arrose de tems en tems, mais toujours légerement, fur-tout avant qu'elles foient enracinées. Quand les plantes commencent à pouffer des branches, on leur donne beaucoup d'air, en foulevant les vitrages, pour les empêcher de filer & de s'affoiblir. Après qu'elles ont pouffé de fortes racines, il eft néceffaire de les accoutumer à l'air par dégrés, & de les mettre enfuite dans la ferre chaude à demeure, en les plaçant près des vitrages, qu'il faut toujours ouvrir dans les tems chauds, afin qu'elles puiffent jouïr de l'air libre, & être cependant à l'abri du froid & de l'humidité.

Pendant l'été, ces plantes doivent être fouvent arrofées, mais toujours légerement, comme nous l'avons déja dit ; & en hiver, on proportionne la quantité d'eau qu'on leur donne, à la chaleur de la ferre ; car, fi l'air en eft toujours fort échauffé, elles auront befoin d'être plus fouvent rafraîchies, fans quoi leurs branches fe rétréciroient ; mais fi la ferre eft tempérée, on les arrofe très peu, pour éviter l'humidité, qui leur eft funefte dans cette faifon.

Ces plantes profitent mieux en hiver, au dégré de chaleur tempérée marqué fur les thermometres de Botanique, que fi elles étoient tenues plus chaudement ; car dans une trop grande chaleur . leurs branches deviennent très-tendres, foibles & défagréables à la vue. Les efpeces qui pouffent naturellement droites, doivent être foutenues avec des bâtons, pour empêcher qu'elles ne foient brifées par leur propre poids.

Bien des perfonnes expofent ces plantes au-dehors pendant l'été ; mais elles réuffiffent beaucoup mieux en les tenant continuellement dans la ferre chaude, pourvu que les vitrages en foient toujours ouverts, & qu'elles y aient beaucoup d'air. Quand on les met en plein air, les groffes pluies qui furviennent fouvent en été, comme dans la température variable de notre climat, diminuent beaucoup leur beauté, & retardent leurs progrès ; quelquefois même, dans des étés humides, elle fe rempliffent de tant d'humidité, qu'elles fe pourriffent fouvent en hiver : d'ailleurs les efpeces tendres qui font trop expofées au plein air, ne produifent pas autant de fleurs & de fruits que celles qu'on tient conftamment dans la ferre chaude.

ORANGER *V.* Aurantium. *L.*

ORANGERIE. Maison de Verdure.

La grande quantité de plantes curieufes exotiques qui ont été apportées depuis peu en Angleterre, a donné lieu à la conftruction d'un grand nombre d'Orangeries ; ce qui a rendu des gens très-habiles, foit dans la culture de ces plantes, foit dans la conftruction, l'ordonnance & la difpofition de ces bâtimens. Comme il y

a plusieurs choses à observer pour rendre ces Orangeries u-tiles & agréables, je vais expliquer, le mieux que je pourrai, la planche ci-jointe.

La longueur de ce bâtiment doit être proportionnée à la quantité de plantes qu'on veut y mettre : elle peut varier encore suivant l'idée du propriétaire ; mais sa profondeur ne doit jamais excéder sa hauteur. Dans des Orangeries ordinaires, cette profondeur ne passe jamais seize ou dix-huit pieds : dans les plus grandes, elle peut aller à vingt ou vingt-quatre, & cette proportion est la plus belle ; car, si ce bâtiment est long & trop étroit, il aura mauvaise apparence, soit en-dedans, soit en-dehors : il ne contiendra pas autant de plantes, & il n'y aura pas assez d'espace pour y ménager un passage sur le devant, & pour mettre dans le fond, des gradins propres à y placer des plantes ; au-lieu qu'une profondeur de vingt-quatre pieds au moins contiendra plus de rangées de plantes, & donnera plus d'aisance pour les arroser & les nettoyer. La trop grande profondeur est cependant plus nuisible qu'une médiocre.

Les fenêtres de la façade doivent commencer à un pied & demi au-dessus du pavé, & s'élever jusqu'au plafond, où elles seront couronnées par la corniche du bâtiment. Cette élévation étant considérable, il seroit difficile que leur largeur y répondît ; car si les châssis étoient faits comme ceux des plus grands bâtimens qui

ont sept pieds ou sept pieds & demi de largeur, il seroient trop lourds pour pouvoir les soulever & les descendre : d'ailleurs les battans des volets, en s'ouvrant, se trouvant plus larges que les trumeaux, ils ôteroient beaucoup de jour aux plantes. Les trumeaux ne doivent avoir que la largeur nécessaire pour soutenir le bâtiment : c'est-pourquoi l'on doit préférer de les faire en pierres de tailles ou en briques bien cuites ; car, si on les construisoit avec des briques mal cuites, ils exigeroient plus d'épaisseur, & le bâtiment seroit moins solide, sur-tout s'il y a des logemens au-dessus de l'Orangerie : ce qui est très-utile pour la garantir des gelées pendant les hivers rudes. Si l'on construit ces trumeaux en pierres de taille, je conseille de leur donner la forme d'une colonne cylindrique, & de les faire de deux pieds & demi de diametre, parce que des trumeaux de cette forme ne parent point les royans du soleil comme ils le feroient par leurs angles, s'ils étoient quarrés. Si on les construit en briques, il sera nécessaire de leur donner trois pieds de largeur, afin qu'ils aient la solidité nécessaire, en observant de les évaser en-dedans, pour admettre plus aisément le soleil.

On pratiquera derriere l'Orangerie un bâtiment qui pourra servir à serrer les outils, & à plusieurs autres commodités, ainsi qu'à empêcher le froid d'y pénétrer. Au moyen de cela, le mur du fond pourra

n'être

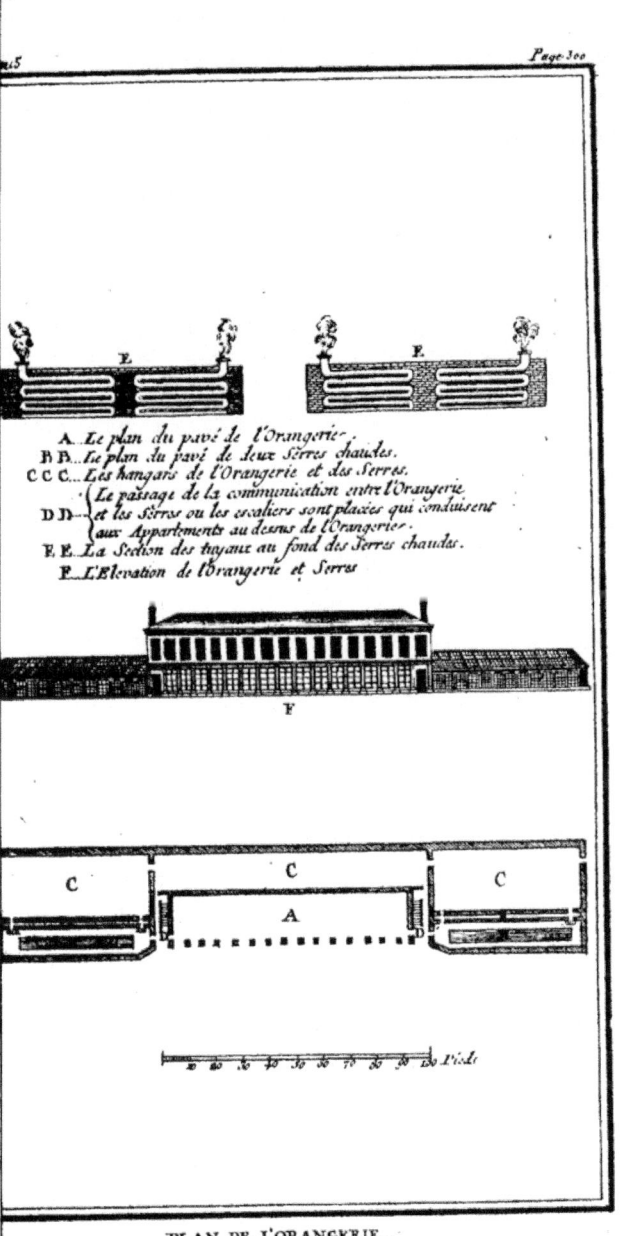

A. *Le plan du pavé de l'Orangerie.*
B.B. *Le plan du pavé de deux Serres chaudes.*
C.C.C. *Les hangars de l'Orangerie et des Serres.*
      Le passage de la communication entre l'Orangerie
D.D. *et les Serres ou les escaliers sont placées qui conduisent*
      aux Appartemens au dessus de l'Orangerie.
E.E. *La Section des tuyaux au fond des Serres chaudes.*
  F. *L'Elevation de l'Orangerie et Serres.*

PLAN DE L'ORANGERIE.

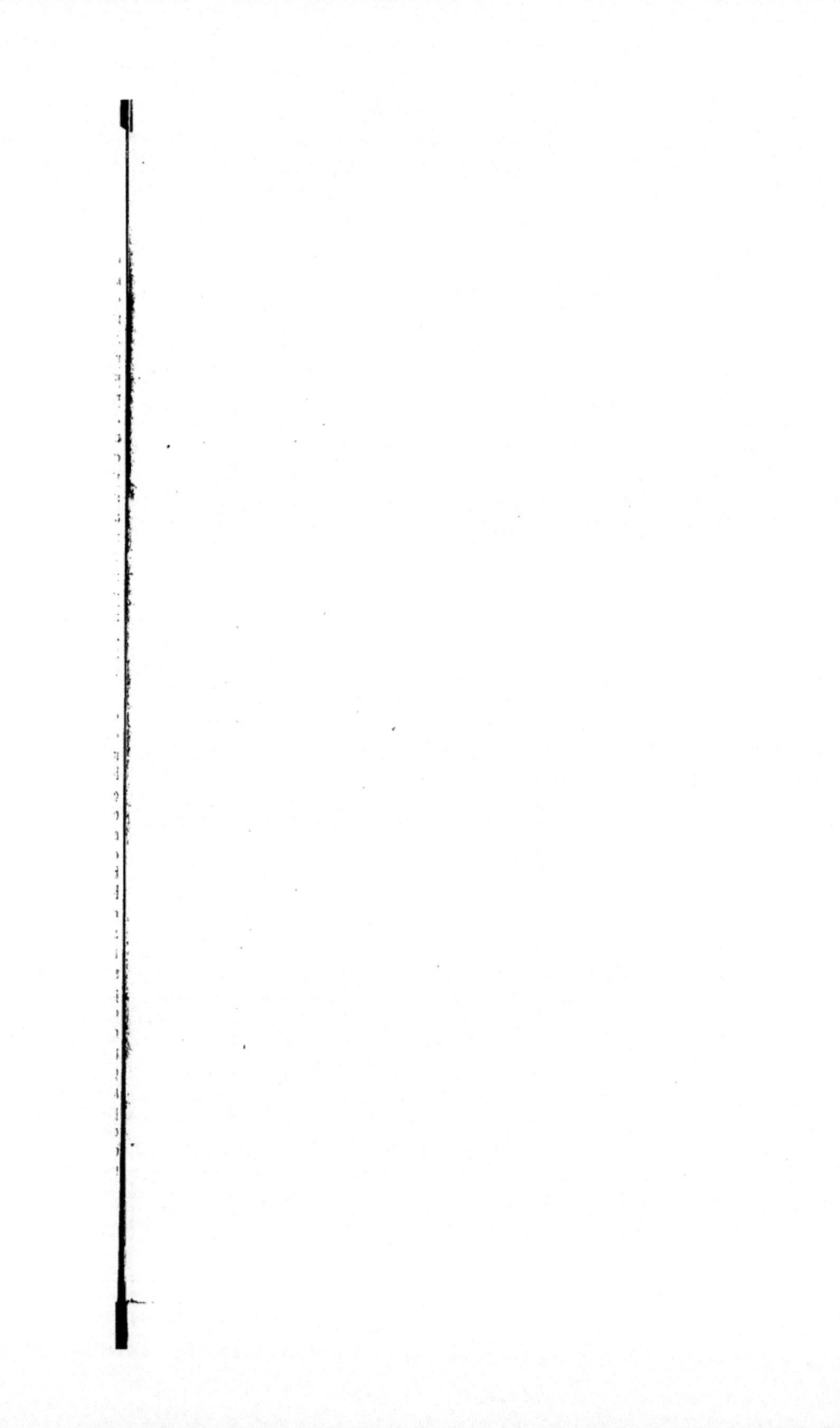

n'être que de deux briques & demie d'épaisseur ; mais si au contraire ce mur est exposé à l'air , il lui faudra au moins trois briques ou trois briques & demie. Si l'on a dessein de faire un beau bâtiment, & de construire des appartemens au-dessus , on pratiquera les escaliers & les passages par derrière au-dessus du magasin des outils , pour ne pas passer dans l'Orangerie : alors on aura au-dessus vingt-cinq ou trente pieds de largeur, & une longueur proportionnée au-dessous des escaliers. On doit pratiquer une porte de dégagement, qui communiquera à l'Orangerie , & par laquelle les Jardiniers pourront y entrer dans les tems des grandes gelées, pendant lesquels on ne doit point ouvrir les vitrages de la façade.

Le pavé de l'Orangerie doit être fait avec des carreaux de pierre ou de brique : on l'élèvera de deux pieds au-dessus de la surface sur laquelle l'édifice est placé ; ce qui sera suffisant dans un terrein sec : mais si le sol est humide, il sera nécessaire de l'élever de trois pieds , & même davantage au-dessus du niveau. Si l'on pratiquoit une voûte au-dessous , on auroit encore moins à craindre de l'humidité , qui est toujours très-nuisible aux plantes, sur-tout après les grands dégels , pendant lesquels le tems est trop froid pour y introduire l'air extérieur. Je pense qu'il sera utile de pratiquer sous le pavé, à un pied de la façade, un tuyau d'un pied de largeur , & de

deux de profondeur , qui regnera dans toute la longueur de l'Orangerie, & qui retournera vers la muraille du fond, où il se terminera en entonnoir dans le magasin des outils , pour en laisser sortir la fumée , après néanmoins qu'il aura tourné trois fois autour de l'Orangerie. Le fourneau peut être placé à une des deux extrémités , de façon que son ouverture, ainsi que la grille des cendres , soit dans le magasin des outils , & qu'on ne l'apperçoive point dans l'Orangerie. La provision du bois sera aussi dans ce magasin , pour être plus à portée.

Plusieurs personnes seront sans doute étonnées de me voir donner le conseil de pratiquer des tuyaux sous une Orangerie ; car depuis long-tems on ne s'en sert plus , & quelques Jardiniers même en ont cru l'usage dangereux. En effet, il a été quelquefois ; mais c'étoit absolument la faute de ceux qui s'en servoient, en tenant toujours les fourneaux allumés, sans faire attention à la température de l'air. Je sais bien qu'on passe souvent deux ou trois ans sans les employer ; mais comme à ces hivers doux succedent quelquefois des hivers très-froids, je ne connois point alors de moyens plus simples & moins dispendieux pour empêcher la gelée de pénétrer dans les Orangeries.

L'intérieur des croisées de l'Orangerie sera garni de bons & fort volets, qui doivent être brisés , pour pouvoir se replier de très-près sur la lar-

geur du trumeau, afin qu'ils ne faffent point d'ombrage : il fuffit de leur donner un pouce & demi d'épaiffeur, ou un peu plus, pour qu'ils foient propres à garantir d'un froid ordinaire. Si le froid devient affez violent pour endommager les plantes, on allume du feu dans le fourneau ; car fans cette commodité, il feroit très-difficile de conferver dans l'Orangerie une température néceffaire, & l'on feroit forcé, comme l'ont pratiqué plufieurs perfonnes, de clouer des nattes fur les fenêtres, ou de remplir de paille le vuide qui fe trouve entre les volers & les châffis, expédient très-nuifible, puifqu'il ôte entierement l'air aux plantes, & qu'il les prive de quelques rayons du foleil, qui paroît fouvent pendant deux ou trois heures dans les gelées les plus fortes, & dont la chaleur & la lumiere font très-utiles aux plantes. On ouvre les volets avec facilité, lorfque le foleil commence à luire, & on les referme, lorfqu'il fe couvre de nuages : quand les fenêtres au contraire font couvertes de paillaffons, on emploie un tems confidérable à les ôter & à les remettre tour-à-tour, & fouvent le foleil eft caché avant que cet ouvrage fatiguant ne foit fini : d'ailleurs, lorfqu'il faut autant de précaution pour fe garantir de la gelée, on peut compter fur l'exactitude de bien peu de Jardiniers. Si le Jardinier n'aime pas fon état, & ne partage pas le goût de fon maître pour les plantes étrangeres, ce long

travail l'effraiera & le rebutera. En fuppofant même qu'il fe donne la peine de couvrir les croifées de paillaffons, penfet-on qu'il les ôtera dès que le foleil commencera à paroître ? Il y a au contraire lieu de croire que les plantes refteront enfermées tant que les grands froids dureront.

On a fait auffi ufage de baffines remplies de charbon, qu'on mettoit dans les Orangeries pendant les grandes gelées. Cette pratique eft dangereufe, & pour ceux qui foignent ce feu, par le rifque qu'ils courent d'en être fuffoqués, & pour les plantes elles-mêmes : auffi les inconvéniens de ces baffines les ont fait généralement profcrire. On doit donner la préférence aux tuyaux de fourneau, qui fe font à peu de frais, en conftruifant l'Orangerie.

Le mur du fond doit être blanchi, & bien enduit de plâtre ou de mortier, pour empêcher les fortes gelées de pénétrer à travers, fur-tout lorfque la gelée eft accompagnée d'un vent fort, tel qu'on en éprouve fouvent dans les hivers rudes.

Quelques perfonnes font la dépenfe de boifer leurs Orangeries ; il eft néceffaire alors d'enduire la muraille avec de la chaux & de la bourre derriere les boiferies, pour arrêter le froid : il n'en faut pas moins blanchir cette boiferie, ainfi que le plafond, parce que cette couleur réfléchit la lumiere en plus grande quantité qu'aucune autre ; qu'elle

ORA

eſt par-là plus avantageuſe aux plantes pendant l'hiver, ſur tout lorſque l'Orangerie eſt exactement fermée. J'ai remarqué que les plantes ont perdu la plus grande partie de leurs feuilles dans des Orangeries peintes en noir ou en couleur brune.

Quand on ne veut point pratiquer des appartemens au-deſſus de l'Orangerie, il faut chercher à la garantir du froid qui peut venir par la toîture : on y parvient en garniſſant le haut de roſeaux, de bruyeres, de genêts ou de mouſſe, placés entre les tuiles & le plafond ; de façon cependant que la charpente du plafond n'en ſoit pas ſurchargée : on entaſſe ces différentes matieres juſqu'à l'épaiſſeur d'un pied au moins ; on en égaliſe la ſurface autant qu'il eſt poſſible, on les fixe avec des lattes, pour empêcher que rien ne ſe ſouleve, & l'on couvre enſuite le tout avec de la chaux mêlée de bourre. Au moyen de cette précaution, le froid ne pourra pénétrer, & les ſouris & autres animaux ne s'y logeront point. On voit pluſieurs Orangeries qui, faute de ce ſoin, ne ſont point à l'abri de la gelée dans les hivers rudes ; ce que l'on attribue quelquefois aux vitrages, pendant que le dommage ne vient que du plafond. Si le bâtiment n'eſt couvert que de tuiles ou d'ardoiſes, les moindres gelées ſe feront ſentir dans l'intérieur.

On place enſuite dans l'Orangerie des gradins mobiles, ſur leſquels on arrange régu-

ORA 303

lierement pluſieurs rangs de pots ſur de petites caiſſes, pour les empêcher de s'entremêler. Le plus petit doit être poſé à quatre pieds des fenêtres : cet eſpace ſervira de paſſage, & à la libre circulation de l'air. On diſpoſe les autres gradins de maniere qu'ils aillent toujours en montant, & que le ſecond ſoit au-deſſus du premier : on ménage derriere, dans la largeur de la muraille du fond, une allée de cinq pieds pour l'arroſement des plantes ; on laiſſe auſſi un eſpace ſur les côtés, afin qu'il y ait toujours un courant d'air pour diſſiper l'humidité que la tranſpiration des plantes occaſionne. On ne doit pas les ſerrer trop, de peur que les jeunes rejettons ne ſe moiſiſſent ; ce qui les feroit périr. On ne place point les *Sedum*, les *Euphorbis*, les *Cierges*, & autres plantes délicates, graſſes & ſucculentes, avec les *Orangers*, les *Myrtes*, & autres arbres toujours verts. En 1729, j'ai fait une expérience déciſive, qui vient à l'appui du conſeil que je viens de donner. Un *Sedum* placé dans le milieu d'arbres toujours verts, & que je n'arroſai point du tout en hiver, augmentoit tous les jours de poids. Cette augmentation de peſanteur ne pouvoit être attribuée qu'à l'humidité de l'air chargé des vapeurs qu'exhaloient ces arbres. Les feuilles pâlirent bien-tôt, ſe fanerent enſuite, & tomberent quelques jours après. J'ai vu en général arriver la même choſe à toutes les plantes

V 2

succulentes, conservées dans une Orangerie remplie d'arbres toujours verts, qui demandent des arrosemens continuels.

Pour éviter cet inconvénient, & séparer ces especes de plantes, je pense qu'il seroit à propos de pratiquer deux aîles à chaque bout de l'Orangerie ; elles lui serviroient d'ornement, & l'on pourroit y entretenir plus de chaleur. La face de l'Orangerie doit être exactement exposée au midi, un des côtés au sud, & l'autre au sud-ouest. De cette maniere, tout le bâtiment jouira du soleil, depuis le lever de cet astre jusqu'à son coucher. On place au-devant de sa façade les plantes exotiques qui peuvent supporter l'air pendant l'été ; & au printems, les plates-bandes de cette place peuvent être garnies d'*Anemones*, de *Renoncu es*, *Tulippes printannieres*, &c., dont les fleurs seront passées, & les racines enlevées, lorsque l'on commence à sortir les plantes ; ce qui contribuera à l'agrément de ce lieu.

On doit pratiquer dans le milieu de ce parterre un petit bassin qui l'ornera beaucoup, & sera fort utile pour l'arrosement des plantes ; l'eau, échauffée par la réverbération du bâtiment sera meilleure pour les arrosemens qu'une eau froide & crue.

Les deux aîles des extrémités seront construites de façon qu'elles puissent contenir des plantes plus ou moins délicates ; ce qui peut s'effectuer par la situation, les fourneaux, &

la maniere d'en conduire les tuyaux. J'entrerai dans un détail particulier à ce sujet à l'article des serres chaudes. J'observerai cependant ici que le côté du sud-est doit toujours être réservé pour la serre la plus chaude, parce que, dans cette situation, elle jouira des premiers rayons du soleil, qui seul peut vivifier les plantes, en les échauffant dès le matin, après les longues nuits de l'hiver.

Ces aîles, dont le plan est ci-joint, ont soixante pieds de longueur ; elles peuvent être divisées chacune dans le milieu par des cloisons en vitrages, avec des communications pour aller de l'une à l'autre. On observera de donner à chaque division un fourneau avec des tuyaux qui s'éleveront contre la muraille du fond, pour le passage de la fumée, & qui se replieront sur eux-mêmes autant que la hauteur pourra le permettre : car plus la fumée séjournera, plus elle donnera de chaleur avec moins de bois ; ce qui doit entrer en considération, sur-tout dans les pays où les matieres combustibles sont cheres : par ce moyen, on pourra séparer les plantes qui exigent différens dégrés de chaleur, ainsi que je l'expliquerai plus amplement dans l'article des serres chaudes.

La seconde aîle du bâtiment qui fait face au sud-ouest, sera aussi divisée de la même maniere, & les tuyaux seront conduits dans toute la longueur des deux parties. On en fera usage suivant la saison & les

especes de plantes qui y feront renfermées ; de sorte qu'il y aura dans les deux ailes quatre divisions, & dans chacune un dégré de chaleur différent ; ce qui, avec l'Orangerie, sera suffisant pour contenir les plantes de tous les pays : il seroit impossible, sans cela, de conserver une quantité de plantes qu'on nous apporte annuellement de l'Afrique & de l'Amérique ; car lorsque ces plantes, qui viennent de différens pays, & sous des climats divers, se trouvent rassemblés dans la même serre, les unes périssent par trop de chaleur, & d'autres faute d'en avoir assez, comme on le voit souvent arriver dans les jardins, où l'on conserve une grande collection de plantes.

En construisant ces serres chaudes, si l'on n'y pratique par-derriere, dans toute leur longueur, une gallerie ou hangar, le mur doit avoir au moins trois briques d'épaisseur, pour empêcher le froid d'y pénétrer, & pour contenir la chaleur des tuyaux dans l'intérieur des serres. Le toît de ces galleries ou hangars doit aussi être garni de roseaux en-dessous, comme nous l'avons préscrit pour l'Orangerie ; ce qui arrêtera le froid & épargnera beaucoup de bois. Les vitrages inclinés, ainsi que ceux du front, doivent aussi être couverts de volets ou de paillassons, pour conserver la chaleur. Lorsque ces bâtimens sont bien conditionnés, on évite beaucoup de dépenses pour l'avenir.

Les vitrages inclinés se font en coulisse, afin de pouvoir les ouvrir plus ou moins dans les tems chauds, & donner de l'air aux plantes. Les vitrages à plomb de la façade sont mobiles sur des gonds, & s'ouvrent comme des portes ; & ceux de dessus seront faits aussi de maniere qu'on puisse les tirer en coulisse. Au moyen de cette disposition, on introduira dans la serre autant d'air qu'on voudra.

Outre toutes ces serres, il sera nécessaire de construire encore des couches profondes à châssis, telles qu'on en a pour élever des plantes annuelles au printems : on y renfermera en hiver les plantes qui viennent de la Caroline, de la Virginie, de l'Espagne, &c., lorsqu'elles sont encore trop jeunes pour être plantées en plein air, ainsi que plusieurs autres especes de l'Espagne, &c., qui exigent seulement d'être abritées des grandes gelées, & qui ont cependant besoin de beaucoup d'air dans les tems doux. On enleve aisément ces vitrages pendant le jour, on les remet pour la nuit ; & pendant les fortes gelées, on les couvre avec des nattes, de la paille ou des chaumes de pois, pour les préserver du froid, qui feroit périr les tiges, & détruiroit entierement les plantes mêmes qui cependant ne craignent point les hivers ordinaires. Ces couches doivent être enfoncées d'un pied & même davantage au-dessous du niveau, à moins que le sol ne soit humide : dans ce cas, il

faudra les tenir fur la furface de la terre. Les côtés de ces couches feront conftruits en briques, & les fommets couvert en madriers, en y pratiquant des gouttieres vis-à-vis les traverfes qui foutiennent les vitrages. Le mur du fond peut avoir quatre pieds de hauteur, & une brique & demie d'épaiffeur; le mur de face, un pied & demi de haut; & l'intérieur de la couche, fix pieds environ de largeur: fa longueur doit être proportionnée au nombre de plantes que l'on veut y renfermer.

ORCANETTE. *Voyez* ANCHUSA TINCTORIA. *L.*

ORCHIS. *Tourn. Inft. R. H.* 431. *tab.* 248. 249. *Lin. Gen. Plant.* 900. d'ὄρχις, un tefticule, parce que la racine de cette plante reffemble aux tefticules d'un homme; ou de ὀρέγειν, avoir appétit, à caufe qu'elle échauffe & excite à l'amour: elle eft auffi appelée κυνοσόρχις, de κύων, un chien, & d'ὄρχις, un tefticule. [*Satyrion, or Fool Stones.*] Satyrion.

Caractères. Cette plante a une tige fimple, avec une fpathe en forme de gaîne: la fleur n'a point de calice; la corolle a cinq pétales, dont trois fe jettent en-dehors, & deux en-dedans, & qui s'élevent & fe joignent en forme d'étendard. Le nectaire eft formé par une feuille fixée à côté du receptacle, entre les divifions des pétales; la levre fupérieure eft courte & érigée, l'inférieure eft groffe, large & étendue; le tube eft fufpendu en forme de corne, & déborde au dos:

la fleur a deux étamines courtes, minces, poftées fur le piftile, & terminées par des antheres ovales, érigées & fixées à la levre fupérieure du nectaire. Elle a un germe oblong & tors fur la corolle, avec un ftyle court, fixé à la levre fupérieure du nectaire, & couronné par un ftigmat obtus & comprimé; ce germe fe change dans la fuite en une capfule oblongue, à une cellule à trois petites valves, en forme de quille, qui s'ouvrent de trois côtés, mais qui font jointes au fommet & au bas: elle eft remplie de petites femences femblables à de la pouffiere.

Ce genre de plantes eft rangé dans la premiere fection de la vingtieme claffe de LINNÉE, qui comprend celles dont les fleurs ont deux étamines jointes au ftyle.

Les efpeces font:

1°. *Orchis morio, bulbis indivifis, nectarii labio quadrifido, crenulato, cornu obtufo. Act. Ups.* 1740. *p.* 8. *Mat. Med.* 195. *Fl. Suec.* 724. 694. *Dalib. Paris.* 273. *Crantz. Auftr. p.* 499. *Pollich. pal. n.* 843. *Gmel. tub. p.* 270. *Mattufch. Sil.* 657; Orchis avec des bulbes non divifées, la levre du nectaire découpée en quatre parties crénelées, & une corne obtufe.

Orchis morio fœmina. C. B. P. 82; Orchis femelle & commune. Satyrion femelle.

Triorchis Serapias mas. Fuchs. Hift 559.

2°. *Orchis mafcula, bulbis indivifis, nectarii labio quadrilobo, crenulato, cornu obtufo, petalis dorfalibus reflexis. Flor.*

Suec. 795. Crantʒ Auſtr. p. 500. Pollich. pal. n. 844. Flor. Dan. t. 457; Orchis avec des bulbes non divisées, dont la levre du nectaire a quatre lobes, ayant une corne obtuſe, & le dos des pétales réfléchi.

Palmata major. Riv hex. f. 10. Segu. ver. t. 15. f. 6.

Orchis morio mas, foliis maculatis. C. B. P. 81.; Orchis mâle avec des feuilles tachetées.

Teſticulus IV. Cam. Epit. 624.
Satyrium mas. Blakw. t. 53; Satyrion mâle.

3°. Orchis bi-folia, bulbis indiviſis, nectarii labio lanceolato, integerrimo, cornu longiſſimo, petalis patentibus. Act. Ups. 1740. p. 5. Fl. Suec. 723. 793. Dalib. Paris. 273. Mat. Med. 195. Flor. Dan. f. 235. Crantʒ. Auſtr. p. 304. Pollich. pal. n. 841. Mattuſch. Sil. n 655; Orchis avec des bulbes non divisées, la levre du nectaire entiere & en forme de lance, une corne très longue & des pétales fort étendus.

Orchis alba bi-folia minor, calcari oblongo. C. B. P. 83; le plus petit Orchis blanc à deux feuilles, avec un éperon oblong, ou Orchis en papillon, ou Orchis mouche.

Satyrium flore albo Riv f. 12; Satyrion à fleurs blanches.

Teſticuli species V. Cam. Epit. 625.

4°. Orchis militaris, bulbis indiviſis, nectarii labio quinque-fido, punctis ſcabro, cornu obtuſo, petalis confluentibus. Act. Ups. 1740. p. 11. Fl. Suec. 725. 798. Dalib. Paris. 271. Gmel. tub. p. 272. Pollich. pal. n. 446. Mattuſch. Sil. n. 658. Crantʒ.

Auſtr. p. 501; Orchis avec des bulbes non divisées, une levre à cinq pointes au nectaire, marquée de points rudes, une corne obtuſe, & des pétales coulans enſemble.

Orchis latifolia, hiante cucullo, major. Tourn. Inſt. R. H. 432; Orchis à figure d'homme.

Cynoſorchis lati-folia, hiante cucullo, major. Bauh. Pin. 80.

5°. Orchis pyramidalis, bulbis indiviſis, nectarii labio tri-fido, æquali, integerrimo, cornu longo, petalis ſub lanceolatis. Act. Ups. 1740. Fl. Suec. 2. n. 798. Jacq. vind. 292. Auſtr. t. 266. Crantʒ. Auſtr. p. 506; Orchis avec des bulbes non divisées, la levre du nectaire découpée en trois parties égales, une longue corne, & des pétales presque en forme de lance.

Cynoſorchis militaris montana. Bauh. Pin. 81. Prodr. 28.

Orchis militaris montana, ſpicâ rubente, conglomeratâ. Tourn. Inſt. R. H. 432; Orchis militaire de montagne, avec un épi rougeâtre & arrondi.

6°. Orchis lati-folia, bulbis ſub palmatis, rectis, nectarii cornu conico, labio trilobo, lateribus reflexo, bracteis flore longioribus. Act. Ups. 1740. p. 15. Fl. Suec. 728. 801. Dalib. Paris. 274. Gmel. Sib. 1. p. 240 Crantʒ. Auſtr. 493; Orchis avec des bulbes droites & presque en forme de main, une corne conique, la levre du nectaire découpée en trois lobes réfléchis ſur les côtés, & des bractées plus longues que la fleur.

Orchis palmata pratenſis, latifolia, longis calcaribus. C. B. P. 85. Vaill. Paris. t 31. f. 1. 2. 3.

4. 5 ; Orchis à larges feuilles des prés, dont les bulbes sont en forme de main ouverte, ayant de longs éperons.

7°. *Orchis maculata, bulbis palmatis, patentibus, nectarii cornu germinibus breviori, labio plano, petalis dorsalibus patulis. Act. Ups. 1740. p. 14. Fl. Suec. 729. 800. Dalib. Paris. 274. Gmel. Sib. 1. p. 23. Crantz. Austr. p. 492. Gmel. tub. 274. Pollich. n. 349* ; Orchis avec des bulbes en forme de main, & étendues, la corne du nectaire plus courte que le germe, une levre unie, & le dos des pétales étendu.

Palmata maculata, non maculata & angusti-folia maculata. Riv. Hexap. t. 8. & 11.

Orchis palmata pratensis maculata. C. B. P. 85 ; Orchis des prés en forme de main, & tachetée. Orchis ordinaire ou fétide.

Satyrium basilicum fæmina. Dod. Pempt. 240.

8°. *Orchis conopica, bulbis palmatis, nectarii cornu setaceo germinibus longiori, labio tri-fido, petalis duobus patentissimis. Act. Ups. 1740. p. 13. Fl. Suec. 727. 799. Dalib. Paris. 275* ; Orchis avec des bulbes en forme de main, la corne du nectaire hérissée & plus longue que le germe, & la levre divisée en trois parties.

Orchis conopsea. Lin. Syst. Plant. tom. 4. p. 14. Sp. 27.

Satyrium basilicum mas. Fuchs. Hist. 712.

Orchis palmata minor, calcaribus oblongis. C. B. P. 85. Vaill. Paris. t. 30. f. 8. Rudb. Elys. 2. p. 212. f. 5 ; le plus petit

Orchis en forme de main, ayant un long éperon à la fleur.

9°. *Orchis abortiva, bulbis fasciculatis, fili-formibus, nectarii labio ovato, integerrimo. Act. Ups. 1740. p. 17. Dalib. Paris. 275. Gouan. Monsp. 471. Jacq. Austr. f. 193* ; Orchis avec des bulbes en forme de filets & en faisceaux, ayant la levre du nectaire ovale & entiere.

Epipactis aphylla, calcare longo, labello ovato, lanceolato. Hall. Helv. n. 1288. f. 36.

Pseudo-Limodorum Austriacum. Clus. Hist. 270 ; Nid d'Oiseau pourpre d'Autriche.

Limodorum. Hall.

Morio. La premiere espece croît naturellement dans des pâturages de plusieurs parties de l'Angleterre ; elle a une racine double & bulbeuse, avec quelques fibres qui sortent du sommet ; elle pousse quatre ou six feuilles oblongues couchées sur la terre & réfléchies ; sa tige, qui s'eleve à la hauteur de neuf ou dix pouces, est embrassée par quatre ou six feuilles, & terminée par un épi court & lâche de fleurs, dont le nectaire a une levre à quatre segmens dentelés, & une corne obtuse. Ces fleurs sont d'un pourpre pâle, & marquées de taches d'un pourpre plus foncé ; elles paroissent dans le mois de Mai.

Mascula. La seconde espece se trouve dans les bois & dans les lieux ombragés de plusieurs parties de l'Angleterre ; sa racine est double, bulbeuse, de la forme d'une olive médiocre, & à-peu-près de la même

grosseur ; elle a six ou huit feuilles longues, larges, de la forme de celles du Lys, & marquées sur leur surface supérieure de plusieurs taches noires ; sa tige est ronde, d'un pied de hauteur, & embrassée par une ou deux feuilles plus petites : ses fleurs sont disposées en épi long sur le sommet de la tige ; elles sont de couleur pourpre, marquées de taches d'un pourpre plus foncé, & d'une odeur agréable : elles paroissent à la fin d'Avril.

Bi-folia. La troisieme naît spontanément sous les arbustes, dans les haies & dans les clôtures des pâturages de plusieurs parties de l'Angleterre ; elle a une racine composée de deux bulbes oblongues & en forme de lance, de laquelle sortent trois ou quatre feuilles semblables à celles du *Lys*, d'un vert pâle, & marquées foiblement de quelques taches: sa tige s'éleve à près d'un pied de hauteur ; elle est mince, sillonnée, embrassée par très-peu de feuilles, & terminée par un épi lâche de fleurs blanches, d'une odeur agréable, & qui ressemblent à un papillon ayant les aîles étendues. Cette plante fleurit dans le mois de Juin.

Militaris. La quatrieme espece a été trouvée sur les montagnes de Cawsham, & dans d'autres endroits dont le sol est sec, & de la nature de la craie ; ses racines sont composées de deux bulbes, desquelles sortent quatre ou cinq feuilles oblongues ; la tige a environ neuf pouces de hauteur,

& soutient un épi lâche de fleurs d'une odeur agréable, dont chacune est suspendue a un long pédoncule ; elle ont une corne courte & obtuse, une houpe & des aîles de couleur de cendre en-dehors, rougeâtre en-dedans, & marquées de lignes plus foncées : la levre est oblongue, divisée en cinq parties, & hérissée de pointes rudes. Elle fleurit dans le mois de Juin.

Pyramidalis. La cinquieme espece croît sans culture sur des montagnes de craie, dans plusieurs parties de l'Angleterre : sa racine est composée de deux bulbes oblongues, desquelles sortent trois ou quatre feuilles étroites & oblongues ; sa tige s'éleve à un pied de hauteur, & a trois ou quatre feuilles étroites & érigées, qui l'embrassent : ses feuilles sont produites en épis épais & ronds au sommet ; elles sont rougeâtres, & ont de longs éperons, & des aîles à pointe aiguë : elles paroissent dans le mois de Juin.

Lati-folia. La sixieme se trouve dans des prairies humides de plusieurs parties de l'Angleterre : sa racine est composée de deux bulbes charnues, divisées en quatre ou cinq doigts, & semblables à une main ouverte ; la tige s'éleve à neuf pouces ou un pied de hauteur ; elle est garnie dans toute sa longueur de feuilles de trois ou quatre pouces de longueur sur un de largeur, & qui embrassent la tige de leur bâse ; elles sont sans tache, & sont terminées en pointe aiguë ;

ses fleurs sortent en un épi au sommet de la tige, avec des bractées entr'elles plus longues que les fleurs. Les éperons ont six lignes de longueur, & s'étendent en arriere; la levre du nectaire est large, & divisée en trois lobes, dont les deux latéraux sont réfléchis : les fleurs & les bractées sont de couleur tirant sur le pourpre, & tachetées d'un pourpre foncé : elles paroissent dans le mois de Mai. Il y a deux variétés de cette espece qui different par la couleur de leurs fleurs, & une autre à feuilles étroites.

Maculata. La septieme espece croît naturellement sur des prés humides, dans plusieurs parties de l'Angleterre : sa racine est composée de deux bulbes grosses, charnues, & divisées en quatre doigts étendus ; sa tige, qui s'éleve à la hauteur d'un pied & demi, est très-forte, de couleur tirant sur le pourpre, & garnie de feuilles dans toute sa longueur ; celles du bas ont six pouces de longueur sur un & demi de largeur, & embrassent la tige de leur bâse : les fleurs sont recueillies en un épi clair ou lâche au sommet de la tige ; elles sont d'un pourpre pâle : l'éperon a environ quatre lignes de longueur ; la levre du nectaire est unie, divisée en trois parties, & tachetée d'un pourpre foncé ; sous chaque pédoncule est placée une bractée de couleur tirant sur le pourpre ; les feuilles & les tiges ont plusieurs taches foncées. Cette plante

fleurit dans le mois de Juin : on en connoît deux variétés, qui different par la couleur de leurs fleurs.

Cornopica. La huitieme, qu'on rencontre encore sur des prés humides en Angleterre, a une racine doublement en forme de main ouverte : la partie qui soutient la tige diminue & perit ; mais l'autre reste pleine, grosse & succulente ; les bulbes en forme de main, qui composent cette racine, sont longues, & s'étendent à une certaine distance : les feuilles du bas ont six ou sept pouces de longueur ; elles sont étroites, d'un vert pâle, & sans aucune tache ; la tige, qui s'éleve à la hauteur d'un pied, est garnie de quelques feuilles courtes & étroites, qui l'embrassent en forme de gaîne, & terminée par un bel épi de fleurs rouges de six pouces de longueur. Ces fleurs sont sans tache, & ont des éperons longs, minces, hérissés comme une griffe d'oiseau, & courbés; la levre du nectaire est dentelée sur ses bords. Cette plante fleurit dans le mois de Juin.

Abortiva. La neuvieme espece croît à l'ombre des bois dans plusieurs cantons de l'Angleterre, & particulierement en Sussex & Hampshire, où je l'ai trouvée plusieurs fois : sa racine est composée de plusieurs fibres épaisses, obliques, longues & charnues ; sa tige, qui s'éleve à près de deux pieds de hauteur, est enveloppée de feuilles en forme de gaîne, & de couleur pourpre : ses fleurs naissent en thyrse lâche au

sommet de la tige ; elles font de couleur pourpre, & ont une levre ovale & entiere au nectaire : la houpe eſt terminée en corne. Cette eſpece fleurit dans le mois de Juin.

Culture. Quoique toutes ces eſpeces d'*Orchis* croiſſent ſauvages dans pluſieurs cantons de l'Angleterre, cependant leur figure extrêmement biſarre, & la beauté de leurs fleurs doivent leur faire donner une place dans tous les beaux jardins. Si l'on ne les y cultive pas, ce n'eſt que la difficulté de les tranſplanter qui s'y oppoſe. On peut cependant y parvenir en les remarquant tandis qu'elles ſont en fleurs, & en ne les enlevant qu'après que leurs feuilles ſont flétries ; ce qu'on peut faire alors avec ſûreté, ainſi qu'on le pratique pour la plupart des eſpeces de plantes à racines bulbeuſes & charnues, qui ſurvivent rarement quand on les tranſplante avant la chûte de leurs feuilles, quoiqu'on les enleve avec une groſſe motte de terre à leurs racines : car l'extrémité de leurs fibres s'étendant à une grande profondeur dans la terre, pour y puiſer leur nourriture ; ſi l'on vient à les rompre ou à les endommager en les enlevant, les plantes profitent rarement après. Il eſt vrai que, dans ce cas, elles ſubſiſtent encore une ou deux années ; mais elles vont toujours en dépériſſant, & meurent enſuite. La même choſe arrive auſſi aux *Tulippes*, aux *Fritillaires* & autres racines bulbeuſes, quand on les

tranſplante lorſqu'elles ont déja pouſſé des tiges. Après avoir enlevé ces racines dans le tems convenable, on les plante dans un ſol, & à une expoſition qui reſſemble, le plus qu'il eſt poſſible, au lieu où elles ont été priſes, ſans quoi elles ne profiteroient pas : ainſi, elles ne peuvent être toutes placées dans la même plate-bande ; car les ûnes croiſſent ſur les montagnes de craie, d'autres dans des prairies humides, & quelques-unes à l'ombre des bois. En donnant à chacune le ſol qui lui eſt propre, elles réuſſiront, ſubſiſteront pluſieurs années, & produiront, pendant tout le tems qu'elles ſeront en fleurs, une variété auſſi agréable que quelque plante que ce ſoit. On trouvera les autres eſpeces qui ne ſont point rappelées ici, dans les articles *Ophrys*, *Satyrium* & *Serapias*.

 OREILLE D'HOMME ou CABARET. *Voyez* ASARUM.

OREILLE DE LIEVRE ou LA PERCE-FEUILLE. *Voy.* BUPLEVRUM ROTUNDI-FOLIUM. *L.*

OREILLE D'OURS. *Voy.* AURICULA URSI. *J. B.*

OREILLE D'OURS DE VIRGINIE ou DODECATHEON. *Voyez* MEADIA.

OREILLE DE RAT ou LA PILOSELLE. *V.* AURICULA MURIS.

OREILLE DE SOURIS. *Voyez* CERASTIUM REPENS. *L.* ou MYOSOTIS.

OREOSELINUM. *Voy.* ATHAMANTHA OREOSELINUM. *L.*

ORGE. *Voy.* HORDEUM. *L.*

ORIGAN. *Voyez* ORIGANUM ; SATUREIA ORIGANOIDES.

ORIGAN SAUVAGE. *Voy.*
ORIGANUM VULGARE. *L.*

ORIGANUM. *Lin. Gen.
Plant.* 645. *Tourn. Inst. R. H.*
108. *tab.* 94. ὀρίγανον, de ὄρος,
une montagne, & γάνος, plaifir ; c'eft-à-dire , une plante
qui fe plait fur les montagnes.
[*Origany, or Pot Marjoram.*]
Origan. Marjolaine.

Caracteres. La fleur eft labiée ; elle a un tube cylindrique & comprimé ; la levre
fupérieure eft unie , érigée ,
obtufe & dentelée ; l'inférieure
eft divifée en trois parties à-
peu-près égales : les fleurs font
difpofées en épis, & compofées de feuilles ovales, colorées & placées les unes fur
les autres en écailles de poiffon. Ces fleurs ont quatre
étamines minces, dont deux
font auffi longues que la corolle , & les deux autres plus
longues , & qui font toutes
terminées par des antheres fimples. Son germe eft quarré ,
& foutient un ftyle mince ,
incliné à la levre fupérieure ,
& couronné par un ftigmat
divifé en deux parties : il fe
change dans la fuite en quatre
femences renfermées dans le
calice de la fleur.

Ce genre de plantes eft rangé dans la premiere fection de
la quatorzieme claffe de LIN-
NÉE , avec celles dont les fleurs
ont deux étamines longues &
deux plus courtes , & qui font
remplacées par des femences
nues.

LINNÉE a ajouté à ce genre
la *Majorana* de TOURNEFORT ,
& le *Dictamnus* de BOERHAA-
VE. La premiere a fes fleurs

difpofées en têtes quarrées &
écailleufes , & l'autre a les
fiennes en têtes lâches & écailleufes , qui fortent entre les
feuilles.

Les efpeces font :

1º. *Origanum vulgare , fpicis
fub-rotundis , paniculatis , conglomeratis , bracteis calyce longioribus , ovatis, Lin. Sp. Plant.*
590. *Mat. Med.* 151. *Gmel. Sib.*
3. *p.* 244. *Deneck. Gallob. p.*
259. *Crantz. Auftr. p.* 282. *Pall.
it.* 1. *p.* 64. 72. *Scop. carn. ed.*
2. *n.* 740 ; Origan fauvage
avec des épis prefque ronds,
en panicules , recueillis en
grappes , & des bractées ovales plus longues que le calice.

*Origanum foliis ovatis , fpicis
laxis , erectis , confertis , paniculatis. Hort. Cliff.* 305. *Fl. Suec.*
480. 5341 *Roy. Lugd.-B.* 323.

*Origanum vulgare fpontaneum.
J. B.* 2. 236 ; Origan commun
& fauvage.

2º. *Origanum Heracleoticum ,
fpicis longis , pedunculatis , aggregatis , bracteis longitudine calycum. Lin. Syft. Plant. tom.* 3.
p. 77. *Sp.* 6 ; Origan avec des
épis longs , dont les fleurs ,
poftées fur des pédoncules ,
font rapprochées en paquets ,
& garnies de bractées de la
longueur des calices.

*Origanum Heracleoticum , Cu-
nila gallinacea Plinii. C. B. P.*
223 ; Marjolaine douce d'hiver.

*Origa , Origanum Heracleoti-
cum , Cunila. Lob. ic.* 492.

3º. *Origanum lati-folium , fpicis
oblongis , paniculatis , conglomeratis , foliis ovatis , glabris ;* Origan
avec des épis oblongs de fleurs
en panicules & rapprochées ,
& des feuilles unies & ovales,

ORI

. *Origanum humilius, lati-foüum, glabrum. Tourn. Inst. R. H.* 199.; Origan bas, uni, & à larges feuilles.

4°. *Origanum humile, caule repente, spicis oblongis, conglomeratis, bracteis florum longioribus*; Origan avec une tige rampante, des épis oblongs de fleurs en grappes, & des bractées plus longues que les fleurs.

Origanum sylvestre, humile. C. B. P. 223. *Prod.* 109; Origan bas & sauvage.

5°. *Origanum Orientale, caule erecto, ramoso, foliis ovatis, rugosis, spicis sub-rotundis, conglomeratis, bracteis calycum brevioribus*; Origan avec une tige érigée & branchue, des feuilles raboteuses & ovales, & des épis presque ronds de fleurs rapprochées, & garnies de bractées plus longues que les calices.

Origanum Orientale Prunellæ folio glauco, flore purpureo. Boerrh. Ind. Alt. 1. 179; Origan oriental, avec une feuille de couleur vert-de-mer de Sanicle, & une fleur pourpre.

6°. *Origanum Creticum, spicis aggregatis longis, prismaticis, rectis, bracteis membranaceis, calyce duplò longioribus. Lin. Sp. Plant.* 589. *Mat. Med.* 151. *Fabric. Helmst. p.* 110. *Hall. Helv. n.* 234; Origan avec des épis longs, droits, en forme de prisme, & disposés en grappes, ayant des bractées membraneuses, deux fois plus longues que les calices.

Origanum Creticum. C. B. P. 223; Origan de Crète.

7°. *Origanum Majorana, foliis ovalibus obtusis, spicis sub-* rotundis, compactis, pubescentibus. *Hort. Cliff.* 304. *Hort. Ups.* 161. *Mat. Med.* 151. *Roy. Lugd.-B.* 324. *Blackw. f.* 319. *Regn. Bot.*; Origan avec des feuilles ovales, & obtuses, des épis presque ronds, comprimés & velus.

Origanum vulgare. C. B. P. 224; Marjolaine commune ou Marjolaine douce.

Amaracus vulgatior. Lob. ic. 498.

8°. *Origanum Ægyptiacum, foliis carnosis, tomentosis, spicis nudis. Lin. Sp. Plant.* 822; Origan avec des feuilles charnues & cotonneuses, & des épis nuds.

Majorana rotundi-folia scutellata, exotica. H. R. Par.; Marjolaine étrangere à feuilles rondes & en forme de cuiller.

Origano cognata Zatarhendi. B. P. 223; Marjolaine de Coq.

9°. *Origanum Smyrnæum, foliis ovatis, acutè serratis, spicis congestis, umbellatim fastigietis. Hort. Cliff.* 304. *Roy. Lugd.-B.* 324; Origan avec des feuilles ovales & sciées, à dents aiguës, ayant des épis de fleurs rapprochées, & disposées en ombelle.

Origanum Smyrnæum. Wheel. Raii Hist. 450; Origan de Smyrne.

Majorana Cretica, Origani foliis, villosa Satureja odore, corymbis majoribus albis. Tourn. cor. 13.

10°. *Origanum Dictamnus, foliis omnibus tomentosis, spicis nutantibus*; Origan dont toutes les feuilles sont cotonneuses, produisant des épis de fleurs penchés.

Dictamnus Creticus. C. B. P. 222; Dictamne de Crète.

Dictamnum Cretense. Cam. E-pit. 472.

11°. *Origanum Sipyleum , foliis omnibus glabris , spicis nutantibus. Hort. Cliff.* 304. *Roy. Lugd. B.* 325 ; Origan dont toutes les feuilles font unies , produifant des épis de fleurs penchés.

Dictamnus montis Sipyli , Origani foliis. Flor. Bat. 2. 72 ; Dictamne de la montagne de Sipyle , à feuilles d'Origan.

12°. *Origanum hybridinum , foliis inferioribus tomentofis , fpicis nutantibus. Hort. Cliff.* 304 ; Origan dont les feuilles du bas font cotonneufes , & dont les épis de fleurs font penchés.

Origanum Dictamnus. Lin. Syft. Plant. tom. 3. *p.* 76. *Sp.* 2.

Origanum Dictamni Cretici facie , folio craffo , nunc villofo , nunc glabro. Tourn. Cor. 13 ; Origan qui reffemble au Dictamne de Crète , avec des feuilles épaiffes , dont les unes font velues & d'autres unies.

13°. *Origanum Onites , fpicis oblongis , aggregatis , hirfutis , foliis cordatis , tomentofis. Lin. Sp. Plant.* 590 ; Origan avec des épis oblongs , velus , & rapprochés , ayant des feuilles cotonneufes & en forme de cœur.

Origanum lignofum Syracufanum perenne , umbellâ ampliffimâ brevi , lato & nervofo folio. Bocc. mus. 2. *p.* 45. *tab.* 38 ; Origan ligneux & vivace de Syracufe, avec une ombelle de fleurs courte & très large , & une feuille très-large & nerveufe.

Vulgare. La première efpece croit naturellement dans les brouffailles & les buiffons de quelques parties de l'Angleterre : fa racine eft vivace , & compofée de plufieurs petites fibres ligneufes ; fes tiges font quarrées , de près de deux pieds de hauteur , ligneufes , & garnies de feuilles ovales & placées par paires , & des aîles defquelles fortent de chaque côté trois ou quatre feuilles plus petites , qui reffemblent à celles de la *Marjolaine* , & font feffiles à la tige. Ces feuilles ont une odeur aromatique : les fleurs font produites en épis ronds , & croiffent en panicules au fommet des tiges. Plufieurs de ces épis font recueillis en une tête ; ces fleurs font de couleur de chair, & paroiffent au-deffus de leurs enveloppes écailleufes ; leur levre fupérieure eft découpée en deux parties érigées , & l'inférieure eft divifée en trois fegmens qui pendent vers le bas ; les étamines s'étendent en-dehors un peu au-delà de la corolle , & font d'une couleur tirant fur le pourpre. Cette plante fleurit dans les mois de Juin & de Juillet , & fes femences mûriffent en automne. On la cultive quelquefois dans les jardins , & quelques-uns lui donnent le nom de *Marjolaine du Pot*, parce que l'on en met généralement dans les potages.

Cette efpece fe reproduit abondamment par fes femences écartées , & l'on peut encore la multiplier en divifant fes racines. Le meilleur tems pour faire cette opération eft l'automne : on plante ces racines dans tous les fols qui ne font

pas trop humides, & elle profitent à toutes les expositions; elles n'exigent aucune autre culture que d'être tenues nettes de mauvaises herbes. Il y en a une variété à fleurs blanches, avec des tiges d'un vert clair, & une autre à feuilles panachées (1).

Heracleoticum. La seconde espece, à laquelle on donne communément le nom de Marjolaine douce d'hiver, étoit autrefois connue sous celui de *Marjolaine de Pot*; elle a une racine vivace, de laquelle sortent plusieurs tiges branchues, quarrées, d'un pied & demi de hauteur, velues, tirant sur le pourpre, & garnies de feuilles ovales, tirant sur le pourpre, qui ressemblent beaucoup à celles de la *Marjolaine*, & sont postées par paires sur de courts pétioles : ses fleurs sont disposées en épis de deux pouces environ de longueur, & plusieurs s'élevent ensemble

des divisions de la tige. Ces fleurs sont petites, blanches, & sortent au-dessus de leurs enveloppes écailleuses ; elles paroissent dans le mois de Juillet, & leurs semences mûrissent en automne. Cette plante croît naturellement en Grece & dans les parties chaudes de l'Europe ; mais elle est assez dure pour profiter en plein air en Angleterre : on la cultive principalement pour en former des bouquets. Comme elle fleurit plutôt que la *Marjolaine*, on l'emploie aux mêmes usages, jusqu'à ce que l'autre soit parvenue à sa maturité. Il y a une variété de cette espece à feuilles panachées. On la multiplie ordinairement en divisant ses racines en automne ; elle exige un sol sec, où elle réussir très-bien : sa culture est la même que celle de la précédente.

Lati-folium. La troisieme, qu'on rencontre dans la France & en Italie, a une racine vivace, de laquelle sortent plusieurs tiges minces, d'un pied environ de hauteur, & garnies de feuilles ovales, unies, & postées sur de longs pétioles : ses fleurs sont disposées en épis oblongs, qui croissent en panicules & en grappes ; elles sont petites, de couleur tirant sur le pourpre, & paroissent au-dessus de leurs enveloppes écailleuses. Cette espece fleurit dans le mois de Juin : on peut la multiplier en divisant ses racines, comme on le pratique pour la précédente.

Humile. La quatrieme espece se trouve en abondance aux

---

(1) L'Origan jouit à-peu-près des mêmes propriétés médicinales que le Thym, le Serpolet, la Marjolaine, &c. auxquels on peut le substituer dans toutes les circonstances où ces simples sont indiqués : il est apéritif, incisif, anti-hystérique, carminatif, stomachique, utérin, emmenagogue, &c. On s'en sert avec quelque succès en infusion froide vineuse, dans les engorgemens catharreux, l'asthme pituiteux, les vices de digestion, la suppression des regles, & enfin, dans toutes les affections morbifiques qui tiennent au relâchement des solides.

L'Origan entre dans le syrop d'armoise & dans l'électuaire des baies de Laurier.

environs d'Orléans ; elle a une racine vivace, de laquelle s'élevent plusieurs tiges quarrées d'environ six pouces de hauteur, inclinées vers la terre, & garnies de feuilles oblongues, velues & sessiles : ses fleurs croissent en épis oblongs, & en paquets aux extrémités des tiges ; elles ont entr'elles des bractées colorées. Ces fleurs sont les unes blanchâtres, & les autres de couleur pourpre dans les mêmes épis ; elles sont petites, & sortent hors de leurs enveloppes écailleuses. Cette plante fleurit en Juin, & peut être multipliée de la même maniere que la précédente.

Orientale. La cinquieme espece est une plante vivace, qui croit spontanément dans le Levant ; ses tiges s'élevent à la hauteur de deux pieds, & poussent dans toute leur longueur des branches de couleur pourpre, & garnies de feuilles ovales, raboteuses, & semblables à celles du *Sanicle*, mais plus petites : ses fleurs, qui sortent en épis ronds & en paquets, ont des bractées courtes & de couleur pourpre : elles paroissent en Juin, mais elles ne produisent point de semences ici. On multiplie cette plante en divisant ses racines, comme on le pratique pour la précédente : elle exige un sol sec.

Creticum. La sixieme espece est l'*Origan de Candie*, dont on fait usage en Médecine ; mais les Botanistes ont bien de la peine à la distinguer. Cette plante s'éleve à la hauteur d'un pied & demi, avec des tiges quarrées & garnies de feuilles ovales, blanches, & d'une odeur forte & aromatique : ses fleurs croissent en épis longs, érigés, & en paquets aux extrémités des tiges ; elles ont entr'elles des bractées membraneuses, & deux fois plus longues que les calices ; ces fleurs sont petites, blanches, & semblables à celles de l'*Origan* sauvage. Cette plante fleurit en Juillet ; mais elle perfectionne rarement ses semences en Angleterre : on la multiplie en divisant ses racines comme celles de la précédente ; elle exige un sol sec & une situation chaude ; mais elle ne subsisteroit pas ici pendant l'hiver en plein air.

Majorana. La septieme espece est la *Marjolaine* commune, qui est si connue, qu'elle n'a pas besoin d'être décrite. On la regarde en Angleterre comme une plante annuelle, quoique ses racines subsistent souvent pendant les hivers doux, ou quand elles sont placées dans une Serre ; mais je crois que dans les pays chauds, elle n'est que bis-annuelle.

On la multiplie par ses semences, que l'on apporte toujours en Angleterre de la France méridionale ou de l'Italie ; car elles ne mûrissent pas souvent ici : on les seme sur une plate-bande chaude vers la fin de Mars. Quand les plantes ont atteint la hauteur d'un pouce, on les transplante sur des planches d'une terre riche, à six pouces de distance de chaque côté, & on les arrose constamment

constamment, jusqu'à ce qu'elles aient formé de nouvelles racines : après quoi elles n'exigeront plus aucun autre soin que d'être tenues nettes de mauvaises herbes. Ces plantes s'étendront & couvriront bientôt la terre. Elles commenceront à fleurir en Juillet, qui est le tems de les couper pour l'usage : on leur donne alors le nom de *Marjolaine* nouée, parce que leurs fleurs sont recueillies en têtes rondes, & serrées comme des nœuds.

Ægyptiacum. La huitieme est originaire d'Afrique. Cette plante est vivace, & a une tige basse d'arbrisseau d'environ un pied & demi de hauteur, garnie de feuilles rondes, épaisses, cotonneuses, creusées en forme de cuiller, & semblables à celles de la *Marjolaine* commune, mais d'une substance plus épaisse & cotonneuse, ayant presque la même odeur : ses fleurs, qui sont disposées en épis ronds & fort rapprochés aux extrémités des tiges & des petites branches latérales, sont d'une couleur de chair pâle, & paroissent audessus de leurs enveloppes écailleuses. Cette plante fleurit dans les mois de Juillet & Août ; mais elle ne perfectionne pas ses semences en Angleterre.

On la multiplie par boutures, qui prennent aisément racine, si on les met sur une plate-bande de bonne terre, dans quelque mois de l'été que ce soit, si on les tient à l'ombre, & si on les arrose constamment ; on peut ensuite les enlever, les planter dans de petits pots remplis d'une terre légere de jardin potager, les tenir à l'ombre jusqu'à ce qu'elles aient poussé de nouvelles racines, & les placer après dans une situation plus ouverte, où elles pourront rester jusqu'à la fin d'Octobre, qui est le tems de les mettre à couvert, parce qu'elles ne peuvent pas résister en plein air durant cette saison ; mais si on les met sous un châssis de couche chaude, où elles puissent être à l'abri des fortes gelées, & avoir autant d'air qu'il est possible dans les tems doux, elles réussiront mieux qu'étant traitées plus délicatement.

La dixieme espece est le *Dictamne de Crète*, dont on fait usage en Médecine ; elle croît naturellement sur le mont Ida en Candie. Cette plante est annuelle ; ses tiges sont velues, de plus de neuf pouces de hauteur, & de couleur pourpre ; elles poussent de petites branches de côté par paires, & garnies de feuilles rondes, épaisses, cotonneuses & très-blanches. La plante entiere a une odeur pénétrante & aromatique, avec un goût piquant : ses fleurs sont recueillies en têtes feuillées, penchées vers le bas ; elles sont petites, de couleur pourpre, & ont des étamines qui s'étendent au-delà de la corolle, & dont deux sont plus longues que les autres. Cette plante fleurit dans les mois de Juin &. de Juillet ; & dans les années chaudes, les semences mûrissent quelquefois en automne.

Tome V.   X

On peut aifément la multi-plier, en la plantant de bou-tures pendant tout l'été, ou dans des pots, ou fur une plate-bande à l'ombre; on les couvre exactement avec une cloche à melon, pour en ex-clurre l'air, & on les arrofe de tems en tems, fans leur don-ner trop d'humidité. Quand elles ont pris racine, on les enleve avec précaution; on les plante chacune féparément dans de petits pots remplis de terre légere; on les place à l'om-bre, pour qu'elles puiffent pouffer de nouvelles fibres, & on les met enfuite dans une fituation ouverte, ou on les laiffera jufqu'à l'automne, pour les mettre alors à l'abri des gelées, fous un vitrage de couche, où l'on puiffe leur don-ner de l'air dans les tems doux. Au printems, on peut tirer des pots quelques-unes des plantes, & les mettre dans une plate-bande chaude, con-tre une muraille bien expofée, & dans un fol fec, où elles fubfifteront fans aucun abri pendant les hivers ordinaires: mais comme elles font fujettes à être détruites par les fortes gelées, il fera prudent d'en garder quelques-unes dans des pots, qu'on mettra à couvert pendant la mauvaife faifon, pour en conferver l'efpece.

Sipyleum. La onzieme fe trouve fur le mont Sipyle, près de Magnefia; où elle a été découverte par le Chevalier GEORGE WHEELER, qui en a envoyé les femences au jardin d'Oxford, où les plantes ont été élevées: elle a une racine

vivace & une tige annuelle; fa racine eft compofée de plu-fieurs fibres minces & ligneu-fes; fes feuilles font ovales, unies & grifes; fes tiges min-ces, quarrées, liffes & pour-pâtres, s'élevent à la hauteur d'environ deux pieds, & pouf-fent des branches minces, op-pofées, & terminées par des épis minces & oblongs de fleurs pourpâtres, qui paroiffent au-deffus de leurs enveloppes écailleufes. Ces fleurs font petites, & femblables à celles de la dixieme, leurs étamines s'allongent hors de la corolle, à une longueur confidérable. Les feuilles qui garniffent le bas de la tige, font prefque auffi larges que celles de l'*Ori-gan* fauvage; mais celles qui couvrent fon fommet, ainfi que les branches, font fort petites & feffiles. Cette plante fleurit dans les mois de Juin & Juillet, & dans les années chaudes; fes femences mûrif-fent ici en automne. On la multiplie par boutures, com-me le *Dictamne* de Crète, & elle exige le même traitement.

Hybridinum. La douzieme eft indubitablement une variété qui a été produite par le mé-lange de la pouffiere fécon-dante du *Dictamne* de Crète, avec celle du mont Sipyle; car les plantes de cette efpece, qui font à préfent dans le jar-din de Chelféa, ont été acci-dentellement produites des fe-mences d'une efpece qui s'eft trouvée voifine de l'autre dans le jardin de JOHN BROWNING, Efq. de LINCOLN'S-INN; les femences étant tombées de la

plante dans la plate-bande entre les deux efpeces, il eft incertain de laquelle des deux elle provenoit : mais comme fes tiges & fes têtes de fleurs reffemblent davantage à celles du *Dictamne* du mont Sipyle, nous pouvons fuppofer qu'elle provient de celle-ci, dont les femences ont été imprégnées de la pouffiere fécondante du *Dictamne* de Crète, qui fe trouvoit dans le voifinage ; fes feuilles baffes font rondes, velues, d'une texture épaiffe, & fi reffemblantes à celles du *Dictamne* de Crète, qu'il eft difficile de les en diftinguer : fes tiges, qui font auffi hautes que celles du *Dictamne* de Sipyle, mais plus chargées de branches dans toute leur longueur, font pourpre & velues ; les feuilles du bas des tiges font beaucoup plus larges que celles du mont Sipyle, velues, & femblables à celles du *Dictamne* de Crète, mais moins épaiffes & moins cotonneufes : celles du haut font unies, & reffemblent à celles de l'autre efpece ; mais elles font plus larges, & les épis de fleurs & les feuilles écailleufes qui les couvrent, font plus larges & d'un pourpre plus foncé.

J'ai auffi des échantillons fecs d'une autre variété qui a été élevée de femences dans le jardin de Leyde. Ses graines avoient été envoyées de Paris fous le nom que TOURNEFORT a donné à celle qu'il a trouvée dans le Levant, & je l'ai jointe à la variété ci-deffus. Ses feuilles font auffi grandes que celles du *Dictamne* de Crète, mais moins épaiffes & moins cotonneufes ; fes tiges font de la hauteur de celles du *Dictamne* du mont Sipyle ; les branches qui en garniffent le fommet font plus étendues : fes fleurs croiffent en grappes plus ferrées, & ne penchent pas vers le bas ; elles font petites, & de la même forme que celles de la précédente ; elles paroiffent dans le même tems.

D'après le nom que LINNÉ a donné au *Dictamne* de Crète, il eft à croire qu'il n'a pas vu la véritable efpece ; car fa dénomination convient mieux à la variété à laquelle je l'ai appliquée. Toutes les feuilles du véritable *Dictamne* font fort épaiffes & cotonneufes, même celles qui font fituées immédiatement au-deffous des fleurs, au-lieu que celles du bas font comme celles décrites ici.

Onites. La treizieme efpece, qu'on trouve à Syracufe, a des tiges ligneufes & vivaces, qui s'élevent à la hauteur d'environ un pied & demi, & fe divifent en plufieurs petites branches, garnies de feuilles en forme de cœur, velues, & un peu plus grandes que celles de la *Marjolaine* : fes fleurs font difpofées en épis oblongs, touffus & velus ; elles font petites, blanches, & fortent au-deffus de leurs enveloppes écailleufes. Elles paroiffent en Juillet ; mais elles perfectionnent rarement leurs femences en Angleterre. On multiplie cette plante par boutures, comme la dixieme, & elle exige le même traitement.

X 2

Culture. Les premiere & sixieme especes font d'usage en Médecine. La premiere étant originaire de ce pays, on la substitue souvent à l'autre, qui est assez rare en Angleterre, & qu'on ne nous apporte pas souvent. Quand on se sert de la premiere, on doit préférer les plantes qui croissent sur une terre seche & stérile, parce qu'elles contiennent plus de principes actifs que celles que l'on récolte sur une bonne terre, ou dans les jardins.

On se sert aussi en Médecine du *Dictamne* de Crète; mais comme cette herbe est fort dessechée, quand elle arrive en Angleterre, & que l'emballage lui a fait perdre beaucoup de sa vertu, on emploie de préférence les plantes de cette espece qui croissent dans nos jardins.

ORME. *Voy.* ULMUS. *L.*

ORME A TROIS FEUILLES. *V.* PTELEA TRI-FOLIATA.

ORMIN. *Voy.* HORMINUM.

ORMIN SAUVAGE. *Voy.* SALVIA VERTICILLATA. *L.* ou HORMINUM VERTICILLATUM.

ORNITHOGALUM. *Tourn. Inst. R. H.* 378. *tab.* 203. *Lin. Gen. Plant.* 377. du Grec, Ὄρνις un oiseau, & de γάλα, du lait; c'est à dire, une plante dont les fleurs sont aussi blanches que les plumes blanches des oiseaux. [*Star of Bethleem.*] Etoile de Bethléem ou Jacinthe du Pérou.

Caracteres. La fleur n'a point de calice: la corolle est composée de six pétales, dont les onglets sont érigés, étendus au sommet, & persistans; la fleur a six étamines érigées; à-peu-près de la moitié de la longueur des pétales, & couronnées par des antheres simples: son germe est angulaire, & soutient un style en forme d'alène, persistant, & terminé par un stigmat obtus. Ce germe devient ensuite une capsule ronde, angulaire, & à trois cellules remplies de semences rondes.

Ce genre de plantes est rangé dans la premiere section de la sixieme classe de LINNÉE, avec celles dont les fleurs ont six étamines & un style.

Les especes sont:

1°. *Ornithogalum Pyrenaïcum, racemo longissimo, filamentis lanceolatis, pedunculis floriferis, patentibus, æqualibus, fructiferis, scapo approximatis. Lin. Sp. Plant.* 440. *Jacq. Austr. t. 103. Gouan. Monsp.* 309. *Illustr.* 26. *Gmel. It. 2. p. 196. Mattusch. Sil. n.* 239; Ornithogalon, avec un fort long épi de fleurs, dont les filamens sont en forme de lance, les pédoncules de fleurs égaux & étendus, & ceux à fruits couchés sur la tige.

Phalangium longissimè spicatum, filamentis latis, lanceolatis. Hall. Helv. n. 1210.

Ornithogalum angusti-folium majus, floribus ex albo virescentibus. C. B. P. 70; la plus grande Etoile, dont les fleurs sont d'un blanc verdâtre, & les feuilles étroites.

Stachyoides. Reneal. Spec. 93. *f.* 90.

2°. *Ornithogalum pyramidale, racemo conico, floribus numerosis adscendentibus. Prod. Leyd.* 32;

ORN

Ornithogalon avec une branche ou tige à fleurs coniques, dont beaucoup sont placées l'une au-dessus de l'autre.

Ornithogalum lacteum maximum. Bejf. *Eyjt. Vern.* 5. *t.* 14. *f.* 2.

Ornithogalum angusti-folium, spicatum, maximum. C. P. B. 70. Rudb. Elys. 2. *p.* 134 *f.* 4 ; la plus grande fleur à étoiles, en épis & à feuilles étroites.

3°. *Ornithogalum lati-folium, racemo longiffimo, foliis lanceolatis, cnfi-formibus.* Lin. Sp. Plant. 307 ; Ornithogalon avec la plus longue tige à fleurs, & des feuilles en forme d'épée & de lance.

Ornithogalum lati-folium & maximum. C. B. P. 70 ; la plus grande fleur à étoiles, à larges feuilles, appelée la fleur à étoiles, d'Alexandrie.

4°. *Ornithogalum nutans, floribus fœcundis, pendulis, nectario flamineo, campani-formi.* Lin. Sp. Plant. 308. Jacq. Auftr. *f.* 301. Scholl. Barb. n. 277. Mœnch. n. 285 ; Ornithogalon à fleurs fructueufes & pendantes, avec un nectaire campanulé

Ornithogalum Neapolitanum. Clus. App. 2. *p.* 9 ; Fleur à étoiles, de Naples.

5°. *Ornithogalum luteum, fcapo angulofo, diphyllo, pedunculis umbellatis, fimplicibus.* Flor. Suec. 270 ; Ornithogalon avec une tige angulaire, garnie de deux feuilles, & des pédoncules en ombelle fimple.

Phalangium radice bulbofi, ftipulis maximis, hirfutis, floribus umbellatis, petiolis unifloris. Hall. Helv. n. 1213.

Ornithogalum luteum. C. B. P. 71 ; Fleur à étoiles, jaune.

ORN 321

Pyrrochiton. Reneal. Spec. 91. *f.* 90.

6°. *Ornithogalum minimum, fcapo angulato, diphyllo, pedunculis umbellatis, ramofis.* Flor. Suec. 271 ; Ornithogalon avec une tige angulaire, & garnie de deux feuilles, dont les pédoncules font branchus & en ombelle.

Ornithogalum luteum minus. C. B. P. 71 ; la plus petite fleur à étoiles, jaune.

Phalangium radice bulbofa, ftipulis maximis, hirfutis, floribus umbellatis, petiolis multifloris. Hall. Helv. n. 1214.

Ornithogalum pallido flore. Bauh. Hift. 2. *p.* 624.

Hypoxis. Reneal. Spec. 92.

Ornithogalum bulbiferum minimum. Colum. Ecphr. 323. 324. Rudb. Elys. 139 ; Variété.

7°. *Ornithogalum umbellatum, floribus corymbofis, pedunculis fcapo altioribus, filamentis emarginatis.* Hort. Cliff. 124. Hort. Ups. 84. Roy. Lugd.-B. 22. Gron. Orient. III. Jacq. Auftr. *f.* 343. Scop. Carn. ed. 2. *n.* 403 ; Ornithogalon dont les fleurs croiffent en corymbe, avec des pédoncules plus longs que la tige, & des filamens échancrés.

Ornithogalum umbellatum, medium, angufti-folium. C. B. P. 70 ; Fleur à étoiles moyenne, à ombelles, avec des feuilles étroites. Dame d'onze heures.

Bulbus Leucanthemos minor, five Ornithogalum. Dodon. Cor. 183. Hift. 221.

Eliocarmos. Reneal. Spec. 89. *f.* 87.

8°. *Ornithogalum Arabicum, floribus corymbofis, pedunculis*

X 3

ſcapo humilioribus, *filamentis emarginatis.* Prod. Leyd. 32 ; Ornithogalon dont les fleurs croiſſent en corymbe , avec des pédoncules plus bas que la tige , & des filamens dentelés.

Ornithogalum umbellatum maximum. Bauh. Pin. 69. Rudb. Elys. 2. p. 130. f. 1.

Ornithogalum Arabicum. Cluſ. Hiſt. 11. p. 186 ; Fleur à étoiles , d'Arabie.

Melenomphale. Reneal. Spec. 89. f. 90.

9°. *Ornithogalum Capenſe* , *foliis cordato-ovatis.* Prod. Leyd. 31. It. Scan. 73 ; Ornithogalon avec des feuilles ovales & en forme de cœur.

Ornithogalum Africanum Plantaginis Roſæ folio , radice tuberoſâ. Hort. Amſt. 2. p. 175 ; Fleur à étoiles , d'Afrique , à feuille de Roſe de Plantain, avec une racine tubéreuſe.

Ornithogalo affinis radice tuberoſâ , Cyclaminis folio , flore pallidè cœruleo. Breyn. cent. f. 41. Rudb. Elys. 1. p. 138. f. 14.

10°. *Ornithogalum tuberoſum, racemo breviſſimo , foliis teretibus, fiſtuloſis* ; Ornithogalon avec une tige de fleurs fort courte, & des feuilles cylindriques & fiſtuleuſes.

Ornithogalum Africanum , luteum , odoratum , foliis Cepaceis, radice tuberoſâ. H. L. ; Fleur à étoiles , d'Afrique , avec des fleurs jaunes & odorantes , des feuilles d'Oignon , & une racine tubéreuſe.

Pyrenaïcum. La premiere eſpece croit naturellement près de Briſtol , & près de Chicheſter en Suſſex , & dans quelques autres parties de l'Angleterre ;

elle a une racine groſſe & bulbeuſe, de laquelle ſortent pluſieurs feuilles longues , en forme de quille , & couchées ſur la terre ; du milieu de ces feuilles ſort une tige ſimple , nue , & de deux pieds de longueur , qui porte un épi long & lâche de fleurs d'un vert jaunâtre, placées ſur de longs pédoncules , qui s'étendent bien au-delà de la tige principale. Les pétales des fleurs ont une odeur agréable : elles paroiſſent en Mai ; & quand les capſules ſont formées , les pédoncules qui les ſoutiennent ſe dreſſent & ſe rapprochent de la tige : ſes ſemences mûriſſent en Août.

Pyramidale. La ſeconde eſpece ſe trouve ſur les montagnes en Portugal & en Eſpagne : on la cultive depuis long-tems dans les jardins Anglois , ſous le nom d'*Etoile de Béthléem* ; elle a une racine fort groſſe , ovale & bulbeuſe , de laquelle ſortent pluſieurs feuilles en forme de carène , & d'un vert foncé ; du centre de ces feuilles ſort une tige nue , de trois pieds environ de hauteur , terminée par un épi long & mince de fleurs blanches , placées ſur de longs pédoncules : elles s'élevent l'une ſur l'autre dans un épi érigé , & paroiſſent dans le mois de Juin ; elles ſont remplacées par des capſules rondes & à trois cellules , remplies de ſemences rondes, qui mûriſſent en Août.

Lati-folium. La troiſieme eſt originaire de l'Arabie ; elle a une racine fort groſſe & bul-

beufe, de laquelle fortent plufieurs feuilles larges en forme d'épée, & couchées fur la terre ; fa tige eft épaiffe, forte, haute de deux à trois pieds, & terminée par de longs épis de fleurs groffes & blanches, poftées fur de longs pédoncules : elles font compofées de fix pétales, qui s'étendent en forme d'étoile ; elles paroiffent en Juin, & ne perfectionnent pas leurs femences en Angleterre.

Nutans. La quatrieme efpece qui naît en grande abondance dans le royaume de Naples, eft à préfent prefque auffi commune en Angleterre ; car elle fe multiplie prodigieufement par les rejettons de fa racine, & par fes femences, & elle devient embarraffante dans les jardins, & par-tout où on les a jettés fur des tas de fumier. Cette efpece a une racine forte, comprimée & bulbeufe, de laquelle fortent plufieurs feuilles longues, étroites, en forme de carène, & d'un vert foncé : fes tiges font fort épaiffes, fucculentes, d'environ un pied de hauteur, & terminées par un épi lâche de dix à douze fleurs, qui font fufpendues chacune à un pédoncule d'un pouce de longueur : elles font compofées de fix pétales blancs en-dedans, d'un vert grifâtre en-dehors, & fans aucune odeur ; dans l'intérieur des pétales eft placé un nectaire campanulé, compofé de fix feuilles, duquel fortent fix étamines terminées par des antheres jaunes. Ces fleurs paroiffent en Avril,

& font fuivies par des capfules larges, rondes, à trois angles, & remplies de femences rondes. Ces capfules font fi lourdes, qu'elles font pencher les tiges jufqu'à terre.

Minimum. La fixieme a des racines bulbeufes de la groffeur d'un pois, qui pouffent une ou deux feuilles de cinq pouces de longueur, & d'une couleur grifâtre : la tige eft angulaire, de quatre pouces de hauteur, & garnie de deux feuilles étroites en forme de carène, précifément au-deffous des fleurs, qui font difpofées en ombelles fur des pédoncules branchus. Ces fleurs font jaunes en-dedans, & d'un vert pourpâtre au-dehors ; elles paroiffent en Mai, & produifent de petites capfules triangulaires, remplies de femences rondes & inégales. Cette efpece croît au bord des champs cultivés, en France & en Allemagne.

Umbellatum. La feptieme efpece croît naturellement dans plufieurs parties de l'Europe : mais on la trouve rarement en Angleterre, fi ce n'eft dans des vergers, ou dans d'autres endroits dans lefquels leurs racines peuvent avoir été jettées avec les ordures des jardins ; fa bulbe, qui eft auffi groffe qu'un petit oignon, produit plufieurs petits rejettons : fes feuilles font longues, étroites, en forme de carène, & couchées fur la terre ; elles ont un trait blanc dans la longueur du fond de la carène : fa tige s'eleve à la hauteur d'environ fix pouces,

& foutient une ombelle de fleurs blanches en-dedans, & rayées de traits verts au-dehors des pétales ; elles font poftées fur de longs pédoncules, qui s'élevent au-deffus de la tige principale. Cette plante produit en Avril & en Mai des fleurs qui font remplacées par des capfules triangulaires, & remplies de femences angulaires, qui mûriffent en Juillet.

Arabicum. La huitieme efpece eft originaire de l'Arabie ; elle a une racine groffe & bulbeufe, de laquelle fortent plufieurs feuilles longues & en forme de carène, qui s'embraffent l'une l'autre à leur bâfe : elles font d'un verr foncé & érigées. Je n'ai point encore vu de fleurs de cette efpece, quoique j'aie effayé toutes fortes de moyens pour la faire fleurir ; fes racines fe multiplient confidérablement, & ne font jamais endommagées par les gelées, quoique fes feuilles pouflent avant l'hiver. On apporte fouvent d'Italie ces racines en Angleterre pour les vendre ; mais je n'ai point entendu dire qu'aucune ait jamais fleuri, & CLUSIUS dit qu'il n'en a vu fleurir qu'une feule qui venoit de Conftantinople.

Capenfe. La neuvieme naît fpontanément au Cap de Bonne Efpérance ; fa racine, qui eft irréguliere & tubéreufe, varie beaucoup en forme & en groffeur : elle eft couverte d'une peau d'un brun foncé, & produit plufieurs feuilles ovales en forme de cœur, poftées fur de longs périoles, & fortifiées par plufieurs veines longitudinales,

comme celles du Plantain ; fes tiges font minces, nues, & d'environ un pied de hauteur : elles foutiennent plufieurs petites fleurs d'un blanc verdâtre, en forme d'épis lâches, & poftées fur des pédoncules minces : elles fe montrent en Novembre, ont peu d'apparence, & ne produifent point de femences en Angleterre.

Tuberofum. La dixieme efpece croît fur des rochers arides au Cap de Bonne-Efpérance ; elle a une racine enfoncée, bulbeufe, auffi groffe que le poing d'un homme, & couverte d'une peau inégale & brune ; elle pouffe plufieurs feuilles creufes, cylindriques, & de neuf à dix pouces de longueur, du milieu defquelles s'élève une tige nue d'un pied de hauteur, & terminée par un épi lâche de fleurs jaunes d'une odeur douce & agréable. Cette plante fleurit en Mai, & ne produit point de femences en Angleterre.

Culture. Les trois premieres efpeces ci deffus font cultivées dans les jardins anglois comme plantes d'ornement. On les multiplie au moyen des rejettons que leurs racines produifent communément en grande abondance. Le meilleur tems pour les tranfplanter, eft dans les mois de Juillet & d'Août, quand leurs feuilles font tombées ; car fi on les enleve tard en automne, elles auront pouflé leurs tiges, & feront fort fujettes à fouffrir. Elles exigent un fol léger, fablonneux, & mêlé de peu de fumier. Elles peuvent être entre-

ORN

mêlées avec d'autres fleurs à racines bulbeuses dans les plates-bandes du parterre, où elles procureront une très-agréable variété. Leurs racines n'ont besoin d'être transplantées que tous les deux ans ; car si on les enlevoit tous les ans, elles ne se multiplieroient pas si fort ; si au contraire on les laisse trop long-tems sans être remuées, elles auront poussé tant de rejettons, que leurs racines à fleurs en seront affoiblies. On peut aussi les multiplier par leurs graines, qu'on seme & qu'on traite comme celles de la plupart des autres especes à racines bulbeuses : les plantes qui en proviennent donnent des fleurs trois ou quatre ans après.

La quatrieme mérite à peine une place dans les jardins ; mais comme elle réussit dans toutes les situations, & même sous des arbres, on peut en mettre quelques plantes dans des endroits couverts, pour la variété.

Comme la cinquieme n'est pas non plus fort remarquable, il suffit d'en avoir quelques racines. Il en est de même des sixieme & septieme especes : ces deux dernieres profitent à l'ombre ; mais la cinquieme exige une situation ouverte.

La huitieme se multiplie si prodigieusement par ses rejettons, qu'elle devient embarrassante dans un jardin ; car chaque petite partie de racine croît ; & en deux ans, elle en produit vingt ou trente de plus ; de maniere que les pla-

ORN   325

tes-bandes s'en trouvent bientôt couvertes, à moins qu'on n'enleve les grosses racines chaque année, pour les débarrasser de leurs rejettons.

La neuvieme étant trop délicate pour supporter en plein air le froid de notre climat, il faut planter ses racines dans des pots remplis de terre légere, & les placer en automne sous un vitrage de couche, où elles puissent être à l'abri des gelées, & jouir de l'air dans les tems doux : ses feuilles paroissent en automne, & continuent de pousser pendant tout l'hiver. Ainsi, il ne faut pas les exposer à la gelée, ni les laisser filer ; car elles fleuriroient peu, & leurs fleurs ne seroient pas grosses. Il sera bon de les exposer de tems en tems, pendant l'hiver, à une pluie douce & légere, sans cependant leur laisser prendre trop d'humidité dans cette saison. Les feuilles & tiges périssent vers le commencement de Juillet ; alors on peut enlever les racines, pour les tenir dans un endroit sec, jusqu'à la fin d'Août, qui est le tems où elles doivent être replantées.

Les autres especes, qui étoient autrefois comprises dans ce genre, se trouvent à présent à l'article Scilla.

ORNITHOPODIUM. *Voyez* Ornithopus.

ORNITHOPUS. *Lin. Gen. Plant.* 790. *Ornithopodium. Tourn. Inst. R. H.* 400. *tab.* 224. [*Birds - foot.*] Pied d'Oiseau.

Caracteres. Le calice de la

ORN

fleur est persistant, & formé par une feuille tubulée & découpée en cinq segmens égaux sur ses bords : la corolle est papillonnacée ; l'étendard est en forme de cœur & entier ; les aîles sont ovales, érigées, & presque aussi larges que l'étendard ; la carène est petite & comprimée : la fleur a dix étamines, dont neuf sont jointes, & l'autre séparée, & qui sont toutes terminées par des antheres simples ; le germe est étroit, & soutient un style hérissé, droit & terminé par un stigmat piqué, il se change dans la suite en un légume cylindrique, courbé, & à plusieurs nœuds ou articles joints ensemble, qui se séparent en murisant, & dont chacun renferme une semence oblongue.

Ce genre de plantes est rangé dans la troisieme section de la dix-septieme classe de Linnée, & fait partie de celles dont les fleurs ont dix étamines jointes en deux corps.

Les especes sont :

1°. *Ornithopus perpusillus, foliis pinnatis, leguminibus compressis, sub-arcuatis. Hort. Upsal.* 234. *Sauv. Monsp.* 236. *De Neck. Gallob. p.* 309. *Pollich. pal. n.* 692. *Flor. Dan.* 730 ; Pied d'Oiseau à feuilles aîlées, avec des légumes comprimés & un peu courbés.

Ornithopodium Majus. C. B. P. 350 ; le plus grand Pied d'Oiseau.

2°. *Ornithopus nodosus, foliis pinnatis, leguminibus confertis, pedunculatis* ; Pied d'Oiseau à feuilles aîlées, avec des légu-

mes rapprochés, & sur des pédoncules.

Ornithopodium radice tuberculis nodosâ. C. B. P. 350 ; Pied d'Oiseau avec des racines noueuses & en tubercules.

3°. *Ornithopus compressus, foliis pinnatis, leguminibus compressis, rugosis. Hort. Cliff.* 364. *Roy. Lugd.-B.* 383. *Sauv. Monsp.* 236 ; Pied d'Oiseau à feuilles linéaires & aîlées, avec des légumes comprimés, rudes, & placés par paires.

Ornithopodium Scorpioïdes, siliquâ compressâ. Tourn. Inst. 400 ; Pied d'Oiseau, qui ressemble à une chenille, avec des légumes plats.

Ornithopodio affinis, hirsuta, Scorpioïdes. B. P. 350.

Scorpioïdes leguminosa. Dalech. Hist. 493.

4°. *Ornithopus Scorpioïdes, foliis ternatis, sub-sessilibus, impari maximo. Hort. Cliff.* 364. *Hort. Ups.* 234. *Roy. Lugd-B.* 383. *Scop. Carn. ed* 2. *n.* 914 ; Pied d'Oiseau à feuilles à trois lobes, sessiles à la tige, dont celui du milieu est fort large.

Telephium Dioscoridis, sive Scorpioïdes. Bauh. Pin. 287.

Ornithopodium Portulacæ folio. Tourn. Inst. 400 ; Pied d'Oiseau à feuilles de Pourpier.

Scorpioïdes Matthioli. Dod. Pempt. 71. *Riv. Tetr.* 210.

Perpusillus. La premiere espece croît naturellement dans la France Méridionale, en Espagne & en Italie ; elle est annuelle, & pousse plusieurs tiges trainantes, d'un pied & demi de longueur, desquelles sortent quelques branches latérales, garnies de feuilles lon-

gues, ailées, & compofées de dix-huit paires environ de petits lobes ovales, & terminés par un lobe impair ; ces lobes font quelquefois oppofés, & quelquefois alternes & velus : fes fleurs naiffent en petits paquets fur des pédoncules qui fortent des aiffelles des tiges, qui ont à-peu-près trois pouces de longueur, & font garnis d'une feuille ailée, dont une partie eft au bas des fleurs, & l'autre au deffus ; de forte qu'elles paroiffent fortir de la côte du milieu de la feuille. Ces fleurs font d'une couleur d'or foncée, & de la figure d'un papillon ; elles paroiffent en Juillet, & produifent des légumes plats, étroits, de trois pouces environ de longueur, & tournés en-dedans au fommet, comme la griffe d'un oifeau ; ils font noueux, un peu velus, & renferment dans chaque nœud une fimple femence, qui mûrit en automne : alors les nœuds fe féparent, & tombent à quelque diftance l'un de l'autre.

Nodofus. La feconde efpece fe trouve fur des terres feches & couvertes de bruyere, dans plufieurs parties de l'Angleterre ; fa racine eft compofée de deux ou trois fibres fortes, auxquelles pendent quelques petites tubercules ou nœuds, comme des grains : elle produit plufieurs tiges minces de quatre ou huit pouces de longueur, couchées fur la terre, & garnies de petites feuilles ailées, velues, & compofées de fix ou fept paires de lobes étroits, & terminés par un

lobe impair : fes fleurs font placées fur des pédoncules longs & minces, qui fortent à chaque nœud de la tige ; elles font petites & jaunes, & font remplacées par des paquets de légumes courts, & un peu courbés au fommet. Cette plante fleurit & perfectionne fes femences vers le même tems que la précédente.

Compreffus. La troifieme efpece croît en abondance aux environs de Meffine & de Naples, fa racine pénetre profondément dans la terre, & pouffe quelques petites fibres fur les côtés ; fes tiges ont environ fix pouces de longueur, & ne s'étendent pas fur la terre comme les autres ; fes feuilles font velues, & compofées de dix ou douze paires de lobes étroits, placés dans la longueur de la côte du milieu, & terminés par un lobe impair : fes fleurs fortent en petits paquets fur le fommet des branches ; elles font jaunes, & produifent généralement deux légumes plats d'un pouce au plus de longueur, & tournés en-dedans comme la griffe d'un oifeau. Cette plante fleurit & perfectionne fes femences en même tems que la précédente.

Scorpioïdes. La quatrieme efpece, qu'on rencontre parmi les bleds en Efpagne & en Italie, a plufieurs tiges liffes & branchues, de près de deux pieds de hauteur, & garnies vers le fommet de feuilles à trois lobes, ovales, feffiles, & ornées de deux petites appendices ou oreilles : les feuil-

les du bas font fouvent fimples & de couleur grifâtre ; le lobe du milieu eft deux fois plus large que les deux latéraux : fes fleurs font poftées fur de minces pédoncules ; elles font jaunes , & produifent des légumes cylindriques de deux pouces de longueur , & femblables à la griffe d'un oifeau. Cette plante fleurit & donne des femences mûres dans le même tems que la précédente.

Culture. On multiplie toutes ces efpeces , en les femant au printems fur une planche de terre fraîche & légere , où elles doivent refter ; car elles ne réuffiffent jamais bien quand elles font tranfplantées. Lorfque les plantes pouffent , on les débarraffe foigneufement de toutes les mauvaifes herbes qui s'y trouvent ; & fi elles font trop ferrées , on en arrache quelques-unes, de maniere qu'elles reftent à dix pouces environ de diftance. Ces plantes fleuriffent en Juin, & leurs femences mûriffent en Août. Comme elles ne font pas fort belles , on ne les conferve dans quelques jardins curieux que par rapport à la fingularité de leurs filiques noueufes. On peut les femer en paquets dans les platesbandes, en diftinguant les efpeces ; on les éclaircit & on n'en laiffe que deux dans chaque touffe, après quoi elles n'exigent plus aucun foin. Ces plantes augmentent la variété, fur-tout celles dont les filiques font en forme de limaçon & de chenille. Toutes ces plantes font annuelles, & périf-

fent auffi-tôt que leurs femences font mûres.

OROBANCHE Rave sauvage a Genet *ou* Ballet. [*Broom Rape.*] L'Orobanche.

On connoît fix ou fept efpeces de ce genre , dont deux croiffent naturellement fur des terres feches dans plufieurs parties de l'Angleterre ; mais comme toutes ne peuvent pas être cultivées , on ne les admet pas dans les jardins. Linnée les range dans la feconde fection de fa quatorzieme claffe , intitulée : *Didynamia Angyofpermia* , qui comprend celles qui ont deux étamines longues & deux plus courtes , & dont les femences font renfermées dans une capfule.

OROBE. *Voy.* Orobus.

OROBUS. *Tourn. Inft. R. H.* 393. *tab.* 214. *Lin. Gen. Plant.* 780 ὄραβος , de βοῦς , un bœuf , & de ἱέπτω , manger ; c'eft-à-dire , herbe dont on nourrit les bœufs , parce que les anciens engraiffoient ces animaux avec ce fourrage. [*Bitter Vetch.*] Orobe.

Caracteres. Le calice de la fleur eft tubulé , & d'une feuille dont la bâfe eft obtufe , & le bord oblique & découpé en cinq parties, dont les trois inférieures font aiguës , & les deux fupérieures plus courtes & obtufes : la corolle eft papilionnacée ; l'étendard eft en forme de cœur ; les deux ailes font prefque auffi longues que l'étendard , & jointes enfemble ; la carène eft divifée en deux fegmens à pointe aiguë & élevés vers le haut ; les bords font com-

ORO

primés, & le corps gonflé. Cette fleur a dix étamines, dont neuf font jointes, & l'autre eft féparée, & qui font toutes érigées, & terminées par des antheres rondes : fon germe eft cylindrique & comprimé ; il foutient un ftyle courbé, élevé, & couronné par un ftigmat étroit, laineux, & attaché par le bord, en-dedans, au milieu, à la pointe du ftyle. Ce germe devient enfuite une filique longue, terminée en pointe aiguë, & à une cellule qui renferme plufieurs femences rondes.

Ce genre de plantes eft rangé dans la troifieme fection de la dix-feptieme claffe, qui comprend celles dont les fleurs ont dix étamines jointes en deux corps.

Les efpeces font :

1°. *Orobus vernus, foliis pinnatis, ovatis, ftipulis femi-fagittatis, integerrimis, caule fimplici. Lin. Sp. Plant.* 728. *Crantz. Auftr.* p. 373. *De Neck. Gallob.* p. 304. *Scop. Carn. ed.* 2. n. 882. *Pallas. It.* 1. p. 160. *Pollich. pal.* n. 672. *Mattufch. Sil.* n. 522. *Dœrr. Naff.* p. 167. *Riv.* t. 58 ; Orobe avec des feuilles ovales & ailées, des ftipules entieres, & en demi-pointe de flèche, & une tige fimple.

Orobus caule fimpliciffimo, foliolis pluribus, ovatis, acutis. Hort. Cliff. 366. *Fl. Suec.* 595. 641. *Roy. Lugd.-B.* 366. *Hall. Helv.* n. 416.

Orobus fylvaticus, purpureus, vernus. C. B. P. 351 ; Orobe des bois, pourpre & printanier.

2°. *Orobus tuberofus, foliis pinnatis, lanceolatis, ftipulis fe-*

ORO 329

mi-fagittatis, integerrimis, caule fimplici. Lin. Sp. Plant. 728. *Scop. Carn. ed.* 2. n. 885. *Pollich. pal.* n. 673 ; Orobe avec des feuilles en forme de lance & ailées, des ftipules très-entieres, à moitié en pointe de flèche, & une tige fimple.

Orobus fylvaticus, foliis oblongis, glabris. Tourn. Inft. R. H. 393 ; Orobe des bois, avec des feuilles unies & oblongues.

Orobus radice tuberofâ. Riv. f. 59.

Aftragalus fylvaticus, foliis oblongis, glabris. Bauh. Pin. 351.

Aftragalus fylvaticus. Thal. Herc. 7. f. 1.

Lathyrus angufti-folius, radice tuberofâ. Læs. Pruff. 183. f. 37.

3°. *Orobus fylvaticus, caulibus decumbentibus, hirfutis, ramofis. Cent. Pl.* 67. *For. Angl.* 275. *Syft. Veg.* 661. *Sp.* 6. ed. 14 ; Orobe avec des tiges velues, branchues & tombantes.

Orobus caule ramofo, hirfuto, decumbente, foliis fub-feptemjugis. Amœn. Acad. 4. p. 284. n. 66.

Orobus fylvaticus noftras. Raii Syn. 324 ; Orobe de bois.

4°. *Orobus niger, caule ramofo, foliis fex-jugis, ovato-oblongis. Hort. Cliff.* 366. *Flor. Suec.* 597. 643. *Roy. Lugd.-B.* 366. *Crantz. Auftr.* p. 372. *Scop. Carn. ed.* 2. n. 884 ; Orobe à tige branchue, avec des feuilles compofées de fix paires de lobes oblongs & ovales.

Orobus caule ramofo, foliis ovatis, duodenis. Hall. Helv. n. 418.

Orobus fylvaticus, foliis Viciæ. C. B. P. 352 ; Orobe des bois, à feuilles de Vefce.

Orobus Pannonicus. 2. Clus.
Hist. 2. p. 230.

5°. Orobus Pyrenaïcus, caule ramoso, foliis bi-jugis, lanceolatis, nervosis, stipulis sub-spinosis. Lin. Sp. 1029. Scop. Carn. ed. 2. n. 885 ; Orobe à tige branchue, avec des feuilles composées de deux paires de lobes, étroites & en forme de lance.

Orobus Pyrenaïcus, foliis nervosis. Tourn. Inst. 393 ; Orobe des Pyrénées, à feuilles nerveuses.

Orobus caule ramoso, foliis tri-jugis, lanceolatis, nervosis. Sauv. Monsp. 235.

Orobus Pyrenaïcus, lati-folius, nervosus. Pluk. Phyt. 210. f. 2.

6°. Orobus Lathyroïdes, foliis conjugatis, sub-sessilibus, stipulis dentatis. Hort. Upsal. 220. Gmel. Sib. 4. p. 12. Pall. It. 2. p. 559 ; Orobe à feuilles conjuguées & sessiles, avec des stipules dentelées.

Lathyroïdes erecta, folio ovato, acuminato, cœruleis Viciæ floribus & siliquis, Sibirica. Amœn. Ruth. 151 ; Gesse érigée & bâtarde, avec une feuille ovale & à pointe aiguë, des fleurs bleues, & des siliques de Vesce.

7°. Orobus luteus, foliis pinnatis, ovato-oblongis, stipulis rotundato-lunatis, dentatis, caule simplici. Lin. Sp. Plant. 728 ; Orobe à feuilles ovales, oblongues & aîlées, avec des stipules rondes en croissant, dentelées, & une tige simple.

Orobus Sibiricus perennis. Gmel ; Orobe vivace de Sibérie.

Galega montana Dalechampi. Bauh. Hist. 2. p. 343.

8°. Orobus Venetus, foliis pinnatis, ovatis, acutis, quatuorjugatis, caule simplici. ; Orobe à feuilles ovales, à pointe aiguë, & aîlées, avec quatre paires de lobes & une tige simple.

Orobus Venetus. Clus. Hist. 322 ; Orobe de Venise.

9°. Orobus Americanus, foliis pinnatis, lineari-lanceolatis, infernè tomentosis, caule ramosissimo, frutescente ; Orobe à feuilles linéaires, en forme de lance, aîlées & cotonneuses endessous, avec une tige d'arbrisseau très-branchue.

Orobus Americanus erectus, foliorum pinnis angustioribus & subtùs incanis, siliquis glabris. Houst. MSS. ; Orobe d'Amérique érigé, dont les lobes des feuilles sont très-étroits, & blancs en dessous, avec des siliques unies.

10°. Orobus argenteus, foliis pinnatis, oblongo-ovatis, infernè sericeis, caule erecto, tomentoso, floribus spicatis terminalibus ; Orobe à feuilles oblongues, ovales, aîlées, & soyeuses en-dessous, avec une tige droite & cotonneuse, terminée par des épis de fleurs.

Orobus Americanus, lati-folius, argenteus, flore purpureo. Houst. MSS. ; Orobe d'Amérique à feuilles larges & argentées, avec une fleur pourpre.

11°. Orobus procumbens, foliis pinnatis, foliolis exterioribus majoribus tomentosis, caule procumbente ; Orobe à feuilles aîlées, dont les plus grands lobes extérieurs sont cotonneux, avec une tige tombante.

Orobus Americanus procumbens

ORO

& hirſutus , flore purpureo. Houſt. MSS. ; Orobe tombant & velu d'Amérique , qui produit une fleur pourpre.

12ᵉ. *Orobus coccineus , foliis pinnatis , foliolis linearibus , villoſis , caule procumbente , floribus alaribus & terminalibus* ; Orobe à feuilles ailées , dont les lobes ſont velus & linéaires , avec une tige tombante , & des fleurs ſur les côtés & aux extrémités des branches.

Orobus procumbens minimus , flore coccineo. Houſt. MSS. ; le plus petit Orobe traînant d'Amérique , à fleur écarlate.

Vernus. La première eſpece croît naturellement dans les forêts d'Allemagne & de la Suiſſe ; ſa racine eſt vivace , & compoſée de pluſieurs fortes fibres ; ſes tiges s'élevent à la hauteur d'un pied , & ſont garnies de feuilles ailées , formées par deux paires de lobes ovales & à pointe aiguë ; à la bâſe du pétiole eſt placée une ſtipule en pointe de flèche , diviſée en deux par le milieu , & qui embraſſe la tige ; les lobes des feuilles ont environ un pouce & demi de longueur , & près d'un pouce de largeur , & ſont terminés en pointe aiguë : les fleurs naiſſent ſur des pédoncules qui ſortent des aiſſelles de la tige ; ils ont environ trois pouces de longueur , & ſoutiennent ſix ou ſept fleurs rangées en forme d'épi. Ces fleurs ſont papillonnacées , d'abord de couleur pourpre , & enſuite bleues. Elles paroiſſent au commencement du printems , & produiſent des légumes cylindriques

d'un pouce & demi de longueur , & à une cellule dans laquelle ſont renfermées quatre ou cinq ſemences oblongues & ameres , qui mûriſſent en Juin. Il y a une variété de cette eſpece à fleurs pâles que l'on conſerve dans quelques jardins.

Tuberoſus. La ſeconde croît naturellement dans les bois & autres lieux couverts de pluſieurs parties de l'Angleterre : elle a une racine vivace & rampante , de laquelle s'élevent des tiges angulaires de neuf ou dix pouces de longueur , & garnies à chaque nœud d'une feuille ailée , compoſée de quatre paires de lobes unis & en forme de lance ; à la bâſe de chaque feuille eſt ſituée une ſtipule ſemblable à celles de la précédente. Les pédoncules ſortent aux aiſſelles des tiges ; ils ont environ quatre pouces de longueur , & ſoutiennent chacun deux ou trois fleurs d'un rouge pourpâtre , qui change en un pourpre foncé , avant que les fleurs ſoient fanées : elles paroiſſent en Avril , & ſont remplacées par des légumes longs & cylindriques , qui contiennent chacun ſix ou ſept ſemences rondes. Ces ſemences mûriſſent au commencement de Juin : on leur donne le nom de *Pois de Bois* ou de *Bruyeres.*

Sylvaticus. La troiſieme eſpece ſe trouve en Cumberland & dans le pays de Galles : ſa racine vivace & ligneuſe pouſſe pluſieurs tiges velues d'un pied & demi de hauteur , &

garnies à chaque nœud d'une feuille ailée, & composée de dix ou douze paires de lobes étroits, & très-rapprochés sur la côte du milieu, à la bâse de laquelle est placée une stipule aiguë, qui embrasse la tige : les fleurs sont disposées en épis serrés, & postées sur des pédoncules de trois pouces de longueur, qui sortent des ailes des feuilles ; elles sont de couleur pourpre, & produisent des légumes courts & plats, qui renferment deux ou trois semences. Cette plante fleurit au commencement de Juin, & ses semences mûrissent en Juillet.

Niger. La quatrieme espece croît spontanément sur les montagnes, en Allemagne & en Suisse ; elle a une racine forte, ligneuse & vivace, de laquelle s'élevent plusieurs tiges branchues de deux pieds de hauteur, & garnies à chaque nœud d'une feuille ailée, & composée de cinq ou six paires de lobes oblongs, ovales, & rangés le long de la côte du milieu ; leurs pédoncules sont fort longs, & sortent des aisselles de la tige ; ils soutiennent à leur sommet quatre, cinq ou six fleurs pourpre, qui paroissent dans le mois de Mai, & produisent des légumes comprimés, & d'un pouce & demi de longueur, dans lesquels sont renfermées quatre ou cinq semences oblongues, qui mûrissent au commencement de Juillet. Ses tiges périssent en automne, & les nouvelles poussent au printems.

Pyrenaïcus. La cinquieme espece, qui est originaire des Pyrénées, a une racine vivace, de laquelle sortent plusieurs tiges unies, branchues, d'un pied & demi de hauteur, & garnies de feuilles ailées, composées de quatre paires de lobes, en forme de lance, fortifiées par trois veines longitudinales ; à la bâse des feuilles est placée une stipule qui embrasse la tige, comme dans la premiere espece : les fleurs naissent sur de longs pédoncules, qui sortent des ailes des feuilles ; vers le haut de la tige elles sont rangées en épis lâches ; elles sont d'une couleur de pourpre, & paroissent en Mai ; à ces fleurs succedent des légumes comprimés, de deux pouces de longueur, qui renferment chacun trois ou quatre semences qui mûrissent en Juillet.

Lathyroïdes. La sixieme espece se trouve en Sibérie ; elle a une racine vivace, de laquelle sortent trois ou quatre tiges branchues, & d'un pied environ de hauteur ; ses feuilles, qui naissent par paires opposées dans la longueur des tiges, sont sessiles, & ont à leur bâse une stipule dentelée ; elles sont unies, roides, & d'un vert luisant : les fleurs croissent en épis serrés sur de courts pédoncules, qui sortent des ailes des feuilles au sommet des tiges, où sont généralement trois ou quatre de ces épis rapprochés.

Ces fleurs sont d'un beau bleu, & ont belle apparence ; elles paroissent en Juin, & produisent

ORO

produisent des légumes courts & plats, renfermant chacun trois ou quatre semences qui mûrissent en Août.

Luteus. La septieme espece est aussi originaire de la Sibérie; elle a une racine vivace, de laquelle sortent plusieurs tiges herbacées, d'un pied & demi de hauteur, & garnies de feuilles ailées, composées de quatre ou cinq paires de lobes ovales & oblongs, ayant à leur bâse une stipule ronde en forme de croissant, qui embrasse la tige: ses fleurs sortent des ailes des feuilles sur de courts pédoncules; elles sont larges, & de couleur pourpre; elles paroissent en Avril, & produisent des légumes gonflés, à-peu près de deux pouces de longueur: ils renferment quatre ou cinq semences qui mûrissent en Juin.

Venetus. La huitieme naît sans culture en Italie; sa racine est vivace, & pousse plusieurs tiges d'un pied environ de hauteur, & garnies de feuilles ailées, composées de quatre paires de lobes ovales, terminés en pointe aiguë, unis, d'un vert pâle, & placés à une bonne distance sur la côte du milieu: ses fleurs sortent sur de minces pédoncules, qui s'élevent aux ailes des feuilles au nombre de quatre ou cinq sur chacun; elles sont de couleur pourpre, paroissent en Mars, & sont remplacées par des légumes gonflés & d'un pouce & demi de longueur. Ces légumes renferment chacun trois ou quatre semences rondes qui mûrissent en Mai.

Tome V.

ORO   333

Americanus. La neuvieme espece a été découverte à la Jamaique, d'où le feu Docteur Houstoun en a envoyé les semences en 1731: elle s'éleve avec une tige fort branchue & ligneuse, à la hauteur d'environ trois pieds; ses branches sont garnies de feuilles ailées, composées de cinq ou six paires de lobes étroits & cotonneux en-dessous: ses fleurs croissent en épis lâches aux extrémités des branches; elles sont d'un pourpre pâle, & donnent des légumes unis, comprimés, & d'un pouce & demi de longueur, qui contiennent chacun cinq ou six semences rondes.

Argenteus. La dixieme a aussi été découverte par le Docteur Houstoun, à la Vera-Cruz, d'où il en a envoyé les semences en Angleterre: elle s'éleve avec une tige d'arbrisseau de cinq ou six pieds de hauteur, qui se divise en plusieurs branches minces, couvertes d'une écorce brune & cotonneuse, & garnies de feuilles molles, satinées & ailées; celles des jeunes branches sont composées de quatre paires de lobes ovales, obtus, d'un vert brunâtre, velus en-dessous, argentés & soyeux en-dessus: les feuilles qui garnissent les branches du haut sont composées de sept ou huit paires de lobes oblongs, ovales, & de la même couleur & consistance que celles du bas: ses fleurs sont disposées en épis longs & érigés aux extrémités des branches; elles sont d'un pourpre foncé, & produisent des légu-

Y

mes longs, cotonneux & comprimés, qui renferment chacun quatre ou cinq semences.

Procumbens. La onzieme espece a encore été envoyée de la Vera-Cruz, par le Docteur HOUSTOUN, en 1730. Cette plante est basse, & pousse des tiges inclinées vers la terre, qui ont rarement plus de six ou huit pouces de longueur, & produisent quelques branches courtes & latérales, garnies de feuilles ailées, & composées de quatre ou cinq paires de petits lobes oblongs, ovales, cotonneux, & terminés par un lobe impair; les lobes supérieurs sont beaucoup plus larges que les inférieurs: les fleurs sortent en petits paquets sur de courts pédoncules qui s'élevent des aisselles de la tige; elles sont petites, d'un pourpre brillant, & donnent des légumes comprimés de deux pouces environ de longueur, qui renferment chacun six ou sept semences rondes & plates.

Coccineus. La douzieme a aussi été découverte dans le même pays que la précédente, par le Docteur HOUSTOUN: elle a une racine épaisse & ligneuse, de laquelle sortent plusieurs tiges minces d'un pied & demi de longueur, qui trainent sur la terre, & sont garnies de feuilles ailées, composées de trois ou quatre paires de lobes étroits, blancs, & de six lignes environ de longueur: les fleurs sortent de côté & aux extrémités des tiges, au nombre de trois ou quatre ensemble, sur un court

pédoncule; elles sont petites, de couleur écarlate, & sont suivies par des légumes courts & cylindriques, contenant chacun trois ou quatre semences petites & rondes.

Culture. Les huit premieres especes ont des racines vivaces & des tiges annuelles, qui périssent en automne. Plusieurs de celles-ci peuvent être multipliées par la division de leurs racines. Le meilleur tems pour faire cette opération, est l'automne, afin que les plantes puissent être bien établies avant le printems; car plusieurs commençant à pousser leurs tiges de fort bonne heure, si on les remuoit alors, elles ne fleuriroient pas aussi bien, ou s'affoibliroient beaucoup. La plupart de ces plantes se plaisent à l'ombre & dans un sol marneux.

On les multiplie aussi par leurs graines: mais il faut les semer en automne; car si l'on attend jusqu'au printems, plusieurs de ces especes ne poussent point, & celles qui réussissent, ne germent souvent que dans l'année suivante. Je n'ai jamais pu faire lever la quatrieme espece, en la semant au printems, quoique je l'aie essayé plusieurs fois dans differentes situations; mais les graines qui se sont répandues elles-mêmes pendant l'été, ont bien poussé au printems suivant, ainsi que celles qui ont été semées en Septembre. Quand les plantes paroissent, on arrache avec soin toutes les herbes inutiles qui croissent avec elles, & on les

ORO   ORT   335

éclaircit dans les endroits où elles font trop ferrées, de façon qu'elles puiffent avoir affez d'efpace pour croître jufqu'à l'automne ; alors on les tranfplante dans les places où elles doivent refter. Si leurs racines font fortes, elles fleuriront fort bien au printems fuivant ; mais comme les plus foibles ne produiront pas de fleurs avant la feconde année, il faut les planter à l'ombre dans une plate-bande, à quatre ou cinq pouces de diftance, & on ne les placera à demeure qu'au bout d'une année : tout le foin qu'elles exigent, confifte à labourer la terre entr'elles pendant l'hiver & à les tenir nettes de mauvaifes herbes en été.

Les quatre dernieres efpeces étant originaires des pays chauds, font tendres, & doivent être confervées dans la ferre chaude, fans quoi elles ne peuvent fubfifter en Angleterre. On les multiplie par leurs graines, qu'on répand au commencement du printems dans de petits pots remplis d'une terre riche & légere : on les plonge dans une couche chaude de tan, & on les arrofe fouvent. Quand les plantes ont pouffé, on les enleve avec précaution : on les met chacune féparément dans de petits pots remplis d'une terre riche, & on les replonge dans une couche de tan, où on les tient à l'ombre, jufqu'à ce qu'elles aient repris racine ; après quoi on leur donne de l'air chaque jour dans les tems chauds, & on les arrofe fouvent. Au

moyen de ce traitement, ces plantes feront de grands progrès. Lorfque quelques unes feront devenues trop hautes pour pouvoir être contenues dans la couche, on les enlevera, & on les placera dans la couche de tan de la ferre, où elles auront plus d'efpace pour croître, fur tout les neuvieme & dixieme efpeces ; mais les deux autres étant d'un crû plus bas, peuvent refter dans la premiere couche jufqu'à la Saint-Michel ; alors les nuits commençant à devenir froides, il fera néceffaire de les enfermer dans la ferre, & de les plonger dans la couche de tan ; on les traite enfuite comme les autres plantes exotiques & tendres : par ce moyen, elles réfifteront à l'hiver ; & l'été fuivant, elles produiront des fleurs. Comme ces plantes font vivaces, fi elles ne perfectionnent pas leurs femences, on pourra les conferver pendant plufieurs années.

ORPIN VRAI. *Voyez* TELEPHIUM. *L.*

ORPIN *ou* REPRISE, *ou* JOUBARBE DES VIGNES. *V.* SEDUM TELEPHIUM. *L.*

ORPIN-ROSE. *Voy.* RHODIOLA ROSEA. *L.*

ORPIN BATARD. *Voy.* ANDRACHNE & PORTULACA ANACAMPSEROS. *L.*

ORTEGIA HISPANICA. Cette plante eft nommée par CLUSIUS *Juncaria Salmantica ;* elle eft baffe & trainante, & pouffe des tiges femblables à celles du *Jonc,* qui produifent à chaque nœud quelques petites fleurs prefque invifibles ;

Y 2

336 O R Y

ce qui fait que cette plante est rarement cultivée, si ce n'est dans les jardins de botanique, pour la variété.

ORTIE. *Voyez* URTICA. *L.*

ORTIE BLANCHE *ou* L'ARCHANGE *V.* LAMIUM ALBUM. *L.*

ORTIE GRANDE. *Voy.* URTICA DIOÏCA. *L.*

ORTIE MORTE, *la plus grande. Voyez* MELITTIS.

ORTIE MORTE A FLEURS JAUNES. *Voyez* GALEOPSIS GALEOBDOLON. *L.*

ORTIE MORTE ET PUANTE. *Voyez* GALEOPSIS. *L.*

ORTIE ROMAINE. *Voyez* URTICA PILULIFERA. *L.*

ORTIE DE HAIE EN ARBRISSEAU. *Voyez* PRASIUM. *L.*

ORVALE *ou* TOUTE-BONNE. *Voyez* SCLAREA.

ORYZA. *Tourn. Inst. R. H.* 513. *tab.* 296. [*Rice.*] Riz.

Caractères. La bale est petite & à pointe aiguë, avec deux petites valves à peu-près égales, qui renferment une simple fleur : la corolle a deux valves creuses, comprimées, en forme de bateau, & terminées par une barbe ou cosse ; elle a un nectaire à deux feuilles, & six étamines velues de la longueur de la corolle, & terminées par des antheres, dont les bâses sont divisées en deux parties : son germe, qui est turbiné, soutient deux styles velus, réfléchis, & couronnés par des stigmats plumacés ; ce germe se change dans la suite en une semence grosse, oblongue, comprimée, sillonnée par deux canelures de chaque côté, & postée sur le pétale de la fleur.

Ce genre de plantes est rangé dans la seconde section de la sixieme classe de LINNÉE, qui comprend celles dont les fleurs ont six étamines & deux styles.

Nous n'avons qu'une espece de ce genre.

Oryza sativa. Matth. 403. *Bauh. Pin.* 24. *Theatr.* 479. *Cam. Epit.* 192. *Dod. Pempt.* 559. *Catesb. Car. 1. p.* 14. *f.* 14. *Hort. Cliff.* 137. *Mat. Med.* 97. *Roy. Lugd.-B.* 58 ; le Riz.

Cette plante forme la principale culture des pays Orientaux, dont les habitans font de sa graine la bâse de leur nourriture : on en transporte une grande quantité en Angleterre & dans d'autres pays de l'Europe, chaque année : on l'emploie parmi nous dans la préparation des puddings, &c. Cette plante est trop délicate pour pouvoir réussir dans nos pays septentrionaux, sans le secours d'une chaleur artificielle ; mais quelques-unes de ses graines, qui ont été autrefois envoyées dans la Caroline Méridionale, y en ont produit abondamment, & il s'est trouvé qu'elle y réussit aussi bien que dans son pays originaire ; ce qui est très-avantageux pour nos établissemens Américains.

Cette plante croît sur des terres humides, que l'on peut inonder quand elle a poussé ; de sorte que, quand on veut la cultiver en Angleterre par curiosité, il faut la semer sur une couche chaude ; & quand les plantes ont poussé, les transplanter dans des pots rem-

OSM

plis d'une terre riche & légere, & les plonger dans des terrines remplies d'eau, que l'on place dans une couche chaude ; à mesure que l'eau diminue, on la renouvelle de tems en tems, de maniere qu'elle furnage toujours au-deſſus des pots, ſans quoi les plantes ne profiteroient pas : on les tient dans une ſerre chaude pendant tout l'été ; & vers la fin du mois d'Août, elles produiſent des graines qui mûriſſent aſſez bien, pourvu que l'automne ſoit favorable.

OSEILLE. *Voyez* ACETOSA.

OSEILLE DES PRÉS. *Voyez* ACETOSA PRATENSIS.

OSEILLE , PATIENCE. *Voyez* RUMEX. *L.*

OSEILLE RONDE. *Voyez*. ACETOSA SCUTATA. *L.*

OSEILLE SAUVAGE *ou* ALLELUIA. *Voyez* OXALIS. *L.*

OSIER. *Voyez* SALIX VITELLINA.

OSMUNDA. [*Oſmund Royal,* or *Flowering Fern.*] Oſmonde royale, *ou* Fougere fleuriſſante.

C'eſt une des eſpeces de Fougeres qui eſt diſtinguée des autres, en ce qu'elle produit des fleurs ſur le ſommet de ſes feuilles ; au-lieu que la plupart des autres les produiſent ſur le dos de leurs feuilles.

Il n'y a qu'une eſpece de cette plante en Angleterre ; mais on en trouve quelques autres en Amérique. Comme on les cultive rarement dans les jardins, je n'en ferai pas mention ici.

L'eſpece commune croît dans nos marais. Si quelqu'un étoit curieux de la tranſplanter dans

OST

un jardin, il faudroit la placer à l'ombre & dans une ſituation humide, ſans quoi elle ne profiteroit point.

OSMONDE *ou* FOUGERE FLEURIE. *Voyez* OSMUNDA. *L.*

OSTEOSPERMUM. *Lin.* Gen. *Plant.* 887. *Moniliſera.* Vaill. *Aɑ. Par.* 1720. *Chryſanthemoïdes.* Tourn. *Aɑ. Par.* 1705. [*Hard-ſeeded Chryſanthemum.*] Chryſanthemum *ou* Marguerite à ſemences dures.

Caraɑeres. Le calice de la fleur eſt hémiſphérique, ſimple & découpé en pluſieurs ſegmens ; la fleur eſt compoſée de pluſieurs fleurons hermaphrodites placés dans le diſque, tubulés & découpés en cinq parties ſur leurs bords ; ils ſont environnés de pluſieurs fleurons femelles, qui forment le rayon. Ces derniers ont chacun une langue longue, étroite, & découpée au ſommet en trois parties. Les fleurons hermaphrodites ont cinq étamines courtes, minces, & terminées par des antheres cylindriques, avec un petit germe qui ſoutient un ſtyle mince , & couronné par un ſtigmat uſé. Ceux-ci ſont ſtériles. Les fleurons femelles ont chacun un germe globulaire qui ſoutient un ſtyle mince & couronné par un ſtigmat découpé. Ce germe devient enſuite une ſemence ſimple & dure.

Ce genre de plantes eſt rangé dans la quatrieme ſection de la dix-neuvieme claſſe de LINNÉE, intitulée, *Syngeneſia Polygamia neceſſaria* ; ſes fleurs étant compoſées de fleurons hermaphrodites & ſtériles dans

338 OST

le difque, & de fleurons femelles fructueux.

Les efpeces font :

1°. *Ofteofpermum Monili-ferum, foliis ovalibus, ferratis, petiolatis, fub-decurrentibus.* Lin. Hort. Cliff. 423. Roy. Lugd-B. 179. Berg. Cap. 331 ; Marguerite à femences dures, avec des feuilles ovales, fciées, & placées fur des pétioles coulans.

Chryfanthemoïdes Afrum Populi albæ foliis. Hort. Elth. 80. tab. 68. f. 79 ; Chryfanthème d'Afrique, à feuilles de Peuplier blanc.

Molini-fera frutefcens bacci-fera, folio fub-rotundo, crenato. Vaill. Act. 1720. p. 573.

2°. *Ofteofpermum pifi-ferum, foliis lanceolatis, acutè dentatis, caule fruticofo.* Ofteofpermum à feuilles en forme de lance, fcié à dents aiguës, avec une tige d'arbriffeau.

Ofteofpermum fruticans, lanuginofum, foliis oblongis, dentatis. Burm. Afr. 171. t. 61. f. 2.

3°. *Ofteofpermum fpinofum, fpinis ramofis.* Lin. Hort. Cliff. 424. Roy. Lugd.-B. 179. Berg. Cap. 327 ; Ofteofpermum avec des épines branchues.

Chryfanthemoïdes Ofteofpermum Africanum, odoratum, fpinofum & vifcofum. Hort. Amft. 2. p. 85 ; Chryfanthème d'Afrique, à femences dures, épineux, vifqueux, & d'une bonne odeur.

Chryfanthemum Africanum frutefcens, fpinofum. Volk. Narib. 105. f. 105. Moris. Hift. 3. S. 6. t. 3. f. 56.

*Ofteofpermum polygaloïdes, foliis lanceolatis, imbricatis, feffi-*libus. Flor. Leyd. Prod. 179 ; Ofteofpermum avec des feuilles en forme de lance, imbriquées & feffiles.

Chryfanthemum fruticofum, Polygoni foliis, Africanum, caulibus fcabris, flore minore. Pluk. Mant. 47. f. 382.

Monili-fera Polygalæ foliis. Vaill. Act. Par. 1720; Monilifera à feuilles de Renouée, que l'on appelle communément Traînaffe.

Monili-ferum. La premiere efpece, qui croît naturellement au Cap de Bonne-Efpérance, eft, depuis plufieurs années, cultivée dans les jardins anglois : elle s'éleve à la hauteur de fix à fept pieds, avec une tige d'arbriffeau, couverte d'une écorce unie & grife ; fa tige fe divife en plufieurs branches garnies de feuilles ovales, inégalement dentelées fur leurs bords, alternes, d'une confiftance épaiffe, & couvertes d'un duvet blanc, qui tombe & fe détache des vieilles feuilles : fes fleurs font difpofées en paquets aux extrémités des branches ; elles fortent fix ou huit enfemble fur des pédoncules d'un pouce & demi de longueur ; elles font jaunes & femblables à celles du *Ragwort*, ou l'*Herbe en chiffons*.

Othonna. Les rayons font compofés d'environ dix demi-fleurons qui s'étendent ; le difque eft formé par des fleurons tubulés, découpés en cinq parties fur leurs bords, & ftériles : mais les demi-fleurons du rayon ont chacun une femence dure. Cette plante fleurit rarement : quand cela arrive, c'eft

dans les mois de Juillet & Août.

Pisi-ferum. La seconde espece croit comme la premiere ; mais ses feuilles sont plus pointues, de couleur verte, & divisées en dentelures aiguës sur leurs bords : les pétioles des feuilles sont profondément veinés. Cette plante produit des touffes de fleurs jaunes aux extrémités de ses branches, depuis le printems jusqu'à l'automne, & perfectionne souvent ses semences.

Spinosum. La troisieme est une plante basse en arbrisseau, qui s'éleve rarement au dessus de trois pieds de hauteur, & se divise en plusieurs branches, dont les extrémités sont garnies d'épines vertes & branchues : ses feuilles sont fort gluantes, sur-tout pendant les tems chauds ; elles sont longues, étroites, & placées sans ordre : ses fleurs naissent simples aux extrémités des branches ; elles sont jaunes, & paroissent dans les mois de Juillet & Aout.

Culture. Ces trois especes étant trop tendres pour pouvoir subsister en plein air en Angleterre, il faut les placer dans une serre au mois d'Octobre, & les traiter comme les *Myrtes* & autres plantes dures de la serre, qui exigent beaucoup d'air dans les tems chauds. Au commencement du mois de Mai, on peut les placer en plein air, dans une situation abritée. Les seconde & troisieme especes exigent beaucoup d'arrosemens.

On multiplie ces especes par boutures, qu'on peut planter dans tous les tems de l'été sur une planche de terre légere : on les arrose, & on les tient à l'ombre, jusqu'à ce qu'elles aient pris racine ; ce qui arrive cinq ou six semaines après : alors on les enleve, & on les met dans des pots ; car si on les laissoit long-tems en pleine terre, elles pousseroient des branches fortes & vigoureuses, & il seroit alors difficile de les transplanter, sur-tout les premiere & seconde especes ; mais il n'y a pas tant de risque pour la troisieme, qui est moins vigoureuse, & qui ne prend pas si aisément racine que les autres. Pendant l'été, il faut souvent changer les pots de place, pour empêcher que les racines des plantes ne passent par les trous du fond, & ne pénetrent dans la terre ; ce qui leur arrive souvent : dans ce cas, leurs branches deviennent vigoureuses & succulentes ; mais quand, en enlevant les pots, on vient à déchirer leurs racines, ces branches périssent, & les plantes entieres sont souvent détruites.

Polygaloïdes. La quatrieme espece se trouve encore au Cap de Bonne-Espérance ; elle a une tige d'arbrisseau de quatre pieds environ de hauteur, qui se divise en plusieurs branches garnies de feuilles petites, oblongues, sessiles & imbriquées sur quelques branches du haut : ses fleurs sont produites aux extrémités des branches, séparément, sur des pédoncules d'un pouce de longueur ; les demi-fleurons qui composent les rayons, sont en pointe aiguë, & étendus : le

diique est composé de fleurons stériles. On multiplie cette espece par boutures, comme les autres, & elle exige le même traitement.

OSYRIS. *Lin. Gen. Plant.* 978. *Cafia. Tourn. Inst. R. H.* 664. *tab.* 488. [*Poets Cafia.*] Cafie des Poëtes.

Caracteres. Les fleurs font mâles & femelles fur différentes plantes : le calice est formé par une feuille divifée en trois fegmens aigus ; la corolle n'a point de pétales : les fleurs des plantes mâles ont trois courtes étamines, & celles des femelles ont un germe, qui devient enfuite une baie globulaire, dans laquelle est renfermée une fimple femence.

Nous ne connoiffons qu'une efpece de ce genre.

Ofyris alba, frutescens, bacci-fera. C. B. P. 212 ; Ofyris en arbriffeau, produifant des baies, nommée par quelques-uns Cafie en arbriffeau, produifant des baies rouges.

Ofyris. Roy. Lugd.-B. 202. *Sauv. Monfp.* 56. *Gouan. Monfp.* 502. *Gron. Orient.* 398. *Scop. Carn. ed.* 2. *n.* 1215.

Cafia Poëtica Monfpeliensium. Cam. Epit. 26. *Lob. ic.* 432.

Cafia Latinorum. Alp. Exot. 41.

Cafia Monfpelii dicta, Gefn. Epit. 50.

Cet arbriffeau s'éleve rarement au-deffus de deux pieds de hauteur, & produit des branches ligneufes, & garnies de feuilles longues, étroites, & de couleur brillante. Ses fleurs paroiffent dans le mois de Juin ; elles font d'une couleur jaunâtre, & produifent

des baies d'abord vertes, enfuite d'un rouge brillant, & qui reffemblent beaucoup à celles de *l'Afperge.*

Cette plante fe trouve dans la France Méridionale, en Efpagne, & dans quelques parties de l'Italie, fur les bords des routes & dans les crevaffes des rochers : mais on la tranfplante difficilement dans les jardins ; car elle ne profite pas après avoir été déplacée ; de forte que la feule méthode pour fe la procurer, est de femer fes baies dans les places qui leur font deftinées ; & comme elles reftent communément un an dans la terre avant que les plantes paroiffent, & quelquefois même deux ou trois ans, il ne faut pas les remuer avant que ce tems foit paffé, fi les plantes n'ont pas paru plutôt. On fe procure ces femences des pays où cette plante croît naturellement ; car celles qu'elle donne dans nos jardins ne font jamais fructueufes : de maniere qu'on ne peut conferver cette efpece qu'avec peine.

OTHONNA. *Lin. Gen. Plant.* 888. *Doria. Raii Meth. Plant.* 33. *Jacobæa. Tourn. Inst. R. H.* 485. *tab.* 276. [*Ragwort.*] Herbe à chiffons, efpece de Jacobée.

Caracteres. La fleur est radiée, & compofée de fleurons hermaphrodites, qui forment le difque, & de demi-fleurons femelles, qui font les rayons ou bordures ; ils font renfermés dans un calice commun, fimple, & formé par une feuille découpée en huit ou dix ter-

OTH

gmens. Les fleurs hermaphrodites sont tubulées, & divisées au sommet en cinq parties ; les demi-fleurons femelles sont étendus au-dehors en forme de langue, dont l'extrémité a trois dents réfléchies. Les fleurons hermaphrodites ont des étamines courtes, & terminées, par des antheres cylindriques, & un germe oblong, qui soutient un style mince, & couronné par un stigmat simple. Les demi-fleurons femelles ont un germe oblong avec un style mince surmonté par un large stigmat, divisé en deux, & réfléchi. Les fleurons hermaphrodites sont rarement fructueux ; mais les demi-fleurons femelles ont une semence oblongue, quelquefois nue, & quelquefois couronnée de duvet, & placée dans le calice, qui persiste.

Ce genre de plantes est rangé dans la quatrieme section de la dix-neuvieme classe de LINNÉE, qui comprend celles à fleurs composées, dont les fleurons femelles sont fructueux, & les hermaphrodites stériles.

Les especes sont :

1°. *Othonna Coronopi-folia*, *foliis infimis lanceolatis, integerrimis, superioribus sinuato-dentatis. Hort. Cliff.* 419. *Roy. Lugd.-B.* 179 ; Othonna dont les feuilles du bas sont entieres & en forme de lance, & celles du haut dentelées & sinuées.

Jacobæa Africana frutescens Coronopi folio. Hort. Amst. 2. p. 139. f. 70. *Raii Suppl.* 175 ; Jacobée d'Afrique en arbrisseau, à feuilles de Corne de cerf.

2°. *Othonna Calthoïdes*, *foliis cunei-formibus, integerrimis, sessilibus, caule fruticoso procumbente, pedunculis longissimis* ; Othonna à feuilles entieres, sessiles, & en forme de coin, avec une tige d'arbrisseau tombante & de très-longs pédoncules.

Calthoïdes Africana, Glasti folio. Jun. ; Souci bâtard d'Afrique, à feuilles de Gaude.

3°. *Othonna pectinata*, *foliis pinnati-fidis, laciniis linearibus, parallelis. Hort. Cliff.* 419. *Roy. Lugd.-B.* 179 ; Othonna à feuilles ailées & pointues, dont les lobes sont paralleles & linéaires.

Jacobæa Africana frutescens, *foliis Absinthii umbelli-feri incanis. Hort. Amst.* 2. p. 137. tab. 69 ; Jacobée d'Afrique en arbrisseau, à feuilles blanches, comme celles de l'Absinthe ombellifere.

Jacobæa Absinthites, tomentosis Cinerariæ foliis, Æthiopica, calyce integro, summis oris dentato. Pluk. Alm. 100 ; Variété à feuilles cotonneuses, avec un calice entier & dentelé au sommet.

4°. *Othonna Abrotani-folia*, *foliis multi-fido-pinnatis, linearibus. Flor. Leyd. Prod.* 380 ; Othonna à feuilles fort étroites, & terminées en plusieurs pointes ailées.

Cineraria Abrotani folia. Berg. cap. 292.

Astero-platy-carpos Africana frutescens, Chrithmi marini-foliis. Comm. Hort. 2. p. 63. f. 32.

Jacobæa Africana frutescens, *foliis Abrotani. Sc. Chrithmi majoris & minoris. Volk. Norib.* 225 ; Jacobée d'Afrique en arbrisseau, à feuilles d'Aurone, ou

342 OTH

de grande & petite Crifte-marine.

5°. *Othonna bulbofa*, *foliis ovato-cunei-formibus, dentatis. Lin. Sp. Plant.* 926 ; Othonna à feuilles ovales, dentelées, & en forme de coin.

Solidago, *foliis oblongis, dentatis, glabris, floribus magnis. Burm. Afr.* 164. *f.* 59 ; Verge à feuilles d'or, oblongues, dentelées & unies, produifant de groffes fleurs.

Jacobæa affinis planta tuberofa, Capitis Bonæ-Spei. Breyn. cent. 1. *t.* 66. *Moris. Hift.* 3. *p. III. S.* 7. *t.* 18. *f.* 33.

Coronopi-folia. La premiere croît naturellement en Ethiopie ; elle s'éleve avec une tige d'arbriffeau à la hauteur de quatre ou cinq pieds, & fe divife en plufieurs branches garnies de feuilles grifâtres, & placées fans ordre ; celles du bas font étroites & entieres, & les autres dentelées fur leurs bords, comme celles de l'*Aubépine* : fes fleurs naiffent en ombelles lâches aux extrémités des branches ; elles font jaunes, & produifent des femences couvertes de duvet.

Calthoides. La feconde a été découverte par le feu Docteur SHAW, près de Tunis, en Afrique, d'où il en a apporté les femences : elle pouffe plufieurs tiges ligneufes, qui s'étendent en-dehors de tous côtés, & font inclinées vers la terre ; ces tiges font garnies de feuilles grifâtres, feffiles, étroites à leur bâfe, mais larges à leur extrémité, où elles font arrondies : fes fleurs font placées fur des pédoncules

longs, épais & fucculens aux extrémités des branches ; elles font jaunes, leurs rayons font terminés en pointe aiguë, & de la même longueur que le calice : fes femences font couronnées d'un duvet long.

Pectinata. La troifieme efpece fe trouve au Cap de Bonne-Efpérance, d'où fes femences ont été apportées en Hollande. Ses graines ont produit des plantes dans le jardin d'Amfterdam en 1699 ; elle s'éleve avec une tige d'arbriffeau d'un pouce de diametre, à la hauteur de deux ou trois pieds, & fe divife en plufieurs branches couvertes d'un duvet blanc, & garnies de feuilles blanches d'environ trois pouces de longueur, fur un de largeur, & découpées en plufieurs fegmens étroits, prefque jufqu'à la côte du milieu. Ces fegmens font égaux, paralleles, & divifées fur leurs bords en deux ou trois pointes : fes fleurs naiffent fur de longs pédoncules qui fortent des aiffelles des tiges vers l'extrémité des branches ; elles ont des rayons larges & jaunes, avec un difque de fleurons, & font remplacées par des femences oblongues, de couléur pourpre, & couronnées de duvet.

Abrotani-folia. La quatrieme efpece, qui croit fpontanément fur les montagnes, près du cap de Bonne-Efpérance, a été élevée de femences dans le jardin d'Amfterdam : elle a une tige baffe d'arbriffeau, & branchue ; fes feuilles font épaiffes comme celles de la *Crifte-marine*, & découpées en

plufieurs fegmens étroits : fes fleurs croiffent fur de courts pédoncules aux extrémités des branches ; elles font jaunes, de la même forme que celles des autres efpeces de ce gen-re, & produifent des femen-ces brunes couronnées d'un duvet mou.

Bulbofa. La cinquieme fe trouve encore au Cap de Bon-ne-Efpérance ; elle a une tige épaiffe d'arbriffeau, qui fe divife en plufieurs branches, & s'éleve à la hauteur de cinq ou fix pieds ; fes feuilles for-tent en paquets du même nœud, & s'étendent de tous côtés ; elles font unies, étroites à leur bâfe, mais plus larges vers leur extrémité, où elles font arrondies : leurs bords font armés de dentelures ai-guës, comme celles du *Houx*; les pédoncules qui foutiennent les fleurs fortent du centre des paquets de feuilles ; ils ont cinq ou fix pouces de lon-gueur, & fe divifent en plu-fieurs autres plus petits, qui s'érendent en-dehors, & fup-portent chacun une fleur jau-ne, radiée, & femblable à celles de la précédente : à ces fleurs fuccedent des fe-mences minces & couronnées de duvet.

Culture. On conferve tou-tes ces efpeces dans les ferres pendant l'hiver, elles n'exi-gent aucune chaleur artificiel-le, & il fuffit de les tenir à couvert de la gelée. en leur donnant beaucoup d'air dans les tems doux : en été, il faut les placer en plein air, dans une fituation abritée, parmi les autres plantes exotiques du-res, où elles augmenteront la variété, & donneront des fleurs durant une grande partie de la belle faifon. On peut les multiplier toutes par boutures pendant tous les mois de l'été, en les plantant fur une vieille couche chaude couverte de vitrages, & en les tenant à l'ombre pendant la chaleur du jour. Quand ces boutures ont pouffé des racines, on les met chacune féparément dans des pots remplis d'une terre molle & marneufe : on les tient à l'ombre jufqu'à ce qu'elles aient formé de nouvelles fibres, & on les place enfuite dans une fituation abritée, où on les laiffera jufqu'à l'automne. El-les exigent le même traitement que les vieilles plantes.

La feconde efpece pourroit fupporter le plein air fous no-tre climat, fi on la plaçoit à une expofition chaude & dans un fol fec. Quelques-unes de ces plantes ont fubfifté ainfi pendant plus de vingt ans dans le jardin de Chelféa, fans au-cun abri. Cette efpece peut être multipliée par boutu-res auffi facilement que les autres.

OXALIS. *Lin. Gen. Plant.* 515. *Oxys. Tourn. Inft. R. H.* 88. tab. 19. [*Wood-forrel.*] Ofeille fauvage, Alleluia.

Caracteres. Le calice de la fleur eft court, perfiftant, & découpé en cinq fegmens ai-gus ; la corolle eft monopéta-le, & divifée prefque jufqu'au fond en cinq parties obtufes : la fleur a dix étamines éri-gées, velues, & terminées

OXA

par des antheres rondes & fillonnées; son germe a cinq angles, & soutient cinq styles
minces, & couronnés par des
stigmats obtus: ce germe se
change ensuite en une capsule
à cinq angles & à cinq cellules, qui s'ouvrent longitudinalement aux angles, & renferment des semences rondes,
qui, étant mûres, sont lancées au loin avec élasticité,
aussi-tôt qu'on les touche.

Ce genre de plantes est rangé dans la cinquieme section
de la dixieme classe de LINNÉE,
avec celles dont les fleurs ont
deux étamines & cinq styles.

Les especes sont:

1°. *Oxalis Acetosella, scapo
uni-floro, foliis ternatis, radice
squamosâ, articulatâ* Hort. Cliff.
175. Fl. Suec. 385. 406. Roy.
Lugd.-B. 458. Mat. Med. p. 118.
Reyg. Ged. t. p. 121. De Neck.
Gallob. 200. Pollich. pall. n.
434. Dærr. Nass. p. 169. Mattusch. Sil. n. 324. Sabb. Hort.
1. f. 32; Oseille sauvage,
avec une seule fleur sur chaque pédoncule, des feuilles à
trois lobes, & une racine
écailleuse & noueuse.

Oxys flore albo. Tourn. Inst.
88; Oseille sauvage à fleurs
blanches.

Oxys Acetosella. Scop. carn.
ed. 2. n. 561.

*Oxys tri-folium acetosum, flo-
ribus lacteis.* Tabern. 525.

Luiula. Blackw. f. 303. L'Alleluia.

Tri-folium acetosum vulgare.
Bauh. Pin. 330.

Tri-folium acetosum. Dod. Pempt.
578.

Oxys flore sub-cæruleo. Tourn.

OXA

Inst. 88; Variété à fleurs presque bleues.

Oxys flore purpurascente Tourn.
Inst. 88; autre variété à fleurs
pourpâtres.

2°. *Oxalis corniculata, caule
ramoso diffuso, pedunculis um-
belli-feris.* Hort. Cliff. 175. Hort.
Ups. 116. Roy. Ludg.-B. 158.
Sauv. Monsp. 173. Pollich. pall.
n. 435. Mattusch. Sil. n. 325.
Dærr. Nass. p. 170. Sabb. Hort.
1. f. 33. Oseille sauvage, avec
une tige branchue, & diffuse,
ayant des pédoncules en forme
d'ombelle.

Oxys lutea. J. B.; Oseille
sauvage jaune, ou Oxis garnie
de cornes.

*Tri-folium acetosum, cornicu-
latum.* Bauh. Pin. 330.

3°. *Oxalis stricta, caule ramoso, erecto, pedunculis umbelli-
feris.* Flor. Virg. 161; Oseille
sauvage avec une tige branchue & érigée, & des pédoncules en ombelle.

*Tri-folium acetosum, cornicu-
latum, luteum, majus, rectum,
Indicum, sivè, Virgineum.* Moris.
Hist. 2. p. 184. S. 2. t. 17. f. 3.

*Oxys lutea, Americana, erec-
tior.* Tourn. Inst. R. H. 88; Oseille sauvage droite & jaune d'A-
mérique.

4°. *Oxalis incarnata, caule
sub-ramoso bulbi-fero, pedunculis
unifloris, foliis passim verticillatis, foliolis ob-cordatis.* Lin.
Sp. 622; Oseille sauvage,
avec des tiges branchues, portant bulbes, une seule fleur
sur chaque pédoncule, des
feuilles généralement verticillées, & des lobes en forme de
cœur.

Oxys bulbosa Æthyopica mi-

OXA

nor , *folio cordato, flore ex albido purpurascente. Tourn. Inst. 89. Comm. Hort. 1. p. 43. f. 22* ; la plus petite Oseille sauvage bulbeuse d'Ethiopie, avec une feuille en forme de cœur, & une fleur d'un blanc tirant sur le pourpre.

5°. *Oxalis purpurea , scapo uni-floro , foliis ternatis , radice bulbosâ. Hort. Cliff.* 175. *Roy. Lugd.-B.* 458. *Ehret. Pict. t.* 10. *f.* 2.; Oseille sauvage avec un pédoncule soutenant une seule fleur, des feuilles à trois lobes, & une racine bulbeuse.

Oxalidi affinis planta bulbosa Africana , flore purpureo magno. Breyn. cent. 102. *f.* 46.

Oxys bulbosa Africana, rotundi-folia , caulibus & floribus purpureis amplis. Hort. Amst. 1. *p.* 41. *tab.* 21 ; Oseille sauvage bulbeuse d'Afrique, avec des feuilles rondes, des tiges & de grosses fleurs pourpre.

Oxys bulbosa tri-folia, hirsuta, flore albo. Burm. Afr. 67. *t.* 27. *f.* 3.

6°. *Oxalis pes-capræ, scapo umbelli-fero , foliis ternatis , bi-partitis. Lin. Sp. Plant.* 434; Oseille sauvage, avec une tige en ombelle , & des feuilles à trois lobes, divisées en deux parties.

Oxalis bulbosa , pentaphylla & hexaphylla , floribus magnis luteis & copiosis. Burm. Afr. 80. *tab.* 28 & 29. *f.* 3 ; Oseille sauvage bulbeuse avec cinq ou six feuilles, & une quantité de grandes fleurs jaunes.

7°. *Oxalis frutescens , caule erecto , fruticoso , foliis ternatis , impari maximo* ; Oseille sauvage , avec une tige droite d'arbris-

OXA   345

seau , & des feuilles à trois lobes , dont celui du milieu est le plus grand.

Oxys lutea frutescens , Tri-folii bituminosi-facie. Plum. Cat. 2. *ic.* 213. *f.* 1 ; Oseille sauvage jaune en arbrisseau, ayant l'apparence du Treffle bitumineux.

8°. *Oxalis Barrelieri , caule ramoso , erecto, pedunculis bi-fidis , racemi-feris. Lin. Sp.* 624 ; Oseille sauvage, avec une tige érigée & branchue, & des pédoncules divisés en deux parties, qui se sous-divisent encore en plusieurs branches.

Oxys Roseo flore erectior. Feuill. Peru. 2. *p.* 773. *f.* 23.

Tri-folium acetosum Americanum , rubro flore. Barrel. Rar. 64. *f.* 1139. *Bocc. Mus.* 2. *p.* 63. *f.* 51; Oseille sauvage à trois feuilles, de l'Amérique , avec une fleur rouge.

Acetosella. La première espece croît naturellement dans des terres humides , à l'ombre des bois, & dans les haies fourrées de presque toute l'Angleterre ; mais on la cultive rarement dans les jardins. Cependant ceux qui aiment les herbes acides dans les salades, ne peuvent en trouver une plus agréable : ses racines sont composées de plusieurs nœuds écailleux, qui se multiplient en grande abondance ; ses feuilles sortent immédiatement des racines sur des pétioles simples & longs; elles sont composées de trois lobes en forme de cœur, qui se rencontrent dans un centre, où ces lobes se joignent au pétiole & sont d'un vert pâle , & velues: entr'elles sortent les fleurs , portées sur de

longs pédoncules, qui en foutiennent chacun une groffe, blanche, & en forme de cloche ouverte ; elles paroiffent dans les mois d'Avril & de Mai, & font remplacées par des capfules oblongues, à cinq angles & à cinq cellules, qui renferment des femences petites & brunâtres. Ces capfules s'ouvrent en mûriffant, & lancent leurs graines à une diftance confiderable, au moindre attouchement.

Cette efpece eft comprife dans la *Pharmacopée* au nombre des plantes d'ufage en Médecine ; mais ceux qui fourniffent de ces herbes les marchés, y portent généralement en place la troifieme efpece, qui eft à préfent devenue trèscommune dans les jardins. Cette derniere, ayant très-peu d'acide, ne peut cependant la remplacer efficacement, ni produire les mêmes effets : mais comme elle s'éleve avec une tige droite & branchue, elle eft plutôt cueillie & mife en paquet ; au-lieu que les feuilles de la premiere, qui croiffent fimples de la racine, exigent plus de temps pour être ramaffées. Il y a une variété de la premiere efpece à fleurs pourpre, qu'on rencontre dans le nord de l'Angleterre ; mais comme elle n'en differe au moins que par cette couleur, je ne l'ai pas mife au nombre des efpeces (1).

Corniculata. La feconde eft une plante annuelle, qui croît naturellement à l'ombre & dans les bois, en Italie & en Sicile : fa racine eft longue, mince & fibreufe ; fes tiges traînent fur la terre, & s'étendent au-dehors à huit ou neuf pouces de chaque côté, & fe divifent en petites branches : fes feuilles font poftées fur de longs périoles, & compofées de trois lobes en forme de cœur, dont les extrémités font plus profondément découpées que celles de la précédente : fes fleurs font jaunes, & croiffent en ombelles fur des pédoncules longs & minces, qui fortent des parties latérales des branches ; elles paroiffent dans les mois de Juin & de Juillet, & produifent des capfules à-peu-près d'un pouce de longueur, qui s'ouvrent avec élafticité, & jettent au loin leurs femences.

Stricta. La troifieme efpece eft originaire de la Virginie,

_____

(1) Le feul principe digne d'attention qu'on decouvre dans cette plante, au moyen de l'analyfe, eft un fel acide concret, fufcepti-

ble de cryftallifation, & dans lequel réfide toute la vertu de l'Ofeille. Cette plante eft en conféquence très-anti-putride, rafraîchiffante, tempérante, diurétique, &c. ; elle produit des effets falutaires dans les fievres ardentes & bilieufes, putrides & malignes, dans le fcorbut, les hémorrhagies, les ébullitions, l'ardeur d'urine, l'Ictere, la manie, la foif ardente, &c. On emploie le fel concret qu'on fait diffoudre dans une liqueur convenable en affez grande quantité, pour qu'elle foit d'une acidité agreable, ou bien on fe fert du fuc exprimé des feuilles, qu'on clarifie avant d'en faire ufage.

& de quelques autres parties de l'Amérique Septentrionale, d'où ses semences ont été apportées en Europe: mais partout où cette plante est introduite & mûrit ses semences, elle y devient une herbe sauvage & commune; elle est annuelle, & s'élève avec une tige branchue & herbacée à la hauteur de huit ou neuf pouces: ses feuilles sont postées sur de fort longs pétioles, & ont la même forme que celles de la seconde espece, ses fleurs sont jaunes, & disposées en espece d'ombelles sur des pédoncules longs, minces & érigés; ses capsules & ses semences ressemblent à celles de la précédente.

Culture. Ces trois especes n'exigent point de culture particuliere; en plaçant les racines de la premiere à l'ombre dans une plate-bande humide, elles y feront des progrès, & se multiplieront considerablement: il suffit de les tenir nettes de mauvaises herbes. En semant les deux autres sur une plate-bande ouverte, les plantes leveront aisément, & n'exigeront aucun soin; & si on leur permet d'écarter leurs semences, elles se multiplieront beaucoup.

Incarnata. La quatrieme espece a une racine ronde & bulbeuse, de laquelle sortent des tiges minces de six pouces de hauteur, & divisées en branches, disposées par paires; les pétioles qui sortent de leurs divisions, sont longs, minces, & soutiennent une feuille composée de trois petits lobes ronds & en forme de cœur; les pédoncules des fleurs sont longs, minces, & naissent des divisions des tiges: ils soutiennent chacun une fleur de couleur tirant sur le pourpre, de la même forme & de la même grosseur à-peu-près que celles de la premiere espece. Celle-ci fleurit dans les mois de Mai, Juin & Juillet, & produit quelque-fois des semences, qui mûrissent en Angleterre. Cette plante, étant originaire du Cap de Bonne-Espérance, est trop délicate pour pouvoir subsister en plein air pendant l'hiver dans ce pays; mais elle n'exige point d'autre abri que d'être placée sous un châssis de couche en hiver, pour être à couvert des fortes gelées: elle se multiplie en abondance par les rejettons de sa racine, ainsi que par les bulbes qui naissent à côté des tiges.

Purpurea. La cinquieme espece se trouve au Cap de Bonne-Espérance; elle s'y multiplie en si grande quantité, que la terre dans laquelle étoient les plantes apportées de ce pays, en étoit remplie: sa racine est ronde, bulbeuse, & couverte d'une peau brune; elle pousse de fortes fibres, qui pénetrent profondément dans la terre; ses feuilles sont composées de trois lobes ronds, larges, velus, & postées sur des pétioles longs & minces, qui sortent d'une tige courte, épaisse, & adhérente à la racine; les pédoncules s'élevent entre les feuilles de la tige, & soutiennent chacun une

OXA

grosse fleur pourpre : ces fleurs paroissent dans les mois de Janvier & Février ; mais elles produisent rarement des semences ici : ses racines poussent des rejettons en grande abondance, qui servent à la multiplier. Comme cette espece ne peut subsister en plein air en hiver, il faut planter ses racines dans des pots, afin de pouvoir les placer sous un châssis ordinaire pendant la mauvaise saison, de maniere qu'on puisse leur donner autant d'air qu'il est possible dans les tems doux, sans quoi leurs feuilles fileroient, & deviendroient foibles ; car elles commencent à pousser en Octobre, & continuent à croitre jusqu'en Mai, qui est le tems où elles se fannent & périssent. Les racines de cette espece peuvent être transplantées en tout tems, quand les feuilles sont tombées, & jusqu'à ce qu'elles commencent à repousser.

Pes-capræ. La sixieme espece est originaire des mêmes contrées que la précédente : ses racines sont bulbeuses ; ses feuilles sont postées sur des pétioles longs & minces, qui sortent d'une tige courte ; elles sont composées de trois lobes, dont la plupart sont divisés en deux parties presque jusqu'à leur bâse. Les pédoncules ont cinq ou six pouces de longueur, & soutiennent plusieurs fleurs grosses & jaunes, en forme d'ombelle : elles paroissent en Mars, & sont quelquefois suivies de semences ici. Cette plante exige le même traitement que la précédente.

Frutescens. La septieme, qui a été découverte par PLUMIER, dans quelques isles françoises des Indes Occidentales, a été trouvée depuis en abondance à la Vera-Cruz par le Docteur HOUSTOUN, qui l'a envoyée en Angleterre. Elle s'éleve en tiges d'arbrisseau à la hauteur d'un pied & demi, & pousse plusieurs branches minces, garnies de petites feuilles à trois. lobes, dont celui du milieu est deux fois plus large que les latéraux ; elles sont placées par paires opposées, & quelquefois par trois autour de la tige, sur de courts pétioles. Les pédoncules s'élevent des aisselles des tiges ; ils ont près de deux pouces de longueur, & soutiennent chacun quatre ou cinq fleurs jaunes, dont les corolles ne sont pas plus longues que les calices ; chaque fleur est supportée par un pédoncule plus petit, & courbé de maniere qu'elle penche vers le bas.

Cette espece, étant beaucoup plus tendre qu'aucune des précédentes, exige une place dans une serre de chaleur modérée, pendant l'hiver : on la multiplie par ses graines qu'on répand dans des pots, & qu'on plonge dans une couche tempérée ; quand les plantes ont poussé, on les met chacune séparément dans un pot rempli d'une terre légere & sablonneuse ; on les plonge dans une nouvelle couche chaude, & on les tient à l'ombre jusqu'à ce qu'elles aient formé de nouvelles racines, après quoi on les traite de la même maniere

OXA

niere que les autres plantes tendres du même pays

Barrelieri. La huitieme espece croît naturellement dans le Brésil; car j'ai vu pousser une grande quantité de ces plantes dans une caisse venue de ce pays : elle s'élève rarement à plus de trois ou quatre pouces de hauteur ; ses tiges sont droites ; ses feuilles sont composées de trois beaux lobes larges, velus, & postés sur de longs pétioles : ses fleurs sortent immédiatement de la racine, sur des pédoncules de la même longueur que les pé-

OXY

tioles des feuilles ; ils sont divisés en deux parties, & soutiennent deux fleurs belles, grandes & rouges, qui sont remplacées par des capsules oblongues & remplies de semences brunes.

Cette espece peut être multipliée par les rejettons de sa racine, ou par semences ; elle exige les mêmes abris que la sixieme.

OXYACANTHA. *Voy.* BER-BERIS.

OXYCEDRE. *Voyez* JUNI-PERUS OXYCEDRUS. *L.*

OXYS. *Voyez* OXALIS.

P

PAD

PADUS. *Lin Gen. edit. prior* 476. *edit.* 5. *Prunus.* 546. *Cerasus & Lauro-Cerasus. Tourn. Inst. R. H.* 625. 627. *tab.* 401. 403. [*The bird-Cherry,* or *Cherry Laurel.*] le Cerisier-d'oiseau ou Laurier-Cerise.

Caracteres. Le calice de la fleur est en cloche, & formé par une feuille découpée à l'extremité en cinq parties, qui s'étendent & s'ouvrent : la corolle est composée de cinq pétales larges, ronds, étendus, & insérés dans le calice ; elle a depuis vingt jusqu'à trente étamines en forme d'alène, fixées dans le calice, & terminées par des antheres ron-

Tome V.

PAD

des, le germe est rond, & soutient un style mince, couronné par un stigmat obtus & entier ; ce germe se change dans la suite en un fruit rond, qui renferme un noyau à pointe ovale, & sillonné par quatre rainures.

LINNÉE, dans les premieres éditions de sa méthode, a séparé ce genre de plantes de celui des *Cerisiers* auquel il avoit été joint avant, parce que les noyaux de ce genre font obtus, & que ceux des *Cerises* sont aigus : mais il y a une différence plus sensible entr'eux, c'est que les fleurs du *Padus* sont rangées en grap-

Z

PAD

pes, & que celles du *Cerifier* ont leurs pédoncules sur un nœud.

Le même Auteur, dans sa derniere édition, a réuni ce genre à celui des *Cerifiers*, *Abricotiers*, *Pruniers*, dont il n'a fait que des especes, en quoi je pense qu'il a excédé les bornes de la Nature. En effet, quoique le *Padus* & le *Cerafus* puissent être réunis dans le même genre, ils ne doivent en aucune maniere être rapprochés des *Pruniers*; car il est certain que la greffe du *Cerifier* ne prend point sur la tige du *Prunier*, ni celle du *Prunier* sur une tige de *Cerifier*. Cependant les arbres du même genre, quoique souvent très-différens dans leur forme extérieure, réussissent toujours les uns sur les autres par la greffe.

Cet arbre est rangé dans la premiere section de la douzieme classe de LINNÉE, qui comprend les plantes dont les fleurs ont un style, & plus de douze étamines insérées dans le calice ou dans les pétales.

Les especes sont :

1°. *Padus avium, glandulis duabus bafi foliorum fubjeĉis. Hort. Cliff.* 185. *Fl. Suec.* 396. 431. *Roy. Lugd.-B.* 269 ; Padus avec deux glandes à la base des feuilles.

Prunus Padus. Lin. Syfl. Plant. tom. 2. *p.* 483. *Sp.* 1 ; Cerifier d'oiseau.

Cerafus racemofa fylveftris, fruĉu non eduli. C. B. P. 451. *Dut. m. Arb.* 3; Cerifier sauvage a grappes, ayant un fruit

qui n'est pas bon à manger, communément appelé arbre ou bois de Sainte-Lucie, que l'on confond avec le *Cerafus Mahaleb.*

2°. *Padus rubra, foliis lanceolato-ovatis, deciduis, petiolis bi-glandulofis. tab.* 196. *fol.* 2 ; Padus avec des feuilles en forme de lance, ovales, & tombantes, ayant deux glandes à chaque pétiole.

Cerafus racemofa fylveftris, fruĉu non eduli, rubro. H. R. Par. ; Cerifier-d'oifeau sauvage, branchu, avec un fruit rouge qui n'est pas bon à manger.

3°. *Padus Virginiana, foliis oblongo-ovatis, ferratis, acuminatis, deciduis, bafi-anticè glandulofis;* Laurier-Cerife à feuilles oblongues, ovales, fciées, à pointe aiguë, & tombantes, ayant des glandes sur le devant des pétioles.

Cerafus latiori folio, fruĉu racemofo purpureo majore. Catefb. Car. 2. *p.* 94. *f* 64.

Cerafi fimilis arbufcula Mariana, Padi folio, flore albo, parvo, racemofo. Pluk. Mant. 43. *Catefb. Car.* 1. *p.* 28 ; Laurier-Cerife d'Amérique.

Prunus Virginiana. Lin. Syfl. Plant. tom. 2. *p.* 483. *Sp.* 2. 1

4°. *Padus Lauro-Cerafus, foliis femper virentibus, lanceolato-ovatis. Hort. Cliff.* 42 ; Laurier-Cerife, à feuilles ovales, en forme de lance, & toujours vertes.

Cerafus folio Laurino. Bauh. Pin. 450.

Lauro-Cerafus. Clus. Hifl. 1. *p.* 4 ; Laurier-commun, Laurier-Cerife.

PAD

Prunus Lauro-Cerasus. Lin. Syst. Plant. t. 2. p. 485. Sp. 5.

5°. *Padus Lusitanica, foliis oblongo-ovatis, semper virentibus, eglandulosis* ; Laurier-Cerise à feuilles oblongues , ovales , toujours vertes, & dénuées de glandes.

Padus foliis glandulâ destitutis. Vir. Cliff. 41. *Hort. Ups.* 126. *Roy. Lugd. - B.* 269.

Lauro - Cerasus Lusitanica minor. Tourn. Inst. 628 ; le plus petit Laurier de Portugal , appelé Azarero par les Portugais.

Prunus Lusitanica. Lin. Syst. Plant. tom. 2. *p.* 484. *Sp.* 4.

6°. *Padus Caroliniana, foliis lanceolatis, acutè denticulatis semper virentibus* ; Laurier-Cerise toujours vert, à feuilles en forme de lance , avec de petites dentelures aiguës, appelé en Amérique Mahagony bâtard.

Avium. La première espece croît naturellement dans les hales du Duché d'Yorck, & dans plusieurs parties Septentrionales de l'Angleterre, ainsi que dans quelques endroits près de Londres. On la cultive dans les pépinieres, comme un arbrisseau à fleurs, pour en faire commerce ; elle s'éleve à la hauteur de dix à douze pieds , avec plusieurs tiges ligneuses ; dont quelques-unes croissent jusqu'à la grosseur de neuf à dix pouces de diametre : mais comme la forme des jardins a souvent été changée depuis cinquante ou soixante ans , il y en a peu dans lesquels on ait laissé subsister des arbres à fleurs & d'ornement aussi long-tems. Les branches

PAD

de cet arbre sont grosses , écartées, couvertes d'une écorce pourpre , & garnies de feuilles ovales, en forme de lance , alternes , légerement sciées sur leurs bords, & pourvues de deux petites protubérances ou glandes à leur bâse : ses fleurs naissent sur les côtès des branches en paquets longs & serrés ; elles ont cinq pétales ronds , blancs , beaucoup plus petits que ceux du *Cerisier ,* & insérés dans la bordure du calice ; la corolle renferme un grand nombre d'étamines , aussi insérées dans le calice. Ces fleurs sont chacune sur un court pédoncule , & rangées alternativement: dans la longueur du pédoncule principal ; elles ont une odeur que beaucoup de gens trouvent désagréable : elles paroissent en Mai , & produisent des fruits ronds , petits , d'abord verds ; ensuite rouges , enfin noirs ; lorsqu'ils sont mûrs, & qui renferment un noyau rond & sillonné.

Rubra. La seconde espece est originaire de l'Amérique , d'où ses semences m'ont été envoyées ; elle a été depuis plusieurs années multipliées dans les pépinieres , près de Londres, ou elle est géneralement connue sous le nom de *Cerisier* de Cornwall ; celle-ci a été souvent confondue avec la première. Plusieurs Botanistes modernes les ont regardées comme ne formant qu'une seule espece ; mais je les ai élevées toutes deux de semences, & j'ai toujours remarqué que les jeunes plantes conservoient

Z 2

leurs différences. Celle-ci s'é-
leve avec une tige croite &
érigée à plus de vingt pieds de
hauteur ; fes branches font plus
courtes, & croiffent plus rapprochées que celles de la premiere ; elle a naturellement
une tête réguliere ; fes feuilles
font plus courtes, plus larges,
& moins rudes que celles de
la précédente ; fes fleurs font
difpofées en épis plus courts,
plus ferrés & plus érigés ; fon
fruit eft plus gros, & rouge lorfqu'il eft mûr. Cet arbre fleurit
un peu après la premiere efpece.

Virginiana. La troifieme,
qu'on rencontre dans la Virginie & dans d'autres parties
de l'Amérique Septentrionale,
s'éleve avec une groffe tige à
la hauteur de dix jufqu'à trente
pieds & fe divife en plufieurs
branches couvertes d'une écorce d'un pourpre foncé, & garnies de feuilles ovales, alternes, fupportées par de courts
pétioles, d'un vert luifant,
légerement fciées fur leurs
bords, & qui confervent leur
verdure en automne plus long-
tems qu'aucune des autres efpeces d'arbres : fes fleurs fortent en grappe comme celles
de la feconde, & font remplacées par de gros fruits,
qui deviennent noirs en mûriffant, & font bientôt dévorés par les oifeaux. Le bois de
cet arbre eft agréablement vei-
né en noir & en blanc ; il fe
polit parfaitement, & fert fouvent à faire des meubles,
comme celui de la premiere
efpece qui eft d'un grand ufage en France fous le nom de
Bois de Sainte-Lucie.

Lauro-Cerafus. La quatrieme
eft le *Laurier commun*, qui eft
à préfent fi bien connu ; qu'il
n'eft pas néceffaire d'en donner aucune defcription : elle
croit fans culture aux environs
de Trébifonde, proche la mer
Noire, d'où elle a été apportée
en Europe vers l'année 1576 ;
mais à préfent, elle eft devenue fort commune, fur-tout
dans les parties les plus chaudes de l'Europe.

Lufitanica. La cinquieme a
été apportée du Portugal en
Angleterre ; mais il eft difficile
de déterminer fi elle eft originaire de ce pays, ou fi elle
y a été introduite de quelques autres endroits. Les Portugais l'appellent *Aferaro* ou
Azeraro : on la regarda d'abord
comme un arbriffeau bas &
toujours vert ; mais l'expérience a appris que, lorfqu'elle
fe trouve dans un fol qui lui
eft propre, elle parvient à une
groffeur confidérable. Il y a
à préfent en Angleterre quelques-uns de ces arbres, dont
les troncs ont plus d'un pied
de diametre, & douze ou quinze pieds de hauteur, quoiqu'ils ne foient pas fort âgés ;
ils font bien garnis de branches couvertes d'une écorce
rougeâtre, lorfqu'elles font jeunes : fes feuilles font plus courtes que celles du *Laurier* commun, prefque ovales, de la
même confiftançe, & d'un vert
luifant ; ces feuilles font, avec
les branches qui font rouges,
un très-bel effet : fes fleurs
naiffent en épis longs & ferrés
fur les côtés des branches ;
elles font blanches, & de la

même forme que celles du *Laurier* ordinaire ; elles paroiſſent en Juin, & ſont ſuivies par des baies ovales, plus petites que celles du *Laurier* commun, d'abord vertes, enſuite rouges & noires, lorſqu'elles ſont mûres ; elles renferment un noyau ſemblable à celui d'une *Ceriſe.*

Caroliniana. Les ſemences de la ſixieme eſpece ont été envoyées de la Caroline ſous le titre de *Mahagony bâtard.* Si nous en jugeons par ſon accroiſſement dans ce pays, elle paroit être un peu plus forte qu'un arbriſſeau ; ſa tige, qui ne s'élève guere qu'à trois pieds de hauteur, pouſſe des branches latérales, qui s'étendent à chaque côté, & ſont couvertes d'une écorce brune, & garnies de feuilles en forme de lance de deux pouces environ de longueur ſur neuf lignes de largeur, ſciées en petites dentelures aiguës ſur leurs bords, alternes, ſupportées par des pétioles fort courts, & d'un vert luiſant : ces feuilles conſervent leur verdure toute l'année. Cette eſpece n'ayant pas encore montré ſes fleurs en Angleterre, je ne puis en donner aucune deſcription ; mais d'après celle qui m'a été envoyée, & d'après l'inſpection de ſes ſemences, je puis aſſurer qu'elle appartient à ce genre.

Cette plante ſubſiſtera en plein air dans ce pays, ſi elle eſt placée à une expoſition chaude, & à l'abri des fortes gelées, qui lui ſeroient très-funeſtes, ſur-tout tandis qu'elle

eſt encore jeune ; mais lorſqu'elle a acquis aſſez de force, elle peut certainement profiter en plein air dans une ſituation abritée. On la multiplie par ſes baies comme le *Laurier* de Portugal ; ſes branches pouſſent auſſi des racines lorſqu'elles ſont marcottées : mais l'expérience m'a appris qu'elle ne réuſſit pas de boutures.

Culture. On multiplie aiſément les trois premieres eſpeces par ſemences ou par marcottes ; on met leurs baies en terre en automne ; car ſi on les conſerve juſqu'au printems, elles pouſſent rarement avant la ſeconde année. On peut les ſemer ſur une plate-bande de terre fertile, comme on le pratique pour les noyaux de Ceriſes deſtinés à former des tiges, & l'on traite les jeunes plantes de la même maniere ; on les tranſplante dans une pépiniere, où on les laiſſe deux ans, pour leur faire acquérir de la force, & on les met enſuite dans les places qui leur ſont deſtinées : on les mêle ordinairement avec d'autres arbriſſeaux à fleurs dans les boſquets, ou d'autres endroits à l'écart, où elles ſerviront beaucoup à la variété.

Si on les multiplie par marcottes, il faut coucher les jeunes branches en automne ; une année après, elles auront acquis de bonnes racines : alors on pourra les ſéparer des vieilles plantes, les tranſplanter dans une pépiniere pour un ou deux ans, afin de leur laiſſer acquérir de la force, & les placer enſuite à demeure.

354 PAD

La troisieme espece devient un grand arbre, quand elle est plantée dans un sol humide ; mais dans un terrein sec, elle s'éleve tout au plus à la hauteur de vingt pieds. Nous avons, depuis quelques années, plusieurs plantes qui ont été élevées avec des semences apportées de la Caroline ; elles ressemblent beaucoup à la troisieme, mais elles sont d'un crû plus bas. Si cette différence vient de ce qu'elles sont originaires d'un climat plus chaud, elles ne pourront supporter le froid de nos hivers, aussi bien que la troisieme ; mais peut-être sont-elles une espece distincte. Au surplus, je ne puis le déterminer, parce qu'elles n'ont point encore produits des fruits dans ce pays.

Le *Laurier* se multiplie aisément par boutures : le meilleur tems pour cette opération est le mois de Septembre, aussi-tôt que les pluies de l'automne commencent à humecter la terre : on prend ces boutures sur des branches de l'année ; & si on leur laisse une petite partie de bois de l'année précédente, elles réussiront plus certainement, & pousseront de plus fortes racines : il faut les planter dans un sol mou & marneux, à six pouces environ de profondeur, & bien presser la terre contre leurs tiges. Si l'on suit exactement toutes ces instructions, & que ces boutures soient placées dans une bonne terre, elles réussiront presque toutes, & auront acquis de bonnes racines pour l'automne suivant : alors on les

met en pépiniere, où on les laisse deux ans, pour qu'elles puissent se fortifier : mais après ce tems, on pourra les transplanter à demeure. On tenoit autrefois ces plantes dans des pots ou caisses, pour les renfermer dans la serre, & les mettre à l'abri des froids de l'hiver ; depuis, on les a plantées contre des murailles chaudes, qui suffisent pour les garantir des fortes gelées : ensuite on les a taillées en piramides & en globes ; mais en les traitant ainsi, leurs feuilles, qui se trouvoient coupées dans le milieu, les rendoient fort désagréables à la vue : depuis quelques années, on les a disposées plus convenablement, en les plantant sur les bords des bosquets & des endroits déserts, où peu d'autres plantes font un aussi bel effet, parce qu'elles croissent sous l'égout des arbres & à l'ombre ; de plus, comme leurs branches s'abaissent jusqu'à terre, elles forment une espece de taillis, & leurs feuilles, qui sont larges & d'un vert éclatant, servent d'ornement dans les bois & autres plantations, pendant l'hiver, quand les autres arbres sont entierement dépouillés ; & en été, elles font une variété agréable avec la verdure de ces autres arbres. Ces plantes sont quelquefois endommagées par les hivers rudes, surtout quand elles sont isolées & fort exposées ; mais lorsqu'elles sont placées au milieu des taillis, & abritées par d'autres, les gelées leur font rarement du tort ; car dans ces

PAD

endroits, il n'y a que leurs jeunes rejettons qui fouffrent ; & comme il en repouffe d'autres immédiatement au-deffus, pour les remplacer, le dommage fe trouve réparé dans l'année. Toutes les fois qu'il furvient des hivers durs, on ne doit tailler ces arbres qu'à la Saint-Jean fuivante ; car ce n'eft qu'alors qu'on peut connoître les branches qui font entierement mortes, pour les retrancher jufqu'aux endroits où pouffent les nouveaux rejettons ; mais fi on les taille au printems, le hâle pénetre dans les jeunes branches, & leur fait plus de tort que les gelées.

Ces arbres fervent auffi d'ornement dans des bofquets, quand ils font entrêmelés avec d'autres toujours verts : on s'en fert pour cacher des objets défagréables, parce que, leurs feuilles étant très-larges, elles mafquent très-bien tout ce qui ne doit pas être vu ; ils font également utiles pour garantir les autres plantes des efforts des vents ; de forte qu'étant plantés parmi des arbriffeaux à fleurs, on peut les dreffer de maniere qu'ils rempliffent les vuides au milieu des plantations ; qu'ils fervent d'abri en hiver, & qu'ils empêchent en toutes faifons la vue de percer à travers les arbriffeaux : ils font auffi très-propres à fervir d'ornement dans plufieurs autres circonftances.

Dans les pays chauds, cet arbre parvient à une groffeur confidérable ; de forte que, dans quelques parties de l'Italie, il y a des bois très-éten-

PAD 355

dus qui en font plantées : mais nous ne pouvons pas efpérer qu'ils aient jamais d'auffi groffes têtes en Angleterre ; car en émondant ces arbres pour former leurs tiges, la gelée leur eft beaucoup plus nuifible que fi leur branches tomboient naturellement vers la terre. Cependant, fi on les plantoit fort près les uns des autres dans de grands bofquets, & fi on les laiffoit croître en liberté, ils s'abriteroient mutuellement de la gelée, & parviendroient à une hauteur confidérable.

On en voit à préfent un exemple dans la belle plantation d'arbres toujours verts du Duc DE BEDFORD, à Wooburn-Abbey, où il y a une montagne étendue entierement couverte de *Lauriers*. Dans quelques parties de cette même plantation, ces arbres, qui font entremélés en grand nombre avec d'autres arbres toujours verts, font déja parvenus à une groffeur confidérable, & produifent un très-bel effet.

Quelques perfonnes multiplient ces arbres par leurs baies. Cette méthode eft certainement la meilleure pour fe procurer de belles plantes ; car celles qui viennent de femences s'élevent naturellement droites, au lieu que toutes celles qu'on obtient par boutures ou par marcottes, pouffent plus volontiers horifontalement, & produifent un plus grand nombre de branches latérales. Quand on veut multiplier cet arbre par femences, il faut

Z 4

préferver fes baies des atteintes des oifeaux, qui, fans cela, les dévoreroient toutes, avant qu'elles fuffent parvenues à leur maturité ; ce qui a rarement lieu avant la fin de Septembre ou le commencement d'Octobre, parce qu'il faut les laiffer fur l'arbre jufqu'à ce que leur chair extérieure foit tout-à-fait noire. On feme ces baies auffi-tôt qu'elles font recueillies ; car fi on les gardoit hors de la terre jufqu'au printems, elles feroient fujettes à manquer ; au-lieu qu'il n'y a point de danger de les femer en automne, pourvu que ce foit dans un fol fec. Quand l'hiver eft rude, on couvre la plate-bande où elles font placées, avec du tan pourri, de la paille, du chaume de pois, ou quelqu'autre litiere légere pour empêcher la gelée de pénétrer dans la terre. La meilleure méthode eft de femer ces baies en rangs éloignés de fix pouces, & à un pouce entr'elles : on donne à-peu-près trois pouces de profondeur aux rigoles ; on écarte bien les baies, & on les couvre de terre : au moyen de cela, elles ne peuvent pas manquer. Les plantes paroîtront au printems fuivant ; alors on les tiendra nettes de mauvaifes herbes ; & fi la faifon eft feche, on les arrofera ; elles feront des progrès fi rapides, qu'elles feront bonnes à être tranfplantées pour l'automne fuivant : alors on les enlevera pour les mettre en pépiniere, en rangs éloignés de trois pieds, & à

un pied de diftance entr'elles. On pourra les laiffer ainfi deux ans ; mais après ce tems, elles feront en état d'être tranfplantées à demeure.

La meilleure faifon pour faire cette tranfplantation, eft l'automne, auffi-tôt que les pluies ont préparé la terre. On la fait fouvent au printems ; mais alors ces arbres ne reprennent pas fi bien, & ne font pas tant de progrès que les autres, fur-tout quand ils ont été élevés dans une terre légere, qui ne refte jamais après leurs racines : mais fi l'on peut les enlever en mottes, & fi on les place à une petite diftance, cette opération peut être faite fans danger au printems, pourvu que ce foit avant que ces arbres commencent à pouffer : car leur féve fe met en mouvement de très-bonne heure au printems ; &, fi on les enleve après qu'ils ont commencé à produire des rejettons, cette opération les fait fouvent périr tout-à-fait.

Quelques perfonnes ont, depuis peu, rejeté ces arbres de leurs jardins, parce que leur eau diftillée ayant été funefte à quelques animaux à qui on en a fait prendre, elles les ont regardées comme vénéneufes. Cependant des expériences réitérées très fouvent, ont démontré que leurs feuilles & leurs fruits n'ont aucune qualité nuifible ; de forte qu'on ne peut attribuer l'effet funefte de leur eau diftillée, qu'à l'huile qui peut être emportée dans la diftillation,

PAD

On a long-tems fait ufage des baies qu'on mettoit dans l'eau-de-vie , pour en faire une efpece de ratafia : on employoit auffi leurs feuilles dans les flans , pour leur donner un goût agreable. Cependant , quoiqu'on s'en foit fervi de cette maniere pendant plufieurs années , elles n'ont jamais occafionné le moindre accident : quant aux baies, j'en ai vu manger en grande quantité , fans qu'elles aient fait aucun tort.

On a quelquefois greffé le *Laurier* fur des tiges de *Cerifier*, pour rendre ces arbres plus larges ; mais quoiqu'ils prennent très-bien l'un fur l'autre , cependant ils font rarement beaucoup de progrès , foit que l'on greffe le *Laurier* fur le *Cerifier*, ou le *Cerifier* fur le *Laurier* ; de forte que cette expérience ne peut jamais être qu'un objet de pure curiofité. Je confeille à ceux qui voudront l'effayer , de greffer le *Laurier* fur le *Cerifier* de Cornwall , plutôt que fur aucune autre efpece , parce que la greffe prend mieux fur celle-ci ; & comme ces arbres font grands & réguliers , les greffes deviendront plus fortes.

Lufitanica. La cinquieme , appelée le *Laurier de Portugal*, peut être multipliée de la même maniere que le *Lauro Cerafus* ou *Laurier commun*, par boutures , par marcottes ou par femences. Les boutures prennent fort aifément racine , en les plantant dans la même faifon , & de la même maniere qui a été prefcrite pour le

PAD   357

Lauro-Cerafus ou *Laurier commun*. En marcottant les jeunes branches en automne , elles pouffront des racines dans l'efpace d'une année : on pourra les tenir enfuite un ou deux ans en pépiniere , pour leur faire acquérir de la force , & après , les tranfplanter à demeure.

Quoique ces deux méthodes foient très-promptes pour multiplier ces plantes , cependant je recommanderai toujours de les élever de baies , furtout quand on en veut faire de grands arbres à plein vent , parce que celles qui font multipliées par boutures ou par marcottes , pouffent plus de branches latérales , fe forment en buiffon , & ne s'élevent jamais auffi droits que celles qui viennent de femences. Il y a à préfent plufieurs de ces arbres dans les jardins anglois, qui produifent des baies chaque année. En les mettant à l'abri des oifeaux jufqu'à leur maturité , on peut s'en procurer affez pour élever un grand nombre de plantes , fans être obligé de les multiplier d'une autre manière. On feme ces baies en automne , & on les traite comme celles du *Lauro-Cerafus*.

Cet arbre fe plait dans un fol léger & marneux , ni trop humide ni trop fec , quoiqu'il croiffe dans prefque tous les terreins , même dans les terres fort humides : on le tranfplante dans le même tems que le *Laurier commun*.

Celui-ci eft beaucoup plus dur que le *Lauro-Cerafus* ; car,

358   P Æ O

dans la forte gelée de 1740, beaucoup de *Lauro-Cerasus* ont été entierement détruits, la plupart ont perdu leurs feuilles; & celui-ci a conservé sa verdure, sans souffrir le moindre dommage: ce qui le rend plus estimable; & comme, par l'apparence de quelques arbres qui croissent dans les jardins, ils doivent devenir très-grands, c'est un de nos plus beaux ornemens en arbres verts.

PÆONIA. *Tourn. Inst. R. H.* 273. *tab.* 146. *Lin. Gen. Plant.* 600; ainsi appelée de PEON le Médecin, parce qu'on dit qu'il a guéri Pluton avec cette herbe, quand il a été blessé par Hercule. [*The Peony.*] Pivoine.

Caractères. La fleur a un calice persistant, & formé par cinq feuilles concaves, réfléchies, & inegales dans leur largeur & leur position: la corolle a cinq pétales larges, ronds & concaves, qui s'étendent & s'ouvrent; la fleur a un grand nombre d'étamines courtes, velues & terminées par des antheres larges, oblongues & quarrées, avec deux, trois ou quatre germes ovales, érigés, velus & placés dans le centre; ils n'ont point de styles, mais seulement des stigmats oblongs, comprimés, obtus, & colorés; ces germes se changent ensuite en autant de capsules ovales, oblongues, velues, réfléchies, sans cellule, qui s'ouvrent longitudinalement, & renferment plusieurs semences ovales, luisantes, colorées, & fixées au sillon.

Ce genre de plantes est rangé dans la seconde section de la treizieme classe de LINNÉE, dans laquelle sont comprises celles dont les fleurs ont plusieurs étamines, & deux germes.

Les especes sont:

1°. *Pæonia mascula, foliis lobatis, ex ovato-lanceolatis.* Haller. Helv. 311. *ed. prim.*; Pivoine avec des feuilles à lobes, ovales, & en forme de lance.

Pæonia folio nigricanti, splendido, quæ mas. C. B. P. 323; Pivoine avec des feuilles tirant sur le noir, & luisantes, ou Pivoine mâle.

Pæonia mas. Lob. Ic. 684. *Blackw. f.* 245.

2°. *Pæonia fœmina, foliis difformiter lobatis.* Haller. Helvet. 311; Pivoine avec des feuilles irrégulieres & à lobes.

Pæonia communis, vel fœmina. C. B. P. 323; Pivoine commune ou femelle.

Pæonia fœmina. Fuch. Hist. 202. Lob. Ic. 602. Blackw. f. 65. Matth. 915. Lob. Ic. 682.

3°. *Pæonia peregrina, foliis difformiter lobatis, lobis incisis, petalis florum rotundioribus;* Pivoine avec des feuilles irrégulieres, à lobes découpés, & des pétales plus ronds à la fleur.

Pæonia peregrina, flore saturatè rubente. C. B. P. 324; Pivoine étrangere, à fleur d'un rouge foncé.

4°. *Pæonia hirsuta, foliis lobatis, lobis lanceolatis, integerrimis;* Pivoine avec des feuilles à lobes entiers & en forme de lance.

Pæonia fœmina flore pleno ru-

PÆO

bro , majore. C. B. P. 324 ; Pivoine femelle, produifant de groffes fleurs rouges & doubles.

5°. *Pæonia Tartarica , foliis difformiter lobatis, pubefcentibus ;* Pivoine avec des feuilles irrégulieres à lobes , & couverte de duvet.

6°. *Pæonia Lufitanica , foliis lobatis , lobis ovatis , infernè incanis ;* Pivoine avec des feuilles à lobes , qui font ovales & blancs en-deffus.

Pæonia Lufitanica , flore fimplici odoro. Juff. ; Pivoine de Portugal, à fleurs fimples & odorantes.

Mafcula. La premiere efpece eft la *Pivoine* mâle & commune , qui croît naturellement dans les bois & fur les montagnes de la Suiffe ; fa racine eft compofée de plufieurs nœuds oblongs, femblables à des mammelles de vache , & fufpendus par des filets à la tette principale : fes tiges s'élevent à la hauteur d'environ deux pieds & demi ; elles font garnies de feuilles compofées de plufieurs lobes ovales, dont quelques-uns font découpés en deux ou trois fegmens ; elles font d'un vert luifant endeffus, & blanches en-deffous ; les tiges font terminées par des fleurs groffes & fimples , compofées de cinq ou fix pétales larges , ronds & rouges, qui renferment un grand nombre d'étamines , terminées par des antheres oblongues & jaunes : dans le centre font fitués deux , trois , & quelquefois cinq germes, qui fe joignent enfemble à leur bâfe ; ils font couverts d'un duvet blanc & ve-

PÆO 359

lu, s'étendent enfuite à quelque diftance , & s'ouvrent longitudinalement , en laiffant a découvert des femences rondes , qui font d'abord rouges, enfuite pourpre, & qui deviennent noires, quand elles font parfaitement mûres : fes fleurs paroiffent dans le mois de Mai , & fes femences mûriffent en automne.

Il y a une variété de cette efpece à fleurs pâles , une autre à fleurs blanches , & une troifieme dont les feuilles ont de larges lobes : mais comme toutes ces variétés ne font qu'accidentelles , je ne les ai point mifes au nombre des efpeces (1).

Fæmina. La feconde, qu'on appelle *Pivoine* femelle, a des racines compofées de plufieurs

---

(1) On emploie en Médecine les racines & les fleurs de Pivoine ; les racines n'ont qu'une odeur foible , leur faveur eft mucilagineufe & un peu aftringente ; elles ne contiennent aucun principe volatil & vaporeux qui puiffe agir fur les nerfs ; on y découvre feulement une fubftance fine , terreufe , & un principe réfineux & gommeux , foluble , qui ne peut opérer avec beaucoup d'activité.

Cette racine fortifie & refferre médiocrement ; elle peut par conféquent convenir dans quelques circonftances, telles que les écoulemens féreux , la diabete , les fleurs blanches, la gonorrhée fimple , quelques efpeces d'hémorrhagies , &c. ; mais quant à fa vertu anti fpafmodique & anti-épileptique, elle eft au moins fort douteufe , & on ne doit pas trop s'y fier.

Cette racine entre dans la compofition de la poudre de Guttete.

nœuds ronds , épais ou tubé-
reux , & fuspendus l'un fur
l'autre par des cordons ; fes
tiges font vertes , & s'élevent
à-peu-près à la même hauteur
que celles de la précédente ;
elles font garnies de feuilles
compofées de plufieurs lobes
inégaux , & différemment découpés en plufieurs fegmens ;
ces lobes font d'un vert plus
pâle que ceux de la premiere
efpece , & velus en deffous :
les fleurs font plus petites , &
d'une couleur de pourpre plus
foncée. Cette plante fleurit en
même tems que la précédente.

Il y a plufieurs variétés de
celle-ci à fleurs doubles , que
l'on cultive dans les jardins
anglois : elles different des autres fleurs par leur groffeur &
leur couleur ; mais on croit
qu'elles on été accidentelle-
ment obtenues de femences.

Peregrina. La troifieme efpece
eft originaire du Levant ; fes
racines font compofées de
nœuds ronds comme celles de
la feconde ; fes feuilles font
auffi femblables , mais d'une
fubftance plus épaiffe ; fes tiges ne s'élevent pas fi haut,
& fes fleurs ont un grand nombre de pétales. Cette efpece
fleurit un peu après les autres. Je crois que la *Pivoine*
large , double , & pourpre , eft
une variété de celle-ci.

Hirfuta. La quatrieme a des
racines femblables à celles de
la feconde ; fes tiges font plus
hautes , de couleur de pourpre ; fes feuilles , beaucoup
plus longues ; leurs lobes
font en forme de lance , &
entiers : fes fleurs font larges ,

& d'un rouge foncé. Cette efpece fleurit en même tems que
les deux premieres.

Tartarica. Les femences de la
cinquieme ont été apportées
du Levant ; elles ont donné
des plantes , dont plufieurs produifent des fleurs fimples , &
d'autres des fleurs doubles de
la même forme , groffeur &
couleur ; fes racines font compofées de nœuds oblongs , char-
nus , de couleur pâle , & fufpendus par des cordons , comme les autres efpeces ; fes tiges s'élevent à la hauteur d'environ deux pieds ; elles font
d'un vert pâle , & garnies de
feuilles compofées de plufieurs
lobes irréguliers dans leur forme & leur grandeur ; quelques-
unes n'ont que fix lobes , d'autres en ont huit ou dix en forme de lance : quelques-uns de
ces lobes font découpés en
deux fegmens , & les autres
en trois ; ils font d'un vert pâle , & couverts de duvet en-
deffus ; les tiges font terminées par des fleurs d'un rouge brillant , un peu plus petites
que celles de la *Pivoine* femelle , avec moins de pétales ;
elles ont un grand nombre d'é-
tamines , avec deux germes, &
quelquefois trois , comme celles
de la *Pivoine* femelle , mais plus
courts & plus blancs. Cette
efpece fleurit un peu plus tard
que la *Pivoine* commune.

Lufitanica. Les femences de
la fixieme m'ont été en-
voyées au jardin de Chelféa
par M. DE JUSSIEU , qui le
avoit apportées de Portugal ,
où cette plante croît fans culture. Sa racine n'eft pas com-

posée de nœuds ronds ; mais elle en a deux ou trois longs & cylindriques, avec des griffes fourchues comme des doigts ; sa tige s'éleve au deffus de la hauteur d'un pied ; elle est garnie de feuilles composées de trois ou quatre lobes ovales, pâles en-deffus, & blancs en-deffous ; la tige est terminée par une fleur simple, d'un rouge brillant, plus petite qu'aucune des précédentes, & d'une odeur agréable. Cette plante fleurit vers le même tems que la *Pivoine* commune.

Culture. La premiere de ces especes est principalement multipliée par ses racines, qui sont d'usage en Médecine ; & ses fleurs, étant simples, ne sont pas si agréables que les doubles, & ne conservent pas si long-tems leur beauté.

Toutes celles à fleurs doubles sont cultivées pour leur beauté, qui augmente la variété, quand elles sont entremêlées avec les autres plantes de pareille groffeur dans les platts bandes des jardins. Ces fleurs font un très-bel ornement dans les appartemens, lorsqu'on les arrange dans des urnes ou des pots.

Elles font toutes extrèmement dures, & croiffent dans presque tous les fols & à toutes les expositions ; ce qui les fait rechercher ; elles profitent encore à l'ombre des arbres, où elles conservent beaucoup plus long-tems leur beauté.

On les multiplie en divisant leurs racines, qui s'étendent confidérablement. La meilleure faison pour les tranfplanter,

est vers la fin d'Août ou au commencement de Septembre ; car si l'on attend pour cela qu'elles aient pouffé de nouvelles fibres, elles fleuriffent très-rarement dans l'été fuivant.

En divifant ces racines, il faut toujours obferver de conferver le jet fur la couronne de chaque rejetton, autrement elles tombent à rien : on ne doit pas non plus féparer ces racines en trop petit volume, fur-tout fi l'on veut avoir des fleurs tout-de-fuite ; car lorfque leurs rejettons font foibles, elles ne fleuriffent point dans la faifon fuivante, ou au moins elles ne produifent qu'une fleur fur chacune ; mais quand on veut les multiplier en quantité, on peut les divifer auffi petites qu'il eft poffible, pourvu qu'il y ait un jet à chaque nœud : alors on les plante dans une planche en pépiniere, pour une faifon ou deux, afin qu'elles puiffent acquérir de la force avant d'être mifes dans le jardin à fleurs.

Les efpeces fimples peuvent être multipliées au moyen des graines qu'elles produifent généralement en grande quantité, quand on laiffe les fleurs fans les couper : on les feme en automne, auffi-tôt qu'elles font mûres, fur une planche de terre fraiche & légere, & on les recouvre d'environ un demi-pouce d'épaiffeur de la même terre légere. Les plantes paroîtront au printems fuivant ; alors on les nettoie exactement, & on les arrofe dans les tems fort fecs, pour hâter leur accroiffement. Elles

doivent refter deux ans dans le femis, avant d'être tranfplantées; mais en automne, lorfque leurs feuilles font flétries, on y répand de la terre fraiche & riche, jufqu'à l'épaiffeur d'un pouce, & on les tient conftamment nettes de mauvaifes herbes. On les enleve au mois de Septembre; alors on prépare quelques planches de terre fraiche & légere, qu'on laboure & qu'on débarraffe de toutes racines & herbes nuifibles. On y plante les racines à fix pouces de diftance entr'elles, & à trois pouces environ de profondeur. On les laiffera ainfi jufqu'à ce qu'elles fleuriffent, mais on les placera enfuite à demeure. Il eft très-poffible qu'en fuivant cette méthode, on obtienne quelques nouvelles variétés, comme il arrive communément pour la plupart des autres plantes qu'on éleve de femences. Celles qui produifent de belles fleurs, peuvent être placées dans le jardin à fleurs; mais celles dont les fleurs font fimples ou de mauvaife couleur, doivent être mifes à part dans des planches, où elles fe multiplieront pour l'ufage de la Médecine.

La *Pionne* ou *Pivoine* de Portugal fe multiplie auffi par femences, ou en divifant fes racines, comme on le pratique pour les autres efpeces; mais elle exige un fol plus léger, & une fituation plus chaude: fes fleurs font fimples, mais d'une odeur fort agréable; ce qui doit lui faire donner une place dans tous les jardins.

PAIN.DE POURCEAU. *V.* Cyclamen.

PAIN DE SINGE. *Voyez* Adansonia & Crescentia.

PALETUVIER DE MONTAGNE. *Voy.* Clusia venosa. *L.*

PALIURE. *Voyez* Paliurus.

PALIURUS. *Tourn. Inft. R. H.* 616. *tab.* 387. *Rhamnus. Lin. Gen. Plant.* 235. [*Chrift's Thorn.*] Paliure, Epine de Chrift, Porte-chapeau.

Caractetes. La fleur n'a point de calice, la corolle eft compofée de cinq pétales rangés circulairement, & terminés en pointe aiguë; elle a cinq étamines inférées dans des écailles au-deffous des pétales, & terminées par de petites antheres; fon germe eft rond, divifé en trois parties, & foutient trois ftyles courts, couronnés par des ftigmats obtus. Ce germe fe change dans la fuite en une noix de la forme d'un bouclier, & divifée en trois cellules, qui contiennent chacune une femence.

Ce genre de plantes a été joint par Linnée au *Rhamnus,* qui eft rangé dans la premiere fection de la cinquieme claffe, laquelle comprend celles dont les fleurs ont cinq étamines & un ftyle: mais comme les fleurs de ce genre-ci ont trois ftyles, il devroit être placé dans la troifieme fection.

Nous ne connoiffons qu'une efpece de ce genre.

Paliurus, Spina Chrifti. Dod. Pempt. 848; Epine de Chrift.

Rhamnus aculeis geminatis, inferiore reflexo, floribus trigynis. Linn. Hort. Cliff. 69. *Hort. Ups.*

PAL

47. *Roy. Lugd.-B.* 224. *Sauv. Monsp.* 306 ; Nerprun épineux, avec de doubles épines, dont les inférieures sont réfléchies, produisant des fleurs à trois germes.

Rhamnus, sivè Paliurus folio Jujubino. Bauh. Hist. 1. *p.* 35.

Cette plante croît naturellement dans les haies de la Palestine ; elle s'éleve avec une tige d'arbriffeau flexible, à la hauteur de huit ou dix pieds, & pouffe plufieurs branches, foibles, minces, & garnies de feuilles ovales, alternes, poftées fur des pétioles d'un pouce environ de longueur ; elles ont trois veines longitudinales, & font d'un vert pâle : fes fleurs naiffent en paquets aux aiffelles de la tige, &. dans prefque toute la longueur des jeunes branches ; elles font d'un jaune verdâtre, paroiffent en Juin, & produifent des capfules larges, rondes, en forme de bouclier, avec des bords comme le contour d'un chapeau : leurs pédoncules font fixés au milieu ; elles renferment trois cellules, qui contiennent chacune une femence.

On croit que c'eft avec cette plante qu'à été faite la couronne d'épines mife fur la tête de notre Sauveur.

Plufieurs Voyageurs, dignes de foi, affûrent que cet arbriffeau eft un des plus communs dans la Judée, & la flexibilité de fes branches fuffit pour donner de la probabilité à cette affertion.

Cet arbriffeau croît fauvage dans plufieurs parties du Levant, en Italie, en Efpagne, en portugal, & dans la France Méridionale, fur-tout près de Montpellier, d'où l'on peut s'en procurer les graines, parce qu'elles ne mûriffent point en Angleterre. Il faut les femer le plutôt qu'il eft poffible, lorfqu'on les reçoit, fur une planche de terre légere, où les plantes poufferont au printems fuivant ; mais fi on les conferve jufqu'à ce tems fans les mettre en terre, elles ne germent que dans l'année fuivante, & fouvent même elles manquent tout-à-fait. Ces plantes de femences peuvent être transplantées la faifon fuivante dans une pépiniere, pour qu'elles puiffent acquérir de la force, avant d'être placées à demeure.

On les multiplie auffi en marcottant leurs branches au printems. Ces marcottes prendront racine dans l'efpace d'une année, fi l'on a foin de les arrofer dans les tems fecs : on peut enfuite les féparer des vieilles plantes, & les placer à demeure.

Le meilleur tems pour transplanter cet arbriffeau eft l'automne, dès que fes feuilles font tombées, ou au commencement d'Avril, précifément avant qu'elles commencent à pouffer : on répand du terreau autour de fes racines, pour les empêcher de fecher, & on l'arrofe légerement de tems en tems, jufqu'à ce qu'il ait pouffé de nouvelles fibres ; mais après cela il n'exige plus aucun foin. Cette plante eft fort dure, & s'éleve à la hauteur de

dix ou douze pieds, si elle est plantée dans un sol chaud & à une exposition favorable. Elle a peu de beauté; mais on la cultive dans les jardins, par curiosité.

PALMA. *Plum. Gen.* 1. *Raii Meth. Plant.* 135. [*The Palm-tree.*] Palmier.

Caractères. Quelques especes de ce genre ont des fleurs mâles & des fleurs femelles sur le même arbre; & dans d'autres, sur des pieds différens. Le calice des fleurs mâles est divisé en trois parties; les corolles de quelques especes ont trois petales, & les fleurs six étamines, terminées par des antheres oblongues, avec un germe usé, qui soutient trois styles courts, stériles, & couronnés par des stigmats aigus: les fleurs femelles ont une enveloppe commune, mais sans calice, six petales courts, & un germe ovale, avec un style en forme d'alène, couronné par un stigmat divisé en trois parties. Ce germe se change dans la suite en un fruit qui varie, pour la forme & la grosseur, dans les différentes especes.

M. RAY range ce genre avec les arbres & arbrisseaux qui ont des fleurs mâles placées à une grande distance du fruit, quelquefois sur le même arbre, & d'autres fois sur des tiges différentes. LINNÉE a séparé les especes sous les genres suivans: les *Chamærops*, *Borassus*, *Corypha*, *Cocos*, *Phænix*, *Areca*, & *Elate*. Il les a ainsi rangés dans son Appendix.

Les especes sont:

1°. *Palma Dactyli-fera, frondibus pinnatis, foliosis angustioribus, aculeis terminalibus;* Palmier à feuilles aîlées, dont les lobes sont étroits, & terminés par des épines.

Phænix Dactyli-fera. Lin. Syst. Plant. tom. 4. p. 634. Mat. Med. 232. Kniph. cent. 2. n. 55. sem.

Palma major. C. B. P. 566; le plus grand Palmier ou le Dattier.

Phænix frondibus pinnatis, foliolis alternis, ensi-formibus, base complicatis, stipitibus compressis, dorso rotundatis. Hort. Cliff. 482. *Hort. Ups.* 306. *Fl. Zeyl.* 390. *Roy. Lugd.-B.* 5.

Dactylis palma. Blackw. f. 202.

Palma dactyli-fera major vulgaris. Sloan. Jam. 174.

Palma. Bauh. Hist. 1. *p.* 351.

Dod. Pempt. Raii. Hist. 1352.

Palma hortensis mas. Kæmpf. Amæn. 688. *tom.* 4. 2. *f.* 1. 2.

Palma hortensis fæmina. Kæmpf. Exot. 668. 686. *tom.* 1. 2. *f.* 2. 16. 17.

2°. *Palma cocos, frondibus pinnatis, foliolis replicatis, spadicibus alaribus, fructu maximo anguloso;* Palmier à feuilles aîlées, dont les lobes se plient en arriere, avec des périoles qui sortent des côtés des branches, & un fruit gros & angulaire.

Palma Indica, cocci-fera, angulosa. C. B. p. 502; Palmier des Indes, à fruit angulaire, ordinairement appelé *Noix de Coco.* [Cet arbre est le même qui a été décrit au mot COCOS.]

Cocus frondibus pinnatis, foliolis ensi-formibus, margine villosis. Hort.Cliff. 483. *Flor.Zeyl.* 391.

Palma

Palma Indica nuci-fera. Bauh. Hift. 1. p. 375.

Calappa. Rumph. 1. p. 1. t. 1. 2.

Tenga. Rheed. Mal. 1. p. 1. t. 1. 2. 3. 4.

Cocos nuci-fera. Lin. Syft. plant. tom. 4. p. 633. Jacq. Amer. 277. f. 169.

3°. Palma spinosa, frondibus pinnatis, ubiquè aculeatis, aculeis, nigricantibus, fructu majore; Palmier à feuilles ailées, & armées par—tout d'épines noires, avec un gros fruit.

Palma caudice aculeatissimo, pinnis ad margines spinosis, fructibus majusculis. Brown. Jam. 343.

Palma tota spinosa major, fructu Pruni-formi. Sloan. Cat. Jam. 1. p. 120; le plus grand Palmier, entierement épineux, avec un fruit en forme de prune, ordinairement appelé grand arbre de Macaw.

Elais Guineensis. Jacq. Amer. 280. t. 172. Lin. Syft. Plant. tòm. 4. p. 635.

4°. Palma altissima, frondibus pinnatis, caudice æquali, fructu minore; Palmier à feuilles ailées, avec un tronc égal, & un plus petit fruit.

Palma altissima, non spinosa, fructu pruni-formi minore, sparso. Sloan. Cat. Jam. 176; le plus grand Palmier sans épines, produisant un plus petit fruit en forme de Prune, croissant épars en paquets longs, ordinairement appelé Arbre à Chou, ou Chou Palmiste.

Areca oleracea. Jacq. Amer. 278. f. 170. Lin. Syft. Plant. tom. 4. p 636.

5°. Palma Gracilis, frondibus pinnatis, caudice tereti acu-

Tome V.

leato, fructu minore ; Palmier à feuilles ailées, avec une tige épineuse & cylindrique, & avec un plus petit fruit.

Bactris minor, fructibus subrotundis. Jacq. Amer. 279. t. 171. f. 1.

Palma spinosa minor, caudice gracili, fructu pruni-formi minimo, rubro. Sloan. Cat. Jam. 178 ; le plus petit Palmier épineux, avec une tige mince, & un plus petit fruit rouge en forme de prune, appelé Pole épineux.

6°. Palma oleosa, frondibus pinnatis, foliolis linearibus, planis, stipitibus spinosis; Palmier à feuilles ailées, ayant des lobes étroits & unis, avec les côtes du milieu piquantes.

Cocos Guineensis. Lin. Syft. Plant. tom. 4. p. 634.

Palma foliorum pediculis spinosis, fructu pruni-formi luteo, oleoso. Sloan. Cat. Jam. 175 ; Palmier avec des pédicules aux feuilles, & un fruit jaune en forme de prune, & huileux, ordinairement appelé Palmier huileux.

7° Palma pruni-fera, frondibus pinnato-palmatis, plicatis, caudice squamato ; Palmier avec des feuilles ailées, en forme de main, plissées, & une tige écailleuse.

Palma Brasiliensis pruni-fera, folio plicatili, seu flabelli-formi, caudice squammato. Raii Hift. 1368; Palmier du Brésil, produisant un fruit semblable à une prune, avec des feuilles plissées en éventail, & une tige écailleuse, appelée Palmetto ou Thatch.

8°. Palma Polypodi-folia, fron

A a

PAL

dibus pinnatis, foliolis lineari-lanceolatis, petiolis spinosis. Hort. Cliff. 482 ; Palmier à feuilles ailées, dont les lobes sont linéaires & en forme de lance, avec des pétioles piquans.

Palma Japonica, spinosis pediculis, Polypodii folio. Boërrh. Ind. Alt. 2. 170 ; Palmier du Japon, avec des pétioles épineux, & une feuille de Polypode, ou l'Arbre du Sagou.

9°. Palma pumila, fructu clavato, polypyreno ; Palmier avec un fruit en forme de massue, qui renferme plusieurs semences.

Palma Americana, foliis Polygonati brevioribus, læviter serratis, & non nihil spinosis, trunco crasso. Pluk, Phyt. tab. 103. fig. 2. & tab. 309. fig. 5 ; Palmier d'Amérique, avec des feuilles plus courtes du Sceau de Salomon, légerement sciées, un peu épineuses, & un gros tronc.

Zamia pumila, frondibus pinnatis, foliolis sub oppositis, lanceolatis. Lin. Syst. Veg. 778. ed. 13.

Palma Americana, crassis rigidisque foliis. Herm. Par. 210. f. 210.

Palmi folia fœmina. Trew. Ehret. 5. f. 26.

10°. Palma Americana, frondibus pinnatis, foliolis lanceolatis, plicatis, geminatis, sparsis ; Palmier à feuilles ailées, dont le lobes sont en forme de lance, plissés, sortant par paires de la même pointe, & épars dans la longueur de la côte du milieu.

Palma altissima non spinosa, fructu oblongo. Houst. MSS. ;

PAL

le plus haut Palmier, sans épines, à fruit oblong.

11°. Palma Draco, foliis simplicibus, ensi-formibus, integerrimis, flaccidis ; Palmier à feuilles simples en forme d'épée, entieres & flasques.

Palma Prunifera foliis Yuccæ, fructu in racemis congestis cerasiformi, duro, cinereo, pisi magnitudine, cujus lachryma Sanguis Draconis est dicta. Comm. Cat. Amst. ; Palmier qui produit des Prunes, avec des feuilles semblables à celles de l'Yucca, des fruits rapprochés en longs paquets, de la forme des Cerises, durs, de couleur de cendre, & de la grosseur d'un Pois, dont les larmes sont connues sous le nom de sang de Dragon ; ordinairement appelé Arbre de Dragon.

Dracœna Draco. Syst. Veg. 275. Asparagus Draco. Lin. Sp. Plant. 451. Cordyline. Roy. Lugd.-B. 22.

Dactyli-fera. La premiere espece est l'arbre qui produit des Dattes ; il croît en abondance dans l'Afrique & dans quelques contrées de l'Orient, d'où l'on apporte son fruit en Angleterre : il s'éleve à une très-grande hauteur dans les pays chauds ; ses tiges sont généralement couvertes de nœuds rudes & raboteux, qui ne sont autres que les cicatrices que les feuilles ont laissées en tombant, parce que les troncs de ces arbres ne sont pas solides comme ceux des autres ; mais que leur centre est rempli d'une moëlle gluante, autour de laquelle il y a une écorce coriace de

PAL

fortes fibres, lorfqu'ils font jeunes : mais quand ils deviennent vieux, cette écorce fe durcit, & devient ligneufe ; les feuilles font attachées très-près du tronc : elles fortent du centre, fort repliées & applanies ; mais lorfqu'elles fe font élevées au-deffus de l'enveloppe qui les entoure, elles s'étendent fort loin tout au tour de la tige ; les anciennes feuilles déclinent & périffent : alors la tige augmente d'autant dans fa hauteur. Lorfque les feuilles de ces arbres croiffent affez pour fupporter le fruit, elles font de fix ou huit pieds de longueur, étroites, longues, terminées en pointe, & placées alternativement ; les lobes, vers la bâfe, ont trois pieds de longueur fur un peu plus d'un pouce de largeur ; ils font fort rapprochés, d'abord liés & garottés par des fibres brunes, qui tombent à mefure que les feuilles fe développent & s'étendent : elles ne s'ouvrent jamais tout-à-fait ; mais elles font creufées en nacelle avec une côte tranchante fur le dos ; elles font très-fermes dans leur naiffance, d'un beau vert, & terminées par une épine noire. Ces arbres ont des fleurs mâles fur des pieds différens de ceux qui produifent le fruit, & il eft néceffaire que quelques arbres mâles croiffent dans le voifinage des plantes femelles, pour les rendre fécondes & fructueufes, en impregnant l'ovaire de la pouffiere fécondante des étamines, fans laquelle les noyaux qui font renfermés dans

PAL

le fruit ne feroient point fufceptibles de germer. La plupart des anciens Auteurs qui ont parlé de ces arbres, affûrent que les femelles de *Palmiers* portant fruit, ont befoin de l'affiftance du mâle, pour être fertiles. S'il ne s'en trouve point dans le voifinage, les habitans vont couper des grappes de fleurs mâles, lorfqu'elles font épanouies, & les placent fur les plantes femelles, près des fleurs à fruits, pour les féconder, fans quoi elles feroient ftériles, & ne produiroient point. Le Pere LABAT, dans fa defcription de l'Amérique, fait mention d'un arbre de cette efpece, placé près d'un couvent dans l'Ifle de la Martinique ; il donnoit une grande quantité de fruits qui parvenoient à une maturité fuffifante pour les manger ; mais comme il n'y avoit point d'autres arbres de cette efpece dans l'Ifle, les habitans, défirant de le multiplier, planterent un grand nombre de noyaux pendant plufieurs années, mais fans fuccès ; car aucuns ne pousferent : & après plufieurs épreuves inutiles, ils furent obligés d'envoyer en Afrique, où ces plantes font en abondance, pour avoir quelques-uns de ces fruits, dont les noyaux leur procurerent plufieurs arbres ; de-là ils imaginerent que le premier ne pouvoit probablement avoir produit des fruits que par le voifinage & à l'aide de quelqu'autre efpece d'arbres qui l'avoient rendu capable de les mûrir, fans cependant avoir

A a 2

368 PAL

pu rendre les germes prolifiques, comme on le remarque
dans le produit des animaux
de différentes especes.

Les fleurs de ces deux genres sortent en grosses grappes
de la tige entre les feuilles ;
elles sont recouvertes d'une
spathe ou voile qui s'ouvre &
se fanne ; celles du mâle ont
six courtes étamines, avec quatre antheres quarrées ou à
quatre faces, & couvertes de
poussiere ; la fleur femelle n'est
garnie que d'un germe ou pistile rougeâtre, qui se change
ensuite en un fruit petit, ovale, & couvert d'une pulpe
épaisse, qui contient un noyau
dur & oblong, dans un côté
duquel il y a un sillon profond
& longitudinal. Les grappes de
ce fruit sont quelquefois fort
grosses.

Cette espece de *Palmier* est
appelée par Linnée, *Phœnix*,
qui est un mot grec, & il en
fait un genre particulier : elle
offre quelques variétés ; mais
ce qui nous empêche de connoître les différences qui distinguent ces arbres des pays
chauds, c'est qu'en Angleterre
nous ne pouvons pas esperer
de les voir dans leur perfection.

Ces fruits, pourvu qu'ils
soient frais, donnent aisément
des plantes : on les seme dans
des pots remplis d'une terre
grasse & légere ; on les plonge
dans une couche de tan de
chaleur modérée, & on les
rafraîchit souvent par des arrosemens.

Lorsque les plantes ont pous-
fé, on les met chacune sépa-

PAL

rément dans de petits pots remplis d'une même terre légere ;
on les replonge dans une couche chaude ; on les arrose,
& on leur donne de l'air à
proportion de la chaleur de la
saison, & de la couche dans
laquelle elles croissent. On les
laisse dans cette couche pendant tout l'été ; & au commencement du mois d'Août,
on leur donne beaucoup d'air,
pour les endurcir, & les accoutumer de bonne heure à
supporter les froids de l'hiver ;
car si on les force trop, elles
deviendront si délicates, qu'on
ne pourra les conserver dans
cette saison sans beaucoup de
difficulté, sur-tout si l'on n'a
point de serre chaude où l'on
puisse les renfermer. Au commencement d'Octobre, il faut
les retirer de la couche, & les
placer dans la serre, de façon
qu'elles y jouissent d'une chaleur modérée. Ces plantes sont
délicates pendant leur jeunesse ;
& quoiqu'on puisse les conserver dans une situation moins
chaude, cependant elles en
font fort retardées, & ont
peine à reprendre leur force
l'été suivant. Il ne faut pas se
donner la peine de les élever
de semences, quand on n'a
point de serre chaude pour
avancer leur accroissement ;
car, sans ce secours, elles ne
parviendront pas à une certaine grandeur pendant vingt
ans.

Quand on change ces plan-
tes de pots, ce qu'il faut faire
une fois par année, on doit
avoir grand soin de ne pas
couper ni blesser leurs racines ;

mais on enleve toutes les petites fibres, qui, si on les laissoit, se pourriroient tôt ou tard, empêcheroient les nouvelles de pousser, & retarderoient beaucoup par-là l'accroissement des plantes.

La terre propre à ces arbres doit être composée de la maniere suivante ; savoir, de terre fraîche & légere, prise dans une prairie, pour moitié, & de terre sablonneuse & de fumier pourri, ou d'écorce de tan, par portions égales : on mêle le tout exactement, & on le tient en monceaux pendant trois ou quatre mois au moins, avant de s'en servir. Pendant ce tems, on le retourne souvent pour l'adoucir, & détruire les mauvaises herbes qui peuvent y naître.

Les pots doivent être proportionnés à la grosseur des plantes, sans cependant qu'ils soient trop grands ; car, dans ce cas, ils leur seroient plus nuisibles que s'ils étoient trop petits. En été, elles ont besoin d'être fréquemment arrosées, mais toujours légerement. Pendant l'hiver on ne leur donne que très-peu d'eau, sur-tout si elles ne sont pas placées dans la serre chaude.

Elles croissent très-lentement, même dans leur pays natal, quoiqu'elles s'y élevent à une très-grande hauteur. Plusieurs anciens habitans de ces pays ont observé qu'elles ne croissoient que de deux pieds en dix ans. Ainsi, quand on les apporte en Europe, on ne peut pas espérer de les voir avancer promptement, sur-tout lors-

qu'on n'a pas soin de leur procurer le dégré de chaleur qui leur est nécessaire en hiver. Malgré la lenteur avec laquelle ces arbres poussent dans leur pays natal, cependant nous parvenons en Europe à les avancer, en plongeant les pots qui les contiennent dans les couches de tan, dont il faut renouveler la chaleur autant de fois qu'il est nécessaire, & les y tenir toute l'année, hiver & été, ayant soin de les changer de pots à mesure qu'ils avancent dans leur accroissement, ainsi que de les arroser à propos. Avec ce traitement, j'en ai eu plusieurs qui ont crû fort promptement. J'ai observé que leurs racines sont fort sujettes à pousser dans le tan, si l'on y laisse les pots pendant un tems considérable, sans les changer, & quand elles y rencontrent une chaleur douce : l'humidité qui s'éleve de la fermentation du tan, conserve leurs fibres pleines & vigoureuses : mais quoique leurs feuilles deviennent grandes en peu d'années, en les traitant bien, cependant il se passe beaucoup de tems avant qu'elles produisent des tiges On voit à présent dans les jardins de Chelsea quelques-unes de ces plantes, dont les feuilles ont sept pieds de longueur ; elles ont été élevées de semences il y a plus de vingt ans, & leurs tiges n'ont pas deux pieds de hauteur. Un de ces arbres a produit quelques pétites grappes de fleurs mâles.

Cocos. La seconde espece produit la Noix de *Cocos* qu'on

nous apporte souvent en Angleterre, & dont quelques-uns sont très-gros. Les branches ou feuilles de cet arbre sont aîlées comme celles de la première espece ; mais leurs lobes ou petites feuilles sont trois fois plus larges ; elles s'ouvrent tout-à-fait, & leurs bords se jettent en arriere ; elles sont d'un vert plus clair que celles de la précédente ; la feuille entiere a souvent douze ou quatorze pieds de longueur : les fleurs mâles croissent en différens endroits de l'arbre ; elles sortent du tronc entre les feuilles, disposées en grappes longues, ainsi que les fleurs femelles ; les noix se montrent de même en gros paquets, couverts d'une enveloppe de fibres très-serrées. Ces Noix sont grosses, ovales, & ont leur coque percée de trois ouvertures vers le sommet ; l'amande est ferme, dure, blanche en-dedans, & la coque contient une grande quantité de jus pâle, que l'on nomme *Lait*.

La Noix de *Cocos* est cultivée dans tous les pays habités des deux Indes ; mais on croit qu'elle vient des Maldives & des Isles desertes des Indes Orientales, d'où elle a été transportée dans toutes les contrées chaudes de l'Amérique ; car on ne l'y trouve point dans aucun endroit éloigné des habitations. Cet arbre est un des plus utiles que les habitans de l'Amérique possedent ; car il leur sert à beaucoup d'usage. L'écorce de la Noix est une espece de filasse

dont on fait des cordages ; la coque sert à faire des vâses, l'amande leur fournit une nourriture saine & agréable, & la liqueur laiteuse que la coque contient, forme une boisson rafraîchissante. On emploie encore les feuilles à couvrir les maisons, à faire des paniers, & plusieurs autres ouvrages qu'on envoie en Europe.

On multiplie cet arbre en plantant les Noix, qui germent au bout de six semaines ou deux mois ; quand elles sont fraîches & bien mûres ; mais il est rare d'en trouver de telles parmi celles qu'on apporte en Angleterre, parce qu'on les cueille toujours avant leur maturité, pour les conserver pendant le passage. La meilleure méthode pour envoyer en Europe des Noix propres à être plantées, seroit de les choisir entierement mûres, & de les mettre dans du table bien sec, pour les garantir des insectes : ces fruits germeroient souvent dans la traversée ; ce qui seroit déja un avantage, parce qu'on pourroit les planter tout de suite dans des pots, que l'on plongeroit dans la couche de tan.

Dans les Isles chaudes de l'Amérique, ces plantes font un progrès considérable dans leur accroissement : on en trouve quelques-unes d'une très-grande hauteur ; mais en Europe, où elles produisent beaucoup de fleurs, il leur faut plusieurs années avant de parvenir à une certaine grandeur. Comme leurs jeunes feuilles

PAL

font bien larges, elles font un bel effet parmi les autres plantes tendres & exotiques, au bout de deux ou trois ans. On les conferve dans quelques jardins en Angleterre, pour la variété. Cette efpece doit être placée dans la couche de tan d'une ferre chaude, & traitée comme la précédente, en obfervant, à chaque fois qu'on la tranfplante, de ne pas couper les fortes racines ; ce qui occafionne toujours la deftruction de la plupart des efpeces de *Palmiers*. Comme ces plantes font peu de progrès, quand leurs racines font gênées, lorfque leurs pots en font remplis, on les met dans des caiffes d'une grandeur médiocre, afin de leur donner le moyen de s'étendre : mais ces caiffes doivent être tenues conftamment dans la couche de tan, fans quoi les plantes ne profiteroient point. La méthode d'élever ces plantes, au moyen de leurs Noix, quand on les met en terre avant qu'elles foient germées, étant amplement détaillée à l'article qui traite de la maniere de foigner les femences des plantes exotiques, je prie le Lecteur d'y avoir recours, pour éviter la répétition.

Spinofa. La troifieme efpece eft ordinairement appelée *Macaw-Tree* par les habitans des Ifles Britanniques en Amérique. Cet arbre s'élève à la hauteur de trente ou quarante pieds. Sa tige eft fouvent plus groffe vers le fommet qu'en-bas ; fes branches ou plutôt

PAL 371

fes feuilles font ailées ; les lobes ou petites feuilles font longues & fort larges ; la tige & les feuilles font fortement armées d'épines noires de toute grandeur ; les fleurs mâles & femelles naiffent fur le même arbre, & fortent de la même maniere que la Noix de *Cocos* ; le fruit eft à-peu-près de la groffeur d'une *Pomme* médiocre, & renfermé dans une coque fort dure.

Ce *Palmier* eft très commun dans les Ifles Caraïbes, où les Negres percent ces fruits, pour en extraire une liqueur agréable, qu'ils aiment beaucoup. Le corps de l'arbre produit un bois dur & folide, avec lequel on fait des javelots & des flèches. Quelques-uns le regardent comme une efpece d'ébene ; il croît très-lentement, & exige de la chaleur pendant l'hiver.

Altiffima. La quatrieme, auquel on donne communément le nom de *Cabbage-Tree* ou *Arbre à Chou*, dans les Indes Occidentales, s'élève à une très-grande hauteur dans fon pays natal LIGON, dans fon Hiftoire des Barbades, dit qu'il y avoit alors de ces arbres de deux-cents pieds d'élevation, & qu'on lui affura qu'ils étoient deux-cents ans à croître, pour parvenir à leur hauteur, & être en état de produire des femences. Les tiges de ces arbres font rarement plus groffes que la cuiffe d'un homme ; elles font plus unies que celles de la plupart des autres efpeces ; car les feuilles tombent naturellement.

A a 4

& ne laiſſent que la marque de leur place.

Ces feuilles ou branches ont douze ou quatorze pieds de longueur ; les lobes ont près d'un pied de long ſur un demi-pouce de large, avec pluſieurs ſillons dans leur longueur ; ils ſont terminés par des pointes tendres & aiguës, alternes, & moins fermes que ceux de la premiere eſpece ; les fleurs naiſſent au-deſſous des feuilles en grappes claires, & ſont ſoutenues par des pédoncules ſéparés, de près de quatre pieds de longueur, ſur leſquels elles ſont éparſes ; les fleurs femelles ſont remplacées par un fruit de la groſſeur d'une noiſette, couvert d'une peau jaunâtre, & fixé fortement aux cordons ſur le pied principal de la tige.

Comme les feuilles du centre entourent les boutons qui doivent pouſſer plus exactement que dans les autres eſpeces, on diſtingue celle-ci par le nom de *Cabbage-Tree* ou *Arbre à Chou*, parce que le centre pouſſe avant d'être expoſé à l'air ; ce qui en rend les feuilles blanches, fort tendres, & ſemblables à celles des autres plantes que l'on fait blanchir. Les habitans ont coutume de les couper pour les manger ; ils les marinent, & les envoient en Angleterre ſous le nom de *Choux marinés* ou *Cabbages* : mais quand ces rejettons ſont coupés, les plantes ſe fannent, & périſſent ; ce qui eſt cauſe qu'il y a peu de ces arbres dans les Iſles près des Etabliſſemens,

& ceux qui reſtent, ſervent d'ornement.

Gracilis. La cinquieme eſpece eſt connue ſous le nom de *Pole épineux* à la Jamaïque, où elle croît naturellement ; on la trouve en grand nombre dans les petits bois ou haliers ; ſa tige eſt mince, & n'a guere plus de cinq ou ſix pouces de diametre : mais elle s'éleve à la hauteur de quarante pieds, & eſt fortement armée de longues épines ; ſes feuilles ſont placées circulairement au ſommet, comme dans la plupart des autres eſpeces ; elles ſont ailées ; mais les lobes ſont plus verts, plus courts que ceux des autres, & fort garnis d'épines : ſes fleurs ſortent de la même maniere que celles de la Noix de *Cocos*, ſur des pédoncules longs & branchus ; elles ſont plus groſſes que les plus gros *Pois gris*, applaties au ſommet, & couvertes d'une peau rouge. Les habitans de la Jamaïque font des baguettes pour nettoyer les fuſils avec les tiges de ces arbres. Le bois en eſt fort rude & ſouple ; mais je ne crois pas qu'on faſſe uſage d'aucune autre de ſes parties.

Oleoſa. La ſixieme eſpece eſt appelée, dans les Indes Occidentales, *Palmier huileux*, & par quelques-uns, *Huile des Negres.* Le fruit de cet arbre a été originairement apporté de l'Afrique en Amérique, par les Negres ; il croît en grande quantité ſur la Côte de Guinée & aux iſles de Cap-verd ; mais depuis qu'il a été apporté dans les

PAL

Colonies de l'Amérique, il s'y est fort multiplié, & les Negres en ont grand soin.

Les feuilles de cet arbre sont ailées, les lobes sont longs, étroits, & moins fermes que ceux de la plupart des autres; les pétioles des feuilles sont larges à leur bâse, où ils embrassent la tige, & plus étroits vers l'extrémité; ils sont armés d'épines jaunâtres, fortes, émoussées, & plus larges à leur bâse: ses fleurs sortent au sommet de la tige entre les feuilles; quelques grappes n'ont que des fleurs mâles, & d'autres des femelles, auxquelles succedent des baies ovales, plus grosses que les plus fortes Olives d'Espagne, & de la même forme. Ces fruits croissent en très-grosses grappes; & quand ils sont mûrs, ils deviennent d'une couleur jaunâtre.

Les habitans tirent une huile de ce fruit de la même manière qu'on tire l'huile des Olives; le corps de l'arbre leur fournit une liqueur, qui, dans sa fermentation, a une qualité vineuse qui enivre; avec les feuilles de cet arbre, les Negres font des nattes sur lesquelles ils couchent.

Pruni-fera. La septieme espece est appelée *Palmette* ou *Thatch* par les Habitans de la Jamaïque, où cet arbre croit sur les rochers de *Honey-Comb*, en grande abondance; il s'éleve à la hauteur de dix à douze pieds, avec une tige mince, nue & unie; son sommet est garni de plusieurs feuilles en forme d'éventail, pla-

PAL 373

cées circulairement sur des pétioles de deux ou trois pieds de longueur, armés de quelques épines fortes, vertes & courbées; les lobes se rapprochent tous dans un centre, près du pétiole, & sont joints ensemble jusqu'à la troisieme partie de leur bâse; ils sont d'abord plissés très-près l'un de l'autre, mais ensuite ils s'étendent en forme d'éventail; leur extrémité est pliante, & souvent penchant vers le bas: entre ces feuilles pendent de longs fils; les fleurs & les fruits sortent des ailes des feuilles: le fruit est de la forme & de la grosseur d'une petite *Olive*. Les feuilles de cet arbre servent à couvrir les maisons dans toutes les Indes Occidentales.

Polypodi-folia. La huitieme espece se trouve au Japon & au Malabar, sur des montagnes seches & remplies de rochers. Cet arbre, avec le tems, s'éleve à la hauteur de quarante pieds; sa tige est marquée dans toute sa longueur d'un grand nombre d'empreintes circulaires, occasionnées par la chûte des feuilles, qui sont toujours placées circulairement autour du tronc, & qui, en se détachant, laissent des vestiges dans l'endroit où leur bâse étoit fixée; l'arbre se termine en un cône obtus, & précisément au-dessous sortent des feuilles longues de huit à neuf pieds sur les grands arbres, mais plus petites sur ceux d'une taille médiocre. La plus considérable que j'aie vu ne surpassoit pas deux pieds; la bâse

374 PAL

du pétiole, dont la plus grande partie embraffe la tige, eft large, triangulaire, armée à chaque côté de courtes épines à l'endroit où font les lobes. Ces lobes font longs, étroits, entiers, d'un vert clair fur le haut, placés par paires fur toute la longueur de la côte du milieu, & fort rapprochés les uns des autres : les fleurs & les fruits naiffent en gros paquets aux pétioles des feuilles ; le fruit eft ovale, à peu-près de la groffeur d'une groffe *Prune*, & prefque de la même forme ; la peau en eft d'abord jaune, & devient rouge en mûriffant ; fous fa chair, qui eft d'un goût douçâtre, eft une coque dure & brune, qui renferme une amande blanche, dont la faveur approche beaucoup de celle de la *Chataigne*.

Le *Sagou* fe fait avec la moëlle renfermée dans le tronc : on le pulvérife d'abord ; & après en avoir formé une pâte, on le réduit en grains.

Pumila. La neuvieme efpece, que le feu Docteur HOUSTOUN a découverte dans les fables près de l'ancienne Vera-Cruz en Amérique, a une tige épaiffe, qui s'éleve rarement au-deffus de deux pieds de hauteur : fes feuilles pouffent fur le haut de la tige ; elles ont des pétioles d'un pied & demi de longueur ; elles font ailées, & leurs lobes, qui ont environ cinq pouces de long fur un & demi de large au milieu, font pointus aux deux extrémités, entiers, fermes, unis, fillonnés par quelques petites crene-

lures à leurs pointes, alternes, & d'un vert pâle. Il y a quatorze ou quinze de ces lobes rangés dans la longueur de la côte du milieu. Le fruit pouffe à côté de la tige fur un pédoncule court & épais ; il fe tient érigé, & a la forme d'une maffue : il renferme plufieurs femences rouges, à-peu-près femblables à de gros *Pois*, & placées féparément dans des cellules autour du centre auquel elles font fixées. Ces plantes ont leurs mâles fur des pieds différens de ceux qui produifent le fruit ; car toutes celles qui ont fleuri en Angleterre, font mâles ; elles perdent leurs feuilles annuellement avant la maturité du fruit. La premiere fois que le Docteur HOUSTOUN vit ces plantes croître à la Vera-Cruz, elles étoient en pleines feuilles ; mais trois mois après, à fon retour dans le même endroit, le fruit fe trouva mûr, & toutes les feuilles étoient tombées. Il obferva la même chofe l'année fuivante.

Americana. La dixieme efpece a encore été découverte par le Docteur HOUSTOUN, dans l'Amérique Efpagnole. Cet arbre s'éleve à une grande hauteur avec une tige nue, & garnie au fommet de feuilles longues & ailées, dont les lobes font pliffés, en forme de lance, & d'une texture plus douce que celles d'aucune autre efpece ; ils fortent par paires du même point, & fe tiennent érigés au même côté de la côte du milieu : elles ont deux lobes à chaque côté,

PAL

dont l'un eſt placé un peu au-deſſous de l'autre ; mais il y a un grand eſpace entre cha-que quatre lobes : les fleurs ſortent en paquets ou grappes longues entre les feuilles ; les fleurs mâles pendent, & ſont fixées à un chaton mince ; mais le fruit, qui eſt à-peu-près de la groſſeur d'une Prune mé-diocre, eſt rapproché en grof-ſes grappes.

Draco. La onzieme eſpece ſe trouve dans les Iſles du Cap-Verd, d'où j'en ai reçu une plante, & ſes ſemences m'ont été envoyées de Madere.

On donne à cette plante le nom d'*Arbre de Dragon*, parce que ſa ſève, qui a une odeur d'épices, ſe réduit en une pou-dre rouge, ſemblable au *Sang de Dragon oriental* : on en fait uſage dans les boutiques, au lieu du *Sang de Dragon* ; mais l'arbre duquel on tire le vrai *Sang de Dragon*, eſt d'un genre tout différent.

Le Docteur VAN-ROYEN, dans ſa *Deſcription du Jardin de Leyde*, a mis cette eſpece au nombre des *Yuccas*, à cauſe de la reſſemblance de ces deux genres ; mais le fruit de cet arbre ayant une graine ou baie ſemblable à celles du *Lau-rier*, & les graines de l'*Yucca* croiſſant dans des capſules à trois cellules, ils ne peuvent être du même genre. Comme nous n'avons eu aucune bonne deſcription des caracteres de cette plante, nous ne pouvons en déterminer le genre. LIN-NÉE, d'après des informations faites par LŒFLING, ſon élève, l'a rangée dans celui d'*Aſpara-*

gus, avec lequel elle paroit avoir quelques rapports ; mais comme pluſieurs Auteurs moder-nes l'ont décrite ſous ce titre, j'ai cru devoir faire de même.

Cet arbre s'éleve avec une tige groſſe, unie, & égale dans toute ſa longueur ; l'in-térieur de cette tige eſt rem-pli d'une moëlle gluante, & environnée d'un cercle de for-tes fibres, & l'extérieur eſt tendre & mou ; elle s'éleve à douze ou quatorze pieds de hauteur, & conſerve à-peu près un diametre de huit à dix pou-ces dans toute ſa longueur, avec des marques circulaires, de diſtance en diſtance, aux endroits où ſe trouvoient les feuilles qui embraſſoient de leur bâſe la moitié de ſa cir-conférence. Le ſommet de l'ar-bre eſt garni d'une groſſe touf-fe de feuilles, qui ſortent une à une tout autour ; elles ont la forme de celles de l'*Iris* com-mun ; mais elles ſont beau-coup plus larges, & ſouvent de quatre à cinq pieds de lon-gueur ſur un pouce & demi de largeur à leur bâſe, où elles embraſſent la tige. Cette largeur diminue par dégrés juf-qu'à leur extrémité, qui eſt terminée en pointe. Les feuil-les ſont ſouples, & penchent vers le bas autour de la tige ; elles ſont entieres, d'un vert foncé, unies ſur les deux fa-ces, & reſſemblent beaucoup à celles de l'*Iris* jaune com-mun. Cette plante n'ayant pas montré ſes fleurs en Angle-terre, je ne puis en donner aucune deſcription ; mais, au-tant que je puis en juger,

d'après les baies que j'ai reçues, elle peut être rangée parmi les *Palmiers*.

Culture. Toutes ces efpeces de *Palmiers* fe multiplient par leurs fruits, qu'il faut traiter fuivant la méthode qui a été prefcrite pour la premiere. Les plantes qui en proviennent, exigent auffi la même culture, avec la différence que celles qui viennent des pays chauds, demandent plus de chaleur. Les feconde, troifieme, quatrieme, cinquieme, fixieme, feptieme, huitieme & neuvieme efpeces doivent être renues conftamment dans la couche de tan de la terre chaude, fans quoi elles ne feront pas de grands progrès en Angleterre; mais lorfqu'elles réuffiffent bien, elles deviennent trop grandes pour pouvoir être contenues dans nos ferres : or ne peut pas même efpérer de leur voir produire des fruits dans notre climat. Les curieux ne les confervent qu'à caufe de la fingularité de leur feuillage, qui differe extrèmement de celui de nos arbres Européens, & les rend dignes du foin qu'on leur donne.

Les autres efpeces peuvent être tenues dans des ferres chaudes, feches en hiver, à une température modérée ; & pendant les chaleurs de l'été, on peut les expofer en plein air fous un abri, à une fituation chaude. On les y laiffe pendant l'efpace de trois mois ; mais il faut les remettre dans la ferre chaude avant les froides matinées de l'automne, & les tenir à un dégré de chaleur modéré. On leur donne peu d'eau

pendant l'hiver ; & en été ; lorfqu'elles font expofées en plein air, il ne faut pas les arrofer fouvent, à moins que la faifon ne foit fort feche & chaude ; car trop d'humidité les détruit bientôt. Leur culture eft à-peu-près la même que celle des *Palmiers-Dattiers*. On ne doit jamais couper leurs principales racines, quand on les change de pots ou de caiffes, & ne pas trop les gêner ; mais à mefure que ces plantes s'élevent en hauteur, on les change annuellement, & on leur donne de plus grands pots. La terre dans laquelle on les plante, doit être légere, afin que l'humidité fe diffipe aifément ; car fi elle eft forte & retient l'eau, les tendres fibres des racines fe pourriffent bientôt.

PALME DE CHRIST ou RICIN *Voyez* RICINUS.

PALMETTO ou THALCH. *Voyez* PALMA PRUNI-FERA.

PALMIER. *Voyez* PALMA.

PALMIER NAIN ou PALMETTO *V.* CHAMÆROPS. *L.*

PAMPAYE ou CONCOMBRE D'EGYPTE. *Voyez* LUFFA.

PAMPELMOUSE ou CHADOCK. *Voyez* AURANTIUM DE-CUMANA.

PANACÉE. *V.* HERACLEUM PANACES. *L.*

PANACHÉ, ou *bigarré*, [*Variegated.*] fignifie rayé ou marqué de plufieurs couleurs. On trouve dans les jardins des curieux une grande quantité de plantes panachées de différens genres, dont les feuilles font marquées de jaune & de blanc. Celles qui font marquées de quelques-unes de ces cou-

PAN

leurs au milieu de la feuille, font appelées tachetées; mais celles qui font bordées de ces deux couleurs, font appelées panachées ou rayées.

Les plantes dont les feuilles ne font que tachetées, deviennent ordinairement unies, & perdent ces taches dans un bon terrein, à tout le moins dans le tems de leur croiffance. Ces deux couleurs ne paroiffent que très-peu; mais les feuilles panachées ou rayées deviennent rarement unies, fur-tout fi les raies font larges, & qu'elles pénetrent les deux furfaces, quoiqu'elles ne paroiffent pas autant dans la faifon où elles croiffent, que dans les autres tems de l'année.

Toutes les différentes efpeces de plantes panachées ne font que des produits accidentels, & font l'effet d'une maladie qu'on entretient, autant qu'on le veut, en appauvriffant le terrein dans lequel elles croiffent: on rend ainfi leurs panaches plus beaux & plus durables.

Cependant je n'ai jamais vu qu'on puiffe faire panacher artificiellement des plantes unies, quoique certains Jardiniers prétendent en avoir le fecret. Il eft vrai qu'on peut rendre panachés les arbres & arbriffeaux ligneux, en y inférant, par la fente ou l'écuffon, un bouton d'une plante panachée. Quoique, dans ce cas, la greffe ne pouffe point, cependant, fi elle conferve fa fraîcheur pendant huit ou dix jours, elle communique fa maladie à la fève de l'arbre fur lequel elle eft appliquée; & en peu de

PAN 377

tems, on voit paroître des taches fur les feuilles les plus voifines de la greffe, & enfuite fur toutes les autres parties de l'arbre: mais dans les plantes herbacées fur lefquelles on ne peut pas pofer des greffes, on ne peut point opérer ce changement.

Cette maladie fe communique fouvent aux femences dans quelques efpeces de plantes; de forte que celles qu'on recueille fur des plantes panachées, produifent ordinairement plufieurs plantes femblables, comme on peut l'obferver dans les *Pois ailes*, *panachés*, le *grand Erable*, &c. On emploie ce moyen pour fe procurer des plantes panachées.

Il eft bien certain que ces panaches ne font occafionnés que par la foibleffe des plantes; car on remarque toujours que les plantes panachées ne produifent pas des feuilles auffi grandes qu'auparavant, & qu'elles font moins capables de fupporter le froid; ce qui eft caufe que plufieurs efpeces de plantes, affez dures par elles-mêmes pour croître en plein air dans notre climat, exigent d'être abritées en hiver, dès qu'elles font devenues panachés, & durent rarement auffi long-tems. Il eft d'autant plus évident que cette variété n'eft qu'une maladie, que toutes les fois que ces plantes deviennent plus vigoureufes, le panache eft moins vifible, & difparoît quelquefois entierement, furtout, comme nous l'avons déja obfervé, quand les feuilles ne font que tachetées, & auffi quand elles font bordées de

jaune, parce que cette couleur ne dure pas tant que le blanc, qui eft le plus eftimé dans les panaches, & qui, lorfqu'il eft bien établi, eft prefque ineffaçable; de plus, le venin de cette matiere morbifique, ne tache pas feulement les feuilles, mais auffi l'écorce & les fruits des arbres qui en font infectés, comme on peut le voir dans l'*Oranger*, le *Poirier*, &c., dont l'écorce & les fruits font panachés de la même maniere que les feuilles.

Les différentes couleurs qui paroiffent dans les fleurs, viennent auffi de la même caufe, quoique ce foit dans un moindre dégré que les feuilles & les branches qn font infectées; car les différentes couleurs que nous obfervons dans les mêmes fleurs, font occafionnées par la féparation des fucs nourriciers des plantes, ou par quelque changement arrivé dans leurs parties, au moyen duquel des corpufcules de forme différente, étant conduits jufqu'à la furface des pétales, réfléchiffent les rayons de lumiere en différentes proportions.

Pour rendre cette théorie plus intelligible, il eft à propos de dire quelque chofe fur les phénomenes des couleurs, d'après les principes établis par le grand NEWTON.

1°. On peut confidérer la couleur de deux manieres; premierement, comme une qualité qui réfide dans un corps, qu'on dit être coloré d'une certaine façon, ou qui modifie la lumiere de telle ou telle maniere; fecondement, plutôt comme la lumiere elle-même, qui, étant ainfi modifiée, frappe l'organe de la vue, & produit la fenfation que nous appelons couleur.

2°. La *Couleur* eft définie, une propriété effentielle à la nature de la lumiere, dont les parties réfléchies excitent, fuivant leur grandeur ou leur force, des vibrations différentes dans les fibres du nerf optique, & produifent dans le cerveau telle ou telle fenfation, d'où réfulte l'idée des couleurs.

3°. On peut définir encore la Couleur, une fenfation de l'âme, occafionnée par l'action de la lumiere fur la rétine de l'œil, laquelle fenfation eft différente, fuivant que la lumiere differe dans le dégré de fa réfraction, & dans la grandeur de fes parties conftituantes.

4°. Selon la premiere définition, la lumiere eft le fujet de la couleur; fuivant la derniere, elle en eft l'agent.

5°. Ainfi, le mot *Lumiere* fignifie quelquefois cette fenfation excitée dans l'efprit, à la vue de quelque corps lumineux; & quelquefois cette propriété des corps, qui les met en état d'occafionner cette fenfation en nous.

6°. Les auteurs anciens & modernes, & auffi les différentes fectes de philofophes, different dans leurs opinions fur la nature & l'origine des couleurs.

7°. Les *Péripatéticiens* difent, que les couleurs font des qua-

PAN

lités réelles & inhérentes dans les corps lumineux, & ils suppofent que la lumiere ne fait que les découvrir, fans les produire.

8°. PLATON penfoit que la couleur étoit une efpece de flamme, compofée de particules très-déliées, & proportionnées aux pores de nos yeux, & que l'objet nous envoyoit.

9°. Quelques modernes veulent que la couleur foit une efpece de lumiere interne des parties les plus luifantes de l'objet obfcurci, & par conféquent altérée par les différens mélanges des parties les moins lumineufes.

10°. D'autres, avec quelques anciens Anatomiftes, foutiennent que la couleur n'eft pas un courant de matiere lumineufe, mais feulement une émanation des corps.

11. Les autres expliquent toutes les couleurs par les mélanges différens de lumiere & d'obfcurité, & les Chymiftes prétendent qu'elles viennent quelquefois du foufre, quelquefois du fel qui entre dans la compofition des corps, & quelquefois auffi d'un troifieme principe hypoftatique, c'eft-à-dire, du mercure.

12°. L'opinion populaire eft celle d'ARISTOTE, qui prétend que la couleur eft une propriété inhérente aux corps, & que fon exiftence ne dépend pas de la lumiere.

13. Les *Cartéfiens* difent, que la fenfation de la lumiere eft une impulfion faite fur l'œil par quelques globules folides, mais très-petits, qui

PAN

pénetrent aifément les pores de l'air, & que les corps tranfparens tirent leur couleur des différentes proportions du mouvement direct de ces globules à leur mouvement autour de leur centre, par le moyen duquel ils font en état de frapper le nerf optique, fuivant des manieres diftinctes & différentes ; & par-là, produifent la fenfation de différentes couleurs.

14°. Ils avouent que les corps colorés n'étant pas immédiatement appliqués à l'organe, pour occafionner la fenfation de la vue, comme les autres corps qui frappent les fens par un contact immédiat, ils n'excitent pas eux-mêmes la fenfation des couleurs, mais ne la produifent qu'en remuant un milieu intermédiaire qui frappe lui-même l'organe de la vue.

15°. Ils ajoûtent que, comme on trouve que les corps n'affectent pas le fens dans l'obfcurité, la lumiere feule occafionne la fenfation de la couleur, en frappant l'organe, & que les corps colorés ne font que réflechir la lumiere d'une certaine maniere ; la différence de couleurs vient de celle qui exifte dans la texture de leurs parties, laquelle les rend propres à réflechir la lumiere de telle ou telle maniere.

16°. Le Docteur HOOK, dans fa *Micographie*, dit que, le fantôme des couleurs étant caufé par la fenfation d'une pulfion oblique ou inégale de la lumiere, cette fenfation

n'admet que deux variétés, qui proviennent des deux côtés de la pulsion oblique ; de sorte que, dans la realité , il n'y a que deux couleurs simples , le jaune & le bleu , du mélange desquelles , avec une proportion convenable de noir & de blanc , c'est-à-dire , de l'obscurité & de la lumiere , naissent toutes les autres couleurs intermédiaires.

17º. Ce phénomene de la Nature ayant long-tems embarrassé les Philosophes , qui ne savoient comment l'expliquer , l'incomparable NEWTON a enfin trouvé , par des expériences , faites avec un prisme , qu'il y a une grande variété dans les rayons de la lumiere ; & que par là l'origine des couleurs peut être expliquée.

La Doctrine qu'il a établie , d'après ces expériences , est renfermée dans les propositions suivantes :

1º. La lumiere est composée d'une infinité de rayons dirigés en ligne droite , & parallele , qui se réfractent diversement dans des différens milieux qu'ils traversent.

2º. Chaque rayon , suivant son dégré de retrangibilité , quand il est ainsi réfracté , paroît aux yeux de différentes couleurs.

3º. Les rayons les moins réfrangibles paroissent d'une couleur écarlate , les rayons les plus réfrangibles d'un bleu violet ; les nuances intermédiaires varient de l'écarlate au jaune , & du vert brillant au bleu.

4º. Les couleurs qui proviennent de différens dégrès de réfrangibilité de la lumiere , ne sont pas seulement les principales couleurs rouge , jaune , vert & bleu ; mais aussi toutes les couleurs intermédiaires du rouge au jaune , & du jaune au vert.

5º. La blancheur , telle que la lumiere du soleil paroît être , renfermant des rayons de toute espece de réfrangibilité , est conséquemment composée de toutes les couleurs.

6º. Les couleurs simples ou homogènes sont celles qui sont produites par des rayons de lumiere , qui ont le même dégré de réfrangibilité , & les couleurs mêlées procedent des rayons d'une réfrangibilité différente.

7º. Les rayons de la même réfrangibilité produisent les mêmes couleurs , lorsqu'elles ne changent pas par des refractions répétées ; mais elles sont seulement plus fortes ou plus foibles , suivant que les rayons sont plus unis ou plus divisés.

8º. Tous les corps paroissent de telle ou telle couleur , suivant que leur surface n'est propre qu'à réfléchir les rayons d'une telle couleur , ou au moins à les réfléchir en plus grande quantité.

On reconnoît , par des expériences , que les rayons de la lumiere sont composés de particules héterogènes , c'est-à-dire , que quelques-unes sont composées de particules plus épaisses , & d'autres de particules plus petites ; car un rayon

rayon de lumiere tombant fur la furface d'un corps dans un endroit obfcur, n'eft pas entierement réfraélé, mais divifé & féparé en plufieurs petits rayons, dont quelques-uns font réfraélés jufqu'aux extrémités, & les autres aux points intermédiaires ; c'eft-à-dire, que ces particules de lumiere, qui font les plus petites, font détournées plus facilement & plus confidérablement que toutes les autres de leur ligne droite, par l'aélion de la furface réfraélante ; de maniere que plus ces particules de lumiere excedent les autres en grandeur, moins le corps réfraélant a de force pour les détourner de leur ligne de direélion.

Or chaque rayon de lumiere, à proportion qu'il differe d'un autre dans fon dégré de réfrangibilité, en differe auffi en couleur, ainfi que l'expérience le démontre.

Les particules qui font le plus réfraélées, forment la couleur violette; c'eft-à-dire, vraifemblablement, que la plus petite particule de lumiere, ainfi féparément chaffée, excite dans la rétine la plus courte vibration, qui, étant communiquée au cerveau par le fecours du nerf optique, donne la fenfation de la couleur violette, qui eft la plus obfcure, & la plus foible de toutes les couleurs.

De plus, les particules les moins réfraélées forment un petit rayon rouge; c'eft-à-dire, que les plus groffes particules de lumiere, excitant les plus

Tome V.

longues vibrations dans la rétine, produifent la fenfation du rouge, qui eft la plus brillante & la plus vive de toutes les couleurs. On peut remarquer que les plantes en pleine crûe changent de tems en tems de couleur, à proportion que les vaiffeaux des jeunes branches deviennent plus gros : leurs feuilles font d'un jaune pâle, quand elles commencent à naitre, & d'un vert clair, & quelquefois rouge, quand elles font dans leur état moyen ; mais lorfque leurs vaiffeaux font parvenus à toute leur groffeur, elles deviennent d'un vert obfcur, & enfuite, vers l'automne, elles prennent une couleur terne ou de feuille morte, qui provient de la maturité de leur fuc, après quoi elles fe putréfient & fe diffolvent en terre, qui eft leur premier principe.

PANAIS. *Voyez* Pastinaca sativa.

PANAIS ÉPINEUX. *Voyez* Echinophora. L.

PANAIS SAUVAGE. *Voyez* Heracleum. L.

PANAX. *Lin. Gen. Plant.* 1031. *Panacea. Mitch. Gen.* 26 *Araliaftrum. Vaill.* 6. [*Ginfeng, or Ninfeng.*] Ginfeng & Ninzin.

Caraéteres. Ce genre a des fleurs mâles & hermaphrodites fur des plantes différentes ; les fleurs mâles ont des ombelles fimples & globulaires, compofées de plufieurs rayons égaux ; l'enveloppe extérieure a plufieurs petites feuilles en forme de lance ; la corolle eft compofée de cinq pétales étroits, oblongs, émouffés,

B b

382 **PAN**

réfléchis, & poftés fur le calice : la fleur a cinq étamines ob-longues, minces, inférées dans le calice, & terminées par des antheres fimples; les ombelles hermaphrodites font fimples, égales, & fort rap-prochées; l'enveloppe eft pe-tite, perfiftante, & compofée de plufieurs feuilles en forme d'alène; le calice eft petit & perfiftant; les corolles ont cinq pétales oblongs, égaux & re-courbés : les fleurs ont cinq étamines courtes, & terminées par des antheres fimples, qui tombent avec un germe rond placé fous le calice, & qui foutient deux ftyles petits, érigés, & couronnés par des ftigmats fimples. Ce germe fe change dans la fuite en une baie, avec un nombril formé par le calice, & a deux cel-lules, qui renferment chacune une femence fimple, unie, convexe, & en forme de cœur.

Ce genre de plantes eft rangé dans la feconde fection de la vingt-troifieme claffe de LIN-NÉE, avec celles dont les fleurs mâles font fur des pieds dif-férens de ceux des fleurs fe-melles & hermaphrodites.

Les efpeces font :

1°. *Panax Quinque-folium*, fo-*liis ternis, quinatis. Flor. Virg.* 147. *Mat. Med.* 222. *Kalm. It.* 3. p. 334. *Blakw. f.* 513. *Regn. Bot.*; Panax avec des feuilles à trois & à cinq lobes.

Aureliana Canadenfis. Lafit. Gens. 51. *f.* 1. *Catefb. Car* 3. p. 16. *f.* 16.

*Araliaftrum Quinque-folii folio, majus Ninzin vocatum. D. Sar-*rafin. *Vaill. Gen.* 43 ; le plus

PAN

grand Aralia à cinq feuilles, appelé Ninzin.

*Araliaftrum foliis ternis quin-*que-partitis. Genzeng, *fivè Ninzin officinarum. Trew. Ehret. t. 6. f.* 1.

2°. *Panax Tri-folium*, *foliis ternis, ternatis. Flor. Virg.* 35 ; Panax avec trois feuilles à trois lobes.

Araliaftrum Fragariæ folio, minus. *Vaill. Gen.* 43 ; le plus petit Aralia bâtard, à feuilles de Fraifier.

*Araliaftrum foliis ternis, tri-*partitis & quadri-partitis. *Trew. Ehret. t.* 6. *f.* 2.

Nafturtium Marianum, *Ane-*mones fylvaticæ foliis, *ennea-*phyllon, floribus exiguis. *Pluk. Mant.* 135. *t.* 435. *f.* 7.

Quinque-folium. Ces deux plan-tes croiffent naturellement dans l'Amérique Septentrionale. La premiere a toujours été prife pour le *Genfeng* de Tartarie. Les figures & la defcription de cette plante, qui ont été envoyées en Europe par les Miffionnaires, s'accordent par-faitement avec la plante d'Amé-rique.

Elle a une racine charnue, conique, auffi groffe que le doigt, noueufe, & divifée vers le bas en deux plus petites fibres : fa tige, qui s'éleve à la hauteur d'environ un pied & demi, eft nue au fommet, où elle fe divife généralement en trois pétioles, qui foutien-nent chacun une feuille com-pofée de cinq lobes en forme de lance, fciés fur leurs bords, d'un vert pâle, & un peu ve-lus : les fleurs, qui naiffent à la divifion des pétioles des feuilles, font rangées en une

petite ombelle ; elles font d'un jaune herbacé , compofées de cinq petits pétales recourbés : elles paroiffent au commencement de Juin , & produifent des baies comprimées ₊ en forme de cœur , d'abord vertes , & enfuite rouges , renfermant deux femences comprimées & en forme de cœur, qui mûriffent au commencement d'Août.

Les Chinois , d'après le rapport des Miffionnaires , font beaucoup de cas de cette plante. Le P. Jartoux dit dans fes Lettres, que les plus fameux Médecins de la Chine ont écrit des volumes fur les vertus de cette plante ; ils la font entrer dans prefque tous les remedes qu'ils adminiftrent à la Nobleffe , parce qu'elle eft d'un trop grand prix pour la donner au menu peuple. Ils affûrent qu'elle eft un remede fouverain dans toutes les foibleffes occafionnées par les grandes fatigues , foit du corps , foit de l'efprit ; qu'elle guérit les maladies de poumon & les pleuréfies ; qu'elle arrête le vomiffement, fortifie l'eftomac, donne de l'appétit , ranime les efprits vitaux, augmente la lymphe dans le fang ; enfin , qu'elle eft bonne pour guérir le vertige, l'affoibliffement de la vue , & qu'elle prolonge la vie des vieillards.

Ce Pere dit auffi qu'il a fait des effais de cette plante fur lui-même , & qu'une heure après avoir pris la moitié d'une racine, il avoit été entierement rétabli d'une très-grande fatigue , s'étoit fenti beaucoup plus vigoureux , & en état de fupporter le travail beaucoup mieux qu'auparavant.

Il rapporte auffi, que l'Empereur avoit employé dix-mille Tartares en l'année 1709, pour recueillir cette plante dans les déferts, où elle croit naturellement, les avoit fait garder par une troupe de Mandarins , qui campoient fous des tentes , dans des endroits convenables à la fubfiftance de leurs chevaux , qui de-là envoyoient des détachemens de troupes , pour veiller à cet ouvrage ; & que, quand la récolte fut complette , ils retournerent avec leur charge à la ville.

Des racines de cette plante , recueillies en Amérique , & apportées en Angleterre, ayant été autrefois envoyées à la Chine , produifirent d'abord un revenu confidérable ; mais la grande quantité qu'on y en porta enfuite ayant rendu cette marchandife trop commune , elle y perdit beaucoup de fon prix.

Cette plante a été introduite dans les jardins anglois, où on la cultive à l'ombre & dans un fol léger ; elle y a profité & produit des fleurs : fes femences y mûriffent même chaque année ; mais aucune n'a germé ; car j'en ai femé pendant plufieurs années après leur maturité , fans aucun fuccès. J'en ai auffi femé plufieurs fois, dans différentes fituations , de celles qui m'avoient été envoyées de l'Amérique , & je n'ai pas été plus heureux. Il paroit que les Miffionnaires, d'après leur propre récit, n'ont

384 **PAN**

pas eu un meilleur fuccès ; car quoiqu'ils aient fouvent femé ces graines à la Chine même, ils n'ont jamais pu obtenir aucune plante. D'après cela, je crois qu'il eft néceffaire qu'il y ait des plantes mâles près des hermaphrodites, pour rendre les femences prolifiques ; car toutes celles que j'ai vues, & que j'ai cultivées, ne produifoient que des fleurs hermaphrodites : & , quoique leurs femences aient paru mûrir parfaitement, cependant aucune n'a réuffi, quoiqu'on les ait laiffées trois ans en terre, fans les remuer.

Tri-folium. La feconde eft originaire des mêmes contrées ; mais j'ignore fi fes propriétés font femblables à celles de la premiere. Je n'ai jamais vu en Angleterre qu'une plante de cette efpece, qui m'avoit été envoyée du Maryland, il y a quelques années ; mais qui a péri dans le premier été, parce qu'il s'eft trouvé fort fec, & qu'elle étoit placée dans un fol aride : fa tige étoit fimple, de cinq pouces de hauteur au plus, & divifée en trois pétioles, qui foutenoient chacun une feuille à trois lobes, plus longs, plus étroits, dentelés beaucoup plus profondément fur leurs bords que ceux de la premiere efpece. Les pédoncules fortoient des divifions des pétioles ; mais comme la plante a péri avant l'épanouiffement des fleurs, je ne puis en donner aucune defcription.

PANCRATIUM. *Dill. Hort. Elth.* 221. *fol.* 289. *Lin. Gen.*

Plant. 365. *Narciffus. Tourn. Inft. R. H.* 353. [*Sea Daffodil.*] Afphodele maritime.

Caraêteres. Les fleurs font renfermées dans une gaîne oblongue, qui s'ouvre fur le côté, & fe fanne : elles ont un neêtaire d'une feuille cylindrique en forme d'entonnoir, & étendue au fommet ; les corolles ont fix pétales en forme de lance, inférés audehors du neêtaire & au deffus de fa bâfe : les fleurs ont fix étamines longues, fixées dans le bord du neêtaire, & terminées par des antheres oblongues & penchées ; le germe, qui eft oblong, a trois angles ; & , placé fous la fleur, foutient un ftyle long, mince, & couronné par un ftigmat obtus. Ce germe devient enfuite une capfule ronde, à trois angles, & à trois cellules remplies de femences globulaires.

Ce genre de plantes eft rangé dans la premiere feêtion de la fixieme claffe de LINNÉE, où fe trouvent celles dont les fleurs ont fix étamines & un ftyle.

Les efpeces font :

1°. *Pancratium maritimum, fpathâ multi florâ, petalis planis, foliis lingulatis. Lin. Sp.* 291 ; Afphodele maritime, dont la fpathe renferme plufieurs fleurs, avec des pétales unis, & des feuilles en forme de langue.

Lilio-Narciffus albus maritimus minor. Moris. Hift. 2. *p.* 365. *S.* 4. *t.* 10. *f.* 28.

Narciffus maritimus. C. B. P. 540 ; Afphodele ou Narciffe maritime.

2°. *Pancratium Illyricum,*

PAN

fpathâ multi-florâ, foliis enfi-for-mibus, ftaminibus nectario lon-gioribus. Flor. Leyd. Prod. 34 ; Afphodele d'Efclavonie, avec plufieurs fleurs dans une fpa-the, en forme d'épée, & des étamines plus longues que le nectaire.

Pancratium, enfi-formibus foliis fpathâ multi-flora, floribus magnis, candidis, fragrantibus. Trew. Ehret. f. 27

Narciffus Illyricus Liliaceus. C. B. P. 55 ; Afphodele ou Narciffe d'Efclavonie, ou Nar-ciffe de Matthiole.

Lilio-Narciffus Hemerocallidis facie. Beft. Eyft. Vern. 3. t. 16. f. 1.

3°. *Pancratium Zeylanicum, fpathâ uni-florâ, petalis reflexis. Flor. Zeyl.* 126 ; Afphodele de Céylan, avec une fleur dans chaque fpathe, & des pétales réfléchis.

Narciffus Zeylanicus, flore al-bo, hexagono, odorato. H. L. 691 ; Afphodele ou Narciffe de Céylan, à fleurs blanches, à fix angles, & d'une bonne odeur.

Lilium Indicum. Rumph. Amb. 6. p. 161. t. 70. f. 2.

Catulli-Pola. Rheed. Mal. 11. f. 40.

4°. *Pancratium Mexicanum, fpathâ bi-florâ. Hort. Cliff.* 133. *Roy. Lugd.-B.* 34 ; Afphodele du Mexique, avec deux fleurs dans une fpathe.

Pancratium Mexicanum, flore gemello candido. Dill. Elth. 299. t. 222. f. 289 ; Afphodele du Mexique, qui produit deux fleurs blanches (1).

[1] *Nota.* Cette efpece a été mife mal-à-propos dans le Diction-

PAN 385

5°. *Pancratium Amboïnenfe, fpathâ multi-florâ, foliis ovatis, nervofis. Linn. Sp. Plant.* 291 ; Narciffe d'Amboine, avec plu-fieurs fleurs dans une fpathe, & des feuilles ovales & ner-veufes.

Narciffus Amboïnenfis, folio latiffimo fub-rotundo. Comm. Hort. t. p. 77. t. 39 ; Narciffe d'Am-boine, avec une feuille très-large, & prefque ronde.

6°. *Pancratium Carolinianum, fpathâ multi-florâ, foliis lineari-bus, ftaminibus nectarii longitu-dine. Lin. Sp. Plant.* 291 ; Nar-ciffe de la Caroline, avec plu-fieurs feuilles dans une fpa-the, des fleurs linéaires, & des étamines de la longueur du nectaire.

Lilio-Narciffus polyanthos, flore albo. Catesb. Car. 3. p. 5 ; Narciffe ou Lys, avec plu-fieurs fleurs blanches.

7°. *Pancratium Americanum, fpathâ multi-florâ, foliis carina-tis, anguftioribus* ; Narciffe de l'Amérique, avec plufieurs fleurs dans une fpathe, & des feuilles très-étroites & en forme de carène.

Narciffus Americanus, flore multiplici albo, odore Balfami Peruviani. Tourn. Inft. R. H. 358 ; Narciffe d'Amérique, avec plufieurs fleurs blanches, qui répandent une odeur fem-blable à celle du Baume du Pérou.

8°. *Pancratium lati-folium, fpathâ multi-florâ, foliis cari-natis latioribus* ; Narciffe Ca-

naire de Miller, tous la dénomi-nation de *Caribæum*, qu'il faut rapporter au n°. 8. *Lati-folium.*

Bb 3

raïbe, avec plufieurs fleurs dans une fpathe, & des feuilles plus larges & en forme de carène.

Narciffus totus albus, lati-folius, pólyanthos, major, odoratus, ftaminibus fex è tubi ampli margine extantibus. Sloan. Cat. Jam. 115 ; Narciffe à larges feuilles, avec plufieurs fleurs groffes, blanches & odorantes, & un large tube, du bord duquel fortent fix étamines.

Pancratium Caribæum. Lyn. Syft. Veg. Murray. ed. 14. *Sp.* 3. *Syft. Plant. tom.* 2. *p.* 22. *Sp.* 3.

Narciffus Americanus, flore multiplici albo, hexagono, odorato. Comm. Hort. 2. *p.* 173. *f.* 87.

9°. *Pancratium ovatum, foliis ovatis, acuminatis, petiolatis, fpathá multiflorá, floribus minoribus, candidis, fragrantibus. Trew. Ehret. t.* 28 ; Pancratium avec des feuilles ovales, terminées en pointes aiguës & pétiolées, ayant plufieurs fleurs dans une fpathe, qui font plus petites, blanches, & d'une très-bonne odeur. Variété du Pancratium Amboïnenfe. *Lin. Syft. Plant. Sp.* 7. *p.* 23. *tom.* 2.

Cepa fylveftris. Rumph. Amb. 6. *p.* 160. *t.* 70. *f.* 1.

Maritimum. La premiere efpece, qui croît naturellement fur les côtes de la mer en Efpagne, & dans la France Méridionale, a une racine groffe, bulbeufe, d'une forme oblongue, & enveloppée par une peau brune : fes feuilles font en forme de langue, de plus d'un pied de longueur fur un pouce de largeur, d'un vert foncé, & fortant fix ou fept enfemble de la même racine,

ayant leur bâfe enveloppée d'une gaine ; la tige s'élève entre les feuilles à la hauteur d'un pied & demi ; elle eft nue, & foutient fix ou huit fleurs blanches, enveloppées d'une fpathe qui fe fanne & s'ouvre fur le côté, pour laiffer fortir les fleurs : au fommet de la tige font fitués les germes, très-près defquels s'élevent les tubes des fleurs, qui ont trois pouces de longueur, font fort étroits, & fe gonflent au haut, où le godet ou nectaire eft fitué ; à l'extérieur du nectaire font fixées les fix pétales, qui font étroits, & s'étendent beaucoup au-delà du nectaire : du bord du nectaire s'élevent fix étamines longues, minces, & terminées par des antheres oblongues & penchées, & du centre fort un ftyle auffi long que les étamines, & terminé par un ftigmat obtus. Les fleurs de cette efpece ne paroiffent en Angleterre qu'à la fin d'Août, & n'y produifent point de femences ; fes feuilles font vertes pendant tout l'hiver, & fe fannent au printems : ainfi, il faut tranfplanter fes racines dans le mois de Juin, auffi-tôt que les feuilles font tombées, les planter dans une plate-bande chaude, & les abriter des fortes gelées, qui les feroient périr fans cela.

Illyricum. La feconde efpece naît fans culture en Efclavonie & en Sicile ; elle a une racine groffe, bulbeufe, couverte & environnée d'une peau brune ; de cette racine fortent plufieurs fibres fortes & épaiffes, qui pénetrent profondément dans la terre ; fes feuilles font en forme

PAN

d'épée, d'un pied & demi de
longueur fur deux pouces de
largeur, & d'une couleur grifâtre ; fes tiges font épaiffes,
fucculentes, hautes d'environ
deux pieds, & terminées par
fix ou fept fleurs blanches de
la même forme que celles de
la précédente, mais dont le
tube eft plus court, & les étamines beaucoup plus longues.
Cette plante fleurit en Juin,
& produit fréquemment des
femences qui mûriffent en Septembre.

Cette efpece eft dure, &
peut réfter tout l'hiver en
pleine terre ; car elle n'eft jamais endommagée par les gelées les plus fortes : & en
couvrant la furface de la terre
avec du tan, des cendres de
charbon de terre, de la paille
ou du chaume de pois, pour
empêcher la gelée de pénétrer, fes racines ne courront
aucun rifque. On multiplie
cette efpece par les rejettons
de fes racines ou par femences. La première méthode eft
plus prompte, parce que les
rejettons fleuriffent très - bien
dès la feconde année ; au-lieu
que les plantes élevées de femences ne produifent guere
de fleurs avant quatre ou cinq
ans.

Les racines de cette plante
ne doivent être enlevées de
terre que chaque trois ans, fi
l'on veut qu'elles fleuriffent
beaucoup. Le meilleur tems
pour les tranfplanter, eft
le commencement d'Octobre,
auffi tôt que leurs feuilles font
flétries. Il ne faut pas les tenir
long-tems hors de la terre ; car

PAN 387

comme elles ne perdent point
leurs fibres lorfqu'on les enleve, fi ces fibres fe deffé-
choient, elles s'affoibliroient
beaucoup. Cette plante exige
un fol léger & fablonneux, &
une fituation abritée. On place
fes racines à neuf pouces, ou
un pied de diftance les unes
des autres, & on les enfonce
à cinq pouces dans la terre.

Quand on veut les multiplier
par femences, on les répand
dans des pots remplis de terre
légere, auffi-tôt qu'elles font
mûres : on tient ces pots fous
un châffis de couche, pendant
l'hiver, pour les abriter de la
gelée, & on a foin d'ôter les
vitrages dans les tems doux.
Comme le refte de leur traitement ne diffère point de celui
qui a été prefcrit pour les *Nar-*
ciffes, je ne répéterai point ici
ce que j'ai déja dit ailleurs ;
j'obferverai feulement que les
jeunes racines exigent un peu
d'abri pendant l'hiver, jufqu'à
ce qu'elles aient acquis de la
force.

Zeylanicum. La troifième efpece, qu'on rencontre dans
l'Ifle de Céylan, a une racine
affez groffe & bulbeufe ; fes
feuilles font longues, étroites,
d'une couleur grifâtre, paffa-
blement épaiffes & érigées ; la
tige, qui s'éleve du milieu des
feuilles jufqu'à la hauteur d'un
pied & demi, eft nue, & foutient une fleur dont les pétales
font inclinés en arrière : le nectaire eft large, & divifé fur fes
bords en plufieurs fegmens aigus : les étamines font longues,
& tournées l'une vers l'autre
à leur pointe, en quoi celle-ci

Bb 4

388 **PAN**

differe des autres efpeces. Cette fleur répand une odeur agréable, mais elle eft de peu de durée, & on la trouve très-rarement aujourd'hui dans les jardins anglois.

Mexicanum. La quatrieme eft originaire de la Vera-Cruz, d'où le Docteur HOUSTOUN en a apporté quelques racines; fes feuilles, qui ont environ un pied de longueur fur prefque deux pouces de largeur, font fillonnées par trois rainures longitudinales; la tige s'éleve à-peu près à un pied de hauteur, & fe divife enfuite comme une fourche en deux pédoncules étrcits, verts, & enveloppés d'abord d'une fpathe mince, qui fe fanne & s'ouvre pour laiffer fortir les fleurs. Ces fleurs font blanches, & de la même forme que celles des autres efpeces, mais fans odeur.

Amboinenfe. La cinquieme efpece fe trouve à Amboine & dans les Ifles de l'Amérique; fa racine eft oblongue, blanche, & garnie de plufieurs fibres épaiffes & charnues, qui s'enfoncent dans la terre: fes feuilles font foutenues par des pétioles fort longs; quelques-unes font fort longues, & d'autres en forme de cœur; elles ont à-peu-près fept pouces de longueur fur cinq de largeur, font terminées en pointe, & ont plufieurs fillons profonds, qui s'étendent fur toute leur longueur; elles font d'un vert clair, & leurs bords font tournés en dedans: la tige, qui eft épaiffe, ronde & fucculente, s'éleve à la hauteur d'environ

PAN

deux pieds, & foutient au fommet plufieurs fleurs blanches, qui ont la même forme que celles des autres efpeces, mais dont les pétales font plus larges, le tube plus court, & les étamines moins longues que les pétales; elles ont des fpathes minces, qui fe fendent dans la longueur, pour laiffer paffer les fleurs.

Carolinianum. La fixieme efpece croît fpontanément fur des terres humides & des fondrieres, en Géorgie, où M. CATESBY l'a découverte: fa racine, qui eft ronde, bulbeufe, & couverte d'une peau d'un brun clair, pouffe plufieurs feuilles étroites, d'un vert foncé, & d'un pied de longueur; entre les feuilles fort une tige épaiffe de neuf pouces environ de longueur, qui foutient fix ou fept fleurs blanches, avec des pétales fort étroits, & un grand nectaire en forme de cloche, profondément découpé fur fes bords; les étamines ne s'élevent pas beaucoup au-deffus du nectaire, & font terminées par des antheres jaunes.

Americanum. La feptieme, qu'on rencontre dans les Ifles de l'Amérique, où on lui donne le nom de *Lys blanc*, a une racine affez groffe, bulbeufe, un peu applatie au fommet, & couverte d'une peau brune; fes feuilles ont près d'un pied & demi de longueur fur un peu plus d'un pouce de largeur; elles font d'un vert foncé, & concaves dans le milieu en forme de carène: les tiges s'élevent à la

hauteur d'environ deux pieds ; elles font épaiffes, fucculentes , nues , & foutiennent huit ou dix fleurs blanches de la même forme que celles de la premiere efpece , mais d'un blanc plus pur , & d'une odeur forte & douce , comme celle du *Baume du Pérou* ; les étamines font fort longues , & s'étendent beaucoup de chaque côté ; le ftyle eft de la même longueur , & placé au milieu du nectaire. Les fleurs font de peu de durée , & confervent rarement leur beauté plus de trois ou quatre jours ; elles fe fannent encore plutôt dans un tems chaud. Lorfqu'elles font paffées , les germes qui font au bas des tubes fe changent en autant de bulbes oblongues , de forme ir—réguliere , qui tombent fur la terre lorfqu'elles font mûres , pouffent des racines , & deviennent de nouvelles plantes.

Ces efpeces étrangeres portent , pour la plupart , fi ce n'eft pas toutes , des bulbes ; au-lieu que les deux premieres ont des capfules à trois cellules , qui renferment des femences rondes & noires ; & quoique toutes s'accordent par les caracteres de leurs fleurs , cependant elles different entr'elles confidérablement par cette particularité.

Lati-folium. La huitieme efpece qui croît naturellement dans les Indes Occidentales , n'eft pas fort différente de la précédente. Mais comme j'ai fouvent multiplié ces deux plantes par les bulbes qui fuccedent aux fleurs , j'ai tou-jours trouvé que celles qu'on éleve ainfi , confervent leurs différences , & je ne doute nullement que ce ne foit des efpeces diftinctes. Celle ci differe de la précédente par fes feuilles , qui font beaucoup plus longues & plus larges ; car elles ont près de deux pieds de longueur fur plus de trois pouces de largeur , & font creufées en forme de carène : fes fleurs font larges , les pétales font plus longs , & leur odeur eft moins forte que celle de la précédente ; fes racines donnent des fleurs dans toutes les faifons de l'année : elle paroît être celle qui a été défignée par le Docteur TREW , dans la vingt-feptieme table de fes *Decades de Plantes Rares* ; mais fi c'eft la même , les feuilles de fa figure font trop plates.

Ovatum. La neuvieme efpece eft originaire des Ifles de l'Amérique ; elle a une racine groffe , ronde & bulbeufe , de laquelle fortent plufieurs feuilles ovales , d'un pied environ de longueur fur fix pouces de largeur au milieu , terminées en pointe aux deux côtés , d'un vert foncé , & fillonnées dans toute leur longueur : la tige eft épaiffe , fucculente , nue , d'un pied & demi de hauteur , & foutient à fon fommet fix ou huit fleurs blanches d'une odeur douce & agréable , de la même forme que celles de la feptieme efpece , mais plus petites , avec des pétales étroits , & des tubes plus courts , ainfi que les fpathes.

PAN

Culture. Les six dernieres especes étant trop délicates pour pouvoir profiter en Angleterre sans le secours d'une chaleur artificielle, la meilleure méthode pour les avoir dans leur perfection, est de plonger les pots qui les contiennent dans la couche de tan de la serre chaude, où elles réussiront & fleuriront très-bien ; car quoiqu'elles puissent être conservées dans des serres seches, cependant elles n'y profitent pas aussi-bien, leurs feuilles ne deviennent pas si fortes que lorsqu'elles sont plongées dans la couche de tan, & elles ne fleurissent qu'une fois l'année : au-lieu que dans la couche les mêmes racines produisent souvent des fleurs deux fois par an. J'ai eu plusieurs de ces especes en fleurs dans toutes les saisons de l'année, & il ne se passoit pas un mois sans que quelques-unes n'en donnassent de nouvelles.

On les multiplie par les rejettons de leurs racines, ainsi que par les bulbes qui succedent aux fleurs. En plantant ces bulbes dans de petits pots remplis de terre légere de jardin potager, & en les plongeant dans une couche chaude tempérée, elles pousseront bientôt des racines & des feuilles. Si elles sont bien traitées, elles deviendront des racines bulbeuses dans un an ; & si on les tient constamment dans la couche de tan de la serre chaude, elles produiront des rejettons, & profiteront aussibien que dans leur pays natal.

PAN

PANICAUT, AMÉTISTE. *V.* ERINGIUM AMETISTINUM. *L.*

PANICAUT, CHARDON A CENT TÊTES *ou* CHARDON ROLAND. *Voyez* ERYNGIUM CAMPESTRE. *L.*

PANICAUT DE MER. *Voy.* ERYNGIUM MARITIMUM. *L.*

PANICULE. On nomme ainsi une tige étendue, & divisée en plusieurs pédoncules qui soutiennent des fleurs ou des fruits, comme dans l'*Avoine*, &c.

PANICUM. *Tourn. Inst. R. H. 515. tab. 298. Lin. Gen. Plant. 70.* [*Panic.*] Panis.

Caracteres. Cette plante a une fleur dans chaque bâle, qui s'ouvre en trois petites valves ovales, & terminées en pointe aiguë ; la corolle est composée de deux petites valves ovales, & aussi à pointe aiguë : les fleurs ont trois étamines courtes, semblables à des poils, & terminées par des antheres oblongues ; le germe est rond, & soutient deux styles comme des poils, & couronnés par des stigmats plumacés. Ce germe se change dans la suite en une semence ronde, & fixée aux pétales fannés.

Le genre de cette plante est rangé dans la seconde section de la troisieme classe de LINNÉE, qui comprend celles dont les fleurs ont trois étamines & deux styles.

Les especes sont :

1°. *Panicum Germanicum, spicâ simplici cernuâ, setis brevioribus, pedunculo hirsuto ;* Panis avec un épi simple, & penché, des poils ou soies courtes, & un pédoncule hérissé.

2°. *Panicum Italicum*, *spicâ compositâ*, *spiculis glomeratis*, *setis immixtis*, *pedunculo hirsuto*. *Lin. Sp. Plant.* 56 ; Panis avec un épi composé, dont les plus petits sont en paquets ronds, des barbes mêlées parmi, & un pédoncule hérissé.

Panicum spicâ compositâ, *aristis flosculo brevioribus*. *Virid. Cliff.* 7. *Hort. Ups.* 19. *Roy. Lugd. - B.* 54. *Gron. Virg.* 134.

Panicum Italicum, *sivè paniculâ majore*. *C. B. P.* 27 ; Panis d'Italie, avec un plus gros épi, ou la Germanie.

Panicum. Rumph. Amb. 5. *p.* 202. *t.* 75. *f.* 2.

3°. *Panicum Indicum*, *spicâ simplici longissimâ*, *setis hispidis*, *pedunculo hirsuto* ; Panis avec un épi simple & très-long, des barbes piquantes, & un pédoncule hérissé.

Panicum Indicum spicâ longissimâ. C. B. P. 27 ; Panis des Indes avec un très-long épi.

Panicum Alopecurodeum, *spicâ tereti*, *involucellis bi-floris*, *fasciculato-pilosis*. *Flor. Zeyl.* 44 ; Panis avec un épi cylindrique, deux fleurs dans chaque enveloppe, & des barbes en paquet.

Panicum Indicum altissimum, *spicis simplicibus*, *mollibus*, *in foliorum alis longissimis pediculis insidentibus*. *Tourn. Inst.* 515 ; le plus grand Panis des Indes, avec des épis mous & simples, qui sortent des ailes des feuilles sur de très-longs pédoncules.

5°. *Panicum cæruleum*, *spicâ simplici*, *æquali*, *pedunculis bi-floris*. *Prod. Leyd.* 54 ; Panis avec un épi simple & égal, & deux fleurs sur chaque pédoncule.

Panicum Indicum, *spicâ obtusâ*,

cæruleâ. C. B. P. 7 ; Panis des Indes, à épis, bleus & obtus.

Germanicum. Il y a plusieurs autres especes de ce genre, dont quelques-unes croissent naturellement en Angleterre ; mais comme on ne les cultive pas, ce seroit augmenter inutilement le volume de cet Ouvrage, que de les insérer ici.

La premiere espece se trouve en Allemagne & en Hongrie : on en connoît trois variétés ; l'une à graines jaunes, l'autre à graines blanches, & la troisieme, à graines pourpre. On la cultivoit autrefois pour en faire du pain dans quelques pays Septentrionaux ; elle s'éleve avec une tige noueuse en forme de Jonc, garnie à chaque nœud d'une feuille semblable à celle de l'herbe commune, d'un pied & demi de longueur sur un pouce environ de largeur à sa bâse, & terminée en pointe aiguë. Les feuilles sont rudes au toucher ; elles embrassent la tige de leur bâse, & penchent vers le bas dans la moitié de leur longueur ; les tiges sont terminées par des épis comprimés, à-peu-près de l'épaisseur du doigt à leur bâse, coniques à leur extrémité, de huit ou neuf pouces de longueur, & fortement garnis de petites graines semblables à celles du *Millet.* Cette plante est annuelle, & périt aussi-tôt que ses semences sont mûres.

Italicum. La seconde espece, qu'on cultive souvent en Italie & dans d'autres contrées Méridionales, s'éleve à la hauteur de quatre pieds, avec une tige en forme de Jonc, & est beau-

coup plus épaisse que celle de la première ; les feuilles sont aussi plus larges, mais de la même forme : les épis ont un pied de longueur, & sont deux fois plus gros que ceux de la précédente, mais moins rapprochés ; ils sont composés de plusieurs petits épis ronds & en paquets ; leur graine est aussi plus grosse, mais de la même forme. Il y a deux ou trois variétés de cette espece, qui ne different que par la couleur de leurs graines. Cette plante est annuelle.

Indicum. La troisieme espece croît naturellement dans les deux Indes ; elle a une tige en forme de *Jonc*, aussi grosse que le pouce, & de plus de cinq pieds de hauteur : ses feuilles ont deux pieds de longueur sur deux pouces de largeur, & sont de la même forme que celles de l'espece précédente ; les épis, qui sont placés au sommet, ont un pied & demi de long, sont tort comprimés, plus épais que le pouce à la base, & cylindriques au sommet : leur graine est beaucoup plus grosse que celle des autres especes ; quelques-unes sont blanches, & d'autres jaunes.

Alopecurodeum. La quatrieme, qui se trouve aussi dans les deux Indes, a une tige forte, en forme de *Jonc*, de six ou sept pieds de hauteur, & garnie de feuilles de plus de trois pieds de longueur sur près de trois pouces de large à leur base, mais terminées en pointe ; leur surface est unie ; les épis s'élevent des aisselles de la tige ; ils sont simples, mais moins comprimés ou rapprochés que ceux de la précédente, & armés de barbes molles ; ils ont environ six pouces de longueur, & sont postés sur de fort longs pédoncules ; leur graine est passablement grosse.

Cæruleum. La cinquieme est originaire du Pérou ; elle s'éleve avec une tige en forme de *Jonc*, à la hauteur de six pieds, & pousse deux ou trois branches latérales, garnies de feuilles longues, & de deux pouces de largeur à leur base. Ces tiges sont de couleur de pourpre ; les feuilles sont aussi presque de la même couleur : les épis sortent des aisselles des tiges, & aux extrémités des branches ; ils ont environ quatre ou cinq pouces de longueur, sont plus épais que le pouce, & leur extremité est presque égale à la base ; ils sont d'un bleu pâle, ainsi que les barbes & les graines, qui sont plus grosses & plus rondes que celles des autres especes.

Culture. Dans quelques cantons de l'Europe, on cultive les deux premieres especes en pleine campagne, comme le *Bled*, pour la nourriture des habitans ; mais elles ne sont pas si estimées que le *Millet.* Cependant on en fait souvent usage dans quelques contrées de l'Allemagne & de l'Italie, pour des gâteaux & du pain. Ces graines sont moins bonnes en Allemagne qu'en Italie ; mais comme elles mûrissent mieux dans les pays froids, on les y cultive dans les en-

PAN

droits où de meilleures graines ne réussiroient pas.

On les seme au printems, en même tems que l'*Orge*, & elles exigent le même traitement ; mais il ne faut pas les répandre si épaisses ; car les graines étant fort petites, & les plantes très-grosses, elles ont besoin d'un plus grand espace.

L'espece d'Allemagne ne s'éleve qu'à la hauteur de trois pieds, à moins qu'elle ne soit semée sur une terre fort riche, où elle croît jusqu'à quatre pieds ; mais comme ses feuilles sont fort larges, & les tiges fort épaisses, il faut laisser entr'elles quatre ou cinq pouces de distance, sans quoi elles fileroient, & deviendroient trop foibles.

On les rend beaucoup plus vigoureuses, en houant la terre entre les rangs, & en la tenant nette de mauvaises herbes. Cette graine mûrit en Août, alors on la coupe pour la faire sécher, & la mettre ensuite à couvert.

Comme le *Panis* d'Italie est bien plus large que celui d'Allemagne, & qu'il produit des épis beaucoup plus gros, il a besoin de plus d'espace pour croître ; mais comme il mûrit plus tard, il n'est pas aussi propre pour des pays froids.

Les autres especes croissent naturellement dans des contrées très-chaudes, où les habitans en font du pain ; elles deviennent fort grosses, & ne mûrissent point dans ce pays-ci, à moins que les étés ne soient très-chauds : on les se-

PAP 393

me à la fin de Mars ou au commencement d'Avril sur une couche de chaleur tempérée ; on les enleve quand elles sont parvenues à une certaine grosseur ; on les plante sur une planche de terre riche & légere, à une exposition chaude, en rangs éloignés de trois pieds, & on les tient nettes de mauvaises herbes. Quand ces plantes ont atteint une certaine hauteur, on les soutient avec des piquets, pour les fortifier contre les efforts du vent ; & quand la graine commence à mûrir, on en écarte les oiseaux, qui la dévoreroient bientôt. On conserve ces especes dans quelques jardins curieux, pour la variété ; mais elles ne valent pas la peine d'être cultivées, pour l'usage, en Angleterre. Les deux dernieres mûrissent rarement dans ce pays.

PANICUM. *Voyez* CYNOSURUS COROCANUS. *L.* HOLCUS SPICATUS. *L.*

PANIS. *Voyez* PANICUM. *L.*

PAPAVER. *Tourn. Inst. R. H.* 2. *tab.* 119. *Lin. Gen. Plant.* 573. [*Poppy.*] Pavot, Coquelicot.

Caracteres. Le calice de la fleur est ovale, dentelé, & composé de deux feuilles presque ovales, concaves, obtuses, & qui tombent ; la corolle a quatre pétales larges, ronds & étendus ; la fleur a un grand nombre d'étamines semblables à des pois, & terminées par des antheres oblongues, comprimées & droites ; dans le centre est placé un germe gros, rond & sans style, mais cou-

ronné par un ftigmat uni , radié , & en forme de bou-clier ; le germe devient en-fuite une capfule groffe & cou-ronnée par un ftigmat uni , avec une cellule qui s'ouvre en plufieurs endroits au fom-met, fous la couronne , & qui eft remplie de petites fe-mences.

Ce genre de plantes eft ran-gé dans la premiere fection de la treizieme claffe de LINNÉE, avec celles dont les fleurs ont plufieurs étamines , & un germe.

Les efpeces font :

1°. *Papaver Rhæas , capfulis glabris , globofis , caule pilofo, multi-floro , foliis pinnati-fidis, in-cifis.* L'n. *Sp. Plant.* 507. *Mat. Med.* 134. *Pollich. pal. n.* 507. *Crantz. Auftr. p.* 137. *n.* 1. *Regn. bot. Pallas. It.* 3. *p.* 546. *Scop. Carn.* 2. *n.* 638. *Mattufch. Sil. n.* 377 ; Coquelicot ou Pavot rouge à capfules unies & gio-bulaires , avec une tige velue, qui produit plufieurs fleurs , & des feuilles découpées en forme d'ailes.

Papaver erraticum , rubrum , campeftre. J. B. 3. 395 ; Coque-licot ou Pavot rouge des champs.

Papaver erraticum , pleno flore. Bauh. Pin. 171. *Knorr. Del.* 1. *f. R.* 12 , 13 ; Variété à fleurs doubles.

Papaver erraticum minus. Bauh. Pin. 171 ; Variété plus pe-tite.

2°. *Papaver hybridum , cap-fults fub-globofis , torofis , hif-pidis , caule foliofo , multi floro. Lin. Sp. Plant.* 506 ; Pavot avec des capfules globulaires ,

fillonnées & épineufes, qui produit une tige feuillée & à plufieurs fleurs.

Argemone capitulo breviori , hif-pido. J. B. 396 ; Argemone avec des têtes plus petites & épineufes.

3°. *Papaver Argemone , cap-fulis clavatis , hifpidis , caule foliofo , multi-floro. Lin. Sp. Plant.* 506 ; Pavot avec des têtes épineufes & en forme de maffue , & une tige feuillée qui produit plufieurs fleurs.

Papaver erraticum , capite lon-giori hispido. Tourn. Inft. 238 ; Pavot des champs , à têtes plus longues , & épineufes.

Argemone capitulo tenuiori , lon-giori , hirfuto. Moris. Hift. 2. *p.* 278. *S.* 3. *t.* 14. *f.* 10.

4°. *Papaver Alpinum , cap-fulâ hifpidâ , fcapo uni-floro , nudo , hifpido , foliis bi-pinnatis. Lin. Sp. Plant.* 507. *Jacq. Auftr. t.* 83. *Scop. Carn.* 2. *n.* 637 ; Pavot avec des têtes épineufes, une tige nue & épineufe, qui porte une feule fleur , & des feuilles à doubles ailes.

Papaver Burferi. Crantz. Auftr. p. 138. *t.* 6. *f.* 4.

Argemone Alpina , Coriandri folio. C. B. P. 172 ; Argemone des Alpes , à feuilles de Cori-andre.

5°. *Papaver Cambricum , cap-fulis glabris , oblongis , caule multi-floro , lævi , foliis pinnatis , in-cifis. Lin. Sp. Plant.* 508 ; Pavot avec des têtes unies & oblon-gues , une tige liffe , portant plufieurs fleurs , & des feuilles découpées en forme d'ailes.

Papaver luteum perenne, la-ciniato folio , Cambro-Britannicum. Raii Syn. ed. 3. *p.* 309 ; Pavot

jaune & vivace de Galle , avec une feuille découpée.

Argemone Cambro – Britannica lutea , capite longiori glabro. Moris. Hift. 2. p.297. S. 3. t. 14. f. 12.

6°. *Papaver nudi-caule , capfulis hifpidis , fcapo uni-floro , nudo , hifpido , foliis fimplicibus pinnato-finuatis. Hort. Ups. 136. Gmel. Sib. 4. p. 180. Gunn. Norv. n. 578. Kniph. cent. 10. n. 68. Flor. Dan. t. 41* ; Pavot avec des têtes épineufes , une tige nue & rude , portant une feule fleur , & des feuilles fimples & découpées en forme d'aîles.

Papaver erraticum , luteo flore , capite oblongo , hifpido. Amman. Rhut. 61 ; Pavot des champs à fleurs jaunes.

7°. *Papaver Orientale , capfulis glabris , caulibus uni floris , fcabris , foliofis , foliis pinnatis , ferratis. Hort. Ups. 136. Knorr. Del. t. R. 14. a.* ; Pavot à têtes unies , avec des tiges rudes , feuillées ; & portant une feule fleur , & des feuilles fciées & aîlées.

Papaver foliis pinnatis , fructu globofo. Roy. Lugd. · B. 279.

Papaver Orientale hirfutiffimum , flore magno. Tourn. Cor. 17. Itin. 3. p. 127. t. 127. Comm. rar. 34. f. 34 ; Pavot d'Orient, très-velu , avec une groffe fleur. Le grand Pavot du Levant.

8°. *Papaver fomni-ferum , calycibus capfulifque glabris , foliis amplexi-caulibus , incifis. Lin. Sp. Plant. 508. Mat. Med. p. 134. Hall. Helv. n. 1065* ; Pavot avec des calices & des capfules unies , & des feuilles amplexicaules & découpées.

Papaver nigrum & album officinale. Crantz. Auftr. p. 138. Blackw. t. 482. 483. Regn. bot.

Papaver hortenfe nigro femine, fylveftre Diofcoridis , nigrum Plinii C. B. P. 176 ; Pavot de jardin , à femences noires.

9°. *Papaver album , capfulis ovatis , glabris , foliis latioribus , amplexi caulibus , marginibus incifo-ferratis* ; Pavot avec des têtes ovales & unies, & des feuilles plus larges & amplexicaules , dont les bords font découpés en forme de fcie.

Papaver hortenfe , femine albo, fativum Diofcoridis , album Plinii. C. B. P. 170 ; Pavot de jardin à femences blanches , ordinairement appelé Pavot blanc.

Rhœas. La première efpece eft le *Coquelicot* ou *Pavot rouge commun* , qui croît naturellement fur les terres labourées , dans la plus grande partie de l'Angleterre. On extrait des feuilles de cette plante une eau fimple , une teinture , un fyrop , & une conferve pour l'ufage de la Médecine : elle eft annuelle ; fes racines produifent plufieurs tiges rudes & branchues , qui s'élevent à un pied & demi de hauteur , & font garnies de feuilles velues de cinq ou fix pouces de longueur , & profondément découpées jufqu'à la côte du milieu : l'incifion du bas des feuilles eft plus profonde que les autres ; les lobes font oppofés & réguliers , comme ceux des feuilles aîlées : au fommet de chaque tige font poftées les fleurs , qui ont des calices ovales & velus qui s'ouvrent en

PAP

deux petites valves , & qui tombent bientôt après. Ces fleurs font compofées de quatre pétales larges , ronds , étroits à leur bâle , & étendus au-dehors dans un ordre circulaire ; elles font d'une belle couleur écarlate , & tombent en peu de tems ; elles paroiffent en Juin, & produifent des têtes oblongues , unies , & couronnées par des ftigmats plats en forme de bouclier , percés dans plufieurs endroits au fommet ; & remplies de petites femences de couleur pourpre. Il y a dans cette efpece plufieurs variétés à fleurs doubles, qu'on cultive dans les jardins , & dont quelques-unes font blanches , d'autres rouges , bordées de blanc , & quelques-unes panachées : mais comme ces variétés proviennent toutes de femences de l'efpece commune , elles n'en doivent point être féparées (1).

_____

(1) Les fleurs de Coquelicot font regardées comme fudoriſiques, bechiques , & légerement calmantes : on les donne en infufion cherſorme dans les fluxions de poitrine, les rhumes opiniatres, l'afthme , l'efquinancie, &c. : le Syrop de Coquelicot , qu'on trouve dans les boutiques , fert auffi aux mêmes ufages : mais on doit peu compter fur les vertus de ces fleurs , & ne point leur donner la préférence fur d'autres remedes , dans les maladies qui exigent de prompts fecours. Les têtes qui contiennent la graine , font plus efficaces : on peut s'en fervir comme d'un doux calmant dans toutes les circonftances où les legers narcotiques font indiqués. L'infufion d'une douzaine de ces têtes fuffit pour une dofe.

PAP

Hybridum. La feconde efpece fe trouve parmi les bleds dans prefque toute l'Angleterre ; fes feuilles font beaucoup plus petites que celles de la premiere , & ont des fegmens plus déliés ; fes tiges font minces , d'un peu plus d'un pied de hauteur , & moins branchues que celles de la précédente : fes fleurs font moins larges , de couleur pourpre foncé , tombent en peu de tems , & durent rarement plus d'un jour ; elles font remplacées par des têtes oblongues , épineufes , & remplies de femences petites & noires. Cette plante fleurit dans le mois de Juin.

Argemone. La troifieme croît auffi parmi les bleds dans quelques parties de l'Angleterre , mais en moins grande abondance que les précédentes ; fes feuilles font plus agréablement découpées , & plus petites que celles de la premiere ; mais elles font moins belles que celles de la feconde ; fes tiges font moins élevées que celles des précédentes , & ont rarement beaucoup de branches : fes fleurs ne font pas moitié auffi larges que celles des précédentes , & font de couleur de cuivre ; elles paroiffent dans le mois de Mai , & tombent en peu d'heures ; elles font remplacées par des têtes longues , minces , épineufes , canelées , & remplies de femences petites & ridées.

Alpinum. La quatrieme efpece fe trouve fur les Alpes , parmi les rochers ; fes feuilles font liffes , & doublement ailees , & leurs fegmens font joliment

liment découpés : les tiges s'é-
levent à la hauteur d'un pied ,
& soutiennent une petite fleur
jaune ou de couleur de cuivre , à laquelle succede une tête
ronde , épineule , & remplie
de petites semences. Cette plante fleurit à peu près dans le même
tems que l'espece précédente.

Cambricum. La cinquieme ,
qui a une racine vivace , croît
naturellement dans le pays de
Galle , ainsi que dans quelques
parties septentrionales de l'Angleterre. Je l'ai rencontrée en
abondance près de Kirby Lons-
dale , en Westmoreland. TOUR-
NEFORT l'a aussi trouvée sur
les montagnes des Pyrénées ;
ses feuilles sont ailées , & leurs
lobes sont profondément découpés sur leurs bords ; ses
tiges , qui s'élevent à la hauteur d'un pied , sont lisses , &
garnies de quelques petites
feuilles de la même forme que
celle du bas. La partie haute
de la tige est nue , & soutient
une fleur grosse & jaune, qui
paroît en Juin , & à laquelle
succede une capsule oblongue,
unie & remplie de petites semences de couleur tirant sur
le pourpre.

Nudi-caule. La sixieme espece naît spontanément sur
les confins de la Russie , près
de la Tartarie ; ses feuilles
sont simples , rudes , velues ,
& découpées presque jusqu'à
la côte du milieu en forme de
feuilles ailées : la tige s'éleve
à la hauteur d'environ deux
pieds ; elle est mince , nue ,
& soutient une fleur composée de quatre pétales ronds ,
d'un jaune pâle , & d'une cou-

Tome V.

leur plus foncée dans le bas.
Ces fleurs ont une odeur agréable , mais elles sont de peu
de durée : elles paroissent dans
le mois de Juin , & produisent
des capsules rondes , rudes , &
remplies de petites semences.

Orientale. La septieme croît
naturellement dans le Levant ,
d'où M. DE TOURNEFORT a envoyé ses semences au jardin
Royal à Paris , d'où elles se sont
répandues ensuite dans les Jardins curieux de l'Angleterre &
de la Hollande ; sa racine est
compotée de deux ou trois fibres
fortes , aussi grosses que le petit
doigt , d'un pied & demi de
long, d'un brun foncé en-dehors , & remplies d'un suc
laiteux , âcre & très-amer ;
ses feuilles sont ailées , sciées
sur leurs bords , d'un pied de
longueur , & fortement couvertes de poils blancs & hérissés ; ses tiges , qui s'élevent
à la hauteur de deux pieds &
demi , sont fort rudes , velues ,
& garnies vers leurs parties
basses de feuilles semblables
aux feuilles radicales , mais
plus petites : la partie haute
des tiges est nue , & supporte
à son extrémité une fleur fort
grosse , & de la même couleur
que le *Coquelicot* ou *Pavot rouge
commun.* Cette plante fleurit
dans le mois de Mai , & produit
des capsules ovales , unies , &
remplies de semences pourpre.

Il y a deux ou trois variétés de cette espece , qui ne
different que par la couleur de
leurs fleurs. J'ai appris qu'il y
en a aussi une à fleurs doubles ,
mais je ne l'ai jamais vue.
TOURNEFORT assure que les

C c

398 **PAP**

Turcs mangent les têtes vertes de ce *Pavot*, malgré leur amertume & leur âcreté.

Somniferum. La huitieme espece est le *Pavot noir commun*, dont on vend les semences dans les boutiques sous le nom de *Maw-Seed*. Le Pavot à fleurs simples croit naturellement dans les contrées chaudes de l'Europe. Il est annuel ; ses tiges, qui s'élevent à la hauteur de trois pieds, sont lisses, & divisées en plusieurs branches garnies de feuilles larges, unies, profondément découpées sur leurs bords, & qui embrassent les tiges de leur bâse : les fleurs croissent au sommet des tiges , & sont composées de quatre pétales larges, ronds, & de couleur pourpre, un peu foncée aux onglets. A ces fleurs succedent des capsules ovales, unies, & remplies de semences noires. Cette plante fleurit en Juin, & perfectionne ses semences à la fin d'Août.

Il y a beaucoup de variétés dans les fleurs de cette espece ; quelques-unes sont grosses, doubles, & panachées de plusieurs couleurs ; d'autres sont rouges & blanches, & quelques-unes sont agréablement tachetées, comme les *Œillets* ; de sorte que, tandis qu'elles sont épanouies, il n'y a guere de plantes qui paroissent aussi belles : mais leur odeur est désagréable. Les feuilles de cette espece entrent dans la composition des onguens rafraichissans, & ses têtes dans le *Syrupus è Melonio* ; mais la derniere Pharmacopée les en a exclues.

Album. La neuvieme espe-

PAP

ce est le *Pavot blanc commun* qu'on cultive dans les jardins pour ses têtes, dont on fait usage en Médecine ; ses tiges sont grosses, lisses, de cinq ou six pieds de hauteur, & divisées en plusieurs plus petites, garnies de feuilles larges & grisâtres, dont la bâse embrasse les tiges, & qui sont régulierement découpées sur leurs bords : les fleurs qui terminent les tiges penchent vers le bas, tandis qu'elles sont encore renfermées dans le calice ; mais elles se redressent avant de s'épanouir. Le calice est composé de deux feuilles larges, ovales, de couleur grisâtre comme les autres, & qui se séparent & tombent en peu de tems. La fleur a quatre pétales ronds, blancs, & de peu de durée ; elle est remplacée par une tête ronde, aussi grosse qu'une *Orange*, applatie aux deux extrémités, surmontée d'une couronne dentelée, & remplie de semences petites & blanches. Cette plante fleurit en Juin, & ses semences mûrissent en automne (1).

───────────

(1) Les graines, & sur-tout les têtes de ce Pavot sont d'un usage assez fréquent en Médecine. Les graines n'ont aucune propriété narcotique ; mais elles contiennent un mucilage & une huile douce, qui les rendent propres à former d'excellentes émulsions dans les différentes maladies où les autres sont indiquées, mais particuliérement dans l'enrouement, la toux âcre, la dyssenterie, l'érosion des conduits, la néphrétique, la dysurie, &c. L'huile qu'on tire par expression de ces graines est plus connue dans la cuisine que dans les

PAP

Il y a plusieurs variétés de cette espece, qui different par

Pharmacies; on l'ajoute cependant aux autres ingrédiens qui entrent dans la composition des onguens anodins. La tête de ce Pavot, ou l'enveloppe qui contient la graine, est très-narcotique, & a les mêmes propriétés que l'Opium, dont il va être question. La décoction d'une de ces têtes est un très-bon calmant; mais on ne doit en user qu'avec beaucoup de prudence; elles entrent comme principal ingrédient dans la composition du syrop de Diacode, dont la dose est depuis une demi-once jusqu'a une once. C'est de ce Pavot, ou au moins d'une espece très-voisine, que les Orientaux tirent l'Opium. Cette substance est une gomme-resine, d'un vert-brun, d'une saveur amere, & d'une odeur forte & noseuse. Outre la gomme & la résine, dont la proportion n'est point constante, l'Opium contient encore un principe vaporeux, très-mobile & très-actif, dans lequel résident toutes les propriétés de cette drogue.

Les effets de cette substance singuliere sont très-différens, suivant la dose à laquelle on l'a prise, le plus ou moins d'habitude qu'on a d'en faire usage, son degré de pureté, le tempérament de celui qui en use, &c.; mais en général, lorsqu'on a pris de l'Opium à une dose modérée, on éprouve une espece d'ivresse, toutes les fibres motrices acquierent une nouvelle activité, une agitation vive & rapide les fonctions du corps s'operent avec plus de force & d'énergie; la circulation est accélérée; on éprouve une gaieté singuliere, un bien-être ravissant; l'imagination apperçoit des fantômes bizarres; les objets ne paroissent plus les mêmes; on est affecté d'une maniere nouvelle. Quelques personnes entrent en délire, en fureur, & se

PAP 399

la couleur de leurs fleurs & le nombre des pétales, celles qui ont les plus belles fleurs servent d'ornement dans les jardins; mais les fleurs simples ne sont cultivées que pour l'usage. On extrait des semences de cette espece une émulsion rafraichissante, qu'on administre avec succès dans les fievres & les maladies inflammatoires, ainsi que dans la strangurie & l'ardeur d'urine. On fait avec les têtes seches, infusées & bouillies dans l'eau, le *Diacode des Boutiques.*

précipitent avec intrépidité au-devant des dangers. Cette premiere effervescence se calme peu-a-peu; à l'agitation succede l'inertie, l'imagination s'eteint, toutes les parties du corps tombent dans le relâchement, & le sommeil, qui survient bientôt, ramene le calme & l'équilibre dans toutes les fonctions.

L'Opium est un excellent cordial, & le meilleur de tous les calmans, il appaise les douleurs les plus vives, procure un relâchement salutaire, lorsqu'il est question de faciliter la descente d'une pierre à travers les uretres, calme la toux convulsive, ainsi que les paroxismes hystériques & hypocondriaques, relâche les parties convulsées, &c.

La connoissance de la maniere d'agir des narcotiques, peut étendre l'usage de l'Opium à une infinité d'autres circonstances; mais les bornes de cet Ouvrage ne me permettent point d'entrer à ce sujet dans de plus longs détails, qui d'ailleurs seroient déplacés ici. J'observerai seulement encore, que les meilleurs remedes contre l'Opium, pris à forte dose, sont d'abord l'émétique, & ensuite les acides végétaux, tels que le vin aigre, le suc de Citron, &c.

Cc 2

On a toujours été dans l'o-
pinion que l'on tiroit l'*Opium*
des têtes de *Pavot* de cette
espece ; mais j'ai une tête dont
on avoit tiré l'*Opium* en Turquie, qui est bien différente
de celles de cette espece.

Culture. On multiplie tous
les Pavots par semences, à
l'exception des cinquieme &
septieme especes, qui ont des
racines vivaces, & qui peuvent aussi être multipliées par
les rejettons. Le meilleur tems
pour les semer, est le mois
de Septembre. Ces graines
réussissent alors plus certainement qu'au printems. Les plantes annuelles de l'automne sont
plus fortes, & fleurissent mieux
que celles du printems. La
bonne méthode est de semer
les especes annuelles en place, & d'en éclaircir les plantes quand elles sont trop serrées. Celles des grosses especes doivent être éloignées d'un
pied & demi au moins l'une
de l'autre, & les plus petites
de huit ou neuf pouces ; après
quoi elles n'exigent plus que
d'être tenues nettes de toutes
mauvaises herbes.

Quand on veut avoir de
bons *Pavots*, il faut examiner
soigneusement les plantes, lorsqu'elles commencent à fleurir,
& arracher toutes celles dont
les fleurs ne sont pas bien
doubles & bien marquées,
avant qu'elles s'épanouissent,
pour empêcher leur poussiere
fécondante de faire dégénérer
les plus belles fleurs, ce qui
arrive souvent quand on ne
prend pas cette précaution ;
& après cela, on s'en prend

mal-à-propos à la nature du
sol.

Le *Pavot jaune de Galle* exige une situation fraîche &
ombrée, où les plantes profitent & produisent annuellement une grande abondance
de semences, qui poussent
beaucoup mieux, quand on
les laisse écarter, qu'en les
semant à la main : mais quand
on veut les semer, on doit
toujours le faire en automne ;
car celles qu'on ne met en
terre qu'au printems, réussissent rarement.

Le meilleur tems pour tranfplanter & diviser les racines
de cette plante, est l'automne,
afin qu'elles puissent être bien
établies dans la terre avant les
secheresses du printems.

Le *Pavot d'Orient* profite ou
au soleil ou à l'ombre ; car
j'ai placé plusieurs plantes de
cette espece sous des arbres,
où elles ont réussi pendant
plusieurs années, & y ont
fleuri tout aussi bien que si
elles avoient été dans une situation plus ouverte, si ce n'est
qu'elles paroissoient plus tard
dans la saison.

Comme elle se multiplie considérablement par ses racines,
il n'est pas nécessaire de la
semer, à moins qu'on ne
veuille se procurer de nouvelles variétés. On la tranfplante dans la même saison
que la précédente, & elle
doit être semée en même tems,
pour les raisons que nous
avons exposées.

PAPAVER CORNICULA-
TUM. *Voyez* CHELIDONIUM
GLAUCIUM. *I.*

PAR

PAPAVER SPINOSUM. *Voy.* ARGEMONE.

PAPAYA *ou* PAPAYER. *Voy.* CARICA PAPAYA.

PAPILIONNACÉE. Une fleur est papilionnacée ou légumineuse, lorsqu'elle ressemble en quelque manière à un papillon ayant les ailes étendues ; elle a toujours un étendard ou *vexillum*, qui est un segment ou pétale large & érigé, *alas*, des ailes, qui forment les côtés, & une carène, qui est un segment ou pétale concave, semblable à la partie basse d'un bateau. Cette carène est quelquefois entière, & quelquefois composée de deux segmens ou pétales fort rapprochés. Les *Pois*, les *Féves*, les *Haricots*, les *Vesces*, & autres plantes légumineuses sont de cette espece.

PAPPUS se dit des semences couronnées d'un duvet qui adhere à leur partie haute, & sert à les transporter dans l'air à une grande distance. Les *Laiterons*, l'*Herbe à Faucon*, le *Pissenlit*, &c, sont de cette espece.

PAQUERETTE *ou* PETITE MARGUERITE. *Voy.* BELLIS FERENNIS. *L.*

PARASITES. On nomme ainsi les plantes qui s'attachent aux troncs des arbres, aux branches & aux autres plantes, & qui en tirent leur nourriture, sans pouvoir croître sur la terre, comme le *Guy*, &c.

PARELLE *ou* PATIENCE DES MARAIS. *Voy.* RUMEX AQUATICUS. *L.*

PARIETAIRE. *Voyez* PARIETARIA. *L.*

PAR

PARIÉTAIRE D'ESPAGNE. *V.* ANTHEMIS PYRETHRUM. *L.*

PARIETARIA. *Tourn. Inst. R. H.* 509. *tab.* 289. *Lin. Gen. Plant.* 1020 ; ainsi appelée de *Paries*, muraille, parce qu'elle croît sur les vieux murs. [*Pellitory.*] Pariétaire.

Caracteres. Cette plante a des fleurs femelles & hermaphrodites sur le même pied ; elle a deux fleurs hermaphrodites renfermées dans une enveloppe à six feuilles ; le calice est formé par une feuille unie & divisée en quatre parties, & de la moitié de la grosseur de l'enveloppe : les fleurs sont apétales, & n'ont que quatre étamines en forme d'alêne, qui persistent, & sont plus longues que le calice ; elles sont terminées par des antheres jumelles ; le germe est ovale, & soutient un style mince, coloré, & couronné par un stigmat en forme de pinceau. Ce germe se change dans la suite en une semence ovale, & renfermée dans le calice.

Les fleurs femelles n'ont point d'étamine ; mais elles ressemblent pour le reste aux hermaphrodites.

Ce genre de plantes est rangé dans la premiere section de la vingt-troisieme classe de LINNÉE, avec celles dont les fleurs sont femelles & hermaphrodites sur la même racine.

Les especes sont :

1°. *Parietaria officinalis*, foliis lanceolato-ovatis, alternis. *Hort. Ups.* 302. *Hort. Cliff.* 496. *Roy. Lugd.-B.* 210. *Dalib. Paris.* 505. *Scop. Carn. ed.* 2. *n.* 1242 ; Pariétaire à feuilles

ovales, en forme de ance, & alternes.

Parietaria Officinarum & Dioscoridis. C. B. P. *121* ; Pariétaire des Boutiques & de Dioscoride, ou l'Herbe Notre-Dame.

Parietaria foliis hirsutis, ellipticolanceolatis. Hall. Helv. n. 1612.

Parietaria Helxine. Tabern. p. 550. Blackw. f. 136.

Helxine. Cam. Epit. 849.

2°. *Parietaria Judaïca, foliis ovatis, caulibus erectiusculis, calycibus tri-floris, corollis hermaphroditis, defloratis, elongatocylindricis.* Lin. Sp. 1492 ; Pariétaire à feuilles ovales, avec des tiges érigées, trois fleurs dans chaque calice, des corolles hermaphrodites, d'une forme cylindrique allongée.

Parietaria minor Ocymi folio. C. B. P. 121 ; la plus petite pariétaire à feuilles de Basilic.

Officinalis. La premiere espece croît naturellement en Allemagne & en Hollande, mais point en Angleterre, où je l'ai apportée en 1727.

On croit que c'est celle qui est recommandée par les anciens pour l'usage de la Médecine ; sa racine est épaisse, vivace, & composée de fibres charnues, rougeâtres ; elle pousse plusieurs tiges d'un pied & demi de hauteur, garnies de feuilles ovales, velues, en forme de lance, de deux pouces environ de longueur sur un de largeur au milieu, & traversées par plusieurs nervures : ses fleurs naissent en petits paquets sur les côtés des tiges ; elles sont petites, herbacées, & n'ont

point d'apparence ; elles se succedent, & continuent à s'épanouïr pendant tout l'été ; leurs semences se perfectionnent dans la même proportion, & sont lancées à une certaine distance, quand elles sont mûres (1).

Judaïca. La seconde croît en abondance sur de vieux murs, & sur les bords des bancs secs dans plusieurs cantons de l'Angleterre ; elle diffère de la premiere en ce que ses tiges sont plus courtes, que ses feuilles sont ovales & plus petites, & que ses fleurs naissent en plus petits paquets : mais ces deux plantes se ressemblent pour le reste.

Une seule plante suffit pour couvrir bientôt un terrein considérable d'une grande quantité de jeunes tiges, par ses semences qui s'écartent, mais qui sont très-difficiles à recueillir, parce que leurs capsules élastiques les lancent au loin aussitôt qu'elles sont mûres.

Il y a trois ou quatre autres especes de ce genre ; mais comme elles ont peu de beauté, & ne sont d'aucun usage, on ne les cultive point dans les jardins.

---

(1) La Pariétaire passe pour être apéritive, émolliente & résolutive ; on l'emploie en infusion ou en substance, après l'avoir réduite en poudre, dans l'ardeur d'urine, la néphrétique, la toux, les obstructions des viscères, l'hydropisie, &c. On l'applique aussi en cataplasmes, comme les autres plantes émollientes. La Pariétaire sert à la composition du syrop de Guimauve de Fernel.

PAR

PARIS. *Lin. Gen. Plant.* 449. *Herba Paris. Tourn. Inst. R. H.* 233. *tab.* 117. [*True-Love*, or *One-berry*.] Véritable Amour. Raisin de Renard.

Caractères. Le calice de la fleur est persistant, & composé de quatre feuilles placées en forme de croix; la corolle a aussi quatre pétales étendus de la même manière, & persistans. Dans le centre de la fleur est placé un germe rond à quatre angles, qui soutient quatre styles étendus, & couronnés par des stigmats simples : ce germe est ensuite accompagné par huit étamines, terminées chacune par une anthere oblongue, & fixée par des filets sur chaque côté des étamines; il se change dans la suite en une baie ronde & à quatre cellules remplies de semences.

Ce genre de plantes est rangé dans la quatrieme section de la huitieme classe de LINNÉE, dans laquelle sont comprises toutes celles dont les fleurs ont huit étamines & quatre styles.

Nous n'avons qu'une espece de ce genre.

Paris quadri-folia, foliis quaternis. Flor. Lapp. 155. *Flor. Suec.* 325. 346. *Hort. Cliff.* 153. *Roy. Lugd.-B.* 461. *Hall. Helv. n.* 1006. *Gmel. Sib.* 4. *p.* 176. *Reyg. Ged.* 2. *p.* 76. *De Neck. Gallob. p.* 188. *Scop. Carn. ed.* 2. *n.* 472. *Pollich. pal. n.* 389. *Mattusch. Sil. n.* 285. *Blackw. t.* 286. *Flor. Dan. t.* 139. *Kniph. Cent.* 12. *n.* 73. *Darr. Nass. p.* 171. *Regn. bot.*; Raisin de Renard.

Solanum quadri-folium bacciferum. Bauh. Pin. 167.

Herba Paris. Matt. 1193. *Bauh. Hist.* 3. *p.* 613.

Aconitum saluti-ferum. Tabern. Hist. 720.

Cette plante croît spontanément à l'ombre des bois humides dans differens cantons de l'Angleterre, sur-tout dans les contrées septentrionales; mais il est difficile de la conserver dans les jardins : la seule maniere de se la procurer est d'enlever les plantes du lieu même où elles naissent, en conservant une bonne motte de terre à leurs racines, de les placer à l'ombre dans une plate-bande humide, & les y laisser sans y toucher. Elles subsisteront ainsi quelques années : mais comme elles ont peu de beauté, on se donne rarement cette peine.

PARKINSONIA. *Plum. Nov. Gen.* 25. *tab.* 3. *Lin. Gen. Plant.* 460. [*Parkinsonia.*] Genêt épineux Caraïbe; la Parkinion.

Caractères. Le calice de la fleur est étendu, & formé par une feuille découpée au sommet en cinq parties; la corolle a cinq pétales presque égaux, & placés circulairement; les quatre supérieurs sont ovales, & l'inférieur est en forme de rein. La fleur a dix étamines inclinées, & terminées par des antheres oblongues; son germe est long, cylindrique, & presque sans style, mais avec un stigmat obtus & élevé. Ce germe devient ensuite un legume ou silique longue & conique, divisée en nœuds gonflés, qui renferment chacun

une femence oblongue.

Ce genre de plantes eft rangé dans la première fection de la dixieme claffe de LINNÉE, qui comprend celles dont les fleurs ont dix étamines & un ftyle.

Nous n'avons qu'une efpece de ce genre.

Parkinfonia aculeata. Hort. Cliff. 157. t. 13. Hort. Ups. 99. Roya-B. 465. Brown. Jam. 221. Plum. Amer. 121. f. 80; la Parkinfon.

Parkinfonia aculeata, foliis minus, utricofiæ adnexis. Plum. Nov. Gen. 25 ; la Parkinfon épineufe, avec de très-petites feuilles fixées à la côte du milieu; Genêt épineux Caraïbe.

Cette plante a été découverte dans l'Amérique par le P. PLUMIER, qui lui a donné ce nom en l'honneur de M. JEAN PARKINSON, Auteur d'une *Hiftoire univerfelle des Plantes*, en 1640.

Elle eft très-commune dans l'Amérique Efpagnole, d'où elle a été portée depuis quelques années dans les établiffemens Anglois, à caufe de fa beauté & de la bonne odeur de fes fleurs : elle s'élève dans fon pays natal, à plus de vingt pieds de hauteur, & produit des fleurs jaunes, difpofées en paquets longs & minces, comme celles du *Laburnum*, & d'une odeur douce & agréable, qui parfume l'air à une grande diftance; ce qui engage les habitans de l'Amérique à la planter aux environs de leurs maifons. Quoique ces arbres n'aient été portés que depuis peu dans les Colonies Angloifes, ce-

pendant ils s'y font multipliés à un tel point, qu'on en voit autour de toutes les maifons. Les plantes de deux ans produifent des fleurs, & une grande abondance de femences, qui les rendent très-communes dans tous les pays chauds ; mais en Europe on ne peut les conferver fans le fecours d'une ferre chaude.

Culture. On multiplie cette plante par fes graines, qu'il faut femer de bonne heure au printems, dans de petits pots remplis d'une terre fraiche & légere; on les plonge dans une couche chaude de tan, où les plantes paroîtront au bout de trois femaines ou d'un mois : on les tient alors nettes de mauvaifes herbes, & on les arrofe fréquemment, mais toujours en petite quantité à la fois : elles feront en état d'être tranfplantées en peu de tems; ce qu'il faut faire avec beaucoup de précaution, pour ne pas endommager leurs racines : on les met chacune féparément dans des pots de la valeur d'un fou, remplis d'une terre fraiche & légere : on les replonge dans la couche de tan qu'on remue bien auparavant, & auquel on en ajoute de nouveau, pour en renouveler la chaleur, fi elle eft bien diminuée ; on tient enfuite ces plantes à l'ombre, jufqu'à ce qu'elles aient formé de nouvelles racines; après quoi on leur donne chaque jour de l'air frais, à proportion de la chaleur de la faifon. Au moyen de cette méthode, ces plantes croîtront fi promptement, que les pots feront rem-

plis de racines au commence-ment de Juillet; alors on leur en donne d'autres un peu plus larges, & on les replonge dans la couche de tan, pour leur faire pousser de nouvelles racines; on les accoutume ensuite par dégrés à supporter le plein air, afin qu'elles puissent s'endurcir avant l'hiver; car si l'on étoit obligé de les tenir trop chaude-ment durant cette saison, elles se fletriroient au printems suivant.

La seule méthode qui m'ait réussi pour conserver ces plan-tes pendant l'hiver, a été de les accoutumer, en Juillet & en Août, à supporter le plein air, & de les placer en Sep-tembre sur les tablettes de la serre chaude seche, le plus loin du feu qu'il étoit possi-ble, de maniere qu'elles ne fussent exposées qu'à une cha-leur très—modérée : par ce moyen, elles se sont entrete-nues en bon état, & ont con-servé leurs feuilles pendant tout l'hiver; au-lieu que celles de la serre chaude ou de l'Oran-gerie ont été entierement dé-truites. Cependant les premieres survivent rarement au second hiver.

PARNASSIA. *Tourn. Inst. R. H.* 246. *tab.* 127. *Lin. Gen. Plant.* 345. [*Grass of Parnassus.*] Herbe ou Gazon du Parnasse.

Caracteres. Le calice est per-sistant, étendu & découpé en cinq parties; la corolle a cinq pétales ronds, concaves, & tout-à-fait ouverts, avec cinq nectaires en forme de cœur, & concaves : la fleur a cinq étamines, terminées par des

antheres penchées; son germe est gros, & sans style; mais à la place de ce dernier sont quatre stigmats et & per-sistans; le germe se change, quand la fleur est passée, en une capsule ovale, à quatre angles, & à une cellule qui renferme plusieurs semences ovales.

Ce genre de plantes est rangé dans la quatrieme section de la cinquieme classe de LINNÉE, avec celles dont les fleurs ont cinq étamines & quatre styles.

Les especes sont :

1°. *Parnassia palustris; Her-*be du Parnasse.

Parnassia palustris & vulgaris. Tourn. Inst. R. H.; Herbe, ou Gazon du Parnasse de marais, commune.

Parnassia. Fl. Lapp. 108. *Fl. Suec.* 252. 363. *Hort. Cliff.* 113. *Mat. Med. p.* 90. *Roy. Lugd-B.* 420. *Dalib. Paris.* 96. *Gmel. Sib.* 4. *p.* 91. *Hall. Helv. n.* 832. *Reyg. Ged.* 1. *p.* 93. *Scop. Carn.* 2. *n.* 378. *Pollich. pal. n.* 316. *Mattusch. Sil. n.* 220. *Fl. Dan. t.* 584. *Kniph. Cent.* 7. *n.* 70. *Dœrr. Nass. p.* 172.

Gramen Parnassi albo simplici flore. Bauh. Pin. 309. *Lob. ic* 603.

Hepatica alba. Cord. Hist. 53.

Pyrola rotundi-folia palustris, flore unico campliore. Moris. Hist. 3. *p.* 505. *S.* 12. *t.* 10. *f.* 3.

2°. *Parnassia pleno flore, vul-garis;* Herbe du Parnasse, commune, à doubles fleurs, Variété de l'espece à fleurs simples.

Palustris. La premiere de ces especes croit naturellement dans des prés humides de plu-sieurs parties de l'Angleterre,

& principalement dans le Nord; mais on n'en voit point dans le voisinage de Londres, ni en aucun endroit plus près qu'à l'autre côté de Watford, dans les prairies basses, près de Cassioberry, où l'on en trouve en grande quantité.

Pleno flore. La seconde est une variété accidentelle de la premiere, qui a été trouvée sauvage, & transplantée dans des jardins; mais elle est rare à présent, & on ne la voit que dans peu de jardins.

Culture. On peut enlever ces plantes des endroits où elles croissent, en conservant une bonne motte de terre à leurs racines, les mettre dans des pots remplis d'une terre fraîche, forte, & sans fumier, & les placer à l'ombre, où elles profiteront fort bien, & fleuriront chaque été, si l'on a soin de les arroser exactement dans les tems secs; mais si l'on met ces plantes en pleine terre, il faut que ce soit dans une plate-bande fort humide, & à l'ombre, sans quoi elles seroient bientôt détruites: elles exigent d'être arrosées copieusement dans les tems secs, ainsi que celles des pots, pour leur faire produire de belles fleurs.

On peut les multiplier, en divisant leurs racines, dans le mois de Mars, avant qu'elles poussent de nouvelles feuilles; mais il ne faut pas les diviser en trop petites parties: car cela les empêcheroit de donner des fleurs dans l'été suivant. Ces racines doivent toujours être plantées dans une terre fraîche & forte; mais elles ne profiteroient pas dans un sol riche & léger: il faut, au printems, les arroser constamment, si le tems est sec, sans quoi elles ne fleurissent point. On ne doit les diviser que tous les trois ans, afin qu'elles restent toujours fortes & vigoureuses. Ces plantes fleurissent dans le mois de Juillet, & perfectionnent leurs semences à la fin d'Août. Elles tirent leur nom du Mont-Parnasse, sur lequel on imagine qu'elles se trouvent; & comme le bétail s'en nourrit, on les nomme encore *Herbe* ou *Gazon,* quoiqu'elles n'aient aucune ressemblance avec l'herbe commune; car leurs fleurs approchent de celles de la *Renoncule,* & leurs feuilles sont assez larges, oblongues & unies.

PARONYCHIA. *Voyez* IL-LECEBRUM.

PARTERRE. *Voyez* JARDIN, PARTERRE.

PARTHENIUM. *Lin. Gen. Plant.* 939 *Partheniastrum Nissol. Act. Par.* 1711. *Dill. Gen.* 13. [*Bastard Feverfew.*] Matricaire bâtarde.

Caractéres. La fleur est composée de fleurons hermaphrodites, & de demi-fleurons femelles, renfermés dans un calice à cinq feuilles étendues; les fleurs hermaphrodites qui forment le disque, ont un pétale tubulé, découpé en cinq parties sur ses bords; elles ont cinq étamines semblables à des poils, aussi longues que le tube, & terminées par des antheres épaisses; le germe, qui est placé au-dessous de la

fleur, eft à peine vifible ; il
foutient un ftyle mince fans
ftigmat. Ces fleurons font ftériles. Les femelles, qui compofent les rayons ou bordures, s'étendent au-dehors fur
un côté en forme de langue ;
elles ont un germe gros, comprimé, & en forme de cœur,
avec un ftyle mince & couronné par deux ftigmats étendus : à ces fleurs fuccede une
femence applatie & en forme
de cœur.

Ce genre de plantes eft rangé dans la cinquieme fection
de la vingt-unieme claffe de
Linnée, qui comprend celles
qui ont des fleurons femelles
& hermaphrodites fur la même plante, & dont les fleurs
mâles ont cinq étamines.

Les efpeces font :

1°. *Parthenium hyfterophorus*,
foliis compofito-multi-fidis. Lin.
Hort. Cliff. 442. Hort. Ups.
285. Roy. Lugd.-B. 86. Brown.
Jam. 340 ; Matricaire bâtarde,
avec des feuilles compofées
de plufieurs lobes.

Abfynthium Eryfimi folio,
ad achoavarum Alpinum quodam
modo accedens. Pluk. Alm. 8.
t. 45. f. 3.

Partheniaftrum Americanum,
Ambrofiæ folio. Niff. Act. 1711.
p. 423. t. 13. f. 2.

Partheniaftrum Artemifiæ fo-
lio, flore albo. Hort. Elth. 152 ;
Matricaire bâtarde, à feuilles
d'Armoife ; Abfinthe d'Amérique.

2°. *Parthenium integri-folium,*
foliis ovatis, crenatis. Lin. Hort.
Cliff. 442. Gron. Virg. 147 ;
Matricaire bâtarde à feuilles
ovales & crenelées.

Ptarmica Virginiana, Scabio-
fæ Auftriacæ foliis diffectis. Pluk.
Alm. 308. t. 53. f. 5. & t.
219. f. 1.

Partheniaftrum Helenii folio.
Hort. Elth. 302 ; Matricaire
bâtarde, à feuilles d'Enule
Campane ou d'Aunée.

Hyfterophorus. La premiere
efpece croît naturellement en
grande abondance dans l'Ifle
de la Jamaïque, & dans quelques autres établiffemens de
l'Amerique, où on lui donne
le nom d'*Abfinthe fauvage*,
& où on la regarde comme
vulnéraire.

Integri-folium. La feconde fe
trouve en abondance dans plufieurs parties de l'Amérique
Efpagnole, d'où fes femences
ont été apportées en Europe.

La premiere eft une plante
annuelle qu'on feme fur une
couche chaude, dans le commencement du printems. Quand
les plantes pouffent, on les
tranfplante fur une autre couche chaude, à cinq ou fix
pouces entr'elles ; on les arrofe, & on les tient à l'ombre, jufqu'à ce qu'elles aient
repris racine, enfuite on leur
donne beaucoup d'air dans les
tems chauds, en foulevant les
vitrages de la couche chaque
jour, & on les arrofe convenablement au moins chaque
deux jours.

Lorfque ces plantes com-
mencent à fe toucher, on les
enleve avec précaution, en
confervant une motte de terre
à leurs racines ; on les met
chacune féparément dans des
pots remplis de terre riche &
légere, & on les place dans

une ferre chaude , pour leur faire prendre facilement racine ; mais fi l'on n'a pas cette commodité , il faut les mettre dans une fituation chaude & abritée , & les tenir à l'ombre , jufqu'à ce qu'elles aient pris racine ; après quoi on les expofe avec d'autres plantes dures & annuelles à une expofition chaude , où elles fleuriront en Juillet , & donneront des femences mûres en Septembre : mais fi la faifon devient froide & humide , il fera prudent de conferver une ou deux plantes dans une ferre chaude ou fous un châffis élevé , pour en avoir des femences , & en conferver l'efpece, en cas que celles du dehors viennent à manquer.

La feconde eft une plante vivace , qui périt jufque fur terre en automne , & repouffe au printems fuivant. Ses femences m'ont été envoyées par le Docteur THOMAS DALE, qui l'a découverte dans la Caroline Méridionale. On peut la multiplier en divifant fes racines en automne ; elle fupporte très-bien le froid de nos hivers ordinaires : elle fleurit en Juillet ; mais elle produit rarement de bonnes femences en Angleterre.

Comme ces plantes ont peu d'apparence , on ne les cultive guere que pour la variété.

PAS D'ASNE ou TUSSILAGE. *Voy.* TUSSILAGO FARFARA. *L.*

PASSERAGE, GRANDE ou POIVRÉE. *Voy.* LEPIDIUM LATI-FOLIUM. *L.*

PASSERINA. *Lin. Gen. Plant.* 440. *Thymelæa. Tourn. Inft. R.*

H. 594. *Pluck. Sanamunda. Clus.* [*Sparrow - wort.*] l'Herbe à l'Hirondelle ; Sanamonde.

Caractères. La fleur n'a point de calice ; elle a un petale fanné , avec un tube cylindrique gonflé au-deffous du milieu , & divifé au fommet en quatre parties étendues ; elle a huit étamines velues, auffi longues que le limbe ou partie fupérieure de la corolle , placées fur le haut du tube , & terminées par des antheres prefque ovales : elle a un germe ovale fixé fous le tube , avec un ftyle mince , qui s'éleve fur un côté du fommet du germe , & qui eft couronné par un ftigmat à tête , & garni de poils aigus fur chaque côté. Le germe fe change dans la fuite en une femence ovale à chaque bout, & renfermée dans une capfule épaiffe, ovale, & à une cellule.

Ce genre de plantes eft rangé dans la premiere fection de la huitieme claffe de LINNÉE , qui comprend celles dont les fleurs ont huit étamines & un ftyle.

Les efpeces font:

1°. *Pafferina fili-formis, foliis linearibus , convexis , quadrifariam imbricatis , ramis tomentofis. Lin. Sp. Plant.* 559. *Berg. Cap.* 130 ; Herbe à l'Hirondelle , avec des feuilles linéaires , convexes , imbriquées en quatre manieres , & des branches garnies de duvet.

Thymelæa Æthiopica fruticofa , foliis in longum ftriatis , furculis valdè tomentofis. Pluk. Alm. 180.

Thymelæa Ethyopica , Paffe—

PAS 409

rina foliis. Bryn. cent. 10. *fig.*
6 ; Laurier Epurge d'Ethiopie
à feuilles d'Herbe à l'Hirondelle.
2°. *Paſſerina hirſuta*, *foliis
carnoſis exius glabris*, *caulibus
tomentoſis*, *Lin. Sp. Plant.* 559 ;
Herbe à l'Hirondelle, à feuilles charnues, & unies en-deſ-
fus, avec des tiges cotonneuſes.
*Thymelæa tomentoſa, foliis Sedi
minoris. Bauh. Pin.* 461.
Sanamunda, 3 *Cluſ. Hiſt.* 1.
p. 89 ; la troiſieme Sanamonde
de Cluſius.
Seſamoides parvum Dalechampi.
Bauh. Hiſt. 1. *p.* 595.
3°. *Paſſerina ciliata*, *foliis
lanceolatis*, *ſub-ciliatis*, *erectis*,
ramis nudis. Lin. Sp. Plant.
559 ; Herbe à l'Hirondelle,
avec des feuilles en forme de
lance, érigées & garnies de
petits poils, avec des branches
nues.
*Paſſerina foliis lanceolatis. Hort.
Cliff.* 146. *Roy. Lugd.-b.* 208.
Gron. Orient. 126.
*Thymelæa foliis Chamæleæ mi-
noribus hirſutis. Bauh. Pin.* 463.
*Thymelæa foliis oblongis, acu-
tis, ad oras fimbriatis. Burm.
Afric.* 129. *t.* 47. *f.* 2.
Sanamunda, 1. *Cluſ. Hiſt.* 88 ;
la premiere Sanamonde de Cluſius.
Erica Africana, *Ruſci folio.
Seb. Muſ.* 2. *p.* 15. *t.* 12. *f.* 9.
4°. *Paſſerina uni-flora*, *foliis
linearibus*, *oppoſitis*, *floribus terminalibus ſolitariis*, *ramis glabris*,
Lin. Sp. Plant. 560. *Bert. Cap.*
128 ; Herbe à l'Hirondelle,
avec des feuilles linéaires &
oppoſées, des fleurs ſolitaires
qui terminent les branches, &
des tiges unies.
Thymelæa foliis triquetris cra-

Thymelæa ramoſa, *lineari ius
cium oppoſitis*, *flore ſericeo. Burm.
Afr.* 132. *t.* 48. *f.* 2.
Thymelæa ramoſa, *linearibus
foliis anguſtis*, *flore ſolitario.
Burm. Afr.* 131. *tab.* 48. *fig.* 1 ;
Lauréole branchue, avec des
feuilles étroites & linéaires, &
des fleurs ſolitaires.
Fili-ſarnis. La premiere eſpece croît naturellement au
Cap de Bonne-Eſpérance, d'où
elle a été d'abord apportée dans
les jardins de la Hollande ; elle
s'éleve, avec une tige d'arbriſſeau, à la hauteur de cinq
ou ſix pieds, & pouſſe dans
toute ſa longueur, des branches érigées, tandis qu'elles
ſont jeunes, mais qui, à meſure qu'elles grandiſſent, déclinent, & prennent une po-
ſition horizontale, ſur-tout lorſque les petits rejettons de l'extrémité ſont chargés de fleurs
& de capſules, dont la peſanteur les fait encore pencher
davantage : les branches ſont
couvertes d'un duvet blanc,
ſemblable à de la farine, &
ſont très-garnies de feuilles fort
étroites, convexes, & placées
les unes ſur les autres en quatre
rangs, comme des écailles de
poiſſon ; de maniere que les
jeunes branches paroiſſent être
quarrées : les fleurs ſortent aux
extrémités des jeunes rejettons
entre les feuilles, & ſur chaque côté ; elles ſont petites,
blanches, & peu apparentes :
à ces fleurs ſuccedent de petites
capſules qui ſemblent être te-
ches & frangées. Les fleurs paroiſſent dans les mois de Juin
& Juillet, & les ſemences mûriſſent en automne.
Cette eſpece peut être mul-

tipliée par boutures, qu'on plante en été sur une planche de terre marneuse ; on les couvre de cloches, pour en exclure l'air ; on les tient à l'ombre, & on les arrose de tems en tems: avec ce traitement, elles auront poussé des racines au bout de deux mois : alors on les enlevera ; on les mettra chacune séparément dans de petits pots remplis d'une terre molle & marneuse, & on les tiendra à l'ombre, jusqu'à ce qu'elles aient produit de nouvelles fibres ; on les placera ensuite dans une situation abritée, où on les laissera jusqu'au mois d'Octobre, pour les transporter alors dans la serre; car elles ne subsisteroient pas ici en plein air pendant l'hiver. Elles n'exigent point d'autres soins que ceux qu'on donne aux *Myrtes* & autres plantes dures de la serre. Comme cette plante conserve son feuillage pendant toute l'année, elle fait une belle variété dans la serre en hiver.

On peut aussi la multiplier par ses graines, qu'on seme en automne aussi-tôt qu'elles sont mûres, parce qu'elles réussissent alors plus certainement qu'en aucune autre saison de l'année : on les répand dans de petits pots remplis de terre légere ; & si on les tient dans une vieille couche de tan, sous un châssis ordinaire pendant l'hiver, les plantes pousseront au printems : alors on les traitera comme les plantes de boutures ; mais celles de semences deviennent plus droites, & paroissent plus belles que ces dernieres.

Hirsuta. La seconde espece croit naturellement en Espagne & en Portugal ; elle a des tiges d'arbrisseau qui s'élevent à une plus grande hauteur que celles de la précédente ; ses branches sont plus étendues, & couvertes d'un duvet farineux ; elles sont garnies de feuilles courtes, epaisses, succulentes, & disposées l'une sur l'autre comme des écailles de poisson. Ces feuilles sont unies & vertes en-dessus, mais couvertes de duvet en-dessous ; ses fleurs sont petites & blanches comme celles de la précédente, & paroissent à-peu-près dans le même tems.

Cette plante résiste en plein air dans les hivers ordinaires, si elle est placée dans un sol sec, & à une exposition chaude ; mais comme elle est souvent détruite par les fortes gelées, il est nécessaire d'en conserver deux plantes dans des pots, & de les abriter en hiver, pour en perpétuer l'espece. On peut la multiplier par boutures, comme la précédente.

Ciliata. La troisieme se trouve encore en Espagne & en Portugal ; ainsi qu'au Cap de Bonne-Espérance ; elle a une tige d'arbrisseau, qui s'éleve à la hauteur de cinq ou six pieds, & qui pousse à son extrémité plusieurs branches nues & garnies de feuilles oblongues, érigées & velues, ses fleurs sont petites, blanches, & sortent entre les feuilles aux extrémités des branches ; elles paroissent en Juin : mais elles ne produisent point de semences en Angleterre. On peut

multiplier cette espece par boutures comme les deux précédentes, & elle exige le même traitement.

Uni-flora. La quatrieme, qui est originaire du Cap de Bonne-Espérance, a une tige basse d'arbrisseau, d'un pied au plus de hauteur, qui se divise en plusieurs branches minces, lisses, étendues au dehors de tous côtés, & garnies de feuilles fort étroites, opposées, d'un vert foncé, & semblables à celles du *Sapin*, mais plus étroites : ses fleurs sont solitaires aux extrémités des branches, & sont plus grosses que celles de la précédente ; le haut du pétale est étendu, & presque plat. Ces fleurs sont de couleur pourpre, & paroissent vers le même tems que celles des précédentes. Cette espece peut être multipliée comme les autres, & elle exige le même traitement que la première.

PASSE-ROSE, MAUVE-ROSE D'OUTRE MER ou DE TREMIER. *Voy.* ALCEA ROSEA.

PASSE-VELOURS ou AMA-RANTHE. *Voy.* AMARANTHUS CAUDATUS. L.

PASSIFLORA. *Lin. Gen. Plant.* 910. *Granadilla. Tourn. Inst. R. H.* 240. [*Passion-flower.*] Fleur de la Passion ; Grenadille.

Caractères. Le calice de la fleur est uni, coloré, & a cinq feuilles ; la corolle a cinq demi-pétales en forme de lance, larges, unis & obtus ; le nectaire a une triple couronne ; le dehors, qui est plus long, est attaché en-dedans des pétales ; mais il est plus large,

& serré au-dessus : la fleur a cinq étamines fixées par leur base à la colonne du style qui est annexé au germe ; elles s'étendent audehors horisontalement, & sont terminées par des antheres oblongues, obtuses & penchées. Le style forme une colonne érigée & cylindrique, sur le sommet de laquelle est un germe ovale, avec trois petits styles qui s'étendent en-dehors, & sont couronnés par des stigmats à tête ; le germe se change dans la suite en un fruit ovale, charnu, & a une cellule placée à l'extrémité du style, & remplie de semences ovales, fixées longitudinalement à la peau.

Ce genre de plantes est rangé dans la quatrieme section de la vingtieme classe de LINNÉE, avec celles dont les parties mâles & femelles sont jointes ensemble, & dont les fleurs ont cinq étamines.

Les especes sont :

1°. *Passi-flora incarnata, foliis trilobis, serratis. Amœn. Acad.* vol. 1. p. 230. *Hort. Ups.* 278. *Gron. Virg.* 140 ; Fleur de la passion, avec des feuilles à trois lobes, sciées.

Granadilla Hispanis, flos Passionis Italis, Hern. Mex. 888. t. 888. *Raii Hist.* 649 ; la Grenadille des Espagnols, & la fleur de la Passion des Italiens, ordinairement appelée Fleur de la Passion, à trois feuilles.

Clematis tri-folia, flore Roseo clavato. Bauh. Pin. 301.

Clematis tri-folia, sive Flos passionalis, flore viridi. Moris. Hist. 2. p. 5. S. 1. t. 1. f. 9.

412 **PAS**

Balsamina Indica , repens , triphylla , sivè folio hastato. Ambr. Phyt. 89. f. 90.

Murucuia Mali-formis alia. Maregr. Bras. 71.

2°. *Passi-flora cærulea , foliis palmatis , integerrimis. Amœn. Acad. vol.* 1. p. 23. f. 20. *Hort. Ups.* 278. *Gouan. Monsp.* 476. *Fabric. Helmst.* 350. *Knorr. Del.* 1. t. P *Kniph. cent.* 2. n. 50; Fleur de la Passion , avec des feuilles entieres & en forme de main ouverte.

Clematis quinque-folia Americana , sivè j.os Passionis. Rob. Ic.

Granadilla pentaphyllos , flore cæruleo magno. Boerrh. Ind. Alt. 2. p 81; Fleur de la Passion ou Grenadille à cinq feuilles , avec une fleur grosse & bleue , ou la Fleur de la Passion commune.

Flos Passionis major pentaphyllis. Sloan. Jam. 104. *Hist.* 1. p. 229. *Raii Suppl.* 339.

3°. *Passi-flora lutea , foliis trilobis , cordatis , æquaiibus , obtusis , glabris , integerrimis. Amœn. Acad. v.* 1 p. 224. f. 13. *Gron. Virg.* 140; Fleur de la Passion , avec des feuilles en forme de cœur , & à trois lobes égaux , unis , obtus & entiers.

Passi-flora foliis cordatis , trilobis , integerrimis , glabris , lateribus ungulatis. Horr. Cliff. 431. *Roy. Lugd.-B.* 261.

Passi-flora foliis tri lobis , integerrimis , laciniis semi-ovatis , acutis , integerrimis , glabris. Gron. Virg. 1 p. 112.

Clematis passionalis triphyllos , flore luteo. Moris. Hist. 1. p. 7. S. 1. t 2. f.

Flos Passionis minor , folio in tres lacinias , non serratas , minus profundas diviso. *Sloan. Jam.* 104. *Hist.* 1. p. 231.

Granadilla folio tri-cuspidi, flore parvo flavescente. Tourn. Inst. R. H. 240 ; Grenadille avec une feuille à trois pointes , & une petite fleur jaune.

4°. *Passi-flora glabra , foliis-trilobis , integerrimis , lobis sublanceolatis , intermedio productiore. Amœn. Acad. vol.* 1. p. 229 ; Fleur de la Passion , avec des feuilles à trois lobes , entieres , un peu en forme de lance , & dont le lobe du milieu est un peu plus long que les autres.

Flos Passionis minor , folio in tres lacinias non serratas profundius diviso , flore luteo. Sloan. Cat. Jam. 104 ; Fleur de la Passion , plus petite , avec une feuille profondément divisée en trois segmens non sciés , & une fleur jaune.

Clematis Indica , folio angusto , tri-fido , fructu Olivæ-formi, Plum. Amer. 70. f. 85.

5°. *Passi-flora sub-erosa , foliis tri-lobis , integerrimis , glabris , cortice sub eroso. Amœn. Acad.* 1. 226. *Jacq. Hort. t.* 163 ; Fleur de la Passion , avec des feuilles à trois lobes , unies & entieres , & une écorce semblable à du Liége.

Passi-floræ affinis , Hederæ folio , Americana. Pluk. Alm. 202. t. 210. f. 4.

Flos Passionis Curassavicus , folio glabro , tri-lobato , & angusto , flore flavescente & omnium minimo. Par. Bat. Pluk. Alm. 282 ; Fleur de la Passion de Curaçao , avec une feuille unie , à trois lobes , & étroite , qui produit une fleur jaune,

PAS

jaune , & la plus petite de toutes.

6°. *Paffi-flora olivæ-formis , foliis haftatis , glabris , petalis florum anguftioribus ;* Fleur de la Paffion, avec des feuilles unies , & en forme de hallebarde , ayant des pétales étroits aux fleurs.

Granadilla folio amplo tri-cuf- pidi , fructu olivæ-formi. Tourn. Inft. R. H. 240 ; Grenadille avec une large feuille à trois pointes , & un fruit comme une Olive.

7°. *Paffi-flora fœtida , foliis tri-lobis , cordatis , pilofis , invo- lucris multi-fido capillaribus Amœn. Acad.* 1. *p.* 228. *f.* 17 ; Fleur de la Paffion , avec des feuilles à trois lobes en forme de cœur , & velues , dont l'enveloppe de la fleur eft compofée de plufieurs folioles ca- pillaires.

Flos Paffionis albus , reticu- latus Herm. Par. 173. *f.* 173.

Granadilla fœtida , folio tri- cuspidi , villoso , flore albo. Tourn. Inft. R. H. 240 ; Grenadille fétide , avec une feuille à trois lobes velus , & une fleur blanche.

Clematis Indica , hirfuta , fœ- tida. Plum. Amer. 71. *f.* 86.

8°. *Paffi-flora variegata , fo- liis haftatis , pilofis , ampliori- bus , involucris multi-fido-capilla- ribus ;* Fleur de la Paffion , avec de larges feuilles , cou- vertes de poils , & à pointe de hallebarde , dont les enve- loppes font compofées de plu- fieurs fegmens capillaires.

Paffi-flora veficaria hederacea , foliis lanuginofis , odore tetro , filamentis florum ex albo & pur-

Tome V.

pureo variegatis. Pluk. Alm. 382. *t.* 104. *f.* 1.

Granadilla fœtida , folio tri- cufpidi , villofo , flore purpureo , variegato. Tourn. Inft. R. H. 241 ; Grenadille avec une feuille à trois lobes velus , & une fleur panachée de pourpre.

9°. *Paffi-flora holofericea , fo- liis tri-lobis , tomentofis , bafi utrin- què denticulo reflexo. Amœn. Acad.* 1. *p.* 229. *f.* 15 ; Fleur de la Paffion , avec des feuilles à trois lobes , cotonneufes , & un peu dentelées à chaque côté de la bâfe , qui eft réfléchie.

Paffi-flora foliis cordato-tri-lo- bis , integerrimis , bafi utrinquè denticulo reflexo. Hort. Cliff. 432. *Roy. Lugd.-B.* 261.

Granadilla folio haftato , ho- lofericeo , petalis candicantibus , fimbriis ex purpureo & luteo va- riis. Martyn. Dec. 51 ; Grena- dille avec une feuille foyeufe , & à pointe de hallebarde , & des fleurs dont les pétales font blancs , & panachés de pour- pre & de jaune.

10°. *Paffi-flora capfularis , fo- liis bi-lobis , cordatis , oblongis , petiolatis. Lin. Sp. Plant.* 957 ; Fleur de la Paffion avec des feuilles oblongues , en forme de cœur , à deux lobes , & poftés fur des pétioles.

Granadilla flore fuavè rubente , folio bi-corni. Tourn. Inft. R. H. 241 ; Grenadille à fleur d'un rouge tendre , ayant une feuil- le à deux cornes.

11°. *Paffi-flora vefpertilio , fo- liis bi-lobis , bafi rotundatis , bi- glandulofis , lobis acutis , diva- ricatis , fubtùs punctatis. Amœn. Acad.* 1. *p.* 223. *f.* 11 ; Fleur de la Paffion , avec des feuil-

414 PAS

les à deux lobes, ayant deux glandes globulaires à leur bâfe, & dont les lobes font aigus, éloignés l'un de l'autre, & ponctués en-deffous.

Granadilla bi-cornis, flore candido, filamentis intortis. Hort. Elth. 164. tab. 137. f. 164; Grenadille avec une feuille à deux cornes, une fleur blanche, & des filamens ou vrilles torfes.

12°. *Paffi-flora normalis, foliis bi-lobis, bafi emarginatis, lobis linearibus, obtufis, divaricatis, intermedio obfoleto mucronato. Amœn. Acad. 5. p. 248. Brown. Jam. 328. n. 11*; Fleur de la Paffion, avec des feuilles à deux lobes, linéaires & obtus, dont la bâfe eft échancrée, qui font éloignés l'un de l'autre, & qui ont leur milieu comme ufé & en pointe.

Coanenepilli feu Contrayerva. Hernand. Mex. 301; Grenadille, appelée Coanenepille ou Contrayerva par Hernandès.

13°. *Paffi-flora bi-corna, foliis bi-lobis, glabris, rigidis, bafi indivifis*; Fleur de la Paffion, avec des feuilles roides, unies, & à deux lobes, qui ne font point divifés à leur bâfe.

Granadilla folio bi-corni, glabro, rigido, flore albo. Houft. MSS.; Grenadille avec une feuille à deux cornes, roide & unie, & une fleur blanche.

14°. *Paffi-flora Murucuïa, foliis bi-lobis, tranfverfis, amplexicaulibus. Amœn. Acad. 1. p. 222. f. 8*; Fleur de la Paffion, avec des feuilles à deux lobes, tranfverfales & amplexicaules.

PAS

Paffi-flora perfoliata. Lin. Syft. Plant. tom. 4. p. 49. Sp. 9.

Murucuïa folio lunato. Tourn. Inft. R. H. 251; Murucuïa à feuilles en forme de croiffant.

Flos Paffionis perfoliatus five Periclymeni perfoliati folio. Sloan. Jam. 104. Hift. 1. p. 230. t. 142. f. 3. 4. Raii Suppl. 342.

15°. *Paffi-flora Mali-formis, foliis indivifis, cordato-oblongis, integerrimis, petiolis bi-glandulofis, involucris integerrimis. Amœn. Acad. 1. p. 220. f. 5*; Fleur de la Paffion, avec des feuilles non-divifées, en forme de cœur, oblongues, entieres, ayant deux glandes aux pétioles, & des enveloppes entieres aux fleurs.

Clematis Indica lati-folia, flore clavato, fructu Mali-formi. Tourn. Inft. R. H. 241; Grenadille à larges feuilles, avec un fruit en forme de Pomme, ordinairement appelée Grenadille dans les Indes Occidentales.

16°. *Paffi-flora Lauri-folia, foliis indivifis, ovatis, integerrimis, petiolis bi-glandulofis, involucris dentatis. Amœn. Acad. 1. p. 220. f. 6. Jacq. Obs. 1. p. 35. Hort. t. 162*; Fleur de la Paffion, avec des feuilles non-divifées, ovales & entieres, dont les pétioles ont deux glandes, & les enveloppes des fleurs font dentelées.

Clematis Indica, fructu Citriformi, foliis oblongis. Plum. Amer. 64. f. 80. Raii Suppl. 341.

Granadilla fructu Citri-formi, foliis oblongis. Tourn. Inft. R. H. 241; Grenadille avec un fruit en forme de Citron, & des feuilles oblongues, ordinairement appelée dans les Indes

PAS

Occidentales , *Limon aquatique* ou *Pomme de Lianne.*

Marquiaas. Mer. Surin. 21. *t.* 21.

17°. *Paſſi-flora cuprea, foliis indiviſis , ovatis , integerrimis , petiolis æqualibus. Amæn. Acad. vol.* 1. *p.* 219. *f.* 3 ; Fleur de la Paſſion, avec des feuilles non-diviſées, ovales & entieres , & des pétioles égaux.

Granadilla Americana , fructu ſub-rotundo , corollâ floris erectâ, petalis amœnè fulvis , foliis integris. Martyn. cent. 1. *p.* 37. *f.* 37 ; Grenadille d'Amérique, avec un fruit preſque rond, les corolles des fleurs érigées, les pétales d'une belle couleur de cuivre, & des feuilles entieres.

Granadilla flore cupreo , flore olivi-formi. Dill. Elth. 165 , *t.* 138. *f.* 165.

18°. *Paſſi-flora ſerrati-folia, foliis indiviſis , ſerratis. Amæn. Acad.* 1. *p.* 217. *f.* 1. *Jacq. Hort. f.* 10 ; Fleur de la Paſſion , avec des feuilles non diviſées, & ſciées.

Granadilla Americana , folio oblongo, læviter ſerrato, petalis ex viridi rubeſcentibus. Mart. cent. 1. *p.* 36. *f.* 36 ; Grenadille d'Amérique, avec des feuilles oblongues , légerement ſciées, & des pétales d'un rouge verdâtre.

19°. *Paſſi-flora multi-flora , foliis indiviſis , oblongis, integerrimis , floribus conſertis. Amæn. Acad.* 1. *p.* 221. *f.* 7 ; Fleur de la Paſſion, avec des feuilles non - diviſées , oblongues & entieres , & des fleurs raſſemblées en paquets.

PAS 415

Granadilla , flore minore corymboſo. Plum. Spec. 7.

Clematis Indica polyanthos , odoratiſſima. Plum. Amer. 75. *tab.* 90. *Raii Suppl.* 343 ; Clematite des Indes , à pluſieurs fleurs très-odorantes.

20°. *Paſſi-flora quadrangulatis , foliis indiviſis, ſub-cordatis , integerrimis , petiolis ſex-glanduloſis , caule membranaceo, tetragono. Lin. Sp. Plant.* 1356. *Jacq. Amer.* 231. *f.* 143 ; Fleur de la Paſſion , avec des feuilles non-diviſées, preſque en forme de cœur, & entieres, ayant ſix glandes aux pétioles, & une tige quarrée & membraneuſe.

Paſſi-flora foliis amplioribus , cordatis, petiolis glandulis ſex , caule quadrangulo , alato. Brown. Jam. 327 ; Fleur de la Paſſion , avec des feuilles très-grandes & en forme de cœur , ſix glandes aux pétioles , & une tige quadrangulaire & aîlée.

Incarnata. La premiere eſpece croît naturellement en Virginie , & dans d'autres parties de l'Amérique Septentrionale. Toutes les eſpeces ont d'abord été connues en Europe ; mais elles n'ont été communes dans les jardins anglois que depuis quelques années. La racine de celle-ci eſt vivace ; mais ſa tige eſt annuelle. Dans l'Amérique Septentrionale, elle périt juſques ſur terre chaque hiver , ainſi qu'en Angleterre , à moins qu'elle ne ſoit conſervée dans une ſerre chaude ; ſes tiges minces , & de quatre ou cinq pieds de hauteur , ſont garnies à chaque nœud de vrilles qui s'attachent à toutes les plantes voiſines, & leur fourniſſent un

D d 2

PAS

soutien. A chaque nœud sort
une feuille sur un pétiole
court. Ces feuilles ont, pour
la plupart, trois lobes oblongs,
qui se joignent à leur bâse;
mais les deux latéraux sont
quelquefois divisés, dans une
partie de leur longueur, en
deux segmens étroits, qui leur
donnent l'apparence de feuilles
à cinq lobes; elles sont minces, d'un vert clair, & légérement dentelées sur leurs bords:
les fleurs naissent aux nœuds
de la tige, près des pétioles
des feuilles; elles sont soutenues sur des pédoncules longs
& déliés, & se succedent à
mesure que les tiges s'élevent
en été. L'enveloppe de la fleur
est composée de cinq feuilles
oblongues, terminées en pointe
émoussée, d'un vert pâle, &
qui laissent voir, en s'ouvrant,
cinq autres feuilles ou pétales
blancs, avec une frange ou
un double cercle en rayons de
couleur pourpre placé autour du
style. Le rang du bas est le
plus long: dans le centre,
s'éleve un style en forme de
colonne, avec un germe rond
au sommet, entouré vers le
bas, où le style est fixé, par
cinq étamines plates, écartées
de tous côtés, & qui soutiennent chacune une anthere
oblongue, suspendue vers le
bas, & couverte en–dessous
d'une poussiere jaune. Les fleurs
ont une odeur agréable, mais
elles durent peu de tems; car
elles s'ouvrent le matin, & se
fannent le soir, pour ne plus
reparoître; elles sont remplacées par d'autres, qui sortent
des nœuds de la tige au dessus

des premieres. Quand la fleur
est fannée, le germe, qui est
rond, se gonfle, & devient un
fruit aussi gros qu'une *Pomme*
ordinaire, d'une couleur d'O-
range pâle, lorsqu'il est mûr,
& qui renferme plusieurs semences rudes & oblongues dans
une chair d'une saveur douce.

Cette espece se multiplie
communément par ses graines,
que l'on apporte de l'Amérique; car elles ne mûrissent pas
souvent en Angleterre, quoique j'aie eu quelquefois plusieurs fruits parfaitement mûrs
sur des plantes qui avoient été
plongées dans une couche de
tan, sous un châssis profond:
mais celles qui restent exposées en plein air, ne produisent
point de fruits ici: on les seme
sur une couche de chaleur
modérée, qui fera pousser les
plantes beaucoup plutôt que si
on les laissoit en plein air;
elles ont aussi plus de tems
pour acquérir de la force avant
l'hiver.

Quand les plantes ont poussé
jusqu'à la hauteur de deux ou
trois pouces, on les enleve
avec précaution; on les met
chacune séparément dans de
petits pots remplis d'une bonne
terre de jardin potager; on les
plonge dans une couche de
chaleur tempérée, pour leur
faire prendre racine, & les
faire avancer, & on les accoutume ensuite par dégrés au
plein air, auquel on les expose entiérement pendant l'été;
mais en automne, on les place
sous des vitrages, de maniere
qu'elles soient à l'abri des gelées, & qu'on puisse les expo-

er au plein air dans les tems doux. Au printems fuivant, on peut enlever des pots quelques-unes de ces plantes, & les mettre dans une plate-bande chaude, où chaque hiver on aura foin de les couvrir avec du tan. Elles fubfifterout ainfi plufieurs années ; leurs tiges périront en automne ; mais au printems fuivant, leurs racines en poufferont de nouvelles, qui fleuriront très-bien dans les années chaudes. Celles qui font en pots, doivent être placées fur une couche de tan ; quelques-unes pourront produire du fruit. On multiplie ici cette efpece, en marcottant, au commencement de Juin, les branches des plantes qui font en pots : elles prendront racine vers la fin d'Août.

Cœrulea. Quoique la feconde n'ait été apportée que depuis peu en Angleterre, elle y eft cependant aujourd'hui la plus commune de toutes. Elle croit naturellement dans le Bréfil ; mais elle eft affez dure pour profiter en plein air, fans être endommagée, fi ce n'est dans les hivers durs, qui détruifent communément fes branches, & quelquefois fes racines. Elle s'élève en peu d'années à une grande hauteur, lorfqu'on lui fournit un foutien.

J'ai vû quelques-unes de ces plantes dont les branches avoient plus de quarante pieds de longueur. Les tiges de cette efpece deviennent prefque auffi groffes que le bras, & font couvertes d'une écorce de couleur pourpre, fans être fort ligneufes. Les rejettons de ces tiges croiffent fouvent de douze ou quinze pieds dans un été ; ils font très-minces, & fe penchent jufqu'à terre, fi l'on ne les fixe point ; ils s'entremêlent alors, & paroiffent fort défagréables à la vue. Ces branches font garnies à chaque nœud d'une feuille en forme de main, compofée de cinq lobes unis & entiers, dont celui du milieu, qui eft le plus grand, a prefque quatre pouces de longueur fur un de largeur au milieu ; les autres diminuent par dégrés, & les deux extérieurs font fouvent divifés fur leurs côtés en deux plus petits fegmens ; leurs pétioles ont près de deux pouces de longueur, & font accompagnés de deux petites feuilles ou oreilles qui embraffent les tiges de leur bâfe ; du même nœud fort une vrille longue, qui s'entortille autour des plantes voifines, pour fupporter les tiges : fes fleurs naiffent aux mêmes nœuds qui produifent les feuilles ; elles font foutenues par des pédoncules de trois pouces de longueur : leur enveloppe extérieure eft compofée de trois feuilles concaves, ovales, d'un vert plus pâle que les feuilles de la plante, & un peu plus longues que la moitié du calice ou godet, qui eft formé par cinq feuilles émouffées, oblongues, & d'un vert très-pâle. Ce calice renferme cinq pétales à-peu-près de la taille & de la grandeur de celles du calice, & poftées alternativement entr'elles ; du centre de la fleur s'élève une efpece de colonne femblable à une maf-

D d 3

418 PAS

sue épàisse, & d'un pouce environ de longueur, au sommet
de laquelle est un germe ovale ;
à sa bâse s'étendent horisontalement cinq étamines en forme
d'alène, & terminées par des
antheres oblongues, larges,
fixées par le milieu à l'étamine,
inclinées vers le bas, qui peuvent se mouvoir sans se déta-
tacher, & couvertes d'une
poussiere jaune : à côté du
germe s'élevent trois styles
minces, de couleur tirant sur
le pourpre, d'un pouce environ de longueur, écartés l'un
de l'autre, & terminés par des
stigmats émoussés ; autour du
bas de la colonne sont deux
rangs de rayons, dont l'intérieur, qui est le plus petit,
est dirigé en-dehors & vers le
haut de la colonne ; ils ont
presque la moitié de la longueur des pétales, & s'étendent à plat au-dessus. Ces rayons
sont composés d'un grand nombre de filamens minces, d'une
couleur pourpre au fond, &
bleue au-dehors. Les fleurs ont
une odeur agréable ; elles ne
durent qu'un jour, & se fa-
nent ensuite ; le germe, qui
est placé au sommet de la colonne, se gonfle, & devient
un fruit gros, large & ovale,
de la forme & de la grosseur
d'une Prune de Mogul, & d'un
jaune pâle, quand il est mùr.
Ce fruit contient une pulpe
un peu douce, mais désagréable, dans laquelle se trouvent
des semences oblongues. Cette
plante commence à fleurir dans
les premiers jours du mois de
Juillet, & continue à produire
des fleurs journellement, jus-

PAS

qu'à ce que les gelées d'automne les arrêtent.

On peut la multiplier par
ses graines, qui doivent être
semées comme celles de la
premiere espece. On traite
aussi de même les plantes qui
en proviennent, jusqu'au printems suivant ; alors on les retire des pots, pour les placer
contre une muraille bien exposée, & assez hautes pour
qu'elles puissent y étendre leurs
jets, qui, sans cela, retom-
beroient, s'entremêleroient,
& auroient une mauvaise apparence.

Cette plante peut servir à
couvrir des murs de bâtimens :
quand elle est bien enracinée,
il suffit de palisser ses jets
exactement, pour les empêcher de tomber ; & si l'hiver
devient dur, de répandre du
terreau sur leurs racines : la
gelée n'y pénetre pas.

On couvre aussi leurs branches avec des nattes, du chaume de Pois, de la paille,
ou quelqu'autre litiere légere,
pour les préserver des rigueurs
du froid. On doit ôter ces couvertures dans des tems doux,
sans quoi les branches se moisiroient ; ce qui leur feroit
plus de tort que la gelée. On
taille ces plantes au printems,
en retranchant toutes les branches foibles, & en raccourcissant les plus fortes à quatre ou cinq pieds, pour leur
faire pousser des jets vigoureux, qui fleurissent dans l'année suivante.

On multiplie aussi cette espece, en marcottant ses branches, qui, dans une année,

auront pouſſé d'aſſez fortes racines pour être ſéparées des
anciennes plantes , & tranſplantées à demeure. Leurs
boutures prennent auſſi racine , ſi on les plante dans un
ſol marneux & pas trop ferme , au printems , avant qu'elles aient commencé à pouſſer.
En les couvrant de cloches
pour en exclure l'air , elles
réuſſiſſent beaucoup mieux que
de toute autre maniere. Quand
elles pouſſent des rejettons ,
on leur donne de l'air , pour
les empêcher de s'affoiblir , &
on les traite enſuite comme
les jeunes marcottes.

Les plantes de marcottes ou
de boutures ne produiſent point
de fruits auſſi abondamment
que celles qu'on éleve de ſemences : & j'ai remarqué qu'a-
près les avoir multipliées ainſi
deux ou trois fois de ſuite ,
elles deviennent preſque toujours ſtériles ; ce qui arrive
auſſi à toutes les autres plantes.

Lorſque , dans un hiver rigoureux , les tiges de ces plantes périſſent juſques ſur terre ,
leurs racines en pouſſent ſouvent de nouvelles dans l'été
ſuivant : c'eſt pourquoi il ne
faut pas les déranger ; cependant ces racines courent moins
de riſque quand on les couvre
avec du terreau , quoique leurs
tiges ſoient également détruites.

Il y a une variété de cette
eſpece , dont les lobes des
feuilles ſont beaucoup plus
étroits , & diviſés preſque juſqu'au bas : ſes fleurs paroiſſent
plus tard dans l'été , & leurs
pétales ſont plus étroits , &
d'un blanc plus pur : mais

comme je ne la regarde que
comme une variété féminale
de la ſeconde , je n'en donnerai pas une plus longue deſcription.

Lutea. La troiſieme eſpece ,
qui eſt originaire de la Virginie & de la Jamaïque , a une
racine vivace & rampante ,
de laquelle ſortent pluſieurs
tiges foibles de trois ou quatre pieds de haut , & garnies
de feuilles fort reſſemblantes
à celles du Lierre , & preſque
auſſi larges , mais d'un vert
clair ou pâle , & de peu de
conſiſtance : ſes fleurs ſortent
des aiſſelles des tiges ſur des
pétioles minces , & d'un pouce & demi de longueur ; elles ont à leur bâſe des vrilles
très-minces , qui s'attachent aux
ſupports voiſins : ſes fleurs
ſont d'un jaune pâle , & de
la largeur d'une piece de douze ſous , lorſqu'elles ſont épa-
noüies , mais de peu d'apparence. Cette plante peut être
multipliée par ſes racines rampantes , que l'on diviſe en
Avril , pour les mettre dans
les places qui leur ſont deſtinées. On les plante dans une
plate-bande chaude , & on les
traite comme celles de la premiere eſpece. J'en ai conſervé quelques unes au jardin de
Chelſéa , pendant pluſieurs années , dans une plate-bande ,
à l'expoſition du ſud-oueſt ;
mais elles furent toutes détruites par la gelée de 1740.

Glabra. La quatrieme ſe
trouve à la Jamaïque ; elle a
une racine vivace , de laquelle
ſortent pluſieurs tiges minces
de quatre ou cinq pieds de

D d 4

haut, dont les nœuds font à quatre ou cinq pouces l'un de l'autre ; de chacun de ces nœuds fort une feuille, une vrille, & une fleur : les feuilles ont trois lobes, dont celui du centre a trois pouces de longueur fur un de largeur au milieu, & les deux latéraux ont environ deux pouces fur neuf lignes de large ; ils font minces, & d'un vert clair : les fleurs font plus petites que celles des précédentes, de couleur verdâtre, & produifent un fruit ovale, de la groffeur d'une petite Olive, & de couleur de pourpre, lorfqu'il eft mûr.

Suberofa. La cinquieme, qui nait fpontanément dans la plupart des Ifles des Indes Occidentales, s'éleve, avec une tige foible, à vingt pieds de hauteur. Quand les tiges vieilliffent, leur écorce devient épaiffe & fpongieufe comme celle du Liége, & fe fend de la même maniere. Les plus petites branches ont une écorce unie, & font garnies à chaque nœud, de feuilles liffes & poftées fur des pétioles fort courts ; elles ont trois lobes, dont celui du milieu eft beaucoup plus large que ceux de côté ; de maniere qu'elles reffemblent à une pointe de hallebarde : fes fleurs fon petites, d'un jaune verdâtre, & produifent un fruit petit, ovale, & d'un pourpre foncé, quand il eft mûr.

Olivæ formis. La fixieme efpece croit, fans culture, en Amérique ; elle a une racine vivace, de laquelle fortent

plufieurs tiges minces de huit ou dix pieds de hauteur, & garnies de feuilles vertes, liffes. foutenues fur de minces pétioles, & légerement découpées en trois lobes terminés en pointe aiguë, & en forme de hallebarde : le lobe du milieu eft pofté obliquement fur le pétiole : les fleurs naiffent aux ailes des feuilles fur des pédoncules fort courts, & font d'un jaune pâle ; leurs pétales font fort étroits, & plus longs que ceux des deux efpeces précedentes : le fruit eft plus petit, d'une forme ovale, & d'un pourpre foncé, quand il eft mûr.

Fœtida. La feptieme croit naturellement dans la plupart des Ifles des Indes Occidentales, où les Colons Anglois l'appellent *Love in a Mift* ; c'eft-à-dire, *Amout dans un brouillard.* Sa racine eft annuelle ; fes tiges s'élevent à la hauteur de huit ou dix pieds, quand on leur fournit un foutien ; elles font canelées & velues ; fes feuilles font en forme de cœur, & divifées en trois lobes, dont celui du milieu a trois pouces de longueur fur un & demi de largeur, & ceux de côté font plus courts, mais plus larges, & couverts d'un poil brun & court. Les vrilles fortent des mêmes nœuds que les feuilles, ainfi que les fleurs, dont les pédoncules ont deux pouces de longueur, font velus, & affez forts : le calice de la fleur eft compofé de filamens minces & veloutés, de la forme à-peu-près

d'un filet, plus longs que les pétales, & qui s'élevent autour, de maniere que les fleurs ne font pas fort vifibles à une certaine diftance ; elles font blanches, & de peu de durée : leur ftructure eft la même que celles des autres efpeces ; elles font remplacées par un fruit ovale, brunàtre, de la groffeur d'une petite *Pomme de Pepin d'or*, d'un vert jaunàtre, & renfermé dans le calice à filet.

On multiplie cette plante par fes graines, qu'on répand fur une couche chaude, dans le commencement du printems ; quand les plantes font en état d'être enlevées, on les met chacune féparément dans de petits pots remplis d'une terre légere de jardin potager ; on le replonge dans une autre couche chaude : on les tient à l'abri du foleil, jufqu'à ce qu'elles aient formé de nouvelles racines, & on les traite enfuite comme les autres plantes qui viennent des mêmes contrées, en obfervant de leur donner des pots plus larges, à mefure que les racines augmentent. Quand ces plantes font devenues trop grandes pour pouvoir être contenues fous les vitrages de la couche, on les met dans une caiffe de vitrage aërée, où elles feront à l'abri du froid, & où l'on pourra leur donner de l'air dans les tems chauds. Ces plantes fleuriffent dans cette fituation pendant le mois de Juillet, & leurs femences mùriffent en automne. Toutes les parties de cette plante

répandent une odeur défagréable, quand on les touche.

Il y a une variété de celle-ci, à moins que ce ne foit une efpece diftincte, à feuilles veloutées, & moins larges que celles de la précédente : la feuille entiere reffemble plus à une hallebarde, & celles qui croiffent vers le haut des tiges ont de très-petites dentelures ; de forte qu'elles paroiffent être fimples & fans lobe : les fleurs font auffi plus petites, mais de la même forme, & les racines font d'une plus courte durée, ce qui me porte à croire que c'eft une efpece diftincte.

Variegata. La huitieme ayant quelque reffemblance, avec la feptieme, plufieurs perfonnes l'ont regardée comme n'en étant qu'une variété accidentelle ; mais il n'eft pas douteux qu'elle ne foit une efpece différente. Les tiges de celle-ci s'élevent à plus de vingt pieds de haut, & fubfiftent deux ou trois ans ; les feuilles font plus larges, mais de la même forme, & veloutées ; fes vrilles font très-longues, ainfi que les pédoncules des fleurs, qui font unies fans être veloutées comme les premieres : le calice de la fleur eft en filet, mais moins long que celui de l'efpece précédente ; les fleurs font plus larges, & leurs rayons font d'un bleu plus clair ; le fruit eft beaucoup plus petit & plus rond ; & quand il eft mùr, il devient d'un jaune foncé.

Holofericea. La neuvieme, que le Docteur HOUSTOUN a

422 PAS

trouvée à la Vera-Cruz, est une plante vivace ; ses tiges s'élevent à vingt pieds de hauteur, & se divisent en plusieurs branches minces, & couvertes d'un coton doux & velouté ; ses feuilles sont en forme de hallebarde, de trois pouces de longueur sur un & demi de large à leur bâse, d'un vert clair, molles, soyeuses au toucher, & placées obliquement sur leurs pétioles : ses fleurs naissent aux ailes des feuilles comme celles de la précédente, & ne sont pas à moitié aussi larges que celles de la seconde, quoique de la même forme. Les pétales sont blancs, & les rayons ou filamens sont d'un pourpre mêlé de jaune ; le fruit est petit, rond & jaune, lorsqu'il est mûr.

Capsularis. La dixieme, qu'on rencontre à la Jamaïque, est aussi vivace ; ses tiges sont minces, de vingt pieds de hauteur, quand on leur fournit un soutien, & divisées en plusieurs branches foibles ; les feuilles, les fleurs & les vrilles sortent des mêmes nœuds : les feuilles ont quatre pouces de longueur sur trois de large ; elles sont arrondies à leur bâse en forme de cœur, mais terminées à leur extrémité par deux cornes, qui, dans quelques feuilles, sont plus aiguës que dans d'autres, & dont plusieurs semblent être un peu creusées au sommet, comme celles du *Tulipier* ; elles ont trois veines longitudinales, qui se joignent à la bâse du pétiole, & dont les deux latérales se jettent vers les bords

de la feuille au milieu, & se retirent en-dedans au sommet. Ces feuilles sont d'un vert foncé en-dessus, pâles en-dessous, & placées sur de courts pétioles ; les pédoncules sont très-minces, de couleur tirant sur le pourpre, & d'un pouce & demi de longueur : les fleurs sont de la même forme que celles des autres especes ; mais quand elles sont épanouïes, elles n'ont qu'un pouce & demi de diametre ; elles sont d'un rouge léger, & ont peu d'odeur : le fruit est petit, ovale ; & lorsqu'il est mûr, il devient d'une couleur de pourpre.

Vespertilio. La onzieme à été découverte par M. ROBERT MILLAR, près de Carthagène, dans l'Amérique Méridionale ; elle a des tiges minces, canelées, d'un rouge brunâtre, & divisées en plusieurs branches minces, & garnies de feuilles en forme d'ailes de chauve-souris quand elles sont étendues, d'environ sept pouces de longueur, & de deux pouces & demi, depuis leur bâse jusqu'à l'extrémité. Les pétioles qui les supportent, ont un demi-pouce de longueur, & de leur extrémité partent trois nervures, dont deux s'étendent de chaque côté vers les deux pointes étroites de la feuille, & l'autre s'éleve vers son extrémité. La figure de cette feuille est fort singuliere : les fleurs sortent des nœuds de la tige comme celles des autres, sur des pédoncules courts & minces ; elles ont à-peu-près trois pouces de diametre, quand elles sont épanouïes ; leurs pé-

tales & leurs rayons font blancs; les rayons font minces, entrelacés, & étendus au-delà des pétales. Je n'ai pas encore vu un fruit entier de cette efpece.

Normalis. La douzieme a été découverte par le Docteur HOUSTOUN, à la Vera-Cruz, dans la Nouvelle Efpagne; elle a des tiges minces & angulaires, qui s'élevent à la hauteur de vingt pieds, & pouffent plufieurs branches garnies de feuilles en forme de croiffant, avec deux lobes émouffés, qui s'étendent à chaque côté, de maniere qu'elles ont la forme d'une demi-lune; les fleurs & les vrilles fortent des mêmes nœuds fur les tiges: les fleurs font petites, & de couleur pâle, mais de la même forme que celles des autres efpeces; elles produifent un fruit ovale, pourpre, & de la groffeur d'une petite graine de Raifin.

Bi-corna. La treizieme reffemble un peu à la douzieme; mais fes tiges font plus rondes, & deviennent ligneufes; fes feuilles font prefque auffi fermes que celles du *Laurier*, & divifées moins profondément que celles de la précédente: fes fleurs font poftées fur de longs pédoncules placés horifontalement; elles font petites, blanches, & de la même forme que celles de la premiere efpece: les fruits qui leur fuccedent, font ovales, petits, de couleur pourpre, & ferrés près des petales des fleurs, qui font perfiftans. Cette plante a été découverte par le Docteur

HOUSTOUN, à Carthagène de l'Amérique, où elle croît naturellement.

Murucuïa. La quatorzieme croît fpontanément dans la plupart des Ifles des Indes Occidentales. TOURNEFORT l'a féparée de ce genre, & l'a nommée *Murucuia*, qui eft fon nom Bréfilien; elle a des tiges minces, grimpantes & cannelées, qui pouffent des vrilles à leurs nœuds, au moyen defquelles elles s'attachent aux plantes voifines, & s'élevent ainfi à la hauteur de dix à douze pieds. Ces tiges font garnies de feuilles découpées en deux lobes à leur bâfe, mais feulement un peu creufées au fommet entre chaque pointe; de maniere que la partie oppofée au pétiole eft faillante; la bâfe des deux lobes s'étend & fe rencontre de façon qu'elles paroiffent avoir embraffé la tige; mais quand on les examine de près, on les trouve divifées vers le pétiole, qui eft court & courbé; ce qui fe rencontre rarement. La feuille a deux veines, dont la couleur tire fur le pourpre, qui s'élevent du pétiole, & s'étendent à chaque côté vers les pointes des lobes; les feuilles font d'un vert luifant en-deffus, & pâle en-deffous. Les vrilles qui fortent avec les feuilles font longues, coriaces, & de couleur pourpre: les fleurs naiffent aux extrémités des branches, & fortent par paires de chaque côté; leurs pédoncules font de couleur pourpre, d'un pouce & demi de longueur, & foutiennent à leur extrémité une fleur

dont le calice est composé de cinq feuilles pourpre, qui renferment, dans une espece de tube, cinq pétales pourpre & fort étroits : la colonne du centre de la fleur est de la même longueur que les pétales ; mais les étamines s'élevent à un pouce au-deffus. Quand ces fleurs sont fannées, le germe se gonfle & se change en un fruit ovale de couleur pourpre, & de la groffeur d'une groseille rouge ; il contient une chair molle, dans laquelle sont renfermées les femences.

Mali-formis. La quinzieme espece se trouve dans les Indes Occidentales, dont les habitans lui donnent le nom de *Grenadille*, & font fervir son fruit fur leur table ; elle a une tige épaisse, grimpante, herbacée & triangulaire, dont chaque nœud produit une vrille mince, qui s'attache aux arbustes, aux haies, &c. ; au moyen de quoi elle s'éleve à la hauteur de quinze ou vingt pieds. Cette tige est garnie à chaque nœud d'une feuille large, ovale, en forme de cœur, de fix pouces de longueur fur quatre de largeur au milieu, & dentelée à fa bâfe, où elle est supportée par un court pétiole qui fort des branches ; elle est ronde au fommet, & terminée en pointe aiguë ; elle a deux larges stipules ou oreilles, jointes aux tiges qui entourent les pédoncules, les pétioles & la bâfe de la vrille. Ces feuilles sont d'un vert vif, d'une texture mince, & fortifiées dans leur longueur par

une côte robuste, de laquelle forrent plusieurs petites veines qui se divergent ou coulent vers les côtés, & se tournent ensuite vers le fommet. Les fleurs sont postées fur des pédoncules affez longs, & garnis de deux petites glandes au milieu. L'enveloppe de la fleur est composée de trois feuilles molles, veloutées, d'un rouge pâle, & rayées d'un rouge vif ; les pétales sont blancs, & les rayons bleus. Ces fleurs sont larges, & ont une belle apparence, lorsqu'elles sont tout-à-fait ouvertes ; mais elles sont d'une courte durée, comme cèlles des autres especes : elles se fuccedent pendant quelque tems fur les mêmes plantes ; quand elles sont paffées, le germe se gonfle, & devient un fruit rond, de la groffeur d'une groffe pomme, de couleur jaune, lorsqu'il est mûr, couvert d'une écorce plus épaisse qu'aucun de ceux des autres especes, & qui renferme une chair douce, dans laquelle se trouvent plusieurs femences plates, oblongues, d'une couleur brunâtre, & un peu rudes au toucher.

Lauri-folia. La seizieme espece qui est auffi originaire des Indes Occidentales, a des tiges grimpantes & coriaces, qui pouffent des vrilles comme les autres, & s'attachent aux arbres & haies voisines pour se foutenir : elles s'élevent ainfi au-deffus de vingt pieds de hauteur, & produisent plusieurs branches latérales ; les feuilles, qui ont quatre ou cinq pouces de longueur fur deux de lar-

geur, font d'une confiftance affez épaiffe, d'un vert vif en-deffus, & pâle en deffous. Les fleurs fortent des nœuds des tiges fur des pédoncules d'un pouce & demi de longueur ; les boutons des fleurs, font auffi gros qu'un œuf de pigeon, avant qu'ils foient épanouïs ; l'enveloppe de la fleur eft com-pofée de trois feuilles larges, vertes, ovales, dentelées fur leurs bords, & creufées en forme de cuiller ; au-dedans eft le godet ou corolle de la fleur, compofée de cinq pé-tales oblongs, d'un vert pâle à l'extérieur, & blanchâtres en-dedans ; elles ont environ un pouce & demi de longueur fur un demi-pouce de largeur ; les pétales font blancs, & pla-cés alternativement avec les feuilles du calice ; mais ils n'ont que la moitié de leur largeur, & font marqués de plufieurs petites taches d'un rouge brunâtre : les rayons de la fleur font violets ; la co-lonne du centre eft jaunâtre, ainfi que le germe, qui eft rond ; mais les trois ftyles font de couleur pourpre. Ces fleurs ont une odeur agréable ; & lorfqu'elles font fannées, le germe fe gonfle jufqu'à la grof-feur d'un œuf de poule, & devient jaune lorfqu'il eft mûr ; fon écorce eft molle, épaiffe, d'un goût acide & agréable ; il étanche la foif, diminue la chaleur de l'efto-mac, donne de l'appétit, & rend de l'activité à l'efprit. On le donne ordinairement dans les fievres ; fes femences font brunâtres & en forme de cœur.

Les habitans de la Martinique appellent ce fruit *Pomme de Lianne.*

Cuprea. La dix-feptieme ef-pece croît naturellement dans les Ifles de Bahama, d'où le feu M. CATESBY a envoyé fes femences en Angleterre ; elle a des tiges minces, grimpan-tes & triangulaires, qui pouf-fent à chaque nœud des vrilles, au moyen defquelles elles s'at-tachent à tout ce qui les avoifine. Ces tiges, qui s'éle-vent à douze ou quatorze pieds de hauteur, font garnies de feuilles ovales, oblongues, de deux pouces environ de longueur fur un de largeur, d'un vert clair, & entieres : leurs pétioles font minces, & d'un pouce de longueur ; ils donnent origine à trois veines longitudinales, dont l'une reg-ne dans le milieu de la feuille, & les deux autres fe divergent fur les côtés, & fe rappro-chent l'une de l'autre à la pointe. Les fleurs naiffent aux aiffelles de la tige fur des pé-doncules minces, & d'un pouce de longueur ; le calice de la fleur eft compofé de cinq feuil-les pourpre, oblongues & étroites, qui renferment cinq pétales étroits de la même couleur, & qui fe tournent en arriere quelque temps après qu'ils font épanouis. La colonne du milieu de la fleur eft fort longue, mince, & foutient un germe rond, de la bâfe duquel fortent cinq étamines minces, & terminées par des antheres penchées & oblongues : du haut du germe s'élevent trois ftyles minces, qui s'étendent

féparément, & font couronnés par des ftigmats ronds.
Quand les feuilles font flétries,
le germe fe gonfle, & de-
vient un fruit ovale de la groffeur d'un œuf de moineau, de
couleur pourpre, lorfqu'il eft
mûr, & rempli de femences
oblongues, renfermées dans
une chair molle.

Multi-flora. La dix-neuvieme
efpece a été découverte par le
Docteur HOUSTOUN à la Vera-
Cruz dans la Nouvelle-Efpag-
ne, où elle croit naturellement, & d'où il a envoyé en
Angleterre, en 1731, fes femences qui ont réuffi dans
plufieurs jardins. Elle a des
tiges minces & grimpantes,
qui pouffent plufieurs petites
branches, & s'élevent à la
hauteur de vingt-cinq ou trente
pieds, quand elles rencontrent
des foutiens dans le voifinage,
auxquels elles s'attachent par
leurs vrilles. Les tiges de cette
efpece deviennent ligneufes
vers le bas en vieilliffant: leurs
nœuds ne font pas éloignés
les uns des autres ; les feuilles
ont des pétioles courts & minces ; leur longueur eft de trois
pouces & demi, & leur largeur de deux ; elles font terminées en une pointe au fommet, unies, entieres, & d'un
vert vif : les fleurs fortent aux
ailes des feuilles fur de longs
pédoncules ; leur calice eft compofé de cinq feuilles oblongues, vertes en-dehors, &
blanchâtres en-dedans ; la corolle a cinq pétales blancs,
oblongs, placés alternativement
avec les feuilles du calice qui
s'ouvrent ; les rayons font

d'une couleur pourpre bleuâ-
tre, & tirant fur le rouge vers
le bas ; la colonne du centre
eft courte & épaiffe ; le germe,
qui eft placé fur le fommet, eft
ovale ; & quand la fleur fe fane,
il fe gonfle jufqu'à la groffeur
d'un œuf de poule, & de-
vient d'un jaune pâle, quand il
eft mûr ; il renferme plufieurs
femences oblongues, éparfes
dans une chair molle. Les fleurs
de cette efpece ont une odeur
agréable, mais elles font d'une
courte durée ; car elles reftent rarement ouvertes plus de
vingt quatre heures : elles fe
fuccedent fur les mêmes plantes
depuis le mois de Juin jufqu'en
Septembre, & quelquefois leurs
fruits mûriffent ici.

Quadrangularis. La vingtieme
efpece reffemble beaucoup à
la quinzieme, par fa tige &
fes feuilles ; mais les tiges de
celle-ci ont quatre angles,
au-lieu que celles de la quinzieme n'en ont que trois ; fes
feuilles font moins creufées à
leur bâfe, & prefque en forme de cœur : fa fleur eft beaucoup plus large, quoique fort
femblable, par fa couleur, à
celles de la quinzieme, & le
fruit eft prefque deux fois plus
gros, & d'un goût fort agréable.

Cette efpece, étant traitée
comme la quinzieme, produira
des fleurs, & perfectionnera
fouvent fon fruit en Angleterre.
Quelques perfonnes l'ont confondu avec cette derniere, &
l'ont fait paffer pour la *Grenadille.*

Culture. Toutes ces efpeces
vivaces étant originaires des

parties méridionales de l'Amérique, ne peuvent être conservées ici fans le fecours d'une ferre chaude. Ce moyen eft le feul qu'on puiffe employer pour leur faire faire des progrès, & pour leur faire produire des graines ; car quoique plufieurs puiffent être confervées en plein air pendant les mois les plus chauds de l'été, cependant elles y profitent peu, & ne produifent pas beaucoup de fleurs. Ainfi, il eft néceffaire de les tenir dans des pots, de les plonger dans la couche de tan de la ferre chaude, & de les dreffer contre un efpalier. La meilleure méthode pour les avoir en perfeétion, eft d'élever une bordure de terre au dos de la couche de tan, que l'on fépare par des planches, pour empêcher la terre de fe mêler avec le tan. Lorfque les plantes font affez fortes, on les fort des pots, pour les planter dans cette bordure, près de laquelle il faut élever un treillage jufqu'au haut de la ferre chaude, pour y palifter les tiges de ces plantes. A mefure qu'elles avanceront, elles formeront une haie qui cachera le mur de la ferre chaude. Comme la plupart de leurs feuilles confervent leur fraîcheur toute l'année, lorfqu'elles font entremêlées avec les fleurs qui pouffent en abondance en été, elles font un effet très-agréable.

La terre n'étant féparée du tan que par une planche, fe confervera chaude, & aidera beaucoup à faire pouffer les racines. Cette bordure ne doit

pas avoir moins de deux pieds de largeur fur trois de profondeur, qui eft celle que l'on donne ordinairement à la foffe du tan. Par tout où l'on veut avoir de ces plates-bandes, on doit donner au moins huit ou neuf pieds de largeur aux couches, de maniere qu'il leur refte encore fix ou feptpieds, déduétion faite de celles de la plate-bande. Il eft néceffaire de la féparer de la couche avec des planches fortes, enduites d'une compofition faite avec de la poix fondue, de la pouffiere de briques pilées, & de l'huile, pour les conferver longtems. On change chaque année cette terre, en obfervant de retirer l'ancienne avec foin d'entre les racines. Par ce traitement, j'ai vu des plantes parvenir à une grande perfeétion ; mais quand on n'a point cette facilité, on tire les plantes hors des pots, & on les met dans le tan, lorfqu'il eft à moitié pourri ; elles y poufferont beaucoup de racines, & profiteront pendant deux ou trois années auffi-bien qu'on puiffe le défirer : mais comme, après ce tems, les racines s'étendent à une grande diftance dans la couche, la fermentation qui s'empare du nouveau tan qu'on eft obligé d'y mettre, les brûle, & détruit bientôt les plantes.

On multiplie ces *Grenadilles* par leurs graines, qu'on feme au printems fur une bonne couche chaude. Quand elles font en état d'être enlevées, on les met chacune féparément dans de petits pots remplis d'une

bonne terre de jardin potager; on les plonge dans une couche de tan: on les tient a l'ombre jufqu'à ce qu'elles aient formé de nouvelles racines, & on les traite enfuite comme les autres plantes délicates qui viennent des mêmes contrées; lorfqu'elles font devenues trop hautes pour pouvoir refter fous les vitrages de la couche, on les enleve hors des pots, pour les planter dans la ferre chaude, comme il a été dit ci-deffus.

Comme ces efpeces ne perfectionnent pas fouvent leurs femences ici, on peut encore les multiplier, en marcottant leurs branches. Si l'on fait cette opération en Avril, elles auront des racines vers le milieu d'Août ou de Septembre; alors on pourra les féparer des vieilles plantes, & les placer dans des pots, pour leur faire acquérir de la force, ou les mettre à demeure dans la bordure de la ferre chaude.

On multiplie auffi quelquesunes de ces efpeces par boutures; on les plante dans des pots vers le milieu ou la fin du mois de Mars; on les plonge dans une couche de chaleur modérée; on les tient à l'ombre, & on les arrofe légérement autant de fois que la terre l'exige: lorfqu'elles auront pouffé des racines; ce qui aura lieu environ fix femaines après, on les traitera comme les plantes de femences.

PASTEL ou GUEDE. Voyez ISATIS TINCTORIA. L.

PASTENADE ou PANAIS. Voyez PASTINACA.

PASTEQUE ou MELON D'EAU. Voy. ANGURIA CITRULLUS DICTA.

PASTINACA. Tourn. Inft. R. H. 319. tab. 170, de Paftus; nourri, parce que la racine de cette plante eft bonne à manger. [Parfnep.] Panais ou Paftenade.

Caracteres. La fleur eft difpofée en ombelle; l'ombelle eft formée par plufieurs autres plus petites, qui font auffi compofées de plufieurs rayons; elles n'ont point d'enveloppe; & le calice eft à peine vifible: l'ombelle eft uniforme; les corolles ont cinq pétales courbés en forme de lance; les fleurs ont cinq étamines femblables à des poils, & terminées par des antheres rondes. Le germe, qui eft placé fous la fleur, foutient deux ftyles réfléchis, & couronnés par des ftigmats obtus; il fe change dans la fuite en un fruit elliptique, uni, comprimé, & divifé en deux parties, qui forment deux femences bordées & elliptiques.

Le genre de cette plante eft rangé dans la feconde fection de la cinquieme claffe de LINNÉE, qui comprend celles dont les fleurs ont cinq étamines & deux ftyles.

Les efpeces font:

1°. Paftinaca fylveftris; foliis fimpliciter pinnatis, hirfutis; Panais à feuilles fimples, ailées & velues.

Paftinaca fylveftris lati-folia. C. B. P. 155; Panais fauvage à larges feuilles.

Selinum Paftinaca. Crantz. Auftr. 161.

2°. Paftinaca fativa, foliis fimpliciter

PAS

simpliciter pinnatis , glabris ; Panais à feuilles ailées simplement , & unies.

Pastinaca sativa , lati-folia. C. B. P. 155 ; Panais cultivé , à larges feuilles.

3°. *Pastinaca opopanax , foliis decompositis , pinnatis. Hort. Cliff. 105. Mat. Med. 84. Roy. Lugd.-B. 114 ;* Panais à feuilles ailées & décomposées.

Pastinaca sylvestris altissima. Tourn. Inst. 319 ; le plus grand Panais sauvage , nommé par GASPAR BAUHIN : *Panax Costinum. Pin. 156.*

Panax Heracleum. Moris. Hist. 3. p. 305. S. 9. t. 17. f. 2.

Sylvestris. La premiere espece croît naturellement à côté des digues , & sur des terres seches en Angleterre. Cette plante est bis-annuelle ; la premiere année, elle pousse des feuilles qui s'étendent sur la terre. Ces feuilles sont velues , ailées simplement , & leurs lobes sont irrégulierement découpés. L'année suivante , on voit paroître des tiges de quatre ou cinq pieds de hauteur, canelées , velues , & garnies de feuilles ailées comme celles du bas , mais plus petites. Ces tiges se divisent vers le haut en branches , dont chacune est terminée par une grande ombelle de fleurs jaunes : à ces fleurs succedent des fruits comprimés , dont chacun forme deux femences plates & bordées. Cette plante fleurit en Juin , & ses femences mûrissent en Août.

Quoique les racines & les femences de cette espece soient quelquefois d'usage en Médecine , on ne la cultive cependant guere dans les jardins ; mais on va recueillir dans les champs les plantes qu'on apporte au marché. Les Droguistes vendent ordinairement les graines de l'espece cultivée , en place de celles-ci , parce qu'ils les achetent à bas prix , quand elles sont trop vieilles pour être femées ; mais alors elles n'ont plus aucune vertu. (1)

Sativa. La feconde espece a des feuilles unies , & d'un vert clair ou jaunâtre , en quoi elle differe de la premiere : ses tiges , qui s'élevent aussi à une hauteur plus considérable , font canelées plus profondément ; les pédoncules des ombelles font beaucoup plus longs , & les fleurs font d'un jaune plus foncé. Ces deux especes n'ont été regardées que comme des variétés. On croyoit que le *Panais* de jardin ne différoit du *Panais* sauvage que par la culture ; mais après les avoir cultivés l'un & l'autre pendant plufieurs années, j'ai reconnu qu'ils ne varient jamais , les femences de chaque plante ayant toujours produit les mêmes ; de maniere que je suis certain qu'elles forment deux especes distinctes & séparées.

On cultive cette feconde espece dans les jardins potagers ; ses racines font gros-

_____

(1) Les graines de Panais, comme celles de la plupart des autres plantes de cette classe , font carminatives & diurétiques : on peut les employer à la même dose que celles du Daucus & de l'Anis.

Tome V.

ses, douces, & fort nourris-
santes : on la multiplie par ses
graines , qu'il faut semer en
Fevrier ou Mars, dans un sol
riche , meuble , & bien labouré,
afin que les racines puissent
s'enfoncer ; car leur plus grande
qualité dépend de leur longueur & de leur grosseur. On
peut les semer seules , ou avec
des *Carottes* , suivant l'usage des
Jardiniers de Londres ; quelques-uns y mêlent aussi des
Porreaux , des *Ognons* & de
la *Laitue* : mais je n'approuve
point cette méthode ; car il n'est
pas possible que tant de différentes especes puissent bien
profiter ensemble , à moins
qu'on né laisse entre chaque
plante une distance considérable , & alors il est égal de
les semer séparément. Cependant les *Panais* & les *Carottes*
peuvent bien être semés ensemble , sur-tout si l'on destine
les *Carottes* à être mangées jeunes , parce que les *Panais* s'é-
tendent , & grossissent presque
toujours vers la fin de l'été ,
tems auquel les *Carottes* sont enlevées. On obtient ainsi une double récolte sur le même terrein.

Quand ces plantes ont poussé , il faut les houer , leur
donner dix pouces ou un pied
de distance , & détruire en
même tems toutes les mauvaises
herbes qui s'y rencontrent ;
car si on les laissoit croître ,
elles s'étendroient bientôt sur
les plantes , & les étoufferoient.
Ce travail doit être répété trois
ou quatre fois dans le printems , suivant la poussée des
mauvaises herbes ; mais à la
fin de l'été , lorsque les plantes

seront assez fortes pour couvrir
la terre , elles détruiront elles-
mêmes toutes les herbes nuisibles , & n'exigeront plus aucun
soin.

Lorsque les feuilles commencent à se flétrir, on peut enlever
les racines pour l'usage : mais
avant ce tems , elles sont rarement de bon goût ; elles ne
sont pas même bonnes au printems, quand elles ont repoussé :
de sorte que la seule maniere
de garder ces racines pour l'u-
sage au printems, est de les
enlever au commencement de
Fevrier , & de les mettre dans
du sable dans un lieu sec, où
elles se conserveront bonnes
jusqu'au milieu d'Avril, & même
plus tard. Lorsqu'on veut se
procurer de bonnes femences
de cette plante , on choisit parmi les racines quelques-unes des
plus longues, des plus droites &
des plus grosses ; & on les plante
à la distance de deux pieds ,
dans un endroit où elles puissent être à l'abri des vents du
sud & de l'ouest , parce que
leurs tiges, qui s'élevent ordinairement à une grande hauteur , sont fort sujettes à être
brisées par les grands vents ,
si elles y sont exposées. On les
tient constamment nettes de
mauvaises herbes ; & si le tems
est fort sec, on les arrose deux
fois par semaine , pour leur
faire produire une grande quantité de femences, & les rendre
plus fortes. Vers la fin d'Août
ou au commencement de Septembre, lorsque leurs femences
seront parvenues à leur maturité , on coupera les ombelles
avec soin ; on les tiendra éten-

PAS

dues pendant deux ou trois jours sous une toile, pour les faire secher, on les battra ensuite, & on les conservera pour l'usage : mais il ne faut jamais se fier sur celles qui ont plus d'une année ; car, après ce tems, elles sont rarement susceptibles de germer.

Il est dangereux de manier les feuilles du *Panais*, sur-tout le matin, quand elles sont encore couvertes de rosée ; car alors elles occasionnent des ampoules aux personnes qui ont la peau délicate. J'ai vu des Jardiniers qui, pour avoir arraché dans la matinée des *Carottes* mêlées parmi les *Panais*, ayant les manches de leurs chemises retroussées jusqu'aux épaules, pour ne point les mouiller dans la rosée, avoient les bras couverts de larges ampoules remplies d'une liqueur brûlante, qui les a beaucoup incommodés pendant plusieurs jours.

Opopanax. La troisieme espece s'éleve à la hauteur de sept à huit pieds, avec une tige verte, rude, & garnie de feuilles ailées, décomposées, fort rudes au toucher, d'un vert foncé, & remplies d'une seve fort jaune, qui s'écoule par toutes les blessures qu'elles reçoivent. Les tiges sont divisées vers le haut en plusieurs branches horisontales, & terminées chacune par une grande ombelle : les fleurs sont jaunes ; elles paroissent en Juillet, & sont remplacées par des semences unies, bordées, & un peu convexes au milieu, qui mû-

PAT 431

rissent en automne. On croit que l'*Opopanax* des boutiques est le suc épaissi de cette plante.

PATAGONE. *Voy.* BOERHAAVIA DIFFUSA. *L.*

PATATE *ou* BATATE. *Voy.* CONVOLVULUS BATATAS.

PATIENCE *ou* RHUBARBE DES MOINES. *Voyez* RUMEX PATIENTIA. *L.*

PATIENCE ROUGE *ou* SANG DRAGON. *Voy.* RUMEX SANGUINEUS. *L.*

PATIENCE DES MARAIS *ou* LA PARELLE. *Voy.* RUMEX AQUATICUS. *L.*

PATURE. Il y a deux especes de terre en pâturage ; l'une fait des prairies basses, & souvent inondées, & l'autre des prairies seches sur des terres élevées. La premiere produit une bien plus grande quantité de foin que la derniere, & n'exige pas autant d'engrais ; mais le foin récolté sur les prairies hautes, est bien préférable, & les bestiaux qui en sont nourris sont d'un plus grand prix, quoique ceux qui ont été élevés sur des prés bas, soient plus gros & plus gras, comme on peut en faire la remarque sur le bétail que l'on amene des terres basses & fertiles de la Province de Lincoln.

Mais quand on est un peu délicat sur le choix des viandes, on paie plus cher les animaux nourris sur les dunes ou dans les hauts prés, dont l'herbe est courte, que ceux qui sont beaucoup plus gras : d'ailleurs les pâtures seches ont un avantage sur les prairies basses, en ce qu'elles peuvent

nourrir pendant tout l'hiver, & qu'elles ne font pas auffi fujettes à être piétinées & gâtées dans les tems humides ; en outre, comme elles produifent moins de mauvaifes herbes, on eft en quelque forte dédommagé par-là du peu d'abondance des récoltes.

J'ai déja parlé des avantages des prairies baffes & inondées ; j'ai indiqué les moyens d'en faire écouler les eaux & de les améliorer, fous l'article *Terres* ou *Champs* : ainfi, je ne les répéterai pas ici ; je me bornerai a donner quelques inftructions fur la maniere d'améliorer les hauts pâturages.

La premiere amélioration pour une prairie, dans un terrein élevé, eft de l'enclorre, & de la divifer en petits cantons de quatre, cinq, fix, huit ou dix âcres chacun, en plantant des arbres de charpente dans les rangées de haies, pour mettre l'herbe à l'abri, & la préferver du hâle de Mars, qui l'empêcheroit de pouffer, fi elle reftoit en grande piece ouverte, fur-tout quand le mois d'Avril eft froid & fec ; au-lieu que dans les endroits abrités, l'herbe commence à croître dans le commencement de Mars, & couvre la terre bientôt après : ce qui empêche le foleil de brûler les racines de l'herbe, & la fait pouffer de maniere à produire une affez bonne récolte, quand même le printems feroit fec : mais en faifant ces enclos, il faut obferver, comme on l'a déja dit, de ne pas les faire trop petits, fur-

tout lorfque les rangs de haies font plantés en arbres, qui, lorfqu'ils font parvenus à une hauteur confidérable, couvrent la prairie, & rendent l'herbe très-aigre, quand ils font trop rapprochés : alors, au-lieu d'être avantageux, ils nuifent beaucoup au pâturage.

Le fecond dégré de perfection d'une prairie élevée, confifte à en rendre le gazon d'une bonne qualité ; fouvent même l'herbe y eft détruite par des *Joncs*, des buiffons ou des taupinieres ; ce qui n'eft occafionné que par la mauvaife qualité du fól, ou faute de foin. Si cela provient de ce que la fuperficie de la terre eft formée par une glaife froide, on peut y remédier, en l'enlevant, & en la brûlant, comme il a été dit dans l'art. *Terres* ou *Champs* ; fi au contraire la terre eft chaude & fablonneufe, il faut y répandre de la craie, de la chaux, de la marne ou de la glaife, qui font de fort bons engrais pour de pareils fols : mais il eft indifpenfable d'employer une grande quantité de ces matieres, fi l'on veut en tirer quelque avantage.

Quand la terre eft couverte de *Joncs* ou de *Buiffons*, on arrache exactement toutes ces plantes vers la fin de l'été ; on les brûle lorfqu'elles font feches, & l'on en répand les cendres fur le terrein, avant les pluies de l'automne : alors on nivelle la terre, & on y feme la graine ; l'herbe pouffera bientôt, fi cette opération eft faite dans le commen-

PAT

cement de l'automne, & couvrira la terre au printems suivant. Si le terrein est rempli de taupinieres, on les abat, on les brûle de même ; & après avoir rendu la surface du sol égale, on y répand aussi la semence au commencement des pluies de l'automne.

Il y a aussi quelques prairies qui sont remplies de fourmilieres ; ce qui est non seulement désagréable à la vue, mais empêche aussi de faucher l'herbe, où elles sont fort nombreuses. Dans ce cas, on divise en trois parties, avec une bêche, le gason qui croît au-dessus ; on l'écarte, & on le renverse à chaque côté ; on creuse ensuite le milieu, qu'on répand sur la terre, & on laisse les trous ouverts pendant tout l'hiver, pour détruire les fourmis ; au printems, on remet le gason ; & lorsque les racines ont repris, on le roule pour l'établir & le rendre uni.

Par-tout où la terre a été ainsi travaillée, il est bon de serrer le gason dans les mois de Fevrier & Mars avec un rouleau de bois fort lourd, & toujours dans un tems humide, afin que le rouleau fasse plus d'effet ; la surface du terrein en deviendra plus unie, & beaucoup plus aisée à faucher, & le gason se trouvera garni de maniere que la prairie sera, comme on le dit ordinairement, un bon fonds ; l'herbe en sera plus douce ; elle prendra bientôt le dessus, & finira par détruire les mauvaises herbes.

Un moyen d'améliorer en-

PAT 433

core beaucoup les prairies hautes, est de les faire pâturer chaque deux ans ; car, sans cela, il sera nécessaire d'y répandre des engrais ; il faut au moins le faire tous les trois ans. Quand un Fermier a beaucoup de terre labourables, il n'est pas disposé à employer ses engrais dans les prairies : c'est pourquoi chaque Fermier devroit proportionner ses pâturages aux terres qu'il a à cultiver, sur-tout dans les cantons où le fumier est rare, sans quoi il s'en ressentira bientôt ; car la pâture est le fonds de tout le bénéfice qu'il peut tirer des terres labourables. Quand on veut réparer les pâturages avec des engrais, dans des endroits élevés, il faut avoir égard à la nature du sol, pour que l'engrais qu'on emploie y soit propre. Par exemple, toutes les terres chaudes & sablonneuses exigent des engrais frais, tels que le fumier de vaches & de porcs, la marne & la glaise ; mais les terres froides demandent des engrais chauds, comme le fumier de cheval, les cendres, le sable, &c. que l'on répand en automne, avant que les pluies aient commencé à détremper la terre, & à la rendre trop molle, pour pouvoir y passer avec des charriots : l'on étend ces engrais avec soin, & on brise toutes les mottes autant qu'il est possible ; au printems, on herse la terre avec des épines, pour faire pénétrer l'engrais jusqu'aux racines de l'herbe : en employant ces engrais en au-

E e 3

tomne, les pluies de l'hiver font pénétrer les sels de maniere qu'au printems suivant, l'herbe en reçoit toute l'influence.

Il faut aussi avoir grand soin de détruire les mauvaises herbes dans les prairies, au printems & en automne ; sans quoi leurs semences, qui se répandront sur la terre, les multiplieront en si grande abondance, qu'elles surmonteroient la bonne herbe, l'affoibliroient, la détruiroient, & deviendroient elles mêmes très-difficiles à déraciner, sur-tout l'*Année-Bœuf*, la *Dent-de-Lion*, & quelques autres dont les semences sont garnies de duvet.

Les hauts prés font rarement degénérer l'herbe qui y est semée, si la terre en est un peu bonne ; au-lieu que les prairies basses & inondées pendant l'hiver, se remplissent en peu d'années d'herbes rudes & de *Joncs*. Les terres élevées continuent à produire une herbe de bonne qualité pendant plusieurs années, sans avoir besoin d'être renouvelées. Il n'y a point de partie de l'Agriculture dans laquelle les Fermiers aient fait moins de progrès que dans celle qui concerne les pâturages : la plupart d'entr'eux sont dans l'opinion qu'un vieux pâturage labouré ne peut plus faire une bonne prairie ; aussi ont-ils l'habitude, quand ils ont labouré de pareils terrains, d'en tirer trois ou quatre récoltes de *Bled*, & d'y semer ensuite de l'*Orge* avec de l'herbe ou du *Trefle rouge*, qu'ils laissent

sur la terre après la récolte ; ou du *Trefle* mêlé avec du *Seigle*. Comme ces plantes ne sont que bis-annuelles, & que leurs racines périssent aussi-tôt que leurs graines sont mûres, ils labourent de nouveau le terrein pour y semer du *Bled*. Cette méthode est celle des meilleurs Fermiers. Je n'en ai jamais connu aucun qui ait eu l'idée de continuer leurs terreins en prairies ; aussi les denrées qu'ils sement sont-elles adaptées à cette manœuvre.

Malgré l'antiquité de cet usage, je prouverai qu'il est possible de semer de l'herbe dans un champ qui a été labouré, de maniere que le gazon en soit aussi bon, & même meilleur, & d'une aussi longue durée que si l'herbe y étoit venue naturellement : ce qu'on n'a jamais pu obtenir en suivant la méthode ordinaire ; car, en semant du *Bled* avec l'herbe, si le premier réussit, l'herbe est mauvaise & étouffée, de façon que, si la terre n'est pas bien bonne, elle mérite à peine qu'on la conserve ; elle ne produit que très-peu de fourrage la seconde année ; & la troisieme, la récolte est réduite à rien, soit pour le fourrage, soit pour les semences : il n'est pas possible d'en obtenir d'avantage, parce que la terre ne peut produire deux récoltes à la fois, quand même le sol n'auroit aucun défaut ; car le *Bled* poussant le premier, & plus vigoureusement, arrête les progrès de l'herbe, qui reste très-foible & fort claire. Celle qui pousse au printems est détruite

par le *Bled* ; de maniere que par-tout où le *Bled* étend ses racines, il ne peut y avoir que très-peu d'herbe. Quand la terre n'est pas assez fertile pour lui fournir une abondante nourriture, après que le *Bled* est enlevé, on ne peut espérer qu'une très-foible récolte de *Trefle* ; & comme les racines de cette plante sont bis-annuelles, plusieurs des plus fortes périssent bientôt après qu'elles ont été coupées ; & les plus foibles, qui n'ont fait que peu de progrès jusques là, restant pour l'année suivante, occupent le terrein presque à pure perte.

Ainsi, lorsqu'une terre est préparée pour en faire une prairie, il n'y faut semer aucune autre espece de graines avec celles de l'herbe ; le terrein doit être bien labouré & débarrassé de toutes les plantes inutiles, qui, si on les laissoit, pousseroient les premieres, deviendroient si fortes, quelles étoufferoient les autres, & finiroient par détruire entierement la prairie. Le meilleur tems pour semer l'herbe sur une terre seche, est vers le milieu d'Août, s'il tombe de la pluie ; car la terre étant échauffée, s'il survient quelques ondées, les semences pousseront bientôt ; & comme elles auront le tems de s'enraciner avant l'hiver, il n'y aura aucun risque que la gelée les déterre, sur-tout si le terrein est bien roulé avant que les gelées commencent, & si le sol est bien affermi & fixé sur les racines. Sans cette précaution,

la gelée dessere souvent la terre de façon que l'air pénetre jusqu'aux racines ; ce qui endommage beaucoup les plantes. Cet inconvénient est cause qu'on ne fait jamais ce semis en automne : on a cependant tort ; car quand on suit exactement ce qui vient d'être prescrit, on peut semer sans danger ces plantes en automne, à moins que le tems ne soit fort sec : en effet, si l'herbe pousse bien, si la terre est bien roulée au milieu ou à la fin d'Octobre, & si l'on recommence la même opération au commencement de Mars, le gason sera bien établi & l'on pourra en attendre une bonne récolte de foin pour le premier été. Dans des terres fort ouvertes, exposées & froides, il est prudent de semer l'herbe plutôt qu'on ne vient de le dire, afin qu'elle ait le tems de pousser de grosses racines avant que le froid n'arrête son accroissement ; car comme la végétation finit de bonne heure en automne, si l'herbe est encore foible, elle pourra être détruite par la gelée : mais en la semant au commencement d'Août, s'il survient quelques pluies qui la fassent bien pousser, elle réussira beaucoup mieux qu'au printems, ainsi que je l'ai éprouvé pendant plusieurs années en Angleterre, dans des endroits très-exposés ; mais lorsque la terre ne peut être préparée assez tôt pour semer dans ce tems, on peut le faire au milieu ou à la fin de Mars, suivant que la saison est plus ou moins avancée, lorsque le printems est tardif. J'ai

souvent femé de l'herbe avec succès au milieu d'Avril, dans des terres froides ; mais lorsqu'on fait cette opération fort tard, on s'expose aux dangers des fecherefles, qui font d'autant plus nuifibles, que la terre eft plus feche & plus légere. J'ai vu quelquefois dans cette faifon la furface entiere d'un terrein être enlevée par les grands vents, de maniere que toutes les femences étoient ramaflées fur un côté du champ : c'eft pourquoi, toutes les fois que l'on a femé tard au printems, il eft prudent de bien rouler la terre aufli-tôt que les femences font répandues, pour en fixer la furface, & empêcher qu'elle ne foit enlevée.

Les meilleures efpeces de femences font celles que l'on a recueillies fur des près hauts, & les plus nets de mauvaises herbes. On crible ces femences, pour en ôter l'ordure, & on en emploie trois ou quatre boifleaux pour un âcre de terre. Le *Tri-folium pratenfe album*, ordinairement connu fous le nom de *Treffle blanc de Hollande*, ou *Herbe de Chevre-feuille blanc*, eft aufli une très-bonne efpece d'herbe ; huit livres de fes graines fuffifent pour un âcre de terre : il faut d'abord femer celles de l'herbe, & enfuite celles du *Treffle* de Hollande ; mais on ne doit pas les mêler enfemble pour les répandre, parce que les femences du *Treffle* étant les plus lourdes, tomberoient au fond, & la terre ne feroit pas femée également.

Après avoir répandu la fe-

mence, on herfe légerement la terre, pour l'enfoncer; mais la herfe qu'on emploie doit avoir des dents courtes, fans quoi ces graines fe trouveroient enterrées trop profondément. Deux ou trois jours après, fi la furface de la terre eft feche, on la roule avec un rouleau, pour en brifer les mottes, la rendre unie, la fixer, & empêcher que le vent ne puiffe déplacer les femences.

Si, après que les graines ont poufflé, la terre produit beaucoup de mauvaifes herbes, il faut les arracher avant qu'elles deviennent affez hautes pour couvrir les bonnes ; car fi l'on néglige cette précaution, elles s'établiffent tellement dans la terre, qu'elles privent les autres de leur nourriture ; & fi on laiffe mûrir leurs femences, la terre s'en trouve fi remplie, que toute la prairie eft perdue : aufli le foin d'arracher les mauvaifes herbes eft-il un des plus utiles dans la pratique de l'Agriculture.

En roulant la terre deux ou trois fois après que l'herbe a poufflé, on la renfonce, & on la rend plus épaiffe par le bas : pour ce qui eft du Treffle Hollandois, comme il pouffe des racines à chaque nœud des branches qui font près de la terre, fi l'on enfonce ces tiges, leurs racines s'entrelacent fi étroitement enfemble, qu'elles forment un Gazon très-épais, qui couvre toute la furface de la terre, & produifent un tapis de verdure, qui fe trouve en état de réfifter aux

PAT

fecherefles ; car fi l'on examine en été les pâturages ordinaires, dans lefquels il y a toujours des paquets de cette herbe, on verra que toute la verdure qui s'y trouve n'eft plus compofée que de ce *Treffle blanc* ; & quoique tous les Fermiers conviennent que cette efpece eft la plus propre à la nourriture du bétail, cependant ils n'ont jamais eu l'idée de la multiplier par femences, fi ce n'eft depuis quelques années. Il n'y a pas long-tems que l'on a introduit cette pratique en Angleterre ; elle eft due à quelques curieux, qui ont fait venir depuis peu cette femence du Brabant, où elle eft cultivée depuis long tems. Avant ce tems, on n'en trouvoit point dans notre Ifle ; mais à préfent plufieurs perfonnes recueillent de ces graines, qui réuffiffent auffi-bien que celles qu'on tire de chez l'étranger.

Comme le *Treffle blanc* eft une plante durable, elle eft certainement la meilleure efpece qu'on puiffe employer pour former un pâturage perpétuel. Les femences de *Foin* qu'on recueille dans les meilleures prairies, renferment différentes efpeces d'herbes, dont les unes font annuelles, & d'autres bis-annuelles ; & quand elles périffent, il refte plufieurs parties de terres nues : alors, s'il n'y a pas une quantité fuffifante de *Treffle blanc* qui s'étende & couvre ces endroits, on ne peut jamais efpérer d'avoir une prairie bien verte & bien garnie. Dans la plupart

PAT 437

des prairies naturelles, c'eft cette plante qui forme la verdure ; elle convient également aux terres feches & aux terres humides : elle croit fpontanément fur le gravier & dans la glaife, dans prefque toute l'Angleterre ; ce qui prouve qu'il eft aifé de la cultiver avec beaucoup d'avantage dans la plupart des terres de ce Royaume.

Ainfi, ce qui eft caufe que la méthode ordinaire ne procure pas de bonnes prairies dans les terres de labour, c'eft que les Fermiers ne diftinguent pas les *Herbes* annuelles de celles qui font vivaces. Les *Herbes* annuelles ou bis-annuelles périffent ordinairement bientôt après que leurs femences font mûres ; de maniere que, s'il ne tombe pas quelques graines pour les remplacer, on ne peut plus efpérer de cette terre que ce qui y croîtra naturellement. Cette méthode, jointe à l'ufage mal entendu de vouloir tirer une récolte de *Bled* avec ce fourrage, a empêché jufqu'à préfent l'accroiffement des pâturages dans plufieurs parties de l'Angleterre, où ils commencent cependant à devenir une des parties les plus précieufes de l'Agriculture.

Après que la terre a été femée, fuivant les principes que nous venons d'établir, & lorfqu'elle a produit un bon gazon, on l'entretient, en y paffant conftamment des rouleaux pefans pendant le printems & l'automne, comme il a été dit plus haut. Cette méthode n'eft

point celle que les Fermiers emploient ordinairement; mais ceux qui en font ufage en retirent un bénéfice confidérable, par le prodigieux accroiffement de l'*Herbe*. On doit auffi avoir grand foin d'arracher les *Ofeilles* ou *Patiences*, les *Dents de Lion*, *Arrête-Bœufs*, & toutes autres mauvaifes plantes, avec leurs racines, au printems & en automne; car cette feule attention peut augmenter beaucoup la quantité des bonnes *Herbes*, & conferver les pâturages dans leur beauté. C'eft auffi une très-bonne pratique de recommencer cette opération chaque trois ans; car, fans cela, on ne peut efpérer que la terre continue à donner de bonnes récoltes. En outre il fera néceffaire de changer le tems où l'on fauche, & de ne pas faucher le même pré tous les ans, mais d'y faire une récolte une année, & de le faire pâturer la fuivante; car fi l'on y coupe l'*Herbe* chaque année, on fera forcé d'y mettre conftamment des engrais, comme on le fait pour les prairies artificielles des environs de Londres, fans quoi la terre feroit bientôt épuifée.

Depuis quelques années, il regne une grande émulation, fur-tout parmi les Gentilshommes, pour l'amélioration des pâturages. On a femé beaucoup d'efpeces d'*Herbes*, & quelques perfonnes peu habiles dans ces matieres en ont impofé à des ignorans, en leur vendant des plantes étrangeres, comme ayant des qualités particulie-

res: mais après en avoir effayé la cuiture, on s'eft apperçu qu'elles n'étoient propres à rien; ce qui a fait perdre une année ou deux, & a mis ces hommes crédules dans le cas de recommencer leur ouvrage: c'eft pourquoi je confeille de ne pas trop fe fier à de pareils praticiens, qui, fur une légere expérience, hafardent des confeils pernicieux; car, après avoir femé, en différens tems, plus de cent efpeces de graines d'*Herbes* différentes, apportées de l'Amérique, je n'en ai trouvé aucune qui fût égale au *Gramen*, qui croit naturellement en Angleterre, pour la durée & la verdure: auffi cette efpece, & fix ou fept autres, font-elles celles qui méritent le plus d'être cultivées: mais on a tant de peine à recueillir leurs femences en grande quantité, qu'il eft rare qu'on l'entreprenne; & comme les femences d'*Herbes* qu'on achete, font toujours mêlées de plus de mauvaifes efpeces que de bonnes, cette raifon m'a engagé à femer le *Trefle blanc* de Hollande feul, au-lieu de le mêler avec de la *Fénaffe*, comme je l'ai recommandé au commencement de cet article. L'*Herbe* ordinaire pouffe bientôt d'elle-même, & fe mêle avec le *Trefle*. En farclant, en roulant, & en dreffant la prairie, toutes les mauvaifes *Herbes* feront bientôt détruites, & l'on obtiendra une verdure belle & durable; au-lieu que la *Pimprenelle*, & plufieurs autres plantes qui ont été exaltées comme un excel-

lent fourrage d'hiver, font de peu de durée, & peu propres à améliorer les pâturages. Il n'y. a point de meilleures plantes, en fait de fourrages, que la *Luferne* & le *Sainfoin* ; car lorfqu'elles fe trouvent fur une terre fertile & bien cultivée, elles donnent des récoltes bien plus abondantes fur une même étendue de terrein, que toute autre plante vivace : c'eft pourquoi j'exhorte les perfonnes qui défirent avoir de l'excellent fourrage pour leur bétail, de ne s'attacher qu'à la culture de ces deux plantes, fans s'engager mal-à-propos dans des expériences incertaines.

PATTE D'OIE. *Voy.* CHE-NOPODIUM.

PAULLINIA. *Lin. Gen. Plant.* 446. *Serjana. Plum. Nov. Gen.* 34. *tab.* 35. *Cururu. Plum. Nov. Gen.* 34. *tab.* 35.

Caractères. Le calice de la fleur eft étendu, perfiftant, & compofé de quatre feuilles petites & ovales ; la corolle a quatre pétales oblongs, ovales, & deux fois plus larges que le calice ; la fleur a huit étamines courtes, & terminées par de petites antheres, avec un germe turbiné, & à trois angles obtus, qui foutient trois ftyles courts, minces, & couronnés par des ftigmats étendus. Ce germe fe change dans la fuite en une groffe capfule à trois angles & à trois cellules, qui renferment chacune une femence prefque ovale. La capfule du *Serjana* de PLU-MIER a des femences attachées à fa bâfe, & celles

du *Cururu* croiffent au fommer.

Ce genre de plantes eft rangé dans la troifieme fection de la huitieme claffe de LINNÉE, avec celles dont les fleurs ont huit étamines & trois ftyles.

Les efpeces font :

1°. *Paullinia Serjana , foliis ternatis , petiolis teretiufculis , foliolis ovato-oblongis. Lin. Sp. Plant.* 365. *Jacq. Obs.* 3. *p.* 11. *t.* 61. *f.* 2 ; Paullinie avec des feuilles ternées , des pétioles cylindriques , & des lobes oblongs & ovales.

Serjana fcandens triphylla & racemofa. Plum. Nov. Gen. 34. *Ic.* 113. *f.* 2 ; Serjana grimpant , branchu , & à trois feuilles.

2°. *Paullinia Mexicana , foliis bi-ternatis , petiolis marginatis , foliolis ovatis , integris. Lin. Sp. Plant.* 366. *Jacq. L. c. t.* 61. *f.* 5 ; Paullinie avec des feuilles à fix lobes , ovales & entieres, & des pétioles ailés.

Serjana fcandens enneaphylla & racemofa. Plum Nov. Gen. 34. *Ic.* 113. *f.* 1 ; Serjana grimpant & branchu, avec des feuilles à neuf lobes.

Quauhmeati. Hern. Mex. 289.

3°. *Paullinia cururu , foliis ternatis , foliolis cunei-formibus , obtufis , fubdentatis. Lin. Sp. Plant.* 365 ; Paullinie avec des feuilles à trois lobes , en forme de coin , obtus , & un peu dentelés.

Cururu fcandens triphylla. Plum. Nov. Gen. 34. *Ic.* 3. *f.* 2 ; Cururu grimpant & à trois feuilles.

4°. *Paullinia Curaffavica , foliis bi-ternatis, foliolis ovatis. Lin.*

440 PAU

Sp. Plant. 366 ; Paullinie avec des feuilles à six lobes, ovales.

Cururu fcandens enneaphylla, fructu racemofo, rubro. Plum. Nov. Gen. 34 ; Cururu grimpant à neuf feuilles, avec un fruit rouge & branchu.

Cordis-Indi folio & facie frutefcens Curaffavica, latifolia. Pluk. Alm. 120. *t.* 168. *f.* 6.

5°. *Paullinia pinnata, foliis pinnatis, foliolis incifis, petiolis marginatis. Hort. Cliff.* 52. *Roy. Lugd.-B.* 464 ; Paullinie avec des feuilles ailées, dont les lobes font découpés, & les pétioles bordés.

Clematis pentaphylla, pediculis alatis, fructu racemofo tricocco, & coccineo. Plum. Amer. 76.

Pifum cordatum non veficarium. Sloan. Jam. III.

Cururu-Ape. Marcgr. Braf. 22. *Pif. Braf.* 114. *Raii. Hift.* 1347.

Cururu fcandens pentaphylla. Plum. Nov. Gen. 37 ; Cururu grimpant, & à cinq feuilles.

6°. *Paullinia tomentofa, foliis pinnatis, tomentofis, foliolis ovatis, incifis, petiolis marginatis ;* Paullinie avec des feuilles ailées & cotonneufes, à lobes ovales & decoupés fur leurs bords, & poftés fur des pétioles bordés.

Cururu fcandens, pentaphylla & villofa, fructu racemofo, rubro. Houft. MSS. ; Cururu grimpant, avec cinq feuilles velues, & des fruits rouges & rapprochés en paquet.

Toutes ces plantes croiffent naturellement en Amérique, où l'on en trouve encore plufieurs autres dont je ne fais

pas mention ici ; elles ont des tiges grimpantes, & garnies à chaque nœud de vrilles, au moyen defquelles elles s'attachent aux arbres voifins, & s'élevent à la hauteur de trente ou quarante pieds. A chaque nœud de fes feuilles fort auffi une feuille, qui, dans quelques efpeces, eft compofée de trois lobes, comme ceux du *Treffle ;* & dans d'autres de cinq ; quelques-uns en ont neuf, & d'autres un nombre plus ou moins grand. Dans quelques efpeces, ces lobes font entiers ; dans d'autres, ils font découpés à la pointe, & quelques-uns le font fur leurs bords. Dans quelques-unes, leurs furfaces font unies, & dans d'autres velues ; les fleurs naiffent en paquets ou grappes longues, comme celles des *Grofeillers ;* elles font petites, blanches, & de peu d'apparence ; à ces fleurs fuccedent des capfules à trois angles & à trois cellules, qui, dans le *Cururu* de PLUMIER, contiennent des femences rondes ; mais celles du *Serjana* ont des femences ailées, comme celles de l'*Erable ;* elles font renverfées, fixées à l'extrémité de la capfule, & pendent vers le bas.

Ces plantes font trop tendres pour pouvoir fubfifter en hiver dans ce pays fans le fecours d'une ferre chaude : elles exigent beaucoup de pluie ; mais on les cultive rarement en Europe, fi ce n'eft dans des jardins de Botanique ; car leurs fleurs ont peu de beauté.

On les multiplie par leurs graines, qu'il faut se procurer des pays où elles croissent naturellement, parce qu'elles n'en produisent point en Angleterre : on les seme dans de petits pôts remplis de terre légere, aussi-tôt qu'on les reçoit, & l'on place ces pots dans une couche de tan de chaleur modérée ; si c'est en automne, on les plonge dans la couche de la serre, & alors il est possible que les plantes poussent au printems suivant : mais si ces semences n'arrivent pas avant le printems, elles ne pousseront pas dans la même année. Ainsi, les pots dans lesquels elles sont placées, doivent être plongés dans une couche de chaleur modérée, sous un vitragé où elles puissent rester pendant tout l'été : en automne, on les met dans la serre, où on les laisse pendant tout l'hiver ; on les arrose très-légerement de tems en tems, quand la terre est seche. Au printems suivant, on ôte ces pots de la serre, pour les plonger dans une nouvelle couche chaude sous un vitrage, où les plantes pousseront dans l'espace de six semaines, si les semences sont bonnes. Quand les plantes sont en état d'être enlevées, on les met chacune séparément dans de petits pots remplis de terre légere : on les plonge dans une couche chaude de tan ; on les tient à l'ombre jusqu'à ce qu'elles aient formé de nouvelles racines ; on leur donne ensuite de l'air tous les jours, à pro-portion de la chaleur de la saison, & on les met en automne dans la couche de tan de la serre, où on les tiendra constamment, & où on les traitera comme les autres plantes tendres.

PAVIA. *Boerrh. Ind. Alt.* 2. *p.* 260. *Æsculus. Lin. Gen. Plant.* 420. [*Scarlet flowering Horse Chesnut.*] Marronier d'Inde, écarlate & à fleurs.

Caracteres. Le calice de la fleur est petit, gonflé, & formé par une feuille découpée au sommet en cinq parties : la corolle a cinq pétales ronds, ondés, plissés sur leurs bords, & étroits à leur bâse, où ils sont insérés dans le calice ; la fleur a huit étamines penchées, aussi longues que les pétales, & terminées par des antheres érigées : son germe est rond, placé sur un style en forme d'alêne, & couronné par un stigmat pointu ; ce germe devient ensuite une capsule ovale, turbinée, coriace, & à trois cellules, qui renferment une & quelquefois deux semences presque angulaires.

Ce genre de plantes devroit être rangé dans la premiere section de la huitieme classe de LINNÉE, avec celles dont les fleurs ont huit étamines & un style ; mais il l'a placé, avec le *Marronier d'Inde*, sous le titre d'ÆSCULUS, dans la septieme classe. Comme les fleurs de celle-ci ont huit étamines, & que celles du *Marronier d'Inde* n'en ont que sept, la capsule de celle-ci étant unie, & celle du *Marronier d'Inde* épi-

plante est une fort mauvaise nourriture pour le bétail ; & quand elle se trouve en grande quantité parmi le *Foin*, ce fourrage a peu de valeur.

Comme les semences de cette plante mûrissent toujours dans le tems de la fenaison, lorsqu'on recueille la graine des herbes pour la semer, il faut avoir soin que celle-ci n'y soit point mêlée. Je n'importunerai point le Lecteur par la description des autres espèces.

PEGANUM. *Lin. Gen. Plant.* 530. *Harmala. Tourn. Inst. R. H.* 257. *tab.* 133. [*Wild Assyrian Rue.*] Rhue sauvage d'Assyrie.

Caractères. Le calice de la fleur est persistant, & composé de cinq feuilles étroites, érigées, & aussi longues que les pétales ; la corolle a cinq pétales ovales & oblongs, qui s'étendent en s'ouvrant : la fleur a quinze étamines en forme d'alêne, de moitié moins longues que les pétales, & dont les bases s'étendent dans un nectaire sous le germe ; elles sont terminées par des antheres oblongues & érigées: son germe est long, a trois angles ; &, placé à la base de la fleur, il soutient un style mince, triangulaire de la longueur des antheres, & a trois stigmats, plus longs que le style. Ce germe devient ensuite une capsule ronde, triangulaire, & à trois cellules remplies de semences ovales, & à pointe aiguë.

Ce genre de plantes est rangé dans la première section de la onzieme classe de LINNÉE, qui comprend celles dont les fleurs ont depuis onze jusqu'à dix-neuf étamines & un style.

Nous n'avons qu'une espece de ce genre dans les jardins anglois.

Peganum Harmala, foliis multi-fidis. Hort. Ups. 144. *Gron. Orient.* 165. *Gmel. Sib.* 4. p. 177. *n.* 96. *Fabric. Helm.* p. 228. *Blackw. f.* 310 ; l'Armel, avec des feuilles à plusieurs pointes.

Harmala. Dod. Pempt. 121.

Ruta sylvestris, flore magno albo. Bauh. Pin. 336 ; Rhue sauvage, à grande fleur blanche.

Cette plante croît naturellement en Espagne & en Syrie : sa racine, qui est aussi grosse que le petit doigt, devient ligneuse avec l'âge : ses tiges périssent en automne, & les nouvelles poussent au printems ; elles s'élevent à la hauteur d'un pied, & se divisent en plusieurs petites branches garnies de feuilles oblongues, épaisses, découpées en plusieurs segmens étroits, d'un vert foncé, gluantes, & d'un goût amer : les fleurs naissent aux extrémités des branches, & sont très-étroitement placées entre les feuilles ; elles sont composées de cinq pétales blancs & ronds, qui s'ouvrent comme une rose, & de quinze étamines en forme d'alêne, terminées par des antheres oblongues. Dans le centre est situé un germe rond & à trois angles, qui soutient un style triangulaire de la longueur des étamines, avec trois stigmats plus longs que le style. Ce germe se change dans la suite en une capsule ronde, à trois angles & trois cellules, qui renferment

PEL

ment plusieurs semences ovales & à pointe aiguë. Cette plante fleurit en Juillet ; & , dans les étés chauds, ses semences mûrissent en automne.

On la multiplie par ses graines, qu'il faut semer clair sur une plate-bande de terre légere, au commencement du mois d'Avril. Lorsque les plantes poussent, on les tient constamment nettes de mauvaises herbes : c'est en cela que consiste toute leur culture jusqu'à la fin d'Octobre ou au commencement de Novembre, que les tiges périssent ; alors on couvre la plate-bande avec du tan, des cendres, de la sciûre, ou quelque autre chose, pour empêcher la gelée d'y pénétrer. Cette méthode est la plus sûre pour conserver les racines, qui, lorsqu'elles sont jeunes, sont un peu délicates. On peut les enlever au mois de Mars suivant, & les transplanter dans un sol sec & à une exposition chaude, où elles subsisteront pendant plusieurs années. Cette plante est quelquefois d'usage en Médecine.

PEIGNE *ou* L'AIGUILLE DE VENUS. *Voy.* SCANDIX PECTEN. *L.*

PELECINUS. *Voyez* BISERRULA.

PELOTTE DE NEIGE , OBIER A FLEURS DOUBLES *ou* ROSE DE GUELDRE. *Voy.* VIBURNUM OPULUS.

PELTARIA. *Jacq. Aust. t.* 123. *Lin. Gen. Plant.* 806. [*Mountain Treacle Mustard.*] Thlaspi de montagne.

Caracteres. Le calice de la fleur est composé de quatre

Tome V.

PEL 445

feuilles concaves , colorées , & qui tombent ; la corolle a quatre pétales placés en forme de croix, dont les onglets sont plus courts que le calice. La fleur a six étamines en forme d'alêne , dont deux sont plus courtes que le calice , & qui sont toutes terminées par des antheres simples ; son germe est rond, & soutient un style court , & couronné par un stigmat obtus ; il se change dans la suite en un légume comprimé, & à une cellule qui renferme une semence ronde.

Ce genre de plantes est rangé dans la premiere section de la quinzieme classe de LINNÉE , intitulée : *Tetradynamia Siliculosa,* avec celles dont les fleurs ont quatre étamines longues & deux courtes, & des semences renfermées dans des siliques courtes.

Nous n'avons qu'une espece de ce genre, qui est la *Peltaria alliacea. Jacq. Vind.* 260. *Lin. Sp. Plant.* 910. *Jacq. Austr. t.* 123. *Scop. Carn. ed.* 2. *n.* 784 ; Peltaria ou Thlaspi de montagne.

Bohadschia foliis radicalibus cordatis, caulinis amplexicaulibus, lanceolatis. Crantz. Austr. p. 5. *t.* 1. *f.* 1.

Clypeola perennis , foliis inferioribus petiolatis , cordato-angulatis superioribus amplexicaulibus, lanceolatis , siliculis uni-locularibus. Ard. Spec. 26. *f.* 6.

Thlaspi montanum , Glasti folio , majus. C. B. P. 106 ; Thlaspi de montagne , à feuilles de Gaude.

Thlaspi montanum. 1. *Cluf. Hist.* 2. *p.* 130.

F f

446 **PEN**

Cette plante croît naturellement sur les montagnes de l'Autriche & de l'Istrie ; elle est bis-annuelle, & périt généralement auſſi-tôt que ſes ſemences ſont mûres : elle s'élève à la hauteur d'environ un pied, avec une tige droite, branchue, & garnie de feuilles unies & en forme de cœur, qui l'embraſſent de leur bâſe ; les tiges ſont terminées par des grappes de fleurs blanches qui croiſſent en ombelles : chaque fleur a quatre pétales placés en forme de croix, & produit une ſilique ronde & comprimée, qui produit une ſemence de la même forme. Cette plante fleurit en Mai, & ſes ſemences mûriſſent en Juillet.

On la multiplie aiſément par ſes graines, qu'on peut ſemer en petites touffes dans les plates-bandes du parterre, au commencement d'Avril. Quand les plantes ont pouſſé, on en laiſſe quatre ou cinq dans chaque touffe, & on arrache les autres, afin de leur donner aſſez d'eſpace pour croître, après quoi elles n'exigeront plus aucun ſoin que celui d'être tenues nettes de mauvaiſes herbes.

PENSÉE ou HERBE DE LA TRINITÉ. *V.* VIOLA TRICOLOR.

PENSTEMON. *Voy.* ASARINA ERECTA.

PENTAPETÈS. *Lin. Gen. Plant.* 757. *Alcea. Raii Supp.* 523. [*Indian Mellow.*] eſpece de Mauve des Indes.

Caractères. La plupart des fleurs ont un double calice, dont l'extérieur eſt petit, &

PEN

compoſé de trois feuilles, & l'intérieur eſt découpé en cinq parties réfléchies ; la corolle a cinq pétales oblongs, qui s'étendent en s'ouvrant : la fleur a quinze étamines étroites, & terminées par cinq antheres, longues & colorées ; elle a un germe rond, avec un ſtyle cylindrique de la longueur des étamines, & couronné par un ſtigmat épais. Ce germe devient enſuite une capſule ovale, & à cinq cellules remplies de ſemences oblongues.

Ce genre de plantes eſt rangé dans la quatrieme ſection de la ſeizieme claſſe de LINNÉE, qui renferme celles dont les fleurs ont quinze étamines fixées au ſtyle par leur bâſe, & qui forment avec lui une colonne.

Nous n'avons qu'une eſpece de ce genre dans les jardins angiois.

Pentapetès Phœnicea , foliis haſtato-lanceolatis , ſerratis. Lin. Sp. Plant. 698 ; Pentapetès avec des feuilles ſciées, en pointe de hallebarde, & en forme de lance.

Alceæ Indicæ cognata. Pluk. Alm. 18. *t.* 255. *f.* 3. *Trew. Rar.* 7. *f.* 5.

Alcea indica , lucido haſtato folio , flore Blattariæ phœniceo. Raii Suppl. 523 ; Mauve des Indes, avec une feuille luiſante & en forme de lance, produiſant une fleur écarlate de Bouillon-Blanc.

Alcea fruticoſa Pentaphylloïdes æmula , floribus amœniſſimè rubellis , calice producto. Pluk. Phyt. 126. *f.* 4.

*Blattaria Zeylanica, flore am-
plo coccineo. Comm. Hort. 1. p.
11. f. 6 ; Blattaire de Céylan.
Flos impius. Rumph. Amb. 5.
p. 288. t. 100. f. 1.
Siamin. Rheed. Mal. 10. p. 1. f. 1.*

Cette plante croît naturellement dans l'Inde, d'où tes
semences m'ont été plusieurs
fois envoyées ; elle est annuelle, & périt en automne,
aussi-tôt que ses branches sont
mûres ; sa tige, droite & haute de deux ou trois pieds,
pousse des branches latérales
dans toute sa longueur ; celles du bas sont les plus longues, & les autres diminuent
par dégrés, & forment par-là
une espece de pyramide ; elles
sont garnies de feuilles de différentes formes ; celles du bas
sont les plus larges, & découpées vers la bâse en deux lobes latéraux, courts ; celle du
milieu se prolonge de deux ou
trois pouces au-delà de l'extrémité des autres ; de sorte
qu'elles ressemblent beaucoup
à une pointe de hallebarde ; elles sont légerement sciées sur
leurs bords, d'un vert luisant
en dessus, mais plus pâles en-
dessous, & portées sur des
pétioles assez longs : les feuilles du haut sont beaucoup plus
étroites ; quelques-unes ont de
petites dentelures sur leurs
bords ; celles-ci sont plus
rapprochées des tiges, &
placées alternativement : les
fleurs qui sortent aux ailes des
feuilles, sont, pour la plupart,
solitaires ; quelquefois cependant il en sort deux du même
bouton, sur les côtés des pétioles des feuilles : le pédon-

cule de la fleur est court &
mince ; son calice extérieur est
composé de trois courtes feuilles qui tombent bientôt, & le
calice intérieur est formé par
une feuille découpée au sommet en cinq segmens aigus,
qui s'étendent en s'ouvrant,
& sont presque aussi longs que
le pétale ; la corolle est monopétale, & divisée presque jusqu'au fond en cinq segmens
obtus ; mais comme ils sont
joints & tombent en une piece,
ils ne forment qu'un pétale,
suivant RAY & TOURNEFORT.
Dans le centre de la fleur s'é-
leve une colonne courte &
épaisse, à laquelle adherent
quinze étamines courtes, &
terminées par des antheres longues & érigées, & entre chaque trois étamines est placée
une autre étamine plus large
avec une anthere oblongue,
érigée, & d'un rouge foncé.
Ces cinq larges antheres sont
stériles, & ne sont point pourvues de poussiere fécondante :
entre les étamines est placé un
germe rond, qui soutient un
style de la longueur des étamines, & couronné par un
stigmat épais. Tout cet appa-
reil étant réuni par sa bâse en
une espece de colonne, distingue l'espece de cette plante,
qui est une *malcavée*. Et quoiqu'au premier aspect cette fleur
paroisse beaucoup ressembler à
celle du Bouillon-Blanc, cependant lorsqu'on examine ses
caracteres essentiels, on recon-
noit qu'elle appartient à la classe
des *malcavées*. Ces fleurs sont
d'une belle couleur écarlate ;
elles paroissent en Juillet, &

448 **PEN**

font remplacées par des capfules rondes & à cinq cellules, un peu ligneufes, dont chacune renferme trois ou quatre femences oblongues, qui mûriffent en automne.

On feme les graines de cette efpèce au commencement de Mars, fur une bonne couche chaude. Quand les jeunes plantes font en état d'être enlevées, on plonge dans une nouvelle couche chaude qu'on a préparée d'avance, quelques petits pots remplis de bonne terre de jardin potager; on en plante une dans chacun: on les arrofe légérement, pour fixer la terre à leurs racines; on les tient à l'ombre, jufqu'à ce qu'elles aient produit de nouvelles fibres, & on les traite enfuite comme les autres plantes délicates & exotiques, en obfervant de leur donner de l'air libre chaque jour, fuivant que la faifon eft plus ou moins chaude, & de couvrir les vitrages tous les foirs, pour conferver la chaleur de la couche. Quand ces plantes ont fait affez de progrès pour remplir les pots de leurs racines, il faut leur en donner de plus grands, que l'on remplit avec la même efpece de terre qui a été indiquée ci-deffus, & les plonger dans une autre couche chaude, où on les laiffera auffi long-tems qu'elles pourront y refter, fans toucher les vitrages; mais après cela, on les placera dans une ferre chaude ou dans une caiffe de vitrages, où elles feront à l'abri du froid, & ou l'on pourra leur procu-

PEN

rer de l'air dans les tems chauds. Au moyen de ce traitement, ces plantes commenceront à fleurir dans le commencement de Juillet, & leurs fleurs fe fuccéderont jufqu'à la fin de Septembre. Pendant tout ce tems, elles auront une très-belle apparence. Comme leurs femences mûriffent fucceffivement, il faut cueillir leurs capfules auffi-tôt qu'elles commencent à s'ouvrir au fommet. Quand ces plantes font fortes, on les tire quelquefois des pots, pour les placer dans les plates-bandes à une bonne expofition, où elles fleuriffent affez bien dans les années chaudes: mais comme leurs femences y mûriffent rarement, on ne peut les avoir dans leur perfection, qu'en les traitant comme il vient d'être dit ci-deffus.

PENTHAPHYLLOIDES. *V.* POTENTILLA. *L.*

PENTHORUM. *Gronov. Virg.* 51. *Lin. Gen. Plant.* 580; efpece de Sedum ou Joubarbe. Elle n'a pas de nom anglois.

Caractères. Le calice de la fleur eft perfiftant, & formé par une feuille découpée en cinq fegmens égaux; la corolle a quelquefois cinq pétales étroits, fitués entre les fegmens du calice: la fleur a dix étamines velues, égales, de la longueur du calice, perfiftantes, & terminées par des antheres rondes, qui tombent; elle a un germe coloré avec cinq ftyles de la longueur des étamines, & couronnés par des ftigmats obtus. Ce germe fe

change dans la suite en une capsule simple, conique, à cinq angles & à cinq cellules, remplies de semences petites & applaties.

Ce genre de plantes est rangé dans la quatrieme section de la dixieme classe de LINNÉE, intitulée *Decandria Pentagynia*, qui comprend celles dont les fleurs ont dix étamines & cinq styles.

Nous n'avons qu'une espece de ce genre.

Penthorum sedoïdes. Gron. Virg 54. *Lin. Sp.* 620 ; Penthorum semblable au Sedum ou Semper-Vivum.

Cette plante est bis-annuelle, & croît naturellement dans la Virginie ; ses tiges s'élevent à la hauteur d'un pied, & sont garnies de feuilles oblongues, alternes, & terminées par des grappes de fleurs d'un jaune verdâtre & de peu d'apparence ; à ces fleurs succedent des capsules coniques, à cinq angles, & remplies de semences petites & plates : ses fleurs paroissent à la fin du mois de Juillet, & ses semences mûrissent en automne. Comme cette plante a peu de beauté, on ne la cultive guere que dans les jardins de Botanique ; ceux qui veulent se la procurer doivent la semer à l'ombre & sur une terre humide. Lorsque les plantes ont poussé, elles n'exigent aucune autre culture que d'être éclaircies & tenues nettes de mauvaises herbes.

PÉPINIERE. On appelle ainsi une piece de terre dans laquelle on éleve toutes les especes d'arbres & de plantes, que l'on transporte ensuite dans les jardins & les plantations. Il y a un grand nombre de ces pépinieres dans plusieurs cantons de l'Angleterre, & principalement aux environs de Londres, qui appartiennent à des Jardiniers dont le métier est d'élever des arbres & des fleurs pour en faire commerce. Dans plusieurs de ces pépinieres, on trouve à présent la plus grande partie des plantes qu'on cultive dans toutes les autres contrées de l'Europe. Les pépinieres de France, qui ne sont qu'en petit nombre en comparaison de celles de l'Angleterre, ne contiennent guere que des arbres fruitiers, & c'est de-là que vient le nom de *Pépiniere*. Dans ces dernieres, on ne trouve point d'arbres toujours verts, des arbrisseaux à fleurs, ni les arbres des forêts. En Hollande, les pépinieres sont principalement remplies de fleurs ; mais aux environs de Londres, on trouve tous ces objets réunis, & c'est de-là que presque tous les étrangers les tirent.

Je ne me propose pas de traiter ici de ces pépinieres étendues, ni d'en donner une description ; je me bornerai à celles qui sont absolument nécessaires à tous ceux qui aiment à planter, afin qu'ils aient sous la main de quoi former tout de suite leurs plantations ; car si ces plantations sont vastes, la dépense pour faire venir des arbres d'une grande distance, ne sera pas médiocre, & on courra encore le risque de les voir manquer, ce qui est presque inévitable

s'ils ont été élevés dans une bonne terre, & replantés ensuite dans une médiocre. Il est de la plus grande importance, pour tout homme qui plante, de commencer par établir une pépiniere ; mais il faut avoir soin qu'elle ne soit pas attachée à un certain terrein, c'est-à-dire, qu'il seroit mal d'élever des arbres pendant un certain nombre d'années sur la même place ; car alors la terre en seroit tellement épuisée, qu'elle deviendroit hors d'état de servir de nouveau à cet usage : aussi les bons Jardiniers de pépiniere changent de terrein de tems en tems ; & lorsqu'ils ont arraché les arbres d'un canton, ils y plantent, pendant un an ou deux, des herbes potageres, ou quelques autres plantes. Pendant ce tems, au moyen de l'engrais & des fossés qu'ils y creusent, le terrein se rétablit, & devient propre à recevoir d'autres arbres : mais les Jardiniers sont assujettis, par la nécessité, à planter toujours dans le même endroit, étant bornés à une seule piece de terre ; ce qui n'arrive pas aux Seigneurs qui ont à leur disposition une grande étendue de terrein. Ainsi, je conseille à ces personnes de faire leurs pépinieres sur les lieux mêmes où ils veulent établir leurs plantations ; on y laisse une quantité suffisante d'arbres, & on retire les autres, pour les planter ailleurs. Cette méthode sera la plus avantageuse pour toute espece de gros arbres, & sur-tout pour les bois de charpente ;

car tous les arbres qui viennent de semences, ou qui sont transplantés fort jeunes dans les places qui leur sont destinées, font beaucoup plus de progrès & deviennent plus beaux que ceux que l'on transplante plus tard. Ainsi, il faut éclaircir les pépinieres de bonne heure, en ôtant, dans leur premiere jeunesse, les arbres que l'on destine pour d'autres plantations : au moyen de cela, on évite la dépense des soutiens & des arrosemens : mais quand on fait des pépinieres dans des situations exposées, il faut laisser les arbres sur pied plus long-tems, afin que, croissant serrés, ils puissent s'abriter mutuellement. On les éclaircit par dégrés & en proportion de leur accroissement ; car si l'on en enlevoit beaucoup à la fois, le froid retarderoit le progrès des autres. Cependant on ne doit pas se promettre que les derniers enlevés puissent réussir ; il sera plus prudent de les brûler, que d'essayer de les transplanter, quand ils sont devenus trop grands, parce qu'en voulant les arracher avec de bonnes racines, on endommage beaucoup les racines des arbres qui restent.

Tout ce que nous venons de dire, doit s'entendre des grandes plantations qu'on fait dans les parcs ou dans les bois ; mais les pépinieres qu'on ne destine qu'à élever des arbres toujours verts, des arbrisseaux à fleurs, ou d'autres plantes propres à l'ornement des jardins, peuvent être bornées à un certain espace, &

il n'en faut qu'un petit pour cet ufage. Deux ou trois âcres de terre fuffifent pour les plants les plus étendus, & un âcre pour un d'une médiocre grandeur. On peut élever dans cette piece de terre des plantes étrangeres, plufieurs efpeces de fleurs bis-annuelles & vivaces, qui font deftinées à être tranfplantées dans les plates-bandes du parterre, & y femer des graines de fleurs à racines bulbeufes, afin d'avoir par-là tous les ans une variété de différentes efpeces, ce qui dédommagera de la peine & de la dépenfe, & fournira un amufement agréable à tous ceux qui fe plaifent au jardinage.

Une pareille pépiniere doit être placée de maniere qu'on puiffe y avoir aifément de l'eau. Si ce fecours manque, on ne pourra s'en procurer autant qu'il en faut pendant les féchereffes, fans beaucoup de peine & de dépenfe : il faut auffi que cette pépiniere foit voifine, autant qu'il eft poffible, de l'habitation, afin de pouvoir la vifiter facilement dans tous les tems. Il eft abfolument néceffaire qu'elle foit fous l'œil du maître; & fi elle n'en fait pas le plaifir, il eft à craindre qu'elle ne réuffiffe pas. Le fol de cette pépiniere doit être bon, pas trop lourd ni trop fort. Des terres de cette qualité font moins propres à la plupart des femences; car comme ces terres retiennent l'humidité au printems & en hiver, les femences des plantes les plus délicates, &

fur-tout celles des fleurs y pourriroient, fi l'on n'avoit pas le foin de les femer de bonne heure. Quand donc on eft reftreint à de pareils terreins, il faut y mettre une grande quantité de fable, de cendres, & d'autres engrais légers, afin d'en féparer les parties, & de les pulvérifer. Il fera auffi très-avantageux d'en faire des levées, pour que la gelée s'y introduife en hiver, ainfi que de les remuer fouvent, avant d'y mettre les plantes, & après qu'elles y font.

Les avantages qu'on peut tirer d'une pépiniere, étant fenfibles à tous ceux qui y réfléchiffent, il eft inutile d'en faire mention ici; je prie feulement qu'on me permette de répéter ce que j'ai fouvent recommandé, qui eft de tenir la terre nette de mauvaifes herbes, parce que, fi on les laiffoit croître, elles priveroient les racines des jeunes arbres de leur nourriture. Il eft encore effentiel de bécher la terre entre les jeunes plantes, au moins une fois par an, afin que leurs racines puiffent y pénétrer; & fi la terre eft forte, on fera bien de répéter le labour deux fois par année, en Octobre & en Mars; ce qui avancera beaucoup l'accroiffement des plantes, & préparera les racines à la tranfplantation.

Plufieurs perfonnes aiment affez le jardinage pour vouloir élever elles-mêmes leurs arbres fruitiers, ce que je recommande à tous ceux qui défirent avoir de bons fruits;

car il eſt très-incertain ſi l'eſpece de fruit qu'on ſe procure dans les pépinieres, eſt celle qu'on demande, & pluſieurs Seigneurs, qui en ont beaucoup planté, ſe ſont toujours plaints d'avoir été trompés. En outre, il y a un autre inconvénient, auquel, faute de connoiſſance, on ne fait preſque point attention; c'eſt de prendre des greffes ſur de jeunes arbres de pépiniere, avant qu'ils aient porté du fruit. Après avoir répété cette mauvaiſe pratique pluſieurs fois, les arbres deviennent auſſi luxurieux que des *Saules.* Ces arbres, en deux ou trois ans, pouſſent des rejettons auſſi hauts que les murailles, & ſont très-rarement fertiles, même avec la culture la mieux entendue.

Quand on ſe détermine à former ſoi même des pépinieres d'arbres fruitiers, il faut obſerver les regles ſuivantes:

1°. Le ſol dans lequel on établit une pépiniere, ne doit pas être meilleur que celui où l'on a envie de planter les arbres à demeure. Sans cette précaution, les arbres ſont ſouvent arrêtés dans leur accroiſſement, ou ne ſont que peu de progrès pendant trois ou quatre ans, après avoir été tranſplantés. C'eſt ce qui arrive communément à ces arbres qu'on éleve autour de Londres, &, qui, étant tranſportés dans les parties Septentrionales de l'Angleterre, & placés dans un ſol plus pauvre, & dans une ſituation plus froide, réuſſiſſent rarement.

Par cette raiſon encore, il vaut donc beaucoup mieux, lorſqu'on s'eſt procuré toutes les eſpeces qu'on déſire, établir une pépiniere de toutes les eſpeces de tiges ou de ſujets convenables pour les différents fruits. On les greffe ou en écuſſon ou ſuivant la méthode ordinaire. Ces arbres, ainſi élevés dans le même ſol & au même dégré de chaleur, réuſſiſſent mieux étant tranſplantés, que ceux qu'on apporte d'un lieu éloigné, & qui ont été élevés dans un ſol plus riche.

2°. La terre d'une pépiniere doit être neuve, & non pas épuiſée par des arbres ou de groſſes plantes; car, dans un pareil ſol, les ſujets ne feroient que peu de progrès.

3°. L'emplacement de la pépiniere ne doit être ni trop ſec ni trop humide, quoiqu'un terrein ſec ſoit préférable dans ces deux extrêmes. Les arbres n'y font pas tant de progrès que dans une terre humide; mais ils ſont généralement plus ſains & plus diſpoſés à produire du fruit.

4°. Il faut entourer la pépiniere, afin que les beſtiaux & les animaux ſauvages ne puiſſent y pénétrer; car ils y feroient un dégât affreux, ſurtout en hiver, quand la terre eſt couverte de neige, & qu'ils n'ont rien à manger. Les lievres & les lapins ſont les animaux les plus dangereux pour les pépinieres; ils détruiſent pendant l'hiver tous les jeunes arbres, en rongeant leur écorce, & il eſt abſolument néceſ-

faire de la mettre à l'abri de ces ennemis.

Quand l'enclos est fait, on creuse tout autour un fossé de dix-huit pouces ou deux pieds de profondeur, si le terrein le permet. Cet ouvrage doit être fait en Août ou Septembre, afin que la pépinière soit en état de recevoir les jeunes sujets dans la saison convenable, qui est vers le milieu ou à la fin d'Octobre. En creusant ces fossés, il faut avoir grand soin d'enlever les racines des plantes nuisibles, telles que celles du *Chiendent*, de *l'O seille*, &c., qui, si on les laissoit, pousseroient parmi les racines des arbres, ne pourroient plus être détruites, se répandroient dans tout le terrein, & causeroient un grand préjudice aux jeunes plants.

La terre étant bien labourée à la bêche, & la saison de planter étant venue, on rabaisse les élévations, & on nivelle le terrein aussi exactement qu'il est possible; on divise la piece de terre en carreaux proportionnés à sa grandeur, que l'on peut partager encore en planches, pour y planter des noyaux de fruits.

Les meilleurs sujets pour les *Pêches* & les *Brugnons*, sont ceux qu'on éleve avec les noyaux de *Prunes sauvages* & de *Prunier blanc* de Paris : mais jamais il ne faut planter les rejettons de ces arbres, comme le font quelques personnes ; car ces rejettons ont rarement d'aussi belles tiges, & ne poussent jamais d'aussi bonnes racines : d'ailleurs ils sont très-

sujets à produire eux-mêmes une grande quantité de nouveaux rejettons, qui sont fort incommodes dans les allées & dans les plates-bandes des jardins, & nuisent beaucoup aux arbres. Ainsi, il vaut mieux semer des noyaux de chaque espece tous les ans, ou au moins chaque deux ans, afin de ne jamais manquer de sujets.

On se sert, pour les *Poiriers*, de tiges élevées de pepins de fruits : on prend pour cela les marcs de poirée, ou des pepins de quelque espece de *Poires* d'été, qui produisent ordinairement des tiges fortes & vigoureuses, comme la *Cuisse Madame* ; mais quand on veut faire usage de ces pepins, il faut laisser le fruit sur l'arbre, jusqu'à ce qu'il tombe, & le garder ensuite, jusqu'à ce qu'il soit pourri ; alors on ôte les pepins, que l'on conserve dans du sable, dans un endroit inaccessible aux souris & à l'humidité. On seme ces pépins dans le commencement du printems, sur une planche de terre neuve & légere, dans laquelle ils pousseront au bout de six mois; & si l'on a soin de tenir les plantes nettes de mauvaises herbes, elles seront assez fortes pour être enlevées au mois d'Octobre suivant. On préfere les tiges de *Coignassier* à celles de *Poirier*, pour plusieurs especes de *Poires* d'été & d'automne. On s'en sert aussi pour toutes les especes de *Poires* douces & fondantes ; mais elles ne sont pas bonnes pour les *Poires* cassantes ; les fruits qui viennent sur ces sortes de

tiges font fujets à être pierreux. On multiplie affez fouvent ces fujets, au moyen des rejettons qui pouffent en abondance fur les racines des vieux arbres; mais ils ne font pas, à beaucoup près, auffi bons que ceux qui ont été produits de boutures ou de marcottes, lefquels ont de meilleures racines, & ne donnent pas autant de rejettons que les autres; ce qui eft fort à défirer, parce que ces rejettons ne privent pas feulement les arbres d'une partie de leur nourriture, mais ils gâtent encore tout le jardin, comme on l'a déja dit.

On greffe les *Pommiers* fur des tiges élevées de pepins, que l'on ramaffe dans les preffoirs où l'on fait le cidre, ou de pepins de *Pommes fauvages*. Ces derniers font plus eftimés, parce que les plantes font d'une plus longue durée, & qu'elles font fur-tout propres à former de grands *Pommiers* à haut vent: on les éleve & on les traite comme celles des *Poiriers*. Celles qu'on fe procure de rejettons, ne font point, à beaucoup près, auffi bonnes. On a fort recherché, pour les petits jardins, les tiges de *Pommes de Paradis*, pendant quelques années : comme cet arbre eft bas, les efpeces qu'on greffe deffus, portent du fruit plutôt, & on peut les contenir dans un plus petit efpace; mais ces arbres ne font bons que pour de très-petits jardins, ou pour fatisfaire la curiofité : ils durent peu, & ne s'élevent point affez pour pro-

duire beaucoup de fruits, à moins que la greffe ou l'écuffon ne foit enterré, de maniere qu'il puiffe pouffer des racines, & devenir par-là femblable à des arbres greffés fur des fujets libres, en ne recevant que peu de fève de fon toc.

On fe fert, pour les *Cerifiers*, de fujets élevés avec les noyaux de la *Cerife noire commune* ou de la *Cerife fauvage mielleufe*. Les uns & les autres deviennent forts, & produifent de plus belles tiges.

Pour les *Pruniers*, on prend des noyaux des efpeces qui croiffent le plus vîte. Ces tiges feront bonnes auffi pour les *Abricotiers*, qui y prennent bien plus aifément que les *Péchers & Brugnons*; mais on ne doit point fe fervir de rejettons, pour les raifons que nous avons déja expofées.

Beaucoup de perfonnes recommandent les tiges d'*Amandiers*, pour plufieurs efpeces de *Péches* tendres & délicates, qui prennent beaucoup mieux fur ces tiges que fur celles de *Pruniers*; mais comme elles ont des racines tendres qui pouffent de bonne heure au printems, & qui durent peu, on doit les rejetter. Cependant les efpeces délicates de *Péches* qui ne prennent pas fur des *Pruniers*, peuvent être greffées en écuffon fur des *Abricotiers*. Toutes les efpeces de *Péchers* plantées dans des terres feches, dureront bien plus long-tems, & feront moins fujettes à la nielle, fi elles font greffées fur des *Abricotiers*. On a ob-

fervé que, dans les fo's où les *Péchers* réuffiffent rarement, les *Abricotiers* y viennent bien; ce qui peut provenir de la force des *Abricotiers*, & du tiffu compact de leurs vaiffeaux, qui les met en état de mieux pomper leur nourriture que la tige du *Prunier*, & de fe l'affimiler. Comme les fols fecs ne fourniffent que très peu de fuc nutritif aux greffes, le *Pécher* étant d'une nature délicate & fpongieufe, n'eft pas auffi en état de l'attirer à lui; ce qui occafionne cette foibleffe qu'on obferve communément dans ces arbres, lorfqu'ils font plantés dans des endroits fecs: auffi la pratique ordinaire des Jardiniers de pépinière, eft de greffer en écuffon le *Prunier* avec l'*Abricotier*, ou avec quelques *Péchers* qui croiffent librement; & après une année, d'y greffer encore en écuffon les efpeces délicates de *Péches*. Par ce moyen, plufieurs efpeces qu'on ne peut multiplier autrement, & qui périffent par toute autre méthode, réuffiffent très-bien. Les Jardiniers appellent ces *Péches*, doublement travaillées.

Depuis peu, quelques perfonnes ont greffé des *Cerifiers* en écuffon & en fente fur des tiges de *Cornouiller*, & des *Cerifes de Morelle*: ce qui, à ce qu'ils prétendent, rend les arbres plus fertiles & moins luxurieux; de maniere qu'ils peuvent être contenus dans un petit efpace. Ces tiges produifent le même effet fur les *Cerifiers*, que les tiges de la

Pomme de Paradis fur la *Pomme* ordinaire.

Quand on eft pourvu de jeunes tiges de toutes ces efpeces, qu'on a élevées dans le femis l'année précédente, on les tranfplante dans la pépiniere au mois d'Octobre, en laiffant entre chaque rang au moins trois pieds & demi ou quatre pieds de diftance, fi on les deftine pour des hauts vents, & un pied & demi entr'elles dans les rangs; mais fi l'on veut en faire des arbres nains, trois pieds entre les rangs, & un pied de diftance entr'elles dans les rangs, fuffiront.

En ôtant ces arbres de l'endroit où ils ont été femés, on les enlevera avec une bêche, afin de conferver leurs racines entieres, autant qu'il eft poffible; on coupera enfuite tous les petits chevelus, & on taillera les racines, qui s'enfonceroient perpendiculairement dans la terre. Après les avoir ainfi préparés, on tracera une ligne à travers le terrein dans lequel on veut les planter: on fera avec la bêche une rigole à côté de cette ligne; on les y placera fuivant les diftances que nous venons de donner, en les pofant droits; on preffera enfuite la terre autour de leurs racines, & l'on remplira la rigole, en comprimant doucement la terre avec le pied, fans déplacer les arbres de leurs rangs; ce qui les rendroit défagréables à la vue. Il ne faut pas couper la tête de ces ar-

bres ; car cette opération les affoibliroit, leur feroit pouffer des branches latérales, & les gâteroit.

Si l'hiver fe trouve froid, on rendra un grand fervice à ces jeunes plants, en répandant fur la furface du fol un peu de terre douce ou de terreau, pour empêcher la gelée d'y pénétrer, & de nuire aux tendres fibres qui auront pouffé depuis la tranfplantation : mais il faut avoir foin de ne pas mettre ce terreau trop épais, près des tiges, ni de le laiffer trop long tems fur la terre, de peur qu'il ne communique trop d'humidité aux racines ; ce qui arrive fouvent, quand on n'a pas l'attention de l'ôter auffi-tôt que les gelées font paffées. En été, il faut les houer avec exactitude, & arracher avec foin toutes les mauvaifes herbes, qui affoibliroient beaucoup les tiges, & en retarderoient l'accroiffement, fi on les laiffoit pouffer dans les pépinieres. Dans les années fuivantes, on laboure la terre à la bêche, au printems, entre les rangs, pour la rendre affez légere, de maniere que les fibres puiffent s'étendre de tous côtés ; les mauvaifes herbes, par ce moyen, feront auffi détruites : il faut encore obferver de retrancher les branches latérales, afin de rendre les tiges plus droites & unies.

La feconde année après la tranfplantation, on pourra greffer en écuffon les tiges dont on voudra faire des arbres

nains. Celles qu'on deftine pour les hauts vents, doivent refter fix ou fept ans avant d'être greffées. La méthode de greffer en fente & en écuffon étant fuffifamment expliquée à l'article GREFFE, je n'en parlerai pas ici ; il n'eft pas non plus néceffaire d'expliquer la maniere de traiter ces arbres, après qu'ils font greffés, parce qu'elle eft affez détaillée dans les articles qui traitent des différentes efpeces de fruits ; j'ajouterai feulement que les tiges greffées en écuffon pendant l'été, & qui n'ont point réuffi, peuvent l'être de nouveau au printems fuivant ; mais comme les *Pêches* & les *Brugnons* ne prennent jamais bien en fente, il faut toujours les greffer en écuffon.

Le terrein que l'on deftine à une pépiniere de fleurs, doit être bien expofé au foleil, & abrité de tous les grands vents par des plantations d'arbres ou des bâtimens ; la terre en doit être légere & feche, fur tout pour les fleurs à racines bulbeufes. La culture particuliere qui convient à chaque efpece de fleurs, eft expliquée dans les différens articles qui en traitent.

C'eft dans ces pépinieres qu'il faut planter les rejettons de fleurs à racines bulbeufes, où ils refteront jufqu'à ce qu'ils foient en état de fleurir, pour les placer alors dans le parterre, en les y arrangeant, foit en planches, foit dans les plates-bandes, fuivant la qualité de la fleur & la culture qu'elle exige.

PER

On peut auffi élever par femence, dans ce terrain, des plantes à racines bulbeufes, pour fe procurer de nouvelles variétés ; mais peu de perfonnes ont le courage d'entreprendre cette méthode, à caufe du tems confidérable qu'il faut à ces plantes, pour être en état de fleurir. Cependant quand on a une fois commencé, & que l'on continue à en femer tous les ans, on fe procure une fucceffion continuelle de fleurs, qui rend cette méthode moins ennuyeufe qu'elle ne le paroît d'abord.

L'*Oreille d'Ours*, la *Tubereufe*, les *Renoncules*, les *Anemones*, les *Œillets* qu'on élève de graines, doivent être femés dans cette pépiniere, où ils refteront jufqu'à ce qu'ils fleuriffent ; alors on marquera toutes les plantes qui méritent d'avoir place dans le parterre, pour les y mettre dans une faifon convenable. Il ne faut jamais élever ces plantes dans un parterre ; car, quand elles commencent à fleurir, il y a toujours parmi elles un grand nombre de fleurs fimples, qui font un très-mauvais effet.

PEPO. *Voy.* CUCURBITA. *L.*

PERCE-BOSSE, CHASSE-BOSSE *ou* CORNEILLE. *Voyez* LYSIMACHIA.

PERCE-FEUILLE *ou* OREILLE DE LIEVRE. *Voy.* BUPLEVRUM ROTUNDI-FOLIUM.

PERCE-MOUSSE. *V.* MUSCUS CAPILLACEUS MINOR.

PERCE-NEIGE. *V.* GALANTHUS. *L.*

PERCE-NEIGE. [Grand] *Voy.* LEUCOIUM.

PERCE-OREILLES. [*Earwigs.*] Ce font des infectes fort dangereux dans les jardins, furtout dans ceux où l'on cultive les *Œillets* ; car ils font fi friands de ces fleurs, que, fi l'on ne prend pas un foin particulier pour les mettre à l'abri de leurs atteintes, ils les détruifent totalement, en mangeant la partie fucrée du bas des pétales. Pour prévenir cet accident, prefque tous les Jardiniers dreffent des baguettes autour defquelles ils adaptent un baffin de terre ou de plomb, qu'ils tiennent toujours rempli d'eau. *Voy.* l'article DIANTHUS, ŒILLET.

D'autres mettent fur des baguettes des griffes de homars, & des onglets de moutons dans différentes parties du jardin : ces infectes s'y retirent pendant le jour ; de maniere qu'en les vifitant fouvent, on les détruit fans beaucoup de peine, & l'on préferve ainfi non feulement les fleurs, mais auffi les fruits tendres qu'ils attaquent volontiers.

PERCE-PIERRE, CRISTE-MARINE *ou* FENOUIL MARIN. *Voyez* CRITHMUM MARITIMUM *L.*

PERESKIA. *Plum. Nov. Gen.* 37. *tab.* 26. *Cactus. Lin. Gen. Plant.* 539. [*Goofeberry.*] Grofeiller d'Amérique.

Caracteres. La fleur eft en forme de rofe, & compofée de plufieurs feuilles placées orbiculairement. Quand cette fleur eft paffée, le calice fe change en un fruit mou, charnu, globulaire, & garni de feuilles. Dans le centre de ce

fruit font plufieurs femences plates & rondes, renfermées dans un mucilage.

Nous n'avons qu'une efpece de ce genre.

Pereskia aculeata, flore albo, fructu flavefcente. Plum. Nov. Gen. 57 ; Grofeiller d'Amérique, épineux, à fleurs blanches & à fruits jaunâtres.

Cactus Pereskia. Lin. Syft. Plant. t. 2. p. 472. Sp. 24.

Cactus caule tereti arboreo, fpinofo, foliis lanceolato-ovatis. Lin. Hort. Ups. 122 ; Cactus à tige épineufe, en arbre & en cierge conique, avec des feuilles ovales & en forme de lance.

Cactus farmentofus, foliatus, fpinofus, fpinis geminis recurvis, foliis mollibus, ovatis. Brown. Jam. 237.

Malus Americana fpinofa, Portulacæ folio, fructu foliofo, femine reni-formi, fplendenti. Comm. Hort. 1. p. 145. f. 70.

Portulaca Americana lati-folia, ad foliorum ortum lanugine obducta, longioribus aculeis horrida. Pluk. Alm. 135. t. 215. f. 6.

Groffulariæ fructu majore arbor fpinofa, fructu foliofo, virid.-allicante. Sloan. Jam. 165. Hift. 2. p. 86. Raü Dendr. 27 ; Grofeiller d'Amérique, épineux.

Cette plante croît dans quelques parties de l'Amérique Efpagnole, d'où elle a été portée dans les Colonies Angloifes, où l'on donne à fon fruit le nom de *Grofeille* Les Hollandois l'appelent *Pomme d'Amérique* & *Blad-Appel.* Cette plante a plufieurs branches minces, qui ne peuvent fe foutenir fans le fecours de quelques bâtons, ou qu'en les at-

tachant à toutes les plantes voifines. Les branches, ainfi que les tiges, font garnies d'épines longues & blanchâtres, qui naiffent en paquets ; fes feuilles font rondes, fort épaiffes, & fucculentes : fon fruit, qui eft à peu-près de la groffeur d'une noix, eft orné de petites feuilles en-deffus, & formé par une chair mucilagineufe.

On peut multiplier cette plante par boutures pendant tout l'été : on place fes branches dans de petits pots remplis d'une terre fraîche & légere ; & on les plonge dans une couche de tan de chaleur modérée ; on les tient à l'ombre pendant la chaleur du jour, & on les arrofe chaque trois ou quatre jours. Lorfque ces boutures auront pouffé de bonnes racines, ce qui aura lieu au bout de deux mois, on peut les enlever avec foin, & les mettre chacune féparément dans de petits pots remplis d'une terre fraîche, pour les replonger enfuite dans la couche chaude, où elles peuvent refter pendant l'été ; mais à la Saint-Michel, quand les nuits commencent à être froides, il faut les retirer dans la ferre chaude, & les y plonger dans la couche de tan. En hiver, ces plantes demandent à être tenues chaudement : on les arrofe deux fois par femaine ; mais pendant les froids, on leur donne très-peu d'eau. En été, il faut leur procurer beaucoup d'air, & les arrofer copieufement, mais en les laiffant toujours dans la ferre ; car, quoiqu'elles puiffent fupporter le plein

aif en été, dans une situation chaude, cependant elles n'y font point de progrès, & elles ne réussissent pas si bien dans une terre seche que dans la couche de tan. Ainsi, la meilleure méthode est de les placer au fond de la couche de tan, près d'un treillage contre lequel leurs branches puissent être attachées, pour les empêcher de ramper sur les autres plantes. Elles n'ont point encore produit de fleurs ni de fruits en Angleterre ; mais comme plusieurs ont assez bien prospéré dans différens jardins, nous pouvons espérer d'en voir fleurir quelques-unes dans peu de tems.

PERICLYMENUM. *Tourn. Inst. R. H.* 608. *tab.* 578. *Capri-folium. Tourn. Insl. R. H.* 608. *tab.* 379. *Lonicéra. Lin. Gen. Plant.* 210. [*Honeysuckle.*] Chevrefeuille.

Caractères. Le calice de la fleur est petit, & découpé en cinq parties situées sur le germe : la corolle est monopétale ; elle a un tube oblong, & découpé au sommet en cinq segmens tournés en arriere, & cinq étamines en forme d'alène, presque de la longueur de la corolle, & terminées par des antheres oblongues ; son germe, qui est rond, & placé au-dessous du réceptacle, soutient un style mince & couronné par un stigmat obtus ; ce germe devient ensuite une baie ombelliquée, & à deux cellules qui renferment chacune une semence ronde.

Ce genre de plantes est placé, par LINNÉE, dans la premiere section de sa cinquieme classe, avec celles dont les fleurs ont cinq étamines & un style : il a joint celui-ci au *Lonicera* de PLUMIER, & au *Chamæcerasus* de TOURNEFORT : mais comme les fleurs de ce genre different beaucoup, par leur forme, de celles de ces plantes, j'ai cru devoir l'en séparer.

Les especes sont :

1°. *Periclymenum semper virens, floribus capitatis, terminalibus omnibus connatis, semper virentibus ;* Chevrefeuille avec des fleurs disposées en têtes aux extrémités des branches, & des feuilles toujours vertes jointes autour de la tige.

Periclymenum perfoliatum Virginianum semper virens & florens. H. L. ; Chevrefeuille de Virginie, perfeuillé, toujours vert & fleurissant, ordinairement appelé Chevrefeuille à trompette.

Lonicera semper virens. Lin. Syst. Plant. t. 1. p. 480. *Sp.* 2.

2°. *Periclymenum racemosum, racemis lateralibus oppositis, floribus pendulis, foliis lanceolatis, integerrimis;* Chevrefeuille avec des branches latérales & opposées, des fleurs pendantes, & des feuilles entieres & en forme de lance.

Periclymenum racemosum, flore favescente, fructu niveo. Hort. Elth. 306. *tab.* 228; Chevrefeuille à fleurs jaunes, rapprochées en paquets, avec un fruit aussi blanc que la neige.

Chiococca racemosa, foliis oppositis. Lin. Syst. Plant. tom. 1. p. 479. *Sp.* 1. *Jacq. Amer.* p. 68.

Lonicera, racemis lateralibus simplicibus laxis, floribus oppo-

460 PER

fitis , pendulis , geniculis compreſſis. Lin. Sp. Plant. 1. p. 175. n. 12. Hort. Cliff. 496.

Jaſminum folio Myrtino , acuminato , flore albicante , racemoſo. Sloan. Jam. 196. Hiſt. 2. p. 97. t. 188. f. 3. Raii Dendr 64.

3°. *Periclymenum verticillatum , corymbis terminalibus , foliis ovatis , verticillatis , petiolatis;*Chevrefeuille avec des fleurs en corymbe placées aux extrémités des branches , & des feuilles ovales , verticillées & pétiolées.

Periclymenum aliud arboreſcens , ramulis inflexis , flore Corallino. Plum. Cat. 17 ; autre Chevrefeuille en arbre , avec des branches courbées , & une fleur de Corail.

4°. *Periclymenum Germanicum , capitulis ovatis , imbricatis , terminalibus , foliis omnibus diſtinctis ;* Chevrefeuille avec des têtes ovales & imbriquées , qui terminent les tiges , & des feuilles détachées.

Capri-folium Germanicum. Dod. p. 411 ; Chevrefeuille d'Allemagne.

Lonicera Periclymenum. Lin. Syſt. Plant. tom. 1. p. 481. Sp. 4.

5°. *Periclymenum Italicum , floribus verticillatis , terminalibus , ſeſſilibus , foliis ſummis connatoperfoliatis. Hort. Cliff.* 45 ; Chevrefeuille avec des fleurs verticillées & ſeſſiles , qui terminent les tiges , & dont les feuilles du haut ſont perſeuillées & rapprochées très-étroitement.

Capri-folium Italicum. Dod. p. 411 ; Chevrefeuille d'Italie. *Duham. arb.* 3.

Lonicera Capri-folium. Lin.

Syſt. Plant. t. 1. pag. 480. Sp. 1.

6°. *Periclymenum vulgare , floribus corymboſis , terminalibus , foliis hirſutis , diſtinctis , viminibus tenuioribus ;* Chevrefeuille dont les fleurs ſont en corymbe , & terminent les branches , les feuilles velues & ſéparées , & les branches fort minces , ordinairement appelé Chevrefeuille.

Capri-folium non perfoliatum. Kniph. Cent. 8. n. 61.

7°. *Periclymenum Americanum , floribus verticillatis , terminalibus , ſeſſilibus , foliis connato-perfoliatis , ſemper virentibus , glabris ;* Chevrefeuille avec des fleurs verticillées & ſeſſiles qui terminent les branches , & des feuilles unies , toujours vertes , & diſpoſées en anneau , ou , en autre terme , perfeuillées.

Capri-folium perfoliatum ſemper virens , floribus ſpecioſis. Hort. Chels. ; Chevrefeuille toujours vert , dont les feuilles ſont enfilées dans le diſque , avec des fleurs d'une grande beauté.

Semper virens. La premiere eſpece , qui croît naturellement en Virginie , & dans pluſieurs autres parties de l'Amérique Septentrionale , eſt , depuis long tems , cultivée dans les jardins anglois , ſous le nom de *Chevrefeuille à trompette de Virginie.* Il y en a deux variétés , ſi elles ne ſont point des eſpeces diſtinctes ; l'une eſt beaucoup plus dure que l'autre. L'ancienne eſpece , qui vient de la Virginie , a des branches plus fortes , des feuilles
les

les d'un vert plus brillant, des paquets de fleurs plus gros, & d'une couleur plus foncée que la seconde qui vient de la Caroline. Ces plantes ont l'apparence du *Chevrefeuille commun*; mais leurs branches sont plus foibles que celles d'aucune de celles-ci ; à l'exception de celles de l'espece sauvage appelée *Woodbind*; elles sont lisses, & d'un rouge tirant sur le pourpre ; ses feuilles sont d'une forme oblongue & ovale, d'un vert luisant endessus, & d'un vert pâle endessous ; elles sont renversées, & environnent étroitement la tige : ses fleurs, qui sortent en paquets aux extrémités des branches, ont des tubes longs, minces, élargis au sommet, & divisés en cinq segmens presque égaux ; l'extérieur des fleurs est écarlate & brillant, & l'intérieur est jaune ; elles ressemblent beaucoup à celles du *Chevrefeuille* commun : mais elles ne sont pas si profondément divisées, & leurs segmens sont moins réfléchis ; elles n'ont point d'odeur : mais on conserve cette plante dans la plupart des jardins des Curieux, à cause de la beauté de ses fleurs, de leur longue durée, & de ses feuilles toujours vertes.

Il faut placer ces plantes contre des murailles ou des palissades, auxquelles on fixe leurs branches, pour les soutenir, sans quoi elles tomberoient à terre ; car il n'est pas possible de les tenir en boule, comme plusieurs *Chevrefeuilles*, parce que leurs branches sont trop foibles, coulantes, &

sujettes à être détruites dans les hivers rudes : ainsi, on doit les planter à une exposition chaude, où elles commenceront à fleurir vers la fin de Juin, & continueront à donner de nouvelles fleurs jusqu'à l'automne. On les multiplie, en marcottant leurs jeunes branches, qui prennent aisément racine, & peuvent être traitées comme le *Chevrefeuille* ordinaire.

Racemosum. La seconde espece est originaire de la Jamaique ; elle pousse plusieurs branches minces qui ne peuvent se soutenir, & se répandent sur les buissons voisins ; elles ont huit ou-dix pieds de longueur, sont couvertes d'une écorce brune, & garnies de feuilles en forme de lance, de deux pouces & demi de longueur sur un de largeur au milieu, d'un vert luisant endessus, pâles en-dessous, disposées par paires, & opposées : ses fleurs naissent à chaque nœud sur les parties latérales des branches, & sont rangées, sur chaque côté du pédoncule, en grappes aussi longues que le sont celles des groseilles. Ces grappes sont opposées, & ont trois ou quatre pouces de longueur : les fleurs sont petites, d'un vert jaunâtre, & produisent de petites baies d'une blancheur de neige ; ce qui fait qu'en Amérique on donne à cette espece le nom de *Buisson à baies de neige.*

Verticillatum. La troisieme, qu'on rencontre dans quelques Isles de l'Amérique, s'éleve, avec une tige d'arbrisseau, à

Tome V.

G g

462 PER

la hauteur de douze pieds, &
pousse plusieurs branches minces, couvertes d'une écorce
d'un brun clair, & garnies de
feuilles ovales, de deux pouces à peu-près de longueur sur
trois lignes de largeur, qui
sortent au nombre de quatre à
chaque nœud, & environnent
la tige ; elles sont postées sur
de courts pétioles, & ont une
forte côte, de laquelle sortent
plusieurs veines, qui coulent
de cette côte du milieu jusqu'aux bords. Les fleurs naissent en paquets ronds aux extrémités des branches ; elles
sont d'une couleur de Corail
foncé en-dehors, & d'un rouge
pâle en-dedans. Cette plante a
été trouvée à la Jamaïque par
le feu Docteur Houstoun,
qui l'a apportée en Angleterre.

Ces especes sont trop délicates pour profiter dans ce
pays sans chaleur artificielle.
On les multiplie par leurs graines, qu'il faut se procurer des
contrées où elles croissent
naturellement, parce qu'elles ne
mûrissent point en Angleterre.
On les seme dans des pots,
que l'on plonge dans une couche de chaleur modérée, où
on les laisse jusqu'à l'automne ;
car les plantes poussent rarement dans la premiere année :
on tient ces pots dans la serre
chaude pendant l'hiver ; & au
printems, on les reporte sur une
nouvelle couche chaude, qui
fera paroître les plantes. Quand
elles sont en état d'être enlevées, on les met chacune séparément dans de petits pots
remplis de terre légere ; on les
plonge dans une nouvelle cou-

PER

che chaude, & on les tient à
l'ombre jusqu'à ce qu'elles aient
formé de nouvelles racines ;
après quoi on les traite comme
les autres plantes tendres qui
viennent des mêmes contrées.
A mesure qu'elles se fortifient,
on les conduit moins délicatement, en les exposant au plein
air, dans une situation abritée,
pendant les deux mois les plus
chauds de l'été. On peut les
enfermer en hiver dans une
serre seche, & les tenir à une
chaleur modérée, où elles profiteront, & produiront des fleurs
en automne.

Germanicum. La quatrieme
espece est le *Chevrefeuille commun de Hollande* & d'*Allemagne*,
laquelle a été généralement regardée comme étant la même
que le *Chevrefeuille sauvage d'Angleterre*, nommé *Woodbind* : mais
elle est certainement fort différente ; car ses branches sont
beaucoup plus fortes, peuvent
s'élever en tige, & former des
têtes : ce qu'on ne peut faire
avec l'espece sauvage, dont les
branches sont trop foibles. Les
branches de celle-ci sont lisses,
de couleur pourpre, & garnies
de feuilles oblongues, ovales,
de trois pouces de longueur
sur un pouce trois quarts de
largeur, d'un vert luisant en-
dessus, d'un vert pâle en-dessous, portées sur de fort courts
pétioles, & placées par paires
sans être jointes à leur base :
ses fleurs naissent en paquets
aux extrémités des branches,
& s'élevent au-dessus d'une
enveloppe écailleuse, qui, après
que les fleurs sont fanées,
forme une tête ovale, dont les

PER

écailles font difposées comme celles d'un poiffon, & jaunâtres en-dedans ; ce qui leur donne un coup-d'œil fort agréable. Les fleurs de cette efpece font rougeâtres, & paroiffent dans les mois de Juin, Juillet & Août. Il y a deux variétés de cette plante, l'une appelée *Chevrefeuille à fleurs longues*, & l'autre *Chevrefeuille rouge*.

Italicum. La cinquieme efpece, à laquelle on donne ordinairement le nom de *Chevrefeuille d'Italie*, offre auffi deux ou trois variétés ; l'une eft le *Chevrefeuille blanc printanier*, dont les fleurs paroiffent les premieres, & toujours dans le mois de Mai : fes branches font minces, couvertes d'une écorce d'un vert clair, & garnies de feuilles ovales d'une texture mince, placées par paires, & feffiles aux branches ; mais celles qui couvrent l'extrémité des branches font jointes à leur bâfe, de maniere qu'il femble que les branches & les tiges paffent au travers : les fleurs, qui font difpofées en paquets verticillés aux extrémités des branches, font blanches, & ont une odeur fort agréable ; mais elles durent peu, & font entièrement paffées au bout de quinze jours : bientôt après, les feuilles paroiffent comme fi elles étoient niellées & foyeufes ; ce qui les fait paroître défagréables pendant tout l'été, & a contribué à faire moins rechercher cette efpece que les autres. La feconde variété eft le *Chevrefeuille jaune d'Italie* ; fes branches reffemblent beaucoup à celles des

PER 463

précédentes, mais elles ont une écorce & des feuilles plus foncées en couleur : fes fleurs font d'un rouge jaunâtre, & paroiffent bientôt après les blanches ; elles font d'une courte durée, & produifent des baies rouges, qui contiennent une femence dure, renfermée dans une chair molle, & qui mûriffent en automne.

Vulgare. La fixieme efpece eft le *Chevrefeuille fauvage commun d'Angleterre*, qui croît naturellement dans les haies : fes branches font fort minces & velues ; elles fe répandent fur les buiffons voifins, & s'entortillent autour des branches des arbres : fes feuilles font oblongues, velues, diftinctes, fans être jointes à leur bâfe, & oppofées : fes fleurs fortent en paquets longs aux extrémités des branches. Il y a deux variétés de cette efpece, l'une à fleurs blanches, & l'autre à fleurs d'un rouge jaunâtre; elles paroiffent en Juillet, & fe fuccedent jufqu'à l'automne.

On voit auffi une variété de celle-ci à feuilles panachées, & une autre à feuilles découpées, comme celles du *Chêne* ; mais comme elles ne font que des accidens de femences, je ne les ai point mifes au nombre des efpeces.

Americanum. La feptieme, qu'on regarde comme originaire de l'Amérique Septentrionale, a des branches fortes, couvertes d'une écorce de couleur pourpre, & garnies de feuilles d'un vert luifant, qui embraffent les tiges, & qui confervent leur fraîcheur toute

l'année; ſes fleurs ſont produites en paquets verticillés
aux extrémités des branches :
il y a ſouvent deux, & quelquefois trois de ces paquets
qui ſortent l'un de l'autre ; elles ſont d'un rouge brillant
en-dehors, mais jaunes en-
dedans, & d'un goût fort aromatique. Cette eſpece commence à fleurir en Juin ; &
comme ſes fleurs ſe ſuccedent
juſqu'à ce que les gelées les
détruiſent, on en fait beaucoup plus de cas que des autres.

Culture. Ces eſpeces de *Chevrefeuilles* ſe multiplient par marcottes ou par boutures : pour
les marcotter, on choiſit les
plus jeunes branches, que l'on
couche en automne. Ces branches auront pouſſé des racines
pour la même ſaiſon de l'année
ſuivante ; alors on les ſéparera
des vieilles plantes, pour les
placer à demeure, ou bien on
les mettra en pépiniere, pour
les dreſſer comme on veut les
avoir. Si l'on veut qu'elles
ſoient à plein vent, on attache
leur tige principale contre un
poteau, & l'on retranche toutes les branches latérales juſqu'à ce qu'elles ſoient parvenues à une hauteur convenable ; alors on les arrête, pour
les obliger à former une tête,
qu'il faut tailler, pour empêcher les branches de trop s'allonger. En répétant ſouvent
cette opération, à meſure que
les branches naiſſent, on peut
leur donner la forme d'un arbre à haut vent ; mais ſi l'on
déſire les voir fleurir, il n'eſt
pas poſſible de leur former une

tête réguliere ; car, en taillant
ſouvent les branches, on retranche les boutons, & l'on ne
peut eſpérer que peu de fleurs.
Ainſi, cette forme n'étant pas
naturelle à cette eſpece d'arbre, il faut en avoir peu à
haut vent ; au - lieu qu'en les
plantant près des buiſſons,
leurs jeunes tiges ſe couleront,
& s'entremêleront parmi, fleuriront beaucoup mieux, & auront une apparence plus agréable que ſi elles étoient dreſſées
régulierement. Ainſi, il ſuffira
d'en avoir dans la pépiniere
deux ou trois dreſſées contre
des poteaux, & l'on tiendra les
autres baſſes ; elles ſeront en
état d'être tranſplantées dès
l'automne ſuivant dans les places qui leur ſont deſtinées ; car
quoiqu'on puiſſe les laiſſer plus
long-tems dans les pépinieres,
cependant elles ne profiteroient
pas ſi bien, ſi elles étoient enlevées plus vieilles. Quand on
multiplie ces plantes par boutures, il faut le faire en Septembre, & les planter auſſi-tôt
que la terre eſt humectée par
la pluie : on leur laiſſe quatre
nœuds ou boutons, dont trois
doivent être enfoncés dans la
terre, & le quatrieme ſera deſ-
tiné à produire des branches.
On peut les planter en rangs,
à un pied de diſtance l'un de
l'autre, & à quatre pouces
dans les rangs : on comprime
la terre tout autour, en la
foulant avec les pieds. Comme
le *Chevrefeuille* toujours vert,
& ceux qui fleuriſſent tard,
ſont un peu plus délicats que
les autres, il faut couvrir la
terre où ils ſont plantés avec

du tan ou du terreau, pour empêcher les gelées d'hiver, & les hâles du printems d'y pénétrer. Cette précaution sera très-avantageuse aux boutures, qui prendront certainement racine, si l'on a laissé au bas un petit morceau de bois de deux ans. Les plantes élevées de boutures sont préférables à celles qui sont multipliées par marcottes, parce qu'elles poussent toujours de meilleures racines : elles croissent dans presque tous les sols & à toutes les expositions, à l'exception des dernieres, qui ne profiteroient pas, si elles étoient exposées au froid pendant l'hiver. Elles réussissent plus certainement dans une terre grasse, molle & sablonneuse, & leur verdure s'y conserve mieux que dans un sol sec & graveleux, ou dans les tems chauds ; leurs feuilles se rétrécissent souvent, & pendent d'une maniere désagréable. Les especes qui fleurissent tard en automne, ne conservent pas long-tems leur beauté sur une terre seche, à moins que la saison ne soit froide & humide. Une terre grasse, douce, pas trop ferme, & humide, leur convient mieux.

Il y a peu d'especes d'arbrisseaux qui méritent plus d'être cultivées que la plupart de ceux-ci ; car leurs fleurs sont très belles, & parfument l'air agréablement à une grande distance, sur-tout le matin, le soir, dans les tems couverts, & lorsque l'activité du soleil ne dissipe pas leur odeur ; de sorte qu'on ne peut pas en mettre un trop grand nombre dans les promenades solitaires, en les entremêlant avec d'autres arbustes. J'ai vu de ces plantes produire l'effet le plus agréable dans des haies d'*Aulnes* & de *Lauriers* ; leurs branches étoient rangées de maniere que leurs fleurs se dispersoient depuis le bas de la haie jusqu'au haut, & se mêloient délicieusement avec la verdure des autres plantes. Les meilleures especes qu'on puisse employer à cet usage, sont les *Chevrefeuilles* toujours verts, dont les fleurs conservent plus long tems leur beauté.

On peut aussi multiplier ces plantes par semences ; mais elles ne poussent pas dans la premiere année, à moins qu'elles ne soient mises en terre en automne, aussi-tôt qu'elles sont mûres.

PERIPLOCA. *Tourn. Inst. R. H.* 93. *tab.* 22. *Lin. Gen. Plant.* 267. περιπλοκή, de περί, environ ou autour, & de πλίκω, nouer ou plisser ; parce que cette plante s'entortille sur elle-même autour des autres plantes voisines. [*Virginian Silk.*] Soie de la Virginie ou Apocin, Bourreau des Arbres.

Le calice de la fleur est persistant, petit, & découpé en cinq pointes ; la corolle est monopetale, unie, & divisée en cinq segmens étroits & dentelés à leur extrémité ; elle a un petit nectaire autour de son centre, avec cinq filamens moins longs que le pétale : la fleur a cinq étamines courtes, & terminées par des antheres érigées & réunies en une tête ;

466　PER

son germe est divisé en deux parties ; son style, qui paroît à peine, est couronné par deux stigmats simples. Ce germe se change dans la suite en deux capsules oblongues, gonflées, & à une cellule remplie de semences couronnées de duvet, & disposées les unes sur les autres comme des écailles de poisson.

Ce genre de plantes est rangé dans la seconde section de la cinquieme classe de LINNÉE, qui comprend celles dont les fleurs ont cinq étamines & deux styles.

Les especes sont :

1°. *Periploca Græca, floribus internè hirsutis. Lin. Sp. Plant.* 211 ; Periploca dont les fleurs sont velues en-dedans.

Periploca foliis lanceolato-ovatis. Hort. Cliff. 78. *Roy. Lugd.-B.* 410. *Hort. Angl. fr.* 15. *Duham. Arb.* 2. *f.* 104. *tab.* 21.

Periploca foliis oblongis. Tourn. Inst. R. H. 93 ; Soie de la Virginie, à feuilles oblongues ; Bourreau des Arbres.

Apocynum angusti-folium. Clus. Hist. 1. p. 125.

2°. *Periploca Africana, caule hirsuto. Lin. Sp. Plant.* 211 ; Periploca à tige velue.

Cynanchum caule volubili ramoso, foliis sub-ovatis cum acumine. Hort. Cliff. 79.

Apocynum scandens Africanum, Vineæ Pervincæ foliis, subincanum. Com. Plant. Rar. 18 ; Apocin d'Afrique, grimpant, avec une feuille blanche de Pervenche.

Cynanchum foliis planis, sinuatis, flore pallidè viridi, fructu crasso, glabro, viridi. Burm.

PER

Afr. 34. *t.* 14. *f.* 2 ; Variété.

3°. *Periploca fruticosa, foliis oblongo-cordatis, pubescentibus, floribus alaribus, caule fruticoso scandente ;* Periploca à feuilles oblongues, en forme de cœur, & garnies d'un poil follet, avec des fleurs qui naissent aux aîles des feuilles, & une tige grimpante d'arbrisseau.

Periploca foliis cordatis, holo-sericeis, floribus parvis, albis, campani-formibus. Houst. MSS. ; Périploca avec des feuilles en forme de cœur, & soyeuses, & des fleurs petites, blanches, & en forme de cloche.

Græca. La premiere espece croît naturellement en Syrie ; mais elle est assez dure pour profiter en plein air sous notre climat ; elle a des tiges d'arbrisseau torses, couvertes d'une écorce de couleur foncée, qui se roulent autour de tous les objets voisins, & s'élevent ainsi à plus de quarante pieds de hauteur. Ces tiges poussent latéralement des branches minces, qui s'entrelacent les unes avec les autres ; elles sont garnies de feuilles ovales, en forme de lance, de quatre pouces environ de longueur sur deux de largeur au milieu, d'un vert luisant en-dessus, d'un vert pâle en-dessous, & placées par paires sur de courts pétioles : ses fleurs, qui sortent en paquets aux extrémités des petites branches, sont de couleur pourpre, velues en-dedans, & composées d'un pétale découpé presque jusqu'au fond en cinq segmens, qui s'étendent en forme d'étoile ; en-dedans est situé un nectaire

velu, qui environne cinq courtes étamines, ainsi que le germe. Ce germe se change, quand
la fleur est passée, en une silique double, longue, cylindrique, & remplie de semences
plates, placées en forme d'é-
cailles de poisson, & couronnées au sommet d'un duvet
mou. Cette plante fleurit en
Juillet & Août ; mais elle perfectionne rarement ses semences en Angleterre.

On la multiplie aisément par
marcottes, qui poussent des racines dans l'espace d'une année :
on peut alors les séparer de la
vieille plante, & les mettre
dans les places qui leur sont
destinées ; elles peuvent être
transplantées en automne, lorsque les feuilles tombent, ou
au printems, avant qu'elles
commencent à pousser ; on les
place de façon qu'elles puissent avoir un soutien, sans
quoi elles traîneroient sur la
terre, & s'attacheroient à toutes les plantes du voisinage.

Africana. La seconde espece
croît naturellement en Afrique ; elle pousse plusieurs tiges
minces, qui s'entrelacent les
unes avec les autres, & se
roulent autour des corps voisins ; elles s'élevent à-peu-près
à trois pieds de hauteur, &
poussent quelques petites branches latérales & velues, ainsi
que les feuilles, qui sont ovales, à-peu-près de neuf lignes
de longueur sur six de largeur,
& placées par paires sur de
forts courts pétioles : ses fleurs,
qui sont produites en petits
paquets sur les côtés des tiges,
sont petites, de couleur pourpre

usée, & d'une odeur agréable ; leur corolle est découpée
en cinq segmens étroits presque jusqu'au fond. Cette plante
fleurit pendant tout l'été ; mais
elle ne produit point de semences ici. Il y a une variété de cette espece à feuilles
& à tiges unies, qui vient
des mêmes contrées.

Fruticosa. La troisieme, que
le Docteur Houstoun a dé-
couverte à la Vera-Cruz en
Amérique, s'élève à la hauteur de cinq ou six pieds, avec
une tige forte, ligneuse, &
couverte d'une écorce grise ;
elle pousse plusieurs branches
foibles, qui se roulent autour
des objets voisins, & s'élevent, par ce moyen, à vingt
pieds de hauteur ; ses branches
sont garnies de feuilles en forme de cœur, de trois pouces
de longueur sur deux de largeur près de la bâse, d'un
vert jaunâtre, couvertes d'un
poil soyeux, douces au toucher, & opposées sur des pétioles assez longs : ses fleurs
sortent en petits paquets aux
ailes des feuilles ; elles sont
petites, blanches, en forme
de cloche ouverte, & produisent des siliques gonflées, cylindriques, & remplies de semences, couronnées d'un duvet long & plumacé.

Culture. La seconde espece
est assez dure pour profiter
dans ce pays, pour peu qu'on
la tienne à l'abri de la gelée.
Ces plantes réussissent & fleurissent assez bien, en les plaçant en hiver sous un châssis ordinaire, ou dans une
serre, & en les exposant en

468 **PER**

plein air pendant l'été avec d'autres plantes exotiques & dures : mais comme toutes les plantes de ce genre font remplies d'une féve laiteufe, elles craignent l'humidité, fur-tout dans les tems froids, & fe pourriffent aifément. On les multiplie en marcottant leurs branches, qui, dans une année pouffent d'affez fortes racines pour pouvoir être tranfplantées : on les place dans une terre graffe, légere, fabionneufe & pas trop riche ; leurs pots ne doivent pas être trop grands, parce qu'elles n'y profiteroient pas.

La troifieme efpece eft trop délicate pour réuffir en Angleterre fans le fecours d'une ferre chaude. On peut la multiplier par marcottes, comme la précédente, ou par femences, quand on peut s'en procurer des endroits où elle croit naturellement : on répand ces graines fur une bonne couche chaude ; & quand les plantes ont pouffé, on les traite comme celles qui font tendres & exotiques.

Toutes ces efpeces font de grands progrès, & fleuriffent beaucoup, lorfqu'on les tient conftamment dans la couche de la ferre chaude, qui leur convient mieux que toute autre fituation ; mais il ne faut pas leur donner trop de chaleur en hiver : elles exigent beaucoup d'air en été, parce que, fi elles étoient trop renfermées, leurs feuilles fe couvriroient d'infectes, & les plantes fe moifiroient en peu de tems.

On regarde toutes ces plantes comme nuifibles aux animaux, ainfi que les *Apocins*, auxquels elles reffemblent beaucoup par leurs caracteres & leurs propriétés.

PERSEA. *Plum. Nov. Gen.* 44. *tab.* 20. *Laurus. Lin. Gen. Plant.* 452. [*Avocado*, or *Avogato.*] Poire d'Avocat. C'eft la même que celle qui eft rappelée au mot *Laurus Perfea.*

Caracteres. La fleur n'a point de calice ; mais elle eft compofée de fix pétales terminés en pointe aiguë, & étendus, de fix étamines qui ont environ la moitié de la longueur des pétales, & font terminées par des antheres rondes, & d'un ftyle court, & couronné par un germe pyramidal, qui devient dans la fuite un fruit gros, charnu & pyramidal, dans lequel eft renfermée une femence ovale & à deux lobes.

LINNÉE a joint ce genre de plante à celui du *Laurus*, & l'a placé dans la premiere fection de fa neuvieme claffe, avec celles dont les fleurs ont neuf étamines & un ftyle.

Nous n'avons qu'une efpece de cette plante, qui eft la

Perfea Americana. Clus. Hift.; Poire d'Avocat.

Laurus Perfea, *foliis ovatis, coriaceis, transverfe venofis, perennantibus, floribus corymbofis.* Jacq. *Obs.* 1. *p.* 37. Lin. *Syft. Plant.* tom. 2. *p.* 227. *Sp.* 8.

Laurus foliis oblongo-ovatis, fructu obovato, pericarpio butyraceo. Brown. *Jam.* 214.

Pyro fimilis fructus in novâ Hifpaniá, nucleo magno. Bauh. *Pin.* 439.

PER

Prunifera arbor, fructu maximo pyri-formi, viridi, pericarpio esculento, butyraceo, nucleum unicum maximum, nullo ossiculo tectum, cingente. Sloan. Jam. 132. *Hist.* 2. *p.* 132. *t.* 122. *f.* 2. *Raii Dendr.* 48.

Arbor Americana, amplissimis pergamenis foliis, superficie nitidissimâ, fructu pyri-formi crustaceo, cortice coriato. Pluk. Alm. 39. *t.* 267. *f.* 1.

Cet arbre croit naturellement & en grande abondance dans l'Amérique Espagnole, ainsi qu'à la Jamaïque. Il a été transplanté dans la plupart des établissemens anglois, à cause de son fruit, qui est estimé par les habitans, non-seulement comme bon à manger dans les desserts, mais aussi comme propre à leur fournir une partie considérable de leur nourriture. Ce fruit, en lui-même, est fort insipide ; mais on l'assaisonne généralement avec du jus de *Limon* & du sucre, pour lui donner un goût acide. Il est très-nourrissant, & on le croit un grand aiguillon à l'amour : d'autres le mangent avec du poivre & du vinaigre.

Dans les pays chauds, cet arbre s'élève à la hauteur de trente pieds & plus ; son tronc est aussi gros que ceux de nos *Pommiers* ordinaires ; l'écorce en est lisse, & de couleur cendrée ; ses branches sont garnies de feuilles oblongues, unies, assez larges, comme celles du *Laurier*, d'un vert foncé, & qui restent sur l'arbre pendant toute l'année ; ses fleurs & ses fruits naissent,

pour la plupart, vers l'extrémité des branches ; le fruit est de la grosseur de nos plus grosses *Poires*, & contient une grosse semence à deux lobes, renfermée dans une coque mince.

Cette plante est conservée en Europe comme une curiosité, par les personnes qui prennent plaisir à rassembler les plantes rares & exotiques ; & quoiqu'il y ait peu d'espérance de lui voir produire du fruit dans notre climat, cependant la beauté de son feuillage, qui conserve sa belle verdure pendant toute l'année, doit lui faire donner une place dans toutes les collections curieuses.

M. JACQUIN, Professeur de Botanique, & Directeur des jardins de l'Empereur à Vienne, dans ses *Observations de Botanique*, (part. I., pag. 38,) dit qu'il y a long-tems que cette plante a été portée du continent de l'Amérique, dans les Isles voisines & adjacentes, où on la rencontre dans les villages, les villes, les jardins, & autres lieux cultivés ; elle égale en hauteur les *Poiriers* les plus hauts de l'Europe ; elle est assez garnie de feuilles, & sa forme n'est point désagréable ; son bois est couvert d'une écorce rousfâtre, & ce bois est lui-même presque de cette couleur : elle produit une grande quantité de fleurs petites, blanches & peu odorantes, auxquelles succedent des fruits énormes pour ce genre, plus gros que le poing, de la forme d'un œuf,

470   PER

& dont le côté le plus obtus
se trouve en-bas : ils sont d'a-
bord d'un vert agréable ; mais
ils deviennent d'un rouge brun,
tirant sur le pourpre, lorsqu'ils
sont mûrs. On connoît qu'ils
sont parfaitement mûrs, par
le son que rend le noyau lorsqu'on le remue : on les garde
cependant encore quelques
jours avant de les manger,
parce qu'ils en sont meilleurs
& plus tendres ; la peau de ce
fruit, qui n'est pas d'ailleurs
fort épaisse, en devient plus
mince, & on peut alors la
séparer aisément de la chair,
qui tire sur le vert, mais qui
est d'autant plus blanche, qu'elle est plus éloignée de l'écorce : elle est grasse au toucher,
presque sans odeur, & d'une
consistance butyreuse ; elle a
une saveur qui lui est particuliere . fort agréable, & qui
tient un peu de celle de l'*Ar-
tichaud* & de l'*Aveline*. Il n'y
a cependant point de fruits en
Europe dont le goût puisse lui
être exactement comparé. Le
noyau se trouve dans le centre du fruit, sans y adhérer ;
il est presque rond, inégal à
sa superficie, blanc, & de
plus d'un pouce de diametre :
il n'est point bon à manger ;
il est rempli d'un suc laiteux
& blanc, que le contact de
l'air fait un peu rougir. Ce
noyau, tiré de son fruit, se
couvre, dès le second jour,
d'une pellicule déliée, membra-
neuse, & légerement humide.
On sert journellement le fruit
de cet arbre sur les meilleures
tables. Les François le mangent avec le bouilli, sans aro-

mates, ni sel ni poivre : on
le coupe ordinairement en longueur avec son écorce, autour
du noyau, en morceaux que
l'on offre à chacun des con-
vives. Il fait non-seulement
les délices des hommes, mais,
ce qui lui est peut-être particulier parmi tous les végétaux,
c'est qu'il n'y a point d'animaux qui n'en soient friands,
& qui ne s'en nourrissent. Les
poules, les vaches, les chiens,
les chats l'aiment également.
M. JACQUIN ajoûte, qu'il n'a
point trouvé en Amérique de
fruit qu'il ait plus recherché
que la *Poire d'Avocat*, quoiqu'il ne lui ait pas plu la premiere & la seconde fois ; ce
qui est assez ordinaire aux étrangers qui commencent à en
goûter. Les Espagnols l'appellent, *Peral de Abogado* ; Les
François *Poire d'Avocat* ; les
Anglois *Pear-Tree* ou *Alligator
Pear-Tree*, *Poire de Crocodile*.
Ces fruits n'ont pu soutenir le
transport en Europe.

Culture. On multiplie cette
espece par ses noyaux, qu'il
faut se procurer de son pays
natal, aussi frais qu'il est possible. En les envoyant dans du
sable, on sera plus certain de
leur réussite. On plante ces
noix ou noyaux dans des pots
remplis d'une terre riche &
légere ; on les plonge dans
une bonne couche chaude de
tan, dont il faut conserver la
chaleur : on arrose les pots
aussi souvent que la terre paroît seche, pour en faciliter la
prompte végétation ; mais on
doit avoir soin de ne pas leur
donner trop d'eau à la fois,

PER

de peur de les faire pourrir.
Cinq ou fix femaines après,
ils commenceront à pouffer;
alors on les traitera très-déli-
catement, en confervant à la
couche le dégré de chaleur
qui lui eft néceffaire, & en
foulevant un peu les vitrages,
pour donner de l'air frais aux
plantes, quand le tems eft
chaud. Lorfqu'elles ont acquis
environ quatre pouces de hauteur, on les tranfplante avec
précaution. S'il y en a plufieurs dans un pot, on les fépare, en confervant une motte
de terre à leurs racines; on
les met chacune féparèment
dans un petit pot rempli de
terre riche & légere, & on
les tient plongées dans une
couche chaude de tan, jufqu'à
ce qu'elles aient formé de nouvelles racines; alors on leur
donne de l'air frais à proportion de la chaleur de la faifon; & vers la Saint-Michel,
on les plonge dans la couche
de tan de la ferre chaude. Pendant l'hiver, on leur procure
une chaleur modérée, & on
les arrofe légerement deux fois
la femaine; au printems, on
leur donne de plus grands pots,
& on renouvelle enfuite la
chaleur de la couche avec du
nouveau tan, pour faire pouffer les plantes de bonne heure, & hâter leurs progrès dans
l'été fuivant. On les tient conftamment dans la ferre chaude;
car elles font trop tendres pour
fupporter le plein air dans ce
pays, en quelque faifon que
ce foit; mais dans les tems
chauds, il eft néceffaire de
leur donner beaucoup d'air.

PER   471

PERSICA. *Tourn. Inft. R. H.*
624. tab. 402. *Amygdalus. Lin.*
Gen. 619, ainfi appellée de
la Perfe, Empire d'Afie,
d'où cette efpece de plante a
été apportée dans notre climat.
[*The Peach-tree.*] Le Pêcher.

Caraêteres. Le calice de la
fleur eft tubulé, & formé par
une feuille découpée en cinq
fegmens obtus & étendus; la
corolle eft compofée de cinq
pétales oblongs, ovales, obtus,
& inférés dans le calice : la
fleur a environ trente étamines
minces, érigées, terminées
par des antheres fimples, &
inférées auffi dans le calice;
elle a un germe rond & velu,
qui foutient un ftyle de la longueur des étamines, & couronné par un ftigmat à tête. Ce
germe fe change dans la fuite
en un fruit rond, gros, cotonneux, & bon à manger,
divifé par un fillon longitudinal, & qui renferme un noyau,
dont la coque eft à filet, &
marquée de différens points.

Ce genre de plantes eft rangé dans la premiere feêtion de
la douzieme claffe de LINNÉE,
qui comprend celles dont les
fleurs ont un ftyle, depuis
vingt jufqu'à trente étamines
inférées dans le calice.

Il y a beaucoup de variétés
de cette plante dans les jardins
des Curieux, qui fe plaifent
à raffembler toutes les efpeces
qu'on cultive dans les différentes parties de l'Europe. Je
commencerai d'abord par parler de deux ou trois efpeces
que l'on cultive pour la beauté
de leurs fleurs; après quoi je
donnerai les différentes variétés

472 PER

des meilleurs fruits qui font parvenus à ma connoiffance.

Les efpeces font :

1°. *Perfica vulgaris, flore pleno.* Tourn. *Inft. R. H.* 625 ; Pêcher commun, à fleurs doubles.

Amygdalus Perfica. Lin. *Syft. Plant. tom.* 2. *p.* 481. *Sp.* 1.

2°. *Perfica nana Africana, flore incarnato, fimplici.* Tourn. *Inft. R. H.* 625 ; Amandier nain, à fleurs fimples, couleur de chair.

3°. *Perfica Amygdalus, Africana, nana, flore incarnato, pleno.* Tourn. *Inft. R. H.* 925 ; Amandier nain, à fleurs doubles, couleur de chair.

Amygdalus pumila. Lin. *Syft. Plant. t.* 2. *p.* 482. *Sp.* 3.

Vulgaris. Le premier de ces arbres orne beaucoup les jardins dans le commencement du printems ; fes fleurs font groffes, doubles, & d'un beau rouge ou de couleur pourpre. On peut le planter à plein vent, & en l'entremêlant avec d'autres arbres à fleurs du même crû, parmi lefquels il fera une variété très-agréable : on le met auffi en efpalier contre une muraille du parterre, ou il fera beaucoup mieux que toutes les efpeces de fruits choifis, qui s'y trouveroient expofés au pillage des domeftiques, & ne pourroient jamais acquérir une maturité parfaite. On multiplie cet arbre en le greffant fur *Amandier* ou fur *Prunier*, & on le plante dans un fol frais, de bonne qualité, & pas trop humide.

Nana. Amygdalus. Les deux autres efpeces font d'un crû plus bas, & s'élevent rarement au-deffus de trois ou quatre pieds de hauteur : elles peuvent être greffées fur des *Amandiers*, ou multipliées par marcottes ; leurs greffes prennent auffi fur des *Pruniers* : mais elles font fujettes à fe gâter quatre ou cinq ans après, furtout celle à fleurs doubles, qui eft plus tendre que l'autre, mais qu'on peut multiplier en abondance par les rejettons que fa racine produit.

Ces arbriffeaux font une variété très-agréable parmi les arbres à fleurs, dans les petits quartiers déferts. L'efpece fimple fleurit au commencement d'Avril, & la double ordinairement trois femaines plus tard.

Je vais détailler à préfent les différentes efpeces de bons *Péchers* qui font venus à ma connoiffance : quoiqu'il y en ait peut-être un plus grand nombre dans des catalogues de fruits, je doute fi plufieurs ne font pas les mêmes, répétées fous différentes dénominations ; car, pour déterminer la diftinction des efpeces, il eft néceffaire d'obferver la forme & la groffeur des fleurs, ainfi que les différentes parties du fruit, ce qui fert à caractérifer quelquefois l'efpece, quand le fruit feul ne fuffit pas ; d'ailleurs, il y a une grande différence dans la groffeur & faveur des *Péches* d'une même efpece, fuivant la nature du fol & l'expofition où les arbres, fe trouvent placés ; de forte qu'il eft prefque impoffible, même aux perfonnes les plus exercées en ce genre, de

PER

diftinguer ces fruits , quand ils ont été recueillis dans des jardins d'un fol différent.

Le tranfport de ces arbres de France en Angleterre , a fouvent occafionné de la confufion dans la nomenclature de ces fruits ; car ordinairement les perfonnes qui fe chargent de les tranfporter pour les vendre , ne connoiffent point du tout la différence de leurs efpeces , & ils s'en rapportent à ce qui leur a été dit par les Jardiniers , qui font métier de les multiplier en grande quantité , pour en fournir les marchés de France ; ceux-ci les tranfportent fur des charriots , & les vendent par paquets à ceux qui les portent en Angleterre. Il arrive auffi quelquefois qu'on les donne fous leurs vrais noms ; mais ces noms peuvent fe perdre dans le trajet , & les arbres parviennent à d'autres perfonnes , qui , ne fachant pas le nom véritable du fruit, lui en donnent un nouveau ; ce qui produit une confufion qu'il eft impoffible de rectifier , & a fait placer dans les Catalogues un plus grand nombre d'efpeces qu'il n'en exifte réellement : d'ailleurs , comme la plupart de ces variétés ont été obtenues de femences , & que leur nombre peut être porté à l'infini par ce moyen , je me contenterai de démontrer ici les principales efpeces que nous connoiffons , & qui fuffifent pour former une collection capable de donner des fruits pendant toute la faifon des Pêches.

1. L'avant-Pêche blanche. [The white Nutmeg.] L'arbre a des feuilles fciées , & pouffe toujours très-foiblement , s'il n'eft pas greffé fur *Abricotier* : fes fleurs font larges & ouvertes ; le fruit eft petit & blanc , ainfi que la chair ; le noyau s'en détache aifément. Ce fruit eft un peu mufqué & fucré : il n'eft cependant eftimé que parce qu'il eft le plus précoce , & qu'il mûrit au commencement de Juillet ; mais il devient bientôt farineux.

2. L'avant-Pêche de Troyes. [The red Nutmeg.] L'arbre qui la donne , a des feuilles fciées : fes fleurs font groffes & ouvertes ; fon fruit eft plus gros & plus rond que l'*avant-Pêche blanche* , & d'une couleur de vermillon brillant ; fa chair eft blanche , & fort rouge fur le noyau , dont elle fe détache aifément , & fon goût eft mufqué. Cette Pêche eft fort eftimée , & mûrit vers la fin de Juillet.

3. La *double Pêche* ou *Mignonette de Troyes*. [The early or fmall Mignon.] L'arbre produit de petites fleurs rétrécies ; fon fruit eft d'une groffeur médiocre , rond , & fort rouge fur le côté expofé au foleil ; fa chair eft blanche , & fe fépare du noyau , autour duquel elle eft rouge ; fon jus eft excellent & vineux. Ce fruit mûrit à la fin de Juillet ou au commencement d'Août.

4. L'*Alberge jaune*. [Yellow Alberge.] a des feuilles unies , des fleurs petites & rétrécies , un fruit de groffeur médiocre , & un peu long , une chair

474 PER

jaune, feche, & rarement bien favoureufe. Il a peu de valeur, lorfqu'on le cueille avant fa parfaite maturité. Il mûrit dans le commencement du mois d'Août.

5. La *Magdeleine blanche.* [*White Magdalen.*] a des feuilles fciées & des fleurs groffes & ouvertes; fon bois eft généralement noir autour de la moëlle; fon fruit eft rond, d'une groffeur médiocre; fa chair eft blanche jufqu'au noyau, dont elle fe détache; fon jus eft rarement d'une grande faveur, le noyau eft fort petit. Ce fruit mûrit de bonne heure en Août.

6. La *Pourprée hative.* [*Early Purple.*] a des feuilles unies, des fleurs longues & ouvertes, & un fruit gros, rond, & d'un beau rouge: fa chair eft blanche & fort rouge autour du noyau: il eft rempli d'un jus excellent & vineux. Cette Pèche eft regardée, par tous les connoiffeurs, comme une des meilleures. Elle mûrit avant le milieu d'Août.

7. La *groffe Mignonne françoife* [*Large or french Mignon*] a des feuilles unies, des fleurs groffes & ouvertes, & un fruit un peu oblong, généralement gonflé fur un côté, & d'une belle couleur; fon jus eft fort fucré, & d'un goût relevé; fa chair eft blanche, & fort rouge près du noyau, qui eft petit, & fe détache aifément. Cette Pèche, qui mûrit à la mi-Août, eft regardée comme une des meilleures. Cette efpece eft tendre, & ne réuffit pas fur un

PER

fujet commun. Ainfi, il faut la greffer fur quelques *Péchers* vigoureux ou fur *Abricotiers.* C'eft pour cette raifon que des Jardiniers de pépiniere en augmentent le prix. La meilleure méthode eft de greffer cette Pèche fur quelques vieux Abricotiers bien fains, & plantés à l'expofition du fud ou fud-eft: on coupe l'Abricotier, quand la greffe a pris & pouffé de bonnes branches. J'ai vu des arbres ainfi traités produire des fruits beaucoup plus beaux, de meilleur goût, & en bien plus grande quantité que de toute autre maniere, & les arbres étoient auffi beaucoup plus fains.

8. La *Chevreufe* ou *Belle Chevreufe.* [*The Chevreufe.*] a des feuilles unies, des fleurs petites & rétrécies, un fruit de groffeur médiocre, un peu oblong, d'un beau rouge, & dont la chair eft blanche, mais fort rouge près du noyau, qui fe détache aifément. Ce fruit eft rempli d'un jus excellent & fucré; il mûrit vers la fin d'Août, & eft fort abondant: il peut être rangé parmi les meilleures Pèches.

9. La *Magdelaine rouge,* [*Red Magdalen.*] appelée par les François des environs de Paris, *Magdelaine de Courfon,* a des feuilles profondément fciées, des fleurs groffes & ouvertes, un fruit gros, rond, & d'un beau rouge; la chair blanche, mais fort rouge près du noyau, dont elle fe détache: un jus fort fucré, & d'un goût exquis. Cette Pèche mûrit à la fin d'Août;

PER 475

elle eſt une des meilleures.

10. La *Newington printaniere* [*Smith's Newington.*] reſſemble fort à celle que les François appellent la *Pavie blanche*; elle a des feuilles ſciées , une fleur graſſe & ouverte, un fruit de groſſeur médiocre , & d'un beau rouge ſur le côté expoſé au ſoleil : ſa chair eſt ferme & blanche , mais fort rouge près du noyau , dont elle ſe détache difficilement ; ſon jus eſt ſucré. Ce fruit mûrit à la fin d'Août.

11. La *Montauban* [*Montauban*,] a des feuilles ſciées , des fleurs groſſes & ouvertes, un fruit de groſſeur médiocre , d'un rouge foncé , preſque pourpre ſur le côté du ſoleil , & pâle de l'autre côté; ſa chair eſt fondante & blanche autour du noyau, dont elle ſe détache, & ſon jus eſt excellent. L'arbre eſt très fécond , & ſon fruit mûrit au milieu d'Août.

12. La *Pêche de Malte* [*Malta Peach.*] reſſemble beaucoup à la *Pêche d'Italie* ; ſes feuilles ſont ſciées , ſes fleurs groſſes & ouvertes , & ſon fruit d'une groſſeur médiocre, & d'un beau rouge ſur le côté expoſé au ſoleil ; ſa chair eſt blanche & fondante, mais rouge près du noyau , dont elle ſe détache. Le noyau eſt plat & pointu. L'arbre produit beaucoup de fruits qui mûriſſent à la fin d'Août.

13. La *Plus Noble* [*The Nobleſt.*] a des feuilles ſciées , des fleurs groſſes & ouvertes, & un gros fruit d'un rouge brillant ſur le côté expoſé au ſoleil ; ſa chair eſt blanche ,

fondante , & ſe détache du noyau , où elle eſt d'un rouge pâle. Cette Pêche eſt fort bonne dans les années chaudes ; elle mûrit à la fin d'Août.

14. La *Chanceliere* [*The Chancellor.*] a des feuilles unies , des fleurs petites , reſſerrées , & un fruit qui reſſemble beaucoup, pour la forme , à la *Belle Chevreuſe* , mais plus rond : ſa chair , blanche & fondante, ſe détache du noyau, où elle eſt d'un beau rouge ; ſa peau eſt fort mince , & ſon jus très-exquis. Cette Pêche mûrit vers la fin d'Août ; elle eſt du nombre des meilleures : l'arbre qui la produit eſt fort délicat, & ne réuſſit pas ſur un ſujet ordinaire : ainſi , il faut le greffer deux fois , comme la *Mignonne.* En le greffant ſur un *Abricotier*, comme il a été recommandé pour cette derniere eſpece, il profitera mieux que de toute autre maniere.

15. La *Galande* [*Bellegarde.*] a des feuilles unies , des fleurs petites , reſſerrées , & un fruit très-gros, rond , & d'un pourpre foncé ſur le côté du ſoleil ; ſa chair eſt blanche , fondante, & détachée du noyau, où elle eſt d'un rouge foncé , ſon jus eſt très-exquis. Cette Pêche mûrit au commencement de Septembre ; elle eſt excellente : mais elle n'eſt pas commune.

16. La *petite Violette hâtive* [*The Liſle.*] a des feuilles unies , des fleurs petites & reſſerrées, & un fruit d'une groſſeur médiocre , & d'un beau violet ſur le côté tourné au ſoleil ; ſa chair eſt d'un jaune

476 **PER**

pâle, & fondante, mais adhérente au noyau, où elle est fort rouge ; son jus est très-vineux. Cette Pêche mûrit au commencement de Septembre.

17. La *Bourdine* [*The Bourdine.*] a des feuilles unies, des fleurs petites & resserrées, & un fruit gros, rond, & d'un beau rouge sur le côté exposé au soleil ; sa chair est blanche, fondante, & détachée du noyau, où elle est d'un beau rouge. Le jus de cette Pêche est vineux & exquis ; elle mûrit au commencement de Septembre, & les curieux en font beaucoup de cas. L'arbre en donne en abondance, & il réussiroit aussi fort bien à plein vent.

18. La *Rossanna* [*Rossanna*] a des feuilles unies, des fleurs petites & resserrées, & un fruit gros, & un peu plus long que l'*Alberge* ; sa chair est jaune, & détachée du noyau, où elle est rouge. Le goût de cette Pêche est exquis & vineux ; elle mûrit au commencement de Septembre, & on la regarde comme une des meilleures. C'est la même que celle qu'on appelle la *Pourpre* ou l'*Alberge rouge*, à cause de sa couleur pourpre, dont elle est teinte sur le côté exposé au soleil.

19. L'*Admirable* [*Admirable*] a des feuilles unies, des fleurs petites & resserrées, & un fruit gros, rond & rouge sur le côté exposé au soleil ; sa chair est fondante & détachée du noyau, où elle est d'un beau rouge ; son jus est agréable & sucré. Cette Pêche mûrit au commencement de Septembre. Quel-

PER

ques personnes la nomment l'*Admirable Printaniere* ; mais c'est certainement la même que celle que les François appellent l'*Admirable* : ils en ont encore d'autres de ce nom, qui mûrissent plus tard.

20. La *Vieille Newington* [*Old Newington*] a des feuilles unies, des fleurs grosses & ouvertes, avec un beau fruit, gros, & d'un beau rouge sur le côté du soleil ; sa chair est blanche, fondante, & attachée au noyau, où elle est d'un rouge foncé ; son jus est exquis & vineux. Ce fruit est regardé comme une des meilleures *Pavies* : il mûrit vers le milieu de Septembre.

21. La *Rambouillette*, [*Rambouillet*] qu'on appelle vulgairement la *Rumbullion*, a des feuilles unies, des fleurs grosses & ouvertes, & un fruit d'une grosseur médiocre, plutôt rond que long, profondément divisé par un sillon dans le milieu, d'un beau rouge sur le côté du soleil, mais d'un jaune clair vers la muraille ; sa chair est fondante, d'un jaune brillant, & détachée du noyau, où elle est d'un rouge foncé. Le jus de cette Pêche est exquis, & d'un goût vineux. Elle mûrit au milieu de Septembre, & l'arbre qui la produit, en donne en abondance.

22. La *Belle de Vitry* [*The Bellis.*] a des feuilles sciées, des fleurs petites & resserrées, & un fruit d'une grosseur médiocre, rond, & d'un rouge pâle sur le côté du soleil ; sa chair est blanche & adhérente au noyau, où elle est rouge.

Le

PER

Le jus de cette Pêche est exquis & vineux : elle mûrit au milieu de Septembre.

23. La *Portugal* [*The Portugal*] a des feuilles unies, des fleurs grosses & ouvertes, & un fruit gros, & d'une belle couleur rouge sur le côté tourné au soleil ; sa peau est généralement tachetée ; sa chair est ferme, blanche, & fortement attachée au noyau, où elle est d'un rouge pâle : le noyau est petit, mais profondément sillonné. Le jus de cette Pêche est agréable & vineux : elle mûrit au milieu de Septembre.

24. Le *Tetton de Vénus*, [*Venus' breast*,] ainsi appelé parce que l'extrémité de son fruit est en forme de mammelon, a des feuilles unies, des fleurs petites & resserrées, & un fruit d'une grosseur médiocre, semblable à l'*Admirable*, & d'un rouge pâle sur le côté tourné au soleil ; sa chair est fondante, blanche, & détachée du noyau, où elle est rouge ; son jus est exquis & sucré. Cette Pêche mûrit sur la fin de Septembre.

25. La *Pourprée tardive* [*The Pourprée*] a des feuilles fort larges, & sciées, des branches fortes, des fleurs petites & resserrées, & un fruit gros, rond, & d'un beau pourpre ; sa chair est blanche, fondante, & détachée du noyau, où elle est rouge ; son jus est exquis & sucré. Cette Pêche mûrit tard en Septembre.

26. La *Nivette* [*The Nivette*] a des feuilles sciées, des fleurs petites & resserrées, & un gros

PER 477

fruit, un peu plus long que rond, d'un rouge brillant sur le côté exposé au soleil, & d'un jaune pâle vers la muraille ; sa chair est fondante, pleine d'un jus exquis, & fort rouge près du noyau dont elle se détache. On met cette Pêche au nombre des meilleures : elle mûrit au milieu de Septembre.

27. La *Royale* [*The Royal*] a des feuilles unies, des fleurs petites & resserrées, & un fruit gros, rond, d'un rouge foncé sur le côté du soleil, & plus pâle de l'autre ; sa chair est blanche, fondante, pleine d'un jus exquis, & détachée du noyau, où elle est d'un rouge foncé. Cette Pêche mûrit au milieu de Septembre ; & quand l'automne est favorable, elle devient excellente.

28. La *Persique* [*The Persica*] a des feuilles sciées, des fleurs petites & resserrées, & un fruit gros, oblong, & d'un beau rouge sur le côté exposé au soleil ; sa chair est fondante, pleine d'un jus exquis, & détachée du noyau, autour duquel elle est d'un rouge foncé ; la tige a un petit nœud au-dessus ; l'arbre est beau & bon producteur. Cette Pêche mûrit à la fin de Septembre. Plusieurs Jardiniers la nomment *Nivette*.

29. La *Pavie rouge de Pompone* [*The monstrous Pavy*.] a des feuilles unies, des fleurs grosses, ouvertes, avec un fruit très gros & rond, dont plusieurs ont jusqu'à quatorze pouces de circonférence ; sa chair est blanche, fondante,

Tome V. Hh

& fortement attachée au noyau, où elle eſt d'un rouge foncé ; l'extérieur eſt d'un beau rouge ſur le côté du ſoleil, & de couleur de chair pâle ſur l'autre. Cette Pêche mûrit à la fin d'Octobre, & quand l'automne eſt chaud, elle devient excellente.

30. La *Catherine* [*The Catherine*] a des feuilles unies, des fleurs petites & reſſerrées, & un fruit gros, rond, & d'un rouge foncé ſur le côté tourné au ſoleil ; ſa chair eſt blanche, fondante, pleine d'un jus exquis, & fortement attachée au noyau, autour duquel elle eſt d'un rouge foncé. Cette Pêche, qui mûrit au commencement d'Octobre, eſt excellente dans les années favorables ; mais comme elle mûrit fort tard, il faut la placer dans les meilleures expoſitions.

31. La *Sanguinole* [*The Bloody Peach*] eſt une Pêche d'une groſſeur médiocre, d'un rouge foncé ſur le côté tourné au ſoleil, & dont la chair eſt auſſi d'un rouge foncé juſqu'au noyau ; ce qui la fait nommer, par quelques Jardiniers, *Pêche de Mûrier*. Ce fruit mûrit rarement en Angleterre : c'eſt pourquoi on l'y cultive peu ; mais quand il eſt cuit, il ſe conſerve très-bien. Ainſi, tant pour cet uſage que par curioſité, on peut en planter un ou deux arbres, quand on a une ſuffiſante étendue de murailles.

Il y a encore d'autres eſpeces de *Pêches* qu'on éleve dans quelques pépinieres ; mais celles dont il vient d'être queſtion, ſont les meilleures, &

qui méritent le plus d'être cultivées. On peut choiſir dans ce nombre celles auxquelles on croit devoir donner la préférence. Je vais cependant donner les noms des eſpeces que je n'ai point décrites, pour la ſatisfaction des curieux. Ces dernieres ſont le *Sion*, la *Bourdeaux*, la *Swalch* ou *Hollandoiſe*, la *Carliſle*, l'*Eaton*, la *Pêche de Pau*, l'*Admirable jaune*, la *Double Fleur*. On cultive plutôt cette derniere pour la beauté de ſa fleur, que pour ſon fruit. Il y a quelques années que des arbres de cette eſpece, à plein vent, ont produit une grande abondance de fruits ; mais cette Pêche mûrit tard, & ſon jus eſt froid, aqueux & inſipide. On plante auſſi le Pêcher nain dans quelques endroits par curioſité. Cet arbre eſt délicat ; il pouſſe des branches très foibles & très-garnies de boutons à fleurs ; ſon fruit eſt moins gros qu'une *Muſcade*, & n'eſt pas bon. Cet arbre n'eſt pas d'une longue durée, & ne vaut pas la peine d'être cultivé.

Des trente-une eſpeces ci-deſſus mentionnées, il y en a au plus dix que je conſeillerois de planter, parce qu'étant les meilleures de toutes, il eſt inutile de s'embarraſſer des autres. Voici celles que je préférerois.

La *Pourpre printaniere*, la *Groſſe Mignonne*, la *Belle Chevreuſe*, la *Magdeleine rouge*, la *Chanceliere*, la *Belle-garde*, la *Bourdine*, la *Roſſanna*, la *Rambouillette* & la *Nivette* ; ce ſont, de toutes les meilleures eſpe-

PER

ces, celles qui méritent le plus d'être cultivées ; elles muriffent les unes après les autres, & peuvent fournir la table pendant toute la faifon des *Pêches*. S'il refte encore un peu de place contre quelque muraille, à une expofition très-chaude, il faut y mettre un ou deux arbres de la *Pêche Catherine*, qui, dans les années favorables, devient excellente.

Comme ces onze efpeces fe fuccedent dans leur maturité, elles fuffifent pour fournir des Pêches à toute une famille, quelque confidérable qu'elle foit, fi l'on a une affez grande étendue de murailles bien expofées. Mais comme, dans quelques années, les *Pêches* fe trouvent très-bonnes, & que, dans d'autres, elles font médiocres, je confeillerois, fi l'on a affez de place dans de bonnes expofitions, d'en planter trois ou quatre efpeces, qui, dans quelques années, deviennent excellentes, quoiqu'en général elles ne foient pas auffi bonnes que celles ci-deffus détaillées. Ces dernieres font la *Montauban*, la *L'ifle*, la *Vieille Newington*, le *Tetton de Vénus*, la *Catherine* & la *Perfique*.

Les François diftinguent cette efpece de fruit en *Pêches* & en *Pavies* ; ils nomment *Pêches*, celles qui fe détachent du noyau ; & *Pavies*, celles dont la chair y eft adhérente. Celles-ci font plus eftimées en France que les *Pêches* ; & en Angleterre, quelques perfonnes les préferent auffi.

Les François les diftinguent encore en mâle & en femelle ;

PER 479

ils font les *Pavies* mâles, & les *Pêches* femelles ; mais cette diftinction eft fans fondement, puifque les amandes de ces deux efpeces produifent également des arbres, & que les fleurs des *Pêchers* font généralement hermaphrodites, & renferment en elles toutes les parties de la génération. Il eft vraifemblable que cette diftinction fe perpétue depuis longtems, & qu'elle a été faite avant qu'on ait eu la connoiffance de la différence des fexes dans les plantes, ou qu'on ait fu comment les diftinguer féparément.

Les *Pavies*, que les François appellent *Brugnons*, différent des deux autres efpeces, en ce qu'ils ont une chair dure & ferme, la peau tout-à-fait liffe & fans aucun duvet. Comme j'ai fait mention de cette efpece à l'article *Brugnons-Nectarins*, auquel le Lecteur peut avoir recours, je n'en parlerai point ici.

Je vais indiquer à préfent en quoi confiftent les qualités d'une bonne *Pêche*, afin que chacun foit en état de les apprécier.

Une bonne Pêche doit avoir une chair ferme, une peau mince, d'un rouge foncé ou brillant fur le côté expofé au foleil, & jaunâtre fur le côté oppofé ; fa chair doit être jaune, pleine de jus, & d'un goût exquis, le noyau petit, & la chair fort épaiffe.

Quand une *Pêche* a toutes ces qualités, on peut la regarder comme un excellent fruit.

Toutes ces différentes efpe-

480　**PER**

ces de *Pêches* ont été originairement obtenues de noyaux, comme il arrive à toutes les femences des autres fruits; de forte qu'on fe procure toujours de bonnes efpeces, quand on a affez de place dans un jardin pour élever ces arbres par noyaux : ceux-ci étant acclimatés, font d'ailleurs préférables à ceux qu'on apporte des pays chauds. Il eft vrai que, dans le nombre, il s'en trouve très-peu de bons, comme il arrive pour la plupart des fruits & des fleurs produites par femences, parmi lefquelles il peut y en avoir quelques - unes eftimables & fupérieures à celles fur lefquelles les femences ont été prifes, mais dont le plus grand nombre eft toujours de peu de valeur : mais quand on obtient feulement deux ou trois bonnes efpeces, on eft amplement dédommagé de fes peines. Lorfqu'on eft curieux de planter les noyaux de ces fruits, on doit choifir avec foin les efpeces, & laiffer les fruits fur les arbres, jufqu'à ce qu'ils tombent; alors les amandes, qui feront parfaitement formées, réuffiront plus certainement. Les meilleures efpeces pour en planter les noyaux, font celles qui ont la chair ferme, & qui fe fendent jufqu'au noyau : parmi celles-ci, il faut choifir celles qui mûriffent de bonne heure, & qui ont un jus exquis & vineux.

Il faut planter ces noyaux en automne, fur une planche de terre légere & feche, à trois pouces environ de pro-

PER

fondeur, & à quatre de diftance, & les couvrir en hiver, pour empêcher la gelée d'y pénétrer, & de les détruire. Au printems, lorfque les plantes pouffent, on les débarraffe avec foin de toutes mauvaifes herbes; & fi le printems eft fort fec, on les arrofe de tems en tems, pour hâter leur accroiffement. On les laiffe dans cette planche jufqu'au printems fuivant : alors on les enleve avec précaution, pour les tranfplanter dans une pépiniere en rangs éloignés de trois pieds, & à un pied de diftance entr'elles dans les rangs; on répand un peu de terreau fur la furface de la terre, autour des racines, pour l'empêcher de fe deffecher trop vite; & , fi le printems eft fort fec, on les arrofe légerement une fois par femaine, jufqu'à ce qu'elles aient formé de nouvelles racines: après quoi on les tient conftamment nettes de mauvaifes herbes, & on laboure avec foin la terre à chaque printems, pour la defferrer, & faciliter à leurs tendres fibres le moyen de s'étendre de tous côtés.

On peut laiffer ces plantes dans cette pépiniere un ou deux ans, fuivant le progrès qu'elles auront fait; après ce tems, on les tranfplante à demeure dans les places qui leur font deftinées.

En enlevant ces arbres, on doit avoir foin, s'ils ont des racines qui pouffent vers le bas, de les tailler fort courtes, & de retrancher toutes celles qui font froiffées, ainfi

PER

que toutes les petites fibres, qui se dessechent toujours, & qui, si on les laissoit en plantant les arbres, se moisiroient, nuiroient beaucoup aux nouvelles fibres, & empêcheroient souvent l'accroissement des arbres ; mais il ne faut pas tailler les têtes ; car les plantes produites de noyaux étant toujours d'une texture plus spongieuse, sont plus sujettes à périr, quand on les coupe, que celles qui sont greffées : d'ailleurs ces arbres étant destinés à croître en plein vent, jusqu'à ce qu'on ait reconnu si leurs fruits méritent d'être élevés en espaliers, on doit se borner à retrancher les branches flétries, & celles qui poussent régulierement sur les côtés ; car, en les taillant davantage, on leur nuiroit infiniment.

La meilleure méthode est de planter ces arbres séparément dans différens endroits du jardin potager, où ils profiteront, & produiront beaucoup plus de fruits que s'ils étoient rapprochés en rangs ; & étant ainsi dispersés, ils ne nuisent point aux plantes qu'on a semées aux environs.

Lorsqu'ils ont produit du fruit, on juge bientôt de leur valeur : alors on peut arracher ceux que l'on n'aime point, & multiplier par la greffe ceux qui se trouvent bons. Cette méthode étant la plus ordinaire, je vais en parler plus en détail, en donnant la maniere dont s'y prennent les Jardiniers de pépinieres ; ensuite je proposerai mes pro-

PER 481

pres observations, pour perfectionner leur pratique en faveur des personnes qui veulent se procurer de bons fruits.

D'abord il faut se pourvoir de sujets des meilleures especes de *Pruniers*, pour y greffer les *Pêches* & les *Pavies*, ainsi que de sujets d'*Amandiers* & d'*Abricotiers*, pour quelques especes de Pêches tendres, qui ne prennent pas bien sur le *Prunier*. Ces derniers doivent être élevés de noyaux, comme il a déja été dit à l'article *Pepiniere*, & non pas de rejettons. On transplante ces sujets après un an d'accroissement ; car plus ils sont enlevés jeunes, & mieux ils réussissent : on les empêche par-là de pousser des racines trop profondes ; & en raccourcissant celles qui paroissent y être disposées, on leur en fait pousser d'horisontales. On plante ces sujets suivant les distances que nous avons déja prescrites, c'est-à-dire, en rangs éloignés de trois pieds, & à un pied entr'eux dans les rangs. Les Jardiniers les rapprochent davantage ; mais je dirai bientôt pourquoi je prescris de les tenir plus éloignés.

Après deux ou trois ans de séjour dans la pépiniere, ces sujets seront assez forts pour être greffés. La saison ordinaire pour cette opération, est à la Saint-Jean, ou pendant tout le mois de Juillet, quand l'écorce se sépare aisément du bois : alors on choisit quelques bonnes branches des especes de fruits que l'on veut multiplier, en observant tou-

Hh 3

jours de les prendre fur des arbres fains, & fur ceux qui produifent beaucoup de fruits de bon goût ; car il eft très-certain que, fans cette précaution, toutes les efpeces de fruits dégénerent de façon à ne pouvoir être reconnues : d'ailleurs, toutes les fois qu'un arbre eft mal-fain, les greffes que l'on prend deffus, en confervent le vice plus ou moins, fuivant qu'elles font plus ou moins imprégnées de la féve altérée de l'arbre.

Un *Pécher*, par exemple, ou un *Pavie*, qui aura été fortement attaqué de la nielle, dont les branches font malades & les feuilles frifées, ne fe rétablira qu'en employant beaucoup d'art, & qu'après plufieurs années de foin ; encore, malgré tout cela & quoique les années fuivantes foient favorables, il fe reffentira toujours de cette maladie, au point de faire croire qu'il eft infecté d'une nouvelle nielle, pendant que, dans le vrai, ce n'eft que la fuite de fa premiere qui s'eft étendue & mélée dans toute la féve de l'arbre : alors, fi l'on prend des greffes fur un pareil arbre, il eft certain qu'elles porteront avec elles le germe de cette maladie.

Tout le fuccès dépend donc du foin qu'on apporte dans le choix des greffes : ainfi, quand on eft curieux d'avoir de bons fruits, on ne peut y apporter trop d'attention. Des Jardiniers de pépinieres, qui, en général, font les plus induftrieux pour multiplier les différentes efpeces d'arbres fruitiers, ne fe trompent pas fur les efpeces ; mais il faudroit encore qu'ils euffent l'attention de choifir les arbres les plus fains, fur-tout pour les *Péchers* & les *Pavies.* Si les greffes font prifes fur de jeunes plantes de pépiniere qui n'ont point encore produit de fruits, & dont les branches font toujours très-fortes & vigoureufes, elles auront une difpofition vicieufe, que l'on corrigera difficilement, pour les contenir en bon état. Elles poufferont plus à la maniere des *Saules* que comme des *Péchers*, les boutons fe trouveront à une grande diftance les uns des autres, les branches deviendront très-groffes, & le bois fort moelleux. Ainfi, par-tout où l'on a l'habitude de prendre les greffes fur des arbres de pépiniere, on ne peut pas trop y compter, pour avoir de bons *Péchers*. Je confeille donc de fe procurer des greffes d'arbres âgés, parfaitement fains, & dont le fruit foit très-favoureux ; de ne jamais prendre les rejettons les plus forts & les plus vigoureux ; mais ceux qui font bien conditionnés, & dont les boutons font affez rapprochés : quoiqu'on ne puiffe efpérer de voir produire à ces derniers, dès l'année fuivante, des jets auffi forts que ceux que donnent les greffes prifes fur des branches gourmandes ; cependant les arbres en feront plus difpofés à porter du fruit, & réuffiront beaucoup mieux.

On doit toujours féparer les greffes le matin ou le foir, ou

PER

dans un jour couvert de nuages ; car si on les coupoit, quand le soleil est très-chaud, lorsque les branches transpirent fort, les greffes seroient privées d'humidité, & couroient risque de manquer. Plutôt elles sont placées sur les sujets, après avoir été séparées des arbres, mieux elles réussissent. La maniere de faire cette opération étant bien détaillée dans l'article *Inoculation* ou *Greffe*, je ne la répéterai pas ici. Le traitement que ces arbres exigent dans la pépiniere, étant aussi exactement décrit dans ce dernier article, je vais continuer à donner quelques instructions pour le choix de ces arbres, quand on les prend dans une pépiniere.

Le premier soin doit être de s'adresser à une personne de probité, sur laquelle on puisse compter, non-seulement pour avoir les véritables especes qu'on désire, mais aussi pour être certain que les greffes ont été prises sur des arbres fructueux : alors on va les prendre soi-même, ou l'on envoie quelqu'un de sûr, parce que la plupart des Jardiniers de pépiniere s'arrangent les uns avec les autres, de maniere que, s'ils n'ont pas l'espece qu'on demande, ils vont la chercher chez un autre, qui peut les tromper, s'il n'est pas aussi honnête ni aussi soigneux qu'eux.

Il faut aussi choisir les arbres en automne, avant que les meilleurs soient enlevés des pépinieres ; car ceux qui y vont les premiers, s'ils sont connoisseurs, marquent toujours les plus beaux plants. On doit observer, dans le choix des arbres, les sujets sur lesquels ils ont été greffés ; si ce sont de véritables especes de *Pruniers* ou d'*Abricotiers* ; s'ils sont sains & jeunes ; si ce ne sont pas des sujets greffés l'année précédente & manqués ; s'ils ont été taillés ; s'ils ne sont pas plus gros que le doigt ; car ceux qui sont plus forts sont moins bons : on en ôte la mousse & les chancres. Les greffes ne doivent avoir qu'un an d'accroissement, & ne doivent point avoir été taillées au printems ; ce qui leur fait produire une nouvelle pousse. On ne doit pas non plus choisir ceux dont les branches sont très-fortes & gourmandes, mais plutôt ceux dont les jets sont nets, d'une grosseur médiocre, & dont les boutons ne sont pas trop séparés les uns des autres. Les arbres qui sont hors des rangs & vers l'extrémité, sont généralement les meilleurs ; car lorsqu'ils sont serrés dans la pépiniere, leurs branches filent en longueur, leurs boutons sont beaucoup plus éloignés, & leurs yeux plus plats. C'est ce qui m'a déterminé à conseiller ci-dessus de planter les sujets à une plus grande distance que ne le font ordinairement les Jardiniers de pépiniere. Quand un Jardinier soigneux & intelligent se donne les peines & fait la dépense d'élever ses arbres suivant cette méthode, il mérite qu'on lui

H h 4

prie ses *Pêchers* trois fchelings de plus par piece que ceux qui font élevés fuivant la routine ordinaire ; car dès que l'on a fait la dépenfe de conftruire des murailles pour fe procurer des fruits, ce n'eft pas le cas d'épargner quelques fchelings fur l'achat des arbres, parce que, s'ils font mauvais, ou fi l'on eft trompé fur les efpeces qu'on défire, toute la dépenfe eft en pure perte ; & cet inconvénient eft d'autant plus fâcheux, qu'on ne s'en apperçoit fouvent qu'après trois ou quatre années ; c'eft ce qui décourage bien des perfonnes, & les empêche de planter, dans la crainte d'un auffi mauvais fuccès.

Dès que les arbres font choifis dans la pépiniere, on doit d'abord les faire enlever avec foin, & de maniere que leurs racines ne foient point dommagée ; car ces arbres étant fort fujets à la gomme caffées ou dechirées, & que leur écorce ne foit point endans les endroits où ils font dechirés, on doit en prendre le plus grand foin. S'ils font deftinés à être tranfportés à une certaine diftance, il faut bien envelopper leurs racines avec des bandes de foin, de paille, ou du chaume de Pois, que l'on recouvre avec des nattes, pour empêcher l'air de les deffecher. Si les feuilles des arbres ne font pas tombées quand on les enleve, on les ôte exactement avant de les empaqueter ; car, en les laiffant, elles échaufferoient les arbres dans une longue route, & occafionneroient du moifi, qui nuiroit beaucoup aux branches.

Voyons à préfent comment on prépare la terre dans laquelle on veut planter les arbres. Celle qui convient le mieux aux *Pêchers*, eft celle que l'on prend dans un pâturage, qui n'eft ni trop ferme ni trop humide, ni trop fec, mais d'une nature douce, telle que celle qu'on nomme ordinairement *Marne de Noifetier* ou de *Coudrier* : on enleve cette terre avec le gazon, jufqu'à la profondeur de dix pouces; on la tient en monceaux pendant huit ou dix mois au moins; & celle qui eft préparée un an & plus, avant de s'en fervir, eft encore la meilleure, parce qu'elle a le tems d'être ameublie par les gelées de l'hiver, & par la chaleur du foleil en été. Pendant ce tems, il faut la remuer & la retourner fouvent, afin de confommer les gazons & de brifer les mottes ; de cette maniere, on la rend fort légere, & aifée à labourer. Vers le commencement de Septembre, on la porte dans le jardin, pour en former les plates-bandes, que l'on enleve à une hauteur proportionnée à la fechereffe ou à l'humidité du fol. Si la terre eft fort humide, il fera prudent de mettre dans le fond des plates-bandes quelques décombres, pour en abforber l'humidité, & empêcher les racines de s'enfoncer vers le bas. Dans ce cas auffi, il fera utile de

PER

pratiquer dans le fond des canaux fouterrains, pour faciliter l'écoulement de l'humidité, qui feroit fort préjudiciable, fi elle féjournoit au pied des arbres : on éleve enfuite la terre de la plate-bande à un pied de hauteur, & dans un fol très-humide, à deux pieds au-deffus du niveau du terrein, afin que les racines puiffent toujours être feches ; mais fi le fol eft affez fec, on ne donne à la plate-bande que fix ou huit pouces d'élévation au-deffus du terrein.

Quant à la largeur de cette plate-bande, elle ne peut être trop grande : mais elle doit être au moins de fix ou huit pieds de large ; car lorfqu'elle eft trop étroite, les racines des arbres font gênées après quatre ou cinq ans, & ne font plus enfuite que de très-foibles progrès. La plate-bande doit auffi avoir au plus deux pieds & demi de profondeur ; car lorfque les terres font préparées, les arbres s'enfoncent beaucoup, ce qui eft une des caufes de leur ftérilité. Dès que les racines ont pénétré au-deffous de l'endroit où l'influence des pluies & du foleil fe fait fentir, elles y puifent une grande quantité de fucs cruds, qui ne procurent aux arbres qu'un accroiffement luxurieux, & qui s'oppofent à leur fructification ; d'ailleurs les fruits que donnent de pareils arbres, n'ont jamais une faveur auffi exquife que ceux qu'on recueille fur des arbres dont les racines font près de la furface, & jouiffent de l'influence du foleil, qui corrige & digere toutes les crudités de la terre.

Lorfque le fol d'un jardin eft bas, & que la craie, la glaife ou le gravier eft près de la furface, il faut creufer la terre, & y faire des tranchées, pour recevoir celle que l'on a préparée, & ne pas fe contenter d'y faire des trous, fuivant l'ufage de quelques Jardiniers : en effet cela ne vaudroit pas mieux que de planter les arbres dans des caiffes où leurs racines feroient gênées, car quand ces racines font une fois parvenues aux côtés de ces trous, les arbres fe niellent & périffent. Si c'eft dans la glaife que l'on fait ces trous, l'humidité y féjourne comme dans un baffin, & la terre de la plate-bande devient comme de la boue dans les tems très-humides, ce qui nuit beaucoup aux racines des arbres. Ainfi, toutes les fois que la terre fe trouve avoir quelqu'une de ces mauvaifes qualités, le mieux eft d'élever les plates-bandes au-deffus du niveau jufqu'à une hauteur convenable, plutôt que de l'enfoncer ; car lorfque les racines des arbres font près de la furface, elles s'étendent à une grande diftance pour y chercher leur nourriture ; mais fi au contraire elles pénetrent au-deffous de la terre rapportée, elles ne peuvent y puifer que des fucs aigres & cruds, qui ne font point propres à la végétation.

Les plates-bandes étant préparées, on doit les laiffer ainfi trois femaines ou un mois, pour

486 **PER**

qu'elles puissent bien s'établir ;
& lorsque la saison de planter
est arrivée (ce qui doit être
aussi-tôt que les feuilles commencent à tomber, afin que
les arbres puissent pousser de
nouvelles racines avant les
gelées), la terre étant en
bon état, & les arbres postés
avec soin sur la place, on les
prépare, pour les planter, en
raccourcissant toutes les racines, en coupant celles qui sont
cassées ou froissées, & en retranchant toutes les petites
fibres. Si quelques-unes de ces
racines se croisent ou s'entrelacent, on ôte les plus mauvaises, afin qu'elles ne puissent
se nuire mutuellement.

Les arbres étant bien préparés, on mesure les distances, qui ne doivent jamais
être moindres que de douze
pieds ; & si la terre est très-
bonne, on les marquera à
quatorze. Je suis certain que
bien des personnes trouveront
cette distance trop considérable, sur-tout la pratique actuelle étant fort opposée à celle-
ci ; mais l'expérience prou-
vera que cet intervalle n'est
pas trop grand pour des arbres
bien traités ; car s'ils réussissent, leurs branches garniront en peu d'années tout le
bas des murailles ; ce que l'on
doit principalement rechercher,
en n'attachant pas les branches
en hauteur, comme on le fait
quelquefois ; car, par cette
manœuvre, le bas des arbres
reste destitué de bois, & quelques années après il n'y a plus
de fruit que sur le haut.

La même chose arrive aussi

PER

quand les arbres sont plantés
trop près les uns des autres,
parce que, n'ayant point de
place pour étendre leurs branches latéralement, on est forcé
de les diriger vers le haut ;
ce qui produit le mauvais effet
dont nous venons de parler.

D'autres penseront peut-être
aussi que cet espace est trop
petit pour ces arbres, parce
que les *Pruniers*, les *Cerisiers*,
& la plupart des autres especes d'arbres fruitiers en exigent davantage ; mais il faut
faire attention que les *Pêchers*
& les *Pavies* ne produisent
leurs fruits que sur le bois de
l'année précédente, & non pas
sur les rejettons, comme la
plupart des *Pruniers*, *Cerisiers*
& *Poiriers*, & que les branches
de ces arbres doivent être raccourcies annuellement dans
chaque partie, pour en obtenir du bois productif ; ce qui
est cause qu'on peut les contenir dans un plus petit espace
que toutes les autres especes,
& que toute la muraille peut
être garnie constamment de
branches à fruits ; mais si ces
arbres sont plantés à une grande distance, on est obligé souvent de donner trop de longueur aux branches, & l'on
dégarnit par-là le milieu des
arbres ; car jamais les vieilles
branches de Pêchers ne produisent de bons rejettons.

Je ne puis m'empêcher de
relever ici une autre faute essen-
tielle que l'on commet tous les
jours dans le traitement des
arbres fruitiers en espalier,
qui est de placer des arbres à
hautes tiges entre les autres,

PER 487

afin de couvrir le haut de la muraille, & de se procurer du fruit, en attendant que les arbres du bas soient devenus affez grands pour garnir tout le mur, afin de retrancher alors ceux à hautes tiges : mais on ne considère pas que plus on met d'arbres dans un petit espace, moins ils reçoivent de nourriture, & que conséquemment ils deviennent plus foibles, parce que le même espace de terre qui peut nourrir dix arbres, ne peut pas en entretenir également vingt ; de sorte que, si les arbres à hautes tiges acquierent beaucoup de force, les nains seront à proportion plus foibles : d'ailleurs, comme il est prouvé que la plupart des arbres étendent leurs racines aussi loin sous la terre, que leurs branches au-dessus, il est absolument nécessaire que ces deux proportions soient égales, si l'on veut avoir des arbres sains & vigoureux : c'est aussi pour cette raison qu'il est inutile d'élever les murs d'espaliers trop haut, à moins que ce ne soit pour des *Poiriers* ; car dix ou douze pieds suffisent pour toute autre espece de fruit. J'ai vu des jardins plantés en arbres fruitiers par des personnes réputées très habiles dans cet art : ils avoient placé les *Péchers* & *Pavies* contre un mur exposé à l'Est & à l'Ouest ; mais on ne voyoit jamais sur ces arbres aucuns fruits parvenir en parfaite maturité : ainsi, je conseillerai toujours de ne jamais suivre de pareils exemples, parce qu'il est bien connu

que les dernieres *Péches* mûriffent mal contre les murailles les mieux exposées, & que le seul aspect qui leur convienne est le Sud, en inclinant un peu vers l'Est : quelques especes peuvent aussi réussir au Sud un peu incliné à l'Ouest.

Dans la distribution des arbres, on fera bien de rapprocher les especes de *Péchers* dont les fruits mûriffent à-peu-près dans le même tems ; au moyen de quoi on fera à même de mieux préserver les fruits des attaques des hommes & des insectes, & l'on s'épargnera beaucoup de peine pour les cueillir ; car lorsque l'on est obligé d'aller d'un bout du jardin à l'autre, & d'examiner toute la longueur des espaliers, pour ramasser quelques fruits, on perd beaucoup de tems.

Mais pour revenir à la plantation, après avoir marqué la place de chaque arbre, on fait un trou avec la bêche, affez large pour recevoir ses racines, ensuite on le place, en observant de tourner la greffe au dehors, afin que la coupe ou partie blessée du sujet soit cachée à la vue, & de laisser la tige de l'arbre à quatre ou cinq pouces environ de la muraille : mais on incline son extrêmité vers elle : on remplit ensuite le trou de terre avec les mains, en brisant les mottes de maniere qu'elle s'insinue, & tombe entre les racines, & qu'il n'y reste aucun vuide. Il faut aussi secouer légerement l'arbre avec la main, pour mieux fixer & arranger la terre autour ; mais on ne

doit pas la piétiner trop fort, ce qui eſt ſouvent une très-grande faute ; car la terre étant naturellement ſujete à ſe reſſerrer, en la piétinant trop fort, on la rend ſouvent ſi dure, que les tendres fibres des racines ne peuvent y pénétrer ; l'arbre alors reſte dans le même état, ſans pouſſer pendant quelque tems, & il meurt à la fin, ſi la terre n'eſt pas deſſerrée ; de ſorte que toutes les fois qu'on s'apperçoit que la terre des plates-bandes eſt devenue trop dure, ſoit par les grandes pluies, ſoit par une autre cauſe, il faut la labourer, pour la rendre plus meuble, en choiſiſſant pour cela un tems ſec, ſi c'eſt en hiver ou au printems ; ou un tems humide, ſi l'on fait cette opération en été.

Quoiqu'en donnant des inſtructions pour le choix des arbres de pépiniere, ſuivant la méthode ordinaire de les planter, j'aie recommandé de prendre ceux qui ont pouſſé des branches d'un an, cependant je préférerois ceux qui ont été greſſés l'été précédent, & qui n'ont point encore pouſſé ; car ſi la greffe eſt ſaine & gonflée, & que l'écorce du ſujet ſoit bien ſerrée ſur la greffe, il n'y aura point de danger qu'elle manque. Quand cette greffe a pouſſé, au printems ſuivant, une branche de la longueur de cinq ou ſix pouces, on l'arrête en pinçant l'extrémité, pour lui faire produire des branches latérales, qu'on puiſſe attacher à la muraille : par ce moyen, on ſera diſpenſé de couper la tête,

comme on le fait aux greffes d'une année dans les pépinieres ; car ces amputations ne ſont point favorables à ces eſpeces d'arbres, & ſurtout à quelques-unes des plus délicates. Ainſi, par la méthode de planter ces arbres avant que la greffe ait pouſſé, il n'y aura pas de tems perdu, puiſque les pouſſées d'un an dans les autres doivent être jetées bas, & qu'il eſt d'ailleurs incertain s'ils repouſſeront. La vérité de ce que j'avance m'a été démontrée par une expérience conſtante.

Après avoir ainſi planté les arbres qui ont formé leurs branches dans la pépiniere, il faut attacher leurs têtes au treillage, pour les empêcher d'être ſecouées par le vent, qui dérangeroit leurs racines, & caſſeroit leurs tendres fibres, dès qu'elles auroient commencé à pouſſer ; ce qui cauſeroit un grand préjudice aux arbres. On met auſſi du terreau ſur la ſurface de la terre autour des racines, avant que les gelées commencent à ſe faire ſentir ; car elles ſeroient très-nuiſibles aux racines, & détruiroient peut-être les petites fibres, mais on ne doit pas répandre ce terreau trop tôt, afin qu'il n'empêche pas les pluies d'automne de détremper & d'humecter ces racines.

Toutes ces choſes étant exactement obſervées, la plantation n'exigera plus aucun autre ſoin juſqu'au commencement ou au milieu de Mars, ſuivant que la ſaiſon ſera plus ou moins avancée ; alors on

PER

coupera les têtes des arbres nouvellement plantés, en ne laiſſant que quatre ou cinq boutons au-deſſus de la greffe. Quand on fait cette opération, on doit avoir bien ſoin de ne pas déranger les racines ; &, pour éviter cet accident, on poſe le pied tout près de la tige de l'arbre ; on tient ferme avec une main la partie de la tige qui eſt au-deſſous de la greffe, & de l'autre on jette bas la tête de l'arbre avec une ſerpette tranchante à l'endroit convenable, en laiſſant, comme il vient d'être dit, quatre ou cinq boutons au-deſſus de la greffe. Ceci doit toujours être fait par un tems ſec ; car s'il ſurvenoit beaûcoup de pluie immédiatement après, il y auroit du riſque que l'humidité n'entrât dans la partie bleſſée, & ne fît tort à l'arbre. Il ne faut pas non plus, par la même raiſon, choiſir un tems de gelée, qui pénétreroit dans la bleſſure, & en empêcheroit la guériſon.

Après avoir coupé la tête des arbres, on laboure légerement les plates-bandes, pour en deſſerrer la terre, & faciliter aux fibres le moyen de mieux s'étendre, en obſervant ſoigneuſement de ne pas couper ni froiſſer les nouvelles racines. Lorſque le terreau qu'on a répandu ſur ces racines en automne eſt entierement pourri, on peut l'enterrer dans la plate-bande à quelque diſtance de l'arbre. Dans les tems de hâle ou de ſechereſſe, on prend quelques gazons de pâturage, que l'on met ſur la ſurface de la plate-bande autour des racines, en tournant l'herbe en-deſſous ; ce qui conſervera une légere humidité dans la terre beaucoup mieux qu'aucune eſpece de terreau, ſans attirer les inſectes nuiſibles aux arbres, comme le font la plupart des fumiers ou litieres.

Les arbres que l'on plante en greffe, & avant qu'ils aient pouſſé des branches, doivent avoir leurs têtes coupées préciſément au-deſſus de la greffe, qui pouſſe rarement avant cette opération ; & plus on les coupe près de la greffe, plutôt ils en ſont recouverts : car, quoiqu'il ſoit quelquefois néceſſaire de laiſſer une partie du ſujet au-deſſus de la greffe, afin de pouvoir y attacher les branches qu'elle peut avoir pouſ-fées pour les empêcher d'être briſées par le vent dans les pépinieres ; cependant ceux-ci étant placés auprès d'une muraille contre laquelle on aſ-ſujettit les branches, il eſt inutile d'y laiſſer aucune partie du ſujet.

Quand on arroſe ces arbres nouvellement plantés (ce qui n'a lieu qu'autant que le printems eſt très-ſec), on doit le faire avec une gerbe placée ſur l'arroſoir, pour que l'eau tombe en gouttes ; car ſi elle ſortoit en gros volume, elle ſerreroit trop la terre. Il ſera auſſi très-avantageux d'arroſer la tête de l'arbre. Ces arroſe-mens ne doivent point être répétés trop ſouvent, ni être trop copieux ; car rien n'eſt plus nuiſible aux arbres nouvellement plantés.

Au milieu ou à la fin de Mai, quand les arbres auront poussé plusieurs branches de six ou huit pouces de longueur, on doit les palissader à la muraille, en observant de les diriger horisontalement, & de retrancher tous les rejettons qui poussent en avant, & ceux qui sont foibles, au moyen de quoi ceux qui ont été conservés deviendront beaucoup plus forts : mais si la greffe n'a poussé que deux branches, & que ces branches soient très-fortes, en ce cas, on pince leur extrémité, pour leur faire produire chacune deux nouveaux rejettons, & même un plus grand nombre, pour mieux garnir la muraille. Il faut aussi continuer à les arroser dans les tems secs, pendant toute la saison, sans quoi ils souffriroient, & leurs racines étant encore mal établies dans la terre la premiere année, ils seroient infiniment retardés dans leur accroissement.

Au commencement d'Octobre, lorsque la séve des arbres est arrêtée, il faut les tailler, & raccourcir les branches à proportion de leurs forces. Si elles sont fortes, on peut leur donner huit pouces de longueur ; mais quand elles sont foibles, on les réduit à quatre ou cinq pouces ; on les fixe ensuite au treillage horisontalement, comme il a été dit ci-dessus, de maniere que le milieu des arbres soit sans branches, parce que cette partie se garnira aisément dans la suite ; & si l'on attachoit les branches perpendiculairement,

les plus fortes attireroient la plus grande partie de la séve, qui va toujours en montant ; & les branches latérales, se trouvant privées de nourriture, deviendroient plus foibles, & périroient souvent. C'est ce qui est cause que nous voyons tant de *Péchers* avec une ou deux branches droites dans le milieu, tandis que les côtés sont entierement dégarnis : dans ce cas, le milieu de l'arbre ne peut produire aucun fruit, parce qu'il n'a que du gros bois, qui ne pousse jamais de branches. Les deux côtés ne peuvent pas non plus être régulierement remplis de branches à fruit, quand l'arbre a un pareil défaut. Ainsi, il faut suivre exactement la méthode que je viens de prescrire, en dressant de jeunes arbres ; car lorsque, dans les commencemens, on les laisse pousser en désordre, il est impossible de les réduire ensuite & de les rendre réguliers, le bois en étant trop mou & trop rempli de moëlle, pour pouvoir être taillé comme les autres arbres fruitiers qui repoussent ensuite plus vigoureusement ; au-lieu que les *Péchers* jettent de la gomme par leurs blessures, & périssent entierement en peu d'années.

Durant l'été suivant, lorsque les arbres commencent à pousser des branches, il faut les examiner avec soin, pour en retrancher tous les boutons extérieurs & ceux qui sont mal placés, & palissader horisontalement à la muraille les branches qui doivent rester, dans

PER

leur ordre naturel comme elles font produites ; car c'eſt alors la ſaiſon où l'on peut le mieux arranger les arbres comme on veut les avoir : au-lieu que, s'ils étoient négligés juſqu'à la Saint-Jean, comme il arrive ſouvent, une grande partie de leur nourriture ſe trouveroit abſorbée par les branches qui pouſſent en avant, ou par les rejettons inutiles, que l'on eſt obligé de retrancher enſuite ; de maniere que les autres deviendroient plus foibles, & que quelque partie de la muraille ſe trouveroit peut-être ainſi dégarnie de branches, tandis qu'on auroit pu en faire pouſſer de nouvelles dans le mois de Mai, en arrêtant quelques branches fortes dans le lieu même où l'on auroit voulu en avoir d'autres : on auroit pu alors conduire ces nouvelles branches dans les places vuides, à meſure qu'elles auroient pris de la croiſſance.

C'eſt ainſi que l'on garnit régulierement un eſpalier de bon bois ; ce qui fait la plus grande beauté des arbres. On ne doit point arrêter les branches en été, quand on n'eſt pas dans la néceſſité de remplir un vuide. On ne peut pas faire une plus grande faute que d'augmenter le nombre des branches, au point de les rendre confuſes ; car on ne les multiplie ainſi qu'aux dépens de leur vigueur ; & plus elles ſont foibles, moins elles ſont en état de produire de bons fruits : d'ailleurs quand elles ſont trop rapprochées, la grande quantité de feuilles qui les

PER 491

couvrent, empêche la libre circulation de l'air entr'elles, le fruit ne mûrit jamais bien, & ne devient jamais auſſi bon que celui qu'on recueille ſur des arbres dont les branches jouiſſent de tous les avantages du ſoleil & de l'air.

Après avoir montré comment on dreſſe les jeunes arbres, je vais donner à préſent la maniere de les tailler & de les traiter pour la ſuite ; & elle pourra ſervir auſſi pour les arbres à plein vent.

Quand on taille des *Péchers* & des *Pavies*, qui exigent le même traitement, on doit ſe conformer exactement aux deux regles ſuivantes, qui ſont ;

1°. Que chaque partie de l'arbre ſoit également fournie de bois à fruit.

2°. Que les branches ne ſoient pas trop rapprochées l'une de l'autre, par les raiſons qui viennent d'être dites, & par d'autres que nous ajouterons encore.

Quant à la premiere regle, on doit obſerver que les *Péchers* produiſent leurs fruits ſur le jeune bois, ſoit de l'année précédente, ſoit ſur les branches de deux ans, & qu'après cet âge, ils n'en donnent plus. Ainſi, les branches doivent être raccourcies de maniere à leur en faire pouſſer de nouvelles chaque année dans toutes les parties de l'arbre ; ce qui ne peut s'effectuer par la maniere ordinaire de tailler.

Les Jardiniers négligent leurs arbres dans les ſaiſons convenables, & les plus propres à les tailler. Ces ſaiſons ſont les

492 PER

mois d'Avril, Mai & Juin :
dans ce tems, on pourroit
arrêter les branches gourmandes, en les pinçant, & leur
en faire produire de nouvelles
où il en manque. Ces nou-
velles branches, produites en
bonne saison, auroient assez
de tems pour mûrir & acquérir
de la force avant l'automne ;
au-lieu que toutes celles qui
poussent après le milieu de
Juin, sont foibles & pleines
de moëlle ; & quoiqu'elles puissent quelquefois produire des
fleurs, cependant elles donnent rarement du fruit, & font
par la suite du mauvais bois :
leurs vaisseaux étant trop larges pour perfectionner la séve,
ils donnent passage à une trop
grande quantité de sucs cruds.
Ainsi, lorsqu'on n'examine les
arbres en espaliers qu'en deux
saisons différentes, c'est-à-dire,
pour la taille d'hiver & celle
de la Saint-Jean, il n'est pas
possible de les avoir en bon
état ; car, en laissant toutes
les branches qui naissent au
printems, jusqu'au milieu ou
à la fin de Juin, comme on le
pratique ordinairement, quelques unes des plus vigoureuses
attireront la plus grande partie
de la nourriture des autres,
qui se trouveront trop foibles,
après que les premieres feront
retranchées, pour produire de
beaux fruits : d'ailleurs les arbres seront épuisés, pour avoir
nourri une grande quantité de
branches inutiles. On ne voit
malheureusement que trop d'arbres traités de la sorte : se
se plaint alors de ce qu'ils ne
produisent que peu de fruits,

PER

& on ne fait pas attention à
ces branches gourmandes, qui,
absorbant presque toute la séve, deviennent très-fortes &
ligneuses, au-lieu qu'on pourroit distribuer également cette
séve dans toutes les branches
que leur foiblesse rend stériles.
Il arrive souvent que de pareils arbres périssent avant qu'on
ait retranché ces branches gourmandes, ou qu'au moins ils
deviennent si foibles, qu'ils ne
sont plus en état de produire
du fruit. En affoiblissant ainsi
les branches, on leur fait souvent produire un grand nombre de fleurs, comme on le
voit quelquefois sur des branches d'automne ; mais ces mêmes fleurs les épuisent de façon qu'elles donnent rarement
du fruit, & que la plupart de
ces branches périssent fort souvent bientôt après ; effet qu'on
attribue à la nielle, comme
je l'ai dit ailleurs, tandis qu'il
ne provient que du peu d'intelligence du Jardinier. Il est
par conséquent de la plus grande
conséquence pour ces arbres
en espalier, sur-tout pour les
Péchers, de retrancher toutes
leurs branches irrégulieres deux
ou trois fois dans les mois
d'Avril, Mai & Juin, & de
palissader en bon ordre les
branches que l'on réserve, de
maniere qu'elles jouissent toutes également de l'influence du
soleil & de l'air, qui sont les
deux agents propres à mûrir le
bois, & à le preparer à donner du fruit l'année suivante.

En observant exactement cette
pratique en été, on diminuera
les fréquentes tailles que l'on
ne

PER

ne fait que trop souvent sur les *Péchers*, à leur grand préjudice ; car les jeunes branches à bois étant généralement molles, tendres, & remplies de moëlle, lorsqu'elles sont fortement blessées, elles ne se guériffent pas auffi-tôt que dans d'autres efpeces d'arbres ; l'humidité qui s'infinue dans ces blessures, occafionne des chancres, & détruit souvent les branches ; ce que l'on peut entierement éviter, en pinçant les branches, & en jettant bas avec le doigt les boutons mal placés, comme il a déja été dit : par ce moyen, on ne fait point de blessure aux arbres, & l'on s'épargne un très-grand travail ; car une feule personne exercée repassera une plus grande quantité d'arbres en un jour, que trois ou quatre ne pourroient le faire fur des arbres négligés, dans le même efpace de tems : de forte qu'en laissant croître des arbres naturellement & fans ces précautions pendant tout le printems, ils exigeront fix fois plus d'ouvrage pour les remettre en état, fans compter le tort que l'on fait aux fruits, qui, lorsqu'ils ont crû à l'ombre de ces branches & des feuilles pendant tout le printems jufqu'à la Saint-Jean, & qu'on les expofe alors brufquement au foleil & à l'air, en palissadant les autres branches contre la muraille, font non - feulement retardés dans leur accroissement, mais deviennent mauvais, & acquierent une peau plus dure.

La diftance que l'on donne

aux branches de ces arbres fur le treillage, doit être proportionnée à la grosseur du fruit & à la longueur des feuilles ; car fi nous obfervons comment les branches de ces arbres font naturellement difpofées à croître, nous trouverons qu'elles font toujours placées à une diftance plus ou moins grande, fuivant que leurs feuilles font plus grandes ou plus petites, ainfi que je l'ai déja obfervé à l'article *Feuilles*.

Comme il n'y a point de guide plus sûr, pour un Artifte curieux, que la Nature elle-même, un bon Jardinier doit toujours la confulter dans toutes les parties de fa profession, parce que fon travail ne confifte qu'à l'aider à perfectionner fes productions; & que, pour y parvenir, il doit fe conformer à fes propres principes.

Mais, pour en revenir à la taille de ces arbres, quand leurs branches font palissadées avec foin, comme il a été dit ci-dessus, pour le printems & l'été, il n'eft plus queftion que de la taille d'hiver, qui s'exécute ordinairement en Février ou en Mars ; mais la meilleure faifon eft le mois d'Octobre, quand les feuilles commencent à tomber ; au moyen de quoi les blessures auront le tems de fe guérir avant les gelées, & il ne fera point à craindre qu'ils en foient endommagés. Les branches des arbres étant proportionnées à la force des racines dans cette faifon, toute la fève du printems ne fera employée qu'à nourrir les

Tome V.

parties des branches utiles qu'on a laissées ; au-lieu qu'en ne les taillant qu'au mois de Février, la sève qui est alors en mouvement, comme on peut le voir par le gonflement des bourgeons, se porte aux extrémités des branches, pour nourrir telles fleurs que l'on est obligé de jetter bas. On peut se convaincre de cette vérité, en observant les plus fortes branches dans cette saison : on y verra les boutons des extrémités se gonfler plutôt que ceux du bas ; car n'y ayant point alors de feuilles sur les branches pour arrêter la sève, elle se porte naturellement aussi loin qu'elle peut aller, sans s'arrêter au bas.

Un principe constant parmi les Jardiniers, & fondé sur une longue expérience, c'est qu'il faut tailler les arbres foibles dans le commencement de l'hiver, & les arbres gourmands fort tard au printems, pour arrêter leur trop forte croissance. A présent, il est évident que ce défaut ne vient pas de quelque perte considérable de sève qui se soit faite par les blessures de la taille, si ce n'est dans quelques arbres qui coulent naturellement, quand on les coupe dans cette saison, mais de toute autre cause, suivant les expériences du Docteur H A L E S, qui, en fixant des mesures mercurielles aux tiges d'arbres nouvellement taillés, a trouvé que ces blessures étoient toujours dans un état absorbant, excepté la *Vigne*, dans la saison où sa sève est fort abondante.

C'est pourquoi, quand un arbre foible est taillé dès le commencement de l'hiver, les orifices des vaisseaux de la sève sont refermés long-temps avant le printems ; & conséquemment, lorsqu'au printems ou en été, les chaleurs commencent à se faire sentir, la force attractive des feuilles n'est pas affoiblie par beaucoup d'ouvertures ; mais elle est au contraire ranimée par la sève de la racine ; au-lieu qu'un arbre gourmand étant taillé tard au printems, la force de ses feuilles, pour attirer la sève de la racine, est beaucoup diminuée par les différentes ouvertures de cette taille tardive.

D'ailleurs, quand même ce ne seroit pas un avantage pour les arbres d'être taillés avant l'hiver, je ne crois pas qu'on puisse douter, d'après l'expérience, qu'au moins cette taille réussit aussi bien que celle du printems ; cependant j'avoue qu'il est très-utile de la faire à la Saint-Jean, qui est une saison beaucoup plus commode pour les Jardiniers, que le printems, parce qu'ils ont alors plus de tems pour soigner leurs arbres, & que dans ce tems ils n'ont point d'ouvrages qui exigent d'être exécutés sur le champ ; au-lieu qu'au printems ils ont à soigner le jardin potager & les couches chaudes, & qu'il seroit alors très-avantageux pour eux d'être débarrassés de la taille des arbres, sur-tout quand ils ont une grande étendue d'espaliers : on a aussi l'avantage, en taillant dans cette saison, de pouvoir

PER

labourer & nettoyer les plates-bandes avant le printems, & rendre le jardin propre pour ce tems.

Après avoir bien differté fur les tems propres à la taille des arbres, je vais donner quelques inftruĉtions générales fur la maniere de la faire pour les *Péchers* & *Pavies*, qui exigent un traitement bien différent de celui qui convient aux autres arbres à fruits.

En taillant ces arbres, on doit obferver l'endroit où cette taille doit être faite, & couper toujours les branches au-deffus d'un bouton à bois, qu'on diftingue aifément des boutons à fruits, en ce que ces derniers font plus courts, plus ronds & plus gonflés : car, fi la branche taillée n'eft pas terminée par un bouton à bois, elle eft fort fujette à périr jufqu'au premier endroit où il y en a un ; de maniere que tous les fruits qui fe trouveroient au-deffus, feroient à pure perte. Ainfi, il eft toujours néceffaire que toutes branches taillées fe terminent par un bouton à bois pour attirer la fève ; car il ne fuffit pas d'avoir un bouton à feuilles, comme quelques perfonnes l'ont imaginé ; celui-ci n'attireroit qu'une petite quantité de fève, le grand ufage des feuilles n'étant que de tranfpirer les fucs cruds qui ne font pas propres à nourrir le fruit. La longueur qu'on laiffe à ces branches doit être proportionnée à la force de l'arbre. Dans un arbre fort & fain, on peut les tailler à dix ou douze pouces & plus ; & dans

PER   495

un foible, elles ne doivent pas avoir plus de fix pouces : mais on doit toujours fe conduire, dans cette opération, d'après la pofition du bouton à bois ; car il vaut mieux donner à la branche trois ou quatre pouces de plus, ou deux ou trois de moins, pour fe procurer un de ces boutons qui font abfolument néceffaires pour la réuffite à venir de l'arbre ; il faut auffi retrancher toutes les branches foibles, quoiqu'elles foient chargées de beaucoup de boutons à fleurs, parce qu'elles n'auroient pas affez de force pour nourrir le fruit, & qu'elles affoibliroient les autres parties de l'arbre.

En paliffadant les branches au treillage, il faut avoir foin de les placer à des diftances auffi égales qu'il eft poffible, de maniere que les feuilles en fortant puiffent avoir de la place pour croître fans donner trop d'ombre aux arbres : mais on ne doit jamais les laiffer droites, fi on peut l'éviter ; car alors les boutons du haut pouffent plus vigoureufement, & le bas des branches fe trouve nud.

Rien n'a plus occupé les amateurs du jardinage, que de chercher comment on pouvoit préferver de la nielle du printems les efpeces tendres de *Péchers*, & cependant j'ai lu bien peu de chofe d'utile fur ce fujet. Quelques-uns ont propofé de placer des paillaffons devant les arbres en efpalier pour les préferver de la nielle ; d'autres ont confeillé de fixer aux murailles des abris

horifontalement placés, pour empêcher la rofée ou la pluie perpendiculaire de tomber fur les fleurs des arbres fruitiers, parce qu'ils ont imaginé que cette pluie ou rofée eſt la principale caufe de la nielle : mais toutes ces inventions font bien éloignées de répondre à ce que l'on attendoit de ces Savans qui les ont mifes en pratique.

C'eſt pourquoi je ne crois pas inutile de rappeler ici quelque chofe de ce que j'ai déja dit à ce fujet.

1°. J'ai obfervé que les nielles dont on a tant à fe plaindre, ne provenoient pas tant de quelque caufe externe ou des mauvaifes faifons, que d'une maladie ou foibleſſe dans les arbres ; car, en obfervant dans ce tems les arbres qui font les plus fujets à ce qu'on appelle nielle, on trouve que ce font ceux dont les branches font fort petites, foibles, pas à moitié mûres, & qui font paliſſadées trop près les unes des autres. Ces branches font, pour la plupart, chargées de boutons à fleurs, ce qui provient principalement de foibleſſe. Ces boutons s'ouvrent, & bien des gens peu expérimentés fur la connoiſſance des arbres fruitiers, imaginent que c'eſt l'annonce d'une récolte abondante, tandis que ce n'eſt qu'une preuve de l'épuifement total de ces branches, qui ne font en état que de faire épanouïr ces fleurs. Ces mêmes fleurs tombent enfuite, ainſi que les boutons des feuilles qui n'ont plus de vigueur ; après quoi fouvent une grande par-

tie des branches périt. On appelle cela une grande nielle, tandis qu'en même tems on voit fouvent des arbres de la même efpece ou d'une efpece différente, qui font plus forts & en bon état, quoique placés dans le même fol, à la même expofition, & expofés aux mêmes rigueurs de la faifon,échapper à cette maladie. C'eſt donc une indication certaine que ce mal procede de quelque caufe intérieure, & non pas d'une nielle produite par l'influence de l'air, & qu'on peut y remédier, en fuivant exactement les inſtructions que nous venons de donner fur la taille & le traitement des arbres, en ne furchargeant jamais trop les arbres de branches, & en ne fouffrant pas qu'une feule partie de l'arbre abforbe la totalité des fucs nutritifs que les racines fourniſſent ; ce qui rend le reſte très-foible : mais fi l'on répartit la féve également dans toutes les branches, de maniere qu'il n'y eñ ait point de trop vigoureufes, & fi l'on retranche continuellement les rejettons inutiles, à mefure qu'ils paroiſſent, la force des arbres ne fera pas diminuée par la perte de la nourriture qu'ils auroient dû fournir à ces petites branches, qui doivent être retranchées dans la fuite. C'eſt cependant ce que l'on ne néglige que trop fouvent dans le traitement de ces arbres.

2°. Il arrive quelquefois que les racines de ces arbres font enfevelies trop profondément dans la terre, & qu'elles fe

PER 497

trouvent dans un fol froid & humide, ce qui eſt un des plus grands déſavantages qui puiſſent arriver à ces arbres délicats ; car la fève continue dans les branches, étant miſe en mouvement de bonne heure au printems par la chaleur du foleil, s'épuiſe pour la nourriture des fleurs, & fe perd par les pores des branches à bois ; de maniere qu'elle eſt entiérement diſſipée avant que la chaleur ait pu atteindre aux racines, pour leur procurer une pareille émotion, & les mettre à même de puiſer de nouveaux fucs, qui puiſſent remplacer ceux qui ont été diſperſés : faute de ce fecours, les fleurs tombent, & périſſent ; les branches paroiſſent être malades, juſqu'à ce que la chaleur ait pénétré juſqu'aux racines : alors à cette langueur ſuccede une vigoureuſe végétation ; & avant la fin de l'été, les arbres fe trouvent garnis de branches beaucoup plus fortes que ne ſont celles des arbres qui jouiſſent de tout l'avantage du foleil : mais ces derniers ont conſervé leurs fleurs, & ſont beaucoup plus fains & plus fructueux. Dans ce cas, le défordre ne peut être attribué qu'à la grande quantité de fucs cruds & humides que l'arbre attire, leſquels, quoique propres à produire beaucoup de bois, ſont néanmoins très-nuiſibles aux fruits. Ainſi, on ne peut fe mettre à l'abri de pareils accidens, qu'en foulevant les arbres, s'ils ſont jeunes ; mais s'ils ſont trop vieux pour être dérangés, il n'y a

d'autre remede que de les arracher, de refaire les platesbandes avec une nouvelle terre fraîche, & d'y planter de jeunes arbres : car c'eſt fe donner bien des peines inutiles, & s'occaſionner beaucoup de dépenſes ſuperflues, que de s'aſſujettir à tailler & traiter des arbres, fans avoir jamais la ſatisfaction d'en tirer quelqu'avantage ; ce qui arrive certainement quand ils ſont auſſi mal plantés.

3°. Le mal procede quelquefois de ce que les arbres manquent auſſi de nourriture ; ce qui arrive ſouvent, quand ils ſont plantés dans un fol dur & graveleux : l'uſage commun étant de creuſer les platesbandes de trois ou quatre pieds de largeur, fur trois de profondeur, dans une roche ou dans un fol graveleux, & de remplir ces foſſés avec une bonne terre fraîche, dans laquelle on plante les arbres, ils y profitent aſſez bien pendant deux années ; mais lorſque leurs racines ont atteint le gravier, qu'elles fe trouvent gênées comme ſi elles étoient renfermées dans une caiſſe, & que la nourriture qu'elles y puiſent n'eſt plus ſuffiſante, l'arbre dépérit, & les branches périſſent annuellement. On ne peut remédier à cet inconvénient, ſi les arbres ont déja quelques années de croiſſance, qu'en les enlevant totalement, qu'en ôtant tout le gravier qui gêne les racines, & qu'en mettant en place une plus grande quantité de bonne terre, qui pourra leur procurer un ſup-

Ii 3

498 **PER**

plément de nourriture , pour quelques années de plus ; mais ces arbres , plantés dans un auffi mauvais fol , ne pourront jamais fe conferver fains , quelque art qu'on y emploie.

Si la ftérilité des arbres ne provient d'aucune des caufes ci-deffus , mais qu'elle foit l'ef-fet des mauvaifes faifons & de l'intempérie de l'air , alors la meilleure méthode qui foit con-nue eft , en tems fec & lorf-qu'il tombe peu de rofée , d'ar-rofer légérement avec la gerbe les branches des arbres , auffi-tôt après la faifon des fleurs , & tandis que le jeune fruit eft tendre. Cette opération doit toujours être faite avant midi , afin que l'humidité puiffe fe diffiper avant la nuit ; & fi l'on couvre encore ces arbres pendant la nuit avec des nat-tes , du canevas , &c. on leur fera beaucoup de bien. Ce-pendant quand les arbres font forts & vigoureux , ils ne font pas fi fujets à fouffrir des pe-tites intempéries de l'air , que ceux qui font foibles ; de forte qu'il y a peu d'années dans lefquelles on ne puiffe leur faire produire une quantité modérée de fruits , quand mê-me on ne feroit point ufage de couvertures : car lorfque l'on emploie ce moyen , fi l'on n'y apporte pas beaucoup de foin , & la plus grande intel-ligence , on fait plus de tort aux arbres qu'en les abandon-nant à l'air & aux faifons. Si l'on place ces couvertures trop près des arbres , fi on les y laiffe trop long-tems , ou fi l'on expofe ces arbres trop fubite-

PER

ment à l'air, après les avoir tenus couverts pendant quel-que tems , ils fouffrent beau-coup plus que fi on les avoit livrés à eux-mêmes. Cependant il faut que je rappelle ici ce qui a été dit dans un autre ar-ticle , fur un moyen qui a été généralement fuivi avec fuc-cès : il confifte à pratiquer une efpece d'avant-toit au - deffus des arbres avec des planches de fapin jointes enfemble , par le moyen duquel on garantit les arbres de toute l'humidité qui tombe perpendiculairement ; on attache cet avant-toit à la muraille , au-deffus des arbres , quand ils commencent à fleu-rir , & on le laiffe jufqu'à ce que le fruit foit bien formé ; alors on l'enleve , pour laiffer jouir les feuilles & les bran-ches des rofées & des pluies. Lorfque la muraille a beau-coup d'étendue , & qu'elle eft expofée à un courant de vent , on dreffe quelques haies de rofeaux croifés à quarante pieds de diftance l'une de l'au-tre , & à dix pieds environ de la muraille , pour rompre la force du vent , & l'empêcher de détacher les fleurs ; mais on les ôte auffi-tôt que le dan-ger eft paffé. Par-tout où l'on s'eft fervi de ces moyens , on en a éprouvé d'excellens ef-fets ; & comme il n'y a plus de foins à avoir , quand une fois le toit eft fixé fur les arbres , on n'a plus rien à craindre des négligences qui ont fouvent lieu lorfque l'on eft obligé de répéter fouvent la même opération.

Quand le fruit eft établi , &

PER

qu'il est parvenu à la grosseur d'une petite noix, il faut examiner les arbres, & l'éclaircir, en donnant au moins cinq ou six pouces de distance entre chacun ; car si on laisse ces fruits en paquets, la nourriture, qui ne devroit être employée que pour un certain nombre, se trouvera absorbée par la totalité, dont une grande partie doit être ensuite retranchée ; de sorte que plutôt cette opération est faite, mieux le fruit qui doit rester s'en trouvera. S'il arrive quelquefois qu'une partie des fruits vienne à périr par certains accidens, alors celui qui reste sera beaucoup plus gros & de meilleur goût ; d'ailleurs les arbres en acquerront plus de force. Ainsi, une quantité médiocre de fruits est toujours préférable à une grande récolte. Quand il y a peu de fruits, ils sont plus gros, de meilleur goût, & les arbres sont mieux disposés pour l'année suivante ; au-lieu que, quand ils sont fort chargés, les fruits sont petits, sans saveur, & les arbres sont si affoiblis, qu'ils sont hors d'état d'en produire de deux ou trois ans. Ainsi, une récolte modérée est avantageuse pour le fruit & pour les arbres. On ne doit jamais laisser sur un arbre en pleine vigueur plus de cinq douzaines de *Pêches*, & trois ou quatre douzaines sur un arbre médiocre.

Si la saison est chaude & seche, il sera prudent d'enlever de la terre autour de la tige de chaque arbre, pour former

PER 499

un bassin d'environ six pieds de diametre ; de couvrir la surface de ce bassin avec du terreau, & une fois la semaine ou tous les quinze jours, suivant la chaleur & secheresse de la saison, d'y répandre quinze ou vingt pots d'eau, en se servant d'une gerbe pour disperser l'eau en gouttes fines, comme celle de la pluie. On peut aussi arroser de même les branches des arbres ; ces précautions tiendront les fruits dans un accroissement suivi, & les empêcheront de tomber, comme cela arrive toujours quand on néglige cette précaution ; par-là le fruit recevra une nourriture continuelle, qui lui procurera une saveur beaucoup plus agréable, & conservera les arbres en vigueur. Je puis recommander cette pratique, d'après une longue expérience, comme étant la plus utile ; mais on ne doit en faire usage que tandis que le fruit croît ; car, après ce tems, ces arrosemens seroient très-nuisibles aux arbres & aux fruits. Un automne sec mûrit mieux le bois & le fruit, qu'une arriere saison humide.

Quand les *Pêchers* sont traités avec soin au printems, suivant les regles qui viennent d'être prescrites, toute la sève que les racines peuvent fournir, est employée utilement, tant pour la nourriture des branches qui doivent rester, que pour celle de la quantité de fruits que chaque arbre doit porter : mais, sans ces soins, les arbres dépérissent bientôt, & ne sont jamais proprement

Ii 4

PER

garnis de branches ; les unes font très-foibles , d'autres font trop vigoureufes , & cet affemblage mal afforti rend les arbres fort défagréables à la vue , mal-fains , & les empêche d'être fruêtueux pendant beaucoup d'années. En paliffadant les branches contre la muraille, à mefure qu'elles pouffent , les fruits font toujours expofés à l'air & au foleil : fi au contraire, fuivant la méthode commune , on laiffe les branches telles qu'elles fortent naturellement pendant tout le printems , les fruits fe trouvent privés de ces deux avantages ; en outre , en jettant bas avec le pouce , au printems , les rejettons inutiles & gourmands , on s'épargnera beaucoup de travail : l'on ne fera pas obligé de faire ufage de la ferpette en été ; ce qui nuit beaucoup aux arbres.

Lorfqu'on fuit exaêtement ces regles , & qu'on laiffe une diftance convenable entre les branches , on n'eft pas forcé d'ôter les feuilles des arbres , pour expofer les fruits au foleil , comme on le fait très-fouvent. Si l'on faifoit attention que les feuilles font abfolument néceffaires pour nourrir les boutons à fleurs qui font toujours formés aux ailes des pétioles , on verroit qu'en les ôtant avant qu'elles aient exécuté ce qui leur eft prefcrit par la Nature , on fait beaucoup de tort aux arbres. Ainfi , j'avertis tous Cultivateurs de ne jamais fuivre cette pratique.

L'opinion commune , que les *Pêchers* ne fubfiftent pas long-tems , à prévalu depuis quelques années , même parmi des perfonnes de bon fens , & l'on a prétendu , par cette raifon , qu'il étoit néceffaire de les renouveler tous les vingt ans : mais c'eft une grande erreur ; car j'ai mangé de très-belles Pêches de différentes efpeces , qui avoient été recueillies fur des arbres âgés de plus de cinquante ans ; & je fuis convaincu , par l'expérience , que des arbres greffés fur des fujets convenables , plantés & traités avec foin , fe confervent fains & fruêtueux pendant plus de foixante années. Les fruits de ces vieilles tiges font bien fupérieurs , pour le goût, à ceux qui font produits par de jeunes arbres. Je crois que cette opinion mal fondée nous vient des Francois , qui greffent généralement leurs *Pêchers* fur des tiges d'*Amandiers* , qui font d'une courte durée , & ne fubfiftent guere plus de vingt ans : mais on fuit rarement cet ufage en Angleterre , & nous aurions grand tort de prendre en aucune maniere cette Nation pour modele , *puifque leurs Doêteurs en l'Art du Jardinage font au moins d'un fiecle plus jeunes que les Anglois* , & ils ne paroiffent point à préfent difpofés à vouloir les atteindre ; car ils s'écartent de la Nature dans prefque toutes leurs opérations de jardinage , & ils aiment mieux introduire leurs petites inventions , pour émonder & traiter leurs arbres fruitiers fuivant leurs fantaifies , que de puifer leurs inftruêtions dans la Nature mê-

PER 501

me, qui feule doit nous fervir de guide. Les Jardiniers ne doivent donc guere s'en écarter, fi ce n'eft dans quelque circonftance où l'art vient à leur fecours, pour en tirer plus d'avantages, comme pour fe procurer plufieurs efpeces de plantes & de fruits beaucoup plutôt, & pour leur donner un dégré de perfection auquel ils ne pourroient atteindre autrement; mais en cela les François font encore bien éloignés du but, parce qu'ils fe fient trop à la Nature, & ne font aucun ufage de l'Art.

Dans un Ouvrage d'un des plus célèbres de leurs Auteurs, qui traite particulierement des *Arbres à fruits*, on recommande de planter les *Péchers* à douze pieds de diftance, & les *Poiriers* à neuf ou dix pieds feulement. Cependant il y eft dit qu'un *Poirier* en bon état étend chaque année fes branches à trois pieds de chaque côté. Ainfi, cet Auteur ne leur donne de l'efpace pour croître, que pour deux ans avant de fe rencontrer : il exige auffi pofitivement de ne jamais mettre de fumier fur les plates-bandes des arbres à fruits, parce que, dit-il, cet engrais donne un mauvais goût au fruit. Cette opinion a trop généralement pris en Angleterre, mais elle a été réfutée par un autre auteur du même pays, qui affirme, que depuis plus de vingt ans il a conftamment mis du fumier dans fes plates-bandes, & que fes arbres ont non-feulement produit les fruits les plus délicieux, mais

fe font encore entretenus dans la plus grande vigueur. Le même Auteur cite la pratique des Jardiniers de Montreuil, près de Paris, qui, depuis plufieurs générations, fe font rendus célebres par la culture des *Péchers*, & qui ont grand foin de mettre du fumier chaque deux ans dans leurs plates-bandes, comme on le pratique dans les jardins potagers pour les légumes.

Je puis affurer l'efficacité de cette méthode, d'après une longue expérience ; car j'ai goûté des fruits dans quelques jardins particuliers, où les *Péchers* ont toujours été fumés chaque deux ans. Ainfi, je recommande cette pratique à tous les curieux, en obfervant cependant de ne jamais fe fervir pour cela que de fumier très-confommé, & de l'enterrer en Novembre, afin que les pluies en faffent defcendre les fels avant le printems fuivant. Dans les terres légeres & fablonneufes, on doit donner la préférence au fumier de vache, parce qu'il eft plus frais & plus compact que celui de cheval, & réferver ce dernier pour les terres fortes & froides.

On rendra grand fervice aux arbres, en labourant exactement la terre chaque année, & même deux ou trois fois par an, fi le fol eft fujet à fe ferrer beaucoup : on doit auffi ne pas furcharger les plates-bandes de plantes d'un grand crû, parce qu'elles priveroient les arbres de leur nourriture : c'eft-pourquoi on n'y met que de petites herbes de cuifine,

que l'on enleve de bonne heure au printems, & qui non-seulement ne feront aucun tort aux arbres, mais encore leur procureront des labours plus fréquens, occasionnés par ces petites récoltes, lesquels n'auroient pas été faits, si les plates-bandes étoient restées vuides.

Si, dans le traitement de ces arbres, on se conforme aux regles que nous venons de prescrire, on aura non-seulement de bons fruits, mais les *Péchers* conserveront leur vigueur pendant un grand nombre d'années.

PERSIL COMMUN. *Voyez* Apium. Petroselinum. *L.*

PERSIL BASTARD. *Voyez* Caucalis.

PERSIL DE MACÉDOINE GROS *ou* le Maceron. *Voy.* Smyrnium. *L.*

PERSIL DE MACÉDOINE. *V.* Bubon Macedonicum. *L.*

PERSIL DE MARAIS *ou* Laiteux. *Voy.* Selinum Palustre.

PERSIMON. *Voy.* Diospyros Virginiana.

PERVENCHE. *Voyez* Vinea. *L.*

PERVENCHE DE MADAGASCAR. *V.* Vinea Rosea. *L.*

PERVINCA. *V.* Vinea. *L.*

PESSE *ou* la Pece *ou* Picea. Epicia *ou* faux Sapin. *Voy.* Abies Picea.

PET D'ASNE *ou* Epine Blanche. *Voy.* Onopordum Acantium. *L..*

PET DU DIABLE *ou* Sablier. *Voy.* Hura Crepitans.

PET DE LÉOPARD. *Voy.* Doronicum.

PÉTALE. Ce sont des feuil-les colorées qui composent les parties les plus visibles de la fleur, ou la corolle. On les appelle en latin *petala*, pour les distinguer des feuilles ordinaires, que l'on nomme *folia.*

PETASITE, *Voyez* Tussilago. *L.*

PETIVERIA. *Plum.* Nov. *'Gen.* 50. *tab.* 39. *Lin. Gen. Plant.* 417. [*Guinea Henweed.*] l'Herbe aux Poules de Guinée. La Petiver.

Caracteres. Le calice de la fleur est persistant, & composé de cinq feuilles étroites, obtuses & égales ; la corolle a quatre petits pétales blancs, placés en forme de croix, & qui tombent en peu de tems ; la fleur a six étamines érigées en forme d'alêne, & terminées par des antheres simples. Dans son centre est placé un germe oblong & comprimé, avec quatre styles en forme d'alêne, & couronnés par des stigmats obtus & persistans. Ce germe se change dans la suite en une semence oblongue, conique, étroite par le bas, mais large au dessus, où elle est comprimée, découpée au sommet, semblable a un bouclier recourbé, & armée d'un style aigu & réfléchi.

Ce genre de plantes est rangé dans la quatrieme section de la sixieme classe de Linnée, qui comprend celles dont les fleurs ont six étamines & quatre styles.

Les especes sont :

1°. *Petiveria Alliacea, floribus hexandriis. Hort. Cliff.* 141. *Hort. Ups.* 91. *Act. Stockh.* 1744. *p.* 287. *f.* 7. *Trew. Ehret.* 33. *f.*

PET

67. *Mat. Med.* 100. *Kniph. Cent.*
2. *n.* 53. ; la Petiver , avec six
étamines dans les fleurs.

Petiveria foliis oblongo-ovatis,
spicis longioribus , terminalibus.
Brown. Jam. 274.

Verbenæ aut Scorodoniæ affinis
anon..., flore albido , calice
afi... Allii odore. Sloan. Hist.
1. p. 192 , ordinairement appellé *Herbe aux Poules de Guinée.*

... *Petiveria octandra , flori-*
bus octandris. Lin. Sp. Plant.
... Jacq. Amer. 201.

Petiveria Solani foliis , loculis
spinosis. Plum. Nov. Gen. 50.
Ic. 219. ; la Petiver , avec des
fleurs à huit étamines , des
feuilles de Solanum , & des
godets épineux.

Le titre de ce genre lui a
été donné par le P. PLUMIER ,
en l'honneur de JACQUES PETI-
VER , Apothicaire de Londres
& Botaniste curieux : il l'a découverte en Amérique.

Alliacea. La première espece
est une plante fort commune
à la Jamaïque , à la Barbade ,
& dans la plupart des Isles
des Indes Occidentales , où
elle croît à l'ombre des bois
& dans les prairies découvertes.
Comme cette plante supporte
bien la secheresse , elle se conserve verte , tandis que les autres sont brûlées par l'ardeur
du soleil ; ce qui fait que le
bétail s'en nourrit ; mais , son
odeur étant forte , & son goût
à-peu-près semblable à celui
de l'Ail sauvage , le lait des
vaches qui en mangent , a la
même qualité , & les animaux
qu'on égorge lorsqu'ils s'en sont
rassasiés , ont un goût défagréable , & leur chair ne vaut

PET   503

rien. Les racines de cette espece sont fortes , & pénetrent
profondément dans la terre ;
ses tiges , qui s'élevent à la
hauteur de deux ou trois pieds ,
sont noueuses , deviennent ligneuses vers le bas , & sont
garnies de feuilles oblongues ,
de trois pouces de longueur
sur un & demi de large , d'un
vert foncé , veinées , alternes ,
& placées sur de courts pétioles : ses fleurs naissent en épis
minces aux extrémités des branches ; elles sont fort petites ,
& ont peu d'apparence : elles
paroissent en Juin , & sont
remplacées par de petites capsules en forme de bouclier ,
recourbées , & qui renferment
une semence oblongue , qui
mûrit en automne.

Octandra. La seconde espece
ressemble fort à la premiere ,
dont elle ne differe qu'en ce
qu'elle a une tige plus courte
& plus étroite ; ses fleurs ont
huit étamines : mais ces distinctions ne peuvent être faites
que par un observateur exact ;
car toutes deux peuvent passer
pour la même.

Culture. On conserve ces plantes en Europe dans les Jardins
Botaniques ; mais elles ont peu
de beauté , & d'ailleurs leur
odeur est si forte , quand on
les touche , que c'est une raison de plus pour n'en pas faire
beaucoup de cas. On les multiplie par leurs graines , qu'il faut
semer sur une couche chaude au
commencement du printems.

Quand les plantes ont poussé , on les met chacune séparément dans un pot , que l'on
plonge dans une couche de

chaleur modérée, pour hâter leurs progrès : lorſqu'elles ont acquis beaucoup de force, on les accoutume par dégrés à ſupporter le plein air, auquel on les expoſe vers la fin de Juin, en les plaçant à une expoſition chaude, où elles peuvent reſter juſqu'à l'automne ; alors on les tranſporte dans la ſerre chaude, & on les y tient pendant l'hiver à un dégré de chaleur modérée, ſans quoi elles ne ſubſiſteroient pas dans ce pays.

Elles produiſent des fleurs & des ſemences chaque été, & ſe conſervent pluſieurs années : elles gardent leurs feuilles toute l'année, & peuvent être multipliées par boutures.

PETREA. *Houſt. Gen. Nov. Lin. Gen. Plant.* 682. [*Petrea.*] La Petre.

Caraɛteres. Le calice de la fleur eſt en cloche, & formé par une feuille découpée preſque juſqu'au fond en cinq ſegmens larges, obtus, colorés, étendus & perſiſtans ; la corolle, qui eſt monopétale, a un tube court, & diviſé au ſommet en cinq ſegmens preſque égaux & étendus ; la fleur a quatre étamines courtes & placées dans le tube, dont deux ſont un peu plus longues que les autres, & qui ſont toutes terminées par des antheres ſimples ; elle a quatre germes, qui ſoutiennent un ſtyle mince & couronné par un ſtigmat obtus : ces germes ſe changent dans la ſuite en quatre ſemences renfermées dans une enveloppe à franges.

Ce genre de plantes eſt rangé dans la ſeconde ſeɛtion de la quatorzieme claſſe de LINNÉE, avec celles dont les fleurs ont deux étamines longues & deux courtes, & dont les ſemences ſont renfermées dans une enveloppe.

Le Doɛteur HOUSTOUN a ainſi nommé ce genre en l'honneur du Lord PETRE, grand Amateur de la Botanique, qui poſſédoit une belle colleɛtion de plantes exotiques.

Nous n'avons qu'une eſpece de ce genre.

Petræa volubilis, fruteſcens, foliis lanceolatis, rigidis, flore racemoſo, pendulo ; la Petre en arbriſſeau, avec des feuilles en forme de lance, & des fleurs en paquets longs & pendans.

Petræa. Hort. Cliff. 319. *Jacq. Amer.* 180. *f.* 114.

Cette plante a d'abord été découverte, par le Doɛteur HOUSTOUN, à la Vera-Cruz, dans la Nouvelle-Eſpagne, en 1731 ; mais depuis, elle m'a été envoyée de l'Iſle des Barbades, où elle croît ſans culture : elle s'éleve à la hauteur de quinze ou ſeize pieds, avec une tige d'arbriſſeau couverte d'une écorce d'un gris clair, & de laquelle ſortent pluſieurs branches longues, dont l'écorce eſt plus blanche que celle de la tige, & qui ſont garnies de feuilles à chaque nœud. Sur le bas des branches, ces feuilles ſont placées autour par trois, & plus haut par paire ; elles ont environ cinq pouces de longueur ſur deux pouces & demi de large au milieu, & ſont preſque terminées en pointe à chaque extrémité ; elles ſont roides, d'un vert clair,

& rudes ; leur côte mitoyenne eft groffe , foncée en couleur, & donne origine à plufieurs nervures tranfverfales, qui s'é-tendent vers les bords , qui font entiers. Les fleurs naiffent aux extrémités des branches en pa-quets clairs de neuf ou dix pou-ces de longueur; chaque fleur eft poftée fur un pédoncule mince , d'un pouce environ de long ; le calice eft compofé de cinq feuilles étroites , obtufes , d'à-peu près un pouce de longueur , & d'un beau bleu, ce qui les rend plus vifibles que les pé-tales , qui font blancs, & n'ont que la moitié de la longueur du calice. Quand la fleur eft paffée , les quatre germes du centre fe changent en autant de femences oblongues , qui font renfermées dans une en-veloppe à franges.

Le Docteur HOUSTOUN a trouvé une variété de cette plante à pétales bleus, d'une couleur auffi brillante que celle du calice, & d'une belle ap-parence, & dont chaque bran-che eft terminée par un long cordon de ces fleurs ; ce qui l'a engagé à ranger cet arbre dans la première claffe des plus beaux de l'Amérique.

Autant que j'ai pu en juger , d'après des échantillons deffe-chés , apportés par ce Docteur en Angleterre , il paroît que les fleurs mâles & les fleurs femelles naiffent fur des par-ties différentes du même ar-bre, ou fur différens pieds ; car une grappe de ces fleurs m'a paru être entierement com-pofée de fleurs mâles , & une autre de fleurs femelles ; mais le Docteur n'en a point fait mention dans fon manufcrit.

Culture. On multiplie cette plante par fes graines, qu'il faut faire venir de fon pays natal : mais dans le nombre , il y en a très-peu de bonnes ; car de celles que le Docteur avoit envoyées en Angleterre, on n'a pu en élever que deux plantes , quoique les femences aient été diftribuées à plufieurs perfonnes : c'eft ce qui me con-firme dans l'idée que les grappes de fleurs font de différens fexes , & que les femences recueillies par le Docteur ont été prifes, foit fur des arbres femelles , éloignés des plantes mâles , foit fur des parties de l'arbre écartées des fleurs mâles. On répand ces graines dans une bonne couche chaude. Quand les plantes ont pouffé , on les met chacune féparément dans de petits pots remplis d'une terre légere & marneufe ; on les plonge dans une couche chaude de tan , & on les place enfuite dans celle de la ferre chaude , où on les laiffera conftamment, en les traitant comme les autres plantes qui viennent des mêmes conrrées.

PETROSELINUM. *V.* APIUM.

PEUCEDANUM. *Tourn. Inft. R. H.* 318. *tab.* 169. *Lin. Gen. Plant.* 302. [*Hog's Fennel , or Sulphur—wort.*] Fenouil de Porc ou l'Herbe au foufre ; Queue de Pourceau.

Caractères. Les fleurs font dif-pofées en ombelle ; l'ombelle principale eft compofée de plu-fieurs autres , long·ies , étroites & étendues ; l'enveloppe de la grande ombelle eft formée par

506 **P E U**

plufieurs feuilles linéaires &
réfléchies. Le calice de la fleur
eft petit & découpé en cinq
parties; les corolles de la grande ombelle font uniformes ;
chaque fleur eft compofée de
cinq pétales oblongs, recourbés en-dedans, égaux & entiers ; elles ont chacune cinq
étamines femblables à des poils,
& terminées par des antheres
fimples, avec un germe oblong
placé fous la fleur, & qui foutient deux ftyles couronnés par
des ftigmats obtus. Ce germe
devient enfuite un fruit ovale,
canelé fur chaque côté, & divifé en deux parties, dont chacune eft une femence convexe
d'un côté, & comprimée de
l'autre, à trois fillons érigés,
avec une bordure large, mem-
braneufe, & dentelée à fon extrémité.

Ce genre de plantes eft rangé
dans la feconde fection de la
cinquieme claffe de LINNÉE,
où fe trouvent celles dont les
fleurs ont cinq étamines &
deux ftyles.

Les efpeces font :

1°. *Peucedanum officinale, foliis quinquiès tripartitis, linearibus. Lin. Sp. Plant. 358. Pollich. pal. n. 280* ; Queue de
Porceau, à feuilles divifées en
cinq parties, dont chacune eft
fous-divifée en trois fegmens
linéaires.

*Peucedanum. Bauh. Hift. 3.
p. 36. Raii Hift. 416.*

*Peucedanum Germanicum. C.
B. P. 149.*

2°. *Peucedanum Italicum, foliis tripartitis, fili-formibus, longioribus, umbellis difformibus* ;
Queue de Pourceau d'Italie,

dont les feuilles font divifées
en trois parties minces & plus
longues, produifant des ombelles difformes.

*Peucedanum majus Italicum. C.
B. P. 149.*

3°. *Peucedanum Alpeftre, foliolis linearibus ramofis. Hort. Cliff.
94. Roy. Lugd.-B. 98* ; Queue
de Porceau, à feuilles branchues, dont les lobes font
très-minces.

*Ferula foliis Libanotidis bre-
vioribus, Alpeftris, umbellis ampliffimis. Boerrh. Ind. Alt. 1.
p. 65.*

4°. *Peucedanum minus, foliis
pinnatis, foliolis pinnatifidis,
laciniis linearibus, oppofitis, caule
ramofiffimo, patulo. Fl. Angl.
101* ; Queue de Pourceau, avec
des feuilles aîlées, dont les
divifions font linéaires & oppofées, & une tige étendue
& branchue.

*Selinum montanum pumilum
Clufii, flore albo. Bauh. Hift.
3. p. 17.*

5°. *Peucedanum nodofum, foliolis alternatim multifidis. Hort.
Cliff. 94. Roy. Lugd-B. 98* ;
Queue de Pourceau, avec des
feuilles à plufieurs pointes,
& alternes.

*Silanum, quod Ligufticum Cre-
ticum, folio Fœniculi, caule no-
dofo. Tourn. Cor. 23. Boerrh.
Lugd.-B. 1. p. 51.*

On prétend que la premiere
efpece croît en Angleterre ;
mais je n'ai pas été affez heureux pour l'y trouver, malgré
toutes mes recherches, dans
les endroits indiqués. Je crois
qu'on la rencontre dans les
prairies marécageufes de plufieurs parties de l'Allemagne ;

PEU

fa racine eſt vivace , & diviſée en pluſieurs fortes fibres qui pénetrent profondément dans la terre ; de cette racine ſortent des pétioles nuds , canelés vers le bas , & diviſés à quatre ou cinq pouces de la racine en cinq pétioles plus petits , leſquels ſont ſous-diviſés chacun en trois , qui ſoutiennent chacun trois feuilles étroites , & d'une odeur ſemblable à celle du ſoufre , quand on les froiſſe. Les tiges , qui s'élevent à deux pieds de hauteur , ſont canelées , & diviſées en deux ou trois branches , terminées chacune par une ombelle de fleurs jaunes , régulieres , compoſée de pluſieurs petites ombelles circulaires. Ces fleurs paroiſſent dans le mois de Juin , & produiſent des ſemences comprimées & profondément ſillonnées , qui mûriſſent en automne (1).

Italicum. La ſeconde eſpece ſe trouve en Italie ſur les montagnes , & dans les vallées près des rivages des rivieres ; ſa racine eſt vivace , & pénetre

---

(1) Cette plante eſt apéritive , inciſive , diurétique , emménagogue , &c. On emploie quelquefois ſa racine en poudre ou en décoction dans les engorgemens glaireux de la poitrine , les affections catharrhales , l'aſthme humide , la ſuppreſſion des urines , des regles & des vuidanges , &c. : on s'en ſert auſſi extérieurement pour modifier les plaies & les ulceres.

Cette racine entre dans la compoſition de l'électuaire lithontriptique , & dans la poudre d'Iapraſſu.

PEU 507

plus profondément dans la terre ; les pétioles des feuilles ſont gros & ſillonnes ; ils ſe diviſent en trois petites branches , qui ſe ſous-diviſent en trois autres , terminées par trois lobes longs & étroits , ou petites feuilles beaucoup plus longues que celles de l'eſpece précédente ; les tiges qui ſoutiennent les ombelles , s'élevent à près de deux pieds de hauteur , & ſe diviſent au ſommet en pluſieurs petites branches qui ſoutiennent chacune une ombelle compoſée de pluſieurs plus petites , ou rayons poſtés ſur de forts petits pédoncules , qui s'étendent en-dehors irrégulierement : les fleurs ſont jaunes , & de la même forme que celles de la précédente , mais plus larges ; les ſemences ſont auſſi plus groſſes , quoique ſemblables à celles de la premiere. Cette eſpece fleurit & perfectionne ſes ſemences vers le même tems que la précédente.

Alpeſtre. La troiſieme eſpece croît naturellement dans la forêt de Fontainebleau & dans quelques autres parties de la France ; ſa racine eſt vivace , & pouſſe des pétioles qui ſe diviſent & ſe ſous-diviſent ; chaque ſous-diviſion eſt garnie de cinq feuilles courtes & étroites ; les tiges ſont rondes , & moins profondément canelées que celles des précédentes ; elles ſoutiennent chacune une grande ombelle de fleurs jaunes , ſemblables à celles des premieres ; ſes ſemences ſont plus courtes , mais de la même forme que celle des autres.

508 **PEU**

Cette plante fleurit dans le mois de Juin, & ses semences mûrissent au commencement de Septembre.

Minus. La quatrieme espece croît naturellement sur le rocher de Saint-Vincent, près de Bristol. Cette plante est bis-annuelle, & périt après avoir perfectionné ses semences ; ses feuilles sont courtes, fort étroites, & couchées sur la terre ; ses tiges s'élevent à un pied de hauteur, & se divisent en branches presque depuis le bas : ces branches sont à-peu-près horisontales, & garnies de quelques feuilles courtes, étroites, & d'un vert luisant ; chaque tige est terminée par une petite ombelle de fleurs, d'un jaune herbacé, petites, & qui produisent des semences petites & canelées.

Nodosum. La cinquieme espece, qui est originaire de l'Isle de Candie, n'est pas d'une longue durée en Angleterre, & ses semences n'y mûrissent pas bien : ses tiges s'élevent à un pied & demi de haut ; elles ont des nœuds assez gros, & à chaque jointure sort une feuille découpée en plusieurs divisions ; ses tiges sont terminées par des fleurs en ombelle, qui paroissent au commencement de Juillet, & qui, dans les années favorables, donnent des semences mûres en automne.

Culture. La premiere espece est du nombre des plantes médicinales ; mais elle est à présent de peu d'usage ; ses racines sont la seule partie dont on se serve : on la croit pro-

PHA

pre à débarrasser les poumons des flegmes âcres & gluans ; elle soulage dans les vieilles toux & dans l'asthme ; elle fond aussi les obstructions du foie & de la rate, & dissipe la jaunisse.

On conserve les autres especes dans les jardins de Botanique, pour la variété ; elles se multiplient toutes par semences, qu'on répand en automne, aussi-tôt qu'elles sont mûres : car si on les conserve jusqu'au printems, il est rare qu'elles réussissent ; & quand elles poussent, ce n'est, pour l'ordinaire, qu'après une année. Lorsque les plantes paroissent, il faut les tenir nettes de mauvaises herbes ; &, à l'automne suivant, on peut les transplanter dans les places qui leur sont destinées : elles se plaisent dans un sol humide, & à une situation abritée ; mais elles ne profitent pas sous l'égout des arbres. Les racines des trois premieres especes subsistent pendant plusieurs années, & produisent toujours des fleurs & des semences.

La quatrieme perfectionne rarement ses graines dans un jardin ; j'ai toujours été obligé de les faire venir des endroits ou elles naissent sans culture.

PEUPLIER BLANC, MASLE *ou* FEMELLE. *Voyez* POPULUS ALBA. *L.*

PEUPLIER NOIR, MASLE *ou* FEMELLE. *Voyez* POPULUS NIGRA. *L.*

PHACA. *Lin. Gen. Plant.* 798. *Astragaloïdes. Tourn. Inst. R. H.* 399. *tab.* 223. [*Bastard Milk-Vetch, or Astragaloïdes.*]

Vetce

PHA

Vefce de lait bâtarde *ou* Aftragaloïde.

Caractères. Le calice eft tubulé, & formé par une feuille découpée fur fes bords en cinq petites dentelures; la corolle eft papilionnacée; elle a un étendard large, ovale & érigé, avec deux aîles plus courtes que l'étendard, & obtufes; la carène eft courte & obtufe: la fleur a dix étamines, dont neuf font jointes en un corps, & l'autre eft féparée, & qui font toutes terminées par des antheres érigées. Dans le centre eft placé un germe oblong, qui foutient un ftyle en forme d'alêne, & couronné par un ftigmat fimple: ce germe devient enfuite une filique oblongue & gonflée, dont la future fupérieure eft abaiffée vers l'intérieur; de maniere qu'elle forme prefque deux cellules, qui renferment plufieurs femences en forme de rein.

Ce genre de plantes eft rangé dans la troifieme fection de la dix-feptieme claffe de LINNÉE, qui comprend celles dont les fleurs ont dix étamines jointes en deux corps.

Les efpeces font:

1º. *Phaca Bœtica, caulefcens, erecta, pilofa, leguminibus tereticymbi-formibus.* Lin. Sp. Plant. 755; Vefce de lait, avec une tige droite & velue, & des légumes cylindriques & en forme de bateau.

Phaca leguminibus rectis. Roy. Lugd.-B. 390.

Aftragalus Bœticus lanuginofus, radice ampliffimâ. Bauh. Pin. 351.

Tome V.

PHA 509

Aftragalus Bœticus. Clus. Hift. 2. p. 234.

Aftragaloïdes Lufitanica. Tourn. Inft. R. H. 39; faux Aftragale de Portugal.

2º. *Phaca Alpina, caulefcens, erecta, glabra, leguminibus oblongis, inflatis, fub-pilofis.* Lin. Sp. Plum. 1064; Vefce de lait, avec une tige droite & unie, & des légumes oblongs, velus & gonflés.

Phaca leguminibus pendulis, femi-ovatis. Gmel. Sib. 5. p. 35. f. 14.

Phaca frigida. Fl. Suec. 2. n. 657.

Aftragalus caule erecto, ramofiffimo, foliis ellipticis, hirfutis, filiquis veficariis. Hall. Helv. n. 401.

Aftragaloïdes elatior, erecta, Viciæ foliis, floribus luteis, filiquis pendulis. Amœn. Ruth. 148; Aftragaloïde droit, & très-élevé, avec des feuilles de Vefce, des fleurs jaunes, & des filiques fufpendues.

Bœtica. La premiere efpece croit naturellement en Efpagne & en Portugal; fes racines fubfiftent plufieurs années, & pénetrent très-profondément dans la terre; mais fes tiges périffent chaque automne; elles s'élevent communément à près de quatre pieds de hauteur, & deviennent ligneufes fes fleurs font rapprochées en petits paquets ou épis aux aîles des feuilles; mais elles paroiffent rarement en Angleterre, à moins que l'année ne foit très-chaude: c'eft-pourquoi ces plantes font peu eftimées, avec d'autant plus de raifon, que

Kk

leurs fleurs ne parviennent pas, une fois en sept ans, à une certaine perfection, & qu'elles ne produifent jamais de femences en Angleterre ; de maniere qu'on eft obligé de les faire venir d'ailleurs, quand on eft curieux de multiplier cette efpece.

Alpina. La feconde a des tiges unies & moins hautes que celles de la précédente : fes fleurs font plus petites, & les légumes, qui font beaucoup plus courts, pendent vers le bas.

Culture. Ces deux efpeces fe multiplient par femences : la premiere doit être femée à demeure, parce que fes racines pénetrent très-profondément dans la terre, & qu'il eft par conféquent fort difficile de les tranfplanter avec fûreté, furtout fi elles ont demeuré un tems confidérable dans le femis : on laiffe environ fix pieds de diftance entre ces plantes, afin de pouvoir labourer aifément la terre entr'elles au printems ; on les tient nettes de mauvaifes herbes, & c'eft en cela que confifte toute la culture qu'elles exigent.

PHALANGIUM. *Voy.* ANTHERICUM. *L.*

PHALARIS. *Lin. Gen. Plant.* 74. [*Canary Grafs.*] Bled ou Graine d'Oifeau.

Caracteres. Cette plante eft du nombre de celles dont la fleur eft renfermée dans le calice, & qui ont deux petites valves comprimées & en forme de bateau ; la fleur eft plus petite que le calice ; la valve

extérieure eft oblongue, en pointe, & roulée ; l'intérieure eft plus petite ; elle a trois étamines en forme de poils, terminées par des antheres oblongues, avec un germe rond, qui foutient deux ftyles comme des poils, & couronnés par des ftigmats velus ; fes femences font renfermées dans les pétales de la fleur, qui en contiennent chacun une, elles font pointues à chaque extrémité.

Ce genre de plantes eft rangé dans la feconde fection de la troifieme claffe de LINNÉE, intitulée *Triandria Digynia*, avec celles dont les fleurs ont trois étamines & deux ftyles.

Les efpeces font :

1°. *Phalaris Canarienfis, paniculá fub-ovatá, fpici-formi, carinatis glumis. Lin. Sp. Plant.* 79. *Hort. Ups. 19. Mat. Med. p.* 46 ; Bled d'Oifeau, avec des panicules ovales, en forme d'épis, & une bâle en forme de bateau.

Phalaris radice annuâ. Hort. Cliff. 23. Roy. Lugd.-B. 63. Dalib. Paris. 20.

Phalaris major, femine albo. C. B. P. 28 ; Bled d'Oifeau de Canarie, à femences blanches, Alpifte.

2°. *Phalaris Arundinacea, paniculá oblongá, ventricofá. Lin. Sp. Plant. 80* ; Bled d'Oifeau, en forme de rofeau, avec un panicule oblong & gonflé.

Gramen Arundinaceum, acerofâ glumâ Jerfeianum. D. Sher.

Arundo foliis planis, paniculá fpicatá, fpiculis confertis. Hort. Cliff. 26. Roy. Lugd.-B. 66.

PHA

Il y a plusieurs autres especes de ce genre, que l'on ne cultive jamais pour l'usage: ainsi, il n'est pas nécessaire d'en parler ici.

Canariensis. On multiplie la premiere dans quelques cantons de l'Angleterre, & particulierement dans l'Isle de Thanet en Kent, où on la regarde comme une récolte très-abondante, sur-tout pour ceux qui peuvent faire transporter cette graine par eau jusques sur les marchés de Londres, où elle a beaucoup de débit. On en seme très-peu dans les environs de Londres, encore n'est-ce que quelques curieux qui la cultivent en petite quantité, pour leur amusement. J'ai cultivé, pendant plusieurs années, cette graine, pour essai; mais, comme je n'en ai jamais semé au delà de quelques arpens, je ne puis donner beaucoup d'instructions à ce sujet: cependant je vais rapporter un détail succinct du succès que j'ai eu dans ces essais.

Ma premiere expérience, en semant cette graine sur une grande piece de terre, n'a pas été heureuse, & la récolte en a été mauvaise, pour l'avoir semée trop épaisse, & c'est ce qui arrive ordinairement à presque tous les Fermiers. Ces semences avoient bien poussé; mais les mois de Mai & de Juin s'étant trouvés trop humides, les plantes filerent, & ne produisirent que des tiges tendres & foibles; une forte pluie, qui survint au commencement d'Août, abattit toute la récolte, qui resta couchée,

PHA 511

après des pluies consécutives, & fut totalement perdue.

L'année suivante, je semai une piece de terre avec cette graine, en rangs éloignés d'un pied les uns des autres; mais elle fut semée si épaisse dans les rigoles, que les plantes filerent, & qu'une grande partie en fut abattue par les pluies du mois d'Août; quelques-unes, qui se trouverent hors des rangs, devinrent beaucoup plus fortes que les autres, resterent droites, & produisirent une bonne quantité de semences, qui parvinrent à une parfaite maturité. Cette circonstance m'encouragea à faire un autre essai: depuis, j'ai semé cette graine très-claire dans des rigoles, à un pied de distance. Lorsque les plantes ont poussé, je les ai éclaircies où elles étoient trop serrées, de maniere qu'elles se trouvoient à un pouce l'une de l'autre dans les rangs. La saison ayant été favorable, ces plantes produisirent des tiges fortes, & capables de se soutenir jusqu'à la parfaite maturité des semences: on tint la terre nette, en détruisant les mauvaises herbes, dans les intervalles, par trois houages; & la récolte fut si abondante, que je suis convaincu que cette culture seroit une des plus avantageuses pour les Fermiers, s'ils étoient assurés du débit, qu'on ne peut guere se procurer que dans la ville de Londres; mais elle seroit peu lucrative dans des terres situées à une certaine distance de cette Capitale, où l'on n'a pas la commodité de

l'eau pour en transporter la graine.

J'ai reconnu, d'après plusieurs autres essais, que trois gallons de semences (*mesure qui contient à-peu-près quatre pintes de Paris*) suffisent pour un âcre de terre, & que la meilleure méthode est de se servir d'un semoir dont les ressorts soient bien arrangés, & ne laissent sortir la graine qu'à des distances égales. En arrachant avec soin toutes les mauvaises herbes, on améliorera la récolte, & la terre en sera mieux préparée pour l'avenir.

Quand la graine est mûre, il faut la recueillir sans perdre de tems, sans quoi il en tombe bientôt une grande quantité : on la retourne deux ou trois fois, pour la sécher ; & quand la saison est bonne, elle est en état d'être battue ; ce qu'il faut faire le plutôt possible, afin d'en perdre moins.

PHASEOLOIDES. *V.* GLYCINE. *L.*

PHASEOLUS. *Tourn. Inst. R. H. tab. 232. Lin. Gen. Plant.* 777. Cette plante prend son nom de φασήλος, un bateau oblong & léger, parce que la cosse de cette plante ressemble à un bateau. [*Kidney Bean.*] Haricot, Phaseole.

Caractères. Le calice de la fleur est formé par une feuille à deux levres, dont la supérieure est dentelée au sommet, & l'inférieure est divisée en trois parties ; la fleur est papilionnacée ; l'étendard est en forme de cœur, obtus, penché & réfléchi sur les côtés ; les ailes sont ovales & de la longueur de l'étendard, la carène est étroite, en forme de spirale, torse & opposée au soleil ; la fleur a dix étamines, dont neuf sont jointes en un corps, & l'autre est séparée ; elles sont en spirale endedans du calice, & terminées par des antheres simples ; le germe, qui est oblong, comprimé, & velu, soutient un style mince, en spirale, réfléchi, & couronné par un stigmat obtus & velu. Ce germe se change dans la suite en une silique longue, épaisse, terminée en pointe obtuse, & dans laquelle sont renfermées des semences oblongues, comprimées, & en forme de rein.

Ce genre de plantes est rangé dans la troisieme section de la dix-septieme classe de LINNÉE, qui comprend celles dont les fleurs ont dix étamines jointes en deux corps. Il a divisé les plantes comprises autrefois dans ce genre en deux especes ; les unes sous le nom de *Dolichos*, & les autres sous celui de *Phaseolus*. Ces dernieres different des premieres, en ce que les parties de la génération dans les fleurs sont tordues en spirale.

Il sera inutile de rapporter toutes les variétés de ces plantes ; car l'Amérique nous en fournit chaque année une si grande quantité, qu'il seroit impossible de connoître toutes celles qui nous viennent, tant de ce pays, que d'autres contrées : d'ailleurs comme il n'est pas vraisemblable qu'on les préfere jamais à quelques-unes des anciennes especes, qui

PHA

font bien meilleures que toutes ces nouvelles variétés, pour l'usage du jardin potager, je me contenterai de parler d'abord de quelques nouvelles especes, que l'on cultive pour leurs fleurs, ou par curiosité, & je ferai ensuite mention de celles qui font les plus estimées pour la table.

Les especes font :

1°. *Phaseolus alatus*, *volubilis*, *floribus laxè spicatis*, *alis longitudine vexilli*. *Lin. Sp. Plant.* 1017; Haricot, avec une tige tournante, & des fleurs en épis lâches, dont les ailes font aussi longues que l'étendard.

Phaseolus flore purpureo, *alis amplis longè protensis*. *Hort. Elth.* 314. *tab.* 235. *f.* 303; Haricot à fleur pourpre, dont les ailes font larges & très-étendues en-dehors.

2°. *Phaseolus Caracalla*, *volubilis*, *vexillis carinâque spiraliter convolutis*. *Lin. Sp. Plant.* 1017. *Trew. Rar.* 14. *f.* 10; Haricot avec une tige tournante, dont l'étendard & la carène font en spirale.

Phaseolus radice perenni subrotundâ, *leguminibus folio longioribus*, *teretiusculis*, *glabris*. *Roy. Lugd.-B.* 367.

Phaseolus Indicus, *cochleato flore*. *Triumfet. Obs.* 93. *f.* 94. *Raii Hist.* 1890; Haricot des Indes, avec une fleur en forme d'escargot, ordinairement appelée en Portugal, *Caracalla*. La Caracolle.

3°. *Phaseol. vexillatus*, *vexillis revolutis*, *patulis*, *leguminibus linearibus*, *strictis*. *Lin. Sp. Plant.* 1017. *Jacq. Hort. f.* 102; Haricot avec une tige tour-

PHA 513

nante, un étendard étendu & tordu en arriere, & des filiques étroites & ferrées.

Phaseolus radice annuâ, *leguminibus strictis*, *erectis*, *torosis*, *linearibus*, *caule hirsuto*. *Roy. Lugd.·B.* 367.

Phaseolus flore odorato, *vexillo amplo*, *patulo*. *Hort. Elth.* 313. *t.* 234. *f.* 302. Haricot à fleurs odorantes, dont l'étendard est large & étendu.

4°. *Phaseolus farinosus*, *volubilis*, *pedunculis sub-capitatis*, *seminibus tetragono-cylindricis*, *pulverulentis*. *Hort. Upsal.* 214; Haricot avec une tige tournante, & des fleurs rassemblées en têtes & fur des pédoncules, ayant des femences quarrées, cylindriques, & de couleur de poussiere.

Phaseolus Indicus, *Hederæ folio anguloso*, *femine oblongo*, *lanuginoso*. *Raii Suppl.* 348.

Phaseolus peregrinus, *flore roseo*, *femine tomentoso*. *Niss. Act.* 1730. *p.* 577. *f.* 42; Haricot étranger, dont la fleur est de couleur de rose, & la femence cotonneuse. Le Pois velu.

5°. *Phaseolus vulgaris*, *volubilis*, *floribus racemosis*, *geminis bracteis calyce minoribus*, *leguminibus pendulis*. *Lin. Sp. Plant.* 724. *Mat. Med.* 171; Haricot avec une tige tournante, des fleurs branchues & disposées par paires, des bractées plus courtes que le calice, & des filiques pendantes. Le Haricot.

Phaseolus radice annuâ, *caule volubili*, *leguminibus pendulis*, *compressis*, *torosis*. *Roy. Lugd.-B.* 367. *Hort. Ups.* 213.

Smilax hortensis, *sivè Phaseolus major*. *Bauh. Pin.* 339.

Kk 3

514 PHA

6°. *Phaseolus Indicus , flore coccineo sivè puniceo. Mor. Hist. 2. p. 69* ; Haricot des Indes, à fleurs écarlate ou pourpre, ordinairement appelé Haricot écarlate.

Phaseolus coccineus , volubilis, floribus racemosis geminis , bracteis calyce brevioribus , leguminibus pendulis.Kniph.cent.12.n.75.

Phaseolus puniceo flore. Cornut. Canad. 184.

Alatus. La premiere espece est une plante annuelle, dont les semences ont été apportées de la Caroline, où elle croît sans culture ; ses tiges s'accrochent à tout ce qui les environne, comme celles des *Haricots* ordinaires ; elles sont velues , & s'élevent à la hauteur de quatre ou cinq pieds ; ses feuilles ressemblent à celles des *Haricots* communs, mais elles sont plus étroites : ses fleurs naissent en épis lâches sur de longs pédoncules ; elles sont grosses , & d'une couleur de pourpre , qui se change en bleu avant que les fleurs soient fanées ; elles paroissent en Juillet , & quand l'automne est chaud , elles sont remplacées par des siliques longues & étroites , qui renferment des semences petites & ovales , qui mûrissent en Octobre.

Cette espece doit être plantée dans une plate-bande chaude vers la fin du mois d'Avril ; lorsque les plantes commencent à monter, on les soutient avec des baguettes , ou on les attache à des treillages , pour les empêcher de ramper sur la terre , & on les tient cons-

PHA

tamment nettes de mauvaises herbes. Si elles sont placées contre une muraille ou une haie bien exposée , elles perfectionneront leurs semences en Angleterre ; sans cette précaution , elles manquent fréquemment , quand l'année n'est pas favorable.

Caracalla. La seconde espece croit naturellement au Brésil, d'où ses semences ont été envoyées en Europe. Cette plante est vivace , & a des tiges tournantes , qui s'élevent à la hauteur de douze ou quatorze pieds ; ses feuilles ressemblent à celles des *Haricots* communs, mais elles sont plus courtes : ses fleurs, qui naissent en épis minces , sont de couleur pourpre & d'une odeur agréable ; elles produisent des siliques minces & comprimées , qui renferment plusieurs semences dures & ovales. On multiplie cette plante par ses graines, qu'on répand sur une couche de chaleur modérée, au printems ; lorsque les plantes ont poussé , on les met avec soin dans des pots remplis de terre fraiche & légere; on les plonge dans une couche chaude, pour les aider à prendre racine , & on les accoutume ensuite par dégrés à supporter le plein air, auquel on les exposera entierement au mois de Juin ou au commencement de Juillet , en les plaçant dans une situation abritée ; lorsqu'elles ont fait assez de progrès pour remplir les pots de leurs racines , on leur en donne de plus grands, qu'on remplit avec la même terre fraiche & légere.

PHA

Elles exigent d'être fréquemment arrosées pendant l'été ; mais en hiver, quand elles font retirées dans la ferre , on ne leur donne que très-peu d'eau : elles font tendres & délicates, tandis qu'elles font jeunes, & jufqu'après le premier hiver ; mais enfuite elles n'ont plus befoin que d'être tenues à l'abri des gelées, & de jouir de beaucoup d'air , autant que la faifon le permet ; car , fans cela , leurs feuilles fe moifiroient, & leurs tendres rejettons fe flétriroient. Cette plante produit fes fleurs dans les mois de Juillet & Août ; mais elle perfectionne rarement fes femences en Angleterre : elle eft fort commune en Portugal, où les habitans la cultivent pour couvrir des loges de jardins ; car fes fleurs font belles , & ont une odeur douce & agréable : d'ailleurs ces plantes profitent très - bien en plein air dans ce pays.

Vexillatus. La troifieme efpece croît naturellement en Amérique ; on la cultive dans quelques jardins, pour la variété, quoiqu'elle ne foit pas d'une grande beauté ; on la multiplie par fes graines, qu'on place au printems fur une couche chaude ; lorfque les plantes ont pouffé, on les tranfplante dans des pots, & on les traite comme celles de la précédente. Cette plante produit fes fleurs en Juillet, & perfectionne fes femences en Septembre.

Farinofus. La quatrieme , qui a été apportée de l'Amérique , eft admife dans nos jardins, à

PHA 515

caufe de la durée de fa fleur. Cette plante eft vivace , & veut être traitée comme la précédente ; mais on ne peut la conferver en hiver, qu'en la tenant dans une ferre chaude.

Vulgaris. La cinquieme, qu'on cultive dans les jardins anglois, à caufe de la beauté de fes fleurs écarlate, a une tige rampante, qui, lorfqu'elle eft foutenue, s'éleve à la hauteur de douze à quatorze pieds ; fes feuilles font plus petites que celles des *Haricots* communs : fes fleurs naiffent en gros épis ; elles font beaucoup plus larges que celles des *Haricots* communs, & d'une couleur écarlate foncée ; fes filiques font groffes & rudes, & fes femences font d'un pourpre tacheté de noir.

Celle - ci n'exige point un traitement différent de celui qui convient à l'efpece commune ; mais il faut en foutenir les tiges avec de longs bâtons, fans quoi elles ramperoient fur la terre, & fe pourriroient bientôt.

Quoiqu'on ne cultive cette efpece que pour la beauté de fes fleurs , je penfe cependant qu'elle produit les meilleurs *Haricots* pour manger, & je fuis affuré qu'ils feront préférés à tous les autres par ceux qui en auront fait l'effai.

Coccineus. La fixieme, étant originaire des contrées méridionales de l'Amérique , ne peut profiter en Angleterre fans le fecours d'une ferre chaude ; & comme fa plus grande beauté confifte dans fes femences, moitié de couleur écarlate, & moitié noires, on

K k 4

PHA

doit plutôt se procurer les plantes des pays où elles croissent naturellement, que de les élever ici.

Je vais parler à présent des différentes especes de *Haricots* que l'on cultive pour l'usage de la table : celles-ci sont bien moins nombreuses que les autres ; & quoique plusieurs ne soient pas fort estimées, cependant comme on les multiplie, à cause du peu de soin qu'elles exigent, je les comprendrai dans le nombre des especes nutritives.

Les trois especes que l'on cultive ordinairement pour les récoltes précoces, sont les petites blanches, dont la plante est naine, ainsi que les noires, appelées Fèves des Negres, & les *Haricots* bruns ou couleur de foie. La tige de ces especes n'est jamais longue ; elles peuvent être plantées beaucoup plus serrées que les plus grandes, & elles exigent peu de soutien : ainsi, on les place fur des couches chaudes à vitrage, ou dans des pots que l'on tient dans la serre, pour les faire pousser de bonne heure au printems ; ce qui les fait préférer, pour cet usage ; mais leur saveur n'approche pas de celle de plusieurs autres : cependant comme elles peuvent donner des récoltes beaucoup plus précoces que les autres, on les cultive généralement dans les jardins. Si l'on n'a point de serres ni de châssis pour les élever, on les plante dans des plates-bandes chaudes contre des haies, des murailles, ou des palissades,

PHA

afin de les avoir quinze jours plutôt que les autres.

Les premieres, après celles-ci, sont les *Haricots de Battersea* ou de Cantorbéry, qui s'étendent au loin, & produisent leurs fleurs près de la racine : ainsi ils donnent une récolte qui dure long-tems.

Le *Haricot de Battersea* vient plutôt que celui de Cantorbéry, & ce dernier continue à produire plus long-tems. Ces deux especes ont une meilleure saveur qu'aucune des trois précédentes ; mais elles deviennent filandreuses & coriaces, quand elles commencent à grossir.

Il y a deux ou trois especes d'*Haricots* à tiges droites & érigées, qui n'ont pas besoin de soutien, parce qu'elles ne poussent point de branches rampantes ; ce qui engage les Jardiniers à en cultiver beaucoup : d'ailleurs leur récolte est très-abondante ; mais elles sont inférieures en qualité à toutes les autres, & particuliérement à celles dont les semences sont blanches & noires, parce qu'elles ont un goût fort, & qu'elles deviennent molles & farineuses lorsqu'elles sont bouillies : ainsi, les personnes de bon goût ne doivent pas les multiplier. La meilleure espece pour la table est le *Haricot écarlate*, dont on a parlé ci-dessus, ainsi que le *Haricot blanc*, de la même forme & grosseur, qui paroît être une variété de l'écarlate, parce qu'il n'en differe que par la couleur de ses fleurs & de ses semences, & qu'il lui ressemble par sa grosseur & sa saveur.

PHA

Après ceux-ci, viennent les gros *Haricots de Hollande*, qui croissent aussi haut qu'aucun des précédens ; aussi faut-il les soutenir avec des bâtons, sans quoi leurs tiges traîneroient & se gâteroient.

L'espece à fleurs écarlate est préférable à celle-ci, pour la qualité : elle est aussi plus dure ; & , quoiqu'elle ne pousse pas d'aussi bonne heure que les petites especes , cependant , comme elle continuë à produire jusqu'aux gelées de l'automne, elle mérite le premier rang sur toutes les autres , d'autant plus que ses légumes, quoique vieux , sont rarement filandreux , & ont un meilleur goût que les jeunes légumes des autres especes ; ils deviennent plus verts étant cuits ; & si on les plante à la même exposition que les *Haricots de Battersea*, on ne les attendra pas quinze jours après que les autres auront paru.

Culture. Toutes les especes de *Haricots* se multiplient par semences, qui sont trop tendres pour être mises en terre avant la fin d'Avril, en plein air ; car si le tems devient froid & humide, après qu'elles sont semées, elles se pourrissent en peu de tems, ou s'il survient quelque gelée après qu'elles ont poussé, elles en sont entierement détruites. Ainsi, la meilleure méthode pour avoir des *Haricots* printaniers, quand on n'a pas de couches vitrées pour les élever, est de les semer en rangs très-près les uns des autres , sur des couches d'une chaleur modérée, à la fin

PHA   517

de Mars ou au commencement d'Avril. Si la chaleur de la couche est suffisante pour faire pousser les plantes, elles réussiront. On garnit cette couche de cercles, afin de pouvoir la couvrir de nattes pendant les nuits & les mauvais tems. Les plantes peuvent y rester jusqu'à ce qu'elles aient poussé leurs feuilles à trois lobes ; alors on les enleve avec précaution , & on les transplante dans des plates-bandes chaudes contre des haies, des palissades ou des murailles : si alors la saison est seche, on les arrose légerement , pour les aider à former de nouvelles racines ; on les traite ensuite comme celles qui ont été élevées en pleine terre.

Ces *Haricots*, transplantés , ne croissent jamais aussi bien que ceux qu'on laisse en place, & ils ne produisent pas aussi long-tems ; mais leurs légumes sont bons à manger, au moins quinze jours plutôt que ceux de pleine terre.

Les especes que l'on met en pleine terre pour la premiere récolte, doivent être semées en Avril, dans une terre seche , à une exposition chaude, sans quoi ces semences pourrissent dans la terre ; ou , si le tems est assez favorable pour les faire pousser promptement, les plantes courront risque d'être détruites par les gelées du matin, qui surviennent souvent au commencement du mois de Mai.

Les *Haricots* destinés à la seconde récolte doivent être une des trois grosses especes ci-dessus mentionnées. En les

femant vers le milieu du mois
de Mai, ils commencent à produire avant que les printaniers
foient paffés ; & fi l'on a choifi
les écarlate, ils fourniront de
nouveaux légumes jufqu'aux
gelées de l'automne.

La meilleure façon de les
planter, eft de tracer des fillons
creux avec la houe, à trois
pieds & demi de diftance, dans
lefquels on jette les femences
à-peu-près à deux pouces les
unes des autres ; on les recouvre avec le rateau ; & lorfque les plantes ont pouffé, on
tire doucement la terre auprès,
lorfqu'elle eft feche, pour les
préferver des vents forts, &
les faire tenir droites, mais
cependant fans coûvrir de terre
les feuilles féminales, qui fe
pourriroient, fi on le faifoit,
& retarderoient beaucoup l'accroiffement des plantes : après
cela, elles n'exigent plus aucun foin que d'être foutenues,
lorfqu'elles commencent à monter, & d'être tenues nettes de
mauvaifes herbes, jufqu'à ce
que les légumes paroiffent ;
alors il faut les recueillir trois
fois par femaine ; car fi on
les laiffoit un peu trop long-
tems, les féves deviendroient
trop groffes, & les plantes s'é-
puiferoient beaucoup.

Les grandes efpeces de *Haricots* doivent être plantées en
rangs fort éloignés les uns des
autres : car comme elles s'é-
levent beaucoup, fi les rangs
étoient rapprochés, l'air ne
pourroit point circuler entr'eux, & les rayons du foleil
ne pourroient y pénétrer : ainfi, il faut laiffer au moins quatre

pieds entre chacun. Lorfque
les plantes ont atteint à-peu-
près quatre pouces de hauteur,
on leur fournit des foutiens,
auxquels elles s'attacheront,
& s'éleveront ainfi jufqu'à la
hauteur de huit à dix pieds ;
elles produiront une grande
quantité de fruits depuis le bas
jufqu'au fommet. Les Hollandois & les François confervent,
pour l'hiver, beaucoup de *Féves
de Hollande*, avec lefquelles ils
font de bons ragoûts. Il y a
des perfonnes qui élevent cette
efpece fur des couches chaudes,
pour en obtenir une récolte
printaniere : alors le feul foin
que ces plantes exigent, eft de
leur donner de la place & de
l'air, lorfque le tems eft doux,
& de ne leur procurer qu'une
chaleur modérée, parce que,
fi les couches étoient trop chaudes, elles brûleroient, ou feroient filer les plantes de maniere qu'elles ne feroient plus
propres à rien.

La méthode que l'on doit
fuivre pour établir ces couches chaudes, étant la même
que celle qui a été prefcrite
pour les couches de *Concombres*, je n'en parlerai point ici ;
j'obferverai feulement que,
quand le fumier eft également
placé, il faut le couvrir de
terre de quatre ou cinq pouces
d'épaiffeur, & en laiffer diffiper les vapeurs avant d'y planter les féves. Le tems de faire
ces couches dépend de celui où
l'on défire avoir de ces légumes ; mais le plus favorable
eft dans le commencement de
Février.

On fe procure de bonnes

PHI

femences, en laiffant plufieurs rangs fans y toucher; car fi l'on y prend quelques légumes, ceux qui refteront ne feront ni auffi beaux ni auffi bons. En automne, lorfque les féves font mures, on arrache les plantes dans un tems fec, & on les étend à l'air, pour les faire fecher, après quoi on les bat, pour en tirer les femences, que l'on conferve dans un lieu fec (1).

PHELLANDRIUM. [*Water Hemlock*] Ciguë aquatique.

Nous connoiffons deux efpeces de ce genre, dont l'une croît naturellement dans les eaux ftagnantes & dans les foffés profonds de plufieurs parties de l'Angleterre; & l'autre fe trouve fur les Alpes : mais comme on ne les admet point dans les jardins, il eft inutile de les décrire.

PHILADELPHUS. *Lin. Gen. Plant.* 540. *Syringa. Tourn. Inft. R. H.* 617. *tab.* 389. [*Syringa, Pipe-tree,* or *Mock-Orange.*] Syringa ou Seringa.

Caracteres. Le calice de la fleur eft perfiftant, & formé par une feuille découpée en cinq parties aiguës, & placées fur le germe; la corolle eft compofée de quatre ou cinq pétales ronds, unis & étendus : la fleur a au-delà de vingt étamines, en forme d'alène, inférées dans le calice, & terminées par des antheres érigées, & fillonnées

(1) La farine de ces légumes eft une des quatre réfolutives, dont on fe fert dans les cataplafmes, pour réfoudre, amollir & préparer les tumeurs à la fuppuration.

PHI 519

par quatre rainures; le germe, qui eft placé fous la fleur, foutient un ftyle mince, divifé en quatre parties, dont chacune eft couronnée par un ftigmat fimple. Ce germe fe change dans la fuite en une capfule ovale, à pointe aiguë, & à quatre cellules remplies de femences oblongues.

Ce genre de plantes eft rangé dans la premiere fection de la douzieme claffe de LINNÉE, avec celles dont les fleurs ont environ vingt étamines fixées, foit aux pétales, foit au calice de la fleur.

Les efpeces font :

1°. *Philadelphus coronarius, foliis fub-dentatis. Lin. Sp.* 671. *Hall. Helv. n.* 1100. *Duham. Arb.* 2. *f.* 83. *Kniph. cent.* 5. *n.* 65. *Mænch. Haff. n.* 401; Syringa à feuilles dentelées.

Philadelphus. Hort. Cliff. 188. *Hort. Ups.* 122.

Syringa alba, five Philadelphus Athenæi. C. B. P. 399; Syringa blanc.

Frutex coronarius. Clus. Hift. 1. *p.* 55.

2°. *Philadelphus nanus, foliis ovatis, fub-dentatis, flore folitario, pleno;* Syringa à feuilles ovales & un peu dentelées, & à fleurs folitaires & doubles.

Syringa nana, nunquam florens. Cat. Hort. Angl.; Syringa nain, qui fleurit rarement.

3°. *Philadelphus inodorus, foliis integerrimis. Lin. Sp. Plant.* 672; Syringa à feuilles entieres, avec des fleurs fans odeur.

Philadelphus flore albo majore, inodoro. Catesb. Carol. 1. *p.* 84. *tab.* 84; Syringa à plus grande fleur blanche, & fans odeur.

Coronarius. La premiere espece est, depuis long-tems, cultivée dans les jardins anglois comme un arbrisseau à fleurs ; mais on ne sait pas bien où elle croît naturellement : elle pousse un grand nombre de tiges minces, & couvertes d'une écorce grise, qui produisent un nombre de branches courtes, latérales, & garnies de feuilles ovales & en forme de lance : celles des jeunes branches ont trois pouces & demi de longueur sur deux de large au milieu ; mais elles sont plus étroites vers les deux extrémités, terminées en pointe aiguë, & dentelées sur leurs bords ; leur surface est rude, d'un vert foncé en dessus, & d'un vert pâle en-dessous, & elles ont un goût de *Concombre* frais ; elles naissent opposées sur de fort courts pétioles : les fleurs, qui croissent en paquets lâches sur les parties latérales & aux extrémités des branches, ont chacune un pédoncule court & distinct, quatre pétales ovales, étendus, & un grand nombre d'étamines qui environnent le style. Ces fleurs sont blanches ; elles répandent une odeur forte, & presque semblable à celle de la Fleur d'*Orange*, que bien des gens ont peine à soutenir de près ; elles paroissent à la fin du mois de Mai, & se succedent durant une grande partie du mois de Juin : mais elles produisent rarement de bonnes semences en Angleterre. Cet arbrisseau s'éleve à la hauteur de sept à huit pieds.

Il y a une variété de cette plante, à feuilles panachées, que l'on conserve dans quelques jardins : mais ce panache disparoît généralement, quand les plantes sont en bon état & vigoureuses.

Nanus. La seconde espece s'éleve rarement au-dessus de trois pieds de hauteur ; ses feuilles sont plus courtes que celles de la précédente, presque ovales, & seulement un peu dentelées sur leurs bords : ses fleurs sont solitaires, & sortent sur les parties latérales des branches ; elles ont un double & triple rang de pétales de la même grandeur que ceux des autres, & ont la même odeur : mais comme cette espece fleurit très-rarement, on n'en fait pas beaucoup de cas.

Ces deux especes sont dures, & profitent dans presque tous les sols & à toutes les expositions ; mais elles s'élevent à une hauteur plus considérable dans un sol fertile & léger, que dans une terre forte. On les multiplie ordinairement au moyen de rejettons que leurs racines produisent en grande abondance ; on les sépare des vieilles plantes en automne, & on les met en pépiniere, dans laquelle on les laisse un ou deux ans, pour leur donner le tems d'acquérir de la force, après quoi on les transplante à demeure dans les quartiers déserts, parmi d'autres arbrisseaux du même crû.

Inodorus. La troisieme espece croît naturellement dans la Caroline ; mais elle est encore très-rare en Europe : elle s'é-

leve en tige d'arbriffeau à feize pieds environ de hauteur, & pouffe latéralement des branches minces, oppofées, & garnies de feuilles unies, de la même forme de celles du *Poirier*, entieres, poftées fur des pétioles affez longs, & oppofées : les fleurs fortent aux extrémités des branches ; elles font groffes, & ont chacune quatre pétales ovales & étendus, avec de gros calices, compofés de quatre feuilles à pointe aiguë ; les pétales font blancs, & renferment un grand nombre d'étamines, terminées par des antheres jaunes. Ces fleurs font remplacées par des capfules ovales, remplies de petites femences.

Cet arbriffeau n'eft pas commun en Angleterre, parce qu'il eft difficile de l'élever de femences. J'ai femé deux ou trois fois de ces graines qui m'avoient été envoyées de la Caroline par le Docteur DALE, mais toujours fans aucun fuccès ; ce qui eft auffi arrivé à d'autres perfonnes : mais quand on poffede quelques-unes de ces plantes, on peut les multiplier par marcottes. Le même Docteur DALE m'a auffi envoyé un de ces arbriffeaux, qui a profpéré dans le jardin de Chelféa pendant près de deux ans : quelques-unes des branches qui avoient été marcottées, poufferent des racines ; mais elles furent totalement détruites par le froid de 1740.

PHILLYREA. *Tourn. Inft. R. H.* 596. *tab.* 367. *Lin. Gen. Plant.* 16. [*Phyllirea*, or Mock-

Privet.] Phillyrea. Filaria.

Caractères. Le calice de la fleur eft petit, perfiftant, & formé par une feuille divifée en cinq fegmens fur fes bords ; la corolle, qui eft monopétale, a un tube fort court, découpé en cinq parties, tournées en arriere : la fleur a deux étamines courtes, oppofées, & terminées par des antheres fimples & érigées ; fon germe eft rond, & foutient un ftyle mince, auffi long que les étamines, & couronné par un ftigmat épais. Ce germe fe change dans la fuite en baie globulaire, & à une cellule groffe & ronde.

Ce genre de plantes eft rangé dans la premiere fection de la feconde claffe de LINNÉE, avec celles dont les fleurs ont deux étamines & un ftyle.

Les efpeces font :

1°. *Phillyrea lati-folia, foliis ovato-lanceolatis, integerrimis* ; Filaria avec des feuilles ovales, entieres & en forme de lance.

Phillyrea lati-folia, lævis. C. B. P. 476 ; Filaria à feuilles larges & liffes, ordinairement appelé *le vrai Filaria.*

2°. *Phillyrea media, foliis ovatis, fub-integerrimis. Lin. Sp.* 10 ; Filaria avec des feuilles ovales & prefque entieres.

Phillyrea folio læviter ferrato. C. B. P. ; Filaria à feuilles légerement fciées, appelé *Filaria à feuilles moyennes.*

Phillyrea tertia. Clus. Hift. 1. *p.* 52.

3°. *Phillyrea fpinofa, foliis cordato-ovatis, ferratis. Hort. Cliff.* 4. *Hort. Ups.* 4, *Roy. Lugd.-B.*

398 ; Filaria à feuilles ovales, en forme de cœur, & fciées.

Phillyrea lati-folia fpinofa. C. B. P. 476 ; Filaria épineux, à larges feuilles.

Phillyrea lati-folia. Lin. Syft. Plant. *1. 1. p. 18. Sp. 3.*

4°. *Phillyrea Liguftri-folia, foliis lanceolatis, integerrimis.* Hort. Cliff. 4 ; Filaria avec des feuilles en forme de lance, & entieres.

Phillyrea folio Liguftri. C. B. P. 476 ; Filaria à feuilles de Troëne.

5°. *Phillyrea Oleæ-folia, foliis lanceolato ovatis, integerrimis, floribus confertis axillaribus;* Filaria avec des feuilles ovales, entieres & en forme de lance, & des fleurs raffemblées en paquets fur les côtés des branches.

Phillyrea Oleæ Ephefiacæ folio. Pluk. Alm. 295. Phyt. tab. 310. fig. 3 ; Filaria à feuilles d'Olivier.

6°. *Phillyrea angufti-folia, foliis lineari-lanceolatis, integerrimis, floribus confertis axillaribus.* Hort. Cliff. 4. Roy. Lugd.-B. 398. Kniph. Orig. cent. X. n. 69 ; Filaria à feuilles étroites, entieres, & en forme de lance, avec des fleurs raffemblées en paquets fur les côtés des branches.

Phillyrea angufti-folia prima. C. B. P. 476 ; le premier Filaria, à feuilles étroites.

Phillyrea. 4. 5. Clus. Hift. 1. p. 52.

7°. *Phillyrea Roris marini folio, foliis linearibus, integerrimis;* Filaria avec des feuilles fort étroites & entieres.

Phillyrea angufti-folia fecunda.

C. B. P. 476 ; le fecond Filaria, à feuilles étroites, ordinairement appelé *Filaria à feuilles de Romarin.*

Lati-folia. La premiere efpece eft la plus commune dans les jardins anglois, où elle eft connue fous le nom de *vrai Filaria*, pour la diftinguer de l'*Alaterne*, appelée fimplement par les Jardiniers, *Filaria.* Elle s'éleve à la hauteur de dix-huit ou vingt pieds, avec une tige forte, droite, & divifée en plufieurs branches couvertes d'une écorce liffe & grifâtre, & garnies de feuilles ovales, en forme de lance, oppofées, entieres, fermes, d'un vert clair, d'un pouce & demi de longueur fur un de largeur, & poftées fur de courts pétioles : fes fleurs fortent des aiffelles de la tige à chaque côté ; elles font d'un blanc herbacé, & difpofées en pétits paquets : elles paroiffent dans le mois de Mai ; mais comme elles font petites, elles n'ont pas grande apparence ; elles font remplacées par des baies globulaires, & à une cellule qui renferme une fimple femence de la même forme.

Media. La feconde efpece s'éleve à la même hauteur que la premiere : mais fes branches s'étendent davantage, & font couvertes d'une écorce toncée en couleur ; fes feuilles font ovales, d'un vert plus forcé, longues de plus de deux pouces, larges d'environ un pouce & demi, un peu fciées fur leurs bords, placées fur de courts pétioles, & oppofées : fes fleurs, qui for-

tent des aiſſelles des branches, ſont d'un blanc herbacé ; elles paroiſſent vers le même tems que les précédentes, & produiſent des baies de la même forme.

Spinoſa. La troiſieme s'éleve avec une tige droite à la même hauteur que les deux premieres, & pouſſe pluſieurs branches érigées, couvertes d'une écorce griſe, & garnies de feuilles ovales, en forme de cœur, d'un pouce & demi environ de longueur ſur un de large, fermes, d'un vert luiſant, & diviſées ſur leurs bords en dentelures, dont chacune eſt terminée par une épine. Les fleurs & les ſemences de cette eſpece reſſemblent à celles des deux précédentes.

Liguſtri-folia. La quatrieme eſt d'un crû plus bas qu'aucune des précédentes ; car elle ne s'éleve guere qu'à huit ou dix pieds de hauteur : ſes branches ſont plus foibles, & s'étendent davantage ; elles ſont couvertes d'une écorce brune & liſſe, & ſont garnies de feuilles roides en forme de lance, à peine de deux pouces de longueur ſur ſix lignes de largeur au milieu, terminées en pointe à chaque extremité, d'un vert clair, feſſiles aux branches, & oppoſées : ſes fleurs ſont produites en petits paquets aux aiſſelles des branches ; elles ſont petites, & plus blanches que celles des eſpeces précédentes ; elles paroiſſent vers le même tems, & produiſent de petites baies qui mûriſſent en automne.

Olea-folia. La cinquieme s'éleve à-peu-près à la même hauteur que la quatrieme : ſes branches ſont plus forces, & s'étendent plus loin ; leur écorce eſt d'une couleur plus claire ; les feuilles ſont roides, unies, entieres, poſtées ſur des pétioles fort courts, & oppoſées ; elles ſont d'un vert luiſant, & terminées en pointe : les fleurs, qui ſortent en paquets ſur des pédoncules aſſez longs aux aiſſelles des jeunes branches, ſont petites & blanches ; elles paroiſſent en même tems que celles des autres eſpeces, & produiſent des baies rondes, qui mûriſſent en automne.

Anguſtifolia. La ſixieme s'éleve avec une tige ligneuſe, à la hauteur de dix à douze pieds, & pouſſe des branches oppoſées, couvertes d'une écorce brune, tachetée de blanc, & garnies de feuilles unies, roides, étroites, en forme de lance, entieres, feſſiles, d'un pouce & demi environ de longueur ſur ſix lignes de large au milieu, terminées en pointe à chaque extremité, d'un vert clair, & tournées vers le haut : ſes fleurs ſortent en gros paquets à chaque nœud des branches, auxquelles elles ſont feſſiles ; elles ſont placées comme des fleurs verticillées, & entourent preſque la tige ; elles ſont petites & blanches ; elles paroiſſent en même tems que celles de la précédente, & produiſent de petites baies qui mûriſſent en automne.

Roris-marini folio. La ſeptieme eſt d'un crû plus bas que

les précédentes ; car elle ne s'éleve guere qu'à quatre ou cinq pieds de hauteur : ses branches sont plus minces, oppoſées, & diſpoſées ſéparément ; ſes feuilles ſont d'un vert foncé, roides, entieres, d'un pouce environ de longueur ſur une ligne & demie de largeur, & ſeſſiles aux branches ; ſes fleurs ſont petites, blanches, & diſpoſées en paquets ſur les parties latérales des branches. Les baies de cette eſpece ſont fort petites, & mûriſſent rarement en Angleterre.

Culture. Toutes ces plantes croiſſent naturellement dans la France méridionale, en Eſpagne & en Italie ; mais elles ſont aſſez dures pour profiter en plein air en Angleterre, & ne ſont jamais endommagées, à moins que les hivers ne ſoient très-rudes ; ſouvent alors leurs feuilles tombent, & quelques-unes des branches les plus foibles périſſent : mais leurs racines repouſſent de nouveaux rejettons dans l'été ſuivant ; de ſorte qu'il y a peu d'arbres verts plus durs que ceux-ci, & qui méritent davantage d'être cultivés pour l'agrément.

On plantoit autrefois ces arbres ou contre des murailles, auxquelles on les paliſſadoit, pour les couvrir, ou on les plaçoit à plein vent en taillant leurs têtes en boule ou en pyramide, comme la plupart des arbres toujours verts : mais l'ancien goût étant rejetté, on a généralement aboli les arbres verts, & on n'en a conſervé que quelques eſpeces ; ce qui eſt cauſe que pluſieurs de ces arbres toujours verts ont été preſque entierement perdus en Angleterre pendant pluſieurs années, & qu'on a eu beaucoup de peine à ſe les procurer depuis ; mais, ſuivant la méthode actuelle de diſpoſer les arbres & arbriſſeaux toujours verts, ils ſont un très-bel effet dans les jardins, ſurtout en hiver, quand les autres arbres ſont deſtitués de feuilles.

Il y en a d'autres qui croiſſent naturellement en Eſpagne & en Italie ; mais ceux-ci ſont les ſeuls qu'on cultive dans les jardins anglois, & même pluſieurs de ces derniers ont été regardés comme n'étant que des variétés accidentelles, produites de ſemences. Je ſuis cependant porté à croire qu'ils ſont ſpécifiquement différens ; car je les ai élevés preſque tous avec des graines qui m'avoient été envoyées d'Italie, où les eſpeces avoient été ſoigneuſement diſtinguées en les recueillant, & je ne les ai jamais vu varier ; ce qui me fait croire que les ſemences qui ont produit deux ou trois eſpeces différentes, & que l'on a priſes pour des variétés, avoient été recueillies ſans aucun ſoin ſur différens arbres.

Les trois premieres eſpeces ſont fort propres à être entremêlées avec d'autres arbres toujours verts du même crû, dans des pieces informes de parcs, ou ſur les bords des bois remplis d'arbres qui perdent leurs feuilles ; l'ombrage épais de ces arbres toujours verts y produira

PHI

produira un effet très-agréable, ainsi que le vert de leurs feuilles, plus brillant que celui des autres. En hiver, lorsque tous les arbres seront dépouillés, ceux-ci produiront un bel effet, & fourniront une retraite aux oiseaux. On peut les dresser en tiges de maniere qu'ils soient hors de la portée du bétail : on les éleve dans des endroits enfermés ; & lorsqu'ils sont devenus assez forts, on les place à demeure.

Les autres especes sont d'un crû plus bas, & doivent être placées dans des jardins, ou dans quelqu'autre lieu fermé, pour y être à l'abri des attaques du bétail, des lievres, lapins, &c., qui les détruiroient bientôt.

On multiplie ces plantes par semences ou par marcottes. Cette derniere méthode est la plus prompte & la plus généralement pratiquée en Angleterre. Le meilleur tems pour les marcotter, est l'automne. On laboure la terre autour des tiges destinées à être couchées, pour la rendre meuble : on choisit ensuite une partie lisse de la branche dans laquelle on fait une fente en montant, comme si on marcottoit des *Œillets* ; on la courbe ensuite doucement vers la terre, où l'on fait un creux avec la main pour la recevoir ; on y place la partie entaillée de maniere que la fente soit ouverte ; on l'assujettit avec un bâton fourchu, pour qu'elle ne puisse se déplacer ; on recouvre cette partie de terre jusqu'à trois pouces d'épaisseur, & on tient

Tome V.

PHI 525

l'extrémité érigée. On arrache exactement les mauvaises herbes au printems & durant l'été suivant, parce qu'elles empêcheroient ces marcottes de prendre racine, si on les laissoit croître.

Presque toutes ces plantes seront enracinées pour l'automne suivant ; alors on pourra les enlever & les placer dans une pépiniere, où on les dressera pendant trois ou quatre ans dans la forme qu'on voudra leur donner, en observant de labourer la terre entre les rangs, de couper chaque année les racines autour des plantes, pour leur faire pousser de fortes fibres, & les rendre plus faciles à enlever en mottes : on aura soin aussi de supporter leurs tiges avec des piquets, pour les dresser, & empêcher qu'elles ne se courbent ; car, sans cela, elles deviendroient fort désagréables à la vue.

Quand les plantes ont été ainsi traitées pendant trois ou quatre années, on peut les transplanter à demeure. Le tems le plus favorable pour cette opération, est la fin de Septembre ou le commencement d'Octobre ; mais pour les enlever, il faut creuser la terre autour de leurs racines, & couper toutes celles qui sont fortes, & qui s'étendent à une grande distance, afin de pouvoir mieux les enlever en motte; car, sans cela, elles courent risque de périr. Lorsqu'elles sont placées à demeure, on répand du terreau sur la terre, pour la tenir fraîche, & on

l l

fixe leurs tiges contre des piquets, jusqu'à ce qu'elles foient
bien établies, afin qu'elles ne
foient ni déterrées ni déplacées
par les vents; ce qui romproit
les fibres nouvellement pouf-
fées, & endommageroit beaucoup les plantes. Ces arbres
fe plaifent dans un fol médiocre, ni trop humide, ni trop
ferme, ni trop fec. Ce dernier
eft cependant préférable au
premier, pourvu qu'il foit frais.

Comme les efpeces de mar-
cottes à petites feuilles ne prennent racine qu'au bout de deux
ans, il ne faut pas les remuer
avant ce tems; car on les retarderoit beaucoup, fi on les
fortoit de terre fans racines.

Lorfqu'on veut les multiplier
par femences, il faut mettre
ces graines en terre en automne, auffi-tôt qu'elles font mûres; car elles ne croiffent pas
dans la première année, fi on
ne les feme qu'au printems.
Elles réuffiffent mieux dans des
pots ou des caiffes remplies de
terre légere & marneufe; on
les place fous un châffis de
couche, où elles foient à l'a-
bri de la gelée, & où l'on
puiffe leur procurer de l'air
dans les tems doux. En les femant de bonne heure en automne, les plantes paroitront
au printems; mais fi elles ne
pouffent pas dans ce tems, on
plonge les pots qui les contiennent dans une plate-bande,
à l'expofition du levant, où
elles puiffent jouïr feulement
du foleil du matin, & on les
y laiffera durant tout l'été fuivant: pendant tout ce tems,
on tiendra toujours les pots

nets de mauvaifes herbes; en
automne, on les remettra fous
un châffis, pour les abriter des
froids; & au printems fuivant,
les plantes pousferont certainement, fi les femences font
bonnes. Vers le milieu d'A-
vril, on replonge les pots dans
une plate bande, à l'expofition
du levant, pour empêcher la
terre de fe deffécher; ce qui
arrive ordinairement quand on
fe contente de les pofer fur le
fol; & dans ce cas, on eft
obligé de les arrofer fouvent,
quoique cela foit contraire aux
plantes. A la Saint-Michel fuivante, on tire avec foin ces
plantes hors des pots, & on
les met dans une pépiniere,
dont on couvre la furface avec
du vieux tan, pour les préferver de la gelée. Si l'hiver eft
rude, on les couvre de nattes,
& on les traîte enfuite comme
les plantes de marcottes.

PHILLYREA DU CAP. *V.*
MAUROCENIA PHILLYREA.

PHLOMIS. *Tourn. Inft. R.
H.* 177. *tab.* 82. *Lin. Gen. Plant.*
642. Φλομὶς, ainfi appelée de
Φλέγω, brûler, parce que, dans
l'ancien tems, les payfans fe
fervoient de cette plante pour
s'éclairer. [*The fage-tree, or
Jerufalemfage.*] Arbre de fauge
ou fauge de Jérufalem; Bouillon fauvage *ou* Sauge en arbriffeau.

Caractères. Le calice eft perfiftant, & formé par une feuille; il a un tube oblong & à
cinq angles; la corolle eft monopétale & papillonnacée; fon
tube eft oblong; la levre fupérieure eft ovale, fourchue
& réfléchie; l'inférieure eft

PHL

découpée en trois fegmens, dont celui du milieu eft large & obtus : la fleur a quatre étamines cachées fous la levre inférieure, dont deux font plus longues que les autres, & qui font toutes terminées par des antheres oblongues; fon germe, qui eft divifé en quatre parties, foutient un ftyle auffi long que les étamines, & couronné par un ftigmat aigu; divifé en deux parties : ce germe fe change dans la fuite en quatre femences oblongues, quarrées & renfermées dans le calice.

Ce genre de plantes eft rangé dans la premiere fection de la quatorzieme claffe de Linnée, qui comprend celles dont les fleurs ont deux étamines courtes, & deux plus longues, & dont les femences font nues, & poftées dans le calice.

Les efpeces font :

1°. *Phlomis fruticofa*, *foliis fub-rotundis*, *tomentofis*, *crenatis*, *Involucris lanceolatis*, *caule fruticofo*. *Lin. Sp. 818. Kniph. cent. 1. n. 61. Sabbat. Hort. Rom. 3. f. 15*; Phlomis avec des feuilles prefque rondes, cotonneufes, & crènelées, ayant des enveloppes en forme de lance, & une tige d'arbriffeau.

Verbafcum latis Salviæ foliis. Bauh. Pin. 240.

Phlomis fruticofa, *Salviæ folio latiori & rotundiori. Tourn. Inft. 177.*; Phlomis en arbriffeau, à feuilles de Sauge, plus larges & plus rondes; Bouillon fauvage ou Sauge en arbre.

2°. *Phlomis angufti-folia*, *foliis ovato-lanceolatis*, *tomentofis*, *integerrimis*, *caule fruticofo*; Phlo-

mis avec des feuilles ovales, en forme de lance, entieres & cotonneufes, & une tige d'arbriffeau.

Phlomis fruticofa, *Salviæ folio, longiori & angufliori. Tourn. Inft. 177*; Phlomis à feuilles de Sauge, plus longues & plus étroites.

3°. *Phlomis lati-folia*, *foliis oblongo-ovatis*, *petiolatis*, *tomentofis*, *floribus capitatis*, *caule fruticofo*; Phlomis à feuilles oblongues, ovales, pétiolées & cotonneufes, à fleurs rapprochées en tête, & à tige d'arbriffeau.

Phlomis lati-folia, *capitata*, *lutea*, *grandi-flora. Hort. Elth. 316. t. 237. f. 306*; Phlomis à feuilles larges, avec de grandes fleurs jaunes qui croiffent en têtes.

4°. *Phlomis herba venti*, *involucris fetaceis*, *hifpidis*, *foliis ovato-oblongis*, *fcabris*, *caule herbaceo. Hort. Ups. 171. Sauv. Monfp. 152. Pall. It. 1. p. 154. Sabb. Hort. 3. f. 17*; Phlomis avec des enveloppes garnies de piquans, & velues, des feuilles ovales, oblongues & rudes, & une tige d'arbriffeau.

Phlomis Narbonenfis, *Hormini folio*, *flore purpurafcente. Tourn. Inft. R. H. 178*; Phlomis de Narbone, à feuilles d'Ormin, & à fleur pourpre.

Marrubium nigrum longi-folium. Bauh. Pin. 230.

5°. *Phlomis tuberofa*, *involucris hifpidis*, *fubulatis*, *foliis cordatis*, *fcabris*, *caule herbaceo. Hort. Ups. 171. Gmel. It. 2. præf. p. 6. Pall. It. 1. p. 319. Kniph. cent. 4. n. 61*; Phlomis avec des enveloppes ve-

528 **PHL**

lues & en forme d'alêne, des feuilles rudes & en forme de cœur, & une tige herbacée.

Phlomis Urticæ foliis, glabra. Amæn. Ruth. 49. Phlomis à feuilles d'Ortie, & unie.

Galeopfis maxima, foliis Hormini. Buxb. cent. 1. *p.* 4. *f.* 6.

6°. *Phlomis lychnitis, foliis lanceolatis, tomentofis, floralibus ovatis, involucris fetaceis, lanatis. Lin. Sp. Plant.* 585 ; Phlomis avec des feuilles en forme de lance, & cotonneufes ; les florales étant ovales, & les enveloppes garnies de piquans, & laineufes.

Phlomis lychnitis. Clus. Hift. 27 ; Phlomis à feuilles étroites.

Verbafcum anguftis Salviæ foliis. Bauh. Pin. 240.

7°. *Phlomis purpurea, foliis ovato-lanceolatis, crenatis, fubtùs tomentofis, involucris fetaceis ;* Phlomis avec des feuilles ovales, en forme de lance, crenelées, & cotonneufes deffous, & des enveloppes garnies de piquans.

Verbafcum fub-rotundo Salviæ folio. Bauh. Pin. 240.

Phlomis fruticofa Lufitanica, flore purpurafcente, foliis acutioribus, Tourn. Inft. 178 ; Phlomis de Portugal en arbriffeau, avec une fleur pourpre, & des feuilles à pointe aiguë.

Salvia fruticofa, Cifti-folio haud incano, floribus purpureis. Pluk. Alm. 329. *t.* 57. *f.* 6.

8°. *Phlomis Samia, foliis cordatis, acutis, fubtùs tomentofis, involucris ftriftis, tri-partitis ;* Phlomis avec des feuilles en forme de cœur, terminées en pointe aiguë,&cotonneufes en-deffous

& des enveloppes rapprochées, & divifées en trois parties.

Phlomis involucris & radiis, fubulatis, ftriftis. Hort. Cliff. 315.

Phlomis Samia herbacea, folio Lunariæ. Tourn. Cor. 10 ; Phlomis de Samos, herbacé, à feuilles de Lunaire.

9°. *Phlomis Orientalis, foliis cordatis, rugofis, fubtùs tomentofis, involucris lanatis, caule herbaceo ;* Phlomis avec des feuilles en forme de cœur, rudes & cotonneufes en - deffous, ayant des enveloppes laineufes, & une tige herbacée.

Phlomis Orientalis lutea, herbacea lati-folia verticillata. Phil. Tranf. vol. 34; Phlomis Oriental, jaune & herbacé, avec de larges feuilles, & des fleurs verticillées.

10°. *Phlomis flavefcens, foliis lanceolatis, crenatis, fubtùs tomentofis, involucris lanatis, caule fruticofo ;* Phlomis avec des feuilles crenelées, en forme de lance, & cotonneutes en-deffous, des enveloppes laineufes, & une tige d'arbriffeau.

Phlomis angufti - folia lutea, cymis flavefcentibus. Sherard. Phil. Tranf. n°. 376 ; Phlomis jaune, à feuilles étroites, avec des fommets jaunâtres.

11°. *Phlomis Niffolii, foliis radicalibus cordatis, utrinquè tomentofis, villofis. Lin. Sp. Plant.* 585 ; Phlomis dont les feuilles radicales font en forme de cœur, cotonneufes & velues fur chaque face.

Phlomis Orientalis, foliis auriculatis, incanis, flore luteo. Niffol. ; Phlomis Oriental, avec des feuilles oreillées & velues,

PHL

produifant une fleur jaune.

12°. *Phlomis ferruginea, invo-lucris lanceolatis, foliis cordatis, fubtùs tomentofis, caule fruticofo;* Phlomis avec des enveloppes en forme de lance, des feuil-les en cœur, cotonneufes en-deffous, & une tige d'arbrif-feau.

Phlomis Hifpanica, fruticofa, candidiffima, flore ferrugineo. Tourn. *Inft.* 178 ; Phlomis d'Ef-pagne, en arbriffeau, & très-blanc, avec une fleur couleur de fer.

13°. *Phlomis rotundi-folia, in-volucris fubulatis, foliis cordato-ovatis, fubtùs tomentofis, caule fruticofo* ; Phlomis avec des en-veloppes en forme d'alêne, des feuilles en forme de cœur, ovales, & cotonneufes, & une tige d'arbriffeau.

Phlomis fruticofa, flore purpu-reo, foliis rotundioribus. Tourn. *Inft.* 178 ; Phlomis en arbrif-feau, avec une fleur pourpre, & des feuilles rondes.

14°. *Phlomis laciniata, foliis alternatim pinnatis, foliolis lacl-niatis, calycibus lanatis.* Lin. *Sp. Plant.* 585 ; Phlomis avec des feuilles ailées alternativement, & dont les lobes font décou-pés, & des calices laineux aux fleurs.

Phlomis Orientalis, foliis laci-niatis. Tourn. *Cor.* 10 ; Phlomis Oriental, à feuilles découpées.

Fruticofa. La premiere efpe-ce croît naturellement en Ef-pagne & en Sicile ; elle a une tige d'arbriffeau affez épaiffe, couverte d'une écorce defferrée, de cinq ou fix pieds de hau-teur, & divifée en plufieurs branches irrégulieres, coton-

PHL 529

neufes & quarrées, lorfqu'elles font jeunes, mais qui devien-nent ligneufes enfuite : leurs nœuds font affez éloignés, & fur chacun font deux feuilles rondes, oppofées, poftées fur de courts pétioles, & cotonneu-fes en-deffous : les fleurs for-tent autour des tiges en têtes épaiffes, & verticillées ; elles font jaunes, & ont deux levres, dont la fupérieure eft fourchue & penchée fur celle du bas, qui eft divifée en trois parties, dont celle du milieu eft large, & s'étend au-delà de deux pe-tits fegmens de côté. Ces fleurs paroiffent dans les mois de Juin, Juillet & Août, & produifent rarement des femences en An-gleterre.

Angufti-folia. La feconde ef-pece a une tige d'arbriffeau, comme la premiere ; mais elle ne s'éleve pas auffi haut ; fes branches font plus foibles ; fes feuilles font ovales, en forme de lance, plus longues, plus étroites, & plus rondes aux deux extrémités que celles de la précédente : fes têtes de fleurs font verticillées & plus petites ; mais les fleurs font de la même forme & couleur, & elles pa-roiffent dans le même tems.

Ces deux efpeces font de-puis long-tems, cultivées dans les jardins Anglois fous le nom d'*Arbre de Sauge* ou *Sauge de Jérufalem* On confervoit autre-fois ces plantes dans des pots, que l'on enfermoit en hiver avec les autres plantes exoti-ques ; mais, depuis quelques années, on les met en pleine terre, où elles ne font pas fou-vent endommagées par le froid,

L l 3

à moins que les hivers ne soient fort rudes, & elles augmentent la variété dans les lieux écartés d'un jardin, quand elles y sont entremêlées avec d'autres arbrisseaux du même crû : car, comme elles conservent pendant toute l'année leurs feuilles blanches & velues, elles produisent un bel effet pendant l'hiver ; d'autant plus que leurs fleurs jaunes se succedent durant une grande partie de l'été.

Ces plantes exigent un sol sec & une situation chaude & abritée, sans quoi elles ne subsisteroient pas en plein air. On peut les placer parmi les *Cistes* de différentes especes : le *Treffle* en croissant, le *Ciste* toujours vert, l'*Absinthe* en arbre, & quelques autres arbrisseaux exotiques des mêmes contrées, auxquels il faut une exposition chaude & un sol sec, étant trop tendres pour des lieux ouverts & exposés aux vents froids & impétueux ; & comme ils ne sont pas de longue durée, il vaut mieux les tenir séparés de ceux qui subsistent plusieurs années. Ceux-ci ne durent guere que douze on quatorze ans dans une terre seche, & au plus la moitié de ce tems dans un terrein froid, humide, & peu abrité.

On les multiplie par boutures, qui poussent de bonnes racines dans l'espace de deux mois ou six semaines, si elles sont plantées en Avril, dans une terre légere, avant que la végétation commence dans les plantes sur lesquelles elles ont été prises ; on les tient à l'ombre avec des nattes, & on les arrose légerement quand la terre est seche ; on les enleve ensuite avec précaution, & on les place dans une pépiniere, où elles pourront rester pendant un an, pour être transplantées après ce tems dans les endroits qui leur sont destinés ; car elles ne souffrent pas d'être déplacées, lorsqu'elles sont plus âgées.

Lati-folia. La troisieme espece a une tige d'arbrisseau comme la précédente, mais beaucoup plus basse ; car elle ne s'éleve guere au-dessus de trois pieds & demi de hauteur ; elle pousse de tous côtés des branches garnies de feuilles blanches, plus larges que celles des précédentes, d'une forme oblongue & ovale, postées sur des pétioles assez longs, & plus blanches que celles de la seconde : ses fleurs croissent en grosses têtes, qui terminent toujours les branches ; elles sont plus larges que celles des especes précédentes ; leur levre supérieure est fort velue, & elles paroissent en même tems que les autres. Ces plantes sont également dures, & peuvent être multipliées par boutures, suivant la méthode qui vient d'être prescrite.

Herba venti. La quatrieme espece croît naturellement dans la France Méridionale & en Italie ; elle a une racine vivace, & une tige annuelle qui s'éleve à la hauteur d'environ deux pieds, & périt en automne. Quand les racines sont grosses, elles poussent un

grand nombre de tiges quarrées, couvertes d'un duvet velu, & garnies de feuilles rudes, oblongues, ovales, opposées & seffiies. Les fleurs, qui croissent en têtes autour des tiges, ont des enveloppes velues & piquantes ; elles sont d'un pourpre brillant, & font un bel effet. Elles paroissent dans le même tems que celles de la précédente ; mais elles ne produisent point de semences.

On peut multiplier cette espece, en divisant ses racines en automne, quand les tiges commencent à périr, parce qu'alors elles ont le tems de bien s'établir dans la terre, avant que les gelées commencent ; mais il ne faut les diviser que chaque trois ou quatre ans, si l'on veut avoir beaucoup de fleurs. Cette plante est dure, & peut être placée dans des lieux découverts, pourvu que le sol n'en soit point humide.

Tuberosa. La cinquieme espece est originaire de la Tartarie ; elle a une racine vivace ; ses tiges sont de couleur pourpre à quatre angles, de cinq ou six pieds de hauteur, & garnies de feuilles en forme de lance, opposées, de six pouces de longueur sur trois de large à leur bâse, terminées en pointe aiguë, & profondément dentelées sur leurs bords : ses fleurs croissent en têtes autour des tiges ; leurs enveloppes sont en forme d'alène, & garnies de poils piquans : elles paroissent dans les mois de Juin & Juillet, & produisent des semences qui mûrissent en Septembre : bientôt après, leurs tiges périssent ; mais les racines durent plusieurs années. On multiplie cette espece par ses graines, qu'il faut semer au printems sur une plate-bande à l'exposition de l'est. Quand les plantes, poussent, on les tient nettes de mauvaises herbes ; & en automne, on les place à demeure, où elles produiront des fleurs & des semences dans l'été suivant. Cette plante est fort dure, & profite dans presque tous les sols & à toutes les expositions.

Lychnitis. La sixieme, qui croît naturellement dans la France méridionale, en Espagne & en Italie, pousse de ses racines des feuilles longues, étroites, cotonneuses, en touffes, enveloppées à leur bâse par une couverture commune, douces au toucher, & couchées sur la terre ; ses tiges sont minces, de deux pieds de longueur, & chargées de nœuds éloignés les uns des autres, dont chacun produit deux feuilles ovales & opposées, qui embrassent la tige de leurs bâses : les têtes des fleurs sont entourées de ces feuilles, & en-dedans est placée une enveloppe velue & rayonnée, qui couvre les fleurs. Ces fleurs sont jaunes, & de la même forme que celles des autres especes ; elles paroissent en Juillet, & ne produisent pas souvent des semences en Angleterre. Les tiges périssent en automne ; mais les feuilles radicales se conservent toute l'année. On peut multi-

plier cette plante par boutures au printems ; elle exige un fol fec & une fituation chaude.

Purpurea. La feptieme fe trouve en Portugal & en Efpagne ; elle a une tige d'arbriffeau de quatre ou cinq pieds de hauteur, qui pouffe des branches minces , à quatre angles , couvertes d'une écorce blanche , & garnies de feuilles ovales , en forme de lance , de quatre pouces environ de longueur fur un & demi de large à leur bâfe , crénelées fur leurs bords , cotonneufes en-deffous , & fupportées par de fort courts pétioles : fes fleurs fortent en têtes à chaque nœud ; elles ont des enveloppes couvertes de poils , & font d'un pourpre foncé. Cette plante fleurit dans les mois de Juin & Juillet ; mais les femences ne mûriffent point en Angleterre. On peut la multiplier par boutures , comme les trois premieres , & la traiter de même.

Samia. La huitieme, que M. de TOURNEFORT a découverte dans le Levant , où elle croit naturellement , a une racine vivace & une tige annuelle ; fes feuilles font en forme de cœur ; les radicales ont trois pouces de longueur fur un & demi de large à leur bâfe ; elles font terminées en pointe aiguë , cotonneufes en-deffous , & ont cinq fortes veines ; les tiges s'élevent à la hauteur d'un pied & demi , & font garnies à chaque nœud de deux feuilles oppofées , de la même forme que celles du bas, mais plus petites : les fleurs croiffent

en têtes autour des tiges ; elles font d'un pourpre ufé , & leurs enveloppes font découpées en deux fegmens bien fermés. Cette efpece ne produit jamais de femences en Angleterre ; & comme fes racines ne font que des progrès fort lents , elle eft à préfent fort rare en Europe. Avant le gros hiver de 1740 , ces plantes fubfiftoient en plein air , dans des plates-bandes chaudes , où on les confervoit depuis plus de vingt ans , fans qu'on en gardât aucune dans des pots ; mais le froid rigoureux de cette année les a toutes détruites.

Orientalis. Les femences de la neuvieme ont été envoyées de Smyrne par le Conful SHERARD , au jardin de Chelféa , où elles ont produit des plantes ; elle a une racine vivace & une tige annuelle ; fes feuilles radicales ont près de trois pouces de longueur fur un & demi de large ; elles font poftées fur des pétioles longs & cotonneux , & font rudes en-deffus , cotonneufes en-deffous, en forme de cœur , & entieres ; les tiges s'élevent à la hauteur d'un pied , & font auffi fort cotonneufes : les fleurs, qui croiffent en têtes autour des tiges , ont des calices fort longs , tubulés , & couverts de duvet ; elles font fort larges , d'un jaune brillant , & ont une belle apparence. Cette plante fleurit à la fin de Juin & en Juillet ; mais elle ne perfectionne jamais fes femences en Angleterre. Cette efpece a réfifté pendant plufieurs années en

plein air dans le jardin de Chelséa ; mais en l'année 1740, elle a été entierement détruite.

Flavescens. Les semences de la dixieme, qui ont aussi été envoyées de Smyrne par le même Consul SHERARD, ont produit plusieurs plantes dans le jardin de Chelséa. Cette espece a des tiges d'arbrisseau d'environ trois pieds de hauteur, couvertes d'un duvet jaunâtre, & qui poussent plusieurs branches minces, irrégulieres, & garnies de feuilles étroites en forme de lance, & couvertes d'un duvet jaunâtre en-dessous : les fleurs sont produites en têtes aux extrémités des branches ; leurs enveloppes sont fort cotonneuses, plus petites qu'aucunes des autres, & d'un jaune sale. Cette espece ressemble presque à la seconde ; mais ses feuilles sont beaucoup plus petites, ses branches sont plus minces, & couvertes d'un duvet jaune, sur-tout vers l'extrémité ; les tête des fleurs sont moins grosses, & sortent généralement aux sommets des branches.

On peut la multiplier par boutures, comme les trois premieres especes, & la traiter de même, avec la seule différence que celles-ci exigent une situation chaude, parce qu'elles sont plus sensibles aux froids. J'en ai cependant vu quelques-unes dans le jardin de Chelséa, qui ont subsisté plusieurs années en plein air dans une plate-bande chaude.

Nissolii. La onzieme espece croît naturellement dans les isles de l'Archipel & en Espagne, d'où ses semences m'ont été envoyées. Cette plante a une tige annuelle ; mais sa racine, ainsi que les feuilles du bas, sont vivaces. Ces feuilles ne s'élevent pas immédiatement de la racine, mais elles sont disposées en grappes ou paquets sur des branches courtes, traînantes & laineuses, supportées par des pétioles fort longs & cotonneux, & placées sans ordre ; elles sont en forme de cœur, couvertes de duvet sur les deux faces, & de quatre pouces environ de longueur sur deux de large à leur bâse ; les tiges sont minces, d'un pied de hauteur, & garnies de feuilles ovales, en forme de lance, mais plus étroites par dégrés jusqu'au sommet, où elles n'ont au plus que six lignes de largeur. Ces tiges poussent généralement des branches latérales & opposées vers le bas : depuis la premiere division jusqu'au sommet, elles sont garnies de fleurs jaunes, verticillées, & très-rapprochées les unes des autres, comme dans les autres especes, chaque fleur étant cependant séparée & distincte ; leurs calices sont ovales, laineux & bien fermés. Ces fleurs paroissent dans les mois de Juin & Juillet, & sont rarement suivies de semences en Angleterre. Cette plante peut être multipliée par boutures, comme la sixieme espece, & traitée de la même maniere.

Ferruginea. La douzieme, qu'on rencontre en Espagne & en Portugal, a une tige pres-

que en arbriſſeau, un peu ligneuſe, de deux pieds & demi
de hauteur, & couverte d'un
duvet blanc & épais ; de la
même racine ſortent pluſieurs
tiges ornées de feuilles en forme de cœur, d'environ deux
pouces de longueur ſur un de
largeur vers leur bâſe ; de chaque nœud, qui garniſſent la
bâſe de ces tiges, naiſſent deux
courts rejettons oppoſés, qui
produiſent quatre ou cinq petites feuilles de la même forme
que les autres, mais couvertes
d'un duvet fort blanc : les fleurs,
qui ſont diſpoſées en petites
têtes vers le haut de la tige,
ont des enveloppes cotonneuſes & en forme de lance ;
elles ſont courtes, & de couleur de fer ; elles paroiſſent
dans les mois de Juin & de Juillet : mais elles ne produiſent
point de ſemences en Angleterre.

On multiplie cette eſpece en
diviſant ſes racines rampantes
chaque deux ans. Le meilleur
tems pour faire cette opération
eſt vers le milieu de Septembre,
afin que les rejettons puiſſent
prendre racine avant les gelées;
mais il faut mettre du terreau
autour, pour empêcher la gelée de pénétrer dans la terre :
on la multiplie auſſi de boutures, comme les trois premieres eſpeces, au printems & en
été : elle exige le même traitement que la dixieme; car elle
eſt moins dure que les trois premieres. On pourra la conſerver en couvrant ſes racines
avec du tan en hiver ; &, s'il
ſurvient des froids rigoureux
qui en faſſent périr les tiges,

il en repouſſera de nouvelles
au printems ſuivant.

Rotundi-folia. La treizieme croît
naturellement en Eſpagne &
en Portugal ; elle s'éleve, avec
pluſieurs tiges d'arbriſſeau, à
la hauteur de trois ou quatre
pieds, & ſe diviſe en pluſieurs
branches quarrées, couvertes
d'un duvet laineux, & garnies
de feuilles en forme de cœur
vers le bas des tiges, mais
ovales, en forme de lance,
cotonneuſes vers le haut, oppoſées, & poſtées ſur de courts
pétioles : ces fleurs ſortent en
têtes rondes autour des tiges ;
elles ont des enveloppes en
forme d'alène, terminées en
pointe aiguë, & couvertes de
duvet. Ces fleurs ſont d'un
pourpre brillant ; elles paroiſſent en Juin, & ne produiſent
jamais de ſemences dans ce
pays. On multiplie cette eſpece
par boutures, comme les trois
premieres, & elle doit être
traitée comme la dixieme.

Laciniata. La quatorzieme
a été découverte par M. de
Tournefort, dans le Levant,
d'où il en a envoyé les ſemences au Jardin Royal, à
Paris, où elles ont réuſſi. Cette
eſpece a une racine vivace &
une tige annuelle, qui périt en
automne ; mais les feuilles du
bas ſubſiſtent toute l'année ;
elles ſont ailées alternativement, & les petits lobes ſont
découpés ſur leurs bords ; les
tiges s'élevent à un pied &
demi de haut, & ſont garnies
de feuilles de la même forme
que celles du bas, mais plus
petites : les fleurs ſortent en
têtes rondes des tiges comme

celles des autres efpeces, ayant des calices laineux ; elles font d'un pourpre ufé, & paroif- fent en Juin ; mais leurs fe- mences ne mûriffent jamais ici.

On multiplie cette plante comme la huitieme efpece, au moyen des rejettons que fes racines produifent ; elle pouffe auffi lentement, & exige le même traitement. Cette efpece eft à préfent fort rare en An- gleterre, parce que les fortes gelées de 1740 en ont détruit toutes les plantes, dont plu- fieurs cependant fubfiftoient, depuis plus de vingt ans, en plein air.

Toutes les efpeces de ce genre font un bel ornement dans les jardins, quand elles y font diftribuées avec goût : ainfi, elles méritent d'y occu- per une place, avec d'autant plus de raifon, que leurs fleurs fe fuccedent pendant deux ou trois mois, & que leurs feuil- les, qui font blanches & co- tonneufes, étant entremêlées parmi d'autres plantes à feuilles vertes, font une agréable variété.

Les feuilles des deux pre- mieres efpeces ont été forte- ment recommandées, comme propres à guérir les maux de gorge, en les prenant en in- fufion, comme le *Thé*

PHLOX. *Lin. Gen. Plant.* 197. *Lychnidea. Dill. Hort. Elth.* 166. [*Lychnidea, or Baſtard Lychnis.*] Lychnide ou Lychnis bâtard.

Caraĉteres. Le calice de la fleur eft cylindrique, perfiftant, & formé par une feuille décou- pée au fommet en cinq par- ties aiguës ; la corolle, qui eft infundibuliforme, impropre-

ment dite, ou hypocratérifor- me, a un tube cylindrique, étroit à fa bâfe, où il eft re- courbé, uni au fommet, & divifé en cinq fegmens égaux, ronds & étendus.

Cette fleur a cinq étamines placées au-dedans du tube, dont deux font plus longues que le tube, & qui font terminées par des antheres renfermées dans les cavités de la corolle ; fon ger- me eft conique, & foutient un ftyle mince de la longueur des étamines, & couronné par un ftigmat aigu, divifé en trois parties. Ce germe devient en- fuite une capfule ovale, pla- cée fur le germe, & à trois cellules, dont chacune con- tient une fimple femence.

Ce genre de plantes eft rangé dans la premiere feĉtion de la cinquieme claffe de Lin- née, qui renferme celles dont les fleurs ont cinq étamines & un ftyle ; mais, fans s'ar- rêter au nombre des étamines, il auroit mieux valu la placer parmi les plantes *perfonnées* ou *en mafque*, qui font comprifes dans la feconde feĉtion de la quatorzieme claffe, la forme de la fleur étant la même.

Les efpeces font :

1°. *Phlox glaberrima, foliis lineari-lanceolatis, glabris, acu- minatis, caule ereĉto, ramofo, corymbo terminali. Lin Sp.* 217. *Hort. Cliff.* 63. *Roy. Lugd.-B.* 433. *Gron. Virg.* 21. *Kalm. It.* 3. *p.* 153 ; Phlox avec des feuilles unies, étroites en for- me de lance, & terminées en pointe aiguë, & une tige droite & branchue, terminée par des fleurs en corymbe.

536 PHL

Lychnidea folio Melampyri.
Dill. Elth. 2c3. *t.* 166. *f.* 202.
Lychnidea Virginiana, Holostei
ampliori folio, floribus umbellatis,
purpureis. Rand. Phil. Trans.
vol. 34 ; Lychnis bâtard de la
Virginie, à larges feuilles
d'*Holosteum*, avec des fleurs
pourpre en ombelle.

2°. *Phlox Caroliniana, foliis*
lanceolatis, lævibus, caule sca-
bro, corymbis sub-fastigiatis. Lin.
Sp. 216 ; Phlox avec des
feuilles en forme de lance,
& lisses, une tige rude, & des
fleurs croissant en corymbe,
terminées en pointe.

Lychnidea Caroliniana, floribus
quasi umbellatim dispositis, foliis
lucidis, crassis, acutis. Martyn.
Dec. 1 ; Lychnis bâtard de la
Caroline, avec des fleurs presque disposées en ombelle, &
des feuilles épaisses, luisantes
& aiguës.

3°. *Phlox maculata, foliis*
lanceolatis lævibus, racemo op-
posite corymboso. Lin. Sp. Plant.
216. *Mant.* 335. *Kalm. It.* 3.
p. 153. *Jacq. Hort. f.* 127 ;
Phlox avec des feuilles en
forme de lance, & lisses, &
des fleurs branchues, opposées
& en corymbe.

Lychnoïdes Marylandica, fo-
liis binis, oppositis, basi & au-
riculis caulem utrinquè amplexi-
caulibus. Raii Supp. 489 ; Lychnis bâtard de Maryland,
avec des feuilles par paires,
& opposées, dont les bâses
& les oreilles embrassent la
tige des deux côtés.

4°. *Phlox divaricata, foliis*
lato-lanceolatis, superioribus al-
ternis, caule bifido, pedunculis
geminis. Lin. Sp. Plant. 217 ;

PHL

Phlox avec des feuilles larges
& en forme de lance, dont
celles du haut sont alternes,
ayant une tige fourchue &
des pédoncules à deux branches.

Lychnidea Virginiana, Alsi-
nes aquaticæ foliis, floribus in
ramulis divaricatis. Pluk. Mant.
121 ; Lychnis bâtard de Virginie, à feuilles de Mouron
aquatique, avec des fleurs produites en rameaux écartés les
uns des autres.

5°. *Phlox paniculata, foliis*
lanceolatis, margine scabris,
corymbis paniculatis. Lin. Sp.
Plant. 216 ; Phlox avec des
feuilles en forme de lance,
& à bords rudes, produisant
des fleurs disposées en corym-
bes paniculés.

Lychnidea folio Salicino. Dill.
Elth. 205. *t.* 166. *f.* 203.
Lysimachia Virginiana umbel-
lata, maxima, Lysimachiæ luteæ
floribus amplioribus. Pluk. Mant.
121 ; Lychnis bâtard de Virginie, ombellée & très-grande,
produisant des fleurs plus grandes & jaunes.

6°. *Phlox pilosa, foliis lanceo-*
latis, villosis, caule erecto,
corymbo terminali. Lin. Sp. Plant.
216 ; Phlox à feuilles velues
& en forme de lance, avec
une tige droite, terminée par
un corymbe de fleurs.

Lychnoïdes Marylandica, ca-
lycibus lanuginosis, foliis angustis,
acutis. Raii Supp. 490 ; Lychnis
bâtard de Maryland, avec des
calices laineux aux fleurs, &
des feuilles étroites, & à
pointe aiguë.

Lychnidea umbelli-fera Blat-
tariæ accedens Virginiana, major,

PHL

repens. Pluk. Alm. 133. *t.* 98. *f.* 1.

7°. *Phlox ovata, foliis ovatis, floribus solitariis. Lin. Sp. Plant.* 152 ; Phlox à feuilles ovales & à fleurs folitaires.

Lychnidea fistulosa Marylandica, Clinopodii vulgaris folio, flore amplo singulari. Pluk. Mant. 122. *t.* 348. *f.* 4 ; Lychnis bâtard fistuleux du Maryland, avec une feuille de Basilic des champs, & une grande fleur folitaire.

Glaberrima. La premiere espece, qui croît naturellement dans la Virginie & dans quelques autres parties de l'Amérique Septentrionale, est, depuis plusieurs années, assez commune dans les jardins anglois ; elle a une racine vivace, qui pousse plusieurs tiges en nombre proportionné à sa grosseur ; ces tiges ont près d'un pied & demi de hauteur, & se divisent vers le sommet en trois ou quatre petites branches, terminées par des fleurs en corymbe ; les feuilles du bas des tiges font opposées, de trois pouces environ de longueur sur près d'un pouce & demi de largeur à leur bâfe, terminées en pointe longue & aiguë, unies & sessiles ; les feuilles du haut font alternes : les fleurs croissent au sommet des tiges en corymbe court, ou plutôt en forme d'ombelle ; plusieurs sortent du même point sur de courts pédoncules ; leurs calices font tubulés, à dix angles ou sillons, & découpés au sommet en cinq segmens ronds & étendus. Ces fleurs font d'un pourpre clair ; elles paroissent

en Juin, & ne produisent point de semences en Angleterre, à moins que l'année ne soit très-chaude.

Caroliniana. La seconde espece croît naturellement dans la Caroline ; sa racine est vivace, & pousse plusieurs tiges rudes de deux pieds de hauteur, & garnies de feuilles roides, luisantes, opposées, sessiles, en forme de lance, entieres, & réfléchies sur leurs bords : le sommet de la tige à généralement deux branches latérales, minces, & terminées par une tête de fleurs verticillées autour des tiges, mais placées si près les unes des autres, qu'elles paroissent n'être qu'un corymbe à quelque distance ; le calice de la fleur est court, & profondément découpé en cinq segmens aigus ; le tube de la fleur est long, & divisé au sommet en cinq parties rondes & étendues ; elles font d'un pourpre plus foncé que celles de la précédente, & paroissent quinze jours plus tard.

Maculata. La troisieme espece est originaire du Maryland ; elle a une racine vivace, de laquelle sortent plusieurs tiges droites, de couleur de pourpre, & toutes couvertes de taches blanches, de trois pieds environ de hauteur, & garnies de feuilles blanches, unies, en forme de cœur, de trois pouces environ de longueur sur un de large à leur bâfe, & terminées en pointe aiguë. Vers le haut des tiges naissent de petites branches opposées, de

chacune eſt terminée par un petit paquet de fleurs; mais la tige principale produit à ſon extrémité un épi de fleurs long, lâche, & compoſé de petits paquets qui ſortent des aiſſelles de la tige à chaque nœud; chaque paquet a un pedoncule commun d'un pouce de longueur, & chaque fleur eſt ſoutenue par un pédoncule particulier & court. Ces fleurs ſont d'un pourpre brillant, & paroiſſent ſur la fin de Juillet. Si le ſol dans lequel elles ſe trouvent eſt humide, & ſi la ſaiſon eſt pluvieuſe, elles conſervent leur beauté durant une grande partie du mois d'Août; mais elles ne produiſent pas ſouvent des ſemences en Angleterre.

Divaricata. La quatrieme eſpece, qui naît ſans culture dans l'Amérique Septentrionale, à une racine vivace, de laquelle s'élevent pluſieurs tiges minces, qui ſont ſujettes à pencher vers la terre ſi elles ne ſont pas ſupportées, & diviſées en pluſieurs petites branches écartées les unes des autres; le bas des tiges eſt garni de feuilles larges en forme de lance, alternes & ſeſſiles; mais celles qui naiſſent ſur les petites branches, ſont plus étroites, & oppoſées: les fleurs croiſſent en paquets lâches aux extrémités des branches: elles ont des calices courts, & diviſés en cinq ſegmens étroits & aigus; le tube de la fleur eſt long & mince; les ſegmens du ſommet ſont larges, en forme de cœur, & renverſés. Ces fleurs, qui ſont

d'un vert clair, paroiſſent à la fin du mois de Mai ou au commencement de Juin; mais elles produiſent rarement des ſemences en Angleterre.

Paniculata. La cinquieme eſpece, qu'on rencontre dans l'Amérique Septentrionale, a une racine vivace & une tige annuelle, liſſe, d'un vert tendre, & de deux pieds de hauteur, qui pouſſe quelques branches latérales, garnies de feuilles en forme de lance, oppoſées, de trois pouces environ de longueur ſur un de largeur au milieu, terminées en pointe à chaque extrémité, & ſeſſiles aux tiges; elles ſont d'un vert foncé, & leurs bords un peu rudes: les fleurs, qui ſont diſpoſées en corymbe au ſommet des tiges, ſont compoſées de pluſieurs petits paquets de fleurs qui ont chacune un pédoncule diſtinct; le calice de la fleur eſt court, & découpé preſque juſqu'au fond en cinq ſegmens étroits & aigus; ſon tube eſt long, mince, & diviſé en cinq ſegmens ovales & étendus. Ces fleurs ſont d'un pourpre pâle; elles paroiſſent ſur la fin de Juillet, & produiſent ſouvent des ſemences qui mûriſſent en automne.

Piloſa. La ſixieme eſpece croit ſans culture dans la Virginie; elle a une racine vivace, de laquelle ſortent quelques tiges d'un pied de hauteur, & garnies de feuilles étroites, en forme de lance, terminées en pointe aiguë, ſeſſiles, & un peu velues: ſes fleurs naiſſent en corymbe lâ-

che au sommet de la tige ; leurs calices sont découpés en segmens aigus presque jusqu'au fond ; le tube de la fleur est mince, assez long, & divisé sur ses bords en cinq segmens ovales & étendus. Ces fleurs sont d'un pourpre clair ; elles paroissent à la fin de Juin ; mais elles produisent rarement des semences en Angleterre.

Ovata. La septieme espece se trouve au Maryland & dans d'autres parties de l'Amérique Septentrionale ; elle a une racine vivace, de laquelle sortent deux ou trois tiges minces, de neuf pouces environ de hauteur, & garnies de feuilles ovales, rudes, velues, d'un pouce & demi de long sur environ neuf lignes de large au milieu, postées sur de courts pétioles, & opposées : les fleurs sortent seules sur le sommet de la tige, leurs tubes sont fort minces, & découpés en cinq segmens ronds & étendus ; elles sont d'un pourpre clair ; elles paroissent en Juillet, & ne produisent jamais de semences en Angleterre.

Culture. Toutes ces plantes sont dures, & profitent en plein air en Angleterre : elles se plaisent dans un sol riche, humide, & pas trop ferme, dans lequel elles deviennent hautes, & produisent des paquets de fleurs beaucoup plus gros que dans une terre seche ; car lorsque le sol est sec & mauvais, elles périssent souvent en été, si on ne les arrose pas constamment.

On les multiplie toujours en divisant leurs racines, parce qu'elles produisent rarement des semences dans notre climat. Le meilleur tems pour faire cette opération est l'automne, quand leurs tiges commencent à se flétrir ; mais il ne faut pas les diviser en trop petites parties, si on veut qu'elles fleurissent fortement dans l'été suivant, & l'on ne doit renouveler cette opération que chaque deux ans, parce qu'en les divisant plus souvent, on les affoiblit, & elles ne poussent plus ensuite qu'un petit nombre de tiges si foibles, qu'elles ne peuvent s'élever à leur hauteur ordinaire, & qui ne donnent que de très-petits paquets de fleurs.

Après que ces racines sont transplantées, il est prudent de répandre du vieux tan ou du terreau sur la terre autour des plantes, pour empêcher la gelée d'y pénétrer ; car comme elles poussent de nouvelles fibres avant l'hiver, si le froid est vif, ces fibres périssent : ce qui fait beaucoup de tort à la plante, & souvent même entraîne sa destruction.

Les premiere, seconde & cinquieme especes se multiplient assez promptement par leurs racines rampantes, & les autres très-lentement de cette façon ; de maniere qu'il vaut beaucoup mieux en faire des boutures : si l'on veut se procurer les trois premieres en abondance, on peut se servir de cette méthode. Le meilleur tems pour planter ces boutures, est à la fin d'Avril ou au commencement de Mai. Quand

les rejettons des racines ont environ deux pouces de hauteur, on les coupe tout près de la terre ; on en raccourcit les sommets ; on les plante sur une plate-bande de terre légere & marneuse ; on les tient à l'ombre, jusqu'à ce qu'elles aient pris racine ; & en les mettant très-près les unes des autres, en les couvrant de cloches, & en les tenant à l'abri des rayons du soleil, elles produiront des racines en cinq ou six semaines ; mais quand elles commencent à pousser, il faut soulever les cloches par dégrés, pour leur donner de l'air, sans quoi elles fileroient, & s'affoibliroient bientôt. Quand elles sont bien enracinées, on ôte les cloches qui les couvroient, pour les accoutumer au plein air ; & bientôt après, on les met en pépiniere sur une terre fertile, en les plantant à six pouces de distance entr'elles. Il est nécessaire de les tenir à l'ombre, & de les arroser constamment, jusqu'à ce qu'elles aient formé entierement de nouvelles racines ; après quoi il suffira de les tenir nettes de mauvaises herbes : & en automne, on les placera à demeure dans les plates-bandes du parterre.

On met quelquefois ces plantes dans des pots, & on les tient sous un châssis de couche chaude en hiver : par ce moyen, elles fleurissent fortement dans l'été suivant. On peut les placer autour d'une habitation, pendant qu'elles sont dans toute leur beauté. En les

mêlant avec d'autres fleurs ; elles produisent un très-bel effet.

PHYLICA. *Lin. Gen. Plant.* 236. *Alaternoïdes. Com. Hort. Amst. p. 1.* [*Bastard Alaternus.*] Alaterne bâtard. Apalanchine. Thé du Cap de Bonne-Espérance.

Caracteres. Les fleurs sont recueillies dans un disque, sur un réceptacle commun ; chacune a un calice persistant, composé de trois feuilles étroites & oblongues ; la corolle est monopétale, & percée à travers avec un tube conique, érigé, découpé au bord en cinq parties, ayant une écaille aiguë à chaque division, qui se joignent ensemble en-dedans, & cinq petites étamines insérées sous les écailles, & terminées par des antheres simples ; le germe, qui est placé au fond de la corolle, soutient un style simple, couronné par un stigmat obtus. Ce germe devient, quand la fleur est passée, une capsule ronde à trois lobes & à trois cellules, dont chacune renferme une simple semence ronde, bossue d'un côté, & angulaire sur l'autre.

Ce genre de plante est rangé dans la premiere section de la cinquieme classe de LINNÉE, qui comprend celles dont les fleurs ont cinq étamines & un style.

Les especes sont :

1°. *Phylica Ericoïdes, foliis linearibus, verticillatis. Lin. Sp. Plant.* 195. *Fabric. Helmst.* 233. *Kniph. cent.* 1. n. 62 ; Phylica à feuilles linéaires & verticillées.

lées, qui reſſemble à la Bruyere.

Alaternoïdes Africana, Ericæ foliis, floribus albicantibus & muſcoſis. Hort. Amſt. 2. p. 1. tab. 1; Alaterne bâtard d'Afrique, à feuilles de Bruyere, avec des fleurs blanches & mouſſeuſes. Bruyere du Cap de Bonne-Eſpérance.

2°. *Phylica plumoſa, foliis lineari-ſubulatis, ſummis hirſutis.* Prod. Leyd. 199; Phylica à feuilles linéaires en forme d'alêne, & velues au ſommet.

Ricinus arboreſcens Africanus, tomentoſis capitulis. Seb. Thes. 1. p. 38.

Alaternoïdes Africana, Rorismarini latiori & piloſiori folio. Comm. Prælud. 63. t. 13.

Chamælæa foliis anguſtis, ſubtùs incanis, floribus capitatis, muſcoſis. Burm. Plant. Afr. 117. tab. 43; Camelée à feuilles étroites & blanches en-deſſous, avec des fleurs recueillies en têtes, & mouſſeuſes.

3°. *Phylica Buxi-folia, foliis ovatis, ſparſis.* Lin. Sp. Plant. 195; Phylica à feuilles ovales & éparſes.

Chamælæa Africana, foliis ſub-rotundis. Herm. Afr. 6.

Chamælæa folio ſub-rotundo, ſubtùs incano, floribus in capitulum collectis. Burm. Plant. Afr. 119; Camelée à feuilles preſque rondes, & blanches en-deſſous, & à fleurs rapprochées en têtes.

Ericoïdes. La première eſpece croît naturellement au Cap de Bonne-Eſpérance, d'où elle a d'abord été portée dans les jardins de Hollande; mais on la trouve auſſi dans les environs de Lisbonne, où il y a de vaſtes terreins qui en ſont couverts, comme on voit les Bruyeres en Angleterre. Cette plante, baſſe & en buiſſon, s'élève rarement à plus de trois pieds de hauteur; ſes tiges ſont ligneuſes, irrégulieres, & diviſées en pluſieurs branches étendues, qui ſe ſous-diviſent en d'autres plus petites, dont les plus jeunes ſont fort garnies de feuilles courtes, étroites, terminées en pointe aiguë, & verticillées autour des tiges, auxquelles elles ſont ſeſſiles; elles ſont d'un vert foncé, & ſe conſervent toute l'année; ſes fleurs ſont diſpoſées en petits paquets aux extrémités des branches, & poſtées tout près des feuilles; elles ſont d'un blanc pur, commencent à paroître en automne, ſe ſuccedent pendant tout l'hiver, ce qui fait eſtimer ces plantes: elles périſſent au printems, & elles ne produiſent point de ſemences en Angleterre.

Plumoſa. La ſeconde eſpece eſt auſſi originaire du Cap de Bonne-Eſpérance, d'où elle a été envoyée dans les jardins de Hollande; elle a une tige d'arbriſſeau érigée, d'environ deux pieds de hauteur, couverte d'une écorce de couleur tirant ſur le pourpre, & chargée çà & là d'un duvet blanc; ſes feuilles ſont étroites, courtes, terminées en pointe aiguë, ſeſſiles aux branches, alternes ſur chaque côté, épaiſſes, nerveuſes, d'un vert foncé en-deſſus, & blanches en-deſſous: les fleurs ſont recueillies

en petites têtes aux extrémités des branches ; elles font blanches , laineufes , ornées de franges fur leurs bords , & découpées au fommet en fix fegmens aigus. Elles paroiffent au commencement de l'hiver , & confervent long-tems leur beauté ; mais elles ne produifent jamais de femences en Angleterre.

Buxi-folia. La troifieme efpece fe trouve encore dans le même pays que la précédente ; elle s'éleve droite en tige d'arbriffeau jufqu'à la hauteur de cinq ou fix pieds : quand elle vieillit , elle eft couverte d'une écorce rude de couleur pourpre ; mais les plus jeunes ont un duvet laineux : elle eft garnie de feuilles épaiffes , ovales , de la même largeur que celles du *Buis*, veinées, unies, d'un vert luifant en-deffus , & blanches en-deffous , portées fur de courts pétioles , & placées fans ordre fur les branches : les fleurs , qui font recuillies en petites têtes aux extrémités des branches, font d'une couleur herbacée , & ne font pas fort belles : elles paroiffent en même tems que celles de la précédente.

Culture. Comme ces plantes ne produifent point de femences en Angleterre, elles n'y peuvent être multipliées que par boutures , qui prennent aifément racine quand elles font bien traitées. On peut les planter en deux faifons ; la premiere eft la fin de Mars , avant que les plantes commencent à pouffer : on les met dans des pots , que l'on plonge dans une couche de chaleur très-modérée ; on les couvre exactement avec des cloches ; on les tient à l'abri du foleil au milieu du jour , & on les arrofe légerement. Au moyen de ces précautions , elles prendront racine dans l'efpace de deux mois : on les accoutumera enfuite à fupporter le plein air ; & quand elles auront acquis de la force, on les mettra chacune féparément dans de petits pots remplis d'une terre molle & marneufe : on les tiendra à l'ombre , jufqu'à ce qu'elles aient formé de nouvelles fibres , & on les placera dans une fituation abritée , où on les laiffera jufqu'à l'automne.

La feconde faifon pour planter ces boutures , eft vers le commencement d'Août ; on les place dans des pots , que l'on plonge ou dans une vieille couche chaude , ou en pleine terre : on les couvre exactement avec des cloches , comme il vient d'être dit , & on les traite comme les premieres. Au moyen de cela , elles poufferont des racines dans l'efpace d'environ deux mois : mais comme alors la faifon fera trop avancée pour les tranfplanter , il faudra les laiffer dans les mêmes pots jufqu'au printems fuivant , & les placer fous un châffis de couche en automne , où elles puiffent être à l'abri des gelées , & expofées en plein air dans les tems doux. Au moyen de ce traitement , elles réuffiront mieux que fi elles avoient été foignées plus délicatement.

Ces plantes étant trop tendres pour pouvoir subfifter ici en plein air , il faut les tenir dans des pots , & les mettre à couvert pendant l'hiver ; car , quoique la premiere efpece puiffe fe conferver dans une fituation chaude & abritée dans les années favorables , cependant elle eft infailliblement détruite lorfqu'il furvient de fortes gelées ! mais comme ces plantes n'ont befoin d'aucune chaleur artificielle , on peut les conferver en hiver fous un châffis de couche , tandis qu'elles font encore jeunes ; & quand elles font devenues plus fortes , on les place dans une ferre , de maniere qu'on puiffe leur procurer de l'air dans les tems doux ; on les traitera comme les autres plantes dures qui viennent des mêmes contrées ; & en été , on les tiendra en plein air dans dans une fituation abritée. Avec ces attentions , elles profiteront , & fubfifteront plufieurs années. Comme ces plantes fleuriffent en hiver , elles produifent un bel effet dans la ferre , durant cette faifon.

PHYLLANTHUS. *Lin. Gen. Plant.* 932. [*Sea-fide Laurel.*] Laurier maritime.

Caractères. Cette plante a des fleurs mâles & des fleurs femelles fur le même pied ; le calice de la fleur , dans les deux fexes , eft perfiftant , en cloche & formé par une feuille découpée , en fix parties étendues & colorées. Quelques perfonnes prétendent qu'elles n'ont point de corolle , & d'autres foutiennent qu'elles

n'ont point de calice : les fleurs mâles ont trois étamines courtes réunies à leur bâfe , écartées au fommet , & terminées par des anthères jumelles ; les fleurs femelles ont un nectaire angulaire qui entoure le germe. Ce germe eft rond , & a trois angles ; il foutient trois ftyles étendus , & couronnés par des ftigmats obtus , & devient enfuite une capfule ronde , à trois fillons & à trois cellules qui renferment chacune une femence fimple & ronde.

Ce genre de plantes eft rangé dans la troifieme fection de la vingt-unieme claffe de LINNÉE , dans laquelle font comprifes celles qui ont des fleurs mâles & femelles fur le même pied , & dont les fleurs mâles ont trois étamines.

Les efpeces font :

1°. *Phyllanthus Epiphyllanthus , foliis lanceolatis , ferratis , crenis flori-feris. Hort. Cliff.* 439. *Roy. Lugd.-B.* 200 ; Phyllanthus avec des feuilles en forme de lance , & fciées , fur les bords defquelles croiffent les fleurs.

Phyllanthus Americana planta , flores è fingulis foliorum crenis proferens. Hort. Amft. 1. p. 199. f. 102. *Seb. Thes.* 1. p. 21. t. 13. f. 2. *Catesb. Car.* 2. p. 26. f. 26 ; Phyllanthus d'Amérique , dont les fleurs naiffent fur chaque dentelure des feuilles.

Phyllanthus foliis latioribus , utrinquè acuminatis , apicem verfùs crenatis. Brown. Jam. 188.

Filici-foliâ Hemionitidi affinis Americana Epiphyllanthos , anguftiori & longiori folio , ramofa ,

544 **PHY**

caulescens. Pluk. Alm. 134. *t.* 247. *f.* 4. *& t.* 36. *f.* 7.

Xylophylla lati-folia , foliis lanceolatis , ramis teretibus. Lin. Syst. Plant. tom. 1. *p.* 741. *Sp.* 2. *Syst. Veget. Murray. ed.* 14. *pag.* 296. *Sp.* 2. *Mant.* 251.

2°. *Phyllanthus Niruri , foliis pinnatis , flori-feris , caule herbaceo , erecto. Flor. Zeyl.* 331. *Hort. Ups.* 282 ; Phyllanthus à feuilles ailées , produisant des fleurs, avec une tige droite & herbacée.

Phyllanthus foliis alternis , alternatim pinnatis , floribus dependentibus ex alis foliorum. Hort. Cliff. 440.

Urinaria Indica , erecta , vulgaris. Burm. Zeyl. 230. *t.* 93. *f.* 2.

Herba mœroris alba. Rumph. Amb. 6. *p.* 41. *t.* 17. *f.* 1.

Fruticulus capsularis hexapetalos , Casiæ Poetarum foliis brevioribus. Pluk. Alm. 159. *t.* 183. *f.* 5.

Niruri Barbadense , folio ovali, subtùs glauco , petiolis florum brevissimis. Martyn. Cent. 9. *tab.* 9 ; Phyllanthus des Barbades, dont les feuilles sont ovales, & d'une couleur grise en-dessous , avec de courts pédoncules aux fleurs. Bois à enivrer.

Kirganeli. Rheed. Mal. 10. *p.* 29. *f.* 15. *Raii Dendr.* 29.

3ᵛ. *Phyllanthus Emblica , foliis pinnatis , flori-feris , caule arboreo , fructu baccato. Flor. Zeyl.* 333. *Mat. Med.* 199 ; Phyllanthus avec des feuilles ailées , qui produisent des fleurs , une tige d'arbre , & un fruit en forme de baie.

Myrobalanus Emblica. Bauh. pin. Rheed. 445. *Rumph. Amb.*

PHY

7. *p.* 1. *t.* 1. *Blackw. t.* 400. *Nelli-Camarum. Rheed. Hort. Mal.* 1. *p.* 69. *t.* 38. *Raii Hist.* 1456.

Nellika. Zan. Hist. 159. *t.* 61.

La premiere espece croît naturellement sur des rochers qui bordent la mer dans toutes les Isles des Indes Occidentales , où les habitans lui donnent le nom de *Laurier Maritime.* On ne la trouve pas souvent dans les Campagnes , ce qui fait qu'elle est très-rare en Europe ; ses racines s'enfoncent si profondément dans les crevasses des rochers, qu'il est presque impossible de les transplanter , & il est très-difficile de les multiplier par semences. Si elles ne sont pas mises en terre aussi-tôt qu'elles sont mûres, elles ne levent point , & la plus grande partie avorte. Il y avoit autrefois une de ces plantes dans les jardins de Hampton-Court (1); mais elle y a péri avec beaucoup d'autres, par l'ignorance des Jardiniers. J'en ai aussi vu une dans les jardins d'Amsterdam.

Cet arbre s'éleve , avec une tige ligneuse , à quinze ou seize pieds de hauteur ; ses feuilles sont placées sans ordre ; elles ont cinq ou six pouces de longueur , & sont unies & épaisses : les fleurs sortent sur le bord des feuilles , & principalement vers le haut, où elles sont placées très-près les unes des autres ; de ma—

_____

(1) Ma son Royale à dix milles de Londres.

niere qu'elles font une espece de bordure qui produit, avec le vert luisant des feuilles, un très-bel effet. Ces feuilles conservent leur verdure toute l'année, ce qui rend ces plantes plus estimables & plus précieuses. Il faut les tenir en hiver dans une serre de chaleur modérée, sans quoi on ne pourroit les conserver en Angleterre. On peut les laisser en plein air, pendant l'été, dans un abri chaud: en les traitant ainsi, on les a tenues en grande vigueur dans le jardin de Botanique d'Amsterdam (1).

Niruri. La seconde espece est originaire de la Barbade, où elle est très-commune. Je l'ai vu pousser très-souvent dans des caisses remplies de terre qu'on m'envoyoit de cette Isle: elle est annuelle, & ses semences sont lancées au loin par l'élasticité de leurs capsules: elle se multiplie aussi de cette maniere en Angleterre; car ses semences étant tombées dans quelques pots placés dans la serre chaude, elles pousserent sans aucun soin. Cette espece a une tige herbacée, d'un pied & demi de hauteur, & divisée en plusieurs branches garnies de feuilles longues, aîlées, & composées d'un grand nombre de lobes ovales, d'une couleur grise en-dessous, & d'un vert brillant en-dessus. Ces lobes se rapprochent & se rétrécissent chaque soirée; ils ont leur

partie inférieure tournée en-dehors: les fleurs sont produites en-dessous dans la longueur de la côte du milieu, & penchent vers le bas; quelques-unes sont mâles, & d'autres femelles, & elles sont entremêlées sur la même plante. Ces fleurs sont découpées sur leurs bords en six segmens colorés. Quelques personnes donnent à ces segmens le nom de apétales ou corolles, & d'autres regardent ces fleurs comme apétales. Les fleurs mâles ont chacune trois étamines: les femelles ont un style simple, qui soutient un stigmat divisé en trois parties; elles sont remplacées par des capsules rondes, & à trois cellules, qui renferment chacune une semence. La plante fleurit ordinairement depuis le mois de Juin jusqu'en Octobre, & ses semences mûrissent successivement.

Emblica. La troisieme espece est le *Nelli-Camarum* de l'*Hortus Malabaricus*, & la *Nux Emblica* des Apothicaires; elle croît naturellement sur la côte de Malabar, où elle s'éleve en tige d'arbre à la hauteur de douze ou quatorze pieds; mais en Europe, elle ne parvient qu'à la moitié de cette hauteur; elle pousse plusieurs branches latérales, garnies de feuilles étroites & aîlées; mais comme elle n'a encore produit ni fleurs ni fruits en Angleterre, je ne puis en donner une plus ample description.

Cette troisieme espece se multiplie par semences: quand on peut s'en procurer des contrées où elle croît naturelle-

---

(1) *Nota*. Cette espece est donnée dans le Système Végétal, sous le titre de *Xylophylla lati-folia*.

546 PHY

ment, on les répand fur une couche chaude ; & , lorfque les plantes qui en proviennent font en état d'être enlevées, on les met chacune féparément dans de petits pots remplis de terre légere ; on les plonge dans une couche chaude de tan ; on les tient à l'ombre, & on les arrofe jufqu'à ce qu'elles aient formé de nouvelles racines : après quoi on les tient conftamment dans la couche de tan de la ferre chaude , où on les traite comme les autres plantes qui viennent des mêmes contrées ; au moyen de cela , on les conferve plufieurs années : mais elles font peu de progrès.

Les autres efpeces dont on a fait mention dans plufieurs éditions de cet ouvrage, & qui ont été jointes à ce genre, font actuellement placées fous le titre *Andrachne.*

PHYLLIS. *Lin. Gen. Plant.* 286. *Buplevroïdes. Boerrh. Ind. Alt.* 71. *Valerianella. Dill. Hort. Elth.* 405. [*Simpla Nobla.*] la Belle-Feuille.

Caractères. Le calice eft fort petit , & compofé de deux feuilles poftées fur le germe ; la corolle a cinq pétales obtus , en forme de lance , & tournés en arriere : la fleur a cinq étamines courtes , & femblables à des poils , foibles , & terminées par des antheres oblongues ; le germe , qui eft placé fous la fleur , n'a point de ftyle , mais il eft couronné par deux ftigmats en forme d'alène , réfléchis & velus. Ce germe fe change dans la fuite en un fruit oblong & angulaire , qui contient deux fe

mences paralleles , courbées en-dehors , plates en-dedans , & larges au fommet.

Ce genre de plantes eft rangé dans la feconde fection de la cinquieme claffe de LINNEÉ , avec celles dont les fleurs ont cinq étamines , deux ftyles ou ftigmats.

Nous n'avons à préfent qu'une efpece de ce genre dans les jardins anglois.

Phyllis Nobla , *ftipulis dentatis. Prod. Leyd.* 92. *Hort. Cliff.* 87. *Hort. Ups.* 57. *Kniph. cent.* 5. *n.* 66 ; Belle-Feuille avec des ftipules dentelées.

Valerianella Canarienfis frutefcens , *Simpla Nobla Dicta. Dill. Elth.* 405. *t.* 299. *f.* 386.

Buplevroïdes , *quæ Simpla Nobla Canarienfium. Pluk. Boerrh. Ind. Alt.* 1. *p.* 72. ; Oreille de Lievre *ou* Simpla Nobla des Canaries. La Belle-Feuille.

Buplevroïdes quæ arbor umbellifera. Walth. Hort. 11. *f.* 6.

Cette plante croît naturellement dans les Ifles Canaries, d'où fes femences ont été autrefois envoyées en Angleterre ; elle s'éleve à la hauteur de deux ou trois pieds , avec une tige molle d'arbriffeau , guere plus groffe que le doigt , d'une couleur herbacée , & remplie de nœuds ; vers le fommet de cette tige fortent plufieurs branches latérales , garnies de feuilles en forme de lance , de quatre pouces environ de longueur fur prefque deux de large au milieu , mais plus étroites , & terminées en pointe à chaque extrémité , d'un vert luifant en-deffus , d'un vert pâle en-deffous , & fortifiées par une groffe côte

qui en occupe le milieu, & de laquelle partent plusieurs veines profondes qui coulent vers les bords. Ces feuilles sont, pour la plupart, placées par trois autour des branches, auxquelles elles sont sessiles : les fleurs, qui naissent en panicules lâches, sont petites d'abord, d'une couleur herbacée ; mais d'un brun foncé ou usé avant de se flétrir, découpées en cinq parties à leur bâse ; où elles sont jointes, & qui tombent sans se séparer ; ce qui fait qu'on doit les regarder comme monopétales ; leurs segmens sont réfléchis en arriere, de maniere qu'ils couvrent le germe, qui est placé sous la fleur, & qui devient ensuite un fruit court, turbiné, obtus, angulaire, divisé, lorsqu'il est menu, en deux parties, dont chacune forme une semence convexe en-dehors, & angulaire. Cette plante fleurit en Juin, & ses semences mûrissent en automne.

On la multiplie par ses graines, qu'on répand sur une plate-bande de terre fraiche & légere vers la fin de Mars. Les plantes poussent au commencement de Mai ; & quand elles sont en état d'être enlevées, on les met dans des pots séparés, & on les tient à l'ombre, jusqu'à ce qu'elles aient formé de nouvelles racines, après quoi on les place dans une situation abritée, & exposée au soleil du matin. En été, on les arrose souvent ; & en hiver, on les met à l'abri des gelées, en leur procurant autant d'air qu'il est pos-

sible dans les tems doux. Ces plantes fleurissent au printems de la seconde année. On peut en placer quelques-unes en pleine terre, où elles perfectionnent leurs semences beaucoup mieux que si elles étoient restées en por.

Comme ces plantes ne se conservent guère en bon état que quatre ou cinq ans, il sera prudent d'en élever à tems une une bonne quantité de jeunes, pour remplacer les anciennes. Elles conservent pendant toute l'année leurs feuilles, qui sont larges, d'un vert luisant, & qui ont une belle apparence en hiver. C'est en cela que consiste toute leur beauté ; car leurs fleurs n'ont rien de remarquable.

PHYSALIS. *Lin. Gen. Plant.* 223. *Alkekengi. Tourn. Inst. R. H.* 150. *tab.* 64. [*Winter Cherry.*] Cerise d'hiver. Alkékenge. Coqueret.

Caractères. Le calice de la fleur est petit, gonflé, persistant, & formé par une feuille à cinq angles, & découpée au sommet en cinq pointes aiguës ; la corolle est monopétale, & en forme de roue ; elle a un tube court, avec un large bord à cinq angles, & plissé. La fleur a cinq petites étamines en forme d'alène, jointes ensemble, & terminées par des antheres érigées ; son germe est rond, & soutient un style mince, & couronné par un stigmat obtus. Ce germe se change dans la suite en une baie presque globulaire, à deux cellules, & renfermée dans un calice large & gonflé ; chaque

548 **PHY**

cellule eft remplie de femences applaties & en forme de rein.

Ce genre de plantes eft rangé
dans la premiere fection de la
cinquieme claffe de LINNÉE,
qui renferme celles dont les
fleurs ont cinq étamines & un
ftyle.

Les efpeces font :

1°. *Phyfalis Alkekengi, foliis
geminis, integris, acutis, caule
herbaceo, infernè fub-ramofo. Lin.
Sp. Plant.* 262. *Mat. Med. p.* 65;
Alkékenge avec deux feuilles
à pointe aiguë, placées fur
chaque nœud, & une tige herbacée qui produit des branches
depuis fa bâfe.

*Alkekengi officinarum. Tourn.
Inft. R. H.* 151; Cerifier d'hiver *ou* Alkékenge des boutiques, *ou* Coqueret.

*Phyfalis Halicacabum. Scop.
Carn. ed.* 2. *n.* 286.

Solanum Veficarium. Bauh. Pin.
166. *Dod. Pempt.* 454.

2°. *Phyfalis vifcofa, foliis
geminis, repandis, obtufis, fub-
tomentofis, caule herbaceo, fupernè
paniculato. Lin. Sp.* 261. *Jacq.
Hort. f.* 136; Phyfalis avec des
feuilles placées par paires,
obtufes & un peu velues, &
une tige herbacée, & terminée par un panicule.

*Alkekengi Bonarienfe repens,
baccá turbinatá vifcofá;* Alkékenge de Buenos, rampant,
produifant des baies vifqueufes & turbinées.

3°. *Phyfalis Penfylvanica,
radice perenni, caule procumbente,
foliis ovatis, acutè dentatis, petiolis longiffimis;* Phyfalis avec
une racine vivace, une tige
trainante, & des feuilles ovales, fciées en dentelures ai-

guës, & poffées fur de très-
longs pétioles.

*Alkekengi Virginianum, peren-
ne, majus, flore luteo amplo,
fructu minimo. Rand. Act. Phil.
n.* 399; le plus grand Alkékenge de Virginie, vivace,
avec une une groffe fleur jaune, & un petit fruit.

4°. *Phyfalis Virginiana, caule
herbaceo, foliis ovato-lanceolatis,
acutè dentatis. Tab.* 206. *fig.* 1;
Phyfalis avec une tige herbacée, des feuilles ovales, en
forme de lance, & à dentelures aiguës.

5°. *Phyfalis Curaffavica, caule
fuffruticofo, foliis ovatis, tomentofis, integerrimis. Vir. Cliff.* 16.
Roy. Lugd.- B. 426; Phyfalis
avec une tige d'arbriffeau, &
des feuilles ovales, cotonneufes & entieres.

*Alkekengi Curaffavicum, foliis
Origani, incanis, flore fufco-ful-
phureo, fundo purpureo. Boerrh.
Ind. Alt.* 2. *p.* 66; Alkékenge
de Curaçao, à feuilles blanches d'Origan, avec une fleur
d'une couleur de foufre rouillé, & à fond pourpre.

*Solanum Veficarium Curaf-
favicum, Solano Antiquorum fi-
mile, foliis Origani fub-incanis.
Moris. Hift.* 3. *p.* 527. *Pluk.
Alm.* 352.

6°. *Phyfalis fomni-fera, caule
fruticofo, ramis rectis, floribus
confertis. Lin. Sp. Plant.* 180;
Phyfalis à tige d'arbriffeau,
avec des branches érigées, &
des fleurs raffemblées en paquets.

*Phyfalis caule fruticofo, tereti,
foliis ovatis, integerrimis, floribus confertis. Hort. Cliff.* 62.
Roy. Lugd.- B. 426.

PHY

Alkekengi fructu parvo, verticillato. Tourn. *Inst.* 151 ; Alkékenge à petits fruits verticillés.

Solanum somni-ferum, verticillatum. Bauh. Pin. 166.

Solanum somni-ferum. Clus. Hist. 2. p. 85.

7°. *Physalis flexuosa, caule fruticoso, ramis flexuosis, floribus confertis.* Lin. Sp. Plant. 182 ; Physalis avec une tige d'arbrisseau, des branches flexibles, & des fleurs rassemblées en paquets.

Bacci-fera Indica, floribus ad foliorum exortus, fructu sulcato, decapyreno. Raii Hist. 1632 ; Plante des Indes, produisant des baies avec des fleurs qui sortent des feuilles, & un fruit sillonné, qui renferme dix semences.

Pevetti. Rheed. Mal. 4. p. 113. f. 55.

8°. *Physalis arborescens, foliis ovato-lanceolatis, integerrimis, oppositis, caule fruticoso.*

Physalis à feuilles ovales, en forme de lance, entieres & opposées, & à tige d'arbrisseau.

Alkekengi Americanum arborescens, fructu sphærico rubro, vesicâ atro-purpureâ. Houst. M. S S. ; Alkékenge d'Amérique en arbre, avec un fruit rouge & sphérique, & une vessie d'un pourpre foncé.

9°. *Physalis ramosa ramosissima, foliis villoso-viscosis, floribus pendulis.* Lin. Sp. Plant. ; Physalis le plus branchu, avec des feuilles velues & visqueuses, & des fleurs pendantes.

Physalis pubescens. Lin. Syst. Plant. tom. 1. p. 509. Sp. 10.

PHY   549

Alkekengi Virginianum fructu luteo. Tourn. Inst. 151 ; Alkékenge de Virginie, à fruit jaune.

Solanum Vesicarium Virginianum, procumbens, annuum, folio lanuginoso. Moris. Hist. 3. p. 527. S. 13. t. 3. f. 24.

10°. *Physalis angulata, ramosissima, ramis angulatis, glabris, foliis ovatis, dentatis.* Lin. Sp. Plant. 262. Kniph. cent. 7. n. 72 ; Physalis le plus branchu, avec des branches unies & angulaires, & des feuilles ovales & dentelées.

Solanum Vesicarium Indicum. Bauh. Pin. 166.

Alkekengi Indicum majus. Tourn. Inst. 151 ; grand Alkékenge des Indes.

Halicacabum, sivè, Solanum Indicum. Cam. Hort. 70. t. 17.

11°. *Physalis minima, ramosissima, foliis ovatis, acuminatis, sub-dentatis, petiolis longioribus ;* Physalis fort petit & branchu, avec des feuilles ovales à pointe aiguë, un peu dentelées, & postées sur de plus longs pétioles.

Alkekengi Indicum minimum, fructu virescente. Tourn. Inst. 151 ; le plus petit Alkékenge des Indes, à fruit verdàtre.

Solanum Vesicarium Indicum minimum. Herm. Lugd. - B. 569. f. 571. Sabb. Hort. 2. f. 64.

Pee-Inota-Inodien. Rheed. Mal. 10. t. 140. f. 71.

12°. *Physalis patula, ramosissima patula, ramis angulatis, glabris, foliis lanceolatis, pinnato-dentatis ;* Physalis branchu & étendu, dont les branches sont lisses & angulaires, les feuilles en forme de lance,

550 **PHY**

& découpées en especes de lobes.

13°. *Physalis villosa, ramosissima, ramis villosis, foliis ovatis, acuminatis, serrato-dentatis*; Physalis très-branchu, avec des branches velues, des feuilles ovales, à pointe aiguë, & dentelées en forme de scie.

Alkekengi Americanum annuum, ramosissimum, villosum, fructu rotundo, è luteo virescente. Houst. *MSS.*; Alkékenge d'Amérique, annuel, très-branchu, & velu, avec un fruit rond, & d'un jaune verdâtre.

14°. *Physalis cordata, caule erecto, ramoso, foliis ovatis, serrato-dentatis, petiolis pedunculisque longissimis*; Physalis avec une tige érigée & branchue, des feuilles ovales, dentelées & sciées, des pétioles & des pédoncules très-longs.

Alkekengi Americanum annuum, Lamii folio, fructu cordato. Houst. *MSS.*; Alkékenge d'Amérique, annuel, à feuilles d'Ortie morte, avec un fruit en forme de cœur.

15°. *Physalis maxima, caule erecto, ramoso, foliis ovato-lanceolatis, viscosis, fructu maximo, cordato*; Physalis avec une tige droite & branchue, des feuilles ovales en forme de lance, & visqueuses, & un gros fruit en forme de cœur.

Alkekengi Americanum annuum, maximum, viscosum. Houst. *M. S S.*; Alkékenge d'Amérique, annuel, très-grand, & visqueux.

16°. *Physalis Peruviana, caule erecto, ramoso, ramis angulatis, foliis sinuatis, calycibus acutangulis*; Physalis avec une tige

PHY

érigée & branchue, des branches angulaires, des feuilles sinuées, & des calices à angles aigus.

Alkekengi amplo flore violaceo. Feuill. *Obs.* 724; Alkékenge avec une grosse fleur violette.

La premiere espece est l'*Alkékenge* commun, dont on fait usage en Médecine; elle croît naturellement en Espagne & en Italie, & on la cultive depuis long-tems dans les jardins anglois : ses racines sont vivaces, & rampent à une grande distance au-dessous de la surface de la terre, si elles ne sont pas resserrées; elles poussent au printems plusieurs tiges de plus d'un pied de hauteur, & garnies de feuilles de différentes formes : quelques-unes sont angulaires & obtuses; d'autres sont oblongues & à pointe aiguë, mais elles sont toutes d'un vert foncé : elles sortent généralement par paires du même nœud sur le même côté de la tige, & sont supportées par de longs pétioles. Les fleurs naissent aux aisselles des tiges, sur de minces pédoncules; elles ont un pétale blanc, dont le tube est court & découpé sur ses bords en cinq angles étendus : dans le centre du tube est placé un germe rond, qui soutient un style mince, couronné par un stigmat obtus; ce style est accompagné de cinq étamines de la même longueur & terminées par des antheres jaunes, oblongues, érigées & jointes ensemble. Ces fleurs paroissent en Juillet, & produisent des baies rondes à-peu-près de la même gros-

PHY

feur d'une petite cerife, & renfermées dans une veffie gonflée, qui devient rouge en automne, & qui s'ouvre alors à fon extrémité, & laiffe voir une baie rouge, molle, charnue, & remplie de femences plates & en forme de rein; bientôt après la maturité du fruit, les tiges périffent jufqu'à la racine.

Cette plante fe multiplie aifément par femences, ou par la divifion de fes racines. Cette derniere méthode, quieft la plus prompte, eft généralement pratiquée. Ces racines peuvent être divifées & tranfplantées en tout tems, depuis que les tiges font flétries, jufqu'à ce que les racines commencent à repouffer au printems. Elles doivent être placées à l'ombre, & refferrées; fans quoi elles s'étendroient à une grande diftance de la premiere année: &, lorfque les tiges font éloignées les unes des autres, elles n'ont point d'apparence; elles ne deviennent belles qu'en automne, lorfque leurs veffies & les fruits font devenus rouges.

Les feuilles de ces plantes font rafraîchiffantes, & de la nature de celles de la *Morelle commune*; fes baies font diurétiques, & particulierement propres contre la gravelle & la pierre. Plufieurs obfervations ont prouvé qu'elles ont la propriété de chaffer une grande quantité de gravier, & qu'elles réuffiffent fouvent dans des circonftances où d'autres remedes ont été fans fuccès. Ces baies, bouillies dans du lait avec du fucre, guériffent les chaleurs

PHY 551

d'urine, qu'elles teignent alors en rouge; elles font auffi cicatrifer les ulceres des reins & de la veffie (1).

Vifcofa. La feconde efpece fe trouve à Buenos Ayrès: elle a une racine rampante, au moyen de laquelle elle fe multiplie confidérablement; elle pouffe un grand nombre de tiges liffes, d'un pied environ de hauteur, & divifées vers le fommet en petites branches rampantes, & garnies de feuilles en forme de cœur, ou ovales, de trois pouces environ de longueur fur deux de largeur à leur bâfe, entieres, rudes au toucher, d'un jaune pâle & verdâtre, alternes, & poftées fur des pétioles affez longs & rudes: fes fleurs fortent des aiffelles des tiges vers le fommet, fur des pédoncules longs & minces; elles font d'un jaune fale avec un fond pourpre: elles paroiffent dans les mois de Juin & Juillet, & produifent des baies vifqueufes, à-peu près de la même groffeur que celles de l'efpece commune, d'un jaune herbacé, & renfermées dans des veffies gonflées & d'un vert clair.

Cette plante fe multiplie aifément, en divifant fes racines au printems ou en au-

_____

(1) Les baies de l'Alkékenge font regardées comme apéritives, & comme très-diurétiques: on en prepare un vin qui a une grande réputation pour chaffer les graviers des reins & de la veffie, & pour faire ecouler les eaux des hydropiques.

Ces fruits entrent dans la compofition du Syrop de Chicorée de Charas.

tomne ; mais comme elle eſt trop tendre pour ſubſiſter en plein air pendant l'hiver en Angleterre, il faut la planter dans des pots , & la tenir à l'abri ſous un châſſis de couche en hiver, où elle puiſſe jouir de l'air dans les tems doux.

Penſylvanica. Les ſemences de la troiſieme eſpece m'ont été envoyées de la Virginie, où elle croît naturellement : elle a une racine vivace & une tige annuelle ; mais ſes racines ne rampent pas dans la terre comme celles des deux précédentes : ſes tiges, longues de deux pieds , reſtent couchées ſur la terre , ſi on ne leur fournit pas un ſoutien ; elles ſont garnies de feuilles ovales, de trois pouces de longueur ſur deux & demi de large , alternes, portées par de longs pétioles, d'un vert pâle, & ſciées à dentelures aiguës ſur leurs bords : ſes fleurs ſortent aux aiſſelles de la tige ſur de fort longs pédoncules ; elles ſont plus larges que celles de l'eſpece commune , & d'un jaune pâle : elles ſont remplacées par de fort petites baies jaunâtres, qui mûriſſent en automne , quand l'année eſt chaude ; mais dans des étés froids & humides, elles acquierent rarement toute leur perfection.

On multiplie cette plante par ſes graines, qu'il faut ſemer ſur une plate-bande chaude vers la fin de Mars. Quand les plantes pouſſent , on les éclaircit dans les endroits où elles ſont trop ſerrées, & on les tient nettes de mauvaiſes herbes juſqu'en automne , pour les transplanter alors dans une ſituation chaude , où elles réſiſteront à des hivers doux : mais les fortes gelées les détruiſent , ſi elles ne ſont pas abritées.

Virginiana. Les ſemences de la quatrieme eſpece m'ont été envoyées de Philadelphie par le Docteur BENSIL , qui l'a découverte dans ce pays ; elle a une racine vivace, compoſée de fortes fibres , de laquelle ſortent deux ou trois tiges velues, de neuf ou dix pouces de hauteur, & diviſées en pluſieurs branches garnies de feuilles ovales, en forme de lance , d'un vert pâle, velues, d'environ deux pouces & demi de longueur ſur un & demi de large, ſciées en dentelures aiguës ſur leurs bords , poſtées ſur de courts pétioles , & alternes : les fleurs qui ſortent ſur le côté des branches, à la bâſe des pétioles des feuilles , ont des pédoncules longs & minces ; leurs tubes ſont fort courts, mais plus élargis que ceux de la plupart des autres eſpeces de ce genre : ces fleurs ſont d'une couleur de ſoufre , & marquées d'un pourpre foncé dans le fond ; elles paroiſſent en Juillet, & ſont remplacées, dans les années chaudes, par des baies ovales & jaunâtres, qui mûriſſent en automne. Cette eſpece peut être multipliée par ſemences , comme la troiſieme , & les plantes qui en proviennent, exigent le même traitement.

La cinquieme eſpece, qui eſt originaire de Curaçao , en Amérique , a une racine vivace

PHY

& rampante, qui produit plufieurs tiges minces, d'un pied environ de hauteur, un peu ligneufes, mais qui durent rarement plus de deux ans : fes feuilles font alternes, & placées fur de courts pétioles ; elles ont deux pouces de longueur fur un & demi de large : les fleurs fortent des aiffelles de la tige vers le fommet, fur des pédoncules courts & minces ; les corolles font ovales, cotonneufes, petites, & de couleur de foufre, avec un fond d'un pourpre foncé : elles paroiffent en Juillet & Août, mais elles produifent rarement des baies en Angleterre.

On ne multiplie guere cette plante qu'en divifant fes racines au printems. Comme elle eft trop tendre pour fubfifter pendant l'hiver en Angleterre, fans le fecours d'une chaleur artificielle, il faut la tenir, durant cette faifon, dans une ferre tempérée ; mais pendant les mois de Juillet, Août & Septembre, on peut l'expofer en plein air dans une fituation chaude.

Somni-fera. La fixieme efpece, qui croît naturellement en Candie, en Sicile & en Efpagne, s'éleve, avec une tige d'arbriffeau, à près de trois pieds de hauteur, & fe divife en plufieurs branches érigées, couvertes d'un duvet laineux, & garnies de feuilles ovales, en forme de lance, de trois pouces à-peu-près de longueur fur un & demi de large au milieu, cotonneufes, & poftées fur de courts pétioles : fes fleurs fortent en paquets des parties latérales des branches ; elles

PHY 553

font petites, d'un blanc herbacé, très-feffiles, & produifent des baies prefque auffi groffes que celles de la premiere efpece, & qui deviennent rouges en mûriffant. Cette plante fleurit dans les mois de Juin & Juillet, & fes baies mûriffent en automne.

On la multiplie par fes graines, qu'on peut femer fur une terre légere au commencement d'Avril. Quand les plantes font parvenues à la hauteur de deux ou trois pouces, on les enleve avec précaution ; on les met chacune féparément dans de petits pots remplis d'une terre de jardin potager : on les tient à l'ombre jufqu'à ce qu'elles aient formé de nouvelles racines, & on les place enfuite dans un endroit où elles puiffent refter jufqu'au commencement d'Octobre, pour les renfermer alors dans la ferre ; car elles font trop tendres pour fubfifter en plein air pendant l'hiver. Ainfi, on doit les traiter comme les autres plantes de la ferre, en obfervant de les arrofer très-légerement en hiver. Elles fubfiftent plufieurs années, fi elles ne font pas conduites trop délicatement.

Flexuofa. La feptieme efpece, naît fans culture fur la côte du Malabar & au Cap de Bonne-Efpérance ; elle s'éleve à la hauteur de cinq ou fix pieds, & produit des branches longues, flexibles, couvertes d'une écorce grife, & garnies de feuilles oblongues, ovales, & fouvent oppofées, quelquefois placées par trois autour

des branches, & feffiles : les fleurs fortent en paquets aux ailes des feuilles ; elles font petites, d'un jaune herbacé, & produifent des baies rondes, purpurines, & à dix cellules, qui renferment chacune une femence. Cette plante fleurit dans les mois de Juillet & Août ; mais fes baies ne mûriffent point en Angleterre, à moins que la faifon ne foit très-chaude.

On multiplie cette efpece par fes graines, qu'il faut femer fur une couche de chaleur modérée, en obfervant de les tenir à l'ombre jufqu'à ce qu'elles aient pouffé ; alors on leur donne de l'air chaque jour dans les tems chauds, pour les empêcher de filer, & on les traite comme les autres plantes exotiques. Quand elles ont atteint la hauteur de trois ou quatre pouces, on les enleve avec précaution ; on les met chacune féparément dans de petits pots remplis d'une terre légere & marneufe ; on les place fous un vitrage de vieille couche, & on les tient à l'ombre jufqu'à ce qu'elles aient formé de nouvelles racines ; enfuite on les accoutume par dégrés à fupporter le plein air, auquel on les expofe tout-à-fait en Juillet, en les plaçant dans une fituation abritée, où on les laiffera jufqu'à la fin de Septembre, pour les transporter alors à l'abri. Durant le premier hiver, on les tiendra dans une ferre de chaleur modérée ; mais quand elles ont une fois acquis de la force, elles peuvent fubfifter pendant l'hiver dans une bonne ferre

Arborefcens. La huitieme efpece a été découverte à Campêche par le Doɗteur Houstoun, qui a envoyé fes femences en Angleterre ; elle a une tige d'arbriffeau de douze pieds de hauteur, & divifée vers fon fommet en plufieurs petites branches, couvertes d'une écorce grife & velue, & garnies de feuilles ovales & en forme de lance, dont celles du haut font oppofées, & celles du bas alternes, & de trois à quatre pouces de long fur deux de large au milieu, terminées en pointe à chaque extrémité, d'un vert pâle & cotonneux : fes fleurs fortent des aiffelles des tiges vers l'extrémité des branches, quelquefois feules, & quelquefois deux enfemble fur le même nœud, oppofées, & poftées fur des pédoncules courts & penchés ; elles font petites, d'un jaune pâle, fale & pourpre au fond, & font remplacées par de petites baies rouges, fphériques, & enfermées dans une veffie ovale, & d'un pourpre foncé. Ces fleurs paroiffent dans les mois de Juin & Juillet ; mais elles ne produifent des baies que dans les années très-chaudes.

On peut multiplier cette efpece par femences, comme la précédente, & les plantes exigent le même traitement : mais comme elles font moins dures, il faut les tenir en hiver dans une ferre médiocrement chaude, & au milieu de l'été, les placer dans une fituation abri-

tée, où on les laisse pendant trois mois : car, si on les tenoit toujours renfermées dans la serre, elles fileroient, & ne fleuriroient point. On peut aussi en faire des boutures, qui prennent aisément racine, en les plantant dans des pots au printems & en été, & en les plongeant dans une couche de chaleur tempérée. On les traite comme celles de la sixieme espece.

Ramosa. La neuvieme est une plante annuelle qu'on rencontre dans la Virginie : elle pousse des branches près de la terre, qui s'étendent fort loin, & se couchent souvent ; elles sont angulaires, remplies de nœuds, & sous-divisées en plusieurs autres plus petites, garnies de feuilles velues, visqueuses, presqu'en forme de cœur, postées sur des pétioles assez longs, de trois pouces environ de longueur sur presque deux de large, & hachées en dentelures aiguës sur leurs bords : leurs fleurs sortent des parties latérales des branches sur des pédoncules courts, minces, & penchés ; elles sont d'un jaune herbacé, & ont une tache foncée dans le fond ; à ces fleurs succedent de grosses vessies gonflées, & d'un vert clair, qui renferment des baies aussi grosses que des *Cerises* ordinaires, & jaunâtres lorsqu'elles sont mûres. Cette plante fleurit dans les mois de Juin & Juillet, & ses baies mûrissent en automne.

Quand on permet à ces graines de se répandre, elles produisent au printems des plantes qui n'exigent aucun autre soin que d'être éclaircies & tenues nettes de mauvaises herbes. Si on les seme au printems sur une plate-bande ordinaire, les plantes leveront fort bien, & n'auront besoin d'aucun autre soin.

Angulata. La dixieme espece est aussi une plante annuelle, qui croît sans culture dans les Isles de l'Amérique : elle s'éleve avec une tige droite & branchue à la hauteur de deux ou trois pieds ; ses branches sont lisses, angulaires, & garnies de feuilles en forme de lance, terminées en pointe aiguë, & fortement dentelées sur leurs bords : ses fleurs naissent vers l'extrémité des branches, sur des pédoncules courts & minces ; elles sont fort petites, d'un blanc sale, & produisent des baies aussi grosses que des *Cerises* ordinaires, & renfermées dans une vessie angulaire. Ces baies deviennent d'une couleur jaunâtre en mûrissant.

On multiplie cette espece par ses graines, qu'il faut semer sur une couche de chaleur tempérée ; quand les plantes ont fait assez de progrès, on les transplante sur une nouvelle couche chaude, pour les faire avancer, & on les traite ensuite comme le *Capsicum.* Lorsqu'elles sont devenues fortes & assez dures pour supporter le plein air, on peut les enlever en mottes, & les placer dans une plate-bande chaude, en observant de les arroser, & de les tenir à l'ombre jusqu'à ce qu'elles aient formé

de nouvelles racines ; après quoi elles n'exigeront plus aucun autre soin que d'être tenues nettes de mauvaises herbes.

Minima. La onzieme espece, qui est aussi originaire de l'Amérique, est une plante annuelle, dont les tiges sont fort branchues, & qui s'élevent rarement à plus d'un pied de hauteur ; ses feuilles sont ovales, d'un vert foncé, & postées sur de longs petioles : ses fleurs sont petites, blanches, & placées sur de courts pedoncules ; ses baies sont petites & vertes, lorsqu'elles sont mûres.

Patula. La douzieme a été découverte à la Vera-Cruz par le Docteur HOUSTOUN. Cette plante, qui est basse & annuelle, pousse une tige fort branchue, étendue & garnie de feuilles en forme de lance, cotonneuses, & dentelées profondément sur leurs bords. Ces dentelures sont opposées, & placées régulierement en forme de feuilles ailées ; les branches sont lisses & angulaires ; les fleurs sont petites & blanches, & le fruit petit & jaunâtre, lorsqu'il est mûr.

Villosa. La troisieme espece a été également découverte par le Docteur HOUSTOUN à la Vera-Cruz. Cette plante est annuelle, & pousse une tige fort branchue & velue ; ses feuilles sont ovales, à pointe aiguë, & sciées sur leurs bords ; ces feuilles sont petites & d'un jaune pâle ; le fruit est rond, de la grosseur d'une Cerise, & d'un vert jaunâtre lorsqu'il est mûr.

Cordata. La quatorzieme, que le Docteur HOUSTOUN a aussi découverte à la Vera-Cruz, est une plante annuelle, qui pousse une tige droite, branchue, de deux pieds environ de hauteur, & garnie de feuilles ovales, dentelées sur leurs bords en forme de scie, portées sur de longs pétioles, & de couleur pourpre en automne : ses fleurs sont petites, blanches, & soutenues sur de fort longs pedoncules ; à ces fleurs succedent des baies presque aussi grosses & de la même forme que les grosses *Cerises noires,* appelées *Cerises en cœur*, & d'un vert jaunâtre, avec quelques rayons pourpre.

Maxima. La quinzieme espece a encore été trouvée par le même Docteur HOUSTOUN dans le même pays que la précédente. Cette plante, qui est annuelle, pousse une tige lisse, érigée, branchue, de trois pieds environ de hauteur, & garnie de feuilles ovales, en forme de lance, visqueuses, & supportées par de longs pétioles : ses fleurs sont d'un jaune pâle, & petites ; elles sont remplacées par de gros fruits en forme de cœur, & d'un jaune pâle, lorsqu'ils sont mûrs.

Les cinq dernieres especes ci dessus se multiplient par semences, comme la dixieme, & elles exigent le même traitement.

Peruviana. La seizieme est originaire de la Perse, d'où M. DE JUSSIEU le jeune en a envoyé les semences. Cette plante est annuelle, & s'éleve à la hauteur de

PHY

de quatre ou cinq pieds, avec
une tige forte, angulaire, herbacée, de couleur tirant fur
le pourpre, & divifée en plufieurs branches angulaires, qui
s'étendent beaucoup de tous
côtés, & font garnies de feuilles oblongues, profondément
finuées fur leurs bords, & d'un
vert foncé ; les pédoncules des
fleurs font courts ; le calice eft
large, en forme de cœur, &
profondément découpé en cinq
fegmens : les fleurs, qui font
larges, en forme de cloche
ouverte, & d'un bleu clair,
produifent des baies à peu-près
de la groffeur d'une *Cerife* ordinaire, & renfermées dans
une groffe veffie gonflée,
& à cinq angles aigus. Cette
plante fleurit en Juillet, &
fes femences mûriffent en automne. Si l'on donne à ces
graines le tems de fe répandre, elles produifent des plantes au printems fuivant ; & ,
en les femant au printems fur
une terre riche, elles lèveront
aifément. Ces plantes peuvent
être placées dans les plates-
bandes du parterre, où il faut
laiffer affez d'efpace entr'elles ;
car, fi la terre en eft bonne, elles y deviendront très-
groffes.

Le Pere FEUILLÉE, qui le
premier a trouvé cette plante
au Pérou, en a donné la figure
& defcription, & a beaucoup
vanté fes propriétés ; il dit
que les Indiens font un grand
ufage de fes baies, pour chaffer le gravier & foulager dans
les rétentions d'urine ; il donne
la manière d'en faire ufage ;

Tome V.

PHY 557

elle confifte à broyer quatre
ou cinq baies dans de l'eau
commune ou dans du vin blanc,
que l'on fait boire au malade.
Il affure que le fuccès de ce
remede eft étonnant.

PHYTOLACCA. *Tourn. Inft.
R. H.* 299. *tab.* 154. *Lin. Gen.
Plant.* 521. Cette plante eft
ainfi appelée de φυτὸν, une
plante & *laque* couleur, parce
qu'on en fait un rouge couleur de laque. [*American
Nightshade.*] Morelle *ou* Raifin
d'Amérique.

Caraĉteres. La fleur n'a point
de pétales, fuivant quelques-
uns, & point de calice, fuivant d'autres, parce que l'enveloppe des parties de la génération étant colorée, ces derniers la regardent comme la
corolle. Cette enveloppe eft
compofée de cinq feuilles ou
pétales ronds, concaves, étendus & perfiftans ; la plupart
des fleurs ont dix étamines auffi
longues que les pétales, & terminées par des antheres rondes ;
elles ont dix germes comprimés,
orbiculaires, joints enfemble en-
dedans, mais féparés au dehors,
& fur lefquels font fitués dix
ftyles fort courts, réfléchis &
couronnés par des ftigmats
fimples. Ces germes fe changent, quand la fleur eft paffée, en une baie comprimée,
orbiculaire, fillonnée par dix
rainures profondes, & à dix
cellules qui contiennent chacune une femence fimple, liffe,
& en forme de rein.

Ce genre de plantes eft
rangé dans la cinquieme fection de la dixieme claffe de

N a

LINNÉE, qui renferme celles dont les fleurs ont dix étamines & dix styles.

Les especes font :

1°. *Phytolacca vulgaris, floribus decandriis decagynis.* Hort. *Cliff.* 177. *Hort. Ups.* 117. *Gron. Virg.* 161. *Roy. Lugd.-B* 222. *Mat. Med.* 118 ; Raisin d'Amérique, dont les fleurs ont dix étamines & dix styles.

Phytolacca decandria, Lin. *Syst. Plant.* tom. 2. p. 406. *Sp.* 2.

Phytolacca Americana, majori fructu. Tourn. Inst. 229 ; Morelle d'Amérique avec un gros fruit, ordinairement appelée *Poche de Virginie.*

Solanum racemosum Americanum. Pluk. Alm. 353. *t.* 225. *f.* 3.

2°. *Phytolacca Mexicana, foliis ovato-lanceolatis, floribus sessilibus ;* Phytolacca avec des feuilles ovales, en forme de lance, & des fleurs sessiles.

Phytolacca octandria. Lin. *Syst. Plant.* tom. 2. p. 406. *Sp.* 1.

Phytolacca Mexicana, baccis sessilibus. Hort. Elth. 318 ; Morelle du Mexique, dont les baies font sessiles.

Jamma goba. Kæmph. Amæn. 828 *f.* 829.

3°. *Phytolacca Icosandria, floribus Icosandriis, decagynis.* Lin. *Sp.* 631 ; Phytolacca avec plusieurs étamines fixées au réceptacle.

Phytolacca spicis florum longissimis, radice annua. Raisin d'Amérique, avec les plus longues grappes de fleurs, & une racine annuelle.

4°. *Phytolacca dioïca, floribus dioicis, caule arboreo, ramoso ;* Raisin d'Amérique, avec une tige en arbre, & des fleurs mâles & femelles fur différentes plantes.

Vulgaris. La premiere espece croît naturellement en Virginie, en Espagne & en Portugal ; elle a une racine fort épaisse, charnue, aussi grosse que la jambe, & divisée en plusieurs fibres épaisses & charnues, qui pénètrent profondément dans la terre. Quand elles font devenues grosses, elles poussent trois ou quatre tiges herbacées, de la grosseur d'un gros bâton, de couleur pourpre, de six ou sept pieds de hauteur, & divisées au sommet en plusieurs branches garnies de feuilles de cinq pouces environ de longueur fur deux & demi de large, rondes à leur bâse, terminées en pointe, placées sans ordre fur de courts pétioles, & d'un vert foncé, qui, en automne, se change en une couleur de pourpre ; les pédoncules sortent aux nœuds des branches & à leurs divisions ; ils ont environ cinq pouces de longueur, font nus vers la bâse ; & depuis leur milieu jusqu'à l'extrémité ils soutiennent un nombre de fleurs rangées fur chaque côté, comme celles de la *Groseille* commune : chaque fleur, qui est postée fur un pédoncule particulier d'un demi-pouce de longueur, est composée de cinq pétales de couleur purpurine, qui renferment dix étamines & dix styles. Quand ces fleurs font fanées, le germe devient une baie comprimée à dix sillons & à dix cellules

PHY

remplies de femences liffes. Cette plante fleurit dans les mois de Juillet & Août ; & quand l'année eft favorable, fes baies mûriffent en automne.

On peut la multiplier par fes graines, qu'on feme au printems fur une terre légere. Quand les plantes ont pouffé, on les place dans les plates-bandes des grands jardins, en leur donnant beaucoup d'ef-pace pour croître. On ne doit pas les mettre trop près des plantes, de peur qu'elles ne les étouffent & ne les détrui-fent, parce qu'elles deviennent très-grandes, fur-tout dans une bonne terre. Quand elles ont pris racine, elles n'exigent plus aucun autre foin que d'être débarraffées des mauvaifes herbes : elles fleuriffent & produi-fent leurs fruits en automne ; leurs tiges périffent aux pre-mieres gelées : mais leurs ra-cines fubfiftent dans la terre, repouffent au printems fuivant, & durent plufieurs années, fur-tout quand elles font plan-tées dans un fol fec ; car fi l'humidité de l'hiver féjournoit autour d'elles, elles feroient bientôt attaquées de pourriture. Comme les fortes gelées les détruifent quelquefois, il eft néceffaire de les couvrir avec du terreau, quoique nos hi-vers ordinaires ne leur faffent aucun tort.

PARKINSON prétend que les habitans de l'Amérique Septen-trionale font ufage du jus de cette racine, comme d'un pur-gatif ordinaire, & que deux cuillerées de ce fuc operent & font beaucoup d'effet. Quel-

PHY   559

ques Charlatans, en dernier lieu, vouloient guérir des chancres avec les feuilles de cette plante ; mais je ne les ai jamais vu réuffir. Les habi-tans du nord de l'Amérique font bouillir les jeunes re-jettons de cette efpece, & les mangent en guife d'Epinars. Le jus des baies teint le papier & le linge en une belle cou-leur pourpre ; mais elle ne dure pas long-tems. Cependant, fi l'on trouvoit le moyen de fixer cette teinture, elle pour-roit devenir très-utile. En Por-tugal, les Vignerons ont fait ufage, pendant plufieurs an-nées, du fuc exprimé des baies de cette plante, en le mêlant avec leur vin rouge pendant la vendange, pour lui donner une couleur plus foncée ; mais s'ils y en mettoient une trop grande quantité, le vin pren-droit un goût fort défagréa-ble. On en a porté des plain-tes au Roi de Portugal, qui a ordonné de couper & dé-truire les plantes de *Phytolacca*, avant qu'elles aient produit des baies, afin d'en empêcher l'ufage à l'avenir, & de réta-blir par-là la réputation des vins du pays. J'ai bu de ces vins fans mélange, & les ai trouvés d'un goût plus agréa-ble ; mais je ne puis affurer que cette pratique ait abfolu-ment ceffé dans ce pays.

Mexicana. La feconde efpece naît fans culture dans l'Amé-rique Efpagnole. Le feu Doc-teur HOUSTOUN l'a trouvée en grande abondance à la Vera-Cruz, où les habitans en font ufage pour la table. Cette plan-

PHY

te eſt bis-annuelle, & ne ſubſiſte gue re plus de deux ans.
Quand elle fleurit & produit
une grande quantité de ſemences dans la premiere année,
elle périt ſouvent au printems
ſuivant : elle a une tige herbacée, de deux pieds de hauteur, de la groſſeur d'un doigt,
& diviſée au ſommet en deux
ou trois branches courtes, &
garnies de feuilles ovales en
forme de lance, d'environ ſix
pouces de longueur ſur preſque trois de large, & terminées en pointe à chaque extrémité. Leur côte du milieu eſt
forte, & donne origine à pluſieurs nervures tranſverſales,
qui ſe prolongent juſqu'aux
bords ; elles ſont d'un vert
foncé, portées ſur des pétioles
d'un pouce & demi de lon-
gueur, & placées ſans ordre
ſur la tige. Les pédoncules, qui
ſortent ſur les côtés des branches oppoſés aux feuilles, ont
ſept ou huit pouces de long ;
leur bâſe eſt nue dans la longueur d'environ deux pouces,
& le reſte eſt garni de fleurs
blanches marquées d'un pourpre rougeâtre au milieu, ſeſſiles, & découpées en cinq ſegmens preſque juſqu'au fond :
elles ont depuis huit juſqu'à
quatorze étamines, & dix ſtyles, & ſont remplacées par des
baies plates, ſillonnées par dix
rainures profondes, & diviſées
en autant de cellules qui contiennent chacune une ou deux
ſemences liſſes.

Cette plante fleurit dans les
mois de Juillet & Août, &
ſes ſemences mûriſſent ſur la
fin de l'automne.

Icoſandria. La troiſieme eſpece naît ſpontanément au Malabar, d'où ſes ſemences m'ont
été envoyées. Cette plante eſt
annuelle, & périt toujours
auſſi-tôt que ſes ſemences ſont
mûres ; de ſorte qu'en cela
elle differe beaucoup de la
premiere ; elle s'éleve à la
hauteur de deux ou trois pieds,
avec une tige herbacée, ſil-
lonnée par pluſieurs rainures
dans toute ſa longueur, & qui
prend une couleur de pourpre
ſur la fin de l'été. Cette tige
ſe diviſe au ſommet en trois
ou quatre branches garnies de
feuilles en forme de lance, de
ſix ou ſept pouces de longueur
ſur preſque trois de largeur
au milieu, terminée en pointe
à chaque bout, d'un vert foncé, poſtées ſur de courts pétioles, quelquefois alternes,
d'autres fois oppoſées, & ſouvent placées obliquement ſur
les pétioles. Les pédoncules
ſortent ſur le côté des branches oppoſées aux feuilles ; ils
ont neuf ou dix pouces de
longueur ; leur partie baſſe eſt
nue, comme dans les autres
eſpeces, & le reſte eſt garni
de fleurs plus groſſes que celles des précédentes, blanches
en-dehors, de couleur herbacée ſur leurs bords, purpurines en-dedans, & poſtées ſur
de courts pédoncules ; elles
n'ont pas toutes le même nombre d'étamines ; quelques-unes
n'en ont que huit, & d'autres
neuf ou onze, qui ſont terminées par des antheres rondes. Les fleurs ſont remplacées
par des baies molles, comprimées, orbiculaires, diviſées en-

dehors en dix fillons profonds , & en-dedans en autant de cellules , dont chacune contient une femence liffe & d'un noir luifant. La grappe de fleurs eft fort étroite au fommet,'où elle eft ordinairement inclinée. Cette plante fleurit dans les mois de Juillet & Août ; fes femences mûriffent en automne, & la plante périt bientôt après.

Les baies de cette efpece font fort fucculentes , & leur fuc teint le papier & le linge en une belle couleur de pourpre , mais qui n'eft pas durable.

Ces deux efpeces étant moins dures que la premiere , il faut les femer au printems fur une couche de chaleur tempérée. Quand les plantes font en état d'être enlevées, on les place fur une autre couche chaude, pour les faire avancer ; on les tient à l'ombre jufqu'à ce qu'elles aient formé de nouvelles racines , & on les traite enfuite comme les autres plantes tendres & exotiques. Au commencement de Juillet, on peut les tranfplanter dans une plate-bande chaude ou dans des pots remplis d'une terre riche & légere , & les abriter du foleil jufqu'à ce qu'elles aient pris racine ; après quoi elles n'exigeront que d'être conftamment arrofées dans les tems fecs, & d'être tenues nettes de mauvaifes herbes. Comme ces plantes perfectionnent leurs femences chaque automne, on peut aifément en conferver l'efpece.

Dioïca. La quatrieme fe trouve au Mexique , d'où fes femences ont été envoyées à Paris, il y a quelques années , & depuis plus long-tems en Efpagne, où l'on trouve, dans plufieurs jardins , de ces arbres qui ont à préfent plus de vingt pieds de hauteur , & dont quelques-uns (d'après le rapport de plufieurs perfonnes dignes de foi) produifent des fleurs mâles , & d'autres des fleurs femelles ; mais comme les plantes du jardin de Chelféa n'ont point encore donné de fleurs entierement épanouïes, je n'en puis parler d'après mes propres obfervations.

Cette plante a une tige forte , ligneufe , & auffi groffe que la jambe, qui pouffe plufieurs branches garnies de feuilles ovales, en forme de lance, de fix pouces de longueur fur prefque trois de largeur , féparées dans leur milieu par une forte côte , & de couleur pourpre quand elles font parvenues à leur grandeur ordinaire : fes fleurs font produites à la bâfe des pétioles des feuilles dans un rameau femblable à celui des autres efpeces ; mais comme les plantes du jardin de Chelféa n'ont produit des fleurs que fort tard, elles font tombées avant de s'ouvrir.

On peut multiplier cette efpece par boutures , pendant tout l'été ; il faut les planter dans des pots remplis de terre légere, les plonger dans une couche de chaleur tempérée , les couvrir avec des cloches de verre , pour en exclure l'air , & les tenir à l'ombre : cinq ou fix femaines après, elles auront pouffé des raci-

PIE

nes : alors on les mettra chacune féparèment dans des petits pots , que l'on plongera dans la couche ; on les tiendra à l'ombre jufqu'à ce qu'elles aient formé de nouvelles fibres , & on les accoutumera enfuite par dégrès au plein air , auquel on les expofera entierement jufqu'à la fin de Septembre , pour les remettre alors dans une ferre de chaleur tempérée , où elles doivent paffer l'hiver ; car elles ne réfifteroient pas dans la ferre, à moins qu'elle ne fût bien chaude.

PICEA, LA PESSE ou PECE ou EPICIA ou FAUX SAPIN. *Voy.* ABIES PICEA. *L.*

PIED D'ALOUETTE ou HERBE AUX POUX. *Voy.* DELPHINIUM CONSOLIDA. *L.*

PIED DE CANARD ou POMME DE MAI. *Voy.* PODOPHYLLUM. *L.*

PIED DE CHAT. *Voy.* GNAPHALIUM DIOÏCUM. *L.*

PIED D'ÉLÉPHANT. *Voy.* ELEPHANTOPUS SCABER.

PIED DE GRIFFON ou ELLÉBORE NOIR. *Voy.* HELLEBORUS FŒTIDUS. *L.*

PIED DE LIEVRE. *Voyez* TRI-FOLIUM ARVENSE. *L.*

PIED DE LION DE CANDIE. *Voy.* CATANANCHE. *L.*

PIED DE LION. *Voy.* ALCHEMILLA VULGARIS. *L.* & LEONTICE. *L.*

PIED D'OIE. *Voy.* CHENOPODIUM. *T.*

PIED D'OISEAU. *Voy.* ORNITHOPUS PERPUSILLUS. *L.*

PIED DE PIGEON. *Voy.* GERANIUM ROTUNDIFOLIUM. *L.* GERANIUM PERENNE. *L.*

PIE

PIED DE POULIN ou PAS D'ASNE. *Voy.* TUSSILAGO. *L.*

PIED DE VEAU. *Voy.* ARUM. *L.*

PIERCEA. *Solanoïdes. Tourn. Act. Par. 1706. Rivina. Lin. Gen. Plant. nov. ed. n. 174. Plum. 39.* [*The Percy.*]

Caractere. La fleur n'a point de pétales ; le calice , qui renferme les parties de la génération , eft compofé de quatre feuilles oblongues , ovales , & colorées, auxquelles quelques-uns donnent le nom de pétales : la fleur a quatre étamines érigées , poftées l'une près de l'autre , & terminées par de petites antheres ; dans fon centre eft placé un germe gros & rond , qui foutient un ftyle court , & couronné par un ftigmat obtus. Ce germe fe change dans la fuite en une baie ronde , poftée fur le calice réfléchi , & à une cellule qui renferme une femence rude & de la même forme.

J'ai pris la liberté de donner à ce genre de plantes le nom du Duc de NORTHUMBERLAND , qui non-feulement encourage beaucoup la Botanique , mais qui eft auffi lui-même fort inftruit dans cette fcience.

TOURNEFORT l'a d'abord placé avec le *Phytolacca* , & n'en a fait qu'une efpece de ce genre ; mais comme les fleurs du *Phytolacca* ont cinq pétales ou feuilles au calice , & dix étamines , & que les fleurs de celle-ci n'ont que quatre pétales & huit étamines ; que les baies du *Phytolacca* ont dix cellules , & que celles-ci n'en ont qu'une , el-

PIE

les ne peuvent être réunies. TOURNEFORT en a conſtitué un nouveau genre, ſous le titre de *Solanoïdes*, dont il a publié les caractères dans les *Mémoires de l'Academie*, pour l'année 1706 ; mais comme tous les titres dont la terminaiſon eſt en *oïdes*, ont été changés par les derniers Botaniſtes, je joindrai celui-ci à la premiere ſection de la huitieme claſſe de LINNÉE, qui l'a ſuppoſé être le même que le *Rivina* ; ce qui l'a engagé à appliquer ce titre à cette plante, croyant que PLUMIER s'étoit trompé, quand il l'a deſſinée avec huit étamines : mais le *Rivina* de PLUMIER eſt tout-à-fait différent, & les fleurs de celle-ci ont effectivement huit étamines, comme PLUMIER l'a repréſenté.

Les eſpeces ſont :

1°. *Piercea glabra, foliis ovato-lanceolatis, glabris* ; Piercea avec des feuilles ovales, liſſes, & en forme de lance.

Solanoides Americana, Circææ foliis glabris. Tourn. Act. Par. 1706 ; Solanoïde avec des feuilles liſſes d'Herbe des Magiciennes.

Rivina lævis. Lin. Syſt. Plant. tom. 1. *p.* 346. *Sp.* 2.

2°. *Piercea tomentoſa, foliis cordatis, pubeſcentibus* ; Piercea à feuilles cotonneuſes & en forme de cœur.

Solanum Barbadenſe racemoſum, minus, tinctorium. Pluk. Alm. 333. t. 112. f. 2.

Solanoïdes Americana, Circææ foliis caneſcentibus. Tourn. Act. Par. 1706 ; Solanoïde d'Amérique, à feuilles blanches,

comme celles de l'Herbe des Magiciennes.

Amaranthus bacci-fer, Circææ foliis. Comm. Hort. 1. *p.* 127. *t. 66.*

Rivina humilis. Lin. Syſt. Plant. tom. 1. *p.* 346. *Sp.* 1.

Glabra. Ces plantes croiſſent naturellement dans la plupart des Iſles des Indes Occidentales ; mais la premiere eſpece y eſt la plus commune ; elle s'éleve à trois ou quatre pieds de hauteur, avec une tige mince & herbacée, qui devient un peu ligneuſe vers le bas en vieilliſſant, & ſe diviſe en pluſieurs branches herbacées, angulaires, & garnies de feuilles ovales, en forme de lance, de quatre pouces environ de longueur ſur deux de large au milieu, d'un vert clair, & ſupportées par des pétioles minces, & d'un pouce & demi de longueur. Les pédoncules ſortent de côté ſur les branches, à la baſe des pétioles des feuilles ; ils ont quatre à cinq pouces de longueur, & ſoutiennent un grand nombre de fleurs petites & blanches, rangées des deux côtés vers leur partie haute. Ces fleurs produiſent de petites baies rouges, remplies d'un ſuc de même couleur, & dans chacune deſquelles eſt renfermée une ſemence dure de la même forme.

Les fleurs ſe ſuccedent ſur la même plante durant la plus grande partie de l'année, & ſont remplacées par des baies qui mûriſſent les unes après les autres ; de ſorte que les plantes en ſont raremen

564 **PIE**

deſtituées ; & , quoique les fleurs n'aient que peu d'apparence , cependant leurs longues grappes de baies d'un rouge brillant , qui pendent après toutes les branches durant une grande partie de l'année , font un agréable effet.

Tomentoſa. La ſeconde eſpece s'éleve plus haut que la premiere , & ſes branches ſont plus érigées ; ſes feuilles ſont plus petites , en forme de cœur, & couvertes d'un duvet court & velu : les grappes de fleurs ſont moins longues , & les fleurs ſont plus éloignées les unes des autres ; elles ſont placées ſur de plus longs pédoncules , ſe ſuccedent , & produiſent des fruits mûrs durant la plus grande partie de l'année , comme la précédente.

Culture. On multiplie ces plantes par leurs graines , qu'il faut mettre en terre auſſi-tôt qu'elles ſont mûres ; car ſi on les conſerve long-tems, elles croiſſent rarement dans la même année : on les ſeme dans des pots remplis de terre légere , & on les plonge dans une couche de chaleur modérée. Quand les plantes pouſſent , on les tient nettes de mauvaiſes herbes ; on les arroſe légerement , à meſure que la terre ſe deſſeche ; & lorſqu'elles ont atteint la haureur de deux pouces , on les tranſplante chacune ſéparément dans de petits pots d'un ſou , remplis de terre légere ; on les plonge dans une couche de chaleur modérée ; on les tient à l'ombre juſqu'à ce qu'elles aient formé de nouvelles raci-

PIM

nes , & on les traite enſuite comme les autres plantes exotiques , en leur donnant de l'air chaque jour , à proportion de la chaleur de la ſaiſon , & en les arroſant toutes les fois qu'elles l'exigent. Quand ces plantes ont acquis de la force , on les place ſur les tablettes de la ſerre chaude , & on les y laiſſe conſtamment ; car elles ſont trop tendres pour profiter en plein air dans ce pays, même pendant le tems le plus chaud de l'année.

Le ſuc de ces baies teint le papier & le linge en un rouge brillant. Je m'en ſuis ſervi pluſieurs fois pour colorer des fleurs , ce qui m'a bien réuſſi : j'ai exprimé le ſuc de ces baies , j'y ai ajouté une certaine quantité d'eau commune, & après l'avoir mis dans une phiole & l'avoir bien ſecoué, pour mêler l'eau avec la teinture, j'y ai plongé des tiges de *Tubéreuſes* & de *Narciſſes* blanches & doubles , fraîchement cueillies ; & dans l'eſpace d'une nuit , ces fleurs ont été agréablement panachées de rouge.

PIGNON D'INDE *ou* Ricinoïdes. *Voyez* Jatropha Curcas.

PIGNON D'INDE *ou* Ricin *ou* Medicinier. *Voy.* Ricinus.

PILOSELLA. *Voyez* Hieracium.

PILOSELLE *ou* Oreille de Rat. *Voy.* Auricula muris.

PIMENT *ou* Botrys vulgaire. *Voyez* Chenopodium Botrys.

PIMENT ROYAL *ou* Gale *Voyez* Myrica Gale.

PIMENT *ou* Toute-Epice.

Voyez Caryophyllus. Pi-
menta.

PIMPINELLA. *Lin. Gen.
Plant.* 328. *Tragoſelinum. Tourn.
Inſt. R. H.* 309. *tab.* 163. [*Bur-
net ſaxifrage.*] Pimprenelle ,
Boucage, Bouquetine , Pimprenelle blanche *ou* Pimprenelle-Saxifrage.

Caractères. La fleur eſt en
ombelle ; l'ombelle principale
eſt compoſée de pluſieurs
rayons ou petites ombelles,
dont aucune des fleurs n'a d'enveloppe, & dont les calices ſont
preſque inviſibles : la plus
grande ombelle eſt uniforme ;
les corolles ont cinq pétales
en forme de cœur, courbés ,
& à-peu-près égaux : la fleur
a cinq étamines plus longues
que les pétales, & terminées
par des antheres rondes. Le
germe, qui eſt placé ſous la
fleur , ſoutient deux ſtyles
courts, & couronnés par des
ſtigmats obtus. Ce germe ſe
change dans la ſuite en un
fruit oblong, ovale, diviſé en
deux parties, & dans lequel
ſont renfermées deux ſemences oblongues , unies ſur un
côté , convexes & ſillonnées
ſur l'autre.

Ce genre de plantes eſt rangé
dans la ſeconde ſection de la
cinquieme claſſe de Linnée ,
qui comprend celles dont les
fleurs ont cinq étamines &
deux ſtyles.

Les eſpeces ſont :

1°. *Pimpinella major , ſoliis
pinnatis , foliolis cordatis , ſerratis , ſummis ſimplicibus , trifi-
dis ;* Boucage avec des feuilles
aîlées près de la racine, dont
les lobes ſont en forme de
cœur, & ſciés, & des feuil-

les ſimples à trois pointes au
ſommet de la tige.

*Pimpinella magna. Lin. Syſt.
Plant. t.* 1. *p.* 723. *Sp.* 2.

*Tragoſelinum majus , umbellâ
candidâ. Tourn. Inſt. R. H.* 309 ;
le plus grand Boucage, avec
une ombelle blanche , Perſil
des bois.

Saxi-fraga magna. Dodon. Purg.
494. *Pempt.* 315 ; la grande
Pimprenelle-Saxifrage.

*Pimpinella Saxi-fraga major ,
umbellâ rubente. Bauh. Pin.* 159.
Riv. ſ. 60 ; Pimprenelle-Saxifrage, à fleurs rouges ; variété de cette premiere eſpece.

2°. *Pimpinella Saxi-fraga ,
foliis pinnatis , foliolis radicalibus , ſub-rotundis , ſummis linearibus. Lin. Sp. Plant.* 263.
Pollich. pal. n. 305. *Jacq. Auſtr.*
4. *ſ.* 395. *Gmel. Sib.* 1. *p.* 220.
Mattuſch. Sil. n. 213. *Blackw.
ſ.* 472 ; Boucage avec des
feuilles aîlées, dont les lobes,
dans les feuilles radicales, ſont
preſque ronds , & linéaires
dans celles du haut.

*Tragoſelinum alterum majus.
Tourn. Inſt. R. H.* 309 ; un
autre plus grand Boucage.

3°. *Pimpinella hircina , foliis
pinnatis , foliolis radicalibus pinnati-fidis , ſummis linearibus , tri-
fidis ;* Boucage avec des feuilles aîlées, dont les lobes des
feuilles radicales ſont à pointes aîlées , & ceux des feuilles
du haut linéaires , & diviſés
en trois parties.

*Tragoſelinum minus. Tourn Inſt.
R. H. ;* le plus petit Boucage.

4°. *Pimpinella nigra , foliis
pinnatis , hirſutis , foliolis radicalibus cordatis , inæqualiter
ſerratis , ſummis linearibus quin-
què-fidis ;* Boucage à feuilles

ailées & velues, dont les lobes des feuilles du bas font en forme de cœur, inégaux & fciés, & ceux des feuilles du haut linéaires & à cinq pointes.

Tragofelinum radice nigrâ Germanicum. Juffieu. Hort. Chels. Cat. 100; Boucage d'Allemagne, à racine noire.

5°. *Pimpinella Auftriaca, foliis pinnati-fidis, lucidis, foliolis radicalibus lanceolatis, pinnato-ferratis, fummis linearibus, pinnati-fidis*; Boucage à feuilles luifantes & ailées, dont les lobes des feuilles du bas font en forme de lance, & fciées, & ceux des feuilles du haut linéaires & à pointes ailées.

Tragofelinum Auftriacum maximum, foliis profundiffimè incifis. Boerh. Chels. Cat. 100; le plus grand Boucage d'Autriche, dont les feuilles font profondément découpées.

6°. *Pimpinella peregrina, foliis radicalibus pinnatis, crenatis, fummis cunei-formibus, incifis. Lin. Sp. Plant.* 164. *Jacq. Hort. f.* 131; Boucage dont les feuilles radicales font ailées, & crénelées fur leurs bords, & celles du haut en forme de coin, & découpées.

Anifum foliis radicalibus pinnatis. Hort. Cliff. 107. *Hort. Ups.* 67. *Roy. Lugd. B.* 115. *Sauv. Meth.* 231.

Apium peregrinum, foliis fubrotundis. C. B. P. 153; Perfil étranger, à feuilles prefque rondes.

Daucus tertius Diofcoridis. Colum. Ecphr. 1. p. 128. *f.* 109.

7°. *Pimpinella Anifum, fo-*

liis radicalibus tri-fidis, incifis. *Lin. Sp. Plant.* 264. *Mat. Med.* 86. *Blackw. f.* 374. *Kniph cent.* 2. *n.* 57; Pimprenelle dont les feuilles radicales font découpées en trois parties.

Anifum Herbariis. Bauh. Pin. 159.

Anifum vulgare. Clus. Hift. 2. *p.* 202; Anis commun.

Cuminum femine rotundiori & minori. Bauh. Pin. 146. *Raii Extr.* 63; Variété.

Major. La premiere efpece croît naturellement dans les bois & dans les haies de plufieurs parties de l'Angleterre, fur-tout dans les terreins de craie; fes feuilles radicales font ailées, & compofées de trois paires de lobes en forme de cœur, que termine un lobe impair. Ces lobes font fortement fciés fur leurs bords, & feffiles à la côte du milieu; ceux du bas, qui font les plus grands, ont deux pouces de long fur un & demi de large à leur bâfe, & font d'un vert foncé; les tiges ont plus d'un pied de hauteur, & fe divifent en quatre ou cinq pétioles branchus; la partie baffe de la tige eft garnie de feuilles ailées, femblables à celles du bas, mais plus petites. Celles qui couvrent les branches font courtes & divifées en trois parties. Les branches font terminées par de petites ombelles de fleurs blanches, compofées d'autres ombelles plus petites ou rayons : ces fleurs ont cinq pétales en forme de lance, tournée en-dedans; elles font remplacées par deux femences étroites, oblongues & cane-

lées. Cette plante fleurit en Juillet, & ses graines mûrissent en automne. On en connoît une variété à fleurs rouges, qu'on trouve souvent parmi les autres, & qui provient des mêmes semences.

Saxi-fraga. La seconde espece se trouve aussi en Angleterre dans des pâturages secs ; ses feuilles radicales sont composées de quatre paires de lobes, terminés par un impair ; ils sont ronds : ceux du bas de la feuille ont environ un demi-pouce de longueur sur une largeur égale, & sont dentelés sur leurs bords ; les tiges, qui s'élevent à près d'un pied de hauteur, poussent trois ou quatre branches minces, garnies de feuilles fort étroites ; les ombelles des fleurs sont plus petites que celles de la premiere, ainsi que les fleurs & les semences. Cette plante fleurit aussi dans le même tems.

Hircina. La troisieme, qu'on rencontre dans des pâturages secs & graveleux de plusieurs parties de l'Angleterre, a ses feuilles radicales composées de cinq ou six paires de lobes, terminés par un lobe impair ; ils sont profondément découpés presqu'à la côte du milieu en forme d'ailes : ses tiges sont minces & de plus d'un pied de hauteur ; elles poussent quelques petites branches, garnies à chaque nœud d'une feuille étroite, divisée en trois parties, & sont terminées par de petites ombelles de fleurs blanches, composées de plusieurs rayons, & placées sur des pédoncules aTlez

longs : les fleurs sont petites, & paroissent dans le même tems que celles de la précedente (1).

Nigra. Les semences de la quatrieme m'ont été envoyées de Paris par M. BERNARD DE JUSSIEU ; les feuilles radicales de cette plante sont composées de six ou sept paires de lobes, & d'un lobe impair, en forme de cœur, presque de deux pouces de longueur sur un & demi de large près de la bâse, velus, & d'un vert pâle ; la tige s'élève à près de deux pieds de hauteur, & se divise en plusieurs branches, garnies à chaque nœud d'une

---

(1) On se sert en Médecine des racines de ces trois premieres especes ; mais on emploie de préférence celles de la troisieme, parce que ses principes sont plus actifs. Ces racines n'ont aucune odeur ; mais quand on les froisse, elles exhalent une vapeur très-subtile, qui picote les yeux, & fait couler les larmes ; leur saveur est âcre & irritante. Cette âcreté ne réside pas seulement dans la substance fixe, résineuse & gommeuse ; mais elle se fait beaucoup plus fortement remarquer dans un principe phlogistico-salin qu'on y découvre. La racine de l'impronelle blanche agace, irrite, & divise fortement : elle produit des effets très-marqués & salutaires dans toutes les affections catharrales & pituiteuses, les écrouelles, la fausse squinancie, l'engorgement des glandes salivaires, le relâchement de la luette, les maladies soporeuses, &c. : on la prépare en infusion froide, vineuse ; on s'en sert fréquemment dans les gargarismes, en masticatoires, & on la fait quelquefois entrer dans les lavemens, mais à foible dose.

feuille étroite & à cinq pointes, & terminées par des ombelles de fleurs blanches semblables à celles de la première espece.

Austriaca. J'ai cueilli les graines de la cinquieme espece dans le jardin particulier du célébre BOERHAVE, près de Leyde: il la nomma *Austriaca*, en me disant qu'il avoit reçu les semences de l'Autriche : ses feuilles radicales ont cinq paires de lobes, terminés par un impair, & placés à une plus grande distance les uns des autres, que ceux des autres especes ; leur longueur est d'environ neuf pouces sur neuf lignes de large au milieu : ils sont terminés en pointe à chaque extrémité, & sont divisés profondément en dentelures régulieres & opposées en forme de feuilles ailées ; ces feuilles sont d'un vert luisant, & supportées par de longs pétioles : les tiges s'élevent à la hauteur de deux pieds, & sont divisées au sommet en deux ou trois branches minces, & garnies à chaque nœud d'une feuille étroite, à pointes ailées. Les ombelles des fleurs ressemblent beaucoup à celles de la premiere espece.

Culture. Toutes ces plantes ont des racines vivaces, & se multiplient par semences. Si on met ces graines en terre en automne, elles réussissent plus certainement que si on les conservoit jusqu'au printems. Les plantes qui en proviennent n'exigent aucun autre soin que d'être éclaircies où elles sont trop épaisses, & d'être tenues

nettes de mauvaises herbes ; elles fleurissent, & perfectionnent leurs semences dans la seconde année : leurs racines se conservent plusieurs années, & continuent à produire des fleurs & des graines, si elles croissent dans une mauvaise terre.

La premiere espece est d'usage en Médecine ; mais les Vendeuses d'herbes portent, en place de celle-ci, sur les marchés, la troisieme, ou même le *Boucage de prairie* ; elle entre dans le *Pulvis Ari compositus*, & on la regarde comme ayant la propriété de chasser le gravier des reins.

Anisum. La derniere espece est l'*Anis commun* ; elle est annuelle, & croît naturellement en Egypte : on la cultive aussi à Malte & en Espagne, d'où l'on apporte ses semences en Angleterre. On en extrait, par la distillation, une eau & une huile pour l'usage de la Médecine. Les Pâtissiers en font grand usage, pour donner à leurs ragoûts une odeur & une saveur aromatique. Les feuilles radicales de cette plante sont divisées en trois lobes profondément découpés sur leurs bords ; sa tige s'éleve à la hauteur d'un pied & demi, & se divise en plusieurs branches minces, garnies de feuilles étroites & découpées en trois ou quatre segmens étroits; ces branches sont terminées par des ombelles peu épaisses, assez larges, & composées de plusieurs plus petites ombelles ou rayons postés sur de longs pédoncules ; les fleurs sont

petites, & les femences oblongues & gonflées. Cette plante fleurit dans le mois de Juillet, & perfectionne fes femences en automne, dans les années chaudes.

Les graines de cette efpece doivent être femées à demeure au commencement d'Avril, fur une plate-bande chaude. Lorfque les plantes ont pouffé, on les éclaircit, & on les tient nettes de mauvaifes herbes ; c'eft en cela que confifte toute leur culture : mais comme cette efpece eft trop délicate, on ne peut la cultiver en Angleterre pour le commerce (1).

PIMPRENELLE , BOUCAGE ou BOUQUETINE. *Voy.* SANGUISORBA OFFICINALIS. *L.* PIMPINELLA. *L.* & POTERIUM. *L.*

PIMPRENELLE AQUATIQUE , *à feuilles rondes. Voyez* SAMOLUS. *L.*

---

(1) L'Anis eft la premiere des quatre femences chaudes majeures, & a à peu-près les mêmes propriétés que les graines de Fenouil ; fon activité réfide uniquement dans fon huile effentielle & dans fon principe réfineux , qui font très-abondans. Ces graines ont une odeur agréable & pénétrante, & une faveur chaude , douceâtre & aromatique ; elles font propres à fortifier l'eftomac , à faciliter les digeftions, à diffiper les vents , & à foulager, dans l'afthme humide, la toux invétérée , la néphrétique pituiteufe, &c. On les prefcrit en infufion vineufe , ou en forme de dragées, & on les fait entrer dans la compofition des liqueurs ftomachiques ; elles font auffi un des ingrédiens du fyrop d'Armoife , du fyrop anti-afthmatique de Charas , &c.

PIMPRENELLE D'AFRIQUE ou MELIANTHE. *Voyez* MELIANTHUS.

PIN. *Voy.* PINUS. *L.*

PINASTER. *Voy.* PINUS. *L.*

PINGUICULA. [*Butterwort.*] Graffette , herbe graffe *ou* huileufe.

Cette plante fe trouve fur des terres marécageufes de plufieurs parties de l'Angleterre ; mais comme on ne la cultive jamais dans les jardins, je n'en dirai pas davantage. On lui attribue beaucoup de vertus en Médecine.

PINGUIN. *Voy.* KARATAS.

PINUS. *Tourn. Inft R. H.* 585. *Raii Meth. Plant.* 138. *Lin. Gen. Plant.* 956. [*The Pinetree.*] le Pin.

Caractères. Les fleurs mâles font recueillies dans une grappe écailleufe ; elles n'ont point de pétales , mais feulement plufieurs étamines réunies à leur bâfe , divifées au fommet , & terminées par des antheres érigées ; elles font renfermées dans les écailles qui fervent de pétales & de calice ; les fleurs femelles font auffi recueillies dans un cône ovale & commun , & font placées à quelque diftance des fleurs mâles fur le même arbre. Sous chaque écaille du cône naiffent deux fleurs qui n'ont point de pétales , mais qui font pourvues d'un petit germe qui foutient un ftyle en forme d'alène , couronné par un ftigmat fimple. Ce germe devient enfuite une noix oblongue , ovale , couronnée d'une aîle , & renfermée dans l'écaille rigide du cône.

570 PIN

Ce genre de plantes eſt rangé dans la neuvieme ſection de la vingt − unieme claſſe de LINNÉE, avec celles qui ont des fleurs mâles & femelles ſur le même pied, & dont les étamines ſont jointes en un corps. A ce genre il ajoute le *Larix* & l'*Abies* de TOURNEFORT.

Les eſpeces ſont :

1°. *Pinus ſylveſtris , foliis geminis , primordialibus ſolitariis , glabris.* Hort. Cliff. 450. Flor. Suec 788. 874. Mat. Med. 204. Roy. Lugd.-B. 89. Dalib. Paris. 295. Gmel. Sib. 1. p. 178. Trew. in nov. Act. A. N. C. III. App. p. 452 ; Pin avec deux feuilles dans chaque gaîne, dont les premieres ſont ſolitaires & liſſes.

Pinus foliis geminis , conis pyramidatis , ſquamis oblongis , obtuſis. Du Roi Harbk. 2. p. 13.

Pinus ſylveſtris. C. B. P. 491 ; le Pin ſauvage *ou* le Pin de Geneve.

2°. *Pinus Pinea , foliis geminis , primordialibus ſolitariis , ciliatis.* Hort. Cliff. 450. Hort. Ups. 288. Mat. Med. 205. Roy. Lugd.-B. 89. Gouan. Monſp. 494. Scop. Carn. 2. n. 1197. Regn. Bot. ; Pin avec deux feuilles griſes ſortant de chaque gaîne, mais dont les premieres ſont ſolitaires & ciliées.

Pinus ſativa. C. B. P. 491. Blackw. t. 189. Duham. Arb. 2. f. 27 ; le Pin cultivé, ordinairement appelé *Pin de Pierre.*

Pinus. Cam. Epit. 93.

3°. *Pinus rubra , foliis geminis brevioribus , glaucis , conis parvis mucronatis* ; Pin avec deux plus courtes feuilles griſes ſortant de chaque gaîne,

& un cône petit & pointu.

Pinus ſylveſtris , foliis brevibus , glaucis , conis parvis , albicantibus. Raii Syn. 2. 288. Duham. Arb. 5. Hort. Angl. f. 17 ; Pin ſauvage, avec de courtes feuilles griſes, & de petits cônes blanchâtres, appelé *Pin d'Ecoſſe.*

4°. *Pinus Tartarica , foliis geminis brevioribus , latius culis, glaucis , conis minimis* ; Pin avec deux plus courtes feuilles, larges & griſes, dans chaque gaîne, & de très-petits cônes, communément appelé *Pin de Tartarie.*

5°. *Pinus montana , foliis ſæpius ternis , tenuioribus , viridibus , conis pyramidatis , ſquamis obtuſis* ; Pin ayant ſouvent trois feuilles étroites & vertes dans chaque gaîne, des cônes en pyramide, & des écailles obtuſes.

Pinus ſylveſtris montana altera. C. B. P. 421 ; autre Pin ſauvage de montagne, appelé *Mugho.*

6°. *Pinus Cembra , foliis quinis lævibus.* It. Scan. 32. Lin. Gen. Plant. 1000 ; Pin avec cinq feuilles liſſes dans chaque gaîne.

Pinus foliis quinis , cono erecto , nucleo eduli. Gmel. Sib. 1. p. 179. f. 39. Duham. Arb. 2. t. 32.

Pinus foliis quinis triquetris. Hall. Helv. n. 1659.

Pinus foliis quinis , conis ovatis , erectis , ſquamis ovalibus , concavis , nucibus cunei-formibus , ala membranaceâ deſtitutis. Du Roi Harbk. 2. p. 51.

Pinus ſativa , cortice fiſſo , foliis ſeroſis , ſub−rigidis , ab unâ

PIN

vaginâ quinis. Amm. Ruth. 178.

Larix femper virens, foliis qui-nis, nucleis edulibus. Breyn. E. N. C. cent. 7. *Obs.* 2.

Pinus fylveftris montana tertia. C. B. P. 491 ; le troifieme Pin fauvage de montagne, appelé *Ceinbrot.*

Pinus fylveftris Cembra. Cam. Ep. 42.

Pinafter. Bell. Conif. 19. *Mich. Gen.* 223.

7°. *Pinus maritima, foliis ge-minis longioribus, glabris, conis longioribus tenuioribusque* ; Pin avec deux feuilles plus lon-gues, & unies dans chaque gaîne, & des cônes plus longs & plus minces.

Pinus maritima fecunda. Tabern. Icon. 937 ; fecond Pin maritime.

Pinafter lati-folius, julis vi-refcentibus, fivè pallefcentibus. Bauh. pin. 492.

8°. *Pinus Alepenfis, foliis geminis tenuiffimis, conis obtufis, ramis patulis. Tab.* 208 ; Pin avec deux feuilles étroites dans cha-que gaîne, des cônes obtus, & des branches étendues.

Pinus Alepenfis, foliis tenui-bus, latè viridibus. Rand. Hort. Chels. Cat. 158 ; Pin d'Alep, avec des feuilles fort étroites & d'un vert foncé.

9°. *Pinus Virginiana, foliis geminis brevioribus, conis parvis, fquamis acutis* ; Pin de Virginie avec deux plus courtes feuil-les dans chaque gaîne, de pe-tits cônes, & des écailles ai-guës.

Pinus Virginiana, foliis binis, brevioribus & craffioribus fetis, minori cono fingulis fquamarum ca-pitibus aculeo donatis. Pluk. Alm. 297 ; Pin de Virginie, avec

PIN 571

deux feuilles plus courtes & plus épaiffes dans chaque gaîne, & un plus petit cône, dont chaque écaiile eft terminée par une pointe, communé-ment appelé *Pin de Jerfey.*

10°. *Pinus rigida, foliis ter-nis, conis longioribus, fquamis rigidioribus* ; Pin avec trois feuil-les, & de plus longs cônes, dont les écailles font roides, ordinairement appelé *Pin de Virginie à trois feuilles.*

11°. *Pinus Tæda, foliis lon-gioribus, tenuioribus, ternis, conis maximis, laxis* ; Pin avec trois feuilles plus longues & plus étroites, & des cônes très-gros & defferrés.

Pinus foliis trinis. Gron. Virg. 152.

Pinus foliis ternis, conis py-ramidatis, fquamis oblongis, ob-tufis, reflexis. Du Roi Harbk. 2. *p.* 48.

Pinus foliis longiffimis ex unâ thecâ ternis. Cold. Noveb. 230.

Pinus Virginiana tenui-folia, triplis, fcilicet ternis plerùmque ex uno folliculo fetis, ftrobilis majoribus. Pluk. Alm. 297. *Raii Dendr.* 8 ; Pin à trois feuilles plus longues & plus étroites, & des cônes defferrés & les plus gros de tous, appelé *Ar-bre à Encens.*

12°. *Pinus echinata, Virgi-niana, prælongis foliis, tenuio-ribus, cono echinato, gracili. Pluk. Alm.* 297 ; Pin de Virginie, avec des feuilles très-longues & plus étroites, & un cône mince & piquant, appelé *Pin bâtard,* à trois feuilles.

13°. *Pinus Strobus, foliis quinis fcabris. Lin. Sp. Plant.* 1001 ; Pin avec cinq feuilles

rudes dans chaque gaîne , ordinairement appelé *Pin du Lord
Weymouth.*

14°. *Pinus palustris , foliis
ternis longissimis ;* Pin dont les
feuilles font plus larges , &
qui naiffent par paires dans
chaque gaîne.

*Pinus Americana palustris tri-
folia , foliis longissimis. Duhamel.* ; Pin d'Amérique à trois
feuilles , qui croît dans les marais & dont les feuilles font
les plus longues.

Il y a en Amérique quel-
ques autres efpeces de ce genre,
qui n'ont point été fuffifamment examinées pour en connoître la différence. Il eft probable que plufieurs efpeces de
l'Europe , qui ne font regardées à préfent que comme des
variétés de celles dont il vient
d'être queftion , peuvent en être diftinguées ;
mais comme je n'ai pas eu
occafion de les voir , je n'ai
pu en faire mention ici.

Sylveftris. La première efpece eft le *Pinéaftre* ou *Pin
fauvage* qui croît naturellement
dans les montagnes de l'Italie
& dans la France méridionale ,
où il forme des forèts entieres. Il s'élève à une très-grande
hauteur , quand on lui donne
le tems de croître ; mais dans
la Suiffe , on en coupe une
grande quantité , pour en faire
des lattes dont on fe fert pour
couvrir les maifons , & pour
en extraire la poix. Dans la
France méridionale , on fait ,
avec ces jeunes arbres , des
peffeaux pour les vignes. Cette
efpece s'élève droite , & à une
grande hauteur ; fes branches

s'étendent de tous côtés , à une
diftance fort confidérable ; &
tandis que les arbres font jeunes , ils font entièrement garnis de feuilles , fur-tout quand
ils ne font pas affez ferrés
pour que l'air ne puiffe pas circuler entr'eux : mais à mefure
qu'ils avancent en âge , les
branches deviennent nues , &
toutes celles du bas , après
quelques années , font défagréables à la vue ; ce qui eft
caufe que , depuis quelque
tems , ils ne font pas fort eftimés. On leur a préféré les
Sapins d'Ecoffe , dont le bois
eft meilleur , & dont les branches font généralement mieux
garnies de feuilles ; auffi ces
derniers ont-ils été plus multipliés que le précédent.

Les branches du *Pin fau-
vage* s'étendent à une plus grande diftance que celles du *Pin
d'Ecoffe ,* & font plus hori-
fontales ; fes feuilles font beaucoup plus larges , plus épaiffes , plus longues & plus droites ; elles ont une furface plus
large en-dedans , où elles ont
un fillon qui coule longitudinalement ; elles font d'un vert
plus foncé , & leurs pointes
font obtufes. Les cônes de cette
efpece ont fept ou huit pouces de longueur ; leur forme
eft pyramidale , & ils ont des
écailles pointues : fes femences font oblongues , un peu
applaties fur leurs côtés , &
ornées d'aîles étroites à leur
fommet.

Pinea. La feconde efpece ,
à laquelle on donne généralement le nom de *Pin de Pierre,*
eft fort commune en Italie ;

mais

PIN

mais j'ignore dans quel pays elle croît naturellement ; car, suivant ce que j'ai entendu dire, tous les arbres de cette espece qu'on trouve en Italie, y ont été plantés, ou ont été produits par les semences écartées des autres. J'ai souvent reçu des semences d'un *Pin* de la Chine, dont les cônes ressembloient à ceux de cette espece, de maniere à ne pas les distinguer ; mais elles n'ont jamais réussi, soit parce qu'elles étoient trop vieilles, soit parce qu'elles avoient été dépouillées : car, lorsque ces graines sont conservées dans leurs cônes, elles levent jusqu'à l'âge de dix à douze ans ; au-lieu qu'étant mises à nud, elles se conservent bonnes rarement au-delà de deux années, & quelques especes même ne levent pas après un an. Les feuilles de celle-ci ne sont pas tout-à-faut aussi longues que celles de la précédente ; elles sont d'une couleur grisâtre ou de vert de mer ; les cônes n'ont pas plus de cinq pouces de longueur ; mais ils sont fort épais, ronds, & terminés en pointe obtuse ; leurs écailles sont plates, & les semences sont plus de deux fois plus grosses que celles de la précédente. On sert les amandes de cette espece sur les tables, en Italie, pendant l'hiver. Autrefois on en faisoit usage ici en Médecine ; mais depuis plusieurs années, on leur a généralement substitué les amandes de pistaches. Le bois de cet arbre est blanc, & moins rempli de résine que celui de

Tome V.

PIN 573

plusieurs autres especes : aussi ne le cultive-t-on pas pour son bois, mais seulement pour la beauté de son feuillage, & pour ses fruits, dont on fait beaucoup de cas dans la France Méridionale & en Italie (1).

Rubra. La troisieme espece, que l'on connoît ici sous le nom de *Pin d'Ecosse*, parce qu'elle croît naturellement dans les montagnes de ce pays, est assez commune dans la plupart des contrées de l'Europe. M. Duhamel Du Monceau, de l'Académie Royale des Sciences de Paris, dit avoir reçu des cônes de cet arbre de Saint-Domingue ; d'où il conclut qu'il croît indifféremment dans les zônes torride, glaciale & tempérée. Cette espece a été décrite par Jean Bauhin, sous le nom de *Pinus sylvestris Genevensis vulgaris* ; de sorte qu'on la trouve aussi dans les montagnes qui environnent cette Ville, en Danemarck, dans la Norvége, & en Suede. Le bois de cet arbre est le *Sapin* rouge ou jaune, qui dure plus que celui de toutes les especes jusqu'à présent connues ; ses feuilles sont beaucoup plus courtes que celles

(1) Les Pignons ou fruits du Pin ne different guere des pistaches & des amandes douces, quant à leurs propriétés médicinales ; mais ils sont plus sujets à se rancir, & doivent être employés frais. Au reste, on s'en sert rarement aujourd'hui comme remede ; mais on en fait un fréquent usage pour la cuisine, dans les provinces meridionales de France & en Italie.

Oo

des précédentes, plus larges, d'une couleur grisâtre, roulées & réunies en paires dans chaque gaîne ; ses cônes sont petits, en pyramide, terminés en pointe, & d'une couleur claire ; ses semences sont petites.

Cette espece réuffit aff.z bien sur presque tous les sols. J'ai planté un grand nombre de ces arbres dans des creux de tourbe, où ils ont fait un grand progrès ; j'en ai auffi placé dans des terres-glaifes, où ils ont réuffi au-dela de mon espérance, ainfi que dans le fable, le gravier & la craie. Ils ne croiffent pas auffi vîte dans le gravier & le fable, que fur une terre humide : mais le bois en eft bien meilleur ; car les arbres coupés fur des terreins humides, où ils ont fait de grands progrès, ne donnent que du bois blanc & d'une texture molle ; au-lieu que ceux qui croiffent dans des terreins fecs & remplis de gravier, ont été trouvés à-peu-près auffi bons que les meilleurs *Sapins* étrangers : auffi je ne doute pas que les plantations qui en ont été faites depuis quelques années, ne foient très - profitables, dans un fiècle, à leurs poffeffeurs, & très-avantageufes à la nation. C'eft auffi l'efpece que je confeille de cultiver de préférence fur des terreins ftériles.

Tartarica. La Quatrieme efpece croît naturellement en Tartarie, d'où j'en ai reçu les femences ; elle reffemble beaucoup au *Pin d'Ecoffe* ; mais fes feuilles font plus larges, plus courtes, & leurs pointes font plus obtufes : elles répandent une odeur balfamique, quand elles font froiffées ; les cônes font forts petits, ainfi que les femences, dont quelques-unes étoient noires, & les autres blanches : mais j'ignore fi elles avoient été recueillies fur différens arbres ou fur le même. On a tiré les femences des cônes ; mais dans le paquet, il ne s'en eft pas trouvé un feul entier.

Montana. La cinquieme fe trouve fur les montagnes de la Suiffe ; fes feuilles font fort étroites, vertes, quelquefois difpofées par paires, & d'autres fois au nombre de trois dans chaque gaîne ; elles font généralement érigées ; les cônes font d'une groffeur médiocre, & en pyramide ; les écailles font plates, & ont chacune une petite élévation obtufe ; elles font fort comprimées, jufqu'à ce que la chaleur du foleil les ait fait ouvrir au fecond printems ; leurs femences font beaucoup plus petites que celles du *Pinéaftre*, mais plus groffes que celles du *Pin d'Ecoffe.*

Cembra. La fixieme, qui eft originaire de la Suiffe, eft regardée comme étant la même que celle de la Sibérie, ce dont je doute fort : car les cônes de celle-ci font courts & ronds, & leurs écailles font ferrés ; au-lieu que ceux du *Pin de Sibérie* font longs & plus ferrés ; leurs feuilles fe reffemblent beaucoup, autant que j'ai pu l'obferver fur des échan-

tillons : mais les plantes qui ont été élevées avec des femences envoyées de la Suisse, ont fait un plus grand progrès que celles de Sibérie, qu'on peut à peine conserver en Angleterre : les feuilles de cette espece font longues, étroites, lisses, d'un vert léger, & fortent au nombre de cinq de la même gaine ; fes branches en font fortement garnies ; les cônes ont environ trois pouces de longueur, & leurs écailles font très-serrées ; les femences font assez grosses, & l'on brise aisément leurs enveloppes.

Maritima. La septieme espece, qu'on rencontre dans les parties maritimes de l'Italie & de la France méridionale, a des feuilles longues, lisses, réunies par paires dans chaque enveloppe ; fes cônes font fort longs & minces ; fes femences font à-peu-près de la grosseur de celles du *Pinéaftre.*

Alepensis. La huitieme se trouve dans les environs d'Alep, & dans plusieurs autres parties de la Syrie. Cet arbre est d'un crû médiocre dans son pays natal ; & en Angleterre il ne parvient point à une grande hauteur. La plupart de ceux qui ont été plantés avant 1740, ont été détruites par le froid de ce rude hiver. Les deux plus grands que j'aie vus, se trouvent à Goodwood en Sussex, dans le Parc du Duc de Richmond. Comme ils avoient été transplantés dans l'année qui a précédé ce terrible hiver, qu'ils avoient à peine réparé les torts occasionnés par le changement de situation, & qu'ils n'avoient

point poussé de branches durant l'été, ils ont échappé au froid plus aisément que les arbres en grande vigueur, dont la plupart ont été détruits. Cet arbre poussé de tous côtés, depuis fa racine, des branches d'abord dirigées horisontalement, mais dont les extrémités se tournent ensuite vers le haut ; fes feuilles ont leur surface supérieure lisse & d'un gris foncé ; elles font disposées par paires dans chaque gaine, & répandent une odeur forte & réfineuse, quand elles font froissées ; les cônes fortent fur le côté des branches ; ils ont à peine la moitié de longueur de ceux du *Pinéaftre* ; mais ils font aussi gros à leur bâse ; leurs écailles font applaties, & l'extrémité du cône est obtuse ; leurs femences font beaucoup plus petites que celles du *Pinéaftre*, mais de la même forme.

Virginiana. La neuvieme espece croît sans culture dans la plus grande partie de l'Amerique Septentrionale. Cet arbre ne s'éleve pas à une grande hauteur, & est le moins estimé de ce genre dans ce pays. Lorsque ces arbres font jeunes, ils ont une assez belle apparence ; mais quand ils parviennent à la hauteur de sept à huit pieds, ils se chiffonnent, & font désagréables à la vue : ainsi, ils ne méritent pas d'être cultivés.

Rigida. La dixieme espece naît spontanément en Virginie & dans d'autres parties de l'Amérique Septentrionale, où elle s'éleve à une fort grande

hauteur ; & , autant que nous pouvons en juger par les progrès des arbres qui font à préfent ici , ils paroiffent devoir devenir fort grands en Angleterre. On en voit plufieurs qui croiffent à préfent dans la belle plantation d'arbres toujours verts du Parc du Duc de BEDFORD à Wooburn. Ils ont déja vingt pieds de hauteur, quoiqu'ils ne foient pas plantés depuis long-tems, & croiffent auffi promptement que les autres efpeces de *Pins* & de *Sapins* qui fe trouvent dans la même plantation. Les feuilles de cet arbre font longues, & fortent toujours par trois de la même enveloppe ; leurs cônes, qui naiffent en paquets autour des branches, font auffi longs que ceux du *Pinéaftre* ; leurs écailles font roides, & leurs femences'ailées, & prefque auffi groffes que celles du *Pinéaftre.*

Tæda. La onzieme efpece croit naturellement dans l'A-mérique Septentrionale ; fes feuilles font fort longues & étroites , & fortent par trois de chaque gaine ; les cônes font auffi gros que ceux du *Pin de Pierre* ; mais leurs écailles font plus defferrées , & les cônes plus pointus. Les écailles de cette efpece s'ouvrent horifontalement, & jettent leurs femences. Cet arbre a été envoyé de l'Amérique à M. BALE, d'Exeter , & au Docteur COMP-TON , Evêque-de Londres, fous le titre de *Pin d'Encens.*

Echinata. La douzieme ef-pece eft originaire de la Vir-ginie ; fes cônes ont été por-

tés en Angleterre , il y a quelques années , fous le nom de *Pin bâtard à trois feuilles* ; fes feuilles font longues & étroites ; quelquefois il y en a trois qui croiffent dans chaque gaine , & d'autres fois feulement deux ; les cônes font longs, minces , & leurs écailles font terminées en pointe aiguë ; ils font plus longs que ceux du *Pinéaftre* , mais moins gros.

Strobus. La treizieme efpece fe trouve dans la plus grande partie de l'Amérique Septentrionale , où elle eft connue fous le nom de *Pin blanc.* Cet arbre eft un des plus élevés de ce genre ; car il parvient fouvent à la hauteur de cent pieds , dans fon pays natal ; fon écorce eft fort liffe & tendre , fur-tout lorfqu'il eft jeune ; fes feuilles , longues & étroites , fortent par cinq de chaque enveloppe , & les branches, qui en font affez garnies , ont une belle apparence : les cônes font longs, minces, très-defferrés , & s'ouvrent à la première chaleur du printems ; de forte que s'ils ne font pas recueillis en hiver , les écailles s'ouvrent, & laiffent tomber les femences. Le bois de cette efpece fait de très-bons mâts de vaiffeaux ; on l'appelle , en Angleterre , *Pin du Lord Weymouth* , ou *Pin de la Nouvelle-Angleterre.* Comme ce bois eft d'un grand ufage dans la Marine , on a fait une loi , dans la deuxieme année du regne de la Reine ANNE , pour la confervation de ces arbres , & en encourager la culture en Amérique. On a commencé,

PIN

il y a plus de quarante ans, à en planter beaucoup en Angleterre. Il y en avoit cependans quelques-uns de fort gros, qui avoient été plantés dans deux ou trois endroits, long-temps auparavant, particulierement chez le Lord WEYMOUTH & le Chevalier WYNDHAM KNATCH-BULL, en Kent. Ce font ces arbres dont les femences ont produit la plus grande partie des autres qu'on voit en Angleterre ; car quoiqu'on en apportât annuellement de l'Amérique, cette quantité étoit peu de chofe en comparaifon de celle qui fut recueillie fur les arbres de Kent. Les nouveaux arbres qui en font provenus, produifent eux-mêmes aujourd'hui beaucoup de graines, fur-tout ceux qui fe trouvent dans les jardins du feu Duc D'ARGYLE, à Whitton, qui a généreufement diftribué à tous les curieux un grand nombre de cônes.

Cette efpece & le *Pin d'E-coffe* méritent d'être cultivés, pour la qualité de leur bois, de préférence à tous les autres qui peuvent être plantés ; ils feront un très-bel effet en hiver, par leur feuillage toujours vert.

Culture. On multiplie toutes les efpeces de *Pins*, au moyen des femences que contiennent les cônes durs & ligneux qu'ils produifent. Pour en tirer les femences, on met ces cônes devant un feu léger, qui en fait ouvrir les cellules, & donne la facilité d'en détacher les femences. Comme ces graines confervent leur qualité vé-

gétative pendant plufieurs années, lorfqu'on les laiffe dans les cônes, la meilleure méthode eft de ne les détacher qu'au moment où on veut les mettre en terre. Si l'on tient ces cônes pendant l'été dans un endroit chaud, ils s'ouvriront, & laifferont fortir les femences ; mais en ne les expofant pas à beaucoup de chaleur, ils refteront entiers plufieurs années de fuite, fur-tout ceux qui font ferrés & compactes. J'ai eu de ces cônes dont on n'a ôté les femences qu'au bout de fept ans, & cependant ces graines ont bien réuffi ; de maniere qu'on peut les tranfporter à quelque diftance que ce foit, pourvu que les cônes foient bien mûrs & bien emballés.

Le meilleur tems pour femer les graines des *Pins*, eft vers la fin de Mars ; mais on ne doit point négliger de couvrir la terre où elles fe trouvent, avec des filets, pour en défendre l'accès aux oifeaux, qui détruiroient les fommets des plantes, lorfqu'elles commenceroient à pouffer.

Quand on n'en feme pas beaucoup, il vaut mieux fe fervir de caiffes ou de pots, que l'on remplit d'une terre légere & marneufe ; au moyen de quoi, on peut les tranfporter par-tout où l'on veut, fuivant la chaleur de la faifon : mais lorfque la quantité de femence eft confiderable, & qu'elle exige un grand terrein, on la répand fur une planche de terre, à l'expofition de l'Eft ou du Nord-eft, de

O o 3

maniere qu'elle puisse être abritée du soleil, dont la chaleur est fort nuisible à ces plantes, quand elles commencent à sortir de terre. Celles qui sont semées dans des caisses ou des pots, doivent aussi être placées à l'ombre, mais pas sous des arbres, & on fera bien de tenir les plantes à l'ombre avec des nattes, lorsqu'elles commenceront à pousser.

Presque toutes les especes leveront bien six ou sept semaines après qu'elles auront été semées; mais les graines du *Pin de Pierre* cultivé, & celles de deux ou trois autres dont les coques sont fort dures, restent souvent une année dans la terre; de sorte que, si les plantes ne poussent pas la premiere année, il ne faut pas remuer la terre, mais la tenir nette de mauvaises herbes, & attendre jusqu'au printems suivant. Ce retard a lieu souvent dans les années seches, & quand les graines se trouvent dans des endroits un peu trop exposés au soleil. Pour éviter cet inconvénient, il sera bon de faire tremper les semences dans l'eau pendant vingt-quatre heures, avant de les mettre en terre.

Lorsque les plantes paroissent, on les tient constamment nettes de mauvaises herbes, & on les arrose légerement de tems en tems dans les tems très-secs, mais avec beaucoup de précaution : car, si on les arrosoit avec trop de promptitude, on les déterreroit, & on les coucheroit sur la terre; ce qui seroit souvent pourrir leurs tiges. Le même accident arrive quand on les arrose trop souvent; de sorte qu'il vaut mieux ne pas les arroser du tout, que de le faire sans soin; mais il faut toujours les tenir à l'ombre.

Si les plantes sont trop serrées, il faut les éclaircir au commencement de Juillet : celles qu'on arrache, peuvent être transplantées sur des planches qui doivent être préparées & prêtes à les recevoir ; car elles doivent être placées aussitôt qu'on les enleve, parce que leurs tendres racines sont bientôt desséchées & gâtées dans cette saison de l'année. On fait cet ouvrage, s'il est possible, par un tems couvert & pluvieux, parce qu'alors on enleve les plantes avec de meilleures racines, & qu'elles repoussent bientôt de nouvelles fibres ; mais si le tems est clair & sec, il faut les tenir à l'ombre chaque jour avec des nattes, & les arroser légerement de tems en tems. Lorsque l'on enleve ces plantes, on doit avoir grand soin de ne pas déranger les racines de celles qui restent dans le semis; & quand la terre est dure, on l'arrose copieusement quelque tems avant de faire cette opération, pour la desserrer. Quand ces plantes sont enlevées, on arrose encore une fois celles qui restent dans le semis, pour affermir la terre sur leurs racines; ce qui leur fera beaucoup de bien : mais il faut faire cette opération avec dextérité, pour ne pas déterrer les racines & abattre

PIN

les tiges. Celles que l'on a enlevées doivent être plantées à quatre ou cinq pouces de rang en rang, & à trois pouces dans les rangs.

Ces plantes peuvent rester dans ces planches pendant un an, & jusqu'au printems suivant : alors elles seront en état d'être mises dans les places qui leur sont destinées ; car plus elles sont transplantées jeunes, & mieux elles réussissent. Quelques especes souffrent la transplantation dans un âge beaucoup plus avancé ; mais celles qu'on enleve plus jeunes & dans le même tems, surpasseront les grosses, & les devanceront dans leur accroissement. En les plantant jeunes, on a aussi l'avantage d'épargner les tuteurs, qui leur sont inutiles, & la peine de les arroser ; ce qui est absolument nécessaire aux grosses plantes.

J'ai souvent vu des plantations de plusieurs especes de *Pins*, faites avec des arbres de sept à huit pieds de hauteur, entre lesquels on en plaçoit en même tems d'autres qui n'avoient qu'un pied d'élévation ; & j'ai constamment observé que ces derniers ont toujours formé de meilleurs arbres que les vieux, & que leur accroissement étoit beaucoup plus vigoureux. Cependant, si la terre qui doit les recevoir, ne peut être préparée pour le tems fixé, alors on met ces jeunes plantes en pépiniere, où elles pourront rester deux ans, mais pas audelà ; car il seroit très-dange-

PIN 579

reux de les transplanter plus tard.

La meilleure saison pour transplanter toutes les especes de *Pins*, est vers la fin de Mars, ou au commencement d'Avril, avant qu'ils commencent à pousser ; car, quoiqu'on puisse transplanter en hiver le *Pin d'Ecosse* & quelques autres des plus durs, sur-tout quand ils croissent dans une terre forte, où ils peuvent être enlevés avec des mottes à leurs racines, cependant je ne le conseillerai jamais, en ayant éprouvé souvent de mauvaises suites, d'autant plus que ceux qu'on transplante au printems manquent très-rarement.

Lorsqu'on place ces arbres dans un lieu exposé au vent, il faut les mettre assez près les uns des autres pour qu'ils puissent se protéger réciproquement ; quelques années après, on peut en couper une partie, pour donner de l'air aux autres ; mais il faut le faire par dégrés, de peur qu'en ouvrant la plantation tout d'un coup, l'air n'y entre avec trop de violence, & n'arrête le progrès de ces arbres.

Quoique plusieurs personnes méprisent ces arbres toujours verts, à cause de leur vert foncé pendant l'été, cependant ils font un très-bon effet en hiver, quand ils sont entremêlés avec une quantité d'autres dans le voisinage d'une maison de campagne ; & en été, ils diversifient la décoration, par le mélange de différentes teintes.

Par-tout où l'on veut en

faire des plantations, la meilleure méthode eft d'élever des plantes de femences, foit fur une partie du terrein même où elle doit être placée, ou fur une piece de terre voifine & de la même qualité. Une petite piece de terre fuffira pour élever affez de plantes pour plufieurs âcres ; mais comme elles exigent d'abord quelques foins , il faut s'affurer de quelque Villageois du voifinage, qui ait un petit enclos près de fa cabanne, ou qui en faffe un exprès pour élever les plantes. On peut lui confier les femences, en lui donnant toutes les inftructions néceffaires pour les femer, & la maniere de traiter les plantes jufqu'à ce qu'elles foient en état d'être tranfplantées. Les femmes & les enfans peuvent être employés utilement à cet ouvrage, & en leur promettant de leur payer un certain prix pour chaque plante, quand elles pourront être enlevées, on les encouragera à avoir foin de la plantation, lorfqu'elle fera faite , & on les empêchera de la détruire.

Le *Pin d'Ecoffe*, comme il a été dit ci-deffus, étant le plus dur de tous , & fon bois etant d'un meilleur ufage, mérite d'être cultivé. Cet arbre profite fur les fables les plus ftériles, où à peine le *Genêt* & la *Bruyere* peuvent croitre ; & comme il y a plufieurs miliers d'âcres de pareille nature , fitués convenablement près des rivieres, qui ne font d'aucun rapport à préfent aux

propriétaires , ils pourroient les employer utilement par des plantations de ces arbres , & faire ainfi le bien de la Nation. Le Gouvernement s'eft déja occupé de cet objet ; car il a donné quelques loix pour encourager ces plantations , ainfi que pour leur confervation & leur fûreté ; de maniere qu'on peut efpérer que les poffeffeurs de pareils terreins entreprendront de les planter dans toutes les parties du Royaume , avec courage & émulation : & , quoiqu'ils ne puiffent en tirer beaucoup de bénéfice eux-mêmes , cependant l'idée du grand avantage qui en réfultera pour leurs héritiers , & le plaifir d'embellir ces contrées, qui font affreufes à préfent, doivent les récompenfer en quelque maniere de leurs peines & de leur dépenfe: d'ailleurs ils peuvent occuper les pauvres à ce travail, & diminuer ainfi les frais de main-d'œuvre.

Beaucoup de perfonnes craignent de s'engager dans la dépenfe de ces plantations ; mais la plus forte eft celle de faire des enclos pour en écarter le bêtail, &c.; le furplus eft une bagatelle, parce qu'il ne fera pas néceffaire de préparer la terre pour recevoir les plantes, & la dépenfe pour planter un âcre de terre, n'excédera pas vingt ou trente fchelings ; &, fi le labour eft cher, les plantes pourront valoir quarante fchelings de plus.

J'ai planté de ces arbres dans plufieurs âcres de terre qui

PIN

étoient couvertes de *Bruyere* & de *Genét*, en me contentant d'y faire creufer des trous entre ces plantes inutiles, & j'ai enfuite fait entaffer ces *Bruyeres* & ces *Genêts* qui avoient été coupés, autour des racines des arbres, pour conferver l'humidité de la terre. Peu de ces arbres ont manqué, quoiqu'ils euffent prefque tous quatre années de femence ; on n'a pris aucun foin de nettoyer la terre enfuite ; & malgré cela, ces *Pins* ont fait affez de progrès dans l'efpace de cinq ou fix années, pour furmonter les Bruyeres & les Genêts, qu'ils ont détruits totalement.

La diftance que j'ai conftamment donnée à ces plantes dans toutes les fituations ouvertes, a été de quatre pieds environ, mais toujours irrégulierement, en évitant, autant qu'il eft poffible, de les placer en rangs. Lorfqu'on plante des arbres, il faut avoir grand foin de ne les enlever de la pépiniere qu'à mefure qu'on les met en place, en employant au premier travail affez d'ouvriers, tandis que d'autres font occupés dans la plantation. Il faut auffi faire en forte de ne point déchirer leurs racines, ni bleffer leur écorce en les arrachant. On couvre les racines dès qu'elles font à l'air, de peur qu'elles ne fe deffechent, & on les place dans leur nouvel établiffement le plutôt qu'il eft poffible.

On doit avoir grand foin, en les plantant, de faire les trous affez larges pour que ces racines y foient à l'aife, de brifer les mottes autant qu'il

PIN   581

eft poffible, de mettre toujours auprès de l'arbre la terre la plus meuble, & de preffer enfuite cette terre légerement avec le pied. Si l'on obferve exactement toutes ces précautions, & fi l'on choifit pour cela une faifon convenable, cette plantation réuffira prefque toujours. J'ai vu quelquefois employer des arbres envoyés de loin, qui étoient fi fortement emballés, qu'ils en étoient beaucoup échauffés, & que la plus grande partie de leurs feuilles étoient devenues jaunes ; auffi ont-ils prefque tous manqué : ce qui a dégoûté, mal-à-propos, plufieurs perfonnes de planter de ces arbres.

Quand les plantations font terminées, le feul foin qu'elles exigent pendant cinq ou fix ans, eft de préferver les plantes du bétail, des lievres & des lapins : car, fi ces animaux peuvent en approcher, ils y font un grand dégât en peu de tems, en rongeant les branches ; ce qui en retarde beaucoup les progrès, & quelquefois les détruit.

Cinq ou fix années après qu'ils font plantés, les branches des jeunes arbres fe rencontrent, & s'entremêlent les unes dans les autres ; alors il eft néceffaire de les tailler : mais il faut le faire avec beaucoup de précaution. On fe contente d'abord de couper feulement les branches du bas ; ce qui s'exécute en Septembre, parce qu'alors il n'y a point de danger que les bleffures coulent trop. La téré-

benthine se durcit sur les blessures à mesure que la saison devient froide, & elle empêche l'humidité d'y pénétrer. Ces branches doivent être coupées tout près de la tige des arbres, & on doit prendre garde, en faisant cette opération, de casser aucunes des branches restantes. On recommence cet ouvrage chaque deux ans; & à chaque fois, on ne retranche que la rangée des branches du bas; car si l'on faisoit à ces arbres beaucoup d'entailles, on retarderoit beaucoup leurs progrès, comme il arrive dans pareil cas à tous les autres arbres : mais comme ceux-ci ne poussent jamais aucuns rejettons dans l'endroit de la taille, ils en souffrent davantage.

Dans les parties de la France où il y a des forêts de ces arbres, les propriétaires donnent toujours les fagots à ceux qui taillent leurs jeunes arbres la première fois, pour les payer de leurs peines, & afin qu'il ne leur en coûte point d'argent; à la seconde taille, le propriétaire a un tiers des fagots, & les ouvriers les deux autres tiers; & ensuite, dans les autres émondages, ils se partagent par moitié entre les ouvriers & les propriétaires : mais il faut avoir grand soin qu'ils ne coupent pas au-delà de ce qu'il est nécessaire de jetter bas.

Douze ou quatorze ans après, ces arbres auront besoin d'être encore plus fortement émondés; car les branches supérieures, en privant d'air celles du bas, les font bientôt périr. Quelque tems après, si ces plantes ont fait un grand progrès, il sera peut-être nécessaire de les éclaircir; mais cette opération doit se faire par dégrés. On commence d'abord au milieu de la plantation, en laissant le dehors serré, pour abriter les arbres du centre, & peu-à-peu on parvient à ceux de l'extérieur : par ce moyen, les premiers qui auront été éclaircis auront eu le tems de se fortifier, & n'auront point souffert des gelées. Quand on éclaircit ces plantations, il ne faut pas arracher les arbres, mais les couper tout près du sol; car leurs racines ne repoussent jamais, & périssent en terre : ainsi il ne peut en arriver aucun inconvénient en les laissant, & les plantes restantes n'en seront point endommagées. Les arbres que l'on retranchera seront propres à plusieurs usages; les plus droits serviront à faire de bons boulins pour les Briquetiers, & des montans d'échaffauds; de sorte qu'en les vendant, on pourra s'indemniser, non-seulement de ce qu'il en aura coûté pour les frais de plantation, mais aussi de l'intérêt de l'argent.

Comme le principal mérite de ces arbres consiste dans leurs tiges droites, on doit les laisser assez près les uns des autres, pour qu'ils s'élèvent plus aisément, & qu'ils parviennent à une grande hauteur. J'en ai vu quelques-uns dont les tiges nues avoient

PIN

plus de vingt pieds de haut , & qui étoient aussi droits que des cannes. Un de ces arbres a fourni assez de planches pour parqueter une chambre de près de vingt pieds quarrés. Il suffit de donner à ces arbres huit pieds de distance en tous sens. Ainsi , si l'on en coupe d'abord une quatrieme partie , les autres pourront rester douze à quatorze ans sans y toucher ; alors ils auront acquis une hauteur suffisante pour faire des échelles , des soutiens d'échaffauds , & plusieurs autres choses , & la vente qu'on en fera , paiera non-seulement le restant des frais de la plantation , mais aussi le produit de la terre , avec intérêt. Les arbres qui resteront en place seront un fonds considérable pour les enfans ou héritiers. Tout ceci peut être démontré à la rigueur , par des exemples récens , qui ont prouvé que le bénéfice est toujours plus considérable que nous ne l'avons dit.

La cinquieme espece est connue en Suisse sous le nom de *Pin de Torche*. Les Paysans de ce pays se servent du bois de cet arbre en guise de torches à brûler. Cet arbre , qui s'éleve à une grande hauteur dans son sol naturel , est bien garni de branches ; son bois est fort rempli de résine ; & aux premieres tailles , il est d'une couleur rougeâtre. Les habitans de la Suisse en font usage dans leurs bâtimens.

La sixieme espece de *Pin* croît lentement en Angleterre , excepté sur le sommet des montagnes septentrionales ou dans

PIN 583

les marais. Celle-ci & le *Pin de Sibérie* paroissent y réussir beaucoup mieux que dans aucune autre partie de la Grande-Bretagne ; car ils croissent naturellement dans la neige.

La huitieme ne devient jamais un grand arbre, même dans son pays natal ; & en Angleterre , elle ressemble plus à un arbrisseau qu'à un arbre. Les froids de l'hiver l'endommagent souvent , & quelquefois les fortes gelées la détruisent ; de sorte qu'on ne la conserve dans les jardins anglois que pour la variété.

Les neuvieme & dixieme especes sont employées par les habitans de l'Amérique Septentrionale , pour leurs bâtimens , & aux mêmes usages que toutes les autres especes de *Pins*.

Il y a en Amérique quelques autres variétés de ces arbres , si elles ne sont pas des especes distinctes. Quelques-unes mûrissent leurs cônes dans la premiere année ; d'autres sont deux ans à les perfectionner , & quelques-unes trois ans ; mais comme celles-ci n'ont pas été assez bien observées par les personnes qui résident dans le pays , & que nous avons peu de ces especes assez grandes pour produire des cônes , je ne puis encore rien dire sur les différences qui les distinguent.

Les onzieme & douzieme especes sont , à ce que je crois, indifféremment appelées *Pins rouges* , dans l'Amérique Septentrionale , où leur bois est très-estimé. Les François du Canada ont construit , avec

ce bois feul , un vaiffeau de foixante canons, appelé le *Saint-Laurent*. J'en ai reçu un peu de l'Amérique ; il m'a paru reffembler beaucoup à celui du *Pin d'Ecoffe* ; mais il contient plus de réfine. Il n'y auroit point de mal de faire effai de quelques-unes de ces efpeces dans nos plantations, pour s'affurer fi elles valent la peine d'être multipliées ; car elles réuffiffent fort bien dans quelques endroits : mais elles ne font pas autant de progrès dans une terre feche que fur un fol humide.

La treizieme eft appelée *Pin blanc* dans plufieurs parties du nord de l'Amérique ; je crois qu'il y en a deux variétés, qui ne font point des efpeces diftinctes : mais comme elles n'ont point été examinées par des perfonnes habiles, nous n'en pouvons rien dire, parce que la defcription qu'a donnée M. GAULTIER d'une efpece, eft fort différente de celle du *Pin de Weymouth*. Cependant il a donné à toutes deux le nom de *Pin blanc*.

Cette efpece mérite d'être cultivée pour fa beauté, qui furpaffe celle de tous les autres *Pins* que nous connoiffons en Angleterre. L'écorce des jeunes arbres , ainfi que celle qui couvre leurs branches, eft parfaitement liffe ; les branches font bien garnies de feuilles longues , & d'un vert agréable , qui fe diftingue en été, mais qui n'a pas en hiver meilleure apparence que celui des autres efpeces. Le bois de cet arbre eft d'un bon ufage ,

fur-tout pour des mâts de vaiffeaux. Comme ces arbres croiffent droits , qu'ils s'élevent beaucoup , & qu'ils font d'ailleurs flexibles , ils font moins fujets à être brifés par le vent : c'eft ce qui a engagé le Gouvernement à publier une loi pour la confervation & la culture de cette efpece en Amérique ; mais comme elle réuffit bien en Angleterre, elle peut y être multipliée dans les cantons où le fol lui convient. Elle réuffit mieux dans un terrein léger & humide , fans être cependant trop rempli d'eau , que par-tout ailleurs : elle profpere auffi fur un fol marneux , s'il ne tient pas trop de la glaife. On doit femer les graines de cet arbre avec un peu plus de foin que celle du *Pin d'Ecoffe* , parce que fes tiges étant moins fortes, elles font plus fujettes à s'abattre , tandis qu'elles font jeunes. Ainfi , lorfqu'on les feme en pleine terre, il faut les tenir à l'ombre avec des nattes, & les expofer aux rofées toutes les nuits. On traite les plantes qui en proviennent, fuivant la méthode qui a été prefcrite pour le *Pin d'Ecoffe* ; & pour les conferver, il fera bon de les tranfplanter toutes dans les planches au mois de Juillet ; mais comme ces plantes croiffent plus vite que celles du *Pin d'Ecoffe* , il faut les mettre à une plus grande diftance, en laiffant un intervalle de fix pouces entre les rangs , & de quatre entre-elles dans les rangs : au moyen de cela , elles auront affez de place pour

croître jufqu'au printems de l'année fuivante ; alors on pourra les tranfplanter à demeure ou dans une pépiniere, où on les laiffera deux ans, pour qu'elles puiffent y acquérir de la force : mais plutôt elles font placées dans les lieux qui leur font deftinés, mieux elles réuffiffent, & plus elles font de progrès : car quoiqu'on puiffe les tranfplanter dans un âge plus avancé, cependant, quand elles font déplacées jeunes, elles deviennent plus fortes, & s'élevent davantage.

Le fol dans lequel cette efpece d'arbre profite le mieux, eft une marne molle de Noifetier, pas trop humide, dans laquelle j'ai fouvent vu des branches d'un an pouffer de deux pieds & demi de longueur, & continuer à croître ainfi pendant plufieurs années. Elle exige une fituation abritée ; car j'ai obfervé que les arbres fort expofés au vent du fud-oueft, ne faifoient pas des progrès auffi rapides que ceux qui croiffent dans un emplacement abrité ; & dans les plantations, ceux de ces arbres qui fe trouvent à l'éxtérieur, ne pouffent pas auffi bien que ceux du centre, & leurs feuilles n'y confervent pas une auffi belle verdure.

Paluftris. La quatorzieme efpece croît naturellement fur des marais, dans plufieurs parties de l'Amérique Septentrionale, où j'ai appris qu'elle s'éleve à la hauteur de vingtcinq ou trente pieds : fes feuilles ont un pied & plus de longueur, elles fortent en touffes aux extrémités des branches ; ce qui leur donne une apparence finguliere : mais je n'ai pas entendu dire que fon bois fût bon à d'autre ufage que pour brûler. Il y a ici quelques endroits où cette plante réuffit bien : mais les fortes gelées détruifent fouvent fes jeunes branches, & dans un terrein fec, elles ne profitent pas ; de forte qu'il eft inutile de la planter dans un fol qui ne lui eft pas convenable.

Le *Pin fauvage* ou *Pinéaftre* produit la térébenthine, dont les Maréchaux font un grand ufage : on en diftille auffi l'huile de *Terebenthine* ; la partie la plus fubtile & la plus eftimée fort la premiere, & eft appelée *Efprit* ; ce qui refte au fond de l'alembic eft la réfine commune.

Les amandes du *Pin de Pierre* font d'une nature balfamique & nourriffante ; elles produifent de bons effets dans la confomption, la toux, l'enrouement, & rétabliffent les forces épuifées, après de longues maladies.

PIPER. *Lin. Gen. Plant.* 42. *Saururus. Plum. Nov. Gen.* 51. *tab.* 12. [*Pepper, or Lizards-tail.*] Poivrier ou Queue de Lézard.

Caracteres. Les fleurs font fortement fixées à une fimple tige, & n'ont point de gaine complette ; elles n'ont ni pétales ni étamines, mais feulement deux antheres rondes, & oppofées à la racine du germe ; elles ont un germe gros, ovale, fans ftyle, & couronné par un ftigmat triple & pi-

quant. Ce germe devient dans la suite une baie ronde à une cellule qui renferme une semence angulaire.

Ce genre de plantes est rangé dans la troisieme section de la seconde classe de LINNÉE, qui comprend celles dont les fleurs ont deux parties de géneration mâles, & trois femelles.

Les especes sont :

1°. *Piper obtusi-folium, foliis ob ovatis, enerviis. Lin. Sp. Plant.* 30 ; Poivrier avec des feuilles presque ovales, obverses, & sans veines.

Saururus humilis, folio carnofo, subrotundo. Plum. Cat. 53. *f.* 70 ; Queue de Lézard, avec une feuille presque ronde & charnue.

2°. *Piper pellucidum, foliis cordatis, petiolatis, caule herbaceo. Lin. Sp. Plant.* 30. *Jacq. Obs.* I. *p.* 16. *Kniph. Orig. cent.* 10. *n.* 70 ; Poivrier avec des feuilles en forme de cœur, placées sur des pétioles, & une tige herbacée.

Piper foliis cordatis, caule procumbente. Hort. Cliff. 6. *tab.* 4. *Roy. Lugd.-B.* 8 ; Poivrier avec des feuilles en forme de cœur, & une tige traînante.

Saururus minor procumbens, botryitis, folio crasso, cordato. Plum. Amer. 58. *f.* 72.

3°. *Piper Amalago, foliis lanceolato-ovatis, quinqué-nerviis, rugosis. Lin. Sp. Plant.* 29 ; Poivrier à feuilles rondes, ovales, en forme de lance, & garnies de cinq veines.

Saururus foliis lanceolato-ovatis, quinqué-nerviis, rugosis. Hort. Cliff. 140 ; Queue de Lé-

zard, avec des feuilles rudes ; ovales, & à cinq veines.

Piper longum, arboreum altius, folio nervoso minori, spicâ gracili & breviore. Sloan. Hist. I. *p.* 134. *t.* 87. *f.* 1.

Piper frutex, spicâ longâ gracili. Pluk. Alm. 297. *t.* 215. *f.* 2.

Piper longum. Rumph. Amb. 5. *p.* 333. *t.* 116. *f.* 1

4°. *Piper humile, foliis lanceolatis, nervosis, rigidis, sessilibus* ; Poivrier avec des feuilles roides, en forme de lance, nerveuses & sessiles.

Piper longum humile, fructu è summitate caulis prodeunte. Sloan. Cat. Jam. 45 ; Poivre long & nain, avec un fruit sortant à l'extrémité de la tige.

5°. *Piper peltatum, foliis peltatis, orbiculato-cordatis, obtusis, repandis, spicis umbellatis. Lin. Sp. Plant.* 30 ; Poivrier à feuilles en forme de bouclier, orbiculaires, obtuses, courbées, & en forme de cœur, avec des épis en ombelles.

Saururus arborescens, foliis amplis, rotundis & umbilicatis. Plum. Amer. 56. *f.* 74 ; Queue de Lézard en arbre, avec des feuilles larges, rondes & en forme de nombril.

Lomba Rumph. Amb. 6. *p.* 133. *t.* 59. *f.* 1.

6°. *Piper Lauri-folium, foliis lanceolato-ovatis, nervosis, spicis brevibus.* Poivrier avec des feuilles ovales, nerveuses, & en forme de lance, & des épis courts.

Saururus frutescens, Lauro-Cerasi folio, fructu breviore & crassiore. Houst. MSS. ; Queue de Lézard en arbrisseau, à feuilles de Laurier-Cerise, avec un fruit plus court & plus épais.

PIP

7°. *Piper tomentofum*, *foliis ovato-lanceolatis*, *tomentofis*, *caule arborefcente* ; Poivrier avec des feuilles ovales, en forme de lance, & cotonneufes, & une tige en arbre.

Saururus arborefcens, *lati-folia*, *villofa fruflu gracili*. *Houff. MSS.* ; Queue de Lézard en arbre, avec des feuilles larges & velues, & un fruit mince.

8°. *Piper aduncum*, *foliis ovato-lanceolatis*, *nervis alternis*, *fpicis uncinatis*. *Lin. Sp. Plant.* 29 ; Poivrier avec des feuilles ovales & en forme de lance, des veines alternes, & des épis courbés.

Piper longum, *folio nervofo*, *pallidè viridi*, *humilius*. *Sloan. Hiff.* 1. *p.* 135. *t.* 87. *f.* 2.

Saururus arborefcens, *fruflu adunco*. *Plum. Cat.* 51 ; Queue de Lézard, avec un fruit courbé.

Saururus foliis ovato-lanceolatis, *nervis alternis*. *Hort. Cliff.* 140. *Roy. Lugd. - B.* 8.

9°. *Piper decumanum*, *foliis cordato-ovatis*, *nervofis*, *acuminatis*, *fpicis reflexis* ; Poivrier à feuilles ovales, nerveufes, à pointe aiguë, & en forme de cœur, avec des épis réfléchis.

Saururus frutefcens Plantaginis folio ampliori, *fruflu breviore*, *graciliore*, *adunco*. *Houff. MSS.* ; Queue de Lézard, avec une large feuille de Plantin, & un épi courbé, plus court & plus mince.

Sirum decumanum. *Rumph. Amb.* 5. *p.* 45. *f.* 27.

10°. *Piper Siriboa*, *foliis cordatis*, *fub-fepti-nerviis*, *venofis*. *Flor. Zeyl.* 29 ; Poivrier avec des feuilles en forme de cœur,

veinées, & avec fept nerfs.

Siriboa. *Rumph. Amb.* 5. *p.* 340. *t.* 117. *f.* 2.

Betela, *quem Siri boa vocant* ; Betele, Betre ou Temboul.

11°. *Piper reticulatum*, *foliis cordatis*, *feptem nervis reticulatis*. *Len. Gen. Plant.* 29 ; Poivrier avec des feuilles en forme de cœur, en filets, & à fept nervures.

Saururus botryoïdes major, *arborefcens*, *foliis Plantaginis*. *Plum. Amer.* 57. *f.* 75 ; la plus grande Queue de Lézard en arbre, & à feuilles de Plantin.

Jaborandi. *Marcgr. Bras.* 37. *Pis. Bras.* 97.

12°. *Piper glabrum*, *foliis ovato-lanceolatis*, *acuminatis glabris*, *tri-nerviis* ; Poivrier avec des feuilles ovales, unies, à pointe aiguë, en forme de lance, & à trois veines.

Saururus racemofus, *feu Botryoïdes minor*. *Plum. Cat.* 51.

13°. *Piper racemofum*, *foliis lanceolato-ovatis*, *rugofis*, *nervis alternis* ; Poivrier avec des feuilles ovales, rudes, en forme de lance, & fortifiées par des veines alternes.

Saururus racemofus, *feu Botryoïdes major*. *Plum. Cat.* 51 ; La plus grande Queue de Lézard, branchue.

Obtufi-folium. La premiere efpece croit naturellement dans plufieurs des Ifles de l'Amérique ; fa racine pouffe quelques tiges fucculentes, herbacées, prefque auffi groffes que le petit doigt, noueufes, & divifées en plufieurs branches ; elles ne s'élevent jamais à plus d'un pied de hauteur ; mais elles s'étendent générale-

ment près de la terre ; les feuilles font fort épaiffes, fucculentes, de trois pouces environ de longueur fur deux de large, fort liffes & entieres ; les pédoncules, qui foutiennent les épis, fortent aux extrémités des branches, & font auffi fort fucculens ; ils ont, y compris l'épi, environ fept pouces de longueur : l'épi, qui eft droit, érigé, & de la groffeur à-peu-près d'un tuyau de plume d'oie, eft fortement couvert de petites fleurs qu'on ne peut diftinguer qu'avec une loupe : auffi n'ont-elles point de beau té ; mais l'épi entier reffemble à la queue d'un *Lézard*, ce qui a engagé PLUMIER à lui donner ce nom.

Les épis paroiffent durant une grande partie de l'année ; mais ils produifent rarement quelques femences en Angleterre. Au refte, on multiplie facilement ces plantes, par les rejettons qui fortent de leurs racines. Cette efpece exige une ferre chaude, pour être confervée en Angleterre, & il ne faut lui donner que très-peu d'humidité, fur tout en hiver. En tenant les plantes dans la couche de la ferre, elles poufferont dans le tems des tiges qu'on pourra couper pour en faire de nouvelles plantes.

Pellucidum. La feconde efpece, qui eft originaire de l'Amérique, eft une plante annuelle, dont les tiges font herbacées, fucculentes, & de fept ou huit pouces de hauteur ; fes feuilles font en forme de cœur, de la longueur d'un pouce & demi fur neuf

lignes de large ; les épis qu'elles forment fortent aux extrémités des tiges ; ils font minces, droits, & d'un pouce à-peu-près de longueur ; ces fleurs font fort petites, & feffiles au pédoncule ou axe commun ; elles paroiffent en Juillet, & font fuivies par de fort petites baies, qui renferment chacune une petite femence femblable à de la pouffiere. Si on laiffe tomber ces femences dans des pots placés exprès au-deffous des plantes, elles poufferont fans peine. On peut auffi les recueillir, pour les femer au printems fur une couche chaude, où elles leveront aifément. On met les plantes qui en proviennent, chacune dans un pot féparé, qu'on plonge dans une couche chaude de tan, & on les traite enfuite comme les autres plantes délicates ; mais il ne faut pas leur donner trop d'humidité.

Amalago. La troifieme efpece, qu'on rencontre à la Jamaïque & à la Barbade, a plufieurs tiges courbées, de douze ou quatorze pieds de hauteur, noueufes, creufes, remplies de moëlle, & divifées en plufieurs petites branches, garnies de feuilles ovales, en forme de lance, de trois pouces & demi environ de longueur fur un & demi de large, rudes, & fortifiées par cinq veines longitudinales ; les épis, qui fortent des extrémités des branches, font minces, de trois pouces à-peu-près de longueur, & garnis de plufieurs petites fleurs,

feffiles

seffiles au chaton, & qui produifent de petites baies.

Humile. La quatrieme efpece croît naturellement à la Jamaïque ; fes tiges minces, & fouvent traînantes, pouffent des racines de leurs nœuds, comme celles de la premiere efpece ; elles font garnies de feuilles roides, en forme de lance, de cinq pouces de longueur, fur deux de large au milieu, terminées en pointe à chaque extrémité, fortifiées par une groffe côte au milieu & par plufieurs veines au dos, qui s'étendent depuis cette côte jufqu'aux bords : l'épi de fleurs eft fort mince, de cinq pouces de longueur, & de la même forme que ceux des efpeces précédentes.

Peltatum. La cinquieme, qui fe trouve encore à la Jamaïque, a une tige épaiffe, fpongieufe, de quinze pieds de hauteur, divifée en plufieurs branches noueufes & remplies de moëlle : fes feuilles font prefque rondes, & les pétioles y font attachés en-deffous, de forte que la furface fupérieure eft en forme de nombril dans l'endroit qui fe joint au pétiole : de ce centre partent des veines qui fe prolongent fur les côtés ; ces feuilles ont environ un pied de diametre ; leur partie baffe eft découpée en forme de cœur, & l'autre eft ronde ; la tige eft fixée vers le milieu ; les feuilles ont l'apparence d'un bouclier ; les épis font petits, & croiffent en forme d'ombelles.

Lauri-folium. La fixieme efpece naît fpontanément à la

Tome V.

Vera-Cruz en Amérique ; elle a des tiges noueufes d'arbriffeau, qui s'élevent à neuf ou dix pieds de hauteur, & fe divifent en plus petites branches, garnies de feuilles ovales, en forme de lance, de fept pouces de longueur fur trois de large, & terminées en pointe aiguë, rudes, veinées, & de la même confiftance que celles du *Laurier* : les épis de fleurs fortent des nœuds des branches fur le côté oppofé aux feuilles ; leur longueur eft d'un pouce & demi ; ils font à peu-près de la groffeur d'un petit tuyau de plume, & fortement garnis de fleurs femblables à celles des autres efpeces.

Tomentofum. La feptieme, qui a été découverte par le Doƈteur HOUSTOUN à la Vera-Cruz, a des tiges minces, & remplies de moëlle, qui s'élevent à la hauteur de douze ou quatorze pieds, & fe divifent en plufieurs branches courbées, divifées par des nœuds gonflés, & garnis de feuilles ovales, en forme de lance, de cinq pouces environ de longueur fur trois de large, fortifiées par plufieurs nervures, & couvertes d'un duvet laineux : les épis des fleurs fortent fur le côté des branches, & font oppofés aux feuilles ; ils font minces, de trois pouces de longueur, & tournés vers le bas.

Aduncum. La huitieme efpece croît fans culture à la Jamaïque ; elle a plufieurs tiges creufes, & de cinq pieds environ de hauteur : leurs nœuds

P p

font affez rapprochés les uns des autres, & gonflés ; elles fe divifent en plus petites branches, garnies de feuilles ovales, en forme de lance, de fept pouces de longueur fur trois de large au milieu, rudes & veinées. Ces veines fortent alternativement de la côte du milieu, fe divergent fur les côtés, & fe rejoignent aux bords de la feuille vers le fommet. Les épis de fleurs font produits vers le côté des branches oppofé aux feuilles ; ils font minces, de cinq pouces de longueur, courbés, & fortement garnis de petites fleurs dans toute leur longueur. Cette plante eft appelée *Sureau* dans les Indes Occidentales.

Decumanum. La neuvieme, qui m'a été envoyée de Carthagène par le Docteur GUILLAUME HOUSTOUN, s'éleve, avec quelques tiges d'arbriffeau, à la hauteur de quinze pieds, & fe divife en plufieurs branches minces, avec des nœuds gonflés, & garnies de feuilles ovales, en forme de lance, de cinq pouces de longueur fur trois de large, terminées en pointe aiguë, unies, & fortifiées à leur bâfe par cinq nervures, dont les deux extérieures fe joignent fur les bords au haut des feuilles, & les trois autres coulent au fommet, celle du milieu en ligne droite, & les deux de côté en fe divergeant, pour fe rejoindre vers le haut. Ces feuilles font d'un vert foncé en-deffus, & d'un vert pâle en-deffous ; les épis de fleurs qui fortent aux côtés des branches, font fort

minces, d'un pouce & demi de long, & réfléchis à l'extrémité comme la queue d'un fcorpion.

Siriboa. La dixieme efpece m'a été envoyée par M. ROBERT MILLAR, de Panama, où elle croît naturellement ; elle a des tiges creufes d'arbriffeau, d'environ quatre pieds de hauteur, & divifées en plufieurs petites branches, garnies de feuilles en forme de lance, de cinq pouces environ de longueur fur quatre de large près de leur bâfe, terminées par une pointe longue & aiguë, & fillonnées à leur bâfe par fept nervures, dont les deux extérieures coulent jufqu'aux bords, & les cinq autres s'étendent prefque dans toute fa longueur, en s'écartant de la côte du milieu vers les côtés, & s'uniffant au fommet. Les épis fortent latéralement fur les branches ; ils font minces, de quatre pouces environ de longueur, penchés dans le milieu en forme d'arc, & fortement garnis de petites fleurs herbacées, auxquelles fuccedent de petites baies, qui renferment chacune une petite femence.

Reticulatum. La onzieme efpece, qui croît naturellement à la Jamaïque, s'éleve avec une tige d'arbriffeau moëlleufe, à cinq pieds environ de hauteur, & pouffe plufieurs branches latérales, garnies de nœuds protubérans, & des feuilles en forme de lance, de fix pouces de longueur fur cinq de large près de la bâfe : elles ont cinq veines, qui fortent du pétiole ; celle du milieu monte

à la pointe en ligne droite, & les deux de côté s'écartent fur les bords vers le milieu de la feuille, & fe rejoignent au fommet ; toute leur furface eft remplie d'autres petites veines, entremêlées en forme de filets : les épis font produits fur le côté des branches oppofé aux feuilles ; ils font minces, de cinq pouces environ de longueur, un peu courbés dans le milieu, & fortement garnis de très-petites fleurs herbacées.

Glabrum. La douzieme efpece fe trouve à Campêche, d'où elle m'a été envoyée par le Docteur Houstoun ; elle a plufieurs tiges d'arbriffeau, qui s'élevent à la hauteur d'environ dix pieds, & fe divifent vers le fommet en plufieurs branches courbées, divifées par des nœuds gonflés, & garnies de feuilles ovales en forme de lance, de quatre pouces environ de longueur fur deux & demi de large, terminées en pointe aiguë, liffes, d'un vert luifant, & fortifiées par trois groffes veines longitudinales : la côte du milieu eft droite, & les deux autres s'écartent vers les côtés ; mais elles fe rejoignent à la pointe : les épis font placés fur le côté des tiges oppofé aux feuilles ; ils font longs, minces, & un peu courbés ; les fleurs & les femences reffemblent à celles des autres efpeces.

Racemofum. La treizieme eft originaire de Campêche ; elle a une tige d'arbriffeau de dix ou douze pieds de haut, & divifée vers fon fommet en un grand nombre de petites branches creufes, couvertes de nœuds gonflés, & garnies de feuilles rudes, ovales, en forme de lance, de cinq pouces environ de longueur fur deux & demi de large, dont quelques-unes ont des pétioles longs, & d'autres fort courts ; elles font d'un vert foncé en-deffus, d'un vert pâle en-deffous, & font terminées en pointe aiguë : les épis, qui paroiffent fur le côté des tiges oppofé aux feuilles, font longs, minces, & fortement garnis de petites fleurs femblables à celles des autres.

Culture. Les onze dernieres efpeces font des plantes vivaces, que l'on peut multiplier par leurs graines, qu'il faut fe procurer fraiches, des pays où ces plantes croiffent naturellement : on les répand fur une bonne couche chaude au printems ; & quand les plantes qui en proviennent font en état d'être enlevées, on les met chacune féparément dans de petits pots remplis d'une terre fraiche & légere ; on les plonge dans une couche chaude de tan ; on les tient à l'ombre jufqu'à ce qu'elles aient formé de nouvelles racines, & on les traite enfuite comme les autres plantes tendres & exotiques. On leur donne de l'air tous les jours à proportion de la chaleur de la faifon, pour les empêcher de filer ; & quand les nuits font froides, on couvre les vitrages de la couche avec des nattes, pour en conferver la chaleur. Comme les tiges de la plupart de ces plantes font tendres lorfqu'elles font

592 PIP

jeunes, il ne faut pas leur donner trop d'humidité, qui les pourriroit, & on les arrose avec précaution, pour ne pas les abattre ; car une fois couchées, elles se relevent rarement.

En automne, on place ces plantes dans la couche de tan de la serre chaude, & on leur donne peu d'eau en hiver; elles exigent la même chaleur que

PIR

le *Caffier* ; & pendant les chaleurs de l'été, il faut leur donner beaucoup d'air : mais il faut les tenir constamment dans la serre chaude ; car elles sont trop délicates pour supporter la rigueur de notre climat, même dans la saison la plus chaude de l'année.

PIROLLE. *Voyez* Pyrola ROTUNDI-FOLIA. *L.*

Fin du Tome cinquieme.

Made at Dunstable, United Kingdom
2022-07-14
http://www.print-info.eu/

83131777R00344